SALAS AND HILLE'S
CALCULUS

SEVERAL VARIABLES

SALAS AND HILLE'S
CALCULUS

SEVERAL VARIABLES

SEVENTH EDITION

REVISED BY

GARRET J. ETGEN

JOHN WILEY & SONS, INC.

New York   Chichester   Brisbane   Toronto   Singapore

ACQUISITIONS EDITOR	Ruth Baruth
DEVELOPMENTAL EDITOR	Nancy Perry
MARKETING MANAGER	Susan Elbe
PRODUCTION EDITOR	Charlotte Hyland
DESIGNER	Ann Marie Renzi
MANUFACTURING MANAGER	Susan Stetzer
ILLUSTRATION EDITOR	Sigmund Malinowski
ELECTRONIC ILLUSTRATION	Fine Line
CHAPTER TITLE DESIGN	Carol Grobe
COVER PHOTOGRAPH	Paul Silverman

This book was set in New Times Roman by Progressive Information Technologies and printed and bound by Von Hoffman Press. The cover was printed by Phoenix Color Corp.

Recognizing the importance of preserving what has been written, it is a policy of John Wiley & Sons, Inc. to have books of enduring value published in the United States printed on acid-free paper, and we exert our best efforts to that end.

The paper in this book was manufactured by a mill whose forest management programs include sustained yield harvesting of its timberlands. Sustained yield harvesting principles ensure that the number of trees cut each year does not exceed the amount of new growth.

Library of Congress Cataloging in Publication Data:
Salas, Saturnino L.
    Salas and Hille's calculus: several variables/revised by Garret J. Etgen. — 7th ed.
    p.   cm.
    Rev. ed. of: Calculus: one and several variables. 6th ed. 1990.
    Includes index.
    ISBN 0-471-12366-8 (pbk.: alk. paper)
    1. Calculus. I. Hille, Einar (deceased) II. Salas, Saturnino L. III. Etgen, Garret J., Calculus: several variables. IV. Title.
QA303.S17   1995b
515—dc20   94-30557
                                                                        CIP

Printed in the United States of America

10 9 8 7 6 5 4 3 2

In fond remembrance of
EINAR HILLE

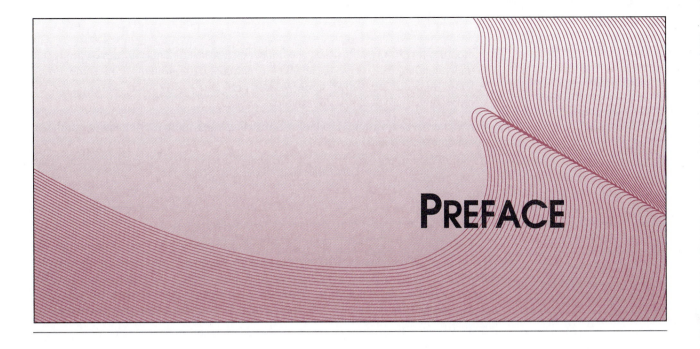

PREFACE

Above all, this is a text on mathematics. The subject is calculus, and the emphasis is on the three basic concepts: limit, derivative, and integral.

This text is designed for a standard introductory multivariable calculus course. Our fundamental goal in preparing the Seventh Edition has been to preserve and enhance the notable strengths that characterized previous editions, including:

- An emphasis on the mathematical exposition—an accurate, understandable treatment of the topics.
- A clear, concise approach. Basic ideas and important points are not obscured by excess verbiage.
- An appropriate level of rigor. Mathematical statements are careful and precise, and all important theorems are proved. This formality is presented in a way that is completely accessible to the beginning calculus student.
- A balance of theory and applications, illustrated by many examples and exercises.

At the same time, we recognize that with the rapid advances in computer technology and the current scrutiny of mathematics education at all levels, the teaching of calculus is undergoing a serious examination. Thus, an equally important and parallel goal of the Seventh Edition has been to incorporate modern technology and current trends without sacrificing the acknowledged strengths of the text.

FEATURES OF THE SEVENTH EDITION

Problem-Solving Skills and Real-World Applications

Over 750 new problems have been added to the Seventh Edition.

- In order to develop students' problem-solving skills, we have significantly increased the number of problems at all levels. A large number of challenging and routine problems are now available in all exercise sets. Many additional medium-level problems are included to assist students in developing the understanding necessary to attack the challenging problems. In some problems, students are called upon to interpret and justify their answers to improve their analytical and communication skills.

- An even wider variety of real-world applications motivates students' study of mathematical topics.
- More illustrations have been added to exercise sets to provide students with visual support as they devise their problem-solving strategies.

Technology

Because the use of graphing calculators and/or computer algebra systems has increased in calculus courses, we have considerably expanded the application of technology in the text. We do not attempt to teach any particular technology and so use a generic approach. Technology problems are clearly designated with an icon (▶) and may be skipped by instructors who prefer that their students not use calculators or computers.

- New technology-based examples appear within the chapter discussions of the material. These support the numerous exercises requiring the use of a graphics calculator or other graphing software located in the end-of-section problems sets.
- "Projects and Explorations Using Technology," a set of problems that requires a combination of approaches involving both analytical and technology skills, ends each chapter. As their title suggests, these problems are also suitable for use by students working in groups. A few of the problems introduce concepts to be developed later in the text, while others explore realistic applications of topics that have already been studied.

Increased Emphasis on Visualization

We recognize the importance of visualization in developing students' understanding of mathematical concepts. For that reason:

- All the artwork from the previous edition has been redrawn for increased clarity and understanding.
- Over 50 new figures have been added.
- Representations in three dimensions are now in full color for increased geometric understanding and include many new computer-generated figures of curves and surfaces in space.

CONTENT AND ORGANIZATION CHANGES IN THE SEVENTH EDITION

In response to the evolutionary state of the current calculus curriculum, many changes have been made in organization and content to meet the needs of today's students and instructors.

Sequences and Series (Chapters 10 and 11)

- The least upper bound axiom now serves as a prelude to sequences, and there is more emphasis on boundedness in the treatment of sequences.
- The treatment of indeterminate forms has been modified: The "other" indeterminate forms—differences, products, exponential forms—are now treated in a separate subsection rather than integrating them with the $0/0$ and ∞/∞ forms.
- The treatment of power series has been expanded slightly; there are some new examples and figures, and the Lagrange form of the remainder is stated explicitly and used to derive bounds on the remainder.

Multivariable Calculus (Chapters 12–17)

- Substantial changes were not necessary in the treatment of these chapters. The major effort in this edition was to upgrade the illustrations and the exercises.
- A large number of computer-generated figures illustrating curves and surfaces in space have been added, and full color has been used where it is most helpful to students' understanding—in three-dimensional figures.

Differential Equations (Chapter 18)

- The material on differential equations has been thoroughly updated and revised to include numerous examples and applications throughout the chapter.
- A new introductory section familiarizes students with the basic terminology and concepts of differential equations.

FEATURES OF THE BOOK

Concise exposition The concepts of calculus are presented clearly and accurately without hand waving.

Theorems and proofs Highlighted theorems direct students to accurate mathematical statements. Most proofs are included to provide a high level of precision.

New examples To facilitate students' understanding, many examples have been revised and new examples have been added.

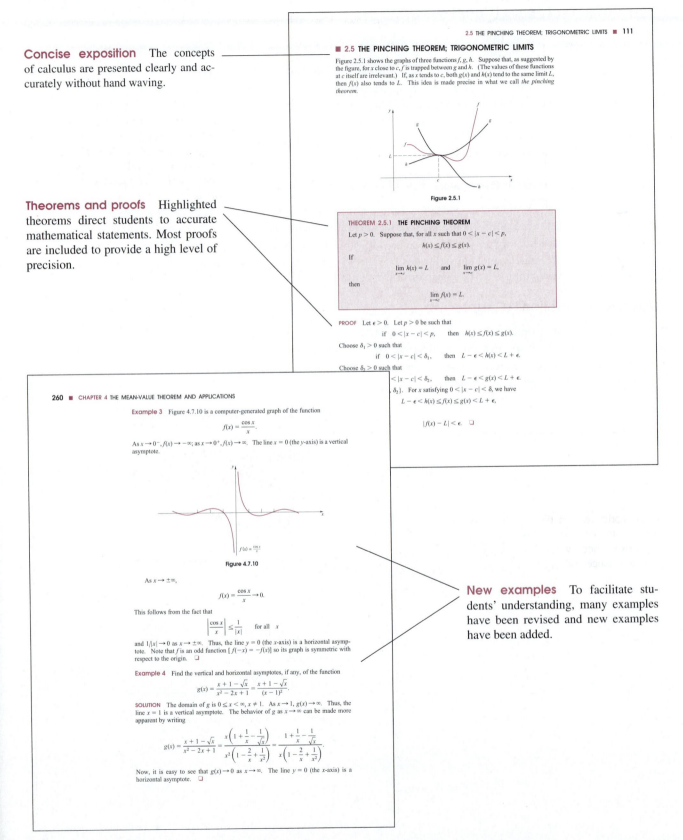

■ 2.5 THE PINCHING THEOREM; TRIGONOMETRIC LIMITS

Figure 2.5.1 shows the graphs of three functions f, g, h. Suppose that, as suggested by the figure, for x close to c, f is trapped between g and h. (The values of these functions at c itself are irrelevant.) If, as x tends to c, both $g(x)$ and $h(x)$ tend to the same limit L, then $f(x)$ also tends to L. This idea is made precise in what we call *the pinching theorem*.

Figure 2.5.1

THEOREM 2.5.1 **THE PINCHING THEOREM**

Let $p > 0$. Suppose that, for all x such that $0 < |x - c| < p$,
$$h(x) \le f(x) \le g(x).$$
If
$$\lim_{x \to c} h(x) = L \quad \text{and} \quad \lim_{x \to c} g(x) = L,$$
then
$$\lim_{x \to c} f(x) = L.$$

PROOF Let $\epsilon > 0$. Let $p > 0$ be such that
$$\text{if} \quad 0 < |x - c| < p, \quad \text{then} \quad h(x) \le f(x) \le g(x).$$
Choose $\delta_1 > 0$ such that
$$\text{if} \quad 0 < |x - c| < \delta_1, \quad \text{then} \quad L - \epsilon < h(x) < L + \epsilon.$$
Choose $\delta_2 > 0$ such that
$$< |x - c| < \delta_2, \quad \text{then} \quad L - \epsilon < g(x) < L + \epsilon.$$
$\delta_2\}$. For x satisfying $0 < |x - c| < \delta$, we have
$$L - \epsilon < h(x) \le f(x) \le g(x) < L + \epsilon,$$

$$|f(x) - L| < \epsilon. \quad \square$$

Example 3 Figure 4.7.10 is a computer-generated graph of the function
$$f(x) = \frac{\cos x}{x}.$$
As $x \to 0^-$, $f(x) \to -\infty$; as $x \to 0^+$, $f(x) \to \infty$. The line $x = 0$ (the y-axis) is a vertical asymptote.

Figure 4.7.10

As $x \to \pm\infty$,
$$f(x) = \frac{\cos x}{x} \to 0.$$
This follows from the fact that
$$\left| \frac{\cos x}{x} \right| \le \frac{1}{|x|} \quad \text{for all} \quad x$$
and $1/|x| \to 0$ as $x \to \pm\infty$. Thus, the line $y = 0$ (the x-axis) is a horizontal asymptote. Note that f is an odd function $[f(-x) = -f(x)]$ so its graph is symmetric with respect to the origin. $\square$

Example 4 Find the vertical and horizontal asymptotes, if any, of the function
$$g(x) = \frac{x + 1 - \sqrt{x}}{x^2 - 2x + 1} = \frac{x + 1 - \sqrt{x}}{(x - 1)^2}.$$

SOLUTION The domain of g is $0 \le x < \infty$, $x \ne 1$. As $x \to 1$, $g(x) \to \infty$. Thus, the line $x = 1$ is a vertical asymptote. The behavior of g as $x \to \infty$ can be made more apparent by writing
$$g(x) = \frac{x + 1 - \sqrt{x}}{x^2 - 2x + 1} = \frac{x\left(1 + \dfrac{1}{x} - \dfrac{1}{\sqrt{x}}\right)}{x^2\left(1 - \dfrac{2}{x} + \dfrac{1}{x^2}\right)} = \frac{1 + \dfrac{1}{x} - \dfrac{1}{\sqrt{x}}}{x\left(1 - \dfrac{2}{x} + \dfrac{1}{x^2}\right)}.$$
Now, it is easy to see that $g(x) \to 0$ as $x \to \infty$. The line $y = 0$ (the x-axis) is a horizontal asymptote. $\square$

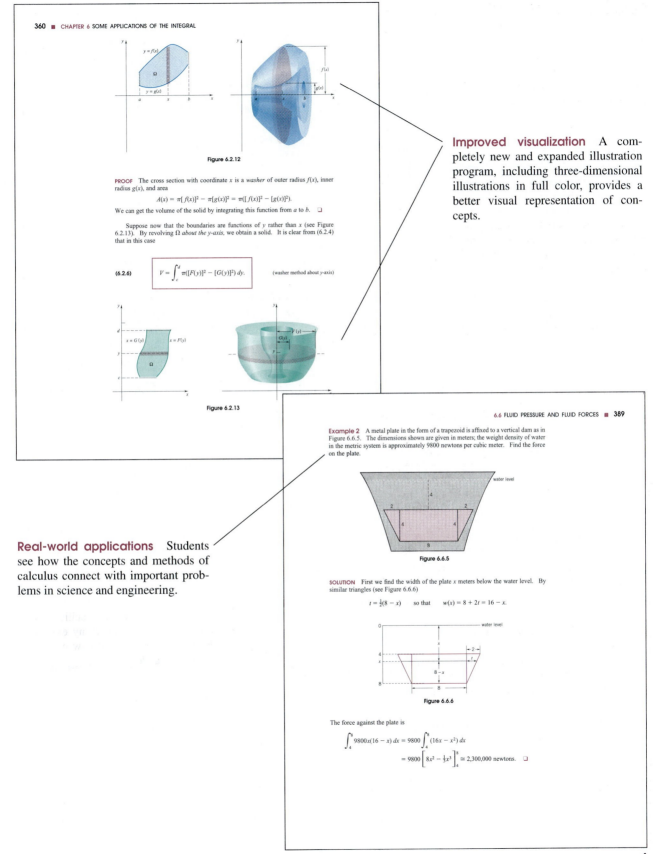

Figure 6.2.12

PROOF The cross section with coordinate x is a *washer* of outer radius $f(x)$, inner radius $g(x)$, and area

$$A(x) = \pi[\,f(x)]^2 - \pi[g(x)]^2 = \pi([\,f(x)]^2 - [g(x)]^2).$$

We can get the volume of the solid by integrating this function from a to b. ❑

Suppose now that the boundaries are functions of y rather than x (see Figure 6.2.13). By revolving Ω *about the y-axis,* we obtain a solid. It is clear from (6.2.4) that in this case

(6.2.6)
$$V = \int_{c}^{d} \pi([F(y)]^2 - [G(y)]^2)\, dy. \qquad \text{(washer method about } y\text{-axis)}$$

Figure 6.2.13

Improved visualization A completely new and expanded illustration program, including three-dimensional illustrations in full color, provides a better visual representation of concepts.

Real-world applications Students see how the concepts and methods of calculus connect with important problems in science and engineering.

Example 2 A metal plate in the form of a trapezoid is affixed to a vertical dam as in Figure 6.6.5. The dimensions shown are given in meters; the weight density of water in the metric system is approximately 9800 newtons per cubic meter. Find the force on the plate.

Figure 6.6.5

SOLUTION First we find the width of the plate x meters below the water level. By similar triangles (see Figure 6.6.6)

$$t = \tfrac{1}{2}(8 - x) \qquad \text{so that} \qquad w(x) = 8 + 2t = 16 - x.$$

Figure 6.6.6

The force against the plate is

$$\int_{4}^{8} 9800x(16 - x)\, dx = 9800 \int_{4}^{8} (16x - x^2)\, dx$$

$$= 9800 \left[8x^2 - \tfrac{1}{3}x^3 \right]_{4}^{8} \cong 2{,}300{,}000 \text{ newtons.} \quad ❑$$

New exercises The exercise sets have been revised and over 1300 new problems added, resulting in an improved balance between drill problems and more challenging exercises involving either theory of applications.

Technology problems Problems marked by the icon ◆ encourage students to use technology as a tool to enhance understanding and problem-solving skills.

26. $f(x) = x + \cos 2x, \quad 0 < x < \pi$.
27. $f(x) = \sin^2 x - \sqrt{3} \sin x, \quad 0 < x < \pi$.
28. $f(x) = \sin^2 x, \quad 0 < x < 2\pi$.
29. $f(x) = \sin x \cos x - 3 \sin x + 2x, \quad 0 < x < 2\pi$.
30. $f(x) = 2 \sin^3 x - 3 \sin x, \quad 0 < x < \pi$.
31. Prove Theorem 4.3.4 by applying Theorem 4.2.3.
32. Prove the validity of the second-derivative test in the case that $f''(c) < 0$.
33. Find the critical numbers and the local extreme values of the polynomial

$$P(x) = x^4 - 8x^3 + 22x^2 - 24x + 4.$$

Then show that the equation $P(x) = 0$ has exactly two real roots, both positive.

34. A function f has derivative f' given by

$$f'(x) = x^3(x-1)^2(x+1)(x-2).$$

At what numbers x, if any, does f have a local maximum? A local minimum?

35. A polynomial function $p(x) = a_n x^n + a_{n-1} x^{n-1} + \cdots + a_1 x + a_0$ has critical numbers at $x = -1, 1, 2$, and 3, and corresponding values $p(-1) = 6$, $p(1) = 1$, $p(2) = 3$, and $p(3) = 1$. Sketch a possible graph for p if:
(a) n is odd, (b) n is even.

36. The quadratic function $f(x) = Ax^2 + Bx + C$ has a local minimum at $x = 2$ and passes through the points $(-1, 3)$ and $(3, -1)$. Find A, B, and C.

37. Determine a and b such that the function $f(x) = ax/(x^2 + b^2)$ has a local minimum at $x = -2$ and $f'(0) = 1$.

38. Let $f(x) = x^p(1-x)^q$, where $p \geq 2$ and $q \geq 2$ are integers.
(a) Show that the critical numbers of f are $x = 0$, $p/(p+q)$, and 1.
(b) Show that if p is even, then f has a local minimum at 0.
(c) Show that if q is even, then f has a local minimum at 1.
(d) Show that f has a local maximum at $p/(p+q)$ for all p and q.

39. Let

$$f(x) = \begin{cases} x^2 \sin(1/x), & x \neq 0 \\ 0, & x = 0. \end{cases}$$

In Exercise 67, Section 3.1, we saw that f is differentiable at 0 and that $f'(0) = 0$. Show that f has neither a local maximum nor a local minimum at 0.

40. Suppose that $C(x)$, $R(x)$, and $P(x)$ are the cost, revenue, functions corresponding to the production and certain item. Suppose, also, that C and functions. Then, since $P = R - C$, it differentiable. Prove that it is possible

to maximize the profit by producing and selling x_0 items, then $C'(x_0) = R'(x_0)$. That is, the marginal cost equals the marginal revenue when the profit is maximized.

41. Let $y = f(x)$ be differentiable and suppose that the graph of f does not pass through the origin. Then the distance D from the origin to a point $P(x, f(x))$ on the graph is given by

$$D = \sqrt{x^2 + [f(x)]^2}.$$

Show that if D has a local extreme value at c, then the line through $(0, 0)$ and $(c, f(c))$ is perpendicular to the tangent line to the graph of f at c.

42. Prove that a polynomial of degree n has at most $n - 1$ local extreme values.

◆ 43. Let $f(x) = x^4 - 2x^2 - 3x + 2$.
(a) Show that f has exactly one critical number c in the interval $(1, 2)$.
(b) Use the bisection method (see Section 2.6) to approximate c to within $\frac{1}{16}$. Does f have a local maximum, a local minimum, or neither a maximum nor a minimum at c?

◆ 44. Let $f(x) = 2 + 20x + 4x^2 - x^4$.
(a) Show that f has exactly one critical number in the interval $(2, 3)$.
(b) Use the bisection method to approximate c to within $\frac{1}{16}$. Does f have a local maximum, a local minimum, or neither a maximum nor a minimum at c?

◆ 45. Let $f(x) = x^4 - 7x^2 + 2x - 3$.
(a) Show that f has exactly one critical number c in the interval $(2, 3)$.
(b) Use the Newton-Raphson method to approximate c; calculate x_3 and round your answer to four decimal places. Does f have a local maximum, a local minimum, or neither a maximum nor a minimum at c?

◆ 46. Let $f(x) = x \cos x$.
(a) Show that f has exactly one critical number in the interval $(0, \pi/2)$.
(b) Use the Newton-Raphson method to approximate c; calculate x_3 and round your answer to four decimal places. Does f have a local maximum, a local minimum, or neither a maximum nor a minimum at c?

◆ 47. Let $f(x) = \sin x + (x^2/2) - 2x$.
(a) Show that f has exactly one critical number in the interval $[2, 3]$.
(b) Use the Newton-Raphson method to approximate c; calculate x_3 and round your answer to four decimal places. Does f have a local maximum, a local minimum, or neither a maximum nor a minimum at c?

◆ In Exercises 48–51, use a graphing utility to graph the function f on the indicated interval. (a) Use the graph to estimate the critical numbers and the local extreme values; and (b) estimate the intervals on which f increases and the intervals on which f decreases. Round off your estimates to three decimal places.

The inverse secant, $y = \sec^{-1} x$, is the inverse of $y = \sec x$, $x \in [0, \frac{1}{2}\pi) \cup (\frac{1}{2}\pi, \pi]$.

graph of $y = \sin^{-1} x$ (p. 462)   graph of $y = \tan^{-1} x$ (p. 465)
graph of $y = \sec^{-1} x$ (p. 468)

$$\frac{d}{dx}(\sin^{-1} x) = \frac{1}{\sqrt{1 - x^2}} \qquad \int \frac{dx}{\sqrt{a^2 - x^2}} = \sin^{-1}\left(\frac{x}{a}\right) + C \quad (a > 0)$$

$$\frac{d}{dx}(\tan^{-1} x) = \frac{1}{1 + x^2} \qquad \int \frac{dx}{a^2 + x^2} = \frac{1}{a}\tan^{-1}\left(\frac{x}{a}\right) + C \quad (a \neq 0)$$

$$\frac{d}{dx}(\sec^{-1} x) = \frac{1}{|x|\sqrt{x^2 - 1}} \qquad \int \frac{dx}{x\sqrt{x^2 - a^2}} = \frac{1}{a}\sec^{-1}\left(\frac{|x|}{a}\right) + C \quad (a > 0)$$

definition of the remaining inverse trigonometric functions (p. 471)

7.9 The Hyperbolic Sine and Cosine

$$\sinh x = \frac{1}{2}(e^x - e^{-x}), \qquad \cosh x = \frac{1}{2}(e^x + e^{-x}),$$

$$\frac{d}{dx}(\sinh x) = \cosh x, \qquad \frac{d}{dx}(\cosh x) = \sinh x.$$

graphs (pp. 475–476)   basic identities (p. 477)

*7.10 The Other Hyperbolic Functions

$$\tanh x = \frac{\sinh x}{\cosh x}, \qquad \coth x = \frac{\cosh x}{\sinh x},$$

$$\operatorname{sech} x = \frac{1}{\cosh x}, \qquad \operatorname{csch} x = \frac{1}{\sinh x}.$$

derivatives (p. 479)   hyperbolic inverses (p. 481)
derivatives of hyperbolic inverses (p. 482)

Chapter Highlights End-of-chapter lists stress important terms, ideas, and theorems.

■ **PROJECTS AND EXPLORATIONS USING TECHNOLOGY**

To do these exercises you will need a graphics calculator or a computer with graphing capability. The majority of these problems are open-ended so different approaches may be used to solve them. You should be aware that different approaches can result in slight variations in the answers. Round your numerical answers to at least four decimal places. The rounding method that your calculator or computer uses also may cause variations in answers.

7.1 The functions $f(x) = a \ln x$, where a is a constant, have a number of applications, one of which will be considered in a later exercise.
(a) Find the values of a for which the graph of f is tangent to the line $y = x$.
(b) For each real number a, how many solutions will there be for $f(x) = x$? What is the value of f' at each solution of $f(x) = x$?
(c) How many solutions are there to $f[f(x)] = x$? What is the value of f' at each of these solutions?
(d) Represent f as a logarithm function in another base.

7.2 Let $A(t)$ denote the area of the rectangle of width $2t$ that has its lower vertices on the x-axis and its upper vertices on the graph of

$$f(x) = e^{(1 - x^4)(2 + x^2)}.$$

See the figure.

Projects and Explorations Using Technology Special problem sets encourage deeper investigation of the material and can be used for cooperative learning activities.

SUPPLEMENTS

Student Aids

Answers to Odd-Numbered Exercises Answers to all the odd-numbered exercises are included at the back of the text.

Student Solutions Manual, Prepared by Garret J. Etgen, University of Houston This manual contains worked-out solutions to all the odd-numbered exercises and is available through your bookstore.

Instructor Aids

Instructor's Manual, by Garret J. Etgen and Sylvain Laroche This manual contains solutions to all the problems in the text.

Test Bank, by Sylvain Laroche A wide range of problems and their solutions are keyed to the text material and exercise sets.

Computerized Test Bank Available in both IBM and Macintosh formats, the Computerized Test Bank allows instructors to create, customize, and print a test containing any combination of questions from the test bank. Instructors can also edit the questions or add their own.

Technology Manuals

Discovering Calculus with Derive, by Jerry Johnson, University of Nevada-Reno, and Benny Evans, Oklahoma State University

- Derive instructions and tutorials
- Solved problems
- Practice problems
- Laboratory exercises

Discovering Calculus with Mathematica, by Cecilia A. Knoll, Florida Institute of Technology, Michael D. Shaw, Florida Institute of Technology, Jerry Johnson, and Benny Evans

- Mathematica introduction and commands
- Solved problems
- Exercises
- Laboratory projects

Discovering Calculus with Maple, by Kent Harris, Western Illinois University, and Robert J. Lopez, Rose-Hulman Institute of Technology

- Maple commands
- Example problems and step-by-step solutions
- Exercises

Discovering Calculus with Graphing Calculators, by Joan McCarter, Arizona State University

- Introductions to various calculators currently on the market (this manual is calculator nonspecific)
- Projects
- Additional exercises
- Critical thinking questions

ACKNOWLEDGMENTS

The revision of a text of this magnitude and stature requires a lot of encouragement and a lot of help. I was fortunate to have an ample supply of both from a variety of sources.

Each edition of this text was developed from those that preceded it. The present book owes much to the people who contributed to the first six editions. I am deeply indebted to them.

The reviewers of the seventh edition supplied detailed criticisms and valuable suggestions. I offer my sincere appreciation to the following individuals:

Linda Becerra
University of Houston

Jay Bourland
Colorado State University

Gary Crown
Wichita State University

Steve Davis
Davidson College

Anthony Dooley
University of New South Wales

William Fox
U.S. Military Academy — West Point

Barbara Gale
Prince Georges Community College

Pamela Gorkin
Bucknell University

Gary Itzkowitz
Rowan College of New Jersey

Harold Jacobs
East Stroudsburg University of Pennsylvania

Adam Lutoborski
Syracuse University

Douglas Mackenzie
University of New South Wales

Katherine Murphy
University of North Carolina-Chapel Hill

Jean E. Rubin
Purdue University

Stuart Smith
University of Toronto

I am especially grateful to Richard D. Byrd, University of Houston, who read the first draft of the manuscript letter by letter, made comments to match, read all the reviewers' comments, and then read the revised manuscript. This was a truly noble effort.

I am equally grateful to John Oman, University of Wisconsin—Oshkosh, who created the Projects and Explorations Using Technology problems.

I also want to thank Jun Ma, University of Houston, who provided detailed solutions for all the new exercises; there were more than 2000 of them.

I am deeply indebted to the editorial staff at John Wiley & Sons: Ruth Baruth, Mathematics Editor, who invited me to undertake this project in the first place and who provided encouragement when I needed encouragement and prodding when I needed prodding; Joan Carrafiello and Nancy Perry, Development Editors—Joan offered invaluable help in the beginning stages of this revision and Nancy continued it, along with exceptional organization and support; Charlotte Hyland, Production Editor, who was patient and understanding as she guided the project through the production stages; Sigmund Malinowski, who carefully directed the entirely new art program; and Ann Marie Renzi, whose creativity produced the attractive interior design as well as the cover. This is an efficient, professional (and patient!) group.

Finally, I want to acknowledge the contributions of my wife, Charlotte. She supplied the sustenance and support that I had to have to complete this work.

Garret J. Etgen

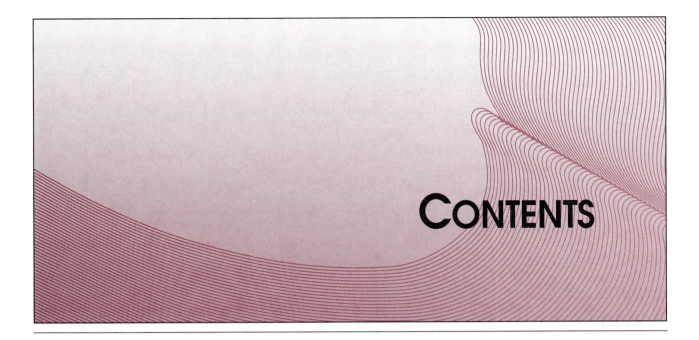

CONTENTS

KEY CONCEPTS FROM SINGLE VARIABLE CALCULUS

■ ANALYTIC GEOMETRY AND THE FUNCTION CONCEPT

Distance and Midpoint Formulas

Let the $P_1(x_1, y_1)$ and $P_2(x_2, y_2)$ be points in the plane.

The distance between P_1 and P_2 is given by

$$d(P_1, P_2) = \sqrt{(x_2 - x_1)^2 + y_2 - y_1)^2}.$$

The coordinates of the mid-point of the line segment $\overline{P_1P_2}$ are

$$\left(\frac{x_1 + x_2}{2}, \frac{y_1 + y_2}{2} \right).$$

Conic Sections

$$Ax^2 + By^2 + Cx + Dy + E = 0$$

Circle: $A = B, A \neq 0$; $\quad Ax^2 + Ay^2 + Cx + Dy + E = 0$ or
$\qquad\qquad\qquad\qquad x^2 + y^2 + C_1x + D_1y + E_1 = 0$ (after dividing by A).
Standard form: $(x - h)^2 + (y - k)^2 = r^2$ *circle of radius r centered at P(h, k).*

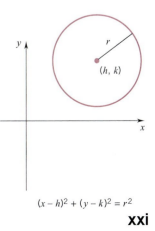

$(x - h)^2 + (y - k)^2 = r^2$

Ellipse: A, B have the same sign, $A \neq B$; $\quad Ax^2 + By^2 + Cx + Dy + E = 0$.
Standard form: $\dfrac{(x - h)^2}{a^2} + \dfrac{(y - k)^2}{b^2} = 1$ *ellipse centered at the point P(h, k).*

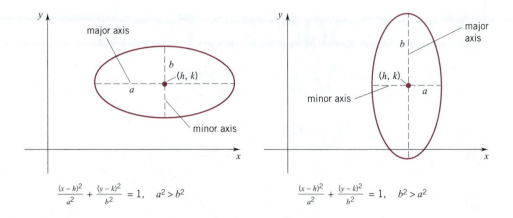

$$\frac{(x-h)^2}{a^2} + \frac{(y-k)^2}{b^2} = 1, \quad a^2 > b^2$$

$$\frac{(x-h)^2}{a^2} + \frac{(y-k)^2}{b^2} = 1, \quad b^2 > a^2$$

Parabola: Either $A = 0$ or $B = 0$; $Ax^2 + Cx + Dy + E = 0$ or
$$By^2 + Cx + Dy + E = 0.$$

Standard forms:
$(x - h)^2 + 4c(y - k)$ or   *parabola, vertex at the point $P(h, k)$ and axis vertical;*
$(y - k)^2 = 4c(x - h)$ *parabola, vertex at the point $P(h, k)$ and axis horizontal.*

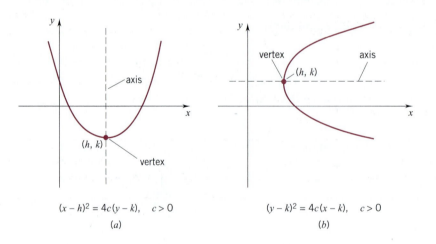

$$(x - h)^2 = 4c(y - k), \quad c > 0$$
$$(a)$$

$$(y - k)^2 = 4c(x - k), \quad c > 0$$
$$(b)$$

[Note: the parabolas open in the other direction if $c < 0$.]

Hyperbola: A, B have opposite sign: $Ax^2 + By^2 + Cx + Dy + E = 0$

Standard Forms:

$$\frac{(x - h)^2}{a^2} - \frac{(y - k)^2}{b^2} = 1 \qquad \text{\textit{hyperbola, center at the point } } P(h, k)\text{\textit{, axis horizontal;}}$$

$$\frac{(y - k)^2}{b^2} - \frac{(x - h)^2}{a^2} = 1 \qquad \text{\textit{hyperbola, center at the point } } P(h, k)\text{\textit{, axis vertical.}}$$

Some Special Limits

1. $\lim\limits_{x \to c} a = a$.

2. $\lim\limits_{x \to c} x = c$.

3. $\lim\limits_{x \to c} P(x) = P(c)$ for any polynomial P.

4. $\lim\limits_{x \to c} \dfrac{P(x)}{Q(x)} = \dfrac{P(c)}{Q(c)}$ for any rational function $R = \dfrac{P}{Q}$, provided $Q(c) \neq 0$.

The Pinching Theorem

Let $p > 0$. Suppose that, for all x such that $0 < |x - c| < p$,

$$h(x) \leq f(x) \leq g(x).$$

If

$$\lim_{x \to c} h(x) = L \quad \text{and} \quad \lim_{x \to c} g(x) = L,$$

then

$$\lim_{x \to c} f(x) = L.$$

Trigonometric Limits

1. $\lim\limits_{x \to c} \sin x = \sin c$ for all real numbers c.

2. $\lim\limits_{x \to c} \cos x = \cos c$ for all real numbers c.

3. $\lim\limits_{x \to 0} \dfrac{\sin x}{x} = 1$.

4. $\lim\limits_{x \to 0} \dfrac{1 - \cos x}{x} = 0$.

Continuity

Let f be defined at least on an open interval $(c - p, c + p), p > 0$. Then f is *continuous* at c iff

$$\lim_{x \to c} f(x) = f(c).$$

Arithmetic of Continuous Functions

> If f and g are continuous at c, then
>
> 1. $f + g$ is continuous at c.
> 2. $f - g$ is continuous at c.
> 3. αf is continuous at c for each real number α.
> 4. $f \cdot g$ is continuous at c.
> 5. f/g is continuous at c provided $g(c) \pm 0$.

Composition of Continuous Functions

> If g is continuous at c and f is continuous at $g(c)$, then the composition $f \circ g$ is continuous at c.

Intermediate Value Theorem

> If f is continuous on $[a, b]$ and C is a number between $f(a)$ and $f(b)$, then there is at least one number $c \in [a, b]$ for which $f(c) = C$.

Boundedness: Extreme-Value Theorem

> Let f be continuous on the closed, bounded interval $[a, b]$.   Then
>
> 1. f is bounded on $[a, b]$.
> 2. f attains both its maximum value M and its minimum value m on $[a, b]$.

■ DIFFERENTIATION

Derivative

> A function f is *differentiable at x* iff
>
> $$\lim_{h \to 0} \frac{f(x + g) - f(x)}{h} \text{ exists.}$$
>
> If this limit exists, it is called the *derivative of f at x,* and is denoted by $f'(x)$. The function f is a *differentiable function* if it is differentiable at each point of its domain.

Geometric Interpretation of the Derivative

$f'(x)$ is the slope of the tangent line to the graph of f at the point $(x, f(x))$.

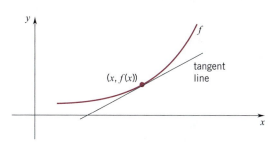

Tangent Lines and Normal Lines

Let (x_0, y_0) be a point on the graph of a function f. If f is differentiable at x_0, then:

1. An equation for the tangent line to the graph of f at (x_0, y_0) is

$$y - y_0 = f'(x_0)(x - x_0).$$

2. An equation for the normal line to the graph of f at (x_0, y_0) is

$$y - y_0 = -\frac{1}{f'(x_0)}(x - x_0) \quad \text{provided } f'(x_0) \neq 0.$$

Differentiability and Continuity

If f is differentiable at x, then f is continuous at x. The converse is false; continuity at x *does not imply* differentiability at x.

Differentiation Rules

Let f and g be differentiable at x. Then

1. $f + g$ is differentiable at x and

$$[f + g]'(x) = f'(x) + g'(x).$$

2. $f - g$ is differentiable at x and

$$[f - g]'(x) = f'(x) - g'(x).$$

3. αf is differentiable at x for each real number α and

$$[\alpha f]'(x) = \alpha f'(x).$$

continues

---

4. $f \cdot g$ is differentiable at x and

$$[f \cdot g]'(x) = f(x) \cdot g'(x) + g(x) \cdot f'(x).$$

5. f/g is differentiable at x and

$$\left[\frac{f}{g}\right]'(x) = \frac{g(x) \cdot f'(x) - f(x) \cdot g'(x)}{[g(x)]^2}, \quad g(x) \neq 0.$$

The Leibniz Notation

Let f be differentiable at x and let $y = f(x)$. Then the derivative $f'(x)$ is also denoted by $\dfrac{dy}{dx}$, called the *Leibniz notation*.

The Chain Rule

If g is differentiable at x and f is differentiable at $g(x)$, then the composition $f \circ g$ is differentiable at x and

$$(f \circ g)'(x) = f'(g(x))g'(x).$$

The Chain Rule in the Leibniz Notation

If $y = f(u)$ and $u = g(x)$ are differentiable functions, then

$$\frac{dy}{dx} = \frac{dy}{du} \cdot \frac{du}{dx}.$$

Differentiation Formulas

Let u be a differentiable function of x.

1. $\dfrac{d}{dx}(u^r) = r\,u^{r-1}\dfrac{du}{dx}.$

2. $\dfrac{d}{dx}(\sin u) = \cos u\,\dfrac{du}{dx}.$ **5.** $\dfrac{d}{dx}(\csc u) = -\csc u \cot u\,\dfrac{du}{dx}.$

3. $\dfrac{d}{dx}(\cos u) = -\sin u\dfrac{du}{dx}.$ **6.** $\dfrac{d}{dx}(\sec u) = \sec u \tan u\,\dfrac{du}{dx}.$

4. $\dfrac{d}{dx}(\tan u) = \sec^2 u\,\dfrac{du}{dx}.$ **7.** $\dfrac{d}{dx}(\cot u) = -\csc^2 u\,\dfrac{du}{dx}.$

Differentials

Let f be a differentiable function and let $h \neq 0$. The difference $f(x + h) - f(x)$ is called the *increment of f from x to x + h*, and is denoted Δf:

$$\Delta f = f(x + h) - f(x).$$

The product $f'(x)h$ is called the *differential of f at x with increment h*, and is denoted df:

$$df = f'(x)h.$$

For small h,

$$\Delta f \cong df.$$

Newton-Raphson Method

$$x_{n+1} = x_n - \frac{f(x_n)}{f'(x_n)}.$$

■ THE MEAN-VALUE THEOREM AND APPLICATIONS

Rolle's Theorem

Let f be differentiable on the open interval (a, b) and continuous on the closed interval $[a, b]$. If $f(a)$ and $f(b)$ are both 0, then there is at least one number c in (a, b) for which

$$f'(c) = 0.$$

The Mean-Value Theorem

If f is differentiable on the open interval (a, b) and continuous on the closed interval $[a, b]$, then there is at least one number c in (a, b) for which

$$f'(c) = \frac{f(b) - f(a)}{b - a}$$

or, equivalently,

$$f(b) - f(a) = f'(c)(b - a).$$

Increasing and Decreasing Functions

A function f is said to:

1. *increase* on the interval I iff for every two numbers x_1, x_2 in I,

$$x_1 < x_2 \quad \text{implies} \quad f(x_1) < f(x_2);$$

2. *decrease* on the interval I iff for every two numbers x_1, x_2 in I,

$$x_1 < x_2 \quad \text{implies} \quad f(x_1) > f(x_2).$$

Theorem

Let f be continuous on an arbitrary interval I and differentiable on the interior of I.

1. If $f'(x) > 0$ for all x in the interior of I, then f increases on I.
2. If $f'(x) < 0$ for all x in the interior of I, then f decreases on I.
3. If $f'(x) = 0$ for all x in the interior of I, then f is constant on I.

Local Extreme Values

A function f is said to have a *local maximum at c* iff

$$f(c) \geq f(x) \quad \text{for all } x \text{ sufficiently close to } c.$$

A function f is said to have a *local minimum at c* iff

$$f(c) \leq f(x) \quad \text{for all } x \text{ sufficiently close to } c.$$

The local maxima and local minima of f are called the *local extreme values of f.*

Theorem

If f has a local maximum or a local minimum at c, then either

$$f'(c) = 0 \quad \text{or} \quad f'(c) \text{ does not exist.}$$

Critical Numbers

Given a function f. The numbers c in the domain of f for which either

$$f'(c) = 0 \quad \text{or} \quad f'(c) \text{ does not exist}$$

are called the *critical numbers of* f.

The First Derivative Test

Suppose that c is a critical number of f and that f is continuous at c. If there is a positive number δ such that:

1. $f'(x) > 0$ for all x in $(c - \delta, c)$ and $f'(x) < 0$ for all x in $(c, c + \delta)$, then $f(c)$ is a local maximum.
2. $f'(x) < 0$ for all x in $(c - \delta, c)$ and $f'(x) > 0$ for all x in $(c, c + \delta)$, then $f(c)$ is a local minimum.
3. If $f'(x)$ keeps constant sign on $(c - \delta, c) \cup (c, c + \delta)$, then $f(c)$ is not a local extreme value.

The Second Derivative Test

Suppose that $f'(c) = 0$ and that $f''(c)$ exists. If $f''(c) > 0$, then $f(c)$ is a local minimum value. If $f''(c) < 0$, then $f(c)$ is a local maximum value.

Endpoint Extreme Values

If c is an endpoint of the domain of f, then f is said to have an *endpoint maximum at* c iff

$$f(c) \geq f(x) \quad \text{for all } x \text{ in the domain of } f \text{ sufficiently close to } c.$$

It is said to have an *endpoint minimum at* c iff

$$f(c) \leq f(x) \quad \text{for all } x \text{ in the domain of } f \text{ sufficiently close to } c.$$

Absolute Extreme Values

A function f is said to have an *absolute maximum at* d iff

$$f(d) \geq f(x) \quad \text{for all } x \text{ in the domain of } f.$$

continues

A function f is said to have an *absolute minimum at d* iff

$$f(d) \le f(x) \quad \text{for all } x \text{ in the domain of } f.$$

These values are called the *absolute extreme values of f.*

To Determine the Absolute Extreme Values of f on (a, b)

1. Find the critical numbers $c_1, c_2, \ldots, c_n$ of f in the open in interval (a, b).
2. Calculate $f(a), f(c_1), f(c_2), \ldots, f(c_n), f(b)$.
3. The largest of the numbers found in step 2 is the absolute maximum of f and the smallest is the absolute minimum of f.

Concavity

Let the function f be differentiable on the open interval I. The graph of f is said to be *concave up* on I iff f' increases on I; it is said to be *concave down* on I iff f' decreases on I.

Point of Inflection

Let the function f be continuous at $x = c$. The point $(c, f(c))$ is called a *point of inflection* iff there exits $\delta > 0$ such that the graph of f is concave in one sense on $(c - \delta, c)$ and concave in the opposite sense on $(c, c + \delta)$.

Theorem

Let f be twice differentiable on an open interval I.

1. If $f''(x) > 0$ for all x in I, then f' increases on I and the graph of f is concave up.
2. If $f''(x) < 0$ for all x in I, then f' decreases on I and the graph of f is concave down.

Vertical Asymptotes

The line $x = c$ is a *vertical asymptote* for a function f if any one of the following conditions holds:

$$f(x) \to \infty \text{ or } -\infty \quad \text{as} \quad x \to c;$$

$$f(x) \to \infty \text{ or } -\infty \quad \text{as} \quad x \to c^-;$$

or

$$f(x) \to \infty \text{ or } -\infty \quad \text{as} \quad x \to c^+.$$

Horizontal Asymptotes

The line $y = L$ is a *horizontal asymptote* for a function f if either one of the following conditions holds:

$$f(x) \to L \text{ as } x \to \infty \quad \text{or} \quad f(x) \to L \text{ as } x \to -\infty.$$

Vertical Tangents and Vertical Cusps

The graph of f has a *vertical tangent* at the point $(c, f(c))$ iff

$$\text{as } x \to c, \quad f'(x) \to \infty, \quad \text{or} \quad f'(x) \to -\infty.$$

The graph of f has a *vertical cusp* at the point $(c, f(c))$ iff

$$\text{as } x \to c^-, \quad f'(x) \to -\infty \quad \text{and} \quad \text{as } x \to c^+, \quad f'(x) \to \infty$$

or

$$\text{as } x \to c^-, \quad f'(x) \to \infty \quad \text{and} \quad \text{as } x \to c^+, \quad f'(x) \to -\infty$$

■ THE TRANSCENDENTAL FUNCTIONS

One-to-One Functions

A function f with domain D is *one-to-one* iff no two distinct points in D have the same image under f, that is,

$$f(x_1) \neq f(x_2) \quad \text{whenever} \quad x_1 \neq x_2, \quad x_1, x_2 \in D.$$

Inverse Functions

Let f be a one-to-one function. The *inverse of f*, denoted f^{-1}, is the unique function that is defined on the range of f and satisfies the equation

$$f(f^{-1}(x)) = x \quad \text{for all } x \text{ in the range of } f.$$

Continuity

Let f be a one-to-one function defined on an interval I. If f is continuous, then its inverse f^{-1} is also continuous.

Differentiability

Suppose that f has an inverse and is differentiable. Let a be a point in the domain of f and let $b = f(a)$. If $f'(a) \neq 0$, then $(f^{-1})'(b)$ exists and is given by

$$(f^{-1})'(b) = \frac{1}{f'(a)}$$

The Natural Logarithm

The *natural logarithm function,* denoted $\ln x$, is defined as the unique function with domain $(0, \infty)$, which satisfies $\ln(0) = 1$ and

$$\frac{d}{dx}(\ln x) = \frac{1}{x}.$$

Properties of ln x

1. $\ln x$ is an increasing function since $\frac{d}{dx}(\ln x)\frac{1}{x} > 0$ on $(0, \infty)$.
2. $\ln x$ is continuous since it is differentiable.
3. $\ln x < 0$ if $0 < x < 1$, $\ln 1 = 0$, $\ln x > 0$ if $x > 1$.
4. The range of $\ln x$ is $(-\infty, \infty)$.
5. There is a unique number $e \cong 2.71828$ such that $\ln e = 1$.

Additional Properties of ln x

1. $\ln xy = \ln x + \ln y \quad (x > 0, y > 0)$.
2. $\ln(1/x) = -\ln x \quad (x > 0)$.
3. $\ln(x/y) = \ln x - \ln y \quad (x > 0, y > 0)$.
4. $\ln x^r = r \ln x \quad (x > 0)$.

The Exponential Function

The exponential function $E(x) = e^x$ is the inverse of the natural logarithm function, $\ln x$:

$$\ln(e^x) = x \quad \text{for all } x, \quad e^{\ln x} = x \quad \text{for all } x > 0.$$

Properties of the Exponential Function

1. $e^x > 0 \quad$ for all real x.
2. $\lim\limits_{x \to \infty} e^x = \infty, \quad \lim\limits_{x \to -\infty} e^x = 0$.
3. $e^{x+y} = e^x \cdot e^y \quad$ for all x and y.
4. $e^{-x} = \dfrac{1}{e^x} \quad$ for all x.
5. $e^{x-y} = \dfrac{e^x}{e^y} \quad$ for all x and y.

Theorem

The exponential function is its own derivative; for all real x

$$\frac{d}{dx}(e^x) = e^x.$$

Exponential Growth and Decay

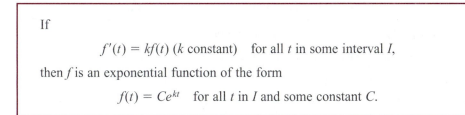

If

$$f'(t) = kf(t) \ (k \text{ constant}) \quad \text{for all } t \text{ in some interval } I,$$

then f is an exponential function of the form

$$f(t) = Ce^{kt} \quad \text{for all } t \text{ in } I \text{ and some constant } C.$$

Derivatives of the Inverse Trigonometric Functions

1. $\dfrac{d}{dx}(\sin^{-1}x) = \dfrac{1}{\sqrt{1-x^2}}.$

2. $\dfrac{d}{dx}(\tan^{-1}x) = \dfrac{1}{1+x^2}.$

3. $\dfrac{d}{dx}(\sec^{-1}x) = \dfrac{1}{|x|\sqrt{x^2-1}}.$

Hyperbolic Sine and Cosine

$$\sinh x = \tfrac{1}{2}(e^x - e^{-x}), \quad \cosh x = \tfrac{1}{2}(e^x + e^{-x}).$$

Derivatives of the Hyperbolic Functions

$$\frac{d}{dx}(\sinh x) = \cosh x, \quad \frac{d}{dx}(\cosh x) = \sinh x.$$

■ INTEGRATION

Upper Sums and Lower Sums

Let f be continuous on $[a, b]$ and let $P = \{a = x_0, x_1, x_2, \ldots, x_n = b\}$ be a partition of $[a, b]$. Let M_i and m_i be the maximum and minimum values of f on the subinterval $[x_{i-1}, x_i]$, $i = 1, 2, \ldots, n$, and let $\Delta x_i = x_i - x_{i-1}$. The number

$$U_f(P) = M_1\,\Delta x_1 + M_2\,\Delta x_2 + \cdots + M_n\,\Delta x_n$$

is called the P *upper sum for f*, and the number

$$L_f(P) = m_1\,\Delta x_1 + m_2\,\Delta x_2 + \cdots + m_n\,\Delta x_n$$

is called the P *lower sum for f.*

The Definite Integral

Let f be continuous on $[a, b]$. The unique number I that satisfies the inequality

$$L_f(P) \leq I \leq U_f(P) \quad \text{for all partitions } P \text{ of } [a, b]$$

is called the *definite integral* (or simply *the integral*) of f *from a to b* and is denoted by

$$\int_a^b f(x)\, dx.$$

Properties of the Definite Integral

Let f be continuous on each of the indicated intervals.

1. $\displaystyle\int_a^c f(x)\, dx + \int_c^b f(x)\, dx = \int_a^b f(x)\, dx.$

2. $\displaystyle\int_b^a f(x)\, dx = -\int_a^b f(x)\, dx.$

3. $\displaystyle\int_c^c f(x)\, dx = 0.$

Theorem

Let f be continuous on $[a, b]$. The function F defined on $[a, b]$ by

$$F(x) = \int_a^x f(t)\, dt$$

is continuous on $[a, b]$, differentiable on (a, b), and has derivative

$$F'(x) = f(x) \quad \text{for all } x \text{ in } (a, b).$$

Antiderivative

Let f be continuous on $[a, b]$. A function G is called an *antiderivative for f* on $[a, b]$ iff

$$G \text{ is continuous on } [a, b] \quad \text{and} \quad G'(x) = f(x) \text{ for all } x \in (a, b).$$

The Fundamental Theorem of Integral Calculus

Let f be continuous on $[a, b]$. If G is any antiderivative for f on $[a, b]$, then

$$\int_a^b f(t)\, dt = G(b) - G(a).$$

Linearity of the Definite Integral

Let f and g be continuous functions and let α and β be constants. Then

$$\int_a^b [\alpha f(x) + \beta g(x)]\, dx = \alpha \int_a^b f(x)\, dx + \beta \int_a^b g(x)\, dx.$$

Area Under a Graph

Let f be continuous and nonnegative on $[a, b]$ and let Ω be the region bounded by the graph of f and the x-axis between $x = a$ and $x = b$. Then

$$\text{area of } \Omega = \int_a^b f(x)\, dx.$$

Area Between Two Curves

Let f and g be continuous functions and suppose that $f(x) \geq g(x)$ on $[a, b]$. Let Ω be the region bounded by the graphs of f and g between $x = a$ and $x = b$. Then

$$\text{area of } \Omega = \int_a^b [f(x) - g(x)]\, dx.$$

The Natural Logarithm as an Integral

The natural logarithm function is given by the integral

$$\ln x = \int_1^x \frac{1}{t}\, dt, \quad x > 0.$$

Integrals of Powers and Trigonometric Functions

$$\int x^r \, dx = \frac{x^{r+1}}{r+1} + C \quad (r \neq -1).$$

$$\int \sin x \, dx = -\cos x + C. \qquad \int \cos x \, dx = \sin x + C.$$

$$\int \sec^2 x \, dx = \tan x + C. \qquad \int \csc^2 x \, dx = -\cot x + C.$$

$$\int \sec x \tan x \, dx = \sec x + C. \qquad \int \csc x \cot x \, dx = -\csc x + C.$$

Integrals Involving the Natural Logarithm and Exponential Functions

$$\int \frac{1}{x} \, dx = \ln |x| + C, \quad x \neq 0. \qquad \int e^x \, dx = e^x + C.$$

$$\int \tan x \, dx = \ln |\sec x| + C. \qquad \int \cot x \, dx = \ln |\sin x| + C.$$

$$\int \sec x \, dx = \ln |\sec x + \tan x| + C. \qquad \int \csc x \, dx = \ln |\csc x - \cot x| + C.$$

Integrals of the Inverse Trigonometric Functions

$$\text{For } a > 0, \quad \int \frac{1}{\sqrt{a^2 - x^2}} \, dx = \sin^{-1}\left(\frac{x}{a}\right) + C.$$

$$\text{For } a \neq 0, \quad \int \frac{1}{a^2 + x^2} \, dx = \frac{1}{a} \tan^{-1}\left(\frac{x}{a}\right) + C.$$

$$\text{For } a > 0, \quad \int \frac{1}{x\sqrt{x^2 - a^2}} \, dx = \frac{1}{a} \sec^{-1}\left(\frac{|x|}{a}\right) + C.$$

Linearity of Indefinite Integrals

$$\int [\alpha f(x) + \beta g(x)] \, dx = \alpha \int f(x) \, dx + \beta \int g(x) \, dx,$$

where α and β are constants.

u-Substitution

If F is an antiderivative of f, then

$$\int f'[g(x)]g'(x)\,dx = F[g(x)] + C.$$

Substitution in Definite Integrals

If g' is continuous on $[a, b]$ and f is continuous on the range of g, then

$$\int_a^b f'[g(x)]g'(x)\,dx = \int_{g(a)}^{g(b)} f(u)\,du.$$

Order Properties of the Definite Integral

Let f and g be continuous functions on $[a, b]$.

1. If $f(x) \geq 0$ for $x \in [a, b]$, then $\displaystyle\int_a^b f(x)\,dx \geq 0$.

2. If $f(x) > 0$ for $x \in [a, b]$, then $\displaystyle\int_a^b f(x)\,dx > 0$.

3. If $f(x) \leq g(x)$ for $x \in [a, b]$, then $\displaystyle\int_a^b f(x)\,dx \leq \int_a^b g(x)\,dx$.

4. If $f(x) < g(x)$ for $x \in [a, b]$, then $\displaystyle\int_a^b f(x)\,dx < \int_a^b g(x)\,dx$.

5. $\left| \displaystyle\int_a^b f(x)\,dx \right| \leq \int_a^b |f(x)|\,dx$.

6. If m is the minimum value of f on $[a, b]$ and M is the maximum value, then

$$m(b - a) \leq \int_a^b f(x)\,dx \leq M(b - a).$$

Symmetry

Let f be continuous on $[-a, a]$.

1. If f is an odd function, then $\displaystyle\int_{-a}^a f(x)\,dx = 0$.

2. If f is an even function, then $\displaystyle\int_{-a}^a f(x)\,dx = 2\int_0^a f(x)\,dx$.

Mean-Value Theorem for Integrals

If f is continuous on $[a, b]$, then there is at least one number c in (a, b) for which

$$\int_a^b f(x)\, dx = f(c)(b - a).$$

The number $f(c)$ is called the *average* (or *mean*) *value of* f on $[a, b]$.

Riemann Sums

Let f be continuous on $[a, b]$ and let $P = \{a = x_0, x_1, x_2, \ldots, x_n = b\}$ be a partition of $[a, b]$. Choose $x_i^* \in [x_{i-1}, x_i]$, $i = 1, 2, \ldots, n$, and let $\Delta x_i = x_i - x_{i-1}$. The sum

$$S^*(P) = f(x_1^*)\, \Delta x_1 + f(x_2^*)\, \Delta x_2 + \cdots + f(x_n^*)\, \Delta x_n$$

is called a *Riemann sum*. If P is any partition of the interval $[a, b]$ and $S^*(P)$ is any corresponding Riemann sum, then

$$L_f(P) \leq S^*(P) \leq U_f(P).$$

Theorem

Let f be continuous on $[a, b]$, let $P = \{a = x_0, x_1, x_2, \ldots, x_n = b\}$ be a partition of $[a, b]$ and let $\|P\| = \max \Delta x_i$. Then

$$\int_a^b f(x)\, dx = \lim_{\|P\| \to 0} [f(x_1^*)\, \Delta x_1 + f(x_2^*)\, \Delta x_2 + \cdots + f(x_n^*)\, \Delta x_n].$$

■ SOME APPLICATIONS OF THE INTEGRAL

Volume by Parallel Cross Sections

Suppose that a solid S lies between $x = a$ and $x = b$. If its cross-sectional area $A(x)$ varies continuously with x on $[a, b]$, then the volume of S is given by

$$V = \int_a^b A(x)\, dx.$$

Solids of Revolution

1. **Disc Method.** Let f be nonnegative and continuous on $[a, b]$, and let Ω be the region bounded by the graph of f and the x-axis. The volume of the solid generated by revolving Ω about the x-axis is given by

$$V = \int_a^b \pi [f(x)]^2 \, dx.$$

2. **Washer Method.** Let f and g be nonnegative continuous functions such that $g(x) \le f(x)$ on $[a, b]$, and let Ω be the region between the graphs of f and g. The volume of the solid generated by revolving Ω around the x-axis is given by

$$V = \int_a^b \pi \{[f(x)]^2 - [g(x)]^2\} \, dx.$$

3. **Shell Method.** Let f be a nonnegative continuous function on $[a, b]$ with $a \ge 0$, and let Ω be the region bounded by the graph of f and the x-axis. The volume of the solid generated by revolving Ω around the y-axis is given by

$$V = \int_a^b 2\pi x f(x) \, dx.$$

Centroid

Let f be a nonnegative continuous function on $[a, b]$ and let Ω be the region bounded by the graph and the x-axis. The coordinates $(\bar{x}, \bar{y})$ of the centroid of Ω are given by

$$\bar{x} A = \int_a^b x f(x) \, dx, \quad \bar{y} A = \int_a^b \tfrac{1}{2} [f(x)]^2 \, dx.$$

where A is the area of Ω.

Pappus's Theorem on Volumes

A plane region is revolved about an axis that lies in its plane. If the region does not cross the axis, then the volume of the resulting solid of revolution is the area of the region multiplied by the circumference of the circle described by the centroid of the region:

$$V = 2\pi \bar{R} A,$$

where A is the area of the region and $\bar{R}$ is the distance from the axis to the centroid of the region.

■ TECHNIQUES OF INTEGRATION

Integration By Parts

$$\int f(x)g'(x)\,dx = f(x)g(x) - \int g(x)f'(x)\,dx,$$

or,

$$\text{if } u = f(x), \quad \text{and} \quad dv = g'(x)\,dx,$$

then

$$du = f'(x)\,dx \quad \text{and} \quad v = g(x),$$

and

$$\int u\,dv = uv - \int v\,du.$$

Trapezoidal Rule

$$T_n = \frac{b-a}{2n}[f(x_0) + 2f(x_1) + 2f(x_2) + \cdots + 2f(x_{n-1}) + f(x_n)].$$

Error:

$$\int_a^b f(x)\,dx - T_n = E_n^T = -\frac{(b-a)^3}{12n^2}f''(c),$$

where c is some number between a and b.

Simpson's Rule

$$S_n = \frac{b-a}{6n}\left\{f(x_0) + f(x_n) + [2f(x_1) + 2f(x_2) + \cdots + 2f(x_{n-1})]\right.$$
$$\left. 4\left[f\left(\frac{x_0 + x_1}{2}\right) + \cdots + f\left(\frac{x_{n-1} + x_n}{2}\right)\right]\right\}.$$

Error:

$$\int_a^b f(x)\,dx - S_n = E_n^S = -\frac{(b-a)^5}{2880n^4}f^{(4)}(c),$$

where c is some number between a and b.

■ POLAR COORDINATES AND PARAMETRIC EQUATIONS

Polar Coordinates

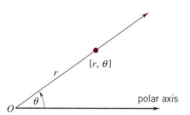

Polar Coordinates and Rectangular Coordinates

$$x = r \cos \theta, \quad y = r \sin \theta.$$

$$r^2 = x^2 + y^2, \quad \tan \theta = \frac{y}{x}.$$

Area in Polar Coordinates

If $r = \rho(\theta) \geq 0$ for $\alpha \leq \theta \leq \beta \leq \alpha + 2\pi$, then the area of the polar region Γ generated by ρ is given by

$$A = \int_{\alpha}^{\beta} \tfrac{1}{2} [\rho(\theta)]^2 \, d\theta.$$

If $\rho_2(\theta) \geq \rho_1(\theta) \geq 0$ for $\alpha \leq \theta \leq \beta \leq \alpha + 2\pi$, then the area of the polar region between the curves ρ_1 and ρ_2 is given by

$$A = \int_{\alpha}^{\beta} \tfrac{1}{2} \{[\rho_2(\theta)]^2 - [\rho_1(\theta)]^2\} \, d\theta.$$

Tangents to Curves Given Parametrically

Let C be a curve that is parametrized by the differentiable functions

$$x = x(t), \quad y = y(t),$$

where x and y are defined on some interval I, and

$$[x'(t)]^2 + [y'(t)]^2 \neq 0.$$

The slope of the tangent line to C at the point

$$(x_0, y_0) = (x(t_0), y(t_0))$$

continues

is given by

$$m = \frac{y'(t_0)}{x'(t_0)},$$

provided $x'(t_0) \neq 0$. If $x'(t_0) = 0$, then $y'(t_0) \neq 0$ and C has a vertical tangent line at (x_0, y_0).

An equation for the tangent line to C at the point (x_0, y_0) is

$$x'(t_0)(y - y_0) - y'(t_0)(x - x_0) = 0.$$

Arc Length

Let C be a curve that is parametrized by the continuously differentiable functions

$$x = x(t), \quad y = y(t), \quad t \in [a, b].$$

The *length of* C, denoted $L(C)$, is given by

$$L(C) = \int_a^b \sqrt{[x'(t)]^2 + [y'(t)]^2} \, dt.$$

If C is the graph of a continuously differentiable function

$$y = f(x), \quad x \in [a, b],$$

then

$$L(C) = \int_a^b \sqrt{1 + [f'(x)]^2} \, dx.$$

If C is the graph of a continuously differentiable polar function

$$r = \rho(\theta), \quad \theta \in [\alpha, \beta],$$

then

$$L(C) = \int_\alpha^\beta \sqrt{[\rho(\theta)]^2 + [\rho'(\theta)]^2} \, d\theta.$$

Surface Area

Let C be a curve in the upper half plane that does not intersect itself and that is parametrized by the continuously differentiable functions

$$x = x(t), \quad y = y(t), \quad t \in [c, d].$$

If C is revolved about the x-axis it generates a *surface of revolution.* The surface area of that surface is given by

$$A = \int_c^d 2\pi y(t)\sqrt{[x'(t)]^2 + [y'(t)]^2} \, dt.$$

Pappus's Theorem on Surface Area

A plane curve is revolved about an axis that lies in its plane. The curve may meet the axis but, if so, only at a finite number of points. If the curve does not cross the axis, then the area of the resulting surface of revolution is the length of the curve multiplied by the circumference of the circle described by the centroid of the curve:

$$A = 2\pi \overline{R} L,$$

where L is the length of the curve and $\overline{R}$ is the distance from the axis to the centroid of the curve.

SEQUENCES; INDETERMINATE FORMS; IMPROPER INTEGRALS

CHAPTER
10

■ 10.1 THE LEAST UPPER BOUND AXIOM

In the next few sections and in the next chapter, we will be talking about limits of sequences. To be able to do that with precision we first need to look a little deeper into the real number system.

We begin with a nonempty set S of real numbers. As indicated in Chapter 1, a number M is called an *upper bound* for S iff

$$x \leq M \qquad \text{for all } x \in S.$$

Note that if M is an upper bound for S, then any number greater than M will also be an upper bound for S. Thus, if a set has *an* upper bound, then it has, in fact, infinitely many upper bounds. Of course, not all sets of real numbers have upper bounds. Those that do are said to be *bounded above.*

For example, the number 3 is an upper bound for the set $\{x: x^2 \leq 1\} = \{x: -1 \leq x \leq 1\}$; so are $\frac{3}{2}$ and 1. This set is bounded above; 1 and every number greater than 1 are upper bounds. The set $\{2, 4, 6, 8, \ldots, 2n, \ldots\}$ is not bounded above.

It is clear that every set that has a largest element has an upper bound: if b is the largest element of S, then

$$x \leq b \qquad \text{for all } x \in S;$$

this makes b an upper bound for S. The converse is false: the sets

$$(-\infty, 0) \quad \text{and} \quad \left\{ \frac{1}{2}, \frac{2}{3}, \frac{3}{4}, \ldots, \frac{n}{n+1}, \ldots \right\}$$

both have upper bounds (2 for instance), but neither has a largest element.

Let's return to the first set, $(-\infty, 0)$. While $(-\infty, 0)$ does not have a largest element, the set of its upper bounds, $[0, \infty)$, does have a least element, 0. We call 0 the *least upper bound of* $(-\infty, 0)$.

637

Now let's reexamine the second set. While the set of quotients

$$\frac{n}{n+1} = 1 - \frac{1}{n+1}$$

does not have a greatest element, the set of its upper bounds, $[1, \infty)$, does have a least element, 1. We call 1 the *least upper bound* of that set of quotients.

In general, we have

DEFINITION 10.1.1   LEAST UPPER BOUND

Let S be a nonempty set of real numbers which is bounded above. A number M is the least upper bound of S iff

  (i) M is an upper bound for S,
  (ii) $M \leq K$, where K is any upper bound for S.

We are ready now to state explicitly one of the key *assumptions* that we make about the real number system. It is called the *least upper bound axiom*. Although we have not made an issue of it up to this point, this axiom has been an implicit assumption throughout the previous nine chapters; it underlies all of calculus.

AXIOM 10.1.2   THE LEAST UPPER BOUND AXIOM

Every nonempty set of real numbers that has an upper bound has a *least* upper bound.

Remark   It is easy to see that if a set of real numbers S is bounded above, then its least upper bound is unique. For if L and M are least upper bounds of S, then $M \leq L$ and $L \leq M$ by property (ii) of Definition 10.1.1, and so $L = M$. Thus, it makes sense to use the term "*the* least upper bound" of S. ❏

To indicate the least upper bound of a set S, we will write lub S. Here are some examples:

1. lub $(-\infty, 0) = 0,$ lub $(-\infty, 0] = 0.$
2. lub $(-4, -1) = -1,$ lub $(-4, -1] = -1.$
3. lub $\left\{ \dfrac{1}{2}, \dfrac{2}{3}, \dfrac{3}{4}, \cdots, \dfrac{n}{n+1}, \cdots \right\} = 1.$
4. lub $\left\{ -\dfrac{1}{2}, -\dfrac{1}{8}, -\dfrac{1}{27}, \cdots, -\dfrac{1}{n^3}, \cdots \right\} = 0.$
5. lub $\{x: x^2 < 3\} =$ lub $\{x: -\sqrt{3} < x < \sqrt{3}\} = \sqrt{3}.$ ❏

The least upper bound of a set has a special property that deserves particular attention. The idea is this: the fact that M is the least upper bound of the set S

does not tell us that M is in S (indeed, it need not be, as illustrated in the preceding examples), but it does tell us that we can approximate M as closely as we wish by elements of S.

THEOREM 10.1.3

If M is the least upper bound of the set S and ϵ is a positive number, then there is at least one number s in S such that

$$M - \epsilon < s \le M.$$

PROOF   Let $\epsilon > 0$.   Since M is an upper bound for S, the condition $s \le M$ is satisfied by all numbers s in S.   All we have to show therefore is that there is some number s in S such that

$$M - \epsilon < s.$$

Suppose on the contrary that there is no such number in S.   We then have

$$x \le M - \epsilon \qquad \text{for all } x \in S.$$

This makes $M - \epsilon$ an upper bound for S.   But this cannot be, for then $M - \epsilon$ is an upper bound for S that is *less* than M, and by assumption, M is the *least* upper bound.   ❑

The theorem we just proved is illustrated in Figure 10.1.1.   Take S as the set of points marked in the figure.   If $M = \text{lub } S$, then S has at least one element in every half-open interval of the form $(M - \epsilon, M]$.

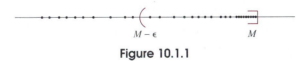

$M - \epsilon$ $\qquad\qquad$ M

Figure 10.1.1

Example 1

(a) Let

$$S = \left\{ \frac{1}{2}, \frac{2}{3}, \frac{3}{4}, \cdots, \frac{n}{n+1}, \cdots \right\}$$

and take $\epsilon = 0.0001$.   Since 1 is the least upper bound of S, there must be a number s in S such that

$$1 - 0.0001 < s \le 1.$$

There is: take, for example, $s = \frac{99999}{100000}$.

(b) Let

$$S = \{1, 2, 3\}$$

and take $\epsilon = 0.00001$. It is clear that 3 is the least upper bound of S. Therefore, there must be a number $s \in S$ such that

$$3 - 0.00001 < s \leq 3.$$

There is: $s = 3$. ❏

We come now to lower bounds. In the first place, a number m is called a *lower bound* for a nonempty set S iff

$$m \leq x \quad \text{for all } x \in S.$$

Sets that have lower bounds are said to be *bounded below.* Not all sets have lower bounds; those that do have *greatest lower bounds.* Paralleling the definition of least upper bound, if a nonempty set S is bounded below, then a number m is the greatest lower bound of S if (i) m is a lower bound, and (ii) $m \geq k$, where k is any lower bound for S.

The existence of a greatest lower bound of a set S which is bounded below does not need to be taken as an axiom. We can prove it as a theorem using the least upper bound axiom.

THEOREM 10.1.4

Every nonempty set of real numbers that has a lower bound has a *greatest* lower bound.

PROOF Suppose that S is nonempty and that it has a lower bound x. Then

$$x \leq s \quad \text{for all } s \in S.$$

It follows that $-s \leq -x$ for all $s \in S$; that is,

$$\{-s: s \in S\} \quad \text{has an upper bound } -x.$$

From the least upper bound axiom we conclude that $\{-s: s \in S\}$ has a least upper bound; call it x_0. Since $-s \leq x_0$ for all $s \in S$, we can see that

$$-x_0 \leq s \quad \text{for all } s \in S,$$

and thus $-x_0$ is a lower bound for S. We now assert that $-x_0$ is the greatest lower bound of the set S. To see this, note that, if there existed a number x_1 satisfying

$$-x_0 < x_1 \leq s \quad \text{for all } s \in S,$$

then we would have

$$-s \leq -x_1 < x_0 \quad \text{for all } s \in S,$$

and thus x_0 would not be the *least* upper bound of $\{-s: s \in S\}$.† ❏

---

† We proved Theorem 10.1.4 by assuming the least upper bound axiom. We could have proceeded the other way. We could have set Theorem 10.1.4 as an axiom, and then proved the least upper bound axiom as a theorem.

As in the case of the least upper bound, the greatest lower bound of a set is unique. Also, the greatest lower bound of a set need not be in the set, but can be approximated as closely as we wish by members of the set. In short, we have the following theorem, the proof of which is left as an exercise.

THEOREM 10.1.5

If m is the greatest lower bound of the set S and ϵ is a positive number, then there is at least one number s in S such that

$$m \leq s < m + \epsilon.$$

The theorem is illustrated in Figure 10.1.2. If $m = $ glb S (that is, if m is the greatest lower bound of the set S), then S has at least one element in every half-open interval of the form $[m, m + \epsilon)$.

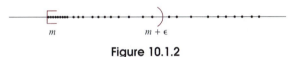

m $m + \epsilon$

Figure 10.1.2

Remark Remember the intermediate-value theorem? It states that a continuous function skips no values. Remember the maximum-minimum theorem? It states that on a bounded closed interval a continuous function takes on both a maximum and a minimum value. We have been using these two results right along, but we have not proved them. Now that you understand least upper bounds and greatest lower bounds, you are in a position to follow proofs of both of these theorems. (See Appendix B.) Better still, try to prove the theorems yourself. ❑

EXERCISES 10.1

In Exercises 1–20, find the least upper bound (if it exists) and the greatest lower bound (if it exists) for the given set.

1. $(0, 2)$.

2. $[0, 2]$.

3. $(0, \infty)$.

4. $(-\infty, 1)$.

5. $\{x: x^2 < 4\}$.

6. $\{x: |x - 1| < 2\}$.

7. $\{x: x^3 \geq 8\}$.

8. $\{x: x^4 \leq 16\}$.

9. $\{2\frac{1}{2}, 2\frac{1}{3}, 2\frac{1}{4}, \ldots\}$.

10. $\{-1, -\frac{1}{2}, -\frac{1}{3}, -\frac{1}{4}, \ldots\}$.

11. $\{0.9, 0.99, 0.999, \ldots\}$.

12. $\{-2, 2, -2.1, 2.1, -2.11, 2.11, \ldots\}$.

13. $\{x: \ln x < 1\}$.

14. $\{x: \ln x > 0\}$.

15. $\{x: x^2 + x - 1 < 0\}$.

16. $\{x: x^2 + x + 2 \geq 0\}$.

17. $\{x: x^2 > 4\}$.

18. $\{x: |x - 1| > 2\}$.

19. $\{x: \sin x \geq -1\}$.

20. $\{x: e^x < 1\}$.

In Exercises 21–24, illustrate the validity of Theorem 10.1.5 taking S and ϵ as given.

21. $S = \{\frac{1}{11}, (\frac{1}{11})^2, (\frac{1}{11})^3, \ldots, (\frac{1}{11})^n, \ldots\}$, $\epsilon = 0.001$.

22. $S = \{1, 2, 3, 4\}$, $\epsilon = 0.0001$.

23. $S = \{\frac{1}{10}, \frac{1}{1000}, \frac{1}{100000}, \ldots, (\frac{1}{10})^{2n-1}, \ldots\}$, $\epsilon = (\frac{1}{10})^k$ $(k \geq 1)$.

24. $S = \{\frac{1}{2}, \frac{1}{4}, \frac{1}{8}, \ldots, (\frac{1}{2})^n, \ldots\}$, $\epsilon = (\frac{1}{4})^k$ $(k \geq 1)$.

25. Prove Theorem 10.1.5 by imitating the proof of Theorem 10.1.3.

26. Let $S = \{a_1, a_2, a_3, \ldots, a_n\}$ be a nonempty, finite set of real numbers.
  (a) Prove that S is bounded.
  (b) Prove that lub S and glb S are elements of S.

27. Suppose that b is an upper bound for a set S of real numbers. Prove that if $b \in S$, then $b =$ lub S.

28. Let S be a bounded set of real numbers and suppose that lub $S =$ glb S. What can you conclude about S?

29. Suppose that S is a nonempty, bounded set of real numbers and that T is a nonempty subset of S.
  (a) Prove that T is bounded.
  (b) Prove that glb $S \leq$ glb $T \leq$ lub $T \leq$ lub S.

30. Let S and T be nonempty sets of real numbers such that $x \leq y$ for all $x \in S$ and all $y \in T$.
  (a) Prove that lub $S \leq y$ for all $y \in T$.
  (b) Prove that lub $S \leq$ glb T.

31. Let c be a positive number. Prove that the set $S = \{c, 2c, 3c, \ldots, nc, \ldots\}$ is not bounded above.

32. Prove that if a and b are any two positive numbers with $a < b$, then there exists a rational number r such that $a < r < b$. HINT: Show that there is a positive integer n such that $na > 1$ and $n(b - a) > 1$. Then show that

there is a positive integer m such that $m > na$ and $m - 1 \leq na$. Let $r = m/n$.

▶ **33.** Let S be the set of irrational numbers

$$S = \{\sqrt{2}, \sqrt{2\sqrt{2}}, \sqrt{2\sqrt{2\sqrt{2}}}, \ldots\}.$$

That is, $S = \{a_1, a_2, a_3, \ldots, a_n, \ldots\}$ where $a_1 = \sqrt{2}$, and for each positive integer n, $a_{n+1} = \sqrt{2a_n}$.
  (a) Calculate the numbers $a_1, a_2, a_3, \ldots, a_{10}$.
  (b) Use mathematical induction to prove that $a_n < 2$ for all n.
  (c) Is 2 the least upper bound of S?
  (d) Choose a positive number other than 2 and repeat this exercise. What can you conclude?

▶ **34.** Let S be the set of irrational numbers

$$S = \{\sqrt{2}, \sqrt{2 + \sqrt{2}}, \sqrt{2 + \sqrt{2 + \sqrt{2}}}, \ldots\}$$

That is, $a_1 = \sqrt{2}$, and for each positive integer n, $a_{n+1} = \sqrt{2 + a_n}$.
  (a) Calculate the numbers $a_1, a_2, a_3, \ldots, a_{10}$.
  (b) Use mathematical induction to show that $a_n < 2$ for all n.
  (c) Is 2 the least upper bound for S?

Choose a positive number other than 2 and repeat this exercise. What can you conclude?

■ 10.2 SEQUENCES OF REAL NUMBERS

So far in our study of calculus, our attention has been fixed on functions defined on an interval or on a union of intervals. Here we study functions defined on the set of positive integers.

DEFINITION 10.2.1 SEQUENCE OF REAL NUMBERS

A real-valued function defined on the set of positive integers is called a *sequence of real numbers*.

The functions defined on the set of positive integers by setting

$$a(n) = n^2, \quad b(n) = \frac{n}{n + 1}, \quad c(n) = \sqrt{\ln n}, \quad d(n) = \frac{e^n}{n}, \quad \text{for } n = 1, 2, 3, \ldots$$

are examples of sequences of real numbers.

The notions developed for functions carry over to sequences. For example, if the functions a and b are sequences, then the linear combination $\alpha a + \beta b$ (α and β are real numbers) and their product ab are also sequences:

$$(\alpha a + \beta b)(n) = \alpha a(n) + \beta b(n) \qquad \text{and} \qquad (ab)(n) = a(n) \cdot b(n).$$

If the sequence b does not take on the value 0, then the reciprocal $1/b$ is a sequence and so is the quotient a/b:

$$\frac{1}{b}(n) = \frac{1}{b(n)} \qquad \text{and} \qquad \frac{a}{b}(n) = \frac{a(n)}{b(n)}.$$

Let a be a sequence. The numbers $a(1), a(2), a(3), \ldots$ are called the *terms* of a. In particular, $a(n)$ is called the *nth term of a*. In discussing sequences it is conventional to use subscript notation a_n rather than functional notation $a(n)$, $n = 1, 2, 3, \ldots$, to denote the terms, and the sequence itself is often written

$$\{a_1, a_2, a_3, \ldots\}$$

or, even more simply,

$$\{a_n\}.$$

For example, the sequence of reciprocals defined by setting

$$a_n = 1/n \qquad \text{for all } n$$

can be written

$$\{1, \tfrac{1}{2}, \tfrac{1}{3}, \ldots\} \qquad \text{or} \qquad \{1/n\}.$$

The sequence defined by setting

$$a_n = 10^{1/n} \qquad \text{for all } n$$

can be written

$$\{10, 10^{1/2}, 10^{1/3}, \ldots\} \qquad \text{or} \qquad \{10^{1/n}\}.$$

In this notation, we can write

$$\alpha\{a_n\} + \beta\{b_n\} = \{\alpha a_n + \beta b_n\}, \qquad \{a_n\}\{b_n\} = \{a_n b_n\}$$

and, provided that none of the b_n are zero,

$$\frac{1}{\{b_n\}} = \{1/b_n\} \qquad \text{and} \qquad \frac{\{a_n\}}{\{b_n\}} = \{a_n/b_n\}.$$

The following assertions will serve to illustrate the notation further.

(1) The sequence $\{1/2^n\}$ multiplied by 5 is the sequence $\{5/2^n\}$:

$$5\{1/2^n\} = \{5/2^n\}.$$

We can also write

$$5\{\tfrac{1}{2}, \tfrac{1}{4}, \tfrac{1}{8}, \tfrac{1}{16}, \ldots\} = \{\tfrac{5}{2}, \tfrac{5}{4}, \tfrac{5}{8}, \tfrac{5}{16}, \ldots\}.$$

(2) The sequence $\{n\}$ plus the sequence $\{1/n\}$ is the sequence $\{n + 1/n\}$:

$$\{n\} + \{1/n\} = \{n + 1/n\}.$$

In expanded form

$$\{1, 2, 3, \ldots\} + \{1, \tfrac{1}{2}, \tfrac{1}{3}, \ldots\} = \{2, 2\tfrac{1}{2}, 3\tfrac{1}{3}, \ldots\}.$$

(3) The sequence $\{n\}$ times the sequence $\{\sqrt{n}\}$ is the sequence $\{n\sqrt{n}\}$:

$$\{n\}\{\sqrt{n}\} = \{n\sqrt{n}\} = \{n^{3/2}\};$$

the sequence $\{2^n\}$ divided by the sequence $\{n\}$ is the sequence $\{2^n/n\}$:

$$\frac{\{2^n\}}{\{n\}} = \{2^n/n\}. \quad \square$$

The concept of boundedness for functions also carries over to sequences. In particular, a sequence a is said to be *bounded above* if $\{a_1, a_2, a_3, \ldots\}$ is bounded above; a is *bounded below* if $\{a_1, a_2, a_3, \ldots\}$ is bounded below. The sequence a is *bounded* if it is bounded above *and* below.

By the least upper bound axiom (10.1.2), we know that if the sequence a is bounded above, then $\{a_1, a_2, a_3, \ldots\}$ has a least upper bound which is called the *least upper bound of the sequence*. Similarly, if a is bounded below, then it has a *greatest lower bound*. The following assertions illustrate these ideas.

(1) The sequence $\{1/n\}$ is bounded above and below:

$$0 \leq 1/n \leq 1 \qquad \text{for } n = 1, 2, 3, \ldots.$$

The number 0 is the greatest lower bound of the set $\{1, \frac{1}{2}, \frac{1}{3}, \ldots\}$ and 1 is the least upper bound.

(2) The sequence $\{2^n\}$ is bounded below. For example, 1 is a lower bound:

$$1 \leq 2^n \qquad \text{for } n = 1, 2, 3, \ldots.$$

The number 2 is the greatest lower bound for this sequence. It is not bounded above: there is no fixed number M that satisfies

$$2^n \leq M \qquad \text{for all } n.$$

(3) The sequence $a_n = (-1)^n 2^n$ can be written

$$\{-2, 4, -8, 16, -32, 64, \ldots\}.$$

It is unbounded below and unbounded above. $\quad \square$

DEFINITION 10.2.2

A sequence $\{a_n\}$ is said to be

(i) *increasing*   iff $a_n < a_{n+1}$ for each positive integer n,

(ii) *nondecreasing*   iff $a_n \leq a_{n+1}$ for each positive integer n,

(iii) *decreasing*   iff $a_n > a_{n+1}$ for each positive integer n,

(iv) *nonincreasing*   iff $a_n \geq a_{n+1}$ for each positive integer n.

If any of these four properties holds, the sequence is said to be *monotonic*.

Remark An increasing sequence is also nondecreasing. However, a nondecreasing sequence will not, in general, be increasing. The sequence $\{1, 1, 1, \ldots\}$ is nondecreasing; it is not increasing. A corresponding relationship holds for decreasing and nonincreasing sequences. $\quad \square$

The sequences

$$\{1, \tfrac{1}{2}, \tfrac{1}{3}, \ldots\}, \qquad \{2, 4, 8, 16, \ldots\}, \qquad \{2, 2, 4, 4, 8, 8, 16, 16, \ldots\}$$

are all monotonic, but the sequences

$$\{1, \tfrac{1}{2}, 1, \tfrac{1}{3}, 1, \tfrac{1}{4}, \ldots\} \qquad \text{and} \qquad \{-2, 4, -8, 16, -32, 64, \ldots\}$$

are not monotonic.

The following examples are less trivial.

Example 1 The sequence $\{a_n\}$ defined by

$$a_n = \frac{n}{n+1}$$

is increasing. It is bounded below by $\tfrac{1}{2}$ (the greatest lower bound) and above by 1 (the least upper bound).

PROOF Since

$$\frac{a_{n+1}}{a_n} = \frac{n+1}{n+2} \cdot \frac{n+1}{n} = \frac{n^2 + 2n + 1}{n^2 + 2n} > 1$$

we have $a_n < a_{n+1}$. This confirms that the sequence is increasing. Since the sequence can be written

$$\{\tfrac{1}{2}, \tfrac{2}{3}, \tfrac{3}{4}, \tfrac{4}{5}, \tfrac{5}{6}, \ldots\}$$

it is easy to see that $\tfrac{1}{2}$ is the greatest lower bound. To show that 1 is the least upper bound, note first that

$$a_n = \frac{n}{n+1} < 1 \qquad \text{for all } n.$$

Thus, the sequence is bounded above and 1 is an upper bound. Now choose any number $K < 1$ and let p be a positive integer such that

$$\frac{1}{p+1} < 1 - K.$$

Then

$$K < 1 - \frac{1}{p+1} = \frac{p}{p+1} = a_p.$$

Thus, no number less than 1 can be an upper bound for the sequence. This means that 1 is the least upper bound. ❑

Example 2 The sequence $\{a_n\}$ defined by

$$a_n = \frac{2^n}{n!} \dagger$$

is nonincreasing. Moreover $a_n > a_{n+1}$ for $n \geq 2$.

PROOF The first two terms are equal:

$$a_1 = \frac{2^1}{1!} = 2 = \frac{2^2}{2!} = a_2.$$

———————
$\dagger$ Recall that $n! = n(n-1)(n-2)\cdots 3 \cdot 2 \cdot 1$. See Section 1.2.

For $n \geq 2$ the sequence decreases:

$$\frac{a_{n+1}}{a_n} = \frac{2^{n+1}}{(n+1)!} \cdot \frac{n!}{2^n} = \frac{2}{n+1} < 1 \qquad \text{if } n \geq 2. \quad \square$$

Remark The sequence in Example 2 is not decreasing since $a_1 = a_2 = 2$, but for $n \geq 2$, the terms *are* decreasing. This suggests an extension of Definition 10.2.2. We will say that a sequence $\{a_n\}$ is *increasing for* $n \geq k$ if $a_n < a_{n+1}$ for all $n \geq k$. The terms *nondecreasing for* $n \geq k$, *decreasing for* $n \geq k$, and *nonincreasing for* $n \geq k$ are defined similarly. $\quad \square$

Example 3 If $c > 1$, the sequence $\{a_n\}$ defined by

$$a_n = c^n$$

increases without bound.

PROOF Choose a number $c > 1$. Then

$$\frac{a_{n+1}}{a_n} = \frac{c^{n+1}}{c^n} = c > 1.$$

This shows that the sequence increases. To show the unboundedness, we take an arbitrary positive number M and show that there exists a positive integer k for which

$$c^k \geq M.$$

A suitable k is one such that

$$k \geq \frac{\ln M}{\ln c},$$

for then

$$k \ln c \geq \ln M, \qquad \ln c^k \geq \ln M, \qquad \text{and thus} \qquad c^k \geq M. \quad \square$$

Since sequences are defined on the set of positive integers and not on an interval, they are not directly susceptible to the methods of calculus. Fortunately, we can sometimes circumvent this difficulty by dealing initially, not with the sequence itself, but with a function of a real variable x that agrees with the given sequence for all positive integers n.

Example 4 The sequence $\{a_n\}$ defined by

$$a_n = \frac{n}{e^n}$$

is decreasing. It is bounded above by $1/e$ and below by 0.

PROOF We will work with the function

$$f(x) = \frac{x}{e^x}. \qquad \text{(See Example 3, Section 7.4.)}$$

Note that $f(1) = 1/e = a_1, f(2) = 2/e^2 = a_2, f(3) = 3/e^3 = a_3$, and so on.

Differentiating f, we get

$$f'(x) = \frac{e^x - xe^x}{e^{2x}} = \frac{1-x}{e^x}.$$

Since $f'(x) < 0$ for $x > 1$, f is decreasing on $[1, \infty)$. Thus $f(1) > f(2) > f(3) > \cdots$, that is $a_1 > a_2 > a_3 > \cdots$, and $\{a_n\}$ is decreasing.

It now follows that $a_1 > a_n$ for all $n \geq 2$, and so $a_1 = 1/e$ is an upper bound for the sequence. In fact, it is the least upper bound. Since all the terms in the sequence are positive, 0 is a lower bound.

Figure 10.2.1 illustrates the graphical relationship between $f(x) = x/e^x$ and $a_n = n/e^n$. The graph of f indicates that $a_n \to 0$ as $n \to \infty$, and so 0 is, in fact, the greatest lower bound. ❏

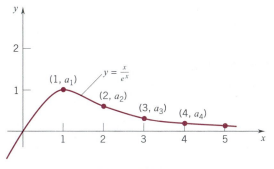

Figure 10.2.1

Example 5 The sequence $\{a_n\}$ defined by

$$a_n = n^{1/n}$$

decreases for $n \geq 3$.

PROOF We could compare a_n with a_{n+1} directly, but it is easier to consider the function

$$f(x) = x^{1/x}$$

instead. Since

$$f(x) = e^{(1/x)\ln x},$$

we have

$$f'(x) = e^{(1/x)\ln x} \frac{d}{dx}\left(\frac{1}{x}\ln x\right) = x^{1/x}\left(\frac{1 - \ln x}{x^2}\right).$$

For $x > e$, $f'(x) < 0$. This shows that f decreases on $[e, \infty)$. Since $3 > e$, f decreases on $[3, \infty)$. It follows that $\{a_n\}$ decreases for $n \geq 3$. ❏

Remark We must be careful when we examine a function of a real variable x in order to analyze the behavior of a sequence. The function $y = f(x)$ and the sequence $\{y_n\} = \{f(n)\}$ may behave differently. For example, the sequence $\{y_n\}$ defined by

$$y_n = f(n) = \frac{1}{n - 4.5}$$

is bounded (by 2 and -2) even though the function

$$f(x) = \frac{1}{x - 4.5}$$

is not bounded; the graph of f has a vertical asymptote at $x = 4.5$. See Figure 10.2.2. ❑

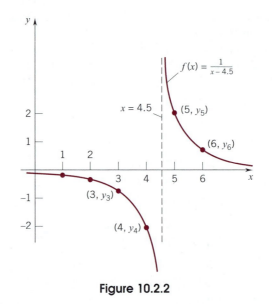

Figure 10.2.2

Clearly, the function $f(x) = \sin \pi x$ is not monotonic on $[0, \infty)$, but the sequence $\{u_n\}$ defined by $u_n = f(n) = \sin n\pi = 0$ for all n is monotonic. Indeed, it is constant. See Figure 10.2.3.

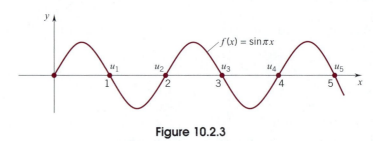

Figure 10.2.3

EXERCISES 10.2

In Exercises 1–8, the first several terms of a sequence $\{a_n\}$ are given. Assuming that the pattern continues as indicated, find an explicit formula for a_n.

1. 2, 5, 8, 11, 14, . . .

2. 2, 0, 2, 0, 2, . . .

3. $1, -\frac{1}{3}, \frac{1}{5}, -\frac{1}{7}, \frac{1}{9}, \dots$

4. $\frac{1}{2}, \frac{3}{4}, \frac{7}{8}, \frac{15}{16}, \frac{31}{32}, \dots$

5. $2, \frac{5}{2}, \frac{10}{3}, \frac{17}{4}, \frac{26}{5}, \dots$

6. $-\frac{1}{4}, \frac{2}{9}, -\frac{3}{16}, \frac{4}{25}, -\frac{5}{36}, \dots$

7. $1, \frac{1}{2}, 3, \frac{1}{4}, 5, \frac{1}{6}, \dots$

8. $1, 2, \frac{1}{9}, 4, \frac{1}{25}, 6, \frac{1}{49}, \dots$

In Exercises 9–44, determine the boundedness and monotonicity of the indicated sequence.

9. $\left\{ \dfrac{2}{n} \right\}$.

10. $\left\{ \dfrac{(-1)^n}{n} \right\}$.

11. $\{\sqrt{n}\}$.

12. $\{(1.001)^n\}$.

13. $\left\{ \dfrac{n + (-1)^n}{n} \right\}$.

14. $\left\{ \dfrac{n-1}{n} \right\}$.

15. $\{(0.9)^n\}$.

16. $\{\sqrt{n^2 + 1}\}$.

17. $\left\{\dfrac{n^2}{n+1}\right\}.$

18. $\left\{\dfrac{2^n}{4^n+1}\right\}.$

19. $\left\{\dfrac{4n}{\sqrt{4n^2+1}}\right\}.$

20. $\left\{\dfrac{n+1}{n^2}\right\}.$

21. $\left\{\dfrac{4^n}{2^n+100}\right\}.$

22. $\left\{\dfrac{n^2}{\sqrt{n^3+1}}\right\}.$

23. $\left\{\dfrac{10^{10}\sqrt{n}}{n+1}\right\}.$

24. $\left\{\dfrac{n^2+1}{3n+2}\right\}.$

25. $\left\{\ln\left(\dfrac{2n}{n+1}\right)\right\}.$

26. $\left\{\dfrac{n+2}{3^{10}\sqrt{n}}\right\}.$

27. $\left\{\dfrac{(n+1)^2}{n^2}\right\}.$

28. $\{(-1)^n\sqrt{n}\}.$

29. $\left\{\sqrt{4-\dfrac{1}{n}}\right\}.$

30. $\left\{\ln\left(\dfrac{n+1}{n}\right)\right\}.$

31. $\{(-1)^{2n+1}\sqrt{n}\}.$

32. $\left\{\dfrac{\sqrt{n+1}}{\sqrt{n}}\right\}.$

33. $\left\{\dfrac{2^n-1}{2^n}\right\}.$

34. $\left\{\dfrac{1}{2n}-\dfrac{1}{2n+3}\right\}.$

35. $\left\{\sin\dfrac{\pi}{n+1}\right\}.$

36. $\{(-\tfrac{1}{2})^n\}.$

37. $\{(1.2)^{-n}\}.$

38. $\left\{\dfrac{n+3}{\ln(n+3)}\right\}.$

39. $\left\{\dfrac{1}{n}-\dfrac{1}{n+1}\right\}.$

40. $\{\cos n\pi\}.$

41. $\left\{\dfrac{\ln(n+2)}{n+2}\right\}.$

42. $\left\{\dfrac{(-2)^n}{n^{10}}\right\}.$

43. $\left\{\dfrac{3^n}{(n+1)^2}\right\}.$

44. $\left\{\dfrac{1-(\tfrac{1}{2})^n}{(\tfrac{1}{2})^n}\right\}.$

45. Show that the sequence $\{5^n/n!\}$ decreases for $n \geq 5$. Is the sequence nonincreasing?

46. Let M be a positive integer. Show that $\{M^n/n!\}$ decreases for $n \geq M$.

47. Show that, if $0 < c < d$, then the sequence

$$a_n = (c^n + d^n)^{1/n}$$

is bounded and monotonic.

48. Show that linear combinations and products of bounded sequences are bounded.

A sequence $\{a_n\}$ is said to be defined *recursively* if, for some $k \geq 1$, the terms $a_1, a_2, \ldots, a_k$ are given and a_n is specified in terms of $a_1, a_2, \ldots, a_{n-1}$ for each $n \geq k$. The formula specifying a_n for $n \geq k$ in terms of some (or all) of its predecessors is called a *recurrence relation*. In Exercises 49–64,

write down the first six terms of the sequence and then give the general term.

49. $a_1 = 1;\quad a_{n+1} = \dfrac{1}{n+1}a_n.$

50. $a_1 = 1;\quad a_{n+1} = a_n + 3n(n+1) + 1.$

51. $a_1 = 1;\quad a_{n+1} = \tfrac{1}{2}(a_n + 1).$

52. $a_1 = 1;\quad a_{n+1} = \tfrac{1}{2}a_n + 1.$

53. $a_1 = 1;\quad a_{n+1} = a_n + 2.$

54. $a_1 = 1;\quad a_{n+1} = \dfrac{n}{n+1}a_n.$

55. $a_1 = 1;\quad a_{n+1} = 3a_n + 1.$

56. $a_1 = 1;\quad a_{n+1} = 4a_n + 3.$

57. $a_1 = 1;\quad a_{n+1} = a_n + 2n + 1.$

58. $a_1 = 1;\quad a_{n+1} = 2a_n + 1.$

59. $a_1 = 1;\quad a_{n+1} = a_n + \cdots + a_1.$

60. $a_1 = 3;\quad a_{n+1} = 4 - a_n.$

61. $a_1 = 2, a_2 = 1, a_3 = 2;\quad a_{n+1} = 6 - (a_n + a_{n-1} + a_{n-2}), n \geq 3.$

62. $a_1 = 1, a_2 = 2;\quad a_{n+1} = 2a_n - a_{n-1}, n \geq 2.$

63. $a_1 = 1, a_2 = 3;\quad a_{n+1} = 2a_n - a_{n-1}, n \geq 2.$

64. $a_1 = 1, a_2 = 3;\quad a_{n+1} = 3a_n - 2n - 1, n \geq 2.$

In Exercises 65–68, use mathematical induction to prove the following assertions for all $n \geq 1$.

65. If $a_1 = 1$ and $a_{n+1} = 2a_n + 1$, then $a_n = 2^n - 1$.

66. If $a_1 = 3$ and $a_{n+1} = a_n + 5$, then $a_n = 5n - 2$.

67. If $a_1 = 1$ and $a_{n+1} = \dfrac{n+1}{2n}a_n$, then $a_n = \dfrac{n}{2^{n-1}}$.

68. If $a_1 = 1$ and $a_{n+1} = a_n - \dfrac{1}{n(n+1)}$, then $a_n = \dfrac{1}{n}$.

69. Let r be a real number, $r \neq 0$. Define a sequence $\{S_n\}$ by

$$S_1 = 1$$
$$S_2 = 1 + r$$
$$S_3 = 1 + r + r^2$$
$$\cdot$$
$$\cdot$$
$$\cdot$$
$$S_n = 1 + r + r^2 + \cdots + r^{n-1}$$
$$\cdot$$
$$\cdot$$
$$\cdot$$

(a) Suppose $r = 1$. What is S_n for $n = 1, 2, 3, \ldots$?

(b) Suppose $r \neq 1$. Find a formula for S_n that does not involve adding up the powers of r. HINT: Calculate $S_n - rS_n$.

70. Let $a_n = \dfrac{1}{n(n+1)}$, $n = 1, 2, 3, \ldots$, and let $\{S_n\}$ be the sequence defined by

$$S_1 = a_1$$
$$S_2 = a_1 + a_2$$
$$S_3 = a_1 + a_2 + a_3$$
$$\cdot$$
$$\cdot$$
$$\cdot$$
$$S_n = a_1 + a_2 + a_3 + \cdots + a_n$$
$$\cdot$$
$$\cdot$$
$$\cdot$$

Find a formula for S_n, $n = 1, 2, 3, \ldots$, that does not involve adding up the terms $a_1, a_2, a_3, \ldots$. HINT: Use partial fractions to write $1/k(k+1)$ as the sum of two fractions.

71. A ball is dropped from a height of 100 feet. Each time it hits the ground it rebounds to 75% of its previous height.
  (a) Let S_n be the distance that the ball travels between the nth and $(n+1)$st bounce, $n = 1, 2, 3, \ldots$. Find a formula for S_n.

(b) Let T_n be the time that the ball is in the air between the nth and $(n+1)$st bounce, $n = 1, 2, 3, \ldots$. Find a formula for T_n.

72. Suppose that the number of bacteria in a culture is growing exponentially (see Section 7.6) and that the number doubles every 12 hours. Find a formula for the number P_n of bacteria in the culture after n hours, given that there are 500 bacteria initially.

▶ 73. Let $\{a_n\}$ be the sequence defined recursively by

$$a_1 = 1; \qquad a_n = 1 + \sqrt{a_{n-1}}, \quad n = 2, 3, 4, \ldots$$

Use mathematical induction to show that:
  (a) $\{a_n\}$ is an increasing sequence.
  (b) $\{a_n\}$ is bounded above.
  (c) Calculate $a_2, a_3, a_4, \ldots , a_{15}$. Can you estimate the least upper bound for $\{a_n\}$?

▶ 74. Let $\{a_n\}$ be the sequence defined recursively by

$$a_1 = 1; \qquad a_n = \sqrt{3a_{n-1}}, \quad n = 2, 3, 4, \ldots$$

Use mathematical induction to show that:
  (a) $\{a_n\}$ is an increasing sequence.
  (b) $\{a_n\}$ is bounded above.
  (c) Calculate $a_1, a_2, a_3, \ldots , a_{15}$. Can you estimate the least upper bound for $\{a_n\}$?

■ 10.3 LIMIT OF A SEQUENCE

The meaning of

$$\lim_{x \to c} f(x) = L$$

is that we can make $f(x)$ as close as we wish to the number L simply by requiring that x be sufficiently close to c. The meaning of

$$\lim_{n \to \infty} a_n = L$$

(read "the limit of a_n as n tends to infinity is L") is that we can make a_n as close as we wish to the number L simply by requiring that n be sufficiently large. For another analogy, recall Section 4.7 where we considered limits of the form

$$\lim_{x \to \infty} f(x)$$

in connection with the problem of finding the horizontal asymptotes for the graph of a function f; if $f(x) \to L$ as $x \to \infty$, then the line $y = L$ is a horizontal asymptote of the graph. The statement

$$\lim_{n \to \infty} a_n = L$$

is the same as saying $a_n \to L$ as $n \to \infty$.

DEFINITION 10.3.1 LIMIT

$$\lim_{n \to \infty} a_n = L$$

iff for each $\epsilon > 0$, there exists a positive integer K such that

$$\text{if} \quad n \geq K, \quad \text{then} \quad |a_n - L| < \epsilon.$$

Example 1

$$\lim_{n \to \infty} \frac{1}{n} = 0.$$

PROOF Let $\epsilon > 0$ and choose an integer $K > 1/\epsilon$. Then $1/K < \epsilon$. Now, for any $n \geq K$, we have

$$0 < \frac{1}{n} \leq \frac{1}{K},$$

and so

$$\left| \frac{1}{n} - 0 \right| = \left| \frac{1}{n} \right| = \frac{1}{n} \leq \frac{1}{K} < \epsilon. \quad \square$$

Example 2

$$\lim_{n \to \infty} \frac{2n - 1}{n} = 2.$$

PROOF Let $\epsilon > 0$. We must show that there exists an integer K such that

$$\left| \frac{2n - 1}{n} - 2 \right| < \epsilon \quad \text{for all } n \geq K.$$

Since

$$\left| \frac{2n - 1}{n} - 2 \right| = \left| \frac{2n - 1 - 2n}{n} \right| = \left| -\frac{1}{n} \right| = \frac{1}{n},$$

again we need only choose $K > 1/\epsilon$. $\quad \square$

The next example justifies the familiar statement

$$\tfrac{1}{3} = 0.333 \ldots.$$

Example 3 The decimal fractions

$$a_n = 0.\overset{n}{\overbrace{33 \ldots 3}}, \quad n = 1, 2, 3, \ldots$$

tend to $\frac{1}{3}$ as a limit:

$$\lim_{n \to \infty} a_n = \tfrac{1}{3}.$$

PROOF Let $\epsilon > 0$. In the first place

$$(1) \qquad \left| a_n - \frac{1}{3} \right| = \left| 0.\overbrace{33 \ldots 3}^{n} - \frac{1}{3} \right| = \left| \frac{0.\overbrace{99 \cdots 9}^{n} - 1}{3} \right| = \frac{1}{3} \cdot \frac{1}{10^n} < \frac{1}{10^n}.$$

Now choose K so that $1/10^K < \epsilon$. If $n \geq K$, then by (1)

$$\left| a_n - \frac{1}{3} \right| < \frac{1}{10^n} \leq \frac{1}{10^K} < \epsilon. \qquad \square$$

Example 4 The limit

$$\lim_{n \to \infty} \frac{n}{\sqrt{n+1}}$$

does not exist.

PROOF Note that

$$\frac{n}{\sqrt{n+1}} = \frac{1}{\dfrac{1}{n}\sqrt{n+1}} = \frac{1}{\sqrt{\dfrac{1}{n} + \dfrac{1}{n^2}}}.$$

Since $1/n \to 0$ and $1/n^2 \to 0$ as $n \to \infty$, the limit does not exist. $\square$

Remark Following the ideas in Section 4.7, we will use the notation $a_n \to \infty$ as $n \to \infty$ to mean that corresponding to any positive number M there is a positive integer K such that $a_n > M$ for all $n \geq K$. Thus, in Example 4, we have

$$\frac{n}{\sqrt{n+1}} \to \infty \quad \text{as } n \to \infty.$$

The notation $a_n \to -\infty$ means that for any negative number Q there is a positive integer K such that $a_n < Q$ for all $n \geq K$. $\square$

Limit Theorems

The limit process for sequences is so similar to the limit process you have already studied that you may find you can prove many of the limit theorems yourself. In any case, try to come up with your own proofs and refer to these only if necessary.

THEOREM 10.3.2 UNIQUENESS OF LIMIT

If $\lim_{n \to \infty} a_n = L$ and $\lim_{n \to \infty} a_n = M$, then $L = M$.

A proof, similar to the proof of Theorem 2.3.1, is given in the supplement at the end of this section.

DEFINITION 10.3.3

A sequence that has a limit is said to be *convergent*. A sequence that has no limit is said to be *divergent*.

Instead of writing

$$\lim_{n \to \infty} a_n = L,$$

we will often write

$$a_n \to L \qquad \text{(read "} a_n \text{ converges to } L\text{")}$$

or more fully

$$a_n \to L \qquad \text{as} \quad n \to \infty.$$

THEOREM 10.3.4

Every convergent sequence is bounded.

PROOF Assume that $a_n \to L$ and choose any positive number: 1, for instance. Using 1 as ϵ, you can see that there must exist a positive integer K such that

$$|a_n - L| < 1 \qquad \text{for all } n \geq K.$$

This means that

$$|a_n| < 1 + |L| \qquad \text{for all } n \geq K$$

and, consequently,

$$|a_n| \leq \max\{|a_1|, |a_2|, \ldots, |a_{K-1}|, 1 + |L|\} \qquad \text{for all } n.$$

This proves that $\{a_n\}$ is bounded. ❏

Since every convergent sequence is bounded, a sequence that is not bounded cannot be convergent; namely,

(10.3.5)

> every unbounded sequence is divergent.

The sequences

$$a_n = \tfrac{1}{2}n, \qquad b_n = \frac{n^2}{n+1}, \qquad c_n = n \ln n$$

are all unbounded. Each of these sequences is therefore divergent.

Boundedness does not imply convergence. As a counterexample, consider the oscillating sequence

$$\{1, 0, 1, 0, \ldots\} = \left\{ \frac{1 + (-1)^{n+1}}{2} \right\}.$$

This sequence is certainly bounded (above by 1 and below by 0), but obviously it does not converge: the limit would have to be arbitrarily close to both 0 and 1 simultaneously.

Boundedness together with monotonicity does imply convergence.

THEOREM 10.3.6

A bounded nondecreasing sequence converges to its least upper bound; a bounded nonincreasing sequence converges to its greatest lower bound.

PROOF Suppose that $\{a_n\}$ is bounded and nondecreasing. If L is the least upper bound of this sequence, then

$$a_n \leq L \qquad \text{for all } n.$$

Now let ϵ be an arbitrary positive number. By Theorem 10.1.3 there exists a_k such that

$$L - \epsilon < a_k.$$

Since the sequence is nondecreasing,

$$a_k \leq a_n \qquad \text{for all } n \geq k.$$

It follows that

$$L - \epsilon < a_n \leq L \qquad \text{for all } n \geq k.$$

This shows that

$$|a_n - L| < \epsilon \qquad \text{for all } n \geq k$$

and proves that

$$a_n \to L.$$

The nonincreasing case can be handled in a similar manner. ❑

Example 5 Take the sequence

$$\{(3^n + 4^n)^{1/n}\}.$$

Since

$$3 = (3^n)^{1/n} < (3^n + 4^n)^{1/n} < (2 \cdot 4^n)^{1/n} = 2^{1/n} \cdot 4 \leq 8,$$

the sequence is bounded. Note that

$$(3^n + 4^n)^{(n+1)/n} = (3^n + 4^n)^{1/n}(3^n + 4^n)$$
$$= (3^n + 4^n)^{1/n}3^n + (3^n + 4^n)^{1/n}4^n.$$

Since

$$(3^n + 4^n)^{1/n} > (3^n)^{1/n} = 3 \qquad \text{and} \qquad (3^n + 4^n)^{1/n} > (4^n)^{1/n} = 4,$$

it follows that

$$(3^n + 4^n)^{(n+1)/n} > 3 \cdot (3^n) + 4 \cdot (4^n) = 3^{n+1} + 4^{n+1}.$$

Taking the $(n + 1)$st root of the left and right sides of this inequality, we have

$$(3^n + 4^n)^{1/n} > (3^{n+1} + 4^{n+1})^{1/(n+1)}.$$

The sequence is decreasing. Being bounded, it must be convergent. (Later you will be asked to show that the limit is 4.) ❑

THEOREM 10.3.7

Let α be a real number. If $a_n \to L$ and $b_n \to M$, then

(i) $a_n + b_n \to L + M$, (ii) $\alpha a_n \to \alpha L$, (iii) $a_n b_n \to LM$.

If, in addition, $M \neq 0$ and $b_n \neq 0$ for all n, then

$$(iv)\ \frac{1}{b_n} \to \frac{1}{M} \qquad \text{and} \qquad (v)\ \frac{a_n}{b_n} \to \frac{L}{M}.$$

Proofs of parts (i) and (ii) are left as exercises. For proofs of parts (iii)–(v), see the supplement at the end of this section.

We are now in a position to handle any rational sequence $\{a_n\}$, where

(2) $$a_n = \frac{\alpha_k n^k + \alpha_{k-1} n^{k-1} + \cdots + \alpha_0}{\beta_j n^j + \beta_{j-1} n^{j-1} + \cdots + \beta_0}, \quad \alpha_k \neq 0, \beta_j \neq 0.$$

To determine the behavior of such a sequence we need only divide both numerator and denominator by the highest power of n that occurs.

Example 6

$$\frac{3n^4 - 2n^2 + 1}{n^5 - 3n^3} = \frac{3/n - 2/n^3 + 1/n^5}{1 - 3/n^2} \to \frac{0}{1} = 0. \quad ❑$$

Example 7

$$\frac{1 - 4n^7}{n^7 + 12n} = \frac{1/n^7 - 4}{1 + 12/n^6} \to \frac{-4}{1} = -4. \quad ❑$$

Example 8

$$\frac{n^4 - 3n^2 + n + 2}{n^3 + 7n} = \frac{1 - 3/n^2 + 1/n^3 + 2/n^4}{1/n + 7/n^3}.$$

Since the numerator tends to 1 and the denominator tends to 0, the sequence is unbounded. Therefore it cannot converge. ❑

In general, if a_n is given by (2), then,

$$\text{as} \quad n \to \infty, \quad a_n \to \begin{cases} 0 & \text{if } k < j \\ \alpha_k/\beta_k & \text{if } k = j \\ \pm\infty & \text{if } k > j. \end{cases}$$

(See Exercise 40.)   The corresponding result for the limit of a rational function $R(x) = P(x)/Q(x)$ as $x \to \infty$ was discussed in Section 4.7.

THEOREM 10.3.8

$$a_n \to L \quad \text{iff} \quad a_n - L \to 0 \quad \text{iff} \quad |a_n - L| \to 0.$$

We leave the proof to you.

THEOREM 10.3.9   THE PINCHING THEOREM FOR SEQUENCES

Suppose that there is a positive integer K such that for all $n \geq K$

$$a_n \leq b_n \leq c_n.$$

If $a_n \to L$ and $c_n \to L$, then $b_n \to L$.

Once again the proof is left to you.

As an immediate and obvious consequence of the pinching theorem we have the following corollary.

(10.3.10)

Suppose that there is a positive integer K such that for all $n \geq K$

$$|b_n| \leq c_n.$$

If $c_n \to 0$, then $|b_n| \to 0$.

Example 9

$$\frac{\cos n}{n} \to 0 \quad \text{since} \quad \left| \frac{\cos n}{n} \right| \leq \frac{1}{n} \quad \text{and} \quad \frac{1}{n} \to 0. \quad \square$$

Example 10

$$\sqrt{4 + \left(\frac{1}{n}\right)^2} \to 2$$

since

$$2 \leq \sqrt{4 + \left(\frac{1}{n}\right)^2} \leq \sqrt{4 + 4\left(\frac{1}{n}\right) + \left(\frac{1}{n}\right)^2} = 2 + \frac{1}{n} \quad \text{and} \quad 2 + \frac{1}{n} \to 2. \quad \square$$

Example 11 Recall from Chapter 7 that there are several ways of defining the number e. Here we complete the argument begun in Theorem 7.5.13.

(10.3.11)

$$\lim_{n\to\infty} \left(1 + \frac{1}{n}\right)^n = e.$$

PROOF You have already seen that, for all positive integers n,

$$\left(1 + \frac{1}{n}\right)^n \le e \le \left(1 + \frac{1}{n}\right)^{n+1}. \qquad \text{(Theorem 7.5.13)}$$

Dividing the right-hand inequality by $1 + 1/n$, we have

$$\frac{e}{1 + 1/n} \le \left(1 + \frac{1}{n}\right)^n.$$

Combining this with the left-hand inequality, we can write

$$\frac{e}{1 + 1/n} \le \left(1 + \frac{1}{n}\right)^n \le e.$$

Since

$$\frac{e}{1 + 1/n} \to \frac{e}{1} = e,$$

we can conclude from the pinching theorem that

$$\left(1 + \frac{1}{n}\right)^n \to e. \quad \square$$

The sequences

$$\left\{\cos\frac{\pi}{n}\right\}, \qquad \left\{\ln\left(\frac{n}{n+1}\right)\right\}, \qquad \{e^{1/n}\}, \qquad \left\{\tan\left(\sqrt{\frac{\pi^2 n^2 - 8}{16 n^2}}\right)\right\}$$

are all of the form $\{f(c_n)\}$ with f a continuous function. Such sequences are frequently easy to deal with. The basic idea is this: When a continuous function is applied to a convergent sequence, the result is itself a convergent sequence. More precisely, we have the following theorem.

THEOREM 10.3.12

Suppose that

$$c_n \to c$$

and that, for each n, c_n is in the domain of f. If f is continuous at c, then

$$f(c_n) \to f(c).$$

PROOF We assume that f is continuous at c and take $\epsilon > 0$. From the continuity of f at c we know that there exists $\delta > 0$ such that

$$\text{if} \quad |x - c| < \delta, \qquad \text{then} \quad |f(x) - f(c)| < \epsilon.$$

Since $c_n \to c$, we know that there exists a positive integer K such that

$$\text{if} \quad n \geq K, \qquad \text{then} \quad |c_n - c| < \delta.$$

It follows therefore that

$$\text{if} \quad n \geq K, \qquad \text{then} \quad |f(c_n) - f(c)| < \epsilon. \quad \square$$

The following examples illustrate the use of Theorem 10.3.12 in calculating limits.

Example 12 Since $\pi/n \to 0$ and the cosine function is continuous at 0,

$$\cos\left(\frac{\pi}{n}\right) \to \cos 0 = 1. \quad \square$$

Example 13 Since

$$\frac{n}{n + 1} = \frac{1}{1 + 1/n} \to 1$$

and the logarithm function is continuous at 1,

$$\ln\left(\frac{n}{n + 1}\right) \to \ln 1 = 0. \quad \square$$

Example 14 Since $1/n \to 0$ and the exponential function is continuous at 0,

$$e^{1/n} \to e^0 = 1. \quad \square$$

Example 15 Since

$$\frac{\pi^2 n^2 - 8}{16 n^2} = \frac{\pi^2 - 8/n^2}{16} \to \frac{\pi^2}{16}$$

and the function $f(x) = \tan \sqrt{x}$ is continuous at $\pi^2/16$,

$$\tan\left(\sqrt{\frac{\pi^2 n^2 - 8}{16 n^2}}\right) \to \tan\left(\sqrt{\frac{\pi^2}{16}}\right) = \tan\frac{\pi}{4} = 1. \quad \square$$

Example 16 Since

$$\frac{2n + 1}{n} + \left(5 - \frac{1}{n^2}\right) \to 7$$

and the square-root function is continuous at 7,

$$\sqrt{\frac{2n + 1}{n} + \left(5 - \frac{1}{n^2}\right)} \to \sqrt{7}. \quad \square$$

Example 17 Since the absolute-value function is everywhere continuous,

$$a_n \to L \quad \text{implies} \quad |a_n| \to |L|. \quad \square$$

Remark For some time now we have asked you to take on faith two fundamentals of integration: that continuous functions do have definite integrals and that these integrals can be expressed as limits of Riemann sums. We could not give you proofs of these assertions because we did not have the necessary tools. Now we do. Proofs are given in Appendix B. $\square$

EXERCISES 10.3

In Exercises 1–36, state whether or not the sequence converges and, if it does, find the limit.

1. $\{2^n\}$.

2. $\left\{\dfrac{2}{n}\right\}$.

3. $\left\{\dfrac{(-1)^n}{n}\right\}$.

4. $\{\sqrt{n}\}$.

5. $\left\{\dfrac{n-1}{n}\right\}$.

6. $\left\{\dfrac{n+(-1)^n}{n}\right\}$.

7. $\left\{\dfrac{n+1}{n^2}\right\}$.

8. $\left\{\sin\dfrac{\pi}{2n}\right\}$.

9. $\left\{\dfrac{2^n}{4^n+1}\right\}$.

10. $\left\{\dfrac{n^2}{n+1}\right\}$.

11. $\{(-1)^n\sqrt{n}\}$.

12. $\left\{\dfrac{4n}{\sqrt{n^2+1}}\right\}$.

13. $\{(-\tfrac{1}{2})^n\}$.

14. $\left\{\dfrac{4^n}{2^n+10^6}\right\}$.

15. $\left\{\tan\dfrac{n\pi}{4n+1}\right\}$.

16. $\left\{\dfrac{10^{10}\sqrt{n}}{n+1}\right\}$.

17. $\left\{\dfrac{(2n+1)^2}{(3n-1)^2}\right\}$.

18. $\left\{\ln\left(\dfrac{2n}{n+1}\right)\right\}$.

19. $\left\{\dfrac{n^2}{\sqrt{2n^4+1}}\right\}$.

20. $\left\{\dfrac{n^4-1}{n^4+n-6}\right\}$.

21. $\{\cos n\pi\}$.

22. $\left\{\dfrac{n^5}{17n^4+12}\right\}$.

23. $\{e^{1/\sqrt{n}}\}$.

24. $\left\{\sqrt{4-\dfrac{1}{n}}\right\}$.

25. $\{(0.9)^{-n}\}$.

26. $\left\{\dfrac{2^n-1}{2^n}\right\}$.

27. $\{\ln n - \ln(n+1)\}$.

28. $\left\{\dfrac{1}{n}-\dfrac{1}{n+1}\right\}$.

29. $\left\{\dfrac{\sqrt{n+1}}{2\sqrt{n}}\right\}$.

30. $\{(0.9)^n\}$.

31. $\left\{\left(1+\dfrac{1}{n}\right)^{2n}\right\}$.

32. $\left\{\left(1+\dfrac{1}{n}\right)^{n/2}\right\}$.

33. $\left\{\dfrac{2^n}{n^2}\right\}$.

34. $\left\{\dfrac{(n+1)\cos\sqrt{n}}{n(1+\sqrt{n})}\right\}$.

35. $\left\{\dfrac{\sqrt{n}\sin(e^n\pi)}{n+1}\right\}$.

36. $\{2\ln 3n - \ln(n^2+1)\}$.

37. Prove that, if $a_n \to L$ and $b_n \to M$, then $a_n + b_n \to L + M$.

38. Let α be a real number. Prove that, if $a_n \to L$, then $\alpha a_n \to \alpha L$.

39. Prove that

$$\left(1+\dfrac{1}{n}\right)^{n+1} \to e \quad \text{given that} \quad \left(1+\dfrac{1}{n}\right)^n \to e.$$

40. Determine the convergence or divergence of a rational sequence

$$a_n = \dfrac{\alpha_k n^k + \alpha_{k-1}n^{k-1} + \cdots + \alpha_0}{\beta_j n^j + \beta_{j-1}n^{j-1} + \cdots + \beta_0}$$

$$\text{with } \alpha_k \neq 0, \beta_j \neq 0,$$

given that: (a) $k = j$; (b) $k < j$; (c) $k > j$. Justify your answers.

41. Prove that a bounded nonincreasing sequence converges to its greatest lower bound.

42. Let $\{a_n\}$ be a sequence of real numbers. Let $\{e_n\}$ be the sequence of even terms:

$$e_n = a_{2n}$$

and let $\{o_n\}$ be the sequence of odd terms:

$$o_n = a_{2n-1}.$$

Show that

$$a_n \to L \quad \text{iff} \quad e_n \to L \quad \text{and} \quad o_n \to L.$$

43. Prove the pinching theorem for sequences.

44. Let $\{a_n\}$ and $\{b_n\}$ be sequences such that $a_n \to 0$ and $\{b_n\}$ is bounded. Prove that $a_n b_n \to 0$.

45. Let $\{a_n\}$ be a convergent sequence with limit L. Prove that if $a_n \le M$ for all n, then $L \le M$.

46. According to Example 17, if $a_n \to L$, then $|a_n| \to |L|$. Is the converse true? That is, if $|a_n| \to |L|$, does it follow that $a_n \to L$? Prove or give a counter-example.

47. Let f be a continuous function on $(-\infty, \infty)$ and let r be a real number. Define the sequence $\{a_n\}$ as follows:

$$a_1 = r, \ a_2 = f(r), \ a_3 = f[f(r)],$$
$$a_4 = f\{f[f(r)]\}, \ \ldots.$$

Prove that if $a_n \to L$, then $f(L) = L$; that is, L is a fixed point of f.

48. Show that

$$\frac{2^n}{n!} \to 0.$$

HINT: First show that

$$\frac{2^n}{n!} = \frac{2}{1} \cdot \frac{2}{2} \cdot \frac{2}{3} \cdot \cdots \cdot \frac{2}{n} \le \frac{4}{n}.$$

49. Prove that $(1/n)^{1/p} \to 0$ for all positive integers p.

50. Prove Theorem 10.3.8.

In Exercises 51–58, the sequences are defined recursively.† Determine in each case whether the sequence converges and, if so, find the limit. Start each sequence with $a_1 = 1$.

51. $a_{n+1} = \frac{1}{e} a_n.$ **52.** $a_{n+1} = 2^{n+1} a_n.$

53. $a_{n+1} = \frac{1}{n+1} a_n.$ **54.** $a_{n+1} = \frac{n}{n+1} a_n.$

55. $a_{n+1} = 1 - a_n.$ **56.** $a_{n+1} = -a_n.$

57. $a_{n+1} = \frac{1}{2} a_n + 1.$ **58.** $a_{n+1} = \frac{1}{3} a_n + 1.$

In Exercises 59–66, evaluate numerically the limit of each sequence as $n \to \infty$. Some of these sequences converge more rapidly than others. Determine for each sequence the least value of n for which the nth term differs from the limit by less than 0.001.

59. $\left\{\dfrac{1}{n^2}\right\}.$ **60.** $\left\{\dfrac{1}{\sqrt{n}}\right\}.$

61. $\left\{\dfrac{n}{10^n}\right\}.$ **62.** $\left\{\dfrac{n^{10}}{10^n}\right\}.$

---
† The notion was introduced in Exercises 10.2.

63. $\left\{\dfrac{1}{n!}\right\}.$ **64.** $\left\{\dfrac{2^n}{n!}\right\}.$

65. $\left\{\dfrac{\ln n}{n^2}\right\}.$ **66.** $\left\{\dfrac{\ln n}{n}\right\}.$

67. (a) Find the exact value of the limit of the sequence $\{a_n\}$ given in Exercise 73, Section 10.2. HINT: Suppose $a_n \to L$, then $a_{n-1} \to L$.
(b) Find the exact value of the limit of the sequence $\{a_n\}$ given in Exercise 74, Section 10.2.

68. Let $\{a_n\}$ be the sequence defined recursively by

$$a_1 = 1, \quad a_n = \sqrt{6 + a_{n-1}}, \quad n = 2, 3, 4, \ldots.$$

(a) Approximate a_2, a_3, a_4, a_5, a_6. Round your answers to six decimal places.
(b) Use mathematical induction to show that $a_n \le 3$ for all n.
(c) Show that $\{a_n\}$ is an increasing sequence. HINT: $a_{n+1}^2 - a_n^2 = (3 - a_n)(2 + a_n).$
(d) What is the limit of this sequence?

69. Let $\{a_n\}$ be the sequence defined recursively by

$$a_1 = 1, \quad a_n = \cos a_{n-1}, \quad n = 2, 3, 4, \ldots.$$

(a) Approximate $a_2, a_3, a_4, \ldots, a_{10}$. Round your answers to six decimal places.
(b) Assuming that $a_n \to L$, approximate L to six decimal places and interpret your result geometrically. HINT: Use Exercise 47.

70. Let $\{a_n\}$ be the sequence defined recursively by

$$a_1 = 1, \quad a_n = a_{n-1} + \cos a_{n-1}, \quad n = 2, 3, 4, \ldots.$$

(a) Approximate $a_2, a_3, a_4, \ldots, a_{10}$. Round your answers to six decimal places.
(b) Assuming that $a_n \to L$, approximate L to six decimal places and interpret your result geometrically.

71. Let R be a positive number. Approximations to $\sqrt{R}$ are generated by the sequence defined recursively by

$$a_1 = 1, \quad a_n = \frac{1}{2}\left(a_{n-1} + \frac{R}{a_{n-1}}\right), n = 2, 3, 4, \ldots.$$

Let $R = 3$.

(a) Approximate $a_2, a_3, \cdots, a_8$. Round your answers to six decimal places.
(b) Assuming that $a_n \to L$, prove that $L = \sqrt{3}$.
(c) Show that the recursion relation given above is simply the sequence generated by the Newton-Raphson method applied to the function $f(x) = x^2 - R$ (see Section 3.9).

72. The Newton-Raphson method (Section 3.9) applied to a differentiable function f generates a sequence $\{x_n\}$ that, under certain conditions, converges to a zero of f. The recursion formula for the sequence is given by

$$x_{n+1} = x_n - \frac{f(x_n)}{f'(x_n)}, \quad n = 1, 2, 3, \ldots .$$

Determine whether the following sequences converge, and if so, give the limit. HINT: Identify the function that generates the sequence.

(a) $x_{n+1} = x_n - \dfrac{x_n^3 - 8}{3x_n^2}; \quad x_1 = 1.$

(b) $x_{n+1} = x_n - \dfrac{\sin x_n - 0.5}{\cos x_n}; \quad x_1 = 0.$

(c) $x_{n+1} = x_n - \dfrac{\ln x_n - 1}{1/x_n} = x_n[2 - \ln x_n]; \quad x_1 = 1.$

---

*SUPPLEMENT TO SECTION 10.3

PROOF OF THEOREM 10.3.2

If $L \neq M$, then

$$\tfrac{1}{2}|L - M| > 0.$$

The assumption that $\lim_{n \to \infty} a_n = L$ and $\lim_{n \to \infty} a_n = M$ gives the existence of K_1 such that

$$\text{if} \quad n \geq K_1, \quad \text{then} \quad |a_n - L| < \tfrac{1}{2}|L - M|$$

and the existence of K_2 such that

$$\text{if} \quad n \geq K_2, \quad \text{then} \quad |a_n - m| < \tfrac{1}{2}|L - M|.†$$

For $n \geq \max\{K_1, K_2\}$ we have

$$|a_n - L| + |a_n - M| < |L - M|.$$

By the triangle inequality we have

$$|L - M| = |(L - a_n) + (a_n - M)| \leq |L - a_n| + |a_n - M| = |a_n - L| + |a_n - M|.$$

Combining the last two statements, we have

$$|L - M| < |L - M|.$$

The hypothesis $L \neq M$ has led to an absurdity. We conclude that $L = M$. ❑

PROOF OF THEOREM 10.3.7 (iii)–(v)

To prove (iii), we set $\epsilon > 0$. For each n,

$$|a_n b_n - LM| = |(a_n b_n - a_n M) + (a_n M - LM)|$$
$$\leq |a_n||b_n - M| + |M||a_n - L|.$$

Since $\{a_n\}$ is convergent, $\{a_n\}$ is bounded; that is, there exists $Q > 0$ such that

$$|a_n| \leq Q \quad \text{for all } n.$$

Since $|M| < |M| + 1$, we have

(1) $\qquad |a_n b_n - LM| \leq Q|b_n - M| + (|M| + 1)|a_n - L|.††$

Since $b_n \to M$, we know that there exists K_1 such that

$$\text{if} \quad n \geq K_1, \quad \text{then} \quad |b_n - M| < \frac{\epsilon}{2Q}.$$

---

† We can reach these conclusions from Definition 10.3.1 by taking $\tfrac{1}{2}|L - M|$ as ϵ.

†† Soon we will want to divide by the coefficient of $|a_n - L|$. We have replaced $|M|$ by $|M| + 1$ because $|M|$ can be zero.

Since $a_n \to L$, we know that there exists K_2 such that

$$\text{if} \quad n \geq K_2, \quad \text{then} \quad |a_n - L| < \frac{\epsilon}{2(|M| + 1)}.$$

For $n \geq \max\{K_1, K_2\}$ both conditions hold, and consequently

$$Q|b_n - M| + (|M| + 1)|a_n - L| < \frac{\epsilon}{2} + \frac{\epsilon}{2} = \epsilon.$$

In view of (1), we can conclude that

$$\text{if} \quad n \geq \max\{K_1, K_2\}, \quad \text{then} \quad |a_n b_n - LM| < \epsilon.$$

This proves that

$$a_n b_n \to LM. \quad \square$$

To prove (iv), once again we set $\epsilon > 0$. In the first place

$$\left| \frac{1}{b_n} - \frac{1}{M} \right| = \left| \frac{M - b_n}{b_n M} \right| = \frac{|b_n - M|}{|b_n||M|}.$$

Since $b_n \to M$ and $|M|/2 > 0$, there exists K_1 such that

$$\text{if} \quad n \geq K_1, \quad \text{then} \quad |b_n - M| < \frac{|M|}{2}.$$

This tells us that for $n \geq K_1$ we have

$$|b_n| > \frac{|M|}{2} \quad \text{and thus} \quad \frac{1}{|b_n|} < \frac{2}{|M|}.$$

Thus for $n \geq K_1$ we have

(2)
$$\left| \frac{1}{b_n} - \frac{1}{M} \right| \leq \frac{2}{|M|^2} |b_n - M|.$$

Since $b_n \to M$ there exists K_2 such that

$$\text{if} \quad n \geq K_2, \quad \text{then} \quad |b_n - M| < \frac{\epsilon|M|^2}{2}.$$

Thus for $n \geq K_2$ we have

$$\frac{2}{|M|^2} |b_n - M| < \epsilon.$$

In view of (2), we can be sure that

$$\text{if} \quad n \geq \max\{K_1, K_2\}, \quad \text{then} \quad \left| \frac{1}{b_n} - \frac{1}{M} \right| < \epsilon.$$

This proves that

$$\frac{1}{b_n} \to \frac{1}{M}. \quad \square$$

The proof of (v) is now easy:

$$\frac{a_n}{b_n} = a_n \cdot \frac{1}{b_n} \to L \cdot \frac{1}{M} = \frac{L}{M}. \quad \square$$

■ 10.4 SOME IMPORTANT LIMITS

Our purpose here is to familiarize you with some limits that are particularly important in calculus and to give you more experience with limit arguments.

(10.4.1)

> If $x > 0$, then
>
> $$x^{1/n} \to 1 \quad \text{as} \quad n \to \infty.$$

PROOF Fix any $x > 0$. Note that

$$\ln(x^{1/n}) = \frac{1}{n} \ln x \to 0 \quad \text{as} \quad n \to \infty.$$

Since the exponential function is continuous at 0, it follows from Theorem 10.3.12 that

$$x^{1/n} = e^{(1/n)\ln x} \to e^0 = 1. \quad \square$$

(10.4.2)

> If $|x| < 1$, then
>
> $$x^n \to 0 \quad \text{as} \quad n \to \infty.$$

PROOF The result clearly holds if $x = 0$. Fix any x with $|x| < 1$ and observe that $\{|x|^n\}$ is a decreasing sequence:

$$|x|^{n+1} = |x|\,|x|^n < |x|^n.$$

Now let $\epsilon > 0$. By (10.4.1)

$$\epsilon^{1/n} \to 1 \quad \text{as} \quad n \to \infty.$$

Thus there exists an integer $k > 0$ such that

$$|x| < \epsilon^{1/k}. \qquad\qquad \text{(explain)}$$

Obviously, then, $|x|^k < \epsilon$. Since $\{|x|^n\}$ is a decreasing sequence,

$$|x^n| = |x|^n < \epsilon \quad \text{for all } n \geq k. \quad \square$$

(10.4.3)

> For each $\alpha > 0$
>
> $$\frac{1}{n^\alpha} \to 0 \quad \text{as} \quad n \to \infty.$$

PROOF Since $\alpha > 0$, there exists an odd positive integer p such that $1/p < \alpha$. Then

$$0 < \frac{1}{n^\alpha} = \left(\frac{1}{n}\right)^\alpha \leq \left(\frac{1}{n}\right)^{1/p}.$$

Since $1/n \to 0$ and $f(x) = x^{1/p}$ is continuous at 0, we have

$$\left(\frac{1}{n}\right)^{1/p} \to 0 \quad \text{and thus by the pinching theorem} \quad \frac{1}{n^{\alpha}} \to 0. \quad \square$$

(10.4.4)

> For each real x
>
> $$\frac{x^n}{n!} \to 0 \qquad \text{as} \quad n \to \infty.$$

PROOF Fix any real number x and choose an integer k such that $k > |x|$. For $n > k + 1$,

$$\frac{k^n}{n!} = \left(\frac{k^k}{k!}\right)\left[\frac{k}{k+1}\frac{k}{k+2}\cdots\frac{k}{n-1}\right]\left(\frac{k}{n}\right) < \left(\frac{k^{k+1}}{k!}\right)\left(\frac{1}{n}\right).$$

$$\text{the middle term is less than 1} \longrightarrow$$

Since $k > |x|$, we have

$$0 < \frac{|x|^n}{n!} < \frac{k^n}{n!} < \left(\frac{k^{k+1}}{k!}\right)\left(\frac{1}{n}\right).$$

Since k is fixed and $1/n \to 0$, it follows from the pinching theorem that

$$\frac{|x|^n}{n!} \to 0 \qquad \text{and thus} \qquad \frac{x^n}{n!} \to 0. \quad \square$$

(10.4.5)

> $$\frac{\ln n}{n} \to 0 \qquad \text{as} \quad n \to \infty.$$

PROOF A routine proof can be based on L'Hospital's rule (10.6.1), but that is not available to us yet. We will appeal to the pinching theorem and base our argument on the integral representation of the logarithm:

$$0 \le \frac{\ln n}{n} = \frac{1}{n}\int_1^n \frac{dt}{t} \le \frac{1}{n}\int_1^n \frac{dt}{\sqrt{t}} = \frac{2}{n}(\sqrt{n} - 1)$$

$$= 2\left(\frac{1}{\sqrt{n}} - \frac{1}{n}\right) \to 0. \quad \square$$

(10.4.6)

> $$n^{1/n} \to 1 \qquad \text{as} \quad n \to \infty.$$

PROOF We know that

$$n^{1/n} = e^{(1/n)\ln n}.$$

Since

(11.1.1) $\qquad\qquad\qquad (1/n)\ln n \to 0 \qquad\qquad\qquad$ (10.4.5)

and the exponential function is continuous at 0, it follows from Theorem 10.3.12 that

$$n^{1/n} \to e^0 = 1. \quad \square$$

(10.4.7)

> For each real x
>
> $$\left(1 + \frac{x}{n}\right)^n \to e^x \qquad \text{as} \quad n \to \infty.$$

PROOF For $x = 0$, the result is obvious. For $x \neq 0$,

$$\ln\left(1 + \frac{x}{n}\right)^n = n \ln\left(1 + \frac{x}{n}\right) = x\left[\frac{\ln(1 + x/n) - \ln 1}{x/n}\right].$$

The crux here is to recognize that the bracketed expression is a difference quotient for the logarithm function. Once we see this, we let $h = x/n$ and write

$$\lim_{n \to \infty}\left[\frac{\ln(1 + x/n) - \ln 1}{x/n}\right] = \lim_{h \to 0}\left[\frac{\ln(1 + h) - \ln 1}{h}\right] = 1.\dagger$$

It follows that

$$\ln\left(1 + \frac{x}{n}\right)^n \to x \quad \text{and therefore} \quad \left(1 + \frac{x}{n}\right)^n = e^{\ln(1 + x/n)^n} \to e^x. \quad \square$$

$\dagger$ For each $t > 0$

$$\lim_{h \to 0}\frac{\ln(t + h) - \ln t}{h} = \frac{d}{dt}(\ln t) = \frac{1}{t}.$$

EXERCISES 10.4

In Exercises 1–36, state whether or not the sequence converges as $n \to \infty$; if it does, find the limit.

1. $\{2^{2/n}\}$.

2. $\{e^{-\alpha/n}\}$.

3. $\left\{\left(\dfrac{2}{n}\right)^n\right\}$.

4. $\left\{\dfrac{\log_{10} n}{n}\right\}$.

5. $\left\{\dfrac{\ln(n + 1)}{n}\right\}$.

6. $\left\{\dfrac{3^n}{4^n}\right\}$.

7. $\left\{\dfrac{x^{100n}}{n!}\right\}$.

8. $\{n^{1/(n+2)}\}$.

9. $\{n^{\alpha/n}\}, \quad \alpha > 0$.

10. $\left\{\ln\left(\dfrac{n + 1}{n}\right)\right\}$.

11. $\left\{\dfrac{3^{n+1}}{4^{n-1}}\right\}$.

12. $\left\{\displaystyle\int_{-n}^{0} e^{2x}\, dx\right\}$.

13. $\{(n + 2)^{1/n}\}$.

14. $\left\{\left(1 - \dfrac{1}{n}\right)^n\right\}$.

15. $\left\{\displaystyle\int_{0}^{n} e^{-x}\, dx\right\}$.

16. $\left\{\dfrac{2^{3n-1}}{7^{n+2}}\right\}$.

17. $\left\{\displaystyle\int_{-n}^{n} \dfrac{dx}{1 + x^2}\right\}$.

18. $\left\{\displaystyle\int_{0}^{n} e^{-nx}\, dx\right\}$.

19. $\{(n + 2)^{1/(n + 2)}\}$.

20. $\{n^2 \sin n\pi\}$.

21. $\left\{\dfrac{\ln n^2}{n}\right\}$.

22. $\left\{\displaystyle\int_{-1 + 1/n}^{1 - 1/n} \dfrac{dx}{\sqrt{1 - x^2}}\right\}$.

23. $\left\{ n^2 \sin \dfrac{\pi}{n} \right\}.$

24. $\left\{ \dfrac{n!}{2n} \right\}.$

25. $\left\{ \dfrac{5^{n+1}}{4^{2n-1}} \right\}.$

26. $\left\{ \left(1 + \dfrac{x}{n}\right)^{3n} \right\}.$

27. $\left\{ \left(\dfrac{n+1}{n+2}\right)^{n} \right\}.$

28. $\left\{ \displaystyle\int_{1/n}^{1} \dfrac{dx}{\sqrt{x}} \right\}.$

29. $\left\{ \displaystyle\int_{n}^{n+1} e^{-x^2}\, dx \right\}.$

30. $\left\{ \left(1 + \dfrac{1}{n^2}\right)^{n} \right\}.$

31. $\left\{ \dfrac{n^n}{2^{n^2}} \right\}.$

32. $\left\{ \displaystyle\int_{0}^{1/n} \cos e^x\, dx \right\}.$

33. $\left\{ \left(1 + \dfrac{x}{2n}\right)^{2n} \right\}.$

34. $\left\{ \left(1 + \dfrac{1}{n}\right)^{n^2} \right\}.$

35. $\left\{ \displaystyle\int_{-1/n}^{1/n} \sin x^2\, dx \right\}.$

36. $\left\{ \left(t + \dfrac{x}{n}\right)^{n} \right\}, \quad t > 0, \quad x > 0.$

37. Show that $\lim\limits_{n\to\infty} (\sqrt{n+1} - \sqrt{n}) = 0.$

38. Show that $\lim\limits_{n\to\infty} (\sqrt{n^2 + n} - n) = \frac{1}{2}.$

39. (a) Show that a regular polygon of n sides inscribed in a circle of radius r has perimeter $p_n = 2rn \sin(\pi/n)$.
(b) Find

$$\lim_{n\to\infty} p_n$$

and give a geometric interpretation of your result.

40. Show that

$$\text{if} \quad 0 < c < d, \quad \text{then} \quad (c^n + d^n)^{1/n} \to d.$$

In Exercises 41–43, find the indicated limit.

41. $\lim\limits_{n\to\infty} \dfrac{1 + 2 + \cdots + n}{n^2}.$

HINT: $1 + 2 + \cdots + n = \dfrac{n(n+1)}{2}.$

42. $\lim\limits_{n\to\infty} \dfrac{1^2 + 2^2 + \cdots + n^2}{(1+n)(2+n)}.$

HINT: $1^2 + 2^2 + \cdots + n^2 = \dfrac{n(n+1)(2n+1)}{6}.$

43. $\lim\limits_{n\to\infty} \dfrac{1^3 + 2^3 + \cdots + n^3}{2n^4 + n - 1}.$

HINT: $1^3 + 2^3 + \cdots + n^3 = \dfrac{n^2(n+1)^2}{4}.$

44. A sequence $\{a_n\}$ is said to be a *Cauchy sequence*† iff

(10.4.8)
> for each $\epsilon > 0$ there exists a positive integer K such that
> $$|a_n - a_m| < \epsilon \qquad \text{for all } m, n \geq K.$$

Show that

(10.4.9)
> every convergent sequence is a Cauchy sequence.

It is also true that every Cauchy sequence is convergent, but this is more difficult to prove.

45. (*Arithmetic means*) For a given sequence $\{a_n\}$, let

$$m_n = \frac{1}{n}(a_1 + a_2 + \cdots + a_n).$$

(a) Prove that if $\{a_n\}$ is increasing, then $\{m_n\}$ is increasing.
(b) Prove that if $a_n \to 0$, then $m_n \to 0$.
HINT: Choose an integer $j > 0$ such that, if $n \geq j$, then $a_n < \epsilon/2$. Then for $n \geq j$,

$$|m_n| < \frac{|a_1 + a_2 + \cdots + a_j|}{n} + \frac{\epsilon}{2}\left(\frac{n-j}{n}\right).$$

46. (a) Let $\{a_n\}$ be a convergent sequence. Prove that

$$\lim_{n\to\infty} (a_n - a_{n-1}) = 0.$$

(b) What can you say about the converse? That is, suppose that $\{a_n\}$ is a sequence such that

$$\lim_{n\to\infty} (a_n - a_{n-1}) = 0.$$

Does $\{a_n\}$ necessarily converge? Prove or give a counter-example.

47. Let a and b be positive numbers with $b > a$. Define two sequences $\{a_n\}$ and $\{b_n\}$ as follows:

$$a_1 = \frac{a+b}{2} \text{ (the arithmetic mean of } a \text{ and } b),$$

$$b_1 = \sqrt{ab} \text{ (the geometric mean of } a \text{ and } b),$$

---

† After the French baron Augustin-Louis Cauchy (1789–1857), one of the most prolific mathematicians of all time.

and

$$a_n = \frac{a_{n-1} + b_{n-1}}{2},$$

$$b_n = \sqrt{a_{n-1}b_{n-1}}, \quad n = 2, 3, 4, \ldots$$

(a) Use mathematical induction to show that

$$a_{n-1} > a_n > b_n > b_{n-1} \quad \text{for } n = 2, 3, 4, \ldots$$

(b) Prove that $\{a_n\}$ and $\{b_n\}$ are convergent sequences and that $\lim_{n\to\infty} a_n = \lim_{n\to\infty} b_n$. The value of these limits is known as the *arithmetic-geometric mean* of a and b.

48. You have seen that for all real x

$$\lim_{n\to\infty} \left(1 + \frac{x}{n}\right)^n = e^x.$$

However, the rate of convergence is different for different x. Verify that at $n = 100$, $(1 + 1/n)^n$ is within 1% of its limit, while $(1 + 5/n)^n$ is still about 12% from its limit. Give comparable accuracy estimates for these two sequences at $n = 1000$.

49. Evaluate

$$\lim_{n\to\infty} \left(\sin \frac{1}{n}\right)^{1/n}$$

numerically and justify your answer by other means.

50. We have stated that

$$\lim_{n\to\infty} (\sqrt{n^2 + n} - n) = \tfrac{1}{2}. \quad \text{(Exercise 38)}$$

Evaluate numerically

$$\lim_{n\to\infty} [(n^3 + n^2)^{1/3} - n].$$

Formulate a conjecture about

$$\lim_{n\to\infty} [(n^k + n^{k-1})^{1/k} - n], \quad k = 1, 2, 3, \ldots$$

and prove that your conjecture is valid.

51. The *Fibonacci sequence* is defined recursively by

$$a_{n+2} = a_{n+1} + a_n, \quad \text{with} \quad a_1 = a_2 = 1.$$

(a) Calculate $a_3, a_4, a_5, \ldots, a_{10}$.
(b) Now define the sequence $\{r_n\}$ by

$$r_n = \frac{a_{n+1}}{a_n}$$

Calculate $r_1, r_2, \ldots, r_6$.
(c) Assuming that $r_n \to L$, find L. HINT: Show that

$$r_n = 1 + \frac{1}{r_{n-1}}.$$

52. Let $\{a_n\}$ be the sequence defined by

$$a_n = \frac{1}{n^2} + \frac{2}{n^2} + \frac{3}{n^2} + \cdots + \frac{n}{n^2}.$$

Show that a_n is a Riemann sum for $\int_0^1 x\, dx$ for each $n \geq 1$. Does the sequence $\{a_n\}$ converge? If so, to what?

■ 10.5 THE INDETERMINATE FORM (0/0)

Recall that the quotient rule for evaluating the limit of a quotient $f(x)/g(x)$ fails when $f(x)$ and $g(x)$ both tend to zero (see Section 2.3). Some important examples of this behavior were

$$\lim_{x\to0} \frac{\sin x}{x} \quad \text{and} \quad \lim_{x\to0} \frac{1 - \cos x}{x}.$$

Also, the definition of the derivative:

$$\lim_{h\to0} \frac{f(x + h) - f(x)}{h}$$

typically leads to a limit of this type. We called such limits *indeterminates of the form* 0/0. In this section we present a method for handling this kind of limit when elementary methods fail or are difficult to apply.

THEOREM 10.5.1 L'HOSPITAL'S RULE (0/0)†

Suppose that f and g are differentiable functions and that

$$f(x) \to 0 \qquad \text{and} \qquad g(x) \to 0$$

as $x \to c^+, x \to c^-, x \to c, x \to \infty$, or $x \to -\infty$.

$$\text{If} \quad \frac{f'(x)}{g'(x)} \to L, \qquad \text{then} \quad \frac{f(x)}{g(x)} \to L.$$

If $\dfrac{f'(x)}{g'(x)} \to \infty$ or $-\infty$, \qquad then $\dfrac{f(x)}{g(x)} \to \infty$ or $-\infty$, respectively.

We will prove the validity of L'Hospital's rule later in the section. First we demonstrate its usefulness.

Example 1 Find

$$\lim_{x \to \pi/2} \frac{\cos x}{\pi - 2x}.$$

SOLUTION As $x \to \pi/2$, both the numerator $f(x) = \cos x$ and the denominator $g(x) = \pi - 2x$ tend to zero, but it is not at all obvious what happens to the quotient

$$\frac{f(x)}{g(x)} = \frac{\cos x}{\pi - 2x}.$$

Therefore we test the quotient of derivatives:

$$\frac{f'(x)}{g'(x)} = \frac{-\sin x}{-2} = \frac{\sin x}{2} \to \frac{1}{2} \qquad \text{as} \quad x \to \frac{\pi}{2}.$$

It follows from L'Hospital's rule that

$$\frac{\cos x}{\pi - 2x} \to \frac{1}{2} \qquad \text{as} \quad x \to \frac{\pi}{2}.$$

We can express all this on just one line using ∗ to indicate the differentiation of numerator and denominator:

$$\lim_{x \to \pi/2} \frac{\cos x}{\pi - 2x} \overset{*}{=} \lim_{x \to \pi/2} \frac{-\sin x}{-2} = \lim_{x \to \pi/2} \frac{\sin x}{2} = \frac{1}{2}. \quad \square$$

---

† Named after a Frenchman G. F. A. L'Hospital (1661–1704). The result was actually discovered by his teacher, Jakob Bernoulli (1654–1705).

Example 2 Find

$$\lim_{x\to 0^+} \frac{x}{\sin\sqrt{x}}.$$

SOLUTION As $x \to 0^+$, both numerator and denominator tend to 0. Since

$$\frac{f'(x)}{g'(x)} = \frac{1}{(\cos\sqrt{x})(1/2\sqrt{x})} = \frac{2\sqrt{x}}{\cos\sqrt{x}} \to 0 \qquad \text{as} \quad x \to 0^+,$$

it follows from L'Hospital's rule that

$$\frac{x}{\sin\sqrt{x}} \to 0 \qquad \text{as} \quad x \to 0^+.$$

For short we can write

$$\lim_{x\to 0^+} \frac{x}{\sin\sqrt{x}} \overset{*}{=} \lim_{x\to 0^+} \frac{2\sqrt{x}}{\cos\sqrt{x}} = 0. \quad ❑$$

Remark It is important to understand that Theorem 10.5.1 does not apply to quotients in general; you should verify first that the numerator and denominator both tend to zero. For example,

$$\lim_{x\to 0} \frac{x}{x + \cos x} = \frac{0}{1} = 0,$$

but a blind application of L'Hospital's rule would lead to

$$\lim_{x\to 0} \frac{x}{x + \cos x} \overset{*}{=} \lim_{x\to 0} \frac{1}{1 - \sin x} = 1.$$

This is, of course, incorrect. ❑

Sometimes it is necessary to differentiate numerator and denominator more than once. The next problem gives such an instance.

Example 3 Find

$$\lim_{x\to 0} \frac{e^x - x - 1}{x^2}.$$

SOLUTION As $x \to 0$, both numerator and denominator tend to 0. Here

$$\frac{f'(x)}{g'(x)} = \frac{e^x - 1}{2x}.$$

Since both numerator and denominator still tend to 0 as $x \to 0$, we differentiate again:

$$\frac{f''(x)}{g''(x)} = \frac{e^x}{2}.$$

Since this last quotient tends to $\frac{1}{2}$, we can conclude that

$$\frac{e^x - 1}{2x} \to \frac{1}{2} \qquad \text{and therefore} \qquad \frac{e^x - x - 1}{x^2} \to \frac{1}{2} \qquad \text{as} \quad x \to 0.$$

For short we can write

$$\lim_{x \to 0} \frac{e^x - x - 1}{x^2} \overset{*}{=} \lim_{x \to 0} \frac{e^x - 1}{2x} \overset{*}{=} \lim_{x \to 0} \frac{e^x}{2} = \frac{1}{2}. \quad \square$$

In the next example we use L'Hospital's rule to find the limit of a sequence.

Example 4   Given the sequence

$$\left\{ \frac{e^{2/n} - 1}{1/n} \right\}.$$

Find

$$\lim_{n \to \infty} \frac{e^{2/n} - 1}{1/n}.$$

SOLUTION   This is an indeterminate of the form $0/0$. To apply the methods of this section, we replace the integer variable n by the real variable x and examine the behavior of

$$\frac{e^{2/x} - 1}{1/x} \quad \text{as} \quad x \to \infty.$$

For an analogous example, see Example 4, Section 10.2. Applying L'Hospital's rule, we have

$$\lim_{x \to \infty} \frac{e^{2/x} - 1}{1/x} \overset{*}{=} \lim_{x \to \infty} \frac{e^{2/x}(-2/x^2)}{(-1/x^2)} = 2 \lim_{x \to \infty} e^{2/x} = 2.$$

This shows that

$$\frac{e^{2/x} - 1}{1/x} \to 2 \quad \text{and so} \quad \frac{e^{2/n} - 1}{1/n} \to 2 \quad \text{as} \quad n \to \infty. \quad \square$$

To derive L'Hospital's rule, we need a generalization of the mean-value theorem.

THEOREM 10.5.2   THE CAUCHY MEAN-VALUE THEOREM†

Suppose that f and g are differentiable on (a, b) and continuous on $[a, b]$. If g' is never 0 in (a, b), then there is a number c in (a, b) for which

$$\frac{f'(c)}{g'(c)} = \frac{f(b) - f(a)}{g(b) - g(a)}.$$

PROOF   We can prove this by applying Rolle's theorem (4.1.3) to the function

$$G(x) = [g(b) - g(a)][f(x) - f(a)] - [g(x) - g(a)][f(b) - f(a)].$$

---

† Another contribution of A. L. Cauchy, after whom Cauchy sequences were named.

Since

$$G(a) = 0 \quad \text{and} \quad G(b) = 0,$$

there exists (by Rolle's theorem) a number c in (a, b) for which

$$G'(c) = 0.$$

Since, in general,

$$G'(x) = [g(b) - g(a)]f'(x) - g'(x)[f(b) - f(a)],$$

we must have

$$[g(b) - g(a)]f'(c) - g'(c)[f(b) - f(a)] = 0,$$

and thus

$$[g(b) - g(a)]f'(c) = g'(c)[f(b) - f(a)].$$

Since g' is never 0 in (a, b),

$$g'(c) \neq 0 \quad \text{and} \quad g(b) - g(a) \neq 0.$$

$\underset{\text{explain}}{\uparrow\text{\textemdash}}$

We can therefore divide by these numbers and obtain

$$\frac{f'(c)}{g'(c)} = \frac{f(b) - f(a)}{g(b) - g(a)}. \qquad \square$$

Now we prove L'Hospital's rule for the case $x \to c^+$. The case $x \to c^-$ is handled in the same way, and the two cases together can be used to prove the rule for the case $x \to c$.

Let f and g be defined on an interval (c, b) for some $b > c$. We assume that, as $x \to c^+$,

$$f(x) \to 0, \quad g(x) \to 0, \quad \text{and} \quad \frac{f'(x)}{g'(x)} \to L$$

and show that

$$\frac{f(x)}{g(x)} \to L.$$

PROOF The fact that

$$\frac{f'(x)}{g'(x)} \to L \qquad \text{as} \qquad x \to c^+$$

assures us that both f' and g' exist on a set of the form $(c, c + h]$ and that g' is not zero there. By setting $f(c) = 0$ and $g(c) = 0$, we ensure that f and g are both continuous on $[c, c + h]$. We can now apply the Cauchy mean-value theorem and conclude that there exists a number c_h between c and $c + h$ such that

$$\frac{f'(c_h)}{g'(c_h)} = \frac{f(c + h) - f(c)}{g(c + h) - g(c)} = \frac{f(c + h)}{g(c + h)}.$$

The result is now obtained by letting $h \to 0^+$. Since the left side tends to L, the right side tends to L. $\quad \square$

Here is a proof of L'Hospital's rule for the case $x \to \infty$.

PROOF   The key here is to set $x = 1/t$:

$$\lim_{x \to \infty} \frac{f'(x)}{g'(x)} = \lim_{t \to 0^+} \frac{f'(1/t)}{g'(1/t)} = \lim_{t \to 0^+} \frac{-t^{-2}f'(1/t)}{-t^{-2}g'(1/t)} = \lim_{t \to 0^+} \frac{f(1/t)}{g(1/t)} = \lim_{x \to \infty} \frac{f(x)}{g(x)}. \quad \square$$

by L'Hospital's rule for the case $t \to 0^+$

EXERCISES 10.5

In Exercises 1–32, find the indicated limit.

1. $\lim\limits_{x \to 0^+} \dfrac{\sin x}{\sqrt{x}}$.

2. $\lim\limits_{x \to 1} \dfrac{\ln x}{1 - x}$.

3. $\lim\limits_{x \to 0} \dfrac{e^x - 1}{\ln(1 + x)}$.

4. $\lim\limits_{x \to 4} \dfrac{\sqrt{x} - 2}{x - 4}$.

5. $\lim\limits_{x \to \pi/2} \dfrac{\cos x}{\sin 2x}$.

6. $\lim\limits_{x \to a} \dfrac{x - a}{x^n - a^n}$.

7. $\lim\limits_{x \to 0} \dfrac{2^x - 1}{x}$.

8. $\lim\limits_{x \to 0} \dfrac{\tan^{-1} x}{x}$.

9. $\lim\limits_{x \to 1} \dfrac{x^{1/2} - x^{1/4}}{x - 1}$.

10. $\lim\limits_{x \to 0} \dfrac{e^x - 1}{x(1 + x)}$.

11. $\lim\limits_{x \to 0} \dfrac{e^x - e^{-x}}{\sin x}$.

12. $\lim\limits_{x \to 0} \dfrac{1 - \cos x}{3x}$.

13. $\lim\limits_{x \to 0} \dfrac{x + \sin \pi x}{x - \sin \pi x}$.

14. $\lim\limits_{x \to 0} \dfrac{a^x - (a + 1)^x}{x}$.

15. $\lim\limits_{x \to 0} \dfrac{e^x + e^{-x} - 2}{1 - \cos 2x}$.

16. $\lim\limits_{x \to 0} \dfrac{x - \ln(x + 1)}{1 - \cos 2x}$.

17. $\lim\limits_{x \to 0} \dfrac{\tan \pi x}{e^x - 1}$.

18. $\lim\limits_{x \to 0} \dfrac{\cos x - 1 + x^2/2}{x^4}$.

19. $\lim\limits_{x \to 0} \dfrac{1 + x - e^x}{x(e^x - 1)}$.

20. $\lim\limits_{x \to 0} \dfrac{\ln(\sec x)}{x^2}$.

21. $\lim\limits_{x \to 0} \dfrac{x - \tan x}{x - \sin x}$.

22. $\lim\limits_{x \to 0} \dfrac{xe^{nx} - x}{1 - \cos nx}$.

23. $\lim\limits_{x \to 1^-} \dfrac{\sqrt{1 - x^2}}{\sqrt{1 - x^3}}$.

24. $\lim\limits_{x \to 0} \dfrac{2x - \sin \pi x}{4x^2 - 1}$.

25. $\lim\limits_{x \to \pi/2} \dfrac{\ln(\sin x)}{(\pi - 2x)^2}$.

26. $\lim\limits_{x \to 0^+} \dfrac{\sqrt{x}}{\sqrt{x} + \sin \sqrt{x}}$.

27. $\lim\limits_{x \to 0} \dfrac{\cos x - \cos 3x}{\sin(x^2)}$.

28. $\lim\limits_{x \to 0} \dfrac{\sqrt{a + x} - \sqrt{a - x}}{x}$.

29. $\lim\limits_{x \to \pi/4} \dfrac{\sec^2 x - 2 \tan x}{1 + \cos 4x}$.

30. $\lim\limits_{x \to 0} \dfrac{x - \sin^{-1} x}{\sin^3 x}$.

31. $\lim\limits_{x \to 0} \dfrac{\tan^{-1} x}{\tan^{-1} 2x}$.

32. $\lim\limits_{x \to 0} \dfrac{\sin^{-1} x}{x}$.

In Exercises 33–36, find the limit of the sequence.

33. $\lim\limits_{n \to \infty} \dfrac{(\pi/2 - \tan^{-1} n)}{1/n}$.

34. $\lim\limits_{n \to \infty} \dfrac{\ln(1 - 1/n)}{\sin(1/n)}$.

35. $\lim\limits_{n \to \infty} \dfrac{1}{n[\ln(n + 1) - \ln n]}$.

36. $\lim\limits_{n \to \infty} \dfrac{\sinh \pi/n - \sin \pi/n}{\sin^3 \pi/n}$.

37. Find the fallacy:

$$\lim_{x \to 0} \frac{2 + x + \sin x}{x^3 + x - \cos x} \overset{*}{=} \lim_{x \to 0} \frac{1 + \cos x}{3x^2 + 1 + \sin x}$$
$$\overset{*}{=} \lim_{x \to 0} \frac{-\sin x}{6x + \cos x} = \frac{0}{1} = 0.$$

38. Show that, if $a > 0$, then

$$\lim_{n \to \infty} n(a^{1/n} - 1) = \ln a.$$

39. Find values for a and b such that

$$\lim_{x \to 0} \frac{\cos ax - b}{2x^2} = -4.$$

40. Find values for a and b so that

$$\lim_{x \to 0} \frac{\sin 2x + ax + bx^3}{x^3} = 0$$

41. Given that f is continuous, use L'Hospital's rule to determine

$$\lim_{x \to 0} \left(\frac{1}{x} \int_0^x f(t)\, dt. \right).$$

42. Let f be a twice differentiable function and fix a value of x.

(a) Prove that

$$\lim_{h \to 0} \frac{f(x+h) - f(x-h)}{2h} = f'(x).$$

(b) Prove that

$$\lim_{h \to 0} \frac{f(x+h) - 2f(x) + f(x-h)}{h^2} = f''(x).$$

43. Let $A(b)$ be the area of the region bounded by the parabola $y = x^2$ and the horizontal line $y = b$ ($b > 0$), and let $T(b)$ be the area of the triangle AOB (see the figure). Find $\lim\limits_{b \to 0^+} T(b)/A(b)$.

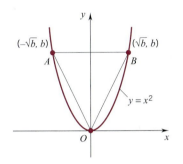

44. Choose an angle θ, $0 < \theta < \pi/2$, in standard position as shown in the figure. Let $T(\theta)$ be the area of the triangle ABC, and let $S(\theta)$ be the area of the segment of the circle formed by the chord AB. Find $\lim\limits_{\theta \to 0^+} T(\theta)/S(\theta)$.

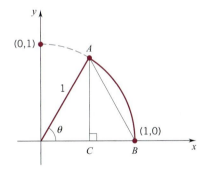

▷ 45. Let

$$f(x) = \frac{x^2 - 16}{\sqrt{x^2 + 9} - 5}.$$

(a) Use a graphing utility to graph f. What is the behavior of the graph as $x \to \infty$ and as $x \to -\infty$?

(b) What is the behavior of f as $x \to 4$? Confirm your answer using L'Hospital's rule.

▷ 46. Let

$$f(x) = \frac{x - \sin x}{x^3}.$$

(a) Use a graphing utility to graph f. What is the behavior of the graph as $x \to \infty$ and as $x \to -\infty$?

(b) What is the behavior of f as $x \to 0$? Confirm your answer using L'Hospital's rule.

▷ 47. Let $f(x) = \dfrac{2^{\sin x} - 1}{x}$.

(a) Use a graphing utility to graph f. Estimate

$$\lim_{x \to 0} f(x).$$

(b) Use L'Hospital's rule to confirm your estimate in part (a).

▷ 48. Let $g(x) = \dfrac{3^{\cos x} - 3}{x^2}$

(a) Use a graphing utility to graph g. Estimate

$$\lim_{x \to 0} g(x).$$

(b) Use L'Hospital's rule to confirm your estimate in part (a).

■ 10.6 THE INDETERMINATE FORM (∞/∞); OTHER INDETERMINATE FORMS

We come now to limits of quotients $f(x)/g(x)$ where numerator and denominator both tend to ∞. Such limits are called *indeterminates of the form* ∞/∞.

THEOREM 10.6.1   L'HOSPITAL'S RULE (∞/∞)

Suppose that f and g are differentiable, and that

$$f(x) \to \pm\infty \qquad \text{and} \qquad g(x) \to \pm\infty$$

as $x \to c^+$, $x \to c^-$, $x \to c$, $x \to \infty$, or $x \to -\infty$.

$$\text{If} \quad \frac{f'(x)}{g'(x)} \to L, \qquad \text{then} \quad \frac{f(x)}{g(x)} \to L.$$

$$\text{If} \quad \frac{f'(x)}{g'(x)} \to \infty \text{ or } -\infty, \qquad \text{then} \quad \frac{f(x)}{g(x)} \to \infty \text{ or } -\infty, \text{ respectively}$$

While the proof of L'Hospital's rule in this setting is a little more complicated than it was in the (0/0) case,† the application of the rule is much the same.

Example 1   Let α be any positive number.   Show that

(10.6.2)
$$\lim_{x \to \infty} \frac{\ln x}{x^\alpha} = 0.$$

SOLUTION   Both numerator and denominator tend to ∞ as $x \to \infty$.   L'Hospital's rule gives

$$\lim_{x \to \infty} \frac{\ln x}{x^\alpha} \overset{*}{=} \lim_{x \to \infty} \frac{1/x}{\alpha x^{\alpha-1}} = \lim_{x \to \infty} \frac{1}{\alpha x^\alpha} = 0. \quad \square$$

For example,

$$\frac{\ln x}{x^{0.01}} \to 0 \qquad \text{and} \qquad \frac{\ln x}{x^{0.001}} \to 0$$

as $x \to \infty$.

Example 2   Let k be any positive integer.   Show that

(10.6.3)
$$\lim_{x \to \infty} \frac{x^k}{e^x} = 0.$$

SOLUTION   Here we differentiate numerator and denominator k times:

$$\lim_{x \to \infty} \frac{x^k}{e^x} \overset{*}{=} \lim_{x \to \infty} \frac{kx^{k-1}}{e^x} \overset{*}{=} \lim_{x \to \infty} \frac{k(k-1)x^{k-2}}{e^x} \overset{*}{=} \cdots \overset{*}{=} \lim_{x \to \infty} \frac{k!}{e^x} = 0. \quad \square$$

---

† We omit the proof.

For example,

$$\frac{x^{100}}{e^x} \to 0 \qquad \text{and} \qquad \frac{x^{1000}}{e^x} \to 0$$

as $x \to \infty$.

Remark   The limits (10.6.2) and (10.6.3) tell us that $\ln x$ tends to infinity *more slowly than* any positive power of x and that e^x tends to infinity *faster than* any positive integral power of x. In the Exercises you are asked to show that e^x tends to infinity faster than *any* positive power of x and that *any* positive power of $\ln x$ tends to infinity more slowly than x. That is, for any positive number α,

$$\lim_{x \to \infty} \frac{x^\alpha}{e^x} = 0 \qquad \text{and} \qquad \lim_{x \to \infty} \frac{[\ln x]^\alpha}{x} = 0.$$

Comparisons of logarithmic and exponential growth were also given in Exercises 17 and 18 in Section 7.6.   ❏

Example 3   Find the limit as $n \to \infty$ of the sequence $\{a_n\}$ given by

$$a_n = \frac{2^n}{n^2}.$$

SOLUTION   To use the methods of calculus, we investigate

$$\lim_{x \to \infty} \frac{2^x}{x^2}.$$

Since both numerator and denominator tend to ∞ with x, we try L'Hospital's rule:

$$\lim_{x \to \infty} \frac{2^x}{x^2} \overset{*}{=} \lim_{x \to \infty} \frac{2^x \ln 2}{2x} \overset{*}{=} \lim_{x \to \infty} \frac{2^x (\ln 2)^2}{2} = \infty.$$

Therefore, the limit of the sequence must also be ∞.   ❏

Other Indeterminate Forms: $0 \cdot \infty$, $\infty - \infty$, 0^0, 1^∞, ∞^0

If f tends to 0 and g tends to ∞ (or $-\infty$) as x approaches some number c (or $\pm\infty$), then it is not clear what the product $f \cdot g$ will do. For example, as $x \to 1$,

$$(x - 1)^3 \to 0, \qquad \frac{1}{(x - 1)^2} \to \infty, \qquad \text{and} \qquad (x - 1)^3 \cdot \frac{1}{(x - 1)^2} = (x - 1) \to 0.$$

On the other hand,

$$\lim_{x \to 1} \left[(x - 1)^3 \cdot \frac{1}{(x - 1)^4} \right] = \lim_{x \to 1} \left[\frac{1}{x - 1} \right] \qquad \text{does not exist.}$$

A limit of this type is called an *indeterminate of the form* $0 \cdot \infty$. We handle these indeterminates by writing the product $f \cdot g$ as a quotient

$$\frac{f}{1/g} \qquad \text{or} \qquad \frac{g}{1/f}.$$

In the first case, the result will be an indeterminate of the form $0/0$, and in the second it will have the form ∞/∞.

Example 4 Find

$$\lim_{x \to 0^+} \sqrt{x} \ln x.$$

SOLUTION As $x \to 0^+$, $\sqrt{x} \to 0$ and $\ln x \to -\infty$. Thus the given limit is an indeterminate of the form $0 \cdot \infty$. Rewriting the product as a quotient, we have

$$\lim_{x \to 0^+} \sqrt{x} \ln x = \lim_{x \to 0^+} \frac{\ln x}{1/\sqrt{x}} \stackrel{*}{=} \lim_{x \to 0^+} \frac{1/x}{-\frac{1}{2}x^{-3/2}} = \lim_{x \to 0^+} -2\sqrt{x} = 0.$$

Therefore, $\lim_{x \to 0^+} \sqrt{x} \ln x = 0$.

Of course, we could have chosen to write $\sqrt{x} \ln x$ as the quotient

$$\frac{\sqrt{x}}{1/\ln x}.$$

Try to evaluate the limit using this quotient. ❏

If f and g both tend to ∞, or if both tend to $-\infty$, as x tends to c (or $\pm\infty$), then $\lim(f - g)$ is called an *indeterminate of the form* $\infty - \infty$. Like the preceding case, indeterminates of this type are handled by converting the difference to a quotient.

Example 5 Find

$$\lim_{x \to (\pi/2)^-} (\tan x - \sec x).$$

SOLUTION Both $\tan x$ and $\sec x$ tend to ∞ as x tends to $\pi/2$ from the left. We first rewrite the difference as a quotient:

$$\tan x - \sec x = \frac{\sin x}{\cos x} - \frac{1}{\cos x} = \frac{\sin x - 1}{\cos x}.$$

Now

$$\lim_{x \to (\pi/2)^-} \frac{\sin x - 1}{\cos x}$$

is an indeterminate of the form $0/0$, and

$$\lim_{x \to (\pi/2)^-} \frac{\sin x - 1}{\cos x} \stackrel{*}{=} \lim_{x \to (\pi/2)^-} \frac{\cos x}{-\sin x} = \frac{0}{-1} = 0.$$

Thus, $\lim_{x \to (\pi/2)^-} (\tan x - \sec x) = 0.$ ❏

Limits involving exponential expressions $[f(x)]^{g(x)}$ are indeterminate when: (1) f and g both tend to 0; (2) f tends to 1 and g tends to $\pm\infty$; and (3) f tends to $\pm\infty$ and g tends to 0. These cases are called *indeterminates of the forms* 0^0, 1^∞, and ∞^0, respectively. Exponential indeterminate forms are treated by taking natural logarithms:

$$\text{if} \quad y = [f(x)]^{g(x)}, \quad \text{then} \quad \ln y = g(x)\ln[f(x)].$$

Now, $\lim \ln y = \lim g(x)\ln[f(x)]$ will be an indeterminate of the form $0 \cdot \infty$

Example 6 Show that

(10.6.4)

$$\lim_{x \to 0^+} x^x = 1.$$

SOLUTION Here we are dealing with an indeterminate of the form 0^0. Our first step is to take the logarithm of x^x. Then we apply L'Hospital's rule:

$$\lim_{x \to 0^+} \ln(x^x) = \lim_{x \to 0^+} (x \ln x) = \lim_{x \to 0^+} \frac{\ln x}{1/x} \overset{*}{=} \lim_{x \to 0^+} \frac{1/x}{-1/x^2} = \lim_{x \to 0^+} (-x) = 0.$$

Since $\ln(x^x) \to 0$ as $x \to 0^+$, $x^x = e^{\ln(x^x)} \to e^0 = 1$. ❑

Example 7 Find

$$\lim_{x \to 0^+} (1 + x)^{1/x}.$$

SOLUTION Here we are dealing with an indeterminate of the form 1^∞: as $x \to 0^+$, $1 + x \to 1$ and $1/x$ increases without bound. Taking the logarithm and then applying L'Hospital's rule, we have

$$\lim_{x \to 0^+} \ln(1 + x)^{1/x} = \lim_{x \to 0^+} \frac{\ln(1 + x)}{x} \overset{*}{=} \lim_{x \to 0^+} \frac{1}{1 + x} = 1.$$

Since $\ln(1 + x)^{1/x} \to 1$ as $x \to 0^+$, $(1 + x)^{1/x} = e^{\ln(1 + x)^{1/x}} \to e^1 = e$. Note that if we set $x = 1/n$, we have the familiar result: $[1 + (1/n)]^n \to e$ as $n \to \infty$. ❑

Example 8 Show that

$$\lim_{x \to \infty} (x^2 + 1)^{1/\ln x} = e^2.$$

SOLUTION Here we have an indeterminate of the form ∞^0. Taking the logarithm and then applying L'Hospital's rule, we find that

$$\lim_{x \to \infty} \ln[(x^2 + 1)^{1/\ln x}] = \lim_{x \to \infty} \frac{\ln(x^2 + 1)}{\ln x} \overset{*}{=} \lim_{x \to \infty} \frac{2x/(x^2 + 1)}{1/x} = \lim_{x \to \infty} \frac{2x^2}{x^2 + 1} = 2.$$

It follows that

$$\lim_{x \to \infty} (x^2 + 1)^{1/\ln x} = e^2. \quad ❑$$

Concluding Remarks Suppose that $\lim(f/g)$ is an indeterminate form (either $0/0$ or ∞/∞). Both versions of L'Hospital's rule (Theorems 10.5.1 and 10.6.1) tell us that if

$$\lim \frac{f'}{g'} = L \text{ (or } \pm\infty), \qquad \text{then} \qquad \lim \frac{f}{g} = L \text{ (or } \pm\infty).$$

However, the rules do not provide any information when $\lim(f'/g')$ fails to exist; $\lim(f/g)$ may or may not exist. For example,

$$\lim_{x \to \infty} \frac{x + \cos x}{x}$$

is an indeterminate of the form ∞/∞. It is easy to show that this limit is 1 (divide numerator and denominator by x). On the other hand, if we try to apply Theorem 10.6.1, we consider the limit

$$\lim_{x\to\infty} \frac{f'(x)}{g'(x)} = \lim_{x\to\infty} \frac{1 - \sin x}{1},$$

and this limit does not exist.

Finally, as noted in Section 10.5, you should always check first to make sure that a given limit actually involves an indeterminate form before trying to apply the methods of these sections. In the case of a quotient, L'Hospital's rule does not apply when either the numerator or denominator has a finite nonzero limit. For example,

$$\lim_{x\to 0^+} \frac{1 + x}{\sin x} = \infty,$$

but a misapplication of L'Hospital's rule would lead to the limit

$$\lim_{x\to 0^+} \frac{1}{\cos x} = 1,$$

and the *incorrect conclusion* that

$$\lim_{x\to 0^+} \frac{1 + x}{\sin x} = 1. \quad \square$$

EXERCISES 10.6

In Exercises 1–36, find the indicated limit.

1. $\displaystyle\lim_{x\to-\infty} \frac{x^2 + 1}{1 - x}$.

2. $\displaystyle\lim_{x\to\infty} \frac{20x}{x^2 + 1}$.

3. $\displaystyle\lim_{x\to\infty} \frac{x^3}{1 - x^3}$.

4. $\displaystyle\lim_{x\to\infty} \frac{x^3 - 1}{2 - x}$.

5. $\displaystyle\lim_{x\to\infty} \left(x^2 \sin \frac{1}{x} \right)$.

6. $\displaystyle\lim_{x\to\infty} \frac{\ln x^k}{x}$.

7. $\displaystyle\lim_{x\to\pi/2^-} \frac{\tan 5x}{\tan x}$.

8. $\displaystyle\lim_{x\to 0} (x \ln|\sin x|)$.

9. $\displaystyle\lim_{x\to 0^+} x^{2x}$.

10. $\displaystyle\lim_{x\to\infty} \left(x \sin \frac{\pi}{x} \right)$.

11. $\displaystyle\lim_{x\to 0} [x(\ln|x|)^2]$.

12. $\displaystyle\lim_{x\to 0^+} \frac{\ln x}{\cot x}$.

13. $\displaystyle\lim_{x\to\infty} \left(\frac{1}{x} \int_0^x e^{t^2}\, dt \right)$.

14. $\displaystyle\lim_{x\to\infty} \frac{\sqrt{1 + x^2}}{x}$.

15. $\displaystyle\lim_{x\to 0} \left[\frac{1}{\sin^2 x} - \frac{1}{x^2} \right]$.

16. $\displaystyle\lim_{x\to 0} |\sin x|^x$.

17. $\displaystyle\lim_{x\to 1} x^{1/(x-1)}$.

18. $\displaystyle\lim_{x\to 0^+} x^{\sin x}$.

19. $\displaystyle\lim_{x\to\infty} \left(\cos \frac{1}{x} \right)^x$.

20. $\displaystyle\lim_{x\to\pi/2} |\sec x|^{\cos x}$.

21. $\displaystyle\lim_{x\to 0} \left[\frac{1}{\ln(1 + x)} - \frac{1}{x} \right]$.

22. $\displaystyle\lim_{x\to\infty} (x^2 + a^2)^{(1/x)^2}$.

23. $\displaystyle\lim_{x\to 0} \left(\frac{1}{x} - \cot x \right)$.

24. $\displaystyle\lim_{x\to\infty} \ln \left(\frac{x^2 - 1}{x^2 + 1} \right)^3$.

25. $\displaystyle\lim_{x\to\infty} (\sqrt{x^2 + 2x} - x)$.

26. $\displaystyle\lim_{x\to\infty} \frac{1}{x} \int_0^x \sin \left(\frac{1}{t + 1} \right) dt$.

27. $\displaystyle\lim_{x\to\infty} (x^3 + 1)^{1/\ln x}$.

28. $\displaystyle\lim_{x\to\infty} (e^x + 1)^{1/x}$.

29. $\displaystyle\lim_{x\to\infty} (\cosh x)^{1/x}$.

30. $\displaystyle\lim_{x\to\infty} (x^4 + 1)^{1/\ln x}$.

31. $\displaystyle\lim_{x\to 0} (e^x + x)^{1/x}$.

32. $\displaystyle\lim_{x\to\infty} \left(1 + \frac{1}{x} \right)^{3x}$.

33. $\displaystyle\lim_{x\to 0} \left(\frac{1}{\sin x} - \frac{1}{x} \right)$.

34. $\displaystyle\lim_{x\to 0} (e^x + 3x)^{1/x}$.

35. $\displaystyle\lim_{x\to 1} \left(\frac{1}{\ln x} - \frac{x}{x - 1} \right)$.

36. $\displaystyle\lim_{x\to 0} \left(\frac{1 + 2^x}{2} \right)^{1/x}$.

In Exercises 37–44, find the limit of the sequence.

37. $\displaystyle\lim_{n\to\infty} \left(\frac{1}{n} \ln \frac{1}{n} \right)$.

38. $\displaystyle\lim_{n\to\infty} \frac{n^k}{2^n}$.

39. $\lim\limits_{n\to\infty} (\ln n)^{1/n}$.

40. $\lim\limits_{n\to\infty} \dfrac{\ln n}{n^p}$, $(p > 0)$.

41. $\lim\limits_{n\to\infty} (n^2 + n)^{1/n}$.

42. $\lim\limits_{n\to\infty} n^{\sin(\pi/n)}$.

43. $\lim\limits_{n\to\infty} \dfrac{n^2 \ln n}{e^n}$.

44. $\lim\limits_{n\to\infty} (\sqrt{n} - 1)^{1/\sqrt{n}}$.

In Exercises 45–50, sketch the curve, specifying all vertical and horizontal asymptotes.

45. $y = x^2 - \dfrac{1}{x^3}$.

46. $y = \sqrt{\dfrac{x}{x-1}}$.

47. $y = xe^x$.

48. $y = xe^{-x}$.

49. $y = x^2 e^{-x}$.

50. $y = \dfrac{\ln x}{x}$.

The graphs of two functions $y = f(x)$ and $y = g(x)$ are said to be *asymptotic as $x \to \infty$* iff

$$\lim_{x\to\infty} [\,f(x) - g(x)] = 0;$$

they are said to be *asymptotic as $x \to -\infty$* iff

$$\lim_{x\to-\infty} [\,f(x) - g(x)] = 0.$$

51. Show that the hyperbolic arc $y = (b/a)\sqrt{x^2 - a^2}$ is asymptotic to the line $y = (b/a)x$ as $x \to \infty$.

52. Show that the graphs of $y = \cosh x$ and $y = \sinh x$ are asymptotic.

53. Give an example of a function the graph of which is asymptotic to the parabola $y = x^2$ as $x \to \infty$ and crosses the graph of the parabola twice.

54. Give an example of a function the graph of which is asymptotic to the line $y = x$ as $x \to \infty$ and crosses the graph of the line infinitely often.

55. Find the fallacy:

$$\lim_{x\to 0^+} \frac{x^2}{\sin x} \overset{*}{=} \lim_{x\to 0^+} \frac{2x}{\cos x} \overset{*}{=} \lim_{x\to 0^+} \frac{2}{-\sin x} = -\infty.$$

56. Let α be a positive number. Show that

$$\lim_{x\to\infty} \frac{x^\alpha}{e^x} = 0.$$

57. (a) Show by induction that, for each positive integer k,

$$\lim_{x\to\infty} \frac{(\ln x)^k}{x} = 0.$$

(b) Show that, for each positive number α,

$$\lim_{x\to\infty} \frac{(\ln x)^\alpha}{x} = 0.$$

58. (a) Try to evaluate

$$\lim_{x\to\infty} \frac{x}{\sqrt{x^2 + 1}}$$

using L'Hospital's rule and see what happens.

(b) Evaluate this limit by some other method.

59. (a) Try to evaluate

$$\lim_{x\to 0} \frac{e^{-1/x^2}}{x}$$

using L'Hospital's rule and see what happens. Then rewrite the quotient in an equivalent form and show that the limit is 0.

(b) Define the function f by

$$f(x) = \begin{cases} e^{-1/x^2} & \text{if } x \neq 0 \\ 0 & \text{if } x = 0. \end{cases}$$

Show that f is differentiable at 0. What is $f'(0)$?

60. The differential equation governing the velocity of an object of mass m dropped from rest under the influence of gravity with air resistance directly proportional to the velocity is

$$(*) \qquad m\frac{dv}{dt} + kv = mg,$$

where $k > 0$ is the constant of proportionality, g is the gravitational constant and $v(0) = 0$. See Exercise 37, Section 7.6. The velocity of the object at time t is given by

$$v(t) = (mg/k)(1 - e^{-(k/m)t}).$$

(a) Fix t and find $\lim\limits_{k\to 0^+} v(t)$.

(b) Set $k = 0$ in equation $(*)$ and solve

$$m\frac{dv}{dt} = mg, \quad v(0) = 0.$$

Does this result agree with the result found in part (a)?

61. Let $f(x) = (1 + x)^{1/x}$ and $g(x) = (1 + x^2)^{1/x}$ on $(0, \infty)$.

(a) Use a graphing utility to graph f and g in the same coordinate system. Estimate

$$\lim_{x\to 0^+} g(x).$$

(b) Use L'Hospital's rule to confirm your estimate in part (a).

62. Let $f(x) = \sqrt{x^2 + 3x + 1} - x$.

(a) Use a graphing utility to graph f. Then use your graph to estimate

$$\lim_{x\to\infty} f(x)$$

(b) Use L'Hospital's rule to confirm your estimate. HINT: "Rationalize."

63. Let $g(x) = \sqrt[3]{x^3 - 5x^2 + 2x + 1} - x$.

(a) Use a graphing utility to graph g. Then use your graph to estimate

$$\lim_{x \to \infty} g(x).$$

(b) Use L'Hospital's rule to confirm your estimate.

64. Exercises 62 and 63 can be generalized as follows: Let n be a positive integer and let P be the polynomial

$$P(x) = x^n + b_1 x^{n-1} + b_2 x^{n-2} + \cdots + b_{n-1}x + b_n.$$

Prove that

$$\lim_{x \to \infty} \left([P(x)]^{1/n} - x\right) = \frac{b_1}{n}.$$

■ 10.7 IMPROPER INTEGRALS

In all of our work involving the theory and applications of the definite integral

$$\int_a^b f(x)\, dx,$$

it has been assumed that the interval $[a, b]$ is finite and that the function f is bounded on $[a, b]$. In the context to be developed here, such integrals are said to be *proper*. In this section, we will use a limit process to calculate integrals in cases where the interval is infinite or where the function is unbounded. Such integrals are called *improper integrals*.

Integrals over Infinite Intervals

We begin with a function f continuous on an unbounded interval $[a, \infty)$. For each number $b > a$ we can form the definite integral

$$\int_a^b f(x)\, dx.$$

If, as b tends to ∞, this integral tends to a finite limit L,

$$\lim_{b \to \infty} \int_a^b f(x)\, dx = L,$$

then we write

$$\int_a^\infty f(x)\, dx = L$$

and say that

the improper integral $\int_a^\infty f(x)\, dx$ converges to L.

Otherwise, we say that

the improper integral $\int_a^\infty f(x)\, dx$ diverges.

In a similar manner, if f is continuous on the unbounded interval $(-\infty, b]$, then for each number $a < b$, we can form the definite integral

$$\int_a^b f(x)\, dx$$

and calculate

$$\lim_{a \to -\infty} \int_a^b f(x)\, dx.$$

If this limit exists and equals L, then

the improper integral $\quad \displaystyle\int_{-\infty}^b f(x)\, dx \quad$ converges to L;

otherwise,

the improper integral $\quad \displaystyle\int_{-\infty}^b f(x)\, dx \quad$ diverges.

Example 1

(a) $\displaystyle\int_0^\infty e^{-2x}\, dx = \tfrac{1}{2}.$

(b) $\displaystyle\int_1^\infty \frac{dx}{x} \quad$ diverges.

(c) $\displaystyle\int_1^\infty \frac{dx}{x^2} = 1.$

(d) $\displaystyle\int_{-\infty}^1 \cos \pi x\, dx \quad$ diverges.

VERIFICATION

(a) $\displaystyle\int_0^\infty e^{-2x}\, dx = \lim_{b \to \infty} \int_0^b e^{-2x}\, dx = \lim_{b \to \infty} \left[-\frac{e^{-2x}}{2} \right]_0^b = \lim_{b \to \infty} \left(\frac{1}{2} - \frac{1}{e^{2b}} \right) = \frac{1}{2}.$

(b) $\displaystyle\int_1^\infty \frac{dx}{x} = \lim_{b \to \infty} \int_1^b \frac{dx}{x} = \lim_{b \to \infty} \ln b = \infty.$

(c) $\displaystyle\int_1^\infty \frac{dx}{x^2} = \lim_{b \to \infty} \int_1^b \frac{dx}{x^2} = \lim_{b \to \infty} \left[-\frac{1}{x} \right]_1^b = \lim_{b \to \infty} \left(1 - \frac{1}{b} \right) = 1.$

(d) Note first that

$$\int_a^1 \cos \pi x\, dx = \left[\frac{1}{\pi} \sin \pi x \right]_a^1 = -\frac{1}{\pi} \sin \pi a.$$

As a tends to $-\infty$, $\sin \pi a$ oscillates between -1 and 1. Therefore the integral oscillates between $1/\pi$ and $-1/\pi$ and does not converge. ❑

The usual formulas for area and volume are extended to the unbounded case by means of improper integrals.

Example 2 Let p be a positive number. If Ω is the region between the x-axis and the graph of

$$f(x) = \frac{1}{x^p}, \qquad x \geq 1 \qquad\qquad \text{(Figure 10.7.1)}$$

then

$$\text{area of } \Omega = \begin{cases} \dfrac{1}{p-1}, & \text{if } p > 1 \\[2mm] \infty, & \text{if } p \leq 1. \end{cases}$$

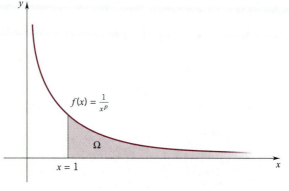

Figure 10.7.1

This comes about from setting

$$\text{area of } \Omega = \lim_{b \to \infty} \int_1^b \frac{dx}{x^p} = \int_1^\infty \frac{dx}{x^p}.$$

For $p \neq 1$,

$$\int_1^\infty \frac{dx}{x^p} = \lim_{b \to \infty} \int_1^b \frac{dx}{x^p} = \lim_{b \to \infty} \frac{1}{1-p}(b^{1-p} - 1) = \begin{cases} \dfrac{1}{p-1} & \text{if } p > 1 \\ \infty & \text{if } p < 1. \end{cases}$$

For $p = 1$,

$$\int_1^\infty \frac{dx}{x^p} = \int_1^\infty \frac{dx}{x} = \infty,$$

as you have seen already. ❏

Remark It is easy to verify that if $p \leq 0$ and Ω is the region between the x-axis and the graph of $f(x) = 1/x^p = x^{-p}$, $-p \geq 0$, then area of $\Omega = \infty$. Thus the conclusion in Example 2 actually holds for all real numbers p. ❏

Example 3 From the last example you know that the region below the graph of

$$f(x) = \frac{1}{x}, \qquad x \geq 1$$

has infinite area. Suppose that this region with infinite area is revolved about the x-axis (see Figure 10.7.2). What is the volume V of the resulting solid? It may surprise you somewhat, but the volume is not infinite. In fact, it is π. Using the disc method to calculate the volume (see Section 6.2), we have

$$V = \int_1^\infty \pi [f(x)]^2 \, dx = \pi \int_1^\infty \frac{dx}{x^2} = \pi \lim_{b \to \infty} \int_1^b \frac{dx}{x^2}$$

$$= \pi \lim_{b \to \infty} \left[\frac{-1}{x} \right]_1^b = \pi \cdot 1 = \pi. \quad ❏$$

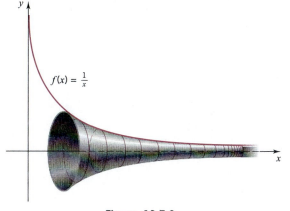

Figure 10.7.2

For future reference we record the following:

(10.7.1)

$$\int_1^\infty \frac{dx}{x^p} \text{ converges for } p > 1 \text{ and diverges for } p \le 1.$$

It is often difficult to determine the convergence or divergence of a given improper integral by direct methods, that is, by calculating the definite integral and evaluating the limit. In such cases we can sometimes gain information by comparison with integrals of known behavior.

(10.7.2)

(*A comparison test*) Suppose that f and g are continuous and

$$0 \le f(x) \le g(x) \qquad \text{for all } x \in [a, \infty).$$

(i) If $\displaystyle\int_a^\infty g(x)\, dx$ converges, then $\displaystyle\int_a^\infty f(x)\, dx$ converges.

(ii) If $\displaystyle\int_a^\infty f(x)\, dx$ diverges, then $\displaystyle\int_a^\infty g(x)\, dx$ diverges.

A similar result holds for integrals from $-\infty$ to b. Figure 10.7.3 illustrates the comparison test (10.7.2). The proof of the result is left to you as an exercise.

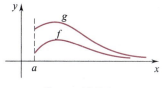

Figure 10.7.3

Example 4 The improper integral

$$\int_1^\infty \frac{dx}{\sqrt{1 + x^3}}$$

converges since

$$\frac{1}{\sqrt{1 + x^3}} < \frac{1}{x^{3/2}} \quad \text{for } x \in [1, \infty) \qquad \text{and} \qquad \int_1^\infty \frac{dx}{x^{3/2}} \text{ converges.}$$

In contrast, if we tried to evaluate

$$\lim_{b \to \infty} \int_1^b \frac{dx}{\sqrt{1 + x^3}}$$

directly, we would have to calculate the integral

$$\int \frac{dx}{\sqrt{1 + x^3}},$$

and this cannot be done by any of the methods we have developed so far. ❏

Example 5 The improper integral

$$\int_1^\infty \frac{dx}{\sqrt{1 + x^2}}$$

diverges since

$$\frac{1}{1 + x} \leq \frac{1}{\sqrt{1 + x^2}} \quad \text{for } x \in [1, \infty) \quad \text{and} \quad \int_1^\infty \frac{dx}{1 + x} \text{ diverges.}$$

This result can also be obtained by evaluating

$$\int_1^b \frac{dx}{\sqrt{1 + x^2}}$$

and then calculating the limit as $b \to \infty$. Try it. ❏

Suppose now that f is continuous on $(-\infty, \infty)$. The *improper integral*

$$\int_{-\infty}^\infty f(x)\, dx$$

is said to *converge* iff

$$\int_{-\infty}^0 f(x)\, dx \quad \text{and} \quad \int_0^\infty f(x)\, dx$$

both converge. We then set

$$\int_{-\infty}^\infty f(x)\, dx = L + M,$$

where

$$\int_{-\infty}^0 f(x)\, dx = L \quad \text{and} \quad \int_0^\infty f(x)\, dx = M.$$

Example 6 Determine whether the improper integral

$$\int_{-\infty}^\infty \frac{e^x}{1 + e^{2x}}\, dx$$

converges or diverges. If it converges, give its value.

SOLUTION   First consider the indefinite integral

$$\int \frac{e^x}{1 + e^{2x}}\, dx.$$

Let $u = e^x$ and $du = e^x\, dx$.   Then

$$\int \frac{e^x}{1 + e^{2x}}\, dx = \int \frac{1}{1 + u^2}\, du = \tan^{-1}u + C = \tan^{-1}(e^x) + C.$$

Now,

$$\int_{-\infty}^{0} \frac{e^x}{1 + e^{2x}}\, dx = \lim_{a \to -\infty} \int_{a}^{0} \frac{e^x}{1 + e^{2x}}\, dx = \lim_{a \to -\infty} \left[\tan^{-1}(e^x) \right]_{a}^{0}$$

$$= \tan^{-1}(1) - \lim_{a \to -\infty} \tan^{-1}(e^a) = \frac{\pi}{4} - 0 = \frac{\pi}{4}$$

and

$$\int_{0}^{\infty} \frac{e^x}{1 + e^{2x}}\, dx = \lim_{b \to \infty} \int_{0}^{b} \frac{e^x}{1 + e^{2x}}\, dx = \lim_{b \to \infty} \left[\tan^{-1}(e^x) \right]_{0}^{b}$$

$$= \lim_{b \to \infty} \tan^{-1}(e^b) - \tan^{-1}(1) = \frac{\pi}{2} - \frac{\pi}{4} = \frac{\pi}{4}.$$

Therefore the improper integral

$$\int_{-\infty}^{\infty} \frac{e^x}{1 + e^{2x}}\, dx$$

converges.   Its value is $\frac{1}{4}\pi + \frac{1}{4}\pi = \frac{1}{2}\pi$.   ❏

Remark   You might be wondering why we did not define

$$\int_{-\infty}^{\infty} f(x)\, dx$$

in terms of the limit

$$\lim_{b \to \infty} \int_{-b}^{b} f(x)\, dx.$$

In fact, it can be shown that if the integral from $-\infty$ to ∞ exists in the sense of the original definition, then the limit just given also exists and they are equal.   On the other hand, if we let $f(x) = x$, then

$$\lim_{b \to \infty} \int_{-b}^{b} x\, dx = \lim_{b \to \infty} \left[\frac{x^2}{2} \right]_{-b}^{b} = \lim_{b \to \infty} \left[\frac{b^2}{2} - \frac{b^2}{2} \right] = 0,$$

while the calculation of

$$\int_{-\infty}^{\infty} x\, dx$$

using the definition will lead to an indeterminate of the form $\infty - \infty$.   ❏

Integrals of Unbounded Functions

Improper integrals can also arise on bounded intervals. Suppose that f is continuous on the half-open interval $[a, b)$ but is unbounded there. See Figure 10.7.4. For each number $c < b$, we can form the definite integral

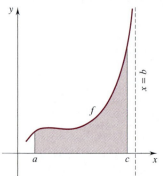

Figure 10.7.4

$$\int_a^c f(x)\, dx.$$

If

$$\lim_{c \to b^-} \int_a^c f(x)\, dx = L$$

exists, then we say that

the improper integral $\quad \displaystyle\int_a^b f(x)\, dx \quad$ converges to L.

Otherwise, *the improper integral diverges.*

In a similar manner, if f is continuous on $(a, b]$ and unbounded at a, then we consider the limit

$$\lim_{c \to a^+} \int_c^b f(x)\, dx.$$

If this limit exists and has the value L, then

the improper integral $\quad \displaystyle\int_a^b f(x)\, dx \quad$ converges to L.

Otherwise, *the improper integral diverges.*

Example 7

(a) $\displaystyle\int_0^1 (1-x)^{-2/3}\, dx = 3.$ **(b)** $\displaystyle\int_0^2 \frac{dx}{x}$ diverges.

VERIFICATION

(a) $\displaystyle\int_0^1 (1-x)^{-2/3}\, dx = \lim_{c \to 1^-} \int_0^c (1-x)^{-2/3}\, dx$

$$= \lim_{c \to 1^-} \left[-3(1-x)^{1/3} \right]_0^c = \lim_{c \to 1^-} \left[-3(1-c)^{1/3} + 3 \right] = 3.$$

(b) $\displaystyle\int_0^2 \frac{dx}{x} = \lim_{c \to 0^+} \int_c^2 \frac{dx}{x} = \lim_{c \to 0^+} [\ln 2 - \ln c] = \infty.$ ❑

Now suppose that f is continuous on an interval $[a, b]$ except at some point c in (a, b) where $f(x) \to \pm\infty$ as $x \to c^-$ or as $x \to c^+$. We say that the *improper integral*

$$\int_a^b f(x)\, dx$$

converges iff *both* of the integrals

$$\int_a^c f(x)\, dx \qquad \text{and} \qquad \int_c^b f(x)\, dx$$

converge. If

$$\int_a^c f(x)\,dx = L \qquad \text{and} \qquad \int_c^b f(x)\,dx = M,$$

then

$$\int_a^b f(x)\,dx = L + M.$$

Example 8 To evaluate

(∗)
$$\int_1^4 \frac{dx}{(x-2)^2}$$

we need to calculate

$$\lim_{c\to 2^-} \int_1^c \frac{dx}{(x-2)^2} \qquad \text{and} \qquad \lim_{c\to 2^+} \int_c^4 \frac{dx}{(x-2)^2}.$$

As you can verify, neither of these limits exists and thus improper integral (∗) diverges.

Notice that, if we ignore the fact that integral (∗) is improper, then we are led to the *incorrect conclusion* that

$$\int_1^4 \frac{dx}{(x-2)^2} = \left[\frac{-1}{x-2}\right]_1^4 = -\frac{3}{2}. \qquad \text{(see Figure 10.7.5)} \quad ❑$$

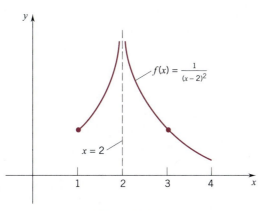

Figure 10.7.5

Example 9 Evaluate

$$\int_{-2}^1 \frac{dx}{x^{4/5}}.$$

SOLUTION Since $1/x^{4/5} \to \infty$ as $x \to 0^-$ and as $x \to 0^+$, the given integral is improper. Therefore, we need to calculate

$$\int_{-2}^0 \frac{dx}{x^{4/5}} \qquad \text{and} \qquad \int_0^1 \frac{dx}{x^{4/5}}.$$

Now

$$\int_{-2}^{0} \frac{dx}{x^{4/5}} = \lim_{c \to 0^-} \int_{-2}^{c} \frac{dx}{x^{4/5}} = \lim_{c \to 0^-} \left[5x^{1/5} \right]_{-2}^{c} = \lim_{c \to 0^-} [5c^{1/5} - 5(-2)^{1/5}] = 5(2^{1/5})$$

and

$$\int_{0}^{1} \frac{dx}{x^{4/5}} = \lim_{c \to 0^+} \int_{c}^{1} \frac{dx}{x^{4/5}} = \lim_{c \to 0^+} \left[5x^{1/5} \right]_{c}^{1} = \lim_{c \to 0^+} [5 - 5c^{1/5}] = 5.$$

Thus, the improper integral converges and

$$\int_{-2}^{1} \frac{dx}{x^{4/5}} = 5 + 5(2^{1/5}) \cong 10.74. \quad \square$$

EXERCISES 10.7

In Exercises 1–34, evaluate the improper integrals that converge.

1. $\int_{1}^{\infty} \frac{dx}{x^2}.$

2. $\int_{0}^{\infty} \frac{dx}{1 + x^2}.$

3. $\int_{0}^{\infty} \frac{dx}{4 + x^2}.$

4. $\int_{0}^{\infty} e^{-px} \, dx, \quad p > 0.$

5. $\int_{0}^{\infty} e^{px} \, dx, \quad p > 0.$

6. $\int_{0}^{1} \frac{dx}{\sqrt{x}}.$

7. $\int_{0}^{8} \frac{dx}{x^{2/3}}.$

8. $\int_{0}^{1} \frac{dx}{x^2}.$

9. $\int_{0}^{1} \frac{dx}{\sqrt{1 - x^2}}.$

10. $\int_{0}^{1} \frac{dx}{\sqrt{1 - x}}.$

11. $\int_{0}^{2} \frac{x}{\sqrt{4 - x^2}} \, dx.$

12. $\int_{0}^{a} \frac{dx}{\sqrt{a^2 - x^2}}.$

13. $\int_{e}^{\infty} \frac{\ln x}{x} \, dx.$

14. $\int_{e}^{\infty} \frac{dx}{x \ln x}.$

15. $\int_{0}^{1} x \ln x \, dx.$

16. $\int_{e}^{\infty} \frac{dx}{x(\ln x)^2}.$

17. $\int_{-\infty}^{\infty} \frac{dx}{1 + x^2}.$

18. $\int_{2}^{\infty} \frac{dx}{x^2 - 1}.$

19. $\int_{-\infty}^{\infty} \frac{dx}{x^2}.$

20. $\int_{1/3}^{3} \frac{dx}{\sqrt[3]{3x - 1}}.$

21. $\int_{1}^{\infty} \frac{dx}{x(x + 1)}.$

22. $\int_{-\infty}^{0} x e^x \, dx.$

23. $\int_{3}^{5} \frac{x}{\sqrt{x^2 - 9}} \, dx.$

24. $\int_{1}^{4} \frac{dx}{x^2 - 4}.$

25. $\int_{-3}^{3} \frac{dx}{x(x + 1)}.$

26. $\int_{1}^{\infty} \frac{x}{(1 + x^2)^2} \, dx.$

27. $\int_{-3}^{1} \frac{dx}{x^2 - 4}.$

28. $\int_{0}^{\infty} \sinh x \, dx.$

29. $\int_{0}^{\infty} \cosh x \, dx.$

30. $\int_{1}^{4} \frac{dx}{x^2 - 5x + 6}.$

31. $\int_{0}^{\infty} e^{-x} \sin x \, dx.$

32. $\int_{0}^{\infty} \cos^2 x \, dx.$

33. $\int_{0}^{1} \frac{e^{\sqrt{x}}}{\sqrt{x}} \, dx.$

34. $\int_{0}^{\pi/2} \frac{\cos x}{\sqrt{\sin x}} \, dx.$

35. The integral

$$\int_{0}^{1} \sin^{-1} x \, dx$$

is a "proper" definite integral. But the integration technique (integration by parts) will lead to an improper integral. Evaluate this integral.

36. (a) For what values of r is

$$\int_{0}^{\infty} x^r e^{-x} \, dx$$

convergent?
(b) Use mathematical induction to show that

$$\int_{0}^{\infty} x^n e^{-x} \, dx = n!, \quad n = 1, 2, 3, \ldots.$$

37. The integral

$$\int_{0}^{\infty} \frac{1}{\sqrt{x}(1 + x)} \, dx$$

is improper for both of the reasons discussed in this section: the interval is infinite and the integrand is unbounded. If we rewrite the integral as

$$\int_0^\infty \frac{1}{\sqrt{x}(1+x)}\,dx = \int_0^1 \frac{1}{\sqrt{x}(1+x)}\,dx + \int_1^\infty \frac{1}{\sqrt{x}(1+x)}\,dx$$

then we have two improper integrals, the first having an unbounded integrand and the second an infinite interval. If each of these integrals converges with values L_1 and L_2, respectively, then the given integral converges and has the value $L_1 + L_2$. Evaluate the given integral.

38. Evaluate

$$\int_1^\infty \frac{1}{x\sqrt{x^2-1}}\,dx$$

using the method given in Exercise 37.

39. Let Ω be the region bounded by the coordinate axes, the graph of $y = 1/\sqrt{x}$, and the line $x = 1$. (a) Sketch Ω. (b) Show Ω has finite area and find it. (c) Show that if Ω is revolved about the x-axis, the solid obtained does not have finite volume.

40. Let Ω be the region between the graph of $y = 1/(1+x^2)$ and the x-axis, $x \ge 0$. (a) Sketch Ω. (b) Find the area of Ω. (c) Find the volume of the solid obtained by revolving Ω about the x-axis. (d) Find the volume of the solid obtained by revolving Ω about the y-axis.

41. Let Ω be the region bounded by the curve $y = e^{-x}$ and the x-axis, $x \ge 0$. (a) Sketch Ω. (b) Find the area of Ω. (c) Find the volume of the solid obtained by revolving Ω about the x-axis. (d) Find the volume obtained by revolving Ω about the y-axis. (e) Find the lateral surface area of the solid in part (c).

42. What point would you call the centroid of the region in Exercise 41? Does Pappus's theorem work in this instance?

43. Let Ω be the region bounded by the curve $y = e^{-x^2}$ and the x-axis, $x \ge 0$. (a) Show that Ω has finite area. (The area is actually $\frac{1}{2}\sqrt{\pi}$, as you will see in Chapter 16.) (b) Calculate the volume generated by revolving Ω about the y-axis.

44. Let Ω be the region bounded below by $y(x^2+1)=x$, above by $xy = 1$, and to the left by $x = 1$. (a) Find the area of Ω. (b) Show that the solid generated by revolving Ω about the x-axis has finite volume. (c) Calculate the volume generated by revolving Ω about the y-axis.

45. Let Ω be the region bounded by the curve $y = x^{-1/4}$ and the x-axis, $0 < x \le 1$. (a) Sketch Ω. (b) Find the area of Ω. (c) Find the volume of the solid obtained by revolving Ω about the x-axis. (d) Find the volume of the solid obtained by revolving Ω about the y-axis.

46. Prove the validity of the comparison test (10.7.2).

In Exercises 47–52, use the comparison test (10.7.2) to determine which of the integrals converge.

47. $\displaystyle\int_1^\infty \frac{x}{\sqrt{1+x^5}}\,dx.$

48. $\displaystyle\int_1^\infty 2^{-x^2}\,dx.$

49. $\displaystyle\int_0^\infty (1+x^5)^{-1/6}\,dx.$

50. $\displaystyle\int_\pi^\infty \frac{\sin^2 2x}{x^2}\,dx.$

51. $\displaystyle\int_1^\infty \frac{\ln x}{x^2}\,dx.$

52. $\displaystyle\int_e^\infty \frac{dx}{\sqrt{x+1}\,\ln x}.$

53. Calculate the arc distance from the origin to the point $(x(\theta_1),\ y(\theta_1))$ along the exponential spiral $r = ae^{c\theta}$. (Take $a > 0$, $c > 0$.)

54. The function

$$f(x) = \frac{1}{\sqrt{2\pi}} \int_{-\infty}^x e^{-t^2/2}\,dt$$

is important in statistics. Prove that the integral on the right converges for all real x.

Exercises 55–58: **Laplace transforms.** Let f be continuous on $[0,\infty)$. The *Laplace transform* of f is the function F defined by

$$F(s) = \int_0^\infty e^{-sx}f(x)\,dx.$$

The domain of F is the set of all real numbers s such that the improper integral converges. Find the Laplace transform F of each of the following functions and give the domain of F.

55. $f(x) = 1.$ **56.** $f(x) = x.$

57. $f(x) = \cos 2x.$ **58.** $f(x) = e^{ax}.$

Exercises 59–62: **Probability density functions.** A nonnegative function f defined on $(-\infty, \infty)$ is a *probability density function* if

$$\int_{-\infty}^\infty f(x)\,dx = 1.$$

59. Show that the function f defined by

$$f(x) = \begin{cases} 6x/(1+3x^2)^2 & x \ge 0 \\ 0 & x < 0 \end{cases}$$

is a probability density function.

60. Show that the function f defined by

$$f(x) = \begin{cases} ke^{-kx} & x \ge 0 \\ 0 & x < 0, \quad k > 0, \end{cases}$$

is a probability density function. It is called the *exponential density function*.

61. If f is a probability density function, then its *mean* μ is given by

$$\mu = \int_{-\infty}^\infty x f(x)\,dx.$$

Calculate the mean for the exponential density function.

62. If f is a probability density function, then its *standard deviation* σ is given by

$$\sigma = \left[\int_{-\infty}^{\infty} (x - \mu)^2 f(x)\, dx \right]^{1/2}$$

where μ is the mean. Calculate the standard deviation for the exponential density function.

63. (*Useful later*) Let f be a continuous, positive, decreasing function on $[1, \infty)$. Show that

$$\int_1^{\infty} f(x)\, dx \quad \text{converges} \quad \text{iff} \quad \left\{ \int_1^{n} f(x)\, dx \right\} \text{ converges.}$$

■ CHAPTER HIGHLIGHTS

10.1 The Least Upper Bound Axiom

Upper bound, bounded above, least upper bound (p. 637)
Least upper bound axiom (p. 638)
Lower bound, bounded below, greatest lower bound (p. 640)

10.2 Sequences of Real Numbers

Sequence (p. 642)
Bounded above, bounded below, bounded (p. 644)
Increasing, nondecreasing, decreasing, nonincreasing (p. 644)
Recurrence relation (p. 649)

It is sometimes possible to obtain useful information about a sequence $y_n = f(n)$ by applying the techniques of calculus to the function $y = f(x)$.

10.3 Limit of a Sequence

Limit of a sequence (p. 651)   Uniqueness of the limit (p. 652)
Convergent, divergent (p. 653)   Pinching theorem (p. 656)

Every convergent sequence is bounded (p. 653); thus, every unbounded sequence is divergent.

A bounded, monotonic sequence converges. (p. 654)

Suppose that $c_n \to c$ as $n \to \infty$, and all the c_n are in the domain of f. If f is continuous at c, then $f(c_n) \to f(c)$. (p. 657)

10.4 Some Important Limits

For $x > 0$, $\quad \lim_{n \to \infty} x^{1/n} = 1$ $\qquad\qquad$ For $|x| < 1$, $\quad \lim_{n \to \infty} x^n = 0$

For each $\alpha > 0$, $\quad \lim_{n \to \infty} \dfrac{1}{n^\alpha} = 0$ $\qquad\qquad$ For each real x, $\quad \lim_{n \to \infty} \dfrac{x^n}{n!} = 0$

$\lim_{n \to \infty} \dfrac{\ln n}{n} = 0$ $\qquad\qquad$ $\lim_{n \to \infty} n^{1/n} = 1$

For each real x, $\quad \lim_{n \to \infty} \left(1 + \dfrac{x}{n} \right)^n = e^x$ $\qquad$ Cauchy sequence (p. 680)

10.5 The Indeterminate Form (0/0)

L'Hospital's rule (0/0) (p. 668)   Cauchy mean-value theorem (p. 670)

10.6 The Indeterminate Form (∞/∞)

L'Hospital's rule (∞/∞) (p. 674)

$\lim_{x \to \infty} \dfrac{\ln x}{x^\alpha} = 0$ $\qquad$ $\lim_{x \to \infty} \dfrac{x^k}{e^x} = 0$ $\qquad$ $\lim_{x \to \infty} x^x = 1$

Other indeterminate forms: $0 \cdot \infty$, $\infty - \infty$, 0^0, 1^∞, ∞^0 (p. 675)

10.7 Improper Integrals

Integrals over infinite intervals (p. 680)   convergent, divergent (p. 680)

$$\int_1^\infty \frac{dx}{x^p} \text{ converges for } p > 1 \text{ and diverges for } p \le 1.$$

A comparison test (p. 683)

Integrals of unbounded functions (p. 686)   convergent, divergent (p. 686)

■ PROJECTS AND EXPLORATIONS USING TECHNOLOGY

To do these exercises you will need a graphics calculator or a computer with graphing capability. The majority of these problems are open-ended so different approaches may be used to solve them. You should be aware that different approaches can result in slight variations in the answers. Round your numerical answers to at least four decimal places. The rounding method that your calculator or computer uses also may cause variations in answers.

10.1 It is easy to show using the comparison test that the improper integral

$$\int_1^\infty \frac{1}{x^3 + 1}\, dx$$

converges. However, computer programs may have trouble evaluating this integral.
  (a) Use the comparison test to show that this integral converges.
  (b) Use the method of partial fractions to evaluate this integral.
  (c) The function

$$F(x) = \int_1^x \frac{1}{t^3 + 1}\, dt$$

  is an increasing function on $[1, \infty)$. Explain why.
  (d) Evaluate F at a sequence of x values (for example, $x = 10$, $x = 100$, $x = 1000$, $x = 10,000$, and so on) using a computer program. Does F start to decrease as x increases? Can you explain why?
  (e) Set up a procedure for checking the accuracy of the values that you compute.

10.2 This problem investigates the convergence of the improper integral

$$\int_1^\infty 1000xe^{-0.1\sqrt{x}}\, dx.$$

  (a) Let $\{I_k\}$ be the sequence defined by

$$I_k = \int_1^k 1000xe^{-0.1\sqrt{x}}\, dx.$$

  Calculate I_k for $k = 10$, 100, 1000, $10,000$, and so on. Can you determine $L = \lim_{k \to \infty} I_k$?
  (b) The difference $E(k) = |L - I_k|$ is the "error." The limit L can also be calculated using integration by parts or by software (be careful that the software gives reasonable answers). Sketch the graph of E to see how quickly the values I_k approach L.
  (c) Let

$$a_k = \int_{10^{k-1}}^{10^k} 1000xe^{-0.1\sqrt{x}}\, dx.$$

Show that

$$I_{10^k} = a_0 + a_1 + a_2 + \cdots + a_k.$$

and compare the values for I_k obtained this way with the values calculated directly in part (a).

(d) Explain why a_k might have advantages over I_k in terms of approximating the integral.

10.3 When calculating compound interest, we work with the functions

$$f_r(n) = \left(1 + \frac{r}{n} \right)^n,$$

where r is the annual interest rate and n is the number of compounding periods per year.

(a) Let $r = 0.1$. Find and discuss the meaning of $f_r(4)$, $f_r(12)$, $f_r(365)$, $f_r(8760)$, $f_r(525,600)$, and $f_r(31,536,600)$.

(b) Repeat the calculations in part (a) using $r = 0.01$, 0.05, 0.075, and 0.12.

(c) Mathematically, f_r can be considered for all real numbers and not just for positive integers n. Based on your results in parts (a) and (b), does it appear that

$$\lim_{x \to \infty} f_r(x)$$

exists when r is a given number?

(d) Now let

$$g_n(r) = \left(1 + \frac{r}{n} \right)^n.$$

For $n = 4$, 12, 365, 8760, 525,600, and 31,536,000, respectively, approximate the following limits

$$\lim_{x \to 0^+} g_n(x).$$

What is the meaning of these limits?

(e) For the values of n and r given in parts (a) and (b), approximate $f_r'(n)$ and $g_n'(r)$. What conclusions can you draw? What are the relationships between these derivatives for a given n and r?

(f) Use a graphing utility to graph the functions $f_r(x)$ and $g_n(x)$ for the values of n and r given in parts (a) and (b). Are these functions increasing or decreasing? Do they have inverse functions?

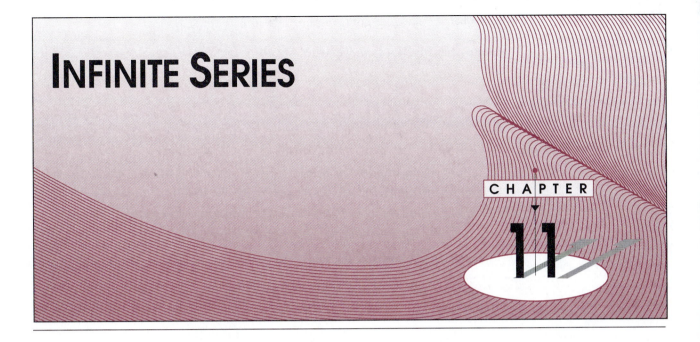

INFINITE SERIES

CHAPTER 11

■ 11.1 SIGMA NOTATION

In Section 10.2 we defined a sequence as a real-valued function whose domain is the set of positive integers. For example, to indicate the sequence

$$\{1, \tfrac{1}{2}, \tfrac{1}{4}, \tfrac{1}{8}, \ldots\}$$

we would set $a_n = (\tfrac{1}{2})^{n-1}$, $n = 1, 2, 3, \ldots$, and write

$$\{a_1, a_2, a_3, a_4, \ldots\}.$$

In this chapter, however, it will often be convenient to begin a sequence with an index other than 1. So, continuing with the example, we can also set $b_n = (\tfrac{1}{2})^n$, $n = 0, 1, 2, \ldots$, and write

$$\{b_0, b_1, b_2, b_3, \ldots\},$$

thereby beginning with the index 0. In general, we can set $c_n = (\tfrac{1}{2})^{n-p}$, where p is an integer and $n = p, p + 1, p + 2, \ldots$, and write

$$\{c_p, c_{p+1}, c_{p+2}, c_{p+3}, \ldots\},$$

so that we begin with the index p.

The symbol Σ is the capital Greek letter "sigma." We write

(1)
$$\sum_{k=0}^{n} a_k$$

(read "the sum of the a sub k from k equals 0 to k equals n") to indicate the sum

$$a_0 + a_1 + \cdots + a_n.$$

More generally, if $n \geq m$, we write

(2)
$$\sum_{k=m}^{n} a_k$$

to indicate the sum

$$a_m + a_{m+1} + \cdots + a_n.$$

In (1) and (2) the letter "k" is being used as a "dummy" variable. That is, it can be replaced by any letter not already engaged. For instance,

$$\sum_{i=3}^{7} a_i, \quad \sum_{j=3}^{7} a_j, \quad \sum_{k=3}^{7} a_k$$

can all be used to indicate the sum

$$a_3 + a_4 + a_5 + a_6 + a_7.$$

Translating

$$(a_0 + \cdots + a_n) + (b_0 + \cdots + b_n) = (a_0 + b_0) + \cdots + (a_n + b_n),$$
$$\alpha(a_0 + \cdots + a_n) = \alpha a_0 + \cdots + \alpha a_n,$$
$$(a_0 + \cdots + a_m) + (a_{m+1} + \cdots + a_n) = a_0 + \cdots + a_n$$

into the Σ-notation, we have

$$\sum_{k=0}^{n} a_k + \sum_{k=0}^{n} b_k = \sum_{k=0}^{n} (a_k + b_k), \quad \alpha \sum_{k=0}^{n} a_k = \sum_{k=0}^{n} \alpha a_k,$$
$$\sum_{k=0}^{m} a_k + \sum_{k=m+1}^{n} a_k = \sum_{k=0}^{n} a_k.$$

At times it is convenient to change indices. In this connection note that

$$\sum_{k=j}^{n} a_k = \sum_{i=0}^{n-j} a_{i+j}. \qquad \text{(set } i = k - j)$$

Both expressions are abbreviations for $a_j + a_{j+1} + \cdots + a_n$.

You can familiarize yourself further with this notation by doing the exercises below, but first one more remark. If all the a_k are equal to some fixed number x, then

$$\sum_{k=0}^{n} a_k \quad \text{can be written} \quad \sum_{k=0}^{n} x.$$

Obviously then

$$\sum_{k=0}^{n} x = \overbrace{x + x + \cdots + x}^{n+1} = (n+1)x.$$

In particular

$$\sum_{k=0}^{n} 1 = n + 1.$$

EXERCISES 11.1

In Exercises 1–12, evaluate the given expression.

1. $\displaystyle\sum_{k=0}^{2} (3k + 1)$.

2. $\displaystyle\sum_{k=1}^{4} (3k - 1)$.

3. $\displaystyle\sum_{k=0}^{3} 2^k$.

4. $\displaystyle\sum_{k=0}^{3} (-1)^k 2^k$.

5. $\displaystyle\sum_{k=0}^{3} (-1)^k 2^{k+1}$.

6. $\displaystyle\sum_{k=2}^{5} (-1)^{k+1} 2^{k-1}$.

7. $\displaystyle\sum_{k=1}^{4} \frac{1}{2^k}$.

8. $\displaystyle\sum_{k=2}^{5} \frac{1}{k!}$.

9. $\displaystyle\sum_{k=3}^{5} \frac{(-1)^k}{k!}$.

10. $\displaystyle\sum_{k=2}^{4} \frac{1}{3^{k-1}}$.

11. $\displaystyle\sum_{k=0}^{3} (\tfrac{1}{2})^{2k}$.

12. $\displaystyle\sum_{k=1}^{3} (-1)^{k+1}(\tfrac{1}{2})^{2k-1}$.

In Exercises 13–28, express in sigma notation.

13. $1 + 3 + 5 + 7 + \cdots + 21$.

14. $1 - 3 + 5 - 7 + \cdots - 19$.

15. $2 \cdot 1 + 2 \cdot 2 + 2 \cdot 3 + \cdots + 2 \cdot 25$.

16. $1 \cdot 2 + 2 \cdot 3 + 3 \cdot 4 + \cdots + 35 \cdot 36$.

17. $1 - \sqrt{2} + \sqrt{3} - 2 + \sqrt{5} - \cdots + 9$.

18. $\dfrac{\tan 1}{2} + \dfrac{\tan 2}{5} + \dfrac{\tan 3}{10} + \dfrac{\tan 4}{17} + \cdots + \dfrac{\tan 10}{101}$.

19. The lower sum $m_1\Delta x_1 + m_2\Delta x_2 + \cdots + m_n\Delta x_n$.

20. The upper sum $M_1\Delta x_1 + M_2\Delta x_2 + \cdots + M_n\Delta x_n$.

21. The Riemann sum
$f(x_1^*)\Delta x_1 + f(x_2^*)\Delta x_2 + \cdots + f(x_n^*)\Delta x_n$.

22. $a^5 + a^4b + a^3b^2 + a^2b^3 + ab^4 + b^5$.

23. $a^5 - a^4b + a^3b^2 - a^2b^3 + ab^4 - b^5$.

24. $a^n + a^{n-1}b + \cdots + ab^{n-1} + b^n$.

25. $a_0x^4 + a_1x^3 + a_2x^2 + a_3x + a_4$.

26. $a_0x^n + a_1x^{n-1} + \cdots + a_{n-1}x + a_n$.

27. $1 - 2x + 3x^2 - 4x^3 + 5x^4$.

28. $3x - 4x^2 + 5x^3 - 6x^4$.

In Exercises 29–32, write the given sums as $\displaystyle\sum_{k=3}^{10} a_k$ and $\displaystyle\sum_{i=0}^{7} a_{i+3}$:

29. $\dfrac{1}{2^3} + \dfrac{1}{2^4} + \cdots + \dfrac{1}{2^{10}}$.

30. $\dfrac{3^3}{3!} + \dfrac{4^4}{4!} + \cdots + \dfrac{10^{10}}{10!}$.

31. $\dfrac{3}{4} - \dfrac{4}{5} + \cdots - \dfrac{10}{11}$.

32. $\dfrac{1}{3} + \dfrac{1}{5} + \dfrac{1}{7} + \cdots + \dfrac{1}{17}$.

In Exercises 33–36, verify by a change of indices that the two sums are identical.

33. $\displaystyle\sum_{k=2}^{10} \frac{k}{k^2 + 1}$; $\displaystyle\sum_{n=-1}^{7} \frac{n + 3}{n^2 + 6n + 10}$.

34. $\displaystyle\sum_{n=2}^{12} \frac{(-1)^n}{n - 1}$; $\displaystyle\sum_{k=1}^{11} \frac{(-1)^{k+1}}{k}$.

35. $\displaystyle\sum_{k=4}^{25} \frac{1}{k^2 - 9}$; $\displaystyle\sum_{n=7}^{28} \frac{1}{n^2 - 6n}$.

36. $\displaystyle\sum_{k=0}^{15} \frac{3^{2k}}{k!}$; $81\displaystyle\sum_{n=-2}^{13} \frac{3^{2n}}{(n + 2)!}$.

37. (a) (*Important*) Show that for $x \neq 1$
$$\sum_{k=0}^{n} x^k = \frac{1 - x^{n+1}}{1 - x}.$$

(b) Determine whether the sequence $a_n = \displaystyle\sum_{k=0}^{n} \frac{1}{3^k}$ converges and, if it does, find the limit.

38. Express $\displaystyle\sum_{k=1}^{n} \frac{a_k}{10^k}$ as a decimal fraction, given that each a_k is an integer from 0 to 9.

39. Let p be a positive integer. Show that, as $n \to \infty$,
$$a_n \to L \qquad \text{iff} \qquad a_{n-p} \to L.$$

40. Show that
$$\sum_{k=1}^{n} \frac{1}{\sqrt{k}} \geq \sqrt{n}.$$

In Exercises 41–44, verify by induction.

41. $\displaystyle\sum_{k=1}^{n} k = \tfrac{1}{2}(n)(n + 1)$.

42. $\displaystyle\sum_{k=1}^{n} (2k - 1) = n^2$.

43. $\displaystyle\sum_{k=1}^{n} k^2 = \tfrac{1}{6}(n)(n + 1)(2n + 1)$.

44. $\displaystyle\sum_{k=1}^{n} k^3 = \left(\sum_{k=1}^{n} k\right)^2$.

In Exercises 45–48, evaluate the sum.

45. $\displaystyle\sum_{k=1}^{10} (2k + 3)$.

46. $\displaystyle\sum_{k=1}^{10} (2k^2 + 3k)$.

47. $\displaystyle\sum_{k=1}^{8} (2k - 1)^2$.

48. $\displaystyle\sum_{k=1}^{n} k(k^2 - 5)$.

■ 11.2 INFINITE SERIES

Introduction; Definitions

While it is possible to add two numbers, three numbers, a hundred numbers, or even a million numbers, how can we attach meaning to the sum of an infinite number of numbers? The theory of infinite series arose from attempts to answer this question. As you might expect, a limit process is involved.

To form an infinite series we begin with an infinite sequence of real numbers: $a_0, a_1, a_2, \ldots$. We can't form the sum of all the a_k (there is an infinite number of them), but we can form the *partial sums*

$$s_0 = a_0 = \sum_{k=0}^{0} a_k,$$

$$s_1 = a_0 + a_1 = \sum_{k=0}^{1} a_k,$$

$$s_2 = a_0 + a_1 + a_2 = \sum_{k=0}^{2} a_k,$$

$$s_3 = a_0 + a_1 + a_2 + a_3 = \sum_{k=0}^{3} a_k,$$

$$\vdots$$

$$s_n = a_0 + a_1 + a_2 + a_3 + \cdots + a_n = \sum_{k=0}^{n} a_k$$

$$\vdots$$

Continuing in this way, we are led to consider the "infinite sum" $\sum_{k=0}^{\infty} a_k$ which is called an *infinite series*. The corresponding sequence $\{s_n\}$ is called the *sequence of partial sums* of the series.

DEFINITION 11.2.1

Given the infinite series $\sum_{k=0}^{\infty} a_k$. If the sequence of partial sums $\{s_n\}$ converges to a finite limit L, then the series $\sum_{k=0}^{\infty} a_k$ is said to *converge* to L, written

$$\sum_{k=0}^{\infty} a_k = L.$$

The number L is called the *sum* of the series. If the sequence of partial sums diverges, then the series $\sum_{k=0}^{\infty} a_k$ *diverges*.

Remark It is important to note that the sum of a series is not a sum in the ordinary sense. It is a limit. ❏

Here are some examples.

Example 1 We begin with the series

$$\sum_{k=0}^{\infty} \frac{1}{(k+1)(k+2)}.$$

To determine whether or not this series converges we must examine the partial sums. Since

$$\frac{1}{(k+1)(k+2)} = \frac{1}{k+1} - \frac{1}{k+2}, \qquad \text{(partial fraction decomposition, see Section 8.6)}$$

you can see that

$$s_n = \frac{1}{1 \cdot 2} + \frac{1}{2 \cdot 3} + \cdots + \frac{1}{n(n+1)} + \frac{1}{(n+1)(n+2)}$$

$$= \left(\frac{1}{1} - \frac{1}{2}\right) + \left(\frac{1}{2} - \frac{1}{3}\right) + \cdots + \left(\frac{1}{n} - \frac{1}{n+1}\right) + \left(\frac{1}{n+1} - \frac{1}{n+2}\right)$$

$$= 1 - \frac{1}{2} + \frac{1}{2} - \frac{1}{3} + \cdots + \frac{1}{n} - \frac{1}{n+1} + \frac{1}{n+1} - \frac{1}{n+2}.$$

Since all but the first and last terms occur in pairs with opposite signs, the sum collapses to give

$$s_n = 1 - \frac{1}{n+2}.$$

Obviously, as $n \to \infty$, $s_n \to 1$. This means that the series converges to 1:

$$\sum_{k=0}^{\infty} \frac{1}{(k+1)(k+2)} = 1. \quad ❏$$

Remark Infinite series with the special property illustrated in Example 1 (that is, except for the first and last term, the terms can be arranged in pairs with opposite signs) are called *telescoping series*. In general,

$$\sum_{k=p}^{n} \{f(k) - f(k+1)\} = f(p) - f(n+1) \qquad \text{and}$$

$$\sum_{k=p}^{n} \{f(k) - f(k-1)\} = f(n) - f(p-1). \quad \text{(verify these)} \quad ❏$$

Example 2 Here we examine two divergent series

$$\sum_{k=0}^{\infty} 2^k \qquad \text{and} \qquad \sum_{k=1}^{\infty} (-1)^k.$$

The partial sums of the first series take the form

$$s_n = \sum_{k=0}^{n} 2^k = 1 + 2 + \cdots + 2^n.$$

The sequence $\{s_n\}$ is unbounded and therefore divergent (10.3.5). This means that the series diverges.

For the second series we have

$$s_n = -1 \quad \text{if } n \text{ is odd} \qquad \text{and} \qquad s_n = 0 \quad \text{if } n \text{ is even.}$$

The sequence of partial sums looks like this:

$$-1, 0, -1, 0, -1, 0, \ldots.$$

The series diverges since the sequence of partial sums diverges. ❏

Remark Example 2 illustrates two types of divergence. In the first case, $s_n \to \infty$ as $n \to \infty$. The notation $\sum_{k=0}^{\infty} a_k = \infty$ is sometimes used to denote this type of divergence. In the second case, s_n oscillates between -1 and 0. ❏

The Geometric Series

Fix a real number x. The sequence $1, x, x^2, x^3, \ldots = \{x^n\}, n = 0, 1, 2, 3, \ldots$, is called a *geometric progression*. Concerning convergence, we know that if $|x| < 1$, then $x^n \to 0$ ("special limit" (10.4.2)). If $x = 1$, then we have the constant sequence $1, 1, 1, \ldots$, which clearly converges to 1. Finally, it is easy to show that $\{x^n\}$ is divergent when $x = -1$ (the sequence oscillates between 1 and -1) and when $|x| > 1$ (the sequence is unbounded).

The sums

$$1, \quad 1 + x, \quad 1 + x + x^2, \quad 1 + x + x^2 + x^3, \ldots$$

generated by numbers in geometric progression are the partial sums of what is known as the *geometric series*:

$$\sum_{k=0}^{\infty} x^k.$$

The geometric series arises in so many contexts that it merits special attention. The following result is fundamental:

(11.2.2)

(i) if $|x| < 1$,   then $\displaystyle\sum_{k=0}^{\infty} x^k = \frac{1}{1-x}$;

(ii) if $|x| \geq 1$,   then $\displaystyle\sum_{k=0}^{\infty} x^k$ diverges.

PROOF The nth partial sum of the geometric series

$$\sum_{k=0}^{\infty} x^k$$

takes the form

(1) $$s_n = 1 + x + \cdots + x^n.$$

Multiplication by x gives

$$xs_n = x + x^2 + \cdots + x^{n+1}.$$

Subtracting the second equation from the first, we find that

$$(1 - x)s_n = 1 - x^{n+1}.$$

For $x \neq 1$, this gives

(2)
$$s_n = \frac{1 - x^{n+1}}{1 - x}.$$

If $|x| < 1$, then $x^{n+1} \to 0$ as $n \to \infty$ and thus, by Equation (2),

$$s_n \to \frac{1}{1 - x}.$$

This proves (i).

Now let's prove (ii). For $x = 1$, we use Equation (1) and deduce that $s_n = n + 1$. Obviously, $\{s_n\}$ diverges. For $x \neq 1$ with $|x| \geq 1$, we use Equation (2). Since in this instance $\{x^{n+1}\}$ diverges, $\{s_n\}$ diverges. ❏

You may have seen (11.2.2) before, written as

$$a + ar + ar^2 + \cdots + ar^n + \cdots = \begin{cases} \dfrac{a}{1 - r}, & |r| < 1 \\ \text{diverges}, & |r| \geq 1, \end{cases} \qquad (a \neq 0).$$

Taking $a = 1$ and $r = \frac{1}{2}$, we have

$$\sum_{k=0}^{\infty} \frac{1}{2^k} = \frac{1}{1 - \frac{1}{2}} = 2.$$

Begin the summation at $k = 1$ instead of at $k = 0$, and you see that

(11.2.3)
$$\boxed{\sum_{k=1}^{\infty} \frac{1}{2^k} = 1.}$$

The partial sums of this series

$$s_1 = \tfrac{1}{2},$$
$$s_2 = \tfrac{1}{2} + \tfrac{1}{4} = \tfrac{3}{4},$$
$$s_3 = \tfrac{1}{2} + \tfrac{1}{4} + \tfrac{1}{8} = \tfrac{7}{8},$$
$$s_4 = \tfrac{1}{2} + \tfrac{1}{4} + \tfrac{1}{8} + \tfrac{1}{16} = \tfrac{15}{16},$$
$$s_5 = \tfrac{1}{2} + \tfrac{1}{4} + \tfrac{1}{8} + \tfrac{1}{16} + \tfrac{1}{32} = \tfrac{31}{32},$$

.

.

.

are illustrated in Figure 11.2.1. Each new partial sum lies halfway between the previous partial sum and the number 1.

Figure 11.2.1

The convergence of the geometric series at $x = \frac{1}{10}$ enables us to assign a precise meaning to infinite decimals. Begin with the fact that

$$\sum_{k=0}^{\infty} \left(\frac{1}{10}\right)^k = \sum_{k=0}^{\infty} \frac{1}{10^k} = \frac{1}{1 - \frac{1}{10}} = \frac{10}{9}.$$

This gives

$$\sum_{k=1}^{\infty} \frac{1}{10^k} = \frac{1}{9}$$

and shows that the partial sums

$$s_n = \frac{1}{10} + \frac{1}{10^2} + \cdots + \frac{1}{10^n}$$

are all less than $\frac{1}{9}$. Now take a series of the form

$$\sum_{k=1}^{\infty} \frac{a_k}{10^k} \qquad \text{with} \qquad a_k = 0, 1, \ldots, \text{ or } 9.$$

Its partial sums

$$t_n = \frac{a_1}{10} + \frac{a_2}{10^2} + \cdots + \frac{a_n}{10^n}$$

are all less than 1:

$$t_n = \frac{a_1}{10} + \frac{a_2}{10^2} + \cdots + \frac{a_n}{10^n} \leq 9\left(\frac{1}{10} + \frac{1}{10^2} + \cdots + \frac{1}{10^n}\right) = 9s_n < 9\left(\frac{1}{9}\right) = 1.$$

Since $\{t_n\}$ is nondecreasing, as well as bounded above, $\{t_n\}$ is convergent; this means that the series

$$\sum_{k=1}^{\infty} \frac{a_k}{10^k}$$

is convergent. The sum of this series is what we mean by the infinite decimal

$$0.a_1 a_2 a_3 \cdots a_n \cdots .$$

Following are two simple examples that lead naturally to geometric series. You will find more in the exercises.

Example 3 An electric fan is turned off, and the blades begin to lose speed. Given that the blades turn N times during the first second of no power and lose at least $\sigma\%$ of their speed with the passing of each ensuing second, show that the blades cannot turn more than $100N\sigma^{-1}$ times after power shutdown.

SOLUTION   The number of turns during the first second of no power is

$$N.$$

The number of turns during the first 2 seconds is at most

$$N + \left(1 - \frac{1}{100}\sigma\right)N;$$

during the first 3 seconds, at most

$$N + \left(1 - \frac{1}{100}\sigma\right)N + \left(1 - \frac{1}{100}\sigma\right)^2 N;$$

and, during the first $n + 1$ seconds, at most

$$N \sum_{k=0}^{n} \left(1 - \frac{1}{100}\sigma\right)^k.$$

The total number of turns after power shutdown cannot exceed the limiting value

$$N \sum_{k=0}^{\infty} \left(1 - \frac{1}{100}\sigma\right)^k.$$

This is a geometric series with $x = 1 - \frac{1}{100}\sigma$. Thus

$$N \sum_{k=0}^{\infty} \left(1 - \frac{1}{100}\sigma\right)^k = N\left[\frac{1}{1 - (1 - \frac{1}{100}\sigma)}\right] = \frac{100N}{\sigma} = 100N\sigma^{-1}. \quad \square$$

Example 4   According to Figure 11.2.2, it is 2 o'clock. At what time between 2 and 3 o'clock will the two hands coincide?

Figure 11.2.2

SOLUTION   We will solve the problem by setting up a geometric series. We will then confirm our answer by approaching the problem from a different perspective.

The hour hand travels one-twelfth as fast as the minute hand. At 2 o'clock the minute hand points to 12 and the hour hand points to 2. By the time the minute hand reaches 2, the hour hand points to $2 + \frac{1}{6}$. By the time the minute hand reaches $2 + \frac{1}{6}$, the hour hand points to

$$2 + \frac{1}{6} + \frac{1}{6 \cdot 12}.$$

By the time the minute hand reaches

$$2 + \frac{1}{6} + \frac{1}{6 \cdot 12},$$

the hour hand points to

$$2 + \frac{1}{6} + \frac{1}{6 \cdot 12} + \frac{1}{6 \cdot 12^2}$$

and so on. In general, by the time the minute hand reaches

$$2 + \frac{1}{6} + \frac{1}{6 \cdot 12} + \cdots + \frac{1}{6 \cdot 12^{n-1}} = 2 + \frac{1}{6} \sum_{k=0}^{n-1} \frac{1}{12^k},$$

the hour hand points to

$$2 + \frac{1}{6} + \frac{1}{6 \cdot 12} + \cdots + \frac{1}{6 \cdot 12^{n-1}} + \frac{1}{6 \cdot 12^n} = 2 + \frac{1}{6} \sum_{k=0}^{n} \frac{1}{12^k}.$$

The two hands coincide when they both point to the limiting value

$$2 + \frac{1}{6} \sum_{k=0}^{\infty} \frac{1}{12^k} = 2 + \frac{1}{6} \left(\frac{1}{1 - \frac{1}{12}} \right) = 2 + \frac{2}{11}.$$

This happens at $2 + \frac{2}{11}$ o'clock, approximately 10 minutes and 55 seconds after 2.

We can confirm this as follows. Suppose that the two hands meet at hour $2 + x$. The hour hand will have moved a distance x and the minute hand a distance $2 + x$. Since the minute hand moves 12 times as fast as the hour hand, we have

$$12x = 2 + x, \qquad 11x = 2, \qquad \text{and thus} \qquad x = \frac{2}{11}. \quad \square$$

We will return to the geometric series later. Right now we turn our attention to series in general.

Some Basic Results

THEOREM 11.2.4

1. If $\displaystyle\sum_{k=0}^{\infty} a_k$ converges and $\displaystyle\sum_{k=0}^{\infty} b_k$ converges, then $\displaystyle\sum_{k=0}^{\infty} (a_k + b_k)$ converges.

   Moreover, if $\displaystyle\sum_{k=0}^{\infty} a_k = L$ and $\displaystyle\sum_{k=0}^{\infty} b_k = M$, then $\displaystyle\sum_{k=0}^{\infty} (a_k + b_k) = L + M$.

2. If $\displaystyle\sum_{k=0}^{\infty} a_k$ converges, then $\displaystyle\sum_{k=0}^{\infty} \alpha a_k$ converges for each real number α.

   Moreover, if $\displaystyle\sum_{k=0}^{\infty} a_k = L$, then $\displaystyle\sum_{k=0}^{\infty} \alpha a_k = \alpha L$.

PROOF Let

$$s_n = \sum_{k=0}^{n} a_k, \qquad t_n = \sum_{k=0}^{n} b_k, \qquad u_n = \sum_{k=0}^{n} (a_k + b_k), \qquad v_n = \sum_{k=0}^{n} \alpha a_k.$$

Note that

$$u_n = s_n + t_n \qquad \text{and} \qquad v_n = \alpha s_n.$$

If $s_n \to L$ and $t_n \to M$, then

$$u_n \to L + M \qquad \text{and} \qquad v_n \to \alpha L. \qquad \text{(Theorem 10.3.7)} \quad \square$$

THEOREM 11.2.5

Let j be a positive integer. The series $\sum\limits_{k=0}^{\infty} a_k$ converges iff the series $\sum\limits_{k=j}^{\infty} a_k$ converges. Moreover, if $\sum\limits_{k=0}^{\infty} a_k = L$, then $\sum\limits_{k=j}^{\infty} a_k = L - (a_0 + a_1 + a_2 + \cdots + a_{j-1})$; or if $\sum\limits_{k=j}^{\infty} a_k = M$, then $\sum\limits_{k=0}^{\infty} a_k = M + (a_0 + a_1 + a_2 + \cdots + a_{j-1})$.

It is important to understand what this theorem means, namely, the convergence (or divergence) of an infinite series is not affected by where you start the summation. In the case of a convergent series, however, the limit (sum) does depend upon where you begin the summation. The proof of the theorem is left to you as an exercise.

THEOREM 11.2.6

The *kth term* of a convergent series tends to 0; namely,

$$\text{if } \sum_{k=0}^{\infty} a_k \text{ converges, then } a_k \to 0 \text{ as } k \to \infty.$$

PROOF To say that the series converges is to say that the sequence of partial sums converges to some number L:

$$s_n = \sum_{k=0}^{n} a_k \to L.$$

Obviously, then, $s_{n-1} \to L$. Since $a_n = s_n - s_{n-1}$, we have $a_n \to L - L = 0$. A change in notation gives $a_k \to 0$. ❑

The next result is an obvious, but important, consequence of Theorem 11.2.6.

THEOREM 11.2.7 A DIVERGENCE TEST

$$\text{If } a_k \not\to 0 \text{ as } k \to \infty, \quad \text{then } \sum_{k=0}^{\infty} a_k \text{ diverges.}$$

Example 5

(a) Since $\dfrac{k}{k+1} \not\to 0$ as $k \to \infty$, the series

$$\sum_{k=0}^{\infty} \frac{k}{k+1} = 0 + \frac{1}{2} + \frac{2}{3} + \frac{3}{4} + \frac{4}{5} + \cdots \quad \text{diverges.}$$

(b) Since $\sin k \nrightarrow 0$ as $k \rightarrow \infty$, the series

$$\sum_{k=0}^{\infty} \sin k = \sin 0 + \sin 1 + \sin 2 + \sin 3 + \cdots \quad \text{diverges.} \quad \square$$

CAUTION Theorem 11.2.6 does *not* say that, if $a_k \rightarrow 0$, then $\sum_{k=0}^{\infty} a_k$ converges. There are divergent series for which $a_k \rightarrow 0$. $\quad \square$

Example 6 In the case of

$$\sum_{k=1}^{\infty} \frac{1}{\sqrt{k}} = \frac{1}{\sqrt{1}} + \frac{1}{\sqrt{2}} + \frac{1}{\sqrt{3}} + \frac{1}{\sqrt{4}} + \cdots$$

we have

$$a_k = \frac{1}{\sqrt{k}} \rightarrow 0 \quad \text{as } k \rightarrow \infty,$$

but, since

$$s_n = \frac{1}{\sqrt{1}} + \frac{1}{\sqrt{2}} + \cdots + \frac{1}{\sqrt{n}} \geq \underbrace{\frac{1}{\sqrt{n}} + \frac{1}{\sqrt{n}} + \cdots + \frac{1}{\sqrt{n}}}_{n \text{ terms}} = \frac{n}{\sqrt{n}} = \sqrt{n},$$

the sequence of partial sums is unbounded, and therefore the series diverges. $\quad \square$

EXERCISES 11.2

In Exercises 1–18, find the sum of the series.

1. $\displaystyle\sum_{k=3}^{\infty} \frac{1}{(k+1)(k+2)}.$

2. $\displaystyle\sum_{k=0}^{\infty} \frac{1}{(k+3)(k+4)}.$

3. $\displaystyle\sum_{k=1}^{\infty} \frac{1}{2k(k+1)}.$

4. $\displaystyle\sum_{k=3}^{\infty} \frac{1}{k^2-k}.$

5. $\displaystyle\sum_{k=1}^{\infty} \frac{1}{k(k+3)}.$

6. $\displaystyle\sum_{k=0}^{\infty} \frac{1}{(k+1)(k+3)}.$

7. $\displaystyle\sum_{k=0}^{\infty} \frac{3}{10^k}.$

8. $\displaystyle\sum_{k=0}^{\infty} \frac{12}{100^k}.$

9. $\displaystyle\sum_{k=0}^{\infty} \frac{67}{1000^k}.$

10. $\displaystyle\sum_{k=0}^{\infty} \frac{(-1)^k}{5^k}.$

11. $\displaystyle\sum_{k=0}^{\infty} \left(\frac{3}{4}\right)^k.$

12. $\displaystyle\sum_{k=0}^{\infty} \frac{3^k+4^k}{5^k}.$

13. $\displaystyle\sum_{k=0}^{\infty} \frac{1-2^k}{3^k}.$

14. $\displaystyle\sum_{k=0}^{\infty} \left(\frac{25}{10^k} - \frac{6}{100^k}\right).$

15. $\displaystyle\sum_{k=3}^{\infty} \frac{1}{2^{k-1}}.$

16. $\displaystyle\sum_{k=0}^{\infty} \frac{1}{2^{k+3}}.$

17. $\displaystyle\sum_{k=0}^{\infty} \frac{2^{k+3}}{3^k}.$

18. $\displaystyle\sum_{k=2}^{\infty} \frac{3^{k-1}}{4^{3k+1}}.$

In Exercises 19–26, write the decimal fraction as an infinite series and express the sum as the quotient of two integers.

19. $0.777\ldots$ . . . .

20. $0.999\ldots$ . . . .

21. $0.2424\ldots$ . . . .

22. $0.8989\ldots$ . . . .

23. $0.112112112\ldots$ . . . .

24. $0.315315315\ldots$ . . . .

25. $0.624545\ldots$ . . . .

26. $0.112019019\ldots$ . . . .

27. Using series, show that every repeating decimal represents a rational number (the quotient of two integers).

28. Prove Theorem 11.2.5.

In Exercises 29 and 30, derive the indicated result from the geometric series.

29. $\displaystyle\sum_{k=0}^{\infty} (-1)^k x^k = \frac{1}{1+x}, \quad |x| < 1.$

30. $\displaystyle\sum_{k=0}^{\infty} (-1)^k x^{2k} = \frac{1}{1+x^2}, \quad |x| < 1.$

In Exercises 31–36, find a series expansion for the given expression.

31. $\dfrac{x}{1-x}$ for $|x| < 1.$

32. $\dfrac{x}{1+x}$ for $|x| < 1.$

33. $\dfrac{x}{1 + x^2}$ for $|x| < 1$.   34. $\dfrac{1}{4 - x^2}$ for $|x| < 2$.

35. $\dfrac{1}{1 + 4x^2}$ for $|x| < \dfrac{1}{2}$.

36. $\dfrac{x^2}{1 - x}$ for $|x| < 1$.

In Exercises 37–40, show that the given series diverges.

37. $1 + \dfrac{3}{2} + \dfrac{9}{4} + \dfrac{27}{8} + \dfrac{81}{16} + \cdots$.

38. $\displaystyle\sum_{k=0}^{\infty} (-1)^k$.   39. $\displaystyle\sum_{k=1}^{\infty} \left(\dfrac{k+1}{k}\right)^k$.

40. $\displaystyle\sum_{k=2}^{\infty} \dfrac{k^{k-2}}{3^k}$.

41. At some time between 4 and 5 o'clock the minute hand is directly above the hour hand. Express this time as a geometric series. What is the sum of this series?

42. Given that a ball dropped to the floor rebounds to a height proportional to the height from which it is dropped, find the total distance traveled by a ball dropped from a height of 6 feet if it rebounds initially to a height of 3 feet.

43. Exercise 42 under the supposition that the ball rebounds initially to a height of 2 feet.

44. In the setting of Exercise 42, to what height does the ball rebound initially if the total distance traveled by the ball is 21 feet?

45. How much money must you deposit at $r\%$ interest compounded annually to enable your descendants to withdraw n_1 dollars at the end of the first year, n_2 dollars at the end of the second year, n_3 dollars at the end of the third year, and so on in perpetuity? Assume that the sequence $\{n_k\}$ is bounded, $n_k \leq N$ for all k, and express your answer as an infinite series.

46. Sum the series you obtained in Exercise 45 setting
(a) $r = 5$, $n_k = 5000(\frac{1}{2})^{k-1}$.
(b) $r = 6$, $n_k = 1000(0.8)^{k-1}$.
(c) $r = 5$, $n_k = N$.

47. Suppose that 90% of each dollar is recirculated into the economy. That is, suppose that when a dollar is put into circulation, 90% of it is spent; then 90% of that is spent, and so on. What is the total economic value of the dollar?

48. Consider the following sequence of steps. First, take the unit interval $[0,1]$ and delete the open interval $(\frac{1}{3}, \frac{2}{3})$. Next delete the two open intervals $(\frac{1}{9}, \frac{2}{9})$ and $(\frac{7}{9}, \frac{8}{9})$ from the two intervals that remain after the first step. For the third step, delete the middle thirds from the four intervals that remain after the second step. Continue on in this

manner. What is the sum of the lengths of the intervals that have been deleted? The set that remains after all of the "middle thirds" have been deleted is called the *Cantor middle third set*. Can you name some points that are in the Cantor set?

49. Start with a square whose sides are four units long. Join the midpoints of the sides of the square to form a second square inside the first. Then join the midpoints of the sides of the second square to form a third square, and so on. See the figure. Find the sum of the areas of the squares.

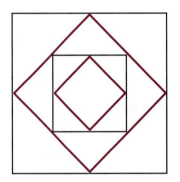

50. (a) Prove that if the series $\Sigma\, a_k$ converges and the series $\Sigma\, b_k$ diverges, then the series $\Sigma\, (a_k + b_k)$ diverges.
(b) Give examples to show that if $\Sigma\, a_k$ and $\Sigma\, b_k$ both diverge, then each of the series
$$\sum (a_k + b_k) \quad \text{and} \quad \sum (a_k - b_k)$$
may either converge or diverge.

51. Let $\Sigma_{k=0}^{\infty}\, a_k$ be a convergent series and let $R_n = \Sigma_{k=n+1}^{\infty}\, a_k$. Prove that $R_n \to 0$ as $n \to \infty$. Note that if s_n is the nth partial sum of the series, then $\Sigma_{k=0}^{\infty}\, a_k = s_n + R_n$; R_n is called the *remainder*.

52. (a) Prove that if $\Sigma_{k=0}^{\infty}\, a_k$ is a convergent series and $a_k \neq 0$ for all k, then $\Sigma_{k=0}^{\infty}\, (1/a_k)$ is divergent.
(b) Suppose that $a_k > 0$ for all k and $\Sigma_{k=0}^{\infty}\, a_k$ diverges. Show by means of examples that $\Sigma_{k=0}^{\infty}\, (1/a_k)$ may either converge or diverge.

53. Let $\{s_n\}$ be the sequence of partial sums of the series $\Sigma_{k=0}^{\infty}\, (-1)^k$. Find a formula for s_n. HINT: Find s_0, s_1, $s_2, \ldots$ and note the pattern.

54. Repeat Exercise 53 for the series $\displaystyle\sum_{k=1}^{\infty} \ln\left(\dfrac{k}{k+1}\right)$.

55. Show that
$$\sum_{k=1}^{\infty} \ln\left(\dfrac{k+1}{k}\right) \quad \text{diverges,} \quad \text{even though}$$
$$\lim_{k \to \infty} \ln\left(\dfrac{k+1}{k}\right) = 0.$$

56. Show that

$$\sum_{k=1}^{\infty} \left(\frac{k}{k+1}\right)^k \quad \text{diverges.}$$

57. (a) Let $\{d_k\}$ be a sequence of real numbers that converges to 0. Show that

$$\sum_{k=1}^{\infty} (d_k - d_{k+1}) = d_1.$$

(b) Sum the following series:

(i) $\sum_{k=1}^{\infty} \dfrac{\sqrt{k+1} - \sqrt{k}}{\sqrt{k(k+1)}}.$ (ii) $\sum_{k=1}^{\infty} \dfrac{2k+1}{2k^2(k+1)^2}.$

58. Show that

$$\sum_{k=1}^{\infty} kx^{k-1} = \frac{1}{(1-x)^2} \qquad \text{for } |x| < 1.$$

HINT: Verify that s_n, the nth partial sum, satisfies the identity

$$(1-x)^2 s_n = 1 - (n+1)x^n + nx^{n+1}.$$

▶ **Speed of Convergence** Suppose that $\Sigma_{k=0}^{\infty} a_k$ is a convergent series with sum L and let $\{s_n\}$ be its sequence of partial sums. It follows from Exercise 51 that $|L - s_n| = |R_n|$. In Exercises 59–62, find the smallest integer N such that $|L - s_N| < 0.0001$.

59. $\sum_{k=0}^{\infty} \dfrac{1}{4^k}.$ **60.** $\sum_{k=0}^{\infty} (0.9)^k.$

61. $\sum_{k=1}^{\infty} \dfrac{1}{k(k+2)}.$ **62.** $\sum_{k=0}^{\infty} \left(\dfrac{2}{3}\right)^k.$

63. Given the geometric series $\Sigma_{k=0}^{\infty} x^k$, with $|x| < 1$, and a positive number ϵ. Determine the smallest positive integer N such that $|L - s_N| < \epsilon$, where L is the sum of the series and s_n is the nth partial sum.

■ 11.3 THE INTEGRAL TEST; COMPARISON THEOREMS

Here and in the next section we direct our attention to *series with nonnegative terms*: $a_k \geq 0$ for all k. The significant feature of a series with nonnegative terms is the fact that its sequence of partial sums is nondecreasing:

$$s_{n+1} = \sum_{k=0}^{n+1} a_k = \sum_{k=0}^{n} a_k + a_{n+1} \geq \sum_{k=0}^{n} a_k = s_n, \qquad n = 0, 1, 2, \ldots.$$

The following theorem is fundamental.

THEOREM 11.3.1

A series with nonnegative terms converges iff the sequence of partial sums is bounded.

PROOF Assume that the series converges. Then the sequence of partial sums is convergent and therefore bounded (Theorem 10.3.4).

Suppose now that the sequence of partial sums is bounded. Since the terms are nonnegative, the sequence is nondecreasing. By being bounded and nondecreasing, the sequence of partial sums converges (Theorem 10.3.6). This means that the series converges. ❑

The convergence or divergence of a series can sometimes be deduced from the convergence or divergence of a closely related improper integral.

THEOREM 11.3.2 THE INTEGRAL TEST

If f is continuous, decreasing, and positive on $[1, \infty)$, then

$$\sum_{k=1}^{\infty} f(k) \quad \text{converges} \quad \text{iff} \quad \int_{1}^{\infty} f(x)\, dx \quad \text{converges.}$$

PROOF In Exercise 63, Section 10.7, you were asked to show that with f continuous, decreasing, and positive on $[1, \infty)$

$$\int_{1}^{\infty} f(x)\, dx \quad \text{converges} \quad \text{iff} \quad \text{the sequence} \quad \left\{ \int_{1}^{n} f(x)\, dx \right\} \quad \text{converges.}$$

We assume this result and base our proof on the behavior of the sequence of integrals. To visualize our argument see Figure 11.3.1.

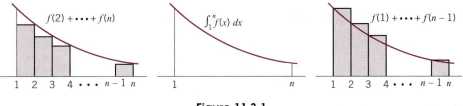

Figure 11.3.1

Since f decreases on the interval $[1, n]$,

$$f(2) + \cdots + f(n) \quad \text{is a lower sum for } f \text{ on } [1, n]$$

and

$$f(1) + \cdots + f(n-1) \quad \text{is an upper sum for } f \text{ on } [1, n].$$

Consequently

$$(1) \quad f(2) + \cdots + f(n) \leq \int_{1}^{n} f(x)\, dx \quad \text{and} \quad \int_{1}^{n} f(x)\, dx \leq f(1) + \cdots + f(n-1).$$

If the sequence of integrals converges, it is bounded. By the first inequality the sequence of partial sums is bounded and the series is therefore convergent.

Suppose now that the sequence of integrals diverges. Since f is positive, the sequence of integrals increases:

$$\int_{1}^{n} f(x)\, dx < \int_{1}^{n+1} f(x)\, dx.$$

Since this sequence diverges, it must be unbounded. By the second inequality, the sequence of partial sums must be unbounded and the series diverges. ❏

Remark The inequalities established in the proof of Theorem 11.3.2 lead to bounds on the sum of the infinite series

$$\sum_{k=1}^{\infty} f(k),$$

where f is continuous, decreasing, and positive on $[1, \infty)$. In particular, it follows from the second inequality in (1) that

$$\int_1^\infty f(x)\, dx \le \sum_{k=1}^\infty f(k),$$

and from the first inequality in (1),

$$\sum_{k=1}^\infty f(k) \le f(1) + \int_1^\infty f(x)\, dx.$$

Combining these two inequalities, we have

$$\int_1^\infty f(x)\, dx \le \sum_{k=1}^\infty f(k) \le f(1) + \int_1^\infty f(x)\, dx.$$

These inequalities also make clear the relation between the convergence of the infinite series and the convergence of the corresponding improper integral. ❏

Applying the Integral Test

Example 1 (*The Harmonic Series*)

(11.3.3)

$$\sum_{k=1}^\infty \frac{1}{k} = 1 + \frac{1}{2} + \frac{1}{3} + \frac{1}{4} + \cdots \quad \text{diverges.}$$

PROOF The function $f(x) = 1/x$ is continuous, decreasing, and positive on $[1, \infty)$. We know that

$$\int_1^\infty \frac{dx}{x} \quad \text{diverges.} \tag{10.7.1}$$

By the integral test

$$\sum_{k=1}^\infty \frac{1}{k} \quad \text{diverges.} ❏$$

The next example gives a more general result.

Example 2 (*The p-series*)

(11.3.4)

$$\sum_{k=1}^\infty \frac{1}{k^p} = 1 + \frac{1}{2^p} + \frac{1}{3^p} + \frac{1}{4^p} + \cdots \quad \text{converges} \quad \text{iff} \quad p > 1.$$

PROOF If $p \le 0$, then each term of the series is greater than or equal to 1. Therefore, the kth term does not have limit 0 and so, by the divergence test (11.2.7) the series cannot converge. (See, also, the remark following Example 2 in Section 10.7.) We assume, therefore, that $p > 0$. The function $f(x) = 1/x^p$ is then continuous, decreasing, and positive on $[1, \infty)$. Thus by the integral test

$$\sum_{k=1}^{\infty} \frac{1}{k^p} \quad \text{converges} \quad \text{iff} \quad \int_1^{\infty} \frac{dx}{x^p} \quad \text{converges}.$$

Earlier you saw that

$$\int_1^{\infty} \frac{dx}{x^p} \quad \text{converges} \quad \text{iff} \quad p > 1. \tag{10.7.1}$$

It follows that

$$\sum_{k=1}^{\infty} \frac{1}{k^p} \quad \text{converges} \quad \text{iff} \quad p > 1. \quad \square$$

Example 3 Here we show that the series

$$\sum_{k=1}^{\infty} \frac{1}{k \ln (k + 1)} = \frac{1}{\ln 2} + \frac{1}{2 \ln 3} + \frac{1}{3 \ln 4} + \cdots$$

diverges. We begin by setting

$$f(x) = \frac{1}{x \ln (x + 1)}.$$

Since f is continuous, decreasing, and positive on $[1, \infty)$, we can use the integral test. Note that

$$\int_1^b \frac{dx}{x \ln (x + 1)} > \int_1^b \frac{dx}{(x + 1) \ln (x + 1)}$$

$$= \left[\ln (\ln (x + 1)) \right]_1^b = \ln (\ln (b + 1)) - \ln (\ln 2).$$

As $b \to \infty$, $\ln (\ln (b + 1)) \to \infty$. This shows that

$$\int_1^{\infty} \frac{dx}{x \ln (x + 1)} \quad \text{diverges}.$$

It follows that the series diverges. $\square$

Remark on Notation You have seen that for each $j \geq 0$

$$\sum_{k=0}^{\infty} a_k \quad \text{converges} \quad \text{iff} \quad \sum_{k=j}^{\infty} a_k \quad \text{converges}$$

(Theorem 11.2.5). This tells you that, in determining whether or not a series converges, it does not matter where we begin the summation. Where detailed indexing would contribute nothing, we will omit it and write $\Sigma \, a_k$ without specifying where the summation begins. For instance, it makes sense to say that

$$\sum \frac{1}{k^2} \quad \text{converges} \quad \text{and} \quad \sum \frac{1}{k} \quad \text{diverges}$$

without specifying where we begin the summation. $\square$

The convergence or divergence of a series with nonnegative terms can sometimes be deduced by comparison with a series of known behavior.

> **THEOREM 11.3.5 THE BASIC COMPARISON TEST**
>
> Let $\Sigma\, a_k$ be a series with nonnegative terms.
>
> **(i)** $\Sigma\, a_k$ converges if there exists a convergent series $\Sigma\, c_k$ with nonnegative terms such that $a_k \le c_k$ for all k sufficiently large;
> **(ii)** $\Sigma\, a_k$ diverges if there exists a divergent series $\Sigma\, d_k$ with nonnegative terms such that $d_k \le a_k$ for all k sufficiently large.

PROOF   The proof is just a matter of noting that, in the first instance, the partial sums of $\Sigma\, a_k$ are bounded and, in the second instance, unbounded.   The details are left to you.   ❑

Example 4

(a) $\sum \dfrac{1}{2k^3 + 1}$ converges by comparison with $\sum \dfrac{1}{k^3}$:

$$\frac{1}{2k^3 + 1} < \frac{1}{k^3} \qquad \text{and} \qquad \sum \frac{1}{k^3} \text{ converges.}$$

(b) $\sum \dfrac{1}{3k + 1}$ diverges by comparison with $\sum \dfrac{1}{3(k + 1)}$:

$$\frac{1}{3(k + 1)} < \frac{1}{3k + 1} \qquad \text{and} \qquad \sum \frac{1}{3(k + 1)} = \frac{1}{3}\sum \frac{1}{k + 1} \qquad \text{diverges}$$

(it's the series $\Sigma 1/k$ with a change of index).

(c) $\sum \dfrac{k^3}{k^5 + 5k^4 + 7}$ converges by comparison with $\sum \dfrac{1}{k^2}$:

$$\frac{k^3}{k^5 + 5k^4 + 7} < \frac{k^3}{k^5} = \frac{1}{k^2} \qquad \text{and} \qquad \sum \frac{1}{k^2} \text{ converges.} ❑$$

Example 5   Show that

$$\sum \frac{1}{\ln(k + 6)} \quad \text{diverges.}$$

SOLUTION   We know that as $k \to \infty$

$$\frac{\ln k}{k} \to 0.$$ (see 10.4.5)

It follows that

$$\frac{\ln(k + 6)}{k + 6} \to 0$$

and therefore that

$$\frac{\ln(k + 6)}{k} = \frac{\ln(k + 6)}{k + 6}\left(\frac{k + 6}{k}\right) \to 0.$$

Thus, for k sufficiently large,

$$\frac{\ln(k + 6)}{k} < 1,$$

and so

$$\ln(k + 6) < k \qquad \text{and} \qquad \frac{1}{k} < \frac{1}{\ln(k + 6)}.$$

Since

$$\sum \frac{1}{k} \quad \text{diverges},$$

we can conclude that

$$\sum \frac{1}{\ln(k + 6)} \quad \text{diverges.} \quad \square$$

Remark Another way to show that $\ln(k + 6) < k$ for large k is to examine the function $f(x) = x - \ln(x + 6)$. At $x = 3$ the function is positive:

$$f(3) = 3 - \ln 9 \cong 3 - 2.197 > 0.$$

Since

$$f'(x) = 1 - \frac{1}{x + 6} > 0 \qquad \text{for all } x > 0,$$

$f(x) > 0$ for all $x \geq 3$. It follows that

$$\ln(x + 6) < x \qquad \text{for all } x \geq 3. \quad \square$$

The basic comparison test is algebraic in nature; it requires that certain inequalities hold. To apply the test to a series $\sum a_k$ you must show that the terms a_k are smaller than the corresponding terms c_k of a known convergent series to establish convergence, or larger than the terms d_k of a known divergent series to establish divergence. However, if the terms a_k are larger than the terms c_k, or smaller than the terms d_k, then the comparison test does not apply. For example, consider the series

$$\sum_{k=2}^{\infty} \frac{1}{k^3 - 1}.$$

It would be natural to compare this series with the convergent series

$$\sum_{k=2}^{\infty} \frac{1}{k^3},$$

but, unfortunately, the inequalities "go the wrong way:"

$$\frac{1}{k^3 - 1} > \frac{1}{k^3} \qquad \text{for all } k \geq 2.$$

Our next theorem is a more sophisticated comparison test. It is analytic in the sense that it involves the evaluation of a limit.

> **THEOREM 11.3.6 THE LIMIT COMPARISON TEST**
>
> Let $\Sigma\, a_k$ and $\Sigma\, b_k$ be series with positive terms. If $a_k/b_k \to L$, where L is some *positive* number, then either both series converge or both series diverge.

PROOF Choose ϵ between 0 and L. Since $a_k/b_k \to L$, we know that for all k sufficiently large (for all k greater than some k_0)

$$\left| \frac{a_k}{b_k} - L \right| < \epsilon.$$

For such k we have

$$L - \epsilon < \frac{a_k}{b_k} < L + \epsilon$$

and thus

$$(L - \epsilon)b_k < a_k < (L + \epsilon)b_k.$$

This last inequality is what we need:

if $\Sigma\, a_k$ converges, then $\Sigma\, (L - \epsilon)b_k$ converges, and thus $\Sigma\, b_k$ converges;
if $\Sigma\, b_k$ converges, then $\Sigma\, (L + \epsilon)b_k$ converges, and thus $\Sigma\, a_k$ converges. ❑

To apply the limit comparison theorem to a series $\Sigma\, a_k$, we must first find a series $\Sigma\, b_k$ of known behavior for which a_k/b_k converges to a positive number. To complete the example started earlier,

$$\sum_{k=2}^{\infty} \frac{1}{k^3 - 1}$$

converges since

$$\sum_{k=2}^{\infty} \frac{1}{k^3} \quad \text{converges} \quad \text{and} \quad \left(\frac{1}{k^3 - 1}\right) \div \left(\frac{1}{k^3}\right) = \frac{k^3}{k^3 - 1} \to 1 \quad \text{as } k \to \infty.$$

Example 6 Determine whether the series

$$\sum \frac{3k^2 + 2k + 1}{k^3 + 1}$$

converges or diverges.

SOLUTION For large k, the terms with the highest powers of k dominate. Here $3k^2$ dominates the numerator and k^3 dominates the denominator. Thus, for large k,

$$\frac{3k^2 + 2k + 1}{k^3 + 1} \quad \text{differs little from} \quad \frac{3k^2}{k^3} = \frac{3}{k}.$$

Since

$$\frac{3k^2 + 2k + 1}{k^3 + 1} \div \frac{3}{k} = \frac{3k^3 + 2k^2 + k}{3k^3 + 3} = \frac{1 + 2/(3k) + 1/(3k^2)}{1 + 1/k^3} \to 1$$

and

$$\sum \frac{3}{k} = 3 \sum \frac{1}{k} \quad \text{diverges,}$$

we know that the series diverges.   ❑

Example 7   Determine whether the series

$$\sum \frac{5\sqrt{k} + 100}{2k^2\sqrt{k} + 9\sqrt{k}}$$

converges or diverges.

SOLUTION   For large values of k, $5\sqrt{k}$ dominates the numerator and $2k^2\sqrt{k}$ dominates the denominator.   Thus, for such k,

$$\frac{5\sqrt{k} + 100}{2k^2\sqrt{k} + 9\sqrt{k}} \quad \text{differs little from} \quad \frac{5\sqrt{k}}{2k^2\sqrt{k}} = \frac{5}{2k^2}.$$

Since

$$\frac{5\sqrt{k} + 100}{2k^2\sqrt{k} + 9\sqrt{k}} \div \frac{5}{2k^2} = \frac{10k^2\sqrt{k} + 200k^2}{10k^2\sqrt{k} + 45\sqrt{k}} = \frac{1 + 20/\sqrt{k}}{1 + 9/2k^2} \to 1 \quad \text{as } k \to \infty$$

and

$$\sum \frac{5}{2k^2} = \frac{5}{2} \sum \frac{1}{k^2} \quad \text{converges,}$$

the series converges.   ❑

Example 8   Determine whether the series

$$\sum \sin \frac{\pi}{k}$$

converges or diverges.

SOLUTION   Recall that

$$\text{as } x \to 0, \quad \frac{\sin x}{x} \to 1. \tag{2.5.5}$$

As $k \to \infty$, $\pi/k \to 0$ and thus

$$\frac{\sin (\pi/k)}{\pi/k} \to 1.$$

Since $\Sigma \ \pi/k$ diverges, $\Sigma \sin (\pi/k)$ diverges.   ❑

Remark   The question of what we can and cannot conclude by limit comparison if $a_k/b_k \to 0$ or if $a_k/b_k \to \infty$ is taken up in Exercises 45 and 46.   ❑

EXERCISES 11.3

In Exercises 1–34, determine whether the series converges or diverges.

1. $\sum \dfrac{k}{k^3 + 1}$.

2. $\sum \dfrac{1}{3k + 2}$.

3. $\sum \dfrac{1}{(2k + 1)^2}$.

4. $\sum \dfrac{\ln k}{k}$.

5. $\sum \dfrac{1}{\sqrt{k + 1}}$.

6. $\sum \dfrac{1}{k^2 + 1}$.

7. $\sum \dfrac{1}{\sqrt{2k^2 - k}}$.

8. $\sum \left(\dfrac{5}{2}\right)^{-k}$.

9. $\sum \dfrac{\tan^{-1}k}{1 + k^2}$.

10. $\sum \dfrac{\ln k}{k^3}$.

11. $\sum \dfrac{1}{k^{2/3}}$.

12. $\sum \dfrac{1}{(k + 1)(k + 2)(k + 3)}$.

13. $\sum \left(\dfrac{3}{4}\right)^{-k}$.

14. $\sum \dfrac{1}{1 + 2\ln k}$.

15. $\sum \dfrac{\ln \sqrt{k}}{k}$.

16. $\sum \dfrac{2}{k(\ln k)^2}$.

17. $\sum \dfrac{1}{2 + 3^{-k}}$.

18. $\sum \dfrac{7k + 2}{2k^5 + 7}$.

19. $\sum \dfrac{2k + 5}{5k^3 + 3k^2}$.

20. $\sum \dfrac{k^4 - 1}{3k^2 + 5}$.

21. $\sum \dfrac{1}{k \ln k}$.

22. $\sum \dfrac{1}{2^{k+1} - 1}$.

23. $\sum \dfrac{k^2}{k^4 - k^3 + 1}$.

24. $\sum \dfrac{k^{3/2}}{k^{5/2} + 2k - 1}$.

25. $\sum \dfrac{2k + 1}{\sqrt{k^4 + 1}}$.

26. $\sum \dfrac{2k + 1}{\sqrt{k^3 + 1}}$.

27. $\sum \dfrac{2k + 1}{\sqrt{k^5 + 1}}$.

28. $\sum \dfrac{1}{\sqrt{2k(k + 1)}}$.

29. $\sum k e^{-k^2}$.

30. $\sum k^2 2^{-k^3}$.

31. $\sum \dfrac{2 + \sin k}{k^2}$.

32. $\sum \dfrac{2 + \cos k}{\sqrt{k + 1}}$.

33. $\sum \dfrac{1}{1 + 2 + 3 + \cdots + k}$.

34. $\sum \dfrac{k}{1 + 2^2 + 3^2 + \cdots + k^2}$.

35. Find the values of p for which the series $\displaystyle\sum_{k=2}^{\infty} \dfrac{1}{k(\ln k)^p}$ converges.

36. Find the values of p for which the series $\displaystyle\sum_{k=2}^{\infty} \dfrac{\ln k}{k^p}$ converges.

37. (a) Prove that $\displaystyle\sum_{k=0}^{\infty} e^{-\alpha k}$ converges for any $\alpha > 0$.

(b) Prove that $\displaystyle\sum_{k=0}^{\infty} ke^{-\alpha k}$ converges for any $\alpha > 0$.

(c) In general, prove that $\displaystyle\sum_{k=0}^{\infty} k^n e^{-\alpha k}$ converges for any nonnegative integer n and any $\alpha > 0$.

38. Let $p > 1$. Use the integral test to show that

$$\frac{1}{(p - 1)(n + 1)^{p-1}} < \sum_{k=1}^{\infty} \frac{1}{k^p} - \sum_{k=1}^{n} \frac{1}{k^p} < \frac{1}{(p - 1)n^{p-1}}.$$

This result gives bounds on the *error* (or remainder) R_n that results from using s_n to approximate L the sum of the convergent *p*-series.

In Exercises 39–40, (a) compute the sum of the first four terms of the given series; use four decimal place accuracy. (b) Use the result in Exercise 38 to give upper and lower bounds on R_4. (c) Use parts (a) and (b) to estimate the sum of the series.

39. $\displaystyle\sum_{k=1}^{\infty} \dfrac{1}{k^3}$.

40. $\displaystyle\sum_{k=1}^{\infty} \dfrac{1}{k^4}$.

In Exercises 41–44, use the error bounds given in Exercise 38.

41. (a) If you were to use s_{100} to approximate $\displaystyle\sum_{k=1}^{\infty} \dfrac{1}{k^2}$ what would be the bounds on your error?

(b) How large would you have to choose n to ensure that R_n is less than 0.0001?

42. (a) If you were to use s_{100} to approximate $\displaystyle\sum_{k=1}^{\infty} \dfrac{1}{k^3}$ what would be the bounds on your error?

(b) How large would you have to choose n to ensure that R_n is less than 0.0001?

43. (a) How many terms of the series $\displaystyle\sum_{k=1}^{\infty} \dfrac{1}{k^4}$ should you use to ensure that R_n is less than 0.0001?

(b) Estimate $\displaystyle\sum_{k=1}^{\infty} \dfrac{1}{k^4}$ to three decimal places.

44. Repeat Exercise 43 for the series $\displaystyle\sum_{k=1}^{\infty} \dfrac{1}{k^5}$.

45. Let Σa_k and Σb_k be series with positive terms and suppose that $a_k/b_k \to 0$.

(a) Show that, if Σb_k converges, then Σa_k converges.

(b) Show that, if Σa_k diverges, then Σb_k diverges.

(c) Show by example that, if $\Sigma\, a_k$ converges, then $\Sigma\, b_k$ may converge or diverge.

(d) Show by example that, if $\Sigma\, b_k$ diverges, then $\Sigma\, a_k$ may converge or diverge.

[Parts (c) and (d) explain why we stipulated $L > 0$ in Theorem 11.3.6.]

46. Let $\Sigma\, a_k$ and $\Sigma\, b_k$ be series with positive terms and suppose that $a_k/b_k \to \infty$.
   (a) Show that if $\Sigma\, b_k$ diverges, then $\Sigma\, a_k$ diverges.
   (b) Show that if $\Sigma\, a_k$ converges, then $\Sigma\, b_k$ converges.
   (c) Show by example that if $\Sigma\, a_k$ diverges, then $\Sigma\, b_k$ may converge or diverge.
   (d) Show by example that if $\Sigma\, b_k$ converges, then $\Sigma\, a_k$ may converge or diverge.

47. Let $\Sigma\, a_k$ be a series with positive terms.
   (a) Prove that if $\Sigma\, a_k$ converges, then $\Sigma\, a_k^2$ converges.
   (b) Suppose that $\Sigma\, a_k^2$ converges. Does $\Sigma\, a_k$ converge or diverge? Prove or give a counterexample.

48. Let $\Sigma\, a_k$ be a series with positive terms. Prove that if $\Sigma\, a_k^2$ converges, then $\Sigma (a_k/k)$ converges.

49. Let f be a continuous, positive, decreasing function on $[1, \infty)$ such that $\int_1^\infty f(x)\, dx$ converges. Then the series $\Sigma_{k=1}^\infty f(k)$ also converges. Prove that

$$0 < L - s_n < \int_n^\infty f(x)\, dx,$$

where L is the sum of the series and s_n is the nth partial sum.

In Exercises 50 and 51, use the result of Exercise 49 to determine the smallest integer N such that the difference between the sum of the given series and the Nth partial sum is less than 0.001.

50. $\displaystyle\sum_{k=1}^\infty \frac{1}{k^2 + 1}$.

51. $\displaystyle\sum_{k=1}^\infty ke^{-k^2}$.

52. All the results of this section were stated for series with nonnegative terms. Corresponding results hold for *series with nonpositive terms*: $a_k \leq 0$ for all k.
   (a) State a comparison theorem analogous to Theorem 11.3.5, this time for series with nonpositive terms.
   (b) As stated, the integral test (Theorem 11.3.2) applies only to series with positive terms. State the equivalent result for series with negative terms.

53. This exercise demonstrates that we cannot always use the same testing series for both the basic comparison test and the limit comparison test.
   (a) Show that

$$\Sigma \frac{\ln n}{n\sqrt{n}} \quad \text{converges by comparison with} \quad \Sigma \frac{1}{n^{5/4}}.$$

   (b) Show that the limit comparison test does not apply.

■ 11.4 THE ROOT TEST; THE RATIO TEST

We continue with our study of series with nonnegative terms. Comparison with geometric series

$$\Sigma\, x^k$$

and with the p-series

$$\Sigma \frac{1}{k^p}$$

leads to two important tests for convergence: the root test and the ratio test.

THEOREM 11.4.1 THE ROOT TEST

Let $\Sigma\, a_k$ be a series with nonnegative terms, and suppose that

$$(a_k)^{1/k} \to \rho \quad \text{as} \quad k \to \infty.$$

(a) If $\rho < 1$, then $\Sigma\, a_k$ converges.

(b) If $\rho > 1$, then $\Sigma\, a_k$ diverges.

(c) If $\rho = 1$, then the test is inconclusive; the series may either converge or diverge.

PROOF   We suppose first that $\rho < 1$ and choose μ so that

$$\rho < \mu < 1.$$

Since $(a_k)^{1/k} \to \rho$, we have

$$(a_k)^{1/k} < \mu \qquad \text{for all } k \text{ sufficiently large.} \qquad \text{(explain)}$$

Thus

$$a_k < \mu^k \qquad \text{for all } k \text{ sufficiently large.}$$

Since $\Sigma \, \mu^k$ converges (a geometric series with $0 < \mu < 1$), we know by the basic comparison theorem that $\Sigma \, a_k$ converges.

We suppose now that $\rho > 1$.   Since $(a_k)^{1/k} \to \rho$, we have

$$(a_k)^{1/k} > 1 \qquad \text{for all } k \text{ sufficiently large.} \qquad \text{(explain)}$$

Thus

$$a_k > 1 \qquad \text{for all } k \text{ sufficiently large.}$$

It now follows that $a_k \nrightarrow 0$ as $k \to \infty$.   Therefore $\Sigma \, a_k$ diverges by the divergence test (11.2.7).

To see the inconclusiveness of the root test when $\rho = 1$, consider the series $\Sigma(1/k^2)$ and $\Sigma(1/k)$.   The first series converges and the second series diverges.   However, in each case we have

$$(a_k)^{1/k} = \left(\frac{1}{k^2} \right)^{1/k} = \left(\frac{1}{k^{1/k}} \right)^2 \to 1^2 = 1 \quad \text{as } k \to \infty,$$

$$(a_k)^{1/k} = \left(\frac{1}{k} \right)^{1/k} = \frac{1}{k^{1/k}} \to 1 \qquad \qquad \text{as } k \to \infty.$$

(Recall that $k^{1/k} \to 1$ as $k \to \infty$, see (10.4.6).)   ❏

Applying the Root Test

Example 1   For the series

$$\Sigma \frac{1}{(\ln k)^k}$$

we have

$$(a_k)^{1/k} = \frac{1}{\ln k} \to 0.$$

The series converges.   ❏

Example 2   For the series

$$\Sigma \frac{2^k}{k^3}$$

we have

$$(a_k)^{1/k} = 2 \left(\frac{1}{k} \right)^{3/k} = 2 \left[\left(\frac{1}{k} \right)^{1/k} \right]^3 = 2 \left[\frac{1}{k^{1/k}} \right]^3 \to 2 \cdot 1^3 = 2 \quad \text{as } k \to \infty.$$

The series diverges.   ❏

Example 3 In the case of

$$\Sigma \left(1 - \frac{1}{k}\right)^k,$$

we have

$$(a_k)^{1/k} = 1 - \frac{1}{k} \to 1.$$

Here the root test is inconclusive. It is also unnecessary: since $a_k = (1 - 1/k)^k$ converges to $1/e$ and not to 0 (10.4.7), the series diverges (11.2.7). ❏

THEOREM 11.4.2 THE RATIO TEST

Let Σa_k be a series with positive terms and suppose that

$$\frac{a_{k+1}}{a_k} \to \lambda \quad \text{as } k \to \infty$$

(a) If $\lambda < 1$, then $\Sigma\, a_k$ converges.
(b) If $\lambda > 1$, then $\Sigma\, a_k$ diverges.
(c) If $\lambda = 1$, then the test is inconclusive; the series may either converge or diverge.

PROOF We suppose first that $\lambda < 1$ and choose μ so that $\lambda < \mu < 1$. Since

$$\frac{a_{k+1}}{a_k} \to \lambda,$$

we know that there exists $k_0 > 0$ such that

$$\text{if } k \geq k_0, \quad \text{then} \quad \frac{a_{k+1}}{a_k} < \mu. \qquad \text{(explain)}$$

This gives

$$a_{k_0+1} < \mu a_{k_0}, \qquad a_{k_0+2} < \mu a_{k_0+1} < \mu^2 a_{k_0},$$

and more generally,

$$a_{k_0+j} < \mu^j a_{k_0}, \qquad j = 1, 2, \ldots .$$

For $k > k_0$ we have

(1)
$$a_k < \mu^{k-k_0} a_{k_0} = \frac{a_{k_0}}{\mu^{k_0}} \mu^k.$$

$$\underset{\text{set } j = k - k_0}{\uparrow\!\!\!_____}$$

Since $\mu < 1$,

$$\Sigma \frac{a_{k_0}}{\mu^{k_0}} \mu^k = \frac{a_{k_0}}{\mu^{k_0}} \Sigma \mu^k \quad \text{converges.}$$

Recalling (1), you can see by the basic comparison theorem that Σa_k converges. The proof of the rest of the theorem is left to the exercises. ❏

Remark Contrary to some people's intuition the root and ratio tests are *not* equivalent. See Exercise 52. ❏

Applying the Ratio Test

Example 4 The ratio test shows that the series

$$\Sigma \frac{1}{k!}$$

converges:

$$\frac{a_{k+1}}{a_k} = \frac{1}{(k+1)!} \cdot \frac{k!}{1} = \frac{1}{k+1} \to 0 \qquad \text{as } k \to \infty. \quad \square$$

Example 5 For the series

$$\Sigma \frac{k}{10^k}$$

we have

$$\frac{a_{k+1}}{a_k} = \frac{k+1}{10^{k+1}} \cdot \frac{10^k}{k} = \frac{1}{10} \frac{k+1}{k} \to \frac{1}{10} \quad \text{as } k \to \infty.$$

The series converges.† $\quad \square$

Example 6 For the series

$$\Sigma \frac{k^k}{k!}$$

we have

$$\frac{a_{k+1}}{a_k} = \frac{(k+1)^{k+1}}{(k+1)!} \cdot \frac{k!}{k^k} = \left(\frac{k+1}{k}\right)^k = \left(1 + \frac{1}{k}\right)^k \to e \quad \text{as } k \to \infty.$$

Since $e > 1$, the series diverges. $\quad \square$

Example 7 For the series

$$\Sigma \frac{1}{2k+1}$$

the ratio test is inconclusive:

$$\frac{a_{k+1}}{a_k} = \frac{1}{2(k+1)+1} \cdot \frac{2k+1}{1} = \frac{2k+1}{2k+3} = \frac{2+1/k}{2+3/k} \to 1 \text{ as } k \to \infty.$$

Therefore, we have to look further. Comparison with the harmonic series shows that the series diverges:

$$\frac{1}{2k+1} \div \frac{1}{k} = \frac{k}{2k+1} \to \frac{1}{2} \qquad \text{and} \qquad \Sigma \frac{1}{k} \text{ diverges.} \quad \square$$

---

† This series can be summed explicitly. See Exercise 41.

Summary on Convergence Tests

In general, the root test is used only if powers are involved. The ratio test is particularly effective with factorials and with combinations of powers and factorials. If the terms are rational functions of k, the ratio test is inconclusive and the root test is difficult to apply. Rational terms are most easily handled by comparison or limit comparison with a p-series, $\Sigma\, 1/k^p$. If the terms have the configuration of a derivative, you may be able to apply the integral test. Finally, keep in mind that, if $a_k \nrightarrow 0$, then there is no reason to apply any special convergence test; the series diverges by Theorem 11.2.7.

EXERCISES 11.4

In Exercises 1–40, determine whether the series converges or diverges.

1. $\Sigma \dfrac{10^k}{k!}$

2. $\Sigma \dfrac{1}{k\,2^k}$.

3. $\Sigma \dfrac{1}{k^k}$.

4. $\Sigma \left(\dfrac{k}{2k+1}\right)^k$.

5. $\Sigma \dfrac{k!}{100^k}$.

6. $\Sigma \dfrac{(\ln k)^2}{k}$.

7. $\Sigma \dfrac{k^2+2}{k^3+6k}$.

8. $\Sigma \dfrac{1}{(\ln k)^k}$.

9. $\Sigma k \left(\dfrac{2}{3}\right)^k$.

10. $\Sigma \dfrac{1}{(\ln k)^{10}}$.

11. $\Sigma \dfrac{1}{1+\sqrt{k}}$.

12. $\Sigma \dfrac{2k+\sqrt{k}}{k^3+\sqrt{k}}$.

13. $\Sigma \dfrac{k!}{10^{4k}}$.

14. $\Sigma \dfrac{k^2}{e^k}$.

15. $\Sigma \dfrac{\sqrt{k}}{k^2+1}$.

16. $\Sigma \dfrac{2^k k!}{k^k}$.

17. $\Sigma \dfrac{k!}{(k+2)!}$.

18. $\Sigma \dfrac{1}{k}\left(\dfrac{1}{\ln k}\right)^{3/2}$.

19. $\Sigma \dfrac{1}{k}\left(\dfrac{1}{\ln k}\right)^{1/2}$.

20. $\Sigma \dfrac{1}{\sqrt{k^3-1}}$.

21. $\Sigma \left(\dfrac{k}{k+100}\right)^k$.

22. $\Sigma \dfrac{(k!)^2}{(2k)!}$.

23. $\Sigma k^{-(1+1/k)}$.

24. $\Sigma \dfrac{11}{1+100^{-k}}$.

25. $\Sigma \dfrac{\ln k}{e^k}$.

26. $\Sigma \dfrac{k!}{k^k}$.

27. $\Sigma \dfrac{\ln k}{k^2}$.

28. $\Sigma \dfrac{k!}{1\cdot 3\cdot\,\cdots\,\cdot(2k-1)}$.

29. $\Sigma \dfrac{2\cdot 4\cdot\,\cdots\,\cdot 2k}{(2k)!}$.

30. $\Sigma \dfrac{(2k+1)^{2k}}{(5k^2+1)^k}$.

31. $\Sigma \dfrac{k!(2k)!}{(3k)!}$.

32. $\Sigma \dfrac{\ln k}{k^{5/4}}$.

33. $\Sigma \dfrac{k^{k/2}}{k!}$.

34. $\Sigma \dfrac{k^k}{(3^k)^2}$.

35. $\Sigma \dfrac{k^k}{3^{(k^2)}}$.

36. $\Sigma(\sqrt{k}-\sqrt{k-1})^k$.

37. $\dfrac{1}{2}+\dfrac{2}{3^2}+\dfrac{4}{4^3}+\dfrac{8}{5^4}+\cdots$.

38. $1+\dfrac{1\cdot 2}{1\cdot 3}+\dfrac{1\cdot 2\cdot 3}{1\cdot 3\cdot 5}+\dfrac{1\cdot 2\cdot 3\cdot 4}{1\cdot 3\cdot 5\cdot 7}+\cdots$.

39. $\dfrac{1}{4}+\dfrac{1\cdot 3}{4\cdot 7}+\dfrac{1\cdot 3\cdot 5}{4\cdot 7\cdot 10}+\dfrac{1\cdot 3\cdot 5\cdot 7}{4\cdot 7\cdot 10\cdot 13}+\cdots$.

40. $\dfrac{2}{3}+\dfrac{2\cdot 4}{3\cdot 7}+\dfrac{2\cdot 4\cdot 6}{3\cdot 7\cdot 11}+\dfrac{2\cdot 4\cdot 6\cdot 8}{3\cdot 7\cdot 11\cdot 15}+\cdots$.

41. Find the sum of the series $\frac{1}{10}+\frac{2}{100}+\frac{3}{1000}+\frac{4}{10000}+\cdots$. HINT: Exercise 58 of Section 11.2.

42. Complete the proof of the ratio test.
    (a) Prove that, if $\lambda > 1$, then $\Sigma\, a_k$ diverges.
    (b) Prove that, if $\lambda = 1$, the ratio test is inconclusive. HINT: Consider $\Sigma\, 1/k$ and $\Sigma\, 1/k^2$.

43. Prove that the sequence $\left\{\dfrac{n!}{n^n}\right\}$ has limit 0. HINT: Consider the series $\Sigma \dfrac{k!}{k^k}$.

44. Let r be a positive number. Prove that the sequence $\left\{\dfrac{r^n}{n!}\right\}$ has limit 0.

45. Let $p\geq 2$ be an integer. Find the values of p (if any) such that $\Sigma \dfrac{(k!)^2}{(pk)!}$ converges.

46. Let r be a positive number. For what values of r (if any) does $\Sigma \dfrac{r^k}{k^r}$ converge?

In Exercises 47–50, find the values of x for which the given series converges.

47. $\displaystyle\sum_{k=1}^{\infty} \frac{|x|^k}{k}$.

48. $\displaystyle\sum_{k=1}^{\infty} \frac{|x|^k}{2^k}$.

49. $\displaystyle\sum_{k=1}^{\infty} \frac{2^k |x|^k}{k!}$.

50. $\displaystyle\sum_{k=2}^{\infty} \frac{|x-2|^k}{k \, 3^k}$.

51. Let $\{a_k\}$ be a sequence of positive numbers and take $r > 0$. Use the root test to show that, if $(a_k)^{1/k} \to \rho$ and $\rho < 1/r$, then $\Sigma a_k r^k$ converges.

52. Consider the series
$\frac{1}{2} + 1 + \frac{1}{8} + \frac{1}{4} + \frac{1}{32} + \frac{1}{16} + \cdots$ formed by rearranging a convergent geometric series. (a) Use the root test to show that the series converges. (b) Show that the ratio test does not apply.

■ 11.5 ABSOLUTE AND CONDITIONAL CONVERGENCE; ALTERNATING SERIES

In this section we consider series that have both positive and negative terms.

Absolute and Conditional Convergence

Let Σa_k be a series with both positive and negative terms. One way to show that Σa_k converges is to show that the series of absolute values, $\Sigma |a_k|$, converges.

THEOREM 11.5.1

If $\Sigma |a_k|$ converges, then Σa_k converges.

PROOF For each k,

$$-|a_k| \le a_k \le |a_k| \quad \text{and therefore} \quad 0 \le a_k + |a_k| \le 2|a_k|.$$

If $\Sigma |a_k|$ converges, then $\Sigma 2|a_k| = 2 \, \Sigma |a_k|$ converges, and therefore, by the basic comparison theorem, $\Sigma (a_k + |a_k|)$ converges. Since

$$a_k = (a_k + |a_k|) - |a_k|,$$

we can conclude that $\Sigma |a_k|$ converges. ❏

DEFINITION 11.5.2 ABSOLUTE CONVERGENCE

A series Σa_k is *absolutely convergent* if the series of absolute values

$$|a_1| + |a_2| + |a_3| + \cdots = \Sigma |a_k|$$

is convergent

The theorem we have just proved says that *an absolutely convergent series is convergent.*

Example 1 Consider the series

$$1 - \frac{1}{2^2} + \frac{1}{3^2} - \frac{1}{4^2} + \frac{1}{5^2} - \frac{1}{6^2} + \cdots = \sum_{k=1}^{\infty} \frac{(-1)^{k+1}}{k^2}.$$

If we replace each term by its absolute value, we obtain the series

$$1 + \frac{1}{2^2} + \frac{1}{3^2} + \frac{1}{4^2} + \frac{1}{5^2} + \frac{1}{6^2} + \cdots = \sum_{k=1}^{\infty} \frac{1}{k^2}.$$

This is a *p*-series with $p = 2$. It is therefore convergent. This means that the initial series is absolutely convergent. ❏

Example 2 Consider the series

$$1 - \frac{1}{2} - \frac{1}{2^2} + \frac{1}{2^3} - \frac{1}{2^4} + \frac{1}{2^5} + \frac{1}{2^6} - \frac{1}{2^7} - \frac{1}{2^8} + \cdots.$$

If we replace each term by its absolute value, we obtain the series

$$1 + \frac{1}{2} + \frac{1}{2^2} + \frac{1}{2^3} + \frac{1}{2^4} + \frac{1}{2^5} + \frac{1}{2^6} + \frac{1}{2^7} + \frac{1}{2^8} + \cdots = \sum_{k=0}^{\infty} \frac{1}{2^k}.$$

This is a convergent geometric series. The initial series is therefore absolutely convergent. ❏

Example 3 As we will see after the next theorem, the series

$$1 - \frac{1}{2} + \frac{1}{3} - \frac{1}{4} + \frac{1}{5} - \frac{1}{6} + \cdots = \sum_{k=0}^{\infty} \frac{(-1)^k}{k+1} †$$

is convergent, but it is not absolutely convergent: if we replace each term by its absolute value, we get the divergent harmonic series

$$1 + \frac{1}{2} + \frac{1}{3} + \frac{1}{4} + \frac{1}{5} + \frac{1}{6} + \cdots = \sum_{k=0}^{\infty} \frac{1}{k+1}. ❏$$

DEFINITION 11.5.3 CONDITIONAL CONVERGENCE

A series Σa_k is *conditionally convergent* if it converges but $\Sigma |a_k|$ diverges.

Thus, the series $\Sigma (-1)^k/(k + 1)$ is conditionally convergent.

Alternating Series

Series in which the consecutive terms have opposite signs are called *alternating series*. For example, the series

$$1 - \frac{1}{2} + \frac{1}{3} - \frac{1}{4} + \frac{1}{5} - \frac{1}{6} + \cdots = \sum_{k=0}^{\infty} \frac{(-1)^k}{k+1}$$

---

† In Section 11.6 we show that the series $1 - \frac{1}{2} + \frac{1}{3} - \frac{1}{4} + \frac{1}{5} - \frac{1}{6} + \cdots$ converges to ln 2.

and

$$-1 + \frac{1}{\sqrt{2}} - \frac{1}{\sqrt{3}} + \frac{1}{\sqrt{4}} - \frac{1}{\sqrt{5}} + \cdots = \sum_{k=1}^{\infty} \frac{(-1)^k}{\sqrt{k}}$$

are alternating series.

The series

$$1 - \frac{1}{2} - \frac{1}{3} + \frac{1}{4} - \frac{1}{5} - \frac{1}{6} + \cdots$$

is not an alternating series because there are consecutive terms with the same sign.

In general, an alternating series will either have the form

$$a_0 - a_1 + a_2 - a_3 + a_4 - \cdots = \sum_{k=0}^{\infty} (-1)^k a_k$$

or the form

$$-a_0 + a_1 - a_2 + a_3 - a_4 + \cdots = \sum_{k=0}^{\infty} (-1)^{k+1} a_k,$$

where $\{a_k\}$ is a sequence of positive numbers. Since the second form is simply the negative of the first, we will focus our attention on the first form.

THEOREM 11.5.4   ALTERNATING SERIES TEST†

Let $\{a_k\}$ be a sequence of positive numbers. If

(a) $a_{k+1} < a_k$ for all k; that is, if the sequence $\{a_k\}$ is decreasing, and
(b) $a_k \to 0 \quad$ as $k \to \infty$,

then $\displaystyle\sum_{k=0}^{\infty} (-1)^k a_k$ converges.

PROOF   First we look at the even partial sums, s_{2m}. Since

$$s_{2m} = (a_0 - a_1) + (a_2 - a_3) + \cdots + (a_{2m-2} - a_{2m-1}) + a_{2m}$$

is the sum of positive numbers, the even partial sums are all positive. Since

$$s_{2m+2} = s_{2m} - (a_{2m+1} - a_{2m+2}) \quad \text{and} \quad a_{2m+1} - a_{2m+2} > 0,$$

we have

$$s_{2m+2} < s_{2m}.$$

This means that the sequence of even partial sums is decreasing. Being bounded below by 0, it is convergent; say,

$$s_{2m} \to L \quad \text{as } m \to \infty.$$

Now

$$s_{2m+1} = s_{2m} - a_{2m+1}.$$

---

† This theorem dates back to Leibniz. He proved the result in 1705.

Since $a_{2m+1} \to 0$ as $m \to \infty$, we also have

$$s_{2m+1} \to L.$$

Since both the even and the odd partial sums tend to L, the sequence of all partial sums tends to L (Exercise 42, Section 10.3). ❑

From this theorem you can see that the following series all converge:

$$1 - \frac{1}{2} + \frac{1}{3} - \frac{1}{4} + \frac{1}{5} - \frac{1}{6} + \cdots, \qquad 1 - \frac{1}{\sqrt{2}} + \frac{1}{\sqrt{3}} - \frac{1}{\sqrt{4}} + \frac{1}{\sqrt{5}} - \frac{1}{\sqrt{6}} + \cdots,$$

$$1 - \frac{1}{2!} + \frac{1}{3!} - \frac{1}{4!} + \frac{1}{5!} - \frac{1}{6!} + \cdots.$$

The first two series converge only conditionally; the third is absolutely convergent.

Remark In the proof of Theorem 11.5.4, we showed that the sequence of even partial sums, $\{s_{2m}\}$, is decreasing and bounded below (by 0). In Exercise 48 you are asked to show that the sequence of odd partial sums $\{s_{2m+1}\}$ is increasing and bounded above. This provides an alternative way to show that the sequence of odd partial sums converges, and since

$$s_{2m+1} - s_{2m} = -a_{2m+1} \to 0 \quad \text{as } m \to \infty,$$

each sequence has the same limit L. This is illustrated in Figure 11.5.1. ❑

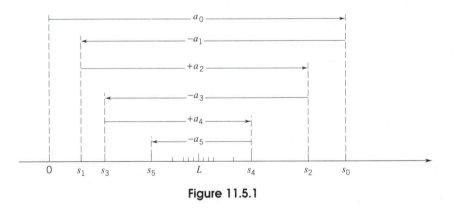

Figure 11.5.1

An Estimate for Alternating Series You have seen that if $\{a_k\}$ is a decreasing sequence of positive numbers that tends to 0, then

$$\sum_{k=0}^{\infty} (-1)^k a_k \quad \text{converges to a sum } L.$$

(11.5.5)

The sum L of a convergent alternating series lies between consecutive partial sums s_n, s_{n+1}, and thus s_n approximates L to within a_{n+1}:

$$|s_n - L| < a_{n+1}.$$

PROOF For all n

$$a_{n+1} > a_{n+2}.$$

If n is odd,

$$s_{n+2} = s_n + a_{n+1} - a_{n+2} > s_n;$$

if n is even,

$$s_{n+2} = s_n - a_{n+1} + a_{n+2} < s_n.$$

The odd partial sums increase toward L; the even partial sums decrease toward L.
For odd n

$$s_n < L < s_{n+1} = s_n + a_{n+1},$$

and for even n

$$s_n - a_{n+1} = s_{n+1} < L < s_n.$$

Thus, for all n, L lies between s_n and s_{n+1}, and s_n approximates L to within a_{n+1}. ❑

Example 5 Both

$$1 - \frac{1}{2} + \frac{1}{3} - \frac{1}{4} + \frac{1}{5} - \frac{1}{6} + \cdots \qquad \text{and} \qquad 1 - \frac{1}{2^2} + \frac{1}{3^2} - \frac{1}{4^2} + \frac{1}{5^2} - \frac{1}{6^2} + \cdots$$

are convergent alternating series. The nth partial sum of the first series approximates the sum of that series within $1/(n + 1)$; the nth partial sum of the second series approximates the sum of the second series within $1/(n + 1)^2$. The second series converges more rapidly than the first series. ❑

Example 6 Approximate the sum L of the alternating series

$$1 - \frac{1}{3!} + \frac{1}{5!} - \frac{1}{7!} + \cdots = \sum_{k=0}^{\infty} \frac{(-1)^k}{(2k + 1)!}$$

within 0.001.

SOLUTION It is easy to verify that the series converges. In fact, it is absolutely convergent. The fourth term of the series, $1/7! \cong 0.0002$, is the first term which is less than 0.001. Thus, we have

$$|L - s_3| < a_4 < 0.001.$$

Now,

$$s_3 = 1 - \frac{1}{3!} + \frac{1}{5!} = 1 - \frac{1}{6} + \frac{1}{120} \cong 0.8417$$

and so $L = 0.842$ with an error of less than 0.001.† ❑

---

† You will see in Section 11.6 that the actual sum of this series is $\sin 1 \cong 0.8415$.

Rearrangements

A *rearrangement* of a series $\Sigma\, a_k$ is a series that has exactly the same terms but in a different order. Thus, for example,

$$1 + \frac{1}{3^3} - \frac{1}{2^2} + \frac{1}{5^5} - \frac{1}{4^4} + \frac{1}{7^7} - \frac{1}{6^6} + \cdots$$

and

$$1 + \frac{1}{3^3} + \frac{1}{5^5} - \frac{1}{2^2} - \frac{1}{4^4} + \frac{1}{7^7} + \frac{1}{9^9} - \cdots$$

are both rearrangements of

$$1 - \frac{1}{2^2} + \frac{1}{3^3} - \frac{1}{4^4} + \frac{1}{5^5} - \frac{1}{6^6} + \frac{1}{7^7} - \cdots.$$

In 1867 Riemann published a theorem on rearrangements of series that underscores the importance of distinguishing between absolute convergence and conditional convergence. According to this theorem all rearrangements of an absolutely convergent series converge absolutely to the same sum. In sharp contrast, a series that is only conditionally convergent can be rearranged to converge to any number we please. It can also be arranged to diverge to $+\infty$, or to diverge to $-\infty$, or even to oscillate between any two bounds we choose.†

Example 7 We have shown that the series

$$\sum_{k=0}^{\infty} \frac{(-1)^k}{k+1}$$

is conditionally convergent, and in the next section you will see that its sum is $\ln 2$. Accepting this fact for now, we have

$$1 - \frac{1}{2} + \frac{1}{3} - \frac{1}{4} + \frac{1}{5} - \frac{1}{6} \cdots = \ln 2$$

and

$$\frac{1}{2} - \frac{1}{4} + \frac{1}{6} - \frac{1}{8} + \frac{1}{10} - \frac{1}{12} + \cdots = \tfrac{1}{2}\ln 2. \qquad \text{(multiply by } \tfrac{1}{2}\text{)}$$

Adding the two series, we get a rearrangement of the given series with a new sum:

$$1 + \frac{1}{3} - \frac{1}{2} + \frac{1}{5} + \frac{1}{7} - \frac{1}{4} + \cdots = \tfrac{3}{2}\ln 2. \quad \square$$

---

† For a complete proof see pp. 138–139, 318–320 in Konrad Knopp's *Theory and Applications of Infinite Series* (Second English Edition), Blackie & Son Limited, London, 1951.

EXERCISES 11.5

In Exercises 1–31, test these series for: (a) absolute convergence, (b) conditional convergence.

1. $1 + (-1) + 1 + \cdots + (-1)^k + \cdots$.

2. $\dfrac{1}{4} - \dfrac{1}{6} + \dfrac{1}{8} - \dfrac{1}{10} + \cdots + \dfrac{(-1)^k}{2k} + \cdots$.

3. $\dfrac{1}{2} - \dfrac{2}{3} + \dfrac{3}{4} - \dfrac{4}{5} + \cdots + (-1)^{k+1}\dfrac{k}{k+1} + \cdots$.

4. $\dfrac{1}{2 \ln 2} - \dfrac{1}{3 \ln 3} + \dfrac{1}{4 \ln 4} - \dfrac{1}{5 \ln 5} + \cdots +$

$\quad (-1)^k \dfrac{1}{k \ln k} + \cdots$.

5. $\Sigma(-1)^k \dfrac{\ln k}{k}$.

6. $\Sigma(-1)^k \dfrac{k}{\ln k}$.

7. $\Sigma\left(\dfrac{1}{k} - \dfrac{1}{k!}\right)$.

8. $\Sigma \dfrac{k^3}{2^k}$.

9. $\Sigma(-1)^k \dfrac{1}{2k+1}$.

10. $\Sigma(-1)^k \dfrac{(k!)^2}{(2k)!}$.

11. $\Sigma \dfrac{k!}{(-2)^k}$.

12. $\Sigma \sin\left(\dfrac{k\pi}{4}\right)$.

13. $\Sigma(-1)^k(\sqrt{k+1} - \sqrt{k})$.

14. $\Sigma(-1)^k \dfrac{k}{k^2 + 1}$.

15. $\Sigma \sin\left(\dfrac{\pi}{4k^2}\right)$.

16. $\Sigma \dfrac{(-1)^k}{\sqrt{k(k+1)}}$.

17. $\Sigma(-1)^k \dfrac{k}{2^k}$.

18. $\Sigma\left(\dfrac{1}{\sqrt{k}} - \dfrac{1}{\sqrt{k+1}}\right)$.

19. $\Sigma \dfrac{(-1)^k}{k - 2\sqrt{k}}$.

20. $\Sigma(-1)^k \dfrac{k+2}{k^2 + k}$.

21. $\Sigma(-1)^k \dfrac{4^{k-2}}{e^k}$.

22. $\Sigma(-1)^k \dfrac{k^2}{2^k}$.

23. $\Sigma(-1)^k k \sin(1/k)$.

24. $\Sigma(-1)^{k+1} \dfrac{k^k}{k!}$.

25. $\Sigma(-1)^k k e^{-k}$.

26. $\Sigma \dfrac{\cos \pi k}{k}$.

27. $\Sigma(-1)^k \dfrac{\cos \pi k}{k}$.

28. $\Sigma \dfrac{\sin(\pi k/2)}{k\sqrt{k}}$.

29. $\Sigma \dfrac{\sin(\pi k/4)}{k^2}$.

30. $\dfrac{1}{2} - \dfrac{1}{3} - \dfrac{1}{4} + \dfrac{1}{5} - \dfrac{1}{6} - \dfrac{1}{7} + \cdots + \dfrac{1}{3k+2} -$

$\quad \dfrac{1}{3k+3} - \dfrac{1}{3k+4} + \cdots$.

31. $\dfrac{2 \cdot 3}{4 \cdot 5} - \dfrac{5 \cdot 6}{7 \cdot 8} + \cdots + (-1)^k \dfrac{(3k+2)(3k+3)}{(3k+4)(3k+5)} + \cdots$.

In Exercises 32–35, estimate the error if the partial sum s_n is used to approximate the sum of the given alternating series.

32. $\displaystyle\sum_{k=1}^{\infty} (-1)^{k+1}\dfrac{1}{k}$; s_{20}.

33. $\displaystyle\sum_{k=0}^{\infty} (-1)^k\dfrac{1}{\sqrt{k+1}}$; s_{80}.

34. $\displaystyle\sum_{k=0}^{\infty} (-1)^k\dfrac{1}{(10)^k}$; s_4.

35. $\displaystyle\sum_{k=1}^{\infty} (-1)^{k+1}\dfrac{1}{k^3}$; s_9.

36. Let s_n be the nth partial sum of the series

$$\sum_{k=0}^{\infty} (-1)^k \dfrac{1}{10^k}.$$

Find the least value of n for which s_n approximates the sum of the series within:
(a) 0.001;   (b) 0.0001.

37. Find the sum of the series in Exercise 36.

In Exercises 38 and 39, find the smallest integer N such that s_N will approximate the sum of the given alternating series to within the indicated accuracy.

38. $\displaystyle\sum_{k=1}^{\infty} (-1)^k\dfrac{(0.9)^k}{k}$;   0.001.

39. $\displaystyle\sum_{k=0}^{\infty} (-1)^k\dfrac{1}{\sqrt{k+1}}$;   0.005.

40. Verify that the series

$$1 - \dfrac{1}{2} + \dfrac{1}{2} - \dfrac{1}{3} + \dfrac{1}{2} - \dfrac{1}{3} - \dfrac{1}{4} + \dfrac{1}{3} -$$

$$\dfrac{1}{4} + \dfrac{1}{3} - \dfrac{1}{4} + \cdots$$

diverges and explain how this does not violate the theorem on alternating series.

41. Let L be the sum of the series

$$\sum_{k=0}^{\infty} (-1)^k \dfrac{1}{k!}$$

and let s_n be the nth partial sum. Find the least value of n for which s_n approximates L within
(a) 0.01.   (b) 0.001.

42. Let $\{a_k\}$ be a nonincreasing sequence of positive numbers that converges to 0. Does the alternating series $\Sigma(-1)^k a_k$ necessarily converge?

43. Can the hypothesis of Theorem 11.5.4 be relaxed to require only that $\{a_{2k}\}$ and $\{a_{2k+1}\}$ be decreasing sequences of positive numbers with limit zero?

44. Prove that if Σa_k is absolutely convergent and $|b_k| \le |a_k|$ for all k, then Σb_k is absolutely convergent.

45. (a) Prove that if Σa_k is absolutely convergent, then Σa_k^2 is convergent.
(b) Show by means of an example that the converse of the result in part (a) is false.

46. Indicate how a conditionally convergent series can be rearranged (a) to converge to an arbitrary real number L; (b) to diverge to $+\infty$; (c) to diverge to $-\infty$. HINT: Collect the positive terms $p_1, p_2, p_3, \ldots$ and also the negative terms $n_1, n_2, n_3, \ldots$ in the order in which they appear in the original series.

47. In Section 11.8 we prove that, if $\Sigma a_k x_1^k$ converges, then $\Sigma a_k x^k$ converges absolutely for $|x| < |x_1|$. Try to prove this now.

48. Let $\displaystyle\sum_{k=0}^{\infty}(-1)^k a_k$ be an alternating series with $\{a_k\}$ a decreasing sequence. Prove that the sequence of odd partial sums $\{s_{2m+1}\}$ is increasing and bounded above.

49. Let a and b be positive numbers and consider the series
$$a - \frac{b}{2} + \frac{a}{3} - \frac{b}{4} + \frac{a}{5} - \frac{b}{6} + \cdots.$$
(a) Express this series in Σ notation.
(b) For what values of a and b is this series absolutely convergent? Conditionally convergent?

■ 11.6 TAYLOR POLYNOMIALS IN x; TAYLOR SERIES IN x

Taylor Polynomials in x

We begin with a function f continuous at 0 and set $P_0(x) = f(0)$. If f is differentiable at 0, the linear function that best approximates f at points close to 0 is the linear function
$$P_1(x) = f(0) + f'(0)x;$$

P_1 has the same value as f at 0 and also the same first derivative (the same rate of change):
$$P_1(0) = f(0), \quad P_1'(0) = f'(0).$$

(See Section 3.9, Exercise 48.) If f is twice differentiable at 0, then we can get a better approximation to f by using the quadratic polynomial
$$P_2(x) = f(0) + f'(0)x + \frac{f''(0)}{2!}x^2;$$

P_2 has the same value as f at 0 and the same first two derivatives:
$$P_2(0) = f(0), \quad P_2'(0) = f'(0), \quad P_2''(0) = f''(0).$$

If f has three derivatives at 0, we can form the cubic polynomial
$$P_3(x) = f(0) + f'(0)x + \frac{f''(0)}{2!}x^2 + \frac{f'''(0)}{3!}x^3;$$

P_3 has the same value as f at 0 and the same first three derivatives:
$$P_3(0) = f(0), \quad P_3'(0) = f'(0), \quad P_3''(0) = f''(0), \quad P_3'''(0) = f'''(0).$$

More generally, if f has n derivatives at 0, we can form the polynomial

$$P_n(x) = f(0) + f'(0)x + \frac{f''(0)}{2!}x^2 + \cdots + \frac{f^{(n)}(0)}{n!}x^n;$$

P_n is the polynomial of degree n that has the same value as f at 0 and the same first n derivatives:

$$P_n(0) = f(0), \quad P'_n(0) = f'(0), \quad P''_n(0) = f''(0), \ldots, P_n^{(n)}(0) = f^{(n)}(0).$$

These approximating polynomials $P_0(x), P_1(x), P_2(x), \ldots, P_n(x)$ are called *Taylor polynomials* after the English mathematician Brook Taylor (1685–1731). In particular, for each nonnegative integer k, $P_k(x)$, is called the *Taylor polynomial of degree k for f* or the *kth Taylor polynomial of f*.

Example 1 The exponential function

$$f(x) = e^x$$

has derivatives

$$f'(x) = e^x, \quad f''(x) = e^x, \quad f'''(x) = e^x, \quad \text{and so on.}$$

Thus

$$f(0) = 1, \quad f'(0) = 1, \quad f''(0) = 1, \quad f'''(0) = 1, \ldots, \quad f^{(n)}(0) = 1.$$

The nth Taylor polynomial takes the form

$$P_n(x) = 1 + x + \frac{x^2}{2!} + \frac{x^3}{3!} + \cdots + \frac{x^n}{n!}.$$

The graphs of $f(x) = e^x$, $P_0(x)$, $P_1(x)$, $P_2(x)$, and $P_3(x)$ are indicated in Figure 11.6.1. ❑

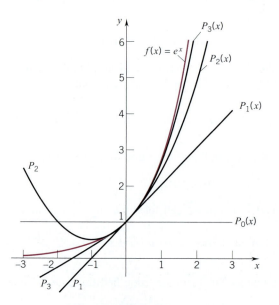

Figure 11.6.1

Example 2 To find the Taylor polynomials that approximate the sine function we write

$$f(x) = \sin x, \quad f'(x) = \cos x, \quad f''(x) = -\sin x, \quad f'''(x) = -\cos x.$$

The pattern now repeats itself:

$$f^{(4)}(x) = \sin x, \quad f^{(5)}(x) = \cos x, \quad f^{(6)}(x) = -\sin x, \quad f^{(7)}(x) = -\cos x.$$

At 0, the sine function and all its even derivatives are 0. The odd derivatives are alternately 1 and -1:

$$f'(0) = 1, \quad f'''(0) = -1, \quad f^{(5)}(0) = 1, \quad f^{(7)}(0) = -1, \quad \text{and so on.}$$

The Taylor polynomials are therefore as follows:

$$P_0(x) = 0$$

$$P_1(x) = P_2(x) = x$$

$$P_3(x) = P_4(x) = x - \frac{x^3}{3!}$$

$$P_5(x) = P_6(x) = x - \frac{x^3}{3!} + \frac{x^5}{5!}$$

$$P_7(x) = P_8(x) = x - \frac{x^3}{3!} + \frac{x^5}{5!} - \frac{x^7}{7!}, \quad \text{and so on.}$$

Only odd powers appear; you should relate this to the fact the $f(x) = \sin x$ is an *odd* function. The graphs of $f(x) = \sin x$, $P_1(x)$, $P_3(x)$, $P_5(x)$, $P_7(x)$ are indicated in Figure 11.6.2. □

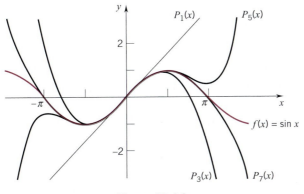

Figure 11.6.2

It is not enough to say that the Taylor polynomials

$$P_n(x) = f(0) + f'(0)x + \frac{f''(0)}{2!}x^2 + \cdots + \frac{f^{(n)}(0)}{n!}x^n$$

approximate $f(x)$. We must describe the accuracy of the approximation.

Our first step is to prove a result known as Taylor's theorem.

THEOREM 11.6.1 TAYLOR'S THEOREM

Suppose that f has $n + 1$ continuous derivatives on an open interval I containing 0. Then, for each $x \in I$,

$$f(x) = f(0) + f'(0)x + \frac{f''(0)}{2!}x^2 + \cdots + \frac{f^{(n)}(0)}{n!}x^n + R_{n+1}(x),$$

where the *remainder* $R_{n+1}(x)$ is given by the formula

$$R_{n+1}(x) = \frac{1}{n!}\int_0^x f^{(n+1)}(t)(x - t)^n \, dt.$$

PROOF Fix x in the interval I. Then

(1)
$$\int_0^x f'(t) \, dt = f(x) - f(0).$$

On the other hand, if we evaluate the integral using integration by parts with

$$u = f'(t) \quad \text{and} \quad dv = dt,$$

then

$$du = f''(t) \, dt \quad \text{and} \quad v = -(x - t) \qquad \text{(verify this expression for } v\text{)}$$

and

(2)
$$\int_0^x f'(t) \, dt = \left[-f'(t)(x - t) \right]_0^x + \int_0^x f''(t)(x - t) \, dt = f'(0)x + \int_0^x f''(t)(x - t) \, dt.$$

Thus, from Equations (1) and (2), we have

$$f(x) = f(0) + f'(0)x + \int_0^x f''(t)(x - t) \, dt.$$

Integrating by parts again [let $u = f''(t)$, $dv = (x - t) \, dt$, then $du = f'''(t) \, dt$, $v = -\frac{1}{2}(x - t)^2$], we get

$$f(x) = f(0) + f'(0)x + \frac{f''(0)}{2!}x^2 + \frac{1}{2!}\int_0^x f'''(t)(x - t)^2 \, dt.$$

If we continue integrating by parts (see Exercise 53), we will get, after n steps,

$$f(x) = f(0) + f'(0)x + \frac{f''(0)}{2!}x^2 + \frac{f'''(0)}{3!}x^3 + \cdots +$$

$$\frac{f^{(n)}(0)}{n!}x^n + \frac{1}{n!}\int_0^x f^{(n+1)}(t)(x - t)^n \, dt.$$

Thus,

$$f(x) = P_n(x) + \frac{1}{n!} \int_0^x f^{(n+1)}(t)(x - t)^n \, dt,$$

and

$$R_{n+1}(x) = \frac{1}{n!} \int_0^x f^{(n+1)}(t)(x - t)^n \, dt. \quad \Box$$

To see how closely

$$P_n(x) = f(0) + f'(0)x + \frac{f''(0)}{2!}x^2 + \cdots + \frac{f^{(n)}(0)}{n!}x^n$$

approximates $f(x)$ we need an estimate for the remainder term $R_{n+1}(x)$. The following corollary to Taylor's theorem gives a more convenient form of the remainder. It was established by Joseph Lagrange in 1797, and it is known as the Lagrange formula for the remainder. The proof is left to you as an exercise.

COROLLARY 11.6.2 LAGRANGE FORMULA FOR THE REMAINDER

Suppose that f has $n + 1$ continuous derivatives on an open interval I containing 0. Let $x \in I$ and let $P_n(x)$ be the nth Taylor polynomial for f. Then

$$R_{n+1}(x) = \frac{f^{(n+1)}(c)}{(n+1)!}x^{n+1},$$

where c is some number between 0 and x.

Remark It is important to understand that the number c indicated in the corollary depends on x and, of course, on f and n. If we rewrite Taylor's theorem using the Lagrange form for the remainder, we have

$$f(x) = f(0) + f'(0)x + \frac{f''(0)}{2!}x^2 + \cdots + \frac{f^{(n)}}{n!}x^n + \frac{f^{(n+1)}(c)}{(n+1)!}x^{n+1}$$

where c is some number between 0 and x. This result is an extension of the mean-value theorem. Indeed, if $n = 0$, we get

$$f(x) = f(0) + f'(c)x \qquad \text{or} \qquad f(x) - f(0) = f'(c)(x - 0),$$

where c is between 0 and x. This is the mean-value theorem for f on the interval $[0, x]$. $\Box$

The following estimate for $R_{n+1}(x)$ is an immediate consequence of Corollary 11.6.2. Let J be the interval joining 0 to x, $x \neq 0$. Then

(11.6.3)
$$|R_{n+1}(x)| \le \left(\max_{t \in J} |f^{(n+1)}(t)| \right) \frac{|x|^{n+1}}{(n+1)!}.$$

Example 3   The Taylor polynomials of the exponential function

$$f(x) = e^x$$

take the form

$$P_n(x) = 1 + x + \frac{x^2}{2!} + \cdots + \frac{x^n}{n!}. \qquad \text{(Example 1)}$$

We will show with our remainder estimate that for all real x

$$R_{n+1}(x) \to 0 \quad \text{as } n \to \infty,$$

and therefore we can approximate e^x as closely as we wish by Taylor polynomials.
   We begin by fixing x and letting M be the maximum value of the exponential function on the interval J that joins 0 to x.   (If $x > 0$, then $M = e^x$; if $x < 0$, $M = e^0 = 1$.)   Since

$$f^{(n+1)}(t) = e^t \qquad \text{for all } n,$$

we have

$$\max_{t \in J} |f^{(n+1)}(t)| = M \qquad \text{for all } n.$$

Thus by (11.6.3)

$$|R_{n+1}(x)| \le M \frac{|x|^{n+1}}{(n+1)!}.$$

By (10.4.4) we know that

$$\frac{|x|^{n+1}}{(n+1)!} \to 0 \quad \text{as } n \to \infty.$$

It follows then that $R_{n+1}(x) \to 0$ as asserted.   ❏

Example 4   We return to the sine function

$$f(x) = \sin x$$

and its Taylor polynomials

$$P_1(x) = P_2(x) = x$$

$$P_3(x) = P_4(x) = x - \frac{x^3}{3!}$$

$$P_5(x) = P_6(x) = x - \frac{x^3}{3!} + \frac{x^5}{5!}, \quad \text{and so on.}$$

The pattern of derivatives was established in Example 2; namely, for all k,

$$f^{(4k)}(x) = \sin x, \quad f^{(4k+1)}(x) = \cos x,$$

$$f^{(4k+2)}(x) = -\sin x, \quad f^{(4k+3)}(x) = -\cos x.$$

Thus, for all n and all real t,

$$|f^{(n+1)}(t)| \le 1.$$

It follows from our remainder estimate (11.6.3) that

$$|R_{n+1}(x)| \leq \frac{|x|^{n+1}}{(n+1)!}.$$

Since

$$\frac{|x|^{n+1}}{(n+1)!} \to 0 \qquad \text{for all real } x,$$

we see that $R_{n+1}(x) \to 0$ for all real x. Thus the sequence of Taylor polynomials converges to the sine function and therefore can be used to approximate $\sin x$ for any real number x as closely as we may wish. ❑

Taylor Series in x

By definition $0! = 1$. By adopting the convention that $f^{(0)} = f$, we can write Taylor polynomials

$$P_n(x) = f(0) + f'(0)x + \frac{f''(0)}{2!}x^2 + \cdots + \frac{f^{(n)}(0)}{n!}x^n$$

in Σ notation:

$$P_n(x) = \sum_{k=0}^{n} \frac{f^{(k)}(0)}{k!}x^k.$$

If f is infinitely differentiable on an open interval I containing 0, then we have

$$f(x) = \sum_{k=0}^{n} \frac{f^{(k)}(0)}{k!}x^k + R_{n+1}(x), \qquad x \in I,$$

for all positive integers n. If, as in the case of the exponential function and the sine function, $R_{n+1}(x) \to 0$ as $n \to \infty$ for each $x \in I$, then

$$\sum_{k=0}^{n} \frac{f^{(k)}(0)}{k!}x^k \to f(x).$$

In this case, we say that $f(x)$ can be expanded as a *Taylor series in x* and write

(11.6.4)
$$f(x) = \sum_{k=0}^{\infty} \frac{f^{(k)}(0)}{k!}x^k.$$

Taylor series in x are sometimes called Maclaurin series after Colin Maclaurin, a Scottish mathematician (1698–1746). In some circles the name Maclaurin remains attached to these series, although Taylor considered them some twenty years before Maclaurin.

From Example 3 it is clear that

(11.6.5)
$$e^x = \sum_{k=0}^{\infty} \frac{x^k}{k!} = 1 + x + \frac{x^2}{2!} + \frac{x^3}{3!} + \cdots \qquad \text{for all real } x.$$

From Example 4 we have

(11.6.6)

$$\sin x = \sum_{k=0}^{\infty} \frac{(-1)^k}{(2k+1)!} x^{2k+1} = x - \frac{x^3}{3!} + \frac{x^5}{5!} - \frac{x^7}{7!} + \cdots \quad \text{for all real } x.$$

Note that $\sin 1 = 1 - \frac{1}{3!} + \frac{1}{5!} - \frac{1}{7!} + \cdots$ as suggested in Section 11.5.
We leave it to you as an exercise to show that

(11.6.7)

$$\cos x = \sum_{k=0}^{\infty} \frac{(-1)^k}{(2k)!} x^{2k} = 1 - \frac{x^2}{2!} + \frac{x^4}{4!} - \frac{x^6}{6!} + \cdots \quad \text{for all real } x.$$

We come now to the logarithm function. Since $\ln x$ is not defined at $x = 0$, we cannot expand $\ln x$ in powers of x. We work instead with $\ln(1 + x)$.

(11.6.8)

$$\ln(1 + x) = \sum_{k=1}^{\infty} \frac{(-1)^{k+1}}{k} x^k = x - \frac{x^2}{2} + \frac{x^3}{3} - \cdots \quad \text{for } -1 < x \leq 1.$$

PROOF†   The function

$$f(x) = \ln(1 + x)$$

is defined on $(-1, \infty)$ and has derivatives

$$f'(x) = \frac{1}{1 + x}, \qquad f''(x) = -\frac{1}{(1 + x)^2}, \qquad f'''(x) = \frac{2}{(1 + x)^3},$$

$$f^{(4)}(x) = -\frac{3!}{(1 + x)^4}, \qquad f^{(5)}(x) = \frac{4!}{(1 + x)^5}, \qquad \text{and so on.}$$

For $k \geq 1$

$$f^{(k)}(x) = (-1)^{k+1} \frac{(k-1)!}{(1 + x)^k}, \qquad f^{(k)}(0) = (-1)^{k+1}(k-1)!, \qquad \frac{f^{(k)}(0)}{k!} = \frac{(-1)^{k+1}}{k}.$$

Since $f(0) = 0$, the nth Taylor polynomial takes the form

$$P_n(x) = \sum_{k=1}^{n} (-1)^{k+1} \frac{x^k}{k} = x - \frac{x^2}{2} + \cdots + (-1)^{n+1} \frac{x^n}{n}.$$

All we have to show therefore is that

$$R_{n+1}(x) \to 0 \qquad \text{for } -1 < x \leq 1.$$

---

† The proof we give here illustrates the methods of this section. A much simpler way of obtaining this series expansion is given in Section 11.9.

Instead of trying to apply our usual remainder estimate (in this case, that estimate is not delicate enough to show that $R_{n+1}(x) \to 0$ for $-1 < x < -\frac{1}{2}$), we write the remainder in its integral form. From Taylor's theorem

$$R_{n+1}(x) = \frac{1}{n!} \int_0^x f^{(n+1)}(t)(x-t)^n \, dt,$$

so that in this case

$$R_{n+1}(x) = \frac{1}{n!} \int_0^x (-1)^{n+2} \frac{n!}{(1+t)^{n+1}}(x-t)^n \, dt = (-1)^n \int_0^x \frac{(x-t)^n}{(1+t)^{n+1}} \, dt.$$

For $0 \le x \le 1$ we have

$$|R_{n+1}(x)| = \int_0^x \frac{(x-t)^n}{(1+t)^{n+1}} \, dt \le \int_0^x (x-t)^n \, dt = \frac{x^{n+1}}{n+1} \to 0.$$

　　　　　　　　　　　　　　└── explain

For $-1 < x < 0$ we have

$$|R_{n+1}(x)| = \left| \int_0^x \frac{(x-t)^n}{(1+t)^{n+1}} \, dt \right| = \int_x^0 \left(\frac{t-x}{1+t} \right)^n \frac{1}{1+t} \, dt.$$

By the First Mean-Value Theorem for Integrals (5.9.1) there exists a number x_n between x and 0 such that

$$\int_x^0 \left(\frac{t-x}{1+t} \right)^n \frac{1}{1+t} \, dt = \left(\frac{x_n - x}{1+x_n} \right)^n \left(\frac{1}{1+x_n} \right)(-x).$$

Since $-x = |x|$ and $0 < 1 + x < 1 + x_n$, we can conclude that

$$|R_{n+1}(x)| < \left(\frac{x_n + |x|}{1+x_n} \right)^n \left(\frac{|x|}{1+x} \right).$$

Since $|x| < 1$ and $x_n < 0$, we have

$$x_n < |x| x_n, \qquad x_n + |x| < |x| x_n + |x| = |x|(1 + x_n)$$

and thus

$$\frac{x_n + |x|}{1 + x_n} < |x|.$$

It now follows that

$$|R_{n+1}(x)| < |x|^n \left(\frac{|x|}{1+x} \right)$$

and, since $|x| < 1$, that $R_{n+1}(x) \to 0$ as $n \to \infty$. ❑

Remark The series expansion for $\ln(1+x)$ that we have just verified for $-1 < x \le 1$ cannot be extended to other values of x. For $x \le -1$ neither side makes sense: $\ln(1+x)$ is not defined, and the series on the right diverges. For $x > 1$, $\ln(1+x)$ is defined, but the series on the right diverges and hence does not represent the function. At $x = 1$, the series gives the intriguing result that was mentioned in Section 11.5:

$$\ln 2 = 1 - \tfrac{1}{2} + \tfrac{1}{3} - \tfrac{1}{4} + \cdots.$$

The graphs of $f(x) = \ln(1 + x)$, $P_1(x)$, $P_2(x)$, and $P_3(x)$ are shown in Figure 11.6.3. Compare these approximations with the Taylor polynomial approximations of e^x and $\sin x$ shown in Figures 11.6.1 and 11.6.2. ❏

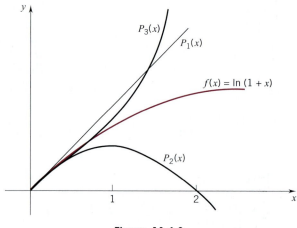

Figure 11.6.3

We want to emphasize again the role played by the remainder term $R_{n+1}(x)$. We can form a Taylor series

$$\sum_{k=0}^{\infty} \frac{f^{(k)}(0)}{k!} x^k$$

for any function f with derivatives of all orders at $x = 0$, but such a series need not converge at any number $x \neq 0$. Even if it does converge, the sum need not be $f(x)$. (See Exercise 55.) The Taylor series converges to $f(x)$ if and only if the remainder term $R_{n+1}(x)$ tends to 0.

Some Numerical Calculations

If the Taylor series converges to $f(x)$, we can use the partial sums (the Taylor polynomials) to calculate $f(x)$ as accurately as we wish. In what follows we show some sample calculations. For ready reference we list some values of $k!$ and $1/k!$ in Tables 11.6.1 and 11.6.2.

■ **Table 11.6.1**

$k!$
$2! = 2$
$3! = 6$
$4! = 24$
$5! = 120$
$6! = 720$
$7! = 5,040$
$8! = 40,320$

■ **Table 11.6.2**

$1/k!$	
$0.16666 < \dfrac{1}{3!} < 0.16667$	$0.00138 < \dfrac{1}{6!} < 0.00139$
$.04166 < \dfrac{1}{4!} < 0.04167$	$0.00019 < \dfrac{1}{7!} < 0.00020$
$0.00833 < \dfrac{1}{5!} < 0.00834$	$0.00002 < \dfrac{1}{8!} < 0.00003$

Example 5 Estimate e within 0.001.

SOLUTION For all x

$$e^x = 1 + x + \frac{x^2}{2!} + \cdots + \frac{x^n}{n!} + \cdots.$$

Taking $x = 1$ we have

$$e = 1 + 1 + \frac{1}{2!} + \cdots + \frac{1}{n!} + \cdots.$$

From Example 3 we know that the nth partial sum of this series, the nth Taylor polynomial of e^x evaluated at $x = 1$,

$$P_n(1) = 1 + 1 + \frac{1}{2!} + \cdots + \frac{1}{n!},$$

approximates e within

$$|R_{n+1}(1)| \leq e \frac{|1|^{n+1}}{(n+1)!} < \frac{3}{(n+1)!}.$$

here $M = e^1 = e$ ⟶ $e < 3$

Since

$$\frac{3}{7!} = \frac{3}{5040} = \frac{1}{1680} < 0.001,$$

we can take $n = 6$ and be sure that

$$P_6(1) = 1 + 1 + \frac{1}{2!} + \frac{1}{3!} + \frac{1}{4!} + \frac{1}{5!} + \frac{1}{6!} = \frac{1957}{720}$$

differs from e by less than 0.001.

Our calculator gives

$$\frac{1957}{720} \cong 2.7180556 \qquad \text{and} \qquad e \cong 2.7182818. \quad \square$$

Example 6 Estimate $e^{0.2}$ within three decimal places (remainder less than 0.0005).

SOLUTION The exponential series at $x = 0.2$ gives

$$e^{0.2} = 1 + 0.2 + \frac{(0.2)^2}{2!} + \cdots + \frac{(0.2)^n}{n!} + \cdots.$$

From Example 3 we know that the nth partial sum of this series, the nth Taylor polynomial of e^x evaluated at $x = 0.2$

$$P_n(0.2) = 1 + 0.2 + \frac{(0.2)^2}{2!} + \cdots + \frac{(0.2)^n}{n!},$$

approximates $e^{0.2}$ within

$$|R_{n+1}(0.2)| \leq e^{0.2} \frac{|0.2|^{n+1}}{(n+1)!} < 3^{0.2} \frac{(0.2)^{n+1}}{(n+1)!} < 1.25 \frac{(0.2)^{n+1}}{(n+1)!}.$$

here $M = e^{0.2}$

Since

$$1.25 \frac{(0.2)^4}{4!} = \frac{1.25(0.0016)}{24} = 0.0000833 \cdots < 0.00009,$$

we can take $n = 3$ and be sure that

$$P_3(0.2) = 1 + 0.2 + \frac{(0.2)^2}{2!} + \frac{(0.2)^3}{3!} = \frac{7.328}{6} \cong 1.22133 \cdots$$

differs from $e^{0.2}$ by less than 0.00009, which is a much better result than the one we asked for. You can verify that $n = 2$ will not yield the desired accuracy.

It now follows that $1.22124 < e^{0.2} < 1.22143$. Our calculator gives $e^{0.2} \cong 1.2214028$. ❏

Example 7 Estimate sin 0.5 within 0.001.

SOLUTION At $x = 0.5$ the sine series gives

$$\sin 0.5 = 0.5 - \frac{(0.5)^3}{3!} + \frac{(0.5)^5}{5!} - \frac{(0.5)^7}{7!} + \cdots.$$

From Example 4 we know that $P_n(0.5)$, where $P_n(x)$ is the nth Taylor polynomial of sin x, approximates sin 0.5 within

$$|R_{n+1}(0.5)| \leq \frac{(0.5)^{n+1}}{(n+1)!}.$$

Since

$$\frac{(0.5)^5}{5!} = \frac{1}{(2^5)(5!)} = \frac{1}{(32)(120)} = \frac{1}{3840} < 0.001,$$

we can be sure that

the coefficient of x^4 is 0

$$P_4(0.5) = P_3(0.5) = 0.5 - \frac{(0.5)^3}{3!} = \frac{23}{48}$$

approximates sin 0.5 within 0.001.

Our calculator gives

$$\frac{23}{48} \cong 0.4791666 \qquad \text{and} \qquad \sin 0.5 \cong 0.4794255. \quad ❏$$

Remark We could have solved the last problem without reference to the remainder estimate derived in Example 4. The series for sin 0.5 is a convergent alternating series with decreasing terms. By (11.5.5) we can conclude immediately that sin 0.5 lies between every two consecutive partial sums. In particular

$$0.5 - \frac{(0.5)^3}{3!} < \sin 0.5 < 0.5 - \frac{(0.5)^3}{3!} + \frac{(0.5)^5}{5!}. \quad ❏$$

Example 8 Estimate ln 1.4 within 0.01.

SOLUTION By (11.6.8)

$$\ln 1.4 = \ln(1 + 0.4) = 0.4 - \tfrac{1}{2}(0.4)^2 + \tfrac{1}{3}(0.4)^3 - \tfrac{1}{4}(0.4)^4 + \cdots .$$

This is a convergent alternating series with decreasing terms. Therefore ln 1.4 lies between every two consecutive partial sums.

The first term less than 0.01 is

$$\tfrac{1}{4}(0.4)^4 = \tfrac{1}{4}(0.0256) = 0.0064.$$

The relation

$$0.4 - \tfrac{1}{2}(0.4)^2 + \tfrac{1}{3}(0.4)^3 - \tfrac{1}{4}(0.4)^4 < \ln 1.4 < 0.4 - \tfrac{1}{2}(0.4)^2 + \tfrac{1}{3}(0.4)^3$$

gives

$$0.335 < \ln 1.4 < 0.341.$$

Within the prescribed limits of accuracy we can take ln $1.4 \cong 0.34$. ❏†

---

† A much more effective tool for computing logarithms is given in the exercises.

EXERCISES 11.6

In Exercises 1–4, find the Taylor polynomial $P_4(x)$ for the given function.

1. $f(x) = x - \cos x$.   **2.** $f(x) = \sqrt{1 + x}$.

3. $f(x) = \ln \cos x$.   **4.** $f(x) = \sec x$.

In Exercises 5–8, find the Taylor polynomial $P_5(x)$ for the given function.

5. $f(x) = (1 + x)^{-1}$.   **6.** $f(x) = e^x \sin x$.

7. $f(x) = \tan x$.   **8.** $f(x) = x \cos x^2$.

9. Determine $P_0(x)$, $P_1(x)$, $P_2(x)$, $P_3(x)$ for
$f(x) = 1 - x + 3x^2 + 5x^3$.

10. Determine $P_0(x)$, $P_1(x)$, $P_2(x)$, $P_3(x)$ for $f(x) = (x + 1)^3$.

In Exercises 11–16, determine the nth Taylor polynomial $P_n(x)$ for the given function.

11. $f(x) = e^{-x}$.   **12.** $f(x) = \sinh x$.

13. $f(x) = \cosh x$.   **14.** $f(x) = \ln(1 - x)$.

15. $f(x) = e^{rx}$, r a real number.

16. $f(x) = \cos bx$, b a real number.

In Exercises 17–24, use Taylor polynomials to estimate the following within 0.01.

17. $\sqrt{e}$.   **18.** $\sin 0.3$.

19. $\sin 1$.   **20.** $\ln 1.2$.

21. $\cos 1$.   **22.** $e^{0.8}$.

23. $\sin 10°$.   **24.** $\cos 6°$.

In Exercises 25–32, find the Lagrange form of the remainder R_{n+1} for the given function and the indicated integer n.

25. $f(x) = e^{2x}$; $n = 4$.

26. $f(x) = \ln(1 + x)$; $n = 5$.

27. $f(x) = \cos 2x$; $n = 4$.

28. $f(x) = \sqrt{x + 1}$; $n = 3$.

29. $f(x) = \tan x$; $n = 2$.

30. $f(x) = \sin x$; $n = 5$.

31. $f(x) = \tan^{-1}x$; $n = 2$.

32. $f(x) = \dfrac{1}{1 + x}$; $n = 4$.

In Exercises 33–36, find the Lagrange form of the remainder R_{n+1} for the given function.

33. $f(x) = e^{-x}$.   **34.** $f(x) = \sin 2x$.

35. $f(x) = \dfrac{1}{1 - x}$.   **36.** $f(x) = \ln(1 + x)$.

37. Let $P_n(x)$ be the nth Taylor polynomial of
$$f(x) = \ln(1 + x).$$

Find the least integer n for which: (a) $P_n(0.5)$ approximates $\ln 1.5$ within 0.01; (b) $P_n(0.3)$ approximates $\ln 1.3$ within 0.01; (c) $P_n(1)$ approximates $\ln 2$ within 0.001.

38. Let $P_n(x)$ be the nth Taylor polynomial of

$$f(x) = \sin x.$$

Find the least integer n for which: (a) $P_n(1)$ approximates $\sin 1$ within 0.001; (b) $P_n(2)$ approximates $\sin 2$ within 0.001; (c) $P_n(3)$ approximates $\sin 3$ within 0.001.

▶ **39.** Let $f(x) = e^x$.
  (a) Find the Taylor polynomial P_n of f of least degree that will approximate $\sqrt{e}$ with four decimal place accuracy. Then evaluate $P_n(1/2)$ to obtain your approximation of $\sqrt{e}$.
  (b) Find the Taylor polynomial P_n of f of least degree that will approximate $1/e$ with three decimal place accuracy. Then evaluate $P_n(-1)$ to obtain your approximation of $1/e$.

▶ **40.** Let $g(x) = \cos x$.
  (a) Find the Taylor polynomial P_n of g of least degree that will approximate $\cos(\pi/30)$ with three decimal place accuracy. Then evaluate $P_n(\pi/30)$ to obtain your approximation of $\cos(\pi/30)$.
  (b) Find the Taylor polynomial P_n of g of least degree that will approximate $\cos 9°$ with four decimal place accuracy. Then evaluate P_n to obtain your approximation of $\cos 9°$. (Remember to convert to radian measure.)

41. Show that a polynomial $P(x) = a_0 + a_1 x + \cdots + a_n x^n$ is its own Taylor series.

42. Show that

$$\cos x = \sum_{k=0}^{\infty} \frac{(-1)^k}{(2k)!} x^{2k} \qquad \text{for all real } x.$$

43. Show that

$$\sinh x = \sum_{k=0}^{\infty} \frac{1}{(2k+1)!} x^{2k+1} \qquad \text{for all real } x.$$

44. Show that

$$\cosh x = \sum_{k=0}^{\infty} \frac{1}{(2k)!} x^{2k} \qquad \text{for all real } x.$$

In Exercises 45–49, derive a series expansion in x for the given function and specify the numbers x for which the expansion is valid. Take $a > 0$.

45. $f(x) = e^{ax}$. HINT: Set $t = ax$ and expand e^t in powers of t.

46. $f(x) = \sin ax$.

47. $f(x) = \cos ax$.

48. $f(x) = \ln(1 - ax)$.

49. $f(x) = \ln(a + x)$.
  HINT: $\ln(a + x) = \ln\{a(1 + x/a)\}$.

50. The series we derived for $\ln(1 + x)$ converges too slowly to be of much practical use. The following logarithm series converges much more quickly:

(11.6.9)
$$\ln\left(\frac{1+x}{1-x}\right) = 2\left(x + \frac{x^3}{3} + \frac{x^5}{5} + \cdots\right)$$
$$\text{for } -1 < x < 1.$$

Derive this series expansion.

51. Set $x = \frac{1}{3}$ and use the first three nonzero terms of (11.6.9) to estimate $\ln 2$.

52. Use the first two nonzero terms of (11.6.9) to estimate $\ln 1.4$.

53. Verify the identity

$$\frac{f^{(k)}(0)}{k!} x^k = \frac{1}{(k-1)!} \int_0^x f^{(k)}(t)(x-t)^{k-1} \, dt$$
$$- \frac{1}{k!} \int_0^x f^{(k+1)}(t)(x-t)^k \, dt$$

by computing the second integral by parts.

54. Prove Corollary 11.6.2 and then derive the remainder estimate (11.6.3).

▶ **55.** (a) Use a graphing utility to draw the graph of the function

$$f(x) = \begin{cases} e^{-1/x^2}, & x \neq 0 \\ 0, & x = 0 \end{cases}.$$

  (b) Use L'Hospital's rule to show that for every positive integer n

$$\lim_{x \to 0} \frac{e^{-1/x^2}}{x^n} = 0.$$

  (c) Use mathematical induction to prove that $f^{(n)}(0) = 0$ for all $n \geq 1$.
  (d) What is the Taylor series of f?
  (e) For what values of x does the Taylor series of f actually represent f?

▶ **56.** Let $f(x) = \cos x$. Use a graphing utility to graph the Taylor polynomials $P_2(x)$, $P_4(x)$, $P_6(x)$, and $P_8(x)$ of f.

▶ **57.** Let $g(x) = \ln(1 + x)$. Use a graphing utility to graph the Taylor polynomials $P_2(x)$, $P_3(x)$, $P_4(x)$, and $P_5(x)$ of g.

58. (*Important*) Show that e is irrational by following these steps.

(1) Take the expansion

$$e = \sum_{k=0}^{\infty} \frac{1}{k!}$$

and show that the qth partial sum

$$s_q = \sum_{k=0}^{q} \frac{1}{k!}$$

satisfies the inequality

$$0 < q!(e - s_q) < \frac{1}{q}.$$

(2) Show that $q!s_q$ is an integer and argue that, if e were of the form p/q, then $q!(e - s_q)$ would be a positive integer less than 1.

■ 11.7 TAYLOR POLYNOMIALS AND TAYLOR SERIES IN $x - a$

So far we have considered series expansions only in powers of x. Here we generalize to expansions in powers of $x - a$, where a is an arbitrary real number. We begin with a more general version of Taylor's theorem.

THEOREM 11.7.1 TAYLOR'S THEOREM

Suppose that g has $n + 1$ continuous derivatives on an open interval I containing the point a. Then, for each $x \in I$,

$$g(x) = g(a) + g'(a)(x - a) + \frac{g''(a)}{2!}(x - a)^2 + \cdots$$

$$+ \frac{g^{(n)}(a)}{n!}(x - a)^n + R_{n+1}(x),$$

where

$$R_{n+1}(x) = \frac{1}{n!} \int_a^x g^{(n+1)}(t)(x - t)^n \, dt.$$

The polynomial

$$P_n(x) = g(a) + g'(a)(x - a) + \frac{g''(a)}{2!}(x - a)^2 + \cdots + \frac{g^{(n)}(a)}{n!}(x - a)^n$$

is call the *nth Taylor polynomial for g in powers of (x − a)*. In this more general setting, the Lagrange formula for the remainder $R_{n+1}(x)$ is given by the following Corollary.

COROLLARY 11.7.2 LAGRANGE FORMULA FOR THE REMAINDER

Suppose that g has $n + 1$ continuous derivatives on an open interval I containing a. Let $x \in I$ and let P_n be the nth Taylor polynomial for g in powers of $(x - a)$. Then

$$R_{n+1}(x) = \frac{g^{(n+1)}(c)}{(n + 1)!}(x - a)^{n+1},$$

where c is some number between a and x.

Now let $x \in I, x \neq a$, and let J be the interval joining a and x. Then an estimate for the remainder can be written:

(11.7.3)
$$|R_{n+1}(x)| \leq \left(\max_{t \in J} |g^{(n+1)}(t)| \right) \frac{|x-a|^{n+1}}{(n+1)!}.$$

If $R_{n+1}(x) \to 0$, then we have

$$g(x) = g(a) + g'(a)(x-a) + \frac{g''(a)}{2!}(x-a)^2 + \cdots + \frac{g^{(n)}(a)}{n!}(x-a)^n + \cdots.$$

In sigma notation we have

(11.7.4)
$$g(x) = \sum_{k=0}^{\infty} \frac{g^{(k)}(a)}{k!}(x-a)^k.$$

This is known as the Taylor expansion of $g(x)$ in powers of $x - a$. The series on the right is called a *Taylor series in* $x - a$.

All this differs from what you saw before only by a translation. Define

$$f(x) = g(x+a).$$

Then obviously

$$f^{(k)}(x) = g^{(k)}(x+a) \quad \text{and} \quad f^{(k)}(0) = g^{(k)}(a).$$

The results of this section as stated for g can be derived by applying the results of Section 11.6 to the function f.

Example 1 Expand $g(x) = 4x^3 - 3x^2 + 5x - 1$ in powers of $x - 2$.

SOLUTION We need to evaluate g and its derivatives at $x = 2$.

$$g(x) = 4x^3 - 3x^2 + 5x - 1$$
$$g'(x) = 12x^2 - 6x + 5$$
$$g''(x) = 24x - 6$$
$$g'''(x) = 24.$$

All higher derivatives are identically 0.

Substitution gives $g(2) = 29$, $g'(2) = 41$, $g''(2) = 42$, $g'''(2) = 24$, and $g^{(k)}(2) = 0$ for all $k \geq 4$. Thus from (11.7.4)

$$g(x) = 29 + 41(x-2) + \frac{42}{2!}(x-2)^2 + \frac{24}{3!}(x-2)^3$$
$$= 29 + 41(x-2) + 21(x-2)^2 + 4(x-2)^3. \quad \square$$

Example 2 Expand $g(x) = x^2 \ln x$ in powers of $x - 1$.

SOLUTION We need to evaluate g and its derivatives at $x = 1$.

$$g(x) = x^2 \ln x$$

$$g'(x) = x + 2x \ln x$$

$$g''(x) = 3 + 2 \ln x$$

$$g'''(x) = 2x^{-1}$$

$$g^{(4)}(x) = -2x^{-2}$$

$$g^{(5)}(x) = (2)(2)x^{-3}$$

$$g^{(6)}(x) = -(2)(2)(3)x^{-4} = -2(3!)x^{-4}$$

$$g^{(7)}(x) = (2)(2)(3)(4)x^{-5} = (2)(4!)x^{-5}, \quad \text{and so on.}$$

The pattern is now clear: for $k \geq 3$

$$g^{(k)}(x) = (-1)^{k+1} 2(k-3)! x^{-k+2}.$$

Evaluation at $x = 1$ gives $g(1) = 0$, $g'(1) = 1$, $g''(1) = 3$ and, for $k \geq 3$,

$$g^{(k)}(1) = (-1)^{k+1} 2(k-3)!.$$

The expansion in powers of $x - 1$ can be written

$$g(x) = (x-1) + \frac{3}{2!}(x-1)^2 + \sum_{k=3}^{\infty} \frac{(-1)^{k+1}(2)(k-3)!}{k!}(x-1)^k$$

$$= (x-1) + \frac{3}{2}(x-1)^2 + 2\sum_{k=3}^{\infty} \frac{(-1)^{k+1}}{k(k-1)(k-2)}(x-1)^k. \quad \square$$

Another way to expand $g(x)$ in powers of $x - a$ is to expand $g(t + a)$ in powers of t and then set $t = x - a$. This is the approach we take when the expansion in t is either known to us, or is easily available.

Example 3 We can expand $g(x) = e^{x/2}$ in powers of $x - 3$ by expanding

$$g(t + 3) = e^{(t+3)/2} \qquad \text{in powers of } t$$

and then setting $t = x - 3$.

Note that

$$g(t + 3) = e^{3/2}e^{t/2} = e^{3/2} \sum_{k=0}^{\infty} \frac{(t/2)^k}{k!} = e^{3/2} \sum_{k=0}^{\infty} \frac{1}{2^k k!} t^k.$$

exponential series ⟵

Setting $t = x - 3$, we have

$$g(x) = e^{3/2} \sum_{k=0}^{\infty} \frac{1}{2^k k!}(x-3)^k.$$

Since the expansion of $g(t + 3)$ is valid for all real t, the expansion of $g(x)$ is valid for all real x. $\square$

Taking this same approach, we can prove:

(11.7.5)

For $a > 0$ and $0 < x \leq 2a$

$$\ln x = \ln a + \frac{1}{a}(x-a) - \frac{1}{2a^2}(x-a)^2 + \frac{1}{3a^3}(x-a)^3 - \cdots$$

$$= \ln a + \sum_{k=1}^{\infty} \frac{(-1)^{k+1}}{k\,a^k}(x-a)^k.$$

PROOF We will expand $\ln(a+t)$ in powers of t and then set $t = x - a$. In the first place

$$\ln(a+t) = \ln\left[a\left(1 + \frac{t}{a}\right)\right] = \ln a + \ln\left(1 + \frac{t}{a}\right).$$

From (11.6.8) it is clear that

$$\ln\left(1 + \frac{t}{a}\right) = \frac{t}{a} - \frac{1}{2}\left(\frac{t}{a}\right)^2 + \frac{1}{3}\left(\frac{t}{a}\right)^3 - \cdots \quad \text{for } -1 < \frac{t}{a} \leq 1 \quad \text{or} \quad -a < t \leq a.$$

Adding $\ln a$ to both sides, we have

$$\ln(a+t) = \ln a + \frac{1}{a}t - \frac{1}{2a^2}t^2 + \frac{1}{3a^3}t^3 - \cdots \qquad \text{for } -a < t \leq a.$$

Setting $t = x - a$, we find that

$$\ln x = \ln a + \frac{1}{a}(x-a) - \frac{1}{2a^2}(x-a)^2 + \frac{1}{3a^3}(x-a)^3 - \cdots$$

for all x such that $-a < x - a \leq a$; that is, for all x such that $0 < x \leq 2a$. ❑

EXERCISES 11.7

In Exercises 1–6, find the Taylor polynomial of the function f for the given values of a and n, and give the Lagrange form of the remainder.

1. $f(x) = \sqrt{x}$; $a = 4$, $n = 3$.
2. $f(x) = \cos x$; $a = \pi/3$, $n = 4$.
3. $f(x) = \sin x$; $a = \pi/4$, $n = 4$.
4. $f(x) = \ln x$; $a = 1$, $n = 5$.
5. $f(x) = \tan^{-1} x$; $a = 1$, $n = 3$.
6. $f(x) = \cos \pi x$; $a = \frac{1}{2}$, $n = 4$.

In Exercises 7–22, expand $g(x)$ as indicated and specify the values of x for which the expansion is valid.

7. $g(x) = 3x^3 - 2x^2 + 4x + 1$ in powers of $x - 1$.
8. $g(x) = x^4 - x^3 + x^2 - x + 1$ in powers of $x - 2$.
9. $g(x) = 2x^5 + x^2 - 3x - 5$ in powers of $x + 1$.

10. $g(x) = x^{-1}$ in powers of $x - 1$.
11. $g(x) = (1 + x)^{-1}$ in powers of $x - 1$.
12. $g(x) = (b + x)^{-1}$ in powers of $x - a$, $a \neq -b$.
13. $g(x) = (1 - 2x)^{-1}$ in powers of $x + 2$.
14. $g(x) = e^{-4x}$ in powers of $x + 1$.
15. $g(x) = \sin x$ in powers of $x - \pi$.
16. $g(x) = \sin x$ in powers of $x - \frac{1}{2}\pi$.
17. $g(x) = \cos x$ in powers of $x - \pi$.
18. $g(x) = \cos x$ in powers of $x - \frac{1}{2}\pi$.
19. $g(x) = \sin \frac{1}{2}\pi x$ in powers of $x - 1$.
20. $g(x) = \sin \pi x$ in powers of $x - 1$.
21. $g(x) = \ln(1 + 2x)$ in powers of $x - 1$.
22. $g(x) = \ln(2 + 3x)$ in powers of $x - 4$.

In Exercises 23–32, expand $g(x)$ as indicated.

23. $g(x) = x \ln x$ in powers of $x - 2$.

24. $g(x) = x^2 + e^{3x}$ in powers of $x - 2$.

25. $g(x) = x \sin x$ in powers of x.

26. $g(x) = \ln(x^2)$ in powers of $x - 1$.

27. $g(x) = (1 - 2x)^{-3}$ in powers of $x + 2$.

28. $g(x) = \sin^2 x$ in powers of $x - \frac{1}{2}\pi$.

29. $g(x) = \cos^2 x$ in powers of $x - \pi$.

30. $g(x) = (1 + 2x)^{-4}$ in powers of $x - 2$.

31. $g(x) = x^n$ in powers of $x - 1$.

32. $g(x) = (x - 1)^n$ in powers of x.

33. (a) Expand e^x in powers of $x - a$.
  (b) Use the expansion to show that $e^{x_1 + x_2} = e^{x_1}e^{x_2}$.
  (c) Expand e^{-x} in powers of $x - a$.

34. (a) Expand $\sin x$ and $\cos x$ in powers of $x - a$.
  (b) Show that both series are absolutely convergent for all real x.
  (c) As noted earlier (Section 11.5), Riemann proved that the order of the terms of an absolutely convergent series may be changed without altering the sum of the series. Use Riemann's discovery and the Taylor expansions of part (a) to derive the addition formulas

$$\sin(x_1 + x_2) = \sin x_1 \cos x_2 + \cos x_1 \sin x_2,$$
$$\cos(x_1 + x_2) = \cos x_1 \cos x_2 - \sin x_1 \sin x_2.$$

35. (a) Determine the Taylor polynomial P_n in powers of $(x - \pi/6)$ and of least degree that will approximate $\sin 35°$ with four decimal place accuracy.
  (b) Evaluate the polynomial that you found in part (a) to obtain your approximation of $\sin 35°$.

36. (a) Determine the Taylor polynomial P_n in powers of $(x - \pi/3)$ and of least degree that will approximate $\cos 57°$ with four decimal place accuracy.
  (b) Evaluate the polynomial that you found in part (a) to obtain your approximation of $\cos 57°$.

37. Choose an appropriate Taylor polynomial of $f(x) = \sqrt{x}$ to approximate $\sqrt{38}$ with three decimal place accuracy.

38. Choose an appropriate Taylor polynomial for $f(x) = \sqrt{x}$ to approximate $\sqrt{61}$ with three decimal place accuracy.

■ 11.8 POWER SERIES

You have become familiar with Taylor series

$$\sum_{k=0}^{\infty} \frac{f^{(k)}(0)}{k!} x^k \quad \text{and} \quad \sum_{k=0}^{\infty} \frac{f^{(k)}(a)}{k!} (x - a)^k.$$

Here we study series of the form

$$\sum_{k=0}^{\infty} a_k x^k \quad \text{and} \quad \sum_{k=0}^{\infty} a_k(x - a)^k$$

without regard to how the coefficients a_k have been generated. Such series are called *power series*. In particular, the first series is a *power series in powers of x*, and the second is a *power series in powers of $x - a$*.

Since a simple translation converts

$$\sum_{k=0}^{\infty} a_k(x - a)^k \quad \text{into} \quad \sum_{k=0}^{\infty} a_k x^k,$$

we can focus our attention on power series of the form

$$\sum_{k=0}^{\infty} a_k x^k.$$

When detailed indexing is unnecessary, we will omit it and write

$$\sum a_k x^k.$$

Note that if we replace the variable x by a real number c, then the result

$$\sum a_k c^k$$

is an infinite series, and so the ideas and methods of Sections 11.2–11.5 can be applied. In particular, our objective here is to determine the set of numbers c for which the resulting infinite series converges. We begin the discussion with a definition.

DEFINITION 11.8.1

A power series $\sum a_k x^k$ is said to converge

  (i) at c iff $\sum a_k c^k$ converges;

  (ii) on the set S iff $\sum a_k x^k$ converges for each $x \in S$.

The following result is fundamental.

THEOREM 11.8.2

If $\sum a_k x^k$ converges at $c \neq 0$, then it converges absolutely for $|x| < |c|$.

If $\sum a_k x^k$ diverges at c, then it diverges for $|x| > |c|$.

PROOF   If $\sum a_k c^k$ converges, then $a_k c^k \rightarrow 0$ as $k \rightarrow \infty$. In particular, for k sufficiently large,

$$|a_k c^k| \leq 1$$

and thus

$$|a_k x^k| = |a_k c^k| \left| \frac{x}{c} \right|^k \leq \left| \frac{x}{c} \right|^k.$$

For $|x| < |c|$, we have

$$\left| \frac{x}{c} \right| < 1.$$

The convergence of $\sum |a_k x^k|$ follows by comparison with the geometric series. This proves the first statement.

    Suppose now that $\sum a_k c^k$ diverges. By a similar argument, there cannot exist x with $|x| > |c|$ such that $\sum a_k x^k$ converges. The existence of such an x would imply the absolute convergence of $\sum a_k c^k$. This proves the second statement.   ❑

    From the theorem we just proved you can see that there are exactly three possibilities for a power series:

Case I. *The series converges only at* $x = 0$. This is what happens with

$$\sum k^k x^k.$$

For if $x \neq 0$, then $\lim_{k \to \infty} k^k x^k \neq 0$, and so the series cannot converge (Theorem 11.2.7).

Case II. The series is absolutely convergent for all real numbers x. This is what happens with the exponential series

$$\sum \frac{x^k}{k!}.$$

Case III. *There exists a positive integer* r *such that the series converges for* $|x| < r$ *and diverges for* $|x| > r$. This is what happens with the geometric series

$$\sum x^k.$$

In this instance, there is absolute convergence for $|x| < 1$ and divergence for $|x| > 1$.

Associated with each case is a *radius of convergence*:

In Case I, we say that the radius of convergence is 0.
In Case II, we say that the radius of convergence is ∞.
In Case III, we say that the radius of convergence is r.

The three cases are pictured in Figure 11.8.1.

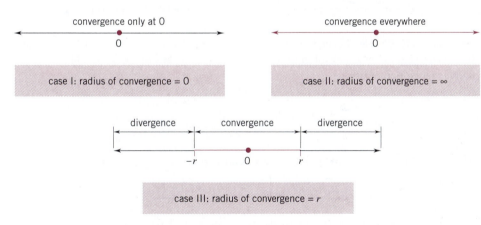

Figure 11.8.1

In general, the behavior of a power series at $-r$ and at r is not predictable. For example, the series

$$\sum x^k, \qquad \sum \frac{(-1)^k}{k} x^k, \qquad \sum \frac{1}{k} x^k, \qquad \sum \frac{1}{k^2} x^k$$

all have radius of convergence 1, but the first series converges only on $(-1, 1)$ (verify this) and, as we show next, the second series converges on $(-1, 1]$, the third on $[-1, 1)$, and the fourth on $[-1, 1]$.

The maximal interval on which a power series converges is called the *interval of convergence*. For a series with infinite radius of convergence, the interval of convergence is $(-\infty, \infty)$. For a series with radius of convergence r, the interval of convergence can be $[-r, r]$, $(-r, r]$, $[-r, r)$, or $(-r, r)$. For a series with radius of convergence 0, the interval of convergence reduces to a point, $\{0\}$.

Example 1 Verify that the series

(1)
$$\sum \frac{(-1)^k}{k} x^k$$

has interval of convergence $(-1, 1]$.

SOLUTION First we show that the radius of convergence is 1 (that the series converges absolutely for $|x| < 1$ and diverges for $|x| > 1$). We do this by forming the series

(2)
$$\sum \left| \frac{(-1)^k}{k} x^k \right| = \sum \frac{1}{k} |x|^k$$

and applying the ratio test.

We set

$$b_k = \frac{1}{k} |x|^k$$

and note that

$$\frac{b_{k+1}}{b_k} = \frac{|x|^{k+1}/(k+1)}{|x|^k/k} = \frac{k}{k+1} \frac{|x|^{k+1}}{|x|^k} = \frac{k}{k+1} |x| \to |x| \quad \text{as } k \to \infty.$$

By the ratio test, series (2) converges for $|x| < 1$ and diverges for $|x| > 1$.† It follows that series (1) converges absolutely for $|x| < 1$ and diverges for $|x| > 1$. The radius of convergence is therefore 1.

Now we test the endpoints $x = -1$ and $x = 1$. At $x = -1$

$$\sum \frac{(-1)^k}{k} x^k \quad \text{becomes} \quad \sum \frac{(-1)^k}{k} (-1)^k = \sum \frac{1}{k}.$$

This is the harmonic series which, as you know, diverges. At $x = 1$

$$\sum \frac{(-1)^k}{k} x^k \quad \text{becomes} \quad \sum \frac{(-1)^k}{k}.$$

This is a convergent alternating series.

We have shown that series (1) converges absolutely for $|x| < 1$, diverges at -1, and converges at 1. The interval of convergence is $(-1, 1]$. ❑

_____

† We could also have used the root test:
$$(b_k)^{1/k} = \left| \frac{1}{k} \right|^{1/k} |x| = \frac{1}{k^{1/k}} |x| \to |x|.$$

Remark   The same arguments can be used to show that the series

$$\sum \frac{1}{k} x^k$$

converges on $[-1, 1)$.   ❑

Example 2   Verify that the series

(1)
$$\sum \frac{1}{k^2} x^k$$

has interval of convergence $[-1, 1]$.

SOLUTION   First we examine the series

(2)
$$\sum \left| \frac{1}{k^2} x^k \right| = \sum \frac{1}{k^2} |x|^k.$$

Here again we use the ratio test.   We set

$$b_k = \frac{1}{k^2} |x|^k$$

and note that

$$\frac{b_{k+1}}{b_k} = \frac{k^2}{(k+1)^2} \frac{|x|^{k+1}}{|x|^k} = \left(\frac{k}{k+1} \right)^2 |x| \rightarrow |x| \quad \text{as } k \rightarrow \infty.$$

By the ratio test, (2) converges for $|x| < 1$ and diverges for $|x| > 1$.†   This shows that (1) converges absolutely for $|x| < 1$ and diverges for $|x| > 1$.   The radius of convergence is therefore 1.
   Now for the endpoints.   At $x = -1$,

$$\sum \frac{1}{k^2} x^k \quad \text{takes the form} \quad \sum \frac{(-1)^k}{k^2} = -1 + \tfrac{1}{4} - \tfrac{1}{9} + \tfrac{1}{16} - \cdots.$$

This is a convergent alternating series.   At $x = 1$,

$$\sum \frac{1}{k^2} x^k \quad \text{becomes} \quad \sum \frac{1}{k^2}.$$

This is a convergent p-series.   The interval of convergence is therefore the entire closed interval $[-1, 1]$.   ❑

Example 3   Find the interval of convergence of

(1)
$$\sum \frac{k}{6^k} x^k.$$

---

† Once again we could have used the root test:

$$(b_k) = \frac{1}{k^{2/k}} |x| \rightarrow |x| \quad \text{as } k \rightarrow \infty.$$

SOLUTION We begin by examining the series

$$(2) \qquad \sum \left| \frac{k}{6^k} x^k \right| = \sum \frac{k}{6^k} |x|^k.$$

We set

$$b_k = \frac{k}{6^k} |x|^k$$

and apply the root test. (The ratio test will also work.) Since

$$(b_k)^{1/k} = \tfrac{1}{6} k^{1/k} |x| \rightarrow \tfrac{1}{6} |x| \quad \text{as } k \rightarrow \infty \qquad \text{(recall } k^{1/k} \rightarrow 1)$$

you can see that (2) converges

$$\text{for} \quad \tfrac{1}{6}|x| < 1, \qquad \text{that is, for} \quad |x| < 6,$$

and diverges

$$\text{for} \quad \tfrac{1}{6}|x| > 1, \qquad \text{that is, for} \quad |x| > 6.$$

Testing the endpoints, we have:

$$\text{at} \quad x = 6, \qquad \sum \frac{k}{6^k} 6^k = \sum k, \qquad \text{which is divergent;}$$

$$\text{at} \quad x = -6, \qquad \sum \frac{k}{6^k}(-6)^k = \sum (-1)^k k, \qquad \text{which is also divergent.}$$

Thus, the interval of convergence is $(-6, 6)$. ❏

Example 4 Find the interval of convergence of

$$(1) \qquad \sum \frac{(2k)!}{(3k)!} x^k.$$

SOLUTION Again, we begin by examining the series

$$(2) \qquad \sum \left| \frac{(2k)!}{(3k)!} x^k \right| = \sum \frac{(2k)!}{(3k)!} |x|^k.$$

Set

$$b_k = \frac{(2k)!}{(3k)!} |x|^k.$$

Since factorials are involved, we will use the ratio test. Note that

$$\frac{b_{k+1}}{b_k} = \frac{\{2(k+1)\}!}{\{3(k+1)\}!} \frac{(3k)!}{(2k)!} \frac{|x|^{k+1}}{|x|^k} = \frac{(2k+2)(2k+1)}{(3k+3)(3k+2)(3k+1)} |x|.$$

Since

$$\frac{(2k+2)(2k+1)}{(3k+3)(3k+2)(3k+1)} \rightarrow 0 \quad \text{as } k \rightarrow \infty,$$

(the numerator is a quadratic in k, the denominator is a cubic), the ratio b_{k+1}/b_k tends to 0 no matter what x is. By the ratio test, (2) converges for all x and therefore (1) converges absolutely for all x. The radius of convergence is ∞ and the interval of convergence is $(-\infty, \infty)$. ❏

Example 5 Find the interval of convergence of

$$\sum (\tfrac{1}{2}k)^k x^k.$$

SOLUTION Since $(\tfrac{1}{2}k)^k x^k \to 0$ only if $x = 0$ (explain), there is no need to invoke the ratio test or the root test. By (11.2.7) the series can converge only at $x = 0$. There it converges trivially, that is, all the terms of the series are 0. ❏

Example 6 Find the interval of convergence of

$$\sum \frac{(-1)^k}{k^2 3^k}(x + 2)^k.$$

SOLUTION We consider the series

$$\sum \left| \frac{(-1)^k}{k^2 3^k}(x + 2)^k \right| = \sum \frac{1}{k^2 3^k}|x + 2|^k.$$

Set

$$b_k = \frac{1}{k^2 3^k}|x + 2|^k$$

and apply the ratio test (the root test will work equally as well):

$$\frac{b_{k+1}}{b_k} = \frac{|x + 2|^{k+1}/(k + 1)^2 3^{k+1}}{|x + 2|^k/k^2 3^k} = \frac{k^2}{3(k + 1)^2}|x + 2| \to \tfrac{1}{3}|x + 2| \quad \text{as } k \to \infty.$$

Thus, the series is absolutely convergent

$$\text{for} \quad \tfrac{1}{3}|x + 2| < 1 \qquad \text{or} \qquad |x + 2| < 3,$$

which is the same as $-5 < x < 1$.
   We now check the endpoints. At $x = -5$:

$$\sum \frac{(-1)^k}{k^2 3^k}(-3)^k = \sum \frac{1}{k^2}.$$

This is a convergent p-series. At $x = 1$:

$$\sum \frac{(-1)^k}{k^2 3^k}(3)^k = \sum \frac{(-1)^k}{k^2},$$

and this is a convergent alternating series. Therefore, the interval of convergence is $[-5, 1]$ ❏

EXERCISES 11.8

In Exercises 1–38, find the interval of convergence.

1. $\sum kx^k.$

2. $\sum \frac{1}{k}x^k.$

3. $\sum \frac{1}{(2k)!}x^k.$

4. $\sum \frac{2^k}{k^2}x^k.$

5. $\sum (-k)^{2k}x^{2k}.$

6. $\sum \frac{(-1)^k}{\sqrt{k}}x^k.$

7. $\sum \frac{1}{k2^k}x^k.$

8. $\sum \frac{1}{k^2 2^k}x^k.$

9. $\sum \left(\frac{k}{100} \right)^k x^k.$

10. $\sum \frac{k^2}{1 + k^2}x^k.$

11. $\sum \frac{2^k}{\sqrt{k}}x^k.$

12. $\sum \frac{1}{\ln k}x^k.$

13. $\sum \frac{k - 1}{k}x^k.$

14. $\sum ka^k x^k.$

15. $\sum \dfrac{k}{10^k} x^k$.

16. $\sum \dfrac{3k^2}{e^k} x^k$.

17. $\sum \dfrac{x^k}{k^k}$.

18. $\sum \dfrac{7^k}{k!} x^k$.

19. $\sum \dfrac{(-1)^k}{k^k} (x-2)^k$.

20. $\sum k!\, x^k$.

21. $\sum (-1)^k \dfrac{2^k}{3^{k+1}} x^k$.

22. $\sum \dfrac{2^k}{(2k)!} x^k$.

23. $\sum (-1)^k \dfrac{k!}{k^3} (x-1)^k$.

24. $\sum \dfrac{(-e)^k}{k^2} x^k$.

25. $\sum \left(\dfrac{k}{k-1}\right) \dfrac{(x+2)^k}{2^k}$.

26. $\sum \dfrac{\ln k}{k} (x+1)^k$.

27. $\sum (-1)^k \dfrac{k^2}{(k+1)!} (x+3)^k$.

28. $\sum \dfrac{k^3}{e^k} (x-4)^k$.

29. $\sum \left(1+\dfrac{1}{k}\right)^k x^k$.

30. $\sum \dfrac{(-1)^k a^k}{k^2} (x-a)^k$.

31. $\sum \dfrac{\ln k}{2^k} (x-2)^k$.

32. $\sum \dfrac{1}{(\ln k)^k} (x-1)^k$.

33. $\sum (-1)^k (\tfrac{2}{3})^k (x+1)^k$.

34. $\sum \dfrac{2^{1/k} \pi^k}{k(k+1)(k+2)} (x-2)^k$.

35. $1 - \dfrac{x}{2} + \dfrac{2x^2}{4} - \dfrac{3x^3}{8} + \dfrac{4x^4}{16} - \cdots$.

36. $\dfrac{(x-1)}{5^2} + \dfrac{4}{5^4}(x-1)^2 + \dfrac{9}{5^6}(x-1)^3 +$
$$\dfrac{16}{5^8}(x-1)^4 + \cdots.$$

37. $\dfrac{3x^2}{4} + \dfrac{9x^4}{9} + \dfrac{27x^6}{16} + \dfrac{81x^8}{25} + \cdots$.

38. $\dfrac{1}{16}(x+1) - \dfrac{2}{25}(x+1)^2 + \dfrac{3}{36}(x+1)^3 -$
$$\dfrac{4}{49}(x+1)^4 + \cdots.$$

39. Let $\sum a_k x^k$ be a power series, and let r be its radius of convergence.
(a) Given that $|a_k|^{1/k} \to \rho$, show that, if $\rho \neq 0$, then $r = 1/\rho$ and, if $\rho = 0$, then $r = \infty$.
(b) Given that $|a_{k+1}/a_k| \to \lambda$, show that, if $\lambda \neq 0$, then $r = 1/\lambda$ and, if $\lambda = 0$, then $r = \infty$.

40. Find the interval of convergence of the series $\sum s_k x^k$ where s_k is the kth partial sum of the series
$$\sum_{n=1}^{\infty} \dfrac{1}{n}.$$

41. Let $\sum a_k x^k$ be a power series and let r, $0 < r < \infty$, be its radius of convergence. Prove that if the series is absolutely convergent at one endpoint of its interval of convergence, then it is absolutely convergent at the other also.

42. Let $\sum a_k x^k$ be a power series and let r, $0 < r < \infty$, be its radius of convergence. Prove that the power series $\sum a_k x^{2k}$ has radius of convergence $\sqrt{r}$.

■ 11.9 DIFFERENTIATION AND INTEGRATION OF POWER SERIES

Suppose that the power series
$$\sum_{k=0}^{\infty} a_k x^k$$
converges on the interval $(-c, c)$. Then, for each $x \in (-c, c)$, the infinite series
$$\sum_{k=0}^{\infty} a_k x^k$$
converges to a number L_x; the sum L_x depends upon x. Let f be the function defined on $(-c, c)$ by $f(x) = L_x$, that is,
$$f(x) = \sum_{k=0}^{\infty} a_k x^k \qquad \text{for } x \in (-c, c).$$

In this section we show that this function is both infinitely differentiable and integrable on $(-c, c)$. We begin with a simple but important result.

THEOREM 11.9.1

If

$$\sum_{k=0}^{\infty} a_k x^k = a_0 + a_1 x + a_2 x^2 + \cdots + a_n x^n + \cdots$$

converges on $(-c, c)$, then

$$\sum_{k=0}^{\infty} \frac{d}{dx}(a_k x^k) = \sum_{k=1}^{\infty} k a_k x^{k-1}$$
$$= a_1 + 2a_2 x + 3a_3 x^2 + \cdots + n a_n x^{n-1} + \cdots$$

also converges on $(-c, c)$.

PROOF   Assume that

$$\sum_{k=0}^{\infty} a_k x^k \quad \text{converges on } (-c, c).$$

By Theorem 11.8.2 the series is absolutely convergent on this interval.

Now let x be some fixed number in $(-c, c)$ and choose $\epsilon > 0$ such that

$$|x| < |x| + \epsilon < c.$$

Since $|x| + \epsilon$ lies within the interval of convergence,

$$\sum_{k=0}^{\infty} |a_k(|x| + \epsilon)^k| \quad \text{converges.}$$

In Exercise 48 you are asked to show that, for all k sufficiently large,

$$|k x^{k-1}| \leq (|x| + \epsilon)^k.$$

It follows that for all such k

$$|k a_k x^{k-1}| \leq |a_k(|x| + \epsilon)^k|.$$

Since

$$\sum_{k=0}^{\infty} |a_k(|x| + \epsilon)^k| \quad \text{converges,}$$

we can conclude that

$$\sum_{k=0}^{\infty} \left| \frac{d}{dx}(a_k x^k) \right| = \sum_{k=1}^{\infty} |k a_k x^{k-1}| \quad \text{converges,}$$

and thus that

$$\sum_{k=0}^{\infty} \frac{d}{dx}(a_k x^k) = \sum_{k=1}^{\infty} k a_k x^{k-1} \quad \text{converges.} \quad ❑$$

Repeated application of the theorem shows that

$$\sum_{k=0}^{\infty} \frac{d^2}{dx^2}(a_k x^k), \qquad \sum_{k=0}^{\infty} \frac{d^3}{dx^3}(a_k x^k), \qquad \sum_{k=0}^{\infty} \frac{d^4}{dx^4}(a_k x^k), \qquad \text{and so on,}$$

all converge on $(-c, c)$.

Example 1 Since the geometric series

$$\sum_{k=0}^{\infty} x^k = 1 + x + x^2 + x^3 + x^4 + x^5 + x^6 + \cdots$$

converges on $(-1, 1)$, the series

$$\sum_{k=0}^{\infty} \frac{d}{dx}(x^k) = \sum_{k=1}^{\infty} kx^{k-1} = 1 + 2x + 3x^2 + 4x^3 + 5x^4 + 6x^5 + \cdots,$$

$$\sum_{k=0}^{\infty} \frac{d^2}{dx^2}(x^k) = \sum_{k=2}^{\infty} k(k-1)x^{k-2} = 2 + 6x + 12x^2 + 20x^3 + 30x^4 + \cdots,$$

$$\sum_{k=0}^{\infty} \frac{d^3}{dx^3}(x^k) = \sum_{k=3}^{\infty} k(k-1)(k-2)x^{k-3} = 6 + 24x + 60x^2 + 120x^3 + \cdots,$$

$$\cdot$$
$$\cdot$$
$$\cdot$$

all converge on $(-1, 1)$. ❏

COROLLARY 11.9.2

If

$$\sum_{k=0}^{\infty} a_k x^k$$

has a radius of convergence r, then each of the series

$$\sum_{k=0}^{\infty} \frac{d}{dx}(a_k x^k), \qquad \sum_{k=0}^{\infty} \frac{d^2}{dx^2}(a_k x^k), \qquad \sum_{k=0}^{\infty} \frac{d^3}{dx^3}(a_k x^k), \qquad \text{and so on,}$$

has radius of convergence r.

Remark Even though $\Sigma\, a_k x^k$ and its "derivative" $\Sigma\, k a_k x^{k-1}$ have the same radius of convergence, their intervals of convergence may be different. For example, the interval of convergence of the series

$$\sum_{k=1}^{\infty} \frac{1}{k^2} x^k$$

is $[-1, 1]$, whereas the interval of convergence of its derivative

$$\sum_{k=1}^{\infty} \frac{1}{k} x^{k-1}$$

is $[-1, 1)$. Endpoints must always be checked separately. Obviously, this remark does not apply in the cases where $r = 0$ or $r = \infty$. ❏

Suppose now that

$$\sum_{k=0}^{\infty} a_k x^k \quad \text{converges on } (-c, c).$$

Then, as we have seen,

$$\sum_{k=0}^{\infty} \frac{d}{dx}(a_k x^k) \quad \text{also converges on } (-c, c).$$

As we noted at the beginning of this section, we can define a function f on $(-c, c)$ by setting

$$f(x) = \sum_{k=0}^{\infty} a_k x^k.$$

Using the second series, we can define a function g on $(-c, c)$ by setting

$$g(x) = \sum_{k=0}^{\infty} \frac{d}{dx}(a_k x^k).$$

The crucial point is that

$$f'(x) = g(x).$$

THEOREM 11.9.3 THE DIFFERENTIABILITY THEOREM

If

$$f(x) = \sum_{k=0}^{\infty} a_k x^k \qquad \text{for all } x \text{ in } (-c, c),$$

then f is differentiable on $(-c, c)$ and

$$f'(x) = \sum_{k=0}^{\infty} \frac{d}{dx}(a_k x^k) \qquad \text{for all } x \text{ in } (-c, c).$$

By applying this theorem to f', you can see that f' is itself differentiable. This in turn implies that f'' is differentiable, and so on. In short, f has derivatives of all orders.

The discussion up to this point can be summarized as follows:

In the interior of its interval of convergence a power series defines an infinitely differentiable function, the derivatives of which can be obtained by differentiating term by term:

$$\frac{d^n}{dx^n}\left(\sum_{k=0}^{\infty} a_k x^k\right) = \sum_{k=0}^{\infty} \frac{d^n}{dx^n}(a_k x^k) \qquad \text{for all } n.$$

For a detailed proof of the differentiability theorem see the supplement at the end of this section. We go on to examples.

Example 2 You know that

$$\frac{d}{dx}(e^x) = e^x.$$

You can see this directly by differentiating the exponential series:

$$\frac{d}{dx}(e^x) = \frac{d}{dx}\left(\sum_{k=0}^{\infty}\frac{x^k}{k!}\right) = \sum_{k=0}^{\infty}\frac{d}{dx}\left(\frac{x^k}{k!}\right) = \sum_{k=1}^{\infty}\frac{x^{k-1}}{(k-1)!} = \sum_{n=0}^{\infty}\frac{x^n}{n!} = e^x. \quad \square$$

set $n = k - 1$

Example 3 You have seen that

$$\sin x = x - \frac{x^3}{3!} + \frac{x^5}{5!} - \frac{x^7}{7!} + \frac{x^9}{9!} - \cdots$$

and

$$\cos x = 1 - \frac{x^2}{2!} + \frac{x^4}{4!} - \frac{x^6}{6!} + \frac{x^8}{8!} - \cdots.$$

The relations

$$\frac{d}{dx}(\sin x) = \cos x, \qquad \frac{d}{dx}(\cos x) = -\sin x$$

can be confirmed by differentiating the series term by term:

$$\frac{d}{dx}(\sin x) = 1 - \frac{3x^2}{3!} + \frac{5x^4}{5!} - \frac{7x^6}{7!} + \frac{9x^8}{9!} - \cdots$$

$$= 1 - \frac{x^2}{2!} + \frac{x^4}{4!} - \frac{x^6}{6!} + \frac{x^8}{8!} - \cdots = \cos x,$$

$$\frac{d}{dx}(\cos x) = -\frac{2x}{2!} + \frac{4x^3}{4!} - \frac{6x^5}{6!} + \frac{8x^7}{8!} - \cdots$$

$$= -x + \frac{x^3}{3!} - \frac{x^5}{5!} + \frac{x^7}{7!} - \cdots$$

$$= -\left(x - \frac{x^3}{3!} + \frac{x^5}{5!} - \frac{x^7}{7!} + \cdots\right) = -\sin x. \quad \square$$

Example 4 We can sum the series

$$\sum_{k=1}^{\infty}\frac{x^k}{k} \qquad \text{for all } x \text{ in } (-1, 1)$$

by setting

$$g(x) = \sum_{k=1}^{\infty}\frac{x^k}{k} \qquad \text{for all } x \text{ in } (-1, 1)$$

and noting that

$$g'(x) = \sum_{k=1}^{\infty} \frac{kx^{k-1}}{k} = \sum_{k=1}^{\infty} x^{k-1} = \sum_{n=0}^{\infty} x^n = \frac{1}{1-x}.$$

the geometric series

With

$$g'(x) = \frac{1}{1-x} \qquad \text{and} \qquad g(0) = 0,$$

we can conclude that

$$g(x) = -\ln(1-x) = \ln\left(\frac{1}{1-x}\right).$$

It follows that

$$\sum_{k=1}^{\infty} \frac{x^k}{k} = \ln\left(\frac{1}{1-x}\right) \qquad \text{for all } x \text{ in } (-1, 1). \quad \square$$

Power series can also be integrated term by term.

THEOREM 11.9.4   TERM-BY-TERM INTEGRATION

If $f(x) = \sum\limits_{k=0}^{\infty} a_k x^k$ converges on $(-c, c)$, then

$$g(x) = \sum_{k=0}^{\infty} \frac{a_k}{k+1} x^{k+1} \quad \text{converges on } (-c, c) \text{ and } \int f(x)\,dx = g(x) + C.$$

PROOF If $\sum\limits_{k=0}^{\infty} a_k x^k$ converges on $(-c, c)$, then $\sum\limits_{k=0}^{\infty} |a_k x^k|$ converges on $(-c, c)$ by Theorem 11.8.2. Since

$$\left| \frac{a_k}{k+1} x^k \right| \leq |a_k x^k| \qquad \text{for all } k,$$

we know by comparison that

$$\sum_{k=0}^{\infty} \left| \frac{a_k}{k+1} x^k \right| \quad \text{also converges on } (-c, c).$$

It follows that

$$x \sum_{k=0}^{\infty} \frac{a_k}{k+1} x^k = \sum_{k=0}^{\infty} \frac{a_k}{k+1} x^{k+1} \quad \text{converges on } (-c, c).$$

With

$$f(x) = \sum_{k=0}^{\infty} a_k x^k \qquad \text{and} \qquad g(x) = \sum_{k=0}^{\infty} \frac{a_k}{k+1} x^{k+1},$$

we know from the differentiability theorem that

$$g'(x) = f(x) \qquad \text{and therefore} \qquad \int f(x) \, dx = g(x) + C. \quad \square$$

Term-by-term integration can be expressed as follows:

(11.9.5)

$$\int \left(\sum_{k=0}^{\infty} a_k x^k \right) dx = \left(\sum_{k=0}^{\infty} \frac{a_k}{k+1} x^{k+1} \right) + C.$$

Remark It follows from Theorem 11.9.4 that if $\sum a_k x^k$ has radius of convergence r, then its "integral" $\sum [1/(k+1)]a_k x^{k+1}$ also has radius of convergence r. As in the case of differentiating power series, convergence at the endpoints (if any) has to be tested separately. ❏

If a power series converges at c and converges at d, then it converges at all numbers in between and

(11.9.6)

$$\int_c^d \left(\sum_{k=0}^{\infty} a_k x^k \right) dx = \sum_{k=0}^{\infty} \left(\int_c^d a_k x^k \, dx \right) = \sum_{k=0}^{\infty} \frac{a_k}{k+1}(d^{k+1} - c^{k+1}).$$

Example 5 You are familiar with the series expansion

$$\frac{1}{1+x} = \frac{1}{1-(-x)} = \sum_{k=0}^{\infty} (-1)^k x^k.$$

It is valid for all x in $(-1, 1)$ and for no other x. Integrating term by term we have

$$\ln(1+x) = \int \left(\sum_{k=0}^{\infty} (-1)^k x^k \right) dx = \left(\sum_{k=0}^{\infty} \frac{(-1)^k}{k+1} x^{k+1} \right) + C$$

for all x in $(-1, 1)$. At $x = 0$ both $\ln(1+x)$ and the series on the right are 0. It follows that $C = 0$ and thus

$$\ln(1+x) = \sum_{k=0}^{\infty} \frac{(-1)^k}{k+1} x^{k+1} = x - \frac{x^2}{2} + \frac{x^3}{3} - \frac{x^4}{4} + \cdots$$

for all x in $(-1, 1)$. ❏

In Section 11.6 we were able to prove that this expansion for $\ln(1 + x)$ was valid on the half-closed interval $(-1, 1]$; this gave us an expansion for $\ln 2$. Term-by-term integration gives us only the open interval $(-1, 1)$. Well, you may say, it's easy to see that the logarithm series also converges at $x = 1$.† True enough, but why to $\ln 2$? This takes us back to consideration of the remainder term, the method of Section 11.6.

---

† An alternating series with $a_k \to 0$.

There is, however, another way to proceed. The great Norwegian mathematician Niels Henrik Abel (1802–1829) proved the following result: suppose that

$$\sum_{k=0}^{\infty} a_k x^k \quad \text{converges on } (-c, c) \text{ and there represents } f(x).$$

If f is continuous at one of the endpoints (c or $-c$) and the series converges there, then the series represents the function at that point. Using Abel's theorem it is evident that the series for $\ln(1+x)$ does represent the function at $x = 1$.

We come now to another important series expansion:

(11.9.7)
$$\tan^{-1} x = x - \frac{x^3}{3} + \frac{x^5}{5} - \frac{x^7}{7} + \cdots \qquad \text{for } -1 \le x \le 1.$$

PROOF For x in $(-1, 1)$

$$\frac{1}{1+x^2} = \frac{1}{1-(-x^2)} = \sum_{k=0}^{\infty} (-1)^k x^{2k}$$

so that, by integration,

$$\tan^{-1} x = \int \left(\sum_{k=0}^{\infty} (-1)^k x^{2k} \right) dx = \left(\sum_{k=0}^{\infty} \frac{(-1)^k}{2k+1} x^{2k+1} \right) + C.$$

The constant C is 0 because the series on the right and the inverse tangent are both 0 at $x = 0$. Thus, for all x in $(-1, 1)$, we have

$$\tan^{-1} x = \sum_{k=0}^{\infty} \frac{(-1)^k}{2k+1} x^{2k+1} = x - \frac{x^3}{3} + \frac{x^5}{5} - \frac{x^7}{7} + \cdots.$$

That the series also represents the function at $x = -1$ and $x = 1$ follows directly from Abel's theorem: at both these points $\tan^{-1} x$ is continuous, and at both of these points the series converges. ❏

Since $\tan^{-1} 1 = \frac{1}{4}\pi$, we have

$$\tfrac{1}{4}\pi = 1 - \tfrac{1}{3} + \tfrac{1}{5} - \tfrac{1}{7} + \tfrac{1}{9} - \cdots.$$

This series was known to the Scottish mathematician James Gregory in 1671. It is an elegant formula for π, but it converges too slowly for computational purposes. A much more effective way of computing π is outlined in the supplement at the end of this section.

Term-by-term integration provides a method of calculating some (otherwise rather intractable) definite integrals. Suppose that you are trying to evaluate

$$\int_a^b f(x) \, dx,$$

but cannot find an antiderivative. If you can expand $f(x)$ in a convergent power series, then you can estimate the integral by forming the series and integrating term by term.

EXAMPLE 6 We will estimate

$$\int_0^1 e^{-x^2}\, dx$$

by expanding the integral in a power series and integrating term by term. Our starting point is the expansion

$$e^x = 1 + x + \frac{x^2}{2!} + \frac{x^3}{3!} + \frac{x^4}{4!} + \frac{x^5}{5!} + \frac{x^6}{6!} + \cdots \qquad \text{for all } x.$$

From this we see that

$$e^{-x^2} = 1 - x^2 + \frac{x^4}{2!} - \frac{x^6}{3!} + \frac{x^8}{4!} - \frac{x^{10}}{5!} + \frac{x^{12}}{6!} - \cdots \qquad \text{for all } x,$$

and therefore

$$\int_0^1 e^{-x^2}\, dx = \left[x - \frac{x^3}{3} + \frac{x^5}{5(2!)} - \frac{x^7}{7(3!)} + \frac{x^9}{9(4!)} - \frac{x^{11}}{11(5!)} + \frac{x^{13}}{13(6!)} - \cdots \right]_0^1$$

$$= 1 - \frac{1}{3} + \frac{1}{5(2!)} - \frac{1}{7(3!)} + \frac{1}{9(4!)} - \frac{1}{11(5!)} + \frac{1}{13(6!)} - \cdots.$$

This is an alternating series with declining terms. Therefore we know that the integral lies between consecutive partial sums. In particular it lies between

$$1 - \frac{1}{3} + \frac{1}{5(2!)} - \frac{1}{7(3!)} + \frac{1}{9(4!)} - \frac{1}{11(5!)}$$

and

$$\left[1 - \frac{1}{3} + \frac{1}{5(2!)} - \frac{1}{7(3!)} + \frac{1}{9(4!)} - \frac{1}{11(5!)} \right] + \frac{1}{13(6!)}.$$

As you can check, the first sum is greater than 0.74673 and the second one is less than 0.74684. It follows that

$$0.74673 < \int_0^1 e^{-x^2}\, dx < 0.74684.$$

The estimate 0.7468 approximates the integral within 0.0001. ❑

The integral of Example 6 was easy to estimate numerically because it could be expressed as an alternating series with decreasing terms. The next example requires more subtlety and illustrates a method more general than that used in Example 6.

EXAMPLE 7 We want to estimate

$$\int_0^1 e^{x^2}\, dx.$$

If we proceed exactly as in Example 6, we find that

$$\int_0^1 e^{x^2}\, dx = 1 + \frac{1}{3} + \frac{1}{5(2!)} + \frac{1}{7(3!)} + \frac{1}{9(4!)} + \frac{1}{11(5!)} + \frac{1}{13(6!)} + \cdots.$$

We now have a series expansion for the integral, but that expansion does not guide us directly to a numerical estimate for the integral. We know that s_n, the nth partial sum of the series, approximates the integral, but we don't know the accuracy of the approximation. We have no handle on the remainder left by s_n.

We start again, this time keeping track of the remainder. For $x \in [0, 1]$

$$0 \le e^x - \left(1 + x + \frac{x^2}{2!} + \cdots + \frac{x^n}{n!}\right) = R_{n+1}(x) \le e\left[\frac{x^{n+1}}{(n+1)!}\right] \le \frac{3}{(n+1)!}.$$

(11.6.3)

If $x \in [0, 1]$, then $x^2 \in [0, 1]$, and therefore

$$0 \le e^{x^2} - \left(1 + x^2 + \frac{x^4}{2!} + \cdots + \frac{x^{2n}}{n!}\right) \le \frac{3}{(n+1)!}$$

Integrating this inequality from $x = 0$ to $x = 1$, we have

$$0 \le \int_0^1 \left[e^{x^2} - \left(1 + x^2 + \frac{x^4}{2!} + \cdots + \frac{x^{2n}}{n!}\right)\right] dx \le \int_0^1 \frac{3}{(n+1)!} \, dx.$$

Carrying out the integration where possible, we see that

$$0 \le \int_0^1 e^{x^2} \, dx - \left[1 + \frac{1}{3} + \frac{1}{5(2!)} + \cdots + \frac{1}{(2n+1)(n!)}\right] \le \frac{3}{(n+1)!}.$$

We can use this inequality to estimate the integral as closely as we wish. Since

$$\frac{3}{7!} = \frac{1}{1680} < 0.0006,$$

we see that

$$\alpha = 1 + \frac{1}{3} + \frac{1}{5(2!)} + \frac{1}{7(3!)} + \frac{1}{9(4!)} + \frac{1}{11(5!)} + \frac{1}{13(6!)}$$

approximates the integral within 0.0006. Arithmetical computation shows that

$$1.4626 \le \alpha \le 1.4627.$$

It follows that

$$1.4626 \le \int_0^1 e^{x^2} \, dx \le 1.4627 + 0.0006 = 1.4633.$$

The estimate 1.463 approximates the integral within 0.0004. ❏

It is time to relate Taylor series

$$\sum_{k=0}^{\infty} \frac{f^{(k)}(0)}{k!} x^k$$

to power series in general. The relation is very simple:

On its interval of convergence a power series is the Taylor series of its sum.

That is, if you have a power series representation of a function f, then the series must be the Taylor series for f. To see this, all you have to do is differentiate

$$f(x) = a_0 + a_1 x + a_2 x^2 + \cdots + a_k x^k + \cdots$$

term by term. Do this and you will find that $f^{(k)}(0) = k!\, a_k$, and therefore

$$a_k = \frac{f^{(k)}(0)}{k!}.$$

The a_k are the Taylor coefficients of f.

We end this section by carrying out a few simple expansions.

Example 8 Expand $\cosh x$ and $\sinh x$ in powers of x.

SOLUTION There is no need to go through the labor of computing the Taylor coefficients

$$\frac{f^{(k)}(0)}{k!}$$

by differentiation. We know that

$$\cosh x = \tfrac{1}{2}(e^x + e^{-x}) \qquad \text{and} \qquad \sinh x = \tfrac{1}{2}(e^x - e^{-x}). \qquad (7.9.1)$$

Since

$$e^x = 1 + x + \frac{x^2}{2!} + \frac{x^3}{3!} + \frac{x^4}{4!} + \frac{x^5}{5!} + \cdots,$$

we have

$$e^{-x} = 1 - x + \frac{x^2}{2!} - \frac{x^3}{3!} + \frac{x^4}{4!} - \frac{x^5}{5!} + \cdots.$$

Thus

$$\cosh x = \frac{1}{2}\left(2 + 2\frac{x^2}{2!} + 2\frac{x^4}{4!} + \cdots\right) = 1 + \frac{x^2}{2!} + \frac{x^4}{4!} + \cdots = \sum_{k=0}^{\infty} \frac{x^{2k}}{(2k)!}$$

and

$$\sinh x = \frac{1}{2}\left(2x + 2\frac{x^3}{3!} + 2\frac{x^5}{5!} + \cdots\right) = x + \frac{x^3}{3!} + \frac{x^5}{5!} + \cdots = \sum_{k=0}^{\infty} \frac{x^{2k+1}}{(2k+1)!}.$$

Both expansions are valid for all real x, since the exponential expansions are valid for all real x. ❏

Example 9 Expand $x^2 \cos x^3$ in powers of x.

SOLUTION

$$\cos x = 1 - \frac{x^2}{2!} + \frac{x^4}{4!} - \frac{x^6}{6!} + \cdots.$$

Thus

$$\cos x^3 = 1 - \frac{(x^3)^2}{2!} + \frac{(x^3)^4}{4!} - \frac{(x^3)^6}{6!} + \cdots = 1 - \frac{x^6}{2!} + \frac{x^{12}}{4!} - \frac{x^{18}}{6!} + \cdots,$$

and

$$x^2 \cos x^3 = x^2 - \frac{x^8}{2!} + \frac{x^{14}}{4!} - \frac{x^{20}}{6!} + \cdots.$$

This expansion is valid for all real x, since the expansion for $\cos x$ is valid for all real x. ❑

ALTERNATIVE SOLUTION Since

$$x^2 \cos x^3 = \frac{d}{dx}\left(\frac{1}{3}\sin x^3\right),$$

we can derive the expansion for $x^2 \cos x^3$ by expanding $\frac{1}{3}\sin x^3$ and then differentiating term by term. ❑

EXERCISES 11.9

In Exercises 1–6, expand f in powers of x, basing your calculations on the geometric series

$$\frac{1}{1-x} = 1 + x + x^2 + \cdots + x^n + \cdots.$$

1. $f(x) = \dfrac{1}{(1-x)^2}$.

2. $f(x) = \dfrac{1}{(1-x)^3}$.

3. $f(x) = \dfrac{1}{(1-x)^k}$.

4. $f(x) = \ln(1-x)$.

5. $f(x) = \ln(1-x^2)$.

6. $f(x) = \ln(2-3x)$.

In Exercises 7 and 8, expand f in powers of x, basing your calculations on the tangent series:

$$\tan x = x + \tfrac{1}{3}x^3 + \tfrac{2}{15}x^5 + \tfrac{17}{315}x^7 + \cdots.$$

7. $f(x) = \sec^2 x$.

8. $f(x) = \ln \cos x$.

In Exercises 9 and 10, find $f^{(9)}(0)$.

9. $f(x) = x^2 \sin x$.

10. $f(x) = x \cos x^2$.

In Exercises 11–22, expand f in powers of x.

11. $f(x) = \sin x^2$.

12. $f(x) = x^2 \tan^{-1} x$.

13. $f(x) = e^{3x^3}$.

14. $f(x) = \dfrac{1-x}{1+x}$.

15. $f(x) = \dfrac{2x}{1-x^2}$.

16. $f(x) = x \sinh x^2$.

17. $f(x) = \dfrac{1}{1-x} + e^x$.

18. $f(x) = \cosh x \sinh x$.

19. $f(x) = x \ln(1+x^3)$.

20. $f(x) = (x^2 + x) \ln(1+x)$.

21. $f(x) = x^3 e^{-x^3}$.

22. $f(x) = x^5 (\sin x + \cos 2x)$.

In Exercises 23–26, evaluate the given limit in two ways: (a) using L'Hospital's rule, and (b) using power series.

23. $\displaystyle\lim_{x\to 0} \frac{1 - \cos x}{x^2}$.

24. $\displaystyle\lim_{x\to 0} \frac{\sin x - x}{x^2}$.

25. $\displaystyle\lim_{x\to 0} \frac{\cos x - 1}{x \sin x}$.

26. $\displaystyle\lim_{x\to 0} \frac{e^x - 1 - x}{x \tan^{-1} x}$.

In Exercises 27–30, find a powers series representation of the improper integral.

27. $\displaystyle\int_0^x \frac{\ln(1+t)}{t}\,dt$.

28. $\displaystyle\int_0^x \frac{1 - \cos t}{t^2}\,dt$.

29. $\displaystyle\int_0^x \frac{\tan^{-1} t}{t}\,dt$.

30. $\displaystyle\int_0^x \frac{\sinh t}{t}\,dt$.

▷ In Exercises 31–36, estimate within 0.01.

31. $\displaystyle\int_0^1 e^{-x^3}\,dx$.

32. $\displaystyle\int_0^1 \sin x^2\,dx$.

33. $\displaystyle\int_0^1 \sin \sqrt{x}\,dx$.

34. $\displaystyle\int_0^1 x^4 e^{-x^2}\,dx$.

35. $\displaystyle\int_0^1 \tan^{-1} x^2\,dx$.

36. $\displaystyle\int_1^2 \frac{1 - \cos x}{x}\,dx$.

▷ In Exercises 37–40, use a power series to estimate the integral within 0.0001.

37. $\displaystyle\int_0^1 \frac{\sin x}{x}\,dx$.

38. $\displaystyle\int_0^{0.5} \frac{1 - \cos x}{x^2}\,dx$.

39. $\displaystyle\int_0^{0.5} \frac{\ln(1+x)}{x}\,dx$.

40. $\displaystyle\int_0^{0.2} x \sin x\,dx$.

In Exercises 41–43, sum the series.

41. $\displaystyle\sum_{k=0}^{\infty} \frac{1}{k!} x^{3k}$.

42. $\displaystyle\sum_{k=0}^{\infty} \frac{1}{k!} x^{3k+1}$.

43. $\sum\limits_{k=1}^{\infty} \dfrac{3k}{k!} x^{3k-1}$.

44. Let $f(x) = \dfrac{e^x - 1}{x}$.

(a) Find a power series representation of f in powers of x.

(b) Differentiate the power series in part (a) and show that

$$\sum_{n=1}^{\infty} \frac{n}{(n+1)!} = 1.$$

45. Let $f(x) = xe^x$.

(a) Find a power series representation of f in powers of x.

(b) Integrate the power series in part (a) and show that

$$\sum_{n=1}^{\infty} \frac{1}{n!(n+2)} = \frac{1}{2}.$$

46. Deduce the differentiation formulas

$$\frac{d}{dx}(\sinh x) = \cosh x, \qquad \frac{d}{dx}(\cosh x) = \sinh x$$

from the expansions of $\sinh x$ and $\cosh x$ in powers of x.

47. Show that, if $\Sigma\, a_k x^k$ and $\Sigma\, b_k x^k$ both converge to the same sum on some interval, then $a_k = b_k$ for each k.

48. Show that, if $\epsilon > 0$, then

$$|kx^{k-1}| < (|x| + \epsilon)^k \qquad \text{for all } k \text{ sufficiently large.}$$

HINT: Take the kth root of the left side and let $k \to \infty$.

49. Suppose that the function f has the power series representation $f(x) = \Sigma_{k=0}^{\infty}\, a_k x^k$.

(a) Show that if f is an even function, then $a_{2k+1} = 0$ for all k.

(b) Show that if f is an odd function, then $a_{2k} = 0$ for all k.

50. Suppose that the function f is infinitely differentiable on an interval containing 0, and suppose that $f'(x) = -2f(x)$ and $f(0) = 1$. Use these properties to find the power series representation of f in powers of x. Do you recognize this function?

In Exercises 51–53, estimate within 0.001 by the method of this section and check your result by carrying out the integration directly.

51. $\displaystyle\int_0^{1/2} x \ln (1 + x)\, dx.$ **52.** $\displaystyle\int_0^1 x \sin x\, dx.$

53. $\displaystyle\int_0^1 xe^{-x}\, dx.$

54. Show that

$$0 \le \int_0^2 e^{x^2}\, dx - \left[2 + \frac{2^3}{3} + \frac{2^5}{5(2!)} + \cdots + \frac{2^{2n+1}}{(2n+1)n!}\right]$$
$$< \frac{e^4 2^{2n+3}}{(n+1)!}.$$

---

*SUPPLEMENT TO SECTION 11.9

PROOF OF THEOREM 11.9.3

Set

$$f(x) = \sum_{k=0}^{\infty} a_k x^k \qquad \text{and} \qquad g(x) = \sum_{k=0}^{\infty} \frac{d}{dx}(a_k x^k) = \sum_{k=1}^{\infty} k a_k x^{k-1}.$$

Select x from $(-c, c)$. We want to show that

$$\lim_{h \to 0} \frac{f(x + h) - f(x)}{h} = g(x).$$

For $x + h$ in $(-c, c)$, $h \ne 0$, we have

$$\left| g(x) - \frac{f(x+h) - f(x)}{h} \right| = \left| \sum_{k=1}^{\infty} k a_k x^{k-1} - \sum_{k=0}^{\infty} \frac{a_k(x+h)^k - a_k x^k}{h} \right|$$
$$= \left| \sum_{k=1}^{\infty} k a_k x^{k-1} - \sum_{k=1}^{\infty} a_k \left[\frac{(x+h)^k - x^k}{h} \right] \right|.$$

By the mean-value theorem

$$\frac{(x+h)^k - x^k}{h} = k(t_k)^{k-1}$$

for some number t_k between x and $x + h$. Thus we can write

$$\left| g(x) - \frac{f(x + h) - f(x)}{h} \right| = \left| \sum_{k=1}^{\infty} ka_k x^{k-1} - \sum_{k=1}^{\infty} ka_k(t_k)^{k-1} \right|$$

$$= \left| \sum_{k=1}^{\infty} ka_k[x^{k-1} - (t_k)^{k-1}] \right|$$

$$= \left| \sum_{k=2}^{\infty} ka_k[x^{k-1} - (t_k)^{k-1}] \right|.$$

By the mean-value theorem

$$\frac{x^{k-1} - (t_k)^{k-1}}{x - t_k} = (k - 1)(p_{k-1})^{k-2}$$

for some number p_{k-1} between x and t_k. Obviously, then,

$$|x^{k-1} - (t_k)^{k-1}| = |x - t_k||(k - 1)(p_{k-1})^{k-2}|.$$

Since $|x - t_k| < |h|$ and $|p_{k-1}| \le |\alpha|$ where $|\alpha| = \max \{|x|, |x + h|\}$,

$$|x^{k-1} - (t_k)^{k-1}| \le |h||(k - 1)\alpha^{k-2}|.$$

Thus

$$\left| g(x) - \frac{f(x + h) - f(x)}{h} \right| \le |h| \sum_{k=2}^{\infty} |k(k - 1)a_k\alpha^{k-2}|.$$

Since the series converges,

$$\lim_{h \to 0} \left(|h| \sum_{k=2}^{\infty} |k(k - 1)a_k\alpha^{k-2}| \right) = 0.$$

This gives

$$\lim_{h \to 0} \left| g(x) - \frac{f(x + h) - f(x)}{h} \right| = 0 \quad \text{and thus} \quad \lim_{h \to 0} \frac{f(x + h) - f(x)}{h} = g(x). \quad \square$$

Calculating π

We base our computation of π on the inverse tangent series (11.9.7)

$$\tan^{-1} x = x - \frac{x^3}{3} + \frac{x^5}{5} - \frac{x^7}{7} + \cdots \quad \text{for } -1 \le x \le 1$$

and the relation

(11.9.8)

$$\boxed{\tfrac{1}{4}\pi = 4 \tan^{-1} \tfrac{1}{5} - \tan^{-1} \tfrac{1}{239}.\dagger}$$

The inverse tangent series gives

$$\tan^{-1} \tfrac{1}{5} = \tfrac{1}{5} - \tfrac{1}{3}(\tfrac{1}{5})^3 + \tfrac{1}{5}(\tfrac{1}{5})^5 - \tfrac{1}{7}(\tfrac{1}{5})^7 + \cdots$$

---

† This relation was discovered in 1706 by John Machin, a Scotsman. It can be verified by repeated applications of the addition formula

$$\tan (A + B) = \frac{\tan A + \tan B}{1 - \tan A \tan B}.$$

First calculate $\tan (2 \tan^{-1} \tfrac{1}{5})$, then $\tan (4 \tan^{-1} \tfrac{1}{5})$, and finally $\tan (4 \tan^{-1} \tfrac{1}{5} - \tan^{-1} \tfrac{1}{239})$.

and

$$\tan^{-1}\tfrac{1}{239} = \tfrac{1}{239} - \tfrac{1}{3}(\tfrac{1}{239})^3 + \tfrac{1}{5}(\tfrac{1}{239})^5 - \tfrac{1}{7}(\tfrac{1}{239})^7 + \cdots.$$

These are alternating series $\Sigma\,(-1)^k a_k$ with a_k decreasing toward 0. Thus we know that

$$\tfrac{1}{5} - \tfrac{1}{3}(\tfrac{1}{5})^3 \le \tan^{-1}\tfrac{1}{5} \le \tfrac{1}{5} - \tfrac{1}{3}(\tfrac{1}{5})^3 + \tfrac{1}{5}(\tfrac{1}{5})^5$$

and

$$\tfrac{1}{239} - \tfrac{1}{3}(\tfrac{1}{239})^3 \le \tan^{-1}\tfrac{1}{239} \le \tfrac{1}{239}.$$

With these inequalities, together with relation (11.9.8), we can show that

$$3.14 < \pi < 3.147.$$

By using six terms of the series for $\tan^{-1}\tfrac{1}{5}$ and still only two of the series for $\tan^{-1}\tfrac{1}{239}$, we can show that

$$3.14159262 < \pi < 3.14159267.$$

Greater accuracy can be obtained by taking more terms into account. For instance, fifteen terms of the series for $\tan^{-1}\tfrac{1}{5}$ and just four terms of the series for $\tan^{-1}\tfrac{1}{239}$ determine π to twenty decimal places:

$$\pi \cong 3.14159\ 26535\ 89793\ 23846.$$

■ 11.10 THE BINOMIAL SERIES

Through a collection of problems we invite you to derive for yourself the basic properties of one of the most celebrated series of all—*the binomial series.*

Start with the binomial $1 + x$. Choose a real number $\alpha \neq 0$ and form the function

$$f(x) = (1 + x)^\alpha.$$

Note that if $\alpha = n$ is a positive integer, then

$$(1 + x)^n = 1 + nx + \frac{n(n-1)}{2!}x^2 + \cdots + nx^{n-1} + x^n$$

is the familiar binomial theorem. The binomial series is a generalization of the binomial theorem. You should also note that if $n = -1$, then $(1 + x)^{-1}$ is the sum of a geometric series

$$\frac{1}{1+x} = 1 - x + x^2 - x^3 + \cdots + (-1)^n x^n + \cdots$$

Problem 1 Show that

$$\frac{f^{(k)}(0)}{k!} = \frac{\alpha[\alpha-1][\alpha-2]\cdots[\alpha-(k-1)]}{k!}.$$

The number you just obtained is the coefficient of x^k in the expansion of $(1 + x)^\alpha$. It is called *the kth binomial coefficient* and is usually denoted by $\binom{\alpha}{k}$:

(11.10.1)
$$\binom{\alpha}{k} = \frac{\alpha[\alpha-1][\alpha-2]\cdots[\alpha-(k-1)]}{k!}.$$

For example, if $\alpha = 7$ and $k = 3$, then

$$\binom{7}{3} = \frac{7 \cdot 6 \cdot 5}{3!} = 35;$$

if $\alpha = 3/2$ and $k = 3$, then

$$\binom{3/2}{3} = \frac{(3/2)[(3/2) - 1][(3/2) - 2]}{3!} = \frac{(3/2)(1/2)(-1/2)}{6} = -\frac{1}{16}. \quad ❏$$

Problem 2 Show that the binomial series

$$\sum \binom{\alpha}{k} x^k$$

has radius of convergence 1.   HINT: Use the ratio test.   ❏

From Problem 2 you know that the binomial series converges on the open interval $(-1, 1)$ and defines there an infinitely differentiable function. The next thing to show is that this function (the one defined by the series) is actually $(1 + x)^\alpha$. To do this, you first need some other results.

Problem 3 Verify the identity

$$(k + 1) \binom{\alpha}{k + 1} + k \binom{\alpha}{k} = \alpha \binom{\alpha}{k}. \quad ❏$$

Problem 4 Use the identity of Problem 3 to show that the sum of the binomial series

$$\phi(x) = \sum_{k=0}^{\infty} \binom{\alpha}{k} x^k$$

satisfies the differential equation

$$(1 + x)\phi'(x) = \alpha\phi(x) \qquad \text{for all } x \text{ in } (-1, 1)$$

together with the side condition $\phi(0) = 1$.   ❏

You are now in a position to prove the main result.

Problem 5 Show that

(11.10.2)
$$(1 + x)^\alpha = \sum_{k=0}^{\infty} \binom{\alpha}{k} x^k \qquad \text{for all } x \text{ in } (-1, 1).$$

You can probably get a better feeling for the series by writing out the first few terms:

(11.10.3)
$$(1 + x)^\alpha = 1 + \alpha x + \frac{\alpha(\alpha - 1)}{2!} x^2 + \frac{\alpha(\alpha - 1)(\alpha - 2)}{3!} x^3 + \cdots . \quad ❏$$

EXERCISES 11.10

In Exercises 1–10, expand f in powers of x up to x^4.

1. $f(x) = \sqrt{1 + x}$.

2. $f(x) = \sqrt{1 - x}$.

3. $f(x) = \sqrt{1 + x^2}$.

4. $f(x) = \sqrt{1 - x^2}$.

5. $f(x) = \dfrac{1}{\sqrt{1 + x}}$.

6. $f(x) = \dfrac{1}{\sqrt[3]{1 + x}}$.

7. $f(x) = \sqrt[4]{1 - x}$.

8. $f(x) = \dfrac{1}{\sqrt[4]{1 + x}}$.

9. $f(x) = (4 + x)^{3/2}$.

10. $f(x) = \sqrt{1 + x^4}$.

11. (a) Use a binomial series to find the Taylor series of $f(x) = 1/\sqrt{1 - x^2}$ in powers of x.
   (b) Use the series for f in part (a) to find the Taylor series for $F(x) = \sin^{-1} x$ and give the radius of convergence.

12. (a) Use a binomial series to find the Taylor series of $f(x) = 1/\sqrt{1 + x^2}$ in powers of x.
   (b) Use the series for f in part (a) to find the Taylor series for $F(x) = \sinh^{-1} x$ and give the radius of convergence.

▶ In Exercises 13–18, estimate by using the first three terms of a binomial expansion, rounding off your answer to four decimal places.

13. $\sqrt{98}$.   HINT: $\sqrt{98} = (100 - 2)^{1/2} = 10(1 - \frac{1}{50})^{1/2}$.

14. $\sqrt[5]{36}$.

15. $\sqrt[3]{9}$.

16. $\sqrt[4]{620}$.

17. $17^{-1/4}$.

18. $9^{-1/3}$.

▶ In Exercises 19–22, approximate each integral to within 0.001.

19. $\displaystyle\int_0^{1/3} \sqrt{1 + x^3}\, dx$.

20. $\displaystyle\int_0^{1/5} \sqrt{1 + x^4}\, dx$.

21. $\displaystyle\int_0^{1/2} \dfrac{1}{\sqrt{1 + x^2}}\, dx$.

22. $\displaystyle\int_0^{1/2} \dfrac{1}{\sqrt{1 - x^3}}\, dx$.

■ CHAPTER HIGHLIGHTS

11.1   Sigma Notation

11.2   Infinite Series

partial sums (p. 696)   convergence, divergence (p. 696)
sum of a series (p. 696)   a divergence test (p. 703)

$$\text{geometric series: } \sum_{k=0}^{\infty} x^k = \begin{cases} \dfrac{1}{1 - x}, & |x| < 1 \\ \text{diverges}, & |x| \geq 1 \end{cases}$$

If $\displaystyle\sum_{k=0}^{\infty} a_k$ converges, then $a_k \to 0$.   The converse is false.

11.3   The Integral Test; Comparison Theorems

integral test (p. 707)   basic comparison (p. 710)
limit comparison (p. 712)

harmonic series: $\displaystyle\sum_{k=1}^{\infty} \dfrac{1}{k}$ diverges p-series: $\displaystyle\sum_{k=1}^{\infty} \dfrac{1}{k^p}$ converges iff $p > 1$

11.4   The Root Test; The Ratio Test

root test (p. 715)   ratio test (p. 717)
summary on convergence tests (p. 719)

11.5   Absolute and Conditional Convergence; Alternating Series

absolutely convergent, conditionally convergent (pp. 720–721)
convergence theorem for alternating series (p. 722)
an estimate for alternating series (p. 723)
rearrangements (p. 725)

11.6 Taylor Polynomials in x; Taylor Series in x

Taylor polynomials in x (p. 728) remainder term $R_{n+1}(x)$ (p. 730)
remainder estimate (p. 731) Lagrange form of the remainder (p. 731)

Taylor series in x (Maclaurin series): $\displaystyle\sum_{k=0}^{\infty} \frac{f^{(k)}(0)}{k!} x^k$

$$e^x = \sum_{k=0}^{\infty} \frac{x^k}{k!}, \quad \text{all real } x \qquad \ln(1+x) = \sum_{k=1}^{\infty} \frac{(-1)^{k+1}}{k} x^k, \quad -1 < x \le 1$$

$$\sin x = \sum_{k=0}^{\infty} \frac{(-1)^k}{(2k+1)!} x^{2k+1}, \quad \text{all real } x \qquad \cos x = \sum_{k=0}^{\infty} \frac{(-1)^k}{(2k)!} x^{2k}, \quad \text{all real } x$$

11.7 Taylor Polynomials and Taylor Series in $x - a$

Taylor series in $x - a$: $\displaystyle\sum_{k=0}^{\infty} \frac{g^{(k)}(a)}{k!}(x-a)^k$

11.8 Power Series

power series (p. 745) radius of convergence (p. 747)
interval of convergence (p. 748)

If a power series converges at $c \ne 0$, then it converges absolutely for $|x| < |c|$; if it diverges at c, then it diverges for $|x| > |c|$.

11.9 Differentiation and Integration of Power Series

$$\tan^{-1} x = \sum_{k=0}^{\infty} \frac{(-1)^k}{2k+1} x^{2k+1}, \quad -1 \le x \le 1$$

$$\cosh x = \sum_{k=0}^{\infty} \frac{x^{2k}}{(2k)!}, \quad \text{all real } x \qquad \sinh x = \sum_{k=0}^{\infty} \frac{x^{2k+1}}{(2k+1)!}, \quad \text{all real } x$$

On the interior of its interval of convergence, a power series can be differentiated and integrated term by term.

On its interval of convergence a power series is the Taylor series of its sum.

11.10 The Binomial Series

$$(1+x)^\alpha = \sum_{k=0}^{\infty} \binom{\alpha}{k} x^k = 1 + \alpha x + \frac{\alpha(\alpha-1)}{2!} x^2 + \cdots, \quad -1 < x < 1$$

■ PROJECTS AND EXPLORATIONS USING TECHNOLOGY

To do these exercises you will need a graphics calculator or a computer with graphing capability. The majority of these problems are open-ended so different approaches may be used to solve them. You should be aware that different approaches can result in slight variations in the answers. Round your numerical answers to at least four decimal places. The rounding method that your calculator or computer uses also may cause variations in answers.

11.1 The integral test is used to determine the convergence of a series of positive terms. In the case of convergent series, it can also be used to obtain reasonable approximations for the sum of the series. Let $\{a_n\}$ be the sequence defined by

$$a_n = \frac{[n^3 + \ln(n)]\ln(n)}{n^{4 + \ln(n)} + 3n^3 + 7}.$$

(a) Find j such that the sequence is decreasing for $n \geq j$.
(b) Use the integral test to show that the series

$$\sum_{n=1}^{\infty} \frac{[n^3 + \ln(n)] \ln(n)}{n^{4 + \ln(n)} + 3n^3 + 7}$$

converges. (You may need to start the series at a value $j > 1$.)
(c) Let

$$I_k = \int_j^k \frac{[x^3 + \ln(x)] \ln(x)}{x^{4 + \ln(x)} + 3x^3 + 7} \, dx,$$

and let

$$S_k = \sum_{n=j}^{k} \frac{[n^3 + \ln(n)] \ln(n)}{n^{4 + \ln(n)} + 3n^3 + 7}.$$

Show that $I_{k-1} < S_k < I_k$.
(d) How large must k be in order to get an estimate for the sum of the series that is accurate to three decimal places.
(e) Since you are approximating I_k, how can you be sure that you have enough accuracy for it so as to provide an accurate approximation of S_k?

11.2 The alternating series $\Sigma(-1)^n a_n$, $(a_n > 0$ for all $n)$, converges if the sequence $\{a_n\}$ is decreasing and has limit zero. Moreover, if the series converges, then the sum S is always between two consecutive partial sums, S_k and S_{k+1}. The following procedure allows even better approximations for the sum. Consider the alternating series

$$\sum_{n=0}^{\infty} \frac{(-1)^n(n^2 + 1)}{n^3 + 1}.$$

(a) Estimate the sum of the first 10, 20, 30, . . . , 100 terms of this series.
(b) Determine k such that the partial sum S_k approximates the sum S with three decimal place accuracy.
(c) Using the values of the partial sums from part (a), estimate the differences between the partial sums and S as a function of n. Add the value given by this function to the partial sum S_{100} to get a new approximation for S. This process is called *extrapolation*.
(d) Now consider the following weighted average of consecutive terms

$$T_n = \frac{a_n S_n + a_{n-1} S_{n-1}}{a_n + a_{n-1}}$$

and calculate T_n for $n = 10, 20, 30, . . . , 100$. Plot T_n versus n and estimate the error as a function of n.
(e) Generate the sequence $\{b_n\}$ where

$$b_n = \frac{(S - S_{n-1})}{(S_n - S_{n-1})}$$

and S is the sum of the series. For our example, what is $\lim_{n\to\infty} b_n$?
(f) Compare the convergence of $\{T_n\}$ with the convergence of the sequence $\{U_n\}$ given by

$$U_n = \frac{S_n + S_{n-1}}{2}.$$

11.3 Consider the function

$$F(z) = \int_0^1 [\cos zx^2 + \sin z^2 x^3] \, dx.$$

(a) Using technology, graph the function F. Based on the graph, make conjectures about the properties of this function. For example, is it continuous? differentiable? Where is it increasing? decreasing? And so on.

(b) Determine the Taylor series in powers of x for each of the functions $\cos zx^2$ and $\sin z^2x^3$. Then approximate $F(z)$ using termwise integration. Can you verify the information that you found in part (a) from this expression for F?

(c) Now assume that the derivative of F with respect to z can be found by differentiating the integrand with respect to z, obtaining a series expansion in powers of x, and then integrating that series termwise. Do you get the same series?

VECTORS

■ 12.1 CARTESIAN SPACE COORDINATES

To introduce a Cartesian coordinate system in three-dimensional space we begin with a plane Cartesian coordinate system O-xy. Through the point O, which we continue to call the origin, we pass a third line, perpendicular to the other two. This third line we call the z-axis. We assign real number coordinates to the z-axis in the usual way, with the z-coordinate 0 being assigned to the origin O.

For later convenience we orient the z-axis so that O-xyz forms a "right-handed" system. That is, if the index finger of the right hand points along the positive x-axis and the middle finger along the positive y-axis, then the thumb will point along the positive z-axis. (See Figure 12.1.1.)

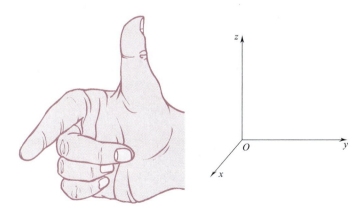

Figure 12.1.1

The planes determined by the x- and y-axes, the x- and z-axes, and the y- and z-axes are called the *coordinate planes*, and are labeled the *xy-plane*, the *xz-plane*, and the *yz-plane*, respectively. If P is an arbitrary point in space, then the plane through P parallel to the yz-plane will intersect the x-axis at a point x_0, the plane through

773

P parallel to the xz-plane will intersect the y-axis at a point y_0, and the plane through P parallel to the xy-plane will intersect the z-axis at a point z_0. We associate P with the ordered triple of numbers (x_0, y_0, z_0) and call (x_0, y_0, z_0) the *Cartesian* (or *rectangular*) *space coordinates* of P, or simply the *space coordinates* of P. In particular, the point on the x-axis with x-coordinate x_0 has space coordinates $(x_0, 0, 0)$; the point on the y-axis with y-coordinate y_0 has space coordinates $(0, y_0, 0)$; and the point on the z-axis with z-coordinate z_0 has space coordinates $(0, 0, z_0)$. (See Figure 12.1.2.)

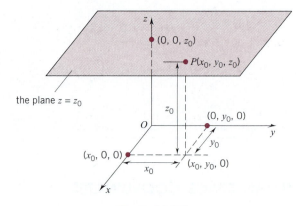

Figure 12.1.2

A point is in the xy-plane iff it is of the form $(x, y, 0)$. Thus, an equation for the xy-plane is $z = 0$. The equation

$$z = z_0$$

represents a plane parallel to the xy-plane, z_0 units above the xy-plane when z_0 is positive, $|z_0|$ units below the xy-plane when z_0 is negative. Similarly, equations of the form $x = x_0$, constant, represent planes parallel to the yz-plane, and equations of the form $y = y_0$, constant, represent planes parallel to xz-plane. (See, for example, Figure 12.1.3. There we have drawn the planes $x = 1$ and $y = 3$.)

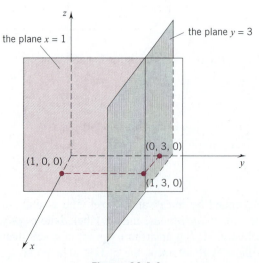

Figure 12.1.3

The Distance Formula

The distance $d(P_1, P_2)$ between two points $P_1(x_1, y_1, z_1)$ and $P_2(x_2, y_2, z_2)$ can be found by applying the Pythagorean theorem twice. With Q and R as in Figure 12.1.4, P_1P_2R and P_1RQ are both right triangles. From the first triangle

$$[d(P_1, P_2)]^2 = [d(P_1, R)]^2 + [d(R, P_2)]^2,$$

and from the second triangle

$$[d(P_1, R)]^2 = [d(Q, R)]^2 + [d(P_1, Q)]^2.$$

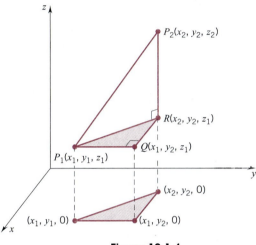

Figure 12.1.4

Combining equations,

$$[d(P_1, P_2)]^2 = [d(Q, R)]^2 + [d(P_1, Q)]^2 + [d(R, P_2)]^2$$
$$= (x_2 - x_1)^2 + (y_2 - y_1)^2 + (z_2 - z_1)^2.$$

Taking square roots, we have the distance formula:

(12.1.1)
$$d(P_1, P_2) = \sqrt{(x_2 - x_1)^2 + (y_2 - y_1)^2 + (z_2 - z_1)^2}.$$

The *sphere* of radius r centered at $P_0(a, b, c)$ is the set of all points $P(x, y, z)$ with $d(P, P_0) = r$. We can obtain an equation of this sphere by using (12.1.1);

(12.1.2)

Equation of a Sphere
$$(x - a)^2 + (y - b)^2 + (z - c)^2 = r^2.$$

See Figure 12.1.5. An equation for the sphere of radius r centered at the origin is

(12.1.3)
$$x^2 + y^2 + z^2 = r^2.$$

See Figure 12.1.6.

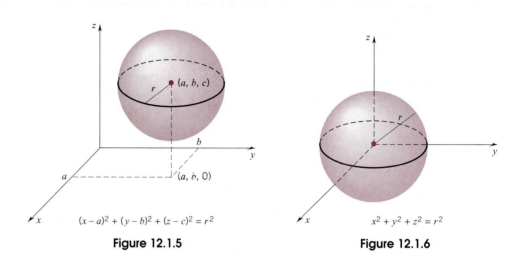

Figure 12.1.5 $(x - a)^2 + (y - b)^2 + (z - c)^2 = r^2$

Figure 12.1.6 $x^2 + y^2 + z^2 = r^2$

Example 1   An equation for the sphere with radius 3 and center $(5, -2, 0)$ is

$$(x - 5)^2 + (y + 2)^2 + z^2 = 9. \quad ❑$$

Example 2   Show that

$$x^2 + y^2 + z^2 + 6x + 2y - 4z = 11$$

is an equation for a sphere.   Find its center and radius.

SOLUTION   We rewrite the equation as

$$(x^2 + 6x) + (y^2 + 2y) + (z^2 - 4z) = 11$$

and complete the squares.   The result is

$$(x^2 + 6x + 9) + (y^2 + 2y + 1) + (z^2 - 4z + 4) = 11 + 9 + 1 + 4 = 25$$

or

$$(x + 3)^2 + (y + 1)^2 + (z - 2)^2 = 25.$$

The sphere has center $(-3, -1, 2)$ and radius $r = 5$.   ❑

Symmetry

You are already familiar with two kinds of symmetry.   Symmetry about a point: for example, the endpoints of a line segment are symmetric about the midpoint of the segment (Figure 12.1.7(a)); the graph of an odd function $y = f(x)$ is symmetric about the origin.   Symmetry about a line (Figure 12.1.7(b)): the graph of an even function $y = f(x)$ is symmetric about the y-axis; if f and f^{-1} are inverse functions, then their graphs are symmetric about the line $y = x$.   In space, we can also speak about symmetry about a plane.   This idea is illustrated in Figure 12.1.7(c).

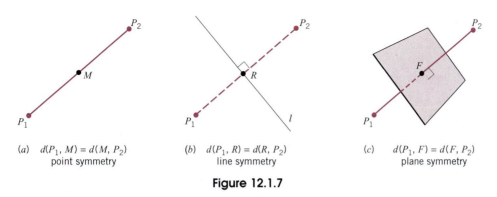

(a) $d(P_1, M) = d(M, P_2)$
point symmetry

(b) $d(P_1, R) = d(R, P_2)$
line symmetry

(c) $d(P_1, F) = d(F, P_2)$
plane symmetry

Figure 12.1.7

The endpoints $P_1(x_1, y_1, z_1)$ and $P_2(x_2, y_2, z_2)$ of the line segment $\overline{P_1P_2}$ are symmetric about the midpoint of the segment, which leads to:

(12.1.4)

Midpoint Formula

$$\left(\frac{x_1 + x_2}{2}, \frac{y_1 + y_2}{2}, \frac{z_1 + z_2}{2}\right).$$

EXERCISES 12.1

In Exercises 1–6, plot points A and B on a right-handed coordinate system. Then calculate the length of the line segment $\overline{AB}$ and find the midpoint.

1. $A(2, 0, 0)$, $B(0, 0, -4)$.
2. $A(0, -2, 0)$, $B(0, 0, 6)$.
3. $A(0, -2, 5)$, $B(4, 1, 0)$.
4. $A(4, 3, 0)$, $B(-2, 0, 6)$.
5. $A(2, 4, 7)$, $B(1, 6, 5)$.
6. $A(-1, 2, 0)$, $B(3, -2, 4)$

In Exercises 7–12, find an equation for the plane through $(3, 1, -2)$ that satisfies the given condition.

7. Parallel to xy-plane.
8. Parallel to xz-plane.
9. Perpendicular to y-axis.
10. Perpendicular to z-axis.
11. Parallel to yz-plane.
12. Perpendicular to x-axis.

In Exercises 13–20, find an equation for the sphere that satisfies the given conditions.

13. Centered at $(0, 2, -1)$ with radius 3.
14. Centered at $(1, 0, -2)$ with radius 4.
15. Centered at $(2, 4, -4)$ and passes through the origin.
16. Centered at the origin and passes through $(1, -2, 2)$.
17. The line segment joining $(0, 4, 2)$ and $(6, 0, 2)$ is a diameter.
18. Centered at $(2, 3, -4)$ and tangent to the xy-plane.
19. Centered at $(2, 3, -4)$ and tangent to the plane $x = 7$.
20. Centered at $(2, 3, -4)$ and tangent to the plane $y = 1$.

In Exercises 21–24, show that the given equation is an equation for a sphere, and find the center and radius.

21. $x^2 + y^2 + z^2 + 4x - 8y - 2z + 5 = 0$.
22. $x^2 + y^2 + z^2 - 2x + y - 4z = 4$.
23. $2x^2 + 2y^2 + 2z^2 + 8x - 4y = -1$.
24. $3x^2 + 3y^2 + 3z^2 - 12x - 6z + 3 = 0$.

The points $P(a, b, c)$ and $Q(2, 3, 5)$ are symmetric in the sense given in Exercises 25–36. Find a, b, c.

25. About the xy-plane.
26. About the xz-plane.
27. About the yz-plane.
28. About the x-axis.
29. About the y-axis.
30. About the z-axis.
31. About the origin.
32. About the plane $x = 1$.
33. About the plane $y = -1$.
34. About the plane $z = 4$.
35. About the point $(0, 2, 1)$.
36. About the point $(4, 0, 1)$.

37. Find an equation for each sphere that passes through the point $(5, 1, 4)$ and is tangent to all three coordinate planes.

38. Find an equation for the largest sphere that is centered at $(2, 1, -2)$ and intersects the sphere $x^2 + y^2 + z^2 = 1$.

39. Is the equation $x^2 + y^2 + z^2 - 4x + 4y + 6z + 20 = 0$ an equation for a sphere? If so, find its center and radius. If not, why isn't it?

40. Find conditions on A, B, C, D such that the equation

$$x^2 + y^2 + z^2 + Ax + By + Cz + D = 0$$

represents a sphere.

41. Show that the points $P(1, 2, 3)$, $Q(4, -5, 2)$, and $R(0, 0, 0)$ are the vertices of a right triangle.

42. The points $(5, -1, 3)$, $(4, 2, 1)$, and $(2, 1, 0)$ are the midpoints of the sides of a triangle PQR. Find the vertices P, Q, and R of the triangle.

In Exercises 43 and 44, the point R lies on the line segment that joins $P(a_1, a_2, a_3)$ and $Q(b_1, b_2, b_3)$.

43. (a) Find the coordinates of R given that

$$d(P, R) = t\, d(P, Q) \qquad \text{where } 0 \le t \le 1.$$

(b) Determine the value of t for which R is the midpoint of $\overline{PQ}$.

44. (a) Find the coordinates of R given that

$$d(P, R) = r\, d(R, Q) \qquad \text{where } r > 0.$$

(b) Determine the value of r for which R is the midpoint of $\overline{PQ}$.

■ 12.2 DISPLACEMENTS, FORCES AND VELOCITIES; VECTORS

Displacements

A *displacement* along a coordinate line can be specified by a real number and depicted by an arrow. For a displacement of a_1 units we use the number a_1 and an arrow that begins at any number x and ends at $x + a_1$. Following convention, if $a_1 > 0$, the arrow will point to the right, and if $a_1 < 0$, the arrow will point to the left. (Figure 12.2.1.) The *magnitude* of the displacement a_1 is defined to be the length of the arrow, which is the same as the distance between the points x and $x + a_1$. Thus, the magnitude of the displacement a_1 is $|a_1|$.

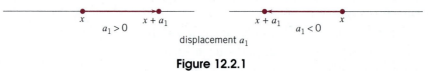

Figure 12.2.1

Displacements in the plane are more interesting. Instead of having two possible directions, there are an infinite number of possible directions. Figure 12.2.2 shows several displacements beginning at the point (x, y). Displacements in the plane are specified by ordered pairs of real numbers. The magnitude of the displacement (a_1, a_2) is $\sqrt{a_1^2 + a_2^2}$, which equals the distance between the points (x, y) and $(x + a_1, y + a_2)$.

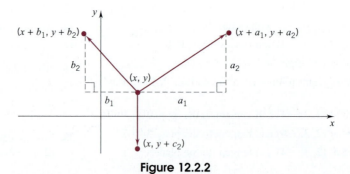

Figure 12.2.2

We now consider displacements in three-dimensional space. Here ordered triples of real numbers come into play. A displacement of a_1 units in the x-coordinate, a_2 units in the y-coordinate, a_3 units in the z-coordinate can be indicated by an arrow that begins at an arbitrary point (x, y, z) and ends at the point $(x + a_1, y + a_2, z + a_3)$. This displacement is represented by the ordered triple (a_1, a_2, a_3). (Figure 12.2.3.) The magnitude of the displacement (a_1, a_2, a_3) is $\sqrt{a_1^2 + a_2^2 + a_3^2}$.

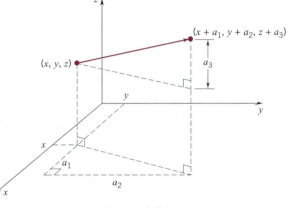

Figure 12.2.3

Remark You have probably noticed that we have ignored displacements of magnitude 0 in the preceding illustrations. In the case of three-dimensional space, the displacement of magnitude 0 is represented by the ordered triple $(0, 0, 0)$, which we will denote by **0**. No direction is associated with this displacement, and consequently we will not use an arrow to depict it. As you will see, the ordered triple $\mathbf{0} = (0, 0, 0)$ arises naturally in the study of vectors, and we will have more to say about it later in this section. ❏

We can follow one displacement by another. A displacement (a_1, a_2, a_3) followed by (b_1, b_2, b_3) results in a total displacement $(a_1 + b_1, a_2 + b_2, a_3 + b_3)$. We can express this by writing

$$(a_1, a_2, a_3) + (b_1, b_2, b_3) = (a_1 + b_1, a_2 + b_2, a_3 + b_3).$$

We can picture the first displacement by an arrow from some point $P(x, y, z)$ to

$$Q(x + a_1, y + a_2, z + a_3),$$

the second displacement by an arrow from $Q(x + a_1, y + a_2, z + a_3)$ to

$$R(x + a_1 + b_1, y + a_2 + b_2, z + a_3 + b_3),$$

and then the resultant displacement by an arrow from $P(x, y, z)$ to

$$R(x + a_1 + b_1, y + a_2 + b_2, z + a_3 + b_3).$$

The three arrows then form a triangular pattern that is easy to remember (Figure 12.2.4).

From a displacement (a_1, a_2, a_3) and a real number α, we can form a new displacement $(\alpha a_1, \alpha a_2, \alpha a_3)$. We view this new displacement as α times the initial displacement and write

$$\alpha(a_1, a_2, a_3) = (\alpha a_1, \alpha a_2, \alpha a_3).$$

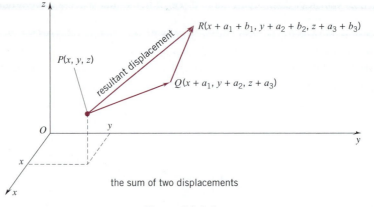

the sum of two displacements

Figure 12.2.4

The effect of multiplying a displacement by a number α is to change the magnitude of the displacement by a factor of $|\alpha|$:

$$\sqrt{(\alpha a_1)^2 + (\alpha a_2)^2 + (\alpha a_3)^2} = \sqrt{\alpha^2(a_1^2 + a_2^2 + a_3^2)} = |\alpha|\sqrt{a_1^2 + a_2^2 + a_3^2},$$

keeping the same direction if $\alpha > 0$, but reversing the direction if $\alpha < 0$. (If $\alpha = 0$, then $0(a_1, a_2, a_3) = (0, 0, 0) = \mathbf{0}$.) The displacement

$$2(a_1, a_2, a_3) = (2a_1, 2a_2, 2a_3)$$

is twice as long as (a_1, a_2, a_3) and has the same direction; the displacement

$$-(a_1, a_2, a_3) = (-a_1, -a_2, -a_3)$$

has the same length as (a_1, a_2, a_3) but the opposite direction; the displacement

$$-\tfrac{3}{2}(a_1, a_2, a_3) = (-\tfrac{3}{2}a_1, -\tfrac{3}{2}a_2, -\tfrac{3}{2}a_3)$$

is one and one-half times as long as (a_1, a_2, a_3) and has the opposite direction. (Figure 12.2.5.)

Forces and Velocities

As we have seen, a (nonzero) displacement is a quantity that has both direction and magnitude. Displacements arise naturally in many settings. For example, in tracking the motion of an object moving along a straight line, its *velocity* at time t can be represented by an arrow that points in the direction of motion. The length of the arrow represents the speed of the object.

Force and *acceleration* have both magnitude† and direction. If a force $\mathbf{F}$ is applied to an object of mass m, then the object will accelerate in the direction of the force (Figure 12.2.6). In this instance, force and acceleration are related by *Newton's Second Law of Motion*

$$\mathbf{F} = m\mathbf{a}.$$

A force $\mathbf{F}$ acting in three-dimensional space is completely determined by its components along the x-, y-, and z-axes. If these components are a_1, a_2, a_3, respectively, then $\mathbf{F}$ can be represented by the ordered triple (a_1, a_2, a_3). See Figure 12.2.7. In

---

† The standard units of measurement for the magnitude of a force $\mathbf{F}$ are: English—*pounds,* and metric—*newtons.*

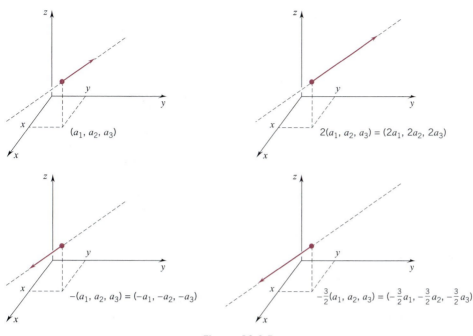

Figure 12.2.5

exactly the same way, the velocity **v** of an object moving in space can be represented by an ordered triple of real numbers.

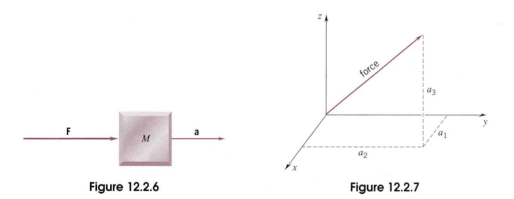

Figure 12.2.6   Figure 12.2.7

It can be verified by physical experiments that if two forces $\mathbf{F}_1 = (a_1, a_2, a_3)$ and $\mathbf{F}_2 = (b_1, b_2, b_3)$ are applied simultaneously at the same point, the effect is the same as that produced by the single force

$$\mathbf{F}_3 = (a_1 + b_1, a_2 + b_2, a_3 + b_3).$$

We call $\mathbf{F}_3$ the *resultant* or *total force* and write $\mathbf{F}_1 + \mathbf{F}_2 = \mathbf{F}_3$. For the ordered triples,

$$(a_1, a_2, a_3) + (b_1, b_2, b_3) = (a_1 + b_1, a_2 + b_2, a_3 + b_3).$$

Pictorially we have the usual force diagram, Figure 12.2.8. It is a parallelogram with the sides representing $\mathbf{F}_1 = (a_1, a_2, a_3)$ and $\mathbf{F}_2 = (b_1, b_2, b_3)$ and the diagonal representing

$$\mathbf{F}_3 = \mathbf{F}_1 + \mathbf{F}_2 = (a_1 + b_1, a_2 + b_2, a_3 + b_3).$$

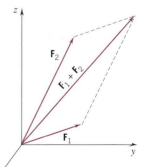

Figure 12.2.8

For any force $\mathbf{F} = (a_1, a_2, a_3)$ and any real number α, the force of $\alpha\mathbf{F}$ is defined by the equation

$$\alpha\mathbf{F} = (\alpha a_1, \alpha a_2, \alpha a_3).$$

Thus, once again we have

$$\alpha(a_1, a_2, a_3) = (\alpha a_1, \alpha a_2, \alpha a_3).$$

Vectors

The algebra of number triples that we have introduced to discuss displacements, forces, and velocities is so prodigiously rich in applications that it has found a firm place in the world of science and engineering and has generated much mathematics. It is to this mathematics that we now turn.

DEFINITION 12.2.1 VECTORS

Ordered triples of real numbers subject to addition defined by

$$(a_1, a_2, a_3) + (b_1, b_2, b_3) = (a_1 + b_1, a_2 + b_2, a_3 + b_3)$$

and multiplication by real numbers (*scalars*†) defined by

$$\alpha(a_1, a_2, a_3) = (\alpha a_1, \alpha a_2, \alpha a_3),$$

are called *vectors*.

Two vectors are called *equal* iff they have exactly the same *components*:

$$(a_1, a_2, a_3) = (b_1, b_2, b_3) \quad \text{iff} \quad a_1 = b_1, a_2 = b_2, a_3 = b_3.$$

Geometrically, we can depict vectors by arrows. To depict the vector (a_1, a_2, a_3) we can choose any initial point $Q = Q(x, y, z)$ and use the arrow

$$\overrightarrow{QR} \quad \text{with} \quad R = R(x + a_1, y + a_2, z + a_3). \qquad \text{(Figure 12.2.9)}$$

Usually we choose the origin as the initial point and use the arrow

$$\overrightarrow{OP} \quad \text{with} \quad P = P(a_1, a_2, a_3).††$$

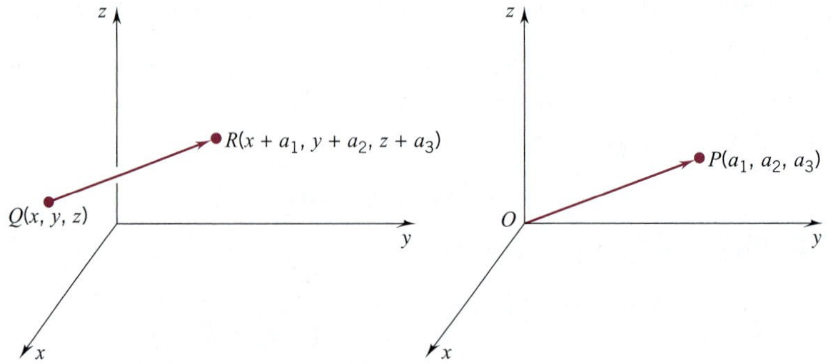

Figure 12.2.9

---

† In discussing vectors, *scalar* is synonymous with *real number*.
†† The ordered triple $(0, 0, 0)$ will not be represented by an arrow but simply by bold $\mathbf{0}$.

Remark Definition 12.2.1 specifically defines vectors in three-dimensional Euclidean space. There are also two-dimensional vectors which are represented by ordered pairs of real numbers (a_1, a_2) and depicted by arrows in the xy-plane (Figure 12.2.10); four-dimensional vectors which are represented by ordered quadruples of real numbers; five-dimensional vectors which are represented by ordered quintuples, and so on. Of course, we cannot draw figures to represent vectors in dimensions higher than three. We will restrict our attention to three-dimensional vectors in the work which follows. A brief discussion of two-dimensional vectors together with some examples involving forces and velocities is given at the end of the section. ❏

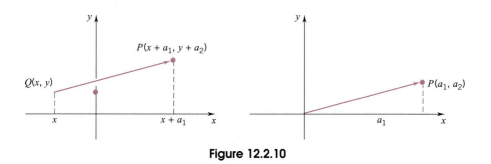

Figure 12.2.10

We use boldface letters to denote vectors. Thus for

$$\mathbf{a} = (a_1, a_2, a_3) \qquad \text{and} \qquad \mathbf{b} = (b_1, b_2, b_3)$$

we have

$$\mathbf{a} + \mathbf{b} = (a_1 + b_1, a_2 + b_2, a_3 + b_3),$$

and, if α is a *scalar* (a real number),

$$\alpha \mathbf{a} = (\alpha a_1, \alpha a_2, \alpha a_3).$$

Addition of vectors satisfies the commutative and associative laws:

$$\mathbf{a} + \mathbf{b} = \mathbf{b} + \mathbf{a}$$

$$\mathbf{a} + (\mathbf{b} + \mathbf{c}) = (\mathbf{a} + \mathbf{b}) + \mathbf{c}.$$

These laws follow immediately from the definition of vector addition and the corresponding properties of real numbers. For the *zero vector* $(0, 0, 0)$ we will use the symbol $\mathbf{0}$. Obviously

$$0\mathbf{a} = \mathbf{0} \qquad \text{for all vectors } \mathbf{a}.$$

By the vector $-\mathbf{b}$ we mean $(-1)\mathbf{b}$; that is,

$$-(b_1, b_2, b_3) = (-b_1, -b_2, -b_3).$$

By $\mathbf{a} - \mathbf{b}$ we mean $\mathbf{a} + (-\mathbf{b})$; that is,

$$(a_1, a_2, a_3) - (b_1, b_2, b_3) = (a_1 - b_1, a_2 - b_2, a_3 - b_3).$$

Example 1 Given that

$$\mathbf{a} = (1, -1, 2), \qquad \mathbf{b} = (2, 3, -1), \qquad \mathbf{c} = (8, 7, 1),$$

find (a) $\mathbf{a} - \mathbf{b}$. (b) $2\mathbf{a} + \mathbf{b}$. (c) $3\mathbf{a} - 7\mathbf{b}$. (d) $2\mathbf{a} + 3\mathbf{b} - \mathbf{c}$.

SOLUTION

(a) $\mathbf{a} - \mathbf{b} = (1, -1, 2) - (2, 3, -1) = (1 - 2, -1 - 3, 2 + 1) = (-1, -4, 3)$.

(b) $2\mathbf{a} + \mathbf{b} = 2(1, -1, 2) + (2, 3, -1) = (2, -2, 4) + (2, 3, -1) = (4, 1, 3)$.

(c) $3\mathbf{a} - 7\mathbf{b} = 3(1, -1, 2) - 7(2, 3, -1)$
$$= (3, -3, 6) - (14, 21, -7) = (-11, -24, 13).$$

(d) $2\mathbf{a} + 3\mathbf{b} - \mathbf{c} = 2(1, -1, 2) + 3(2, 3, -1) - (8, 7, 1)$
$$= (2, -2, 4) + (6, 9, -3) - (8, 7, 1) = (0, 0, 0) = \mathbf{0}. \quad ❑$$

The addition of vectors can be visualized as the "addition" of arrows by the parallelogram law (as in the case of forces). If you picture $\mathbf{a}$, $\mathbf{b}$, and $\mathbf{a} + \mathbf{b}$ all as arrows emanating from the same point, say the origin, then $\mathbf{a} + \mathbf{b}$ acts as the diagonal of the parallelogram generated by $\mathbf{a}$ and $\mathbf{b}$. (See Figure 12.2.11.)

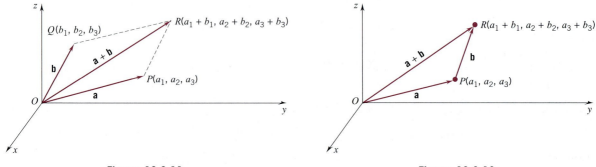

| Figure 12.2.11 | Figure 12.2.12 |

The addition of vectors can also be visualized as the "tail-to-head" addition of arrows (as in the case of displacements). If, instead of starting $\mathbf{b}$ at the origin, you start $\mathbf{b}$ at the tip of $\mathbf{a}$, then $\mathbf{a} + \mathbf{b}$ goes from the tail of $\mathbf{a}$ to the tip of $\mathbf{b}$. (Figure 12.2.12.) These two pictorial representations of vector addition lead to the same result.

DEFINITION 12.2.2 PARALLEL VECTORS

Two nonzero vectors $\mathbf{a}$ and $\mathbf{b}$ are said to be *parallel* provided that

$$\mathbf{a} = \alpha\mathbf{b} \quad \text{for some real number } \alpha.$$

If $\alpha > 0$, $\mathbf{a}$ and $\mathbf{b}$ are said to have the *same direction*; if $\alpha < 0$, they are said to have *opposite directions*.

In the case of $\mathbf{a} = (2, -2, 6)$, $\mathbf{b} = (1, -1, 3)$, $\mathbf{c} = (-1, 1, -3)$ we have

$$\mathbf{a} = 2\mathbf{b} \quad \text{and} \quad \mathbf{a} = -2\mathbf{c}.$$

This tells us that $\mathbf{a}$ and $\mathbf{b}$ are parallel and have the same direction, whereas $\mathbf{a}$ and $\mathbf{c}$, though parallel, have opposite directions. (See Figure 12.2.13.)

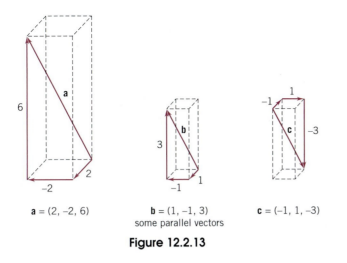

$$\mathbf{a} = (2, -2, 6)$$

$$\mathbf{b} = (1, -1, 3)$$
some parallel vectors

$$\mathbf{c} = (-1, 1, -3)$$

Figure 12.2.13

Definition 12.2.2 did not include the zero vector $\mathbf{0}$. By special convention, $\mathbf{0}$ is said to be *parallel to every vector*.

(Since $\mathbf{0}$ is represented geometrically by a point, there is no geometric meaning to saying that $\mathbf{0}$ is parallel to another vector. However, it simplifies the statement of certain results to maintain that $\mathbf{0}$ is parallel to every vector. Algebraically this is warranted by the fact that $\mathbf{0}$ is a scalar multiple of every vector $\mathbf{b}$: $\mathbf{0} = 0\mathbf{b}$.)

Example 2 Show that, if $\mathbf{a}$ and $\mathbf{b}$ are parallel to $\mathbf{c}$, then every linear combination, $\alpha\mathbf{a} + \beta\mathbf{b}$, of $\mathbf{a}$ and $\mathbf{b}$ is also parallel to $\mathbf{c}$.

SOLUTION Suppose that $\mathbf{a}$ and $\mathbf{b}$ are parallel to $\mathbf{c}$. If $\mathbf{c} = \mathbf{0}$, then every vector is parallel to $\mathbf{c}$ and there is nothing to prove. If $\mathbf{c} \neq \mathbf{0}$, then $\mathbf{a}$ and $\mathbf{b}$ are scalar multiples of $\mathbf{c}$:

$$\mathbf{a} = \alpha_1\mathbf{c}, \qquad \mathbf{b} = \beta_1\mathbf{c}.$$

Then

$$\alpha\mathbf{a} + \beta\mathbf{b} = \alpha(\alpha_1\mathbf{c}) + \beta(\beta_1\mathbf{c}) = (\alpha\alpha_1 + \beta\beta_1)\mathbf{c}$$

is also parallel to $\mathbf{c}$. ❏

DEFINITION 12.2.3 NORM

The *norm* of a vector $\mathbf{a} = (a_1, a_2, a_3)$, denoted by $\|\mathbf{a}\|$, is the number

$$\|\mathbf{a}\| = \sqrt{a_1^2 + a_2^2 + a_3^2}$$

The norm of $\mathbf{a}$ is also called the *length* or *magnitude* of $\mathbf{a}$. As discussed previously, if we represent the vector $\mathbf{a}$ by an arrow, say $\overrightarrow{QR}$, with $Q = Q(x, y, z)$ and $R = R(x + a_1, y + a_2, z + a_3)$, then $\|a\|$ gives the length of $\overrightarrow{QR}$ or, what amounts to the same thing, the distance between Q and R. (Figure 12.2.14).

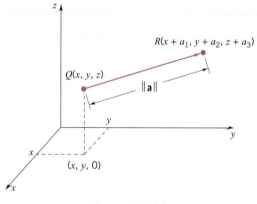

Figure 12.2.14

The norm properties of vectors are very similar to the absolute-value properties of real numbers. In particular

(12.2.4)

> **(1)** $\|\mathbf{a}\| \geq 0$ and $\|\mathbf{a}\| = 0$ iff $\mathbf{a} = 0$.
> **(2)** $\|\alpha\mathbf{a}\| = |\alpha| \|\mathbf{a}\|$.
> **(3)** $\|\mathbf{a} + \mathbf{b}\| \leq \|\mathbf{a}\| + \|\mathbf{b}\|$.   (the triangle inequality)

Property (1) is obvious. Property (2) is easy to verify:

$$\|\alpha\mathbf{a}\| = \sqrt{(\alpha a_1)^2 + (\alpha a_2)^2 + (\alpha a_3)^2} = |\alpha|\sqrt{a_1^2 + a_2^2 + a_3^2} = |\alpha| \|\mathbf{a}\|.$$

We prove Property (3), the triangle inequality, in the next section, where we have "dot products" at our disposal. A proof at this time would be laborious.

You can interpret the triangle inequality as saying that the length of a side of a triangle cannot exceed the sum of the lengths of the other two sides. (See Figure 12.2.15.)

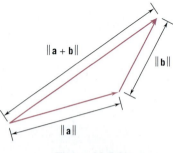

Figure 12.2.15

Example 3 Given that $\mathbf{a} = (1, -2, 3)$ and $\mathbf{b} = (-4, 1, 0)$, calculate

(a) $\|\mathbf{a}\|$.   **(b)** $\|\mathbf{b}\|$.   **(c)** $\|\mathbf{a} + \mathbf{b}\|$.   **(d)** $\|\mathbf{a} - \mathbf{b}\|$.   **(e)** $\|-7\mathbf{a}\|$.
(f) $\|2\mathbf{a} - 3\mathbf{b}\|$.

SOLUTION

(a) $\|\mathbf{a}\| = \sqrt{1^2 + (-2)^2 + (3)^2} = \sqrt{1 + 4 + 9} = \sqrt{14}$.

(b) $\|\mathbf{b}\| = \sqrt{(-4)^2 + 1^2 + 0^2} = \sqrt{16 + 1 + 0} = \sqrt{17}$.

(c) $\|\mathbf{a} + \mathbf{b}\| = \|(-3, -1, 3)\| = \sqrt{(-3)^2 + (-1)^2 + (3)^2} = \sqrt{9 + 1 + 9} = \sqrt{19}$.

(d) $\|\mathbf{a} - \mathbf{b}\| = \|(5, -3, 3)\| = \sqrt{(5)^2 + (-3)^2 + (3)^2} = \sqrt{25 + 9 + 9} = \sqrt{43}$.

(e) $\|-7\mathbf{a}\| = |-7|\|\mathbf{a}\| = 7\sqrt{14}$.

(f) $\|2\mathbf{a} - 3\mathbf{b}\| = \|2(1, -2, 3) - 3(-4, 1, 0)\|$
$$= \|(14, -7, 6)\| = \sqrt{(14)^2 + (-7)^2 + (6)^2}$$
$$= \sqrt{196 + 49 + 36} = \sqrt{281}. \quad \square$$

To multiply a nonzero vector by a nonzero scalar α is to change its length by a factor of $|\alpha|$,

$$\|\alpha\mathbf{a}\| = |\alpha|\|\mathbf{a}\|,$$

keeping the same direction if $\alpha > 0$ and reversing the direction if $\alpha < 0$. The vector obtained from $\mathbf{a}$ simply by reversing its direction is the vector $(-1)\mathbf{a} = -\mathbf{a}$. (See Figure 12.2.16.)

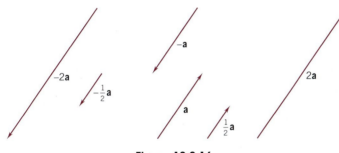

Figure 12.2.16

Since $\mathbf{a} - \mathbf{b} = \mathbf{a} + (-\mathbf{b})$, we can draw the vector $\mathbf{a} - \mathbf{b}$ by drawing $-\mathbf{b}$ and adding it to the vector $\mathbf{a}$. (Figure 12.2.17.)

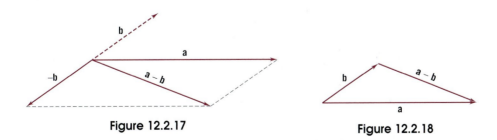

Figure 12.2.17   **Figure 12.2.18**

We can obtain the same result more easily by noting that $\mathbf{a} - \mathbf{b}$ is the vector that we must add to $\mathbf{b}$ to obtain $\mathbf{a}$. (Figure 12.2.18.)

Vectors of norm 1 are called *unit vectors*. If **b** is a nonzero vector, then there is a unit vector $\mathbf{u_b}$ that has the direction of **b**. To find $\mathbf{u_b}$, note that

$$\|\mathbf{u_b}\| = 1 \quad \text{and} \quad \mathbf{u_b} = \alpha\mathbf{b} \quad \text{for some } \alpha > 0.$$

It follows that

$$1 = \|\mathbf{u_b}\| = \|\alpha\mathbf{b}\| = |\alpha|\|\mathbf{b}\| = \alpha\|\mathbf{b}\|.$$

Thus

$$\alpha = \frac{1}{\|\mathbf{b}\|} \quad \text{and consequently} \quad \mathbf{u_b} = \frac{1}{\|\mathbf{b}\|}\mathbf{b} = \frac{\mathbf{b}}{\|\mathbf{b}\|}.$$

While

$$\mathbf{u_b} = \frac{\mathbf{b}}{\|\mathbf{b}\|}$$

is the unit vector in the direction of **b**,

$$-\mathbf{u_b} = -\frac{\mathbf{b}}{\|\mathbf{b}\|}$$

is the unit vector in the opposite direction.

We single out for special attention the vectors

$$\mathbf{i} = (1, 0, 0), \qquad \mathbf{j} = (0, 1, 0), \qquad \mathbf{k} = (0, 0, 1).$$

These vectors all have norm 1 and, if pictured as emanating from the origin, lie along the positive coordinate axes. They are called *the unit coordinate vectors.* (Figure 12.2.19.)

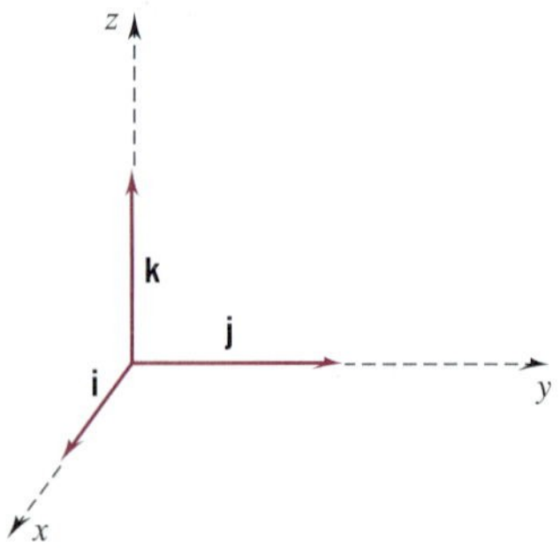

Every vector can be expressed as a linear combination of the unit coordinate vectors:

(12.2.5) | for $\mathbf{a} = (a_1, a_2, a_3)$ we have $\mathbf{a} = a_1\mathbf{i} + a_2\mathbf{j} + a_3\mathbf{k}.$

Figure 12.2.19

PROOF

$$\begin{aligned}(a_1, a_2, a_3) &= (a_1, 0, 0) + (0, a_2, 0) + (0, 0, a_3) \\ &= a_1(1, 0, 0) + a_2(0, 1, 0) + a_3(0, 0, 1) \\ &= a_1\mathbf{i} + a_2\mathbf{j} + a_3\mathbf{k}.\end{aligned}$$

The numbers a_1, a_2, a_3 are called the **i, j, k** components of the vector **a**.

Example 4 Given that

$$\mathbf{a} = 3\mathbf{i} - \mathbf{j} + \mathbf{k} \qquad \text{and} \qquad \mathbf{b} = 2\mathbf{i} + 3\mathbf{j} - \mathbf{k},$$

(1) Express $2\mathbf{a} - \mathbf{b}$ as a linear combination of **i, j, k**.

(2) Calculate $\|2\mathbf{a} - \mathbf{b}\|$.

(3) Find the unit vector $\mathbf{u_c}$ in the direction of $\mathbf{c} = 2\mathbf{a} - \mathbf{b}$.

SOLUTION

(1) $2\mathbf{a} - \mathbf{b} = 2(3\mathbf{i} - \mathbf{j} + \mathbf{k}) - (2\mathbf{i} + 3\mathbf{j} - \mathbf{k})$
$\qquad\qquad = 6\mathbf{i} - 2\mathbf{j} + 2\mathbf{k} - 2\mathbf{i} - 3\mathbf{j} + \mathbf{k} = 4\mathbf{i} - 5\mathbf{j} + 3\mathbf{k}.$

(2) $\|2\mathbf{a} - \mathbf{b}\| = \|4\mathbf{i} - 5\mathbf{j} + 3\mathbf{k}\| = \sqrt{16 + 25 + 9} = \sqrt{50} = 5\sqrt{2}.$

(3) $\mathbf{u_c} = \dfrac{2\mathbf{a} - \mathbf{b}}{\|2\mathbf{a} - \mathbf{b}\|} = \dfrac{1}{5\sqrt{2}}(4\mathbf{i} - 5\mathbf{j} + 3\mathbf{k}).$ ❑

Vectors in the Plane

A vector $\mathbf{a} = a_1\mathbf{i} + a_2\mathbf{j} + a_3\mathbf{k}$ for which $a_3 = 0$ can be written more simply as

(12.2.6)
$$\mathbf{a} = a_1\mathbf{i} + a_2\mathbf{j}$$

and identified with the vector (a_1, a_2) in the xy-plane. In particular, the vectors $\mathbf{i} = (1, 0)$ and $\mathbf{j} = (0, 1)$ are the unit coordinate vectors. See Figure 12.2.20.

It is easy to see that if $\mathbf{a}$ and $\mathbf{b}$ are vectors in the xy-plane and α is a scalar, then $\mathbf{a} + \mathbf{b}$ and $\alpha\mathbf{a}$ are also vectors in the plane. Thus, all of the definitions and results stated previously for vectors in space hold for vectors in the plane as well. In particular, if α is a scalar, and $\mathbf{a} = (a_1, a_2)$ and $\mathbf{b} = (b_1, b_2)$ are vectors in the xy-plane, then:

(1) $\mathbf{a} = \mathbf{b}$ iff $a_1 = b_1$ and $a_2 = b_2$.

(2) $\mathbf{a} + \mathbf{b} = (a_1 + b_1, a_2 + b_2) = (a_1 + b_1)\mathbf{i} + (a_2 + b_2)\mathbf{j}.$

(3) $\alpha\mathbf{a} = \alpha(a_1, a_2) = (\alpha a_1, \alpha a_2) = \alpha a_1\mathbf{i} + \alpha a_2\mathbf{j}.$

(4) $\mathbf{0} = (0, 0) = 0\mathbf{i} + 0\mathbf{j}$ is the zero vector.

(5) $\|\mathbf{a}\| = \sqrt{a_1^2 + a_2^2}$ is the norm, or magnitude, of $\mathbf{a}$.

A nonzero vector $\mathbf{a}$ can also be represented in terms of its magnitude and the angle θ that it makes with the positive x-axis (see Figure 12.2.21).

(12.2.7)
$$\mathbf{a} = \|\mathbf{a}\| \cos\theta\,\mathbf{i} + \|\mathbf{a}\| \sin\theta\,\mathbf{j}$$

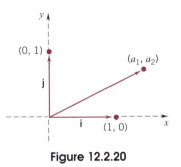

Figure 12.2.20

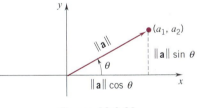

Figure 12.2.21

We will use either ordered pairs (a_1, a_2) or the form (12.2.6) to treat problems that are two-dimensional in nature. The following two examples illustrate analyses of forces and velocities in the plane.

Example 5 Two forces $\mathbf{F}_1$ and $\mathbf{F}_2$ with magnitudes 6 pounds and 4 pounds, respectively, are applied to an object at a point P as shown in Figure 12.2.22(a). Find the resultant force $\mathbf{F}$, give its magnitude, and give its direction in terms of an angle θ.

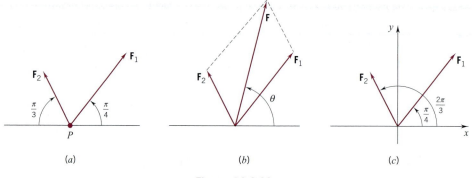

(a)   (b)   (c)

Figure 12.2.22

SOLUTION The resultant force **F** is shown in Figure 12.2.22(b). In Figure 12.2.22(c) we have introduced a coordinate system in order to find $\mathbf{F}_1$ and $\mathbf{F}_2$ in terms of their components:

$$\mathbf{F}_1 = \left(6 \cos \frac{\pi}{4} \right) \mathbf{i} + \left(6 \sin \frac{\pi}{4} \right) \mathbf{j} = 3\sqrt{2}\mathbf{i} + 3\sqrt{2}\mathbf{j}$$

$$\mathbf{F}_2 = \left(4 \cos \frac{2\pi}{3} \right) \mathbf{i} + \left(4 \sin \frac{2\pi}{3} \right) \mathbf{j} = -2\mathbf{i} + 2\sqrt{3}\mathbf{j}.$$

Thus, the resultant force $\mathbf{F} = \mathbf{F}_1 + \mathbf{F}_2$ is given by

$$\mathbf{F} = (3\sqrt{2} - 2)\mathbf{i} + (3\sqrt{2} + 2\sqrt{3})\mathbf{j} \cong 2.243\mathbf{i} + 7.707\mathbf{j}.$$

The magnitude of **F** is

$$\|\mathbf{F}\| = \sqrt{(3\sqrt{2} - 2)^2 + (3\sqrt{2} + 2\sqrt{3})^2} \cong \sqrt{(2.243)^2 + (7.707)^2} \cong 8.026.$$

To find the angle θ, note that

$$\cos \theta = \frac{3\sqrt{2} - 2}{\|\mathbf{F}\|} \cong \frac{2.243}{8.026} \cong 0.279 \qquad \text{so} \qquad \theta \cong 1.29 \text{ radians or } 73.8°. \quad \square$$

Example 6 A man wants to paddle his canoe across a river from a point X to the point Y on the opposite shore directly across from X. If he can paddle the canoe at the rate of 5 miles per hour and the current in the river is 3 miles per hour, in what direction θ should he steer his canoe in order to go straight across the river? See Figure 12.2.23a. Also, what is his resultant speed across the river?

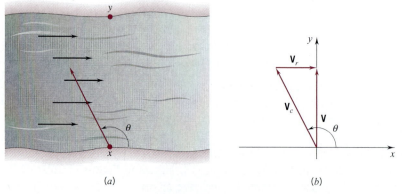

(a)   (b)

Figure 12.2.23

SOLUTION A velocity diagram is shown in Figure 12.2.23(b) where $\mathbf{V}_c$ denotes that velocity of the canoe, $\mathbf{V}_r$ denotes the velocity of the river, and $\mathbf{V}$ denotes the resultant velocity directed straight across the river. These velocities are given by

$$\mathbf{V}_c = 5\cos\theta\mathbf{i} + 5\sin\theta\mathbf{j}, \qquad \mathbf{V}_r = 3\mathbf{i} \qquad \text{and} \qquad \mathbf{V} = \alpha\mathbf{j} \qquad \text{for some number } \alpha.$$

From the diagram, we have

$$\mathbf{V}_c + \mathbf{V}_r = \mathbf{V}$$

so

$$5\cos\theta\mathbf{i} + 5\sin\theta\mathbf{j} + 3\mathbf{i} = \alpha\mathbf{j}$$

and

$$(5\cos\theta + 3)\mathbf{i} + 5\sin\theta\mathbf{j} = \alpha\mathbf{j}.$$

Thus, we must have

$$5\cos\theta + 3 = 0 \qquad \text{or} \qquad \cos\theta = -\frac{3}{5} = -0.6.$$

It follows from this that $\theta \cong 2.214$ radians or $126.87°$.

The relative speed across the river is given by $\|\mathbf{V}\| = \alpha$ and, from Figure 12.2.23(b),

$$\alpha^2 + 3^2 = 5^2 \qquad \text{and so} \qquad \alpha = 4;$$

therefore, the resultant speed is 4 miles per hour. ❑

EXERCISES 12.2

In Exercises 1–6, points P and Q are given. Find the vector $\overrightarrow{PQ}$ and determine its norm.

1. $P(1, -2, 5)$, $Q(4, 2, 3)$.

2. $P(-1, 0, 2)$, $Q(-3, -2, 4)$.

3. $P(4, -2)$, $Q(2, 4)$.

4. $P(\frac{3}{2}, -\frac{1}{4})$, $Q(-\frac{1}{2}, \frac{3}{4})$.

5. $P(3, 2, \frac{1}{2})$, $Q(-1, 4, \frac{5}{2})$.

6. $P(-4, 0, 7)$, $Q(0, 3, -1)$.

In Exercises 7–10, let $\mathbf{a} = (1, -2, 3)$, $\mathbf{b} = (3, 0, -1)$, and $\mathbf{c} = (-4, 2, 1)$, and find:

7. $2\mathbf{a} - \mathbf{b}$.

8. $2\mathbf{b} + 3\mathbf{c}$.

9. $-2\mathbf{a} + \mathbf{b} - \mathbf{c}$.

10. $\mathbf{a} + 3\mathbf{b} - 2\mathbf{c}$.

In Exercises 11–14, simplify the linear combinations.

11. $(2\mathbf{i} - \mathbf{j} + \mathbf{k}) + (\mathbf{i} - 3\mathbf{j} + 5\mathbf{k})$.

12. $(6\mathbf{j} - \mathbf{k}) + (3\mathbf{i} - \mathbf{j} + 2\mathbf{k})$.

13. $2(\mathbf{j} + \mathbf{k}) - 3(\mathbf{i} + \mathbf{j} - 2\mathbf{k})$.

14. $2(\mathbf{i} - \mathbf{j}) + 6(2\mathbf{i} + \mathbf{j} - 2\mathbf{k})$.

In Exercises 15–20, calculate the norm of the given vector.

15. $3\mathbf{i} + 4\mathbf{j}$.

16. $\mathbf{i} - \mathbf{j}$.

17. $2\mathbf{i} + \mathbf{j} - 2\mathbf{k}$.

18. $6\mathbf{i} + 2\mathbf{j} - \mathbf{k}$.

19. $\frac{1}{2}(\mathbf{i} + 4\mathbf{j}) - (\frac{3}{2}\mathbf{i} + \mathbf{k})$.

20. $(\mathbf{i} - \mathbf{j}) + 2(\mathbf{j} - \mathbf{i}) + (\mathbf{k} - \mathbf{j})$.

21. Let

$$\mathbf{a} = \mathbf{i} - \mathbf{j} + 2\mathbf{k}, \qquad \mathbf{b} = 2\mathbf{i} - \mathbf{j} + 2\mathbf{k},$$
$$\mathbf{c} = 3\mathbf{i} - 3\mathbf{j} + 6\mathbf{k}, \qquad \mathbf{d} = -2\mathbf{i} + 2\mathbf{j} - 4\mathbf{k}.$$

    (a) Which vectors are parallel?
    (b) Which vectors have the same direction?
    (c) Which vectors have opposite directions?

22. (*Important*) Prove the following version of the triangle inequality:

$$\big|\|\mathbf{a}\| - \|\mathbf{b}\|\big| \le \|\mathbf{a} - \mathbf{b}\|.$$

    HINT: Apply the previously stated triangle inequality to $\mathbf{a} = (\mathbf{a} - \mathbf{b}) + \mathbf{b}$.

In Exercises 23–28, find the unit vector in the direction of the given vector $\mathbf{a}$.

23. $\mathbf{a} = (3, -4)$.

24. $\mathbf{a} = (-2, 3)$.

25. $\mathbf{a} = (-4, 0, 3)$.

26. $\mathbf{a} = (2, 1, 2)$.

27. $\mathbf{a} = \mathbf{i} - 2\mathbf{j} + 2\mathbf{k}$.

28. $\mathbf{a} = 2\mathbf{i} + 3\mathbf{j} + \mathbf{k}$.

In Exercises 29–30, find the unit vector whose direction is opposite to the direction of **a**.

29. $\mathbf{a} = -\mathbf{i} + 3\mathbf{j} + 2\mathbf{k}$.   **30.** $\mathbf{a} = 2\mathbf{i} - \mathbf{k}$.

31. Label the vectors.

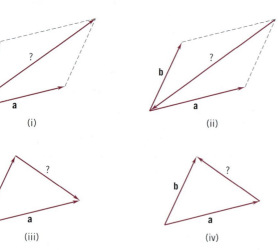

(i)   (ii)

(iii)   (iv)

32. Let $\mathbf{a} = (1, 1, 1)$, $\mathbf{b} = (-1, 3, 2)$, $\mathbf{c} = (-3, 0, 1)$, $\mathbf{d} = (4, -1, 1)$.
   (a) Express $\mathbf{a} + 2\mathbf{b} + 3\mathbf{c} + 4\mathbf{d}$ as a linear combination of **i**, **j**, **k**.
   (b) Find scalars A, B, C such that $\mathbf{d} = A\mathbf{a} + B\mathbf{b} + C\mathbf{c}$.

33. Let $\mathbf{a} = (2, 0, -1)$, $\mathbf{b} = (1, 3, 5)$, $\mathbf{c} = (-1, 1, 1)$, $\mathbf{d} = (1, 1, 6)$.
   (a) Express $\mathbf{a} - 3\mathbf{b} + 2\mathbf{c} + 4\mathbf{d}$ as a linear combination of **i**, **j**, **k**.
   (b) Find scalars A, B, C such that $\mathbf{d} = A\mathbf{a} + B\mathbf{b} + C\mathbf{c}$.

34. Find α given that $3\mathbf{i} + \mathbf{j} - \mathbf{k}$ and $\alpha\mathbf{i} - 4\mathbf{j} + 4\mathbf{k}$ are parallel.

35. Find α given that $3\mathbf{i} + \mathbf{j}$ and $\alpha\mathbf{j} - \mathbf{k}$ have the same length.

36. Find the unit vector in the direction of $\mathbf{i} - 2\mathbf{j} + 2\mathbf{k}$.

37. Find α given that $\|\alpha\mathbf{i} + (\alpha - 1)\mathbf{j} + (\alpha + 1)\mathbf{k}\| = 2$.

38. Find the vector of norm 2 in the direction of $\mathbf{i} + 2\mathbf{j} - \mathbf{k}$.

39. Find the vectors of norm 2 parallel to $3\mathbf{j} + 2\mathbf{k}$.

40. Express **c** in terms of **a** and **b**, given that the tip of **c** bisects the line segment.

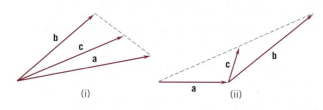

(i)   (ii)

In Exercises 41–46, find all vectors $\mathbf{v} = a\mathbf{i} + b\mathbf{j}$ that have the given properties.

41. Makes an angle of 30° with the positive x-axis and has norm 2.

42. Makes an angle of 135° with the positive x-axis and has norm 4.

43. Makes an angle of $\pi/4$ with the positive x-axis and has norm 1.

44. Makes an angle of $-5\pi/6$ with the positive x-axis and has norm 5.

45. Has norm 3 and the **i**-component is twice the **j**-component.

46. Has norm 5 and the **j**-component is three-fourths the **i**-component.

47. Let $\mathbf{a} = (2, -1)$ and $\mathbf{b} = (1, 2)$. Show that **a** and **b** are perpendicular.   HINT: Assume that **a** and **b** are the sides of a triangle.

48. Repeat Exercise 47 for $\mathbf{a} = (2, 0, 3)$ and $\mathbf{b} = (-6, 2, 4)$.

49. Let **a** and **b** be nonzero vectors such that $\|\mathbf{a} - \mathbf{b}\| = \|\mathbf{a} + \mathbf{b}\|$.
   (a) What can you conclude about the parallelogram generated by **a** and **b**?
   (b) Show that, if $\mathbf{a} = a_1\mathbf{i} + a_2\mathbf{j} + a_3\mathbf{k}$ and $\mathbf{b} = b_1\mathbf{i} + b_2\mathbf{j} + b_3\mathbf{k}$, then
   $$a_1b_1 + a_2b_2 + a_3b_3 = 0.$$

50. (a) Show that, if **a** and **b** have the same direction, then $\|\mathbf{a} + \mathbf{b}\| = \|\mathbf{a}\| + \|\mathbf{b}\|$.
   (b) Does this equation necessarily hold if **a** and **b** are only parallel?

51. Let P and Q be two points in space and let M be the midpoint of the line segment $\overline{PQ}$. Let $\mathbf{p} = \overrightarrow{OP}$, $\mathbf{q} = \overrightarrow{OQ}$, and $\mathbf{m} = \overrightarrow{OM}$.
   (a) Show that $\mathbf{m} = \mathbf{p} + \frac{1}{2}(\mathbf{q} - \mathbf{p})$.
   (b) Derive the midpoint formula (12.1.4).

52. Let P and Q be two points in space and let R be the point on $\overline{PQ}$ which is twice as far from P as it is from Q. Let $\mathbf{p} = \overrightarrow{OP}$, $\mathbf{q} = \overrightarrow{OQ}$, and $\mathbf{r} = \overrightarrow{OR}$. Prove that $\mathbf{r} = \frac{1}{3}\mathbf{p} + \frac{2}{3}\mathbf{q}$.

53. A 200-pound weight is suspended by wires as shown in the following figure. Find the force exerted by each of the wires (called the *tension*) and determine the magnitude of each.   HINT: The resultant of the two tensions must counterbalance the weight.

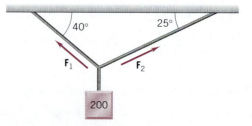

54. The figure below illustrates two tugboats pulling a barge. If tugboat *A* exerts a force of 5000 pounds, what force must tugboat *B* exert if the barge is to move along the straight line *l*?

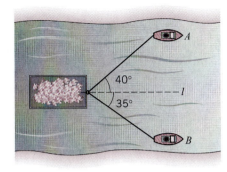

55. An airplane has an airspeed of 600 miles per hour. If the pilot is steering the plane in the direction N30°E and the wind is blowing from the northwest (N45°W) at the rate of 50 miles per hour, find the resultant velocity of the plane. The direction of the resultant is called the *true course* of the plane and the magnitude of the resultant is called the *ground speed.* Find the true course and ground speed of the plane.

56. The true course of an airplane is N40°W and its ground speed is 550 miles per hour (see Exercise 55). What is the airspeed of the plane if the wind is blowing from the east at 70 miles per hour?

A vector **r** emanating from the origin is called a *radius vector.* Each radius vector determines a unique point of space: the point at the tip of the vector. Conversely, each point of space determines a unique radius vector: the radius vector whose tip falls on that point. This one-to-one correspondence between the set of all radius vectors and the set of all points in three-dimensional space enables us to use radius vectors to specify sets in space. Thus, for example, the radius-vector equation $\|\mathbf{r}\| = 3$ can be used to represent the sphere of radius 3 centered at the origin: that sphere consists of the tips of all the radius vectors **r** that satisfy that equation.

57. Write a radius-vector equation or inequality for each of the following sets.
   (a) The sphere of radius 3 centered at $P(a_1, a_2, a_3)$.
   (b) The set of all points on or inside the sphere of radius 2 centered at the origin. (This set is called the *ball* of radius 2 about the origin.)
   (c) The ball of radius 1 about the point $P(a_1, a_2, a_3)$.
   (d) The set of all points equidistant from $P(a_1, a_2, a_3)$ and $Q(b_1, b_2, b_3)$. (This set forms a plane.)
   (e) The set of all points the sum of whose distances from $P(a_1, a_2, a_3)$ and $Q(b_1, b_2, b_3)$ is a constant $k > d(P, Q)$. [Such a set is called an *ellipsoid.* The ellipsoid is a three-dimensional analogue of the ellipse.] What happens if $k = d(P, Q)$? If $k < d(P, Q)$?

■ 12.3 THE DOT PRODUCT

In this section we introduce the first of two products that we define for vectors.

Introduction

We begin with two nonzero vectors

$$\mathbf{a} = a_1\mathbf{i} + a_2\mathbf{j} + a_3\mathbf{k}, \qquad \mathbf{b} = b_1\mathbf{i} + b_1\mathbf{j} + b_3\mathbf{k}.$$

How can we tell from the components of these vectors whether these vectors meet at right angles? To explore this question we draw Figure 12.3.1. By the Pythagorean Theorem **a** and **b** meet at right angles iff

$$\|\mathbf{a}\|^2 + \|\mathbf{b}\|^2 = \|\mathbf{b} - \mathbf{a}\|^2.$$

In terms of components this equation reads

$$(a_1^2 + a_2^2 + a_3^2) + (b_1^2 + b_2^2 + b_3^2) = (b_1 - a_1)^2 + (b_2 - a_2)^2 + (b_3 - a_3)^2,$$

which, as you can readily check, simplifies to

$$a_1b_1 + a_2b_2 + a_3b_3 = 0. \quad \square$$

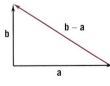

Figure 12.3.1

The expression $a_1b_1 + a_2b_2 + a_3b_3$ is widely used in geometry and in physics. It has a name, the *dot product* of **a** and **b**, and there is a special notation for it, **a · b**. The notion is so important that it deserves a formal definition.

Definition of the Dot Product

DEFINITION 12.3.1

For any two vectors

$$\mathbf{a} = a_1\mathbf{i} + a_2\mathbf{j} + a_3\mathbf{k} \qquad \text{and} \qquad \mathbf{b} = b_1\mathbf{i} + b_2\mathbf{j} + b_3\mathbf{k}$$

we define the *dot product* **a · b** by setting

$$\mathbf{a} \cdot \mathbf{b} = a_1b_1 + a_2b_2 + a_3b_3.$$

Remark Dot products for n-dimensional vectors, $n = 2, 4, 5, \ldots$, are defined in exactly the same way. In particular, for any 2 two-dimensional vectors $\mathbf{a} = a_1\mathbf{i} + a_2\mathbf{j}$ and $\mathbf{b} = b_1\mathbf{i} + b_2\mathbf{j}$

$$\mathbf{a} \cdot \mathbf{b} = a_1b_1 + a_2b_2.$$

The properties of the dot product discussed in this section for three-dimensional vectors also hold in the other dimensions. ❏

Example 1 For

$$\mathbf{a} = 2\mathbf{i} - \mathbf{j} + 3\mathbf{k}, \qquad \mathbf{b} = -3\mathbf{i} + \mathbf{j} + 4\mathbf{k}, \qquad \mathbf{c} = \mathbf{i} + 3\mathbf{j}$$

we have

$$\mathbf{a} \cdot \mathbf{b} = (2)(-3) + (-1)(1) + (3)(4) = -6 - 1 + 12 = 5,$$
$$\mathbf{a} \cdot \mathbf{c} = (2)(1) + (-1)(3) + (3)(0) = 2 - 3 = -1,$$
$$\mathbf{b} \cdot \mathbf{c} = (-3)(1) + (1)(3) + (4)(0) = -3 + 3 = 0.$$

The last equation tells us that **b** and **c** meet at right angles. (Verify this by drawing a figure.) ❏

Because **a · b** is not a vector, but a scalar, it is sometimes called the *scalar product* of **a** and **b**. We will continue to call it the dot product and speak of "dotting **a** with **b**."

Properties of the Dot Product

If we dot a vector with itself, we obtain the square of its norm:

(12.3.2) $$\mathbf{a} \cdot \mathbf{a} = \|\mathbf{a}\|^2.$$

PROOF

$$\mathbf{a} \cdot \mathbf{a} = a_1a_1 + a_2a_2 + a_3a_3 = a_1^2 + a_2^2 + a_3^2 = \|\mathbf{a}\|^2. \quad ❏$$

The dot product of any vector with the zero vector is zero:

(12.3.3)
$$\mathbf{a} \cdot \mathbf{0} = 0, \qquad \mathbf{0} \cdot \mathbf{a} = 0.$$

PROOF

$$(a_1)(0) + (a_2)(0) + (a_3)(0) = 0, \qquad (0)(a_1) + (0)(a_2) + (0)(a_3) = 0. \quad \square$$

The dot product is commutative:

(12.3.4)
$$\mathbf{a} \cdot \mathbf{b} = \mathbf{b} \cdot \mathbf{a},$$

and scalars can be factored:

(12.3.5)
$$\alpha\mathbf{a} \cdot \beta\mathbf{b} = \alpha\beta(\mathbf{a} \cdot \mathbf{b}).$$

PROOF

$$\mathbf{a} \cdot \mathbf{b} = a_1b_1 + a_2b_2 + a_3b_3 = b_1a_1 + b_2a_2 + b_3a_3 = \mathbf{b} \cdot \mathbf{a},$$

and

$$\alpha\mathbf{a} \cdot \beta\mathbf{b} = (\alpha a_1)(\beta b_1) + (\alpha a_2)(\beta b_2) + (\alpha a_3)(\beta b_3)$$
$$= \alpha\beta(a_1b_1 + a_2b_2 + a_3b_3) = \alpha\beta(\mathbf{a} \cdot \mathbf{b}). \quad \square$$

The dot product satisfies the distributive laws:

(12.3.6)
$$\mathbf{a} \cdot (\mathbf{b} + \mathbf{c}) = \mathbf{a} \cdot \mathbf{b} + \mathbf{a} \cdot \mathbf{c}, \qquad (\mathbf{a} + \mathbf{b}) \cdot \mathbf{c} = \mathbf{a} \cdot \mathbf{c} + \mathbf{b} \cdot \mathbf{c}.$$

PROOF

$$\mathbf{a} \cdot (\mathbf{b} + \mathbf{c}) = a_1(b_1 + c_1) + a_2(b_2 + c_2) + a_3(b_3 + c_3)$$
$$= a_1b_1 + a_1c_1 + a_2b_2 + a_2c_2 + a_3b_3 + a_3c_3$$
$$= (a_1b_1 + a_2b_2 + a_3b_3) + (a_1c_1 + a_2c_2 + a_3c_3)$$
$$= \mathbf{a} \cdot \mathbf{b} + \mathbf{a} \cdot \mathbf{c}.$$

The second equation can be verified in a similar manner. $\quad \square$

Example 2 Given that

$$\|\mathbf{a}\| = 1, \qquad \|\mathbf{b}\| = 3, \qquad \|\mathbf{c}\| = 4, \qquad \mathbf{a} \cdot \mathbf{b} = 0, \qquad \mathbf{a} \cdot \mathbf{c} = 1, \qquad \mathbf{b} \cdot \mathbf{c} = -2,$$

find **(a)** $3\mathbf{a} \cdot (\mathbf{b} + 4\mathbf{c})$. **(b)** $(\mathbf{a} - \mathbf{b}) \cdot (2\mathbf{a} + \mathbf{b})$. **(c)** $[(\mathbf{b} \cdot \mathbf{c})\mathbf{a} - (\mathbf{a} \cdot \mathbf{c})\mathbf{b}] \cdot \mathbf{c}$.

SOLUTION

(a) $3\mathbf{a} \cdot (\mathbf{b} + 4\mathbf{c}) = (3\mathbf{a} \cdot \mathbf{b}) + (3\mathbf{a} \cdot 4\mathbf{c}) = 3(\mathbf{a} \cdot \mathbf{b}) + 12(\mathbf{a} \cdot \mathbf{c}) = 12.$

(b) $(\mathbf{a} - \mathbf{b}) \cdot (2\mathbf{a} + \mathbf{b}) = (\mathbf{a} \cdot 2\mathbf{a}) + (\mathbf{a} \cdot \mathbf{b}) + (-\mathbf{b} \cdot 2\mathbf{a}) + (-\mathbf{b} \cdot \mathbf{b})$
$$= 2(\mathbf{a} \cdot \mathbf{a}) + (\mathbf{a} \cdot \mathbf{b}) - 2(\mathbf{b} \cdot \mathbf{a}) - (\mathbf{b} \cdot \mathbf{b})$$
$$= 2\|\mathbf{a}\|^2 + (\mathbf{a} \cdot \mathbf{b}) - 2(\mathbf{a} \cdot \mathbf{b}) - \|\mathbf{b}\|^2$$
$$= 2 + 0 - 2(0) - 9 = -7.$$

(c) $[(\mathbf{b} \cdot \mathbf{c})\mathbf{a} - (\mathbf{a} \cdot \mathbf{c})\mathbf{b}] \cdot \mathbf{c} = [(\mathbf{b} \cdot \mathbf{c})\mathbf{a} \cdot \mathbf{c}] - [(\mathbf{a} \cdot \mathbf{c})\mathbf{b} \cdot \mathbf{c}]$
$$= (\mathbf{b} \cdot \mathbf{c})(\mathbf{a} \cdot \mathbf{c}) - (\mathbf{a} \cdot \mathbf{c})(\mathbf{b} \cdot \mathbf{c}) = 0. \quad ❏$$

Geometric Interpretation of the Dot Product

We begin with a triangle with sides a, b, c. (Figure 12.3.2.) If θ were $\frac{1}{2}\pi$, the Pythagorean Theorem would tell us that $c^2 = a^2 + b^2$. The Law of Cosines,

$$c^2 = a^2 + b^2 - 2ab \cos \theta,$$

is a generalization of the Pythagorean Theorem.

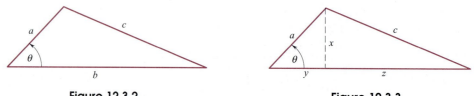

Figure 12.3.2   **Figure 12.3.3**

To derive the Law of Cosines we drop a perpendicular to side b. (Figure 12.3.3.) We then have

$$c^2 = z^2 + x^2 = (b - y)^2 + x^2 = b^2 - 2by + y^2 + x^2.$$

From the figure $y^2 + x^2 = a^2$ and $y = a \cos \theta$. Therefore

$$c^2 = a^2 + b^2 - 2ab \cos \theta,$$

as asserted. (What if the angle θ is obtuse? We leave that case to you.) ❏

Now back to dot products. If neither $\mathbf{a}$ nor $\mathbf{b}$ is zero, we can interpret $\mathbf{a} \cdot \mathbf{b}$ from the triangle of Figure 12.3.4. The lengths of the sides are $\|\mathbf{a}\|, \|\mathbf{b}\|, \|\mathbf{a} - \mathbf{b}\|$. By the law of cosines

$$\|\mathbf{a} - \mathbf{b}\|^2 = \|\mathbf{a}\|^2 + \|\mathbf{b}\|^2 - 2\|\mathbf{a}\| \|\mathbf{b}\| \cos \theta.$$

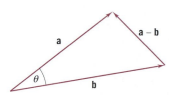

Figure 12.3.4

This gives

$$2\|\mathbf{a}\| \|\mathbf{b}\| \cos \theta = \|\mathbf{a}\|^2 + \|\mathbf{b}\|^2 - \|\mathbf{a} - \mathbf{b}\|^2 = a_1^2 + a_2^2 + a_3^2 + b_1^2 + b_2^2 + b_3^2$$
$$- (a_1 - b_1)^2 - (a_2 - b_2)^2 - (a_3 - b_3)^2 = 2(a_1b_1 + a_2b_2 + a_3b_3) = 2(\mathbf{a} \cdot \mathbf{b}),$$

and thus

(12.3.7)

$$\mathbf{a} \cdot \mathbf{b} = \|\mathbf{a}\| \|\mathbf{b}\| \cos \theta.$$

(By convention, θ, the angle between $\mathbf{a}$ and $\mathbf{b}$, is measured in radians and taken from 0 to π, inclusive.)

From (12.3.7) you can see that the dot product of two vectors depends on the norms of the vectors and on the angle between them. For vectors of a given norm, the dot product measures the extent to which the vectors agree in direction. As the difference in direction increases, the dot product decreases:

If **a** and **b** have the same direction, then $\theta = 0$ and

$$\mathbf{a} \cdot \mathbf{b} = \|\mathbf{a}\| \, \|\mathbf{b}\|; \qquad\qquad (\cos 0 = 1)$$

this is the largest possible value for $\mathbf{a} \cdot \mathbf{b}$.

If **a** and **b** meet at right angles, then $\theta = \frac{1}{2}\pi$ and

$$\mathbf{a} \cdot \mathbf{b} = 0. \qquad\qquad (\cos \tfrac{1}{2}\pi = 0)$$

If **a** and **b** have opposite directions, then $\theta = \pi$ and

$$\mathbf{a} \cdot \mathbf{b} = -\|\mathbf{a}\| \, \|\mathbf{b}\|; \qquad\qquad (\cos \pi = -1)$$

this is the least possible value for $\mathbf{a} \cdot \mathbf{b}$

Two vectors are said to be *perpendicular* iff they lie at right angles or one of the vectors is the zero vector; in other words, two vectors are said to be perpendicular iff their dot product is zero.† In symbols

(12.3.8)

$$\boxed{\quad \mathbf{a} \perp \mathbf{b} \quad\quad \text{iff} \quad\quad \mathbf{a} \cdot \mathbf{b} = 0. \quad}$$

(Figure 12.3.5)

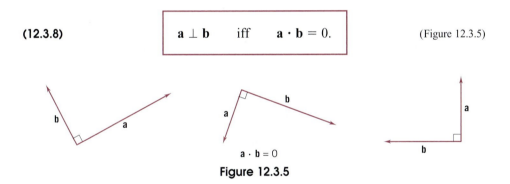

Figure 12.3.5

The unit coordinate vectors are obviously mutually perpendicular:

$$\mathbf{i} \cdot \mathbf{j} = 0, \quad\quad \mathbf{i} \cdot \mathbf{k} = 0, \quad\quad \mathbf{j} \cdot \mathbf{k} = 0.$$

Example 3 Verify that the vectors

$$\mathbf{a} = 2\mathbf{i} + \mathbf{j} + \mathbf{k} \quad\quad \text{and} \quad\quad \mathbf{b} = \mathbf{i} + \mathbf{j} - 3\mathbf{k}$$

are perpendicular.

SOLUTION

$$\mathbf{a} \cdot \mathbf{b} = (2)(1) + (1)(1) + (1)(-3) = 2 + 1 - 3 = 0. \quad \square$$

Example 4 Find the value of α for which

$$(3\mathbf{i} - \alpha\mathbf{j} + \mathbf{k}) \perp (\mathbf{i} + 2\mathbf{j}).$$

---

† This makes the zero vector both parallel and perpendicular to every vector. There is, however, no contradiction since we do not apply the notion of angle to the zero vector.

SOLUTION For the two vectors to be perpendicular their dot product must be zero. Since

$$(3\mathbf{i} - \alpha\mathbf{j} + \mathbf{k}) \cdot (\mathbf{i} + 2\mathbf{j}) = (3)(1) + (-\alpha)2 + (1)(0) = 3 - 2\alpha,$$

α must be $\frac{3}{2}$. ❑

With **a** and **b** both different from zero, we can divide the equation

$$\mathbf{a} \cdot \mathbf{b} = \|\mathbf{a}\| \|\mathbf{b}\| \cos \theta$$

by $\|\mathbf{a}\| \|\mathbf{b}\|$ to obtain

$$\cos \theta = \frac{\mathbf{a} \cdot \mathbf{b}}{\|\mathbf{a}\| \|\mathbf{b}\|} = \left(\frac{\mathbf{a}}{\|\mathbf{a}\|} \cdot \frac{\mathbf{b}}{\|\mathbf{b}\|} \right).$$

Writing

$$\mathbf{u_a} = \frac{\mathbf{a}}{\|\mathbf{a}\|} \quad \text{and} \quad \mathbf{u_b} = \frac{\mathbf{b}}{\|\mathbf{b}\|},$$

we have

(12.3.9)

$$\cos \theta = \mathbf{u_a} \cdot \mathbf{u_b}.$$

The cosine of the angle between two vectors is the dot product of the corresponding unit vectors.

Example 5 Calculate the angle between

$$\mathbf{a} = 2\mathbf{i} + 3\mathbf{j} + 2\mathbf{k} \quad \text{and} \quad \mathbf{b} = \mathbf{i} + 2\mathbf{j} - \mathbf{k}.$$

SOLUTION

$$\mathbf{u_a} = \frac{2\mathbf{i} + 3\mathbf{j} + 2\mathbf{k}}{\|2\mathbf{i} + 3\mathbf{j} + 2\mathbf{k}\|} = \frac{1}{\sqrt{17}} (2\mathbf{i} + 3\mathbf{j} + 2\mathbf{k}),$$

$$\mathbf{u_b} = \frac{\mathbf{i} + 2\mathbf{j} - \mathbf{k}}{\|\mathbf{i} + 2\mathbf{j} - \mathbf{k}\|} = \frac{1}{\sqrt{6}} (\mathbf{i} + 2\mathbf{j} - \mathbf{k}).$$

Therefore

$$\cos \theta = \mathbf{u_a} \cdot \mathbf{u_b} = \frac{1}{\sqrt{17}} \frac{1}{\sqrt{6}} [(2\mathbf{i} + 3\mathbf{j} + 2\mathbf{k}) \cdot (\mathbf{i} + 2\mathbf{j} - \mathbf{k})].$$

Since

$$(2\mathbf{i} + 3\mathbf{j} + 2\mathbf{k}) \cdot (\mathbf{i} + 2\mathbf{j} - \mathbf{k}) = (2)(1) + (3)(2) + (2)(-1) = 6,$$

we have

$$\cos \theta = \frac{6}{\sqrt{17} \sqrt{6}} = \frac{1}{17} \sqrt{102} \cong \frac{10.1}{17} \cong 0.594.$$

Thus, $\theta \cong 0.935$ radians, which is about 54 degrees. ❑

Example 6 Show that, if **a** and **b** are both perpendicular to **c**, then every linear combination $\alpha\mathbf{a} + \beta\mathbf{b}$ is also perpendicular to **c**.

SOLUTION Suppose that **a** and **b** are both perpendicular to **c**. Then

$$\mathbf{a} \cdot \mathbf{c} = 0 \qquad \text{and} \qquad \mathbf{b} \cdot \mathbf{c} = 0.$$

It follows that

$$(\alpha\mathbf{a} + \beta\mathbf{b}) \cdot \mathbf{c} = \alpha\underbrace{(\mathbf{a} \cdot \mathbf{c})}_{0} + \beta\underbrace{(\mathbf{b} \cdot \mathbf{c})}_{0} = 0$$

and therefore $\alpha\mathbf{a} + \beta\mathbf{b}$ is perpendicular to **c**. ❑

Projections and Components

If $\mathbf{b} \neq \mathbf{0}$, then every vector **a** can be written in a unique manner as the sum of a vector $\mathbf{a}_{\parallel}$ parallel to **b** and a vector $\mathbf{a}_{\perp}$ perpendicular to **b**:

$$\mathbf{a} = \mathbf{a}_{\parallel} + \mathbf{a}_{\perp}. \hspace{3cm} \text{(Exercise 39)}$$

The idea is illustrated in Figure 12.3.6. If **a** is parallel to **b**, then $\mathbf{a}_{\parallel} = \mathbf{a}$ and $\mathbf{a}_{\perp} = \mathbf{0}$. If **a** is perpendicular to **b**, then $\mathbf{a}_{\parallel} = \mathbf{0}$ and $\mathbf{a}_{\perp} = \mathbf{a}$.

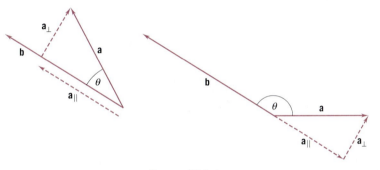

Figure 12.3.6

The vector $\mathbf{a}_{\parallel}$ is called *the projection of* **a** *on* **b** and is denoted by $\text{proj}_\mathbf{b}\mathbf{a}$. Since $\text{proj}_\mathbf{b}\mathbf{a}$ is parallel to **b**, it is a scalar multiple of the unit vector in the direction of **b**:

$$\text{proj}_\mathbf{b}\mathbf{a} = \lambda\mathbf{u}_\mathbf{b}.$$

The scalar λ is called *the component of* **a** *in the direction of* **b** (or more briefly, the **b**-*component* of **a**) and is denoted by $\text{comp}_\mathbf{b}\mathbf{a}$. In symbols,

(12.3.10)

$$\boxed{\text{proj}_\mathbf{b}\mathbf{a} = (\text{comp}_\mathbf{b}\mathbf{a})\mathbf{u}_\mathbf{b}.}$$

The component of **a** in the direction of **b** measures the "advance" of **a** in the direction of **b**. In Figure 12.3.6 we used θ to indicate the angle between **a** and **b**. If $0 \leq \theta < \frac{1}{2}\pi$, the projection and **b** have the same direction and the component is positive. If $\theta = \frac{1}{2}\pi$, the projection is **0** and the component is 0. If $\frac{1}{2}\pi < \theta \leq \pi$, the projection and **b** have opposite directions and, consequently, the component is negative.

Projections and components are closely related to dot products: if $\mathbf{b} \neq \mathbf{0}$, then

(12.3.11)

$$\text{proj}_b\mathbf{a} = (\mathbf{a} \cdot \mathbf{u_b})\mathbf{u_b} \qquad \text{and} \qquad \text{comp}_b\mathbf{a} = \mathbf{a} \cdot \mathbf{u_b}.$$

PROOF The second assertion follows trivially from the first. We will prove the first. We begin with the identity

$$\mathbf{a} = (\mathbf{a} \cdot \mathbf{u_b})\mathbf{u_b} + [\mathbf{a} - (\mathbf{a} \cdot \mathbf{u_b})\mathbf{u_b}].$$

Since the first vector $(\mathbf{a} \cdot \mathbf{u_b})\mathbf{u_b}$ is a scalar multiple of $\mathbf{b}$, it is parallel to $\mathbf{b}$. All we have to show now is that the second vector is perpendicular to $\mathbf{b}$. We do this by showing that its dot product with $\mathbf{u_b}$ is zero:

$$[\mathbf{a} - (\mathbf{a} \cdot \mathbf{u_b})\mathbf{u_b}] \cdot \mathbf{u_b} = (\mathbf{a} \cdot \mathbf{u_b}) - (\mathbf{a} \cdot \mathbf{u_b})(\mathbf{u_b} \cdot \mathbf{u_b}) = 0. \quad \square$$

$$\mathbf{u_b} \cdot \mathbf{u_b} = \|\mathbf{u_b}\|^2 = 1$$

Example 7 Find $\text{comp}_b\mathbf{a}$ and $\text{proj}_b\mathbf{a}$ given that

$$\mathbf{a} = -2\mathbf{i} + \mathbf{j} + \mathbf{k} \qquad \text{and} \qquad \mathbf{b} = 4\mathbf{i} - 3\mathbf{j} + \mathbf{k}.$$

SOLUTION Since

$$\|\mathbf{b}\| = \sqrt{(4)^2 + (-3)^2 + (1)^2} = \sqrt{26},$$

we have

$$\mathbf{u_b} = \frac{\mathbf{b}}{\|\mathbf{b}\|} = \frac{1}{\sqrt{26}}(4\mathbf{i} - 3\mathbf{j} + \mathbf{k}).$$

Thus

$$\text{comp}_b\mathbf{a} = \mathbf{a} \cdot \mathbf{u_b} = (-2\mathbf{i} + \mathbf{j} + \mathbf{k}) \cdot \frac{1}{\sqrt{26}}(4\mathbf{i} - 3\mathbf{j} + \mathbf{k})$$

$$= \frac{1}{\sqrt{26}}[(-2)(4) + (1)(-3) + (1)(1)] = -\frac{10}{\sqrt{26}} = -\frac{5}{13}\sqrt{26}$$

and

$$\text{proj}_b\mathbf{a} = (\text{comp}_b\mathbf{a})\,\mathbf{u_b} = -\frac{5}{13}(4\mathbf{i} - 3\mathbf{j} + \mathbf{k}). \quad \square$$

The following characterization of $\mathbf{a} \cdot \mathbf{b}$ is frequently used in physical applications:

(12.3.12)

$$\text{if } \mathbf{b} \neq \mathbf{0}, \quad \mathbf{a} \cdot \mathbf{b} = (\text{comp}_b\mathbf{a})\|\mathbf{b}\|.$$

PROOF

$$\mathbf{a} \cdot \mathbf{b} = \left(\mathbf{a} \cdot \frac{\mathbf{b}}{\|\mathbf{b}\|}\right)\|\mathbf{b}\| = (\mathbf{a} \cdot \mathbf{u_b})\|\mathbf{b}\| = (\text{comp}_b\mathbf{a})\|\mathbf{b}\|. \quad \square$$

For an arbitrary vector $\mathbf{a} = a_1\mathbf{i} + a_2\mathbf{j} + a_3\mathbf{k}$ we have

$$\text{comp}_{\mathbf{i}}\mathbf{a} = \mathbf{a} \cdot \mathbf{i} = a_1, \qquad \text{comp}_{\mathbf{j}}\mathbf{a} = \mathbf{a} \cdot \mathbf{j} = a_2, \qquad \text{comp}_{\mathbf{k}}\mathbf{a} = \mathbf{a} \cdot \mathbf{k} = a_3.$$

This agrees with our previous use of the term "component" (Section 12.2) and gives the identity

(12.3.13)
$$\mathbf{a} = (\mathbf{a} \cdot \mathbf{i})\mathbf{i} + (\mathbf{a} \cdot \mathbf{j})\mathbf{j} + (\mathbf{a} \cdot \mathbf{k})\mathbf{k}.$$

Direction Angles, Direction Cosines

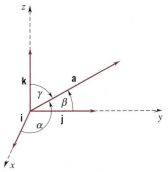

Figure 12.3.7

In Figure 12.3.7 we have portrayed a nonzero vector $\mathbf{a}$. The angles α, β, γ that the vector makes with the unit coordinate vectors are called the *direction angles* of $\mathbf{a}$, and $\cos \alpha$, $\cos \beta$, $\cos \gamma$ are called the *direction cosines*. By (12.3.7)

$$\mathbf{a} \cdot \mathbf{i} = \|\mathbf{a}\| \cos \alpha, \qquad \mathbf{a} \cdot \mathbf{j} = \|\mathbf{a}\| \cos \beta, \qquad \mathbf{a} \cdot \mathbf{k} = \|\mathbf{a}\| \cos \gamma.$$

Thus by (12.3.13)

(12.3.14)
$$\mathbf{a} = \|\mathbf{a}\|(\cos \alpha\, \mathbf{i} + \cos \beta\, \mathbf{j} + \cos \gamma\, \mathbf{k}).$$

Taking the norm of both sides, we have

$$\|\mathbf{a}\| = \|\mathbf{a}\|\sqrt{\cos^2\alpha + \cos^2\beta + \cos^2\gamma}$$

and therefore

(12.3.15)
$$\cos^2\alpha + \cos^2\beta + \cos^2\gamma = 1.$$

The sum of the squares of the direction cosines is always 1.
For a unit vector $\mathbf{u}$, Equation 12.3.14 takes the form

(12.3.16)
$$\mathbf{u} = \cos \alpha\, \mathbf{i} + \cos \beta\, \mathbf{j} + \cos \gamma\, \mathbf{k}.$$

Thus, for a unit vector, the $\mathbf{i}$, $\mathbf{j}$, $\mathbf{k}$ components are simply the direction cosines.

Example 8 Find the unit vector with direction angles

$$\alpha = \tfrac{1}{4}\pi, \qquad \beta = \tfrac{2}{3}\pi, \qquad \gamma = \tfrac{1}{3}\pi.$$

What is the vector of norm 4 with these same direction angles?

SOLUTION The unit vector with these direction angles is

$$\cos \tfrac{1}{4}\pi\, \mathbf{i} + \cos \tfrac{2}{3}\pi\, \mathbf{j} + \cos \tfrac{1}{3}\pi\, \mathbf{k} = \tfrac{1}{2}\sqrt{2}\, \mathbf{i} - \tfrac{1}{2}\mathbf{j} + \tfrac{1}{2}\mathbf{k}.$$

The vector of norm 4 with these direction angles is

$$4(\tfrac{1}{2}\sqrt{2}\, \mathbf{i} - \tfrac{1}{2}\mathbf{j} + \tfrac{1}{2}\mathbf{k}) = 2\sqrt{2}\, \mathbf{i} - 2\mathbf{j} + 2\mathbf{k}. \qquad \square$$

Example 9 Find the direction cosines of

$$\mathbf{a} = 2\mathbf{i} + 3\mathbf{j} - 6\mathbf{k}.$$

What are the direction angles?

SOLUTION Here

$$\|\mathbf{a}\| = \sqrt{2^2 + 3^2 + (-6)^2} = 7$$

so that

$$2 = 7\cos\alpha, \qquad 3 = 7\cos\beta, \qquad -6 = 7\cos\gamma$$

and

$$\cos\alpha = \tfrac{2}{7}, \qquad \cos\beta = \tfrac{3}{7}, \qquad \cos\gamma = -\tfrac{6}{7}.$$

Since angles between vectors are measured in radians and taken from 0 to π,

$$\alpha = \cos^{-1}\tfrac{2}{7} \cong 1.28 \text{ radians}, \qquad \beta = \cos^{-1}\tfrac{3}{7} \cong 1.13 \text{ radians},$$

$$\gamma = \cos^{-1}(-\tfrac{6}{7}) \cong 2.60 \text{ radians}. \quad ❏$$

Proving the Triangle Inequality

By taking absolute values, the relation

$$\mathbf{a} \cdot \mathbf{b} = \|\mathbf{a}\|\,\|\mathbf{b}\|\cos\theta$$

gives

(12.3.17)

$$\boxed{|\mathbf{a} \cdot \mathbf{b}| \le \|\mathbf{a}\|\,\|\mathbf{b}\|.}$$

$(|\cos\theta| \le 1)$

This inequality, called *Schwarz's inequality*, enables us to give a simple proof of the triangle inequality

$$\|\mathbf{a} + \mathbf{b}\| \le \|\mathbf{a}\| + \|\mathbf{b}\|.$$

PROOF

$$\begin{aligned}
\|\mathbf{a} + \mathbf{b}\|^2 &= (\mathbf{a} + \mathbf{b}) \cdot (\mathbf{a} + \mathbf{b}) \\
&= (\mathbf{a} \cdot \mathbf{a}) + (\mathbf{b} \cdot \mathbf{a}) + (\mathbf{a} \cdot \mathbf{b}) + (\mathbf{b} \cdot \mathbf{b}) \\
&= \|\mathbf{a}\|^2 + 2(\mathbf{a} \cdot \mathbf{b}) + \|\mathbf{b}\|^2 \\
&\le \|\mathbf{a}\|^2 + 2|\mathbf{a} \cdot \mathbf{b}| + \|\mathbf{b}\|^2 \qquad (\mathbf{a} \cdot \mathbf{b} \le |\mathbf{a} \cdot \mathbf{b}|)\\
&\le \|\mathbf{a}\|^2 + 2\|\mathbf{a}\|\,\|\mathbf{b}\| + \|\mathbf{b}\|^2 = (\|\mathbf{a}\| + \|\mathbf{b}\|)^2
\end{aligned}$$

by Schwarz's inequality ⬑

and therefore

$$\|\mathbf{a} + \mathbf{b}\| \le \|\mathbf{a}\| + \|\mathbf{b}\|. \quad ❏$$

It is worth remarking that Schwarz's inequality, and hence the triangle inequality, can be proved purely algebraically. (See Exercise 50.)

EXERCISES 12.3

In Exercises 1–6, find $\mathbf{a} \cdot \mathbf{b}$.

1. $\mathbf{a} = (2, -3, 1)$, $\mathbf{b} = (-2, 0, 3)$.

2. $\mathbf{a} = (4, 2, -1)$, $\mathbf{b} = (-2, 2, 1)$.

3. $\mathbf{a} = (2, -4)$, $\mathbf{b} = (1, \frac{1}{2})$.

4. $\mathbf{a} = (-2, 5)$, $\mathbf{b} = (3, 1)$.

5. $\mathbf{a} = 2\mathbf{i} + \mathbf{j} - 2\mathbf{k}$, $\mathbf{b} = \mathbf{i} + \mathbf{j} + 2\mathbf{k}$.

6. $\mathbf{a} = 2\mathbf{i} + 3\mathbf{j} + \mathbf{k}$, $\mathbf{b} = \mathbf{i} + 4\mathbf{j}$.

In Exercises 7–10, simplify the given expression.

7. $(3\mathbf{a} \cdot \mathbf{b}) - (\mathbf{a} \cdot 2\mathbf{b})$.

8. $\mathbf{a} \cdot (\mathbf{a} - \mathbf{b}) + \mathbf{b} \cdot (\mathbf{b} + \mathbf{a})$.

9. $(\mathbf{a} - \mathbf{b}) \cdot \mathbf{c} + \mathbf{b} \cdot (\mathbf{c} + \mathbf{a})$.

10. $\mathbf{a} \cdot (\mathbf{a} + 2\mathbf{c}) + (2\mathbf{b} - \mathbf{a}) \cdot (\mathbf{a} + 2\mathbf{c}) - 2\mathbf{b} \cdot (\mathbf{a} + 2\mathbf{c})$.

11. Taking

$$\mathbf{a} = 2\mathbf{i} + \mathbf{j}, \qquad \mathbf{b} = 3\mathbf{i} - \mathbf{j} + 2\mathbf{k}, \qquad \mathbf{c} = 4\mathbf{i} + 3\mathbf{k}$$

calculate

(a) the three dot products $\mathbf{a} \cdot \mathbf{b}$, $\mathbf{a} \cdot \mathbf{c}$, $\mathbf{b} \cdot \mathbf{c}$;

(b) the cosines of the angles between these vectors;

(c) the component of $\mathbf{a}$ (i) in the $\mathbf{b}$ direction, (ii) in the $\mathbf{c}$ direction;

(d) the projection of $\mathbf{a}$ (i) in the $\mathbf{b}$ direction, (ii) in the $\mathbf{c}$ direction.

12. Exercise 11 with $\mathbf{a} = \mathbf{j} + 3\mathbf{k}$, $\mathbf{b} = 2\mathbf{i} - \mathbf{j} + 2\mathbf{k}$, $\mathbf{c} = 3\mathbf{i} - \mathbf{k}$.

13. Find the unit vector with direction angles $\frac{1}{3}\pi$, $\frac{1}{4}\pi$, $\frac{2}{3}\pi$.

14. Find the vector of norm 2 with direction angles $\frac{1}{4}\pi$, $\frac{1}{4}\pi$, $\frac{1}{2}\pi$.

15. Find the angle between the vectors $3\mathbf{i} - \mathbf{j} - 2\mathbf{k}$ and $\mathbf{i} + 2\mathbf{j} - 2\mathbf{k}$.

16. Find the angle between the vectors $2\mathbf{i} - 3\mathbf{j} + \mathbf{k}$ and $-3\mathbf{i} + \mathbf{j} + 9\mathbf{k}$.

17. Find the direction angles of the vector $\mathbf{i} - \mathbf{j} + \sqrt{2}\mathbf{k}$.

18. Find the direction angles of the vector $\mathbf{i} - \sqrt{3}\mathbf{k}$.

▶C In Exercises 19–22, estimate the angle between the given vectors. Express your answers in radians rounded to the nearest hundredth of a radian, and in degrees to the nearest tenth of a degree.

19. $\mathbf{a} = (3, 1, -1)$, $\mathbf{b} = (-2, 1, 4)$.

20. $\mathbf{a} = (-2, -3, 0)$, $\mathbf{b} = (-6, 0, 4)$.

21. $\mathbf{a} = -\mathbf{i} + 2\mathbf{k}$, $\mathbf{b} = 3\mathbf{i} + 4\mathbf{j} - 5\mathbf{k}$.

22. $\mathbf{a} = -3\mathbf{i} + \mathbf{j} - \mathbf{k}$, $\mathbf{b} = \mathbf{i} - \mathbf{j}$.

▶C In Exercises 23–26, find the direction cosines and direction angles of the given vector. Express the angles in degrees rounded to the nearest tenth of a degree.

23. $\mathbf{a} = (1, 2, 2)$.

24. $\mathbf{a} = (2, 6, -1)$.

25. $\mathbf{a} = 3\mathbf{i} + 12\mathbf{j} + 4\mathbf{k}$.

26. $\mathbf{a} = 3\mathbf{i} + 5\mathbf{j} - 4\mathbf{k}$.

27. Show that

(a) $\text{proj}_{\mathbf{b}}\alpha\mathbf{a} = \alpha \, \text{proj}_{\mathbf{b}}\mathbf{a}$ for all real α, and

(b) $\text{proj}_{\mathbf{b}}(\mathbf{a} + \mathbf{c}) = \text{proj}_{\mathbf{b}}\mathbf{a} + \text{proj}_{\mathbf{b}}\mathbf{c}$.

28. Show that

(a) $\text{proj}_{\beta\mathbf{b}}\mathbf{a} = \text{proj}_{\mathbf{b}}\mathbf{a}$ for all real $\beta \neq 0$, but

(b) $\text{comp}_{\beta\mathbf{b}}\mathbf{a} = \begin{cases} \text{comp}_{\mathbf{b}}\mathbf{a}, & \text{for } \beta > 0 \\ -\text{comp}_{\mathbf{b}}\mathbf{a}, & \text{for } \beta < 0. \end{cases}$

29. (a) (*Important*) Let $\mathbf{a} \neq \mathbf{0}$. Show that $\mathbf{a} \cdot \mathbf{b} = \mathbf{a} \cdot \mathbf{c}$ does not necessarily imply that $\mathbf{b} = \mathbf{c}$, but only that $\mathbf{b}$ and $\mathbf{c}$ have the same projection on $\mathbf{a}$. Draw a figure illustrating this for $\mathbf{b}$ and $\mathbf{c}$ different from $\mathbf{0}$.

(b) Show that if $\mathbf{u} \cdot \mathbf{b} = \mathbf{u} \cdot \mathbf{c}$ for all unit vectors $\mathbf{u}$, then $\mathbf{b} = \mathbf{c}$. HINT: Consider the unit coordinate vectors.

30. What can you conclude about $\mathbf{a}$ and $\mathbf{b}$ given that

(a) $\|\mathbf{a}\|^2 + \|\mathbf{b}\|^2 = \|\mathbf{a} + \mathbf{b}\|^2$?

(b) $\|\mathbf{a}\|^2 + \|\mathbf{b}\|^2 = \|\mathbf{a} - \mathbf{b}\|^2$?

HINT: Draw figures.

31. (a) Show that

$$4(\mathbf{a} \cdot \mathbf{b}) = \|\mathbf{a} + \mathbf{b}\|^2 - \|\mathbf{a} - \mathbf{b}\|^2.$$

(b) Use (a) to verify that

$$\mathbf{a} \perp \mathbf{b} \qquad \text{iff} \qquad \|\mathbf{a} + \mathbf{b}\| = \|\mathbf{a} - \mathbf{b}\|.$$

(c) Show that, if $\mathbf{a}$ and $\mathbf{b}$ are nonzero vectors such that

$$(\mathbf{a} + \mathbf{b}) \perp (\mathbf{a} - \mathbf{b}) \quad \text{and} \quad \|\mathbf{a} + \mathbf{b}\| = \|\mathbf{a} - \mathbf{b}\|,$$

then the parallelogram generated by $\mathbf{a}$ and $\mathbf{b}$ is a square.

32. Under what conditions does $|\mathbf{a} \cdot \mathbf{b}| = \|\mathbf{a}\| \, \|\mathbf{b}\|$?

33. Given two vectors $\mathbf{a}$ and $\mathbf{b}$, prove the *parallelogram law*:

$$\|\mathbf{a} + \mathbf{b}\|^2 + \|\mathbf{a} - \mathbf{b}\|^2 = 2\|\mathbf{a}\|^2 + 2\|\mathbf{b}\|^2.$$

Geometric interpretation: The sum of the squares of the lengths of the diagonals of a parallelogram determined by $\mathbf{a}$ and $\mathbf{b}$ equals the sum of the squares of the lengths of the four sides. See the following figure.

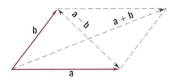

34. Suppose that the direction angles of a vector are equal. What are the angles?

35. What are the direction angles of $-\mathbf{a}$ if the direction angles of $\mathbf{a}$ are α, β, γ?

36. Let θ be the angle between $\mathbf{a}$ and $\mathbf{b}$, and let β be a negative number. Use the dot product to compute the angle between $\mathbf{a}$ and $\beta\mathbf{b}$ in terms of θ. Draw a figure to verify your answer geometrically.

37. Show that, if $\mathbf{a} \perp \mathbf{b}$ and $\mathbf{a} \perp \mathbf{c}$, then $\mathbf{a} \perp (\alpha\mathbf{b} + \beta\mathbf{c})$ for all real α, β.

38. Show that, if $\mathbf{a} \parallel \mathbf{b}$ and $\mathbf{a} \parallel \mathbf{c}$, then $\mathbf{a} \parallel (\alpha\mathbf{b} + \beta\mathbf{c})$ for all real α, β. ($\mathbf{a}\parallel\mathbf{b}$ means $\mathbf{a}$ is parallel to $\mathbf{b}$.)

39. (*Important*) Show that, if $\mathbf{b}$ is a nonzero vector, then every vector $\mathbf{a}$ can be written in a unique manner as the sum of a vector $\mathbf{a}_\parallel$ parallel to $\mathbf{b}$ and a vector $\mathbf{a}_\perp$ perpendicular to $\mathbf{b}$:

$$\mathbf{a} = \mathbf{a}_\parallel + \mathbf{a}_\perp.$$

40. Find all the numbers x for which
$$(x\mathbf{i} + 11\mathbf{j} - 3\mathbf{k}) \perp (2x\mathbf{i} - x\mathbf{j} - 5\mathbf{k}).$$

41. Find all the numbers x for which the angle between $\mathbf{c} = x\mathbf{i} + \mathbf{j} + \mathbf{k}$ and $\mathbf{d} = \mathbf{i} + x\mathbf{j} + \mathbf{k}$ is $\frac{1}{3}\pi$.

42. Set $\mathbf{a} = \mathbf{i} + x\mathbf{j} + \mathbf{k}$ and $\mathbf{b} = 2\mathbf{i} - \mathbf{j} + y\mathbf{k}$. Compute all values of x and y for which $\mathbf{a} \perp \mathbf{b}$ and $\|\mathbf{a}\| = \|\mathbf{b}\|$.

43. Find the angle between the diagonal of a cube and one of its edges.

44. Find the angle between the diagonal of a cube and the diagonal of one of its faces.

45. (a) Show that $\frac{1}{4}\pi, \frac{1}{6}\pi, \frac{2}{3}\pi$ cannot be the direction angles of a vector.
    (b) Show that, if $\mathbf{a} = a_1\mathbf{i} + a_2\mathbf{j} + a_3\mathbf{k}$ has direction angles $\alpha, \frac{1}{4}\pi, \frac{1}{4}\pi$, then $a_1 = 0$.

46. Let $r = f(\theta)$ be the polar equation of a curve in the plane and let
$$\mathbf{u}_r = (\cos\theta)\mathbf{i} + (\sin\theta)\mathbf{j} \qquad \mathbf{u}_\theta = (-\sin\theta)\mathbf{i} + (\cos\theta)\mathbf{j}$$

    (a) Show that $\mathbf{u}_r$ and $\mathbf{u}_\theta$ are unit vectors and that they are perpendicular.
    (b) Let $P(r, \theta)$ be a point on the curve. Show that $\mathbf{u}_r$ has the same direction as the vector $\overrightarrow{OP}$ and that the direction of $\mathbf{u}_\theta$ is 90° counterclockwise from $\mathbf{u}_r$.

47. Find the unit vectors $\mathbf{u}$ that are perpendicular to both $\mathbf{i} + 2\mathbf{j} + \mathbf{k}$ and $3\mathbf{i} - 4\mathbf{j} + 2\mathbf{k}$.

48. Find two mutually perpendicular unit vectors that are perpendicular to $2\mathbf{i} + 3\mathbf{j}$.

49. Two points on a sphere are called *antipodal* iff they are opposite endpoints of a diameter. Show that, if P_1 and P_2 are antipodal points on a sphere and Q is any other point on the sphere, then $\overrightarrow{P_1Q} \perp \overrightarrow{P_2Q}$.

50. Give an algebraic proof of Schwarz's inequality $|\mathbf{a} \cdot \mathbf{b}| \le \|\mathbf{a}\| \, \|\mathbf{b}\|$. HINT: If $\mathbf{b} = \mathbf{0}$, the inequality is trivial, so assume $\mathbf{b} \ne \mathbf{0}$. Note that for any number λ we have $\|\mathbf{a} - \lambda\mathbf{b}\|^2 \ge 0$. First expand this inequality using the fact that $\|\mathbf{a} - \lambda\mathbf{b}\|^2 = (\mathbf{a} - \lambda\mathbf{b}) \cdot (\mathbf{a} - \lambda\mathbf{b})$. After collecting terms, make the special choice $\lambda = (\mathbf{a} \cdot \mathbf{b})/\|\mathbf{b}\|^2$ and see what happens.

Work

In Section 6.5, we defined the *work* W done by a force F in moving an object along the x-axis from $x = a$ to $x = b$ by

$$W = F(b - a).$$

In words, *work equals force times displacement.* This definition can be extended as follows: If a constant force $\mathbf{F}$ is applied to an object moving in a straight line throughout a displacement $\mathbf{r}$ (see the following figure), then the work done by $\mathbf{F}$ is given by

$$W = (\text{comp}_\mathbf{r}\, \mathbf{F})\, \|\mathbf{r}\|.$$

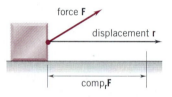

If $\mathbf{F}$ is measured in pounds and distance in feet, then the work is measured in *foot-pounds*; if $\mathbf{F}$ is measured in newtons and distance in meters, then the work is measured in *newton-meters* or *joules*.

51. Let the force $\mathbf{F}$ be applied throughout a displacement $\mathbf{r}$.
    (a) Express the work done by $\mathbf{F}$ as a dot product.
    (b) What is the work done by $\mathbf{F}$ if $\mathbf{F} \perp \mathbf{r}$?
    (c) Show that the work done by $\mathbf{F} = \|\mathbf{F}\|\mathbf{i}$ applied throughout the displacement $(b - a)\mathbf{i}$ reduces to the definition given in Section 6.5.

52. Find the work done by the force $\mathbf{F} = 2\mathbf{i} + \mathbf{j} + 3\mathbf{k}$ in moving an object from the point $P(1, 2, 0)$ to the point $Q(3, 5, 2)$.

53. (a) A sled is pulled along level ground by exerting a force of 15 Newtons on a rope that makes an angle of 35° with the ground. Find the work done by the force in pulling the sled 50 meters. See Figure A.

Figure A   Figure B

(b) Suppose the same sled is pulled 50 meters up a hill that makes an angle of 15° with level ground. Find the work done by the force in this case. See Figure B.

▶ 54. A wooden crate is pulled along a level floor by a rope that makes an angle of 40° with the floor. If the force of friction (which acts in a direction opposite to the motion) between the carton and the floor is 50 pounds, what is the minimum force that must be applied to the rope to move the crate?

55. Two forces of the same magnitude, $\mathbf{F}_1$ and $\mathbf{F}_2$, are applied throughout a displacement $\mathbf{r}$ at angles θ_1 and θ_2, respectively. Compare the work done by $\mathbf{F}_1$ to that done by $\mathbf{F}_2$ if
 (a) $\theta_1 = -\theta_2$.
 (b) $\theta_1 = \pi/3$ and $\theta_2 = \pi/6$.

56. What is the total work done by a constant force $\mathbf{F}$ if the object to which it is applied moves around a triangle? Justify your answer.

■ 12.4 THE CROSS PRODUCT

Although all of the focus in the preceding sections was on three-dimensional vectors, we did remark on several occasions that corresponding definitions and properties also held for vectors in other dimensions. In this section we study the *cross product* of two vectors, and we emphasize at the outset that this product is defined for three-dimensional vectors only. Throughout this section the term "vector" will mean "three-dimensional vector."

Definition of the Cross Product

While the dot product $\mathbf{a} \cdot \mathbf{b}$ is a scalar (and as such is sometimes called the scalar product of $\mathbf{a}$ and $\mathbf{b}$), the cross product $\mathbf{a} \times \mathbf{b}$ is a vector (sometimes called the *vector product* of $\mathbf{a}$ and $\mathbf{b}$). What is the vector $\mathbf{a} \times \mathbf{b}$? We could directly write down a formula that gives the components of $\mathbf{a} \times \mathbf{b}$ in terms of the components of $\mathbf{a}$ and $\mathbf{b}$, but at this stage that would reveal little. Instead we will begin geometrically. We will define $\mathbf{a} \times \mathbf{b}$ by giving its direction and its magnitude.

The Direction of $\mathbf{a} \times \mathbf{b}$ If the vectors $\mathbf{a}$ and $\mathbf{b}$ are not parallel, they determine a plane. The vector $\mathbf{a} \times \mathbf{b}$ is perpendicular to this plane and is directed in such a way that (like $\mathbf{i}$, $\mathbf{j}$, $\mathbf{k}$) the vectors $\mathbf{a}$, $\mathbf{b}$, $\mathbf{a} \times \mathbf{b}$ form a right-handed triple. (Figure 12.4.1.) (If the index finger of the right hand points along $\mathbf{a}$ and the middle finger points along $\mathbf{b}$, then the thumb will point in the direction of $\mathbf{a} \times \mathbf{b}$. NOTE: In saying that $\mathbf{a}$, $\mathbf{b}$, $\mathbf{a} \times \mathbf{b}$ form a right-handed triple, we do not require the vectors $\mathbf{a}$ and $\mathbf{b}$ to be perpendicular (like $\mathbf{i}$ and $\mathbf{j}$); the general notion of a "right-handed triple" does not require the vectors involved to be perpendicular to each other.)

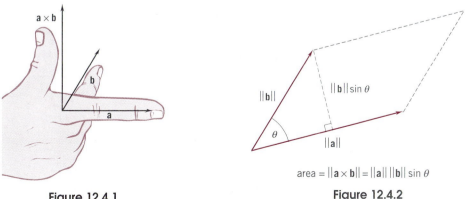

Figure 12.4.1

area $= \|\mathbf{a} \times \mathbf{b}\| = \|\mathbf{a}\| \, \|\mathbf{b}\| \sin \theta$

Figure 12.4.2

The Magnitude of a × b. If **a** and **b** are not parallel, they form the sides of a parallelogram. (Figure 12.4.2.) The magnitude of **a** × **b** is the area of this parallelogram: $\|\mathbf{a}\| \|\mathbf{b}\| \sin \theta$. (Recall that the area of a parallelogram with base b and height h is $A = bh$.)

One more point. What if **a** and **b** are parallel? Then there is no parallelogram and we define **a** × **b** = **0**.

We summarize all this below.

DEFINITION 12.4.1

If **a** and **b** are not parallel, then **a** × **b** is the vector with the following properties:

1. **a** × **b** is perpendicular to the plane of **a** and **b**.
2. **a**, **b**, **a** × **b** form a right-handed triple.
3. $\|\mathbf{a} \times \mathbf{b}\| = \|\mathbf{a}\| \|\mathbf{b}\| \sin \theta$, where θ is the angle between **a** and **b**, $0 < \theta < \pi$.

If **a** and **b** are parallel ($\theta = 0$ or $\theta = \pi$), then **a** × **b** = **0**.

Properties of Right-Handed Triples

I. Note first of all that if (**a**, **b**, **c**) form a right-handed triple, then (**b**, **c**, **a**) and (**c**, **a**, **b**) also form right-handed triples. (Figure 12.4.3.) To maintain right-handedness we don't have to keep the vectors in the same order, but we do have to keep them in the *same cyclic order:*

$$\mathbf{a} \rightarrow \mathbf{b} \rightarrow \mathbf{c}$$

Alter the cyclic order and you reverse the orientation.

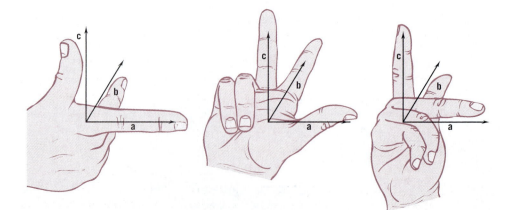

Figure 12.4.3

II. A triple $(\mathbf{a}, \mathbf{b}, \mathbf{c})$ is right-handed iff $\mathbf{c}$ and $\mathbf{a} \times \mathbf{b}$ lie on the same side of the plane determined by $\mathbf{a}$ and $\mathbf{b}$. (Figure 12.4.4.) This means that $(\mathbf{a}, \mathbf{b}, \mathbf{c})$ is right-handed iff $(\mathbf{a} \times \mathbf{b}) \cdot \mathbf{c} > 0$. (Explain)

III. It follows from II that, if $(\mathbf{a}, \mathbf{b}, \mathbf{c})$ is right-handed, then $(\mathbf{a}, \mathbf{b}, -\mathbf{c})$ is not right-handed. Similarly $(-\mathbf{a}, \mathbf{b}, \mathbf{c})$ and $(\mathbf{a}, -\mathbf{b}, \mathbf{c})$ are also not right-handed. However, multiplication by positive scalars does maintain right-handedness: if $(\mathbf{a}, \mathbf{b}, \mathbf{c})$ is right-handed and α, β, γ are positive, then $(\alpha\mathbf{a}, \beta\mathbf{b}, \gamma\mathbf{c})$ is also right-handed.

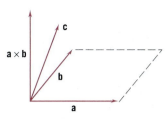

Figure 12.4.4

Properties of the Cross Product

The cross product is *anticommutative*:

(12.4.2)
$$\mathbf{b} \times \mathbf{a} = -(\mathbf{a} \times \mathbf{b}).$$

To see this, note that both vectors are perpendicular to the plane determined by $\mathbf{a}$ and $\mathbf{b}$ and both have the same norm. Thus $\mathbf{b} \times \mathbf{a} = \pm(\mathbf{a} \times \mathbf{b})$. That the minus sign holds and not the plus sign follows from observing that $\mathbf{b}, \mathbf{a}, \mathbf{b} \times \mathbf{a}$ is a right-handed triple and that $\mathbf{b}, \mathbf{a}, \mathbf{a} \times \mathbf{b}$ is not right-handed since $\mathbf{a}, \mathbf{b}, \mathbf{a} \times \mathbf{b}$ is right-handed. ❏

Scalars can be factored:

(12.4.3)
$$\alpha\mathbf{a} \times \beta\mathbf{b} = \alpha\beta(\mathbf{a} \times \mathbf{b}).$$

If α or β is zero, the result is obvious. We will assume that α and β are both nonzero. In this case the two vectors are perpendicular to $\mathbf{a}$, perpendicular to $\mathbf{b}$, and have the same norm. Thus $\alpha\mathbf{a} \times \beta\mathbf{b} = \pm\alpha\beta(\mathbf{a} \times \mathbf{b})$. That the positive sign holds comes from noting that $\alpha\mathbf{a}, \beta\mathbf{b}, \alpha\beta(\mathbf{a} \times \mathbf{b})$ is a right-handed triple. This is obvious if α and β are both positive. If not, two of the three coefficients α, β, $\alpha\beta$ are negative and the other is positive. In this case, the first minus sign reverses the orientation but the second one restores it. ❏

Finally, there are two distributive laws, the verification of which we postpone for a moment.

(12.4.4)

$$\mathbf{a} \times (\mathbf{b} + \mathbf{c}) = (\mathbf{a} \times \mathbf{b}) + (\mathbf{a} \times \mathbf{c}), \qquad (\mathbf{a} + \mathbf{b}) \times \mathbf{c} = (\mathbf{a} \times \mathbf{c}) + (\mathbf{b} \times \mathbf{c}).$$

The Scalar Triple Product

Earlier we saw that $\mathbf{a}, \mathbf{b}, \mathbf{c}$ is a right-handed triple iff $(\mathbf{a} \times \mathbf{b}) \cdot \mathbf{c} > 0$. The expression $(\mathbf{a} \times \mathbf{b}) \cdot \mathbf{c}$ is called a *scalar triple product*. The absolute value of this number (it is a number and not a vector) has geometric significance. To describe it we refer to Figure

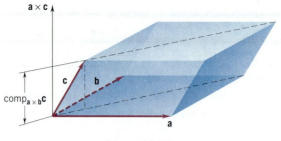

Figure 12.4.5

12.4.5. There you see a parallelepiped with edges **a**, **b**, **c**. The absolute value of the scalar triple product gives the volume of that parallelepiped:

(12.4.5)

$$V = |(\mathbf{a} \times \mathbf{b}) \cdot \mathbf{c}|.$$

PROOF The area of the base is $\|\mathbf{a} \times \mathbf{b}\|$. The height is $|\text{comp}_{\mathbf{a} \times \mathbf{b}} \mathbf{c}|$. Therefore

$$V = |\text{comp}_{\mathbf{a} \times \mathbf{b}} \mathbf{c}| \, \|\mathbf{a} \times \mathbf{b}\| = |(\mathbf{a} \times \mathbf{b}) \cdot \mathbf{c}|. \quad \square$$

$$\underset{\text{(12.3.12)}}{\uparrow}$$

Of course we could have formed the same parallelogram using a different base (for example, using the vectors **c** and **a**) with a correspondingly different height ($\text{comp}_{\mathbf{c} \times \mathbf{a}} \mathbf{b}$). Therefore

$$|(\mathbf{a} \times \mathbf{b}) \cdot \mathbf{c}| = |(\mathbf{c} \times \mathbf{a}) \cdot \mathbf{b}| = |(\mathbf{b} \times \mathbf{c}) \cdot \mathbf{a}|.$$

Since the **a**, **b**, **c** appear in the same cyclic order, the expressions inside the absolute value signs all have the same sign. (Property II of right-handed triples) Therefore

(12.4.6)

$$(\mathbf{a} \times \mathbf{b}) \cdot \mathbf{c} = (\mathbf{c} \times \mathbf{a}) \cdot \mathbf{b} = (\mathbf{b} \times \mathbf{c}) \cdot \mathbf{a}.$$

Verification of the Distributive Laws

We will verify the first distributive law,

$$\mathbf{a} \times (\mathbf{b} + \mathbf{c}) = (\mathbf{a} \times \mathbf{b}) + (\mathbf{a} \times \mathbf{c}).$$

The second follows readily from this one. The argument is left to you as an exercise.

Take an arbitrary vector **r** and form the dot product $[\mathbf{a} \times (\mathbf{b} + \mathbf{c})] \cdot \mathbf{r}$. We can then write

$$[\mathbf{a} \times (\mathbf{b} + \mathbf{c})] \cdot \mathbf{r} = (\mathbf{r} \times \mathbf{a}) \cdot (\mathbf{b} + \mathbf{c}) \qquad (12.4.6)$$

$$= [(\mathbf{r} \times \mathbf{a}) \cdot \mathbf{b}] + [(\mathbf{r} \times \mathbf{a}) \cdot \mathbf{c}] \qquad (12.3.6)$$

$$= [(\mathbf{a} \times \mathbf{b}) \cdot \mathbf{r}] + [(\mathbf{a} \times \mathbf{c}) \cdot \mathbf{r}] \qquad (12.4.6)$$

$$= [(\mathbf{a} \times \mathbf{b}) + (\mathbf{a} \times \mathbf{c})] \cdot \mathbf{r}. \qquad (12.3.6)$$

Since this holds true for all vectors **r**, it holds true for **i**, **j**, **k** and proves that

$$\mathbf{a} \times (\mathbf{b} + \mathbf{c}) = (\mathbf{a} \times \mathbf{b}) + (\mathbf{a} \times \mathbf{c}). \quad \square$$

The Components of a × b

You have learned a lot about cross products, but you still have not seen $\mathbf{a} \times \mathbf{b}$ expressed in terms of the components of $\mathbf{a}$ and $\mathbf{b}$. To derive the formula that does this, we need to observe one more fact, one which is obvious from the very definition of cross product:

(12.4.7)

$$\mathbf{i} \times \mathbf{j} = \mathbf{k}, \quad \mathbf{j} \times \mathbf{k} = \mathbf{i}, \quad \mathbf{k} \times \mathbf{i} = \mathbf{j}.$$

(Figure 12.4.6)

One way to remember these products is to arrange $\mathbf{i}$, $\mathbf{j}$, $\mathbf{k}$ in cyclic order, $\mathbf{i} \rightarrow \mathbf{j} \rightarrow \mathbf{k}$, and note that

[each coordinate unit vector] × [the next one] = [the third one].

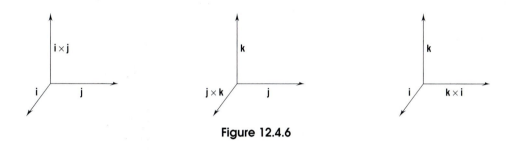

Figure 12.4.6

THEOREM 12.4.8

For vectors $\mathbf{a} = a_1\mathbf{i} + a_2\mathbf{j} + a_3\mathbf{k}$ and $\mathbf{b} = b_1\mathbf{i} + b_2\mathbf{j} + b_3\mathbf{k}$

$$\mathbf{a} \times \mathbf{b} = (a_2b_3 - a_3b_2)\mathbf{i} - (a_1b_3 - a_3b_1)\mathbf{j} + (a_1b_2 - a_2b_1)\mathbf{k}.$$

Determinants provide an especially convenient notational device for representing this formula for $\mathbf{a} \times \mathbf{b}$. If you are not already familiar with determinants, a brief treatment is given in Appendix A-2. Theorem 12.4.8 stated in terms of determinants is:

THEOREM 12.4.8′

For vectors $\mathbf{a} = a_1\mathbf{i} + a_2\mathbf{j} + a_3\mathbf{k}$ and $\mathbf{b} = b_1\mathbf{i} + b_2\mathbf{j} + b_3\mathbf{k}$

$$\mathbf{a} \times \mathbf{b} = \begin{vmatrix} \mathbf{i} & \mathbf{j} & \mathbf{k} \\ a_1 & a_2 & a_3 \\ b_1 & b_2 & b_3 \end{vmatrix} = \begin{vmatrix} a_2 & a_3 \\ b_2 & b_3 \end{vmatrix} \mathbf{i} - \begin{vmatrix} a_1 & a_3 \\ b_1 & b_3 \end{vmatrix} \mathbf{j} + \begin{vmatrix} a_1 & a_2 \\ b_1 & b_2 \end{vmatrix} \mathbf{k}.$$

(The 3×3 determinant with $\mathbf{i}$, $\mathbf{j}$, $\mathbf{k}$ in the top row is there only as a mnemonic device.)

PROOF The hard work has all been done. With what you know about cross products now, the proof is just a matter of simple algebra:

$$\mathbf{a} \times \mathbf{b} = (a_1\mathbf{i} + a_2\mathbf{j} + a_3\mathbf{k}) \times (b_1\mathbf{i} + b_2\mathbf{j} + b_3\mathbf{k})$$

$$= a_1b_2(\mathbf{i} \times \mathbf{j}) + a_1b_3(\mathbf{i} \times \mathbf{k}) + a_2b_1(\mathbf{j} \times \mathbf{i}) + a_2b_3(\mathbf{j} \times \mathbf{k}) + a_3b_1(\mathbf{k} \times \mathbf{i}) + a_3b_2(\mathbf{k} \times \mathbf{j})$$

$$\uparrow\!\!\rule{2em}{0pt} \mathbf{i} \times \mathbf{i} = \mathbf{j} \times \mathbf{j} = \mathbf{k} \times \mathbf{k} = 0$$

$$= a_1b_2\mathbf{k} - a_1b_3\mathbf{j} - a_2b_1\mathbf{k} + a_2b_3\mathbf{i} + a_3b_1\mathbf{j} - a_3b_2\mathbf{i}$$

$$= (a_2b_3 - a_3b_2)\mathbf{i} - (a_1b_3 - a_3b_1)\mathbf{j} + (a_1b_2 - a_2b_1)\mathbf{k}$$

$$= \begin{vmatrix} a_2 & a_3 \\ b_2 & b_3 \end{vmatrix}\mathbf{i} - \begin{vmatrix} a_1 & a_3 \\ b_1 & b_3 \end{vmatrix}\mathbf{j} + \begin{vmatrix} a_1 & a_2 \\ b_1 & b_2 \end{vmatrix}\mathbf{k}. \quad ❏$$

Examples

Example 1 Calculate $\mathbf{a} \times \mathbf{b}$ given that

$$\mathbf{a} = \mathbf{i} - 2\mathbf{j} + 3\mathbf{k} \qquad \text{and} \qquad \mathbf{b} = 2\mathbf{i} + \mathbf{j} - \mathbf{k}.$$

SOLUTION

$$\mathbf{a} \times \mathbf{b} = \begin{vmatrix} \mathbf{i} & \mathbf{j} & \mathbf{k} \\ 1 & -2 & 3 \\ 2 & 1 & -1 \end{vmatrix} = \begin{vmatrix} -2 & 3 \\ 1 & -1 \end{vmatrix}\mathbf{i} - \begin{vmatrix} 1 & 3 \\ 2 & -1 \end{vmatrix}\mathbf{j} + \begin{vmatrix} 1 & -2 \\ 2 & 1 \end{vmatrix}\mathbf{k}$$

$$= -\mathbf{i} + 7\mathbf{j} + 5\mathbf{k}. \quad ❏$$

Example 2 Calculate $\mathbf{a} \times \mathbf{b}$ given that

$$\mathbf{a} = \mathbf{i} - \mathbf{j} \qquad \text{and} \qquad \mathbf{b} = \mathbf{i} + \mathbf{k}.$$

SOLUTION

$$\mathbf{a} \times \mathbf{b} = \begin{vmatrix} \mathbf{i} & \mathbf{j} & \mathbf{k} \\ 1 & -1 & 0 \\ 1 & 0 & 1 \end{vmatrix} = \begin{vmatrix} -1 & 0 \\ 0 & 1 \end{vmatrix}\mathbf{i} - \begin{vmatrix} 1 & 0 \\ 1 & 1 \end{vmatrix}\mathbf{j} + \begin{vmatrix} 1 & -1 \\ 1 & 0 \end{vmatrix}\mathbf{k}$$

$$= -\mathbf{i} - \mathbf{j} + \mathbf{k}. \quad ❏$$

In Examples 1 and 2 we calculated some cross products using Theorem 12.4.8′. Of course, we can obtain the same results just by applying the distributive laws. For example, for $\mathbf{a} = \mathbf{i} - 2\mathbf{j} + 3\mathbf{k}$ and $\mathbf{b} = 2\mathbf{i} + \mathbf{j} - \mathbf{k}$, we have

$$\mathbf{a} \times \mathbf{b} = (\mathbf{i} - 2\mathbf{j} + 3\mathbf{k}) \times (2\mathbf{i} + \mathbf{j} - \mathbf{k})$$

$$= (\mathbf{i} \times \mathbf{j}) - (\mathbf{i} \times \mathbf{k}) - 4(\mathbf{j} \times \mathbf{i}) + 2(\mathbf{j} \times \mathbf{k}) + 6(\mathbf{k} \times \mathbf{i}) + 3(\mathbf{k} \times \mathbf{j})$$

$$= \mathbf{k} + \mathbf{j} + 4\mathbf{k} + 2\mathbf{i} + 6\mathbf{j} - 3\mathbf{i} = -\mathbf{i} + 7\mathbf{j} + 5\mathbf{k}.$$

Example 3 Show that the scalar triple product can be written as a determinant:

(12.4.9)

$$(\mathbf{a} \times \mathbf{b}) \cdot \mathbf{c} = \begin{vmatrix} a_1 & a_2 & a_3 \\ b_1 & b_2 & b_3 \\ c_1 & c_2 & c_3 \end{vmatrix}.$$

SOLUTION

$(\mathbf{a} \times \mathbf{b}) \cdot \mathbf{c} = \mathbf{c} \cdot (\mathbf{a} \times \mathbf{b})$

$$= (c_1\mathbf{i} + c_2\mathbf{j} + c_3\mathbf{k}) \cdot \left(\begin{vmatrix} a_2 & a_3 \\ b_2 & b_3 \end{vmatrix} \mathbf{i} - \begin{vmatrix} a_1 & a_3 \\ b_1 & b_3 \end{vmatrix} \mathbf{j} + \begin{vmatrix} a_1 & a_2 \\ b_1 & b_2 \end{vmatrix} \mathbf{k} \right)$$

$$= c_1 \begin{vmatrix} a_2 & a_3 \\ b_2 & b_3 \end{vmatrix} - c_2 \begin{vmatrix} a_1 & a_3 \\ b_1 & b_3 \end{vmatrix} + c_3 \begin{vmatrix} a_1 & a_2 \\ b_1 & b_2 \end{vmatrix}.$$

This is the expansion of

$$\begin{vmatrix} a_1 & a_2 & a_3 \\ b_1 & b_2 & b_3 \\ c_1 & c_2 & c_3 \end{vmatrix}$$

by the bottom row. (Exercise 45.)   ❏

A SUGGESTION: Vectors were defined as ordered triples, and many of the early proofs were done by "breaking up" vectors into their components. This may give you the impression that the method of "breakup" and working with the components is the first thing to try when confronted with a problem that involves vectors. If it is a *computational* problem, this method may give good results. But if you have to *analyze* a situation involving vectors, particularly one in which geometry plays a role, then the "breakup" strategy is seldom the best. Think instead of using the *operations* we have defined on vectors: addition, subtraction, scalar multiplication, dot product, cross product. Being geometrically motivated, these operations are likely to provide greater understanding than breaking up everything in sight into components.

Example 4   Let $\mathbf{a}, \mathbf{b}, \mathbf{c}$ be nonzero vectors that do not lie in the same plane. Find all the vectors $\mathbf{d}$ for which

(∗) $\mathbf{d} \cdot \mathbf{a} = \mathbf{d} \cdot \mathbf{b} = \mathbf{d} \cdot \mathbf{c}.$

SOLUTION   We could begin by writing

$$\mathbf{d} = d_1\mathbf{i} + d_2\mathbf{j} + d_3\mathbf{k}, \qquad \mathbf{a} = a_1\mathbf{i} + a_2\mathbf{j} + a_3\mathbf{k}, \qquad \text{and so on.}$$

Equation (∗) would then take the form

$$d_1a_1 + d_2a_2 + d_3a_3 = d_1b_1 + d_2b_2 + d_3b_3 = d_1c_1 + d_2c_2 + d_3c_3,$$

and we would be faced with finding all d_1, d_2, d_3 that satisfied these equations. This is a messy task.

Here is a better approach. The vectors $\mathbf{d}$ that satisfy (∗) are the vectors $\mathbf{d}$ for which

$$\mathbf{d} \cdot (\mathbf{a} - \mathbf{b}) = 0 \qquad \text{and} \qquad \mathbf{d} \cdot (\mathbf{b} - \mathbf{c}) = 0.$$

These are the vectors $\mathbf{d}$ that are perpendicular to both $\mathbf{a} - \mathbf{b}$ and $\mathbf{b} - \mathbf{c}$. One such vector is $(\mathbf{a} - \mathbf{b}) \times (\mathbf{b} - \mathbf{c})$. The vectors $\mathbf{d}$ that satisfy (∗) are the scalar multiples of that cross product.   ❏

Example 5   Verify that

$$\|\mathbf{a} \times \mathbf{b}\|^2 + (\mathbf{a} \cdot \mathbf{b})^2 = \|\mathbf{a}\|^2 \|\mathbf{b}\|^2.$$

SOLUTION   We could begin by writing

$$\|\mathbf{a} \times \mathbf{b}\|^2 = (a_2b_3 - a_3b_2)^2 + (a_1b_3 - a_3b_1)^2 + (a_1b_2 - a_2b_1)^2$$

$$(\mathbf{a} \cdot \mathbf{b})^2 = (a_1b_1 + a_2b_2 + a_3b_3)^2$$

$$\|\mathbf{a}\|^2 \|\mathbf{b}\|^2 = (a_1^2 + a_2^2 + a_3^2)(b_1^2 + b_2^2 + b_3^2).$$

but this would take us into a morass of arithmetic.   It is much more fruitful to proceed as follows:

$$\|\mathbf{a} \times \mathbf{b}\| = \|\mathbf{a}\| \|\mathbf{b}\| \sin \theta \qquad \text{and} \qquad \mathbf{a} \cdot \mathbf{b} = \|\mathbf{a}\| \|\mathbf{b}\| \cos \theta.$$

Therefore

$$\|\mathbf{a} \times \mathbf{b}\|^2 + (\mathbf{a} \cdot \mathbf{b})^2 = \|\mathbf{a}\|^2 \|\mathbf{b}\|^2 \sin^2 \theta + \|\mathbf{a}\|^2 \|\mathbf{b}\|^2 \cos^2 \theta$$

$$= \|\mathbf{a}\|^2 \|\mathbf{b}\|^2 (\sin^2 \theta + \cos^2 \theta) = \|\mathbf{a}\|^2 \|\mathbf{b}\|^2. \quad ❑$$

Three Important Identities

It may be tempting to think that $\mathbf{a} \times (\mathbf{b} \times \mathbf{c}) = (\mathbf{a} \times \mathbf{b}) \times \mathbf{c}$.   This is in general false:

$$\mathbf{i} \times (\mathbf{i} \times \mathbf{j}) = \mathbf{i} \times \mathbf{k} = -\mathbf{j} \qquad \text{but} \qquad (\mathbf{i} \times \mathbf{i}) \times \mathbf{j} = \mathbf{0} \times \mathbf{j} = \mathbf{0}.$$

What is true instead is that

(12.4.10)
$$\mathbf{a} \times (\mathbf{b} \times \mathbf{c}) = (\mathbf{a} \cdot \mathbf{c})\mathbf{b} - (\mathbf{a} \cdot \mathbf{b})\mathbf{c},$$
$$(\mathbf{a} \times \mathbf{b}) \times \mathbf{c} = (\mathbf{c} \cdot \mathbf{a})\mathbf{b} - (\mathbf{c} \cdot \mathbf{b})\mathbf{a}.$$

There is one more identity that we want to mention:

(12.4.11)
$$(\mathbf{a} \times \mathbf{b}) \cdot (\mathbf{c} \times \mathbf{d}) = (\mathbf{a} \cdot \mathbf{c})(\mathbf{b} \cdot \mathbf{d}) - (\mathbf{a} \cdot \mathbf{d})(\mathbf{b} \cdot \mathbf{c}).$$

The proof of this, as well as the proof of (12.4.10), is left to you in the exercises.

Remark   Dot products and cross products are almost ubiquitous in physics and engineering.   Work is a dot product.   So is the power expended by a force.   Every line integral is a dot product.   Torque and angular momentum are cross products.   Turn on a television set and watch the dots on the screen.   How they move is determined by a cross product.   Take a look at any text on electromagnetism.   It is all based on Maxwell's four equations.   Two of them are dot products, two of them are cross products.   ❑

EXERCISES 12.4

In Exercises 1–12, find the indicated quantity.

1. $(\mathbf{i} + \mathbf{j}) \times (\mathbf{i} - \mathbf{j})$.

2. $(\mathbf{i} - \mathbf{j}) \times (\mathbf{j} - \mathbf{i})$.

3. $(\mathbf{i} - \mathbf{j}) \times (\mathbf{j} - \mathbf{k})$.

4. $\mathbf{j} \times (2\mathbf{i} - \mathbf{k})$.

5. $(2\mathbf{j} - \mathbf{k}) \times (\mathbf{i} - 3\mathbf{j})$.

6. $\mathbf{i} \cdot (\mathbf{j} \times \mathbf{k})$.

7. $\mathbf{j} \cdot (\mathbf{i} \times \mathbf{k})$.

8. $(\mathbf{j} \times \mathbf{i}) \cdot (\mathbf{i} \times \mathbf{k})$.

9. $(\mathbf{i} \times \mathbf{j}) \times \mathbf{k}$.

10. $\mathbf{k} \cdot (\mathbf{j} \times \mathbf{i})$.

11. $\mathbf{j} \cdot (\mathbf{k} \times \mathbf{i})$.   12. $\mathbf{j} \times (\mathbf{k} \times \mathbf{i})$.

In Exercises 13–20, find the indicated quantity.

13. $(\mathbf{i} + 3\mathbf{j} - \mathbf{k}) \times (\mathbf{i} + \mathbf{k})$.

14. $(3\mathbf{i} - 2\mathbf{j} + \mathbf{k}) \times (\mathbf{i} - \mathbf{j} + \mathbf{k})$.

15. $(\mathbf{i} + \mathbf{j} + \mathbf{k}) \times (2\mathbf{i} + \mathbf{k})$.

16. $(2\mathbf{i} - \mathbf{k}) \times (\mathbf{i} - 2\mathbf{j} + 2\mathbf{k})$.

17. $[2\mathbf{i} + \mathbf{j}] \cdot [(\mathbf{i} - 3\mathbf{j} + \mathbf{k}) \times (4\mathbf{i} + \mathbf{k})]$.

18. $[(-2\mathbf{i} + \mathbf{j} - 3\mathbf{k}) \times \mathbf{i}] \times [\mathbf{i} + \mathbf{j}]$.

19. $[(\mathbf{i} - \mathbf{j}) \times (\mathbf{j} - \mathbf{k})] \times [\mathbf{i} + 5\mathbf{k}]$.

20. $[\mathbf{i} - \mathbf{j}] \times [(\mathbf{j} - \mathbf{k}) \times (\mathbf{j} + 5\mathbf{k})]$.

21. Find two unit vectors which are perpendicular to the vectors $\mathbf{a} = (1, 3, -1)$ and $\mathbf{b} = (2, 0, 1)$.

22. Repeat Exercise 21 for the vectors $\mathbf{a} = (1, 2, 3)$ and $\mathbf{b} = (2, 1, 1)$.

In Exercises 23–26, find a vector $\mathbf{N}$ that is perpendicular to the plane determined by the points P, Q, R, and find the area of the triangle PQR.

23. $P(0, 1, 0)$, $Q(-1, 1, 2)$, $R(2, 1, -1)$.

24. $P(1, 2, 3)$, $Q(-1, 3, 2)$, $R(3, -1, 2)$.

25. $P(1, -1, 4)$, $Q(2, 0, 1)$, $R(0, 2, 3)$.

26. $P(2, -1, 3)$, $Q(4, 1, -1)$, $R(-3, 0, 5)$.

In Exercises 27 and 28, find the volume of the parallelepiped with the given edges.

27. $\mathbf{i} + \mathbf{j}$, $2\mathbf{i} - \mathbf{k}$, $3\mathbf{j} + \mathbf{k}$.

28. $\mathbf{i} - 3\mathbf{j} + \mathbf{k}$, $2\mathbf{j} - \mathbf{k}$, $\mathbf{i} + \mathbf{j} - 2\mathbf{k}$.

29. Find the volume of the parallelepiped whose edges are $\overrightarrow{OP}$, $\overrightarrow{OQ}$, and $\overrightarrow{OR}$, where $O(0, 0, 0)$, $P(1, 2, 3)$, $Q(1, 1, 2)$, and $R(2, 1, 1)$.

30. Find the volume of the parallelepiped whose edges are $\overrightarrow{PQ}$, $\overrightarrow{PR}$, and $\overrightarrow{PS}$, where $P(1, -1, 4)$, $Q(2, 0, 1)$, $R(0, 2, 3)$, and $S(3, 5, 7)$.

31. Express $(\mathbf{a} + \mathbf{b}) \times (\mathbf{a} - \mathbf{b})$ as a scalar multiple of $\mathbf{a} \times \mathbf{b}$.

32. Earlier we verified that $\mathbf{a} \times (\mathbf{b} + \mathbf{c}) = (\mathbf{a} \times \mathbf{b}) + (\mathbf{a} \times \mathbf{c})$. Show now that

$$(\mathbf{a} + \mathbf{b}) \times \mathbf{c} = (\mathbf{a} \times \mathbf{c}) + (\mathbf{b} \times \mathbf{c}).$$

33. Suppose that $\mathbf{a} \times \mathbf{i} = \mathbf{0}$ and $\mathbf{a} \times \mathbf{j} = \mathbf{0}$. What can you conclude about $\mathbf{a}$?

34. Let $\mathbf{a} = a_1\mathbf{i} + a_2\mathbf{j}$ and $\mathbf{b} = b_1\mathbf{i} + b_2\mathbf{j}$ be nonzero vectors in the xy-plane. Show that $\mathbf{a} \times \mathbf{b}$ is parallel to $\mathbf{k}$.

35. Express $(\alpha\mathbf{a} + \beta\mathbf{b}) \times (\gamma\mathbf{a} + \delta\mathbf{b})$ as a scalar multiple of $\mathbf{a} \times \mathbf{b}$.

36. (a) Let $\mathbf{a}, \mathbf{b}, \mathbf{c}$ be distinct nonzero vectors. Show that

$$\mathbf{a} \times \mathbf{b} = \mathbf{a} \times \mathbf{c} \quad \text{iff} \quad \mathbf{a} \text{ and } \mathbf{b} - \mathbf{c} \text{ are parallel.}$$

(b) Sketch a figure depicting all those vectors $\mathbf{c}$ that satisfy $\mathbf{a} \times \mathbf{b} = \mathbf{a} \times \mathbf{c}$.

37. Which of the following dot products are equal?

$\mathbf{a} \cdot (\mathbf{b} \times \mathbf{c})$ $\mathbf{a} \cdot (\mathbf{c} \times \mathbf{b})$ $(\mathbf{a} \times \mathbf{b}) \cdot \mathbf{c}$ $(\mathbf{c} \times \mathbf{a}) \cdot \mathbf{b}$

$(\mathbf{b} \times \mathbf{c}) \cdot \mathbf{a}$ $\mathbf{c} \cdot (\mathbf{b} \times \mathbf{a})$ $(-\mathbf{a} \times \mathbf{b}) \cdot \mathbf{c}$ $(\mathbf{a} \times -\mathbf{c}) \cdot \mathbf{b}$.

38. Show that $(\mathbf{a} \times \mathbf{b}) \cdot \mathbf{b} = 0$ for all vectors $\mathbf{a}$ and $\mathbf{b}$.

39. Show that the vectors $\mathbf{a}, \mathbf{b},$ and $\mathbf{c}$ are coplanar iff $(\mathbf{a} \times \mathbf{b}) \cdot \mathbf{c} = 0$.

40. If $\mathbf{a}, \mathbf{b},$ and $\mathbf{c}$ are mutually perpendicular, show that $\mathbf{a} \times (\mathbf{b} \times \mathbf{c}) = \mathbf{0}$.

41. Let $\mathbf{a} \neq \mathbf{0}$. Show that if $\mathbf{a} \times \mathbf{b} = \mathbf{a} \times \mathbf{c}$ and $\mathbf{a} \cdot \mathbf{b} = \mathbf{a} \cdot \mathbf{c}$ then $\mathbf{b} = \mathbf{c}$.

42. (a) Show that

$$\mathbf{a} \times (\mathbf{b} \times \mathbf{c}) = (\mathbf{a} \cdot \mathbf{c})\mathbf{b} - (\mathbf{a} \cdot \mathbf{b})\mathbf{c}.$$

HINT: Verify that the $\mathbf{i}$ components of the two sides agree. By analogy, it will be clear that the same holds true for the $\mathbf{j}$ and $\mathbf{k}$ components.

(b) Show that

$$(\mathbf{a} \times \mathbf{b}) \times \mathbf{c} = (\mathbf{c} \cdot \mathbf{a})\mathbf{b} - (\mathbf{c} \cdot \mathbf{b})\mathbf{a}.$$

HINT: $(\mathbf{a} \times \mathbf{b}) \times \mathbf{c} = -[\mathbf{c} \times (\mathbf{a} \times \mathbf{b})]$.

(c) Finally, show that

$$(\mathbf{a} \times \mathbf{b}) \cdot (\mathbf{c} \times \mathbf{d}) = (\mathbf{a} \cdot \mathbf{c})(\mathbf{b} \cdot \mathbf{d}) - (\mathbf{a} \cdot \mathbf{d})(\mathbf{b} \cdot \mathbf{c}).$$

HINT: Set $\mathbf{c} \times \mathbf{d} = \mathbf{r}$ and use (12.4.6).

43. Let $\mathbf{a}, \mathbf{b}$ be nonzero vectors with $\mathbf{a} \perp \mathbf{b}$, and set $\mathbf{c} = \mathbf{a} \times \mathbf{b}$. Express $\mathbf{c} \times \mathbf{a}$ in terms of $\mathbf{b}$.

44. Given that $\mathbf{u}$ is a unit vector, show that each vector $\mathbf{a}$ can be decomposed as follows into a part parallel to $\mathbf{u}$ and a part perpendicular to $\mathbf{u}$:

$$\mathbf{a} = \underbrace{(\mathbf{a} \cdot \mathbf{u})\mathbf{u}}_{\substack{\text{parallel} \\ \text{to } \mathbf{u}}} + \underbrace{(\mathbf{u} \times \mathbf{a}) \times \mathbf{u}}_{\substack{\text{perpendicular} \\ \text{to } \mathbf{u}}}.$$

HINT: Use Exercise 42(b).

45. Verify the conclusion given in Example 3.

Torque. Suppose that a rigid body is free to rotate about a fixed point O. If a force $\mathbf{F}$ acts on the body at a point P, then the body will tend to rotate about an axis through O. This effect is measured by the *torque* vector $\boldsymbol{\tau}$ which is given by

$$\boldsymbol{\tau} = \mathbf{r} \times \mathbf{F},$$

where $\mathbf{r}$ is the position vector $\overrightarrow{OP}$. The straight line through O determined by $\boldsymbol{\tau}$ is the axis of rotation. By Definition 12.4.1, the vectors **r**, **F**, and $\boldsymbol{\tau}$ form a right-handed system, and the magnitude of $\boldsymbol{\tau}$ is

$$\|\boldsymbol{\tau}\| = \|\mathbf{r}\| \, \|\mathbf{F}\| \sin \theta$$

where θ is the angle between the position and force vectors. See the following figure.

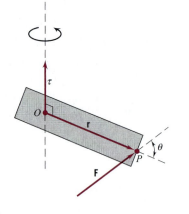

46. Find the magnitude of the torque exerted at the origin by the force $\mathbf{F} = \mathbf{i} + 2\mathbf{j} + \mathbf{k}$ at the point $P(1, 1, 1)$.

▶ 47. A bolt is being tightened by a 20-pound force applied to a 10-inch wrench as shown in the figure. Find the magnitude of the torque. Assuming that the wrench and the force are in the plane of the paper, in what direction will the bolt move?

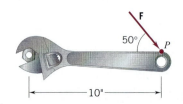

▶ 48. Repeat Exercise 47 if the 20-pound force is applied as shown in the figure.

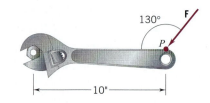

■ 12.5 LINES

Vectors used to specify position are called *position vectors*. Position vectors that emanate from the origin are known as *radius vectors*. In this section we use radius vectors to characterize lines.

Vector Parametrizations

We begin with the idea that two distinct points determine a line. In Figure 12.5.1 we have marked two points, P and Q, and the line l that they determine. To obtain a vector characterization of l, we choose the vectors $\mathbf{r}_0$ and $\mathbf{d} = \overrightarrow{PQ}$ as in Figure 12.5.2. Since we began with two distinct points, the vector $\mathbf{d}$ is nonzero.

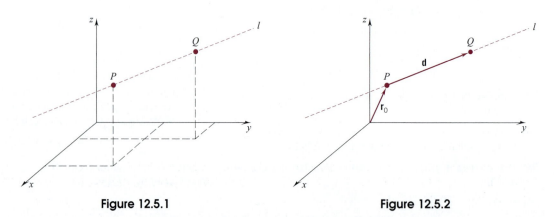

Figure 12.5.1   **Figure 12.5.2**

In Figure 12.5.3 we have drawn an additional vector $\mathbf{r}$. The vector that begins at the tip of $\mathbf{r}_0$ and ends at the tip of $\mathbf{r}$ is $\mathbf{r} - \mathbf{r}_0$. Therefore, the tip of $\mathbf{r}$ will fall on l iff

$$\mathbf{r} - \mathbf{r}_0 \quad \text{and} \quad \mathbf{d} \quad \text{are parallel;}$$

this in turn will happen iff

$$\mathbf{r} - \mathbf{r}_0 = t\mathbf{d} \qquad \text{for some real number } t,$$

or equivalently, iff

$$\mathbf{r} = \mathbf{r}_0 + t\mathbf{d} \qquad \text{for some real number } t.$$

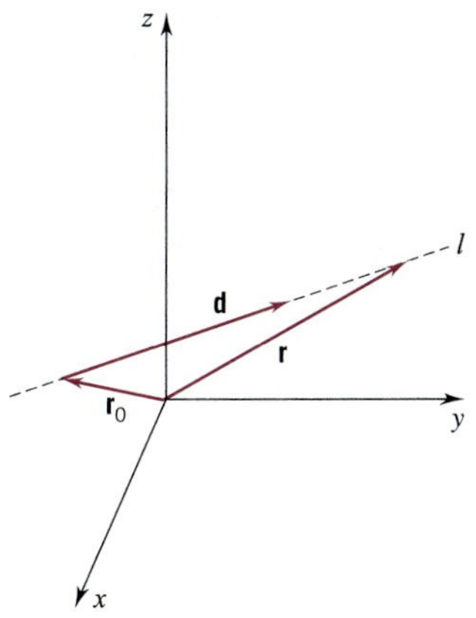

The vector equation

(12.5.1)
$$\boxed{\mathbf{r}(t) = \mathbf{r}_0 + t\mathbf{d}, \qquad t \text{ real}}$$

parametrizes the line l: by varying t, we vary the vector $\mathbf{r}(t)$, but its tip remains on l; as t ranges over the set of real numbers, the tip $\mathbf{r}(t)$ traces out the line l.

Now set

$$\mathbf{r}_0 = x_0\mathbf{i} + y_0\mathbf{j} + z_0\mathbf{k}, \qquad \mathbf{d} = d_1\mathbf{i} + d_2\mathbf{j} + d_3\mathbf{k}$$

and denote the tip of $\mathbf{r}_0$ by $P(x_0, y_0, z_0)$. The line l given by

(12.5.2)
$$\boxed{\mathbf{r}(t) = \mathbf{r}_0 + t\mathbf{d} = (x_0 + td_1)\mathbf{i} + (y_0 + td_2)\mathbf{j} + (z_0 + td_3)\mathbf{k}}$$

passes through the point $P(x_0, y_0, z_0)$ and is parallel to $\mathbf{d}$ (Figure 12.5.4). The vector $\mathbf{d}$, which by definition is not zero, is called a *direction vector* for l, and its components d_1, d_2, d_3 are called *direction numbers.*

Figure 12.5.3

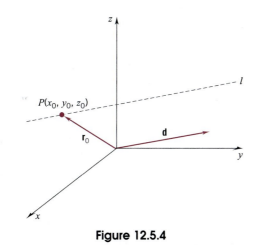

Figure 12.5.4

Remark A direction vector for the line l determined by two distinct points $P(x_0, y_0, z_0)$ and $Q(x_1, y_1, z_1)$ is $\mathbf{d} = \overrightarrow{PQ} = (x_1 - x_0)\mathbf{i} + (y_1 - y_0)\mathbf{j} + (z_1 - z_0)\mathbf{k}$; a set of direction numbers for l is: $(x_1 - x_0), (y_1 - y_0), (z_1 - z_0)$. ❑

Example 1 Find a vector equation that parametrizes the line that passes through the point $P(1, -1, 2)$ and is parallel to the vector $2\mathbf{i} - 3\mathbf{j} + \mathbf{k}$.

SOLUTION Here we can set

$$\mathbf{r}_0 = \mathbf{i} - \mathbf{j} + 2\mathbf{k} \qquad \text{and} \qquad \mathbf{d} = 2\mathbf{i} - 3\mathbf{j} + \mathbf{k}.$$

As a vector parametrization for the line we have

$$\mathbf{r}(t) = (\mathbf{i} - \mathbf{j} + 2\mathbf{k}) + t(2\mathbf{i} - 3\mathbf{j} + \mathbf{k}),$$

which we can rewrite as

$$\mathbf{r}(t) = (1 + 2t)\mathbf{i} - (1 + 3t)\mathbf{j} + (2 + t)\mathbf{k}. \qquad \square$$

As a direction vector for a given line we can take any nonzero vector that is parallel to the line. Thus, if $\mathbf{d}$ is a direction vector for l, so is $\alpha\mathbf{d}$, provided that $\alpha \neq 0$. If d_1, d_2, d_3 are direction numbers for l, so are $\alpha d_1, \alpha d_2, \alpha d_3$, provided again that $\alpha \neq 0$.

The line that passes through the origin with direction vector $\mathbf{d}$ can be parametrized by the vector equation

(12.5.3)
$$\mathbf{r}(t) = t\mathbf{d} = td_1\mathbf{i} + td_2\mathbf{j} + td_3\mathbf{k}.$$
(Figure 12.5.5)

There are, however, other ways to parametrize this line.

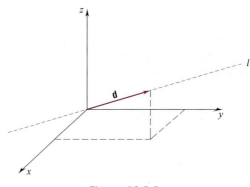

Figure 12.5.5

Example 2 Find all vector parametrizations (12.5.1) for the line through the origin with direction vector $\mathbf{d}$.

SOLUTION Since $\mathbf{d}$ is a direction vector, so is every vector $\alpha\mathbf{d}$ with $\alpha \neq 0$. We therefore write

$(*)$ $$\mathbf{r}(t) = \mathbf{r}_0 + t\alpha\mathbf{d}.$$

Since the line passes through the origin, it may appear at first glance that $\mathbf{r}_0$ has to be $\mathbf{0}$, but that is not true. From the fact that the line passes through the origin we can conclude only that $\mathbf{r}_0 + t\alpha\mathbf{d}$ is $\mathbf{0}$ for some value of t. Call this value t_0. Then

$$\mathbf{r}_0 + t_0\alpha\mathbf{d} = \mathbf{0} \qquad \text{and thus} \qquad \mathbf{r}_0 = -t_0\alpha\mathbf{d}.$$

Substitution in (∗) gives

$$\mathbf{r}(t) = -t_0\alpha\mathbf{d} + t\alpha\mathbf{d} = (t - t_0)\alpha\mathbf{d}.$$

All the desired parametrizations can be written

$$\mathbf{r}(t) = (t - t_0)\alpha\mathbf{d} \qquad \text{with } \alpha \text{ and } t_0 \text{ real}, \ \alpha \neq 0,$$

and all equations of this form parametrize that same line. ❑

Scalar Parametric Equations

It follows from equation (12.5.2) that the line that passes through the point $P(x_0, y_0, z_0)$ with direction numbers d_1, d_2, d_3 can be parametrized by three scalar equations:

(12.5.4)
$$x(t) = x_0 + d_1 t, \qquad y(t) = y_0 + d_2 t, \qquad z(t) = z_0 + d_3 t.$$

These equations are the **i**, **j**, **k** components of the vector $\mathbf{r}(t) = \mathbf{r}_0 + t\mathbf{d}$.

Example 3 Write scalar parametric equations for the line that passes through the point $P(-1, 4, 2)$ with direction numbers 1, 2, 3.

SOLUTION In this case the scalar equations

$$x(t) = x_0 + d_1 t, \qquad y(t) = y_0 + d_2 t, \qquad z(t) = z_0 + d_3 t$$

take the form

$$x(t) = -1 + t, \qquad y(t) = 4 + 2t, \qquad z(t) = 2 + 3t. \quad ❑$$

Example 4 What direction numbers are displayed by the parametric equations

$$x(t) = 3 - t, \qquad y(t) = 2 + 4t, \qquad z(t) = 1 - 5t?$$

What other direction numbers could be used for the same line?

SOLUTION The direction numbers displayed are $-1, 4, -5$. Any triple of the form

$$-\alpha, \quad 4\alpha, \quad -5\alpha \qquad \text{with } \alpha \neq 0$$

could be used as a set of direction numbers for that same line. ❑

Symmetric Form

If the direction numbers are all nonzero, then each of the scalar parametric equations can be solved for t:

$$t = \frac{x(t) - x_0}{d_1}, \qquad t = \frac{y(t) - y_0}{d_2}, \qquad t = \frac{z(t) - z_0}{d_3}.$$

Eliminating the parameter t, we obtain three equations:

$$\frac{x - x_0}{d_1} = \frac{y - y_0}{d_2}, \qquad \frac{y - y_0}{d_2} = \frac{z - z_0}{d_3}, \qquad \frac{x - x_0}{d_1} = \frac{z - z_0}{d_3}.$$

Obviously, any two of these equations suffice; the third is redundant and can be discarded. Rather than decide which equation to discard, we simply write

(12.5.5)

$$\frac{x - x_0}{d_1} = \frac{y - y_0}{d_2} = \frac{z - z_0}{d_3}.$$

These are the equations of a line written in *symmetric form.* They can be used only if d_1, d_2, d_3 are all different from zero.

Example 5 Write equations in symmetric form for the line that passes through the points $P(x_0, y_0, z_0)$ and $Q(x_1, y_1, z_1)$. Under what conditions are the equations valid?

SOLUTION As direction numbers we can take the triple

$$x_1 - x_0, \qquad y_1 - y_0, \qquad z_1 - z_0.$$

We can base our calculations on $P(x_0, y_0, z_0)$ and write

$$\frac{x - x_0}{x_1 - x_0} = \frac{y - y_0}{y_1 - y_0} = \frac{z - z_0}{z_1 - z_0},$$

or we can base our calculations on $Q(x_1, y_1, z_1)$ and write

$$\frac{x - x_1}{x_1 - x_0} = \frac{y - y_1}{y_1 - y_0} = \frac{z - z_1}{z_1 - z_0}.$$

Both sets of equations are valid provided that $x_1 \neq x_0, y_1 \neq y_0, z_1 \neq z_0$. ❑

Equations 12.5.5 can be used only if the direction numbers are all different from zero. If one of the direction numbers is zero, then one of the coordinates is constant. As you will see, this simplifies the algebra. Geometrically, it means that the line lies on a plane that is parallel to one of the coordinate planes.

Suppose, for example, that $d_3 = 0$. Then the scalar parametric equations take the form

$$x(t) = x_0 + d_1 t, \qquad y(t) = y_0 + d_2 t, \qquad z(t) = z_0.$$

Eliminating t, we are left with two equations:

$$\frac{x - x_0}{d_1} = \frac{y - y_0}{d_2}, \qquad z = z_0.$$

The line lies on the horizontal plane $z = z_0$, and its projection onto the xy-plane (see Figure 12.5.6) is the line l' with equation

$$\frac{x - x_0}{d_1} = \frac{y - y_0}{d_2}.$$

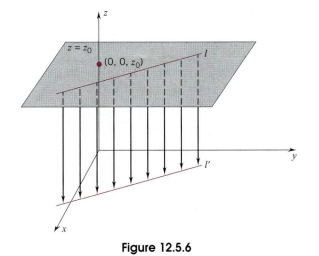

Figure 12.5.6

Intersecting Lines, Parallel Lines

Two distinct lines

$$l_1: \mathbf{r}(t) = \mathbf{r}_0 + t\mathbf{d}, \qquad l_2: \mathbf{R}(u) = \mathbf{R}_0 + u\mathbf{D}$$

intersect iff there are numbers t and u at which

$$\mathbf{r}(t) = \mathbf{R}(u).$$

Example 6 Find the point at which the lines

$$l_1: \mathbf{r}(t) = (\mathbf{i} - 6\mathbf{j} + 2\mathbf{k}) + t(\mathbf{i} + 2\mathbf{j} + \mathbf{k}), \qquad l_2: \mathbf{R}(u) = (4\mathbf{j} + \mathbf{k}) + u(2\mathbf{i} + \mathbf{j} + 2\mathbf{k})$$

intersect.

SOLUTION We set

$$\mathbf{r}(t) = \mathbf{R}(u)$$

and solve for t and u:

$$(\mathbf{i} - 6\mathbf{j} + 2\mathbf{k}) + t(\mathbf{i} + 2\mathbf{j} + \mathbf{k}) = (4\mathbf{j} + \mathbf{k}) + u(2\mathbf{i} + \mathbf{j} + 2\mathbf{k}),$$

$$\mathbf{i} - 6\mathbf{j} + 2\mathbf{k} + t\mathbf{i} + 2t\mathbf{j} + t\mathbf{k} = 4\mathbf{j} + \mathbf{k} + 2u\mathbf{i} + u\mathbf{j} + 2u\mathbf{k},$$

and therefore

$$(1 + t - 2u)\mathbf{i} + (-10 + 2t - u)\mathbf{j} + (1 + t - 2u)\mathbf{k} = 0.$$

This tells us that

$$1 + t - 2u = 0,$$
$$-10 + 2t - u = 0,$$
$$1 + t - 2u = 0.$$

Note that the first and third equations are the same. Solving the first two equations simultaneously, we obtain $t = 7, u = 4$. As you can verify,

$$\mathbf{r}(7) = 8\mathbf{i} + 8\mathbf{j} + 9\mathbf{k} = \mathbf{R}(4).$$

The two lines intersect at the tip of this vector, which is the point $P(8, 8, 9)$. ❑

Remark To give a physical interpretation of the result in Example 6, think of the parameters t and u as representing time, and think of particles moving along the lines l_1 and l_2. At time $t = u = 0$, the particle on l_1 is at the point $(1, -6, 2)$ and the particle on l_2 is at the point $(0, 4, 1)$. The particle on l_1 passes through the point $P(8, 8, 9)$ at time $t = 7$ while the particle on l_2 passes through P at time $u = 4$; both particles pass through the same point P, but at different times! ❏

In the setting of plane geometry we can think of two lines as parallel iff they do not intersect. This point of view is not satisfactory in three-dimensional space. (See Figure 12.5.7.)

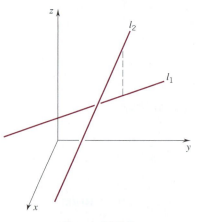

Figure 12.5.7

The lines l_1 and l_2 marked in Figure 12.5.7 do not intersect, and yet we would hesitate to call them parallel. We can avoid this difficulty by using direction vectors: we call two lines *parallel* iff their direction vectors are parallel. Nonparallel, nonintersecting lines are said to be *skew*.

If two lines l_1, l_2 intersect, we can find the angle between them by finding the angle between their direction vectors, **d** and **D**. Depending on our choice of direction vectors, there are two such angles, each the supplement of the other (Figure 12.5.8). We choose the smaller of the two angles, the one with nonnegative cosine:

Figure 12.5.8

(12.5.6)

$$\cos \theta = |\mathbf{u_d} \cdot \mathbf{u_D}|.$$

Example 7 Earlier we verified that the lines

$$l_1: \mathbf{r}(t) = (\mathbf{i} - 6\mathbf{j} + 2\mathbf{k}) + t(\mathbf{i} + 2\mathbf{j} + \mathbf{k}), \qquad l_2: \mathbf{R}(u) = (4\mathbf{j} + \mathbf{k}) + u(2\mathbf{i} + \mathbf{j} + 2\mathbf{k})$$

intersect at $P(8, 8, 9)$. What is the angle between these lines?

SOLUTION As direction vectors we can take

$$\mathbf{d} = \mathbf{i} + 2\mathbf{j} + \mathbf{k} \qquad \text{and} \qquad \mathbf{D} = 2\mathbf{i} + \mathbf{j} + 2\mathbf{k}.$$

Then, as you can check,

$$\mathbf{u_d} = \tfrac{1}{6}\sqrt{6}(\mathbf{i} + 2\mathbf{j} + \mathbf{k}) \qquad \text{and} \qquad \mathbf{u_D} = \tfrac{1}{3}(2\mathbf{i} + \mathbf{j} + 2\mathbf{k}).$$

It follows that

$$\cos \theta = |\mathbf{u_d} \cdot \mathbf{u_D}| = \tfrac{1}{3}\sqrt{6} \cong 0.816 \qquad \text{and} \qquad \theta \cong 0.62 \text{ radian.} \quad \square$$

Two intersecting lines are said to be *perpendicular* iff their direction vectors are perpendicular.

Example 8 Let l_1 and l_2 be the lines of the last exercise. These lines intersect at $P(8, 8, 9)$. Find a vector parametrization for the line l_3 that passes through $P(8, 8, 9)$ and is perpendicular to both l_1 and l_2.

SOLUTION We are given that l_3 passes through $P(8, 8, 9)$. All we need to parametrize this line is a direction vector $\mathbf{d_3}$. We require that $\mathbf{d_3}$ be perpendicular to the direction vectors of l_1 and l_2; namely, we require that

$$\mathbf{d_3} \perp \mathbf{d} \quad \text{and} \quad \mathbf{d_3} \perp \mathbf{D}, \qquad \text{where} \quad \mathbf{d} = \mathbf{i} + 2\mathbf{j} + \mathbf{k} \quad \text{and} \quad \mathbf{D} = 2\mathbf{i} + \mathbf{j} + 2\mathbf{k}.$$

Since $\mathbf{d} \times \mathbf{D}$ is perpendicular to both $\mathbf{d}$ and $\mathbf{D}$, we can set

$$\mathbf{d_3} = \mathbf{d} \times \mathbf{D} = \begin{vmatrix} \mathbf{i} & \mathbf{j} & \mathbf{k} \\ 1 & 2 & 1 \\ 2 & 1 & 2 \end{vmatrix} = \begin{vmatrix} 2 & 1 \\ 1 & 2 \end{vmatrix} \mathbf{i} - \begin{vmatrix} 1 & 1 \\ 2 & 2 \end{vmatrix} \mathbf{j} + \begin{vmatrix} 1 & 2 \\ 2 & 1 \end{vmatrix} \mathbf{k} = 3\mathbf{i} - 3\mathbf{k}.$$

As a parametrization for l_3 we can write

$$\gamma_3(t) = (8\mathbf{i} + 8\mathbf{j} + 9\mathbf{k}) + t(3\mathbf{i} - 3\mathbf{k}). \quad \square$$

Example 9

(a) Find a vector parametrization for the line

$$l: y = mx + b \qquad \text{in the } xy\text{-plane.}$$

(b) Show by vector methods that

$$l_1: y = m_1 x + b_1 \perp l_2: y = m_2 x + b_2 \qquad \text{iff} \qquad m_1 m_2 = -1.$$

SOLUTION

(a) We seek a parametrization of the form

$$\mathbf{r}(t) = \mathbf{r_0} + t\mathbf{d}.$$

Since $P(0, b)$ lies on l, we can set

$$\mathbf{r_0} = 0\mathbf{i} + b\mathbf{j} = b\mathbf{j}.$$

To find a direction vector, we take $x_1 \neq 0$ and note that the point $Q(x_1, mx_1 + b)$ also lies on l. (See Figure 12.5.9.) As direction numbers we can take

$$x_1 - 0 = x_1 \qquad \text{and} \qquad (mx_1 + b) - b = mx_1$$

or, more simply, 1 and m. This choice of direction numbers gives us the direction vector $\mathbf{d} = \mathbf{i} + m\mathbf{j}$. The vector equation

$$\mathbf{r}(t) = b\mathbf{j} + t(\mathbf{i} + m\mathbf{j})$$

parametrizes the line l.

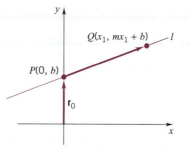

Figure 12.5.9

(b) As direction vectors for l_1 and l_2 we have

$$\mathbf{d}_1 = \mathbf{i} + m_1\mathbf{j} \qquad \text{and} \qquad \mathbf{d}_2 = \mathbf{i} + m_2\mathbf{j}.$$

Since

$$\mathbf{d}_1 \cdot \mathbf{d}_2 = (\mathbf{i} + m_1\mathbf{j}) \cdot (\mathbf{i} + m_2\mathbf{j}) = 1 + m_1 m_2,$$

you can see that

$$\mathbf{d}_1 \cdot \mathbf{d}_2 = 0 \qquad \text{iff} \qquad m_1 m_2 = -1. \quad \square$$

Distance from a Point to a Line

In Figure 12.5.10 we have drawn a line l and a point P not on l. We are interested in finding the distance $d(P, l)$ between P and l.

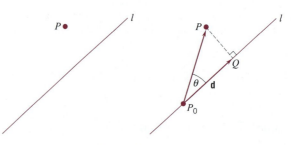

Figure 12.5.10

With P_0 and Q as in the figure, we can take $\mathbf{d} = \overrightarrow{P_0Q}$ as a direction vector for l. From the figure you can see that

$$d(P, l) = \|\overrightarrow{QP}\| = \|\overrightarrow{P_0P}\| \sin \theta.$$

Since $\|\overrightarrow{P_0P} \times d\| = \|\overrightarrow{P_0P}\|\,\|\mathbf{d}\| \sin \theta$, we have

(12.5.7)

$$d(P, l) = \frac{\|\overrightarrow{P_0P} \times \mathbf{d}\|}{\|\mathbf{d}\|}.$$

This elegant little formula gives the distance from a point P to any line l in terms of any point P_0 on l and any direction vector $\mathbf{d}$ for l. ❑

Computations based on this formula are left to the exercises.

EXERCISES 12.5

1. Which of the points $P(1, 2, 0)$, $Q(-5, 1, 5)$, $R(-4, 2, 5)$ lie on the line

$$l: \mathbf{r}(t) = (\mathbf{i} + 2\mathbf{j}) + t(6\mathbf{i} + \mathbf{j} - 5\mathbf{k})?$$

2. Determine which of the lines are parallel:

$$l_1: \mathbf{r}_1(t) = (\mathbf{i} + 2\mathbf{k}) + t(\mathbf{i} - 2\mathbf{j} + 3\mathbf{k}),$$

$$l_2: \mathbf{r}_2(u) = (\mathbf{i} + 2\mathbf{k}) + u(\mathbf{i} + 2\mathbf{j} - 3\mathbf{k}),$$

$$l_3: \mathbf{r}_3(v) = (6\mathbf{i} - \mathbf{j}) - v(2\mathbf{i} - 4\mathbf{j} + 6\mathbf{k}),$$

$$l_4: \mathbf{r}_4(w) = (\tfrac{1}{2} + \tfrac{1}{2}w)\mathbf{i} - w\mathbf{j} + (1 + \tfrac{3}{2}w)\mathbf{k}.$$

In Exercises 3–6, find a vector parametrization for the line that satisfies the given conditions.

3. Passes through $P(3, 1, 0)$ and is parallel to the line $\mathbf{r}(t) = (\mathbf{i} - \mathbf{j}) + t\mathbf{k}$.

4. Passes through $P(1, -1, 2)$ and is parallel to the line $\mathbf{r}(t) = t(3\mathbf{i} - \mathbf{j} + \mathbf{k})$.

5. Passes through the origin and $Q(x_1, y_1, z_1)$.

6. Passes through $P(x_0, y_0, z_0)$ and $Q(x_1, y_1, z_1)$.

In Exercises 7–10, find a set of scalar parametric equations for the line that satisfies the given conditions.

7. Passes through $P(1, 0, 3)$ and $Q(2, -1, 4)$.

8. Passes through $P(x_0, y_0, z_0)$ and $Q(x_1, y_1, z_1)$.

9. Passes through $P(2, -2, 3)$ and is perpendicular to the xz-plane.

10. Passes through $P(1, 4, -3)$ and is perpendicular to the yz-plane.

11. Give a vector parametrization for the line that passes through $P(-1, 2, -3)$ and is parallel to the line $2(x + 1) = 4(y - 3) = z$.

12. Write equations in symmetric form for the line that passes through the origin and the point $P(x_0, y_0, z_0)$.

In Exercises 13–16, find the point where l_1 and l_2 intersect and find the angle between l_1 and l_2.

13. $l_1: \mathbf{r}_1(t) = \mathbf{i} + t\mathbf{j}, \qquad l_2: \mathbf{r}_2(u) = \mathbf{j} + u(\mathbf{i} + \mathbf{j})$.

14. $l_1: \mathbf{r}_1(t) = (\mathbf{i} - 4\sqrt{3}\mathbf{j}) + t(\mathbf{i} + \sqrt{3}\mathbf{j})$,
$l_2: \mathbf{r}_2(u) = (4\mathbf{i} + 3\sqrt{3}\mathbf{j}) + u(\mathbf{i} - \sqrt{3}\mathbf{j})$.

15. $l_1:\quad x_1(t) = 3 + t, \qquad y_1(t) = 1 - t, \qquad z_1(t) = 5 + 2t$.
$l_2:\quad x_2(u) = 1, \qquad y_2(u) = 4 + u, \qquad z_2(u) = 2 + u$.

16. $l_1:\quad x_1(t) = 1 + t, \quad y_1(t) = -1 - t, \quad z_1(t) = -4 + 2t$.
$l_2:\quad x_2(u) = 1 - u, \qquad y_2(u) = 1 + 3u, \qquad z_2(u) = 2u$.

17. Where does the line

$$\frac{x - x_0}{d_1} = \frac{y - y_0}{d_2} = \frac{z - z_0}{d_3}$$

intersect the xy-plane?

18. What can you conclude about the lines

$$\frac{x - x_0}{d_1} = \frac{y - y_0}{d_2} = \frac{z - z_0}{d_3}, \quad \frac{x - x_0}{D_1} = \frac{y - y_0}{D_2} = \frac{z - z_0}{D_3}$$

given that $d_1 D_1 + d_2 D_2 + d_3 D_3 = 0$?

19. What can you conclude about the lines

$$\frac{x - x_0}{d_1} = \frac{y - y_0}{d_2} = \frac{z - z_0}{d_3}, \quad \frac{x - x_1}{D_1} = \frac{y - y_1}{D_2} = \frac{z - z_1}{D_3}$$

given that $d_1/D_1 = d_2/D_2 = d_3/D_3$?

20. Let P_0, P_1 be two distinct points and let $\mathbf{r}_0$, $\mathbf{r}_1$ be the radius vectors that they determine:

$$\mathbf{r}_0 = \overrightarrow{OP_0}, \qquad \mathbf{r}_1 = \overrightarrow{OP_1}.$$

As t ranges over the set of real numbers, $\mathbf{r}(t) = \mathbf{r}_0 + t(\mathbf{r}_1 - \mathbf{r}_0)$ traces out the line determined by P_0 and P_1. Restrict t so that $\mathbf{r}(t)$ traces out only the line segment $\overline{P_0 P_1}$.

21. Find a vector parametrization for the line segment that begins at $(2, 7, -1)$ and ends at $(4, 2, 3)$.

22. Restrict t so that the equations

$$x(t) = 7 - 5t, \qquad y(t) = -3 + 2t, \qquad z(t) = 4 - t$$

parametrize the line segment that begins at $(12, -5, 5)$ and ends at $(-3, 1, 2)$.

23. Determine a unit vector $\mathbf{u}$ and the values of t for which the equation

$$\mathbf{r}(t) = (6\mathbf{i} - 5\mathbf{j} + \mathbf{k}) + t\mathbf{u}$$

parametrizes the line segment that begins at $P(0, -2, 7)$ and ends at $Q(-4, 0, 11)$.

24. Suppose that the lines

$$l_1: \mathbf{r}(t) = \mathbf{r}_0 + t\mathbf{d}, \qquad l_2: \mathbf{R}(u) = \mathbf{R}_0 + u\mathbf{D}$$

intersect at right angles. Show that the point of intersection is the origin iff $\mathbf{r}(t) \perp \mathbf{R}(u)$ for all real numbers t and u.

25. Find scalar parametric equations for all the lines that are perpendicular to the line

$$x(t) = 1 + 2t, \qquad y(t) = 3 - 4t, \qquad z(t) = 2 + 6t$$

and intersect that line at the point $P(3, -1, 8)$.

26. Suppose that $\mathbf{r}(t) = \mathbf{r}_0 + t\mathbf{d}$ and $\mathbf{R}(u) = \mathbf{R}_0 + u\mathbf{D}$ both parametrize the same line. (a) Show that $\mathbf{R}_0 = \mathbf{r}_0 + t_0\mathbf{d}$ for some real number t_0. (b) Then show that, for some real number α, $\mathbf{R}(u) = \mathbf{r}_0 + (t_0 + \alpha u)\mathbf{d}$ for all real u.

In Exercises 27 and 28, find the distance from $P(1, 0, 2)$ to the indicated line.

27. The line through the origin parallel to $2\mathbf{i} - \mathbf{j} + 2\mathbf{k}$.

28. The line through $P_0(1, -1, 1)$ parallel to $\mathbf{i} - 2\mathbf{j} - 2\mathbf{k}$.

In Exercises 29–31, find the distance from the point to the line.

29. $P(1, 2, 3), \quad l: \mathbf{r}(t) = \mathbf{i} + 2\mathbf{k} + t(\mathbf{i} - 2\mathbf{j} + 3\mathbf{k})$.

30. $P(0, 0, 0), \quad l: \mathbf{r}(t) = \mathbf{i} + t\mathbf{j}$.

31. $P(1, 0, 1), \quad l: \mathbf{r}(t) = 2\mathbf{i} - \mathbf{j} + t(\mathbf{i} + \mathbf{j})$.

32. Find the distance from the point $P(x_0, y_0, z_0)$ to the line $y = mx + b$ in the xy-plane.

33. What is the distance from the origin (a) to the line that joins $P(1, 1, 1)$ and $Q(2, 2, 1)$? (b) To the line segment that joins these same points? [For (b) find the point of the line segment $\overline{PQ}$ closest to the origin and calculate its distance from the origin.]

34. Let l be the line

$$\mathbf{r}(t) = \mathbf{r}_0 + t\mathbf{d}.$$

(a) Find the scalar t_0 for which $\mathbf{r}(t_0) \perp l$.
(b) Find the parametrizations $\mathbf{R}(t) = \mathbf{R}_0 + t\mathbf{D}$ for l in which $\mathbf{R}_0 \perp l$ and $\|\mathbf{D}\| = 1$. These are the *standard vector parametrizations* for l.

In Exercises 35 and 36, find the standard vector parametrizations (Exercise 34) for the specified line.

35. The line through $P(0, 1, -2)$ parallel to $\mathbf{i} - \mathbf{j} + 3\mathbf{k}$.

36. The line through $P(\sqrt{3}, 0, 0)$ parallel to $\mathbf{i} + \mathbf{j} + \mathbf{k}$.

37. Let A, B, C be the vertices of a triangle in the xy-plane. Given that $0 < s < 1$, determine the values of t for which the tip of the radius vector

$$\overrightarrow{OA} + s\,\overrightarrow{AB} + t\,\overrightarrow{BC}$$

lies inside the triangle. HINT: Draw a diagram.

■ 12.6 PLANES

Ways of Specifying a Plane

How can we specify a plane? There are a number of ways of doing so. For example, by giving three distinct points on it, so long as they are not all on the same line; by giving two distinct lines on it; or by giving a line on it and a point on it, so long as the point does not lie on the line. There is still another way to specify a plane, and that is to give a point on the plane and a nonzero vector perpendicular to the plane

Scalar Equation of a Plane

Figure 12.6.1 shows a plane. On it we have marked a point $P(x_0, y_0, z_0)$ and starting at that point, a nonzero vector $\mathbf{N} = A\mathbf{i} + B\mathbf{j} + C\mathbf{k}$ perpendicular to the plane. We call $\mathbf{N}$ a *normal vector*. We can obtain an equation for the plane in terms of the coordinates of P and the components of $\mathbf{N}$.

To find such an equation we take an arbitrary point $Q(x, y, z)$ in space and form the vector

$$\overrightarrow{PQ} = (x - x_0)\mathbf{i} + (y - y_0)\mathbf{j} + (z - z_0)\mathbf{k}.$$

The point Q will lie on the given plane iff

$$\mathbf{N} \cdot \overrightarrow{PQ} = 0,$$

which is to say, iff

(12.6.1) $$A(x - x_0) + B(y - y_0) + C(z - z_0) = 0.$$

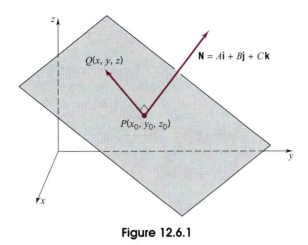

Figure 12.6.1

This is an equation in x, y, z for the plane that passes through $P(x_0, y_0, z_0)$ and has normal vector $\mathbf{N} = A\mathbf{i} + B\mathbf{j} + C\mathbf{k}$.

Remark If $\mathbf{N}$ is normal to a given plane, then so is every nonzero scalar multiple of $\mathbf{N}$. Suppose we had chosen $-2\mathbf{N}$ as our normal. Then Equation (12.6.1) would have read

$$-2A(x - x_0) - 2B(y - y_0) - 2C(z - z_0) = 0.$$

Cancelling the -2, we would have the same equation we had before. It does not matter which normal we choose. They all give equivalent equations. ❑

We can write Equation (12.6.1) in the form

$$Ax + By + Cz + D = 0$$

simply by setting $D = -Ax_0 - By_0 - Cz_0$.

Example 1 Write an equation for the plane that passes through the point $P(1, 0, 2)$ and has normal vector $\mathbf{N} = 3\mathbf{i} - 2\mathbf{j} + \mathbf{k}$.

SOLUTION The general equation

$$A(x - x_0) + B(y - y_0) + C(z - z_0) = 0$$

becomes

$$3(x - 1) + (-2)(y - 0) + (1)(z - 2) = 0,$$

which simplifies to

$$3x - 2y + z - 5 = 0. \quad ❑$$

Example 2 Find an equation for the plane p that passes through $P(-2, 3, 5)$ and is perpendicular to the line l whose scalar parametric equations are $x = -2 + t$, $y = 1 + 2t$, $z = 4$.

SOLUTION We can take $\mathbf{N} = \mathbf{i} + 2\mathbf{j}$ as a direction vector for l. Since p and l are perpendicular, $\mathbf{N}$ is a normal vector for p. Thus, an equation for p is

$$1(x + 2) + 2(y - 3) + 0(z - 5) = 0,$$

which simplifies to

$$x + 2y - 4 = 0.$$

The last equation looks very much like the equation of a line in the xy-plane. If the context of our discussion were the xy-plane, then the equation $x + 2y - 4 = 0$ would represent a line. In this case, however, our context is three-dimensional space. Hence, the equation $x + 2y - 4 = 0$ represents the set of all points $Q(x, y, z)$, where $x + 2y - 4 = 0$ and z is unrestricted. This set forms a vertical plane that intersects the xy-plane in the line just mentioned (Figure 12.6.2). ❏

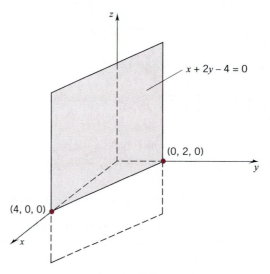

Figure 12.6.2

Example 3 Show that every equation of the form

$$ax + by + cz + d = 0 \qquad \text{with} \quad \sqrt{a^2 + b^2 + c^2} \neq 0$$

represents a plane in space.

SOLUTION Since $\sqrt{a^2 + b^2 + c^2} \neq 0$, the numbers a, b, c are not all zero, and therefore there exist numbers x_0, y_0, z_0 such that

$$ax_0 + by_0 + cz_0 + d = 0.†$$

The equation

$$ax + by + cz + d = 0$$

can now be written

$$(ax + by + cz + d) - (ax_0 + by_0 + cz_0 + d) = 0,$$

and so, after factoring, we have

$$a(x - x_0) + b(y - y_0) + c(z - z_0) = 0.$$

This equation (and hence the initial equation) represents the plane through the point $P(x_0, y_0, z_0)$ with normal $\mathbf{N} = a\mathbf{i} + b\mathbf{j} + c\mathbf{k}$. The initial assumption that $\sqrt{a^2 + b^2 + c^2} \neq 0$ guarantees that $\mathbf{N} \neq \mathbf{0}$. ❏

---

† Would such numbers necessarily exist if $\sqrt{a^2 + b^2 + c^2}$ were zero?

Vector Equation of a Plane

We can write the equation of a plane entirely in vector notation. With

$$\mathbf{N} = A\mathbf{i} + B\mathbf{j} + C\mathbf{k}$$

and

$$\mathbf{r}_0 = x_0\mathbf{i} + y_0\mathbf{j} + z_0\mathbf{k}, \qquad \mathbf{r} = x\mathbf{i} + y\mathbf{j} + z\mathbf{k},$$

Equation (12.6.1) reads

(12.6.2)
$$\mathbf{N} \cdot (\mathbf{r} - \mathbf{r}_0) = 0.$$

This vector equation represents the plane that passes through the tip of $\mathbf{r}_0$ and has normal $\mathbf{N}$. (See Figure 12.6.3.) If the plane passes through the origin, we can take $\mathbf{r}_0 = \mathbf{0}$. Equation (12.6.2), then takes the form

(12.6.3)
$$\mathbf{N} \cdot \mathbf{r} = 0.$$

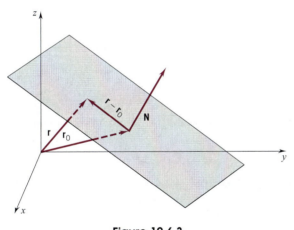

Figure 12.6.3

Collinear Vectors, Coplanar Vectors

Collinear points are points that lie on the same line; *coplanar points* are points that lie on the same plane. The terms ''collinear'' and ''coplanar'' are also applied to vectors: by definition, two vectors $\mathbf{a}$ and $\mathbf{b}$ are said to be *collinear* iff there exist scalars s and t not both 0 such that

$$s\mathbf{a} + t\mathbf{b} = \mathbf{0}.$$

If $s \neq 0$, then $\mathbf{a} = -(t/s)\mathbf{b}$; if $t \neq 0$, then $\mathbf{b} = -(s/t)\mathbf{a}$. Collinear vectors are thus parallel. If we set

$$\mathbf{a} = \overrightarrow{PA} \qquad \text{and} \qquad \mathbf{b} = \overrightarrow{PB},$$

then the points P, A, B all fall on the same line; hence the term ''collinear vectors.''

Three vectors **a, b, c** are said to be *coplanar* iff there exist scalars s, t, u not all zero such that

$$s\mathbf{a} + t\mathbf{b} + u\mathbf{c} = \mathbf{0}.$$

This term, too, is justified:

(12.6.4)

> $\mathbf{a} = \overrightarrow{PA}$, $\mathbf{b} = \overrightarrow{PB}$, $\mathbf{c} = \overrightarrow{PC}$ are coplanar vectors iff the points P, A, B, C all lie on the same plane.

PROOF   Here we show that, if the three vectors are coplanar, then the four points all lie on the same plane. We leave the converse to you (Exercise 45).

Suppose that the three vectors are coplanar. Then we can write

$$s\,\overrightarrow{PA} + t\,\overrightarrow{PB} + u\,\overrightarrow{PC} = \mathbf{0} \qquad \text{with } s, t, u \text{ not all zero.}$$

Without loss of generality, we assume that $s \neq 0$. Then

$$\overrightarrow{PA} = -\frac{t}{s}\,\overrightarrow{PB} - \frac{u}{s}\,\overrightarrow{PC}.$$

Since $\overrightarrow{PB} \times \overrightarrow{PC}$ is perpendicular to both $\overrightarrow{PB}$ and $\overrightarrow{PC}$, we see that

$$(\overrightarrow{PB} \times \overrightarrow{PC}) \cdot \overrightarrow{PA} = (\overrightarrow{PB} \times \overrightarrow{PC}) \cdot \left(-\frac{t}{s}\,\overrightarrow{PB} - \frac{u}{s}\,\overrightarrow{PC} \right)$$

$$= -\frac{t}{s}\,(\overrightarrow{PB} \times \overrightarrow{PC}) \cdot \overrightarrow{PB} - \frac{u}{s}\,(\overrightarrow{PB} \times \overrightarrow{PC}) \cdot \overrightarrow{PC}$$

$$= 0.$$

But $|(\overrightarrow{PB} \times \overrightarrow{PC}) \cdot \overrightarrow{PA}|$ gives the volume of the parallelepiped with edges $\overrightarrow{PA}, \overrightarrow{PB}, \overrightarrow{PC}$. This volume can be zero only if P, A, B, C all lie on the same plane.   ❏

Unit Normals

If **N** is normal to a given plane, then all other normals to that plane are parallel to **N** and hence scalar multiples of **N**. In particular there are only two normals of length 1:

$$\mathbf{u_N} = \frac{\mathbf{N}}{\|\mathbf{N}\|} \qquad \text{and} \qquad -\mathbf{u_N} = -\frac{\mathbf{N}}{\|\mathbf{N}\|}.$$

These are called the *unit normals*.

Example 4   Find the unit normals for the plane

$$3x - 4y + 12z + 8 = 0.$$

SOLUTION   We can take

$$\mathbf{N} = 3\mathbf{i} - 4\mathbf{j} + 12\mathbf{k}.$$

Since

$$\|\mathbf{N}\| = \sqrt{(3)^2 + (-4)^2 + (12)^2} = \sqrt{169} = 13,$$

we have

$$\mathbf{u_N} = \tfrac{1}{13}(3\mathbf{i} - 4\mathbf{j} + 12\mathbf{k}) \qquad \text{and} \qquad -\mathbf{u_N} = -\tfrac{1}{13}(3\mathbf{i} - 4\mathbf{j} + 12\mathbf{k}).   ❏$$

Parallel Planes, Intersecting Planes

Two planes are called *parallel* iff their normals are parallel. If two planes, p_1, p_2 are not parallel, we can find the angle between them by finding the angle between their normals, $\mathbf{N}_1, \mathbf{N}_2$. (See Figure 12.6.4.) Depending on our choice of normals, there are two such angles, each the supplement of the other. We will choose the smaller angle, the one with the nonnegative cosine:

(12.6.5)

$$\cos\theta = |\mathbf{u}_{\mathbf{N}_1} \cdot \mathbf{u}_{\mathbf{N}_2}|.$$

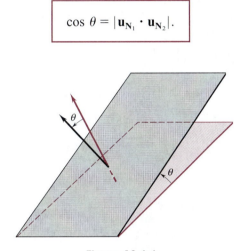

Figure 12.6.4

Example 5 Here are some planes:

$$p_1: 2(x-1) - 3y + 5(z-2) = 0, \qquad p_2: -4x + 6y + 10z + 24 = 0,$$
$$p_3: 4x - 6y - 10z + 1 = 0, \qquad p_4: 2x - 3y + 5z - 12 = 0.$$

(a) Indicate which planes are identical.

(b) Indicate which planes are distinct but parallel.

(c) Find the angle between p_1 and p_2.

SOLUTION

(a) p_1 and p_4 are identical, as you can verify by simplifying the equation of p_1.

(b) p_2 and p_3 are distinct but parallel. The planes are distinct since $P(0, 0, \frac{1}{10})$ lies on p_3 but not on p_2; they are parallel since the normals

$$-4\mathbf{i} + 6\mathbf{j} + 10\mathbf{k} \qquad \text{and} \qquad 4\mathbf{i} - 6\mathbf{j} - 10\mathbf{k}$$

are parallel.

(c) Taking

$$\mathbf{N}_1 = 2\mathbf{i} - 3\mathbf{j} + 5\mathbf{k} \qquad \text{and} \qquad \mathbf{N}_2 = -4\mathbf{i} + 6\mathbf{j} - 10\mathbf{k},$$

we have

$$\mathbf{u}_{\mathbf{N}_1} = \frac{1}{\sqrt{38}}(2\mathbf{i} - 3\mathbf{j} + 5\mathbf{k}) \qquad \text{and} \qquad \mathbf{u}_{\mathbf{N}_2} = \frac{1}{\sqrt{152}}(-4\mathbf{i} + 6\mathbf{j} + 10\mathbf{k}).$$

As you can check,

$$\cos\theta = |\mathbf{u}_{\mathbf{N}_1} \cdot \mathbf{u}_{\mathbf{N}_2}| = \tfrac{6}{19} \cong 0.316 \quad \text{and thus} \quad \theta \cong 1.25 \text{ radians.} \quad \square$$

Example 6 The planes

$$p_1: A_1x + B_1y + C_1z + D_1 = 0, \qquad p_2: A_2x + B_2y + C_2z + D_2 = 0$$

intersect to form a line l. Find a vector parametrization for l.

SOLUTION We need to find a point P_0 on l and a direction vector for l. Finding P_0 is a matter of finding numbers x, y, z that simultaneously satisfy the equations given above for p_1 and p_2. In concrete cases this is not hard, and we will not try to give a general formula for such a P_0. We will just assume that P_0 has been found and focus on finding a direction vector for l.

Since p_1 and p_2 intersect in a line, the normals

$$\mathbf{N}_1 = A_1\mathbf{i} + B_1\mathbf{j} + C_1\mathbf{k}, \qquad \mathbf{N}_2 = A_2\mathbf{i} + B_2\mathbf{j} + C_2\mathbf{k}$$

are not parallel. This guarantees that $\mathbf{N}_1 \times \mathbf{N}_2$ is not $\mathbf{0}$. Since l lies on both p_1 and p_2, the line l, like the vector $\mathbf{N}_1 \times \mathbf{N}_2$, is perpendicular to both $\mathbf{N}_1$ and $\mathbf{N}_2$. This makes l parallel to $\mathbf{N}_1 \times \mathbf{N}_2$. (See Figure 12.6.5.) We can therefore take $\mathbf{N}_1 \times \mathbf{N}_2$ as a direction vector for l and write

$$l: \mathbf{r}(t) = \overrightarrow{OP_0} + t(\mathbf{N}_1 \times \mathbf{N}_2). \quad \square$$

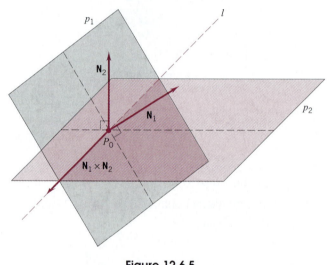

Figure 12.6.5

Example 7 Given the planes

$$p_1: 2x - 3y + 2z = 9 \qquad \text{and} \qquad p_2: x + 2y - z = -4.$$

Show that p_1 and p_2 are not parallel and find scalar parametric equations for the line of intersection l of the two planes.

SOLUTION Normal vectors for p_1 and p_2 are $\mathbf{N}_1 = 2\mathbf{i} - 3\mathbf{j} + 2\mathbf{k}$ and $\mathbf{N}_2 = \mathbf{i} + 2\mathbf{j} - \mathbf{k}$, respectively. Since neither vector is a scalar multiple of the other, the vectors are not parallel and therefore p_1 and p_2 are not parallel.

As shown in Example 6, a direction vector for l is given by

$$\mathbf{N}_1 \times \mathbf{N}_2 = \begin{vmatrix} \mathbf{i} & \mathbf{j} & \mathbf{k} \\ 2 & -3 & 2 \\ 1 & 2 & -1 \end{vmatrix} = -\mathbf{i} + 4\mathbf{j} + 7\mathbf{k}.$$

Now we need a point that lies on l. To find one, we solve the equations p_1 and p_2 simultaneously. If, for example, we set $x = 0$ in the two equations, we get

$$-3y + 2z = 9$$

$$2y - z = -4.$$

Solving this pair of equations for y and z, we find that $y = 1$ and $z = 6$. Thus, the point $(0, 1, 6)$ is on l and scalar parametric equations for l are

$$x = -t, \qquad y = 1 + 4t, \qquad z = 6 + 7t. \quad \square$$

The Plane Determined by Three Noncollinear Points

Suppose now that we are given three noncollinear points P_1, P_2, P_3. These points determine a plane. How can we find an equation for that plane?

First we form the vectors $\overrightarrow{P_1P_2}$, $\overrightarrow{P_1P_3}$. Since P_1, P_2, P_3 are noncollinear, the vectors are not parallel. Therefore their cross product $\overrightarrow{P_1P_2} \times \overrightarrow{P_1P_3}$ can be used as a normal for the plane. We are back in a familiar situation. We have a point of the plane, say P_1, and we have a normal vector, $\overrightarrow{P_1P_2} \times \overrightarrow{P_1P_3}$. A point P will lie on the plane iff

(12.6.6)
$$\boxed{\overrightarrow{P_1P} \cdot (\overrightarrow{P_1P_2} \times \overrightarrow{P_1P_3}) = 0.}$$
(Figure 12.6.6)

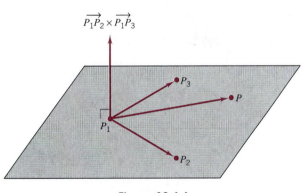

Figure 12.6.6

Example 8 Find an equation in x, y, z for the plane that passes through the points $P_1(0, 1, 1)$, $P_2(1, 0, 1)$, $P_3(1, 1, 0)$.

SOLUTION The point $P = P(x, y, z)$ will lie on this plane iff

$$\overrightarrow{P_1P} \cdot (\overrightarrow{P_1P_2} \times \overrightarrow{P_1P_3}) = 0.$$

Here

$$\overrightarrow{P_1P} = x\mathbf{i} + (y - 1)\mathbf{j} + (z - 1)\mathbf{k}, \qquad \overrightarrow{P_1P_2} = \mathbf{i} - \mathbf{j}, \qquad \overrightarrow{P_1P_3} = \mathbf{i} - \mathbf{k}.$$

As you can check,

$$\overrightarrow{P_1P_2} \times \overrightarrow{P_1P_3} = \mathbf{i} + \mathbf{j} + \mathbf{k}$$

and thus

$$\overrightarrow{P_1P} \cdot (\overrightarrow{P_1P_2} \times \overrightarrow{P_1P_3}) = [x\mathbf{i} + (y-1)\mathbf{j} + (z-1)\mathbf{k}] \cdot [\mathbf{i} + \mathbf{j} + \mathbf{k}]$$
$$= x + (y-1) + (z-1) = x + y + z - 2.$$

An equation for the plane reads

$$x + y + z - 2 = 0. \quad \square$$

The Distance from a Point to a Plane

In Figure 12.6.7, we have drawn a plane p: $Ax + By + Cz + D = 0$ and a point $P_1(x_1, y_1, z_1)$ not on p. The distance between the point P_1 and the plane p is given by the formula

(12.6.7)
$$d(P_1, p) = \frac{|Ax_1 + By_1 + Cz_1 + D|}{\sqrt{A^2 + B^2 + C^2}}.$$

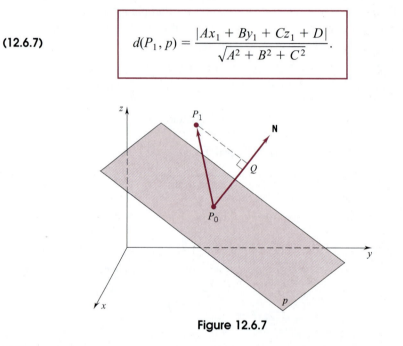

Figure 12.6.7

PROOF Pick any point $P_0(x_0, y_0, z_0)$ in the plane. As a normal to p we can take the vector

$$\mathbf{N} = A\mathbf{i} + B\mathbf{j} + C\mathbf{k}.$$

Then

$$\mathbf{u_N} = \frac{A\mathbf{i} + B\mathbf{j} + C\mathbf{k}}{\sqrt{A^2 + B^2 + C^2}}$$

is the corresponding unit normal. From Figure 12.6.7,

$$d(P_1, p) = d(P_0, Q) = |\text{comp}_\mathbf{N} \, \overrightarrow{P_0P_1}|$$
$$= |\overrightarrow{P_0P_1} \cdot \mathbf{u_N}|$$
$$= \frac{|(x_1 - x_0)A + (y_1 - y_0)B + (z_1 - z_0)C|}{\sqrt{A^2 + B^2 + C^2}}$$
$$= \frac{|Ax_1 + By_1 + Cz_1 - (Ax_0 + By_0 + Cz_0)|}{\sqrt{A^2 + B^2 + C^2}}.$$

Since $P_0(x_0, y_0, z_0)$ lies on p,

$$Ax_0 + By_0 + Cz_0 = -D$$

and

$$d(P_1, p) = \frac{|Ax_1 + By_1 + Cz_1 + D|}{\sqrt{A^2 + B^2 + C^2}}. \quad \square$$

EXERCISES 12.6

1. Which of the points $P(3, 2, 1)$, $Q(2, 3, 1)$, $R(1, 4, 1)$ lie on the plane

$$3(x - 1) + 4y - 5(z + 2) = 0?$$

2. Which of the points $P(2, 1, -2)$, $Q(2, 0, 0)$, $R(4, 1, -1)$, $S(0, -1, -3)$ lie on the plane $\mathbf{N} \cdot (\mathbf{r} - \mathbf{r}_0) = 0$ if $\mathbf{N} = \mathbf{i} - 3\mathbf{j} + \mathbf{k}$ and $\mathbf{r}_0 = 4\mathbf{i} + \mathbf{j} - \mathbf{k}$?

In Exercises 3–9, find an equation for the plane which satisfies the given conditions.

3. Passes through the point $P(2, 3, 4)$ and is perpendicular to $\mathbf{i} - 4\mathbf{j} + 3\mathbf{k}$.

4. Passes through the point $P(1, -2, 3)$ and is perpendicular to $\mathbf{j} + 2\mathbf{k}$.

5. Passes through the point $P(2, 1, 1)$ and is parallel to the plane $3x - 2y + 5z - 2 = 0$.

6. Passes through the point $P(3, -1, 5)$ and is parallel to the plane $4x + 2y - 7z + 5 = 0$.

7. Passes through the point $P(1, 3, 1)$ and contains the line l: $x = t$, $y = t$, $z = -2 + t$.

8. Passes through the point $P(2, 0, 1)$ and contains the line l: $x = 1 - 2t$, $y = 1 + 4t$, $z = 2 + t$.

9. Passes through the point $P_0(x_0, y_0, z_0)$ and is perpendicular to $\overrightarrow{OP_0}$.

In Exercises 10 and 11, find the unit normals to the given plane.

10. $2x - 3y + 7z - 3 = 0$.   11. $2x - y + 5z - 10 = 0$.

12. Show that the plane $x/a + y/b + z/c = 1$ intersects the coordinate axes at $x = a$, $y = b$, $z = c$. This is the equation of a plane in *intercept form*.

In Exercises 13 and 14, write the equation of the given plane in intercept form and find the points where it intersects the coordinate axes.

13. $4x + 5y - 6z = 60$.   14. $3x - y + 4z + 2 = 0$.

In Exercises 15–18, find the angle between the planes.

15. $5(x - 1) - 3(y + 2) + 2z = 0$,
    $x + 3(y - 1) + 2(z + 4) = 0$.

16. $2x - y + 3z = 5$, $5x + 5y - z = 1$.

17. $x - y + z - 1 = 0$, $2x + y + 3z + 5 = 0$.

18. $4x + 4y - 2z = 3$, $2x + y + z = -1$.

In Exercises 19–22, determine whether or not the vectors are coplanar.

19. $4\mathbf{j} - \mathbf{k}$, $3\mathbf{i} + \mathbf{j} + 2\mathbf{k}$, $\mathbf{0}$.

20. $\mathbf{i}$, $\mathbf{i} - 2\mathbf{j}$, $3\mathbf{j} + \mathbf{k}$.

21. $\mathbf{i} + \mathbf{j} + \mathbf{k}$, $2\mathbf{i} - \mathbf{j}$, $3\mathbf{i} - \mathbf{j} - \mathbf{k}$.

22. $\mathbf{j} - \mathbf{k}$, $3\mathbf{i} - \mathbf{j} + 2\mathbf{k}$, $3\mathbf{i} - 2\mathbf{j} + 3\mathbf{k}$.

In Exercises 23–26, find the distance from the point P to the given plane.

23. $P(2, -1, 3)$; $2x + 4y - z + 1 = 0$.

24. $P(3, -5, 2)$; $8x - 2y + z = 5$.

25. $P(1, -3, 5)$; $-3x + 4z + 5 = 0$.

26. $P(1, 3, 4)$; $x + y - 2z = 0$.

In Exercises 27–30, find an equation in x, y, z for the plane that passes through the given points.

27. $P_1(1, 0, 1)$, $P_2(2, 1, 0)$, $P_3(1, 1, 1)$.

28. $P_1(1, 1, 1)$, $P_2(2, -2, -1)$, $P_3(0, 2, 1)$.

29. $P_1(3, -4, 1)$, $P_2(3, 2, 1)$, $P_3(-1, 1, -2)$.

30. $P_1(3, 2, -1)$, $P_2(3, -2, 4)$, $P_3(1, -1, 3)$.

31. Write equations in symmetric form for the line that passes through $P_0(x_0, y_0, z_0)$ and is perpendicular to the plane $Ax + By + Cz + D = 0$.

32. Find the distance between the parallel planes

$$Ax + By + Cz + D_1 = 0$$

and

$$Ax + By + Cz + D_2 = 0.$$

33. Show that the equations of a line in symmetric form

$$\frac{x - x_0}{d_1} = \frac{y - y_0}{d_2} = \frac{z - z_0}{d_3}$$

express the line as an intersection of two planes by finding equations for two such planes.

34. Suppose that we are given equations in x, y, z for two planes that intersect in a line. How can we determine a triple x_0, y_0, z_0 that places $P(x_0, y_0, z_0)$ on the line? One possible procedure is this: select a value for x, say x_0; substitute x_0 in the two equations given; we now have two

linear equations in y and z that we can solve simultaneously. Is this procedure foolproof? If not, identify the circumstances under which the procedure will fail and indicate how the procedure can be made airtight.

In Exercises 35 and 36, find a set of scalar parametric equations for the line formed by the two intersecting planes. HINT: To find a point P_0 on the line, see Exercise 34.

35. $p_1: x + 2y + 3z = 0$, $p_2: -3x + 4y + z = 0$.

36. $p_1: x + y + z + 1 = 0$, $p_2: x - y + z + 2 = 0$.

In Exercises 37 and 38, let l be the line determined by P_1, P_2, and let p be the plane determined by Q_1, Q_2, Q_3. Where, if anywhere, does l intersect p?

37. $P_1(1, -1, 2)$, $P_2(-2, 3, 1)$; $Q_1(2, 0, -4)$, $Q_2(1, 2, 3)$, $Q_3(-1, 2, 1)$.

38. $P_1(4, -3, 1)$, $P_2(2, -2, 3)$; $Q_1(2, 0, -4)$, $Q_2(1, 2, 3)$, $Q_3(-1, 2, 1)$.

39. Let l_1, l_2 be lines that pass through the origin and have direction vectors

$$\mathbf{d} = \mathbf{i} + 2\mathbf{j} + 4\mathbf{k}, \qquad \mathbf{D} = -\mathbf{i} - \mathbf{j} + 3\mathbf{k}.$$

Find an equation for the plane that contains l_1 and l_2.

40. Show that two nonparallel lines $\mathbf{r}(t) = \mathbf{r}_0 + t\mathbf{d}$ and $\mathbf{R}(t) = \mathbf{R}_0 + t\mathbf{D}$ intersect iff the vectors $\mathbf{r}_0 - \mathbf{R}_0$, $\mathbf{d}$, and $\mathbf{D}$ are coplanar.

41. Starting with the plane that contains the point P and has normal $\mathbf{N}$, describe the set of points Q on the plane for which $(\mathbf{N} + \overrightarrow{PQ}) \perp (\mathbf{N} - \overrightarrow{PQ})$. HINT: Draw a figure.

42. Let $\mathbf{a}, \mathbf{b}, \mathbf{c}$ be three nonzero vectors such that the angle between any pair of them is $\frac{1}{2}\pi$. Can these vectors be coplanar?

43. Suppose that $\mathbf{N} = A\mathbf{i} + B\mathbf{j} + C\mathbf{k}$ is a nonzero vector with its initial point on the plane $Ax + By + Cz + D = 0$. Take $P_1(x_1, y_1, z_1)$ as a point of space and set $\alpha = Ax_1 + By_1 + Cz_1 + D$. If $\alpha = 0$, then P_1 lies on the plane. What can you conclude about P_1 if α is positive? If α is negative?

44. Suppose that $\mathbf{a}, \mathbf{b}, \mathbf{c}$ are radius vectors the tips of which are not collinear. Give a geometric interpretation of the equation

$$\begin{vmatrix} x - a_1 & y - a_2 & z - a_3 \\ x - b_1 & y - b_2 & z - b_3 \\ x - c_1 & y - c_2 & z - c_3 \end{vmatrix} = 0.$$

45. Suppose that the points P, A, B, and C all lie on the same plane. Show that the vectors $\mathbf{a} = \overrightarrow{PA}$, $\mathbf{b} = \overrightarrow{PB}$, and $\mathbf{c} = \overrightarrow{PC}$ are coplanar vectors.

Graph Sketching

To sketch the graph of the plane $Ax + By + Cz + D = 0$, find, if possible, the *trace* of the graph in the coordinate

planes. The trace in the xy-plane is found by setting $z = 0$ in the given equation; the result will be the linear equation $Ax + By + D = 0$ which, in general, can be drawn as a line in the xy-plane. Similarly, to find the trace in the xz-plane, set $y = 0$ and sketch the line $Ax + Cz + D = 0$; and to find the trace in the yz-plane, set $x = 0$ and sketch the line $By + Cz + D = 0$. For example, the graph of the plane $3x + 4y + 2z - 12 = 0$ is shown in the following figure.

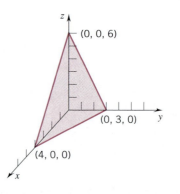

In Exercises 46–49, sketch the graph of the given plane.

46. $x + 2y + 3z - 6 = 0$.   **47.** $5x + 4y + 10z = 20$.

48. $3x + 2y - 6 = 0$.   **49.** $3x + 2z - 12 = 0$.

In Exercises 50–53, find an equation for the plane whose graph is shown in the figure.

50.

51.

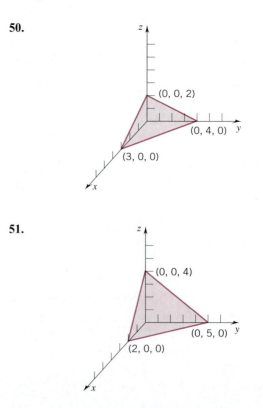

52.

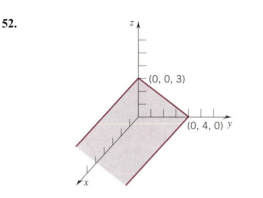

53.

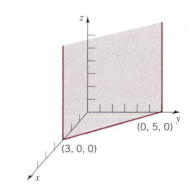

■ 12.7 SOME GEOMETRY BY VECTOR METHODS

Try your hand at proving the following theorems by vector methods. Follow the hints if you like, but you may find it more interesting to disregard them and come up with proofs that are entirely your own.

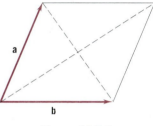

Proposition 1 The diagonals of a parallelogram are perpendicular iff the parallelogram is a rhombus.

HINT FOR PROOF With $\mathbf{a}$ and $\mathbf{b}$ as in Figure 12.7.1, the diagonals are $\mathbf{a} + \mathbf{b}$ and $\mathbf{a} - \mathbf{b}$. Show that $(\mathbf{a} + \mathbf{b}) \cdot (\mathbf{a} - \mathbf{b}) = 0$ iff $\|\mathbf{a}\| = \|\mathbf{b}\|$. ❏

Figure 12.7.1

Proposition 2 Every angle inscribed in a semicircle is a right angle.

HINT FOR PROOF Take $\mathbf{c}$ and $\mathbf{d}$ as in Figure 12.7.2; express $\mathbf{c}$ and $\mathbf{d}$ in terms of $\mathbf{a}$ and $\mathbf{b}$; then show that $\mathbf{c} \cdot \mathbf{d} = 0$. ❏

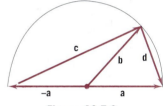

Proposition 3 In a parallelogram the sum of the squares of the lengths of the diagonals equals the sum of the squares of the lengths of the sides.

HINT FOR PROOF With $\mathbf{a}$ and $\mathbf{b}$ as in Figure 12.7.1, the diagonals are $\mathbf{a} + \mathbf{b}$ and $\mathbf{a} - \mathbf{b}$. Show that $\|\mathbf{a} + \mathbf{b}\|^2 + \|\mathbf{a} - \mathbf{b}\|^2 = 2\|\mathbf{a}\|^2 + 2\|\mathbf{b}\|^2$. ❏

Figure 12.7.2

Proposition 4 The three altitudes of a triangle meet at one point.

HINT FOR PROOF As in Figure 12.7.3 assume that the altitudes from P_1 and P_2 intersect at Q. Use the fact that $\overrightarrow{P_1Q} \perp \overrightarrow{OP_2}$ and $\overrightarrow{P_2Q} \perp \overrightarrow{OP_1}$ to show that $\overrightarrow{OQ} \perp \overrightarrow{P_1P_2}$. ❏

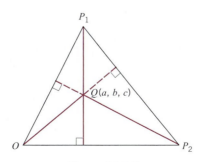

Figure 12.7.3

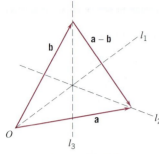

Figure 12.7.4

Proposition 5 The three medians of a triangle meet at one point.

HINT FOR PROOF With l_1, l_2, l_3 as in Figure 12.7.4,

$$l_1: \mathbf{r}_1(t) = t(\mathbf{a} + \mathbf{b}), \qquad l_2: \mathbf{r}_2(u) = \tfrac{1}{2}\mathbf{b} + u(\mathbf{a} - \tfrac{1}{2}\mathbf{b}), \qquad l_3: \mathbf{r}_3(v) = \tfrac{1}{2}\mathbf{a} + v(\mathbf{b} - \tfrac{1}{2}\mathbf{a}).$$

Show that l_1 intersects both l_2 and l_3 at the same point. ❑

Proposition 6 (The Law of Sines) If a triangle has sides $\mathbf{a}$, $\mathbf{b}$, $\mathbf{c}$ and opposite angles A, B, C, then

$$\frac{\sin A}{\|\mathbf{a}\|} = \frac{\sin B}{\|\mathbf{b}\|} = \frac{\sin C}{\|\mathbf{c}\|}.$$

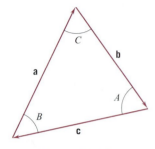

Figure 12.7.5

HINT FOR PROOF With $\mathbf{a}$, $\mathbf{b}$, $\mathbf{c}$ as in Figure 12.7.5, $\mathbf{a} + \mathbf{b} + \mathbf{c} = \mathbf{0}$. Observe then that $\mathbf{a} \times [\mathbf{a} + \mathbf{b} + \mathbf{c}] = \mathbf{0}$ and $\mathbf{b} \times [\mathbf{a} + \mathbf{b} + \mathbf{c}] = \mathbf{0}$. ❑

Proposition 7 If two planes have a point in common, then they have a line in common.

HINT FOR PROOF As equations for the two planes, take $\mathbf{n} \cdot (\mathbf{r} - \mathbf{r}_0) = 0$ and $\mathbf{N} \cdot (\mathbf{R} - \mathbf{R}_0) = 0$. If the point $P(a_1, a_2, a_3)$ lies on both planes, then the vector $\mathbf{a} = a_1\mathbf{i} + a_2\mathbf{j} + a_3\mathbf{k}$ satisfies both equations. In that case we have $\mathbf{n} \cdot (\mathbf{a} - \mathbf{r}_0) = 0$ and $\mathbf{N} \cdot (\mathbf{a} - \mathbf{R}_0) = 0$. Consider the line $\mathbf{r}(t) = \mathbf{a} + t(\mathbf{n} \times \mathbf{N})$. ❑

■ CHAPTER HIGHLIGHTS

12.1 Cartesian Space Coordinates

distance formula (p. 775)   equation of sphere (p. 775)
midpoint formula (p. 777)

12.2 Displacements, Forces and Velocities; Vectors

addition, multiplication by scalars (p. 782)   component (p. 782)
parallel vectors (p. 784)   norm (p. 785)   unit vector (p. 788)

The zero vector $\mathbf{0}$ is parallel to every vector.
Vectors in the plane.

12.3 The Dot Product

$\mathbf{a} \cdot \mathbf{b} = a_1b_1 + a_2b_2 + a_3b_3 = \|\mathbf{a}\| \|\mathbf{b}\| \cos\theta$ $\mathbf{a} \perp \mathbf{b}$ iff $\mathbf{a} \cdot \mathbf{b} = 0$ $\mathbf{a} = \mathbf{a}_\| + \mathbf{a}_\perp$
projections and components (p. 799)   direction angles, direction cosines
Schwarz's inequality (p. 802)   (p. 801)
work (p. 804)

12.4 The Cross Product

definition of $\mathbf{a} \times \mathbf{b}$ (p. 806)   properties of right-handed triples (p. 806)
properties of the cross product (p. 807)   distributive laws (p. 807)
scalar triple product (p. 807)   components of $\mathbf{a} \times \mathbf{b}$ (p. 809)
identities (pp. 812)   torque (p. 813)

12.5 Lines

position vector, radius vector (p. 814)
vector parametrization: $\mathbf{r}(t) = \mathbf{r}_0 + t\mathbf{d}$
direction vector, direction numbers (p. 815)
scalar parametric equations: $x(t) = x_0 + d_1 t, \quad y(t) = y_0 + d_2 t, \quad z(t) = z_0 + d_3 t$

symmetric form (p. 817)
distance from a point to a line (p. 822)

Two lines are parallel iff their direction vectors are parallel; two intersecting lines are perpendicular iff their direction vectors are perpendicular.

12.6 Planes

normal vector (p. 824)
scalar equation: $A(x - x_0) + B(y - y_0) + C(z - z_0) = 0$
vector equation of a plane (p. 827)
collinear vectors, coplanar vectors (p. 827)
parallel planes (p. 829)
angle between intersecting planes (p. 829)
plane determined by three noncollinear points (p. 831)
distance between a point and a plane (p. 832)

12.7 Some Geometry by Vector Methods

■ PROJECTS AND EXPLORATIONS USING TECHNOLOGY

To do these exercises you will need a graphics calculator or a computer with graphing capability. The majority of these problems are open-ended so different approaches may be used to solve them. You should be aware that different approaches can result in slight variations in the answers. Round your numerical answers to at least four decimal places. The rounding method that your calculator or computer uses also may cause variations in answers.

12.1 This exercise uses vectors to look at a geometric problem. To start, consider the triangle T with vertices (0, 0), (3, 0) and (1, 2) in the plane.
   (a) Show that the lengths of the medians of this triangle satisfy the triangle inequality. That is, show that the length of any one of the medians is less than or equal to the sum of the lengths of the other two.
   (b) It follows from the result in part (a) that if we view the medians as vectors, then these vectors form the sides of a triangle M. How does the perimeter of M compare with the perimeter of T? How do the areas of the two triangles compare?
   (c) Compare the angles of T and M.
   (d) What can you say about the triangle formed by the medians of M?
   (e) Repeat parts (a)–(c) for the triangle whose vertices are (0, 0), (3, 0), and (a, b), where (a, b) is an arbitrary point in the first quadrant. Based on these results, conjecture when the triangle formed by the medians is congruent to, or similar to, the original triangle.

12.2 Let P be a point on a circle of radius r that is rolling around the outside of the unit circle. See the figure. The curve traced out by P is called an *epicycloid*.

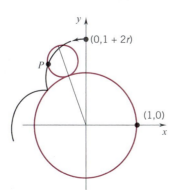

(a) Use vectors to describe the path of P, assuming that the initial position of P is $(0, 1 + 2r)$ and that the circle is rolling at a constant angular velocity ω.

(b) Use a graphing utility to draw the path of P in the case $r = \frac{1}{4}$. Have the graph drawn in such a way that you can watch the path being traced out.

(c) Now use your graphing utility to draw the path of P for several other values of r. Determine conditions on r that will ensure that the path is a closed curve.

(d) Using the graph drawn in part (b), does it appear that the angular velocity of the traced point is constant if the center of the rolling circle is moving at a constant rate.

(e) Repeat parts (a)–(d) for the case where $r < 1$ and the circle is rolling around the inside of the unit circle. The curve in this case is called a *hypocycloid*.

12.3 Consider the triangle R with vertices $A(0, 0, 0)$, $B(4, 0, 0)$, and $C(3, 2, 1)$.

(a) Show that if $0 \le x \le 1$ and $0 \le y \le 1$, then the point $P(4x(1 - y), 2y, y)$ is either inside or on the boundary of R.

(b) Let $f = f(x, y)$ be the sum of the lengths of the vectors from P to the vertices of R. Use a graphing utility to sketch the graph of f.

(c) Based on the graph of f, estimate the absolute minimum value and the absolute maximum value of f.

(d) Let P_m be a point where the absolute minimum value of f occurs. Find the angles between the vectors from P_m to the vertices of R.

(e) Repeat parts (a)–(d) for the triangle S with vertices $A(0, 0, 0)$, $B(4, 0, 0)$, and $C(3, 3, 3)$; and for the triangle T with vertices $A(0, 0, 0)$, $B(1, 0, 0)$, and $C(3, 3, 3)$.

(f) Suppose that the vertices A, B, and C are the locations of retail stores and P is the location of a distribution center. Then the points P_m found in parts (d) and (e) would be the locations of the distribution centers that would minimize transportation costs if the same amount is shipped to each store. Now suppose that twice as much is shipped to B as to A, and three times as much is shipped to C as to A. Use graphs to estimate the minimum transportation costs in the three cases. Find the angles between the vectors from P_m to the vertices of the respective triangles. Is there a relationship between the size of these angles and the corresponding transportation costs?

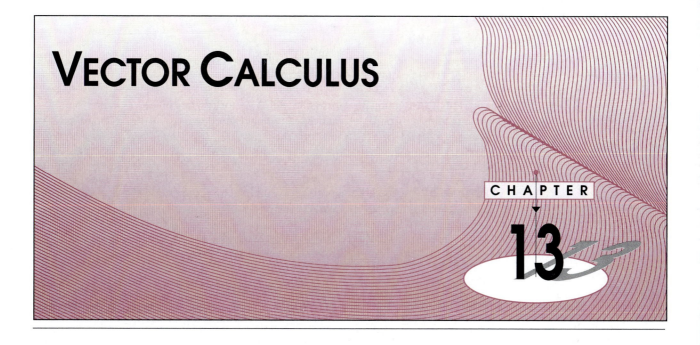

VECTOR CALCULUS

■ 13.1 VECTOR FUNCTIONS

Introduction

If f_1, f_2, f_3 are real-valued functions defined on some interval I, then for each $t \in I$ we can form the vector

$$\mathbf{f}(t) = f_1(t)\mathbf{i} + f_2(t)\mathbf{j} + f_3(t)\mathbf{k}$$

and thereby create a *vector-valued function* $\mathbf{f}$. For short we will call such a function a *vector function*. The real-valued functions f_1, f_2, f_3 are called the *components* of $\mathbf{f}$. A point t is in the *domain* of a vector function $\mathbf{f}$ iff it is in the domain of each of its components. If a domain for $\mathbf{f}$ is not specified explicitly, then we will take the domain to be the common domain of its components.

For example, if we take

$$f_1(t) = x_0 + d_1 t, \qquad f_2(t) = y_0 + d_2 t, \qquad f_3(t) = z_0 + d_3 t$$

we can form the vector function

$$\mathbf{f}(t) = (x_0 + d_1 t)\mathbf{i} + (y_0 + d_2 t)\mathbf{j} + (z_0 + d_3 t)\mathbf{k}.$$

The domain of $\mathbf{f}$ is the set of all real numbers. If d_1, d_2, d_3 are not all 0, then, as we saw in Section 12.5, the radius vector $\mathbf{f}(t)$ traces out the line that passes through the point $P(x_0, y_0, z_0)$ and has direction numbers d_1, d_2, d_3. If d_1, d_2, d_3 are all 0, then we have the constant function

$$\mathbf{f}(t) = x_0\mathbf{i} + y_0\mathbf{j} + z_0\mathbf{k}.$$

Example 1 From the functions

$$f_1(t) = \cos t, \qquad f_2(t) = \sin t, \qquad f_3(t) = 0$$

we can form the vector function

$$\mathbf{f}(t) = \cos t\,\mathbf{i} + \sin t\,\mathbf{j}.$$

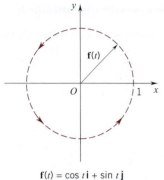

$\mathbf{f}(t) = \cos t\,\mathbf{i} + \sin t\,\mathbf{j}$

Figure 13.1.1

For each t

$$\|\mathbf{f}(t)\| = \sqrt{\cos^2 t + \sin^2 t} = 1.$$

Since the third component is zero, the radius vector $\mathbf{f}(t)$ lies in the xy-plane. Note that the equations $x = \cos t$, $y = \sin t$ are parametric equations of the unit circle. As t increases, the tip of $\mathbf{f}(t)$ traces out the unit circle in a counterclockwise manner, effecting a complete revolution as t increases by 2π. (Figure 13.1.1.) ❑

Example 2 Each real-valued function f defined on some interval $[a, b]$ gives rise to a vector-valued function $\mathbf{f}$ in a natural way. Setting

$$f_1(t) = t, \qquad f_2(t) = f(t), \qquad f_3(t) = 0,$$

we obtain the vector function

$$\mathbf{f}(t) = t\mathbf{i} + f(t)\mathbf{j}.$$

As t ranges from a to b, the radius vector $\mathbf{f}(t)$ traces out the graph of f from left to right. (See Figure 13.1.2.) ❑

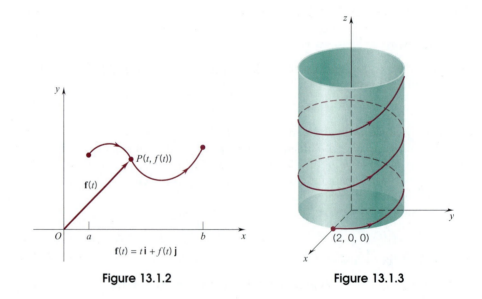

Figure 13.1.2

Figure 13.1.3

Example 3 From the functions

$$f_1(t) = 2\cos t, \qquad f_2(t) = 2\sin t, \qquad f_3(t) = t, \qquad t \geq 0,$$

we can form the vector function

$$\mathbf{f}(t) = 2\cos t\,\mathbf{i} + 2\sin t\,\mathbf{j} + t\mathbf{k}.$$

As indicated, the domain of $\mathbf{f}$ is the set of nonnegative real numbers. At $t = 0$, the radius vector $\mathbf{f}(0)$ has the value $(2, 0, 0)$. As t increases, the tip of $\mathbf{f}(t)$ winds around the circular cylinder $x^2 + y^2 = 4$ (z arbitrary) one revolution for each increase of 2π in t. This curve is called a *circular helix* (Figure 13.1.3). ❑

In general, the *graph* of a vector function $\mathbf{f}(t) = f_1(t)\mathbf{i} + f_2(t)\mathbf{j} + f_3(t)\mathbf{k}$ defined on an interval I is a curve C in the sense that, as t varies over the interval I, the tip of the radius vector $\mathbf{f}$ traces out C. The equations

$$x = f_1(t), \qquad y = f_2(t), \qquad z = f_3(t)$$

corresponding to the components of $\mathbf{f}$ are the parametric equations of C. If one of the components is identically 0 on I, for example, if $\mathbf{f}$ has the form $\mathbf{f}(t) = f_1(t)\mathbf{i} + f_2(t)\mathbf{j}$, then C is said to be a *plane curve*; otherwise C is a *space curve*.

The Limit Process

DEFINITION 13.1.1 LIMIT OF A VECTOR FUNCTION

Let the vector function $\mathbf{f}$ be defined on some interval I containing the point t_0, except possibly at t_0 itself, and let $\mathbf{L}$ be a vector. Then

$$\lim_{t \to t_0} \mathbf{f}(t) = \mathbf{L} \qquad \text{iff} \qquad \lim_{t \to t_0} \|\mathbf{f}(t) - \mathbf{L}\| = 0.$$

Note that for each t, $\|\mathbf{f}(t) - \mathbf{L}\|$ is a real number, and therefore the limit on the right is the limit of a real-valued function. Thus we are still in familiar territory.

The first thing we show is that

(13.1.2)

$$\text{if} \quad \lim_{t \to t_0} \mathbf{f}(t) = \mathbf{L}, \qquad \text{then} \quad \lim_{t \to t_0} \|\mathbf{f}(t)\| = \|\mathbf{L}\|.$$

PROOF By Exercise 22, Section 12.2,

$$0 \le \big| \|\mathbf{f}(t)\| - \|\mathbf{L}\| \big| \le \|\mathbf{f}(t) - \mathbf{L}\|.$$

Thus, by the pinching theorem,

$$\text{if} \quad \lim_{t \to t_0} \|\mathbf{f}(t) - \mathbf{L}\| = 0, \qquad \text{then} \quad \lim_{t \to t_0} \big| \|\mathbf{f}(t)\| - \|\mathbf{L}\| \big| = 0. \quad \square$$

Remark The converse of (13.1.2) is false, that is $\lim_{t \to t_0} \|\mathbf{f}(t)\| = \|\mathbf{L}\|$ does not imply that $\lim_{t \to t_0} \mathbf{f}(t) = \mathbf{L}$. For example, let $\mathbf{a}$ be a nonzero vector; set $\mathbf{f}(t) = \mathbf{a}$ for all t and set $\mathbf{L} = -\mathbf{a}$. Then, for any real number t_0,

$$\lim_{t \to t_0} \|\mathbf{f}(t)\| = \lim_{t \to t_0} \|\mathbf{a}\| = \|-\mathbf{a}\| = \|\mathbf{L}\|$$

but

$$\lim_{t \to t_0} \mathbf{f}(t) = \lim_{t \to t_0} \mathbf{a} = \mathbf{a} \ne -\mathbf{a} = \mathbf{L}. \quad \square$$

We can indicate that $\lim_{t \to t_0} \mathbf{f}(t) = \mathbf{L}$ by writing

$$\text{as } t \to t_0, \quad \mathbf{f}(t) \to \mathbf{L}.$$

We will state the limit rules in this form. As you will see later, there are no surprises.

THEOREM 13.1.3 LIMIT RULES

Let $\mathbf{f}$ and $\mathbf{g}$ be vector functions and let u be a real-valued function. Suppose that, as $t \to t_0$,

$$\mathbf{f}(t) \to \mathbf{L}, \qquad \mathbf{g}(t) \to \mathbf{M}, \qquad u(t) \to A \qquad \text{(A a real number)}$$

Then

$$\mathbf{f}(t) + \mathbf{g}(t) \to \mathbf{L} + \mathbf{M}, \qquad \alpha\mathbf{f}(t) \to \alpha\mathbf{L} \qquad \text{(α real)}$$
$$u(t)\mathbf{f}(t) \to A\mathbf{L}, \qquad \mathbf{f}(t) \cdot \mathbf{g}(t) \to \mathbf{L} \cdot \mathbf{M}, \qquad \mathbf{f}(t) \times \mathbf{g}(t) \to \mathbf{L} \times \mathbf{M}.$$

Each of these limit rules is easy to verify. We will verify the last one. To do this, we have to show that

$$\text{as } t \to t_0, \quad \|[\mathbf{f}(t) \times \mathbf{g}(t)] - [\mathbf{L} \times \mathbf{M}]\| \to 0.$$

We do this as follows:

$$\|[\mathbf{f}(t) \times \mathbf{g}(t)] - [\mathbf{L} \times \mathbf{M}]\| = \|[\mathbf{f}(t) \times \mathbf{g}(t)] - [\mathbf{L} \times \mathbf{g}(t)] + [\mathbf{L} \times \mathbf{g}(t)] - [\mathbf{L} \times \mathbf{M}]\|$$
$$= \|[(\mathbf{f}(t) - \mathbf{L}) \times \mathbf{g}(t)] + [\mathbf{L} \times (\mathbf{g}(t) - \mathbf{M})]\| \qquad (12.4.4)$$
$$\leq \|(\mathbf{f}(t) - \mathbf{L}) \times \mathbf{g}(t)\| + \|\mathbf{L} \times (\mathbf{g}(t) - \mathbf{M})\|$$

(triangle inequality) _____↑

$$\leq \|\mathbf{f}(t) - \mathbf{L}\|\|\mathbf{g}(t)\| + \|\mathbf{L}\|\|\mathbf{g}(t) - \mathbf{M}\|.$$

explain _____↑

As $t \to t_0$, $\|\mathbf{g}(t)\| \to \|\mathbf{M}\|$. [This follows from (13.1.2).] Therefore, as $t \to t_0$,

$$\|\mathbf{f}(t) - \mathbf{L}\| \|\mathbf{g}(t)\| + \|\mathbf{L}\| \|\mathbf{g}(t) - \mathbf{M}\| \to (0)\|\mathbf{M}\| + \|\mathbf{L}\|(0) = 0.$$

If follows from the pinching theorem that $\|[\mathbf{f}(t) \times \mathbf{g}(t)] - [\mathbf{L} \times \mathbf{M}]\| \to 0$. ❏

The limit process can be carried out component by component. Let $\mathbf{f}(t) = f_1(t)\mathbf{i} + f_2(t)\mathbf{j} + f_3(t)\mathbf{k}$ and let $\mathbf{L} = L_1\mathbf{i} + L_2\mathbf{j} + L_3\mathbf{k}$. Then

(13.1.4)

$$\lim_{t \to t_0} \mathbf{f}(t) = \mathbf{L} \qquad \text{iff}$$
$$\lim_{t \to t_0} f_1(t) = L_1, \qquad \lim_{t \to t_0} f_2(t) = L_2, \qquad \lim_{t \to t_0} f_3(t) = L_3.$$

PROOF

$$\lim_{t \to t_0} \mathbf{f}(t) = \mathbf{L} \qquad \text{iff} \qquad \lim_{t \to t_0} \|\mathbf{f}(t) - \mathbf{L}\| = 0$$
$$\text{iff} \qquad \lim_{t \to t_0} \sqrt{[f_1(t) - L_1]^2 + [f_2(t) - L_2]^2 + [f_3(t) - L_3]^2} = 0$$
$$\text{iff} \qquad \lim_{t \to t_0} f_1(t) = L_1, \quad \lim_{t \to t_0} f_2(t) = L_2, \quad \lim_{t \to t_0} f_3(t) = L_3. \quad ❏$$

Remark As you will see, being able to handle the limit of a vector function in terms of corresponding limits of its component functions means that the other processes of calculus, namely differentiation and integration, can also be carried out component-wise. As a result, all that we have learned about the calculus of real-valued functions can be used in the development of the calculus of vector functions. ❏

Example 4   Find $\lim\limits_{t \to 0} \mathbf{f}(t)$ given that

$$\mathbf{f}(t) = \cos(t + \pi)\mathbf{i} + \sin(t + \pi)\mathbf{j} + e^{-t^2}\mathbf{k}.$$

SOLUTION

$$\lim_{t \to 0} \mathbf{f}(t) = \lim_{t \to 0} [\cos(t + \pi)\mathbf{i} + \sin(t + \pi)\mathbf{j} + e^{-t^2}\mathbf{k}]$$

$$= \left[\lim_{t \to 0} \cos(t + \pi)\right]\mathbf{i} + \left[\lim_{t \to 0} \sin(t + \pi)\right]\mathbf{j} + \left[\lim_{t \to 0} e^{-t^2}\right]\mathbf{k}$$

$$= (-1)\mathbf{i} + (0)\mathbf{j} + (1)\mathbf{k} = -\mathbf{i} + \mathbf{k}. \quad \square$$

Continuity and Differentiability

As you would expect, $\mathbf{f}$ is said to be *continuous* at t_0 iff

$$\lim_{t \to t_0} \mathbf{f}(t) = \mathbf{f}(t_0).$$

Thus, by (13.1.4), $\mathbf{f}$ is continuous at t_0 iff each component of $\mathbf{f}$ is continuous at t_0.

The derivative of a vector function is defined as the limit of a *vector difference quotient*:

DEFINITION 13.1.5   DERIVATIVE OF A VECTOR FUNCTION

The vector function $\mathbf{f}$ is *differentiable* at t iff

$$\lim_{h \to 0} \frac{\mathbf{f}(t + h) - \mathbf{f}(t)}{h} \qquad \text{exists.}$$

If this limit exists, it is called the *derivative of $\mathbf{f}$ at t* and is denoted $\mathbf{f}'(t)$. The vector function $\mathbf{f}$ is *differentiable* if it is differentiable at each point of its domain.

Differentiation can be carried out component by component. If $\mathbf{f}(t) = f_1(t)\mathbf{i} + f_2(t)\mathbf{j} + f_3(t)\mathbf{k}$ is differentiable at t, then

$$\mathbf{f}'(t) = f_1'(t)\mathbf{i} + f_2'(t)\mathbf{j} + f_3'(t)\mathbf{k}.$$

PROOF

$$\mathbf{f}'(t) = \lim_{h \to 0} \frac{\mathbf{f}(t + h) - \mathbf{f}(t)}{h}$$

$$= \lim_{h \to 0} \left[\frac{f_1(t + h) - f_1(t)}{h}\mathbf{i} + \frac{f_2(t + h) - f_2(t)}{h}\mathbf{j} + \frac{f_3(t + h) - f_3(t)}{h}\mathbf{k}\right]$$

$$= \left[\lim_{h \to 0} \frac{f_1(t + h) - f_1(t)}{h}\right]\mathbf{i} + \left[\lim_{h \to 0} \frac{f_2(t + h) - f_2(t)}{h}\right]\mathbf{j}$$

$$+ \left[\lim_{h \to 0} \frac{f_3(t + h) - f_3(t)}{h}\right]\mathbf{k}$$

$$= f_1'(t)\mathbf{i} + f_2'(t)\mathbf{j} + f_3'(t)\mathbf{k}. \quad \square$$

As with real-valued functions, if **f** is differentiable at t, then **f** is continuous at t. (Exercise 53.)

Interpretations of the vector derivative and applications of vector differentiation are introduced later in the chapter. Here we limit ourselves to computation.

Example 5 Given that

$$\mathbf{f}(t) = t\mathbf{i} + \sqrt{t + 1}\,\mathbf{j} - e^t\mathbf{k},$$

find

(1) The domain of **f**.   (2) **f**(0).
(3) **f**′(t).   (4) **f**′(0).
(5) ‖**f**(t)‖.   (6) **f**(t) · **f**′(t).

SOLUTION

(1) For a number to be in the domain of **f**, it is necessary only that it be in the domain of each of the components. The domain of **f** is $[-1, \infty)$.

(2) $\mathbf{f}(0) = 0\mathbf{i} + \sqrt{0 + 1}\,\mathbf{j} - e^0\mathbf{k} = \mathbf{j} - \mathbf{k}$.

(3) $\mathbf{f}'(t) = \mathbf{i} + \dfrac{1}{2\sqrt{t + 1}}\,\mathbf{j} - e^t\mathbf{k}$.

(4) $\mathbf{f}'(0) = \mathbf{i} + \dfrac{1}{2\sqrt{0 + 1}}\,\mathbf{j} - e^0\mathbf{k} = \mathbf{i} + \dfrac{1}{2}\mathbf{j} - \mathbf{k}$.

(5) $\|\mathbf{f}(t)\| = \sqrt{t^2 + (\sqrt{t + 1})^2 + (-e^t)^2} = \sqrt{t^2 + t + 1 + e^{2t}}$.

(6) $\mathbf{f}(t) \cdot \mathbf{f}'(t) = (t\mathbf{i} + \sqrt{t + 1}\,\mathbf{j} - e^t\mathbf{k}) \cdot \left(\mathbf{i} + \dfrac{1}{2\sqrt{t + 1}}\,\mathbf{j} - e^t\mathbf{k}\right)$

$$= (t)(1) + (\sqrt{t + 1})\left(\dfrac{1}{2\sqrt{t + 1}}\right) + (-e^t)(-e^t) = t + \dfrac{1}{2} + e^{2t}. \quad \square$$

If **f**′ is itself differentiable, we can form the second derivative **f**″ and so on.

Example 6 Find **f**″(t) for

$$\mathbf{f}(t) = t\sin t\,\mathbf{i} + e^{-t}\mathbf{j} + t\mathbf{k}.$$

SOLUTION

$$\mathbf{f}'(t) = (t\cos t + \sin t)\mathbf{i} - e^{-t}\mathbf{j} + \mathbf{k},$$

$$\mathbf{f}''(t) = (-t\sin t + \cos t + \cos t)\mathbf{i} + e^{-t}\mathbf{j} = (2\cos t - t\sin t)\mathbf{i} + e^{-t}\mathbf{j}. \quad \square$$

Integration

Since we can differentiate vector functions component by component, we can define integration component by component. For $\mathbf{f}(t) = f_1(t)\mathbf{i} + f_2(t)\mathbf{j} + f_3(t)\mathbf{k}$ continuous on $[a, b]$, we set

(13.1.6)
$$\int_a^b \mathbf{f}(t)\,dt = \left(\int_a^b f_1(t)\,dt\right)\mathbf{i} + \left(\int_a^b f_2(t)\,dt\right)\mathbf{j} + \left(\int_a^b f_3(t)\,dt\right)\mathbf{k}.$$

Example 7 Find

$$\int_0^1 \mathbf{f}(t)\, dt \qquad \text{for} \qquad \mathbf{f}(t) = t\mathbf{i} + \sqrt{t+1}\,\mathbf{j} - e^t\mathbf{k}.$$

SOLUTION

$$\int_0^1 \mathbf{f}(t)\, dt = \left(\int_0^1 t\, dt\right)\mathbf{i} + \left(\int_0^1 \sqrt{t+1}\, dt\right)\mathbf{j} + \left(\int_0^1 (-e^t)\, dt\right)\mathbf{k}$$

$$= \left[\tfrac{1}{2}t^2\right]_0^1 \mathbf{i} + \left[\tfrac{2}{3}(t+1)^{3/2}\right]_0^1 \mathbf{j} + \left[-e^t\right]_0^1 \mathbf{k}$$

$$= \tfrac{1}{2}\mathbf{i} + \tfrac{2}{3}(2\sqrt{2} - 1)\mathbf{j} + (1 - e)\mathbf{k}. \quad \square$$

We can calculate indefinite integrals.

Example 8 Find $\mathbf{f}(t)$ given that

$$\mathbf{f}'(t) = 2\cos t\,\mathbf{i} - t\sin t^2\mathbf{j} + 2t\mathbf{k} \qquad \text{and} \qquad \mathbf{f}(0) = \mathbf{i} + 3\mathbf{k}.$$

SOLUTION By integrating $\mathbf{f}'(t)$, we find that

$$\mathbf{f}(t) = (2\sin t + C_1)\mathbf{i} + (\tfrac{1}{2}\cos t^2 + C_2)\mathbf{j} + (t^2 + C_3)\mathbf{k},$$

where C_1, C_2, C_3 are constants to be determined. Since

$$\mathbf{i} + 3\mathbf{k} = \mathbf{f}(0) = C_1\mathbf{i} + (\tfrac{1}{2} + C_2)\mathbf{j} + C_3\mathbf{k},$$

you can see that

$$C_1 = 1, \qquad C_2 = -\tfrac{1}{2}, \qquad C_3 = 3.$$

Thus

$$\mathbf{f}(t) = (2\sin t + 1)\mathbf{i} + (\tfrac{1}{2}\cos t^2 - \tfrac{1}{2})\mathbf{j} + (t^2 + 3)\mathbf{k}. \quad \square$$

(Integration can also be carried out without direct reference to components. See Exercise 54.)

Properties of the Integral

It is easy to see that

(13.1.7)

$$\int_a^b [\mathbf{f}(t) + \mathbf{g}(t)]\, dt = \int_a^b \mathbf{f}(t)\, dt + \int_a^b \mathbf{g}(t)\, dt$$

and

(13.1.8)

$$\int_a^b [\alpha\mathbf{f}(t)]\, dt = \alpha\int_a^b \mathbf{f}(t)\, dt \qquad \text{for every scalar } \alpha.$$

It is also true that

(13.1.9)

$$\int_a^b [\mathbf{c} \cdot \mathbf{f}(t)] \, dt = \mathbf{c} \cdot \left(\int_a^b \mathbf{f}(t) \, dt \right) \qquad \text{for every constant vector } \mathbf{c}$$

and

(13.1.10)

$$\left\| \int_a^b \mathbf{f}(t) \, dt \right\| \le \int_a^b \| \mathbf{f}(t) \| \, dt.$$

The proof of (13.1.9) is left as an exercise. (See Exercise 56.) Here we prove (13.1.10). It is an important inequality.

PROOF Set

$$\mathbf{r} = \int_a^b \mathbf{f}(t) \, dt$$

and note that

$$\|\mathbf{r}\|^2 = \mathbf{r} \cdot \mathbf{r} = \mathbf{r} \cdot \int_a^b \mathbf{f}(t) \, dt = \int_a^b [\mathbf{r} \cdot \mathbf{f}(t)] \, dt \le \int_a^b \|\mathbf{r}\| \, \|\mathbf{f}(t)\| \, dt = \|\mathbf{r}\| \int_a^b \|\mathbf{f}(t)\| \, dt.$$

by (13.1.9) ⟶ ⟵ by Schwarz's inequality (12.3.17)

If $\mathbf{r} \neq \mathbf{0}$, we can divide by $\|\mathbf{r}\|$ and conclude that

$$\|\mathbf{r}\| \le \int_a^b \|\mathbf{f}(t)\| \, dt.$$

If $\mathbf{r} = \mathbf{0}$, the result is obvious in the first place. ❏

EXERCISES 13.1

In Exercise 1–8, differentiate the given vector-valued function.

1. $\mathbf{f}(t) = (1 + 2t)\mathbf{i} + (3 - t)\mathbf{j} + (2 + 3t)\mathbf{k}$.

2. $\mathbf{f}(t) = 2\mathbf{i} - \cos t \, \mathbf{k}$.

3. $\mathbf{f}(t) = \sqrt{1 - t}\,\mathbf{i} + \sqrt{1 + t}\,\mathbf{j} + (1 - t)^{-1}\mathbf{k}$.

4. $\mathbf{f}(t) = e^t\mathbf{i} + \ln t \, \mathbf{j} + \tan^{-1} t \, \mathbf{k}$.

5. $\mathbf{f}(t) = \sin t \, \mathbf{i} + \cos t \, \mathbf{j} + \tan t \, \mathbf{k}$.

6. $\mathbf{f}(t) = e^t(\mathbf{i} + t\mathbf{j} + t^2\mathbf{k})$.

7. $\mathbf{f}(t) = \ln(1 - t)\mathbf{i} + \cos t \, \mathbf{j} + t^2\mathbf{k}$.

8. $\mathbf{f}(t) = \dfrac{t + 1}{t - 1}\mathbf{i} + te^{2t}\mathbf{j} + \sec t \, \mathbf{k}$.

In Exercises 9–12, calculate the second derivative of the given vector-valued function.

9. $\mathbf{f}(t) = 4t\mathbf{i} + 2t^3\mathbf{j} + (t^2 + 2t)\mathbf{k}$.

10. $\mathbf{f}(t) = t \sin t \, \mathbf{i} + t \cos t \, \mathbf{k}$.

11. $\mathbf{f}(t) = e^t \cos t \, \mathbf{i} + e^t \sin t \, \mathbf{j} + 4t\mathbf{k}$.

12. $\mathbf{f}(t) = \sqrt{t}\,\mathbf{i} + t\sqrt{t}\,\mathbf{j} + \ln t \, \mathbf{k}$.

In Exercises 13–18, Calculate the given integral.

13. $\displaystyle \int_1^2 \mathbf{f}(t) \, dt \quad \text{for} \quad \mathbf{f}(t) = \mathbf{i} + 2t\mathbf{j}$.

14. $\displaystyle \int_0^\pi \mathbf{r}(t) \, dt \quad \text{for} \quad \mathbf{r}(t) = \sin t \, \mathbf{i} + \cos t \, \mathbf{j} + t\mathbf{k}$.

15. $\int_0^1 \mathbf{g}(t) \, dt$ for $\mathbf{g}(t) = e^t \mathbf{i} + e^{-t} \mathbf{k}$.

16. $\int_0^1 \mathbf{h}(t) \, dt$ for $\mathbf{h}(t) = e^{-t}[t^2 \mathbf{i} + \sqrt{2} \, t \mathbf{j} + \mathbf{k}]$.

17. $\int_0^1 \mathbf{f}(t) \, dt$ for $\mathbf{f}(t) = \dfrac{1}{1 + t^2} \mathbf{i} + \sec^2 t \mathbf{j}$.

18. $\int_1^3 \mathbf{F}(t) \, dt$ for $\mathbf{F}(t) = \dfrac{1}{t} \mathbf{i} + \dfrac{\ln t}{t} \mathbf{j} + e^{-2t} \mathbf{k}$.

In Exercises 19–24, find $\lim\limits_{t \to 0} \mathbf{f}(t)$ if it exists.

19. $\mathbf{f}(t) = \dfrac{\sin t}{2t} \mathbf{i} + e^{2t} \mathbf{j} + \dfrac{t^2}{e^t} \mathbf{k}$.

20. $\mathbf{f}(t) = 3(t^2 - 1)\mathbf{i} + \cos t \mathbf{j} + \dfrac{t}{|t|} \mathbf{k}$.

21. $\mathbf{f}(t) = t^2 \mathbf{i} + \dfrac{1 - \cos t}{3t} \mathbf{j} + \dfrac{t}{t + 1} \mathbf{k}$.

22. $\mathbf{f}(t) = 3t \mathbf{i} + (t^2 + 1)\mathbf{j} + e^{2t} \mathbf{k}$.

23. $\mathbf{f}(t) = \dfrac{1}{t} \mathbf{i} + \cos t \mathbf{j} + \dfrac{e^t}{t} \mathbf{k}$.

24. $\mathbf{f}(t) = (t^2 - 1)\mathbf{i} + t \sin\left(\dfrac{1}{t}\right)\mathbf{j} + \sqrt{t} \mathbf{k}$.

In Exercises 25–34, sketch the curve represented by the given vector-valued function and indicate the orientation.

25. $\mathbf{r}(t) = 2t \mathbf{i} + t^2 \mathbf{j}, \quad t \geq 0$.

26. $\mathbf{r}(t) = t^3 \mathbf{i} + 2t \mathbf{j}, \quad t \geq 0$.

27. $\mathbf{r}(t) = 2 \sinh t \mathbf{i} + 2 \cosh t \mathbf{j}, \quad t \geq 0$.

28. $\mathbf{r}(t) = 3 \cos t \mathbf{i} + 3 \sin t \mathbf{k}, \quad 0 \leq t \leq 2\pi$.

29. $\mathbf{r}(t) = 2 \cos t \mathbf{i} + 3 \sin t \mathbf{j}, \quad 0 \leq t \leq 2\pi$.

30. $\mathbf{r}(t) = 2t \mathbf{i} + (5 - 2t)\mathbf{j} + 3t \mathbf{k}, \quad t \geq 0$.

31. $\mathbf{r}(t) = (t^2 + 1)\mathbf{i} + t \mathbf{j} + 4 \mathbf{k}, \quad -2 \leq t \leq 2$.

32. $\mathbf{r}(t) = t \mathbf{i} + t^2 \mathbf{j} + t^3 \mathbf{k}, \quad 0 \leq t \leq 4$.

33. $\mathbf{r}(t) = 2 \cos t \mathbf{i} + 2 \sin t \mathbf{j} + (2\pi - t)\mathbf{k}, \quad 0 \leq t \leq 2\pi$.

34. $\mathbf{r}(t) = 3 \sin t \mathbf{i} + 4 \cos t \mathbf{j} + e^{-t} \mathbf{k}, \quad t \geq 0$.

▶ In Exercises 35–37, use a graphing utility to sketch the curve represented by the given vector-valued function and indicate the orientation.

35. $\mathbf{r}(t) = 2 \cos (t^2)\mathbf{i} + (2 - \sqrt{t})\mathbf{j}$.

36. $\mathbf{r}(t) = e^{\cos 2t} \mathbf{i} + e^{-\sin t} \mathbf{j}$.

37. $\mathbf{r}(t) = (2 - \sin 2t)\mathbf{i} + (3 + 2 \cos t)\mathbf{j}$.

▶ 38. The vector-valued function

$$\mathbf{r}(t) = (A \cos at + B \cos bt)\mathbf{i} + A \sin at \, \mathbf{j} + B \sin bt \, \mathbf{k}$$

where A, B, a, and b are constants, describes the motion of an object on a torus. Use a graphing utility that can plot three-dimensional graphs to sketch the curve generated by letting:

(a) $A = 2, B = 1, a = 1, b = 1$.
(b) $A = 1, B = 2, a = 2, b = 1$.

In Exercises 39–44, find a vector-valued function $\mathbf{f}$ that traces out the given curve in the indicated direction.

39. $4x^2 + 9y^2 = 36$ (a) Counterclockwise. (b) Clockwise.

40. $(x - 1)^2 + y^2 = 1$ (a) Counterclockwise. (b) Clockwise.

41. $y = x^2$ (a) From left to right. (b) From right to left.

42. $y = x^3$ (a) From left to right. (b) From right to left.

43. The directed line segment from $(1, 4, -2)$ to $(3, 9, 6)$.

44. The directed line segment from $(3, 2, -5)$ to $(7, 2, 9)$.

45. Set $\mathbf{f}(t) = t \mathbf{i} + f(t)\mathbf{j}$ and calculate

$$\mathbf{f}'(t_0), \qquad \int_a^b \mathbf{f}(t) \, dt, \qquad \int_a^b \mathbf{f}'(t) \, dt$$

given that

$$f'(t_0) = m, \quad f(a) = c, \quad f(b) = d, \quad \int_a^b f(t) \, dt = A.$$

In Exercises 46–49, find $\mathbf{f}(t)$ from the following information.

46. $\mathbf{f}'(t) = t \mathbf{i} + t(1 + t^2)^{-1/2} \mathbf{j} + t e^t \mathbf{k}$ and $\mathbf{f}(0) = \mathbf{i} + 2\mathbf{j} + 3\mathbf{k}$.

47. $\mathbf{f}'(t) = \mathbf{i} + t^2 \mathbf{j}$ and $\mathbf{f}(0) = \mathbf{j} - \mathbf{k}$.

48. $\mathbf{f}'(t) = 2\mathbf{f}(t)$ and $\mathbf{f}(0) = \mathbf{i} - \mathbf{k}$.

49. $\mathbf{f}'(t) = \alpha \mathbf{f}(t)$ with α a real number and $\mathbf{f}(0) = \mathbf{c}$.

50. No ϵ, δ's have surfaced so far, but they are still there at the heart of the limit process. Give an ϵ, δ characterization of

$$\lim_{t \to t_0} \mathbf{f}(t) = \mathbf{L}.$$

51. (a) Show that, if $\mathbf{f}'(t) = \mathbf{0}$ for all t in an interval I, then $\mathbf{f}$ is a constant vector on I.
(b) Show that, if $\mathbf{f}'(t) = \mathbf{g}'(t)$ for all t in an interval I, then $\mathbf{f}$ and $\mathbf{g}$ differ by a constant vector on I.

52. Assume that, as $t \to t_0$, $\mathbf{f}(t) \to \mathbf{L}$ and $\mathbf{g}(t) \to \mathbf{M}$. Show that

$$\mathbf{f}(t) \cdot \mathbf{g}(t) \to \mathbf{L} \cdot \mathbf{M}.$$

53. Show that, if $\mathbf{f}$ is differentiable at t, then $\mathbf{f}$ is continuous at t.

54. A vector-valued function $\mathbf{G}$ is called an *antiderivative* for $\mathbf{f}$ on $[a, b]$ iff (i) $\mathbf{G}$ is continuous on $[a, b]$ and (ii) $\mathbf{G}'(t) = \mathbf{f}(t)$ for all $t \in (a, b)$. Show that, if $\mathbf{f}$ is continuous on $[a, b]$ and $\mathbf{G}$ is an antiderivative for $\mathbf{f}$ on $[a, b]$, then

$$\int_a^b \mathbf{f}(t) \, dt = \mathbf{G}(b) - \mathbf{G}(a).$$

(This is a vector version of the fundamental theorem of integral calculus.)

55. Is it always true that

$$\int_a^b [\mathbf{f}(t) \cdot \mathbf{g}(t)] \, dt = \left[\int_a^b \mathbf{f}(t) \, dt \right] \cdot \left[\int_a^b \mathbf{g}(t) \, dt \right]?$$

56. Prove that, if $\mathbf{f}$ is continuous on $[a, b]$, then for each vector $\mathbf{c}$

$$\int_a^b [\mathbf{c} \cdot \mathbf{f}(t)] \, dt = \mathbf{c} \cdot \int_a^b \mathbf{f}(t) \, dt \quad \text{and}$$

$$\int_a^b [\mathbf{c} \times \mathbf{f}(t)] \, dt = \mathbf{c} \times \int_a^b \mathbf{f}(t) \, dt.$$

57. Let $\mathbf{f}$ be a differentiable vector-valued function. Show that if $\|\mathbf{f}(t)\| \neq 0$, then

$$\frac{d}{dt}(\|\mathbf{f}(t)\|) = \frac{\mathbf{f}(t) \cdot \mathbf{f}'(t)}{\|\mathbf{f}(t)\|}.$$

58. Let $\mathbf{f}$ be a differentiable vector-valued function. Show that if $\|\mathbf{f}(t)\| \neq 0$, then

$$\frac{d}{dt}\left(\frac{\mathbf{f}(t)}{\|\mathbf{f}(t)\|} \right) = \frac{\mathbf{f}'(t)}{\|\mathbf{f}(t)\|} - \frac{\mathbf{f}(t) \cdot \mathbf{f}'(t)}{\|\mathbf{f}(t)\|^3} \mathbf{f}(t).$$

■ 13.2 DIFFERENTIATION FORMULAS

Vector functions with a common domain can be combined in many ways to form new functions. From $\mathbf{f}$ and $\mathbf{g}$ we can form the sum $\mathbf{f} + \mathbf{g}$:

$$(\mathbf{f} + \mathbf{g})(t) = \mathbf{f}(t) + \mathbf{g}(t).$$

We can form scalar multiples $\alpha\mathbf{f}$ and thus linear combinations $\alpha\mathbf{f} + \beta\mathbf{g}$:

$$(\alpha\mathbf{f})(t) = \alpha\mathbf{f}(t), \qquad (\alpha\mathbf{f} + \beta\mathbf{g})(t) = \alpha\mathbf{f}(t) + \beta\mathbf{g}(t).$$

We can form the dot product $\mathbf{f} \cdot \mathbf{g}$:

$$(\mathbf{f} \cdot \mathbf{g})(t) = \mathbf{f}(t) \cdot \mathbf{g}(t).$$

We can also form the cross product $\mathbf{f} \times \mathbf{g}$:

$$(\mathbf{f} \times \mathbf{g})(t) = \mathbf{f}(t) \times \mathbf{g}(t).$$

Of course, these operations on vector functions are simply the pointwise application of the algebraic operations on vectors that we introduced in Chapter 12.

There are two ways of bringing *scalar functions* (real-valued functions) into play. If a scalar function u has the same domain as $\mathbf{f}$, we can form the scalar product $u\mathbf{f}$:

$$(u\mathbf{f})(t) = u(t)\mathbf{f}(t).$$

If $u(t)$ is in the domain of $\mathbf{f}$ for each t in some interval, then we can form the composition $\mathbf{f} \circ u$:

$$(\mathbf{f} \circ u)(t) = \mathbf{f}(u(t)).$$

Example 1 Let $u(t) = t^2$ and $\mathbf{f}(t) = e^t\mathbf{i} + \sin 2t\mathbf{j} + \sqrt{t^2 + 1}\mathbf{k}$. Then

$$(u\mathbf{f})(t) = u(t)\mathbf{f}(t) = t^2e^t\mathbf{i} + t^2\sin 2t\mathbf{j} + t^2\sqrt{t^2 + 1}\mathbf{k}$$

and

$$(\mathbf{f} \circ u)(t) = \mathbf{f}(u(t)) = e^{t^2}\mathbf{i} + \sin 2t^2\mathbf{j} + \sqrt{t^4 + 1}\mathbf{k}. \quad \square$$

From Theorem 13.1.3 it is clear that, if $\mathbf{f}$, $\mathbf{g}$, and u are continuous on a common domain, then $\mathbf{f} + \mathbf{g}$, $\alpha\mathbf{f}$, $\mathbf{f} \cdot \mathbf{g}$, $\mathbf{f} \times \mathbf{g}$, and $u\mathbf{f}$ are all continuous on that same set. We have yet to show the continuity of $\mathbf{f} \circ u$. The verification of that is left to you. (Exercise 36.) What interests us here is that, if $\mathbf{f}$, $\mathbf{g}$, and u are differentiable, then the newly constructed functions are also differentiable and their derivatives satisfy the following rules:

(13.2.1)

(1) $(\mathbf{f} + \mathbf{g})'(t) = \mathbf{f}'(t) + \mathbf{g}'(t).$

(2) $(\alpha\mathbf{f})'(t) = \alpha\mathbf{f}'(t),$ (α constant).

(3) $(u\mathbf{f})'(t) = u(t)\mathbf{f}'(t) + u'(t)\mathbf{f}(t).$

(4) $(\mathbf{f} \cdot \mathbf{g})'(t) = [\mathbf{f}(t) \cdot \mathbf{g}'(t)] + [\mathbf{f}'(t) \cdot \mathbf{g}(t)].$

(5) $(\mathbf{f} \times \mathbf{g})'(t) = [\mathbf{f}(t) \times \mathbf{g}'(t)] + [\mathbf{f}'(t) \times \mathbf{g}(t)].$

(6) $(\mathbf{f} \circ u)'(t) = \mathbf{f}'(u(t))u'(t) = u'(t)\mathbf{f}'(u(t))$ (the chain rule).

Rules (3), (4), (5) are all "product" rules and should remind you of the rule for differentiating the product of ordinary functions. Keep in mind, however, that the cross product is not commutative and therefore the order in Rule (5) is important.

In Rule (6) we first wrote the scalar part $u'(t)$ on the right so that the formula would look like the chain rule for ordinary functions. In general, $\mathbf{a}\alpha$ has the same meaning as $\alpha\mathbf{a}$.

Example 2 Taking

$$\mathbf{f}(t) = 2t^2\mathbf{i} - 3\mathbf{j}, \qquad \mathbf{g}(t) = \mathbf{i} + t\mathbf{j} + t^2\mathbf{k}, \qquad u(t) = \tfrac{1}{3}t^3$$

we have

$$\mathbf{f}'(t) = 4t\mathbf{i}, \qquad \mathbf{g}'(t) = \mathbf{j} + 2t\mathbf{k}, \qquad u'(t) = t^2.$$

Therefore

(a) $(\mathbf{f} + \mathbf{g})'(t) = \mathbf{f}'(t) + \mathbf{g}'(t) = 4t\mathbf{i} + (\mathbf{j} + 2t\mathbf{k}) = 4t\mathbf{i} + \mathbf{j} + 2t\mathbf{k};$

(b) $(2\mathbf{f})'(t) = 2\mathbf{f}'(t) = 2(4t\mathbf{i}) = 8t\mathbf{i};$

(c) $(u\mathbf{f})'(t) = u(t)\mathbf{f}'(t) + u'(t)\mathbf{f}(t) = \tfrac{1}{3}t^3(4t\mathbf{i}) + t^2(2t^2\mathbf{i} - 3\mathbf{j}) = \tfrac{10}{3}t^4\mathbf{i} - 3t^2\mathbf{j};$

(d) $(\mathbf{f} \cdot \mathbf{g})'(t) = [\mathbf{f}(t) \cdot \mathbf{g}'(t)] + [\mathbf{f}'(t) \cdot \mathbf{g}(t)]$
$= [(2t^2\mathbf{i} - 3\mathbf{j}) \cdot (\mathbf{j} + 2t\mathbf{k})] + [4t\mathbf{i} \cdot (\mathbf{i} + t\mathbf{j} + t^2\mathbf{k})] = -3 + 4t;$

(e) $(\mathbf{f} \times \mathbf{g})'(t) = [\mathbf{f}(t) \times \mathbf{g}'(t)] + [\mathbf{f}'(t) \times \mathbf{g}(t)]$
$= [(2t^2\mathbf{i} - 3\mathbf{j}) \times (\mathbf{j} + 2t\mathbf{k})] + [4t\mathbf{i} \times (\mathbf{i} + t\mathbf{j} + t^2\mathbf{k})]$
$= (2t^2\mathbf{k} - 4t^3\mathbf{j} - 6t\mathbf{i}) + (4t^2\mathbf{k} - 4t^3\mathbf{j}) = -6t\mathbf{i} - 8t^3\mathbf{j} + 6t^2\mathbf{k}$

while

$(\mathbf{g} \times \mathbf{f})'(t) = [\mathbf{g}(t) \times \mathbf{f}'(t)] + [\mathbf{g}'(t) \times \mathbf{f}(t)]$
$= [(\mathbf{i} + t\mathbf{j} + t^2\mathbf{k}) \times 4t\mathbf{i}] + [(\mathbf{j} + 2t\mathbf{k}) \times (2t^2\mathbf{i} - 3\mathbf{j})]$
$= (-4t^2\mathbf{k} + 4t^3\mathbf{j}) + (-2t^2\mathbf{k} + 4t^3\mathbf{j} + 6t\mathbf{i})$
$= 6t\mathbf{i} + 8t^3\mathbf{j} - 6t^2\mathbf{k} = -(\mathbf{f} \times \mathbf{g})'(t);$

(f) $(\mathbf{f} \circ u)'(t) = \mathbf{f}'(u(t))u'(t)$
$= [4u(t)\mathbf{i}]u'(t) = [4(\tfrac{1}{3}t^3)\mathbf{i}]\,t^2 = \tfrac{4}{3}t^5\mathbf{i}.$ ❑

The differentiation formulas that we have given can all be derived component by component, and they can all be derived in a component-free manner. Take for example formula (3):

$$(u\mathbf{f})'(t) = u(t)\mathbf{f}'(t) + u'(t)\mathbf{f}(t).$$

COMPONENT-BY-COMPONENT DERIVATION   Set

$$\mathbf{f}(t) = f_1(t)\mathbf{i} + f_2(t)\mathbf{j} + f_3(t)\mathbf{k}.$$

Then

$$(u\mathbf{f})(t) = u(t)\mathbf{f}(t) = u(t)f_1(t)\mathbf{i} + u(t)f_2(t)\mathbf{j} + u(t)f_3(t)\mathbf{k}$$

and

$$\begin{aligned}
(u\mathbf{f})'(t) &= [u(t)f_1'(t) + u'(t)f_1(t)]\mathbf{i} + [u(t)f_2'(t) + u'(t)f_2(t)]\mathbf{j} \\
&\quad + [u(t)f_3'(t) + u'(t)f_3(t)]\mathbf{k} \\
&= u(t)[f_1'(t)\mathbf{i} + f_2'(t)\mathbf{j} + f_3'(t)\mathbf{k}] + u'(t)[f_1(t)\mathbf{i} + f_2(t)\mathbf{j} + f_3(t)\mathbf{k}] \\
&= u(t)\mathbf{f}'(t) + u'(t)\mathbf{f}(t). \quad \square
\end{aligned}$$

COMPONENT-FREE DERIVATION   We find $(u\mathbf{f})'(t)$ by taking the limit as $h \to 0$ of the difference quotient

$$\frac{u(t+h)\mathbf{f}(t+h) - u(t)\mathbf{f}(t)}{h}.$$

By adding and subtracting $u(t+h)\mathbf{f}(t)$, we can rewrite this quotient as

$$\frac{u(t+h)\,\mathbf{f}(t+h) - u(t+h)\,\mathbf{f}(t) + u(t+h)\,\mathbf{f}(t) - u(t)\,\mathbf{f}(t)}{h}$$

which is equal to

$$u(t+h)\frac{\mathbf{f}(t+h) - \mathbf{f}(t)}{h} + \frac{u(t+h) - u(t)}{h}\mathbf{f}(t).$$

As $h \to 0$,

$$u(t+h) \to u(t), \qquad \text{(differentiable functions are continuous)}$$

$$\frac{\mathbf{f}(t+h) - \mathbf{f}(t)}{h} \to \mathbf{f}'(t), \qquad \text{(definition of derivative for vector functions)}$$

$$\frac{u(t+h) - u(t)}{h} \to u'(t). \qquad \text{(definition of derivative for scalar functions)}$$

It follows from the limit rules (Theorem 13.1.3) that

$$u(t+h)\frac{\mathbf{f}(t+h) - \mathbf{f}(t)}{h} \to u(t)\mathbf{f}'(t), \qquad \frac{u(t+h) - u(t)}{h}\mathbf{f}(t) \to u'(t)\mathbf{f}(t)$$

and therefore

$$u(t+h)\frac{\mathbf{f}(t+h) - \mathbf{f}(t)}{h} + \frac{u(t+h) - u(t)}{h}\mathbf{f}(t) \to u(t)\mathbf{f}'(t) + u'(t)\mathbf{f}(t). \quad \square$$

In Leibniz's notation the formulas take the following form:

(13.2.2)

$$(1) \quad \frac{d}{dt}(\mathbf{f} + \mathbf{g}) = \frac{d\mathbf{f}}{dt} + \frac{d\mathbf{g}}{dt}.$$

$$(2) \quad \frac{d}{dt}(\alpha\mathbf{f}) = \alpha\frac{d\mathbf{f}}{dt}. \qquad (\alpha \text{ constant})$$

$$(3) \quad \frac{d}{dt}(u\mathbf{f}) = u\frac{d\mathbf{f}}{dt} + \frac{du}{dt}\mathbf{f}. \qquad (u = u(t))$$

$$(4) \quad \frac{d}{dt}(\mathbf{f} \cdot \mathbf{g}) = \left(\mathbf{f} \cdot \frac{d\mathbf{g}}{dt}\right) + \left(\frac{d\mathbf{f}}{dt} \cdot \mathbf{g}\right).$$

$$(5) \quad \frac{d}{dt}(\mathbf{f} \times \mathbf{g}) = \left(\mathbf{f} \times \frac{d\mathbf{g}}{dt}\right) + \left(\frac{d\mathbf{f}}{dt} \times \mathbf{g}\right).$$

$$(6) \quad \frac{d\mathbf{f}}{dt} = \frac{d\mathbf{f}}{du}\frac{du}{dt}. \qquad (\mathbf{f} = \mathbf{f}(u(t))) \quad \text{(chain rule)}$$

We come now to a couple of results that will prove useful as we go on.

Example 3 Let $\mathbf{r}$ be a differentiable vector function of t and set $r = \|\mathbf{r}\|$. Show that r is differentiable wherever it is not 0 and

(13.2.3)

$$\mathbf{r} \cdot \frac{d\mathbf{r}}{dt} = r\frac{dr}{dt}.$$

SOLUTION If $\mathbf{r}$ is differentiable, then

$$\mathbf{r} \cdot \mathbf{r} = \|\mathbf{r}\|^2 = r^2$$

is differentiable. Let's assume now that $r \neq 0$. Since the square-root function is differentiable at all positive numbers and r^2 is positive, we can apply the square-root function to r^2 and conclude by the chain rule that r is itself differentiable.

To obtain the formula we differentiate the identity $\mathbf{r} \cdot \mathbf{r} = r^2$:

$$\mathbf{r} \cdot \frac{d\mathbf{r}}{dt} + \frac{d\mathbf{r}}{dt} \cdot \mathbf{r} = 2r\frac{dr}{dt}$$

$$2\mathbf{r} \cdot \frac{d\mathbf{r}}{dt} = 2r\frac{dr}{dt}$$

$$\mathbf{r} \cdot \frac{d\mathbf{r}}{dt} = r\frac{dr}{dt}. \quad \square$$

Example 4 Let $\mathbf{r}$ be a differentiable vector function of t and set $r = \|\mathbf{r}\|$. Show that where $r \neq 0$

(13.2.4)

$$\frac{d}{dt}\left(\frac{\mathbf{r}}{r}\right) = \frac{1}{r^3}\left[\left(\mathbf{r} \times \frac{d\mathbf{r}}{dt}\right) \times \mathbf{r}\right].$$

SOLUTION This is a little tricky:

$$\frac{d}{dt}\left(\frac{\mathbf{r}}{r}\right) = \frac{1}{r}\frac{d\mathbf{r}}{dt} - \frac{1}{r^2}\frac{dr}{dt}\mathbf{r}$$

$$= \frac{1}{r^3}\left[r^2\frac{d\mathbf{r}}{dt} - r\frac{dr}{dt}\mathbf{r}\right]$$

$$= \frac{1}{r^3}\left[(\mathbf{r}\cdot\mathbf{r})\frac{d\mathbf{r}}{dt} - \left(\mathbf{r}\cdot\frac{d\mathbf{r}}{dt}\right)\mathbf{r}\right] = \frac{1}{r^3}\left[\left(\mathbf{r}\times\frac{d\mathbf{r}}{dt}\right)\times\mathbf{r}\right]. \quad \square$$

$$(\mathbf{a}\times\mathbf{b})\times\mathbf{c} = (\mathbf{c}\cdot\mathbf{a})\mathbf{b} - (\mathbf{c}\cdot\mathbf{b})\mathbf{a}$$

EXERCISES 13.2

In Exercises 1–12, find $\mathbf{f}'(t)$ and $\mathbf{f}''(t)$.

1. $\mathbf{f}(t) = \mathbf{a} + t\mathbf{b}$.   **2.** $\mathbf{f}(t) = \mathbf{a} + t\mathbf{b} + t^2\mathbf{c}$.

3. $\mathbf{f}(t) = e^{2t}\mathbf{i} - \sin t\mathbf{j}$.

4. $\mathbf{f}(t) = [(t^2\mathbf{i} - \mathbf{j})\cdot(\mathbf{i} - t^2\mathbf{j})]\,\mathbf{i}$.

5. $\mathbf{f}(t) = [(t^2\mathbf{i} - 2t\mathbf{j})\cdot(t\mathbf{i} + t^3\mathbf{j})]\,\mathbf{j}$.

6. $\mathbf{f}(t) = [(3t\mathbf{i} - t^2\mathbf{j} + \mathbf{k})\cdot(\mathbf{i} + t^3\mathbf{j} - 2t\mathbf{k})]\,\mathbf{k}$.

7. $\mathbf{f}(t) = (e^t\mathbf{i} + t\mathbf{k})\times(t\mathbf{j} + e^{-t}\mathbf{k})$.

8. $\mathbf{f}(t) = (t\mathbf{i} - t^2\mathbf{j} + \mathbf{k})\times(\mathbf{i} + t^3\mathbf{j} + 5t\mathbf{k})$.

9. $\mathbf{f}(t) = (\cos t\mathbf{i} + \sin t\mathbf{j} + \mathbf{k})\times(\sin 2t\mathbf{i} + \cos 2t\mathbf{j} + t\mathbf{k})$.

10. $\mathbf{f}(t) = t\mathbf{g}(t^2)$.   **11.** $\mathbf{f}(t) = t\mathbf{g}(\sqrt{t})$.

12. $\mathbf{f}(t) = (e^{2t}\mathbf{i} + e^{-2t}\mathbf{j} + \mathbf{k})\times(e^{2t}\mathbf{i} - e^{-2t}\mathbf{j} + \mathbf{k})$.

In Exercises 13–20, find the indicated derivative.

13. $\dfrac{d}{dt}[e^{\cos t}\mathbf{i} + e^{\sin t}\mathbf{j}]$.

14. $\dfrac{d^2}{dt^2}[e^t\cos t\mathbf{i} + e^t\sin t\mathbf{j}]$.

15. $\dfrac{d^2}{dt^2}\left[(e^t\mathbf{i} + e^{-t}\mathbf{j})\cdot(e^t\mathbf{i} - e^{-t}\mathbf{j})\right]$.

16. $\dfrac{d}{dt}\left[(\ln t\mathbf{i} + t\mathbf{j} - (t^2 + 1)\mathbf{k})\times\left(\dfrac{1}{t}\mathbf{i} + t^2\mathbf{j} - t\mathbf{k}\right)\right]$.

17. $\dfrac{d}{dt}[(\mathbf{a} + t\mathbf{b})\times(\mathbf{c} + t\mathbf{d})]$.

18. $\dfrac{d}{dt}[(\mathbf{a} + t\mathbf{b})\times(\mathbf{a} + t\mathbf{b} + t^2\mathbf{c})]$.

19. $\dfrac{d}{dt}[(\mathbf{a} + t\mathbf{b})\cdot(\mathbf{c} + t\mathbf{d})]$.

20. $\dfrac{d}{dt}[(\mathbf{a} + t\mathbf{b})\cdot(\mathbf{a} + t\mathbf{b} + t^2\mathbf{c})]$.

In Exercises 21–24, find $\mathbf{r}(t)$ given that:

21. $\mathbf{r}'(t) = \mathbf{b}$ for all real t, $\mathbf{r}(0) = \mathbf{a}$.

22. $\mathbf{r}''(t) = \mathbf{c}$ for all real t, $\mathbf{r}'(0) = \mathbf{b}$, $\mathbf{r}(0) = \mathbf{a}$.

23. $\mathbf{r}''(t) = \mathbf{a} + t\mathbf{b}$ for all real t, $\mathbf{r}'(0) = \mathbf{c}$, $\mathbf{r}(0) = \mathbf{d}$.

24. $\mathbf{r}''(t) = \cos 2t\mathbf{i} + \sin 2t\mathbf{j}$ for all real t,
$\mathbf{r}'(0) = 2\mathbf{i} - \frac{1}{2}\mathbf{j}$, $\mathbf{r}(0) = \frac{3}{4}\mathbf{i} + \mathbf{j}$.

25. Show that, if $\mathbf{r}(t) = \sin t\mathbf{i} + \cos t\mathbf{j}$, then $\mathbf{r}(t)$ and $\mathbf{r}''(t)$ are parallel. Is there a value of t for which $\mathbf{r}(t)$ and $\mathbf{r}''(t)$ have the same direction?

26. Show that, if $\mathbf{r}(t) = e^{kt}\mathbf{i} + e^{-kt}\mathbf{j}$, then $\mathbf{r}(t)$ and $\mathbf{r}''(t)$ are parallel.

27. Calculate $\mathbf{r}(t)\cdot\mathbf{r}'(t)$ and $\mathbf{r}(t)\times\mathbf{r}'(t)$ given that $\mathbf{r}(t) = \cos t\mathbf{i} + \sin t\mathbf{j}$.

In Exercises 28–30, assume the rule for differentiating a cross product and show the following.

28. $(\mathbf{g}\times\mathbf{f})'(t) = -(\mathbf{f}\times\mathbf{g})'(t)$.

29. $\dfrac{d}{dt}[\mathbf{f}(t)\times\mathbf{f}'(t)] = \mathbf{f}(t)\times\mathbf{f}''(t)$.

30. $\dfrac{d}{dt}[u_1(t)\mathbf{r}_1(t)\times u_2(t)\mathbf{r}_2(t)] = u_1(t)u_2(t)\dfrac{d}{dt}[\mathbf{r}_1(t)\times\mathbf{r}_2(t)]$
$+ [\mathbf{r}_1(t)\times\mathbf{r}_2(t)]\dfrac{d}{dt}[u_1(t)u_2(t)]$.

31. Prove that $\{\mathbf{f}(t)\cdot[\mathbf{g}(t)\times\mathbf{h}(t)]\}' = \mathbf{f}'(t)\cdot[\mathbf{g}(t)\times\mathbf{h}(t)]$
$+ \mathbf{f}(t)\cdot[\mathbf{g}'(t)\times\mathbf{h}(t)] + \mathbf{f}(t)\cdot[\mathbf{g}(t)\times\mathbf{h}'(t)]$.

32. Suppose that $\mathbf{f}(t)$ is parallel to $\mathbf{f}''(t)$ for all t. Prove that $\mathbf{f}\times\mathbf{f}'$ is constant.   HINT: See Exercise 29.

33. Assume the rule for differentiating a dot product and show that

$\|\mathbf{r}(t)\|$ is constant   iff $\mathbf{r}(t)\cdot\mathbf{r}'(t) = 0$ identically.

34. Derive the formula

$$(\mathbf{f} \cdot \mathbf{g})'(t) = [\mathbf{f}(t) \cdot \mathbf{g}'(t)] + [\mathbf{f}'(t) \cdot \mathbf{g}(t)]$$

(a) by appealing to components; (b) without appealing to components.

35. Derive the formula

$$(\mathbf{f} \times \mathbf{g})'(t) = [\mathbf{f}(t) \times \mathbf{g}'(t)] + [\mathbf{f}'(t) \times \mathbf{g}(t)]$$

without appealing to components.

36. (a) Show that, if u is continuous at t_0 and $\mathbf{f}$ is continuous at $u(t_0)$, then the composition $(\mathbf{f} \circ u)$ is continuous at t_0.

(b) Derive the chain rule for vector functions:

$$\frac{d\mathbf{f}}{dt} = \frac{d\mathbf{f}}{du}\frac{du}{dt}.$$

■ 13.3 CURVES

Introduction

Earlier we explained how every linear vector function

$$\mathbf{r}(t) = \mathbf{r}_0 + t\mathbf{d} \qquad \text{with} \qquad \mathbf{d} \neq \mathbf{0}$$

parametrizes a line. More generally, every differentiable vector function parametrizes a curve.

Suppose that

$$\mathbf{r}(t) = x(t)\mathbf{i} + y(t)\mathbf{j} + z(t)\mathbf{k}$$

is differentiable on some interval I. (At the endpoints, if there are any, we require only continuity.) For each number $t \in I$, the tip of the radius vector $\mathbf{r}(t)$ is the point $P(x(t), y(t), z(t))$. As t ranges over I, the point P traces out some path C. (Figure 13.3.1) We call C a *differentiable curve* and say that C is *parametrized* by $\mathbf{r}$. It is also important to understand that a parametrized curve C is an *oriented* curve in the sense that as t increases on I, the tip of the radius vector traces out C in a certain direction. For example, the unit circle parametrized by

$$\mathbf{r}(t) = \cos t\,\mathbf{i} + \sin t\,\mathbf{j}, \qquad t \in [0, 2\pi]$$

is traversed in the counterclockwise direction starting at the point $(1, 0)$ (Figure 13.3.2). The orientation of the elliptical helix parametrized by

$$\mathbf{r}(t) = t\mathbf{i} + 2\cos t\,\mathbf{j} + 3\sin t\,\mathbf{k}, \qquad t \geq 0,$$

is indicated by the arrows in Figure 13.3.3.

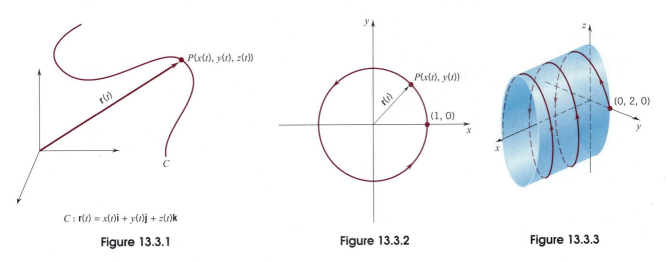

$C : \mathbf{r}(t) = x(t)\mathbf{i} + y(t)\mathbf{j} + z(t)\mathbf{k}$

Figure 13.3.1

Figure 13.3.2

Figure 13.3.3

Tangent Vector, Tangent Line

Let's try to interpret the derivative

$$\mathbf{r}'(t) = x'(t)\mathbf{i} + y'(t)\mathbf{j} + z'(t)\mathbf{k}$$

geometrically. First of all, for fixed t

$$\mathbf{r}'(t) = \lim_{h \to 0} \frac{\mathbf{r}(t + h) - \mathbf{r}(t)}{h}.$$

If $\mathbf{r}'(t) \neq \mathbf{0}$, then we can be sure that for $t + h$ close enough to t, the vector

$$\mathbf{r}(t + h) - \mathbf{r}(t)$$

will not be $\mathbf{0}$. (Explain.) Consequently, we can think of the vector $\mathbf{r}(t + h) - \mathbf{r}(t)$ as pictured in Figure 13.3.4.

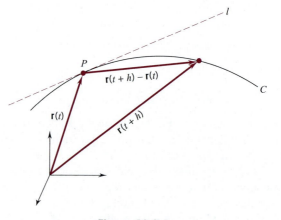

Figure 13.3.4

Let's agree that the line marked l in the figure corresponds to our intuitive notion of tangent line at the point P. As h tends to zero, the vector

$$\mathbf{r}(t + h) - \mathbf{r}(t)$$

comes increasingly closer to serving as a direction vector for that tangent line. It may be tempting therefore to take the limiting case

$$\lim_{h \to 0} [\mathbf{r}(t + h) - \mathbf{r}(t)]$$

and call that a direction vector for the tangent line. The trouble is that this limit vector is $\mathbf{0}$ and $\mathbf{0}$ has no direction.

We can circumvent this difficulty by replacing $\mathbf{r}(t + h) - \mathbf{r}(t)$ by a vector which, for small h, has greater length: the difference quotient

$$\frac{\mathbf{r}(t + h) - \mathbf{r}(t)}{h}.$$

For each real number $h \neq 0$, the vector $[\mathbf{r}(t + h) - \mathbf{r}(t)]/h$ is parallel to $\mathbf{r}(t + h) - \mathbf{r}(t)$, and therefore its limit,

$$\mathbf{r}'(t) = \lim_{h \to 0} \frac{\mathbf{r}(t + h) - \mathbf{r}(t)}{h},$$

which by assumption is not $\mathbf{0}$, can be taken as a direction vector for the tangent line.

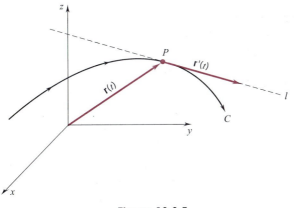

Figure 13.3.5

Finally, an obvious question is: In which of the two possible directions on l does $\mathbf{r}'(t)$ point? You are asked to show in Exercise 48 that $\mathbf{r}'(t)$ points in the direction of increasing t (see Figure 13.3.5). Hence the following definition:

DEFINITION 13.3.1 TANGENT VECTOR

Let

$$C: \quad \mathbf{r}(t) = x(t)\mathbf{i} + y(t)\mathbf{j} + z(t)\mathbf{k}$$

be a differentiable curve. The vector

$$\mathbf{r}'(t) = x'(t)\mathbf{i} + y'(t)\mathbf{j} + z'(t)\mathbf{k},$$

if not $\mathbf{0}$, is said to be *tangent* to the curve C at the point $P(x(t), y(t), z(t))$, and $\mathbf{r}'(t)$ points in the direction of increasing t.

At each point of a line

$$l: \quad \mathbf{r}(t) = \mathbf{r}_0 + t\mathbf{d}$$

the tangent vector $\mathbf{r}'(t)$ is parallel to the line itself:

$$\mathbf{r}'(t) = \mathbf{d} \qquad \text{and } \mathbf{d} \text{ is parallel to } l. \qquad\qquad \text{(Figure 13.3.6)}$$

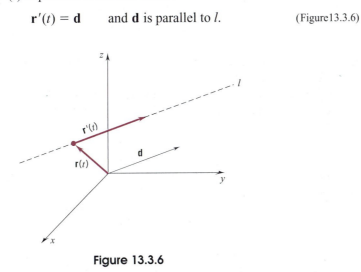

Figure 13.3.6

In the case of a circle

$$C: \quad \mathbf{r}(t) = a \cos t\mathbf{i} + a \sin t\mathbf{j}, \qquad a > 0,$$

the tangent vector $\mathbf{r}'(t)$ is perpendicular to the radius vector $\mathbf{r}(t)$:

$$\mathbf{r}'(t) \cdot \mathbf{r}(t) = (-a \sin t\mathbf{i} + a \cos t\mathbf{j}) \cdot (a \cos t\mathbf{i} + a \sin t\mathbf{j})$$
$$= -a^2\sin t \cos t + a^2\cos t \sin t = 0. \qquad \text{(Figure 13.3.7)}$$

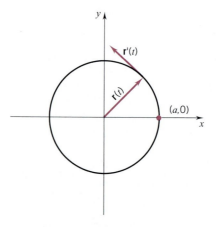

Figure 13.3.7

Example 1 Find the point P on the curve

$$\mathbf{r}(t) = (1 - 2t)\mathbf{i} + t^2\mathbf{j} + 2e^{2(t-1)}\mathbf{k}$$

at which the tangent vector $\mathbf{r}'(t)$ is parallel to the radius vector $\mathbf{r}(t)$.

SOLUTION

$$\mathbf{r}'(t) = -2\mathbf{i} + 2t\mathbf{j} + 4e^{2(t-1)}\mathbf{k}.$$

For $\mathbf{r}'(t)$ to be parallel to $\mathbf{r}(t)$ there must exist a scalar α such that

$$\mathbf{r}(t) = \alpha\mathbf{r}'(t).$$

This vector equation holds iff

$$1 - 2t = -2\alpha, \qquad t^2 = 2\alpha t, \qquad 2e^{2(t-1)} = 4\alpha e^{2(t-1)}.$$

The last scalar equation requires that $\alpha = \frac{1}{2}$. The only value of t that satisfies all three equations with $\alpha = \frac{1}{2}$ is $t = 1$. Therefore the only point at which $\mathbf{r}'(t)$ is parallel to $\mathbf{r}(t)$ is the tip of $\mathbf{r}(1)$. This is the point $P(-1, 1, 2)$. ❏

If $\mathbf{r}'(t_0) \neq \mathbf{0}$, then $\mathbf{r}'(t_0)$ is tangent to the curve at the tip of $\mathbf{r}(t_0)$. The *tangent line* at this point can be parametrized by setting

(13.3.2)

$$\boxed{\mathbf{R}(u) = \mathbf{r}(t_0) + u\mathbf{r}'(t_0).}$$

Example 2 Find a vector tangent to the *twisted cubic*

$$\mathbf{r}(t) = t\mathbf{i} + t^2\mathbf{j} + t^3\mathbf{k} \qquad \text{(Figure 13.3.8)}$$

at the point $P(2, 4, 8)$, and then parametrize the tangent line.

SOLUTION Here

$$\mathbf{r}'(t) = \mathbf{i} + 2t\mathbf{j} + 3t^2\mathbf{k}.$$

Since $P(2, 4, 8)$ is the tip of $\mathbf{r}(2)$, the vector

$$\mathbf{r}'(2) = \mathbf{i} + 4\mathbf{j} + 12\mathbf{k}$$

is tangent to the curve at the point $P(2, 4, 8)$. The vector function

$$\mathbf{R}(u) = (2\mathbf{i} + 4\mathbf{j} + 8\mathbf{k}) + u(\mathbf{i} + 4\mathbf{j} + 12\mathbf{k})$$

parametrizes the tangent line. ❑

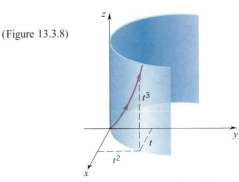

the twisted cubic, $t > 0$

Figure 13.3.8

Intersecting Curves

Two curves

$$C_1: \quad \mathbf{r}_1(t) = x_1(t)\mathbf{i} + y_1(t)\mathbf{j} + z_1(t)\mathbf{k}, \qquad C_2: \quad \mathbf{r}_2(u) = x_2(u)\mathbf{i} + y_2(u)\mathbf{j} + z_2(u)\mathbf{k}$$

intersect iff there are numbers t and u for which

$$\mathbf{r}_1(t) = \mathbf{r}_2(u).$$

The angle between two intersecting curves (which, by definition, is the angle between the corresponding tangent lines) can be obtained by examining the tangent vectors at the point of intersection.

Example 3 Show that the circles

$$C_1: \quad \mathbf{r}_1(t) = \cos t\,\mathbf{i} + \sin t\,\mathbf{j}, \qquad C_2: \quad \mathbf{r}_2(u) = \cos u\,\mathbf{j} + \sin u\,\mathbf{k}$$

intersect at right angles at $P(0, 1, 0)$ and $Q(0, -1, 0)$.

SOLUTION Since $\mathbf{r}_1(\pi/2) = \mathbf{j} = \mathbf{r}_2(0)$, the curves meet at the tip of $\mathbf{j}$, which is $P(0, 1, 0)$. Also, since $\mathbf{r}_1(3\pi/2) = -\mathbf{j} = \mathbf{r}_2(\pi)$, the curves meet at the tip of $-\mathbf{j}$, which is $Q(0, -1, 0)$. Differentiation gives

$$\mathbf{r}_1'(t) = -\sin t\,\mathbf{i} + \cos t\,\mathbf{j} \qquad \text{and} \qquad \mathbf{r}_2'(u) = -\sin u\,\mathbf{j} + \cos u\,\mathbf{k}.$$

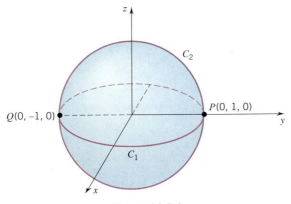

Figure 13.3.9

Since $\mathbf{r}_1'(\pi/2) = -\mathbf{i}$ and $\mathbf{r}_2'(0) = \mathbf{k}$, we have

$$\mathbf{r}_1'(\pi/2) \cdot \mathbf{r}_2'(0) = 0.$$

This tells us that the curves are perpendicular at $P(0, 1, 0)$. Since $\mathbf{r}_1'(3\pi/2) = \mathbf{i}$ and $\mathbf{r}_2'(\pi) = -\mathbf{k}$, we have

$$\mathbf{r}_1'(3\pi/2) \cdot \mathbf{r}_2'(\pi) = 0.$$

This tells us that the curves are perpendicular at $Q(0, -1, 0)$. The curves appear in Figure 13.3.9 on page 857. ❏

The Unit Tangent, the Principal Normal, the Osculating Plane

Suppose now that the curve

$$C: \quad \mathbf{r}(t) = x(t)\mathbf{i} + y(t)\mathbf{j} + z(t)\mathbf{k}$$

is twice differentiable and $\mathbf{r}'(t)$ is never zero. Then at each point $P(x(t), y(t), z(t))$ of the curve, there is a *unit tangent vector*:

(13.3.3)

$$\mathbf{T}(t) = \frac{\mathbf{r}'(t)}{\|\mathbf{r}'(t)\|}.$$

Since $\|\mathbf{r}'(t)\| > 0$, $\mathbf{T}(t)$ points in the same direction as $\mathbf{r}'(t)$, that is, in the direction of increasing t along the curve. Also, since $\|\mathbf{T}(t)\| = 1$, we have $\mathbf{T}(t) \cdot \mathbf{T}(t) = 1$ and differentiation gives

$$\mathbf{T}(t) \cdot \mathbf{T}'(t) + \mathbf{T}'(t) \cdot \mathbf{T}(t) = 0.$$

Since the dot product is commutative, we have

$$2[\mathbf{T}'(t) \cdot \mathbf{T}(t)] = 0 \quad \text{and thus} \quad \mathbf{T}'(t) \cdot \mathbf{T}(t) = 0.$$

At each point of the curve *the vector $\mathbf{T}'(t)$ is perpendicular to $\mathbf{T}(t)$*.

The vector $\mathbf{T}'(t)$ measures the rate of change of $\mathbf{T}(t)$ with respect to t. Since the norm of $\mathbf{T}(t)$ is constantly 1, $\mathbf{T}(t)$ can change only in direction. The vector $\mathbf{T}'(t)$ measures this change in direction.

If the unit tangent vector is not changing in direction (as in the case of a straight line), then $\mathbf{T}'(t) = \mathbf{0}$. If $\mathbf{T}'(t) \neq \mathbf{0}$, then we can form what is called the *principal normal vector*:

(13.3.4)

$$\mathbf{N}(t) = \frac{\mathbf{T}'(t)}{\|\mathbf{T}'(t)\|}.$$

This is the unit vector in the direction of $\mathbf{T}'(t)$. The *normal line* at P is the line through P parallel to the principal normal.

Figure 13.3.10 shows a curve on which we have marked several points. At each point we have drawn the unit tangent and the principal normal. The plane determined by these two vectors is called the *osculating plane* (literally, the "kissing plane"). This is the plane of greatest contact with the curve at the point in question.

Other Changes of Parameter

Not all changes of parameter change the succession of points. Suppose that

$$\mathbf{r} = \mathbf{r}(t), \qquad t \in I$$

is a differentiable curve, and let ϕ be a function that maps some interval J onto the interval I (domain J, range I). Now set

$$\mathbf{R}(u) = \mathbf{r}(\phi(u)) \qquad \text{for all } u \in J.$$

If $\phi'(u) > 0$ for all $u \in J$, then $\mathbf{r}$ and $\mathbf{R}$ are said to differ by a *sense-preserving change of parameter*. In this case, $\mathbf{r}$ and $\mathbf{R}$ take on exactly the same values in exactly the same order. In other words, they produce exactly the same oriented curve (Exercise 44).

If, on the other hand, $\phi'(u) < 0$ for all $u \in J$, then the change in parameter is said to be *sense-reversing*. In that case, $\mathbf{r}$ and $\mathbf{R}$ still take on exactly the same values but in opposite order. The paths are the same, but they are traversed in opposite directions. You have already seen one example of this:

$$\mathbf{R}(u) = \mathbf{r}(a + b - u).$$

Example 5 The parametrization

$$\mathbf{r}(t) = a \cos t\, \mathbf{i} + a \sin t\, \mathbf{j} + bt\, \mathbf{k}, \qquad t \in [0, 2\pi]$$

gives one "spiral" of the circular helix with the orientation indicated by the arrows (Figure 13.3.12). If we let $\varphi(u) = \pi u$, $u \in [0, 2]$, then φ maps the interval $J = [0, 2]$ onto the interval $I = [0, 2\pi]$, and $\varphi'(u) = \pi > 0$. Thus,

$$\mathbf{R}(u) = \mathbf{r}(\varphi(u)) = a \cos(\pi u)\mathbf{i} + a \sin(\pi u)\mathbf{j} + b\pi u\mathbf{k}, \qquad u \in [0, 2]$$

is precisely the same curve with the same orientation.

On the other hand, if we let $\psi(u) = (2 - u)\pi$, $u \in [0, 2]$, then ψ also maps the interval $J = [0, 2]$ onto the interval $I = [0, 2\pi]$, but $\psi'(u) = -\pi < 0$. Thus,

$$\mathbf{R}(u) = \mathbf{r}(\psi(u)) = a \cos(2 - u)\pi\mathbf{i} + a \sin(2 - u)\pi\mathbf{j} + b(2 - u)\pi\mathbf{k},$$

and this gives the same curve but with the opposite orientation. (See Figure 13.3.13.)

$(a, 0, 0)$

Figure 13.3.12

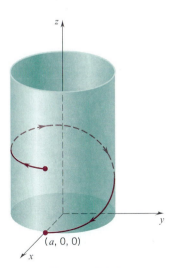

$(a, 0, 0)$

Figure 13.3.13

EXERCISES 13.3

In Exercises 1–8, find the tangent vector $\mathbf{r}'(t)$ and an equation for the tangent line at the indicated point.

1. $\mathbf{r}(t) = \cos \pi t \mathbf{i} + \sin \pi t \mathbf{j} + t \mathbf{k}$ at $t = 2$.

2. $\mathbf{r}(t) = e^t \mathbf{i} + e^{-t} \mathbf{j} - \ln t \mathbf{k}$ at $t = 1$.

3. $\mathbf{r}(t) = \mathbf{a} + t \mathbf{b} + t^2 \mathbf{c}$ at $t = -1$.

4. $\mathbf{r}(t) = (t + 1)\mathbf{i} + (t^2 + 1)\mathbf{j} + (t^3 + 1)\mathbf{k}$ at $P(1, 1, 1)$.

5. $\mathbf{r}(t) = 2t^2 \mathbf{i} + (1 - t)\mathbf{j} + (3 + 2t^2)\mathbf{k}$ at $P(2, 0, 5)$.

6. $\mathbf{r}(t) = 3t\mathbf{a} + \mathbf{b} - t^2 \mathbf{c}$ at $t = 2$.

7. $\mathbf{r}(t) = 2 \cos t \mathbf{i} + 3 \sin t \mathbf{j} + t \mathbf{k}; \; t = \pi/4$.

8. $\mathbf{r}(t) = t \sin t \mathbf{i} + t \cos t \mathbf{j} + 2t \mathbf{k}; \;\; t = \pi/2$.

9. Show that $\mathbf{r}(t) = at \mathbf{i} + bt^2 \mathbf{j}$ parametrizes a parabola. Find an equation in x and y for this parabola.

10. Show that $\mathbf{r}(t) = \frac{1}{2}a(e^{\omega t} + e^{-\omega t})\mathbf{i} + \frac{1}{2}a(e^{\omega t} - e^{-\omega t})\mathbf{j}$ parametrizes the right branch $(x > 0)$ of the hyperbola $x^2 - y^2 = a^2$.

11. Find (a) the points on the curve $\mathbf{r}(t) = t\mathbf{i} + (1 + t^2)\mathbf{j}$ at which $\mathbf{r}(t)$ and $\mathbf{r}'(t)$ are perpendicular; (b) the points at which they have the same direction; (c) the points at which they have opposite directions.

12. Find the curve given that $\mathbf{r}'(t) = \alpha \mathbf{r}(t)$ for all real t and $\mathbf{r}(0) = \mathbf{i} + 2\mathbf{j} + 3\mathbf{k}$.

13. Suppose that $\mathbf{r}'(t)$ and $\mathbf{r}(t)$ are parallel for all t.   Show that, if $\mathbf{r}'(t)$ is never $\mathbf{0}$, then the tangent line at each point passes through the origin.

▶ In Exercises 14–16, the given curves intersect at the indicated point.   Find the angle of intersection.   Express your answer in radians rounded to the nearest hundredth, and in degrees rounded to the nearest tenth.

14. $\mathbf{r}_1(t) = t\mathbf{i} + t^2 \mathbf{j} + t^3 \mathbf{k}$,
$\mathbf{r}_2(u) = \sin 2u \mathbf{i} + u \cos u \mathbf{j} + u \mathbf{k}; \;\; P(0, 0, 0)$.

15. $\mathbf{r}_1(t) = (e^t - 1)\mathbf{i} + 2 \sin t \mathbf{j} + \ln(t + 1)\mathbf{k}$,
$\mathbf{r}_2(u) = (u + 1)\mathbf{i} + (u^2 - 1)\mathbf{j} + (u^3 + 1)\mathbf{k}; \; P(0, 0, 0)$.

16. $\mathbf{r}_1(t) = e^{-t}\mathbf{i} + \cos t \mathbf{j} + (t^2 + 4)\mathbf{k}$,
$\mathbf{r}_2(u) = (2 + u)\mathbf{i} + u^4 \mathbf{j} + 4u^2 \mathbf{k}; \;\; P(1, 1, 4)$.

17. Find the point at which the curves

$$\mathbf{r}_1(t) = e^t \mathbf{i} + 2 \sin(t + \tfrac{1}{2}\pi)\mathbf{j} + (t^2 - 2)\mathbf{k},$$

$$\mathbf{r}_2(u) = u\mathbf{i} + 2\mathbf{j} + (u^2 - 3)\mathbf{k}$$

intersect and find the angle of intersection.

18. Consider the vector function $\mathbf{f}(t) = t\mathbf{i} + f(t)\mathbf{j}$ formed from a differentiable real-valued function f.   The vector function $\mathbf{f}$ parametrizes the graph of f.
(a) Parametrize the tangent line at $P(t_0, f(t_0))$.

(b) Show that the parametrization obtained in (a) reduces to the usual equation for the tangent line:

$$y - f(t_0) = f'(t_0)(x - t_0) \quad \text{if } f'(t_0) \neq 0;$$

$$y = f(t_0) \quad \text{if } f'(t_0) = 0.$$

19. Define a vector function $\mathbf{r}$ on the interval $[0, 2\pi]$ that satisfies the initial condition $\mathbf{r}(0) = a\mathbf{i}$ and, as t increases to 2π, traces out the ellipse $b^2x^2 + a^2y^2 = a^2b^2$.
(a) Once in a counterclockwise manner.
(b) Once in a clockwise manner.
(c) Twice in a counterclockwise manner.
(d) Three times in a clockwise manner.

20. Exercise 19 with $\mathbf{r}(0) = b\mathbf{j}$.

In Exercises 21–26, sketch the plane curve determined by the given vector-valued function $\mathbf{r}$ and indicate the orientation. Find $\mathbf{r}'(t)$ and draw the position vector and the tangent vector for the indicated value of t, placing the tangent vector at the tip of the position vector.

21. $\mathbf{r}(t) = \frac{1}{4}t^4 \mathbf{i} + t^2 \mathbf{j}; \;\; t = 2$.

22. $\mathbf{r}(t) = 2t\mathbf{i} + (t^2 + 1)\mathbf{j}; \;\; t = 4$.

23. $\mathbf{r}(t) = e^{2t}\mathbf{i} + e^{-4t}\mathbf{j}; \;\; t = 0$.

24. $\mathbf{r}(t) = \sin t \mathbf{i} - 2 \cos t \mathbf{j}; \;\; t = \pi/3$.

25. $\mathbf{r}(t) = 2 \cos t \mathbf{i} + 3 \sin t \mathbf{j}; \;\; t = \pi/6$.

26. $\mathbf{r}(t) = \sec t \mathbf{i} + \tan t \mathbf{j}, \;\; |t| < \pi/2; \;\; t = \pi/4$.

In Exercises 27–30, find a vector parametrization for the curve.

27. $y^2 = x - 1, \;\; y \geq 1$.

28. $r = 1 - \cos \theta, \;\;\; \theta \in [0, 2\pi]$.   (Polar coordinates.)

29. $r = \sin 3\theta, \;\;\; \theta \in [0, \pi]$.   (Polar coordinates.)

30. $y^4 = x^3, \;\; y \leq 0$.

31. Find an equation in x and y for the curve $\mathbf{r}(t) = t^3\mathbf{i} + t^2\mathbf{j}$.   Draw the curve.   Does the curve have a tangent vector at the origin?   If so, what is the unit tangent vector?

32. (a) Show that the curve

$$\mathbf{r}(t) = (t^2 - t + 1)\mathbf{i} + (t^3 - t + 2)\mathbf{j} + (\sin \pi t)\mathbf{k}$$

intersects itself at $P(1, 2, 0)$ by finding numbers $t_1 < t_2$ for which P is the tip of both $\mathbf{r}(t_1)$ and $\mathbf{r}(t_2)$.
(b) Find the unit tangents at $P(1, 2, 0)$ first taking $t = t_1$, then taking $t = t_2$.

33. Find the point(s) at which the twisted cubic

$$\mathbf{r}(t) = t\mathbf{i} + t^2\mathbf{j} + t^3\mathbf{k}$$

intersects the plane $4x + 2y + z = 24$.   What is the angle of intersection between the curve and the plane?

34. (a) Find the unit tangent and the principal normal at an arbitrary point of the ellipse

$$\mathbf{r}(t) = a \cos t\mathbf{i} + b \sin t\mathbf{j}.$$

(b) Write vector equations for the tangent line and the normal line at the tip of $\mathbf{r}(\frac{1}{4}\pi)$.

In Exercises 35–42, find the unit tangent vector, the principal normal vector, and an equation in x, y, z for the osculating plane at the point on the curve corresponding to the indicated value of t.

35. $\mathbf{r}(t) = \mathbf{i} + 2t\mathbf{j} + t^2\mathbf{k}; \quad t = 1.$

36. $\mathbf{r}(t) = t\mathbf{i} + t^2\mathbf{j} + 2t^2\mathbf{k}; \quad t = 1.$

37. $\mathbf{r}(t) = \cos 2t\mathbf{i} + \sin 2t\mathbf{j} + t\mathbf{k} \quad$ at $t = \frac{1}{4}\pi.$

38. $\mathbf{r}(t) = t\mathbf{i} + 2t\mathbf{j} + t^2\mathbf{k} \quad$ at $t = 2.$

39. $\mathbf{r}(t) = t\mathbf{i} + t^2\mathbf{j} + t^3\mathbf{k} \quad$ at $t = 1.$

40. $\mathbf{r}(t) = \cos 3t\mathbf{i} + t\mathbf{j} - \sin 3t\mathbf{k} \quad$ at $t = \frac{1}{3}\pi.$

41. $\mathbf{r}(t) = e^t\sin t\mathbf{i} + e^t\cos t\mathbf{j} + e^t\mathbf{k}; \quad t = 0.$

42. $\mathbf{r}(t) = (\cos t + t \sin t)\mathbf{i} + (\sin t - t \cos t)\mathbf{j} + 2\mathbf{k};$ $t = \frac{1}{4}\pi.$

43. Let $\mathbf{r} = \mathbf{r}(t), t \in [a, b]$ and set

$$\mathbf{R}(u) = \mathbf{r}(a + b - u), \quad u \in [a, b].$$

Show that this change of parameter changes the sign of the unit tangent but does not alter the principal normal. HINT: Let P be the tip of $\mathbf{R}(u) = \mathbf{r}(a + b - u)$. At that point, $\mathbf{R}$ produces a unit tangent $\mathbf{T}_1(u)$ and a principal normal $\mathbf{N}_1(u)$. At that same point, $\mathbf{r}$ produces a unit tangent $\mathbf{T}(a + b - u)$ and a principal normal $\mathbf{N}(a + b - u)$.

44. Show that two vector functions that differ by a sense-preserving change of parameter take on exactly the same values in exactly the same order. That is, set

$$\mathbf{r} = \mathbf{r}(t), \quad t \in I.$$

Assume that ϕ maps an interval J onto the interval I and that $\phi'(u) > 0$ for all $u \in J$. Set

$$\mathbf{R}(u) = \mathbf{r}(\phi(u)),$$

and show that $\mathbf{R}$ and $\mathbf{r}$ take on exactly the same values in exactly the same order.

45. Show that the unit tangent vector, the principal normal vector, and the osculating plane are left invariant (left unchanged) by every sense-preserving change of parameter.

In Exercises 46 and 47, let $\mathbf{r}$ be the vector-valued function defined by

$$\mathbf{r}(t) = 2 \cos t\mathbf{i} + 2 \sin t\mathbf{j} + 4t\mathbf{k} \quad \text{for } 0 \le t \le 2\pi.$$

The graph of $\mathbf{r}$ is one revolution on a circular helix, starting at the point $(2, 0, 0)$ and ending at the point $(2, 0, 8\pi)$.

46. Let $\varphi(u) = u^2$ for $0 \le u \le \sqrt{2\pi}$, and let

$$\mathbf{R}(u) = \mathbf{r}[\varphi(u)] = 2 \cos u^2\mathbf{i} + 2 \sin u^2\mathbf{j} + 4u^2\mathbf{k}.$$

(a) Show that φ determines a sense-preserving change of parameter on $[0, \sqrt{2\pi}]$.

(b) Show that the unit tangent and principal normal vectors for $\mathbf{r}$ at the point $t = \frac{1}{4}\pi$ are the same as the unit tangent and principal normal vectors for $\mathbf{R}$ at $u = \frac{1}{2}\sqrt{\pi}.$

47. Let $\psi(v) = (2\pi - v^2)$ for $0 \le v \le \sqrt{2\pi}$, and let

$$\mathbf{R}(v) = \mathbf{r}[\psi(v)] = 2 \cos(2\pi - v^2)\mathbf{i} + 2 \sin(2\pi - v^2)\mathbf{j} + 4(2\pi - v^2)\mathbf{k}.$$

(a) Show that ψ determines a sense-reversing change of parameter on $[0, \sqrt{2\pi}]$.

(b) Show that principal normal vector for $\mathbf{r}$ at $t = \pi/4$ is the same as the principal normal for $\mathbf{R}$ at $v = \frac{1}{2}\sqrt{7\pi}$, and show that the unit tangent vector for $\mathbf{r}$ at $\pi/4$ is the negative of the unit tangent vector for $\mathbf{R}$ at $v = \frac{1}{2}\sqrt{7\pi}.$

48. Let $\mathbf{r} = \mathbf{r}(t)$ be a differentiable vector-valued function. Prove that the tangent vector to the graph of $\mathbf{r}$ points in the direction of increasing t.

■ 13.4 ARC LENGTH

In Section 9.9, we treated arc length for plane curves in an intuitive manner. We showed that if a curve C is defined parametrically by the continuously differentiable functions

$$x = x(t), \quad y = y(t), \quad t \in [a, b],$$

then the length of C, $L(C)$, is given by the arc length formula

$$L(C) = \int_a^b \sqrt{[x'(t)]^2 + [y'(t)]^2} \, dt.$$

For a space curve C parametrized by the continuously differentiable functions

$$x = x(t), \qquad y = y(t), \qquad z = z(t), \qquad t \in [a, b],$$

the formula becomes

$$L(C) = \int_a^b \sqrt{[x'(t)]^2 + [y'(t)]^2 + [z'(t)]^2} \, dt.$$

In vector notation, both formulas can be written

$$L(C) = \int_a^b \| \mathbf{r}'(t) \| \, dt.$$

We will prove this result in this form, but first we give a precise definition of arc length.

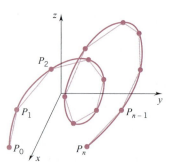

Figure 13.4.1

In Figure 13.4.1, we have sketched a space curve C which is parametrized by the continuously differentiable functions $x = x(t)$, $y = y(t)$, $z = z(t)$, $t \in [a, b]$. Before proceeding, we need to recall the underlying assumption about the nature of the parametrization of C that we made in Section 9.9. In particular, we assume throughout this discussion that the point $(x(t), y(t), z(t))$ traces out the curve C only once as t increases from a to b.

Now, to decide what should be meant by the length of C, we approximate C by the union of a finite number of line segments as we did in treating plane curves.

Choosing a finite number of points in $[a, b]$,

$$a = t_0 < t_1 < \cdots < t_{i-1} < t_i < \cdots < t_{n-1} < t_n = b,$$

we obtain a finite number of points $P_0, P_1, \ldots, P_{i-1}, P_i, \ldots, P_{n-1}, P_n$, on C, where for each k, $0 \le k \le n$, P_k denotes the point $P(x(t_k), y(t_k), z(t_k))$. Join these points consecutively by line segments and call the resulting path

$$\gamma = \overline{P_0 P_1} \cup \cdots \cup \overline{P_{i-1} P_i} \cup \cdots \cup \overline{P_{n-1} P_n},$$

Figure 13.4.2

a *polygonal path* inscribed in C (Figure 13.4.2).

The length of such a polygonal path is the sum of the distances between consecutive vertices:

$$L(\gamma) = d(P_0, P_1) + \cdots + d(P_{i-1}, P_i) + \cdots + d(P_{n-1}, P_n).$$

The path γ serves as an approximation to the curve C, but obviously a better approximation can be obtained by adding more vertices to γ. We now ask ourselves exactly what should we require of the number that we shall call the length of C. Certainly we require that

$$L(\gamma) \le L(C) \qquad \text{for each } \gamma \text{ inscribed in } C.$$

But that is not enough. There is another requirement that seems reasonable. If we can choose γ to approximate C as closely as we wish, then we should be able to choose γ so that $L(\gamma)$ approximates the length of C as closely as we wish; namely, for each positive number ϵ there should exist a polygonal path γ such that

$$L(C) - \epsilon < L(\gamma) \le L(C).$$

In Section 10.1, we introduced the concept of least upper bound of a set of real numbers. Theorem 10.1.3 tells us that we can achieve the result we want by defining the length of the curve C to be the least upper bound of the set of all $L(\gamma)$. This is in fact what we do.

DEFINITION 13.4.1 ARC LENGTH

$$L(C) = \begin{cases} \text{the least upper bound of the set of all} \\ \text{lengths of polygonal paths inscribed in } C. \end{cases}$$

We are now ready to establish the arc length formula.

THEOREM 13.4.2 ARC LENGTH FORMULA

The length of a continuously differentiable curve

$$C: \quad \mathbf{r} = \mathbf{r}(t), \qquad t \in [a, b]$$

is given by the formula

$$L(C) = \int_a^b \|\mathbf{r}'(t)\| \, dt.$$

PROOF First we show that

$$L(C) \le \int_a^b \|\mathbf{r}'(t)\| \, dt.$$

To do this, we begin with an arbitrary partition P of $[a, b]$:

$$P = \{a = t_0, \; \ldots, t_{i-1}, t_i, \; \ldots, t_n = b\}.$$

Such a partition gives rise to a finite number of points of C:

$$\mathbf{r}(a) = \mathbf{r}(t_0), \; \ldots, \mathbf{r}(t_{i-1}), \mathbf{r}(t_i), \; \ldots, \mathbf{r}(t_n) = \mathbf{r}(b)$$

and thus to an inscribed polygonal path of total length

$$L_P = \sum_{i=1}^n \|\mathbf{r}(t_i) - \mathbf{r}(t_{i-1})\|. \qquad \text{(Figure 13.4.3)}$$

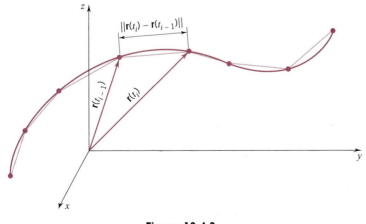

Figure 13.4.3

For each i,

$$\mathbf{r}(t_i) - \mathbf{r}(t_{i-1}) = \int_{t_{i-1}}^{t_i} \mathbf{r}'(t)\, dt.$$

This gives

by (13.1.10)

$$\|\mathbf{r}(t_i) - \mathbf{r}(t_{i-1})\| = \left\| \int_{t_{i-1}}^{t_i} \mathbf{r}'(t)\, dt \right\| \le \int_{t_{i-1}}^{t_i} \|\mathbf{r}'(t)\|\, dt$$

and thus

$$L_P = \sum_{i=1}^{n} \|\mathbf{r}(t_i) - \mathbf{r}(t_{i-1})\| \le \sum_{i=1}^{n} \int_{t_{i-1}}^{t_i} \|\mathbf{r}'(t)\|\, dt = \int_{a}^{b} \|\mathbf{r}'(t)\|\, dt.$$

From the arbitrariness of P, we know that the inequality

$$L_P \le \int_{a}^{b} \|\mathbf{r}'(t)\|\, dt$$

must hold for all the L_P. This makes the integral on the right an upper bound for all the L_P. Since $L(C)$ is the *least* upper bound of all the L_P, we can conclude right now that

$$L(C) \le \int_{a}^{b} \|\mathbf{r}'(t)\|\, dt.$$

To show equality, we need to know that arc length is additive: that is, with P, Q, R as in Figure 13.4.4, the arc length from P to Q, plus the arc length from Q to R, equals the arc length from P to R.† A proof of the additivity of arc length is given in the supplement at the end of this section. From this point on we will assume that arc length is additive.

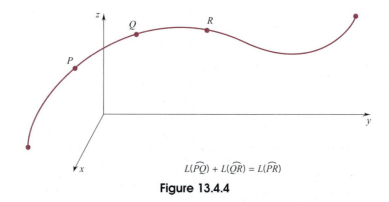

$$L(\widehat{PQ}) + L(\widehat{QR}) = L(\widehat{PR})$$

Figure 13.4.4

In Figure 13.4.5 we display the initial vector $\mathbf{r}(a)$, a general radius vector $\mathbf{r}(t)$, and a nearby vector $\mathbf{r}(t + h)$. Now set

$$s(t) = \text{length of the curve from } \mathbf{r}(a) \text{ to } \mathbf{r}(t),$$

$$s(t + h) = \text{length of the curve from } \mathbf{r}(a) \text{ to } \mathbf{r}(t + h).$$

---

† It is obvious that we want arc length to have this property. What we have to verify is that arc length *as we have defined it* has this property.

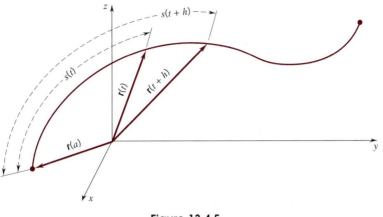

Figure 13.4.5

Then, by the additivity of arc length (remember we are assuming that),

$$s(t + h) - s(t) = \text{length of the curve from } \mathbf{r}(t) \text{ to } \mathbf{r}(t + h).$$

From what we have shown already, you can see that

$$\|\mathbf{r}(t + h) - \mathbf{r}(t)\| \le s(t + h) - s(t) \le \int_t^{t+h} \|\mathbf{r}'(u)\|\, du.$$

Dividing this inequality by h (which we are taking as positive), we get

$$\left\| \frac{\mathbf{r}(t + h) - \mathbf{r}(t)}{h} \right\| \le \frac{s(t + h) - s(t)}{h} \le \frac{1}{h}\int_t^{t+h} \|\mathbf{r}'(u)\|\, du.$$

As $h \to 0^+$, the left-hand side tends to $\|\mathbf{r}'(t)\|$ and, by the First Mean-Value Theorem for Integrals (Theorem 5.9.1), so does the right-hand side:

$$\frac{1}{h}\int_t^{t+h} \|\mathbf{r}'(u)\|\, du = \frac{1}{h}\|\mathbf{r}'(c_h)\|(t + h - t) = \|\mathbf{r}'(c_h)\| \to \|\mathbf{r}'(t)\|.$$

$$c_h \in (t, t+h) \underline{\hspace{2cm}}\uparrow$$

Therefore

$$\lim_{h \to 0^+} \frac{s(t + h) - s(t)}{h} = \|\mathbf{r}'(t)\|.$$

By taking $h < 0$ and proceeding in a similar manner, one can show that

$$\lim_{h \to 0^-} \frac{s(t + h) - s(t)}{h} = \|\mathbf{r}'(t)\|.$$

It follows that

$$s'(t) = \lim_{h \to 0} \frac{s(t + h) - s(t)}{h} = \|\mathbf{r}'(t)\|.$$

Recall now that $s(t) = $ length of the curve from $\mathbf{r}(a)$ to $\mathbf{r}(t)$. Obviously, then, $s(a) = 0$. This, together with the fact that $s'(t) = \|\mathbf{r}'(t)\|$, implies that

$$s(t) = \text{length of the curve from } \mathbf{r}(a) \text{ to } \mathbf{r}(t) = \int_a^t \|\mathbf{r}'(u)\|\, du.$$

The total length of C is therefore

$$s(b) = \int_a^b \|\mathbf{r}'(t)\| \, dt. \quad \square$$

Example 1 Find the length of the curve

$$\mathbf{r}(t) = 2 \cos t\mathbf{i} + 2 \sin t\mathbf{j} + t^2\mathbf{k} \quad \text{from } t = 0 \text{ to } t = \pi/2$$

and compare it to the straight-line distance between the endpoints of the curve. (See Figure 13.4.6.)

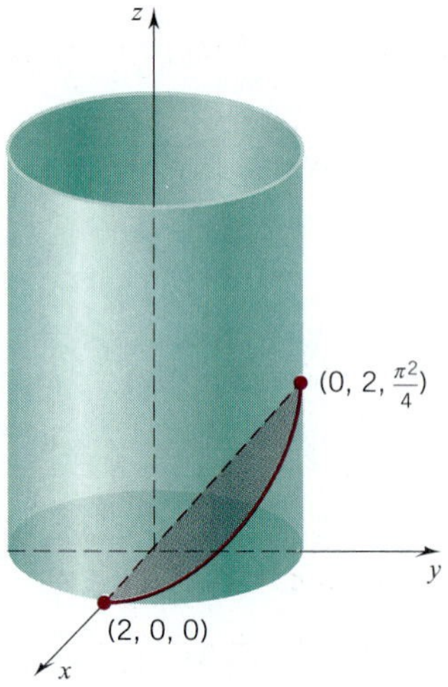

SOLUTION

$$\mathbf{r}'(t) = -2 \sin t\mathbf{i} + 2 \cos t\mathbf{j} + 2t\mathbf{k}.$$

$$\|\mathbf{r}'(t)\| = \sqrt{4 \sin^2 t + 4 \cos^2 t + 4t^2} = 2\sqrt{\sin^2 t + \cos^2 t + t^2} = 2\sqrt{1 + t^2}.$$

$$L(C) = \int_0^{\pi/2} \|\mathbf{r}'(t)\| \, dt = \int_0^{\pi/2} 2\sqrt{1 + t^2} \, dt = \left[t\sqrt{1 + t^2} + \ln(t + \sqrt{1 + t^2}) \right]_0^{\pi/2}$$

$$= \frac{\pi}{2}\sqrt{1 + \frac{\pi^2}{4}} + \ln\left[\frac{\pi}{2} + \sqrt{1 + \frac{\pi^2}{4}} \right] \cong 4.158.$$

The curve begins at $\mathbf{r}(0) = 2\mathbf{i}$ and ends at $\mathbf{r}(\pi/2) = 2\mathbf{j} + (\pi^2/4)\mathbf{k}$. The distance between these two points is

$$\|\mathbf{r}(\pi/2) - \mathbf{r}(0)\| = \sqrt{2^2 + 2^2 + \frac{\pi^4}{16}} \cong 3.753.$$

The curve is about 11 percent longer than the straight-line distance between the endpoints of the curve. ❑

(0, 2, $\frac{\pi^2}{4}$)

(2, 0, 0)

Figure 13.4.6

EXERCISES 13.4

In Exercises 1–14, find the length of the given curve.

1. $\mathbf{r}(t) = t\mathbf{i} + \frac{2}{3}t^{3/2}\mathbf{j}$, from $t = 0$ to $t = 8$.

2. $\mathbf{r}(t) = (\frac{1}{3}t^3 - t)\mathbf{i} + t^2\mathbf{j}$, from $t = 0$ to $t = 2$.

3. $\mathbf{r}(t) = a \cos t\mathbf{i} + a \sin t\mathbf{j} + bt\mathbf{k}$ from $t = 0$ to $t = 2\pi$.

4. $\mathbf{r}(t) = t\mathbf{i} + \frac{2}{3}\sqrt{2}\,t^{3/2}\mathbf{j} + \frac{1}{2}t^2\mathbf{k}$ from $t = 0$ to $t = 2$.

5. $\mathbf{r}(t) = t\mathbf{i} + \ln(\sec t)\mathbf{j} + 3\mathbf{k}$ from $t = 0$ to $t = \frac{1}{4}\pi$.

6. $\mathbf{r}(t) = \tan^{-1}t\mathbf{i} + \frac{1}{2}\ln(1 + t^2)\mathbf{j}$ from $t = 0$ to $t = 1$.

7. $\mathbf{r}(t) = t^3\mathbf{i} + t^2\mathbf{j}$ from $t = 0$ to $t = 1$.

8. $\mathbf{r}(t) = t\mathbf{i} + \mathbf{j} + (\frac{1}{6}t^3 + \frac{1}{2}t^{-1})\mathbf{k}$ from $t = 1$ to $t = 3$.

9. $\mathbf{r}(t) = e^t[\cos t\mathbf{i} + \sin t\mathbf{j}]$ from $t = 0$ to $t = \pi$.

10. $\mathbf{r}(t) = 3t \cos t\mathbf{i} + 3t \sin t\mathbf{j} + 4t\mathbf{k}$ from $t = 0$ to $t = 4$.

11. $\mathbf{r}(t) = 2t\mathbf{i} + (t^2 - 2)\mathbf{j} + (1 - t^2)\mathbf{k}$ from $t = 0$ to $t = 2$.

12. $\mathbf{r}(t) = t^2\mathbf{i} + (t^2 - 2)\mathbf{j} + (1 - t^2)\mathbf{k}$ from $t = 0$ to $t = 2$.

13. $\mathbf{r}(t) = (\ln t)\mathbf{i} + 2t\mathbf{j} + t^2\mathbf{k}$, from $t = 1$ to $t = e$.

14. $\mathbf{r}(t) = (t \sin t + \cos t)\mathbf{i} + (t \cos t - \sin t)\mathbf{j} + 2\mathbf{k}$, from $t = 0$ to $t = 2$.

15. (*Important*) Let $\mathbf{r}(t) = x(t)\mathbf{i} + y(t)\mathbf{j} + z(t)\mathbf{k}$, $t \in [a, b]$ be a continuously differentiable curve. Show that, if s is the length of the curve from the tip of $\mathbf{r}(a)$ to the tip of $\mathbf{r}(t)$, then

(13.4.3)
$$\frac{ds}{dt} = \sqrt{\left(\frac{dx}{dt}\right)^2 + \left(\frac{dy}{dt}\right)^2 + \left(\frac{dz}{dt}\right)^2}.$$

16. Use vector methods to show that, if $y = f(x)$ has a continuous first derivative, then the length of the graph from $x = a$ to $x = b$ is given by the integral

$$\int_a^b \sqrt{1 + [f'(x)]^2} \, dx.$$

17. (*Important*) Let $y = f(x)$, $x \in [a, b]$ be a continuously differentiable function. Show that, if s is the length of the graph from $(a, f(a))$ to $(x, f(x))$, then

(13.4.4)
$$\frac{ds}{dx} = \sqrt{1 + \left(\frac{dy}{dx}\right)^2}.$$

18. Let C_1 be the curve

$$\mathbf{r}(t) = (t - \ln t)\mathbf{i} + (t + \ln t)\mathbf{j}, \qquad 1 \leq t \leq e$$

and let C_2 be the graph of

$$y = e^x, \qquad 0 \leq x \leq 1.$$

Find a relation between the length of C_1 and the length of C_2.

19. Show that the length of a continuously differentiable curve is left invariant (left unchanged) by a sense-preserving (or a sense-reversing) change of parameter. That is, set

$$\mathbf{r} = \mathbf{r}(t), \qquad t \in [a, b].$$

Assume that ϕ maps $[c, d]$ onto $[a, b]$ and that ϕ' is positive (or negative) and continuous on $[a, b]$. Set $\mathbf{R}(u) = \mathbf{r}(\phi(u))$ and show that the length of the curve as computed from $\mathbf{R}$ is the length of the curve as computed from $\mathbf{r}$.

20. Let $\mathbf{r}(t) = x(t)\mathbf{i} + y(t)\mathbf{j} + z(t)\mathbf{k}$ be a differentiable vector-valued function such that $\mathbf{r}'(t) \neq 0$ for all $t \geq 0$.
 (a) Show that the arc length function s defined by

$$s(t) = \int_0^t \sqrt{\left(\frac{dx}{dt}\right)^2 + \left(\frac{dy}{dt}\right)^2 + \left(\frac{dz}{dt}\right)^2}\, dt, \quad t \geq 0$$

has an inverse, $t = \varphi(s)$.

(b) Let $\mathbf{R}(s) = \mathbf{r}[\varphi(s)]$. Show that $\|\mathbf{R}'(s)\| = 1$.

21. Consider the circular helix
$\mathbf{r}(t) = 3 \cos t\mathbf{i} + 3 \sin t\mathbf{j} + 4t\mathbf{k}$ for $t \geq 0$.
 (a) Determine the arc length s as a function of t by evaluating the integral

$$s = \int_0^t \sqrt{\left(\frac{dx}{dt}\right)^2 + \left(\frac{dy}{dt}\right)^2 + \left(\frac{dz}{dt}\right)^2}\, dt.$$

(b) Use the relation found in (a) to express t as a function of s, $t = \varphi(s)$, and set

$$\mathbf{R}(s) = \mathbf{r}[\varphi(s)] = 3 \cos \varphi(s)\mathbf{i} + 3 \sin \varphi(s)\mathbf{j} + 4\varphi(s)\mathbf{k}.$$

(c) Find the coordinates of the point Q on the helix such that the arc length from $P(3, 0, 0)$ to Q is 5π.
 (d) Show that $\|\mathbf{R}'(s)\| = 1$.

22. Repeat Exercise 21 (a), (b), (d) for the vector-valued function

$$\mathbf{r}(t) = (\sin t - t \cos t)\mathbf{i} + (\cos t + t \sin t)\mathbf{j} + \tfrac{1}{2}t^2\mathbf{k}$$

for $t \geq 0$.

▶ In Exercises 23–26, use a computer or calculator and Simpson's rule to estimate the length of the given curve. Obtain answers which are accurate to four decimal places.

23. $\mathbf{r}(t) = \frac{2}{5}t^{5/2}\mathbf{j} + t\mathbf{k}$ from $t = 0$ to $t = \frac{1}{2}$.

24. $\mathbf{r}(t) = t\mathbf{i} + \frac{1}{3}t^3\mathbf{j}$ from $t = 0$ to $t = 2$.

25. $\mathbf{r}(t) = 3 \cos t\mathbf{i} + 4 \sin t\mathbf{j} + 2\mathbf{k}$ from $t = 0$ to $t = 2\pi$.

26. $\mathbf{r}(t) = t\mathbf{i} + t^2\mathbf{j} + (\ln t)\mathbf{k}$ from $t = 1$ to $t = 4$.

---

*SUPPLEMENT TO SECTION 13.4

The Additivity of Arc Length

We wish to show that with P, Q, R as in Figure 13.4.3,

$$L(\widehat{PQ}) + L(\widehat{QR}) = L(\widehat{PR}).$$

Let γ_1 be an arbitrary polygonal path inscribed in $\widehat{PQ}$ and γ_2 an arbitrary polygonal path inscribed in $\widehat{QR}$. Then $\gamma_1 \cup \gamma_2$ is a polygonal path inscribed in $\widehat{PR}$. Since

$$L(\gamma_1) + L(\gamma_2) = L(\gamma_1 \cup \gamma_2) \qquad \text{and} \qquad L(\gamma_1 \cup \gamma_2) \leq L(\widehat{PR}),$$

we have

$$L(\gamma_1) + L(\gamma_2) \leq L(\widehat{PR}) \qquad \text{and thus} \qquad L(\gamma_1) \leq L(\widehat{PR}) - L(\gamma_2).$$

From the arbitrariness of γ_1 we can conclude that $L(\widehat{PR}) - L(\gamma_2)$ is an upper bound for the set of all lengths of polygonal paths inscribed in $\widehat{PQ}$. Since $L(\widehat{PQ})$ is the *least* upper bound of this set, we have

$$L(\widehat{PQ}) \leq L(\widehat{PR}) - L(\gamma_2).$$

It follows that

$$L(\gamma_2) \le L(\widehat{PR}) - L(\widehat{PQ}).$$

Arguing as we did with γ_1, we can conclude that

$$L(\widehat{QR}) \le L(\widehat{PR}) - L(\widehat{PQ}).$$

This gives

$$L(\widehat{PQ}) + L(\widehat{QR}) \le L(\widehat{PR}).$$

We now set out to prove that $L(\widehat{PR}) \le L(\widehat{PQ}) + L(\widehat{QR})$. To do this, we need only take $\gamma = \overline{T_0T_1} \cup \cdots \cup \overline{T_{n-1}T_n}$ as an arbitrary polygonal path inscribed in $\widehat{PR}$ and show that

$$L(\gamma) \le L(\widehat{PQ}) + L(\widehat{QR}).$$

If Q is one of the T_i, say $Q = T_k$, then

$$\gamma_1 = \overline{T_0T_1} \cup \cdots \cup \overline{T_{k-1}T_k} \quad \text{is inscribed in } \widehat{PQ}$$

and

$$\gamma_2 = \overline{T_kT_{k+1}} \cup \cdots \cup \overline{T_{n-1}T_n} \quad \text{is inscribed in } \widehat{QR}.$$

Moreover, $L(\gamma) = L(\gamma_1) + L(\gamma_2)$, so that

$$L(\gamma) \le L(\widehat{PQ}) + L(\widehat{QR}).$$

If Q is none of the T_i, then Q lies between two consecutive points T_k and T_{k+1}. Set

$$\gamma' = \overline{T_0T_1} \cup \cdots \cup \overline{T_kQ} \cup \overline{QT_{k+1}} \cup \cdots \cup \overline{T_{n-1}T_n}.$$

Since

$$d(T_k, T_{k+1}) \le d(T_k, Q) + d(Q, T_{k+1}),$$

we have

$$L(\gamma) \le L(\gamma').$$

Proceed as before and you will see that

$$L(\gamma') \le L(\widehat{PQ}) + L(\widehat{QR}),$$

and once again

$$L(\gamma) \le L(\widehat{PQ}) + L(\widehat{QR}). \quad ❑$$

■ 13.5 CURVILINEAR MOTION; VECTOR CALCULUS IN MECHANICS

Introduction

The tools we have just developed find their premier application in Newtonian mechanics, the study of bodies in motion subject to Newton's laws. The heart of Newton's mechanics is his Second Law of Motion:

$$\text{force} = \text{mass} \times \text{acceleration}.$$

We have worked with Newton's second law, but only in a very restricted context: motion along a coordinate line under the influence of a force directed along that same line. In that special setting Newton's law was written as a scalar equation: $F = ma$. In general, objects do not move along straight lines (they move along curved paths) and the forces on them vary in direction. What happens to Newton's second law then? It becomes the vector equation

$$\mathbf{F} = m\mathbf{a}.$$

This is Newton's Second Law of Motion in its full glory. To understand this law and be able to use it effectively, you must learn to look at curvilinear motion from a vector viewpoint.

Curvilinear Motion from a Vector Viewpoint

We can describe the position of a moving object at time t by a radius vector $\mathbf{r}(t)$. As t ranges over a time interval I, the object traces out some path

$$C: \quad \mathbf{r}(t) = x(t)\mathbf{i} + y(t)\mathbf{j} + z(t)\mathbf{k}, \quad t \in I.$$

If $\mathbf{r}$ is twice differentiable, we can form $\mathbf{r}'(t)$ and $\mathbf{r}''(t)$. In this context these vectors have special names and special significance: $\mathbf{r}'(t)$ is called the *velocity* of the object at time t, and $\mathbf{r}''(t)$ is called the *acceleration*. In symbols, we have

(13.5.1)

$$\mathbf{r}'(t) = \mathbf{v}(t) \quad \text{and} \quad \mathbf{r}''(t) = \mathbf{v}'(t) = \mathbf{a}(t).$$

There should be nothing surprising about this. As before, velocity is the time rate of change of position and acceleration the time rate of change of velocity.

Since $\mathbf{v}(t) = \mathbf{r}'(t)$, the velocity vector, when not $\mathbf{0}$, is tangent to the path of the motion at the tip of $\mathbf{r}(t)$. (Section 13.3.) The direction of the velocity vector at time t thus gives the direction of the motion at time t. (Figure 13.5.1.)

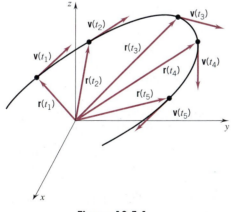

Figure 13.5.1

The magnitude of the velocity vector is called the *speed* of the object:

(13.5.2)

$$\|\mathbf{v}(t)\| = \text{the speed at time } t.$$

The reasoning is as follows: during a time interval $[t_0, t]$ the object moves along its path from $\mathbf{r}(t_0)$ to $\mathbf{r}(t)$ for a total distance

$$s(t) = \int_{t_0}^{t} \|\mathbf{r}'(u)\| \, du. \qquad \text{(Section 13.4.)}$$

Differentiating with respect to t, we have

$$s'(t) = \|\mathbf{r}'(t)\|.$$

The magnitude of the velocity vector is thus *the rate of change of arc distance with respect to time.* This is why we call it the speed of the object.

Speed is commonly denoted by the letter v. In symbols

(13.5.3)

$$s'(t) = \|\mathbf{v}(t)\| = v(t).$$

Motion Along a Straight Line

The position at time t is given by a function of the form

$$\mathbf{r}(t) = \mathbf{r}_0 + f(t)\mathbf{d}, \qquad \mathbf{d} \neq 0.$$

For convenience we take $\mathbf{d}$ as a unit vector.

The velocity and acceleration vectors are both directed along the line of the motion:

$$\mathbf{v}(t) = f'(t)\mathbf{d} \qquad \text{and} \qquad \mathbf{a}(t) = f''(t)\mathbf{d}.$$

The speed is $|f'(t)|$:

$$v(t) = \|f'(t)\mathbf{d}\| = |f'(t)| \|\mathbf{d}\| = |f'(t)|,$$

and the magnitude of the acceleration is $|f''(t)|$:

$$\|\mathbf{a}(t)\| = \|f''(t)\mathbf{d}\| = |f''(t)| \|\mathbf{d}\| = |f''(t)|.$$

Circular Motion About the Origin

The position function can be written

$$\mathbf{r}(t) = r[\cos \theta(t)\mathbf{i} + \sin \theta(t)\mathbf{j}].$$

Here $\theta'(t)$ gives the time rate of change of the central angle θ. If $\theta'(t) > 0$, the motion is counterclockwise; if $\theta'(t) < 0$, the motion is clockwise. We call $\theta'(t)$ the *angular velocity* and $|\theta'(t)|$ the *angular speed.*

Uniform circular motion is circular motion with constant angular speed ω ($\omega > 0$). The position function for uniform circular motion in the counterclockwise direction can be written

$$\mathbf{r}(t) = r(\cos \omega t \, \mathbf{i} + \sin \omega t \, \mathbf{j}).$$

Differentiation gives

$$\mathbf{v}(t) = r\omega(-\sin \omega t \, \mathbf{i} + \cos \omega t \, \mathbf{j}),$$

$$\mathbf{a}(t) = -r\omega^2(\cos \omega t \, \mathbf{i} + \sin \omega t \, \mathbf{j}) = -\omega^2 \mathbf{r}(t).$$

The acceleration is directed along the line of the radius vector toward the center of the circle and is therefore perpendicular to the velocity vector, which, as always, is tangential. As you can verify, the speed is $r\omega$ and the magnitude of acceleration is $r\omega^2$. ❑

Motion along a circular helix is a combination of circular motion and motion along a straight line.

Example 1 A particle moves along a circular helix (see Figure 13.3.11) with position at time t given by the function

$$\mathbf{r}(t) = a \cos \omega t \,\mathbf{i} + a \sin \omega t \,\mathbf{j} + b\omega t \,\mathbf{k}. \qquad (a > 0, \, b > 0, \, \omega > 0)$$

For each time t, find

(a) the velocity of the particle;　　**(b)** the speed;　　**(c)** the acceleration;
(d) the magnitude of acceleration;
(e) the angle between the velocity vector and the acceleration vector.

SOLUTION

(a) Velocity: $\mathbf{v}(t) = \mathbf{r}'(t) = -a\omega \sin \omega t \,\mathbf{i} + a\omega \cos \omega t \,\mathbf{j} + b\omega \,\mathbf{k}$.
(b) Speed $v(t) = \|\mathbf{v}(t)\| = \sqrt{a^2\omega^2\sin^2\omega t + a^2\omega^2\cos^2\omega t + b^2\omega^2}$
$$= \sqrt{a^2\omega^2 + b^2\omega^2} = \omega\sqrt{a^2 + b^2}.$$
(The speed is thus constant.)
(c) Acceleration: $\mathbf{a}(t) = \mathbf{v}'(t) = -a\omega^2\cos \omega t \,\mathbf{i} - a\omega^2\sin \omega t \,\mathbf{j}$
$$= -a\omega^2(\cos \omega t \,\mathbf{i} + \sin \omega t \,\mathbf{j}).$$
(Since the speed is constant, the acceleration comes entirely from the change in direction.)
(d) Magnitude of acceleration: $\|\mathbf{a}(t)\| = a\omega^2$.
(e) Angle between $\mathbf{v}(t)$ and $\mathbf{a}(t)$:

$$\cos\theta = \frac{\mathbf{v}(t)}{\|\mathbf{v}(t)\|} \cdot \frac{\mathbf{a}(t)}{\|\mathbf{a}(t)\|}$$

$$= \left[\frac{-a\omega \sin \omega t \,\mathbf{i} + a\omega \cos \omega t \,\mathbf{j} + b\omega \,\mathbf{k}}{\omega\sqrt{a^2 + b^2}}\right] \cdot \left[\frac{-a\omega^2(\cos \omega t \,\mathbf{i} + \sin \omega t \,\mathbf{j})}{a\omega^2}\right]$$

$$= \frac{a(\sin \omega t \cos \omega t - \cos \omega t \sin \omega t)}{\sqrt{a^2 + b^2}} = 0;$$

Therefore $\theta = \tfrac{1}{2}\pi$. (At each point the acceleration vector is perpendicular to the velocity vector.) ❑

An Introduction to Vector Mechanics

We are now ready to work with Newton's Second Law of Motion in its vector form: $\mathbf{F} = m\mathbf{a}$. Since at each time t we have $\mathbf{a}(t) = \mathbf{r}''(t)$, Newton's law can be written

(13.5.4)
$$\mathbf{F}(t) = m\mathbf{a}(t) = m\mathbf{r}''(t).$$

This is a second-order differential equation in t. In Chapter 18 we give an introduction to the general theory of differential equations. Second-order differential equations of the type that we encounter here are treated in Sections 18.5 and 18.6. Our approach in this section is intuitive; we will search for solutions of equation (13.5.4) in particular situations.

When objects are moving, certain quantities (positions, velocities, and so on) are continually changing. This can make a situation difficult to grasp. In these circumstances it is particularly satisfying to find quantities that do not change. Such quantities are said to be *conserved*. (These conserved quantities are called the *constants of the motion*.) Mathematically we can determine whether or not a quantity is conserved by looking at its derivative with respect to time (the time derivative): *The quantity is conserved (is constant) iff its time derivative remains zero.* A *conservation law* is the assertion that in such and such a context a certain quantity does not change.

Momentum

We start with the idea of momentum. The *momentum* $\mathbf{p}$ of an object is the mass of the object times the velocity of the object:

$$\mathbf{p} = m\mathbf{v}.$$

To indicate the time dependence we write

(13.5.5)
$$\mathbf{p}(t) = m\mathbf{v}(t) = m\mathbf{r}'(t).$$

Assume that the mass of the object is constant. Then differentiation gives

$$\mathbf{p}'(t) = m\mathbf{r}''(t) = \mathbf{F}(t).$$

Thus, *the time derivative of the momentum of an object is the net force on the object.* If the net force on an object is continually zero, the momentum $\mathbf{p}(t)$ is constant. This is the law of *conservation of momentum*:

(13.5.6)
> If the net force on an object is continually zero, then the momentum of the object is conserved.

Angular Momentum

The angular momentum of an object about any given point is a vector quantity that is intended to measure the extent to which the object is circling about that point. If the position of the object at time t is given by the radius vector $\mathbf{r}(t)$, then the object's *angular momentum about the origin* is defined by the formula

(13.5.7)
$$\mathbf{L}(t) = \mathbf{r}(t) \times \mathbf{p}(t) = \mathbf{r}(t) \times m\mathbf{v}(t).$$

At each time t of a motion, $\mathbf{L}(t)$ is perpendicular to $\mathbf{r}(t)$, perpendicular to $\mathbf{v}(t)$, and oriented so that $\mathbf{r}(t)$, $\mathbf{v}(t)$, $\mathbf{L}(t)$ form a right-handed triple. The magnitude of $\mathbf{L}(t)$ is given by the relation

$$\|\mathbf{L}(t)\| = \|\mathbf{r}(t)\|\,\|m\mathbf{v}(t)\|\sin\theta(t)$$

where $\theta(t)$ is the angle between $\mathbf{r}(t)$ and $\mathbf{v}(t)$. (All this, of course, comes simply from the definition of cross product.)

If $\mathbf{r}(t)$ and $\mathbf{v}(t)$ are not zero, then we can express $\mathbf{v}(t)$ as a vector parallel to $\mathbf{r}(t)$ plus a vector perpendicular to $\mathbf{r}(t)$:

$$\mathbf{v}(t) = \mathbf{v}_{\|}(t) + \mathbf{v}_{\perp}(t). \qquad \text{(See Figure 13.5.2.)}$$

Thus

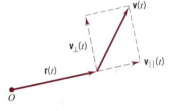

$$\begin{aligned}
\mathbf{L}(t) &= \mathbf{r}(t) \times m\mathbf{v}(t) \\
&= \mathbf{r}(t) \times m[\mathbf{v}_{\|}(t) + \mathbf{v}_{\perp}(t)] \\
&= \underbrace{[\mathbf{r}(t) \times m\mathbf{v}_{\|}(t)]}_{\mathbf{0}} + [\mathbf{r}(t) \times m\mathbf{v}_{\perp}(t)] \\
&= \mathbf{r}(t) \times m\mathbf{v}_{\perp}(t).
\end{aligned}$$

Figure 13.5.2

The component of velocity that is parallel to the radius vector contributes nothing to angular momentum. *The angular momentum comes entirely from the component of velocity that is perpendicular to the radius vector.*

Example 2 In uniform circular motion about the origin,

$$\mathbf{r}(t) = r(\cos\omega t\,\mathbf{i} + \sin\omega t\,\mathbf{j}),$$

the velocity vector $\mathbf{v}(t)$ is always perpendicular to the radius vector $\mathbf{r}(t)$. In this case all of $\mathbf{v}(t)$ contributes to the angular momentum.

We can calculate $\mathbf{L}(t)$ as follows:

$$\begin{aligned}
\mathbf{L}(t) &= \mathbf{r}(t) \times m\mathbf{v}(t) \\
&= [r(\cos\omega t\,\mathbf{i} + \sin\omega t\,\mathbf{j})] \times [mr(-\omega\sin\omega t\,\mathbf{i} + \omega\cos\omega t\,\mathbf{j})] \\
&= mr^2\omega(\cos^2\omega t + \sin^2\omega t)\mathbf{k} = mr^2\omega\mathbf{k}.
\end{aligned}$$

The angular momentum is constant and is perpendicular to the xy-plane. If the motion is counterclockwise (if $\omega > 0$), then the angular momentum points up from the xy-plane. If the motion is clockwise (if $\omega < 0$), then the angular momentum points down from the xy-plane. (This is the right-handedness of the cross product coming in.) ❑

Example 3 In uniform straight-line motion with constant velocity $\mathbf{d}$,

$$\mathbf{r}(t) = \mathbf{r}_0 + t\mathbf{d},$$

we have

$$\mathbf{L}(t) = \mathbf{r}(t) \times m\mathbf{v}(t) = (\mathbf{r}_0 + t\mathbf{d}) \times m\mathbf{d} = m(\mathbf{r}_0 \times \mathbf{d}).$$

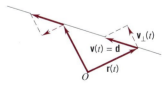

Here again the angular momentum is constant, but all is not quite so simple as it looks. In Figure 13.5.3 you can see the radius vector $\mathbf{r}(t)$, the velocity vector $\mathbf{v}(t) = \mathbf{d}$, and the part of the velocity vector that gives rise to angular momentum, the part perpendicular to $\mathbf{r}(t)$. As before we have called this $\mathbf{v}_{\perp}(t)$. While $\mathbf{v}(t)$ is constant,

Figure 13.5.3

$\mathbf{v}_{\perp}(t)$ is not constant. What happens here is that $\mathbf{r}(t)$ and $\mathbf{v}_{\perp}(t)$ vary in such a way that the cross product

$$\mathbf{L}(t) = \mathbf{r}(t) \times m\mathbf{v}_{\perp}(t)$$

remains constant. If, of course, the line of motion passes through the origin, then $\mathbf{v}(t)$ is parallel to $\mathbf{r}(t)$, $\mathbf{v}_{\perp}(t)$ is zero, and the angular momentum is zero. ❑

Torque

How the angular momentum of an object changes in time depends on the force acting on the object and on the position of the object relative to the origin: since $\mathbf{L}(t) = \mathbf{r}(t) \times m\mathbf{r}'(t)$,

$$\mathbf{L}'(t) = [\mathbf{r}(t) \times m\mathbf{r}''(t)] + \underbrace{[\mathbf{r}'(t) \times m\mathbf{r}'(t)]}_{0} = \mathbf{r}(t) \times \mathbf{F}(t).$$

The cross product

(13.5.8)

$$\boxed{\boldsymbol{\tau}(t) = \mathbf{r}(t) \times \mathbf{F}(t)}$$

is called the *torque* about the origin.†

Since $\mathbf{L}'(t) = \boldsymbol{\tau}(t)$, we have the following conservation law:

(13.5.9)

> If the net torque on an object is continually zero,
> then the angular momentum of the object is conserved.

A force $\mathbf{F}(t)$ is called a *central force* (or a *radial force*) iff $\mathbf{F}(t)$ is always parallel to $\mathbf{r}(t)$. (Gravitational force, for example, is a central force.) For a central force, the cross product $\mathbf{r}(t) \times \mathbf{F}(t)$ is always zero. Thus a central force produces no torque about the origin. As you will see, this places severe restrictions on the kind of motion possible under a central force.

THEOREM 13.5.10

If an object moves under a central force and the angular momentum $\mathbf{L}$ is different from zero, then:

1. The object is confined to the plane that passes through the origin and is perpendicular to $\mathbf{L}$.

2. The radius vector of the object sweeps out equal areas in equal times.

(This theorem plays an important role in astronomy and is embodied in Kepler's three celebrated laws of planetary motion. We will study Kepler's laws in Section *13.6.)

---

† The symbol τ is the Greek letter tau. The word *torque* comes from the Latin word *torquere,* "to twist."

PROOF OF THEOREM 13.5.10 The first assertion is easy to verify: all the radius vectors pass through the origin and (by the very definition of angular momentum) they are all perpendicular to the constant vector **L**.

To verify the second assertion we introduce a right-handed coordinate system O-xyz setting the xy-plane as the plane of the motion, the positive z-axis pointing along **L**. On the xy-plane we introduce polar coordinates r, θ. Thus, at time t, the object has some position $[r(t), \theta(t)]$.

Let's denote by $A(t)$ the area swept out by the radius vector from some fixed time t_0 up to time t. Our task is to show that $A'(t)$ is constant.

The area swept out during the time interval $[t, t + h]$ is simply

$$A(t + h) - A(t).$$

Assuming that the motion takes place in the direction of increasing polar angle θ (see Figure 13.5.4), we have, with obvious notation,

$$\underbrace{\tfrac{1}{2} \min [r(t)]^2 \cdot [\theta(t + h) - \theta(t)]}_{\text{area of inner sector}} \le A(t + h) - A(t) \le \underbrace{\tfrac{1}{2} \max [r(t)]^2 \cdot [\theta(t + h) - \theta(t)]}_{\text{area of outer sector}}.$$

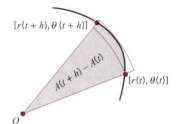

Divide through by h, take the limit as h tends to 0, and you will see that

$$(*) \qquad\qquad A'(t) = \tfrac{1}{2}[r(t)]^2 \theta'(t).$$

Now

$$\mathbf{r}(t) = r(t)[\cos \theta(t)\mathbf{i} + \sin \theta(t)\mathbf{j}],$$

$$\mathbf{v}(t) = r'(t)[\cos \theta(t)\mathbf{i} + \sin \theta(t)\mathbf{j}] + r(t)\theta'(t)[-\sin \theta(t)\mathbf{i} + \cos \theta(t)\mathbf{j}]$$
$$= [r'(t)\cos \theta(t) - r(t)\theta'(t)\sin \theta(t)]\mathbf{i} + [r'(t)\sin \theta(t) + r(t)\theta'(t)\cos \theta(t)]\mathbf{j}.$$

A calculation that you can carry out yourself shows that

$$\mathbf{L} = \mathbf{r}(t) \times m\mathbf{v}(t) = mr^2(t)\theta'(t)\mathbf{k}.$$

Since **L** is constant, $r^2(t)\theta'(t)$ is constant. Thus, by $(*)$, $A'(t)$ is constant:

$$A'(t) = L/2m \qquad \text{where} \quad L = \|\mathbf{L}\|. \quad \square$$

Figure 13.5.4

Initial-Value Problems

In physics one tries to make predictions about the future on the basis of current information and a knowledge of the forces at work. In the case of an object in motion, the task can be to determine $\mathbf{r}(t)$ for all t given the force and some "initial conditions." Frequently the initial conditions give the position and velocity of the object at some time t_0. The problem then is to solve the differential equation

$$\mathbf{F} = m\mathbf{r}''$$

subject to conditions of the form

$$\mathbf{r}(t_0) = \mathbf{r}_0, \qquad \mathbf{v}(t_0) = \mathbf{v}_0.$$

Such problems are known as initial-value problems. We have considered initial-value problems in several different contexts earlier in the text. For example, see Sections 5.6 and 7.6.

By far the simplest problem of this sort concerns a *free particle,* an object on which there is no net force.

Example 4 At time t_0 a free particle has position $\mathbf{r}(t_0) = \mathbf{r}_0$ and velocity $\mathbf{v}(t_0) = \mathbf{v}_0$. Find $\mathbf{r}(t)$ for all t.

SOLUTION Since there is no net force on the object, the acceleration is zero and the velocity is constant. Since $\mathbf{v}(t_0) = \mathbf{v}_0$,

$$\mathbf{v}(t) = \mathbf{v}_0 \qquad \text{for all } t.$$

Integration with respect to t gives

$$\mathbf{r}(t) = t\mathbf{v}_0 + \mathbf{c}$$

where $\mathbf{c}$, the constant of integration, is a vector that we can determine from the initial position. The initial position $\mathbf{r}(t_0) = \mathbf{r}_0$ gives

$$\mathbf{r}_0 = t_0\mathbf{v}_0 + \mathbf{c} \qquad \text{and therefore} \qquad \mathbf{c} = \mathbf{r}_0 - t_0\mathbf{v}_0.$$

Using this value for $\mathbf{c}$ in our equation for $\mathbf{r}(t)$, we have

$$\mathbf{r}(t) = t\mathbf{v}_0 + (\mathbf{r}_0 - t_0\mathbf{v}_0),$$

which we write as

$$\mathbf{r}(t) = \mathbf{r}_0 + (t - t_0)\mathbf{v}_0.$$

This is the equation of a straight line with direction vector $\mathbf{v}_0$. Free particles travel in straight lines with constant velocity. (We have tacitly assumed that $\mathbf{v}_0 \neq \mathbf{0}$. If $\mathbf{v}_0 = \mathbf{0}$, the particle remains at rest at $\mathbf{r}_0$.) ❏

Example 5 An object of mass m is subject to a force of the form

$$\mathbf{F}(t) = -m\omega^2\mathbf{r}(t) \qquad \text{with } \omega > 0.$$

Find $\mathbf{r}(t)$ for all t given that

$$\mathbf{r}(0) = a\mathbf{i} \qquad \text{and} \qquad \mathbf{v}(0) = \omega a\mathbf{j} \qquad \text{with } a > 0.$$

SOLUTION The force is a vector version of the restoring force exerted by a linear spring (Hooke's law). Since the force is central, the angular momentum of the object, $\mathbf{L}(t) = \mathbf{r}(t) \times m\mathbf{v}(t)$, is conserved. So $\mathbf{L}(t)$ is constantly equal to the value it had at time $t = 0$: for all t

$$\mathbf{L}(t) = \mathbf{L}(0) = \mathbf{r}(0) \times m\mathbf{v}(0) = a\mathbf{i} \times m\omega a\mathbf{j} = ma^2\omega\mathbf{k}.$$

From our earlier discussion (Theorem 13.5.10) we can conclude that the motion takes place in the plane that passes through the origin and is perpendicular to $\mathbf{k}$. This is the xy-plane. Thus we can write

$$\mathbf{r}(t) = x(t)\mathbf{i} + y(t)\mathbf{j}.$$

Since $\mathbf{F}(t) = m\mathbf{r}''(t)$, the force equation can be written $\mathbf{r}''(t) = -\omega^2\mathbf{r}(t)$. In terms of components we have

$$x''(t) = -\omega^2 x(t), \qquad y''(t) = -\omega^2 y(t).$$

These are the equations of simple harmonic motion. We have already seen that functions of the form

$$x(t) = A_1\sin(\omega t + \phi_1), \qquad y(t) = A_2\sin(\omega t + \phi_2)$$

are solutions of these equations (see Exercises 61 and 62, Section 3.6). In Section 18.5, we show that *all* solutions have this form. To evaluate the constants we use the initial conditions. That $\mathbf{r}(0) = a\mathbf{i}$ means that $x(0) = a$ and $y(0) = 0$. So

(∗) $\qquad\qquad\qquad A_1 \sin \phi_1 = a, \qquad A_2 \sin \phi_2 = 0.$

That $\mathbf{v}(0) = \omega a\mathbf{j}$ means that $x'(0) = 0$ and $y'(0) = \omega a$. Since

$$x'(t) = \omega A_1 \cos(\omega t + \phi_1) \qquad \text{and} \qquad y'(t) = \omega A_2 \cos(\omega t + \phi_2),$$

we have

(∗∗) $\qquad\qquad\qquad \omega A_1 \cos \phi_1 = 0, \qquad \omega A_2 \cos \phi_2 = \omega a.$

Conditions (∗) and (∗∗) are met by setting $A_1 = a$, $A_2 = a$, $\phi_1 = \frac{1}{2}\pi$, $\phi_2 = 0$. Thus

$$x(t) = a \sin(\omega t + \tfrac{1}{2}\pi) = a \cos \omega t, \qquad y(t) = a \sin \omega t.$$

The vector equation reads

$$\mathbf{r}(t) = a \cos \omega t\, \mathbf{i} + a \sin \omega t\, \mathbf{j}.$$

The object moves in a circle of radius a about the origin with constant angular velocity ω. ❑

Example 6 A particle of charge q in a magnetic field $\mathbf{B}$ is subject to the force

$$\mathbf{F}(t) = \frac{q}{c}[\mathbf{v}(t) \times \mathbf{B}(t)]$$

where c is the speed of light and $\mathbf{v}$ is the velocity of the particle. Given that $\mathbf{r}(0) = \mathbf{r}_0$ and $\mathbf{v}(0) = \mathbf{v}_0$, find the path of the particle in the constant magnetic field $\mathbf{B}(t) = B_0\mathbf{k}$, $B_0 \neq 0$.

SOLUTION There is no conservation law that we can conveniently appeal to here. Neither momentum nor angular momentum are conserved: the force is not zero and it is not central. We start directly with Newton's $\mathbf{F} = m\mathbf{r}''$.
 Since $\mathbf{r}'' = \mathbf{v}'$, we have

$$m\mathbf{v}'(t) = \frac{q}{c}[\mathbf{v}(t) \times B_0\mathbf{k}],$$

which we can write as

$$\mathbf{v}'(t) = \frac{qB_0}{mc}[\mathbf{v}(t) \times \mathbf{k}].$$

To simplify notation we set $qB_0/mc = \omega$. We then have

$$\mathbf{v}'(t) = \omega[\mathbf{v}(t) \times \mathbf{k}].$$

Placing $\mathbf{v}(t) = v_1(t)\mathbf{i} + v_2(t)\mathbf{j} + v_3(t)\mathbf{k}$ in this last equation and working out the cross product, we find that

$$v_1'(t)\mathbf{i} + v_2'(t)\mathbf{j} + v_3'(t)\mathbf{k} = \omega[v_2(t)\mathbf{i} - v_1(t)\mathbf{j}].$$

This gives the scalar equations

$$v_1'(t) = \omega v_2(t), \qquad v_2'(t) = -\omega v_1(t), \qquad v_3'(t) = 0.$$

The last equation is trivial. It says that v_3 is constant:

$$v_3(t) = C.$$

The equations for v_1 and v_2 are linked together. We can get an equation that involves only v_1 by differentiating the first equation:

$$v_1''(t) = \omega v_2'(t) = -\omega^2 v_1(t).$$

As we know from our earlier work, this gives

$$v_1(t) = A_1 \sin(\omega t + \phi_1).$$

Since $v_1'(t) = \omega v_2(t)$, we have

$$v_2(t) = \frac{v_1'(t)}{\omega} = \frac{A_1 \omega}{\omega} \cos(\omega t + \phi_1) = A_1 \cos(\omega t + \phi_1).$$

Therefore

$$\mathbf{v}(t) = A_1 \sin(\omega t + \phi_1)\mathbf{i} + A_1 \cos(\omega t + \phi_1)\mathbf{j} + C\mathbf{k}.$$

A final integration with respect to t gives

$$\mathbf{r}(t) = \left[-\frac{A_1}{\omega} \cos(\omega t + \phi_1) + D_1 \right]\mathbf{i} + \left[\frac{A_1}{\omega} \sin(\omega t + \phi_1) + D_2 \right]\mathbf{j} + [Ct + D_3]\mathbf{k}$$

where D_1, D_2, D_3 are constants of integration. All six constants of integration—A_1, ϕ_1, C, D_1, D_2, D_3—can be evaluated from the initial conditions. We will not pursue this. What is important here is that the path of the particle is a circular helix with axis parallel to $\mathbf{B}$, in this case parallel to $\mathbf{k}$. You should be able to see this from the equation for $\mathbf{r}(t)$: the z-component of $\mathbf{r}$ varies linearly with t from the value D_3, while the x and y components represent uniform motion with angular velocity ω in a circle of radius $|A_1/\omega|$ around the center (D_1, D_2). ❏

(Physicists express the behavior just found by saying that charged particles *spiral around* the magnetic field lines. Qualitatively, this behavior still holds even if the magnetic field lines are "bent," as is the case with the earth's magnetic field. Many charged particles become trapped by the earth's magnetic field. They keep spiraling around the magnetic field lines that run from pole to pole.)

EXERCISES 13.5

1. A particle moves in a circle of radius r at constant speed v. Find the angular speed and the magnitude of the acceleration.

2. A particle moves so that

   $$\mathbf{r}(t) = (a \cos \pi t + bt^2)\mathbf{i} + (a \sin \pi t - bt^2)\mathbf{j}.$$

   Find the velocity, speed, acceleration, and the magnitude of the acceleration all at time $t = 1$.

3. A particle moves so that $\mathbf{r}(t) = at\mathbf{i} + b \sin at\mathbf{j}$. Show that the magnitude of the acceleration of the particle is proportional to the distance of the particle from the x-axis.

4. A particle moves so that $\mathbf{r}(t) = 2\mathbf{i} + t^2\mathbf{j} + (t-1)^2\mathbf{k}$. At what time is the speed a minimum?

In Exercises 5–8, sketch the curve. Then compute and sketch the acceleration vector at the indicated points.

5. $\mathbf{r}(t) = (t/\pi)\mathbf{i} + \cos t\mathbf{j}$, $t \in [0, 2\pi]$; at $t = \frac{1}{4}\pi, \frac{1}{2}\pi, \pi$.

6. $\mathbf{r}(t) = t^3\mathbf{i} + t\mathbf{j}$, t real; at $t = -\frac{1}{2}, \frac{1}{2}, 1$.

7. $\mathbf{r}(t) = \sec t\mathbf{i} + \tan t\mathbf{j}$, $t \in [\frac{1}{4}\pi, \frac{1}{2}\pi)$; at $t = -\frac{1}{6}\pi, 0, \frac{1}{3}\pi$.

8. $\mathbf{r}(t) = \sin \pi t + t\mathbf{j}$, $t \in [0, 2]$; at $t = \frac{1}{2}, 1, \frac{5}{4}$.

9. An object moves so that

$$\mathbf{r}(t) = x_0\mathbf{i} + [y_0 + (\alpha \cos \theta)t]\mathbf{j} \\ + [z_0 + (\alpha \sin \theta)t - 16t^2]\mathbf{k}, \qquad t \geq 0.$$

Find (a) the initial position, (b) the initial velocity, (c) the initial speed, (d) the acceleration throughout the motion. Finally, (e) identify the curve.

10. A particle moves so that $\mathbf{r}(t) = 2 \cos 2t\mathbf{i} + 3 \cos t\mathbf{j}$.
(a) Show that the particle oscillates on an arc of the parabola $4y^2 - 9x = 18$. (b) Draw the path. (c) What are the acceleration vectors at the points of zero velocity? (d) Draw these vectors at the points in question.

11. An object of mass m moves so that

$$\mathbf{r}(t) = \tfrac{1}{2}a(e^{\omega t} + e^{-\omega t})\mathbf{i} + \tfrac{1}{2}b(e^{\omega t} - e^{-\omega t})\mathbf{j}.$$

(a) What is the velocity at $t = 0$? (b) Show that the acceleration vector is a constant positive multiple of the radius vector. (This shows that the force is central and repelling.) (c) What does (b) imply about the angular momentum and the torque? Verify your answers by direct calculation.

12. (a) An object moves so that

$$\mathbf{r}(t) = a_1 e^{bt}\mathbf{i} + a_2 e^{bt}\mathbf{j} + a_3 e^{bt}\mathbf{k}.$$

Show that, if $b > 0$, the object experiences a repelling central force.
(b) An object moves so that

$$\mathbf{r}(t) = \sin t\mathbf{i} + \cos t\mathbf{j} + (\sin t + \cos t)\mathbf{k}.$$

Show that the object experiences an attracting central force.
(c) Compute the angular momentum $\mathbf{L}(t)$ for the motion in (b).

13. A constant force of magnitude α directed upward from the xy-plane is continually applied to an object of mass m. Given that the object starts at time 0 at the point $P(0, y_0, z_0)$ with initial velocity $2\mathbf{j}$, find: (a) the velocity of the object t seconds later; (b) the speed of the object t seconds later; (c) the momentum of the object t seconds later; (d) the path followed by the object, both in vector form and in Cartesian coordinates.

14. Show that, if the force on an object is always perpendicular to the velocity of the object, then the *speed* of the object is constant. (This tells us that the speed of a charged particle in a magnetic field is constant.)

15. Find the force required to propel a particle of mass m so that $\mathbf{r}(t) = t\mathbf{j} + t^2\mathbf{k}$.

16. Show that for an object of constant velocity the angular momentum is constant.

17. At each point $P(x(t), y(t), z(t))$ of its motion an object of mass m is subject to a force

$$\mathbf{F}(t) = m\pi^2[a \cos \pi t\mathbf{i} + b \sin \pi t\mathbf{j}]. \quad (a > 0, b > 0)$$

Given that $\mathbf{v}(0) = -\pi b\mathbf{j} + \mathbf{k}$ and $\mathbf{r}(0) = b\mathbf{j}$, find the following at time $t = 1$:
(a) The velocity. (b) The speed.
(c) The acceleration. (d) The momentum.
(e) The angular momentum. (f) The torque.

18. If an object of mass m moves with velocity $\mathbf{v}(t)$ subject to a force $\mathbf{F}(t)$, the scalar product

(13.5.11) $$\mathbf{F}(t) \cdot \mathbf{v}(t)$$

is called the *power* (expended by the force) and the number

(13.5.12) $$\tfrac{1}{2}m[v(t)]^2$$

is called the *kinetic energy* of the object. Show that the time rate of change of the kinetic energy of an object is the power expended on it:

(13.5.13) $$\frac{d}{dt}(\tfrac{1}{2}m[v(t)]^2) = \mathbf{F}(t) \cdot \mathbf{v}(t).$$

19. Two particles of equal mass m, one with constant velocity $\mathbf{v}$ and the other at rest, collide elastically (i.e., the kinetic energy of the system is preserved) and go off in different directions. Show that the two particles go off at right angles.

20. (*Elliptic harmonic motion*) Show that, if the force on a particle of mass m is of the form

$$\mathbf{F}(t) = -m\omega^2\mathbf{r}(t),$$

then the path of the particle may be written

$$\mathbf{r}(t) = \cos \omega t\mathbf{A} + \sin \omega t\mathbf{B}$$

with $\mathbf{A}$ and $\mathbf{B}$ constant vectors. Give the physical significance of $\mathbf{A}$ and $\mathbf{B}$ and specify conditions on $\mathbf{A}$ and $\mathbf{B}$ that restrict the particle to a circular path.
HINT: The solutions of the differential equation

$$x''(t) = -\omega^2 x(t)$$

can be written in the form

$$x(t) = A \cos \omega t + B \sin \omega t. \quad \text{(Exercise 61, Section 3.6)}$$

21. A particle moves with constant acceleration $\mathbf{a}$. Show that the path of the particle lies entirely in some plane. Find a vector equation for this plane.

22. In Example 6 we stated that the path

$$\mathbf{r}(t) = \left[-\frac{A_1}{\omega} \cos(\omega t + \phi_1) + D_1 \right] \mathbf{i}$$
$$+ \left[\frac{A_1}{\omega} \sin(\omega t + \phi_1) + D_2 \right] \mathbf{j} + [Ct + D_3]\mathbf{k}$$

was a circular helix. Set $\omega = -1$ and show that, if $\mathbf{r}(0) = a\mathbf{i}$ and $\mathbf{v}(0) = a\mathbf{j} + b\mathbf{k}$, then the path takes the form

$$\mathbf{r}(t) = a \cos t\mathbf{i} + a \sin t\mathbf{j} + bt\mathbf{k},$$

the circular helix described in Section 13.3.

23. A charged particle in a constant electric field $\mathbf{E}$ experiences the force $q\mathbf{E}$, where q is the charge of the particle. Assume that the field has the constant value $\mathbf{E} = E_0\mathbf{k}$ and find the path of the particle given that $\mathbf{r}(0) = \mathbf{i}$ and $\mathbf{v}(0) = \mathbf{j}$.

24. (*Important*) A wheel is rotating about an axle with angular speed ω. Let $\boldsymbol{\omega}$ be the *angular velocity vector,* the vector of length ω that points along the axis of the wheel in such a direction that, observed from the tip of $\boldsymbol{\omega}$, the wheel rotates counterclockwise. Take the origin as the center of the wheel and let $\mathbf{r}$ be the vector from the origin to a point P on the rim of the wheel. Express the velocity $\mathbf{v}$ of P in terms of $\boldsymbol{\omega}$ and $\mathbf{r}$.

25. Solve the initial-value problem

$$\mathbf{F}(t) = m\mathbf{r}''(t) = t\mathbf{i} + t^2\mathbf{j}, \qquad \mathbf{r}_0 = \mathbf{r}(0) = \mathbf{i},$$
$$\mathbf{v}_0 = \mathbf{v}(0) = \mathbf{k}.$$

26. Solve the initial-value problem

$$\mathbf{F}(t) = m\mathbf{r}''(t) = -m\beta^2 z(t)\mathbf{k}, \qquad \mathbf{r}_0 = \mathbf{r}(0) = \mathbf{k},$$
$$\mathbf{v}_0 = \mathbf{v}(0) = \mathbf{0}.$$

[Here $z(t)$ is the z-component of $\mathbf{r}(t)$.]

27. An object of mass m moves subject to the force

$$\mathbf{F}(\mathbf{r}) = 4r^2\mathbf{r}$$

where $\mathbf{r}$ is the position of the object. Suppose $\mathbf{r}(0) = \mathbf{0}$ and $\mathbf{v}(0) = 2\mathbf{u}$, where $\mathbf{u}$ is a unit vector. Show that at each time t the speed v of the object satisfies the relation

$$v = \sqrt{4 + \frac{2}{m}r^4}.$$

HINT: Examine the quantity $\frac{1}{2}mv^2 - r^4$. (This is the *energy* of the object, a notion we will take up in Chapter 17.)

28. A moving particle traces a path that at each point has a unit tangent vector $\mathbf{T}(t)$ and a principal normal $\mathbf{N}(t)$. In anticipation of Section 13.7, (a) write the velocity vector as a scalar multiple of $\mathbf{T}(t)$, and then (b) write the acceleration vector as a linear combination of $\mathbf{T}(t)$ and $\mathbf{N}(t)$. (Incidentally, this will show that the acceleration vector lies in the osculating plane.)

---

*■ 13.6 Planetary Motion

Tycho Brahe, Johannes Kepler

In the middle of the sixteenth century the arguments on planetary motion persisted. Was Copernicus right? Did the planets move in circles about the sun? Obviously not. Was not the earth the center of the universe?

In 1576, with the generous support of his king, the Danish astronomer Tycho Brahe built an elaborate astronomical observatory on the isle of Hveen and began his painstaking observations. For more than twenty years he looked through his telescopes and recorded what he saw. He was a meticulous observer, but he could draw no definite conclusions.

In 1599 the German astronomer–mathematician Johannes Kepler began his study of Brahe's voluminous tables. For a year and a half Brahe and Kepler worked together. Then Brahe died and Kepler went on wrestling with the data. His persistence paid off. By 1619 Kepler had made three stupendous discoveries, known today as *Kepler's Laws of Planetary Motion:*

I. Each planet moves in a plane, not in a circle, but in an elliptic orbit with the sun at one focus.

II. The radius vector from sun to planet sweeps out equal areas in equal times.

III. The square of the period of the motion varies directly as the cube of the major semiaxis, and the constant of proportionality is the same for all the planets.

What Kepler formulated, Newton was able to explain. Each of these laws, Newton showed, was deducible from his laws of motion and his law of gravitation.

Newton's Second Law of Motion for Extended Three-Dimensional Objects

Imagine an object that consists of n particles with masses $m_1, m_2, \ldots, m_n$ located at $\mathbf{r}_1, \mathbf{r}_2, \ldots, \mathbf{r}_n$.† The total mass M of the object is the sum of the masses of the constituent particles:

$$M = m_1 + \cdots + m_n.$$

The center of mass of the object is by definition the point $\mathbf{R}_M$ where

$$M\mathbf{R}_M = m_1\mathbf{r}_1 + \cdots + m_n\mathbf{r}_n.$$

The total force $\mathbf{F}_{\text{TOT}}$ on the object is by definition the sum of all the forces that act on the particles that constitute the object:

$$\mathbf{F}_{\text{TOT}} = \mathbf{F}_1 + \cdots + \mathbf{F}_n.$$

Since $\mathbf{F}_1 = m_1\mathbf{r}_1'', \ldots, \mathbf{F}_n = m_n\mathbf{r}_n''$, we have

$$\mathbf{F}_{\text{TOT}} = m_1\mathbf{r}_1'' + \cdots + m_n\mathbf{r}_n'',$$

which we can write as

$$\mathbf{F}_{\text{TOT}} = M\mathbf{R}_M''.$$

The total force on an extended object is thus the total mass of the object times the acceleration of the center of mass.

We can simplify this still further. The forces that act between the constituent particles, the so-called "internal forces," cancel in pairs: if particle 23 tugs at particle 71 in a certain direction with a certain strength, then particle 71 tugs at particle 23 in the opposite direction with the same strength. (Newton's third law: to every action there is an equal reaction.) Therefore, in calculating the total force on our object, we can disregard the internal forces and simply add up the external forces. $\mathbf{F}_{\text{TOT}} = \mathbf{F}_{\text{TOT}}^{(\text{Ext})}$ and Newton's second law takes the form

(13.6.1)

$$\mathbf{F}_{\text{TOT}}^{(\text{Ext})} = M\mathbf{R}_M''.$$

---

† The case of a continuously distributed mass is taken up in Chapter 16.

The total external force on an extended three-dimensional object is thus the total mass of the object times the acceleration of the center of mass.

When an external force is applied to an extended object, we cannot predict the reaction of all the constituent particles, but we can predict the reaction of the center of mass. The center of mass will react to the force as if it were a particle with all the mass concentrated there. Suppose, for example, that a bomb is dropped from an airplane. The center of mass, "feeling" only the force of gravity (we are neglecting air resistance), falls in a parabolic arc toward the ground even if the bomb explodes at a thousand meters and individual pieces fly every which way. The forces of explosion are internal and do not affect the center of mass.

Some Preliminary Comments About the Planets

Roughly speaking, a planet is a massive object in the form of a ball with the center of mass at the center. In what follows, when we refer to the position of a planet, you are to understand that we really mean the position of the center of mass of the planet. Two other points require comment. First, we will write our equations as if the sun affected the motion of the planet, but not vice versa: we will assume that the sun stays put and that the planet moves. Really each affects the other. Our viewpoint is justified by the immense difference in mass between the planets and the sun. In the case of the earth and the sun, for example, a reasonable analogy is to imagine a tug of war in space between someone who weighs three pounds and someone who weighs a million pounds: to a good approximation, the million-pound person does not move. The second point is that the planet is affected not only by the pull from the sun, but also by the gravitational pulls from the other planets and all the other celestial bodies. But these forces are much smaller, and they tend to cancel. We will ignore them. (Our results are only approximations, but they prove to be very good approximations.)

A Derivation of Kepler's Laws from Newton's Laws of Motion and His Law of Gravitation

The gravitational force exerted by the sun on a planet can be written in vector form as follows:

$$(*) \qquad \mathbf{F}(\mathbf{r}) = -G\,\frac{mM}{r^3}\,\mathbf{r}.$$

Here m is the mass of the planet, M is the mass of the sun, G is the gravitational constant, $\mathbf{r}$ is the vector from the center of the sun to the center of the planet, and r is the magnitude of $\mathbf{r}$. (Thus we are placing the sun at the origin of our coordinate system; namely, we are using what is known as a *heliocentric* coordinate system, from *Hēlios,* the Greek word for sun.)

Let's make sure that equation $(*)$ conforms to our earlier characterization of gravitational force. First of all, because of the minus sign, the direction is toward the origin, where the sun is located. Taking norms we have

$$\|\mathbf{F}(\mathbf{r})\| = \frac{GmM}{r^3}\,\|\mathbf{r}\| = \frac{GmM}{r^2}.$$

$$\underset{\|\mathbf{r}\| = r}{\uparrow}$$

Thus the magnitude of the force is as expected: it does vary directly as the product of the masses and inversely as the square of the distance between them. So we are back in familiar territory.

We will derive Kepler's laws in a somewhat piecemeal manner. Since the force on the planet is a central force, **L**, the angular momentum of the planet, is conserved. (Section 13.5.) If **L** were zero, we would have

$$\|\mathbf{L}\| = m\|\mathbf{r}\|\,\|\mathbf{v}\|\sin\theta = 0.$$

This would mean that either $\mathbf{r} = \mathbf{0}$, $\mathbf{v} = \mathbf{0}$, or $\sin\theta = 0$. The first equation would place the planet at the center of the sun. The second equation could hold only if the planet stopped. The third equation could hold only if the planet moved directly toward the sun or directly away from the sun. We can be thankful that none of those things happen.

Since the planet moves under a central force and **L** is not zero, *the planet stays on the plane that passes through the center of the sun and is perpendicular to* **L**, and the radius vector does sweep out equal areas in equal times. (We know all this from Theorem 13.5.10.)

Now we go on to show that the path is an ellipse with the sun at one focus.† The equation of motion of a planet of mass m can be written

$$m\mathbf{a}(t) = -GmM\frac{\mathbf{r}(t)}{[r(t)]^3}.$$

Clearly m drops out of the equation. Setting $GM = \rho$ and suppressing the explicit dependence on t, we have

$$\mathbf{a} = -\rho\frac{\mathbf{r}}{r^3},$$

which gives

(∗∗)
$$\frac{\mathbf{r}}{r^3} = -\frac{1}{\rho}\mathbf{a} = -\frac{1}{\rho}\frac{d\mathbf{v}}{dt}.$$

From (13.2.4) we know that in general

$$\frac{d}{dt}\left(\frac{\mathbf{r}}{r}\right) = (\mathbf{r}\times\mathbf{v})\times\frac{\mathbf{r}}{r^3}.$$

Since $\mathbf{L} = \mathbf{r}\times m\mathbf{v}$, we have

$$\frac{d}{dt}\left(\frac{\mathbf{r}}{r}\right) = \frac{\mathbf{L}}{m}\times\frac{\mathbf{r}}{r^3}.$$

Inserting (∗∗) we see that

$$\frac{d}{dt}\left(\frac{\mathbf{r}}{r}\right) = \frac{\mathbf{L}}{m}\times\left(-\frac{1}{\rho}\frac{d\mathbf{v}}{dt}\right) = \frac{d\mathbf{v}}{dt}\times\frac{\mathbf{L}}{m\rho} = \frac{d}{dt}\left(\mathbf{v}\times\frac{\mathbf{L}}{m\rho}\right)$$

$$\mathbf{L}, m, \rho \text{ are all constant} \longrightarrow$$

and therefore

$$\frac{d}{dt}\left[\left(\mathbf{v}\times\frac{\mathbf{L}}{m\rho}\right) - \frac{\mathbf{r}}{r}\right] = \mathbf{0}.$$

---

† If some of the steps seem unanticipated to you, you should realize that we are discussing one of the most celebrated physical problems in history and there have been three hundred years to think of ingenious ways to deal with it. The argument we give here follows the lines of the excellent discussion that appears in Harry Pollard's *Celestial Mechanics,* Mathematical Association of America (1976).

Integration with respect to t gives

(***)
$$\left(\mathbf{v} \times \frac{\mathbf{L}}{m\rho}\right) - \frac{\mathbf{r}}{r} = \mathbf{e}$$

where $\mathbf{e}$ is a constant vector that depends on the initial conditions. Dotting both sides with $\mathbf{r}$, we have

$$\mathbf{r} \cdot \left(\mathbf{v} \times \frac{\mathbf{L}}{m\rho}\right) - \frac{\mathbf{r} \cdot \mathbf{r}}{r} = \mathbf{r} \cdot \mathbf{e}.$$

Since

$$\mathbf{r} \cdot \left(\mathbf{v} \times \frac{\mathbf{L}}{m\rho}\right) = \frac{\mathbf{L}}{m\rho} \cdot (\mathbf{r} \times \mathbf{v}) = \frac{\mathbf{L}}{m\rho} \cdot \frac{\mathbf{L}}{m} = \frac{L^2}{m^2\rho} \qquad \text{and} \qquad \frac{\mathbf{r} \cdot \mathbf{r}}{r} = \frac{r^2}{r} = r,$$

(12.4.6)

we find that

$$\frac{L^2}{m^2\rho} - r = \mathbf{r} \cdot \mathbf{e},$$

which we write as

$$r + (\mathbf{r} \cdot \mathbf{e}) = \frac{L^2}{m^2\rho}. \qquad \text{(Orbit equation.)}$$

If $\mathbf{e} = \mathbf{0}$, the orbit is a circle:

$$r = \frac{L^2}{m^2\rho}.$$

This is a possibility that requires very special initial conditions, conditions not met by any of the planets in our solar system. In our solar system, at least, $\mathbf{e} \neq \mathbf{0}$.

Given that $\mathbf{e} \neq \mathbf{0}$, we can write the orbit equation as

(13.6.2)
$$r(1 + e \cos \theta) = \frac{L^2}{m^2\rho}$$

where $e = \|\mathbf{e}\|$ and θ is the angle between $\mathbf{r}$ and $\mathbf{e}$. We are almost through. Since $\mathbf{v} \times \mathbf{L}$ and $\mathbf{r}$ are both perpendicular to $\mathbf{L}$, we know from (***) that $\mathbf{e}$ is perpendicular to $\mathbf{L}$. Therefore, in the plane of $\mathbf{e}$ and $\mathbf{r}$ (see Figure 13.6.1) we can interpret r and θ as the usual polar coordinates. The orbit equation (13.6.2) is then a polar equation and, as such, represents a conic section with focus at the origin, which is to say, *focus at the sun.* According to 9.4.1, the conic section can be a parabola, a hyperbola, or an ellipse. The repetitiveness of planetary motion rules out the parabola and the hyperbola. *The orbit is therefore an ellipse.*

Finally we will verify Kepler's third law: that the square of the period of the motion varies directly as the cube of the major semi-axis and the constant of proportionality is the same for all planets.

The elliptic orbit has an equation of the form

$$r(1 + e \cos \theta) = ed \qquad \text{with} \quad 0 < e < 1, \quad d > 0.$$

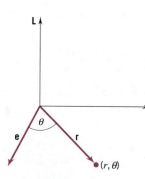

Figure 13.6.1

By Section 9.4 the major and minor semi-axes of the ellipse are given by

$$a = \frac{ed}{1 - e^2}, \qquad b = a\sqrt{1 - e^2}.$$

(See Theorem 9.4.2.) Denote the period of revolution by T. Since the radius vector sweeps out area at the constant rate of $L/2m$ (we know this from the proof of Theorem 13.5.10),

$$\left(\frac{L}{2m}\right) T = \text{area of the ellipse} = \pi ab = \pi a^2 \sqrt{1 - e^2}.$$

Thus

$$T = \frac{2\pi m a^2 \sqrt{1 - e^2}}{L} \qquad \text{and} \qquad T^2 = \frac{4\pi^2 m^2 a^4 (1 - e^2)}{L^2}.$$

From (13.6.2) we know that

$$ed = \frac{L^2}{m^2 \rho} = \frac{L^2}{m^2 GM} \qquad \text{and therefore} \qquad \frac{m^2}{L^2} = \frac{1}{edGM}.$$

It follows that

$$T^2 = \frac{4\pi^2 a^4 (1 - e^2)}{edGM} = \frac{4\pi^2 a^4 (1 - e^2)}{a(1 - e^2)GM} = \frac{4\pi^2}{GM} a^3.$$

$$\underline{\qquad\qquad} ed = a(1 - e^2)$$

T^2 does vary directly with a^3, and the constant of proportionality $4\pi^2/GM$ is the same for all planets. ❏

EXERCISES *13.6

1. Kepler's third law can be stated as follows: for each planet the square of the period of revolution varies directly as the cube of the planet's average distance from the sun, and the constant of proportionality is the same for all planets. A *year* on a given planet is the time taken by the planet to make one circuit around the sun; thus a year on a planet is the period of revolution of that planet. Given that on average Venus is 0.72 times as far from the sun as the earth, how does the length of a "Venus year" compare with the length of an "earth year"?

2. Verify by differentiation with respect to time t that, if the acceleration of a planet is given by

$$\mathbf{a} = -\rho \frac{\mathbf{r}}{r^3},$$

then the energy $E = \frac{1}{2}mv^2 - m\rho r^{-1}$ is constant.

3. Given that a planet moves in a plane, its motion can be described by rectangular coordinates (x, y) or polar coordinates $[r, \theta]$, with origin at the sun. The kinetic energy of a planet is

$$\tfrac{1}{2}mv^2 = \tfrac{1}{2}m\left[\left(\frac{dx}{dt}\right)^2 + \left(\frac{dy}{dt}\right)^2\right].$$

Show that in polar coordinates

$$\tfrac{1}{2}mv^2 = \tfrac{1}{2}m\left[\left(\frac{dr}{dt}\right)^2 + r^2\left(\frac{d\theta}{dt}\right)^2\right].$$

4. We have seen that the energy of a planet

$$E = \tfrac{1}{2}mv^2 - \frac{m\rho}{r}$$

is constant. (Exercise 2.) Setting $dr/dt = \dot{r}$, $d\theta/dt = \dot{\theta}$ and using Exercise 3, we have

$$E = \tfrac{1}{2}m(\dot{r}^2 + r^2 \dot{\theta}^2) - \frac{m\rho}{r}.$$

We also know that the angular momentum $\mathbf{L}$ is constant and that $L = mr^2\dot{\theta}$. Use this fact to verify that

$$E = \frac{L^2}{2m}\left\{\frac{1}{r^2} + \frac{1}{r^4}\left(\frac{dr}{d\theta}\right)^2\right\} - \frac{m\rho}{r}.$$

Since E is a constant, this is a differential equation for r as a function of θ.

5. Show that the function

$$r = \frac{a}{1 + e\cos\theta} \quad \text{with} \quad a = \frac{L^2}{m^2\rho} \quad \text{and} \quad e^2 = \frac{2Ea}{m\rho} + 1$$

satisfies the equation derived in Exercise 4.

■ 13.7 CURVATURE

The Curvature of a Plane Curve

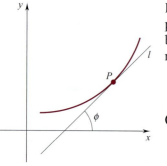

Figure 13.7.1 shows a plane curve which we assume to be twice differentiable. At the point P we have drawn the tangent line l. The angle that l makes with the x-axis has been labeled ϕ. As P moves along the curve, l changes and ϕ changes also. The magnitude k of the change in ϕ per unit of arc length is called the *curvature*:

(13.7.1)
$$k = \left| \frac{d\phi}{ds} \right|.$$

Figure 13.7.1

Calculating Curvature

If the curve is the graph of a twice differentiable function

$$y = y(x),$$

then the curvature can be calculated from the formula

(13.7.2)
$$k = \frac{|y''|}{[1 + (y')^2]^{3/2}}$$

where the primes indicate differentiation with respect to x.

DERIVATION OF FORMULA 13.7.2 We know that $\tan \phi = y'$. Therefore,

$$\phi = \tan^{-1}(y').$$

Differentiating with respect to x, we have

$$\frac{d\phi}{dx} = \frac{1}{1 + (y')^2} \cdot \frac{d}{dx}(y') = \frac{y''}{1 + (y')^2}.$$

Since

$$\frac{d\phi}{dx} = \frac{d\phi}{ds}\frac{ds}{dx} = \frac{d\phi}{ds}\sqrt{1 + (y')^2},$$

chain rule ⟶   ⟵ (13.4.4)

we have

$$\frac{d\phi}{ds}\sqrt{1 + (y')^2} = \frac{y''}{1 + (y')^2} \qquad \text{and therefore} \qquad \left| \frac{d\phi}{ds} \right| = \frac{|y''|}{[1 + (y')^2]^{3/2}}. \quad ❑$$

If the curve is given parametrically by a twice differentiable vector function

$$\mathbf{r}(t) = x(t)\mathbf{i} + y(t)\mathbf{j},$$

then

(13.7.3)
$$k = \frac{|x'y'' - y'x''|}{[(x')^2 + (y')^2]^{3/2}}$$

where the primes now indicate differentiation with respect to t.

We will derive the formula under the assumption that $x' \neq 0$. Actually, the formula holds provided that $(x')^2 + (y')^2 \neq 0$.

DERIVATION OF FORMULA 13.7.3

$$\frac{dy}{dx} = \frac{dy/dt}{dx/dt} = \frac{y'}{x'}.$$

Therefore, as you can verify,

$$\frac{d^2y}{dx^2} = \frac{(dx/dt)(d^2y/dt^2) - (dy/dt)(d^2x/dt^2)}{(dx/dt)^3} = \frac{x'y'' - y'x''}{(x')^3}.$$

(9.8.5) ⟶

Thus

$$k = \frac{|d^2y/dx^2|}{[1 + (dy/dx)^2]^{3/2}} = \left|\frac{x'y'' - y'x''}{(x')^3}\right| \frac{1}{[1 + (y'/x')^2]^{3/2}} = \frac{|x'y'' - y'x''|}{[(x')^2 + (y')^2]^{3/2}}. \quad ❑$$

⟵ (13.7.2)

Example 1 Since a straight line has a constant angle of inclination, we have

$$\frac{d\phi}{ds} = 0 \qquad \text{and thus} \qquad k = 0.$$

(13.7.4) Along a straight line the curvature is constantly zero. ❑

Example 2 For a circle of radius r,

$$\mathbf{r}(t) = r(\cos t\, \mathbf{i} + \sin t\, \mathbf{j}),$$

we have

$$x = r\cos t, \qquad y = r\sin t.$$

Differentiation with respect to t gives

$$x' = -r\sin t, \quad x'' = -r\cos t; \qquad y' = r\cos t, \quad y'' = -r\sin t.$$

Thus

$$k = \frac{|x'y'' - y'x''|}{[(x')^2 + (y')^2]^{3/2}} = \frac{|(-r\sin t)(-r\sin t) - (r\cos t)(-r\cos t)|}{[(-r\sin t)^2 + (r\cos t)^2]^{3/2}} = \frac{r^2}{r^3} = \frac{1}{r}.$$

(13.7.5) Along a circle of radius r the curvature is constantly $1/r$.

Hardly surprising. It is geometrically evident that along a circular path the change in direction takes place at a constant rate. Since a complete revolution entails a change of direction of 2π radians and this change is effected on a path of length $2\pi r$, the change in direction per unit of arc length is $2\pi/2\pi r = 1/r$. ❏

To say that a curve $y = f(x)$ has slope m at a point P is to say that at the point P the curve is rising or falling at the rate of a line of slope m. To say that a plane curve C has curvature $1/r$ at a point P is to say that at the point P the curve is turning at the rate of a circle of radius r. (Figure 13.7.2.) The smaller the circle, the tighter the turn and, thus, the greater the curvature.

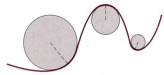

Figure 13.7.2

The reciprocal of the curvature,

$$\rho = \frac{1}{k}, \qquad \text{(for } k \neq 0\text{)}$$

is called the *radius of curvature*. The point at a distance ρ from the curve in the direction of the principal normal is called the *center of curvature*.

Example 3 Take $a > 0$. Show that the point of maximal curvature on the catenary

$$y = \frac{a}{2}(e^{x/a} + e^{-x/a})$$

is the lowest point on the curve. (See Exercise 46, Section 9.9.) What is the radius of curvature at that point?

SOLUTION We will use the formula

$$k = \frac{|y''|}{[1 + (y')^2]^{3/2}}.$$

Here

$$y' = \frac{1}{2}(e^{x/a} - e^{-x/a}) \qquad \text{and} \qquad y'' = \frac{1}{2a}(e^{x/a} + e^{-x/a}) = \frac{y}{a^2} > 0.$$

Observe that

$$[1 + (y')^2]^{3/2} = [1 + \tfrac{1}{4}(e^{x/a} - e^{-x/a})^2]^{3/2},$$

which, as you can check, reduces to give

$$[1 + (y')^2]^{3/2} = \frac{y^3}{a^3}.$$

Thus

$$k = \left(\frac{y}{a^2}\right)\left(\frac{a^3}{y^3}\right) = \frac{a}{y^2}.$$

Since $y > 0$, the curvature is maximal where y is minimal; that is, at the lowest point of the curve.

The value of y at the lowest point of the curve is a. The curvature at that point is $1/a$; the radius of curvature ρ at that point is a. ❏

Example 4 Find the curvature at an arbitrary point (x, y) of the ellipse

$$\frac{x^2}{a^2} + \frac{y^2}{b^2} = 1. \qquad\qquad (a > b > 0)$$

Determine the points of maximal curvature and the points of minimal curvature. What is the radius of curvature at each of these points?

SOLUTION We parametrize the ellipse by setting

$$\mathbf{r}(t) = a \cos t\,\mathbf{i} + b \sin t\,\mathbf{j}$$

and use the fact that

$$k = \frac{|x'y'' - y'x''|}{[(x')^2 + (y')^2]^{3/2}}.$$

Here

$$x = a \cos t, \qquad y = b \sin t$$

and therefore

$$x' = -a \sin t, \quad x'' = -a \cos t; \qquad y' = b \cos t, \quad y'' = -b \sin t.$$

Thus

$$k = \frac{|(-a \sin t)(-b \sin t) - (b \cos t)(-a \cos t)|}{[(-a \sin t)^2 + (b \cos t)^2]^{3/2}} = \frac{ab}{[a^2 \sin^2 t + b^2 \cos^2 t]^{3/2}}.$$

As you can check, the curvature at the point (x, y) can be written

$$\frac{a^4 b^4}{(b^4 x^2 + a^4 y^2)^{3/2}}.$$

To find the points of maximal and minimal curvature, we go back to the parameter t. Observe that

$$a^2\sin^2 t + b^2\cos^2 t = (a^2 - b^2)\sin^2 t + b^2(\sin^2 t + \cos^2 t)$$
$$= (a^2 - b^2)\sin^2 t + b^2.$$

Thus we have

$$k = \frac{ab}{[(a^2 - b^2)\sin^2 t + b^2]^{3/2}}.$$

Since we have assumed that $a > b > 0$, the curvature is maximal when $\sin^2 t = 0$; that is, when $t = 0$ and when $t = \pi$. Thus the points of maximal curvature are the points $P(\pm a, 0)$, the ends of the major axis. The curvature at these points is a/b^2 and the radius of curvature is b^2/a. The curvature is minimal when $\sin^2 t = 1$; that is, when $t = \frac{1}{2}\pi$ and when $t = \frac{3}{2}\pi$. The points of minimal curvature are the points $P(0, \pm b)$, the ends of the minor axis. The curvature at these points is b/a^2 and the radius of curvature is a^2/b. ❑

In Figure 13.7.3 you can see a plane curve. At the point P we have affixed the unit tangent vector $\mathbf{T}$. As P moves along the curve, $\mathbf{T}$ changes, not in length, but in

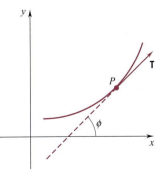

Figure 13.7.3

direction. The curvature of the curve is the magnitude of the change in **T** per unit of arc length:

(13.7.6)

$$k = \left\| \frac{d\mathbf{T}}{ds} \right\|.$$

PROOF Since **T** has length 1, we can write

$$\mathbf{T} = \cos\phi\,\mathbf{i} + \sin\phi\,\mathbf{j}.$$

Note that ϕ represents the angle between the tangent line and the x-axis. Differentiation with respect to s gives

$$\frac{d\mathbf{T}}{ds} = -\sin\phi\frac{d\phi}{ds}\mathbf{i} + \cos\phi\frac{d\phi}{ds}\mathbf{j} = \frac{d\phi}{ds}(-\sin\phi\,\mathbf{i} + \cos\phi\,\mathbf{j}).$$

Taking norms we have

$$\left\| \frac{d\mathbf{T}}{ds} \right\| = \left| \frac{d\phi}{ds} \right| \sqrt{\sin^2\theta + \cos^2\theta} = \left| \frac{d\phi}{ds} \right| = k$$

as asserted. ❑

This characterization of curvature generalizes to space curves.

The Curvature of a Space Curve

A space curve bends in two ways. It bends in the osculating plane (the plane of the unit tangent **T** and the principal normal **N**) and it bends away from that plane. The first form of bending is measured by the rate at which the unit tangent **T** changes direction. The second form of bending is measured by the rate at which the vector **T** × **N** changes direction. We will concentrate here on the first form of bending, the bending in the osculating plane. The measure of this is called *curvature.*

What was a theorem on curvature in the case of a plane curve becomes a definition of curvature in the case of a space curve; namely, in the case of a space curve, we *define* the curvature k by setting

$$k = \left\| \frac{d\mathbf{T}}{ds} \right\|.$$

If the space curve is given in terms of a parameter t, say

$$C: \quad \mathbf{r}(t) = x(t)\mathbf{i} + y(t)\mathbf{j} + z(t)\mathbf{k}, \quad t \in [a, b],$$

then the curvature can be calculated from the formula

(13.7.7)

$$k = \frac{\|d\mathbf{T}/dt\|}{ds/dt}.$$

We arrive at this by noting that

$$\frac{d\mathbf{T}}{ds}\frac{ds}{dt} = \frac{d\mathbf{T}}{dt},$$

dividing through by ds/dt and taking norms. The assumption here is that $ds/dt = \|\mathbf{r}'(t)\|$ remains nonzero.

Example 5 Calculate the curvature of the circular helix

$$\mathbf{r}(t) = r\sin t\mathbf{i} + r\cos t\mathbf{j} + t\mathbf{k}. \qquad (r > 0)$$

SOLUTION We will use the Leibniz notation.

$$\frac{d\mathbf{r}}{dt} = r\cos t\mathbf{i} - r\sin t\mathbf{j} + \mathbf{k}, \qquad \frac{ds}{dt} = \left\|\frac{d\mathbf{r}}{dt}\right\| = \sqrt{r^2 + 1},$$

$$\mathbf{T} = \frac{d\mathbf{r}/dt}{\|d\mathbf{r}/dt\|} = \frac{r\cos t\mathbf{i} - r\sin t\mathbf{j} + \mathbf{k}}{\sqrt{r^2 + 1}}, \qquad \frac{d\mathbf{T}}{dt} = \frac{-r\sin t\mathbf{i} - r\cos t\mathbf{j}}{\sqrt{r^2 + 1}},$$

$$k = \frac{\|d\mathbf{T}/dt\|}{ds/dt} = \frac{1}{\sqrt{r^2 + 1}}\cdot\frac{r}{\sqrt{r^2 + 1}} = \frac{r}{r^2 + 1}. \quad \square$$

Components of Acceleration

In straight-line motion, acceleration is purely tangential; that is, the acceleration vector points in the direction of the motion. (Section 13.5.) In uniform circular motion, the acceleration is normal; the acceleration vector is perpendicular to the tangent vector and points along the line of the normal vector toward the center of the circle. (Also Section 13.5.) In general, acceleration has two components, a tangential component and a normal component. To see this, let's suppose that the position of an object at time t is given by the vector function

$$\mathbf{r}(t) = x(t)\mathbf{i} + y(t)\mathbf{j} + z(t)\mathbf{k}.$$

Since

$$\mathbf{T} = \frac{d\mathbf{r}/dt}{\|d\mathbf{r}/dt\|} = \frac{\mathbf{v}}{ds/dt},$$

we have

$$\mathbf{v} = \frac{ds}{dt}\mathbf{T}.$$

Differentiation gives

$$\mathbf{a} = \frac{d^2s}{dt^2}\mathbf{T} + \frac{ds}{dt}\frac{d\mathbf{T}}{dt}.$$

Observe now that

$$\frac{d\mathbf{T}}{dt} = \left\|\frac{d\mathbf{T}}{dt}\right\|\mathbf{N} = k\frac{ds}{dt}\mathbf{N}.$$

$$(13.3.4) \longrightarrow \qquad \qquad \longleftarrow (13.7.7)$$

Substitution in the previous display gives

(13.7.8)
$$\mathbf{a} = \frac{d^2s}{dt^2}\mathbf{T} + k\left(\frac{ds}{dt}\right)^2\mathbf{N}.$$

The acceleration vector lies in the osculating plane, the plane of $\mathbf{T}$ and $\mathbf{N}$. The tangential component of acceleration

$$a_{\mathbf{T}} = \frac{d^2s}{dt^2}$$

depends only on the change of speed of the object; if the speed is constant, the tangential component of acceleration is zero and the acceleration is directed entirely toward the center of curvature of the path. On the other hand, the normal component of acceleration

$$a_{\mathbf{N}} = k\left(\frac{ds}{dt}\right)^2$$

depends both on the speed of the object and the curvature of the path. At a point where the curvature is zero, the normal component of acceleration is zero and the acceleration is directed entirely along the path of motion. If the curvature is not zero, then the normal component of acceleration is a multiple of the *square* of the speed. This means, for example, that if you are in a car going around a curve at 50 miles per hour, you will feel *four times* the normal component of acceleration that you would feel going around the same curve at 25 miles per hour.

EXERCISES 13.7

In Exercises 1–10, find the curvature of the given curve.

1. $y = e^{-x}$.
2. $y = x^3$.
3. $y = \sqrt{x}$.
4. $y = 1/x$.
5. $y = \ln \sec x$.
6. $y = x - x^2$.
7. $y = \sin x$.
8. $y = \tan x$.
9. $y = x^{-3/2}$.
10. $x^2 - y^2 = a^2$.

In Exercises 11–18, find the radius of curvature at the indicated point.

11. $6y = x^3$; $(2, \frac{4}{3})$.
12. $2y = x^2$; $(0, 0)$.
13. $y^2 = 2x$; $(2, 2)$.
14. $y = 2 \sin 2x$; $(\frac{1}{4}\pi, 2)$.
15. $y = \ln(x + 1)$; $(2, \ln 3)$.
16. $y = \sec x$; $(\frac{1}{4}\pi, \sqrt{2})$.
17. $4x^2 + 9y^2 = 36$; $(3, 0)$.
18. $x^2 - 4y^2 = 9$; $(5, 2)$.

19. Find the point of maximal curvature on the curve $y = \ln x$.

20. Find the curvature of the graph of $y = 3x - x^3$ at the point where the function takes on its local maximum value.

In Exercises 21–28, express the curvature in terms of t.

21. $\mathbf{r}(t) = t\mathbf{i} + \frac{1}{2}t^2\mathbf{j}$.
22. $\mathbf{r}(t) = e^t\mathbf{i} + e^{-t}\mathbf{j}$.
23. $\mathbf{r}(t) = 2t\mathbf{i} + t^3\mathbf{j}$.
24. $\mathbf{r}(t) = t^2\mathbf{i} + t^3\mathbf{j}$.
25. $\mathbf{r}(t) = e^t(\cos t\mathbf{i} + \sin t\mathbf{j})$.
26. $\mathbf{r}(t) = 2 \cos t\mathbf{i} + 3 \sin t\mathbf{j}$.
27. $\mathbf{r}(t) = (t \cos t)\mathbf{i} + (t \sin t)\mathbf{j}$.
28. $\mathbf{r}(t) = (\cos t + t \sin t)\mathbf{i} + (\sin t - t \cos t)\mathbf{j}$, $t > 0$.

29. Find the radius of curvature at the vertices of the hyperbola $xy = 1$.

30. Find the radius of curvature at the vertices of the hyperbola $x^2 - y^2 = 1$.

31. Find the curvature at each (x, y) of the hyperbola $b^2x^2 - a^2y^2 = a^2b^2$.
    HINT: Parametrize the hyperbola by setting $\mathbf{r}(t) = a \cosh t\mathbf{i} + b \sinh t\mathbf{j}$.

32. Find the curvature at the highest point of an arch of the cycloid

$$x(t) = r(t - \sin t), \qquad y(t) = r(1 - \cos t).$$

33. Show that the curvature of the path of a moving object can be written

(13.7.9)

$$k = \frac{\|\mathbf{v} \times \mathbf{a}\|}{(ds/dt)^3}.$$

HINT: $\mathbf{v} = \dfrac{ds}{dt}\mathbf{T}$ and $\mathbf{a} = \dfrac{d^2s}{dt^2}\mathbf{T} + k\left(\dfrac{ds}{dt}\right)^2\mathbf{N}.$

In Exercises 34–39, interpret $\mathbf{r}(t)$ as the position of a moving object at time t. Find the curvature of the path and determine the tangential and normal components of acceleration.

34. $\mathbf{r}(t) = (1 - 2t)\mathbf{i} + (3 + 4t)\mathbf{j} + (2 - 3t)\mathbf{k}.$

35. $\mathbf{r}(t) = e^t\cos t\mathbf{i} + e^t\sin t\mathbf{j} + e^t\mathbf{k}.$

36. $\mathbf{r}(t) = 2\cos t\mathbf{i} + t\mathbf{j} + \sin t\mathbf{k}$

37. $\mathbf{r}(t) = \cos 2t\mathbf{i} + \sin 2t\mathbf{j} + \mathbf{k}.$

38. $\mathbf{r}(t) = t\mathbf{i} + t^2\mathbf{j} + (\ln t)\mathbf{k}$

39. $\mathbf{r}(t) = t\mathbf{i} + \frac{1}{2}t^2\mathbf{j} + \frac{1}{3}t^3\mathbf{k}.$

40. Show that the curvature of a polar curve $r = f(\theta)$ is given by

$$k = \frac{|[f(\theta)]^2 + 2[f'(\theta)]^2 - f(\theta)f''(\theta)|}{([f(\theta)]^2 + [f'(\theta)]^2)^{3/2}}.$$

41. Find the curvature of the logarithmic spiral $r = e^{a\theta}$, $a > 0$.

42. Find the curvature of the spiral of Archimedes $r = a\theta$, $a > 0$.

43. Find the curvature of the cardioid $r = a(1 - \cos \theta)$ in terms of r.

44. (*Transition Curves*) When engineers lay railroad track, they cannot allow abrupt changes in curvature. To join a straightaway that ends at a point P to a curved track that begins at a point Q, they need to lay some transitional track that has curvature zero at P and the curvature of the second piece at Q. Find an arc of the form

$$y = Cx^n, \qquad x \in [0, 1]$$

that joins the arcs C_1 and C_2 of Figure A without any discontinuities in curvature.

45. Let $s(\theta)$ be the arc distance from the highest point of the cycloidal arch

$$x(\theta) = R(\theta - \sin \theta), \quad y(\theta) = R(1 - \cos \theta), \quad \theta \in [0, 2\pi]$$

to the point $(x(\theta), y(\theta))$ of that same arch. (The arch is pictured in Figure 9.11.1.) Let $\rho(\theta)$ be the radius of curvature at the point $(x(\theta), y(\theta))$. (a) Calculate $s(\theta)$. (b) Calculate $\rho(\theta)$. (c) Then find an equation in s and ρ for that arch. (Such an equation is called a *natural equation* for the curve.)

46. Let $s(\theta)$ be the arc distance from the origin to the point $(x(\theta), y(\theta))$ along the exponential spiral $r = ae^{c\theta}$. (Take $a > 0, c > 0$.) Let $\rho(\theta)$ be the radius of curvature at that same point. Find an equation in s and ρ for that curve.

47. Let $s(\theta)$ be the arc distance from the point $(-2a, 0)$ to the point $(x(\theta), y(\theta))$ along the cardioid $r = a(1 - \cos \theta)$. (Take $a > 0$.) Let $\rho(\theta)$ be the radius of curvature at that same point. Find an equation in s and ρ for that curve.

(*The Frenet Formulas*) Figure B shows a space curve. At a point of the curve we have drawn the unit tangent $\mathbf{T}$, the principal normal $\mathbf{N}$, and the vector $\mathbf{B} = \mathbf{T} \times \mathbf{N}$, which, being normal to both $\mathbf{T}$ and $\mathbf{N}$, is called the *binormal*. At each point of the curve, the vectors $\mathbf{T}, \mathbf{N}, \mathbf{B}$ form what is called the *Frenet trihedral*, a set of mutually perpendicular unit vectors that, in the order given, form a local right-handed coordinate system.

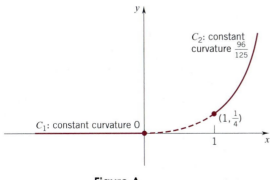

Figure A

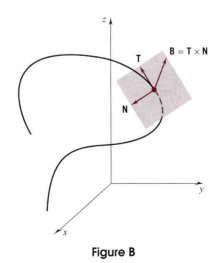

Figure B

48. (a) Show that $d\mathbf{B}/ds$ is parallel to $\mathbf{N}$, and therefore there is a scalar τ for which

$$\frac{d\mathbf{B}}{ds} = \tau\mathbf{N}.$$

HINT: Since $\mathbf{B}$ has constant length one, $d\mathbf{B}/ds \perp \mathbf{B}$. Show that $d\mathbf{B}/ds \perp \mathbf{T}$ by carrying out the differentiation

$$\frac{d\mathbf{B}}{ds} = \frac{d}{ds}(\mathbf{T} \times \mathbf{N}).$$

(b) Now show that

$$\frac{d\mathbf{N}}{ds} = -k\mathbf{T} - \tau\mathbf{B}.$$

HINT: Since $\mathbf{T}, \mathbf{N}, \mathbf{B}$ form a right-handed system of mutually perpendicular unit vectors, we can show that

$$\mathbf{N} \times \mathbf{B} = \mathbf{T} \qquad \text{and} \qquad \mathbf{B} \times \mathbf{T} = \mathbf{N}.$$

You can assume these relations.

(c) The scalar τ is called the *torsion* of the curve. Give a geometric interpretation to $|\tau|$.

The three formulas

$$\textbf{(13.7.10)} \qquad \frac{d\mathbf{T}}{ds} = k\mathbf{N}, \quad \frac{d\mathbf{N}}{ds} = -k\mathbf{T} - \tau\mathbf{B}, \quad \frac{d\mathbf{B}}{ds} = \tau\mathbf{N}$$

were developed in 1847 by the French geometer Frederic Frenet (1816–1900). Another Frenchman, Alfred Serret (1819–1885), discovered these same relations independently in 1850. Accordingly, the formulas are often called the Frenet–Serret formulas.

■ **CHAPTER HIGHLIGHTS**

13.1 Vector Functions

basic definition: $\lim_{t \to t_0} \mathbf{f}(t) = \mathbf{L}$ iff $\lim_{t \to t_0} \|\mathbf{f}(t) - \mathbf{L}\| = 0$

theorem: if $\lim_{t \to t_0} \mathbf{f}(t) = \mathbf{L}$ then $\lim_{t \to t_0} \|\mathbf{f}(t)\| = \|\mathbf{L}\|$

limit rules (p. 842)
limits can be taken component by component (p. 842)
continuity and differentiability (p. 843)   integration (p. 844)
properties of the integral (p. 845)

13.2 Differentiation Formulas

$$\frac{d}{dt}(\mathbf{f} + \mathbf{g}) = \frac{d\mathbf{f}}{dt} + \frac{d\mathbf{g}}{dt}. \qquad\qquad \frac{d}{dt}(\alpha\mathbf{f}) = \alpha\frac{d\mathbf{f}}{dt} \qquad (\alpha \text{ constant}).$$

$$\frac{d}{dt}(u\mathbf{f}) = u\frac{d\mathbf{f}}{dt} + \frac{du}{dt}\mathbf{f} \qquad (u = u(t)). \qquad\qquad \frac{d}{dt}(\mathbf{f} \cdot \mathbf{g}) = \left(\mathbf{f} \cdot \frac{d\mathbf{g}}{dt}\right) + \left(\frac{d\mathbf{f}}{dt} \cdot \mathbf{g}\right).$$

$$\frac{d}{dt}(\mathbf{f} \times \mathbf{g}) = \left(\mathbf{f} \times \frac{d\mathbf{g}}{dt}\right) + \left(\frac{d\mathbf{f}}{dt} \times \mathbf{g}\right). \qquad\qquad \frac{d\mathbf{f}}{dt} = \frac{d\mathbf{f}}{du}\frac{du}{dt} \qquad (\text{chain rule}).$$

13.3 Curves

tangent vector (p. 854)   tangent line (p. 856)
unit tangent (p. 858)   principal normal (p. 858)
normal line (p. 858)   osculating plane (p. 858)
sense-preserving (sense-reversing) change of parameter (p. 861)

13.4 Arc Length

$$L(C) = \int_a^b \|\mathbf{r}'(t)\| \, dt \qquad \frac{ds}{dt} = \sqrt{\left(\frac{dx}{dt}\right)^2 + \left(\frac{dy}{dt}\right)^2 + \left(\frac{dz}{dt}\right)^2}$$

■ PROJECTS AND EXPLORATIONS USING TECHNOLOGY

To do these exercises you will need a graphics calculator or a computer with graphing capability. The majority of these problems are open-ended so different approaches may be used to solve them. You should be aware that different appraoches can result in slight variations in the answers. Round your numerical answers to at least four decimal places. The rounding method that your calculator or computer uses also may cause variations in answers.

13.1 (a) Use your technology to calculate the derivatives of the following vector functions:

$$\text{(i) } \mathbf{f}(t) = t^3\,\mathbf{i} - \sin t\,\mathbf{j} + \ln t\,\mathbf{k} \quad \text{at } t = 1.$$

$$\text{(ii) } \mathbf{f}(t) = \sec t\,\mathbf{i} + e^{3t}\,\mathbf{j} - (3 + \tan t^3)^6\,\mathbf{k} \quad \text{at } t = 0.75.$$

If your technology can do this directly, do so. Otherwise, find an efficient way to use it component-wise.

(b) Use the difference quotient

$$\frac{\mathbf{f}(t + h) - \mathbf{f}(t)}{h}$$

with $h = 0.1, 0.01, 0.001, 0.0001$ to find approximations for the derivatives of the vector functions in part (a).

(c) We are interested in "how good" our approximations are. Use the standard vector norm $\|\cdot\|$, called the *Euclidean norm,* to measure the difference between the approximate derivative and the analytic derivative for each of the values found in part (b). Does the norm approach zero in the same way as it would for a scalar function?

(d) Mathematicians often use norms other than the Euclidean norm to find the lengths of vectors. One such norm, called the "sup norm" and denoted $\|\cdot\|_\infty$, is given by

$$\|a\,\mathbf{i} + b\,\mathbf{j} + c\,\mathbf{k}\|_\infty = \max(|a|, |b|, |c|).$$

For example, $\|2\,\mathbf{i} - 3\,\mathbf{j} + \mathbf{k}\|_\infty = 3$. Discuss the possible advantages of this norm over the Euclidean norm.

(e) Repeat part (c) using the sup norm.

(f) Use the symmetric difference quotient

$$\frac{\mathbf{f}(t + h) - \mathbf{f}(t - h)}{2h}$$

with $h = 0.1, 0.01, 0.001, 0.0001$ to find approximations for the derivatives in part (a).

     (g) Use the Euclidean norm to measure the difference between the approximate derivative and the analytic derivative for each of the values calculated in part (f). Repeat using the sup norm.

13.2 (a) Use your technology to calculate the integrals of the following vector functions:

        (i) $\mathbf{f}(t) = t^3\,\mathbf{i} - \sin t\,\mathbf{j} + \ln t\,\mathbf{k}$ from $t = 1$ to $t = 3$.

        (ii) $\mathbf{f}(t) = \sec t\,\mathbf{i} + e^{3t}\,\mathbf{j} - (3 + \tan t^3)^6\,\mathbf{k}$ from $t = 0.75$ to $t = 1$.

     If your technology can do this directly, do so. Otherwise, determine an efficient way to use it component-wise.

     (b) Approximate the integrals in part (a) by using a variation of the trapezoidal rule modified to deal with vector functions. Let $n = 2, 4, 8, 16, 32$.

     (c) Using the Euclidean norm, the trapezoidal rule, and the number of subintervals $n = 2$, 4, 8, 16, 32, compare the rate of convergence here with that for the one-dimensional case.

     (d) Repeat part (c) using the sup norm introduced in Exercise 1.

     (e) Repeat parts (b), (c), and (d) using a vector version of Simpson's rule.

13.3 Let $\mathbf{r}(t) = x(t)\,\mathbf{i} + y(t)\,\mathbf{j} + z(t)\,\mathbf{k}$, $t \in [a, b]$, be a continuously differentiable curve. Consider using the summation

$$\sum_{i=0}^{n-1} \sqrt{(x_{i+1} - x_i)^2 + (y_{i+1} - y_i)^2 + (z_{i+1} - z_i)^2}$$

as an approximation for arc length, where the points (x_i, y_i, z_i) are spaced along the curve. In particular, let $h = (b - a)/n$, $t_i = a + ih$, $x_i = x(t_i)$, $y_i = y(t_i)$, and $z_i = z(t_i)$ for $i = 0, 1, 2, \ldots, n$. Not only is this method intuitively clear, it may have advantages over approximations of the arc length integral.

     (a) Test this method on the curve $\mathbf{r}(t) = t^5\,\mathbf{i} + t\,\mathbf{j} - \sin t\,\mathbf{k}$, $t \in [0, 1]$, with $n = 1, 2, 4, 8, 16, 32$.

     (b) Use arc length software or technology to calculate the arc length. Plot n versus error and try to fit a curve through these points.

     (c) Using the same number of intervals, approximate the definite integral that represents this arc length by applying either the trapezoidal rule or Simpson's rule. How do these approximations compare with those found in part (a)?

     (d) Our theory claims that the summations should increase toward the arc length. Does this seem to be true?

     (f) The arc length of a curve should be independent of the particular parametrization used for the curve. Find another parametrization for the curve in part (a) and check this result.

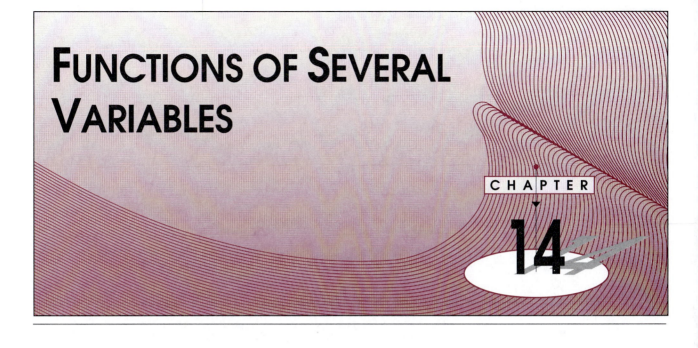

FUNCTIONS OF SEVERAL VARIABLES

■ 14.1 ELEMENTARY EXAMPLES

First a remark on notation. Points $P(x, y)$ of the xy-plane will be written (x, y) and points $P(x, y, z)$ of three-space will be written (x, y, z).

Let D be a nonempty subset of the xy-plane. A rule f that assigns a real number $f(x, y)$ to each point (x, y) in D is called a *real-valued function of two variables*. The set D is called the *domain of f* and the set of all values $f(x, y)$ is called the *range of f*.

Example 1 Take D as the entire xy-plane and to each point (x, y) assign the number

$$f(x, y) = xy. \quad \square$$

Example 2 Take D as the set of all points (x, y) with $y \neq 0$. To each such point assign the number

$$f(x, y) = \tan^{-1}\left(\frac{x}{y}\right). \quad \square$$

Example 3 Take D as the *open unit disc*:

$$D = \{(x, y): x^2 + y^2 < 1\}.$$

This set consists of all points inside the *unit circle* $x^2 + y^2 = 1$; the circle itself is omitted.† To each point (x, y) in D assign the number

$$f(x, y) = \frac{1}{\sqrt{1 - (x^2 + y^2)}}. \quad \square$$

---

† The *closed unit disc* $D = \{(x, y): x^2 + y^2 \leq 1\}$ consists of all points on or inside the unit circle; the circle is included.

Now let D be a nonempty subset of three-space. A rule f that assigns a real number $f(x, y, z)$ to each point (x, y, z) in D is called a *real-valued function of three variables.* The set D is called the *domain of f* and the set of all values $f(x, y, z)$ is called the *range of f.*

Example 4 Take D as all of three-space and to each point (x, y, z) assign the number

$$f(x, y, z) = xyz. \quad ❏$$

Example 5 Take D as the set of all points (x, y, z) with $z \neq x + y$. (Thus D consists of all points not on the plane $x + y - z = 0$.) To each point of D assign the number

$$f(x, y, z) = \cos\left(\frac{1}{x + y - z}\right). \quad ❏$$

Example 6 Take D as the *open unit ball*:

$$D = \{(x, y, z): x^2 + y^2 + z^2 < 1\}.$$

This set consists of all points inside the *unit sphere* $x^2 + y^2 + z^2 = 1$; the sphere itself is omitted.† To each point (x, y, z) in D assign the number

$$f(x, y, z) = \frac{1}{\sqrt{1 - (x^2 + y^2 + z^2)}}. \quad ❏$$

Functions of several variables arise naturally in very elementary settings.

$f(x, y) = \sqrt{x^2 + y^2}$ gives the distance between $P(x, y)$ and the origin;

$f(x, y) = xy$ gives the area of a rectangle of dimensions x, y, and

$f(x, y) = 2(x + y)$ gives the perimeter.

$f(x, y, z) = \sqrt{x^2 + y^2 + z^2}$ gives the distance between $P(x, y, z)$ and the origin;

$f(x, y, z) = xyz$ gives the volume of of rectangular solid of dimensions x, y, z, and

$f(x, y, z) = 2(xy + xz + yz)$ gives the total surface area.

Functions of several variables are also key to many important problems in science, engineering, economics, and so on. Indeed, the mathematical models for "real" problems are much more likely to involve functions of several variables than functions of a single variable. Here are some simple examples.

If interest is compounded continuously, then the accumulation of principal A is a function of the interest rate r and the time t given by

$$A(t, r) = A_0 e^{rt},$$

where A_0 is the initial investment (Section 7.6).

---

† The *closed unit ball* $D = \{(x, y, z): x^2 + y^2 + z^2 \leq 1\}$ consists of all points on or inside the unit sphere; the sphere is included.

The magnitude of the gravitational force exerted by a body of mass M situated at the origin on a body of mass m at the point (x, y, z) is given by

$$F(x, y, z) = \frac{GmM}{x^2 + y^2 + z^2}.$$ (Section 13.6.)

The *ideal gas law* states that the pressure P of a gas is a function of the volume V and the temperature T according to the equation

$$P = \frac{cT}{V},$$

where c is a constant.

The deflection S at the midpoint of a rectangular beam which is supported at both ends and subjected to a uniform load is given by

$$S(L, w, h) = \frac{CL^3}{wh^3},$$

where L is the length, w the width, h the height, and C is a constant.

If the domain of a function of several variables is not explicitly given, it is to be understood that the domain is the set of all points for which the definition makes sense. Thus, in the case of

$$f(x, y) = \frac{1}{x - y},$$

the domain is understood to be all points (x, y) with $y \neq x$. That is, all points of the plane not on the line $y = x$. In the case of

$$g(x, y, z) = \sin^{-1}(x + y + z),$$

the domain is understood to be all points (x, y, z) with $-1 \leq x + y + z \leq 1$. This set is the slab bounded by the parallel planes

$$x + y + z = -1 \qquad \text{and} \qquad x + y + z = 1.$$

To say that a function is *bounded* is to say that its range is bounded. Since the function

$$f(x, y) = \frac{1}{x - y}$$

takes on all values other than 0, its range is $(-\infty, 0) \cup (0, \infty)$. The function is unbounded. In the case of

$$g(x, y, z) = \sin^{-1}(x + y + z)$$

the range is the closed interval $[-\tfrac{1}{2}\pi, \tfrac{1}{2}\pi]$. This function is bounded, below by $-\tfrac{1}{2}\pi$ and above by $\tfrac{1}{2}\pi$.

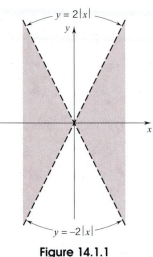

$y = 2|x|$

$y = -2|x|$

Figure 14.1.1

Example 7 Find the domain and range of

$$f(x, y) = \frac{1}{\sqrt{4x^2 - y^2}}$$

SOLUTION A point (x, y) is in the domain of f iff

$$4x^2 - y^2 > 0.$$

This occurs iff

$$y^2 < 4x^2$$

and thus iff

$$-2|x| < y < 2|x|.$$

The domain of f is shaded in Figure 14.1.1. It consists of all the points of the xy-plane that lie between the graph of $y = -2|x|$ and the graph of $y = 2|x|$. On this set $\sqrt{4x^2 - y^2}$ takes on all positive values, and so does its reciprocal $f(x, y)$. The range of f is $(0, \infty)$. ❑

EXERCISES 14.1

In Exercises 1–22, find the domain and range of the given function.

1. $f(x, y) = \sqrt{xy}.$

2. $f(x, y) = \sqrt{1 - xy}.$

3. $f(x, y) = \dfrac{1}{x + y}.$

4. $f(x, y) = \dfrac{1}{x^2 + y^2}.$

5. $f(x, y) = \dfrac{e^x - e^y}{e^x + e^y}.$

6. $f(x, y) = \dfrac{x^2}{x^2 + y^2}.$

7. $f(x, y) = \ln(xy).$

8. $f(x, y) = \ln(1 - xy).$

9. $f(x, y) = \dfrac{1}{\sqrt{y - x^2}}.$

10. $f(x, y) = \dfrac{\sqrt{9 - x^2}}{1 + \sqrt{1 - y^2}}.$

11. $f(x, y) = \sqrt{9 - x^2} - \sqrt{4 - y^2}.$

12. $f(x, y, z) = \cos x + \cos y + \cos z.$

13. $f(x, y, z) = \dfrac{x + y + z}{|x + y + z|}.$

14. $f(x, y, z) = \dfrac{z^2}{x^2 - y^2}.$

15. $f(x, y, z) = -\dfrac{z^2}{\sqrt{x^2 - y^2}}.$

16. $f(x, y, z) = \dfrac{z}{x - y}.$

17. $f(x, y) = \dfrac{2}{\sqrt{9 - (x^2 + y^2)}}.$

18. $f(x, y, z) = \ln(|x + 2y + 3z| + 1).$

19. $f(x, y, z) = \ln(x + 2y + 3z).$

20. $f(x, y, z) = e^{\sqrt{4 - (x^2 + y^2 + z^2)}}.$

21. $f(x, y, z) = e^{-(x^2 + y^2 + z^2)}.$

22. $f(x, y, z) = \dfrac{\sqrt{1 - x^2} + \sqrt{4 - y^2}}{1 + \sqrt{9 - z^2}}.$

23. Let $f(x) = \sqrt{x}$, $g(x, y) = \sqrt{x}$, and $h(x, y, z) = \sqrt{x}$. Determine the domain and range of each function and compare your results. Sketch the graphs of f and g.

24. Let $f(x, y) = \cos \pi x \sin \pi y$ and $g(x, y, z) = \cos \pi x \sin \pi y$. Determine the domain and range of each of these functions and compare your results.

In Exercises 25–30, form the difference quotients

$$\frac{f(x + h, y) - f(x, y)}{h} \quad \text{and} \quad \frac{f(x, y + h) - f(x, y)}{h}, \quad (h \neq 0).$$

Then, assuming that x and y are fixed, calculate the limit as $h \to 0$. Can you make a connection between your results and derivatives?

25. $f(x, y) = 2x^2 - y.$ **26.** $f(x, y) = xy + 2y.$

27. $f(x, y) = 3x - xy + 2y^2.$

28. $f(x, y) = x \sin y.$ **29.** $f(x, y) = \cos(xy).$

30. $f(x, y) = x^2 e^y.$

31. Determine a function f of two variables whose value at (x, y) is:
(a) The volume of a box with a square base of side length x and height y.
(b) The volume of a right circular cylinder whose radius is x and whose height is y.
(c) The area of the parallelogram whose sides are the vectors $2\mathbf{i}$ and $x\mathbf{i} + y\mathbf{j}$.

32. Determine a function f of three variables whose value at (x, y, z) is:
(a) The surface area of a box with no top whose sides have lengths x, y, and z.

(b) The angle between the vectors $\mathbf{i} + \mathbf{j}$ and $x\mathbf{i} + y\mathbf{j} + z\mathbf{k}$.

(c) The volume of the parallelepiped whose sides are the vectors $\mathbf{i}$, $\mathbf{i} + \mathbf{j}$, and $x\mathbf{i} + y\mathbf{j} + z\mathbf{k}$.

33. A closed box is to have a total surface area of 20 square feet. Express the volume V of the box as a function of its length l and height h.

34. An open box is to contain a volume of 12 cubic meters. If the material for the sides of the box costs \$2 per square meter and the material for the bottom costs \$4 per square meter, express the total cost C of the box as a function of the length l and width w.

35. A petrochemical company is designing a cylindrical tank with hemispherical ends to be used in the transportation of its products. See the figure. Express the volume of the tank as a function of its radius r and the length h of the cylindrical portion.

36. A 10-foot section of gutter is to be made from a 12-inch-wide strip of metal by folding up strips of length x on each side so that they make an angle θ with the bottom of the gutter. See the figure. Express the cross-sectional area of the gutter as a function of x and θ.

■ **14.2 A BRIEF CATALOG OF THE QUADRIC SURFACES; PROJECTIONS**

As you know, the graph of an equation in two variables, say x and y, is typically a curve in the xy-plane. We will see in this section, and in the sections which follow, that the graph of an equation in three variables is, in general, a *surface* in three-space.

In this section we examine in a systematic manner the surfaces defined by equations of the form

$$Ax^2 + By^2 + Cz^2 + Dxy + Exz + Fyz + Hx + Iy + Jz + K = 0$$

where A, B, C, . . . , K are constants. Such surfaces are called *quadric surfaces.*

By suitable translations and rotations of the coordinate axes (see Section 1.4 and Appendix A.1) we can simplify such equations and thereby show that the nondegenerate† quadrics fall into nine distinct types:

1. The ellipsoid.

2. The hyperboloid of one sheet.

3. The hyperboloid of two sheets.

4. The quadric cone.

5. The elliptic paraboloid.

6. The hyperbolic paraboloid.

7. The parabolic cylinder.

8. The elliptic cylinder.

9. The hyperbolic cylinder.

---

† We are excluding such degenerate quadrics as

$$x^2 + y^2 + z^2 + 1 = 0 \quad \text{and} \quad x^2 + y^2 + z^2 = 0.$$

The first one has no points and the second consists of only one point, the origin.

As you go on with calculus, you will encounter these surfaces time and time again. Here we give you a picture of each one, together with its equation in standard form and some information about its special properties. These are some of the things to look for:

(a) The *intercepts* (the points at which the surface intersects the coordinate axes).

(b) The *traces* (the intersections with the coordinate planes).

(c) The *sections* (the intersections with planes in general).

(d) The *center* (some quadrics have a center; some do not).

(e) *Symmetry.*

(f) *Boundedness, unboundedness.*

1. The Ellipsoid

$$\frac{x^2}{a^2} + \frac{y^2}{b^2} + \frac{z^2}{c^2} = 1.†$$
(Figure 14.2.1.)

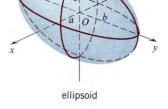

ellipsoid

Figure 14.2.1

This ellipsoid is centered at the origin and is symmetric about the three coordinate planes. It intersects the coordinate axes at six points: $(\pm a, 0, 0)$, $(0, \pm b, 0)$, $(0, 0, \pm c)$. These points are called the *vertices*. The surface is bounded, being contained in the rectangular solid: $|x| \le a$, $|y| \le b$, $|z| \le c$. All three traces are ellipses; thus, for example, the trace in the xy-plane (set $z = 0$) is the ellipse

$$\frac{x^2}{a^2} + \frac{y^2}{b^2} = 1.$$

All sections parallel to the coordinate planes are also ellipses; for example, taking $y = y_0$ we have

$$\frac{x^2}{a^2} + \frac{z^2}{c^2} = 1 - \frac{y_0^2}{b^2}.$$

This ellipse is the intersection of the ellipsoid with the plane $y = y_0$. The numbers a, b, c are called the *semi-axes* of the ellipsoid. If two of the semi-axes are equal, then we have an *ellipsoid of revolution*. (If, for example, $a = c$, then all sections parallel to the xz-plane are circles and the surface can be obtained by revolving the trace in the xy-plane about the y-axis.) If all three semi-axes are equal, the surface is a *sphere*. ❑

2. The Hyperboloid of One Sheet

$$\frac{x^2}{a^2} + \frac{y^2}{b^2} - \frac{z^2}{c^2} = 1.$$
(Figure 14.2.2.)

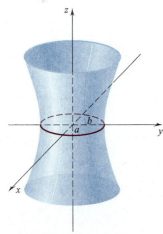

hyperboloid of one sheet

Figure 14.2.2

The surface is unbounded. It is centered at the origin and is symmetric about the three coordinate planes. The surface intersects the coordinate axes at four points: $(\pm a, 0, 0)$, $(0, \pm b, 0)$. The trace in the xy-plane (set $z = 0$) is the ellipse

$$\frac{x^2}{a^2} + \frac{y^2}{b^2} = 1.$$

† Throughout this section we take a, b, c as positive constants.

All sections parallel to the xy-plane are ellipses. The trace in the xz-plane (set $y = 0$) is the hyperbola

$$\frac{x^2}{a^2} - \frac{z^2}{c^2} = 1,$$

and the trace in the yz-plane (set $x = 0$) is the hyperbola

$$\frac{y^2}{b^2} - \frac{z^2}{c^2} = 1.$$

All sections parallel to the xz-plane or yz-plane are hyperbolas. If $a = b$, then all sections parallel to the xy-plane are circles and we have a *hyperboloid of revolution*. ❑

3. The Hyperboloid of Two Sheets

$$\frac{x^2}{a^2} + \frac{y^2}{b^2} - \frac{z^2}{c^2} = -1. \qquad \text{(Figure 14.2.3.)}$$

The surface intersects the coordinate axes only at the two vertices $(0, 0, \pm c)$. The surface consists of two parts: one for which $z \geq c$, another for which $z \leq -c$. We can see this by rewriting the equation as

$$\frac{x^2}{a^2} + \frac{y^2}{b^2} = \frac{z^2}{c^2} - 1$$

and noting that we must have

$$\frac{z^2}{c^2} - 1 \geq 0$$

in order for there to be solutions for x and y. It follows from this that $z^2 \geq c^2$, and so $|z| \geq c$. Each of the parts is unbounded. All sections parallel to the xy-plane are ellipses: for $z = z_0$ with $|z_0| \geq c$, we have

$$\frac{x^2}{a^2} + \frac{y^2}{b^2} = \frac{z_0^2}{c^2} - 1.$$

Sections parallel to the other coordinate planes are hyperbolas; for example, for $y = y_0$ we have

$$\frac{z^2}{c^2} - \frac{x^2}{a^2} = 1 + \frac{y_0^2}{b^2}.$$

The surface is symmetric about the three coordinate planes and is centered at the origin. ❑

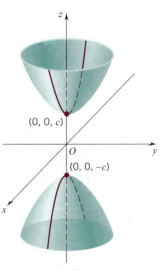

hyperboloid of two sheets

Figure 14.2.3

4. The Quadric Cone

$$\frac{x^2}{a^2} + \frac{y^2}{b^2} = z^2. \qquad \text{(Figure 14.2.4.)}$$

The surface intersects the coordinate axes only at the origin. The surface is unbounded. Once again there is symmetry about the three coordinate planes. The trace in the xz-plane is a pair of intersecting lines: $z = \pm x/a$. The trace in the yz-plane is

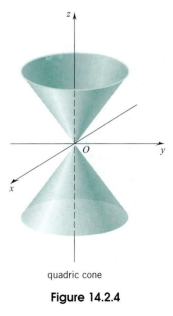

quadric cone

Figure 14.2.4

also a pair of intersecting lines: $z = \pm y/b$. The trace in the xy-plane is just the origin. Sections parallel to the xy-plane are ellipses. If $a = b$, these sections are circles and we have what is commonly called a *double circular cone* or simply a *cone*. The upper and lower portions of the cone are called *nappes*. ❏

We come now to the *paraboloids*. The equations in standard form will involve x^2 and y^2, but then z instead of z^2.

5. The Elliptic Paraboloid

$$\frac{x^2}{a^2} + \frac{y^2}{b^2} = z.$$ (Figure 14.2.5.)

The surface does not extend below the xy-plane; it is unbounded above. The origin is called the *vertex*. Sections parallel to the xy-plane are ellipses; sections parallel to the other coordinate planes are parabolas. Hence the term "elliptic paraboloid." The surface is symmetric about the xz-plane and about the yz-plane. It is also symmetric about the z-axis. If $a = b$, then the surface is a *paraboloid of revolution*. ❏

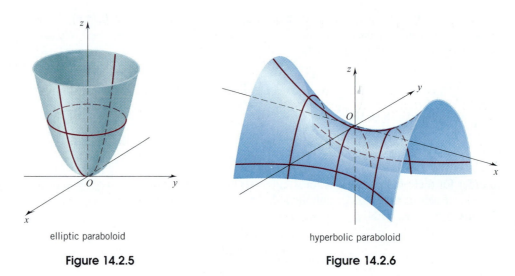

elliptic paraboloid

Figure 14.2.5

hyperbolic paraboloid

Figure 14.2.6

6. The Hyperbolic Paraboloid

$$\frac{x^2}{a^2} - \frac{y^2}{b^2} = z.$$ (Figure 14.2.6.)

Here there is symmetry about the xz-plane and yz-plane. Sections parallel to the xy-plane are hyperbolas; sections parallel to the other coordinate planes are parabolas. Hence the term "hyperbolic paraboloid." The origin is a minimum point for the trace in the xz-plane, but a maximum point for the trace in the yz-plane. The origin is called a *minimax* or *saddle point* of the surface. *Note:* The orientation of the coordinate axes was chosen to enhance the view of the surface. ❏

The rest of the quadric surfaces are *cylinders*. The term deserves definition. Take any plane curve C. All the lines through C that are perpendicular to the plane of C form a surface. Such a surface is called a *cylinder, the cylinder with base curve C*. The perpendicular lines are known as the *generators* of the cylinder.

If the base curve lies in the xy-plane (or in a plane parallel to the xy-plane), then the generators of the cylinder are parallel to the z-axis. In such a case the equation of the cylinder involves only x and y. The z-coordinate is left unrestricted; it can take on all values.

There are three basic types of quadric cylinders. We give you their equations in standard form: base curve in the xy-plane, generators parallel to the z-axis.

7. The Parabolic Cylinder

$$x^2 = 4cy. \qquad \text{(Figure 14.2.7.)}$$

This surface is formed by all lines that pass through the parabola $x^2 = 4cy$ and are perpendicular to the xy-plane. ❏

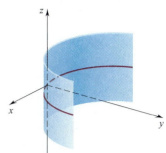

parabolic cylinder

Figure 14.2.7

8. The Elliptic Cylinder

$$\frac{x^2}{a^2} + \frac{y^2}{b^2} = 1. \qquad \text{(Figure 14.2.8.)}$$

The surface is formed by all lines that pass through the ellipse

$$\frac{x^2}{a^2} + \frac{y^2}{b^2} = 1$$

and are perpendicular to the xy-plane. If $a = b$, we have the common *right circular cylinder.* ❏

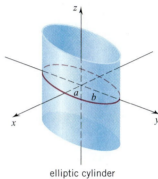

elliptic cylinder

Figure 14.2.8

9. The Hyperbolic Cylinder

$$\frac{x^2}{a^2} - \frac{y^2}{b^2} = 1. \qquad \text{(Figure 14.2.9.)}$$

The surface has two parts, each generated by a branch of the hyperbola

$$\frac{x^2}{a^2} - \frac{y^2}{b^2} = 1. \quad ❏$$

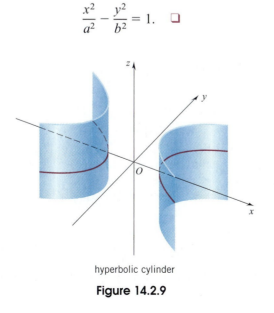

hyperbolic cylinder

Figure 14.2.9

Projections

Suppose that $S_1 : z = f(x, y)$ and $S_2 : z = g(x, y)$ are surfaces in three-space that intersect in a space curve C. See Figure 14.2.10.

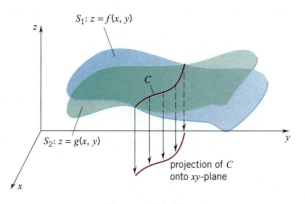

$S_1: z = f(x, y)$

$S_2: z = g(x, y)$

projection of C onto xy-plane

Figure 14.2.10

The curve C is the set of all points (x, y, z) with

$$z = f(x, y) \qquad \text{and} \qquad z = g(x, y).$$

The set of all points (x, y, z) with

$$f(x, y) = g(x, y) \qquad \text{(here } z \text{ is unrestricted)}$$

is the vertical cylinder that passes through C.

The set of all points $(x, y, 0)$ with

$$f(x, y) = g(x, y) \qquad \text{(here } z = 0\text{)}$$

is called the *projection of C onto the xy-plane*. In Figure 14.2.10 it appears as the curve in the xy-plane that lies directly below C.

Example 1 The paraboloid of revolution

$$z = x^2 + y^2$$

and the plane

$$z = 2y + 3$$

intersect in a curve C. See Figure 14.2.11. The projection of this curve onto the xy-plane is the set of all points $(x, y, 0)$ with

$$x^2 + y^2 = 2y + 3.$$

This equation can be written

$$x^2 + (y - 1)^2 = 4.$$

The projection of C onto the xy-plane is the circle of radius 2 centered at $(0, 1, 0)$. ❑

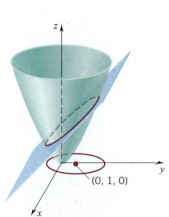

$(0, 1, 0)$

Figure 14.2.11

EXERCISES 14.2

In Exercises 1–12, identify the given surface.

1. $x^2 + 4y^2 - 16z^2 = 0.$

2. $x^2 + 4y^2 + 16z^2 - 12 = 0.$

3. $x - 4y^2 = 0.$ **4.** $x^2 - 4y^2 - 2z = 0.$

5. $5x^2 + 2y^2 - 6z^2 - 10 = 0.$

6. $2x^2 + 4y^2 - 1 = 0.$ **7.** $x^2 + y^2 + z^2 - 4 = 0.$

8. $5x^2 + 2y^2 - 6z^2 + 10 = 0.$

9. $x^2 + 2y^2 - 4z = 0.$ **10.** $2x^2 - 3y^2 - 6 = 0.$

11. $x - y^2 + 2z^2 = 0.$ **12.** $x - y^2 - 6z^2 = 0.$

In Exercises 13–24, sketch the given cylinder.

13. $25y^2 + 4z^2 - 100 = 0.$ **14.** $25x^2 + 4y^2 - 100 = 0.$

15. $y^2 - z = 0.$ **16.** $x^2 - y + 1 = 0.$

17. $y^2 + z = 0.$ **18.** $25x^2 - 9y^2 - 225 = 0.$

19. $x^2 + y^2 = 9.$ **20.** $\dfrac{x^2}{4} + \dfrac{y^2}{9} = 1.$

21. $y^2 - 4x^2 = 4.$ **22.** $z = x^2.$

23. $y = x^2 + 1.$

24. $(x - 1)^2 + (y - 1)^2 = 1.$

In Exercises 25–38, identify the surface and find the traces. Then sketch the surface.

25. $9x^2 + 4y^2 - 36z = 0.$

26. $9x^2 + 4y^2 + 36z^2 - 36 = 0.$

27. $9x^2 + 4y^2 - 36z^2 = 0.$

28. $9x^2 + 4y^2 - 36z^2 - 36 = 0.$

29. $9x^2 + 4y^2 - 36z^2 + 36 = 0.$

30. $9x^2 - 4y^2 - 36z = 0.$

31. $9x^2 - 4y^2 - 36z^2 = 36.$

32. $9x^2 + 4z^2 - 36y^2 - 36 = 0.$

33. $9x^2 + 4y^2 - 36z = 0.$

34. $9x^2 + 4z^2 - 36y^2 = 0.$

35. $9y^2 - 4x^2 - 36z^2 - 36 = 0.$

36. $9y^2 + 4z^2 - 36x = 0.$

37. $x^2 + y^2 - 4z = 0.$

38. $36x^2 + 9y^2 + 4z^2 - 36 = 0.$

39. Identify all possibilities for the surface

$$z = Ax^2 + By^2$$

taking (a) $AB > 0.$ (b) $AB < 0.$ (c) $AB = 0.$

40. Find the planes of symmetry for the cylinder $x - 4y^2 = 0.$

41. Write an equation for the surface obtained by revolving the parabola $4z - y^2 = 0$ about the z-axis.

42. The hyperbola $c^2y^2 - b^2z^2 - b^2c^2 = 0$ is revolved about the z-axis. Find an equation for the resulting surface.

43. (a) The equation

$$\sqrt{x^2 + y^2} = kz \qquad \text{with } k > 0$$

represents the upper nappe of a cone, with vertex at the origin and the positive z-axis as the axis of symmetry. Describe the section in the plane $z = z_0$, $z_0 > 0.$

(b) Let S be one nappe of a cone, with vertex at the origin. Write an equation for S given that

  (i) The negative z-axis is the axis of symmetry and the section in the plane $z = -2$ is a circle of radius 6.

  (ii) The positive y-axis is the axis of symmetry and the section in the plane $y = 3$ is a circle of radius 1.

44. Form the elliptic paraboloid

$$x^2 + \frac{y^2}{b^2} = z.$$

(a) Describe the section in the plane $z = 1.$

(b) What happens to this section as b tends to infinity?

(c) What happens to the paraboloid as b tends to infinity?

In Exercises 45–52, the given surfaces intersect in a space curve C. Determine the projection of C onto the xy-plane.

45. The planes $x + 2y + 3z = 6$ and $x + y - 2z = 6.$

46. The planes $x - 2y + z = 4$ and $3x + y - 2z = 1.$

47. The sphere $x^2 + y^2 + (z - 1)^2 = \frac{3}{2}$ and the hyperboloid $x^2 + y^2 - z^2 = 1.$

48. The sphere $x^2 + y^2 + (z - 2)^2 = 2$ and the cone $x^2 + y^2 = z^2.$

49. The paraboloids $x^2 + y^2 + z = 4$ and $x^2 + 3y^2 = z.$

50. The cylinder $y^2 + z - 4 = 0$ and the paraboloid $x^2 + 3y^2 = z.$

51. The cone $x^2 + y^2 = z^2$ and the plane $y + z = 2.$

52. The cone $x^2 + y^2 = z^2$ and the plane $y + 2z = 2.$

■ 14.3 GRAPHS; LEVEL CURVES AND LEVEL SURFACES

We begin with a function f of two variables defined on a subset D of the xy-plane. By the *graph of f* we mean the graph of the equation

$$z = f(x, y) \qquad (x, y) \in D.$$

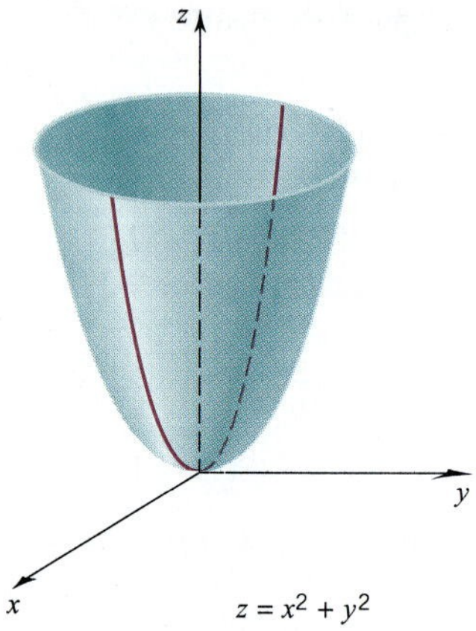

Example 1 In the case of

$$f(x, y) = x^2 + y^2$$

the domain is the entire plane. The graph of f is a paraboloid of revolution:

$$z = x^2 + y^2.$$

This surface can be generated by revolving the parabola

$$z = x^2 \text{ (in the } xz\text{-plane)}$$

$z = x^2 + y^2$

Figure 14.3.1

about the z-axis. See Figure 14.3.1. ❑

Example 2 Let a, b, and c be positive constants. The domain of the function

$$g(x, y) = c - ax - by$$

is also the entire xy-plane. The graph of g is the plane

$$z = c - ax - by \qquad\qquad \text{(Figure 14.3.2.)}$$

with intercepts $x = c/a$, $y = c/b$, $z = c$. ❑

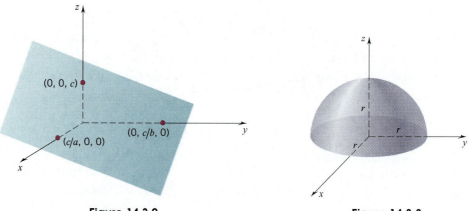

Figure 14.3.2

Figure 14.3.3

Example 3 The function

$$f(x, y) = \sqrt{r^2 - (x^2 + y^2)}, \qquad r \text{ a positive constant,}$$

is defined only on the closed disc $x^2 + y^2 \le r^2$. The graph is the surface

$$z = \sqrt{r^2 - (x^2 + y^2)}, \qquad\qquad \text{(Figure 14.3.3.)}$$

which is the upper half of the sphere

$$x^2 + y^2 + z^2 = r^2. \quad ❑$$

Example 4 The function

$$f(x, y) = xy$$

is simple enough, but its graph, the surface $z = xy$, is quite difficult to draw. It is a "saddle-shaped" surface, a hyperbolic paraboloid: rotate the x and y axes by $\frac{1}{4}\pi$ radians and the equation takes the form

$$\frac{X^2}{2} - \frac{Y^2}{2} = z.†$$

A computer-generated graph of this surface is shown in Figure 14.3.9. ❏

Level Curves

In practice the graph of a function of two variables is usually difficult to visualize and draw. Moreover, if drawn, it is often difficult to interpret. Of course, the current techniques of computer graphics do make it possible to sketch and analyze more complicated surfaces. We will provide some illustrations of this later in the section.

Here we discuss an approach to sketching surfaces in space which we take from the map maker. In mapping mountainous terrain it is a common practice to sketch curves joining points of constant elevation. A collection of such curves, properly labeled, gives a good idea of the altitude variation in a region and suggests the shape of mountains and valleys. (See Figure 14.3.4.)

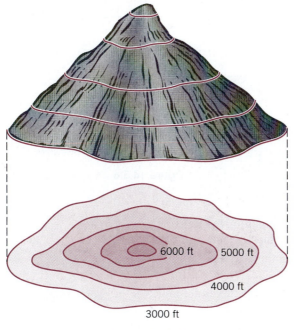

Figure 14.3.4

We can do the same thing to portray functions of two variables. Suppose that f is a nonconstant function defined on some portion of the xy-plane. If c is some value in the range of f, then we can sketch the curve $f(x, y) = c$. Such a curve is called a *level*

---

† To see this, set $\alpha = \frac{1}{4}\pi$ in Formula A.1.2, Appendix A-1.

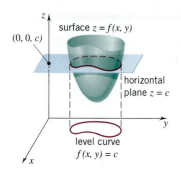

Figure 14.3.5

curve for f. It can be obtained by intersecting the graph of f with the horizontal plane $z = c$ and then projecting this intersection onto the xy-plane. (See Figure 14.3.5.)

The level curve $f(x, y) = c$ lies entirely in the domain of f, and on this curve f is constantly c.

A collection of level curves, properly drawn and labeled, can lead to a good understanding of the overall behavior of a function.

Example 5 We begin with the function $f(x, y) = x^2 + y^2$ (see Figure 14.3.1). The level curves are circles centered at the origin:

$$x^2 + y^2 = c, \qquad c \geq 0 \qquad \text{(Figure 14.3.6.)}$$

The function has the value c on the circle of radius $\sqrt{c}$ centered at the origin. At the origin, the function has the value 0. ❏

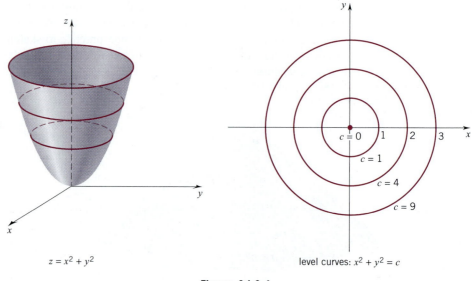

$z = x^2 + y^2$

level curves: $x^2 + y^2 = c$

Figure 14.3.6

Example 6 The graph of the function $g(x, y) = 4 - x - y$ is a plane. The level curves are parallel lines of the form

$$4 - x - y = c.$$

The surface and the level curves are indicated in Figure 14.3.7. ❏

Example 7 Now we consider the function

$$h(x, y) = \begin{cases} \sqrt{x^2 + y^2}, & x \geq 0 \\ |y|, & x < 0. \end{cases}$$

For $x \geq 0$, $h(x, y)$ is the distance from (x, y) to the origin. For $x < 0$, $h(x, y)$ is the distance from (x, y) to the x-axis. The level curves are pictured in Figure 14.3.8. The 0-level curve is the nonpositive x-axis. The other level curves are horseshoe-shaped: pairs of horizontal rays capped on the right by semicircles. ❏

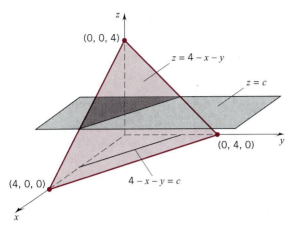
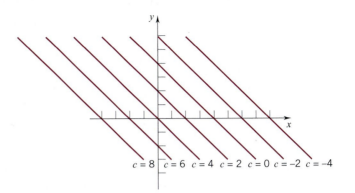

Figure 14.3.7

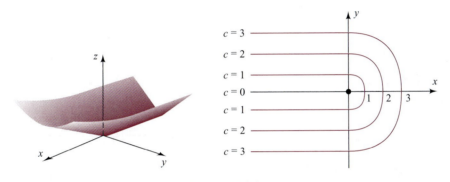

Figure 14.3.8

Example 8 Let's return to the function $f(x, y) = xy$. Earlier we noted that the graph is a "saddled-shaped" surface. You can visualize the surface from the few level curves sketched in Figure 14.3.9. The 0-level curve, $xy = 0$, consists of the two coordinate axes. The other level curves, $xy = c$ with $c \neq 0$, are hyperbolas. ❏

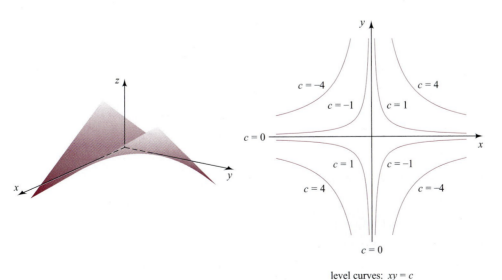

level curves: $xy = c$

Figure 14.3.9

Computer-Generated Graphs

The preceding examples illustrate how difficult it is to sketch an accurate graph of a function of two variables. But powerful help is at hand. Three-dimensional graphing programs for modern computers make it possible to visualize even quite complicated surfaces in three-space. These programs allow the user to view a surface from different perspectives, and they show level curves and the traces of various planes. Examples of computer-generated graphs are shown in Figure 14.3.10 and in the Exercises.

Level Surfaces

While drawing graphs for functions of two variables is quite difficult, drawing graphs for functions of three variables is actually impossible. To draw such figures we would need four dimensions at our disposal; the domain itself is a portion of three-space.

One can try to visualize the behavior of a function of three variables by examining the *level surfaces*. These are the subsets of the domain with equations of the form

$$f(x, y, z) = c.$$

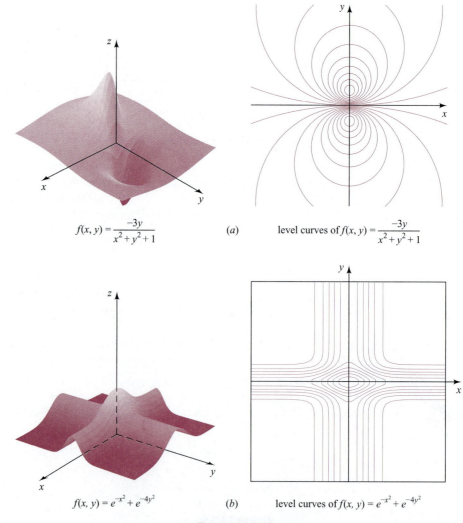

$$f(x, y) = \frac{-3y}{x^2 + y^2 + 1}$$
(a) level curves of $f(x, y) = \dfrac{-3y}{x^2 + y^2 + 1}$

$$f(x, y) = e^{-x^2} + e^{-4y^2}$$
(b) level curves of $f(x, y) = e^{-x^2} + e^{-4y^2}$

Figure 14.3.10

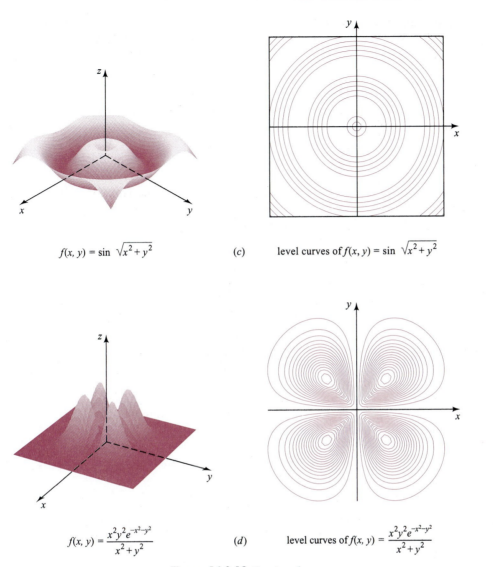

$$f(x, y) = \sin \sqrt{x^2 + y^2}$$

(c)   level curves of $f(x, y) = \sin \sqrt{x^2 + y^2}$

$$f(x, y) = \frac{x^2 y^2 e^{-x^2 - y^2}}{x^2 + y^2}$$

(d)   level curves of $f(x, y) = \frac{x^2 y^2 e^{-x^2 - y^2}}{x^2 + y^2}$

Figure 14.3.10 *Continued*

Level surfaces are usually difficult to draw. Nevertheless, a knowledge of what they are can be helpful. Here we restrict ourselves to a few simple examples.

Example 9   For the function $f(x, y, z) = Ax + By + Cz$ the level surfaces are parallel planes

$$Ax + By + Cz = c. \quad \square$$

Example 10   For the function $g(x, y, z) = \sqrt{x^2 + y^2 + z^2}$ the level surfaces are concentric spheres

$$x^2 + y^2 + z^2 = c^2. \quad \square$$

Example 11   As our final example we take the function

$$f(x, y, z) = \frac{|z|}{x^2 + y^2}.$$

We extend this function to the origin by defining it to be zero there. At other points of the z-axis we leave f undefined.

In the first place note that f takes on only nonnegative values. Since f is zero only when $z = 0$, the 0-level surface is the xy-plane. To find the other level surfaces, we take $c > 0$ and set $f(x, y, z) = c$. This gives

$$\frac{|z|}{x^2 + y^2} = c \quad \text{and thus} \quad |z| = c(x^2 + y^2).$$

(Figure 14.3.11). Each of these surfaces is a double-paraboloid of revolution.† ❏

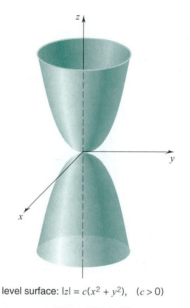

level surface: $|z| = c(x^2 + y^2)$, $(c > 0)$

Figure 14.3.11

---

† Surface 5 of the last section together with its mirror image below the xy-plane.

EXERCISES 14.3

In Exercises 1–18, identify the level curves $f(x, y) = c$ and sketch the curves corresponding to the indicated values of c.

1. $f(x, y) = x - y$; $c = -2, 0, 2$.

2. $f(x, y) = 2x - y$; $c = -2, 0, 2$.

3. $f(x, y) = x^2 - y$; $c = -1, 0, 1, 2$.

4. $f(x, y) = \dfrac{1}{x - y^2}$; $c = -2, -1, 1, 2$.

5. $f(x, y) = \dfrac{x}{x + y}$; $c = -1, 0, 1, 2$.

6. $f(x, y) = \dfrac{y}{x^2}$; $c = -1, 0, 1, 2$.

7. $f(x, y) = x^3 - y$; $c = -1, 0, 1, 2$.

8. $f(x, y) = e^{xy}$; $c = \frac{1}{2}, 1, 2, 3$.

9. $f(x, y) = x^2 - y^2$; $c = -2, -1, 0, 1, 2$.

10. $f(x, y) = x^2$; $c = 0, 1, 4, 9$.

11. $f(x, y) = y^2$; $c = 0, 1, 4, 9$.

12. $f(x, y) = x(y - 1)$; $c = -2, -1, 0, 1, 2$.

13. $f(x, y) = \ln(x^2 + y^2)$; $c = -1, 0, 1$.

14. $f(x, y) = \ln\left(\dfrac{y}{x^2}\right)$; $c = -2, -1, 0, 1, 2$.

15. $f(x, y) = \dfrac{\ln y}{x^2}$; $c = -2, -1, 0, 1, 2$.

16. $f(x, y) = x^2 y^2$; $c = -4, -1, 0, 1, 4$.

17. $f(x, y) = \dfrac{x^2}{x^2 + y^2}$, $c = 0, \frac{1}{2}, \frac{1}{4}$.

18. $f(x, y) = \dfrac{\ln y}{x}$; $c = -2, -1, 0, 1, 2$.

In Exercises 19–24, identify the c-level surface and sketch it.

19. $f(x, y, z) = x + 2y + 3z,\quad c = 0.$

20. $f(x, y, z) = x^2 + y^2,\quad c = 4.$

21. $f(x, y, z) = z(x^2 + y^2)^{-1/2},\quad c = 1.$

22. $f(x, y, z) = x^2/4 + y^2/6 + z^2/9,\quad c = 1.$

23. $f(x, y, z) = 4x^2 + 9y^2 - 72z,\quad c = 0.$

24. $f(x, y, z) = z^2 - 36x^2 - 9y^2,\quad c = 1.$

25. Identify the c-level surfaces of

$$f(x, y, z) = x^2 + y^2 - z^2$$

taking (i) $c < 0$, (ii) $c = 0$, (iii) $c > 0$.

26. Identify the c-level surfaces of

$$f(x, y, z) = 9x^2 - 4y^2 + 36z^2$$

taking (i) $c < 0$, (ii) $c = 0$, (iii) $c > 0$.

In Exercises 27–30, find an equation for the level curve of f that contains the given point P.

27. $f(x, y) = 1 - 4x^2 - y^2;\quad P(0, 1).$

28. $f(x, y) = (x^2 + y^2)e^{xy};\quad P(1, 0).$

29. $f(x, y) = y^2 \tan^{-1}x;\quad P(1, 2).$

30. $f(x, y) = (x^2 + y)\ln[2 - x + e^y];\quad P(2, 1).$

In Exercises 31 and 32, find an equation for the level surface of f that contains the given point P.

31. $f(x, y, z) = x^2 + 2y^2 - 2xyz;\quad P(-1, 2, 1).$

32. $f(x, y, z) = \sqrt{x^2 + y^2} - \ln z;\quad P(3, 4, e).$

33. The magnitude of the gravitational force exerted by a body of mass M situated at the origin on a body of mass m located at the point (x, y, z) is given by

$$F(x, y, z) = \frac{GmM}{x^2 + y^2 + z^2}$$

where G is the universal gravitational constant. If m and M are constants, describe the level surfaces of F. What is the physical significance of these surfaces?

34. The strength E of an electric field at a point (x, y, z) due to an infinitely long charged wire lying along the y-axis is given by

$$E(x, y, z) = \frac{k}{\sqrt{x^2 + z^2}}$$

where k is a positive constant. Describe the level surfaces of E.

35. A thin metal plate is situated in the xy-plane. The temperature T (in $C°$) at the point (x, y) is inversely proportional to the square of the distance from the origin.

(a) Express T as a function of x and y.

(b) Describe the level curves and sketch a representative set. NOTE: The level curves of T are called *isothermals*; all points on an isothermal have the same temperature.

(c) Suppose the temperature at the point $(1, 2)$ is $50°$. What is the temperature at the point $(4, 3)$?

36. The formula

$$V(x, y) = \frac{k}{\sqrt{r^2 - x^2 - y^2}},$$

where k and r are positive constants, gives the electric potential (in volts) at a point (x, y) in the xy-plane. Describe the level curves of V and sketch a representative set. NOTE: The level curves of V are called the *equipotential curves*; all points on an equipotential curve have the same electric potential.

In Exercises 37–42, a function f, together with a set of level curves for f, is given. Figures A–F are the surfaces $z = f(x, y)$ (in some order). Match f and its system of level curves with its graph $z = f(x, y)$.

37. $f(x, y) = y^2 - y^3$.

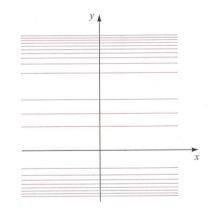

38. $f(x, y) = \sin x,\ 0 \le x \le 2\pi$.

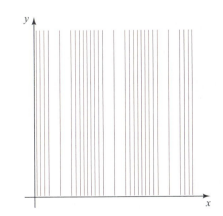

39. $f(x, y) = \cos \sqrt{x^2 + y^2}$, $-10 \le x \le 10$, $-10 \le y \le 10$.

41. $f(x, y) = xy\, e^{-\frac{1}{2}(x^2 + y^2)}$.

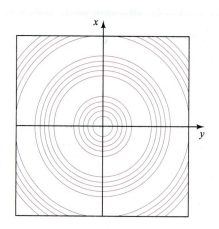

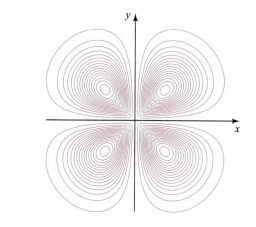

40. $f(x, y) = 2x^2 + 4y^2$.

42. $f(x, y) = \sin x \sin y$.

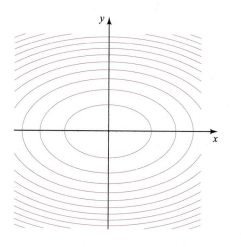

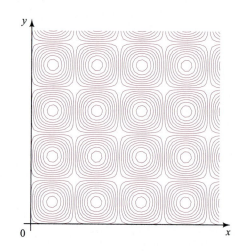

A.

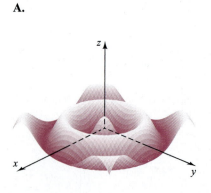

B.

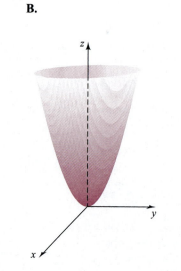

C.

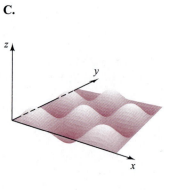

D.   **E.**   **F.**

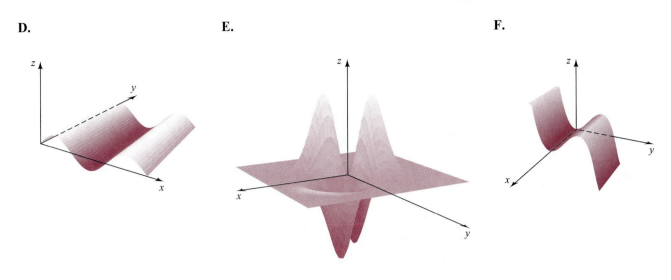

■ 14.4 PARTIAL DERIVATIVES

Functions of Two Variables

Let f be a function of x and y; for example

$$f(x, y) = 3x^2y - 5x \cos \pi y.$$

The *partial derivative of f with respect to x* is the function f_x obtained by differentiating f with respect to x, treating y as a constant; in this case

$$f_x(x, y) = 6xy - 5 \cos \pi y.$$

The *partial derivative of f with respect to y* is the function f_y obtained by differentiating f with respect to y, treating x as a constant; in this case

$$f_y(x, y) = 3x^2 + 5\pi x \sin \pi y.$$

These partial derivatives are formally defined as limits:

DEFINITION 14.4.1 PARTIAL DERIVATIVES OF $f(x, y)$

Let f be a function of two variables. The partial derivatives of f with respect to x and y are the functions f_x and f_y defined by

$$f_x(x, y) = \lim_{h \to 0} \frac{f(x + h, y) - f(x, y)}{h}$$

$$f_y(x, y) = \lim_{h \to 0} \frac{f(x, y + h) - f(x, y)}{h}$$

provided these limits exist.

Example 1 For

$$f(x, y) = x \tan^{-1} xy$$

we have

$$f_x(x, y) = x\frac{y}{1 + (xy)^2} + \tan^{-1} xy = \frac{xy}{1 + x^2y^2} + \tan^{-1} xy$$

and

$$f_y(x, y) = x\frac{x}{1 + (xy)^2} = \frac{x^2}{1 + x^2y^2}. \quad \square$$

In the one-variable case, $f'(x_0)$ gives the rate of change with respect to x of $f(x)$ at $x = x_0$. In the two-variable case, $f_x(x_0, y_0)$ *gives the rate of change with respect to x of $f(x, y_0)$ at $x = x_0$, and $f_y(x_0, y_0)$ gives the rate of change with respect to y of $f(x_0, y)$ at $y = y_0$.*

Example 2 For the function

$$f(x, y) = e^{xy} + \ln (x^2 + y)$$

we have

$$f_x(x, y) = ye^{xy} + \frac{2x}{x^2 + y} \quad \text{and} \quad f_y(x, y) = xe^{xy} + \frac{1}{x^2 + y}.$$

The number

$$f_x(2, 1) = e^2 + \frac{4}{4 + 1} = e^2 + \frac{4}{5}$$

gives the rate of change with respect to x of the function

$$f(x, 1) = e^x + \ln (x^2 + 1) \qquad \text{at } x = 2;$$

the number

$$f_y(2, 1) = 2e^2 + \frac{1}{4 + 1} = 2e^2 + \frac{1}{5}$$

gives the rate of change with respect to y of the function

$$f(2, y) = e^{2y} + \ln (4 + y) \qquad \text{at } y = 1. \quad \square$$

A Geometric Interpretation

In Figure 14.4.1 we have sketched a surface $z = f(x, y)$, which you can take as every-where defined. Through the surface we have passed a plane $y = y_0$ parallel to the xz-plane. The plane $y = y_0$ intersects the surface in a curve, the y_0-section of the surface.

The y_0-section of the surface is the graph of the function

$$g(x) = f(x, y_0).$$

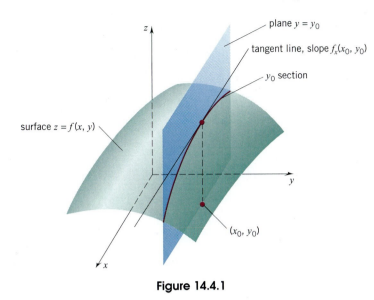

Figure 14.4.1

Differentiating with respect to x, we have

$$g'(x) = f_x(x, y_0)$$

and in particular

$$g'(x_0) = f_x(x_0, y_0).$$

The number $f_x(x_0, y_0)$ is thus the slope of the y_0-section of the surface $z = f(x, y)$ at the point $P(x_0, y_0, f(x_0, y_0))$.

The other partial derivative f_y can be given a similar interpretation. In Figure 14.4.2 you can see the same surface $z = f(x, y)$, this time sliced by a plane $x = x_0$ parallel to the yz-plane. The plane $x = x_0$ intersects the surface in a curve, the x_0-section of the surface.

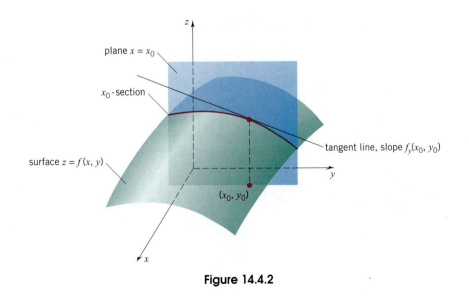

Figure 14.4.2

The x_0-section of the surface is the graph of the function

$$h(y) = f(x_0, y).$$

Differentiating, this time with respect to y, we have

$$h'(y) = f_y(x_0, y)$$

and thus

$$h'(y_0) = f_y(x_0, y_0).$$

The number $f_y(x_0, y_0)$ is the slope of the x_0-section of the surface $z = f(x, y)$ at the point $P(x_0, y_0, f(x_0, y_0))$.

Functions of Three Variables

In the case of a function of three variables, you can look for three partial derivatives: the partial with respect to x, the partial with respect to y, and also the partial with respect to z. These partials

$$f_x(x, y, z), \qquad f_y(x, y, z), \qquad f_z(x, y, z)$$

are defined as follows:

DEFINITION 14.4.2   PARTIAL DERIVATIVES OF $f(x, y, z)$

Let f be a function of three variables. The partial derivatives of f with respect to x, y, and z are the functions f_x, f_y, and f_z defined by

$$f_x(x, y, z) = \lim_{h \to 0} \frac{f(x + h, y, z) - f(x, y, z)}{h},$$

$$f_y(x, y, z) = \lim_{h \to 0} \frac{f(x, y + h, z) - f(x, y, z)}{h},$$

$$f_z(x, y, z) = \lim_{h \to 0} \frac{f(x, y, z + h) - f(x, y, z)}{h},$$

provided these limits exist.

Each partial can be found by differentiating with respect to the subscript variable, treating the other two variables as constants.

Example 3   For the function

$$f(x, y, z) = xy^2z^3$$

the partial derivatives are

$$f_x(x, y, z) = y^2z^3, \qquad f_y(x, y, z) = 2xyz^3, \qquad f_z(x, y, z) = 3xy^2z^2.$$

In particular

$$f_x(1, -2, -1) = -4, \qquad f_y(1, -2, -1) = 4, \qquad f_z(1, -2, -1) = 12. \quad \square$$

Example 4 For
$$g(x, y, z) = x^2 e^{y/z}$$
we have
$$g_x(x, y, z) = 2xe^{y/z}, \qquad g_y(x, y, z) = \frac{x^2}{z} e^{y/z}, \qquad g_z(x, y, z) = -\frac{x^2 y}{z^2} e^{y/z}. \quad ❏$$

Example 5 For a function of the form
$$f(x, y, z) = F(x, y)G(y, z)$$
we can write
$$f_x(x, y, z) = F_x(x, y)G(y, z),$$
$$f_y(x, y, z) = F(x, y)G_y(y, z) + F_y(x, y)G(y, z),$$
$$f_z(x, y, z) = F(x, y)G_z(y, z). \quad ❏$$

The number $f_x(x_0, y_0, z_0)$ gives the rate of change with respect to x of $f(x, y_0, z_0)$ at $x = x_0$; $f_y(x_0, y_0, z_0)$ gives the rate of change with respect to y of $f(x_0, y, z_0)$ at $y = y_0$; and $f_z(x_0, y_0, z_0)$ gives the rate of change with respect to z of $f(x_0, y_0, z)$ at $z = z_0$.

Example 6 The function
$$f(x, y, z) = xy^2 - yz^2$$
has partial derivatives
$$f_x(x, y, z) = y^2, \qquad f_y(x, y, z) = 2xy - z^2, \qquad f_z(x, y, z) = -2yz.$$
The number $f_x(1, 2, 3) = 4$ gives the rate of change with respect to x of the function
$$f(x, 2, 3) = 4x - 18 \qquad \text{at } x = 1;$$
$f_y(1, 2, 3) = -5$ gives the rate of change with respect to y of the function
$$f(1, y, 3) = y^2 - 9y \qquad \text{at } y = 2;$$
$f_z(1, 2, 3) = -12$ gives the rate of change with respect to z of the function
$$f(1, 2, z) = 4 - 2z^2 \qquad \text{at } z = 3. \quad ❏$$

The geometric significance that we attached to the partials of a function of two variables has no analog in the case of three variables. The interpretations shown in Figures 14.4.1 and 14.4.2 do not generalize to three variables.

Other Notations

There is obviously no need to restrict ourselves to the variables x, y, z. Where more convenient we can use other letters.

Example 7 The volume of the frustum of a cone (Figure 14.4.3) is given by the function

$$V(R, r, h) = \tfrac{1}{3} \pi h (R^2 + Rr + r^2).$$

Find the rate of change of the volume with respect to each of its dimensions if the other dimensions are held constant. Determine the values of these rates of change when $R = 8$, $r = 4$, and $h = 6$.

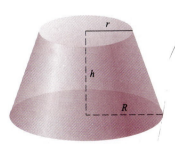

frustum of a cone

Figure 14.4.3

SOLUTION The partial derivatives of V with respect to R, r, and h are as follows:

$$V_R(R, r, h) = \tfrac{1}{3}\pi h(2R + r),$$

$$V_r(R, r, h) = \tfrac{1}{3}\pi h(R + 2r),$$

$$V_h(R, r, h) = \tfrac{1}{3}\pi(R^2 + Rr + r^2).$$

When $R = 8$, $r = 4$, and $h = 6$,

the rate of change of V with respect to R is $V_R(8, 4, 6) = 40\pi$,
the rate of change of V with respect to r is $V_r(8, 4, 6) = 32\pi$,
the rate of change of V with respect to h is $V_h(8, 4, 6) = \tfrac{112}{3}\pi$. ❑

The subscript notation is not the only one used for partial differentiation. A variant of Leibniz's double-d notation is also commonly used. In this notation the partials f_x, f_y, f_z are denoted by

$$\frac{\partial f}{\partial x}, \quad \frac{\partial f}{\partial y}, \quad \frac{\partial f}{\partial z}.$$

Thus, for

$$f(x, y, z) = x^3 y^2 z + \sin xy$$

we have

$$\frac{\partial f}{\partial x}(x, y, z) = 3x^2 y^2 z + y \cos xy,$$

$$\frac{\partial f}{\partial y}(x, y, z) = 2x^3 yz + x \cos xy,$$

$$\frac{\partial f}{\partial z}(x, y, z) = x^3 y^2,$$

or more simply,

$$\frac{\partial f}{\partial x} = 3x^2 y^2 z + y \cos xy,$$

$$\frac{\partial f}{\partial y} = 2x^3 yz + x \cos xy,$$

$$\frac{\partial f}{\partial z} = x^3 y^2.$$

We can also write

$$\frac{\partial}{\partial x}(x^3 y^2 z + \sin xy) = 3x^2 y^2 z + y \cos xy,$$

$$\frac{\partial}{\partial y}(x^3 y^2 z + \sin xy) = 2x^3 yz + x \cos xy,$$

$$\frac{\partial}{\partial z}(x^3 y^2 z + \sin xy) = x^3 y^2.$$

Of course, this notation is not restricted to the letters x, y, z. For instance, we can write

$$\frac{\partial}{\partial r}(r^2\cos\theta + e^{\theta r}) = 2r\cos\theta + \theta e^{\theta r},$$

$$\frac{\partial}{\partial\theta}(r^2\cos\theta + e^{\theta r}) = -r^2\sin\theta + re^{\theta r}.$$

For the function

$$\rho = \sin 2\theta \cos 3\phi$$

we have

$$\frac{\partial\rho}{\partial\theta} = 2\cos 2\theta \cos 3\phi \qquad \text{and} \qquad \frac{\partial\rho}{\partial\phi} = -3\sin 2\theta \sin 3\phi.$$

EXERCISES 14.4

In Exercises 1–28, calculate the partial derivatives of the given function

1. $f(x, y) = 3x^2 - xy + y$. **2.** $g(x, y) = x^2e^{-y}$.

3. $\rho = \sin\phi \cos\theta$. **4.** $\rho = \sin^2(\theta - \phi)$.

5. $f(x, y) = e^{x-y} - e^{y-x}$. **6.** $z = \sqrt{x^2 - 3y}$.

7. $g(x, y) = \dfrac{Ax + By}{Cx + Dy}$. **8.** $u = \dfrac{e^z}{xy^2}$.

9. $u = xy + yz + zx$. **10.** $z = Ax^2 + Bxy + Cy^2$.

11. $f(x, y, z) = z\sin(x - y)$.

12. $g(u, v, w) = \ln(u^2 + vw - w^2)$.

13. $\rho = e^{\theta + \phi}\cos(\theta - \phi)$.

14. $f(x, y) = (x + y)\sin(x - y)$.

15. $f(x, y) = x^2y \sec xy$.

16. $g(x, y) = \tan^{-1}(2x + y)$.

17. $h(x, y) = \dfrac{x}{x^2 + y^2}$. **18.** $z = \ln\sqrt{x^2 + y^2}$.

19. $f(x, y) = \dfrac{x\sin y}{y\cos x}$. **20.** $f(x, y, z) = e^{xy}\sin xz$.

21. $h(x, y) = [f(x)]^2 g(y)$. **22.** $h(x, y) = e^{f(x)g(y)}$.

23. $f(x, y, z) = z^{xy^2}$.

24. $h(x, y, z) = [f(x, y)]^3[g(x, z)]^2$.

25. $h(r, \theta, t) = r^2e^{2t}\cos(\theta - t)$.

26. $u = \ln(x/y) - ye^{xz}$.

27. $f(x, y, z) = z\tan^{-1}(y/x)$.

28. $w = xy\sin z - yz\sin x$.

29. Find $f_x(0, e)$ and $f_y(0, e)$ given that $f(x, y) = e^x \ln y$.

30. Find $g_x(0, \tfrac{1}{4}\pi)$ and $g_y(0, \tfrac{1}{4}\pi)$ given that $g(x, y) = e^{-x}\sin(x + 2y)$.

31. Find $f_x(1, 2)$ and $f_y(1, 2)$ given that $f(x, y) = \dfrac{x}{x + y}$.

32. Find $g_x(1, 2)$ and $g_y(1, 2)$ given that $g(x, y) = \dfrac{y}{x + y^2}$.

In Exercises 33–38, find $f_x(x, y)$ and $f_y(x, y)$ by forming the appropriate difference quotient and taking the limit as h tends to zero (Definition 14.4.1).

33. $f(x\,y) = x^2y$. **34.** $f(x, y) = y^2$.

35. $f(x, y) = \ln(x^2y)$. **36.** $f(x, y) = \dfrac{1}{x + 4y}$.

37. $f(x\,y) = \dfrac{1}{x - y}$. **38.** $f(x, y) = e^{2x+3y}$.

In Exercises 39 and 40, find $f_x(x, y, z), f_y(x, y, z),$ and $f_z(x, y, z)$ by forming the appropriate difference quotient and taking the limit as h tends to zero (Definition 14.4.2).

39. $f(x, y, z) = xy^2z$. **40.** $f(x, y, z) = \dfrac{x^2y}{z}$.

41. The intersection of a surface $z = f(x, y)$ with a plane $y = y_0$ is a curve C in space. The slope of the tangent line to C at the point $P(x_0, y_0, f(x_0, y_0))$ is $f_x(x_0, y_0)$. (See Figure 14.4.1.)

(a) Show that equations for the tangent line can be written in the form:

$$y = y_0, \qquad z - z_0 = f_x(x_0, y_0)(x - x_0).$$

(b) Now let C be the curve formed by intersecting the surface $z = f(x, y)$ with the plane $x = x_0$. Derive equations for the tangent line to C at the point $P(x_0, y_0, f(x_0, y_0))$. (See Figure 14.4.2.)

In Exercises 42 and 43, let $z = x^2 + y^2$ and let C be the curve of intersection of the surface with the given plane. Find equations for the tangent line to the graph of C at the given point.

42. Plane: $y = 3$; point $(1, 3, 10)$.

43. Plane: $x = 2$; point $(2, 1, 5)$.

In Exercises 44 and 45, let

$$z = \frac{x^2}{y^2 - 3}$$

and let C be the curve of intersection of the surface with the given plane. Find equations for the tangent line to C at the given point.

44. Plane: $x = 3$; point $(3, 2, 9)$.

45. Plane: $y = 2$; point $(3, 2, 9)$.

46. The surface $z = \sqrt{4 - x^2 - y^2}$ is a hemisphere of radius 2 centered at the origin.
 (a) Find equations for the tangent line l_1 to the curve of intersection of the hemisphere with the plane $x = 1$ at the point $(1, 1, \sqrt{2})$.
 (b) Find equations for the tangent line l_2 to the curve of intersection of the hemisphere with the plane $y = 1$ at the point $(1, 1, \sqrt{2})$.
 (c) The tangent lines l_1 and l_2 determine a plane. Find an equation for this plane. As you might expect, this plane is tangent to the surface at the point $(1, 1, \sqrt{2})$

In Exercises 47–50, show that the functions u and v satisfy

$$u_x(x, y) = v_y(x, y) \quad \text{and} \quad u_y(x, y) = -v_x(x, y).$$

These equations are called the *Cauchy–Riemann equations*. They arise in the study of functions of a complex variable and are of fundamental importance.

47. $u(x, y) = x^2 - y^2$; $v(x, y) = 2xy$.

48. $u(x, y) = e^x \cos y$; $v(x, y) = e^x \sin y$.

49. $u(x, y) = \frac{1}{2} \ln(x^2 + y^2)$; $v(x, y) = \tan^{-1}\frac{y}{x}$.

50. $u(x, y) = \frac{x}{x^2 + y^2}$; $v(x, y) = \frac{-y}{x^2 + y^2}$.

51. Assume that f is a function which is defined on a domain D in the xy-plane, and assume that its partial derivatives exist throughout D.
 (a) Suppose that $f_x(x, y) = 0$ for all $(x, y) \in D$. What can you conclude about f?
 (b) Suppose that $f_y(x, y) = 0$ for all $(x, y) \in D$. What can you conclude about f?

52. The law of cosines for a triangle can be written

$$a^2 = b^2 + c^2 - 2bc \cos \theta.$$

At time t_0 we have $b_0 = 10$ inches, $c_0 = 15$ inches, $\theta_0 = \frac{1}{3}\pi$ radians.

 (a) Find a_0.
 (b) Find the rate of change of a with respect to b at time t_0 if c and θ remain constant.
 (c) Using the rate found in (b), calculate (by differentials) the approximate change in a if b is decreased by 1 inch.
 (d) Find the rate of change of a with respect to θ at time t_0 if b and c remain constant.
 (e) Find the rate of change of c with respect to θ at time t_0 if a and b remain constant.

53. The area of a triangle is given by the formula

$$A = \tfrac{1}{2}bc \sin \theta.$$

At time t_0 we have $b_0 = 10$ inches, $c_0 = 20$ inches, $\theta_0 = \frac{1}{3}\pi$ radians.
 (a) Find the area of the triangle at time t_0.
 (b) Find the rate of change of the area with respect to b at time t_0 if c and θ remain constant.
 (c) Find the rate of change of the area with respect to θ at time t_0 if b and c remain constant.
 (d) Using the rate found in (c), calculate (by differentials) the approximate change in area if the angle is increased by one degree.
 (e) Find the rate of change of c with respect to b at time t_0 if the area and the angle are to remain constant.

54. Let f be a function of x and y that satisfies a relation of the form

$$\frac{\partial f}{\partial x} = kf, \quad k \text{ a constant.}$$

Show that

$$f(x, y) = g(y)e^{kx},$$

where g is some function of y.

55. Let $z = f(x, y)$ be a surface everywhere defined.
 (a) Find a vector function that parametrizes the y_0-section of the surface. (See Figure 14.4.1.) Find a vector function that parametrizes the line tangent to this section at the point $P(x_0, y_0, f(x_0, y_0))$. See Exercise 41.
 (b) Find a vector function that parametrizes the x_0-section of the surface. (See Figure 14.4.2.) Find a vector function that parametrizes the line tangent to this section at the point $P(x_0, y_0, f(x_0, y_0))$.
 (c) Show that the equation of the plane determined by the tangent lines found in (a) and (b) can be written

$$z - f(x_0, y_0) = (x - x_0)\frac{\partial f}{\partial x}(x_0, y_0)$$
$$+ (y - y_0)\frac{\partial f}{\partial y}(x_0, y_0).$$

(We show in Chapter 15 that, for certain functions of two variables, this plane plays a role similar to the role played by the tangent line for differentiable functions of a single variable.)

56. *(A chain rule)* Let f be a function of x and y, and g be a function of a single variable. Form the composition $h(x, y) = g(f(x, y))$ and show that

$$h_x(x, y) = g'(f(x, y))f_x(x, y) \qquad \text{and}$$
$$h_y(x, y) = g'(f(x, y))f_y(x, y).$$

In Leibniz's notation, setting $u = f(x, y)$, we have

$$\frac{\partial h}{\partial x} = \frac{dg}{du}\frac{\partial u}{\partial x} \qquad \text{and} \qquad \frac{\partial h}{\partial y} = \frac{dg}{du}\frac{\partial u}{\partial y}.$$

57. Let g be a differentiable function of a single variable. Use Exercise 56 to verify the following results.
(a) If $w = g(ax + by)$, then

$$b\frac{\partial w}{\partial x} = a\frac{\partial w}{\partial y}.$$

(b) If m and n are nonzero integers and $w = g(x^m y^n)$, then

$$nx\frac{\partial w}{\partial x} = my\frac{\partial w}{\partial y}.$$

Partial Differential Equations

Differential equations were introduced in Section 7.6. At that stage, the only type of derivative that we had available was an "ordinary" derivative, that is, the derivative of a function of one variable. The differential equations we considered are called *ordinary differential equations*. This term is used to distinguish them from *partial differential equations,* which we introduce here. A *partial differential equation* is a differential equation in which the unknown function is a function of more than one variable. Since the unknown is a function of several variables, the derivatives of the unknown that appear in the equation must be partial derivatives. Partial differential equations play a significant role in the science and engineering disciplines since the descriptions of most natural phenomena have mathematical models that involve functions of several variables.

58. Show that $u = \dfrac{x^2 y^2}{x + y}$
satisfies the partial differential equation

$$x\frac{\partial u}{\partial x} + y\frac{\partial u}{\partial y} = 3u.$$

59. Show that $u = Ax^4 + 2Bx^2y^2 + Cy^4$
satisfies the partial differential equation

$$x\frac{\partial u}{\partial x} + y\frac{\partial u}{\partial y} = 4u.$$

60. Show that $u = x^2y + y^2z + z^2x$
satisfies the partial differential equation

$$\frac{\partial u}{\partial x} + \frac{\partial u}{\partial y} + \frac{\partial u}{\partial z} = (x + y + z)^2.$$

61. Given that $x = r \cos \theta$ and $y = r \sin \theta$, find

$$\frac{\partial x}{\partial r}\frac{\partial y}{\partial \theta} - \frac{\partial x}{\partial \theta}\frac{\partial y}{\partial r}.$$

62. For a gas confined in a container, the ideal gas law states that the pressure P is related to the volume V and the temperature T by an equation of the form

$$P = k\frac{T}{V}$$

where k is a positive constant. Show that

$$V\frac{\partial P}{\partial V} = -P \qquad \text{and} \qquad V\frac{\partial P}{\partial V} + T\frac{\partial P}{\partial T} = 0.$$

■ 14.5 OPEN SETS AND CLOSED SETS

Recall that the fundamental concepts of calculus for a function f of one variable (limit, continuity, derivative) were so-called point concepts, and in the definitions we required f to be defined on some open interval of the form $(x - p, x + p)$ containing x (except possibly at x itself in the case of the limit). Open intervals played a basic role in that development. In this section we extend the notion of open interval to sets in the plane and sets in space. In the next section we consider limits, continuity, and differentiability for functions of several variables.

Points in the domain of a function of several variables can be written in vector notation. In the two-variable case, set

$$\mathbf{x} = (x, y),$$

and, in the three-variable case, set

$$\mathbf{x} = (x, y, z).$$

The vector notation enables us to treat the two cases together.

In this section we introduce five important notions:

(1) Neighborhood of a point.

(2) Interior of a set.

(3) Boundary of a set.

(4) Open set.

(5) Closed set.

For our purposes, the fundamental notion here is "neighborhood of a point." The other four notions can be derived from it.

DEFINITION 14.5.1. NEIGHBORHOOD OF A POINT

A *neighborhood* of a point $\mathbf{x}_0$ is a set of the form

$$\{\mathbf{x}: \|\mathbf{x} - \mathbf{x}_0\| < \delta\}$$

where δ is some number greater than zero.

In the plane, a neighborhood of $\mathbf{x}_0 = (x_0, y_0)$ consists of all the points inside a disc centered at (x_0, y_0). In three-space, a neighborhood of $\mathbf{x}_0 = (x_0, y_0, z_0)$ consists of all the points inside a ball (sphere) centered at (x_0, y_0, z_0). See Figure 14.5.1.

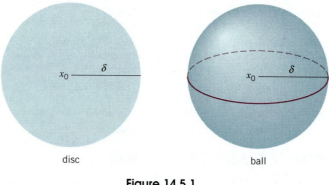

disc   ball

Figure 14.5.1

Remark On the real line, the norm of a number x is simply its absolute value, $\|x\| = |x|$. Thus a neighborhood of a point x_0 on the real line is a set of the form $\{x: |x - x_0| < \delta\}$ for some number $\delta > 0$, which is the same as the open interval $(x_0 - \delta, x_0 + \delta)$. ❏

DEFINITION 14.5.2 THE INTERIOR OF A SET

A point $\mathbf{x}_0$ is said to be an *interior point* of the set S iff the set S contains some neighborhood of $\mathbf{x}_0$. The set of all interior points of S is called the *interior* of S.

Example 1 Let Ω be the plane set shown in Figure 14.5.2. The point marked x_1 is an interior point of Ω because Ω contains a neighborhood of x_1. The point x_2 is not an interior point of Ω because *no* neighborhood of x_2 is completely contained in Ω. (Every neighborhood of x_2 has points that lie outside of Ω.) ❏

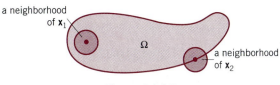

a neighborhood of x_1

Ω

a neighborhood of x_2

Figure 14.5.2

DEFINITION 14.5.3 THE BOUNDARY OF A SET

A point x_0 is said to be a *boundary point* of the set S iff every neighborhood of x_0 contains points that are in S and points that are not in S. The set of all boundary points of S is called the *boundary* of S.

Example 2 The point marked x_2 in Figure 15.4.2 is a boundary point of Ω: each neighborhood of x_2 contains points in Ω and points not in Ω. ❏

DEFINITION 14.5.4 OPEN SET

A set S is said to be *open* iff it contains a neighborhood of each of its points.

Statements which are equivalent to this definition of open set are:

(a) A set S is open iff each of its points is an interior point.

(b) A set S is open iff it contains no boundary points.

DEFINITION 14.5.5 CLOSED SET

A set S is said to be *closed* iff it contains its boundary.

Here are some examples of sets that are open, sets that are closed, and sets that are neither open nor closed:

Two-Dimensional Examples

The sets

$$S_1 = \{(x, y): 1 < x < 2, 1 < y < 2\},$$
$$S_2 = \{(x, y): 3 \le x \le 4, 1 \le y \le 2\},$$
$$S_3 = \{(x, y): 5 \le x \le 6, 1 < y < 2\}$$

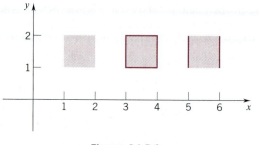

Figure 14.5.3

are displayed in Figure 14.5.3. S_1 is the inside of the first square. S_1 is open because it contains a neighborhood of each of its points. S_2 is the inside of the second square together with the four bounding line segments. S_2 is closed because it contains its entire boundary. S_3 is the inside of the last square together with the two vertical bounding line segments. S_3 is not open because it contains part of its boundary, and it is not closed because it does not contain all of its boundary. ❑

Three-Dimensional Examples

We now examine some three-dimensional sets:

$$S_1 = \{(x, y, z): z > x^2 + y^2\},$$
$$S_2 = \{(x, y, z): z \geq x^2 + y^2\},$$
$$S_3 = \left\{(x, y, z): 1 \geq \frac{x^2 + y^2}{z}\right\}.$$

The boundary of each of these sets is the paraboloid of revolution

$$z = x^2 + y^2. \qquad \text{(Figure 14.5.4)}$$

The first set consists of all points above this surface. This set is open because, if a point is above this surface, then all points sufficiently close to it are also above this surface. Thus the set contains a neighborhood of each of its points. The second set is closed because it contains all of its boundary. The third set is neither open nor closed. It is not open because it contains some boundary points; for example, it contains the point $(1, 1, 2)$. It is not closed because it fails to contain the boundary point $(0, 0, 0)$. ❑

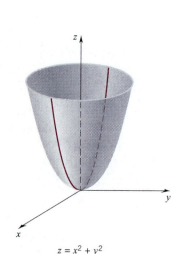

$z = x^2 + y^2$

Figure 14.5.4

A Final Remark A neighborhood of $\mathbf{x}_0$ is a set of the form

$$\{\mathbf{x}: \|\mathbf{x} - \mathbf{x}_0\| < \delta\}.$$

If we remove $\mathbf{x}_0$ from the set, we have the set

$$\{\mathbf{x}: 0 < \|\mathbf{x} - \mathbf{x}_0\| < \delta\}.$$

Such a set is called a *deleted neighborhood* of $\mathbf{x}_0$. (We use deleted neighborhoods in the next section.) ❑

EXERCISES 14.5

In Exercises 1–10, specify the interior and the boundary of the set. State whether the set is open, closed, or neither. Then sketch the set.

1. $\{(x, y): 2 \leq x \leq 4, 1 \leq y \leq 3\}$.

2. $\{(x, y): 2 < x < 4, 1 < y < 3\}$.

3. $\{(x, y): 1 < x^2 + y^2 < 4\}$.

4. $\{(x, y): 1 \leq x^2 \leq 4\}$.

5. $\{(x, y): 1 < x^2 \leq 4\}$.   6. $\{(x, y): y < x^2\}$.

7. $\{(x, y): y \leq x^2\}$.

8. $\{(x, y, z): 1 \leq x \leq 2, 1 \leq y \leq 2, 1 \leq z < 2\}$.

9. $\{(x, y, z): x^2 + y^2 \leq 1, 0 \leq z \leq 4\}$.

10. $\{(x, y, z): (x - 1)^2 + (y - 1)^2 + (z - 1)^2 < \frac{1}{4}\}$.

11. Let $S = \{\mathbf{x}_1, \mathbf{x}_2, \ldots, \mathbf{x}_n\}$ be a non-empty, finite set of points. (a) What is the interior of S? (b) What is the boundary of S? (c) Is S open, closed, or neither?

All the notions introduced in this section can be applied to sets of real numbers: write x for $\mathbf{x}$ and $|x|$ for $\|\mathbf{x}\|$. A *neighborhood* of a number x_0 is then a set of the form

$$\{x: |x - x_0| < \delta\} \qquad \text{with } \delta > 0.$$

This is just an open interval $(x_0 - \delta, x_0 + \delta)$. In Exercises 12–19, specify the interior and boundary of the given set of real numbers. State whether the set is open, closed, or neither.

12. $\{x: 1 < x < 3\}$.   13. $\{x: 1 \leq x \leq 3\}$.

14. $\{x: 1 \leq x < 3\}$.   15. $\{x: x > 1\}$.

16. $\{x: x \leq -1\}$.   17. $\{x: x < -1 \text{ or } x \geq 1\}$.

18. The set of positive integers: $\{1, 2, 3, \ldots, n, \ldots\}$.

19. The set of reciprocals: $\{1, 1/2, 1/3, \ldots, 1/n, \ldots\}$.

20. Let $\varnothing$ be the empty set. Let X be the real line, the entire plane, or, in the three-dimensional case, all of three-space. For each subset A of X, let $X - A$ be the set of all points $\mathbf{x} \in X$ such that $\mathbf{x} \notin A$.

    (a) Show that $\varnothing$ is both open and closed.

    (b) Show that X is both open and closed. (It can be shown that $\varnothing$ and X are the only subsets of X that are both open and closed.)

    (c) Let U be a subset of X. Show that U is open iff $X - U$ is closed.

    (d) Let F be a subset of X. Show that F is closed iff $X - F$ is open.

■ 14.6 LIMITS AND CONTINUITY; EQUALITY OF MIXED PARTIALS

The Basic Notions

The limit process used in taking partial derivatives involved nothing new because in each instance all but one of the variables remained fixed. In this section we take up limits of the form

$$\lim_{(x, y) \to (x_0, y_0)} f(x, y) \qquad \text{and} \qquad \lim_{(x, y, z) \to (x_0, y_0, z_0)} f(x, y, z).$$

To avoid having to treat the two- and three-variable cases separately, we will write instead

$$\lim_{\mathbf{x} \to \mathbf{x}_0} f(\mathbf{x}).$$

This gives us both the two-variable case [set $\mathbf{x} = (x, y)$ and $\mathbf{x}_0 = (x_0, y_0)$] and the three-variable case [set $\mathbf{x} = (x, y, z)$ and $\mathbf{x}_0 = (x_0, y_0, z_0)$].

To take the limit of $f(\mathbf{x})$ as $\mathbf{x}$ tends to $\mathbf{x}_0$ we do not need f to be defined at $\mathbf{x}_0$ itself, but we do need f to be defined at points $\mathbf{x}$ close to $\mathbf{x}_0$. At this stage, we will assume that f is defined at all points $\mathbf{x}$ in some deleted neighborhood of $\mathbf{x}_0$ (f may or may not be defined at $\mathbf{x}_0$). This will guarantee that we can form $f(\mathbf{x})$ for all $\mathbf{x} \neq \mathbf{x}_0$ that are "sufficiently close" to $\mathbf{x}_0$. This approach is consistent with our approach to limits of functions of one variable in Chapter 2.

To say that

$$\lim_{\mathbf{x} \to \mathbf{x}_0} f(\mathbf{x}) = L$$

is to say that for $\mathbf{x}$ sufficiently close to $\mathbf{x}_0$ but different from $\mathbf{x}_0$, the number $f(\mathbf{x})$ is close to L; or, to put it another way, as $\|\mathbf{x} - \mathbf{x}_0\|$ tends to zero but remains different from zero, $|f(\mathbf{x}) - L|$ tends to zero. The ϵ-δ definition is a direct generalization of the ϵ-δ definition in the single-variable case.

DEFINITION 14.6.1   THE LIMIT OF A FUNCTION OF SEVERAL VARIABLES

Let f be a function defined at least on some deleted neighborhood of $\mathbf{x}_0$.

$$\lim_{\mathbf{x} \to \mathbf{x}_0} f(\mathbf{x}) = L \qquad \text{iff}$$

for each $\epsilon > 0$ there exists $\delta > 0$ such that

$$\text{if} \quad 0 < \|\mathbf{x} - \mathbf{x}_0\| < \delta \qquad \text{then} \qquad |f(\mathbf{x}) - L| < \epsilon$$

Example 1   We will show that the function

$$f(x, y) = \frac{xy + y^3}{x^2 + y^2}$$

does not have a limit at $(0, 0)$. Note that f is not defined at $(0, 0)$, but is defined for all $(x, y) \neq (0, 0)$.

Along the obvious paths to $(0, 0)$, the coordinate axes, the limiting value is 0:

Along the x-axis, $y = 0$; thus, $f(x, y) = f(x, 0) = 0$ and $\lim_{x \to 0} f(x, 0) = \lim_{x \to 0} 0 = 0$.

Along the y-axis, $x = 0$; thus, $f(x, y) = f(0, y) = y$ and $\lim_{y \to 0} f(0, y) = \lim_{y \to 0} y = 0$.

Along the line $y = 2x$, however, the limiting value is $\frac{2}{5}$:

$$f(x, y) = f(x, 2x) = \frac{2x^2 + 8x^3}{x^2 + 4x^2} = \frac{2}{5} + \frac{8}{5}x \to \frac{2}{5} \qquad \text{as } x \to 0.$$

There is nothing special about the line $y = 2x$ here. For example, you can verify that $f(x, y) \to -\frac{1}{2}$ as $(x, y) \to (0, 0)$ along the line $y = -x$. See Figure 14.6.1.

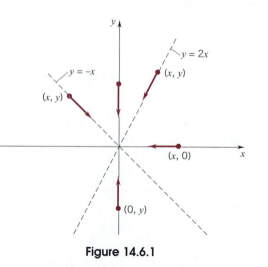

Figure 14.6.1

We have shown that not all paths to (0, 0) yield the same limiting value. It follows that f does not have a limit at (0, 0). ❏

Example 2 Show that the function

$$g(x, y) = \frac{x^2 y}{x^4 + y^2}$$

has limiting value 0 as $(x, y) \to (0, 0)$ along *any* line through the origin, but

$$\lim_{(x,y)\to(0,0)} g(x, y)$$

still does not exist. Note that the domain of g is all $(x, y) \neq (0, 0)$.

SOLUTION As in Example 1, it is easy to verify that $g(x, y) \to 0$ when $(x, y) \to (0, 0)$ along the coordinate axes. If we let $y = mx$, then

$$g(x, y) = g(x, mx) = \frac{mx^3}{x^4 + m^2 x^2} = \frac{mx}{x^2 + m^2} \qquad (x \neq 0)$$

and

$$\lim_{x\to0} g(x, mx) = \lim_{x\to0} \frac{mx}{x^2 + m^2} = 0.$$

Therefore, $g(x, y) \to 0$ as $(x, y) \to (0, 0)$ along any line through the origin.
Now suppose that $(x, y) \to (0, 0)$ along the parabola $y = x^2$. Then we have

$$g(x, y) = g(x, x^2) = \frac{x^4}{x^4 + x^4} = \frac{1}{2}$$

and

$$\lim_{x\to0} g(x, x^2) = \lim_{x\to0} \frac{1}{2} = \frac{1}{2}.$$

Thus, $g(x, y) \to \frac{1}{2}$ as $(x, y) \to (0, 0)$ along $y = x^2$ (Figure 14.6.2).

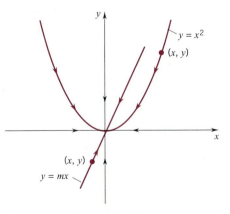

Figure 14.6.2

Since not all paths to (0, 0) yield the same limiting value, we conclude that g does not have a limit at (0, 0). ❏

As in the one-variable case, the limit (if it exists) is unique. Moreover, if

$$\lim_{\mathbf{x} \to \mathbf{x}_0} f(\mathbf{x}) = L \qquad \text{and} \qquad \lim_{\mathbf{x} \to \mathbf{x}_0} g(\mathbf{x}) = M,$$

then

$$\lim_{\mathbf{x} \to \mathbf{x}_0} [f(\mathbf{x}) + g(\mathbf{x})] = L + M, \qquad \lim_{\mathbf{x} \to \mathbf{x}_0} [\alpha f(\mathbf{x})] = \alpha L, \quad \alpha \text{ a real number,}$$

$$\lim_{\mathbf{x} \to \mathbf{x}_0} [f(\mathbf{x})g(\mathbf{x})] = LM, \qquad \text{and} \qquad \lim_{\mathbf{x} \to \mathbf{x}_0} [f(\mathbf{x})/g(\mathbf{x})] = L/M \qquad \text{provided } M \neq 0.$$

These results are not hard to derive. You can do it simply by imitating the comparable arguments in the one-variable case.

Suppose now that $\mathbf{x}_0$ is an interior point of the domain of f. To say that f is *continuous* at $\mathbf{x}_0$ is to say that

(14.6.2)

$$\lim_{\mathbf{x} \to \mathbf{x}_0} f(\mathbf{x}) = f(\mathbf{x}_0)$$

or, equivalently, that

(14.6.3)

$$\lim_{\mathbf{h} \to \mathbf{0}} f(\mathbf{x}_0 + \mathbf{h}) = f(\mathbf{x}_0).$$

For two variables we can write

$$\lim_{(x,y) \to (x_0, y_0)} f(x, y) = f(x_0, y_0)$$

and for three variables

$$\lim_{(x,y,z) \to (x_0, y_0, z_0)} f(x, y, z) = f(x_0, y_0, z_0).$$

To say that f is *continuous on an open set S* is to say that f is continuous at all points of S.

Some Examples of Continuous Functions

Polynomials in several variables, for example,

$$P(x, y) = x^2 y + 3x^3 y^4 - x + 2y \qquad \text{and} \qquad Q(x, y, z) = 6x^3 z - z^3 y + 2xyz,$$

are everywhere continuous. In the two-variable case, that means continuity at each point of the xy-plane; in the three-variable case, continuity at each point of three-space.

Rational functions (quotients of polynomials) are continuous everywhere except where the denominator is zero. Thus

$$f(x, y) = \frac{2x - y}{x^2 + y^2}$$

is continuous at each point of the xy-plane other than the origin $(0, 0)$;

$$g(x, y) = \frac{x^4}{x - y}$$

is continuous except on the line $y = x$;

$$h(x, y) = \frac{1}{x^2 - y}$$

is continuous except on the parabola $y = x^2$;

$$F(x, y, z) = \frac{2x}{x^2 + y^2 + z^2}$$

is continuous at each point of three-space other than the origin $(0, 0, 0)$;

$$G(x, y, z) = \frac{x^5 - y}{ax + by + cz},$$

where a, b, c are constants, is continuous except on the plane $ax + by + cz = 0$.

You can construct more elaborate continuous functions by forming composites: take, for example,

$$f(x, y, z) = \tan^{-1}\left(\frac{xz^2}{x + y}\right), \qquad g(x, y, z) = \sqrt{x^2 + y^4 + z^6}, \qquad h(x, y, z) = \sin xyz.$$

The first function is continuous except along the vertical plane $x + y = 0$. The other two functions are continuous at each point of space. The continuity of such composites follows from a simple theorem that we now state and prove. In the theorem, g is a function of several variables, but f is a function of a single variable.

THEOREM 14.6.4   THE CONTINUITY OF COMPOSITE FUNCTIONS

If g is continuous at the point $\mathbf{x}_0$ and f is continuous at the number $g(\mathbf{x}_0)$, then the composition $f \circ g$ is continuous at the point $\mathbf{x}_0$.

PROOF   We begin with $\epsilon > 0$. We must show that there exists $\delta > 0$ such that

$$\text{if} \quad \|\mathbf{x} - \mathbf{x}_0\| < \delta \quad \text{then} \quad |f(g(\mathbf{x})) - f(g(\mathbf{x}_0))| < \epsilon.$$

From the continuity of f at $g(\mathbf{x}_0)$, we know that there exists $\delta_1 > 0$ such that

$$\text{if} \quad |u - g(\mathbf{x}_0)| < \delta_1 \quad \text{then} \quad |f(u) - f(g(\mathbf{x}_0))| < \epsilon.$$

From the continuity of g at $\mathbf{x}_0$, we know that there exists $\delta > 0$ such that

$$\text{if} \quad \|\mathbf{x} - \mathbf{x}_0\| < \delta \quad \text{then} \quad |g(\mathbf{x}) - g(\mathbf{x}_0)| < \delta_1.$$

This last δ obviously works; namely,

$$\text{if} \quad \|\mathbf{x} - \mathbf{x}_0\| < \delta \quad \text{then} \quad |g(\mathbf{x}) - g(\mathbf{x}_0)| < \delta_1$$

$$\text{and} \quad |f(g(\mathbf{x})) - f(g(\mathbf{x}_0))| < \epsilon \quad \square$$

Continuity in Each Variable Separately

A *continuous function of several variables is continuous in each of its variables separately.* In the two-variable case, this means that, if

$$\lim_{(x,y)\to(x_0,y_0)} f(x,y) = f(x_0, y_0),$$

then

$$\lim_{x\to x_0} f(x, y_0) = f(x_0, y_0) \quad \text{and} \quad \lim_{y\to y_0} f(x_0, y) = f(x_0, y_0).$$

(This is not hard to prove.) The converse is false. *It is possible for a function to be continuous in each variable separately and yet fail to be continuous as a function of several variables.* You can see this in the next example.

Example 3 We set

$$f(x,y) = \begin{cases} \dfrac{2xy}{x^2 + y^2}, & (x,y) \neq (0,0) \\ 0, & (x,y) = (0,0). \end{cases}$$

Since

$$f(x,0) = 0 \quad \text{for all } x \quad \text{and} \quad f(0,y) = 0 \quad \text{for all } y,$$

we have

$$\lim_{x\to 0} f(x,0) = 0 = f(0,0) \quad \text{and} \quad \lim_{y\to 0} f(0,y) = 0 = f(0,0).$$

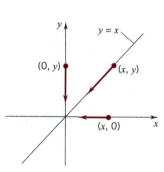

Figure 14.6.3

Thus, at the point $(0,0)$, f is continuous in x and continuous in y. As a function of two variables, however, f is not continuous at $(0,0)$. One way to see this is to note that we can approach $(0,0)$ as closely as we wish by points of the form (t,t) with $t \neq 0$, (that is, along the line $y = x$). (See Figure 14.6.3.) At such points f takes on the value 1:

$$f(t,t) = \frac{2t^2}{t^2 + t^2} = 1.$$

Hence, f cannot tend to $f(0,0) = 0$ as required. ❏

Continuity and Partial Differentiability

For functions of a single variable the existence of the derivative guarantees continuity (Theorem 3.1.4). *For functions of several variables the existence of partial derivatives fails to guarantee continuity.*†

To show this, we can use the same function

$$f(x,y) = \begin{cases} \dfrac{2xy}{x^2 + y^2}, & (x,y) \neq (0,0) \\ 0, & (x,y) = (0,0). \end{cases}$$

Since both $f(x,0)$ and $f(0,y)$ are constantly zero, both partials exist (and are zero) at $(0,0)$, and yet, as you saw, the function is discontinuous at $(0,0)$.

---

† See, however, Exercise 34.

It is not hard to understand how a function can have partial derivatives and yet fail to be continuous. The existence of $\partial f/\partial x$ at (x_0, y_0) depends on the behavior of f only at points of the form $(x_0 + h, y_0)$. Similarly, the existence of $\partial f/\partial y$ at (x_0, y_0) depends on the behavior of f only at points of the form $(x_0, y_0 + k)$. On the other hand, continuity at (x_0, y_0) depends on the behavior of f at points of the more general form $(x_0 + h, y_0 + k)$. More briefly, we can put it this way: *the existence of a partial derivative depends on the behavior of the function along a line segment (two directions), whereas continuity depends on the behavior of the function in all directions.*

Derivatives of Higher Order; Equality of Mixed Partials

Suppose that f is a function of x and y with first partials f_x and f_y. These are again functions of x and y and may themselves possess partial derivatives such as $(f_x)_x$, $(f_x)_y$, $(f_y)_x$, and $(f_y)_y$. These functions are called the *second-order partials*. If $z = f(x, y)$, we use the following notations for second-order partials

$$(f_x)_x = f_{xx} = \frac{\partial}{\partial x}\left(\frac{\partial f}{\partial x}\right) = \frac{\partial^2 f}{\partial x^2} = \frac{\partial^2 z}{\partial x^2},$$

$$(f_x)_y = f_{xy} = \frac{\partial}{\partial y}\left(\frac{\partial f}{\partial x}\right) = \frac{\partial^2 f}{\partial y \partial x} = \frac{\partial^2 z}{\partial y \partial x},$$

$$(f_y)_x = f_{yx} = \frac{\partial}{\partial x}\left(\frac{\partial f}{\partial y}\right) = \frac{\partial^2 f}{\partial x \partial y} = \frac{\partial^2 z}{\partial x \partial y},$$

$$(f_y)_y = f_{yy} = \frac{\partial}{\partial y}\left(\frac{\partial f}{\partial y}\right) = \frac{\partial^2 f}{\partial y^2} = \frac{\partial^2 z}{\partial y^2}.$$

Note that there are two "mixed" partials: f_{xy} (or $\partial^2 f/\partial y \partial x$) and f_{yx} (or $\partial^2 f/\partial x \partial y$). The first of these is obtained by differentiating first with respect to x and then with respect to y. The second is obtained by differentiating first with respect to y and then with respect to x.

Example 4 The function $f(x, y) = \sin x^2 y$ has first partials

$$\frac{\partial f}{\partial x} = 2xy \cos x^2 y \qquad \text{and} \qquad \frac{\partial f}{\partial y} = x^2 \cos x^2 y.$$

The second-order partials are

$$\frac{\partial^2 f}{\partial x^2} = -4x^2 y^2 \sin x^2 y + 2y \cos x^2 y, \qquad \frac{\partial^2 f}{\partial y \partial x} = -2x^3 y \sin x^2 y + 2x \cos x^2 y,$$

$$\frac{\partial^2 f}{\partial x \partial y} = -2x^3 y \sin x^2 y + 2x \cos x^2 y, \qquad \frac{\partial^2 f}{\partial y^2} = -x^4 \sin x^2 y. \quad ❏$$

Example 5 Setting $f(x, y) = \ln(x^2 + y^3)$, we have

$$\frac{\partial f}{\partial x} = \frac{2x}{x^2 + y^3} \qquad \text{and} \qquad \frac{\partial f}{\partial y} = \frac{3y^2}{x^2 + y^3}.$$

The second-order partials are

$$\frac{\partial^2 f}{\partial x^2} = \frac{(x^2 + y^3)2 - 2x(2x)}{(x^2 + y^3)^2} = \frac{2(y^3 - x^2)}{(x^2 + y^3)^2},$$

$$\frac{\partial^2 f}{\partial y \partial x} = \frac{-2x(3y^2)}{(x^2 + y^3)^2} = -\frac{6xy^2}{(x^2 + y^3)^2},$$

$$\frac{\partial^2 f}{\partial x \partial y} = \frac{-3y^2(2x)}{(x^2 + y^3)^2} = -\frac{6xy^2}{(x^2 + y^3)^2},$$

$$\frac{\partial^2 f}{\partial y^2} = \frac{(x^2 + y^3)6y - 3y^2(3y^2)}{(x^2 + y^3)^2} = \frac{3y(2x^2 - y^3)}{(x^2 + y^3)^2}. \quad ❏$$

Perhaps you noticed that in both examples we had

$$\frac{\partial^2 f}{\partial y \partial x} = \frac{\partial^2 f}{\partial x \partial y}.$$

Since in neither case was f symmetric in x and y, this equality of the mixed partials was not due to symmetry. Actually it was due to continuity. It can be proved that

(14.6.5)

$$\frac{\partial^2 f}{\partial y \partial x} = \frac{\partial^2 f}{\partial x \partial y}$$

on every open set U on which f and its partials

$$\frac{\partial f}{\partial x}, \quad \frac{\partial f}{\partial y}, \quad \frac{\partial^2 f}{\partial y \partial x}, \quad \frac{\partial^2 f}{\partial x \partial y}$$

are all continuous.†

In the case of a function of three variables you can look for three first partials

$$\frac{\partial f}{\partial x}, \quad \frac{\partial f}{\partial y}, \quad \frac{\partial f}{\partial z},$$

and nine second partials

$$\frac{\partial^2 f}{\partial x^2}, \quad \frac{\partial^2 f}{\partial x \partial y}, \quad \frac{\partial^2 f}{\partial x \partial z} \quad \frac{\partial^2 f}{\partial y \partial x}, \quad \frac{\partial^2 f}{\partial y^2}, \quad \frac{\partial^2 f}{\partial y \partial z} \quad \frac{\partial^2 f}{\partial z \partial x}, \quad \frac{\partial^2 f}{\partial z \partial y}, \quad \frac{\partial^2 f}{\partial z^2}.$$

Here again, there is equality of the mixed partials

$$\frac{\partial^2 f}{\partial y \partial x} = \frac{\partial^2 f}{\partial x \partial y}, \quad \frac{\partial^2 f}{\partial z \partial x} = \frac{\partial^2 f}{\partial x \partial z}, \quad \frac{\partial^2 f}{\partial y \partial z} = \frac{\partial^2 f}{\partial z \partial y}$$

provided that f and its first and second partials are continuous.

---

† For a proof consult a text on advanced calculus.

Example 6 For

$$f(x, y, z) = xe^y \sin \pi z$$

we have

$$\frac{\partial f}{\partial x} = e^y \sin \pi z, \qquad \frac{\partial f}{\partial y} = xe^y \sin \pi z, \qquad \frac{\partial f}{\partial z} = \pi xe^y \cos \pi z,$$

$$\frac{\partial^2 f}{\partial x^2} = 0, \qquad \frac{\partial^2 f}{\partial y^2} = xe^y \sin \pi z, \qquad \frac{\partial^2 f}{\partial z^2} = -\pi^2 xe^y \sin \pi z,$$

$$\frac{\partial^2 f}{\partial y \partial x} = e^y \sin \pi z = \frac{\partial^2 f}{\partial x \partial y},$$

$$\frac{\partial^2 f}{\partial z \partial x} = \pi e^y \cos \pi z = \frac{\partial^2 f}{\partial x \partial z},$$

$$\frac{\partial^2 f}{\partial y \partial z} = \pi xe^y \cos \pi z = \frac{\partial^2 f}{\partial z \partial y}. \qquad ❏$$

EXERCISES 14.6

In Exercises 1–20, calculate the second-order partial derivatives of the given function. (NOTE: Treat A, B, C, and D as constants.)

1. $f(x, y) = Ax^2 + 2Bxy + Cy^2$.

2. $f(x, y) = Ax^3 + Bx^2y + Cxy^2$.

3. $f(x, y) = Ax + By + Ce^{xy}$

4. $f(x, y) = x^2\cos y + y^2\sin x$.

5. $f(x, y, z) = (x + y^2 + z^3)^2$.

6. $f(x, y) = \sqrt{x + y^2}$.

7. $f(x, y) = \ln\left(\dfrac{x}{x + y}\right)$.

8. $f(x, y) = \dfrac{Ax + By}{Cx + Dy}$.

9. $f(x, y, z) = (x + y)(y + z)(z + x)$.

10. $f(x, y, z) = \tan^{-1}xyz$. **11.** $f(x, y) = x^y$.

12. $f(x, y, z) = \sin(x + z^y)$. **13.** $f(x, y) = xe^y + ye^x$.

14. $f(x, y) = \tan^{-1}(y/x)$. **15.** $f(x, y) = \ln\sqrt{x^2 + y^2}$.

16. $f(x, y) = \sin(x^3y^2)$. **17.** $f(x, y) = \cos^2(xy)$.

18. $f(x, y) = e^{xy^2}$.

19. $f(x, y, z) = xy \sin z - xz \sin y$.

20. $f(x, y, z) = xe^y + ye^z + ze^x$.

21. Show that

$$\text{if} \quad u = \frac{xy}{x + y}$$

$$\text{then} \quad x^2\frac{\partial^2 u}{\partial x^2} + 2xy\frac{\partial^2 u}{\partial x \partial y} + y^2\frac{\partial^2 u}{\partial y^2} = 0.$$

Laplace's Equation†   The partial differential equation

$$\frac{\partial^2 f}{\partial x^2} + \frac{\partial^2 f}{\partial y^2} = 0$$

is known as *Laplace's equation in two dimensions*. It is used to describe potentials and steady-state temperature distributions in the plane. In three dimensions, Laplace's equation is

$$\frac{\partial^2 f}{\partial x^2} + \frac{\partial^2 f}{\partial y^2} + \frac{\partial^2 f}{\partial z^2} = 0$$

and it is satisfied by gravitational and electrostatic potentials, and by steady-state temperature distributions in space. Functions that satisfy Laplace's equation are called *harmonic* functions.

---

† Named after the French mathematician Pierre-Simon Laplace (1749–1827). Laplace wrote two monumental works: one on celestial mechanics, the other on probability theory. He also made major contributions to the theory of differential equations.

In Exercises 22–25, show that the given function is harmonic.

22. $f(x, y) = x^3 - 3xy^2$.　　　　**23.** $f(x, y) = e^x \sin y$.

24. $f(x, y) = \cos x \sinh y + \sin x \cosh y$.

25. $f(x, y) = \ln \sqrt{x^2 + y^2}$.

In Exercises 26 and 27, show that the given function satisfies the three-dimensional Laplace equation.

26. $f(x, y, z) = \dfrac{1}{\sqrt{x^2 + y^2 + z^2}}$.

27. $f(x, y, z) = e^{x+y} \cos \sqrt{2} z$.

The Wave Equation　The partial differential equation

$$\frac{\partial^2 f}{\partial t^2} - c^2 \frac{\partial^2 f}{\partial x^2} = 0,$$

where c is a positive constant, is known as the *wave equation*. It arises in the study of phenomena involving the propagation of waves in a continuous medium. For example, studies of water waves, sound waves, and light waves are all based on this equation. The wave equation is also used in the study of mechanical vibrations such as a vibrating string.

In Exercises 28–32, show that the given function is a solution of the wave equation.

28. $f(x, t) = (Ax + B)(Ct + D)$, where A, B, C, and D are constants.

29. $f(x, t) = \sin(x + ct) \cos(2x + 2ct)$.

30. $f(x, t) = \ln(x + ct)$.

31. $f(x, t) = (Ae^{kx} + Be^{-kx})(Ce^{ckt} + De^{-ckt})$, where c and k are constants.

32. $f(x, t) = g(x + ct) + h(x - ct)$, where g and h are any two, twice differentiable functions. (This is the most general form of a solution of the wave equation.)

33. Verify that

$$\frac{\partial^2 f}{\partial y \partial x} = \frac{\partial^2 f}{\partial x \partial y}$$

given that
(a) $f(x, y) = g(x) + h(y)$ with g and h differentiable.
(b) $f(x, y) = g(x)h(y)$ with g and h differentiable.
(c) $f(x, y)$ is a polynomial in x and y.
　　HINT:　Check each term $x^m y^n$ separately.

34. If a function of several variables has all first partials at a point, then it is continuous in each variable separately at that point. Show, for example, that
if $\dfrac{\partial f}{\partial x}$ exists at (x_0, y_0), then f is continuous in x at (x_0, y_0).

35. Let f be a function of x and y with everywhere continuous second partials. Is it possible that

(a) $\dfrac{\partial f}{\partial x} = x + y$　and　$\dfrac{\partial f}{\partial y} = y - x$?

(b) $\dfrac{\partial f}{\partial x} = xy$　and　$\dfrac{\partial f}{\partial y} = xy$?

36. Let g be a twice-differentiable function of one variable and set

$$h(x, y) = g(x + y) + g(x - y).$$

Show that

$$\frac{\partial^2 h}{\partial x^2} = \frac{\partial^2 h}{\partial y^2}.$$

HINT: The chain rule (Exercise 56, Section 14.4).

37. Let f be a function of x and y with third-order partials

$$\frac{\partial^3 f}{\partial x^2 \partial y} = \frac{\partial}{\partial x} \left(\frac{\partial^2 f}{\partial x \partial y} \right) \quad \text{and} \quad \frac{\partial^3 f}{\partial y \partial x^2} = \frac{\partial}{\partial y} \left(\frac{\partial^2 f}{\partial x^2} \right).$$

Show that, if all the partials are continuous, then

$$\frac{\partial^3 f}{\partial x^2 \partial y} = \frac{\partial^3 f}{\partial y \partial x^2}.$$

38. Show that the following functions do not have a limit at $(0, 0)$:

(a) $f(x, y) = \dfrac{x^2 - y^2}{x^2 + y^2}$.　　(b) $f(x, y) = \dfrac{y^2}{x^2 + y^2}$.

In Exercises 39 and 40, evaluate the limit as (x, y) approaches the origin along:

(a) The x-axis.　　(b) The y-axis.
(c) The line $y = mx$.　　(d) The spiral $r = \theta$, $\theta > 0$.
(e) The differentiable curve $y = f(x)$, given that $f(0) = 0$.
(f) The arc $r = \sin 3\theta$, $\frac{1}{6}\pi < \theta < \frac{1}{3}\pi$.
(g) The path $\mathbf{r}(t) = \dfrac{1}{t}\mathbf{i} + \dfrac{\sin t}{t}\mathbf{j}$, $t > 0$.

39. $\displaystyle \lim_{(x,y)\to(0,0)} \frac{xy}{x^2 + y^2}$.　　**40.** $\displaystyle \lim_{(x,y)\to(0,0)} \frac{xy^2}{(x^2 + y^2)^{3/2}}$.

41. Set

$$g(x, y) = \begin{cases} \dfrac{x^2 y^2}{x^4 + y^4}, & (x, y) \neq (0, 0) \\ 0, & (x, y) = (0, 0). \end{cases}$$

(a) Show that $\partial g/\partial x$ and $\partial g/\partial y$ both exist at $(0, 0)$. What are their values at $(0, 0)$?
(b) Show that $\displaystyle \lim_{(x,y)\to(0,0)} g(x, y)$ does not exist.

42. Set

$$f(x, y) = \frac{x - y^4}{x^3 - y^4}.$$

Determine whether or not f has a limit at $(1, 1)$.
HINT: Let (x, y) tend to $(1, 1)$ along the lines $x = 1$ and $y = 1$.

43. Set

$$f(x, y) = \begin{cases} \dfrac{xy(y^2 - x^2)}{x^2 + y^2}, & (x, y) \neq (0, 0) \\ 0, & (x, y) = (0, 0). \end{cases}$$

It can be shown that some of the second partials are discontinuous at $(0, 0)$. Show that

$$\frac{\partial^2 f}{\partial y \partial x}(0, 0) \neq \frac{\partial^2 f}{\partial x \partial y}(0, 0).$$

44. Let f be a function of x and y which has continuous first and second partial derivatives throughout some set D in the plane. Suppose that $f_{xy}(x, y) = 0$ for all $(x, y) \in D$. What can you conclude about f?

■ CHAPTER HIGHLIGHTS

Points in the domain of a function of several variables can be written in vector notation. In the two-variable case, set $\mathbf{x} = (x, y)$, and, in the three-variable case, set $\mathbf{x} = (x, y, z)$. The vector notation enables us to treat the two cases together simply by writing $f(\mathbf{x})$.

A continuous function of several variables is continuous in each of its variables. The converse is false: it is possible for a function to be continuous in each variable separately and yet fail to be continuous as a function of several variables. (p. 936)

For functions of several variables the existence of partial derivatives fails to guarantee continuity. (p. 936)

The existence of a partial derivative depends on the behavior of the function along a line segment (two directions), whereas continuity depends on the behavior of the function in all directions. (p. 937)

equality of mixed partials (p. 938)

■ PROJECTS AND EXPLORATIONS USING TECHNOLOGY

To do these exercises you will need a graphics calculator or a computer with graphing capability. The majority of these problems are open-ended so different approaches may be used to solve them. You should be aware that different approaches can result in slight variations in the answers. Round your numerical answers to at least four decimal places. The rounding method that your calculator or computer uses also may cause variations in answers.

14.1 A problem about surface area known as Schwarz's paradox involves computing the following limit for the surface area of a cylinder

$$\lim_{n,m\to\infty} 2\pi r \frac{\sin(\pi/m)}{\pi/m}\left[h^2 + \pi^2 r^2\left(\frac{n^2}{m^2}\right)\left(\frac{1-\cos(\pi/m)}{\pi/m}\right)^2 \right]^{1/2},$$

where r and h are positive constants.

(a) Suppose that a cylinder of height h and radius r is subdivided into n small cylindrical bands and each band is divided into m equilateral triangles (see the figure). Show that the sum of the areas of these triangles is the expression just given.

(b) Try to compute the limit by holding n fixed and letting $m\to\infty$, and then letting $n\to\infty$.

(c) Now fix m and let $n\to\infty$, and then let $m\to\infty$.

(d) Set $n=m$ and let $n\to\infty$.

(e) Set $n=m^2$ and let $m\to\infty$.

14.2 One version of the Newton-Raphson iteration method for finding a zero of a function of two variables is

$$x_{n+1} = x_n - \frac{f(x_n, y_n)}{\dfrac{\partial f}{\partial x}(x_n, y_n)}$$

$$y_{n+1} = y_n - \frac{f(x_n, y_n)}{\dfrac{\partial f}{\partial y}(x_n, y_n)}.$$

(a) Let $f(x, y) = \sqrt{x^2 + y^2}$. If this version of Newton's method is used with an initial guess of $(1, 1)$, it does not converge. Does it converge for other initial guesses? Can you explain? Now try the method using the function

$$g(x, y) = e^{xy^2} - \ln(x + y).$$

(b) As a variation of the method, consider

$$x_{n+1} = x_n - \frac{f(x_n, y_n)}{\dfrac{\partial f}{\partial x}(x_n, y_n)}$$

$$y_{n+1} = y_n - \frac{f(x_{n+1}, y_n)}{\dfrac{\partial f}{\partial y}(x_{n+1}, y_n)}.$$

Apply this method to the functions f and g in part (a) using the same initial guesses. Compare the convergence of this method with the method in part (a).

(c) A modification to find the intersecton of two surfaces is given by the solution to the following system:

$$0 = f(x_n, y_n) + (x_{n+1} - x_n)\frac{\partial f}{\partial x}(x_n, y_n) + (y_{n+1} - y_n)\frac{\partial f}{\partial y}(x_n, y_n)$$

$$0 = g(x_n, y_n) + (x_{n+1} - x_n)\frac{\partial g}{\partial x}(x_n, y_n) + (y_{n+1} - y_n)\frac{\partial g}{\partial y}(x_n, y_n).$$

Apply this method to the pair of functions

$$f(x, y) = 1 - \sqrt{x^2 + 2y^4},$$

$$g(x, y) = 1 + x^6 - 5y^2,$$

with an initial guess of $(\frac{1}{2}, \frac{1}{2})$. Does this method seem to converge as quickly as the Newton-Raphson method for functions of one variable?

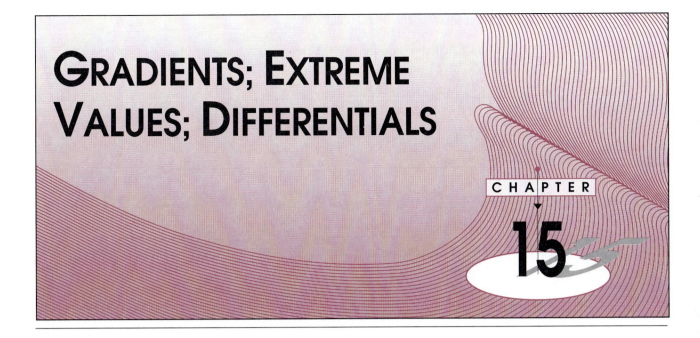

GRADIENTS; EXTREME VALUES; DIFFERENTIALS

CHAPTER

15

■ 15.1 DIFFERENTIABILITY AND GRADIENT

The Notion of Differentiability

Our object here is to extend the notion of differentiability from real-valued functions of one variable to real-valued functions of several variables. Partial derivatives alone do not fulfill this role because they reflect behavior only in particular directions (the coordinate directions).

In the one-variable case we formed the difference quotient

$$\frac{f(x + h) - f(x)}{h}$$

and called f differentiable at x provided that this quotient had a limit as h tended to zero. In the multivariable case we can still form the difference

$$f(\mathbf{x} + \mathbf{h}) - f(\mathbf{x}),$$

but the "quotient"

$$\frac{f(\mathbf{x} + \mathbf{h}) - f(\mathbf{x})}{\mathbf{h}}$$

is not defined because it makes no sense to divide a real number $[f(\mathbf{x} + \mathbf{h}) - f(\mathbf{x})]$ by a vector $\mathbf{h}$.

We can get around this difficulty by going back to an idea introduced in the exercises to Section 3.9. We review the idea here.

Let g be a real-valued function of a single variable which is defined on some open interval containing 0. We say that $g(h)$ is *little-o(h)* (read "little oh of h") and write $g(h) = o(h)$ iff

$$\lim_{h \to 0} \frac{g(h)}{|h|} = 0.$$

For a function of one variable the following statements are equivalent:

$$\lim_{h \to 0} \frac{f(x + h) - f(x)}{h} = f'(x),$$

$$\lim_{h \to 0} \frac{[f(x + h) - f(x)] - f'(x)h}{h} = 0,$$

$$\lim_{h \to 0} \frac{[f(x + h) - f(x)] - f'(x)h}{|h|} = 0,$$

$$[f(x + h) - f(x)] - f'(x)h = o(h),$$

$$[f(x + h) - f(x)] = f'(x)h + o(h).$$

Thus, for a function of one variable, the derivative of f at x is the unique number $f'(x)$ such that

$$f(x + h) - f(x) = f'(x)h + o(h).$$

It is this view of the derivative that inspires the notion of differentiability in the multivariable case.

Paralleling the definition for a function of a single variable, let g be a function of several variables which is defined in some neighborhood of **0**. We will say that $g(\mathbf{h})$ is $o(\mathbf{h})$ iff

$$\lim_{\mathbf{h} \to 0} \frac{g(\mathbf{h})}{\|\mathbf{h}\|} = 0.$$

We will denote by $o(\mathbf{h})$ any function $g(\mathbf{h})$ that has this property.

Now let f be a function of several variables *defined at least in some neighborhood of* **x**. [In the three-variable case, $\mathbf{x} = (x, y, z)$; in the two-variable case, $\mathbf{x} = (x, y)$.]

DEFINITION 15.1.1 DIFFERENTIABILITY

We say that f is *differentiable* at **x** iff there exists a vector **y** such that

$$f(\mathbf{x} + \mathbf{h}) - f(\mathbf{x}) = \mathbf{y} \cdot \mathbf{h} + o(\mathbf{h}).$$

It is not hard to show that, if such a vector **y** exists, it is unique (Exercise 43). We call this unique vector *the gradient of f at* **x** and denote it by $\nabla f(\mathbf{x})$:†

DEFINITION 15.1.2 GRADIENT

Let f be differentiable at **x**. The *gradient of f at* **x** is the unique vector $\nabla f(\mathbf{x})$ such that

$$f(\mathbf{x} + \mathbf{h}) - f(\mathbf{x}) = \nabla f(\mathbf{x}) \cdot \mathbf{h} + o(\mathbf{h}).$$

---

† The notation grad f is sometimes used to denote the gradient of f.

The similarities between the one-variable case,

$$f(x + h) - f(x) = f'(x)h + o(h),$$

and the multivariable case,

$$f(\mathbf{x} + \mathbf{h}) - f(\mathbf{x}) = \nabla f(\mathbf{x}) \cdot \mathbf{h} + o(\mathbf{h}),$$

are obvious. We point to the differences. There are essentially two of them:

(1) While the derivative $f'(x)$ is a number, the gradient $\nabla f(\mathbf{x})$ is a vector.

(2) While $f'(x)h$ is the ordinary product of two real numbers, $\nabla f(\mathbf{x}) \cdot \mathbf{h}$ is the dot product of two vectors.

Calculating Gradients

First we calculate some gradients by applying the definition directly. Then we give a theorem that makes such calculations much easier. Finally, we calculate some gradients with the aid of the theorem. As for notation, in the two-variable case we write

$$\nabla f(\mathbf{x}) = \nabla f(x, y) \qquad \text{and} \qquad \mathbf{h} = (h_1, h_2),$$

and in the three-variable case,

$$\nabla f(\mathbf{x}) = \nabla f(x, y, z) \qquad \text{and} \qquad \mathbf{h} = (h_1, h_2, h_3).$$

Example 1 For the function

$$f(x, y) = x^2 + y^2$$

we have

$$\begin{aligned}
f(\mathbf{x} + \mathbf{h}) - f(\mathbf{x}) &= f(x + h_1, y + h_2) - f(x, y) \\
&= [(x + h_1)^2 + (y + h_2)^2] - [x^2 + y^2] \\
&= [2xh_1 + 2yh_2] + [h_1^2 + h_2^2] \\
&= [2x\mathbf{i} + 2y\mathbf{j}] \cdot \mathbf{h} + \|\mathbf{h}\|^2.
\end{aligned}$$

The remainder $\|\mathbf{h}\|^2$ is $o(\mathbf{h})$:

$$\frac{\|\mathbf{h}\|^2}{\|\mathbf{h}\|} = \|\mathbf{h}\| \to 0 \quad \text{as} \quad \mathbf{h} \to \mathbf{0}.$$

Thus

$$\nabla f(\mathbf{x}) = \nabla f(x, y) = 2x\mathbf{i} + 2y\mathbf{j}. \quad \square$$

Example 2 For the function

$$f(x, y, z) = 2xy - 3z^2$$

we have

$$\begin{aligned}
f(\mathbf{x} + \mathbf{h}) - f(\mathbf{x}) &= f(x + h_1, y + h_2, z + h_3) - f(x, y, z) \\
&= 2(x + h_1)(y + h_2) - 3(z + h_3)^2 - [2xy - 3z^2] \\
&= 2xh_2 + 2yh_1 + 2h_1h_2 - 6zh_3 - 3h_3^2 \\
&= (2y\mathbf{i} + 2x\mathbf{j} - 6z\mathbf{k}) \cdot (h_1\mathbf{i} + h_2\mathbf{j} + h_3\mathbf{k}) + 2h_1h_2 - 3h_3^2 \\
&= (2y\mathbf{i} + 2x\mathbf{j} - 6z\mathbf{k}) \cdot \mathbf{h} + 2h_1h_2 - 3h_3^2.
\end{aligned}$$

It remains to show that the remainder $g(\mathbf{h}) = 2h_1 h_2 - 3h_3^2$ is $o(\mathbf{h})$. We can rewrite g as

$$g(\mathbf{h}) = (2h_2 \mathbf{i} - 3h_3 \mathbf{k}) \cdot (h_1 \mathbf{i} + h_2 \mathbf{j} + h_3 \mathbf{k}) = (2h_2 \mathbf{i} - 3h_3 \mathbf{k}) \cdot \mathbf{h}.$$

Then

$$\frac{|g(\mathbf{h})|}{\|\mathbf{h}\|} = \frac{\|2h_2 \mathbf{i} - 3h_3 \mathbf{k}\| \cdot \|\mathbf{h}\| \, |\cos \theta|}{\|\mathbf{h}\|} \leq \frac{\|2h_2 \mathbf{i} - 3h_3 \mathbf{k}\| \cdot \|\mathbf{h}\|}{\|\mathbf{h}\|} = \|2h_2 \mathbf{i} - 3h_3 \mathbf{k}\|.$$

Now, since $\mathbf{h} \to \mathbf{0}$ iff $h_1 \to 0$, $h_2 \to 0$, and $h_3 \to 0$, it follows that $\|2h_2 \mathbf{i} - 3h_3 \mathbf{k}\| \to 0$ as $\mathbf{h} \to \mathbf{0}$. Therefore, $g(\mathbf{h})/\|\mathbf{h}\| \to 0$ as $\mathbf{h} \to \mathbf{0}$, and

$$\nabla f(\mathbf{x}) = 2y\mathbf{i} + 2x\mathbf{j} - 6z\mathbf{k}. \quad \square$$

Examples 1 and 2 illustrate the use of the definition to calculate gradients. The following theorem provides an alternative method which you will find to be much easier to apply.

THEOREM 15.1.3

If f has continuous first partials in a neighborhood of $\mathbf{x}$, then f is differentiable at $\mathbf{x}$ and

$$\nabla f(\mathbf{x}) = \frac{\partial f}{\partial x}(\mathbf{x})\mathbf{i} + \frac{\partial f}{\partial y}(\mathbf{x})\mathbf{j} \qquad (f \text{ a function of two variables})$$

or

$$\nabla f(\mathbf{x}) = \frac{\partial f}{\partial x}(\mathbf{x})\mathbf{i} + \frac{\partial f}{\partial y}(\mathbf{x})\mathbf{j} + \frac{\partial f}{\partial z}(\mathbf{x})\mathbf{k}. \qquad (f \text{ a function of three variables})$$

The proof is somewhat difficult. A proof of the two-variable case is given in a supplement at the end of this section. ❑

Returning to Examples 1 and 2: for $f(x, y) = x^2 + y^2$, we have $\partial f/\partial x = 2x$ and $\partial f/\partial y = 2y$, and since these functions are continuous, we have

$$\nabla f(\mathbf{x}) = 2x\mathbf{i} + 2y\mathbf{j}.$$

For $f(x, y, z) = 2xy - 3z^2$, we have $\partial f/\partial x = 2y$, $\partial f/\partial y = 2x$ and $\partial f/\partial z = -6z$, and each of these functions is continuous. Therefore,

$$\nabla f(\mathbf{x}) = 2y\mathbf{i} + 2x\mathbf{j} - 6z\mathbf{k}.$$

Example 3 For

$$f(x, y) = xe^y - ye^x$$

we have

$$\frac{\partial f}{\partial x}(x, y) = e^y - ye^x, \qquad \frac{\partial f}{\partial y}(x, y) = xe^y - e^x$$

and therefore

$$\nabla f(x, y) = (e^y - ye^x)\mathbf{i} + (xe^y - e^x)\mathbf{j}. \quad \square$$

When there is no reason to emphasize the point of evaluation, we don't write

$$\nabla f(\mathbf{x}) \quad \text{or} \quad \nabla f(x, y) \quad \text{or} \quad \nabla f(x, y, z)$$

but simply ∇f. Thus for the function

$$f(x, y) = xe^y - ye^x$$

we write

$$\frac{\partial f}{\partial x} = e^y - ye^x, \qquad \frac{\partial f}{\partial y} = xe^y - e^x$$

and

$$\nabla f = (e^y - ye^x)\mathbf{i} + (xe^y - e^x)\mathbf{j}.$$

Example 4   For

$$f(x, y, z) = \sin(xy^2z^3)$$

we have

$$\frac{\partial f}{\partial x} = y^2z^3\cos(xy^2z^3), \qquad \frac{\partial f}{\partial y} = 2xyz^3\cos(xy^2z^3), \qquad \frac{\partial f}{\partial z} = 3xy^2z^2\cos(xy^2z^3)$$

and

$$\nabla f = yz^2\cos(xy^2z^3)[yz\mathbf{i} + 2xz\mathbf{j} + 3xy\mathbf{k}]. \quad ❏$$

Example 5   Let f be the function defined by

$$f(x, y, z) = x \sin \pi y + y \cos \pi z.$$

Evaluate ∇f at $(0, 1, 2)$.

SOLUTION   Here

$$\frac{\partial f}{\partial x} = \sin \pi y, \qquad \frac{\partial f}{\partial y} = \pi x \cos \pi y + \cos \pi z, \qquad \frac{\partial f}{\partial z} = -\pi y \sin \pi z.$$

At $(0, 1, 2)$

$$\frac{\partial f}{\partial x} = 0, \qquad \frac{\partial f}{\partial y} = 1, \qquad \frac{\partial f}{\partial z} = 0 \qquad \text{and thus} \qquad \nabla f(0, 1, 2) = \mathbf{j}. \quad ❏$$

Of special interest for later work are the powers of r where, as usual,

$$r = \|\mathbf{r}\| \qquad \text{and} \qquad \mathbf{r} = x\mathbf{i} + y\mathbf{j} + z\mathbf{k}.$$

We begin by showing that, for $r \neq 0$,

(15.1.4)
$$\nabla r = \frac{\mathbf{r}}{r} \qquad \text{and} \qquad \nabla \left(\frac{1}{r}\right) = -\frac{\mathbf{r}}{r^3}.$$

PROOF

$$\nabla r = \nabla(x^2 + y^2 + z^2)^{1/2}$$

$$= \frac{\partial}{\partial x}(x^2 + y^2 + z^2)^{1/2}\mathbf{i} + \frac{\partial}{\partial y}(x^2 + y^2 + z^2)^{1/2}\mathbf{j} + \frac{\partial}{\partial z}(x^2 + y^2 + z^2)^{1/2}\mathbf{k}$$

$$= \frac{x}{(x^2 + y^2 + z^2)^{1/2}}\mathbf{i} + \frac{y}{(x^2 + y^2 + z^2)^{1/2}}\mathbf{j} + \frac{z}{(x^2 + y^2 + z^2)^{1/2}}\mathbf{k}$$

$$= \frac{1}{(x^2 + y^2 + z^2)^{1/2}}(x\mathbf{i} + y\mathbf{j} + z\mathbf{k}) = \frac{\mathbf{r}}{r}.$$

$$\nabla\left(\frac{1}{r}\right) = \nabla(x^2 + y^2 + z^2)^{-1/2}$$

$$= \frac{\partial}{\partial x}(x^2 + y^2 + z^2)^{-1/2}\mathbf{i} + \frac{\partial}{\partial y}(x^2 + y^2 + z^2)^{-1/2}\mathbf{j} + \frac{\partial}{\partial z}(x^2 + y^2 + z^2)^{-1/2}\mathbf{k}$$

$$= -\frac{x}{(x^2 + y^2 + z^2)^{3/2}}\mathbf{i} - \frac{y}{(x^2 + y^2 + z^2)^{3/2}}\mathbf{j} - \frac{z}{(x^2 + y^2 + z^2)^{3/2}}\mathbf{k}$$

$$= -\frac{1}{(x^2 + y^2 + z^2)^{3/2}}(x\mathbf{i} + y\mathbf{j} + z\mathbf{k}) = -\frac{\mathbf{r}}{r^3}. \quad ❑$$

The formulas we just derived can be generalized. As you are asked to show in the exercises, for each integer n and all $\mathbf{r} \neq \mathbf{0}$,

(15.1.5)

$$\nabla r^n = nr^{n-2}\mathbf{r}.$$

(If n is positive and even, the result also holds at $\mathbf{r} = \mathbf{0}$.)

Differentiability Implies Continuity

As in the one-variable case, differentiability implies continuity; namely,

(15.1.6)

if f is differentiable at $\mathbf{x}$, then f is continuous at $\mathbf{x}$.

To see this, write

$$f(\mathbf{x} + \mathbf{h}) - f(\mathbf{x}) = \nabla f(\mathbf{x}) \cdot \mathbf{h} + o(\mathbf{h})$$

and note that

$$|f(\mathbf{x} + \mathbf{h}) - f(\mathbf{x})| = |\nabla f(\mathbf{x}) \cdot \mathbf{h} + o(\mathbf{h})| \leq |\nabla f(\mathbf{x}) \cdot \mathbf{h}| + |o(\mathbf{h})|. \qquad \text{(triangle inequality)}$$

As $\mathbf{h} \to \mathbf{0}$,

$$|\nabla f(\mathbf{x}) \cdot \mathbf{h}| \leq \|\nabla f(\mathbf{x})\| \, \|\mathbf{h}\| \to 0 \qquad \text{and} \qquad |o(\mathbf{h})| \to 0.$$

Schwarz's Inequality ⎯⎯⎯⎯⎯↑ ↑⎯⎯ Exercise 44

It follows that

$$[f(\mathbf{x} + \mathbf{h}) - f(\mathbf{x})] \to 0 \qquad \text{and therefore} \quad f(\mathbf{x} + \mathbf{h}) \to f(\mathbf{x}). \quad ❑$$

EXERCISES 15.1

In Exercises 1–20, find the gradient.

1. $f(x, y) = xe^{xy}$.

2. $f(x, y) = x^3 + y^2$.

3. $f(x, y) = 3x^2 - xy + y$.

4. $f(x, y) = Ax^2 + Bxy + Cy^2$.

5. $f(x, y) = x^2 y^{-2}$.

6. $f(x, y) = e^{x-y} - e^{y-x}$.

7. $f(x, y, z) = z \sin(x - y)$.

8. $f(x, y, z) = xy^2 e^{-z}$.

9. $f(x, y, z) = xy + yz + zx$.

10. $f(x, y, z) = \sqrt{x + y + z}$.

11. $f(x, y) = (x + y)e^{x-y}$.

12. $f(x, y) = (x + y) \sin(x - y)$.

13. $f(x, y) = e^x \ln y$.

14. $f(x, y, z) = x^2 y + y^2 z + z^2 x$.

15. $f(x, y) = \dfrac{Ax + By}{Cx + Dy}$.

16. $f(x, y) = \dfrac{x - y}{x^2 + y^2}$.

17. $f(x, y, z) = xye^x + ye^z - e^y \sin xz$.

18. $f(x, y, z) = e^{yz^2/x^3}$.

19. $f(x, y, z) = e^{x+2y} \cos(z^2 + 1)$.

20. $f(x, y, z) = \sin(xy) + e^{xyz} + \ln(x^2 z)$.

In Exercises 21–30, find the gradient vector at the indicated point.

21. $f(x, y) = 2x^2 - 3xy + 4y^2$ at $(2, 3)$.

22. $f(x, y) = 2x(x - y)^{-1}$ at $(3, 1)$.

23. $f(x, y) = \ln(x^2 + y^2)$ at $(2, 1)$.

24. $f(x, y) = x \tan^{-1}\left(\frac{y}{x}\right)$ at $(1, 1)$.

25. $f(x, y) = x \sin(xy)$ at $(1, \pi/2)$.

26. $f(x, y) = xye^{-(x^2+y^2)}$ at $(1, -1)$.

27. $f(x, y, z) = e^{-x} \sin(z + 2y)$ at $(0, \frac{1}{4}\pi, \frac{1}{4}\pi)$.

28. $f(x, y, z) = (x - y)\cos \pi z$ at $(1, 0, \frac{1}{2})$.

29. $f(x, y, z) = x - \sqrt{y^2 + z^2}$ at $(2, -3, 4)$.

30. $f(x, y, z) = \cos(xyz^2)$ at $(\pi, \frac{1}{4}, -1)$.

In Exercises 31–34, use Definition 15.1.2 to find the gradient of the function f.

31. $f(x, y) = 3x^2 - xy + y$.

32. $f(x, y) = \frac{1}{2}x^2 + 2xy + y^2$.

33. $f(x, y, z) = x^2 y + y^2 z + z^2 x$.

34. $f(x, y, z) = 2x^2 y - \frac{1}{z}$.

In Exercises 35–38, find a function f whose gradient is the given vector-valued function.

35. $\mathbf{F}(x, y) = 2xy\mathbf{i} + (1 + x^2)\mathbf{j}$.

36. $\mathbf{F}(x, y) = (2xy + x)\mathbf{i} + (x^2 + y)\mathbf{j}$.

37. $\mathbf{F}(x, y) = (x + \sin y)\mathbf{i} + (x \cos y - 2y)\mathbf{j}$.

38. $\mathbf{F}(x, y, z) = yz\mathbf{i} + (xz + 2yz)\mathbf{j} + (xy + y^2)\mathbf{k}$.

39. Calculate: (a) $\nabla(\ln r)$. (b) $\nabla(\sin r)$. (c) $\nabla(e^r)$, where $r = \sqrt{x^2 + y^2 + z^2}$.

40. Derive (15.1.5).

41. Let $f(x, y) = 1 + x^2 + y^2$.
(a) Find the points (x, y), if any, at which $\nabla f(x, y) = \mathbf{0}$.
(b) Sketch the graph of the surface $z = f(x, y)$.
(c) What can you say about the surface at the point(s) found in part (a)?

42. Repeat Exercise 41 for the function $f(x, y) = \sqrt{4 - x^2 - y^2}$.

43. (a) Show that, if $\mathbf{c} \cdot \mathbf{h}$ is $o(\mathbf{h})$, then $\mathbf{c} = \mathbf{0}$.   HINT: First set $\mathbf{h} = h\mathbf{i}$, then set $\mathbf{h} = h\mathbf{j}$, then $\mathbf{h} = h\mathbf{k}$.
(b) Show that, if

$$f(\mathbf{x} + \mathbf{h}) - f(\mathbf{x}) = \mathbf{y} \cdot \mathbf{h} + o(\mathbf{h})$$

and

$$f(\mathbf{x} + \mathbf{h}) - f(\mathbf{x}) = \mathbf{z} \cdot \mathbf{h} + o(\mathbf{h}),$$

then

$$\mathbf{y} = \mathbf{z}.$$

HINT: Verify that $\mathbf{y} \cdot \mathbf{h} - \mathbf{z} \cdot \mathbf{h} = (\mathbf{y} - \mathbf{z}) \cdot \mathbf{h}$ is $o(\mathbf{h})$.

44. Show that, if g is $o(\mathbf{h})$, then

$$\lim_{\mathbf{h} \to \mathbf{0}} g(\mathbf{h}) = 0.$$

45. Set

$$f(x, y) = \begin{cases} \dfrac{2xy}{x^2 + y^2}, & (x, y) \neq (0, 0) \\ 0, & (x, y) = (0, 0). \end{cases}$$

(a) Show that f is not differentiable at $(0, 0)$.
(b) In Section 14.6 you saw that the first partials $\partial f/\partial x$ and $\partial f/\partial y$ exist at $(0, 0)$. Since these partials obviously exist at every other point of the plane, we can conclude from Theorem 15.1.3 that at least one of these partials is not continuous in a neighborhood of $(0, 0)$. Show that $\partial f/\partial x$ is discontinuous at $(0, 0)$.

46. Let $f(x, y) = 2x^2 - y^2 - x^4 + 2$.
(a) Find the points (x, y), if any, at which $\nabla f(x, y) = \mathbf{0}$.
(b) Use a graphing utility to graph the surface $z = f(x, y)$ and investigate the behavior of the surface at the points found in part (a).

47. Repeat Exercise 46 for the function $f(x, y) = 4xye^{-(x^2+y^2)}$.

48. Repeat Exercise 46 for the function $f(x, y) = (x^2 + 4y^2)e^{1-(x^2+y^2)}$.

*SUPPLEMENT TO SECTION 15.1

PROOF OF THEOREM 15.1.3

We prove the theorem in the two-variable case. A similar argument yields a proof in the three-variable case, but there the details are more burdensome.

In the first place,

$$f(\mathbf{x} + \mathbf{h}) - f(\mathbf{x}) = f(x + h_1, y + h_2) - f(x, y).$$

Adding and subtracting $f(x, y + h_2)$, we have

(1) $f(\mathbf{x} + \mathbf{h}) - f(\mathbf{x}) = [f(x + h_1, y + h_2) - f(x, y + h_2)] + [f(x, y + h_2) - f(x, y)].$

By the mean-value theorem for functions of one variable, we know that there are numbers

$$0 < \theta_1 < 1 \qquad \text{and} \qquad 0 < \theta_2 < 1$$

such that

$$f(x + h_1, y + h_2) - f(x, y + h_2) = \frac{\partial f}{\partial x}(x + \theta_1 h_1, y + h_2)h_1$$

and

$$f(x, y + h_2) - f(x, y) = \frac{\partial f}{\partial y}(x, y + \theta_2 h_2)h_2. \quad \text{(Exercise 36, Section 4.1)}$$

By the continuity of $\partial f/\partial x$,

$$\frac{\partial f}{\partial x}(x + \theta_1 h_1, y + h_2) = \frac{\partial f}{\partial x}(x, y) + \epsilon_1(\mathbf{h})$$

where

$$\epsilon_1(\mathbf{h}) \to 0 \quad \text{as} \quad \mathbf{h} \to \mathbf{0}.\dagger$$

By the continuity of $\partial f/\partial y$,

$$\frac{\partial f}{\partial y}(x, y + \theta_2 h_2) = \frac{\partial f}{\partial y}(x, y) + \epsilon_2(\mathbf{h})$$

where

$$\epsilon_2(\mathbf{h}) \to 0 \quad \text{as} \quad \mathbf{h} \to \mathbf{0}.$$

Substituting these expressions in equation (1), we find that

$$
\begin{aligned}
f(\mathbf{x} + \mathbf{h}) - f(\mathbf{x}) &= \left[\frac{\partial f}{\partial x}(x, y) + \epsilon_1(\mathbf{h})\right]h_1 + \left[\frac{\partial f}{\partial y}(x, y) + \epsilon_2(\mathbf{h})\right]h_2 \\
&= \left[\frac{\partial f}{\partial x}(x, y) + \epsilon_1(\mathbf{h})\right](\mathbf{i} \cdot \mathbf{h}) + \left[\frac{\partial f}{\partial y}(x, y) + \epsilon_2(\mathbf{h})\right](\mathbf{j} \cdot \mathbf{h}) \\
&= \left[\frac{\partial f}{\partial x}(x, y)\mathbf{i} + \epsilon_1(\mathbf{h})\mathbf{i}\right] \cdot \mathbf{h} + \left[\frac{\partial f}{\partial y}(x, y)\mathbf{j} + \epsilon_2(\mathbf{h})\mathbf{j}\right] \cdot \mathbf{h} \\
&= \left[\frac{\partial f}{\partial x}(x, y)\mathbf{i} + \frac{\partial f}{\partial y}(x, y)\mathbf{j}\right] \cdot \mathbf{h} + \left[\epsilon_1(\mathbf{h})\mathbf{i} + \epsilon_2(\mathbf{h})\mathbf{j}\right] \cdot \mathbf{h}.
\end{aligned}
$$

$\dagger$

$$\epsilon_1(\mathbf{h}) = \frac{\partial f}{\partial x}(x + \theta_1 h_1, y + h_2) - \frac{\partial f}{\partial x}(x, y) \to 0$$

since, by the continuity of $\partial f/\partial x$,

$$\frac{\partial f}{\partial x}(x + \theta_1 h_1, y + h_2) \to \frac{\partial f}{\partial x}(x, y).$$

To complete the proof of the theorem we need only show that

(2) $$[\epsilon_1(\mathbf{h})\mathbf{i} + \epsilon_2(\mathbf{h})\mathbf{j}] \cdot \mathbf{h} = o(\mathbf{h}).$$

From Schwarz's inequality, $|\mathbf{a} \cdot \mathbf{b}| \leq \|\mathbf{a}\|\,\|\mathbf{b}\|$, we know that

$$|[\epsilon_1(\mathbf{h})\mathbf{i} + \epsilon_2(\mathbf{h})\mathbf{j}] \cdot \mathbf{h}| \leq \|\epsilon_1(\mathbf{h})\mathbf{i} + \epsilon_2(\mathbf{h})\mathbf{j}\|\,\|\mathbf{h}\|.$$

It follows that

$$\frac{|[\epsilon_1(\mathbf{h})\mathbf{i} + \epsilon_2(\mathbf{h})\mathbf{j}] \cdot \mathbf{h}|}{\|\mathbf{h}\|} \leq \|\epsilon_1(\mathbf{h})\mathbf{i} + \epsilon_2(\mathbf{h})\mathbf{j}\| \leq \underset{\underset{\text{by the triangle inequality}}{\uparrow}}{\|\epsilon_1(\mathbf{h})\mathbf{i}\| + \|\epsilon_2(\mathbf{h})\mathbf{j}\|} = |\epsilon_1(\mathbf{h})| + |\epsilon_2(\mathbf{h})|.$$

As $\mathbf{h} \to \mathbf{0}$, the expression on the right tends to 0. This shows that (2) holds and completes the proof of the theorem. ❏

■ 15.2 GRADIENTS AND DIRECTIONAL DERIVATIVES

Some Elementary Formulas

In many respects gradients behave just as derivatives do in the one-variable case. In particular, if $\nabla f(\mathbf{x})$ and $\nabla g(\mathbf{x})$ exist, then $\nabla[f(\mathbf{x}) + g(\mathbf{x})]$, $\nabla[\alpha f(\mathbf{x})]$, and $\nabla[f(\mathbf{x})g(\mathbf{x})]$ all exist, and

(15.2.1)
$$\begin{aligned} \nabla[f(\mathbf{x}) + g(\mathbf{x})] &= \nabla f(\mathbf{x}) + \nabla g(\mathbf{x}), \\ \nabla[\alpha f(\mathbf{x})] &= \alpha \, \nabla f(\mathbf{x}), \\ \nabla[f(\mathbf{x})g(\mathbf{x})] &= f(\mathbf{x}) \, \nabla g(\mathbf{x}) + g(\mathbf{x}) \, \nabla f(\mathbf{x}). \end{aligned}$$

The first two formulas are easy to derive. To derive the third formula, let's assume that $\nabla f(\mathbf{x})$ and $\nabla g(\mathbf{x})$ both exist. Our task is to show that

$$f(\mathbf{x} + \mathbf{h})g(\mathbf{x} + \mathbf{h}) - f(\mathbf{x})g(\mathbf{x}) = [f(\mathbf{x}) \, \nabla g(\mathbf{x}) + g(\mathbf{x}) \, \nabla f(\mathbf{x})] \cdot \mathbf{h} + o(\mathbf{h}).$$

We now sketch how this can be done. We leave it to you to justify each step. The "trick" in proving a product rule is to add and subtract an appropriate expression (for example, see the proof of the product rule, Theorem 3.2.5). Starting from

$$f(\mathbf{x} + \mathbf{h})g(\mathbf{x} + \mathbf{h}) - f(\mathbf{x})g(\mathbf{x})$$

we add and subtract the term $f(\mathbf{x})g(\mathbf{x} + \mathbf{h})$. This gives:

$$\begin{aligned} &[f(\mathbf{x} + \mathbf{h})g(\mathbf{x} + \mathbf{h}) - f(\mathbf{x})g(\mathbf{x} + \mathbf{h})] + [f(\mathbf{x})g(\mathbf{x} + \mathbf{h}) - f(\mathbf{x})g(\mathbf{x})] \\ &= [f(\mathbf{x} + \mathbf{h}) - f(\mathbf{x})]g(\mathbf{x} + \mathbf{h}) + f(\mathbf{x})[g(\mathbf{x} + \mathbf{h}) - g(\mathbf{x})] \\ &= [\nabla f(\mathbf{x}) \cdot \mathbf{h} + o(\mathbf{h})]g(\mathbf{x} + \mathbf{h}) + f(\mathbf{x})[\nabla g(\mathbf{x}) \cdot \mathbf{h} + o(\mathbf{h})] \\ &= g(\mathbf{x} + \mathbf{h}) \, \nabla f(\mathbf{x}) \cdot \mathbf{h} + f(\mathbf{x}) \, \nabla g(\mathbf{x}) \cdot \mathbf{h} + o(\mathbf{h}) \qquad \text{(Exercise 30)} \\ &= g(\mathbf{x}) \, \nabla f(\mathbf{x}) \cdot \mathbf{h} + f(\mathbf{x}) \, \nabla g(\mathbf{x}) \cdot \mathbf{h} + [g(\mathbf{x} + \mathbf{h}) - g(\mathbf{x})] \, \nabla f(\mathbf{x}) \cdot \mathbf{h} + o(\mathbf{h}) \\ &= [g(\mathbf{x}) \, \nabla f(\mathbf{x}) + f(\mathbf{x}) \, \nabla g(\mathbf{x})] \cdot \mathbf{h} + o(\mathbf{h}). \qquad \text{(Exercise 30)} \quad ❏ \end{aligned}$$

In Exercise 43, you are asked to verify this formula using Theorem 15.1.3.

Directional Derivatives

Here we take up an idea that generalizes the notion of partial derivative. Its connection with gradients will be made clear as we go on. First, recall the definitions of the first partial derivatives:

(two variables)   (three variables)

$$\frac{\partial f}{\partial x}(x, y) = \lim_{h \to 0} \frac{f(x + h, y) - f(x, y)}{h}, \qquad \frac{\partial f}{\partial x}(x, y, z) = \lim_{h \to 0} \frac{f(x + h, y, z) - f(x, y, z)}{h},$$

$$\frac{\partial f}{\partial y}(x, y) = \lim_{h \to 0} \frac{f(x, y + h) - f(x, y)}{h}, \qquad \frac{\partial f}{\partial y}(x, y, z) = \lim_{h \to 0} \frac{f(x, y + h, z) - f(x, y, z)}{h},$$

$$\frac{\partial f}{\partial z}(x, y, z) = \lim_{h \to 0} \frac{f(x, y, z + h) - f(x, y, z)}{h}.$$

Expressed in vector notation, these definitions take the form

$(\mathbf{x} = (x, y))$ $(\mathbf{x} = (x, y, z))$

$$\frac{\partial f}{\partial x}(\mathbf{x}) = \lim_{h \to 0} \frac{f(\mathbf{x} + h\mathbf{i}) - f(\mathbf{x})}{h}, \qquad \frac{\partial f}{\partial x}(\mathbf{x}) = \lim_{h \to 0} \frac{f(\mathbf{x} + h\mathbf{i}) - f(\mathbf{x})}{h},$$

$$\frac{\partial f}{\partial y}(\mathbf{x}) = \lim_{h \to 0} \frac{f(\mathbf{x} + h\mathbf{j}) - f(\mathbf{x})}{h}, \qquad \frac{\partial f}{\partial y}(\mathbf{x}) = \lim_{h \to 0} \frac{f(\mathbf{x} + h\mathbf{j}) - f(\mathbf{x})}{h},$$

$$\frac{\partial f}{\partial z}(\mathbf{x}) = \lim_{h \to 0} \frac{f(\mathbf{x} + h\mathbf{k}) - f(\mathbf{x})}{h}.$$

Each partial is thus the limit of a quotient

$$\frac{f(\mathbf{x} + h\mathbf{u}) - f(\mathbf{x})}{h},$$

where $\mathbf{u}$ is one of the unit coordinate vectors $\mathbf{i}$, $\mathbf{j}$, or $\mathbf{k}$. There is no reason to be so restrictive on $\mathbf{u}$. If f is defined in a neighborhood of $\mathbf{x}$, then, for small h, the difference quotient

$$\frac{f(\mathbf{x} + h\mathbf{u}) - f(\mathbf{x})}{h}$$

makes sense for any unit vector $\mathbf{u}$.

DEFINITION 15.2.2 DIRECTIONAL DERIVATIVE

For each unit vector $\mathbf{u}$, the limit

$$f'_{\mathbf{u}}(\mathbf{x}) = \lim_{h \to 0} \frac{f(\mathbf{x} + h\mathbf{u}) - f(\mathbf{x})}{h},$$

if it exists, is called the *directional derivative of f at $\mathbf{x}$ in the direction $\mathbf{u}$.*

It is important to realize that each of the partial derivatives $(\partial f/\partial x)$, $(\partial f/\partial y)$, $(\partial f/\partial z)$ is itself a directional derivative:

(15.2.3)
$$\frac{\partial f}{\partial x}(\mathbf{x}) = f'_{\mathbf{i}}(\mathbf{x}), \qquad \frac{\partial f}{\partial y}(\mathbf{x}) = f'_{\mathbf{j}}(\mathbf{x}), \qquad \frac{\partial f}{\partial z}(\mathbf{x}) = f'_{\mathbf{k}}(\mathbf{x}).$$

As you know, the partials of f give the rates of change of f in the $\mathbf{i}$, $\mathbf{j}$, $\mathbf{k}$ directions. The directional derivative $f'_{\mathbf{u}}$ gives *the rate of change of f in the direction* $\mathbf{u}$.

The geometric interpretation of the directional derivative for a function f of two variables is shown in Figure 15.2.1. Fix a point (x, y) in the domain of f and let $\mathbf{u}$ be a unit vector with initial point (x, y) in the xy-plane. Let C be the curve of intersection of the surface $z = f(x, y)$ and the plane p which contains $\mathbf{u}$ and is perpendicular to the xy-plane. Then $f'_{\mathbf{u}}(\mathbf{x})$ is the slope of the tangent line to C at the point $(x, y, f(x, y))$.

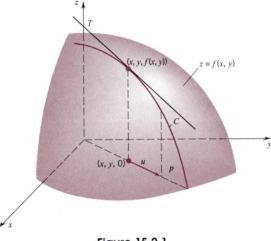

Figure 15.2.1

Remark The definition of the directional derivative of f in the direction $\mathbf{u}$ requires $\mathbf{u}$ to be a unit vector. However, we can extend the definition to arbitrary nonzero vectors as follows: The directional derivative of f at $\mathbf{x}$ in the direction of a nonzero vector $\mathbf{a}$ is $f'_{\mathbf{u}}(\mathbf{x})$ where $\mathbf{u} = \mathbf{a}/\|\mathbf{a}\|$ is the unit vector having the same direction as $\mathbf{a}$. ❏

There is an important connection between the gradient at $\mathbf{x}$ and the directional derivatives at $\mathbf{x}$.

THEOREM 15.2.4

If f is differentiable at $\mathbf{x}$, then f has a directional derivative at $\mathbf{x}$ in every direction $\mathbf{u}$, where $\mathbf{u}$ is a unit vector, and

$$f'_{\mathbf{u}}(\mathbf{x}) = \nabla f(\mathbf{x}) \cdot \mathbf{u}.$$

PROOF We take $\mathbf{u}$ as a unit vector and assume that f is differentiable at $\mathbf{x}$. The differentiability at $\mathbf{x}$ tells us that $\nabla f(\mathbf{x})$ exists and

$$f(\mathbf{x} + h\mathbf{u}) - f(\mathbf{x}) = \nabla f(\mathbf{x}) \cdot h\mathbf{u} + o(h\mathbf{u}).$$

Division by h gives

$$\frac{f(\mathbf{x} + h\mathbf{u}) - f(\mathbf{x})}{h} = \nabla f(\mathbf{x}) \cdot \mathbf{u} + \frac{o(h\mathbf{u})}{h}.$$

Since

$$\left| \frac{o(h\mathbf{u})}{h} \right| = \frac{|o(h\mathbf{u})|}{|h|} = \frac{|o(h\mathbf{u})|}{\|h\mathbf{u}\|} \to 0,$$

you can see that

$$\frac{o(h\mathbf{u})}{h} \to 0 \quad \text{and thus} \quad \frac{f(\mathbf{x} + h\mathbf{u}) - f(\mathbf{x})}{h} \to \nabla f(\mathbf{x}) \cdot \mathbf{u}. \quad \square$$

Earlier (Theorem 15.1.3) you saw that, *if f has continuous first partials in a neighborhood of* $\mathbf{x}$*, then f is differentiable at* $\mathbf{x}$ *and*

$$\nabla f(\mathbf{x}) = \frac{\partial f}{\partial x}(\mathbf{x})\mathbf{i} + \frac{\partial f}{\partial y}(\mathbf{x})\mathbf{j}. \qquad (\mathbf{x} = (x, y))$$

$$\nabla f(\mathbf{x}) = \frac{\partial f}{\partial x}(\mathbf{x})\mathbf{i} + \frac{\partial f}{\partial y}(\mathbf{x})\mathbf{j} + \frac{\partial f}{\partial z}(\mathbf{x})\mathbf{k}. \qquad (\mathbf{x} = (x, y, z))$$

The next theorem shows that this formula for $\nabla f(\mathbf{x})$ holds wherever f is differentiable.

THEOREM 15.2.5

If f is differentiable at $\mathbf{x}$, then all the first partial derivatives of f exist at $\mathbf{x}$ and

$$\nabla f(\mathbf{x}) = \frac{\partial f}{\partial x}(\mathbf{x})\mathbf{i} + \frac{\partial f}{\partial y}(\mathbf{x})\mathbf{j} \qquad (\mathbf{x} = (x, y)),$$

$$\nabla f(\mathbf{x}) = \frac{\partial f}{\partial x}(\mathbf{x})\mathbf{i} + \frac{\partial f}{\partial y}(\mathbf{x})\mathbf{j} + \frac{\partial f}{\partial z}(\mathbf{x})\mathbf{k} \qquad (\mathbf{x} = (x, y, z)).$$

PROOF It is sufficient to prove the theorem for the case $\mathbf{x} = (x, y, z)$. Assume that f is differentiable at $\mathbf{x}$. Then $\nabla f(\mathbf{x})$ exists and we can write

$$\nabla f(\mathbf{x}) = [\nabla f(\mathbf{x}) \cdot \mathbf{i}]\mathbf{i} + [\nabla f(\mathbf{x}) \cdot \mathbf{j}]\mathbf{j} + [\nabla f(\mathbf{x}) \cdot \mathbf{k}]\mathbf{k}. \qquad (12.3.13)$$

The result follows from observing that

$$\begin{array}{cc} (15.2.4) & (15.2.3) \\ \downarrow & \downarrow \end{array}$$

$$\nabla f(\mathbf{x}) \cdot \mathbf{i} = f'_{\mathbf{i}}(\mathbf{x}) = \frac{\partial f}{\partial x}(\mathbf{x}),$$

$$\nabla f(\mathbf{x}) \cdot \mathbf{j} = f'_{\mathbf{j}}(\mathbf{x}) = \frac{\partial f}{\partial y}(\mathbf{x}),$$

$$\nabla f(\mathbf{x}) \cdot \mathbf{k} = f'_{\mathbf{k}}(\mathbf{x}) = \frac{\partial f}{\partial z}(\mathbf{x}). \quad \square$$

Example 1   Find the directional derivative of the function

$$f(x, y) = x^2 + y^2$$

at the point (1, 2) in the direction of the vector $2\mathbf{i} - 3\mathbf{j}$.

SOLUTION   In the first place, $2\mathbf{i} - 3\mathbf{j}$ is not a unit vector; its norm is $\sqrt{13}$. The unit vector in the direction of $2\mathbf{i} - 3\mathbf{j}$ is the vector

$$\mathbf{u} = \frac{1}{\sqrt{13}}[2\mathbf{i} - 3\mathbf{j}].$$

Next

$$\nabla f = 2x\mathbf{i} + 2y\mathbf{j},$$

and therefore

$$\nabla f(1, 2) = 2\mathbf{i} + 4\mathbf{j}.$$

By (15.2.4) we have

$$f_{\mathbf{u}}'(1, 2) = \nabla f(1, 2) \cdot \mathbf{u}$$

$$= (2\mathbf{i} + 4\mathbf{j}) \cdot \frac{1}{\sqrt{13}}[2\mathbf{i} - 3\mathbf{j}] = \frac{-8}{\sqrt{13}} \cong -2.219. \quad \square$$

Example 2   Find the directional derivative of the function

$$f(x, y, z) = x \cos y \sin z$$

at the point $(1, \pi, \tfrac{1}{4}\pi)$ in the direction of the vector $2\mathbf{i} - \mathbf{j} + 4\mathbf{k}$.

SOLUTION   The unit vector in the direction of $2\mathbf{i} - \mathbf{j} + 4\mathbf{k}$ is the vector

$$\mathbf{u} = \frac{1}{\sqrt{21}}[2\mathbf{i} - \mathbf{j} + 4\mathbf{k}].$$

Here

$$\frac{\partial f}{\partial x}(x, y, z) = \cos y \sin z, \qquad \frac{\partial f}{\partial y}(x, y, z) = -x \sin y \sin z,$$

$$\frac{\partial f}{\partial z}(x, y, z) = x \cos y \cos z$$

so that

$$\frac{\partial f}{\partial x}\left(1, \pi, \tfrac{1}{4}\pi\right) = -\frac{1}{2}\sqrt{2}, \qquad \frac{\partial f}{\partial y}\left(1, \pi, \tfrac{1}{4}\pi\right) = 0, \qquad \frac{\partial f}{\partial z}\left(1, \pi, \tfrac{1}{4}\pi\right) = -\frac{1}{2}\sqrt{2}.$$

Therefore

$$\nabla f\left(1, \pi, \tfrac{1}{4}\pi\right) = -\frac{1}{2}\sqrt{2}\,\mathbf{i} - \frac{1}{2}\sqrt{2}\,\mathbf{k}$$

and

$$f_{\mathbf{u}}'\left(1, \pi, \tfrac{1}{4}\pi\right) = \nabla f\left(1, \pi, \tfrac{1}{4}\pi\right) \cdot \mathbf{u}$$

$$= -\frac{1}{2}\sqrt{2}[\mathbf{i} + \mathbf{k}] \cdot \frac{1}{\sqrt{21}}[2\mathbf{i} - \mathbf{j} + 4\mathbf{k}] = -\frac{3\sqrt{2}}{\sqrt{21}} \cong -0.926. \quad \square$$

You know that for each unit vector **u**

$$f'_{\mathbf{u}}(\mathbf{x}) = \nabla f(\mathbf{x}) \cdot \mathbf{u}.$$

Since

$$\nabla f(\mathbf{x}) \cdot \mathbf{u} = \operatorname{comp}_{\mathbf{u}} \nabla f(\mathbf{x}), \qquad (12.3.11)$$

we have

(15.2.6)

$$\boxed{f'_{\mathbf{u}}(\mathbf{x}) = \operatorname{comp}_{\mathbf{u}} \nabla f(\mathbf{x}).}$$

Namely, the directional derivative in a direction **u** *is the component of the gradient vector in that direction.* (Figure 15.2.2)

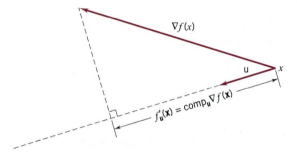

Figure 15.2.2

If $\nabla f(\mathbf{x}) \neq \mathbf{0}$, then

$$f'_{\mathbf{u}}(\mathbf{x}) = \nabla f(\mathbf{x}) \cdot \mathbf{u} = \| \nabla f(\mathbf{x}) \| \, \| \mathbf{u} \| \cos \theta = \| \nabla f(\mathbf{x}) \| \cos \theta$$

$$\underset{(12.3.7)}{\uparrow} \qquad \underset{\| \mathbf{u} \| = 1}{\uparrow}$$

where θ is the angle between $\nabla f(\mathbf{x})$ and **u**. Since $-1 \leq \cos \theta \leq 1$, we have

$$-\| \nabla f(\mathbf{x}) \| \leq f'_{\mathbf{u}}(\mathbf{x}) \leq \| \nabla f(\mathbf{x}) \| \qquad \text{for all directions } \mathbf{u}.$$

If **u** points in the direction of $\nabla f(\mathbf{x})$, then

$$f'_{\mathbf{u}}(\mathbf{x}) = \| \nabla f(\mathbf{x}) \|, \qquad\qquad (\theta = 0, \cos \theta = 1)$$

and, if **u** points in the direction of $-\nabla f(\mathbf{x})$, then

$$f'_{\mathbf{u}}(\mathbf{x}) = -\| \nabla f(\mathbf{x}) \|. \qquad\qquad (\theta = \pi, \cos \theta = -1)$$

Since the directional derivative gives the rate of change of the function in that direction, it is clear that

(15.2.7)

> a differentiable function f increases most rapidly in the direction of the gradient (the rate of change is then $\| \nabla f(\mathbf{x}) \|$) and it decreases most rapidly in the opposite direction (the rate of change is then $-\| \nabla f(\mathbf{x}) \|$).

Example 3 The graph of the function

$$f(x, y) = \sqrt{1 - (x^2 + y^2)}$$

is the upper half of the unit sphere $x^2 + y^2 + z^2 = 1$. The function is defined on the closed unit disc $x^2 + y^2 \le 1$, but differentiable only on the open unit disc.

In Figure 15.2.3 we have marked a point (x, y) and drawn the corresponding radius vector $\mathbf{r} = x\mathbf{i} + y\mathbf{j}$. The gradient

$$\nabla f(x, y) = \frac{-x}{\sqrt{1 - (x^2 + y^2)}}\mathbf{i} + \frac{-y}{\sqrt{1 - (x^2 + y^2)}}\mathbf{j}$$

is a negative multiple of $\mathbf{r}$:

$$\nabla f(x, y) = -\frac{1}{\sqrt{1 - (x^2 + y^2)}}(x\mathbf{i} + y\mathbf{j}) = -\frac{1}{\sqrt{1 - (x^2 + y^2)}}\mathbf{r}.$$

Since $\mathbf{r}$ points from the origin to (x, y), the gradient points from (x, y) to the origin. This means that f increases most rapidly toward the origin. This is borne out by the observation that along the hemispherical surface the path of steepest ascent from the point $P(x, y, f(x, y))$ is the "great circle route to the north pole." ❑

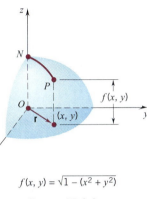

$f(x, y) = \sqrt{1 - (x^2 + y^2)}$

Figure 15.2.3

Example 4 Suppose that the temperature at each point of a metal plate is given by the function

$$T(x, y) = e^x \cos y + e^y \cos x.$$

(a) In what direction does the temperature increase most rapidly at the point $(0, 0)$? What is this rate of increase?

(b) In what direction does the temperature decrease most rapidly at $(0, 0)$?

SOLUTION

$$\nabla T(x, y) = \frac{\partial T}{\partial x}(x, y)\mathbf{i} + \frac{\partial T}{\partial y}(x, y)\mathbf{j}$$

$$= (e^x \cos y - e^y \sin x)\mathbf{i} + (e^y \cos x - e^x \sin y)\mathbf{j}.$$

(a) At $(0, 0)$ the temperature increases most rapidly in the direction of the gradient

$$\nabla T(0, 0) = \mathbf{i} + \mathbf{j}.$$

This rate of increase is

$$\|\nabla T(0, 0)\| = \|\mathbf{i} + \mathbf{j}\| = \sqrt{2}.$$

(b) The temperature decreases most rapidly in the direction of

$$-\nabla T(0, 0) = -\mathbf{i} - \mathbf{j}. \quad ❑$$

Example 5 The mass density (mass per unit volume) of a metal ball centered at the origin is given by the function

$$\lambda(x, y, z) = ke^{-(x^2 + y^2 + z^2)}, \qquad k \text{ a positive constant}.$$

(a) In what direction does the density increase most rapidly at the point (x, y, z)? What is this rate of density increase?

(b) In what direction does the density decrease most rapidly?

(c) What are the rates of density change at (x, y, z) in the $\mathbf{i}$, $\mathbf{j}$, $\mathbf{k}$ directions?

SOLUTION The gradient

$$\nabla \lambda(x, y, z) = \frac{\partial \lambda}{\partial x}(x, y, z)\mathbf{i} + \frac{\partial \lambda}{\partial y}(x, y, z)\mathbf{j} + \frac{\partial \lambda}{\partial z}(x, y, z)\mathbf{k}$$

$$= -2ke^{-(x^2+y^2+z^2)}[x\mathbf{i} + y\mathbf{j} + z\mathbf{k}],$$

and since $\lambda(x, y, z) = ke^{-(x^2+y^2+z^2)}$, we have

$$\nabla \lambda(x, y, z) = -2\lambda(x, y, z)\mathbf{r}.$$

From this, we see that the gradient points from (x, y, z) in the direction opposite to that of the radius vector.

(a) The density increases most rapidly toward the origin. The rate of increase is

$$\|\nabla \lambda(x, y, z)\| = 2\lambda(x, y, z)\|\mathbf{r}\| = 2\lambda(x, y, z)\sqrt{x^2 + y^2 + z^2}.$$

(b) The density decreases most rapidly directly away from the origin.

(c) The rates of density change in the $\mathbf{i}, \mathbf{j}, \mathbf{k}$ directions are given by the directional derivatives

$$\lambda_{\mathbf{i}}'(x, y, z) = \nabla \lambda(x, y, z) \cdot \mathbf{i} = -2x\, \lambda(x, y, z),$$

$$\lambda_{\mathbf{j}}'(x, y, z) = \nabla \lambda(x, y, z) \cdot \mathbf{j} = -2y\, \lambda(x, y, z),$$

$$\lambda_{\mathbf{k}}'(x, y, z) = \nabla \lambda(x, y, z) \cdot \mathbf{k} = -2z\, \lambda(x, y, z).$$

These are just the first partials of λ. ❑

Example 6 Suppose that the temperature at each point of a metal plate is given by the function

$$T(x, y) = 1 + x^2 - y^2.$$

Find the path followed by a heat-seeking particle that originates at $(-2, 1)$.

SOLUTION The particle moves in the direction of the gradient vector

$$\nabla T = 2x\mathbf{i} - 2y\mathbf{j}.$$

We want the curve

$$C: \mathbf{r}(t) = x(t)\mathbf{i} + y(t)\mathbf{j}$$

which begins at $(-2, 1)$ and at each point has its tangent vector in the direction of ∇T. We can satisfy the first condition by setting

$$x(0) = -2, \qquad y(0) = 1.$$

We can satisfy the second condition by setting

$$x'(t) = 2x(t), \qquad y'(t) = -2y(t). \qquad\qquad \text{(explain)}$$

These differential equations, together with initial conditions at $t = 0$, imply that

$$x(t) = -2e^{2t}, \qquad y(t) = e^{-2t}. \qquad\qquad \text{(Section 7.6)}$$

We can eliminate the parameter t by noting that

$$x(t)y(t) = (-2e^{2t})(e^{-2t}) = -2.$$

In terms of just x and y we have

$$xy = -2.$$

The particle moves from the point $(-2, 1)$ along the left branch of the hyperbola $xy = -2$ in the direction of decreasing x (see Figure 15.2.4).

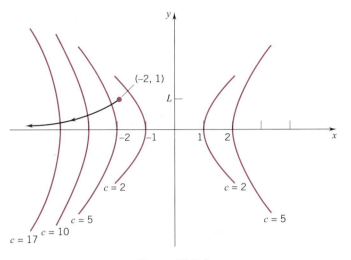

Figure 15.2.4

The level curves (isothermals) of the temperature distribution T are also hyperbolas. You can verify that the path of the particle is perpendicular to each of the level curves $x^2 - y^2 = c - 1$. ❑

Remark The pair of differential equations

$$x'(t) = 2x(t), \qquad y'(t) = -2y(t)$$

can be set as a single differential equation in x and y: the relation

$$\frac{y'(t)}{x'(t)} = -\frac{y(t)}{x(t)}$$

gives

$$\frac{dy}{dx} = -\frac{y}{x}.$$

This equation is readily solved directly (see Section 7.6):

$$\frac{1}{y}\frac{dy}{dx} = -\frac{1}{x}$$

$$\ln|y| = -\ln|x| + C$$

$$\ln|x| + \ln|y| = C$$

$$\ln|xy| = C.$$

Thus xy is constant:

$$xy = k.$$

Since the curve passes through the point $(-2, 1)$, $k = -2$ and once again we have the curve

$$xy = -2.$$

You will be called upon to use this method of solution in some of the exercises. ❑

EXERCISES 15.2

In Exercises 1–14, find the directional derivative at the given point in the direction indicated.

1. $f(x, y) = x^2 + 3y^2$ at $(1, 1)$ in the direction of $\mathbf{i} - \mathbf{j}$.

2. $f(x, y) = x + \sin(x + y)$ at $(0, 0)$ in the direction of $2\mathbf{i} + \mathbf{j}$.

3. $f(x, y) = xe^y - ye^x$ at $(1, 0)$ in the direction of $3\mathbf{i} + 4\mathbf{j}$.

4. $f(x, y) = \dfrac{2x}{x - y}$ at $(1, 0)$ in the direction of $\mathbf{i} - \sqrt{3}\mathbf{j}$.

5. $f(x, y) = \dfrac{ax + by}{x + y}$ at $(1, 1)$ in the direction of $\mathbf{i} - \mathbf{j}$.

6. $f(x, y) = \dfrac{x + y}{cx + dy}$ at $(1, 1)$ in the direction of $c\mathbf{i} - d\mathbf{j}$.

7. $f(x, y) = \ln(x^2 + y^2)$ at $(0, 1)$ in the direction of $8\mathbf{i} + \mathbf{j}$.

8. $f(x, y) = x^2y + \tan y$ at $(-1, \pi/4)$ in the direction of $\mathbf{i} - 2\mathbf{j}$.

9. $f(x, y, z) = xy + yz + zx$ at $(1, -1, 1)$ in the direction of $\mathbf{i} + 2\mathbf{j} + \mathbf{k}$.

10. $f(x, y, z) = x^2y + y^2z + z^2x$ at $(1, 0, 1)$ in the direction of $3\mathbf{j} - \mathbf{k}$.

11. $f(x, y, z) = (x + y^2 + z^3)^2$ at $(1, -1, 1)$ in the direction of $\mathbf{i} + \mathbf{j}$.

12. $f(x, y, z) = Ax^2 + Bxyz + Cy^2$ at $(1, 2, 1)$ in the direction of $A\mathbf{i} + B\mathbf{j} + C\mathbf{k}$.

13. $f(x, y, z) = x \tan^{-1}(y + z)$ at $(1, 0, 1)$ in the direction of $\mathbf{i} + \mathbf{j} - \mathbf{k}$.

14. $f(x, y, z) = xy^2\cos z - 2yz^2\sin \pi x + 3zx^2$ at $(0, -1, \pi)$ in the direction of $2\mathbf{i} - \mathbf{j} + 2\mathbf{k}$.

15. Find the directional derivative of $f(x, y) = \ln \sqrt{x^2 + y^2}$ at $(x, y) \neq (0, 0)$ toward the origin.

16. Find the directional derivative of $f(x, y) = (x - 1)y^2e^{xy}$ at $(0, 1)$ toward the point $(-1, 3)$.

17. Find the directional derivative of $f(x, y) = Ax^2 + 2Bxy + Cy^2$ at (a, b) toward (b, a) (a) if $a > b$; (b) if $a < b$.

18. Find the directional derivative of $f(x, y, z) = z \ln\left(\dfrac{x}{y}\right)$ at $(1, 1, 2)$ toward the point $(2, 2, 1)$.

19. Find the directional derivative of $f(x, y, z) = xe^{y^2 - z^2}$ at $(1, 2, -2)$ in the direction of the path $\mathbf{r}(t) = t\mathbf{i} + 2\cos(t - 1)\mathbf{j} - 2e^{t-1}\mathbf{k}$.

20. Find the directional derivative of $f(x, y, z) = x^2 + yz$ at $(1, -3, 2)$ in the direction of the path $\mathbf{r}(t) = t^2\mathbf{i} + 3t\mathbf{j} + (1 - t^3)\mathbf{k}$.

21. Find the directional derivative of $f(x, y, z) = x^2 + 2xyz - yz^2$ at $(1, 1, 2)$ in a direction parallel to the straight line

$$\frac{x - 1}{2} = y - 1 = \frac{z - 2}{-3}.$$

22. Find the directional derivative of $f(x, y, z) = e^x\cos \pi yz$ at $(0, 1, \frac{1}{2})$ in a direction parallel to the line of intersection of the planes $x + y - z - 5 = 0$ and $4x - y - z + 2 = 0$.

In Exercises 23–26, find a unit vector in the direction in which f increases most rapidly at P and give the rate of change of f in that direction; find a unit vector in the direction in which f decreases most rapidly at P and give the rate of change of f in that direction.

23. $f(x, y) = y^2e^{2x}$; $P(0, 1)$.

24. $f(x, y) = x + \sin(x + 2y)$; $P(0, 0)$.

25. $f(x, y, z) = \sqrt{x^2 + y^2 + z^2}$; $P(1, -2, 1)$.

26. $f(x, y, z) = x^2ze^y + xz^2$; $P(1, \ln 2, 2)$.

27. Let $f = f(x)$ be a differentiable function of one variable. What is the gradient of f at x_0? What is the geometric significance of the direction of this gradient?

28. Suppose that f is differentiable at (x_0, y_0) and $\nabla f(x_0, y_0) \neq \mathbf{0}$. Compute the rate of change of f in the direction of the vector

$$\frac{\partial f}{\partial y}(x_0, y_0)\mathbf{i} - \frac{\partial f}{\partial x}(x_0, y_0)\mathbf{j}.$$

Give a geometric interpretation to your answer.

29. Let

$$f(x, y) = \sqrt{x^2 + y^2}.$$

(a) Show that $\partial f/\partial x$ is not defined at $(0, 0)$.
(b) Is f differentiable at $(0, 0)$?

30. Verify that, if g is continuous at $\mathbf{x}$, then
 (a) $g(\mathbf{x} + \mathbf{h})o(\mathbf{h}) = o(\mathbf{h})$ and
 (b) $[g(\mathbf{x} + \mathbf{h}) - g(\mathbf{x})] \nabla f(\mathbf{x}) \cdot \mathbf{h} = o(\mathbf{h})$.

31. Given the density function $\lambda(x, y) = 48 - \frac{4}{3}x^2 - 3y^2$, find the rate of density change (a) at $(1, -1)$ in the direction of the most rapid density decrease; (b) at $(1, 2)$ in the $\mathbf{i}$ direction; (c) at $(2, 2)$ away from the origin.

32. The intensity of light in a neighborhood of the point $(-2, 1)$ is given by a function of the form $I(x, y) = A - 2x^2 - y^2$. Find the path followed by a light-seeking particle that originates at the center of the neighborhood.

33. Determine the path of steepest descent along the surface $z = x^2 + 3y^2$ from each of the following points: (a) $(1, 1, 4)$; (b) $(1, -2, 13)$.

34. Determine the path of steepest ascent along the hyperbolic paraboloid $z = \frac{1}{2}x^2 - y^2$ from each of the following points: (a) $(-1, 1, -\frac{1}{2})$; (b) $(1, 0, \frac{1}{2})$.

35. Determine the path of steepest descent along the surface $z = a^2x^2 + b^2y^2$ from the point $(a^2, b^2, a^4 + b^4)$.

36. The temperature in a neighborhood of the origin is given by a function of the form
$$T(x, y) = T_0 + e^y \sin x.$$
Find the path followed by a heat-fleeing particle that originates at the origin.

37. The temperature in a neighborhood of the point $(\frac{1}{4}\pi, 0)$ is given by the function
$$T(x, y) = \sqrt{2}\, e^{-y} \cos x.$$
Find the path followed by a heat-seeking particle that originates at the center of the neighborhood.

38. Determine the path of steepest descent along the surface $z = A + x + 2y - x^2 - 3y^2$ from the point $(0, 0, A)$.

39. Set $f(x, y) = 3x^2 + y$.
 (a) Find
$$\lim_{h \to 0} \frac{f(x(2 + h), y(2 + h)) - f(2, 4)}{h}$$
 given that $x(t) = t$ and $y(t) = t^2$. (These functions parametrize the parabola $y = x^2$.)
 (b) Find
$$\lim_{h \to 0} \frac{f(x(4 + h), y(4 + h)) - f(2, 4)}{h}$$

given that $x(t) = \frac{1}{4}(t + 4)$ and $y(t) = t$. (These functions parametrize the line $y = 4x - 4$.)
 (c) Compute the directional derivative of f at $(2, 4)$ in the direction of $\mathbf{i} + 4\mathbf{j}$.
 (d) Notice that $\mathbf{i} + 4\mathbf{j}$ is a direction vector for the line $y = 4x - 4$ and this line is tangent to the parabola $y = x^2$ at $(2, 4)$. Explain then why the computations in (a), (b), and (c) yield different values.

40. According to Newton's law of gravitation, the force exerted on a particle of mass m located at the point (x, y, z) by a particle of mass M located at the origin is given by
$$\mathbf{F}(x, y, z) = \frac{-GMm}{\|\mathbf{r}\|^3}\mathbf{r}$$
where $\mathbf{r} = x\mathbf{i} + y\mathbf{j} + z\mathbf{k}$ and G is the gravitational constant. Show that $\mathbf{F}$ is the gradient of
$$f(x, y, z) = \frac{GMm}{\sqrt{x^2 + y^2 + z^2}}.$$

41. Let $\mathbf{u}$ be a unit position vector (initial point at the origin) in the plane and let θ be the angle measured in the counterclockwise direction from the positive x-axis to $\mathbf{u}$. Let f be a function of two variables.
 (a) Show that $f'_{\mathbf{u}}(x, y) = \dfrac{\partial f}{\partial x} \cos \theta + \dfrac{\partial f}{\partial y} \sin \theta$.
 (b) Let $f(x, y) = x^3 + 2xy - xy^2$ and $\theta = 2\pi/3$. Find $f'_{\mathbf{u}}(-1, 2)$.

42. Refer to Exercise 41. Let $f(x, y) = x^2 e^{2y}$ and $\theta = 5\pi/4$. Find $f'_{\mathbf{u}}(x, y)$ and $f'_{\mathbf{u}}(2, \ln 2)$.

43. Assume that $\nabla f(\mathbf{x})$ and $\nabla g(\mathbf{x})$ exist. Use Theorem 15.1.3 to derive the product rule
$$\nabla[f(\mathbf{x})g(\mathbf{x})] = f(\mathbf{x})\nabla g(\mathbf{x}) + g(\mathbf{x})\nabla f(\mathbf{x}).$$

44. Assume that $\nabla f(\mathbf{x})$ and $\nabla g(\mathbf{x})$ exist, and that $g(\mathbf{x}) \neq 0$. Derive the quotient rule
$$\nabla \left[\frac{f(\mathbf{x})}{g(\mathbf{x})}\right] = \frac{g(\mathbf{x})\nabla f(\mathbf{x}) - f(\mathbf{x})\nabla g(\mathbf{x})}{g^2(\mathbf{x})}.$$

45. Assume that $\nabla f(\mathbf{x})$ exists. Prove that, for any integer n,
$$\nabla f^n(\mathbf{x}) = nf^{n-1}(\mathbf{x})\nabla f(\mathbf{x}).$$
Does this result hold if n is replaced by any real number?

■ 15.3 THE MEAN-VALUE THEOREM; TWO INTERMEDIATE-VALUE THEOREMS

You have seen the important role played by the mean-value theorem in the calculus of functions of one variable. Here we take up the analogous result for functions of several variables. Let $\mathbf{a}$ and $\mathbf{b}$ be points (either in the plane or in three space); by $\overline{\mathbf{ab}}$ we mean the line segment that joins point $\mathbf{a}$ to point $\mathbf{b}$.

THEOREM 15.3.1 THE MEAN-VALUE THEOREM (SEVERAL VARIABLES)

If f is differentiable at each point of the line segment $\overline{\mathbf{ab}}$, then there exists on that line segment a point $\mathbf{c}$ between $\mathbf{a}$ and $\mathbf{b}$ such that

$$f(\mathbf{b}) - f(\mathbf{a}) = \nabla f(\mathbf{c}) \cdot (\mathbf{b} - \mathbf{a}).$$

PROOF As t ranges from 0 to 1, $\mathbf{a} + t(\mathbf{b} - \mathbf{a})$ traces out the line segment $\overline{\mathbf{ab}}$. The idea of the proof is to apply the one-variable mean-value theorem to the function

$$g(t) = f(\mathbf{a} + t[\mathbf{b} - \mathbf{a}]), \qquad t \in [0, 1].$$

To show that g is differentiable on the open interval $(0, 1)$, we take $t \in (0, 1)$ and form

$$
\begin{aligned}
g(t + h) - g(t) &= f(\mathbf{a} + (t + h)[\mathbf{b} - \mathbf{a}]) - f(\mathbf{a} + t[\mathbf{b} - \mathbf{a}]) \\
&= f(\mathbf{a} + t[\mathbf{b} - \mathbf{a}] + h[\mathbf{b} - \mathbf{a}]) - f(\mathbf{a} + t[\mathbf{b} - \mathbf{a}]) \\
&= \nabla f(\mathbf{a} + t[\mathbf{b} - \mathbf{a}]) \cdot h[\mathbf{b} - \mathbf{a}] + o(h[\mathbf{b} - \mathbf{a}]).
\end{aligned}
$$

Since

$$\nabla f(\mathbf{a} + t[\mathbf{b} + \mathbf{a}]) \cdot h(\mathbf{b} - \mathbf{a}) = [\nabla f(\mathbf{a} + t[\mathbf{b} - \mathbf{a}]) \cdot (\mathbf{b} - \mathbf{a})]h$$

and the $o(h[\mathbf{b} - \mathbf{a}])$ term is obviously $o(h)$, we can write

$$g(t + h) - g(t) = [\nabla f(\mathbf{a} + t[\mathbf{b} - \mathbf{a}]) \cdot (\mathbf{b} - \mathbf{a})]h + o(h).$$

Dividing both sides by h, we see that g is differentiable and

$$g'(t) = \nabla f(\mathbf{a} + t[\mathbf{b} - \mathbf{a}]) \cdot (\mathbf{b} - \mathbf{a}).$$

The function g is clearly continuous at 0 and at 1. Applying the one-variable mean-value theorem to g, we can conclude that there exists a number t_0 between 0 and 1 such that

(1) $$g(1) - g(0) = g'(t_0)(1 - 0).$$

Since $g(1) = f(\mathbf{b})$, $g(0) = f(\mathbf{a})$, and $g'(t_0) = \nabla f(\mathbf{a} + t_0[\mathbf{b} - \mathbf{a}]) \cdot (\mathbf{b} - \mathbf{a})$, condition (1) gives

$$f(\mathbf{b}) - f(\mathbf{a}) = \nabla f(\mathbf{a} + t_0[\mathbf{b} - \mathbf{a}]) \cdot (\mathbf{b} - \mathbf{a}).$$

Setting $\mathbf{c} = \mathbf{a} + t_0[\mathbf{b} - \mathbf{a}]$, we have

$$f(\mathbf{b}) - f(\mathbf{a}) = \nabla f(\mathbf{c}) \cdot (\mathbf{b} - \mathbf{a}). \quad \square$$

U is connected

Figure 15.3.1

A nonempty open set U (in the plane or in three-space) is said to be *connected* iff any two points of U can be joined by a polygonal path that lies entirely in U. You can see such a set in Figure 15.3.1.

The set shown in Figure 15.3.2 is the union of two disjoint open sets. The set is open but not connected: it is impossible to join $\mathbf{a}$ and $\mathbf{b}$ by a polygonal path that lies within the set.

In Chapter 4 you saw that, if $f'(x) = 0$ for all x in an open interval I, then f is constant on I. We have a similar result for functions of several variables.

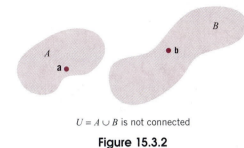

$U = A \cup B$ is not connected

Figure 15.3.2

THEOREM 15.3.2

Let U be an open connected set and let f be a differentiable function on U.

If $\nabla f(\mathbf{x}) = \mathbf{0}$ for all $\mathbf{x}$ in U, then f is constant on U.

PROOF Let $\mathbf{a}$ and $\mathbf{b}$ be any two points in U. Since U is connected, we can join these points by a polygonal path with vertices $\mathbf{a} = \mathbf{c}_0, \mathbf{c}_1, \mathbf{c}_2, \ldots, \mathbf{c}_{n-1}, \mathbf{c}_n = \mathbf{b}$ (see Figure 15.3.3). By the mean-value theorem (15.3.1) there exist points

$\mathbf{c}_1^*$ between $\mathbf{c}_0$ and $\mathbf{c}_1$ such that $f(\mathbf{c}_1) - f(\mathbf{c}_0) = \nabla f(\mathbf{c}_1^*) \cdot (\mathbf{c}_1 - \mathbf{c}_0),$

$\mathbf{c}_2^*$ between $\mathbf{c}_1$ and $\mathbf{c}_2$ such that $f(\mathbf{c}_2) - f(\mathbf{c}_1) = \nabla f(\mathbf{c}_2^*) \cdot (\mathbf{c}_2 - \mathbf{c}_1),$

$\qquad\qquad\qquad\qquad\qquad\qquad\vdots$

$\mathbf{c}_n^*$ between $\mathbf{c}_{n-1}$ and $\mathbf{c}_n$ such that $f(\mathbf{c}_n) - f(\mathbf{c}_{n-1}) = \nabla f(\mathbf{c}_n^*) \cdot (\mathbf{c}_n - \mathbf{c}_{n-1}).$

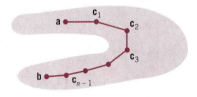

Figure 15.3.3

If $\nabla f(\mathbf{x}) = \mathbf{0}$ for all $\mathbf{x}$ in U, then

$$f(\mathbf{c}_1) - f(\mathbf{c}_0) = 0, \quad f(\mathbf{c}_2) - f(\mathbf{c}_1) = 0, \ldots, f(\mathbf{c}_n) - f(\mathbf{c}_{n-1}) = 0.$$

This shows that

$$f(\mathbf{a}) = f(\mathbf{c}_0) = f(\mathbf{c}_1) = f(\mathbf{c}_2) = \cdots = f(\mathbf{c}_{n-1}) = f(\mathbf{c}_n) = f(\mathbf{b}).$$

Since $\mathbf{a}$ and $\mathbf{b}$ are arbitrary points of U, f must be constant on U. ❏

THEOREM 15.3.3

Let U be an open connected set, and let f and g be differentiable functions on U.

If $\nabla f(\mathbf{x}) = \nabla g(\mathbf{x})$ for all $\mathbf{x}$ in U, then f and g differ by a constant on U.

PROOF If $\nabla f(\mathbf{x}) = \nabla g(\mathbf{x})$ for all $\mathbf{x}$ in U, then

$$\nabla[\,f(\mathbf{x}) - g(\mathbf{x})] = \nabla f(\mathbf{x}) - \nabla g(\mathbf{x}) = \mathbf{0} \qquad \text{for all } \mathbf{x} \text{ in } U.$$

By Theorem 15.3.2, $f - g$ must be constant on U. ❑

You have seen that a function continuous on an interval skips no values (Theorem 2.6.1). There are analogous results for functions of several variables. Here is one of them.

THEOREM 15.3.4 AN INTERMEDIATE-VALUE THEOREM (SEVERAL VARIABLES)

Suppose that f is continuous on an open connected set U and A, B, C are real numbers such that $A < C < B$. If, somewhere in U, f takes on the value A and, somewhere in U, f takes on the value B, then, somewhere in U, f takes on the value C.

PROOF Let $\mathbf{a}$ and $\mathbf{b}$ be points of U for which

$$f(\mathbf{a}) = A \qquad \text{and} \qquad f(\mathbf{b}) = B.$$

We must show that there exists a point $\mathbf{c}$ in U for which $f(\mathbf{c}) = C$.

Since U is connected, there is a polygonal path γ in U that joins $\mathbf{a}$ to $\mathbf{b}$. Let $\mathbf{r} = \mathbf{r}(t)$, $a \le t \le b$, be a continuous parametrization of the path γ with $\mathbf{r}(a) = \mathbf{a}$ and $\mathbf{r}(b) = \mathbf{b}$. Since $\mathbf{r}$ is continuous, the composition

$$g(t) = f(\mathbf{r}(t))$$

is also continuous on $[a, b]$. Since

$$g(a) = f(\mathbf{r}(a)) = f(\mathbf{a}) = A \qquad \text{and} \qquad g(b) = f(\mathbf{r}(b)) = f(\mathbf{b}) = B,$$

we know from Theorem 2.6.1 that there is a number c in $[a, b]$ for which $g(c) = C$. Setting $\mathbf{c} = \mathbf{r}(c)$ we have $f(\mathbf{c}) = C$. ❑

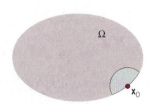

Figure 15.3.4

Continuity on a Closed Region

An open connected set is called an *open region*. If we start with an open region and adjoin to it the boundary, then we have what is called a *closed region*. (A closed region is therefore a closed set, the interior of which is an open region.)

Continuity on a closed region Ω requires continuity at the boundary points of Ω as well as the interior points.

Let the function f be defined on a closed region Ω, and let $\mathbf{x}_0 \in \Omega$.

If $\mathbf{x}_0$ is an interior point of Ω, then all points $\mathbf{x}$ sufficiently close to $\mathbf{x}_0$ are in Ω and, by definition, f is continuous at $\mathbf{x}_0$ iff

$$\text{as } \mathbf{x} \text{ approaches } \mathbf{x}_0, f(\mathbf{x}) \text{ approaches } f(\mathbf{x}_0). \qquad \text{(Figure 15.3.4)}$$

If $\mathbf{x}_0$ is a boundary point of Ω, then we have to modify the definition and say: f is continuous at $\mathbf{x}_0$ iff

$$\text{as } \mathbf{x} \text{ approaches } \mathbf{x}_0 \text{ within } \Omega, f(\mathbf{x}) \text{ approaches } f(\mathbf{x}_0). \qquad \text{(Figure 15.3.5)}$$

Figure 15.3.5

In terms of ϵ-δ, f is continuous at a boundary point $\mathbf{x}_0$ iff for each $\epsilon > 0$ there exists $\delta > 0$ such that

$$\text{if} \quad \|\mathbf{x} - \mathbf{x}_0\| < \delta \quad \text{and} \quad \mathbf{x} \in \Omega, \quad \text{then} \quad |f(\mathbf{x}) - f(\mathbf{x}_0)| < \epsilon.$$

(This is completely analogous to one-sided continuity at an endpoint of a closed interval $[a, b]$ as defined in Section 2.4.)

The intermediate-value result that we just proved for open connected sets can be extended to closed regions.

THEOREM 15.3.5 A SECOND INTERMEDIATE-VALUE THEOREM (SEVERAL VARIABLES)

Suppose that f is continuous on a closed region Ω and A, B, C are real numbers such that $A < C < B$. If, somewhere in Ω, f takes on the value A and, somewhere in Ω, f takes on the value B, then, somewhere in Ω, f takes on the value C.

PROOF Let $\mathbf{a}$ and $\mathbf{b}$ be points of Ω for which

$$f(\mathbf{a}) = A \quad \text{and} \quad f(\mathbf{b}) = B.$$

If $\mathbf{a}$ and $\mathbf{b}$ are both in the interior of Ω, then the result follows from the previous theorem. But one or both of these points could lie on the boundary of Ω. To take care of that possibility, we can proceed as follows. Take ϵ small enough that

$$A + \epsilon < C < B - \epsilon.$$

By continuity there exist points $\mathbf{x}_1$, $\mathbf{x}_2$ in the interior of Ω for which

$$f(\mathbf{x}_1) < A + \epsilon \quad \text{and} \quad B - \epsilon < f(\mathbf{x}_2). \qquad \text{(Exercise 7)}$$

Then

$$f(\mathbf{x}_1) < C < f(\mathbf{x}_2)$$

and the result follows from the previous theorem. ❏

EXERCISES 15.3

1. The function $f(x, y) = x^3 - xy$ is differentiable for all (x, y). Let $\mathbf{a} = (0, 1)$ and $\mathbf{b} = (1, 3)$. Find a point $\mathbf{c}$ on the line segment joining $\mathbf{a}$ and $\mathbf{b}$ such that

$$f(\mathbf{b}) - f(\mathbf{a}) = \nabla f(\mathbf{c}) \cdot (\mathbf{b} - \mathbf{a}).$$

2. The function $f(x, y, z) = 4xz - y^2 + z^2$ is differentiable for all (x, y, z). Let $\mathbf{a} = (0, 1, 1)$ and $\mathbf{b} = (1, 3, 2)$. Find a point $\mathbf{c}$ on the line segment joining $\mathbf{a}$ and $\mathbf{b}$ such that

$$f(\mathbf{b}) - f(\mathbf{a}) = \nabla f(\mathbf{c}) \cdot (\mathbf{b} - \mathbf{a}).$$

3. (a) Find f if $\nabla f(x, y, z) = a_1\mathbf{i} + a_2\mathbf{j} + a_3\mathbf{k}$ for all (x, y, z).

(b) What can you conclude about f and g if
$\nabla f(x, y, z) - \nabla g(x, y, z) = a_1\mathbf{i} + a_2\mathbf{j} + a_3\mathbf{k}$ for all (x, y, z)?

4. (*Rolle's theorem for functions of several variables*) Show that, if f is differentiable at each point of the line segment $\overline{\mathbf{a}\mathbf{b}}$ and $f(\mathbf{a}) = f(\mathbf{b})$, then there exists a point $\mathbf{c}$ between $\mathbf{a}$ and $\mathbf{b}$ for which $\nabla f(\mathbf{c}) \perp (\mathbf{b} - \mathbf{a})$.

5. Let $U = \{\mathbf{x}: \|\mathbf{x}\| \neq 1\}$. Define f on U by setting

$$f(\mathbf{x}) = \begin{cases} 0, & \|\mathbf{x}\| < 1 \\ 1, & \|\mathbf{x}\| > 1. \end{cases}$$

(a) Note that $\nabla f(\mathbf{x}) = \mathbf{0}$ for all $\mathbf{x}$ in U, but f is not constant on U. Explain how this does not contradict Theorem 15.3.2.

(b) Define a function g on U different from f such that $\nabla f(\mathbf{x}) = \nabla g(\mathbf{x})$ for all $\mathbf{x}$ in U and $f - g$ is (i) constant on U, (ii) not constant on U.

(c) The function f takes on the value 0, takes on the value 1, but takes on no value in between. Explain how this does not contradict Theorem 15.3.4.

6. A set of points is said to be *convex* provided that every pair of points in the set can be joined by a line segment that lies entirely within the set. Show that, if $\|\nabla f(\mathbf{x})\| \leq M$ for all $\mathbf{x}$ in some convex set Ω, then

$$|f(\mathbf{x}_1) - f(\mathbf{x}_2)| \leq M\|\mathbf{x}_1 - \mathbf{x}_2\| \qquad \text{for all } \mathbf{x}_1 \text{ and } \mathbf{x}_2 \text{ in } \Omega.$$

7. Justify the assertion made in the proof of Theorem 15.3.5 that there exist points $\mathbf{x}_1$, $\mathbf{x}_2$ in the interior of Ω for which

$$f(\mathbf{x}_1) < A + \epsilon \qquad \text{and} \qquad B - \epsilon < f(\mathbf{x}_2).$$

■ 15.4 CHAIN RULES

For functions of a single variable there is basically only one chain rule: Theorem 3.5.7. For functions of several variables there are many chain rules.

A vector-valued function is said to be *continuous* provided that its components are continuous. If $f = f(x, y, z)$ is a scalar-valued function (a real-valued function), then its gradient ∇f is a vector-valued function. We say that f is *continuously differentiable* on an *open set U* iff f is differentiable on U and ∇f is continuous on U.

If a curve $\mathbf{r}$ lies in the domain of f, then we can form the composition

$$(f \circ \mathbf{r})(t) = f(\mathbf{r}(t)).$$

The composition $f \circ \mathbf{r}$ is a real-valued function of a real variable t. The numbers $f(\mathbf{r}(t))$ are the values taken on by f *along the curve* $\mathbf{r}$.

THEOREM 15.4.1 CHAIN RULE (ALONG A CURVE)

If f is continuously differentiable on an open set U and $\mathbf{r} = \mathbf{r}(t)$ is a differentiable curve that lies entirely in U, then the composition $f \circ \mathbf{r}$ is differentiable and

$$\frac{d}{dt}[f(\mathbf{r}(t))] = \nabla f(\mathbf{r}(t)) \cdot \mathbf{r}'(t).$$

PROOF We will show that

$$\lim_{h \to 0} \frac{f(\mathbf{r}(t + h)) - f(\mathbf{r}(t))}{h} = \nabla f(\mathbf{r}(t)) \cdot \mathbf{r}'(t).$$

For $h \neq 0$ and sufficiently small, the line segment that joins $\mathbf{r}(t)$ to $\mathbf{r}(t + h)$ lies entirely in U. This we know because U is open and $\mathbf{r}$ is continuous. (See Figure 15.4.1.) For such h, the mean-value theorem we proved in Section 15.3 assures us that there exists a point $\mathbf{c}(h)$ between $\mathbf{r}(t)$ and $\mathbf{r}(t + h)$ such that

$$f(\mathbf{r}(t + h)) - f(\mathbf{r}(t)) = \nabla f(\mathbf{c}(h)) \cdot [\mathbf{r}(t + h) - \mathbf{r}(t)].$$

Dividing both sides by h, we have

$$\frac{f(\mathbf{r}(t + h)) - f(\mathbf{r}(t))}{h} = \nabla f(\mathbf{c}(h)) \cdot \left[\frac{\mathbf{r}(t + h) - \mathbf{r}(t)}{h}\right].$$

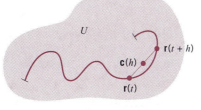

Figure 15.4.1

As h tends to zero, $\mathbf{c}(h)$ tends to $\mathbf{r}(t)$ and by the continuity of ∇f,

$$\nabla f(\mathbf{c}(h)) \to \nabla f(\mathbf{r}(t)).$$

Since

$$\frac{\mathbf{r}(t + h) - \mathbf{r}(t)}{h} \to \mathbf{r}'(t),$$

the result follows. ❑

Example 1   Use the chain rule to find the rate of change of

$$f(x, y) = \tfrac{1}{3}(x^3 + y^3)$$

with respect to t along the ellipse $\mathbf{r}(t) = a \cos t\, \mathbf{i} + b \sin t\, \mathbf{j}$.

SOLUTION   The rate of change of f with respect to t along the curve $\mathbf{r}$ is the derivative

$$\frac{d}{dt}\,[f(\mathbf{r}(t))].$$

By the chain rule (Theorem 15.4.1)

$$\frac{d}{dt}\,[f(\mathbf{r}(t))] = \nabla f(\mathbf{r}(t)) \cdot \mathbf{r}'(t).$$

Here

$$\nabla f = x^2 \mathbf{i} + y^2 \mathbf{j}.$$

With $x(t) = a \cos t$ and $y(t) = b \sin t$, we have

$$\nabla f(\mathbf{r}(t)) = a^2 \cos^2 t\, \mathbf{i} + b^2 \sin^2 t\, \mathbf{j}.$$

Since $\mathbf{r}'(t) = -a \sin t\, \mathbf{i} + b \cos t\, \mathbf{j}$, you can see that

$$\begin{aligned}
\frac{d}{dt}\,[f(\mathbf{r}(t))] &= \nabla f(\mathbf{r}(t)) \cdot \mathbf{r}'(t) \\
&= (a^2 \cos^2 t\, \mathbf{i} + b^2 \sin^2 t\, \mathbf{j}) \cdot (-a \sin t\, \mathbf{i} + b \cos t\, \mathbf{j}) \\
&= -a^3 \sin t \cos^2 t + b^3 \sin^2 t \cos t \\
&= \sin t \cos t (b^3 \sin t - a^3 \cos t). \quad ❑
\end{aligned}$$

Remark   Note that we could have obtained the same result without invoking Theorem 15.4.1 by first forming $f(\mathbf{r}(t))$ and then differentiating with respect to t:

$$f(\mathbf{r}(t)) = f(x(t), y(t)) = \tfrac{1}{3}([x(t)]^3 + [y(t)]^3) = \tfrac{1}{3}(a^3 \cos^3 t + b^3 \sin^3 t)$$

so that

$$\frac{d}{dt}[f(\mathbf{r}(t))] = \tfrac{1}{3}[3a^3(\cos^2 t)(-\sin t) + 3b^3\sin^2 t \cos t]$$

$$= \sin t \cos t \,(b^3\sin t - a^3\cos t). \quad \square$$

Example 2   Use the chain rule to find the rate of change of

$$f(x, y, z) = x^2y + z \cos x$$

with respect to t along the twisted cubic $\mathbf{r}(t) = t\mathbf{i} + t^2\mathbf{j} + t^3\mathbf{k}$.

SOLUTION   Once again we use the relation

$$\frac{d}{dt}[f(\mathbf{r}(t))] = \nabla f(\mathbf{r}(t)) \cdot \mathbf{r}'(t).$$

This time

$$\nabla f = (2xy - z \sin x)\mathbf{i} + x^2\mathbf{j} + \cos x \,\mathbf{k}.$$

With $x(t) = t$, $y(t) = t^2$, $z(t) = t^3$, we have

$$\nabla f(\mathbf{r}(t)) = (2t^3 - t^3\sin t)\mathbf{i} + t^2\mathbf{j} + \cos t\,\mathbf{k}.$$

Since $\mathbf{r}'(t) = \mathbf{i} + 2t\mathbf{j} + 3t^2\mathbf{k}$, we have

$$\frac{d}{dt}[f(\mathbf{r}(t))] = \nabla f(\mathbf{r}(t)) \cdot \mathbf{r}'(t)$$

$$= [(2t^3 - t^3\sin t)\mathbf{i} + t^2\mathbf{j} + \cos t\,\mathbf{k}] \cdot [\mathbf{i} + 2t\mathbf{j} + 3t^2\mathbf{k}]$$
$$= 2t^3 - t^3\sin t + 2t^3 + 3t^2\cos t$$
$$= 4t^3 - t^3\sin t + 3t^2\cos t.$$

You can check this answer by first forming $f(\mathbf{r}(t))$ and then differentiating. $\square$

Another Formulation of (15.4.1)

The chain rule for functions of one variable,

$$\frac{d}{dt}[u(x(t))] = u'(x(t))x'(t),$$

can be written

$$\frac{du}{dt} = \frac{du}{dx}\frac{dx}{dt}.$$

In a similar manner, the relation

$$\frac{d}{dt}[u(\mathbf{r}(t))] = \nabla u(\mathbf{r}(t)) \cdot \mathbf{r}'(t)$$

can be written

(1)
$$\frac{du}{dt} = \nabla u \cdot \frac{d\mathbf{r}}{dt}.$$

With

$$\nabla u = \frac{\partial u}{\partial x}\mathbf{i} + \frac{\partial u}{\partial y}\mathbf{j} + \frac{\partial u}{\partial z}\mathbf{k} \qquad \text{and} \qquad \frac{d\mathbf{r}}{dt} = \frac{dx}{dt}\mathbf{i} + \frac{dy}{dt}\mathbf{j} + \frac{dz}{dt}\mathbf{k},$$

equation (1) takes the form

(15.4.2)
$$\frac{du}{dt} = \frac{\partial u}{\partial x}\frac{dx}{dt} + \frac{\partial u}{\partial y}\frac{dy}{dt} + \frac{\partial u}{\partial z}\frac{dz}{dt}.$$

In the two-variable case, the z-term drops out and we have

(15.4.3)
$$\frac{du}{dt} = \frac{\partial u}{\partial x}\frac{dx}{dt} + \frac{\partial u}{\partial y}\frac{dy}{dt}.$$

Example 3 Find du/dt if

$$u = x^2 - y^2 \qquad \text{and} \qquad x = t^2 - 1, \quad y = 3\sin \pi t.$$

SOLUTION Here we are in the two-variable case

$$\frac{du}{dt} = \frac{\partial u}{\partial x}\frac{dx}{dt} + \frac{\partial u}{\partial y}\frac{dy}{dt}.$$

Since

$$\frac{\partial u}{\partial x} = 2x, \quad \frac{\partial u}{\partial y} = -2y \qquad \text{and} \qquad \frac{dx}{dt} = 2t, \quad \frac{dy}{dt} = 3\pi \cos \pi t,$$

we have

$$\frac{du}{dt} = (2x)(2t) + (-2y)(3\pi \cos \pi t)$$
$$= 2(t^2 - 1)(2t) + (-2)(3\sin \pi t)(3\pi \cos \pi t)$$
$$= 4t^3 - 4t - 18\pi \sin \pi t \cos \pi t.$$

You can obtain this same result by first writing u directly as a function of t and then differentiating:

$$u = x^2 - y^2 = (t^2 - 1)^2 - (3\sin \pi t)^2$$

so that

$$\frac{du}{dt} = 2(t^2 - 1)2t - 2(3\sin \pi t)3\pi \cos \pi t = 4t^3 - 4t - 18\pi \sin \pi t \cos \pi t. \quad \square$$

Example 4 A solid is in the shape of a frustum of a right circular cone (see Figure 15.4.2). If the upper radius x decreases at the rate of 2 inches per minute, the lower radius y increases at the rate of 3 inches per minute, and the height z decreases at the

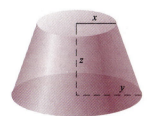

Figure 15.4.2

rate of 4 inches per minute, at what rate is the volume V changing at the instant the upper radius is 10 inches, the lower radius is 12 inches, and the height is 18 inches.

SOLUTION   The formula for the volume of a frustum of a right circular cone is

$$V = \frac{1}{3} \pi z (x^2 + xy + y^2) \qquad \text{(Exercise 35, Section 6.2)}$$

so that

$$\frac{\partial V}{\partial x} = \frac{1}{3} \pi z (2x + y), \qquad \frac{\partial V}{\partial y} = \frac{1}{3} \pi z (x + 2y), \qquad \frac{\partial V}{\partial z} = \frac{1}{3} \pi (x^2 + xy + y^2).$$

Since

$$\frac{dV}{dt} = \frac{\partial V}{\partial x} \frac{dx}{dt} + \frac{\partial V}{\partial y} \frac{dy}{dt} + \frac{\partial V}{\partial z} \frac{dz}{dt},$$

we have

$$\frac{dV}{dt} = \frac{1}{3} \pi z (2x + y) \frac{dx}{dt} + \frac{1}{3} \pi z (x + 2y) \frac{dy}{dt} + \frac{1}{3} \pi (x^2 + xy + y^2) \frac{dz}{dt}.$$

Set

$$x = 10, \quad y = 12, \quad z = 18, \quad \frac{dx}{dt} = -2, \quad \frac{dy}{dt} = 3, \quad \frac{dz}{dt} = -4,$$

and you will find that

$$\frac{dV}{dt} = -\frac{772}{3} \pi \cong -808.4.$$

The volume decreases at the rate of approximately 808 cubic inches per minute.   ❑

Other Chain Rules

In the setting of functions of several variables there are numerous chain rules. Some are stated here, others in the exercises. They can all be deduced from Theorem 15.4.1 and its corollaries, (15.4.2) and (15.4.3).

   If, for example,

$$u = u(x, y) \qquad \text{where} \quad x = x(s, t) \qquad \text{and} \qquad y = y(s, t),$$

then

(15.4.4)
$$\frac{\partial u}{\partial s} = \frac{\partial u}{\partial x} \frac{\partial x}{\partial s} + \frac{\partial u}{\partial y} \frac{\partial y}{\partial s} \qquad \text{and} \qquad \frac{\partial u}{\partial t} = \frac{\partial u}{\partial x} \frac{\partial x}{\partial t} + \frac{\partial u}{\partial y} \frac{\partial y}{\partial t}.$$

   To obtain the first equation, keep t fixed and differentiate u with respect to s according to Formula (15.4.3); to obtain the second equation, keep s fixed and differentiate u with respect to t. These formulas can be viewed as a chain rule on a surface, extending the chain rule along a curve.

In Figure 15.4.3 we have drawn a *tree diagram* for Formula (15.4.4). We construct such a tree by branching at each stage from a function to all the variables that directly determine it. Each path starting at u and ending at a variable determines a product of (partial) derivatives. The partial derivative of u with respect to each variable is the sum of the products generated by all the direct paths to that variable.

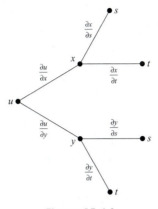

Figure 15.4.3

Example 5 Let $u = x^2 - 2xy + 2y^3$, where $x = s^2 \ln t$ and $y = 2st^3$. Find $\partial u / \partial s$ and $\partial u / \partial t$.

SOLUTION Here u is a function of two variables, x and y, where each of x and y is itself a function of two variables. Thus (15.4.4) applies. Since

$$\frac{\partial u}{\partial x} = (2x - 2y), \qquad \frac{\partial u}{\partial y} = (-2x + 6y^2)$$

and

$$\frac{\partial x}{\partial s} = 2s \ln t, \qquad \frac{\partial y}{\partial s} = 2t^3, \qquad \frac{\partial x}{\partial t} = \frac{s^2}{t}, \qquad \frac{\partial y}{\partial t} = 6st^2,$$

we have

$$\frac{\partial u}{\partial s} = (2x - 2y)(2s \ln t) + (-2x + 6y^2)(2t^3),$$

and

$$\frac{\partial u}{\partial t} = (2x - 2y)\left(\frac{s^2}{t}\right) + (-2x + 6y^2)(6st^2).$$

These results can be expressed entirely in terms of s and t by replacing x by $s^2 \ln t$ and y by $2st^3$:

$$\frac{\partial u}{\partial s} = (2s^2 \ln t - 4st^3)(2s \ln t) + (-2s^2 \ln t + 24s^2t^6)(2t^3),$$

$$\frac{\partial u}{\partial t} = (2s^2 \ln t - 4st^3)\left(\frac{s^2}{t}\right) + (-2s^2 \ln t + 24s^2t^6)(6st^2). \qquad \square$$

Now suppose that u is a function of three variables:

$$u = u(x, y, z) \qquad \text{where} \quad x = x(s, t), \quad y = y(s, t), \quad z = z(s, t).$$

A tree diagram for the partials of u appears in Figure 15.4.4. The partials of u with respect to s and t can be read from the diagram:

(15.4.5)
$$\frac{\partial u}{\partial s} = \frac{\partial u}{\partial x}\frac{\partial x}{\partial s} + \frac{\partial u}{\partial y}\frac{\partial y}{\partial s} + \frac{\partial u}{\partial z}\frac{\partial z}{\partial s}, \qquad \frac{\partial u}{\partial t} = \frac{\partial u}{\partial x}\frac{\partial x}{\partial t} + \frac{\partial u}{\partial y}\frac{\partial y}{\partial t} + \frac{\partial u}{\partial z}\frac{\partial z}{\partial t}.$$

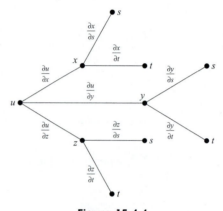

Figure 15.4.4

Example 6 Let $u = x^2 y^3 e^{xz}$, where $x = s^2 + t^2$, $y = 2st$ and $z = s \ln t$. Find $\partial u/\partial s$.

SOLUTION In this case, u is a function of three variables, x, y, z, where each of x, y, and z is a function of s and t. To find $\partial u/\partial s$, we use chain rule (15.4.5):

$$\frac{\partial u}{\partial s} = \frac{\partial u}{\partial x}\frac{\partial x}{\partial s} + \frac{\partial u}{\partial y}\frac{\partial y}{\partial s} + \frac{\partial u}{\partial z}\frac{\partial z}{\partial s}.$$

Now,

$$\frac{\partial u}{\partial x} = 2xy^3 e^{xz} + x^2 y^3 z e^{xz}, \qquad \frac{\partial u}{\partial y} = 3x^2 y^2 e^{xz}, \qquad \frac{\partial u}{\partial z} = x^3 y^3 e^{xz}$$

and

$$\frac{\partial x}{\partial s} = 2s, \qquad \frac{\partial y}{\partial s} = 2t, \qquad \frac{\partial z}{\partial s} = \ln t.$$

Therefore,

$$\frac{\partial u}{\partial s} = (2xy^3 e^{xz} + x^2 y^3 z e^{xz})(2s) + (3x^2 y^2 e^{xz})(2t) + (x^3 y^3 e^{xz})(\ln t).$$

This result can be expressed entirely in terms of s and t by substituting $s^2 + t^2$ for x, $2st$ for y, and $s \ln t$ for z. ❑

Implicit Differentiation

In Section 3.7 we presented a method for finding the derivative of a function $y = y(x)$ defined implicitly by an equation $u(x, y) = 0$. Here we show how the chain rule (15.4.1) can be used to derive a simpler method for finding the derivative of an implicitly defined function.

Let $u = u(x, y)$ be a continuously differentiable function and suppose that the equation

$$u(x, y) = 0$$

defines y implicitly as a differentiable function of x. We introduce a variable t by setting $x = t$. Then we have

$$u = u(x, y) \quad \text{with} \quad x = t \quad \text{and} \quad y = y(t).$$

By (15.4.3),

$$\frac{du}{dt} = \frac{\partial u}{\partial x}\frac{dx}{dt} + \frac{\partial u}{\partial y}\frac{dy}{dt}$$

Now, $u(t, y(t)) = 0$ for all t, so $du/dt = 0$. Also, since $x = t$, we have $dx/dt = 1$ and $dy/dt = dy/dx$. Therefore,

$$0 = \frac{\partial u}{\partial x} + \frac{\partial u}{\partial y}\frac{dy}{dx}.$$

If $\partial u/\partial y \neq 0$, we can solve for dy/dx to obtain

$$\frac{dy}{dx} = -\frac{\partial u/\partial x}{\partial u/\partial y}.$$

This result can be summarized as follows:

(15.4.6)

> Let $u = u(x, y)$ be differentiable and suppose that the equation $u(x, y) = 0$ defines y implicitly as a differentiable function of x. If $\partial u/\partial y \neq 0$, then
> $$\frac{dy}{dx} = -\frac{\partial u/\partial x}{\partial u/\partial y}.$$

Example 7 The equation

$$u(x, y) = 2xy - y^3 + 1 - x - 2y = 0$$

appeared in Example 1(a), Section 3.7. Using the method we have just derived,

$$\frac{\partial u}{\partial x} = 2y - 1, \qquad \frac{\partial u}{\partial y} = 2x - 3y^2 - 2,$$

and

$$\frac{dy}{dx} = -\frac{2y - 1}{2x - 3y^2 - 2} = \frac{1 - 2y}{2x - 3y^2 - 2},$$

which is the same as our previous result. ❏

As illustrated in the next example, this method can be extended to expressions involving more than two variables.

Example 8   Let $u = u(x, y, z)$ be a continuously differentiable function, and suppose that the equation

$$u(x, y, z) = 0$$

defines z implicitly as a differentiable function of x and y.   Show that

(15.4.7) if $\dfrac{\partial u}{\partial z} \neq 0$,   then $\dfrac{\partial z}{\partial x} = -\dfrac{\partial u/\partial x}{\partial u/\partial z}$ and $\dfrac{\partial z}{\partial y} = -\dfrac{\partial u/\partial y}{\partial u/\partial z}$.

SOLUTION   To be able to apply (15.4.5), we write

$$u = u(x, y, z) \quad \text{with} \quad x = s, \quad y = t, \quad z = z(s, t).$$

Then

$$\frac{\partial u}{\partial s} = \frac{\partial u}{\partial x}\frac{\partial x}{\partial s} + \frac{\partial u}{\partial y}\frac{\partial y}{\partial s} + \frac{\partial u}{\partial z}\frac{\partial z}{\partial s}.$$

Since $u(s, t, z(s, t)) = 0$ for all s and t,

$$\frac{\partial u}{\partial s} = 0.$$

Since

$$\frac{\partial x}{\partial s} = 1 \quad \text{and} \quad \frac{\partial y}{\partial s} = 0,$$

we have

$$0 = \frac{\partial u}{\partial x}\cdot 1 + \frac{\partial u}{\partial y}\cdot 0 + \frac{\partial u}{\partial z}\frac{\partial z}{\partial s} = \frac{\partial u}{\partial x} + \frac{\partial u}{\partial z}\frac{\partial z}{\partial s} = \frac{\partial u}{\partial x} + \frac{\partial u}{\partial z}\frac{\partial z}{\partial x}.$$

$$x = s$$

If $\partial u/\partial z \neq 0$, then

$$\frac{\partial z}{\partial x} = -\frac{\partial u/\partial x}{\partial u/\partial z}.$$

The formula for $\partial z/\partial y$ can be obtained in a similar manner.   ❏

EXERCISES 15.4

In Exercises 1–10, find the rate of change of f with respect to t along the given curve.

1. $f(x, y) = x^2 y$, $\mathbf{r}(t) = e^t\mathbf{i} + e^{-t}\mathbf{j}$.

2. $f(x, y) = x - y$, $\mathbf{r}(t) = at\mathbf{i} + b\cos at\mathbf{j}$.

3. $f(x, y) = \tan^{-1}(y^2 - x^2)$, $\mathbf{r}(t) = \sin t\mathbf{i} + \cos t\mathbf{j}$.

4. $f(x, y) = \ln(2x^2 + y^3)$, $\mathbf{r}(t) = e^{2t}\mathbf{i} + t^{1/3}\mathbf{j}$.

5. $f(x, y) = xe^y + ye^{-x}$, $\mathbf{r}(t) = (\ln t)\mathbf{i} + t(\ln t)\mathbf{j}$.

6. $f(x, y, z) = \ln(x^2 + y^2 + z^2)$,
$\mathbf{r}(t) = \sin t\mathbf{i} + \cos t\mathbf{j} + e^{2t}\mathbf{k}$.

7. $f(x, y, z) = xy - yz$, $\mathbf{r}(t) = t\mathbf{i} + t^2\mathbf{j} + t^3\mathbf{k}$.

8. $f(x, y, z) = x^2 + y^2$,
$\mathbf{r}(t) = a\cos\omega t\mathbf{i} + b\sin\omega t\mathbf{j} + b\omega t\mathbf{k}$.

9. $f(x, y, z) = x^2 + y^2 + z$,
$\mathbf{r}(t) = a \cos \omega t \mathbf{i} + b \sin \omega t \mathbf{j} + b \omega t \mathbf{k}$.

10. $f(x, y, z) = y^2 \sin(x + z)$, $\quad \mathbf{r}(t) = 2t \mathbf{i} + \cos t \mathbf{j} + t^3 \mathbf{k}$.

In Exercises 11–18, find du/dt by applying (15.4.2) or (15.4.3).

11. $u = x^2 - 3xy + 2y^2$; $\quad x = \cos t, \quad y = \sin t$.

12. $u = x + 4\sqrt{xy} - 3y$; $\quad x = t^3, \quad y = t^{-1} \quad (t > 0)$.

13. $u = e^x \sin y + e^y \sin x$; $\quad x = \frac{1}{2}t, \quad y = 2t$.

14. $u = 2x^2 - xy + y^2$; $\quad x = \cos 2t, \quad y = \sin t$.

15. $u = e^x \sin y$; $\quad x = t^2, \quad y = \pi t$.

16. $u = z \ln\left(\dfrac{y}{x}\right)$; $\quad x = t^2 + 1, y = \sqrt{t}, z = te^t$.

17. $u = xy + yz + zx$; $\quad x = t^2, y = t(1 - t)$, $z = (1 - t)^2$.

18. $u = x \sin \pi y - z \cos \pi x$; $\quad x = t^2, y = 1 - t$, $z = 1 - t^2$.

19. The radius of a right circular cone is increasing at the rate of 3 inches per second and the height is decreasing at the rate of 2 inches per second. At what rate is the volume of the cone changing at the instant the height is 20 inches and the radius is 14 inches?

20. The radius of a right circular cylinder is decreasing at the rate of 2 centimeters per second and the height is increasing at the rate of 3 centimeters per second. At what rate is the volume of the cylinder changing at the instant the radius is 13 centimeters and the height is 18 centimeters?

21. If the lengths of two sides of a triangle are x and y, and θ is the angle between the two sides, then the area A of the triangle is given by $A = \frac{1}{2}xy \sin \theta$. See the figure. If the sides are each increasing at the rate of 3 inches per second and θ is decreasing at the rate of 0.10 radian per second, how fast is the area changing at the instant $x = 1.5$ feet, $y = 2$ feet, and $\theta = 1$ radian?

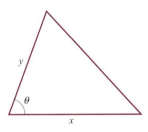

22. An object is moving along the curve of intersection of the paraboloid $z = x^2 + \frac{1}{4} y^2$ and the circular cylinder $x^2 + y^2 = 13$. If the x-coordinate is increasing at the rate of 5 centimeters per second, how fast is the z-coordinate changing at the instant when $x = 2$ centimeters and $y = 3$ centimeters?

In Exercises 23–28, find $\partial u/\partial s$ and $\partial u/\partial t$.

23. $u = x^2 - xy$ where $x = s \cos t, \quad y = t \sin s$.

24. $u = \sin(x - y) + \cos(x + y)$ where $x = st$, $y = s^2 - t^2$.

25. $u = x^2 \tan y$ where $x = s^2 t, \quad y = s + t^2$.

26. $u = z^2 \sec xy$ where $x = 2st, \quad y = s - t^2$, $z = s^2 t$.

27. $u = x^2 - xy + z^2$ where $x = s \cos t$, $y = \sin(t - s), \quad z = t \sin s$.

28. $u = xe^{yz^2}$ where $x = \ln st, \quad y = t^3, \quad z = s^2 + t^2$.

29. An object moves so that at time t it has position $\mathbf{r}(t) = x(t)\mathbf{i} + y(t)\mathbf{j} + z(t)\mathbf{k}$. Show that

$$\frac{d}{dt}[f(\mathbf{r}(t))]$$

is the directional derivative of f in the direction of the motion times the speed of the object.

30. (*Important*) Set $r = \|\mathbf{r}\|$ where $\mathbf{r} = x\mathbf{i} + y\mathbf{j} + z\mathbf{k}$. If f is a continuously differentiable function of r, then

(15.4.8)
$$\nabla[f(r)] = f'(r)\frac{\mathbf{r}}{r}.$$

Derive this formula.

In Exercises 31–33, calculate the following gradients $(r = \sqrt{x^2 + y^2 + z^2})$.

31. (a) $\nabla(\sin r)$. (b) $\nabla(r \sin r)$.

32. (a) $\nabla(r \ln r)$. (b) $\nabla(e^{1 - r^2})$.

33. (a) $\nabla\left(\dfrac{\sin r}{r}\right)$. (b) $\nabla\left(\dfrac{r}{\sin r}\right)$.

34. (a) Draw a tree diagram for du/dt given that

$u = u(x, y)$ where $x = x(s), \quad y = y(s), \quad$ and $\quad s = s(t)$.

(b) Compute du/dt.

35. Set $u = u(x, y, z)$ where

$$x = x(w, t), \quad y = y(w, t), \quad z = z(w, t)$$

$$\text{and} \quad w = w(r, s), \quad t = t(r, s).$$

(a) Draw a tree diagram for the partials of u.
(b) Compute $\partial u/\partial r$ and $\partial u/\partial s$.

36. Set $u = u(x, y, z, w)$ where

$$x = x(r, s, t), \quad y = y(s, t, v), \quad z = z(r, t)$$

$$\text{and} \quad w = w(r, s, t, v).$$

(a) Draw a tree diagram for the partials of u.
(b) Compute $\partial u/\partial r$ and $\partial u/\partial v$.

Higher Derivatives

37. Let $u = u(x, y)$, where $x = x(t)$ and $y = y(t)$, and assume that these functions have continuous second derivatives. Show that

$$\frac{d^2u}{dt^2} = \frac{\partial^2 u}{\partial x^2}\left(\frac{dx}{dt}\right)^2 + 2\frac{\partial^2 u}{\partial x \partial y}\frac{dx}{dt}\frac{dy}{dt} + \frac{\partial^2 u}{\partial y^2}\left(\frac{dy}{dt}\right)^2$$
$$+ \frac{\partial u}{\partial x}\frac{d^2x}{dt^2} + \frac{\partial u}{\partial y}\frac{d^2y}{dt^2}.$$

38. Let $u = u(x, y)$, where $x = x(s, t)$ and $y = y(s, t)$, and assume that these functions have continuous second derivatives. Show that

$$\frac{\partial^2 u}{\partial s^2} = \frac{\partial^2 u}{\partial x^2}\left(\frac{\partial x}{\partial s}\right)^2 + 2\frac{\partial^2 u}{\partial x \partial y}\frac{\partial x}{\partial s}\frac{\partial y}{\partial s} + \frac{\partial^2 u}{\partial y^2}\left(\frac{\partial y}{\partial s}\right)^2$$
$$+ \frac{\partial u}{\partial x}\frac{\partial^2 x}{\partial s^2} + \frac{\partial u}{\partial y}\frac{\partial^2 y}{\partial s^2}.$$

Polar Coordinates

39. Let $u = u(x, y)$ and assume that u is differentiable.
(a) Show that the change of variables to polar coordinates $x = r \cos \theta$ and $y = r \sin \theta$ gives

$$\frac{\partial u}{\partial r} = \frac{\partial u}{\partial x}\cos \theta + \frac{\partial u}{\partial y}\sin \theta,$$

$$\frac{\partial u}{\partial \theta} = -\frac{\partial u}{\partial x}r \sin \theta + \frac{\partial u}{\partial y}r \cos \theta.$$

(b) Express

$$\left(\frac{\partial u}{\partial r}\right)^2 + \frac{1}{r^2}\left(\frac{\partial u}{\partial \theta}\right)^2$$

entirely in terms of $\partial u/\partial x$ and $\partial u/\partial y$.

40. Let w be a function of polar coordinates r and θ. Then w is also a function of rectangular coordinates

$$x = r \cos \theta \quad \text{and} \quad y = r \sin \theta.$$

(a) Using the first part of Exercise 39, verify that

$$\frac{\partial w}{\partial x} = \frac{\partial w}{\partial r}\cos \theta - \frac{1}{r}\frac{\partial w}{\partial \theta}\sin \theta,$$

$$\frac{\partial w}{\partial y} = \frac{\partial w}{\partial r}\sin \theta + \frac{1}{r}\frac{\partial w}{\partial \theta}\cos \theta.$$

(b) Deduce from (a) that

$$\frac{\partial r}{\partial x} = \cos \theta, \qquad \frac{\partial r}{\partial y} = \sin \theta;$$

$$\frac{\partial \theta}{\partial x} = -\frac{1}{r}\sin \theta, \qquad \frac{\partial \theta}{\partial y} = \frac{1}{r}\cos \theta.$$

(c) Find the fallacy:

$$x = r \cos \theta, \qquad r = \frac{x}{\cos \theta}, \qquad \frac{\partial r}{\partial x} = \frac{1}{\cos \theta}.$$

41. *The gradient in polar coordinates*: Let $u = u(x, y)$ be differentiable. Show that if u is written in terms of polar coordinates, then

$$\nabla u = \frac{\partial u}{\partial r}\mathbf{e_r} + \frac{1}{r}\frac{\partial u}{\partial \theta}\mathbf{e_\theta},$$

where

$$\mathbf{e_r} = \cos \theta \mathbf{i} + \sin \theta \mathbf{j} \quad \text{and} \quad \mathbf{e_\theta} = -\sin \theta \mathbf{i} + \cos \theta \mathbf{j}.$$

In Exercises 42 and 43, use the formula in Exercise 41 to express the gradient of the given function in polar coordinates.

42. $u(x, y) = x^2 + y^2.$ **43.** $u(x, y) = x^2 - xy + y^2.$

44. Let $u = u(x, y)$, where $x = r \cos \theta$ and $y = r \sin \theta$, and assume that u has continuous second partial derivatives. Derive a formula for $\partial^2 u/\partial r \partial \theta.$

45. *Laplace's equation in polar coordinates*: Let $u = u(x, y)$ have continuous second partial derivatives. Show that

$$\frac{\partial^2 u}{\partial x^2} + \frac{\partial^2 u}{\partial y^2} = \frac{\partial^2 u}{\partial r^2} + \frac{1}{r^2}\frac{\partial^2 u}{\partial \theta^2} + \frac{1}{r}\frac{\partial u}{\partial r}.$$

In Exercises 46–49, y is defined implicitly as a differentiable function of x by the given equation. Find dy/dx.

46. $x^2 - 2xy + y^4 = 4.$ **47.** $xe^y + ye^x - 2x^2y = 0.$

48. $x^{2/3} + y^{2/3} = a^{2/3}$ (a a constant).

49. $x \cos xy + y \cos x = 2.$

In Exercises 50–53, z is defined implicitly as a differentiable function of x and y by the given equation. Find $\partial z/\partial x$ and $\partial z/\partial y.$

50. $z^4 + x^2z^3 + y^2 + xy = 2.$

51. $\cos xyz + \ln(x^2 + y^2 + z^2) = 0.$

52. $xe^y + ye^z + ze^x = 1.$

53. $z \tan x - xy^2z^3 = 2xyz.$

54. (*A chain rule for vector-valued functions*) Suppose that

$$\mathbf{u}(x, y) = u_1(x, y)\mathbf{i} + u_2(x, y)\mathbf{j}$$

where $x = x(t), \quad y = y(t).$

(a) Show that

(15.4.9)
$$\frac{d\mathbf{u}}{dt} = \frac{\partial \mathbf{u}}{\partial x}\frac{dx}{dt} + \frac{\partial \mathbf{u}}{\partial y}\frac{dy}{dt},$$

where

$$\frac{\partial \mathbf{u}}{\partial x} = \frac{\partial u_1}{\partial x}\mathbf{i} + \frac{\partial u_2}{\partial x}\mathbf{j} \quad \text{and} \quad \frac{\partial \mathbf{u}}{\partial y} = \frac{\partial u_1}{\partial y}\mathbf{i} + \frac{\partial u_2}{\partial y}\mathbf{j}.$$

(b) Let

$$\mathbf{u} = e^x \cos y \, \mathbf{i} + e^x \sin y \, \mathbf{j}$$

where $x = \frac{1}{2}t^2, \quad y = \pi t.$

Compute $d\mathbf{u}/dt$ (i) by applying (15.4.9), (ii) by forming $\mathbf{u}(t)$ directly.

55. Set

$$\mathbf{u}(x, y) = u_1(x, y)\mathbf{i} + u_2(x, y)\mathbf{j}$$

where $x = x(s, t), \quad y = y(s, t).$

Find $\partial \mathbf{u}/\partial s$ and $\partial \mathbf{u}/\partial t$.

56. Set

$$\mathbf{u}(x, y, z) = u_1(x, y, z)\mathbf{i} + u_2(x, y, z)\mathbf{j} + u_3(x, y, z)\mathbf{k},$$

where

$$x = x(t), \quad y = y(t), \quad z = z(t).$$

Derive a formula for $d\mathbf{u}/dt$ analogous to (15.4.9).

57. Homogeneous Functions. A function $f = f(x, y)$ is *homogeneous of degree n*, where n is a positive integer, if $f(tx, ty) = t^n f(x, y)$. Assume that f has continuous second partial derivatives, and is homogeneous of degree n.

(a) Show that f satisfies the equation

$$x\frac{\partial f}{\partial x} + y\frac{\partial f}{\partial y} = nf(x, y).$$

(b) Show that f satisfies the equation

$$x^2\frac{\partial^2 f}{\partial x^2} + 2xy\frac{\partial^2 f}{\partial x \partial y} + y^2\frac{\partial^2 f}{\partial y^2} = n(n-1)f(x, y).$$

■ 15.5 THE GRADIENT AS A NORMAL; TANGENT LINES AND TANGENT PLANES

Functions of Two Variables

We begin with a nonconstant function $f = f(x, y)$ that is continuously differentiable. (*Remember:* That means f is differentiable and its gradient ∇f is continuous.) You have seen that at each point of the domain the gradient vector, if not $\mathbf{0}$, points in the direction of the most rapid increase of f. Here we show that

(15.5.1)	at each point of the domain, the gradient vector, if not $\mathbf{0}$, is perpendicular to the level curve that passes through that point.

PROOF We choose a point (x_0, y_0) in the domain and assume that $\nabla f(x_0, y_0) \neq \mathbf{0}$. The level curve through this point has equation

$$f(x, y) = c \qquad \text{where } c = f(x_0, y_0).$$

Under our assumptions on f, this curve can be parametrized in a neighborhood of (x_0, y_0) by a continuously differentiable vector function

$$\mathbf{r}(t) = x(t)\mathbf{i} + y(t)\mathbf{j}, \qquad t \in I$$

with nonzero tangent vector $\mathbf{r}'(t)$.†

Now take t_0 such that

$$\mathbf{r}(t_0) = x_0\mathbf{i} + y_0\mathbf{j} = (x_0, y_0).$$

---

† This follows from a result of advanced calculus known as the *implicit function theorem*.

We will show that

$$\nabla f(\mathbf{r}(t_0)) \perp \mathbf{r}'(t_0).$$

Since f is constantly c on the curve, we have

$$f(\mathbf{r}(t)) = c \qquad \text{for all } t \in I.$$

For such t

$$\frac{d}{dt}[f(\mathbf{r}(t))] = \nabla f(\mathbf{r}(t)) \cdot \mathbf{r}'(t) = 0.$$

⌐———— (Theorem 15.4.1)

In particular

$$\nabla f(\mathbf{r}(t_0)) \cdot \mathbf{r}'(t_0) = 0,$$

and thus

$$\nabla f(\mathbf{r}(t_0)) \perp \mathbf{r}'(t_0). \quad \square$$

Figure 15.5.1 illustrates the result.

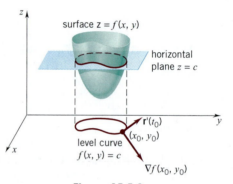

Figure 15.5.1

Example 1 For the function

$$f(x, y) = x^2 + y^2$$

the level curves are concentric circles:

$$x^2 + y^2 = c.$$

At each point $(x, y) \neq (0, 0)$ the gradient vector

$$\nabla f(x, y) = 2x\mathbf{i} + 2y\mathbf{j} = 2\mathbf{r}$$

points away from the origin along the line of the radius vector and is thus perpendicular to the circle in question. At the origin the level curve is reduced to a point and the gradient is simply **0**. See Figure 15.5.2. ❏

Consider now a curve in the xy-plane

$$C: \quad f(x, y) = c.$$

As before we assume that f is nonconstant and continuously differentiable. Let's suppose that (x_0, y_0) lies on the curve and $\nabla f(x_0, y_0) \neq \mathbf{0}$.

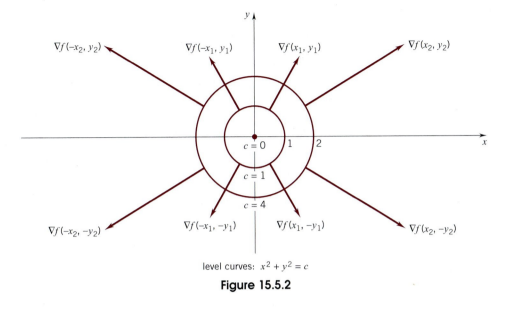

level curves: $x^2 + y^2 = c$

Figure 15.5.2

We can view C as the c-level curve of f and conclude from (15.5.1) that the gradient

(15.5.2)
$$\nabla f(x_0, y_0) = \frac{\partial f}{\partial x}(x_0, y_0)\mathbf{i} + \frac{\partial f}{\partial y}(x_0, y_0)\mathbf{j}$$

is perpendicular to C at (x_0, y_0). We call it a *normal vector*.

The vector

(15.5.3)
$$\mathbf{t}(x_0, y_0) = \frac{\partial f}{\partial y}(x_0, y_0)\mathbf{i} - \frac{\partial f}{\partial x}(x_0, y_0)\mathbf{j}$$

is perpendicular to the gradient:

$$\nabla f(x_0, y_0) \cdot \mathbf{t}(x_0, y_0) = \frac{\partial f}{\partial x}(x_0, y_0)\frac{\partial f}{\partial y}(x_0, y_0) - \frac{\partial f}{\partial y}(x_0, y_0)\frac{\partial f}{\partial x}(x_0, y_0) = 0.$$

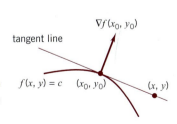

Figure 15.5.3

It is therefore a *tangent vector*.

The line through (x_0, y_0) perpendicular to the gradient is the tangent line. To obtain an equation for the tangent line we refer to Figure 15.5.3. A point (x, y) will lie on the tangent line iff

$$[(x - x_0)\mathbf{i} + (y - y_0)\mathbf{j}] \cdot \nabla f(x_0, y_0) = 0,$$

that is, iff

(15.5.4)
$$\frac{\partial f}{\partial x}(x_0, y_0)(x - x_0) + \frac{\partial f}{\partial y}(x_0, y_0)(y - y_0) = 0.$$

This is an equation for the *tangent line*.

The line through (x_0, y_0) perpendicular to the tangent vector $\mathbf{t}(x_0, y_0)$ is the normal line (Figure 15.5.4). A point (x, y) will lie on the normal line iff

$$[(x - x_0)\mathbf{i} + (y - y_0)\mathbf{j}] \cdot \mathbf{t}(x_0, y_0) = 0,$$

that is, iff

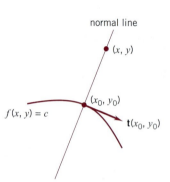

normal line

(x, y)

(x_0, y_0)

$f(x, y) = c$

$\mathbf{t}(x_0, y_0)$

Figure 15.5.4

(15.5.5)
$$\frac{\partial f}{\partial y}(x_0, y_0)(x - x_0) - \frac{\partial f}{\partial x}(x_0, y_0)(y - y_0) = 0.$$

This is an equation for the *normal line*.

Example 2   Choose a point (x_0, y_0) on the hyperbola

$$\frac{x^2}{a^2} - \frac{y^2}{b^2} = 1.$$

To avoid denominators, we write

$$b^2 x^2 - a^2 y^2 = a^2 b^2.$$

This equation is of the form

$$f(x, y) = c \qquad \text{with} \quad f(x, y) = b^2 x^2 - a^2 y^2 \quad \text{and} \quad c = a^2 b^2.$$

Partial differentiation gives

$$\frac{\partial f}{\partial x}(x, y) = 2b^2 x, \qquad \frac{\partial f}{\partial y}(x, y) = -2a^2 y.$$

At (x_0, y_0) the gradient

$$\frac{\partial f}{\partial x}(x_0, y_0)\mathbf{i} + \frac{\partial f}{\partial y}(x_0, y_0)\mathbf{j} = 2b^2 x_0 \mathbf{i} - 2a^2 y_0 \mathbf{j}$$

is normal to the curve and the vector

$$\frac{\partial f}{\partial y}(x_0, y_0)\mathbf{i} - \frac{\partial f}{\partial x}(x_0, y_0)\mathbf{j} = -2a^2 y_0 \mathbf{i} - 2b^2 x_0 \mathbf{j}$$

is tangent to the curve.

The equation of the tangent line can be written

$$2b^2 x_0(x - x_0) + (-2a^2 y_0)(y - y_0) = 0.$$

Dividing by 2 and noting that $b^2x_0^2 - a^2y_0^2 = a^2b^2$,† we can simplify the equation to

$$(b^2x_0)x - (a^2y_0)y = a^2b^2.$$

The equation of the normal line takes the form

$$(-2a^2y_0)(x - x_0) + (-2b^2x_0)(y - y_0) = 0.$$

This simplifies to

$$(a^2y_0)x + (b^2x_0)y = (a^2 + b^2)x_0y_0. \quad \square$$

Functions of Three Variables

Here, instead of level curves, we have level surfaces, but the results are similar. If $f = f(x, y, z)$ is nonconstant and continuously differentiable, then

---

(15.5.6) at each point of the domain, the gradient vector, if not **0**, is perpendicular to the level surface that passes through that point.

PROOF We choose a point $\mathbf{x}_0 = (x_0, y_0, z_0)$ in the domain and assume that $\nabla f(x_0, y_0, z_0) \neq \mathbf{0}$. The level surface through this point has equation

$$f(x, y, z) = c \qquad \text{where } c = f(x_0, y_0, z_0).$$

We suppose now that

$$\mathbf{r}(t) = x(t)\mathbf{i} + y(t)\mathbf{j} + z(t)\mathbf{k}, \qquad t \in I$$

is a differentiable curve that lies on this surface and passes through the point $\mathbf{x}_0 = (x_0, y_0, z_0)$. We choose t_0 so that

$$\mathbf{r}(t_0) = \mathbf{x}_0 = (x_0, y_0, z_0)$$

and suppose that $\mathbf{r}'(t_0) \neq \mathbf{0}$.

Since the curve lies on the given surface, we have

$$f(\mathbf{r}(t)) = c \qquad \text{for all } t \in I.$$

For such t

$$\frac{d}{dt}[f(\mathbf{r}(t))] = \nabla f(\mathbf{r}(t)) \cdot \mathbf{r}'(t) = 0.$$

In particular,

$$\nabla f(\mathbf{r}(t_0)) \cdot \mathbf{r}'(t_0) = 0.$$

The gradient vector

$$\nabla f(\mathbf{r}(t_0)) = \nabla f(\mathbf{x}_0) = \nabla f(x_0, y_0, z_0)$$

is thus perpendicular to the curve in question.

---

† The point (x_0, y_0) lies on the hyperbola.

This same argument applies to *every* differentiable curve that lies on this level surface and passes through the point $\mathbf{x}_0 = (x_0, y_0, z_0)$ with nonzero tangent vector. (See Figure 15.5.5.) Consequently, $\nabla f(\mathbf{x}_0)$ must be perpendicular to the surface itself. ❑

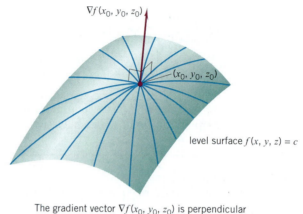

The gradient vector $\nabla f(x_0, y_0, z_0)$ is perpendicular
to the level surface at (x_0, y_0, z_0)

Figure 15.5.5

Example 3 For the function

$$f(x, y, z) = x^2 + y^2 + z^2$$

the level surfaces are concentric spheres:

$$x^2 + y^2 + z^2 = c.$$

At each point $(x, y, z) \neq (0, 0, 0)$ the gradient vector

$$\nabla f(x, y, z) = 2x\mathbf{i} + 2y\mathbf{j} + 2z\mathbf{k} = 2\mathbf{r}$$

points away from the origin along the line of the radius vector and is thus perpendicular to the sphere in question. At the origin the level surface is reduced to a point and the gradient is $\mathbf{0}$. ❑

The *tangent plane* to a surface

$$f(x, y, z) = c$$

at a point $\mathbf{x}_0 = (x_0, y_0, z_0)$ is the plane through $\mathbf{x}_0$ with normal $\nabla f(\mathbf{x}_0)$. See Figure 15.5.6.

The tangent plane at a point $\mathbf{x}_0$ is the plane through $\mathbf{x}_0$ that best approximates the surface in a neighborhood of $\mathbf{x}_0$. (We return to this later.)

A point $\mathbf{x}$ lies on the tangent plane through $\mathbf{x}_0$ iff

(15.5.7)

$$\nabla f(\mathbf{x}_0) \cdot (\mathbf{x} - \mathbf{x}_0) = 0.$$

(Figure 15.5.6)

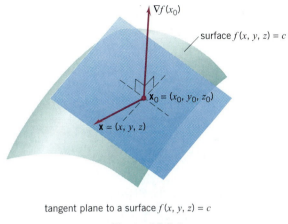

surface $f(x, y, z) = c$

$\nabla f(x_0)$

$\mathbf{x}_0 = (x_0, y_0, z_0)$

$\mathbf{x} = (x, y, z)$

tangent plane to a surface $f(x, y, z) = c$

Figure 15.5.6

This is an equation for the tangent plane in vector notation. In Cartesian coordinates the equation takes the form

(15.5.8)
$$\frac{\partial f}{\partial x}(x_0, y_0, z_0)(x - x_0) + \frac{\partial f}{\partial y}(x_0, y_0, z_0)(y - y_0) + \frac{\partial f}{\partial z}(x_0, y_0, z_0)(z - z_0) = 0.$$

The *normal line* to the surface $f(x, y, z) = c$ at a point $\mathbf{x}_0 = (x_0, y_0, z_0)$ on the surface is the line which passes through (x_0, y_0, z_0) parallel to $\nabla f(\mathbf{x}_0)$. Thus, $\nabla f(\mathbf{x}_0)$ is a direction vector for the normal line and

(15.5.9)
$$\mathbf{r}(t) = \mathbf{r}_0 + \nabla f(\mathbf{x}_0)t \qquad (\mathbf{r}_0 = x_0\mathbf{i} + y_0\mathbf{j} + z_0\mathbf{k})$$

is a vector equation for the line. In scalar parametric form, equations for the normal line are

(15.5.10)
$$x = x_0 + \frac{\partial f}{\partial x}(x_0, y_0, z_0)t,$$
$$y = y_0 + \frac{\partial f}{\partial y}(x_0, y_0, z_0)t,$$
$$z = z_0 + \frac{\partial f}{\partial z}(x_0, y_0, z_0)t.$$

Example 4 Find an equation in x, y, z for the plane tangent to the surface
$$xy + yz + zx = 11 \qquad \text{at the point } (1, 2, 3).$$

SOLUTION The surface is of the form

$$f(x, y, z) = c \qquad \text{with} \quad f(x, y, z) = xy + yz + zx \quad \text{and} \quad c = 11.$$

Observe that

$$\frac{\partial f}{\partial x} = y + z, \qquad \frac{\partial f}{\partial y} = x + z, \qquad \frac{\partial f}{\partial z} = x + y.$$

At the point $(1, 2, 3)$

$$\frac{\partial f}{\partial x} = 5, \qquad \frac{\partial f}{\partial y} = 4, \qquad \frac{\partial f}{\partial z} = 3.$$

The equation for the tangent plane can therefore be written

$$5(x - 1) + 4(y - 2) + 3(z - 3) = 0.$$

This simplifies to

$$5x + 4y + 3z - 22 = 0. \quad \square$$

Example 5 Find an equation for the tangent plane and scalar parametric equations for the normal line to the quadric cone

$$x^2 + 4y^2 = z^2$$

at the point $(3, 2, 5)$ on the cone (Figure 15.5.7).

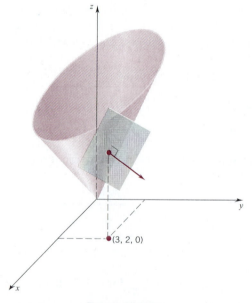

(3, 2, 0)

Figure 15.5.7

SOLUTION The surface is of the form $f(x, y, z) = c$ with

$$f(x, y, z) = x^2 + 4y^2 - z^2 \quad \text{and} \quad c = 0.$$

The partial derivatives of f are

$$\frac{\partial f}{\partial x} = 2x, \qquad \frac{\partial f}{\partial y} = 8y, \qquad \frac{\partial f}{\partial z} = -2z$$

and

$$\nabla f = 2x\mathbf{i} + 8y\mathbf{j} - 2z\mathbf{k}.$$

Now, $\nabla f(3, 2, 5) = 6\mathbf{i} + 16\mathbf{j} - 10\mathbf{k}$ is normal to the cone at the point $(3, 2, 5)$. Note that $\frac{1}{2}\nabla f(3, 2, 5) = 3\mathbf{i} + 8\mathbf{j} - 5\mathbf{k}$ is also normal to the cone and is a little simpler. An equation for the tangent plane is

$$3(x - 3) + 8(y - 2) - 5(z - 5) = 0,$$

which simplifies to

$$3x + 8y - 5z = 0.$$

Note that this plane passes through the origin, as we would expect. Scalar parametric equations for the normal line are

$$x = 3 + 3t, \qquad y = 2 + 8t, \qquad z = 5 - 5t. \quad \square$$

Example 6 The curve

$$\mathbf{r}(t) = \tfrac{1}{2}t^2\mathbf{i} + 4t^{-1}\mathbf{j} + (\tfrac{1}{2}t - t^2)\mathbf{k}$$

intersects the hyperbolic paraboloid $x^2 - 4y^2 - 4z = 0$ at the point $(2, 2, -3)$. What is the angle of intersection?

SOLUTION We want the angle φ between the tangent vector of the curve and the tangent plane of the surface at the point of intersection (Figure 15.5.8).

A simple calculation shows that the curve passes through the point $(2, 2, -3)$ at $t = 2$. Since

$$\mathbf{r}'(t) = t\mathbf{i} - 4t^{-2}\mathbf{j} + (\tfrac{1}{2} - 2t)\mathbf{k},$$

we have

$$\mathbf{r}'(2) = 2\mathbf{i} - \mathbf{j} - \tfrac{7}{2}\mathbf{k}.$$

Now set

$$f(x, y, z) = x^2 - 4y^2 - 4z.$$

This function has gradient $2x\mathbf{i} - 8y\mathbf{j} - 4\mathbf{k}$. At the point $(2, 2, -3)$

$$\nabla f = 4\mathbf{i} - 16\mathbf{j} - 4\mathbf{k}.$$

Now let θ be the angle between $\mathbf{r}'(2)$ and this gradient. By (12.3.9)

$$\cos\theta = \frac{\mathbf{r}'(2)}{\|\mathbf{r}'(2)\|} \cdot \frac{\nabla f}{\|\nabla f\|} = \frac{19}{414}\sqrt{138} \cong 0.539$$

so that $\theta \cong 1.00$ radian. Since the gradient is normal to the tangent plane, the angle ϕ we want is

$$\phi = \tfrac{1}{2}\pi - \theta \cong 1.57 - 1.00 = 0.57 \text{ radian.} \quad \square$$

A surface of the form

$$z = g(x, y)$$

can be written in the form

$$f(x, y, z) = 0$$

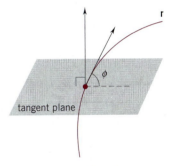

Figure 15.5.8

by setting

$$f(x, y, z) = g(x, y) - z.$$

If g is differentiable, so is f.   Moreover,

$$\frac{\partial f}{\partial x}(x, y, z) = \frac{\partial g}{\partial x}(x, y), \qquad \frac{\partial f}{\partial y}(x, y, z) = \frac{\partial g}{\partial y}(x, y), \qquad \frac{\partial f}{\partial z}(x, y, z) = -1.$$

By (15.5.8), the tangent plane at (x_0, y_0, z_0) has equation

$$\frac{\partial g}{\partial x}(x_0, y_0)(x - x_0) + \frac{\partial g}{\partial y}(x_0, y_0)(y - y_0) + (-1)(z - z_0) = 0,$$

which we can rewrite as

(15.5.11) $$\boxed{(z - z_0) = \frac{\partial g}{\partial x}(x_0, y_0)(x - x_0) + \frac{\partial g}{\partial y}(x_0, y_0)(y - y_0).}$$

If $\nabla g(x_0, y_0) = \mathbf{0}$, then both partials of g are zero at (x_0, y_0) and the equation reduces to

$$\boxed{z = z_0.}$$

In this case the tangent plane is *horizontal.*
  Scalar parametric equations for the normal line to the surface $z = g(x, y)$ at the point (x_0, y_0, z_0) are

(15.5.12) $$\boxed{x = x_0 + \frac{\partial g}{\partial x}(x_0, y_0)t, \qquad y = y_0 + \frac{\partial g}{\partial y}(x_0, y_0)t, \qquad z = z_0 + (-1)t.}$$

Example 7   Find an equation for the tangent plane and symmetric equations for the normal line to the surface

$$z = \ln(x^2 + y^2)$$

at the point $(-2, 1, \ln 5)$ on the surface.

SOLUTION   Set

$$g(x, y) = \ln(x^2 + y^2).$$

The partial derivatives of g are

$$\frac{\partial g}{\partial x}(x, y) = \frac{2x}{x^2 + y^2}, \qquad \frac{\partial g}{\partial y}(x, y) = \frac{2y}{x^2 + y^2},$$

and when $x = -2$ and $y = 1$,

$$\frac{\partial g}{\partial x} = -\frac{4}{5}, \qquad \frac{\partial g}{\partial y} = \frac{2}{5}.$$

Therefore, at the point $(-2, 1, \ln 5)$, the tangent plane has equation

$$z - \ln 5 = -\frac{4}{5}(x + 2) + \frac{2}{5}(y - 1).$$

Symmetric equations for the normal line are

$$\frac{x + 2}{-\frac{4}{5}} = \frac{y - 1}{\frac{2}{5}} = \frac{z - \ln 5}{-1}.$$

A graph of this surface is shown in Figure 15.5.9. ❏

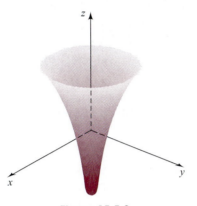

Figure 15.5.9

Example 8 At what points of the surface

$$z = 3xy - x^3 - y^3$$

is the tangent plane horizontal?

SOLUTION The function

$$g(x, y) = 3xy - x^3 - y^3$$

has first partials

$$\frac{\partial g}{\partial x}(x, y) = 3y - 3x^2, \qquad \frac{\partial g}{\partial y}(x, y) = 3x - 3y^2.$$

We set these partial derivatives equal to 0 and solve the resulting system of equations:

$$\begin{array}{ccc} 3y - 3x^2 = 0 & & y - x^2 = 0 \\ & \text{or} & \\ 3x - 3y^2 = 0 & & x - y^2 = 0. \end{array}$$

From the first equation, we get $y = x^2$. Substituting this into the second equation, it follows that

$$x - x^4 = 0$$

or

$$x(1 - x^3) = 0.$$

Therefore, $x = 0$ (which implies $y = 0$) or $x = 1$ (which implies $y = 1$). Thus, these partials are both zero only at $(0, 0)$ and $(1, 1)$. The surface has a horizontal tangent plane only at $(0, 0, 0)$ and $(1, 1, 1)$. A graph of this surface is shown in Figure 15.5.10. ❏

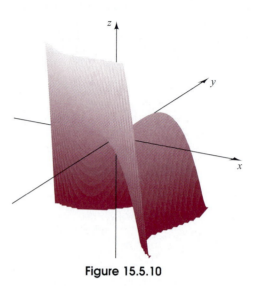

Figure 15.5.10

EXERCISES 15.5

In Exercises 1–8, find a normal vector and a tangent vector at the point indicated. Write an equation for the tangent line and an equation for the normal line.

1. $x^2 + xy + y^2 = 3$; $(-1, -1)$.

2. $(y - x)^2 = 2x$; $(2, 4)$.

3. $(x^2 + y^2)^2 = 9(x^2 - y^2)$; $(\sqrt{2}, 1)$.

4. $x^3 + y^3 = 9$; $(1, 2)$.

5. $xy^2 - 2x^2 + y + 5x = 6$; $(4, 2)$.

6. $x^5 + y^5 = 2x^3$; $(1, 1)$.

7. $2x^3 - x^2y^2 = 3x - y - 7$; $(1, -2)$.

8. $x^3 + y^2 + 2x = 6$; $(-1, 3)$.

9. Find the slope of the curve $x^2y = a^2(a - y)$ at the point $(0, a)$.

In Exercises 10–18, find an equation for the tangent plane and scalar parametric equations for the normal line at the point indicated.

10. $z = (x^2 + y^2)^2$; $(1, 1, 4)$.

11. $x^3 + y^3 = 3xyz$; $(1, 2, \frac{3}{2})$.

12. $xy^2 + 2z^2 = 12$; $(1, 2, 2)$.

13. $z = axy$; $(1, 1/a, 1)$.

14. $\sqrt{x} + \sqrt{y} + \sqrt{z} = 4$; $(1, 4, 1)$.

15. $z = \sin x + \sin y + \sin(x + y)$; $(0, 0, 0)$.

16. $z = x^2 + xy + y^2 - 6x + 2$; $(4, -2, -10)$.

17. $b^2c^2x^2 - a^2c^2y^2 - a^2b^2z^2 = a^2b^2c^2$; (x_0, y_0, z_0).

18. $z = \sin(x \cos y)$; $(0, \frac{1}{2}\pi, 0)$.

In Exercises 19–23, find the point(s) on the surface at which the tangent plane is horizontal.

19. $xy + a^3x^{-1} + b^3y^{-1} - z = 0$.

20. $z = 4x + 2y - x^2 + xy - y^2$.

21. $z = xy$.

22. $x + y + z + xy - x^2 - y^2 = 0$.

23. $z - 2x^2 - 2xy + y^2 + 5x - 3y + 2 = 0$.

24. (a) Find the *upper unit normal* (the unit normal with positive **k** component) for the surface $z = xy$ at the point $(1, 1, 1)$.
    (b) Find the *lower unit normal* (the unit normal with negative **k** component) for the surface $z = 1/x - 1/y$ at the point $(1, 1, 0)$.

25. Let $f = f(x, y, z)$ be continuously differentiable. Write equations in symmetric form for the line normal to the surface $f(x, y, z) = c$ at the point (x_0, y_0, z_0).

26. Show that in the case of a surface of the form $z = xf(x/y)$ with f continuously differentiable, all the tangent planes share a point in common.

27. Given that the surfaces $F(x, y, z) = 0$ and $G(x, y, z) = 0$ intersect at right angles in a curve γ, what condition must be satisifed by the partial derivatives of F and G on γ?

28. Show that, for all planes tangent to the surface $\sqrt{x} + \sqrt{y} + \sqrt{z} = \sqrt{a}$, the sum of the intercepts is the same.

29. Show that all pyramids formed by the coordinate planes and a plane tangent to the surface $xyz = a^3$ have the same volume. What is this volume?

30. Show that, for all planes tangent to the surface $x^{2/3} + y^{2/3} + z^{2/3} = a^{2/3}$, the sum of the squares of the intercepts is the same.

31. The curve $\mathbf{r}(t) = 2t\mathbf{i} + 3t^{-1}\mathbf{j} - 2t^2\mathbf{k}$ and the ellipsoid $x^2 + y^2 + 3z^2 = 25$ intersect at $(2, 3, -2)$. What is the angle of intersection?

32. Show that the curve $\mathbf{r}(t) = \frac{3}{2}(t^2 + 1)\mathbf{i} + (t^4 + 1)\mathbf{j} + t^3\mathbf{k}$ is perpendicular to the ellipsoid $x^2 + 2y^2 + 3z^2 = 20$ at the point $(3, 2, 1)$.

33. The surfaces $x^2y^2 + 2x + z^3 = 16$ and $3x^2 + y^2 - 2z = 9$ intersect in a curve that passes through the point $(2, 1, 2)$. What are the equations of the respective tangent planes to the two surfaces at this point?

34. Show that the sphere $x^2 + y^2 + z^2 - 8x - 8y - 6z + 24 = 0$ is tangent to the ellipsoid $x^2 + 3y^2 + 2z^2 = 9$ at the point $(2, 1, 1)$.

35. Show that the sphere $x^2 + y^2 + z^2 - 4y - 2z + 2 = 0$ is perpendicular to the paraboloid $3x^2 + 2y^2 - 2z = 1$ at the point $(1, 1, 2)$.

36. Show that the following surfaces are mutually perpendicular: $xy = az^2, x^2 + y^2 + z^2 = b, z^2 + 2x^2 = c(z^2 + 2y^2)$.

37. The surface S: $z = x^2 + 3y^2 + 2$ intersects the vertical plane p: $3x + 4y + 6 = 0$ in a space curve C.
  (a) Let C_1 be the projection of C in the xy-plane. Find an equation for C_1.
  (b) Find a parametrization $\mathbf{r}(t) = x(t)\mathbf{i} + y(t)\mathbf{j} + z(t)\mathbf{k}$ for C setting $x(t) = 4t - 2$.
  (c) Find a parametrization $\mathbf{R}(s) = \mathbf{R}_0 + s\mathbf{d}$ for the line l tangent to C at $(2, -3, 33)$.
  (d) Find an equation for the plane p_1 tangent to S at $(2, -3, 33)$.
  (e) Find a parametrization $\mathbf{r}(t) = x(t)\mathbf{i} + y(t)\mathbf{j} + z(t)\mathbf{k}$ for the line l' formed by the intersection of p with p_1 taking $x(t) = t$. What is the relation between l and l'?

■ 15.6 MAXIMUM AND MINIMUM VALUES

In Chapter 4 we discussed the local extreme values of a function of one variable. Here we take up the same subject for functions of several variables. The ideas are very similar.

DEFINITION 15.6.1 LOCAL MAXIMUM AND LOCAL MINIMUM

Let f be a function of several variables and let $\mathbf{x}_0$ be an interior point of the domain:
  f is said to have a *local maximum* at $\mathbf{x}_0$ iff

  $$f(\mathbf{x}_0) \geq f(\mathbf{x}) \qquad \text{for all } \mathbf{x} \text{ in some neighborhood of } \mathbf{x}_0;$$

  f is said to have a *local minimum* at $\mathbf{x}_0$ iff

  $$f(\mathbf{x}_0) \leq f(\mathbf{x}) \qquad \text{for all } \mathbf{x} \text{ in some neighborhood of } \mathbf{x}_0.$$

As in the one-variable case, the local maxima and local minima together comprise the *local extreme values*.

In the one-variable case we know that, if f has a local extreme value at x_0, then

$$\text{either} \quad f'(x_0) = 0 \quad \text{or} \quad f'(x_0) \text{ does not exist.}$$

We have a similar result for functions of several variables.

> **THEOREM 15.6.2**
>
> If f has a local extreme value at $\mathbf{x}_0$, then
>
> $$\text{either} \quad \nabla f(\mathbf{x}_0) = \mathbf{0} \quad \text{or} \quad \nabla f(\mathbf{x}_0) \text{ does not exist.}$$

PROOF We assume that f has a local extreme value at $\mathbf{x}_0$ and that f is differentiable at $\mathbf{x}_0$ [that $\nabla f(\mathbf{x}_0)$ exists]. We need to show that $\nabla f(\mathbf{x}_0) = \mathbf{0}$. For simplicity we set $\mathbf{x}_0 = (x_0, y_0)$. The three-variable case is similar.

Since f has a local extreme value at (x_0, y_0), the function $g(x) = f(x, y_0)$ has a local extreme value at x_0. Since f is differentiable at (x_0, y_0), g is differentiable at x_0 and therefore

$$g'(x_0) = \frac{\partial f}{\partial x}(x_0, y_0) = 0.$$

Similarly, the function $h(y) = f(x_0, y)$ has a local extreme value at y_0 and, being differentiable there, satisfies the relation

$$h'(y_0) = \frac{\partial f}{\partial y}(x_0, y_0) = 0.$$

The gradient is $\mathbf{0}$ since both partials are 0. ❑

Interior points of the domain at which the gradient is zero or the gradient does not exist are called *critical points*. By Theorem 15.6.2 these are the only points at which a local extreme value can occur.

Although the ideas introduced so far are completely general, their application to functions of more than two variables is generally laborious. We restrict ourselves mostly to functions of two variables. Not only are the computations less formidable, but also we can make use of our geometric intuition.

Two Variables

We suppose for the moment that $f = f(x, y)$ is defined on an open set and is continuously differentiable there. The graph of f is a surface

$$z = f(x, y).$$

Where f has a local maximum, the surface has a local high point. Where f has a local minimum, the surface has a local low point. Where f has either a local maximum or a local minimum, the gradient is $\mathbf{0}$ and therefore the tangent plane is horizontal. See Figure 15.6.1.

A zero gradient signals the possibility of a local extreme value; it does not guarantee it. For example, in the case of the saddle-shaped surface of Figure 15.6.2, there is a horizontal tangent plane at the origin and therefore the gradient is zero there, yet the origin gives neither a local maximum nor a local minimum.

Critical points at which the gradient is zero are called *stationary points*. The stationary points that do not give rise to extreme values are called *saddle points*.

Below we test some differentiable functions for extreme values. In each case, our first step is to seek out the stationary points.

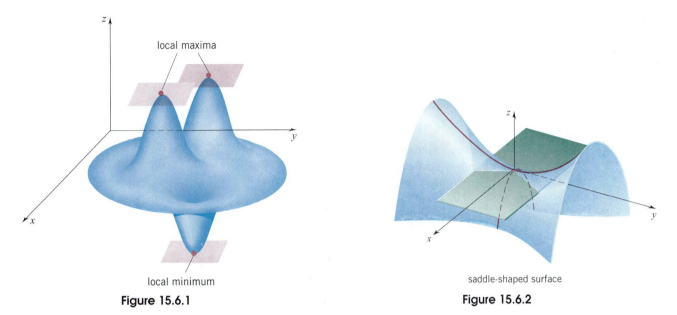

local maxima

local minimum

Figure 15.6.1

saddle-shaped surface

Figure 15.6.2

Example 1 For the function

$$f(x, y) = 2x^2 + y^2 - xy - 7y,$$

we have

$$\nabla f(x, y) = (4x - y)\mathbf{i} + (2y - x - 7)\mathbf{j}.$$

To find the stationary points, we set $\nabla f(x, y) = \mathbf{0}$. This gives

$$4x - y = 0 \qquad \text{and} \qquad 2y - x - 7 = 0.$$

The only simultaneous solution to these equations is $x = 1, y = 4$. The point $(1, 4)$ is therefore the only stationary point.

We now compare the value of f at $(1, 4)$ with the values of f at nearby points $(1 + h, 4 + k)$:

$$f(1, 4) = 2 + 16 - 4 - 28 = -14,$$

$$
\begin{aligned}
f(1 + h, 4 + k) &= 2(1 + h)^2 + (4 + k)^2 - (1 + h)(4 + k) - 7(4 + k) \\
&= 2 + 4h + 2h^2 + 16 + 8k + k^2 - 4 - 4h - k - hk - 28 - 7k \\
&= 2h^2 + k^2 - hk - 14.
\end{aligned}
$$

The difference

$$
\begin{aligned}
f(1 + h, 4 + k) - f(1, 4) &= 2h^2 + k^2 - hk \\
&= h^2 + (h^2 - hk + k^2) \\
&\geq h^2 + (h^2 - 2|h||k| + k^2) \\
&= h^2 + (|h| - |k|)^2 \geq 0.
\end{aligned}
$$

Thus $f(1 + h, 4 + k) \geq f(1, 4)$ for all small h and k (in fact for all real h and k).† It follows that f has a local minimum at $(1, 4)$. This local minimum is -14. ❑

---

† Another way to see that $2h^2 + k^2 - hk$ is nonnegative is to complete the square:

$$2h^2 + k^2 - hk = \tfrac{1}{4}h^2 - hk + k^2 + \tfrac{7}{4}h^2 = (\tfrac{1}{2}h - k)^2 + \tfrac{7}{4}h^2 \geq 0.$$

Example 2 In the case of

$$f(x, y) = y^2 - xy + 2x + y + 1$$

we have

$$\nabla f(x, y) = (2 - y)\mathbf{i} + (2y - x + 1)\mathbf{j}.$$

The gradient is **0** where

$$2 - y = 0 \quad \text{and} \quad 2y - x + 1 = 0.$$

The only simultaneous solution to these equations is $x = 5, y = 2$. The point $(5, 2)$ is the only stationary point.

We now compare the value of f at $(5, 2)$ with the values of f at nearby points $(5 + h, 2 + k)$:

$$f(5, 2) = 4 - 10 + 10 + 2 + 1 = 7,$$

$$\begin{aligned} f(5 + h, 2 + k) &= (2 + k)^2 - (5 + h)(2 + k) + 2(5 + h) + (2 + k) + 1 \\ &= 4 + 4k + k^2 - 10 - 2h - 5k - hk + 10 + 2h + 2 + k + 1 \\ &= k^2 - hk + 7. \end{aligned}$$

The difference

$$d = f(5 + h, 2 + k) - f(5, 2) = k^2 - hk = k(k - h)$$

does not keep a constant sign for small h and k. See Figure 15.6.3: $d < 0$ in region I $(k > 0, k < h)$; $d > 0$ in region II $(k > 0, k > h)$; and so on. Therefore, it follows that $(5, 2)$ is a saddle point. ❑

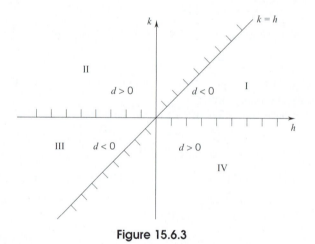

Figure 15.6.3

Example 3 For the function

$$f(x, y) = -xye^{-(x^2 + y^2)/2}$$

we have

$$\frac{\partial f}{\partial x} = -ye^{-(x^2 + y^2)/2} + x^2 ye^{-(x^2 + y^2)/2} = y(x^2 - 1)e^{-(x^2 + y^2)/2},$$

$$\frac{\partial f}{\partial y} = -xe^{-(x^2 + y^2)/2} + xy^2 e^{-(x^2 + y^2)/2} = x(y^2 - 1)e^{-(x^2 + y^2)/2},$$

and

$$\nabla f(x, y) = e^{-(x^2+y^2)/2}[y(x^2 - 1)\mathbf{i} + x(y^2 - 1)\mathbf{j}].$$

Since $e^{-(x^2+y^2)/2} \neq 0$ for all (x, y), $\nabla f(x, y) = \mathbf{0}$ iff

$$y(1 - x^2) = 0 \qquad \text{and} \qquad x(y^2 - 1) = 0.$$

The simultaneous solutions to these equations are $x = y = 0$; $x = 1, y = \pm 1$; $x = -1$, $y = \pm 1$. Thus, $(0, 0)$, $(1, 1)$, $(1, -1)$, $(-1, 1)$, and $(-1, -1)$ are the stationary points.

A computer-generated graph of this function is shown in Figure 15.6.4. From the graph, we see that f has local minima at $(1, 1)$ and $(-1, -1)$, and local maxima at $(1, -1)$ and $(-1, 1)$. The point $(0, 0)$ is a saddle point since

$$f(h, k) - f(0, 0) = f(h, k) = -hke^{-(h^2+k^2)/2}$$

does not keep constant sign for small h and k. ❏

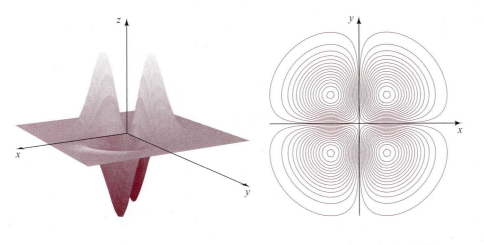

level curves of $f(x, y) = -xye^{-(x^2+y^2)^2}$

Figure 15.6.4

The function in the next example has an infinite number of stationary points.

Example 4   For all points (x, y) with $x + y \neq 0$, set

$$f(x, y) = -\frac{x^2 + 2y^2}{(x + y)^2}.$$

Then

$$\nabla f(x, y) = 2y\frac{2y - x}{(x + y)^3}\mathbf{i} + 2x\frac{x - 2y}{(x + y)^3}\mathbf{j}.$$

The gradient is $\mathbf{0}$ where

$$2y\frac{2y - x}{(x + y)^3} = 0 \qquad \text{and} \qquad 2x\frac{x - 2y}{(x + y)^3} = 0.$$

Solving these two equations simultaneously, we get

$$x = 2y.$$

The stationary points are all the points of the line $x = 2y$ other than the point $(0, 0)$, which is not in the domain of f. At each stationary point (x_0, y_0) the function takes on the value $-\frac{2}{3}$.

$$f(x_0, y_0) = f(2y_0, y_0) = -\frac{(2y_0)^2 + 2y_0^2}{(2y_0 + y_0)^2} = -\frac{6y_0^2}{9y_0^2} = -\frac{2}{3}.$$

In a moment we will show that for all (x, y) in the domain,

$$f(x, y) = -\frac{x^2 + 2y^2}{(x + y)^2} \leq -\frac{2}{3}.$$

It will then follow that the number $-\frac{2}{3}$ is a local maximum value.

The inequality

$$-\frac{x^2 + 2y^2}{(x + y)^2} \leq -\frac{2}{3}$$

can be justified by the following sequence of equivalent inequalities, the last of which is obvious:

$$-3(x^2 + 2y^2) \leq -2(x + y)^2,$$
$$3(x^2 + 2y^2) \geq 2(x + y)^2,$$
$$3x^2 + 6y^2 \geq 2x^2 + 4xy + 2y^2,$$
$$x^2 - 4xy + 4y^2 \geq 0,$$
$$(x - 2y)^2 \geq 0. \quad \square$$

Each function we have considered so far was differentiable on its entire domain. The only critical points were thus the stationary points. Our next example exhibits a function that is everywhere defined but is not differentiable at the origin. The origin is therefore a critical point, though not a stationary point.

Example 5 The function

$$f(x, y) = 1 + \sqrt{x^2 + y^2}$$

is everywhere defined and everywhere continuous. The graph is the upper nappe of a right circular cone. (See Figure 15.6.5.) The number $f(0, 0) = 1$ is obviously a local minimum.

Since the partials

$$\frac{\partial f}{\partial x} = \frac{x}{\sqrt{x^2 + y^2}}, \qquad \frac{\partial f}{\partial y} = \frac{y}{\sqrt{x^2 + y^2}}$$

are not defined at $(0, 0)$, the gradient is not defined at $(0, 0)$ (Theorem 15.2.5). The point $(0, 0)$ is thus a critical point, but not a stationary point. At $(0, 0, 1)$ the surface comes to a sharp point and there is no tangent plane. $\quad \square$

Absolute Extreme Values

In Chapter 2 we saw that a function of one variable that is continuous on a bounded closed interval takes on both an absolute maximum and an absolute minimum on that interval. (Theorem 2.6.2.) More generally, it can be proved that, if f, now a function of one or several variables, is continuous on a *bounded closed set* (a closed set that can

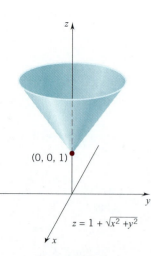

$z = 1 + \sqrt{x^2 + y^2}$

$(0, 0, 1)$

Figure 15.6.5

be contained in a ball of finite radius), then f takes on both an absolute maximum and an absolute minimum on that set.

In the search for local extreme values the critical points are interior points of the domain: the stationary points and the interior points at which the gradient does not exist. In the search for absolute extreme values we must also test the boundary points. This usually requires special methods. One approach is to try to parametrize the boundary by some vector function $\mathbf{r} = \mathbf{r}(t)$ and then work with $f(\mathbf{r}(t))$. This is the approach we take in the example below. Another, altogether different approach is outlined in Section 15.8.

Example 6 The triangular region

$$\Omega = \{(x, y): 0 \le x \le 4, 0 \le y \le 2x\} \qquad \text{(Figure 15.6.6)}$$

is a closed, bounded set. The function

$$f(x, y) = xy - 2x - 3y,$$

being continuous everywhere, is continuous on Ω. Therefore, we know that f takes on an absolute maximum and an absolute minimum on Ω.

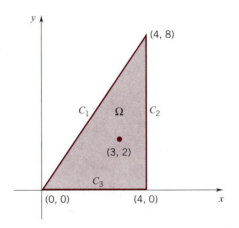

Figure 15.6.6

To see whether either of these values is taken on in the interior of the set, we form the gradient

$$\nabla f(x, y) = (y - 2)\mathbf{i} + (x - 3)\mathbf{j}.$$

The gradient, defined everywhere, is $\mathbf{0}$ only at the point $(3, 2)$, and $(3, 2) \in \Omega$. As you can check,

$$f(3 + h, 2 + k) - f(3, 2) = hk.$$

Since this difference does not keep a constant sign for small h and k, the point $(3, 2)$ does not give rise to an extreme value. It is a saddle point.

Now we will look for extreme values on the boundary by writing each side of the triangle in the form $\mathbf{r} = \mathbf{r}(t)$ and then analyzing $f(\mathbf{r}(t))$. With C_1, C_2, C_3 as in the figure, we have

$$C_1: \quad \mathbf{r}_1(t) = t\mathbf{i} + 2t\mathbf{j}, \quad t \in [0, 4],$$

$$C_2: \quad \mathbf{r}_2(t) = 4\mathbf{i} + t\mathbf{j}, \quad t \in [0, 8],$$

$$C_3: \quad \mathbf{r}_3(t) = t\mathbf{i}, \qquad t \in [0, 4].$$

The values of f on these line segments are given by the functions

$$f_1(t) = f(\mathbf{r}_1(t)) = 2t^2 - 8t, \quad t, \in [0, 4],$$
$$f_2(t) = f(\mathbf{r}_2(t)) = t - 8, \qquad t \in [0, 8],$$
$$f_3(t) = f(\mathbf{r}_3(t)) = -2t, \qquad t \in [0, 4].$$

The maximum value of f is the maximum of the maxima of f_1, f_2, f_3. And, the minimum value of f is the minimum of the minima of f_1, f_2, f_3. In this case, it happens that each of the f_i has a maximum of 0 and a minimum of -8. (Check this out.) It follows that the maximum value of f is 0 and the minimum value of f is -8. ❑

In some cases, physical or geometric considerations may allow us to conclude that an absolute maximum or an absolute minimum exists even if the function is not continuous, or the domain is not bounded or not closed.

Example 7 The rectangle

$$\{(x, y): 0 \le x \le a, -b \le y \le b\}$$

is a bounded closed subset of the plane. The function

$$f(x, y) = 1 + \sqrt{x^2 + y^2}$$

being everywhere continuous, is continuous on this rectangle. Thus we can be sure that f takes on both an absolute maximum and an absolute minimum on this set. The absolute maximum is taken on at the points $(a, -b)$ and (a, b), the points of the rectangle furthest away from the origin. (This should be clear from Figure 15.6.5.) The value at these points is $1 + \sqrt{a^2 + b^2}$. The absolute minimum is taken on at the origin $(0, 0)$. The value there is 1.
  Now let's continue with the same function but apply it instead to the rectangle

$$\{(x, y): 0 < x \le a, -b \le y \le b\}.$$

This rectangle is bounded but not closed. On this set f takes on an absolute maximum (the same maximum as before and at the same points), but it takes on no absolute minimum (the origin is not in the set).
  Finally, on the entire plane (which is closed but not bounded), f takes on an absolute minimum (1 at the origin) but no maximum. ❑

EXERCISES 15.6

In Exercises 1–14, find the stationary points and determine the local extreme values.

1. $f(x, y) = 2x - x^2 - y^2$.
2. $f(x, y) = x^2 + 2y^2 - 4y$.
3. $f(x, y) = 2x + 2y - x^2 + y^2 + 5$.
4. $f(x, y) = x^2 - xy + y^2 + 1$.
5. $f(x, y) = x^2 + xy + y^2 + 3x + 1$.
6. $f(x, y) = x^3 - y^2 + 2y$.

7. $f(x, y) = x^3 - 3x + y$.
8. $f(x, y) = x^5 - 3x^2 - 2y$.
9. $f(x, y) = x^2 + xy + y^2 - 3x - 3y$.
10. $f(x, y) = x^2 - xy + y^2 + 2x + 2y - 6$.
11. $f(x, y) = x^2 + xy + 2x + 2y + 1$.
12. $f(x, y) = 5xy - 7x^2 - y^2 + 3x - 6y$.
13. $f(x, y) = 4x^3y - 4xy^3$.
14. $f(x, y) = x \sin y, \ -\pi < y < \pi$.

In Exercises 15–24, find the absolute extreme values taken on by the function on the set D.

15. $f(x, y) = \dfrac{1}{\sqrt{x^2 + y^2}}$,
$D = \{(x, y): 1 \le x \le 3, 1 \le y \le 4\}$.

16. $f(x, y) = |x| + |y|$,
$D = \{(x, y): -1 \le x \le 1, -1 \le y \le 1\}$.

17. $f(x, y) = 4x^2 - 9y^2$,
$D = \{(x, y): -1 \le x \le 1\}$.

18. $f(x, y) = (x^2 + y^2)e^{-(x^2+y^2)}$,
$D = \{(x, y): x^2 + y^2 \le 25\}$.   HINT: Consider te^{-t}.

19. $f(x, y) = (x - 1)^2 + (y - 1)^2$,
$D = \{(x, y): x^2 + y^2 \le 4\}$.

20. $f(x, y) = y(x - 3)$, $D = \{(x, y): x^2 + y^2 \le 9\}$.

21. $f(x, y) = (x - y)^2$,
$D = \{(x, y): 0 \le x \le 6, 0 \le y \le 12 - 2x\}$.

22. $f(x, y) = (x - 3)^2 + y^2$,
$D = \{(x, y): 0 \le x \le 4, x^2 \le y \le 4x\}$.

23. $f(x, y) = (x - 4)^2 + y^2$,
$D = \{(x, y): 0 \le x \le 2, x^3 \le y \le 4x\}$.

24. $f(x, y) = (x + y - 2)^2$,
$D = \{(x, y): 0 \le x \le 3, x \le y \le 3\}$.

25. Find the point in the plane $2x - y + 2z = 16$ that is closest to the origin, and calculate the distance from the origin to the plane.   Check your answer by using Formula (12.6.7).

26. Find the point in the plane $3x - 4y + 2z + 32 = 0$ that is closest to the point $P(-1, 2, 4)$ and calculate the distance from P to the plane.   Check your answer by using Formula (12.6.7).

27. Find positive numbers x, y, z such that $x + y + z = 18$ and xyz is a maximum.   HINT: Maximize $f(x, y) = 18xy - x^2y - xy^2$ on the triangle formed by the positive x- and y-axes and the line $x + y = 18$.

28. Find positive numbers x, y, z such that $x + y + z = 30$ and xyz^2 is a maximum.   HINT: Maximize $f(y, z) = 30yz^2 - y^2z^2 - yz^3$ over the triangle formed by the positive y- and z-axes and the line $y + z = 30$.

29. A rectangular box without a top is to have a volume of 12 cubic feet.   Find the dimensions for the box that will yield the minimum surface area.

30. Suppose that the material to be used to construct the box in Exercise 29 costs $3 per square foot for the sides and

$4 per square foot for the bottom.   What dimensions will yield the minimum cost?

31. Define $f(x, y) = \frac{1}{4}x^2 - \frac{1}{9}y^2$ on the closed unit disc.   Find (a) the stationary points, (b) the local extreme values, (c) the absolute extreme values.

32. Find the dimensions of the rectangular solid of maximum volume given that the sum of these dimensions is b.

33. Find the point with the property that the sum of the squares of its distances from $P_1(x_1, y_1)$, $P_2(x_2, y_2)$, $P_3(x_3, y_3)$ is an absolute minimum.

34. A manufacturer finds that it can sell in two distinct markets.   If it sells Q_1 units in the first market and Q_2 units in the second market, the revenue functions are

$$R_1 = A_1Q_1 - B_1Q_1^2, \qquad R_2 = A_2Q_2 - B_2Q_2^2$$

and the total cost function is

$$C = A_3 + B_3(Q_1 + Q_2).$$

Given that the A's and B's are known positive constants, find the quantities Q_1 and Q_2 that will maximize profit.

35. Find the distance from the point $(\frac{1}{2}, \frac{1}{2}, \frac{1}{2})$ to the sphere $x^2 + y^2 + z^2 = 1$.

36. Given that $0 < a < b$, find the absolute maximum value taken on by the function

$$f(x, y) = \frac{xy}{(a + x)(x + y)(b + y)}$$

on the open square $\{(x, y): a < x < b, a < y < b\}$.

37. A manufacturer produces razors and blades at a constant average cost of 80 cents per razor and 60 cents per dozen blades.   If the razors are sold at x cents each and the blades at y cents per dozen, the demand of the market each week is

$1000(320 - x)$ razors and $1000(140 - y)$ dozen blades.

Find the selling prices for maximum profit.

▷ In Exercises 38–41, use a graphing utility to graph the surface and locate any stationary points and saddle points.

38. $f(x, y) = 4xy - x^4 - y^4 + 1$.

39. $f(x, y) = 3xy - x^3 - y^3 + 2$.

40. $f(x, y) = (x^2 + 2y^2)e^{-(x^2+y^2)}$.

41. $f(x, y) = \dfrac{-2x}{x^2 + y^2 + 1}$.

■ 15.7 SECOND-PARTIALS TEST

We start with a function g of one variable and suppose that $g'(x_0) = 0$.   According to the second-derivative test, (Theorem 4.3.5), g has

a local minimum at x_0 if $g''(x_0) > 0$,

a local maximum at x_0 if $g''(x_0) < 0$.

We have a similar test for functions of two variables. As one might expect, the test is somewhat more complicated to state and definitely more difficult to prove. We will omit the proof.†

THEOREM 15.7.1 THE SECOND-PARTIALS TEST

Suppose that f has continuous second-order partial derivatives in a neighborhood of (x_0, y_0) and that $\nabla f(x_0, y_0) = \mathbf{0}$. Set

$$A = \frac{\partial^2 f}{\partial x^2}(x_0, y_0), \qquad B = \frac{\partial^2 f}{\partial y \partial x}(x_0, y_0), \qquad C = \frac{\partial^2 f}{\partial y^2}(x_0, y_0)$$

and form the *discriminant* $D = B^2 - AC$.

1. If $D > 0$, then (x_0, y_0) is a saddle point.
2. If $D < 0$, then f has

$$\text{a local minimum at } (x_0, y_0) \quad \text{if} \quad A > 0,$$
$$\text{a local maximum at } (x_0, y_0) \quad \text{if} \quad A < 0.$$

The test is geometrically evident for functions of the form

$$f(x, y) = \tfrac{1}{2}ax^2 + \tfrac{1}{2}cy^2. \qquad\qquad (a \neq 0, c \neq 0)$$

The graph of such a function is a paraboloid:

$$z = \tfrac{1}{2}ax^2 + \tfrac{1}{2}cy^2.$$

The gradient is $\mathbf{0}$ at the origin $(0, 0)$. Moreover

$$A = \frac{\partial^2 f}{\partial x^2}(0, 0) = a, \qquad B = \frac{\partial^2 f}{\partial y \partial x}(0, 0) = 0, \qquad C = \frac{\partial^2 f}{\partial y^2}(0, 0) = c.$$

and $D = B^2 - AC = -ac$. If $D > 0$, then a and c have opposite signs and the surface has a saddle point. (Figure 15.7.1.)

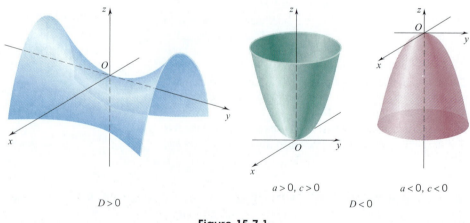

D > 0 $a > 0, c > 0$ $a < 0, c < 0$

D < 0

Figure 15.7.1

---

† You can find a proof in most texts on advanced calculus.

Suppose now that $D < 0$. If $a > 0$, then $c > 0$ and the surface has a minimum point; if $a < 0$, then $c < 0$ and the surface has a maximum point.

In the examples that follow we apply the second-partials test to a variety of functions.

Example 1 For the function

$$f(x, y) = 2x^2 + y^2 - xy - 7y$$

we have

$$\frac{\partial f}{\partial x} = 4x - y, \qquad \frac{\partial f}{\partial y} = 2y - x - 7.$$

Setting both partials equal to zero, we have

$$4x - y = 0, \qquad 2y - x - 7 = 0.$$

The only simultaneous solution to these equations is $x = 1$, $y = 4$. The point $(1, 4)$ is thus the only stationary point.

The second partials are constant:

$$\frac{\partial^2 f}{\partial x^2} = 4, \qquad \frac{\partial^2 f}{\partial y \partial x} = -1, \qquad \frac{\partial^2 f}{\partial y^2} = 2.$$

Thus $A = 4$, $B = -1$, $C = 2$, and

$$D = B^2 - AC = -7 < 0.$$

Since $A > 0$, it follows from the second-partials test that

$$f(1, 4) = 2 + 16 - 4 - 28 = -14$$

is a local minimum. (We obtained this result in Section 15.6 without benefit of the second-partials test; see Example 1.) ❏

Example 2 The function

$$f(x, y) = \frac{x}{y^2} + xy$$

has partial derivatives

$$\frac{\partial f}{\partial x} = \frac{1}{y^2} + y, \qquad \frac{\partial f}{\partial y} = -\frac{2x}{y^3} + x.$$

Setting both of these partials equal to zero, we have

$$\frac{1}{y^2} + y = 0, \qquad x\left(-\frac{2}{y^3} + 1\right) = 0.$$

As you can check, the only simultaneous solution to these equations is $x = 0$, $y = -1$. The point $(0, -1)$ is thus the only stationary point.

The second partials are

$$\frac{\partial^2 f}{\partial x^2} = 0, \qquad \frac{\partial^2 f}{\partial y \partial x} = -\frac{2}{y^3} + 1, \qquad \frac{\partial^2 f}{\partial y^2} = \frac{6x}{y^4}.$$

Evaluating these partials at the point $(0, -1)$, we find that $A = 0$, $B = 3$, $C = 0$, and thus

$$D = B^2 - AC = 9 > 0.$$

By the second-partials test, $(0, -1)$ is a saddle point.

We could have determined that $(0, -1)$ is a saddle point by comparing the value of f at $(0, -1)$ with its value at nearby points $(h, -1 + k)$:

$$f(h, -1 + k) - f(0, -1) = \frac{h}{(k - 1)^2} + h(k - 1) - 0 = h\left[\frac{1 + (k - 1)^3}{(k - 1)^2}\right].$$

For small $k > 0$, this expression is positive if $h > 0$ and negative if $h < 0$. ❏

Example 3 The function

$$f(x, y) = x^3 + y^3 - 3xy$$

has partial derivatives

$$\frac{\partial f}{\partial x} = 3x^2 - 3y, \qquad \frac{\partial f}{\partial y} = 3y^2 - 3x.$$

Setting both partials equal to zero, we have

$$x^2 - y = 0, \qquad y^2 - x = 0.$$

The only simultaneous solutions are $x = 1$, $y = 1$ and $x = 0$, $y = 0$. The points $(1, 1)$ and $(0, 0)$ are the only stationary points.

The second partials are

$$\frac{\partial^2 f}{\partial x^2} = 6x, \qquad \frac{\partial^2 f}{\partial y \partial x} = -3, \qquad \frac{\partial^2 f}{\partial y^2} = 6y.$$

At $(1, 1)$, we have $A = 6$, $B = -3$, $C = 6$, and thus

$$D = B^2 - AC = 9 - 36 = -27 < 0.$$

Since $A > 0$, we know that

$$f(1, 1) = 1 + 1 - 3 = -1$$

is a local minimum. At $(0, 0)$, we have $A = 0$, $B = -3$, $C = 0$. Thus

$$D = B^2 - AC = 9 > 0.$$

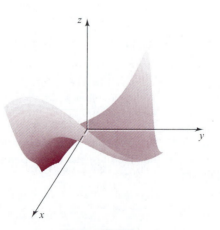

Figure 15.7.2

The origin is a saddle point. A computer-generated graph of this function is shown in Figure 15.7.2. ❑

Example 4 We considered the function

$$f(x, y) = -xye^{-(x^2+y^2)/2}$$

in Example 3 of the preceding section and found the stationary points $(0, 0)$, $(1, 1)$, $(1, -1)$, $(-1, 1)$, and $(-1, -1)$. Here we apply the second-partials test to confirm the results that we found there. You can verify that the second-partial derivatives of f are

$$\frac{\partial^2 f}{\partial x^2} = xy(3 - x^2)e^{-(x^2+y^2)/2}, \qquad \frac{\partial^2 f}{\partial y^2} = xy(3 - y^2)e^{-(x^2+y^2)/2}$$

and

$$\frac{\partial^2 f}{\partial y \partial x} = (x^2 - 1)(1 - y^2)e^{-(x^2+y^2)/2}.$$

The data for the second partials test are recorded in the following table.

Point	A	B	C	D	Result
$(0, 0)$	0	-1	0	1	Saddle point
$(1, 1)$	$2e^{-1}$	0	$2e^{-1}$	$-4e^{-2}$	Loc. min.
$(1, -1)$	$-2e^{-1}$	0	$-2e^{-1}$	$-4e^{-2}$	Loc. max.
$(-1, 1)$	$-2e^{-1}$	0	$-2e^{-1}$	$-4e^{-2}$	Loc. max.
$(-1, -1)$	$2e^{-1}$	0	$2e^{-1}$	$-4e^{-2}$	Loc. min.

❑

Example 5 Here we test the function

$$f(x, y) = \sin x + \sin y + \sin(x + y)$$

on the open square $0 < x < \pi, 0 < y < \pi$.

In the first place

$$\frac{\partial f}{\partial x} = \cos x + \cos(x + y), \qquad \frac{\partial f}{\partial y} = \cos y + \cos(x + y).$$

Setting both of these partials equal to zero, we have

$$\cos x + \cos(x + y) = 0, \qquad \cos y + \cos(x + y) = 0.$$

Together these equations imply that $\cos x = \cos y$. Since both x and y lie between 0 and π, and since the cosine function is one-to-one on $(0, \pi)$, we can conclude that $x = y$. The condition

$$\cos x + \cos(x + y) = 0$$

now gives

$$\cos x + \cos 2x = 0.$$

Into this last equation we substitute the identity $\cos 2x = 2\cos^2 x - 1$ and thereby obtain the relation

$$2\cos^2 x + \cos x - 1 = 0.$$

This is a quadratic in cos x, which can be written in factored form

$$(2 \cos x - 1)(\cos x + 1) = 0$$

Thus

$$\cos x = -1 \quad \text{or} \quad \tfrac{1}{2}$$

We throw out the solution -1 because x has to lie strictly between 0 and π. The other possibility, $\cos x = \tfrac{1}{2}$, gives $x = \tfrac{1}{3}\pi$. Since $x = y$, we also have $y = \tfrac{1}{3}\pi$. This shows that $(\tfrac{1}{3}\pi, \tfrac{1}{3}\pi)$ is the only stationary point.

The second partials are

$$\frac{\partial^2 f}{\partial x^2} = -\sin x - \sin(x + y), \qquad \frac{\partial^2 f}{\partial y \partial x} = -\sin(x + y),$$

$$\frac{\partial^2 f}{\partial y^2} = -\sin y - \sin(x + y).$$

At $(\tfrac{1}{3}\pi, \tfrac{1}{3}\pi)$ we have

$$A = -\sin \tfrac{1}{3}\pi - \sin \tfrac{2}{3}\pi = -\tfrac{1}{2}\sqrt{3} - \tfrac{1}{2}\sqrt{3} = -\sqrt{3}, \qquad B = -\sin \tfrac{2}{3}\pi = -\tfrac{1}{2}\sqrt{3},$$

$$C = -\sin \tfrac{1}{3}\pi - \sin \tfrac{2}{3}\pi = -\tfrac{1}{2}\sqrt{3} - \tfrac{1}{2}\sqrt{3} = -\sqrt{3},$$

and thus

$$D = B^2 - AC = \tfrac{3}{4} - 3 < 0.$$

Since $A < 0$, we can conclude that

$$f(\tfrac{1}{3}\pi, \tfrac{1}{3}\pi) = \sin \tfrac{1}{3}\pi + \sin \tfrac{1}{3}\pi + \sin \tfrac{2}{3}\pi = \tfrac{3}{2}\sqrt{3}$$

is a local maximum. A computer-generated drawing of the graph of $f(x, y) = \sin x + \sin y + \sin(x + y)$ is shown in Figure 15.7.3. ❑

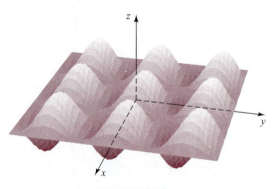

Figure 15.7.3

The second-derivative test for a function of one variable applies to points x_0 where $g'(x_0) = 0$ but $g''(x_0) \neq 0$. If $g''(x_0) = 0$, the second-derivative test provides no conclusive information. The second-partials test suffers from a similar limitation. It applies to points (x_0, y_0) where $\nabla f(x_0, y_0) = \mathbf{0}$ but $D \neq 0$. If $D = 0$, the second-partials test provides no information.

Consider, for example, the functions

$$f(x, y) = x^4 + y^4, \qquad g(x, y) = -(x^4 + y^4), \qquad h(x, y) = x^4 - y^4.$$

Each of these functions has zero gradient at the origin, and, as you can check, in each case $D = 0$. Yet,

(1) for f, $(0, 0)$ gives a local minimum;

(2) for g, $(0, 0)$ gives a local maximum;

(3) for h, $(0, 0)$ is a saddle point.

Statements (1) and (2) are obvious. To confirm (3), note that $h(0, 0) = 0$, but in every neighborhood of $(0, 0)$ the function h takes on both positive and negative values:

$$h(x, 0) > 0 \quad \text{for } x \neq 0 \qquad \text{while} \qquad h(0, y) < 0 \quad \text{for } y \neq 0. \quad \square$$

EXERCISES 15.7

In Exercises 1–24, find the stationary points and the local extreme values.

1. $f(x, y) = x^2 + xy + y^2 - 6x + 2$.

2. $f(x, y) = x^2 + 2xy + 3y^2 + 2x + 10y + 1$.

3. $f(x, y) = 4x + 2y - x^2 + xy - y^2$.

4. $f(x, y) = x^2 - xy + y^2 + 2x + 2y - 1$.

5. $f(x, y) = x^3 - 6xy + y^3$.

6. $f(x, y) = 3x^2 + xy - y^2 + 5x - 5y + 4$.

7. $f(x, y) = x^3 + y^2 - 6xy + 6x + 3y - 2$.

8. $f(x, y) = x^2 + xy + y^2 - 6y + 5$.

9. $f(x, y) = e^x \cos y$.

10. $f(x, y) = x^2 - 2xy + 2y^2 - 3x + 5y$.

11. $f(x, y) = x \sin y$.   **12.** $f(x, y) = y + x \sin y$.

13. $f(x, y) = (x + y)(xy + 1)$.

14. $f(x, y) = xy^{-1} - yx^{-1}$.

15. $f(x, y) = xy + x^{-1} + 8y^{-1}$.

16. $f(x, y) = x^2 - 2xy - y^2 + 1$.

17. $f(x, y) = xy + x^{-1} + y^{-1}$.

18. $f(x, y) = (x - y)(xy - 1)$.

19. $f(x, y) = \dfrac{-2x}{x^2 + y^2 + 1}$.   **20.** $f(x, y) = (x - 3) \ln xy$.

21. $f(x, y) = x^4 - 2x^2 + y^2 - 2$.

22. $f(x, y) = (x^2 + y^2)e^{x^2 - y^2}$.

23. $f(x, y) = \cos x + \cos y - \cos(x + y)$ where $-\pi/2 < x < \pi/2, \ -\pi/2 < y < \pi/2$.

24. $f(x, y) = 8xye^{-(x^2 + y^2)}$.

25. Let $f(x, y) = x^2 + kxy + y^2$, k a constant.
  (a) Show that f has a stationary point at $(0, 0)$ independent of the value of k.
  (b) For what values of k will f have a saddle point at $(0, 0)$?
  (c) For what values of k will f have a local minimum at $(0, 0)$?

  (d) For what values of k is the second-derivative test inconclusive?

26. Repeat Exercise 25 for the function
$$f(x, y) = x^2 + kxy + 4y^2, \ k \text{ a constant}.$$

27. Find the point in the plane $2x - 3y + 6z = 14$ that is closest to the point $(-1, 1, 2)$. Justify your answer using the second-partials test.

28. Use the methods of this section to derive the formula for the distance from the origin to the plane
$$Ax + By + Cz + D = 0.$$

29. Find the maximum volume for a rectangular solid in the first octant $(x \geq 0, y \geq 0, z \geq 0)$ with one vertex at the origin and opposite vertex on the plane $x + y + z = 1$.

30. Find the maximum volume for a rectangular solid in the first octant with one vertex at the origin and the opposite vertex on the plane
$$\frac{x}{a} + \frac{y}{b} + \frac{z}{c} = 1.$$

31. Find the shortest distance from the point $(1, 2, 0)$ to the elliptic cone $z = \sqrt{x^2 + 2y^2}$.   HINT: Minimize the square of the distance.

32. Let n be an integer greater than 2 and set
$$f(x, y) = ax^n + cy^n, \text{ taking } ac \neq 0.$$
  (a) Find the stationary points.
  (b) Find the discriminant at each stationary point.
  (c) Find the local and absolute extreme values given that
    (i) $a > 0, c > 0$.
    (ii) $a < 0, c < 0$.
    (iii) $a > 0, c < 0$.

33. Describe the behavior of the function at the origin.
  (a) $f(x, y) = x^2 - y^3$.
  (b) $f(x, y) = 2 \cos(x + y) + e^{xy}$.

34. Find the maximum volume for a rectangular solid inscribed in the sphere
$$x^2 + y^2 + z^2 = a^2$$

35. A pentagon is composed of a rectangle surmounted by an isosceles triangle (see the figure). Given that the perimeter of the pentagon has a fixed value P, find the dimensions for maximum area.

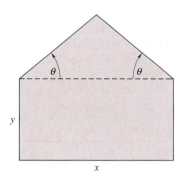

36. A bakery produces two types of bread, one at a cost of 50 cents per loaf, the other at a cost of 60 cents per loaf. Assume that, if the first bread is sold at x cents a loaf and the second at y cents a loaf, then the number of loaves that can be sold each week is given by the formulas

$$N_1 = 250(y - x), \qquad N_2 = 32{,}000 + 250(x - 2y).$$

Determine x and y for maximum profit.

37. Find the distance between the lines $x = \frac{1}{2}y = \frac{1}{3}z$ and $x = y - 2 = z$.

38. Find the absolute maximum value of

$$f(x, y) = \frac{(ax + by + c)^2}{x^2 + y^2 + 1}.$$

39. Find the dimensions of the most economical open-top rectangular crate 96 cubic meters in volume given that the base costs 30 cents per square meter and the sides cost 10 cents per square meter.

40. Let $f(x, y) = ax^2 + bxy + cy^2$, taking $abc \neq 0$.
(a) Find the discriminant D.
(b) Find the stationary points and local extreme values if $D \neq 0$.
(c) Suppose that $D = 0$. Find the stationary points and the local and absolute extreme values given that
(i) $a > 0, c > 0$. (ii) $a < 0, c < 0$.

41. Show that a closed rectangular box of maximum volume having a prescribed surface area S is a cube.

42. If an open rectangular box has a prescribed surface area S, what dimensions yield the maximum volume?

43. (*The method of least squares*) In this exercise we illustrate an important method of fitting a curve to a collection of points. Consider three points

$$(x_1, y_1) = (0, 2), \quad (x_2, y_2) = (1, -5), \quad (x_3, y_3) = (2, 4).$$

(a) Find the line $y = mx + b$ that minimizes the sum of the squares of the vertical distances
$$d_i = |y_i - (mx_i + b)|$$ from these points to the line.

(b) Find the parabola $y = \alpha x^2 + \beta$ that minimizes the sum of the squares of the vertical distances $d_i = |y_i - (\alpha x_i^2 + \beta)|$ from the points to the parabola.

44. Repeat Exercise 43 taking $(x_1, y_1) = (-1, 2)$, $(x_2, y_2) = (0, -1)$, $(x_3, y_3) = (1, 1)$.

45. According to U.S. Postal Service regulations the length plus the girth (the perimeter of a cross section) of a package cannot exceed 108 inches. (See the figure.)

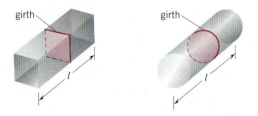

(a) Find the dimensions of the rectangular box of maximum volume that is acceptable for mailing.
(b) Find the dimensions of the cylindrical tube of maximum volume that is acceptable for mailing.

46. A petrochemical company is designing a cylindrical tank with hemispherical ends to be used in transporting its products. (See Exercise 35, Section 14.1.) If the volume of the tank is to be 10,000 cubic meters, what dimensions should be used to minimize the amount of metal required?

47. A 10-foot section of gutter is to be made from a 12-inch-wide strip of metal by folding up strips of length x on each side so that they make an angle θ with the bottom of the gutter. (See the figure.) Determine values for x and θ that will maximize the cross-sectional area of the gutter.

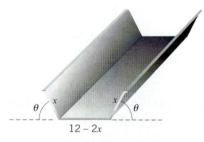

48. Find the volume of the largest rectangular box with edges parallel to the coordinate axes that can be inscribed in the ellipsoid

$$\frac{x^2}{a^2} + \frac{y^2}{b^2} + \frac{z^2}{c^2} = 1.$$

■ 15.8 MAXIMA AND MINIMA WITH SIDE CONDITIONS

When we ask for the distance from a point $P(x_0, y_0)$ to a line l: $Ax + By + C = 0$, we are asking for the minimum value of

$$f(x, y) = \sqrt{(x - x_0)^2 + (y - y_0)^2}$$

with (x, y) subject to the side condition† $Ax + By + C = 0$. When we ask for the distance from a point $P(x_0, y_0, z_0)$ to a plane p: $Ax + By + Cz + D = 0$, we are asking for the minimum value of

$$f(x, y, z) = \sqrt{(x - x_0)^2 + (y - y_0)^2 + (z - z_0)^2}$$

with (x, y, z) subject to the side condition $Ax + By + Cz + D = 0$.

We have already treated these particular problems by special techniques. Our interest here is to present techniques for handling problems of this sort in general. In the two-variable case, the problems will take the form of maximizing (or minimizing) some expression $f(x, y)$ subject to a side condition $g(x, y) = 0$. In the three-variable case, we will seek to maximize (or minimize) some expression $f(x, y, z)$ subject to a side condition $g(x, y, z) = 0$. We begin with two simple examples.

Example 1 Maximize the product xy subject to the side condition $x + y - 1 = 0$.

SOLUTION The condition $x + y - 1 = 0$ gives $y = 1 - x$. Our initial problem can therefore be solved simply by maximizing the product $h(x) = x(1 - x)$. The derivative $h'(x) = 1 - 2x$ is 0 only at $x = \frac{1}{2}$. Since $h''(x) = -2 < 0$, we know from the second-derivative test that $h(\frac{1}{2}) = \frac{1}{2}(1 - \frac{1}{2}) = \frac{1}{4}$ is the desired maximum. ❑

Example 2 Find the maximum volume of a rectangular solid given that the sum of the lengths of its edges is $12a$.

SOLUTION We denote the dimensions of the solid by x, y, z. (Figure 15.8.1.) The volume is given by

$$V = xyz.$$

The stipulation on the edges requires that

$$4(x + y + z) = 12a.$$

Solving this last equation for z, we find that

$$z = 3a - (x + y).$$

Substituting this expression for z in the volume formula, we have

$$V = xy[3a - (x + y)].$$

Since x, y, and $z = 3a - (x + y)$ must all remain positive, our problem is to find the maximum value of V on the interior of the triangle shown in Figure 15.8.2.

The first partials are

$$\frac{\partial V}{\partial x} = xy(-1) + y[3a - (x + y)] = 3ay - 2xy - y^2,$$

$$\frac{\partial V}{\partial y} = xy(-1) + x[3a - (x + y)] = 3ax - x^2 - 2xy.$$

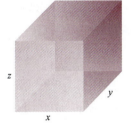

Figure 15.8.1

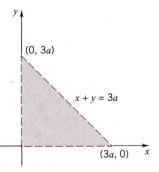

$(0, 3a)$

$x + y = 3a$

$(3a, 0)$

Figure 15.8.2

---

† Side conditions are often called *constraints*.

Setting both partials equal to zero, we have

$$(3a - 2x - y)y = 0 \quad \text{and} \quad (3a - x - 2y)x = 0.$$

Since x and y are assumed positive, we can divide by x and y and get

$$3a - 2x - y = 0 \quad \text{and} \quad 3a - x - 2y = 0.$$

Solving these equations simultaneously, we find that $x = y = a$. The point (a, a), which does lie within the triangle, is the only stationary point. The value of V at that point is a^3. The conditions of the problem make it clear that this is a maximum. (If you are skeptical, you can confirm this by appealing to the second-partials test.) ❑

The last two problems were easy. They were easy in part because the side conditions were such that we could solve for one of the variables in terms of the other(s). In general this is not possible and a more sophisticated approach is required.

The Method of Lagrange

We begin with what looks like a detour. To avoid having to make separate statements for the two- and three-variable cases, we will use vector notation.

Throughout the discussion f will be a function of two or three variables which is continuously differentiable on some open set U. We take

$$C: \quad \mathbf{r} = \mathbf{r}(t), \quad t \in I$$

as a curve that lies entirely in U and has at each point a nonzero tangent vector $\mathbf{r}'(t)$. The basic result is this:

(15.8.1)

> if $\mathbf{x}_0$ maximizes (or minimizes) $f(\mathbf{x})$ on C,
> then $\nabla f(\mathbf{x}_0)$ is perpendicular to C at $\mathbf{x}_0$.

PROOF Choose t_0 so that

$$\mathbf{r}(t_0) = \mathbf{x}_0.$$

The composition $f(\mathbf{r}(t))$ has a maximum (or minimum) at t_0. Consequently, its derivative,

$$\frac{d}{dt}[f(\mathbf{r}(t))] = \nabla f(\mathbf{r}(t)) \cdot \mathbf{r}'(t).$$

must be zero at t_0:

$$0 = \nabla f(\mathbf{r}(t_0)) \cdot \mathbf{r}'(t_0) = \nabla f(\mathbf{x}_0) \cdot \mathbf{r}'(t_0).$$

This shows that

$$\nabla f(\mathbf{x}_0) \perp \mathbf{r}'(t_0).$$

Since $\mathbf{r}'(t_0)$ is tangent to C at $\mathbf{x}_0$, $\nabla f(\mathbf{x}_0)$ is perpendicular to C at $\mathbf{x}_0$. ❑

We are now ready for the side condition problems. Suppose that g is a continuously differentiable function of two or three variables defined on a subset of the domain of f. Lagrange made the following observation:†

(15.8.2)

> if $\mathbf{x}_0$ maximizes (or minimizes) $f(\mathbf{x})$ subject to the side condition $g(\mathbf{x}) = 0$, then $\nabla f(\mathbf{x}_0)$ and $\nabla g(\mathbf{x}_0)$ are parallel. Thus, if $\nabla g(\mathbf{x}_0) \neq \mathbf{0}$, then there exists a scalar λ such that
> $$\nabla f(\mathbf{x}_0) = \lambda \nabla g(\mathbf{x}_0).$$

Such a scalar λ has come to be called a *Lagrange multiplier.*

PROOF OF (15.8.2) Let's suppose that $\mathbf{x}_0$ maximizes (or minimizes) $f(\mathbf{x})$ subject to the side condition $g(\mathbf{x}) = 0$. If $\nabla g(\mathbf{x}_0) = \mathbf{0}$, the result is trivially true: every vector is parallel to the zero vector. We suppose therefore that $\nabla g(\mathbf{x}_0) \neq \mathbf{0}$.

In the two-variable case we have

$$\mathbf{x}_0 = (x_0, y_0) \qquad \text{and} \qquad \text{the side condition} \quad g(x, y) = 0.$$

The side condition defines a curve C that has a nonzero tangent vector at (x_0, y_0).†† Since (x_0, y_0) maximizes (or minimizes) $f(x, y)$ on C, we know from (15.8.1) that $\nabla f(x_0, y_0)$ is perpendicular to C at (x_0, y_0). By (15.5.2), $\nabla g(x_0, y_0)$ is also perpendicular to C at (x_0, y_0). The two gradients are therefore parallel.

In the three-variable case we have

$$\mathbf{x}_0 = (x_0, y_0, z_0) \qquad \text{and} \qquad \text{the side condition} \quad g(x, y, z) = 0.$$

The side condition defines a surface Γ that lies in the domain of f. Now let C be a curve that lies on Γ and passes through (x_0, y_0, z_0) with nonzero tangent vector. We know that (x_0, y_0, z_0) maximizes (or minimizes) $f(x, y, z)$ on C. Consequently, $\nabla f(x_0, y_0, z_0)$ is perpendicular to C at (x_0, y_0, z_0). Since this is true for each such curve C, $\nabla f(x_0, y_0, z_0)$ must be perpendicular to Γ itself. But $\nabla g(x_0, y_0, z_0)$ is also perpendicular to Γ at (x_0, y_0, z_0) (15.5.6). It follows that $\nabla f(x_0, y_0, z_0)$ and $\nabla g(x_0, y_0, z_0)$ are parallel. ❏

We come now to some problems that are susceptible to Lagrange's method. In each case ∇g is not $\mathbf{0}$ where g is 0 and therefore we can focus entirely on those points $\mathbf{x}$ that satisfy the Lagrange condition

$$\nabla f(\mathbf{x}) = \lambda \nabla g(\mathbf{x}). \tag{15.8.2}$$

Example 3 Maximize and minimize

$$f(x, y) = xy \quad \text{on the unit circle} \quad x^2 + y^2 = 1.$$

SOLUTION Since f is continuous and the unit circle is closed and bounded, it is clear that both a maximum and a minimum exist (see Section 15.6).

---

† Another contribution of the French mathematician Joseph Louis Lagrange.

†† $\mathbf{t}(x_0, y_0) = \dfrac{\partial g}{\partial y}(x_0, y_0)\mathbf{i} - \dfrac{\partial g}{\partial x}(x_0, y_0)\mathbf{j} \neq \mathbf{0}.$

To apply Lagrange's method we set

$$g(x, y) = x^2 + y^2 - 1.$$

We want to maximize and minimize

$$f(x, y) = xy \quad \text{subject to the side condition} \quad g(x, y) = 0.$$

The gradients are

$$\nabla f(x, y) = y\mathbf{i} + x\mathbf{j}, \qquad \nabla g(x, y) = 2x\mathbf{i} + 2y\mathbf{j}.$$

Setting

$$\nabla f(x, y) = \lambda \nabla g(x, y),$$

we obtain

$$y = 2\lambda x, \qquad x = 2\lambda y.$$

Multiplying the first equation by y and the second equation by x, we find that

$$y^2 = 2\lambda xy, \qquad x^2 = 2\lambda xy$$

and thus

$$y^2 = x^2.$$

The side condition $x^2 + y^2 = 1$ now implies that $2x^2 = 1$ and therefore that $x = \pm\frac{1}{2}\sqrt{2}$. The only points that can give rise to an extreme value are

$$(\tfrac{1}{2}\sqrt{2}, \tfrac{1}{2}\sqrt{2}), \quad (\tfrac{1}{2}\sqrt{2}, -\tfrac{1}{2}\sqrt{2}), \quad (-\tfrac{1}{2}\sqrt{2}, \tfrac{1}{2}\sqrt{2}), \quad (-\tfrac{1}{2}\sqrt{2}, -\tfrac{1}{2}\sqrt{2}).$$

At the first and fourth points f takes on the value $\frac{1}{2}$. At the second and third points f takes on the value $-\frac{1}{2}$. Clearly then, $\frac{1}{2}$ is the maximum value and $-\frac{1}{2}$ the minimum value. ❏

Example 4 Find the minimum value taken on by the function

$$f(x, y) = x^2 + (y - 2)^2 \quad \text{on the hyperbola} \quad x^2 - y^2 = 1.$$

SOLUTION Note that the expression $f(x, y) = x^2 + (y - 2)^2$ gives the square of the distance between the points $(0, 2)$ and (x, y). Therefore, the problem asks us to minimize the square of the distance from the point $(0, 2)$ to the hyperbola, and this minimum value clearly exists. Note, also, that there is no maximum value. See Figure 15.8.3.

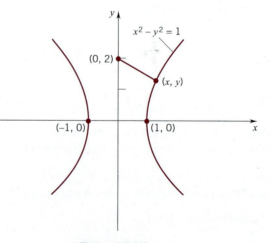

Figure 15.8.3

Now set

$$g(x, y) = x^2 - y^2 - 1.$$

We want to minimize

$$f(x, y) = x^2 + (y - 2)^2 \quad \text{subject to the side condition} \quad g(x, y) = 0.$$

Here

$$\nabla f(x, y) = 2x\mathbf{i} + 2(y - 2)\mathbf{j}, \qquad \nabla g(x, y) = 2x\mathbf{i} - 2y\mathbf{j}.$$

The Lagrange condition $\nabla f(x, y) = \lambda \nabla g(x, y)$ gives

$$2x = 2\lambda x, \qquad 2(y - 2) = -2\lambda y,$$

which we can simplify to

$$x = \lambda x, \qquad y - 2 = -\lambda y.$$

The side condition $x^2 - y^2 = 1$ shows that x cannot be zero. Dividing $x = \lambda x$ by x, we get $\lambda = 1$. This means that $y - 2 = -y$ and therefore $y = 1$. With $y = 1$, the side condition gives $x = \pm\sqrt{2}$. The points to be checked are therefore $(-\sqrt{2}, 1)$ and $(\sqrt{2}, 1)$. At each of these points f takes on the value 3. This is the desired minimum. ❑

Remark The last problem could have been solved more simply by rewriting the side condition as $x^2 = 1 + y^2$ and eliminating x from $f(x, y)$ by substitution. It would then have been only a matter of minimizing the function $h(y) = 1 + y^2 + (y - 2)^2 = 2y^2 - 4y + 5$. ❑

Example 5 Maximize

$$f(x, y, z) = xyz \quad \text{subject to the side condition} \quad x^3 + y^3 + z^3 = 1$$

with $x \geq 0, y \geq 0, z \geq 0$.

SOLUTION The set of all points (x, y, z) that satisfy the side condition can be shown to be closed and bounded. Since f is continuous, we can be sure that the desired maximum exists.

We begin by setting

$$g(x, y, z) = x^3 + y^3 + z^3 - 1$$

so that the side condition becomes $g(x, y, z) = 0$. We seek those triples (x, y, z) that simultaneously satisfy the Lagrange condition

$$\nabla f(x, y, z) = \lambda \nabla g(x, y, z) \quad \text{and the side condition} \quad g(x, y, z) = 0.$$

The gradients are

$$\nabla f(x, y, z) = yz\mathbf{i} + xz\mathbf{j} + xy\mathbf{k}, \qquad \nabla g(x, y, z) = 3x^2\mathbf{i} + 3y^2\mathbf{j} + 3z^2\mathbf{k}.$$

The Lagrange condition $\nabla f(x, y, z) = \lambda \nabla g(x, y, z)$ gives

$$yz = 3\lambda x^2, \qquad xz = 3\lambda y^2, \qquad xy = 3\lambda z^2.$$

Multiplying the first equation by x, the second by y, and the third by z, we get

$$xyz = 3\lambda x^3, \qquad xyz = 3\lambda y^3, \qquad xyz = 3\lambda z^3$$

and consequently

$$\lambda x^3 = \lambda y^3 = \lambda z^3.$$

We can exclude $\lambda = 0$ because, if $\lambda = 0$, then x, y, or z would have to be zero. That would force xyz to be 0, and 0 is obviously not a maximum. Having excluded $\lambda = 0$, we can divide by λ and get $x^3 = y^3 = z^3$ and thus $x = y = z$. The side condition $x^3 + y^3 + z^3 = 1$ now gives

$$x = (\tfrac{1}{3})^{1/3}, \qquad y = (\tfrac{1}{3})^{1/3}, \qquad z = (\tfrac{1}{3})^{1/3}.$$

The desired maximum is $\tfrac{1}{3}$. ❏

Example 6 Show that, of all the triangles inscribed in a fixed circle of radius R, the equilateral triangle has the largest perimeter.

SOLUTION It is intuitively clear that this maximum exists and geometrically clear that the triangle that offers this maximum contains the center of the circle in its interior or on its boundary. As in Figure 15.8.4 we denote by x, y, z the central angles that subtend the three sides. As you can verify by trigonometry, the perimeter of the triangle is given by the function

$$f(x, y, z) = 2R(\sin \tfrac{1}{2}x + \sin \tfrac{1}{2}y + \sin \tfrac{1}{2}z).$$

As a side condition we have

$$g(x, y, z) = x + y + z - 2\pi = 0.$$

To maximize the perimeter we form the gradients

$$\nabla f(x, y, z) = R[\cos \tfrac{1}{2}x\,\mathbf{i} + \cos \tfrac{1}{2}y\,\mathbf{j} + \cos \tfrac{1}{2}z\,\mathbf{k}], \qquad \nabla g(x, y, z) = \mathbf{i} + \mathbf{j} + \mathbf{k}.$$

The Lagrange condition $\nabla f(x, y, z) = \lambda \nabla g(x, y, z)$ gives

$$\lambda = R \cos \tfrac{1}{2}x, \qquad \lambda = R \cos \tfrac{1}{2}y, \qquad \lambda = R \cos \tfrac{1}{2}z$$

and therefore

$$\cos \tfrac{1}{2}x = \cos \tfrac{1}{2}y = \cos \tfrac{1}{2}z.$$

With x, y, z all in $(0, \pi]$, we can conclude that $x = y = z$. Since the central angles are equal, the sides are equal. The triangle is therefore equilateral. ❏

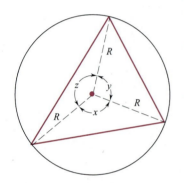

Figure 15.8.4

An Application of the Cross Product

The Lagrange condition can be replaced by a cross-product equation: points that satisfy $\nabla f = \lambda \nabla g$ satisfy

(15.8.3)
$$\nabla f \times \nabla g = \mathbf{0}.$$

If f and g are functions of two variables, then $\nabla f \times \nabla g = 0$ has a simpler form:

$$\nabla f \times \nabla g = \begin{vmatrix} \mathbf{i} & \mathbf{j} & \mathbf{k} \\ f_x & f_y & 0 \\ g_x & g_y & 0 \end{vmatrix} = (f_x g_y - f_y g_x)\mathbf{k}$$

which implies that

(15.8.4)

$$\frac{\partial f}{\partial x}\frac{\partial g}{\partial y} - \frac{\partial f}{\partial y}\frac{\partial g}{\partial x} = 0.$$

Example 7 Maximize and minimize $f(x, y) = xy$ on the unit circle $x^2 + y^2 = 1$.

SOLUTION This problem was solved earlier by means of the Lagrange equation. This time we will use (15.8.4) instead.

As before we set

$$g(x, y) = x^2 + y^2 - 1$$

so that the side condition takes the form $g(x, y) = 0$. Since

$$\frac{\partial f}{\partial x} = y, \quad \frac{\partial f}{\partial y} = x \quad \text{and} \quad \frac{\partial g}{\partial x} = 2x, \quad \frac{\partial g}{\partial y} = 2y,$$

(15.8.4) takes the form

$$y(2y) - x(2x) = 0.$$

This gives $x^2 = y^2$.

As before, the side condition $x^2 + y^2 = 1$ implies that $2x^2 = 1$ and therefore that $x = \pm\frac{1}{2}\sqrt{2}$. The points under consideration are

$$(\tfrac{1}{2}\sqrt{2}, \tfrac{1}{2}\sqrt{2}), \quad (\tfrac{1}{2}\sqrt{2}, -\tfrac{1}{2}\sqrt{2}), \quad (-\tfrac{1}{2}\sqrt{2}, \tfrac{1}{2}\sqrt{2}), \quad (-\tfrac{1}{2}\sqrt{2}, -\tfrac{1}{2}\sqrt{2}).$$

At the first and fourth points f takes on the value $\frac{1}{2}$. At the second and third points f takes on the value $-\frac{1}{2}$. Thus, $\frac{1}{2}$ is the maximum value of f and $-\frac{1}{2}$ is the minimum value of f on the unit circle. ❑

In three variables the computations demanded by the cross-product equation are often quite complicated, and it is usually easier to follow the method of Lagrange.

EXERCISES 15.8

1. Minimize $x^2 + y^2$ on the hyperbola $xy = 1$.

2. Maximize xy on the ellipse $b^2x^2 + a^2y^2 = a^2b^2$.

3. Minimize xy on the ellipse $b^2x^2 + a^2y^2 = a^2b^2$.

4. Minimize xy^2 on the unit circle $x^2 + y^2 = 1$.

5. Maximize xy^2 on the ellipse $b^2x^2 + a^2y^2 = a^2b^2$.

6. Maximize $x + y$ on the curve $x^4 + y^4 = 1$.

7. Maximize $x^2 + y^2$ on the curve $x^4 + 7x^2y^2 + y^4 = 1$.

8. Minimize xyz on the unit sphere $x^2 + y^2 + z^2 = 1$.

9. Maximize xyz on the ellipsoid $x^2/a^2 + y^2/b^2 + z^2/c^2 = 1$.

10. Minimize $x + 2y + 4z$ on the sphere $x^2 + y^2 + z^2 = 7$.

11. Maximize $2x + 3y + 5z$ on the sphere $x^2 + y^2 + z^2 = 19$.

12. Minimize $x^4 + y^4 + z^4$ on the plane $x + y + z = 1$.

13. Maximize the volume of a rectangular solid in the first octant with one vertex at the origin and opposite vertex on the plane $x/a + y/b + z/c = 1$. (Take $a > 0, b > 0, c > 0$.)

14. Show that the square has the largest area of all the rectangles with a given perimeter.

15. Find the distance from the point $(0, 1)$ to the parabola $x^2 = 4y$.

16. Find the distance from the point $(p, 4p)$ to the parabola $y^2 = 2px$.

17. Find the points on the sphere $x^2 + y^2 + z^2 = 1$ that are closest to and furthest from the point $(2, 1, 2)$.

18. Let x, y, and z be the angles of a triangle. Determine the maximum value of $f(x, y, z) = \sin x \sin y \sin z$.

19. Determine the maximum value of $f(x, y, z) = 3x - 2y + z$ on the sphere $x^2 + y^2 + z^2 = 14$.

20. A rectangular box has three of its faces on the coordinate planes and one vertex in the first octant on the paraboloid $z = 4 - x^2 - y^2$. Determine the maximum volume of the box.

21. Use the method of Lagrange to find the distance from the origin to the plane with equation $Ax + By + Cz + D = 0$.

22. Maximize the volume of a rectangular solid given that the sum of the areas of the six faces is $6a^2$.

23. Within a triangle there is a point P such that the sum of the squares of its distances from the sides of the triangle is a minimum. Find this minimum.

24. Show that of all the triangles inscribed in a fixed circle the equilateral one has the largest: (a) product of the lengths of the sides; (b) sum of the squares of the lengths of the sides.

25. The curve $x^3 - y^3 = 1$ is asymptotic to the line $y = x$. Find the point(s) on the curve $x^3 - y^3 = 1$ furthest from the line $y = x$.

26. A plane passes through the point (a, b, c). Find its intercepts with the coordinate axes if the volume of the solid bounded by the plane and the coordinate planes is to be a minimum.

27. Show that, of all the triangles with a given perimeter, the equilateral triangle has the largest area. HINT: Area $= \sqrt{s(s - a)(s - b)(s - c)}$, where s represents the semiperimeter $s = \frac{1}{2}(a + b + c)$.

28. Show that the rectangular box of maximum volume that can be inscribed in the sphere $x^2 + y^2 + z^2 = a^2$ is a cube.

29. (a) Determine the maximum value of $f(x, y) = (xy)^{1/2}$ given that x and y are nonnegative numbers and $x + y = k$, k a constant.
(b) Use the result in (a) to show that if x and y are nonnegative numbers, then
$$(xy)^{1/2} \leq \frac{x + y}{2}.$$
(See Exercise 76, Section 1.3.)

30. (a) Determine the maximum value of $f(x, y, z) = (xyz)^{1/3}$ given that x, y, and z are nonnegative numbers and $x + y + z = k$, k a constant.
(b) Use the result in (a) to show that if x, y, and z are nonnegative numbers, then
$$(xyz)^{1/3} \leq \frac{x + y + z}{3}.$$
Note: $(xyz)^{1/3}$ is the *geometric mean* of x, y, z.

31. Let $x_1, x_2, \ldots, x_n$ be nonnegative numbers such that $x_1 + x_2 + \cdots + x_n = k$, k a constant. Prove that
$$(x_1 x_2 \cdots x_n)^{1/n} \leq \frac{x_1 + x_2 + \cdots + x_n}{n}.$$
In words, the geometric mean of n nonnegative numbers cannot exceed the arithmetic mean of the numbers.

32. The Celsius temperature T at a point (x, y, z) on the sphere $x^2 + y^2 + z^2 = 1$ is given by
$$T(x, y, z) = 10xy^2z.$$
Find the point(s) on the sphere at which the temperature is greatest and the point(s) at which it is least. Give the temperature at each of these points.

33. A soft drink manufacturer wants to design an aluminum can in the shape of a right circular cylinder to hold a given volume V (measured in cubic inches). If the objective is to minimize the amount of aluminum needed (top, sides, and bottom), what dimensions should be used?

In Exercises 34–38, use the Lagrange method to give alternative solutions to the indicated exercises in Sections 15.6 and 15.7.

34. Exercise 28, Section 15.6.

35. Exercise 29, Section 15.6.

36. Exercise 29, Section 15.7.

37. Exercise 45, Section 15.7.

38. Exercise 46, Section 15.7.

39. A manufacturer can produce three distinct products in quantities Q_1, Q_2, Q_3, respectively, and thereby derive a profit $P(Q_1, Q_2, Q_3) = 2Q_1 + 8Q_2 + 24Q_3$. Find the values of Q_1, Q_2, Q_3 that maximize profit if production is subject to the constraint $Q_1^2 + 2Q_2^2 + 4Q_3^2 = 4.5 \times 10^9$.

40. Find the volume of the largest rectangular box that can be inscribed in the ellipsoid
$$4x^2 + 9y^2 + 36z^2 = 36$$
if the edges of the box are parallel to the coordinate axes.

Problems with Two Side Conditions

The Lagrange method can be extended to problems with two side conditions as follows: If $\mathbf{x}_0$ is a maximum (or minimum) of $f(\mathbf{x})$ subject to the two side conditions $g(\mathbf{x}) = 0$ and $h(\mathbf{x}) = 0$, then there exist scalars λ and μ such that
$$\nabla f(\mathbf{x}_0) = \lambda \nabla g(\mathbf{x}_0) + \mu \nabla h(\mathbf{x}_0).$$

41. Find the maximum value of $f(x, y, z) = xy^2z - x^2yz$ subject to the side conditions $x^2 + y^2 = 1$ and $z = \sqrt{x^2 + y^2}$.   HINT: Set $\nabla f = \lambda \nabla g + \mu \nabla h$ and equate components.

42. The planes $x + 2y + 3z = 0$ and $2x + 3y + z = 4$ intersect in a straight line. Find the point on the line that is closest to the origin.   HINT: Minimize $D(x, y, z) = $ $x^2 + y^2 + z^2$ subject to the side conditions $x + 2y + 3z = 0$ and $2x + 3y + z = 4$; set $\nabla f = \lambda \nabla g + \mu \nabla h$ and equate components.

43. The plane $x + y - z + 1 = 0$ intersects the upper nappe of the cone $z^2 = x^2 + y^2$ in an ellipse. Find the points on this ellipse that are closest to and furthest from the origin.

■ 15.9 DIFFERENTIALS

We begin by reviewing the one-variable case. If f is differentiable at x, then, for small h, the increment

$$\Delta f = f(x + h) - f(x)$$

can be approximated by the differential

$$df = f'(x)h.$$

The geometric interpretations of Δf and df are shown in Figure 15.9.1. We write

$$\Delta f \cong df.$$

How good is this approximation?   Good enough that the ratio

$$\frac{\Delta f - df}{|h|}$$

tends to 0 as h tends to 0.

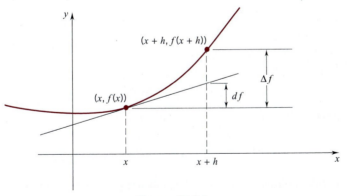

Figure 15.9.1

The differential of a function of several variables, defined in an analogous manner, plays a similar approximating role. Let's suppose that f, now a function of several variables, is differentiable at $\mathbf{x}$. The difference

(15.9.1)

$$\Delta f = f(\mathbf{x} + \mathbf{h}) - f(\mathbf{x})$$

is called the *increment* of f, and the dot product

(15.9.2)
$$df = \nabla f(\mathbf{x}) \cdot \mathbf{h}$$

is called the *differential* (more formally, the *total differential*). As in the one-variable case, for small $\mathbf{h}$, the differential and the increment are approximately equal:

(15.9.3)
$$\Delta f \cong df.$$

How approximately equal are they? Enough so that the ratio

$$\frac{\Delta f - df}{\|\mathbf{h}\|}$$

tends to 0 as $\mathbf{h}$ tends to $\mathbf{0}$. How do we know this? We know that

$$f(\mathbf{x} + \mathbf{h}) - f(\mathbf{x}) = \nabla f(\mathbf{x}) \cdot \mathbf{h} + o(\mathbf{h}).$$

Therefore,

$$[f(\mathbf{x} + \mathbf{h}) - f(\mathbf{x})] - \nabla f(\mathbf{x}) \cdot \mathbf{h} = o(\mathbf{h}).$$

Now, it follows that

$$\frac{\overbrace{[f(\mathbf{x} + \mathbf{h}) - f(\mathbf{x})]}^{\Delta f} - \overbrace{\nabla f(\mathbf{x}) \cdot \mathbf{h}}^{df}}{\|\mathbf{h}\|} \to 0$$

as $\mathbf{h} \to \mathbf{0}$ since $(o(\mathbf{h}))/\|\mathbf{h}\| \to 0$ as $\mathbf{h} \to \mathbf{0}$. (See Section 15.1.)

In the two-variable case we set $\mathbf{x} = (x, y)$ and $\mathbf{h} = (\Delta x, \Delta y)$. The increment $\Delta f = f(\mathbf{x} + \mathbf{h}) - f(\mathbf{x})$ then takes the form

$$\Delta f = f(x + \Delta x, y + \Delta y) - f(x, y),$$

and the differential $df = \nabla f(\mathbf{x}) \cdot \mathbf{h}$ takes the form

$$df = \frac{\partial f}{\partial x}(x, y)\, \Delta x + \frac{\partial f}{\partial y}(x, y)\, \Delta y.$$

By suppressing the point of evaluation, we can write

(15.9.4)
$$df = \frac{\partial f}{\partial x}\, \Delta x + \frac{\partial f}{\partial y}\, \Delta y.$$

The approximation $\Delta f \cong df$ is illustrated in Figure 15.9.2. There we have represented f as a surface $z = f(x, y)$. Through a point $P(x_0, y_0, f(x_0, y_0))$ we have drawn the tangent plane. *The difference $df - \Delta f$ is the vertical separation between this tangent plane and the surface as measured at the point* $(x_0 + \Delta x, y_0 + \Delta y)$.

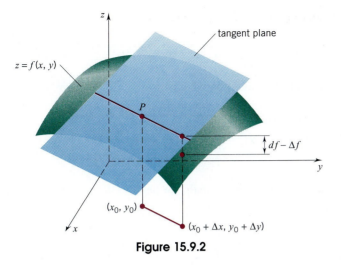

Figure 15.9.2

PROOF The tangent plane at P has equation

$$z - f(x_0, y_0) = \frac{\partial f}{\partial x}(x_0, y_0)(x - x_0) + \frac{\partial f}{\partial y}(x_0, y_0)(y - y_0).$$

The z-coordinate of this plane at the point $(x_0 + \Delta x, y_0 + \Delta y)$ is

$$f(x_0, y_0) + \frac{\partial f}{\partial x}(x_0, y_0)\,\Delta x + \frac{\partial f}{\partial y}(x_0, y_0)\,\Delta y. \qquad \text{(check this)}$$

The z-coordinate of the surface at this same point is

$$f(x_0 + \Delta x, y_0 + \Delta y).$$

The difference between these two:

$$\left[f(x_0, y_0) + \frac{\partial f}{\partial x}(x_0, y_0)\,\Delta x + \frac{\partial f}{\partial y}(x_0, y_0)\,\Delta y \right] - \left[f(x_0 + \Delta x, y_0 + \Delta y) \right]$$

can be written as

$$\left[\frac{\partial f}{\partial x}(x_0, y_0)\,\Delta x + \frac{\partial f}{\partial y}(x_0, y_0)\,\Delta y \right] - \left[f(x_0 + \Delta x, y_0 + \Delta y) - f(x_0, y_0) \right].$$

This is just $df - \Delta f$. ❑†

For the three-variable case we set $\mathbf{x} = (x, y, z)$ and $\mathbf{h} = (\Delta x, \Delta y, \Delta z)$. The increment becomes

$$\Delta f = f(x + \Delta x, y + \Delta y, z + \Delta z) - f(x, y, z),$$

---

† As in the figure, we have been assuming that the tangent plane lies above the surface. If the tangent plane lies below the surface, then $df - \Delta f$ is negative. The vertical separation between the tangent plane and the surface is then $\Delta f - df$.

and the approximating differential becomes

$$df = \frac{\partial f}{\partial x}(x, y, z)\,\Delta x + \frac{\partial f}{\partial y}(x, y, z)\,\Delta y + \frac{\partial f}{\partial z}(x, y, z)\,\Delta z.$$

Suppressing the point of evaluation we have

(15.9.5)
$$df = \frac{\partial f}{\partial x}\Delta x + \frac{\partial f}{\partial y}\Delta y + \frac{\partial f}{\partial z}\Delta z.$$

To illustrate the use of differentials we begin with a rectangle of sides x and y. The area is given by

$$A(x, y) = xy.$$

An increase in the dimensions of the rectangle to $x + \Delta x$ and $y + \Delta y$ produces a change in area

$$\begin{aligned}
\Delta A &= (x + \Delta x)(y + \Delta y) - xy \\
&= (xy + x\,\Delta y + y\,\Delta x + \Delta x\,\Delta y) - xy \\
&= x\,\Delta y + y\,\Delta x + \Delta x\,\Delta y.
\end{aligned}$$

The differential estimate for this change in area is

$$dA = \frac{\partial A}{\partial x}\Delta x + \frac{\partial A}{\partial y}\Delta y = y\,\Delta x + x\,\Delta y. \qquad \text{(Figure 15.9.3)}$$

The error of our estimate, the difference between the actual change and the estimated change, is the difference $\Delta A - dA = \Delta x\,\Delta y$. ❏

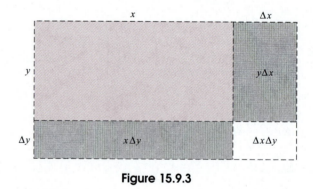

Figure 15.9.3

Example 1   Given that

$$f(x, y) = yx^{2/5} + x\sqrt{y},$$

estimate by a differential the change in f from $(32, 16)$ to $(35, 18)$.

SOLUTION   Since

$$\frac{\partial f}{\partial x} = \frac{2}{5}\left(\frac{1}{x}\right)^{3/5} y + \sqrt{y} \qquad \text{and} \qquad \frac{\partial f}{\partial y} = x^{2/5} + \frac{x}{2\sqrt{y}},$$

we have

$$df = \left[\frac{2}{5} \left(\frac{1}{x} \right)^{3/5} y + \sqrt{y} \right] \Delta x + \left[x^{2/5} + \frac{x}{2\sqrt{y}} \right] \Delta y.$$

At $x = 32$, $y = 16$, $\Delta x = 3$, $\Delta y = 2$,

$$df = \left[\frac{2}{5} \left(\frac{1}{32} \right)^{3/5} 16 + \sqrt{16} \right] 3 + \left[32^{2/5} + \frac{32}{2\sqrt{16}} \right] 2 = 30.4.$$

The change increases the value of f by approximately 30.4. ❏

Example 2   Use differentials to estimate $\sqrt{27} \sqrt[3]{1021}$.

SOLUTION   We know $\sqrt{25}$ and $\sqrt[3]{1000}$. What we need is an estimate for the increase of

$$f(x, y) = \sqrt{x} \sqrt[3]{y} = x^{1/2} y^{1/3}$$

from $x = 25$, $y = 1000$ to $x = 27$, $y = 1021$. The differential is

$$df = \tfrac{1}{2} x^{-1/2} y^{1/3} \Delta x + \tfrac{1}{3} x^{1/2} y^{-2/3} \Delta y.$$

With $x = 25$, $y = 1000$, $\Delta x = 2$, $\Delta y = 21$, df becomes

$$(\tfrac{1}{2} \cdot 25^{-1/2} \cdot 1000^{1/3}) 2 + (\tfrac{1}{3} \cdot 25^{1/2} \cdot 1000^{-2/3}) 21 = 2.35.$$

The change increases the value of the function by about 2.35. It follows that

$$\sqrt{27} \sqrt[3]{1021} \cong \sqrt{25} \sqrt[3]{1000} + 2.35 = 52.35.$$

(Our calculator gives $\sqrt{27} \sqrt[3]{1021} \cong 52.323$.) ❏

Example 3   Estimate by a differential the change in the volume of the frustum of a right circular cone if the upper radius r is decreased from 3 to 2.7 centimeters, the base radius R is increased from 8 to 8.1 centimeters, and the height h is increased from 6 to 6.3 centimeters.

SOLUTION   Since

$$V(r, R, h) = \tfrac{1}{3} \pi h (R^2 + Rr + r^2),$$

we have

$$dV = \tfrac{1}{3} \pi h (R + 2r) \Delta r + \tfrac{1}{3} \pi h (2R + r) \Delta R + \tfrac{1}{3} \pi (R^2 + Rr + r^2) \Delta h.$$

At $r = 3$, $R = 8$, $h = 6$, $\Delta r = -0.3$, $\Delta R = 0.1$, and $\Delta h = 0.3$,

$$dV = (28 \pi)(-0.3) + (38 \pi)(0.1) + \tfrac{1}{3}(97 \pi)(0.3) = 5.1 \pi \cong 16.02.$$

According to our differential estimate, the volume increases by about 16 cubic centimeters. (As you can check, the exact change in volume is 5.015π cubic centimeters, $5.015 \pi \cong 15.76$.) ❏

EXERCISES 15.9

In Exercises 1–12, find the differential df.

1. $f(x, y) = x^3 y - x^2 y^2$.

2. $f(x, y, z) = xy + yz + xz$.

3. $f(x, y) = x \cos y - y \cos x$.

4. $f(x, y, z) = x^2 y e^{2z}$.

5. $f(x, y, z) = x - y \tan z$.

6. $f(x, y) = (x - y) \ln(x + y)$.

7. $f(x, y, z) = \dfrac{xy}{x^2 + y^2 + z^2}$.

8. $f(x, y) = \ln(x^2 + y^2) + xe^{xy}$.

9. $f(x, y) = \sin(x + y) + \sin(x - y)$.

10. $f(x, y) = x \ln\left[\dfrac{1 + y}{1 - y}\right]$.

11. $f(x, y, z) = y^2 e^{xz} + x \ln z$.

12. $f(x, y) = xy e^{-(x^2 + y^2)}$.

13. Compute Δu and du for $u = x^2 - 3xy + 2y^2$ at $x = 2$, $y = -3$, $\Delta x = -0.3$, $\Delta y = 0.2$.

14. Compute du for $u = (x + y)\sqrt{x - y}$ at $x = 6$, $y = 2$, $\Delta x = \frac{1}{4}$, $\Delta y = -\frac{1}{2}$.

15. Compute Δu and du for $u = x^2 z - 2yz^2 + 3xyz$ at $x = 2$, $y = 1$, $z = 3$, $\Delta x = 0.1$, $\Delta y = 0.3$, $\Delta z = -0.2$.

16. Calculate du for

$$u = \frac{xy}{\sqrt{x^2 + y^2 + z^2}}$$

at $x = 1$, $y = 3$, $z = -2$, $\Delta x = \frac{1}{2}$, $\Delta y = \frac{1}{4}$, $\Delta z = -\frac{1}{4}$.

In Exercises 17–20, use differentials to find the approximate value.

17. $\sqrt{125} \sqrt[4]{17}$.

18. $(1 - \sqrt{10})(1 + \sqrt{24})$.

19. $\sin \frac{6}{7}\pi \cos \frac{1}{5}\pi$.

20. $\sqrt{8} \tan \frac{5}{16}\pi$.

In Exercises 21–24, use differentials to approximate the value of f at the given point.

21. $f(x, y) = x^2 e^{xy}$; $(2.9, 0.01)$.

22. $f(x, y, z) = x^2 y \cos \pi z$; $(2.12, 2.92, 3.02)$.

23. $f(x, y, z) = x \tan^{-1} yz$; $(2.94, 1.1, 0.92)$.

24. $f(x, y) = \sqrt{x^2 + y^2}$; $(3.06, 3.88)$.

25. Given that $z = (x - y)(x + y)^{-1}$, use dz to find the approximate change in z if x is increased from 4 to $4\frac{1}{10}$ and y is increased from 2 to $2\frac{1}{10}$. What is the exact change?

26. Estimate by a differential the change in the volume of a right circular cylinder if the height is increased from 12 to 12.2 inches and the radius is decreased from 8 to 7.7 inches.

27. Estimate the change in total surface area for the cylinder of Exercise 26.

28. Use a differential to estimate the change in $T = x^2 \cos \pi z - y^2 \sin \pi z$ from $x = 2$, $y = 2$, $z = 2$ to $x = 2.1$, $y = 1.9$, $z = 2.2$.

29. Estimate the surface area of a closed rectangular box whose dimensions are length $= 9.98$ inches; width $= 5.88$ inches; height $= 4.08$ inches.

30. Estimate the volume of a right circular cone whose dimensions are radius $= 7.2$ centimeters and height $= 10.15$ centimeters.

31. The dimensions of a closed rectangular box change from length $= 12$; width $= 8$; height $= 6$, to length $= 12.02$; width $= 7.95$; height $= 6.03$.
  (a) Use a differential to approximate the change in volume.
  (b) Calculate the exact change in volume.

32. Use the dimensions of the rectangular box in Exercise 31.
  (a) Approximate the change in surface area using a differential.
  (b) Calculate the exact change in the surface area.

33. The function $T(x, y, z) = 100 - x^2 - y^2 - z^2 + 2xyz$ gives the temperature T at the point $P(x, y, z)$ in space. Use differentials to approximate the temperature difference between the points $P(1, 3, 4)$ and $Q(1.15, 2.90, 4.10)$.

34. According to the ideal gas law, the relation between the pressure P, the temperature T, and the volume V of a confined gas is given by the equation $PV = kT$, where k is a constant. If $P = 4$ pounds per square inch when $V = 81$ cubic inches and $T = 300$ K, approximate the change in pressure if the volume is decreased to 75 cubic inches and the temperature is increased to 325 K.

35. As illustrated in the following figure, the side x in the right triangle is increased by Δx and the angle θ is increased by $\Delta\theta$. Use a differential to approximate the change in the area of the triangle. Is the area more sensitive to a change in x or to a change in θ?

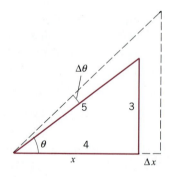

36. Use a differential to approximate the change in the area of the isosceles triangle shown in the following figure if x changes by Δx and θ changes by $\Delta\theta$. Is the area more sensitive to changes in x or to changes in θ? HINT: The area of a triangle with sides a and b and included angle θ is $A = \frac{1}{2} ab \sin \theta$.

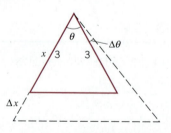

37. The radius of a right circular cylinder of height h is increased from r to $r + \Delta r$.
    (a) Determine the exact change in h that will keep the volume constant. Then estimate this change in h by using a differential.
    (b) Determine the exact change in h that will keep the total surface area constant. Then estimate this change in h by using a differential.

38. The dimensions of a rectangular box with a top are length = 4 feet; width = 2 feet; height = 3 feet. It has a coat of paint $\frac{1}{16}$ inch thick. Estimate the amount of paint (cubic inches) on the box.

Error Estimates

Let $u = u(x, y)$ be differentiable. If the variables x and y are known to be $x_0 \pm \Delta x$ and $y_0 \pm \Delta y$, then the maximum possible error in the calculated value $u(x_0, y_0)$ is

$$\frac{\partial u}{\partial x}(x_0, y_0)(\pm\Delta x) + \frac{\partial u}{\partial y}(x_0, y_0)(\pm\Delta y)$$

39. The legs of a right triangle are measured to be 5 and 12 centimeters with a possible error of ± 15 millimeters in each measurement. What is the maximum possible error in the calculated value of (a) the hypotenuse and (b) the area of the triangle?

40. The radius of a right circular cone is measured to be 5 inches with a possible error of ± 0.2 inch and the height is measured to be 12 inches with a possible error of ± 0.3 inch. What is the maximum possible error in the calculated values of (a) the volume and (b) the lateral surface area of the cone?

41. The specific gravity of a solid is given by the formula $s = A(A - W)^{-1}$ where A is the weight in air and W is the weight in water. What is the maximum possible error in the calculated value of s if A is measured to be 9 pounds (within a tolerance of 0.01 pound) and W is measured to be 5 pounds (within a tolerance of 0.02 pound)?

42. The measurements of a closed rectangular box are length = 5 feet; width = 3 feet; height = 3.5 feet, with a possible error of $\pm\frac{1}{12}$ inch in each measurement. What is the maximum possible error in the calculated value of (a) the volume and (b) the surface area of the box? HINT: Extend the error estimate to three dimensions.

■ 15.10 RECONSTRUCTING A FUNCTION FROM ITS GRADIENT

This section has three parts. In Part 1 we show how to find $f(x, y)$ given its gradient

$$\nabla f(x, y) = \frac{\partial f}{\partial x}(x, y)\mathbf{i} + \frac{\partial f}{\partial y}(x, y)\mathbf{j}.$$

In Part 2 we show that, although all gradients $\nabla f(x, y)$ are expressions of the form

$$P(x, y)\mathbf{i} + Q(x, y)\mathbf{j}$$

(set $P = \partial f/\partial x$ and $Q = \partial f/\partial y$), not all such expressions are gradients. In Part 3 we tackle the problem of recognizing which expressions $P(x, y)\mathbf{i} + Q(x, y)\mathbf{j}$ are actually gradients.

Part 1

Example 1 Find f given that

$$\nabla f(x, y) = (4x^3y^3 - 3x^2)\mathbf{i} + (3x^4y^2 + \cos 2y)\mathbf{j}.$$

SOLUTION The first partial derivatives of f are

$$\frac{\partial f}{\partial x}(x, y) = 4x^3y^3 - 3x^2, \qquad \frac{\partial f}{\partial y}(x, y) = 3x^4y^2 + \cos 2y.$$

Integrating $\partial f/\partial x$ with respect to x, treating y as a constant, we find that

$$f(x, y) = x^4y^3 - x^3 + \phi(y)$$

where ϕ is an unknown function of y. The function ϕ plays the same role as the arbitrary constant C that arises when you integrate a function of one variable. Now, differentiation with respect to y gives

$$\frac{\partial f}{\partial y}(x, y) = 3x^4y^2 + \phi'(y).$$

Equating the two expressions for $\partial f/\partial y$, we have

$$3x^4y^2 + \phi'(y) = 3x^4y^2 + \cos 2y,$$

which implies

$$\phi'(y) = \cos 2y \quad \text{and thus} \quad \phi(y) = \tfrac{1}{2}\sin 2y + C \quad (C \text{ a constant}).$$

This means that

$$f(x, y) = x^4y^3 - x^3 + \tfrac{1}{2}\sin 2y + C. \quad \square$$

Remark The procedure for finding f just illustrated is symmetric in x and y. That is, rather than integrating $\partial f/\partial x$ with respect to x, we could have started by integrating $\partial f/\partial y$ with respect to y, with x held constant, followed by differentiating the result with respect to x:

$$\text{if} \quad \frac{\partial f}{\partial y}(x, y) = 3x^4y^2 + \cos 2y \quad \text{then} \quad f(x, y) = x^4y^3 + \tfrac{1}{2}\sin 2y + \psi(x),$$

where ψ is an unknown function of x. Now, differentiating with respect to x, we have

$$\frac{\partial f}{\partial x}(x, y) = 4x^3y^3 + \psi'(x).$$

Equating the two expressions for $\partial f/\partial x$ gives

$$4x^3y^3 + \psi'(x) = 4x^3y^3 - 3x^2.$$

Therefore,

$$\psi'(x) = -3x^2 \quad \text{which implies} \quad \psi(x) = -x^3 + C \quad (C \text{ a constant}).$$

Thus,

$$f(x, y) = x^4y^3 + \tfrac{1}{2}\sin 2y - x^3 + C$$

as we found before. $\square$

Example 2 Find f given that

$$\nabla f(x, y) = \left(\sqrt{y} - \frac{y}{2\sqrt{x}} + 2x\right)\mathbf{i} + \left(\frac{x}{2\sqrt{y}} - \sqrt{x} + 1\right)\mathbf{j}.$$

SOLUTION Here we have

$$\frac{\partial f}{\partial x}(x, y) = \sqrt{y} - \frac{y}{2\sqrt{x}} + 2x, \quad \frac{\partial f}{\partial y}(x, y) = \frac{x}{2\sqrt{y}} - \sqrt{x} + 1.$$

We will proceed as in the first example. Integrating $\partial f/\partial x$ with respect to x, we have

$$f(x, y) = x\sqrt{y} - y\sqrt{x} + x^2 + \phi(y)$$

with $\phi(y)$ independent of x. Differentiation with respect to y gives

$$\frac{\partial f}{\partial y}(x, y) = \frac{x}{2\sqrt{y}} - \sqrt{x} + \phi'(y).$$

The two equations for $\partial f/\partial y$ can be reconciled only by having

$$\phi'(y) = 1 \quad \text{and thus} \quad \phi(y) = y + C.$$

This means that

$$f(x, y) = x\sqrt{y} - y\sqrt{x} + x^2 + y + C. \quad \square$$

The function

$$f(x, y) = x\sqrt{y} - y\sqrt{x} + x^2 + y + C$$

is the *general solution* of the vector differential equation

$$\nabla f(x, y) = \left(\sqrt{y} - \frac{y}{2\sqrt{x}} + 2x \right) \mathbf{i} + \left(\frac{x}{2\sqrt{y}} - \sqrt{x} + 1 \right) \mathbf{j}.$$

Each *particular* solution can be obtained by assigning a particular value to the constant of integration C.

Part 2

The next example shows that not all linear combinations $P(x, y)\mathbf{i} + Q(x, y)\mathbf{j}$ are gradients.

Example 3   Show that $y\mathbf{i} - x\mathbf{j}$ is not a gradient.

SOLUTION   Suppose on the contrary that it is a gradient. Then there exists a function f such that

$$\nabla f(x, y) = y\mathbf{i} - x\mathbf{j}.$$

Obviously

$$\frac{\partial f}{\partial x}(x, y) = y, \qquad \frac{\partial f}{\partial y}(x, y) = -x$$

$$\frac{\partial^2 f}{\partial y \partial x}(x, y) = 1, \qquad \frac{\partial^2 f}{\partial x \partial y}(x, y) = -1$$

and thus

$$\frac{\partial^2 f}{\partial y \partial x}(x, y) \neq \frac{\partial^2 f}{\partial x \partial y}(x, y).$$

This contradicts (14.6.5): the four partial derivatives under consideration are everywhere continuous and thus, according to (14.6.5), we must have

$$\frac{\partial^2 f}{\partial y \partial x}(x, y) = \frac{\partial^2 f}{\partial x \partial y}(x, y).$$

This contradiction shows that $y\mathbf{i} - x\mathbf{j}$ is not a gradient. $\square$

Part 3

We come now to the problem of recognizing which linear combinations

$$P(x, y)\mathbf{i} + Q(x, y)\mathbf{j}$$

are actually gradients. But first we need to review some ideas and establish some new terminology.

As indicated earlier (Section 15.3), an open set (in the plane or in three-space) is said to be *connected* iff any two points of the set can be joined by a polygonal path that lies entirely within the set. An open connected set is called an *open region*. A curve

$$C: \quad \mathbf{r} = \mathbf{r}(t), \qquad t \in [a, b]$$

is said to be *closed* iff it begins and ends at the same point:

$$\mathbf{r}(a) = \mathbf{r}(b).$$

It is said to be *simple* iff it does not intersect itself:

$$a < t_1 < t_2 < b \qquad \text{implies} \qquad \mathbf{r}(t_1) \neq \mathbf{r}(t_2).$$

These notions are illustrated in Figure 15.10.1.

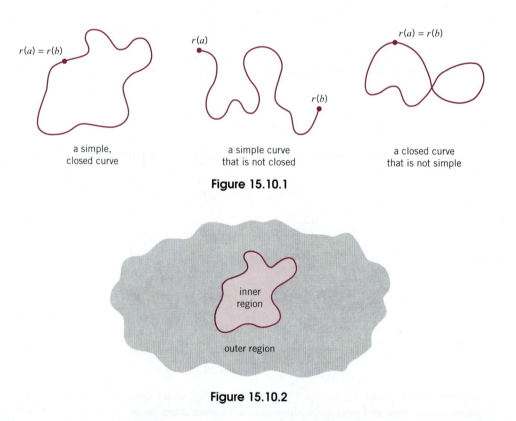

| a simple, closed curve | a simple curve that is not closed | a closed curve that is not simple |

Figure 15.10.1

Figure 15.10.2

As is intuitively clear (Figure 15.10.2), a simple closed curve in the plane separates the plane into two disjoint open connected sets: a bounded inner region consisting of all points surrounded by the curve and an unbounded outer region consisting of all points not surrounded by the curve.†

Finally, we come to the notion we need in our work with gradients:

(15.10.1)

> Let Ω be an open region of the plane, and let C be an arbitrary simple closed curve. Ω is said to be *simply connected* iff
>
> C is in Ω implies   the inner region of C is in Ω.

The first two regions in Figure 15.10.3 are simply connected. The annular region is not; the annular region contains the simple closed curve drawn there, but it does not contain all of the inner region of that curve.

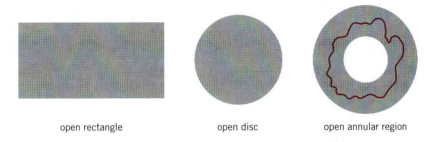

open rectangle   open disc   open annular region

Figure 15.10.3

THEOREM 15.10.2

Let P and Q be functions of two variables, each continuously differentiable on a simply connected open region Ω. The linear combination $P(x, y)\mathbf{i} + Q(x, y)\mathbf{j}$ is a gradient on Ω iff

$$\frac{\partial P}{\partial y}(x, y) = \frac{\partial Q}{\partial x}(x, y) \qquad \text{for all } (x, y) \in \Omega.$$

A complete proof of this theorem for a general region Ω is complicated. We will prove the result under the additional assumption that Ω has the form of an open rectangle with sides parallel to the coordinate axes.

Suppose that $P(x, y)\mathbf{i} + Q(x, y)\mathbf{j}$ is a gradient on this open rectangle Ω, say

$$\nabla f(x, y) = P(x, y)\mathbf{i} + Q(x, y)\mathbf{j}.$$

---

† Add to this the assertion that the curve in question constitutes the total boundary of both regions and you have what is called the Jordan curve theorem, named after the French mathematician Camille Jordan (1838–1922). Jordan was the first to point out that, although all this is apparently obvious, it nevertheless requires proof. In recognition of Jordan, plane simple closed curves are now commonly known as *Jordan curves*.

Since

$$\nabla f(x, y) = \frac{\partial f}{\partial x}(x, y)\mathbf{i} + \frac{\partial f}{\partial y}(x, y)\mathbf{j},$$

we have

$$P = \frac{\partial f}{\partial x} \qquad \text{and} \qquad Q = \frac{\partial f}{\partial y}.$$

Since P and Q have continuous first partials, f has continuous second partials. Thus, according to (14.6.5), the mixed partials are equal and we have

$$\frac{\partial P}{\partial y} = \frac{\partial^2 f}{\partial y \partial x} = \frac{\partial^2 f}{\partial x \partial y} = \frac{\partial Q}{\partial x}.$$

Conversely, suppose that

$$\frac{\partial P}{\partial y}(x, y) = \frac{\partial Q}{\partial x}(x, y) \qquad \text{for all } (x, y) \in \Omega.$$

We must show that $P(x, y)\mathbf{i} + Q(x, y)\mathbf{j}$ is a gradient on Ω. To do this, we choose a point (x_0, y_0) from Ω and form the function

$$f(x, y) = \int_{x_0}^{x} P(u, y_0)\, du + \int_{y_0}^{y} Q(x, v)\, dv, \qquad (x, y) \in \Omega.$$

[If you want to visualize f, you can refer to Figure 15.10.4. The function P is being integrated along the horizontal line segment joining (x_0, y_0) to (x, y_0), and Q is being integrated along the vertical line segment joining (x, y_0) to (x, y). Our assumptions on Ω guarantee that these line segments remain in Ω.]

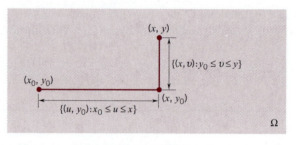

Figure 15.10.4

The first integral is independent of y. Hence

$$\frac{\partial f}{\partial y}(x, y) = \frac{\partial}{\partial y}\left(\int_{y_0}^{y} Q(x, v)\, dv \right) = Q(x, y).$$

The last equality holds because we are differentiating an integral with respect to its upper limit. (See Theorem 5.3.5.) Differentiating f with respect to x we have

$$\frac{\partial f}{\partial x}(x, y) = \frac{\partial}{\partial x}\left(\int_{x_0}^{x} P(u, y_0)\, du \right) + \frac{\partial}{\partial x}\left(\int_{y_0}^{y} Q(x, v)\, dv \right).$$

The first term is $P(x, y_0)$ since once again we are differentiating with respect to the upper limit. In the second term the variable x appears under the sign of integration. It can be shown that, since Q and $\partial Q/\partial x$ are continuous,

$$\frac{\partial}{\partial x} \left(\int_{y_0}^{y} Q(x, v) \, dv \right) = \int_{y_0}^{y} \frac{\partial Q}{\partial x}(x, v) \, dv.\dagger$$

Anticipating this result, we have

$$\frac{\partial f}{\partial x}(x, y) = P(x, y_0) + \int_{y_0}^{y} \frac{\partial Q}{\partial x}(x, v) \, dv = P(x, y_0) + \int_{y_0}^{y} \frac{\partial P}{\partial y}(x, v) \, dv$$

explain ⟶

$$= P(x, y_0) + P(x, y) - P(x, y_0) = P(x, y).$$

We have now shown that

$$P(x, y) = \frac{\partial f}{\partial x}(x, y) \qquad \text{and} \qquad Q(x, y) = \frac{\partial f}{\partial y}(x, y) \qquad \text{for all } (x, y) \in \Omega.$$

It follows that $P(x, y)\mathbf{i} + Q(x, y)\mathbf{j}$ is the gradient of f on Ω. ❏

Example 4 The vector functions

$$\mathbf{F}(x, y) = 2x \sin y \, \mathbf{i} + x^2 \cos y \, \mathbf{j} \qquad \text{and} \qquad \mathbf{G}(x, y) = xy \mathbf{i} + \tfrac{1}{2}(x + 1)^2 y^2 \mathbf{j}$$

are both defined everywhere. The first vector function is the gradient of a function that is defined everywhere since for $P(x, y) = 2x \sin y$ and $Q(x, y) = x^2 \cos y$, we have

$$\frac{\partial P}{\partial y}(x, y) = 2x \cos y = \frac{\partial Q}{\partial x} \qquad \text{for all } (x, y).$$

It is easy to verify that if $f(x, y) = x^2 \sin y + C$, where C is a constant, then $\nabla f(x, y) = \mathbf{F}(x, y)$.

The vector function $\mathbf{G}$ is not a gradient: $P(x, y) = xy$, $Q(x, y) = \tfrac{1}{2}(x + 1)^2 y^2$, and

$$\frac{\partial P}{\partial y}(x, y) = x \qquad \text{and} \qquad \frac{\partial Q}{\partial x}(x, y) = (x + 1)y^2.$$

Thus,

$$\frac{\partial P}{\partial y}(x, y) \neq \frac{\partial Q}{\partial x}(x, y). \quad ❏$$

Example 5 The vector function $\mathbf{F}$ defined on the *punctured disc* $0 < x^2 + y^2 < 1$ by setting

$$\mathbf{F}(x, y) = \frac{y}{x^2 + y^2} \mathbf{i} - \frac{x}{x^2 + y^2} \mathbf{j}$$

satisfies

$$\frac{\partial P}{\partial y}(x, y) = \frac{\partial Q}{\partial x}(x, y) \qquad \text{on the punctured disc.}$$

---

† See Exercise 60, Section 16.4.

(Check this out.)   Nevertheless, as you will see in Chapter 17, **F** is not a gradient on that set.   The punctured disc is not simply connected and therefore Theorem 15.10.2 does not apply.   ❏

EXERCISES 15.10

In Exercises 1–16, determine whether the vector function is the gradient $\nabla f(x, y)$ of a function everywhere defined. If so, find such a function f.

1. $xy^2\mathbf{i} + x^2y\mathbf{j}$

2. $x\mathbf{i} + y\mathbf{j}$.

3. $y\mathbf{i} + x\mathbf{j}$.

4. $(x^2 + y)\mathbf{i} + (y^3 + x)\mathbf{j}$.

5. $(y^3 + x)\mathbf{i} + (x^2 + y)\mathbf{j}$.

6. $(y^2e^x - y)\mathbf{i} + (2ye^x - x)\mathbf{j}$.

7. $(\cos x - y \sin x)\mathbf{i} + \cos x\mathbf{j}$.

8. $(1 + e^y)\mathbf{i} + (xe^y + y^2)\mathbf{j}$.

9. $e^x\cos y^2\mathbf{i} - 2ye^x\sin y^2\mathbf{j}$.

10. $e^x\cos y\mathbf{i} + e^x\sin y\mathbf{j}$.

11. $ye^x(1 + x)\mathbf{i} + (xe^x - e^{-y})\mathbf{j}$.

12. $(e^x + 2xy)\mathbf{i} + (x^2 + \sin y)\mathbf{j}$.

13. $(xe^{xy} + x^2)\mathbf{i} + (ye^{xy} - 2y)\mathbf{j}$.

14. $(y \sin x + xy \cos x)\mathbf{i} + (x \sin x + 2y + 1)\mathbf{j}$.

15. $(1 + y^2 + xy^2)\mathbf{i} + (x^2y + y + 2xy + 1)\mathbf{j}$.

16. $\left[2 \ln(3y) + \dfrac{1}{x} \right]\mathbf{i} + \left[\dfrac{2x}{y} + y^2 \right]\mathbf{j}$.

In Exercises 17–20, find the most general function with the given gradient.

17. $\dfrac{x}{\sqrt{x^2 + y^2}}\mathbf{i} + \dfrac{y}{\sqrt{x^2 + y^2}}\mathbf{j}$.

18. $(x \tan y + \sec^2 x)\mathbf{i} + (\tfrac{1}{2}x^2\sec^2 y + \pi y)\mathbf{j}$.

19. $(x^2\sin^{-1}y)\mathbf{i} + \left(\dfrac{x^3}{3\sqrt{1 - y^2}} - \ln y \right)\mathbf{j}$.

20. $\left(\dfrac{\tan^{-1}y}{\sqrt{1 - x^2}} + \dfrac{x}{y} \right)\mathbf{i} + \left(\dfrac{\sin^{-1}x}{1 + y^2} - \dfrac{x^2}{2y^2} + 1 \right)\mathbf{j}$.

21. Find the general solution of the differential equation $\nabla f(x, y) = f(x, y)\mathbf{i} + f(x, y)\mathbf{j}$.

22. Given that g and its first and second partials are every-where continuous, find the general solution of the differential equation $\nabla f(x, y) = e^{g(x, y)}[g_x(x, y)\mathbf{i} + g_y(x, y)\mathbf{j}]$.

Theorem 15.10.2 has a three-dimensional analog. In particular we can show that, if P, Q, R are continuously differentiable on an open rectangular box S, then the vector function

$$P(x, y, z)\mathbf{i} + Q(x, y, z)\mathbf{j} + R(x, y, z)\mathbf{k}$$

is a gradient on S iff

$$\frac{\partial P}{\partial y} = \frac{\partial Q}{\partial x}, \qquad \frac{\partial P}{\partial z} = \frac{\partial R}{\partial x}, \qquad \frac{\partial Q}{\partial z} = \frac{\partial R}{\partial y} \qquad \text{throughout } S.$$

23. (a) Verify that $2x\mathbf{i} + z\mathbf{j} + y\mathbf{k}$ is the gradient of a function f that is everywhere defined.
  (b) Deduce from the relation $\partial f/\partial x = 2x$ that $f(x, y, z) = x^2 + g(y, z)$.
  (c) Verify then that $\partial f/\partial y = z$ gives $g(y, z) = zy + h(z)$ and finally that $\partial f/\partial z = y$ gives $h(z) = C$.
  (d) What is $f(x, y, z)$?

In Exercises 24–28, determine whether the function is a gradient $\nabla f(x, y, z)$ and, if so, find such a function f.

24. $yz\mathbf{i} + xz\mathbf{j} + xy\mathbf{k}$.

25. $(2x + y)\mathbf{i} + (2y + x + z)\mathbf{j} + (y - 2z)\mathbf{k}$.

26. $(2x \sin 2y \cos z)\mathbf{i} + (2x^2 \cos 2y \cos z)\mathbf{j} - (x^2 \sin 2y \sin z)\mathbf{k}$.

27. $(y^2z^3 + 1)\mathbf{i} + (2xyz^3 + y)\mathbf{j} + (3xy^2z^2 + 1)\mathbf{k}$.

28. $\left[\dfrac{y}{z} - e^z \right]\mathbf{i} + \left[\dfrac{x}{z} + 1 \right]\mathbf{j} - \left[xe^z + \dfrac{xy}{z^2} \right]\mathbf{k}$.

29. Verify that the gravitational force function

$$\mathbf{F}(\mathbf{r}) = -G\frac{mM}{r^3}\mathbf{r} \qquad (\mathbf{r} = x\mathbf{i} + y\mathbf{j} + z\mathbf{k})$$

is a gradient.

30. Verify that every vector function of the form

$$\mathbf{h}(\mathbf{r}) = kr^n\mathbf{r} \qquad (k \text{ constant}, n \text{ an integer})$$

is a gradient.

■ CHAPTER HIGHLIGHTS

15.1 Differentiability and Gradient

gradient of f at $\mathbf{x}$: $\nabla f(\mathbf{x})$ (p. 946)

$$\nabla f(x, y) = \frac{\partial f}{\partial x}\mathbf{i} + \frac{\partial f}{\partial y}\mathbf{j} \qquad \nabla f(x, y, z) = \frac{\partial f}{\partial x}\mathbf{i} + \frac{\partial f}{\partial y}\mathbf{j} + \frac{\partial f}{\partial z}\mathbf{k}$$

If f is differentiable at $\mathbf{x}$, then f is continuous at $\mathbf{x}$.

$$\nabla r^n = nr^{n-2}\mathbf{r}$$

15.2 Gradients and Directional Derivatives

directional derivative: $f'_{\mathbf{u}}(\mathbf{x}) = \nabla f(\mathbf{x}) \cdot \mathbf{u}$ (p. 955)

The directional derivative $f'_{\mathbf{u}}$ gives the rate of change of f in the direction of the unit vector $\mathbf{u}$.

The directional derivative in a direction $\mathbf{u}$ is the component of the gradient vector in that direction. (p. 958)

A differentiable function f increases most rapidly in the direction of the gradient (the rate of change is then $\|\nabla f(\mathbf{x})\|$) and it decreases most rapidly in the opposite direction (the rate of change is then $-\|\nabla f(\mathbf{x})\|$).

15.3 The Mean-Value Theorem; Two Intermediate-Value Theorems

The Mean-Value Theorem (p. 964) open, connected set (p. 964)
intermediate-value theorems (p. 966, 967) open, closed regions (p. 966)

15.4 Chain Rules

chain rule along a curve (p. 968) tree diagram (p. 973)

In the setting of functions of several variables, there are numerous chain rules. They can all be deduced from the chain rule along a curve. If, for example, $u = u(x, y)$ where $x = x(s, t)$ and $y = y(s, t)$, then

$$\frac{\partial u}{\partial s} = \frac{\partial u}{\partial x}\frac{\partial x}{\partial s} + \frac{\partial u}{\partial y}\frac{\partial y}{\partial s} \quad \text{and} \quad \frac{\partial u}{\partial t} = \frac{\partial u}{\partial x}\frac{\partial x}{\partial t} + \frac{\partial u}{\partial y}\frac{\partial y}{\partial t}.$$

If f is a continuously differentiable function of $r = \|\mathbf{r}\|$, then

$$\nabla[f(r)] = f'(r)\frac{\mathbf{r}}{r}.$$

15.5 The Gradient as a Normal; Tangent Lines and Tangent Planes

tangent and normal lines to a curve $f(x, y) = c$ (p. 982)
tangent plane to a surface $f(x, y, z) = c$ (p. 984)
upper and lower unit normals (p. 990)

At each point of the domain of a function, the gradient vector, if not $\mathbf{0}$, is perpendicular to the level curve (level surface) that passes through that point.

15.6 Maximum and Minimum Values

local maximum and local minimum (p. 991) critical points (p. 992)
stationary points, saddle points (p. 992) absolute extreme values (p. 996)

If f has a local extreme value at $\mathbf{x}_0$, then either $\nabla f(\mathbf{x}_0) = \mathbf{0}$ or $\nabla f(\mathbf{x}_0)$ does not exist.

15.7 Second-Partials Test

second-partials test, discriminant (p. 1000) method of least squares (p. 1006)

15.8 Maxima and Minima with Side Conditions

If $\mathbf{x}_0$ maximizes (or minimizes) $f(\mathbf{x})$ subject to the side condition $g(\mathbf{x}) = 0$, then $\nabla f(\mathbf{x}_0)$ and $\nabla g(\mathbf{x}_0)$ are parallel. Thus, if $\nabla g(\mathbf{x}_0) \neq \mathbf{0}$, then there exists a scalar λ, called a Lagrange multiplier, such that $\nabla f(\mathbf{x}_0) = \lambda \nabla g(\mathbf{x}_0)$.

15.9 Differentials

increment: $\Delta f = f(\mathbf{x} + \mathbf{h}) - f(\mathbf{x})$ differential: $df = \nabla f(\mathbf{x}) \cdot \mathbf{h}$

$\Delta f \cong df$ in the sense that $\dfrac{\Delta f - df}{\|\mathbf{h}\|} \to 0$ as $\mathbf{h} \to \mathbf{0}$

two variables: $df = \dfrac{\partial f}{\partial x} \Delta x + \dfrac{\partial f}{\partial y} \Delta y$

three variables: $df = \dfrac{\partial f}{\partial x} \Delta x + \dfrac{\partial f}{\partial y} \Delta y + \dfrac{\partial f}{\partial z} \Delta z$

15.10 Reconstructing a Function from its Gradient

necessary and sufficient conditions for a vector-valued function to be a gradient:

■ PROJECTS AND EXPLORATIONS USING TECHNOLOGY

To do these exercises you will need a graphics calculator or a computer with graphing capability. The majority of these problems are open-ended so different approaches may be used to solve them. You should be aware that different approaches can result in slight variations in the answers. Round your numerical answers to at least four decimal places. The rounding method that your calculator or computer uses also may cause variations in answers.

15.1 (a) A fence of height h is located w feet in front of a tall building. We want to find the length of the shortest ladder that can be leaned over the fence, against the side of the building. Determine the function that is to be minimized and the side condition.

(b) Suppose that the fence is 6 feet tall and 4 feet from the building. Find the length of the shortest ladder using the methods of single-variable calculus.

(c) Solve the same problem using Lagrange multipliers.

(d) Suppose that a moat of width w has walls of heights h_1 and h_2 on either side. Ladders of lengths L_1 and L_2 lean over the walls and meet at a point which is at a height y above the moat. Let x be the distance from the first wall to the point directly below the point where the ladders meet. (See the figure.) We want to minimize the sum of the lengths of the two ladders. Determine the function that is to be minimized and the side conditions. Explain why this problem should have a solution.

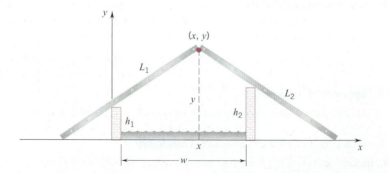

(e) If x and y are given, solve for L_1 and L_2 in terms of x, y, h_1, h_2, and w.

(f) Let $w = 20$, $h_1 = 5$, and $h_2 = 6$. Find L_1, L_2 and $L_1 + L_2$ for the following pairs (x, y): $(5, 8)$, $(7, 10)$, $(10, 12)$, $(12, 15)$.

(g) Now solve the problem using Lagrange multipliers.

15.2 In this Exercise and the next, we are going to use Lagrange multipliers to study municipal recycling. In this problem we will determine where to locate the recycling center. We assume that the garbage is picked up from the homes and businesses, transported to the recycling center, and then sorted at the center. A percentage γ of the garbage is recycled and the remaining $1 - \gamma$ is trucked to a landfill. Let c_1 be the cost per mile of trucking garbage from homes and business to the recycling center and c_2 be the cost per mile of trucking garbage from the recycling center to the landfill. For simplicity, assume that the city is rectangular with vertices $(0, 0)$, $(a, 0)$, (a, b), and $(0, b)$, that the recycling center is at the point (p, q) inside the city limits, and that the landfill is at the point (r, s) outside the city limits. (See the figure.)

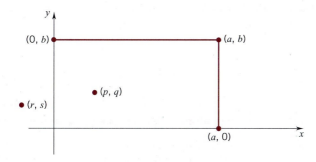

(a) Using the usual distance function for points in the plane

$$d[(x_1, y_1), (x_2, y_2)] = \sqrt{(x_2 - x_1)^2 + (y_2 - y_1)^2},$$

give the integrals that represent the average distance from a home or business to the recycling center. Find the location (p, q) for the recycling center which will minimize the average distance.

(b) Next, assume that the distance between points in the plane is given by

$$d[(x_1, y_1), (x_2, y_2)] = |x_2 - x_1| + |y_2 - y_1|.$$

Give the integrals that represent the average distance from a home to the recycling center and find the location (p, q) for the recycling center which will minimize this average distance.

(c) Using the distance function from part (b), find an expression for the total cost of transporting the garbage to the center and transporting the nonrecycled material to the landfill.

(d) Using the cost function from part (c), find the point (p, q) which minimizes the total cost (as a function of γ, c_1, and c_2) for each of the following cases (i) $r = 0$, $0 < s < b$; (ii) $r < 0$, $0 < s < b$; (iii) $r < 0$, $s < 0$.

15.3 Now we will determine how much sorting to do in order to minimize total costs. In addition to the transportation costs, c_1 and c_2, the costs that we consider are (i) the unit cost c_3 of dumping at the landfill; (ii) p, the unit profit of selling recycled material; and (iii) $c_4\gamma/(1 - \gamma)$, the labor cost of sorting.

(a) Discuss the labor cost function. Determine the total cost function.

(b) Let $a = 2$, $b = 6$, $r = 0$, $s = 1$, and $c_1 = 1$. Find the value of γ that minimizes the total cost in each of the following situations:

(i) $c_2 = 0.5$, $c_3 = 1$, $c_4 = 1$. (ii) $c_2 = 0.5$, $c_3 = 0.5$, $c_4 = 1$.
(iii) $c_2 = 0.5$, $c_3 = 1$, $c_4 = 0.5$. (iv) $c_2 = 0.5$, $c_3 = 0.5$, $c_4 = 0.5$.
(v) $c_2 = 0.25$, $c_3 = 1$, $c_4 = 1$.

    (c) Suppose that the recycling center is built at the point (p, q) found in Exercise 2, part (b). How much is the total cost above the minimum in each of the cases in part (b).

    (d) Use the values from part (c) to approximate the partials of γ with respect to c_2, c_3, and c_4. What is the importance of these partials?

15.4 Here we are going to reexamine a familiar problem—the problem of minimizing the cost of manufacturing a can in the shape of a right circular cylinder that has a given volume. Suppose the cost per unit area of the top and bottom is k times the cost per unit area of the sides. Let r be the radius of the can, h the height, and V the given volume.

    (a) Find the ratio of the height to the radius of the cheapest can as a function of k. Also give the minimal cost as a function of k. Treat this as a single-variable problem.

    (b) Now do the problem using Lagrange multipliers.

    (c) Find the partial derivatives of the minimal cost with respect to k and with respect to V. Is the minimal cost an increasing function of V for all k?

    (d) Suppose that for aesthetic reasons we add the constraint that the height of the can must be at least equal to the diameter of the base, but at most twice the diameter. Use Lagrange multipliers to solve this problem.

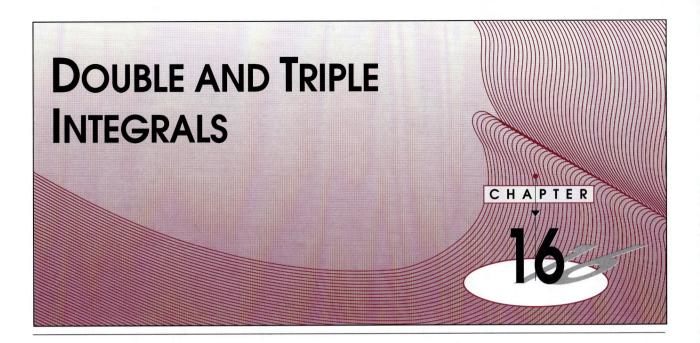

DOUBLE AND TRIPLE INTEGRALS

CHAPTER

16

■ 16.1 MULTIPLE-SIGMA NOTATION

In Chapter 5 we studied ordinary integrals

$$\int_J f(x)\, dx$$

where $J = [a, b]$ is an interval on the real line and

$$\int_J f(x)\, dx = \int_a^b f(x)\, dx.$$

Here we study double integrals

$$\iint_\Omega f(x, y)\, dx\, dy,$$

where Ω is some region in the x,y-plane, and a little later we will study triple integrals

$$\iiint_T f(x, y, z)\, dx\, dy\, dz,$$

where T is a solid in three-dimensional space.

Our first step is to introduce some new notation. In an ordinary sequence $\{a_i\}$, each term a_i is indexed by a single integer. The sum of all the a_i from $i = 1$ to $i = m$ is then denoted by

$$\sum_{i=1}^m a_i.$$

When two indices are involved, say

$$a_{ij} = 2^i 5^j, \qquad a_{ij} = \frac{2i}{5 + j}, \quad \text{or} \quad a_{ij} = (1 + i)^j,$$

then we use double-sigma notation to denote the sum of all the doubly indexed terms. By

(16.1.1)

$$\sum_{i=1}^{m} \sum_{j=1}^{n} a_{ij}$$

we mean *the sum of all the a_{ij} where i ranges from 1 to m and j ranges from 1 to n.* For example,

$$\sum_{i=1}^{3} \sum_{j=1}^{2} 2^i 5^j = 2 \cdot 5 + 2 \cdot 5^2 + 2^2 \cdot 5 + 2^2 \cdot 5^2 + 2^3 \cdot 5 + 2^3 \cdot 5^2 = 420.$$

Since addition is associative and commutative, we can add the terms of (16.1.1) in any order we choose. Usually we set

(16.1.2)

$$\sum_{i=1}^{m} \sum_{j=1}^{n} a_{ij} = \sum_{i=1}^{m} \left(\sum_{j=1}^{n} a_{ij} \right).$$

We can expand the expression on the right by expanding first with respect to i and then with respect to j:

$$\sum_{i=1}^{m} \left(\sum_{j=1}^{n} a_{ij} \right) = \sum_{j=1}^{n} a_{1j} + \sum_{j=1}^{n} a_{2j} + \cdots + \sum_{j=1}^{n} a_{mj}$$

$$= (a_{11} + a_{12} + \cdots + a_{1n}) + (a_{21} + a_{22} + \cdots + a_{2n})$$

$$+ \cdots + (a_{m1} + a_{m2} + \cdots + a_{mn}),$$

or we can expand first with respect to j and then with respect to i:

$$\sum_{i=1}^{m} \left(\sum_{j=1}^{n} a_{ij} \right) = \sum_{i=1}^{m} (a_{i1} + a_{i2} + \cdots + a_{in})$$

$$= (a_{11} + a_{12} + \cdots + a_{1n}) + (a_{21} + a_{22} + \cdots + a_{2n})$$

$$+ \cdots + (a_{m1} + a_{m2} + \cdots + a_{mn}).$$

The results are the same.

For example, we can write

$$\sum_{i=1}^{3} \left(\sum_{j=1}^{2} a_{ij} \right) = \sum_{j=1}^{2} a_{1j} + \sum_{j=1}^{2} a_{2j} + \sum_{j=1}^{2} a_{3j}$$

$$= (a_{11} + a_{12}) + (a_{21} + a_{22}) + (a_{31} + a_{32}),$$

or we can write

$$\sum_{i=1}^{3} \left(\sum_{j=1}^{2} a_{ij} \right) = \sum_{i=1}^{3} (a_{i1} + a_{i2}) = (a_{11} + a_{12}) + (a_{21} + a_{22}) + (a_{31} + a_{32}).$$

Since constants can be factored through single sums, they can also be factored through double sums; namely,

(16.1.3)

$$\sum_{i=1}^{m} \sum_{j=1}^{n} \alpha a_{ij} = \alpha \sum_{i=1}^{m} \sum_{j=1}^{n} a_{ij}.$$

Also,

(16.1.4)

$$\sum_{i=1}^{m} \sum_{j=1}^{n} (a_{ij} + b_{ij}) = \sum_{i=1}^{m} \sum_{j=1}^{n} a_{ij} + \sum_{i=1}^{m} \sum_{j=1}^{n} b_{ij}.$$

The easiest double sums to handle are those where each term a_{ij} appears as a product $b_i c_j$ in which each of the factors bears only one index. In that case, we can express the double sum as the product of two single sums:

(16.1.5)

$$\sum_{i=1}^{m} \sum_{j=1}^{n} b_i c_j = \left(\sum_{i=1}^{m} b_i \right) \left(\sum_{j=1}^{n} c_j \right).$$

PROOF   Set

$$B = \sum_{i=1}^{m} b_i, \qquad C = \sum_{j=1}^{n} c_j.$$

Then

$$\sum_{i=1}^{m} \sum_{j=1}^{n} b_i c_j = \sum_{i=1}^{m} \left(\sum_{j=1}^{n} b_i c_j \right) \overset{\dagger}{=} \sum_{i=1}^{m} b_i \left(\sum_{j=1}^{n} c_j \right) = \sum_{i=1}^{m} b_i C$$

$$= C \sum_{i=1}^{m} b_i = CB = BC. \quad \square$$

For example,

$$\sum_{i=1}^{4} \sum_{j=1}^{2} 2^i 5^j = \left(\sum_{i=1}^{4} 2^i \right) \left(\sum_{j=1}^{2} 5^j \right) = (2 + 2^2 + 2^3 + 2^4)(5 + 5^2) = (30)(30) = 900.$$

Triple-sigma notation is used when three indices are involved. The sum of all the a_{ijk} where i ranges from 1 to m, j from 1 to n, and k from 1 to q can be written

(16.1.6)

$$\sum_{i=1}^{m} \sum_{j=1}^{n} \sum_{k=1}^{q} a_{ijk}.$$

---

† Since b_i is independent of j, it can be factored through the j-summation.

Multiple sums appear in the following sections. We introduced them here so as to avoid lengthy asides later.

EXERCISES 16.1

In Exercises 1–4, evaluate the given sum.

1. $\displaystyle\sum_{i=1}^{3}\sum_{j=1}^{3} 2^{i-1}3^{j+1}.$

2. $\displaystyle\sum_{i=1}^{4}\sum_{j=1}^{2} (1+i)^{j}.$

3. $\displaystyle\sum_{i=1}^{4}\sum_{j=1}^{3} (i^2 + 3i)(j-2).$

4. $\displaystyle\sum_{i=1}^{3}\sum_{j=1}^{2}\sum_{k=1}^{3} \frac{2i}{k+j^2}.$

In Exercises 5–16, let

$P_1 = \{x_0, x_1, \ldots, x_m\}$ be a partition of $[a_1, a_2]$,

$P_2 = \{y_0, y_1, \ldots, y_n\}$ be a partition of $[b_1, b_2]$,

$P_3 = \{z_0, z_1, \ldots, z_q\}$ be a partition of $[c_1, c_2]$,

and let

$$\Delta x_i = x_i - x_{i-1}, \qquad \Delta y_j = y_j - y_{j-1}, \qquad \Delta z_k = z_k - z_{k-1}.$$

Evaluate the following sums.

5. $\displaystyle\sum_{i=1}^{m} \Delta x_i.$

6. $\displaystyle\sum_{j=1}^{n} \Delta y_j.$

7. $\displaystyle\sum_{i=1}^{m}\sum_{j=1}^{n} \Delta x_i\, \Delta y_j.$

8. $\displaystyle\sum_{j=1}^{n}\sum_{k=1}^{q} \Delta y_j\, \Delta z_k.$

9. $\displaystyle\sum_{i=1}^{m} (x_i + x_{i-1})\Delta x_i.$

10. $\displaystyle\sum_{j=1}^{n} \tfrac{1}{2}(y_j^2 + y_j y_{j-1} + y_{j-1}^2)\Delta y_j.$

11. $\displaystyle\sum_{i=1}^{m}\sum_{j=1}^{n} (x_i + x_{i-1})\Delta x_i\, \Delta y_j.$

12. $\displaystyle\sum_{i=1}^{m}\sum_{j=1}^{n} (y_j + y_{j-1})\Delta x_i\, \Delta y_j.$

13. $\displaystyle\sum_{i=1}^{m}\sum_{j=1}^{n} (2\Delta x_i - 3\Delta y_j).$

14. $\displaystyle\sum_{i=1}^{m}\sum_{j=1}^{n} (3\Delta x_i - 2\Delta y_j).$

15. $\displaystyle\sum_{i=1}^{m}\sum_{j=1}^{n}\sum_{k=1}^{q} \Delta x_i\, \Delta y_j\, \Delta z_k.$

16. $\displaystyle\sum_{i=1}^{m}\sum_{j=1}^{n}\sum_{k=1}^{q} (x_i + x_{i-1})\Delta x_i\, \Delta y_j\, \Delta z_k.$

17. Evaluate

$$\sum_{i=1}^{n}\sum_{j=1}^{n}\sum_{k=1}^{n} \delta_{ijk}a_{ijk} \qquad \text{where} \qquad \delta_{ijk} = \begin{cases} 1, & \text{if } i=j=k \\ 0, & \text{otherwise.} \end{cases}$$

18. Show that any sum written in double-sigma notation can also be expressed in ordinary (that is, single) sigma notation.

19. Verify property (16.1.3).

20. Verify property (16.1.4).

■ 16.2 THE DOUBLE INTEGRAL OVER A RECTANGLE

We start with a function f continuous on a rectangle

$$R: \quad a \le x \le b, \quad c \le y \le d.$$

See Figure 16.2.1. Our object is to define the double integral of f over R:

$$\iint_R f(x, y)\, dx\, dy.$$

Recall that our approach to defining the definite integral of a function f of one variable,

$$\int_a^b f(x)\, dx,$$

involved some auxiliary notions, namely: partition P of $[a, b]$, upper sum $U_f(P)$, and lower sum $L_f(P)$ (see Section 5.2). We were then able to define

$$\int_a^b f(x)\, dx$$

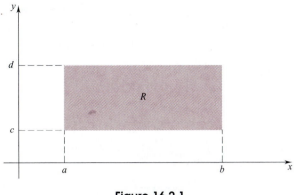

Figure 16.2.1

as the unique number I that satisfies the inequality

$$L_f(P) \le I \le U_f(P) \qquad \text{for all partitions } P \text{ of } [a, b].$$

We will follow exactly the same procedure to define the double integral

$$\iint\limits_{R} f(x, y) \, dx \, dy.$$

First we explain what we mean by a partition of the rectangle R. To do this, we begin with a partition

$$P_1 = \{x_0, x_1, \ldots, x_m\} \quad \text{of} \quad [a, b]$$

and a partition

$$P_2 = \{y_0, y_1, \ldots, y_n\} \quad \text{of} \quad [c, d].$$

The set

$$P = P_1 \times P_2 = \{(x_i, y_j): x_i \in P_1, y_j \in P_2\}\dagger$$

is called a *partition of R* (see Figure 16.2.2); P consists of all the grid points (x_i, y_j).

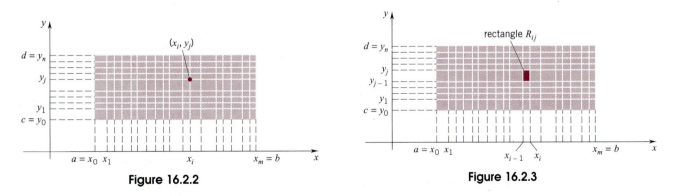

Figure 16.2.2   **Figure 16.2.3**

The partition P breaks up R into $m \times n$ nonoverlapping rectangles

$$R_{ij}: \quad x_{i-1} \le x \le x_i, \quad y_{j-1} \le y \le y_j, \qquad \text{(Figure 16.2.3)}$$

$\dagger$ $P_1 \times P_2$ is the Cartesian product of P_1 and P_2; if A and B are sets, then the *Cartesian product* of A and B is the set $A \times B = \{(a, b): a \in A \text{ and } b \in B\}$.

where $1 \leq i \leq m$, $1 \leq j \leq n$. On each rectangle R_{ij}, the function f takes on a maximum value M_{ij} and a minimum value m_{ij}. We know this because f is continuous and R_{ij} is closed and bounded (Section 15.6). The sum of all the products

$$M_{ij}(\text{area of } R_{ij}) = M_{ij}(x_i - x_{i-1})(y_j - y_{j-1}) = M_{ij}\,\Delta x_i\,\Delta y_j$$

is called the P *upper sum* for f:

(16.2.1)
$$U_f(P) = \sum_{i=1}^{m}\sum_{j=1}^{n} M_{ij}(\text{area of } R_{ij}) = \sum_{i=1}^{m}\sum_{j=1}^{n} M_{ij}\,\Delta x_i\,\Delta y_j.$$

The sum of all the products

$$m_{ij}(\text{area of } R_{ij}) = m_{ij}(x_i - x_{i-1})(y_j - y_{j-1}) = m_{ij}\,\Delta x_i\,\Delta y_j$$

is called the P *lower sum* for f:

(16.2.2)
$$L_f(P) = \sum_{i=1}^{m}\sum_{j=1}^{n} m_{ij}(\text{area of } R_{ij}) = \sum_{i=1}^{m}\sum_{j=1}^{n} m_{ij}\,\Delta x_i\,\Delta y_j.$$

Example 1 Consider the function

$$f(x, y) = x + y - 2$$

on the rectangle

$$R: \quad 1 \leq x \leq 4, \quad 1 \leq y \leq 3.$$

As a partition of $[1, 4]$ take

$$P_1 = \{1, 2, 3, 4\},$$

and as a partition of $[1, 3]$ take

$$P_2 = \{1, \tfrac{3}{2}, 3\}.$$

The partition $P = P_1 \times P_2$ then breaks up the initial rectangle into the six rectangles marked in Figure 16.2.4. On each rectangle R_{ij}, the function f takes on its maximum value M_{ij} at the point (x_i, y_j), the corner farthest from the origin:

$$M_{ij} = f(x_i, y_j) = x_i + y_j - 2.$$

Thus

$$U_f(P) = M_{11}(\text{area of } R_{11}) + M_{12}(\text{area of } R_{12}) + M_{21}(\text{area of } R_{21})$$
$$+ M_{22}(\text{area of } R_{22}) + M_{31}(\text{area of } R_{31}) + M_{32}(\text{area of } R_{32})$$
$$= \tfrac{3}{2}(\tfrac{1}{2}) + 3(\tfrac{3}{2}) + \tfrac{5}{2}(\tfrac{1}{2}) + 4(\tfrac{3}{2}) + \tfrac{7}{2}(\tfrac{1}{2}) + 5(\tfrac{3}{2}) = \tfrac{87}{4}.$$

On each rectangle R_{ij}, f takes on its minimum value m_{ij} at the point (x_{i-1}, y_{j-1}), the corner closest to the origin:

$$m_{ij} = f(x_{i-1}, y_{j-1}) = x_{i-1} + y_{j-1} - 2.$$

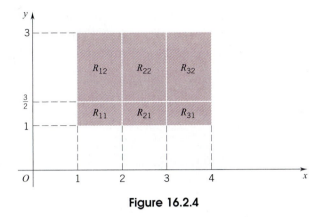

Figure 16.2.4

Thus

$$L_f(P) = m_{11}(\text{area of } R_{11}) + m_{12}(\text{area of } R_{12}) + m_{21}(\text{area of } R_{21})$$
$$+ m_{22}(\text{area of } R_{22}) + m_{31}(\text{area of } R_{31}) + m_{32}(\text{area of } R_{32})$$
$$= 0(\tfrac{1}{2}) + \tfrac{1}{2}(\tfrac{3}{2}) + 1(\tfrac{1}{2}) + \tfrac{3}{2}(\tfrac{3}{2}) + 2(\tfrac{1}{2}) + \tfrac{5}{2}(\tfrac{3}{2}) = \tfrac{33}{4}. \quad \square$$

We return now to the general situation. As in the one-variable case, it can be shown that, if f is continuous, then there exists one and only one number I that satisfies the inequality

$$L_f(P) \le I \le U_f(P) \qquad \text{for all partitions } P \text{ of } R.$$

DEFINITION 16.2.3 THE DOUBLE INTEGRAL OVER A RECTANGLE R

Let f be continuous on a closed rectangle R. The unique number I that satisfies the inequality

$$L_f(P) \le I \le U_f(P) \qquad \text{for all partitions } P \text{ of } R$$

is called the *double integral* of f over R and is denoted by

$$\iint\limits_{R} f(x, y) \, dx \, dy.$$

The Double Integral as a Volume

If f is continuous and nonnegative on the rectangle R, the equation

$$z = f(x, y)$$

represents a surface that lies above R. (See Figure 16.2.5.) In this case the double integral

$$\iint\limits_{R} f(x, y) \, dx \, dy$$

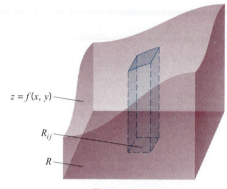

Figure 16.2.5

gives the volume of the solid that is bounded below by R and bounded above by the surface $z = f(x, y)$.

To see this, consider a partition P of R. P breaks up R into subrectangles R_{ij} and thus the entire solid T into parts T_{ij}. Since T_{ij} contains a rectangular solid with base R_{ij} and height m_{ij} (Figure 16.2.6), we must have

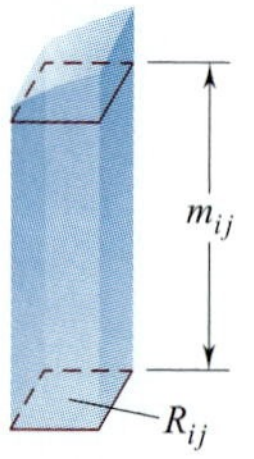

$$m_{ij}(\text{area of } R_{ij}) \leq \text{volume of } T_{ij}.$$

Since T_{ij} is contained in a rectangular solid with base R_{ij} and height M_{ij} (Figure 16.2.7), we must have

$$\text{volume of } T_{ij} \leq M_{ij}(\text{area of } R_{ij}).$$

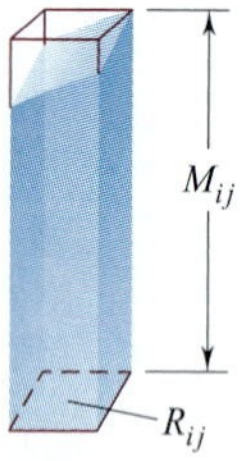

In short, for each pair of indices i and j, we must have

$$m_{ij}(\text{area of } R_{ij}) \leq \text{volume of } T_{ij} \leq M_{ij}(\text{area of } R_{ij}).$$

Adding up these inequalities, we can conclude that

$$L_f(P) \leq \text{volume of } T \leq U_f(P).$$

Since P is arbitrary, the volume of T must be the double integral:

Figure 16.2.6

Figure 16.2.7

(16.2.4)

$$\text{volume of } T = \iint\limits_{R} f(x, y) \, dx \, dy.$$

The double integral

$$\iint\limits_{R} 1 \, dx \, dy = \iint\limits_{R} dx \, dy$$

gives the volume of a solid of constant height 1 erected over R. In square units this is just the area of R:

(16.2.5)

$$\text{area of } R = \iint\limits_{R} dx \, dy.$$

Some Computations

Double integrals are generally computed by techniques that we will take up later. It is possible, however, to evaluate simple double integrals directly from the definition.

Example 2 Evaluate

$$\iint_R \alpha \, dx \, dy$$

where α is a constant and R is the rectangle

$$R: \quad a \le x \le b, \quad c \le y \le d.$$

SOLUTION Here

$$f(x, y) = \alpha \qquad \text{for all } (x, y) \in R.$$

We begin with $P_1 = \{x_0, x_1, \ldots, x_m\}$ as an arbitrary partition of $[a, b]$ and $P_2 = \{y_0, y_1, \ldots, y_n\}$ as an arbitrary partition of $[c, d]$. This gives

$$P = P_1 \times P_2 = \{(x_i, y_j): x_i \in P_1, y_j \in P_2\}$$

as an arbitrary partition of R. On each rectangle R_{ij}, f has constant value α. This gives $M_{ij} = \alpha$ and $m_{ij} = \alpha$ throughout. Thus

$$U_f(P) = \sum_{i=1}^{m} \sum_{j=1}^{n} \alpha \, \Delta x_i \, \Delta y_j = \alpha \left(\sum_{i=1}^{m} \Delta x_i \right) \left(\sum_{j=1}^{n} \Delta y_j \right) = \alpha(b - a)(d - c).$$

Similarly

$$L_f(P) = \alpha(b - a)(d - c).$$

Since

$$L_f(P) \le \alpha(b - a)(d - c) \le U_f(P)$$

and P was chosen arbitrarily, the inequality must hold for all partitions P of R. This means that

$$\iint_R f(x, y) \, dx \, dy = \alpha(b - a)(d - c). \quad \square$$

Remark If $\alpha > 0$,

$$\iint_R \alpha \, dx \, dy = \alpha(b - a)(d - c)$$

gives the volume of the rectangular solid of constant height α erected over the rectangle R (Figure 16.2.8). ◻

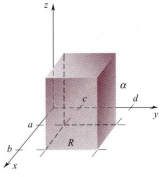

Figure 16.2.8

Example 3 Evaluate

$$\iint_R (y - 2x) \, dx \, dy$$

where R is the rectangle

$$R: \quad 1 \le x \le 2, \quad 3 \le y \le 5.$$

SOLUTION With

$$P_1 = \{x_0, x_1, \ldots, x_m\} \text{ as an arbitrary partition of } [1, 2]$$

and

$$P_2 = \{y_0, y_1, \ldots, y_n\} \text{ as an arbitrary partition of } [3, 5]$$

we have

$$P = P_1 \times P_2 = \{(x_i, y_j): x_i \in P_1, y_j \in P_2\}$$

as an arbitrary partition of R.   On each rectangle

$$R_{ij}: \quad x_{i-1} \le x \le x_i, \quad y_{j-1} \le y \le y_j$$

the function

$$f(x, y) = y - 2x$$

has a maximum

$$M_{ij} = y_j - 2x_{i-1} \qquad (y \text{ is maximized and } x \text{ is minimized})$$

and a minimum

$$m_{ij} = y_{j-1} - 2x_i. \qquad (y \text{ is minimized and } x \text{ is maximized})$$

Thus,

$$U_f(P) = \sum_{i=1}^{m} \sum_{j=1}^{n} (y_j - 2x_{i-1}) \, \Delta x_i \, \Delta y_j \quad \text{and} \quad L_f(P) = \sum_{i=1}^{m} \sum_{j=1}^{n} (y_{j-1} - 2x_i) \, \Delta x_i \, \Delta y_j.$$

For each pair of indices i and j,

$$y_{j-1} - 2x_i \le \tfrac{1}{2}(y_j + y_{j-1}) - (x_i + x_{i-1}) \le y_j - 2x_{i-1}. \qquad \text{(explain)}$$

This means that for arbitrary P we have

$$L_f(P) \le \sum_{i=1}^{m} \sum_{j=1}^{n} [\tfrac{1}{2}(y_j + y_{j-1}) - (x_i + x_{i-1})] \, \Delta x_i \, \Delta y_j \le U_f(P).$$

The double sum in the middle of this set of inequalities can be written

$$\sum_{i=1}^{m} \sum_{j=1}^{n} \tfrac{1}{2}(y_j + y_{j-1}) \, \Delta x_i \, \Delta y_j - \sum_{i=1}^{m} \sum_{j=1}^{n} (x_i + x_{i-1}) \, \Delta x_i \, \Delta y_j.$$

The first double sum reduces to

$$\sum_{i=1}^{m} \sum_{j=1}^{n} \tfrac{1}{2} \Delta x_i (y_j^2 - y_{j-1}^2) = \tfrac{1}{2} \left(\sum_{i=1}^{m} \Delta x_i \right) \left(\sum_{j=1}^{n} (y_j^2 - y_{j-1}^2) \right)$$

$$= \tfrac{1}{2}(2 - 1)(25 - 9) = 8.$$

The second double sum reduces to

$$-\sum_{i=1}^{m} \sum_{j=1}^{n} (x_i^2 - x_{i-1}^2) \, \Delta y_j = -\left(\sum_{i=1}^{m} (x_i^2 - x_{i-1}^2) \right) \left(\sum_{j=1}^{n} \Delta y_j \right)$$

$$= -(4 - 1)(5 - 3) = -6.$$

The sum of these two numbers

$$I = 8 + (-6) = 2$$

satisfies the inequality

$$L_f(P) \leq 2 \leq U_f(P) \qquad \text{for arbitrary } P.$$

The integral is therefore 2:

$$\iint\limits_{R} (y - 2x) \, dx \, dy = 2. \quad \square$$

Remark This last integral should not be interpreted as a volume. The expression $y - 2x$ does not keep a constant sign on

$$R: \quad 1 \leq x \leq 2, \quad 3 \leq y \leq 5. \quad \square$$

EXERCISES 16.2

For Exercises 1–3, take

$$f(x, y) = x + 2y \qquad \text{on} \qquad R: \ 0 \leq x \leq 2, \ 0 \leq y \leq 1,$$

and P as the partition $P = P_1 \times P_2$.

1. Find $L_f(P)$ and $U_f(P)$ if $P_1 = \{0, 1, \frac{3}{2}, 2\}$ and $P_2 = \{0, \frac{1}{2}, 1\}$.
2. Find $L_f(P)$ and $U_f(P)$ if $P_1 = \{0, \frac{1}{2}, 1, \frac{3}{2}, 2\}$ and $P_2 = \{0, \frac{1}{4}, \frac{1}{2}, \frac{3}{4}, 1\}$.
3. (a) Find $L_f(P)$ and $U_f(P)$ if

$$P_1 = \{x_0, x_1, \ldots, x_m\}$$

is an arbitrary partition of $[0, 2]$,

$$P_2 = \{y_0, y_1, \ldots, y_n\}$$

is an arbitrary partition of $[0, 1]$.

(b) Use (a) to evaluate the double integral

$$\iint\limits_{R} (x + 2y) \, dx \, dy$$

and give a geometric interpretation to your answer.

For Exercises 4–6, take

$$f(x, y) = x - y \qquad \text{on} \qquad R: \ 0 \leq x \leq 1, \ 0 \leq y \leq 1,$$

and P as the partition $P = P_1 \times P_2$.

4. Find $L_f(P)$ and $U_f(P)$ if $P_1 = \{0, \frac{1}{2}, \frac{3}{4}, 1\}$ and $P_2 = \{0, \frac{1}{2}, 1\}$.
5. Find $L_f(P)$ and $U_f(P)$ if $P_1 = \{0, \frac{1}{4}, \frac{1}{2}, \frac{3}{4}, 1\}$ and $P_2 = \{0, \frac{1}{3}, \frac{2}{3}, 1\}$.

6. (a) Find $L_f(P)$ and $U_f(P)$ if

$$P_1 = \{x_0, x_1, \ldots, x_m\} \text{ and } P_2 = \{y_0, y_1, \ldots, y_n\}$$

are arbitrary partitions of $[0, 1]$.

(b) Use (a) to evaluate the double integral

$$\iint\limits_{R} (x - y) \, dx \, dy.$$

For Exercises 7–9, take $R: 0 \leq x \leq b, 0 \leq y \leq d$ and let $P = P_1 \times P_2$ where

$$P_1 = \{x_0, x_1, \ldots, x_m\} \text{ is an arbitrary partition of } [0, b],$$

$$P_2 = \{y_0, y_1, \ldots, y_n\} \text{ is an arbitrary partition of } [0, d].$$

7. (a) Find $L_f(P)$ and $U_f(P)$ for $f(x, y) = 4xy$.
   (b) Calculate

$$\iint\limits_{R} 4xy \, dx \, dy.$$

   HINT: $4x_{i-1}y_{j-1} \leq (x_i + x_{i-1})(y_j + y_{j-1}) \leq 4x_i y_j.$

8. (a) Find $L_f(P)$ and $U_f(P)$ for $f(x, y) = 3(x^2 + y^2)$.
   (b) Calculate

$$\iint\limits_{R} 3(x^2 + y^2) \, dx \, dy.$$

   HINT: If $0 \leq s \leq t$, then $3s^2 \leq t^2 + ts + s^2 \leq 3t^2$.

9. (a) Find $L_f(P)$ and $U_f(P)$ for $f(x, y) = 3(x^2 - y^2)$.
   (b) Calculate

$$\iint\limits_{R} 3(x^2 - y^2) \, dx \, dy.$$

10. Let $f = f(x, y)$ be continuous on the rectangle R: $a \leq x \leq b$, $c \leq y \leq d$. Suppose that $L_f(P) = U_f(P)$ for some partition P of R. What can you conclude about f? What is

$$\iint_R f(x, y)\, dx\, dy?$$

11. Let $f(x, y) = \sin(x + y)$ on R: $0 \leq x \leq 1$, $0 \leq y \leq 1$. Show that

$$0 \leq \iint_R \sin(x + y)\, dx\, dy \leq 1.$$

■ 16.3 THE DOUBLE INTEGRAL OVER A REGION

We start with a closed bounded region Ω in the xy-plane (Figure 16.3.1). We assume that Ω is a *basic region*. A *basic region* is a region whose boundary consists of a finite number of continuous arcs $y = \phi(x)$, $x = \psi(y)$. See Figure 16.3.2.

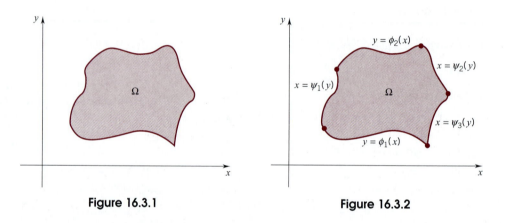

Figure 16.3.1   Figure 16.3.2

Now let f be a continuous function on Ω. We want to define the double integral

$$\iint_\Omega f(x, y)\, dx\, dy.$$

To do this, we enclose Ω by a rectangle R as in Figure 16.3.3. We now extend f to all of R by setting f equal to 0 outside Ω. This extended function, which we continue to call f, is bounded on R, and it is continuous on all of R except possibly at the boundary of Ω. In spite of these possible discontinuities, it can be shown that f is still integrable on R; that is, there still exists a unique number I such that

$$L_f(P) \leq I \leq U_f(P) \qquad \text{for all partitions } P \text{ of } R.$$

(We will not attempt to prove this, but an analogous result is given in Exercises 29 and 30, Section 5.2.) This number I is by definition the double integral

$$\iint_R f(x, y)\, dx\, dy.$$

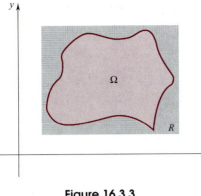

Figure 16.3.3

As you probably guessed by now, we define the double integral over Ω by setting

(16.3.1)

$$\iint_{\Omega} f(x, y) \, dx \, dy = \iint_{R} f(x, y) \, dx \, dy.$$

If f is continuous and nonnegative over Ω, the extended f is nonnegative on all of R. (Figure 16.3.4.) The double integral gives the volume of the solid whose upper boundary is the surface $z = f(x, y)$ and whose lower boundary is the rectangle R. But since the surface has height 0 outside of Ω, the volume outside Ω is 0. Therefore, it follows that

$$\iint_{\Omega} f(x, y) \, dx \, dy$$

gives the volume of the solid T bounded above by $z = f(x, y)$ *and below by* Ω:

(16.3.2)

$$\text{volume of } T = \iint_{\Omega} f(x, y) \, dx \, dy.$$

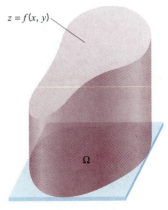

Figure 16.3.4

The double integral

$$\iint_{\Omega} 1 \, dx \, dy = \iint_{\Omega} dx \, dy$$

gives the volume of a solid of constant height 1 over Ω. In square units this is the area of Ω:

(16.3.3)

$$\text{area of } \Omega = \iint_{\Omega} dx\, dy.$$

Below we list four elementary properties of the double integral. They are all analogous to what you saw in the one-variable case. The Ω referred to is a basic region. The functions f and g are assumed to be continuous on Ω.

I. Linearity: the double integral of a linear combination is the linear combination of the double integrals

$$\iint_{\Omega} [\alpha f(x, y) + \beta g(x, y)]\, dx\, dy = \alpha \iint_{\Omega} f(x, y)\, dx\, dy + \beta \iint_{\Omega} g(x, y)\, dx\, dy.$$

II. Order:

$$\text{if } f \geq 0 \text{ on } \Omega, \text{ then } \iint_{\Omega} f(x, y)\, dx\, dy \geq 0;$$

$$\text{if } f \leq g \text{ on } \Omega, \text{ then } \iint_{\Omega} f(x, y)\, dx\, dy \leq \iint_{\Omega} g(x, y)\, dx\, dy.$$

III. Additivity: if Ω is broken up into a finite number of non-overlapping basic regions $\Omega_1, \ldots \Omega_n$, then

$$\iint_{\Omega} f(x, y)\, dx\, dy = \iint_{\Omega_1} f(x, y)\, dx\, dy + \cdots + \iint_{\Omega_n} f(x, y)\, dx\, dy.$$

See, for example, Figure 16.3.5.

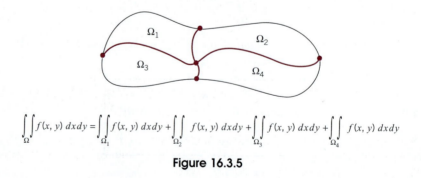

$$\iint_{\Omega} f(x, y)\, dx\, dy = \iint_{\Omega_1} f(x, y)\, dx\, dy + \iint_{\Omega_2} f(x, y)\, dx\, dy + \iint_{\Omega_3} f(x, y)\, dx\, dy + \iint_{\Omega_4} f(x, y)\, dx\, dy$$

Figure 16.3.5

IV. Mean-value condition: There is a point (x_0, y_0) in Ω for which

$$\iint_{\Omega} f(x, y)\, dx\, dy = f(x_0, y_0) \cdot (\text{area of } \Omega).$$

We call $f(x_0, y_0)$ *the average value of f on* Ω.

The notion of average given in (IV) enables us to write

(16.3.4)

$$\iint\limits_{\Omega} f(x, y)\, dxdy = \left(\begin{array}{c}\text{the average value}\\ \text{of } f \text{ on } \Omega\end{array}\right) \cdot (\text{area of } \Omega).$$

This is a powerful, intuitive way of viewing the double integral. We will capitalize on it as we go on.

THEOREM 16.3.5 MEAN-VALUE THEOREM FOR DOUBLE INTEGRALS

Let f and g be continuous functions on the basic region Ω. If g is nonnegative on Ω, then there exists a point (x_0, y_0) in Ω for which

$$\iint\limits_{\Omega} f(x, y)g(x, y)\, dx\, dy = f(x_0, y_0) \iint\limits_{\Omega} g(x, y)\, dx\, dy\dagger$$

NOTE: The value $f(x_0, y_0)$ specified in this theorem is called the *g-weighted average of f on Ω*.

PROOF Since f is continuous on Ω and Ω is closed and bounded, we know that f takes on a minimum value m and a maximum value M. Since g is nonnegative on Ω,

$$mg(x, y) \le f(x, y)g(x, y) \le Mg(x, y) \qquad \text{for all } (x, y) \text{ in } \Omega.$$

Therefore (by property II)

$$\iint\limits_{\Omega} mg(x, y)\, dx\, dy \le \iint\limits_{\Omega} f(x, y)g(x, y)\, dx\, dy \le \iint\limits_{\Omega} Mg(x, y)\, dx\, dy$$

and (by property I)

$$(*) \quad m\iint\limits_{\Omega} g(x, y)\, dx\, dy \le \iint\limits_{\Omega} f(x, y)g(x, y)\, dx\, dy \le M\iint\limits_{\Omega} g(x, y)\, dx\, dy.$$

We know that $\iint_\Omega g(x, y)\, dx\, dy \ge 0$ (again, by Property II). If $\iint_\Omega g(x, y)\, dx\, dy = 0$, then by (*) we have $\iint_\Omega f(x, y)g(x, y)\, dx\, dy = 0$ and the theorem holds for all choices of (x_0, y_0) in Ω. If $\iint_\Omega g(x, y)\, dx\, dy > 0$, then

$$m \le \frac{\iint_\Omega f(x, y)g(x, y)\, dx\, dy}{\iint_\Omega g(x, y)\, dx\, dy} \le M$$

† Property IV is this equation with g constantly 1.

and by the intermediate-value theorem (Theorem 15.3.5) there exists (x_0, y_0) in Ω for which

$$f(x_0, y_0) = \frac{\iint_\Omega f(x, y)g(x, y)\, dx\, dy}{\iint_\Omega g(x, y)\, dx\, dy}.$$

Obviously then

$$f(x_0, y_0) \iint_\Omega g(x, y)\, dx\, dy = \iint_\Omega f(x, y)g(x, y)\, dx\, dy. \quad \square$$

EXERCISES 16.3

1. Let $\phi = \phi(x)$ be a continuous, nonnegative function on $[a, b]$, and set

$$\Omega = \{(x, y): a \le x \le b,\ 0 \le y \le \phi(x)\}.$$

Compare

$$\iint_\Omega dx\, dy \quad \text{to} \quad \int_a^b \phi(x)\, dx.$$

2. Begin with a function f continuous on a closed bounded region Ω. Now surround Ω by a rectangle R as in Figure 16.3.3 and extend f to all of R by defining f to be 0 outside of Ω. Explain how the extended f can fail to be continuous on the boundary of Ω although the original function f, being continuous on all of Ω, was continuous on the boundary of Ω.

3. Suppose that f is continuous on a disc Ω with center at (x_0, y_0) and that

$$\iint_R f(x, y)\, dx\, dy = 0$$

for every rectangle R contained in Ω. Show that $f(x_0, y_0) = 0$.

4. Calculate the average value of $f(x, y) = x + 2y$ on the rectangle R: $0 \le x \le 2, 0 \le y \le 1$. HINT: See Exercise 3, Section 16.2.

5. Calculate the average value of $f(x, y) = 4xy$ on the rectangle R: $0 \le x \le 2, 0 \le y \le 3$. HINT: See Exercise 7, Section 16.2.

6. Calculate the average value of $f(x, y) = x^2 + y^2$ on the rectangle R: $0 \le x \le b, 0 \le y \le d$. Hint: See Exercise 8, Section 16.2.

7. Let f be continuous on a closed bounded region Ω and let (x_0, y_0) be a point in the interior of Ω. Let D_r be a closed disc with center (x_0, y_0) and radius r. Show that

$$\lim_{r \to 0} \frac{1}{\pi r^2} \iint_{D_r} f(x, y)\, dx\, dy = f(x_0, y_0).$$

HINT: Use the mean-value theorem.

■ 16.4 THE EVALUATION OF DOUBLE INTEGRALS BY REPEATED INTEGRALS

The Reduction Formulas

If an integral

$$\int_a^b f(x)\, dx$$

proves difficult to evaluate, it is not because of the interval $[a, b]$ but because of the integrand f. Difficulty in evaluating a double integral

$$\iint_\Omega f(x, y)\, dx\, dy$$

can come from two sources: from the integrand f or from the base region Ω. Even such a simple looking integral as $\iint_\Omega 1 \, dx \, dy$ is difficult to evaluate if Ω is complicated.

In this section we introduce a technique for evaluating double integrals of continuous functions over regions that have special shapes, the shapes shown in Figure 16.4.1. In each case, the region Ω is a basic region and so we know that the double integral exists. The fundamental idea of this section is that double integrals over sets of these shapes can be reduced to a pair of ordinary integrals.

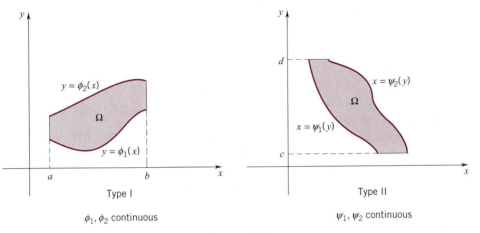

Type I
ϕ_1, ϕ_2 continuous

Type II
ψ_1, ψ_2 continuous

Figure 16.4.1

Type I Region. The *projection* of Ω onto the x-axis is a closed interval $[a, b]$ and Ω consists of all points (x, y) with

$$a \leq x \leq b \quad \text{and} \quad \phi_1(x) \leq y \leq \phi_2(x).$$

Then

(16.4.1)
$$\iint_\Omega f(x, y) \, dx \, dy = \int_a^b \left(\int_{\phi_1(x)}^{\phi_2(x)} f(x, y) \, dy \right) dx.$$

Here we first calculate

$$\int_{\phi_1(x)}^{\phi_2(x)} f(x, y) \, dy$$

by integrating $f(x, y)$ with respect to y from $y = \phi_1(x)$ to $y = \phi_2(x)$. The resulting expression is a function of x alone, which we then integrate with respect to x from $x = a$ to $x = b$.

Type II Region. The *projection* of Ω onto the y-axis is a closed interval $[c, d]$ and Ω consists of all points (x, y) with

$$c \leq y \leq d \quad \text{and} \quad \psi_1(y) \leq x \leq \psi_2(y).$$

Then

(16.4.2)

$$\iint_{\Omega} f(x, y) \, dx \, dy = \int_c^d \left(\int_{\psi_1(y)}^{\psi_2(y)} f(x, y) \, dx \right) dy.$$

This time we first calculate

$$\int_{\psi_1(y)}^{\psi_2(y)} f(x, y) \, dx$$

by integrating $f(x, y)$ with respect to x from $x = \psi_1(y)$ to $x = \psi_2(y)$. The resulting expression is a function of y alone, which we then integrate with respect to y from $y = c$ to $y = d$.

These formulas are easy to understand geometrically.

The Reduction Formulas Viewed Geometrically

Suppose that f is nonnegative and Ω is a region of type I. The double integral over Ω gives the volume of the solid T whose upper boundary is the surface $z = f(x, y)$ and whose lower boundary is the region Ω:

(1)
$$\iint_{\Omega} f(x, y) \, dx \, dy = \text{volume of } T. \qquad \text{(Figure 16.4.2)}$$

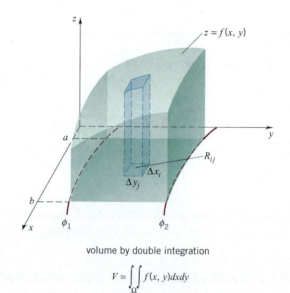

volume by double integration

$$V = \iint_{\Omega} f(x, y) dx dy$$

Figure 16.4.2

We can also compute the volume of T by the method of parallel cross sections (see Section 6.2). As in Figure 16.4.3, let $A(x)$ be the area of that cross section of T that has first coordinate x. Then by (6.2.1)

$$\int_a^b A(x)\, dx = \text{volume of } T.$$

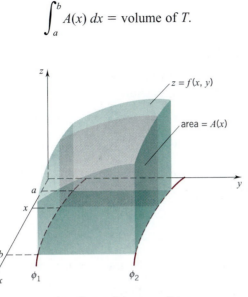

volume by parallel cross sections

$$V = \int_a^b A(x)\, dx = \int_a^b \left(\int_{\phi_1(x)}^{\phi_2(x)} f(x, y)\, dy \right) dx$$

Figure 16.4.3

Since

$$A(x) = \int_{\phi_1(x)}^{\phi_2(x)} f(x, y)\, dy,$$

we have

(2)
$$\int_a^b \left(\int_{\phi_1(x)}^{\phi_2(x)} f(x, y)\, dy \right) dx = \text{volume of } T.$$

Combining (1) with (2), we have the first reduction formula

$$\iint_\Omega f(x, y)\, dx\, dy = \int_a^b \left(\int_{\phi_1(x)}^{\phi_2(x)} f(x, y)\, dy \right) dx.$$

The other reduction formula can be obtained in a similar manner. ❑

Remark Note that our argument was a very loose one and certainly not a proof. How do we know, for example, that the "volume" obtained by double integration is the same as the "volume" obtained by the method of parallel cross sections? Intuitively it seems evident, but actually it is quite difficult to prove. ❑

Computations

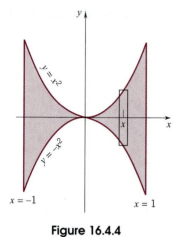

Figure 16.4.4

Example 1 Evaluate

$$\iint_\Omega (x^4 - 2y)\, dx\, dy$$

with Ω as in Figure 16.4.4.

SOLUTION By projecting Ω onto the x-axis we obtain the interval $[-1, 1]$. The region Ω consists of all points (x, y) with

$$-1 \le x \le 1 \quad \text{and} \quad -x^2 \le y \le x^2.$$

This is a region of type I. By (16.4.1)

$$\iint_\Omega (x^4 - 2y)\, dx\, dy = \int_{-1}^{1} \left(\int_{-x^2}^{x^2} [x^4 - 2y]\, dy \right) dx$$

$$= \int_{-1}^{1} \left[x^4 y - y^2 \right]_{-x^2}^{x^2} dx$$

$$= \int_{-1}^{1} [(x^6 - x^4) - (-x^6 - x^4)]\, dx$$

$$= \int_{-1}^{1} 2x^6\, dx = \left[\tfrac{2}{7} x^7 \right]_{-1}^{1} = \tfrac{4}{7}. \quad \square$$

Example 2 Evaluate

$$\iint_\Omega (xy - y^3)\, dx\, dy$$

with Ω as in Figure 16.4.5.

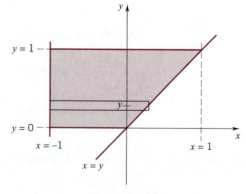

Figure 16.4.5

SOLUTION By projecting Ω onto the y-axis we obtain the interval $[0, 1]$. The region Ω consists of all points (x, y) with

$$0 \le y \le 1 \quad \text{and} \quad -1 \le x \le y.$$

This is a region of type II.   By (16.4.2)

$$\iint_{\Omega} (xy - y^3) \, dx \, dy = \int_0^1 \left(\int_{-1}^y (xy - y^3) \, dx \right) dy$$

$$= \int_0^1 \left[\tfrac{1}{2}x^2 y - xy^3 \right]_{-1}^y dy$$

$$= \int_0^1 [(\tfrac{1}{2}y^3 - y^4) - (\tfrac{1}{2}y + y^3)] \, dy$$

$$= \int_0^1 (-\tfrac{1}{2}y^3 - y^4 - \tfrac{1}{2}y) \, dy$$

$$= \left[-\tfrac{1}{8}y^4 - \tfrac{1}{5}y^5 - \tfrac{1}{4}y^2 \right]_0^1 = -\tfrac{23}{40}.$$

We can also project Ω onto the x-axis and express Ω as a region of type I, but then the lower boundary is defined piecewise (see figure) and the calculations are somewhat more complicated: setting

$$\phi(x) = \begin{cases} 0, & -1 \le x \le 0 \\ x, & 0 \le x \le 1, \end{cases}$$

we have Ω as the set of all points (x, y) with

$$-1 \le x \le 1 \quad \text{and} \quad \phi(x) \le y \le 1;$$

thus

$$\iint_{\Omega} (xy - y^3) \, dx \, dy = \int_{-1}^1 \left(\int_{\phi(x)}^1 (xy - y^3) \, dy \right) dx$$

$$= \int_{-1}^0 \left(\int_{\phi(x)}^1 (xy - y^3) \, dy \right) dx$$

$$+ \int_0^1 \left(\int_{\phi(x)}^1 (xy - y^3) \, dy \right) dx$$

$$= \int_{-1}^0 \left(\int_0^1 (xy - y^3) \, dy \right) dx + \int_0^1 \left(\int_x^1 (xy - y^3) \, dy \right) dx$$

as you can check ⟶ $\stackrel{.}{=} (-\tfrac{1}{2}) + (-\tfrac{3}{40}) = -\tfrac{23}{40}.$ ❑

Repeated integrals

$$\int_a^b \left(\int_{\phi_1(x)}^{\phi_2(x)} f(x, y) \, dy \right) dx \quad \text{and} \quad \int_c^d \left(\int_{\psi_1(y)}^{\psi_2(y)} f(x, y) \, dx \right) dy$$

can be written in more compact form by omitting the large parentheses. From now on we will simply write

$$\int_a^b \int_{\phi_1(x)}^{\phi_2(x)} f(x, y) \, dy \, dx \quad \text{and} \quad \int_c^d \int_{\psi_1(y)}^{\psi_2(y)} f(x, y) \, dx \, dy.$$

Sometimes a region can be expressed either as a type I region: $a \le x \le b$, $\phi_1(x) \le y \le \phi_2(x)$, or as a type II region: $c \le y \le d$, $\psi_1(y) \le x \le \psi_2(y)$. Since

$$\iint_{\Omega} f(x, y)\, dx\, dy = \int_a^b \int_{\phi_1(x)}^{\phi_2(x)} f(x, y)\, dy\, dx$$

and

$$\iint_{\Omega} f(x, y)\, dx\, dy = \int_c^d \int_{\psi_1(y)}^{\psi_2(y)} f(x, y)\, dx\, dy,$$

it follows that

$$\int_a^b \int_{\phi_1(x)}^{\phi_2(x)} f(x, y)\, dy\, dx = \int_c^d \int_{\psi_1(y)}^{\psi_2(y)} f(x, y)\, dx\, dy.$$

Therefore we can, at least in theory, perform the integration in either order. However, there are instances where one order is preferable over the other. This is illustrated at the end of the section.

Example 3 Evaluate

$$\iint_{\Omega} (x^{1/2} - y^2)\, dx\, dy$$

with Ω as in Figure 16.4.6.

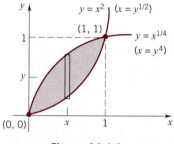

Figure 16.4.6

SOLUTION The projection of Ω onto the x-axis is the closed interval $[0, 1]$, and Ω can be characterized as the set of all (x, y) with

$$0 \le x \le 1 \quad \text{and} \quad x^2 \le y \le x^{1/4}.$$

Thus

$$\iint_{\Omega} (x^{1/2} - y^2)\, dx\, dy = \int_0^1 \int_{x^2}^{x^{1/4}} (x^{1/2} - y^2)\, dy\, dx$$

$$= \int_0^1 \left[x^{1/2} y - \tfrac{1}{3} y^3 \right]_{x^2}^{x^{1/4}} dx$$

$$= \int_0^1 (\tfrac{2}{3} x^{3/4} - x^{5/2} + \tfrac{1}{3} x^6)\, dx$$

$$= \left[\tfrac{8}{21} x^{7/4} - \tfrac{2}{7} x^{7/2} + \tfrac{1}{21} x^7 \right]_0^1 = \tfrac{8}{21} - \tfrac{2}{7} + \tfrac{1}{21} = \tfrac{1}{7}.$$

We can also integrate in the other order. The projection of Ω onto the y-axis is the closed interval $[0, 1]$, and Ω can be characterized as the set of all (x, y) with

$$0 \le y \le 1 \quad \text{and} \quad y^4 \le x \le y^{1/2}.$$

This gives the same result:

$$\iint\limits_{\Omega} (x^{1/2} - y^2)\, dx\, dy = \int_0^1 \int_{y^4}^{y^{1/2}} (x^{1/2} - y^2)\, dx\, dy$$

$$= \int_0^1 \left[\tfrac{2}{3}x^{3/2} - y^2 x \right]_{y^4}^{y^{1/2}} dy$$

$$= \int_0^1 (\tfrac{2}{3}y^{3/4} - y^{5/2} + \tfrac{1}{3}y^6)\, dy$$

$$= \left[\tfrac{8}{21}y^{7/4} - \tfrac{2}{7}y^{7/2} + \tfrac{1}{21}y^7 \right]_0^1 = \tfrac{8}{21} - \tfrac{2}{7} + \tfrac{1}{21} = \tfrac{1}{7}. \quad \square$$

Example 4 Calculate by double integration the area of the region Ω that lies between

$$\sqrt{x} + \sqrt{y} = \sqrt{a} \quad \text{and} \quad x + y = a, \qquad a > 0.$$

SOLUTION The region Ω is pictured in Figure 16.4.7. Its area is given by the double integral

$$\iint\limits_{\Omega} dx\, dy. \qquad\qquad \text{(by 16.3.3)}$$

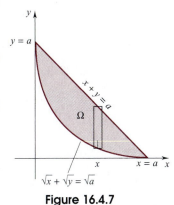

Figure 16.4.7

Here again we can integrate in either order. We can project Ω onto the x-axis and write the boundaries as functions of x:

$$y = (\sqrt{a} - \sqrt{x})^2 \quad \text{and} \quad y = a - x;$$

Ω is then the set of all (x, y) with

$$0 \le x \le a \quad \text{and} \quad (\sqrt{a} - \sqrt{x})^2 \le y \le a - x.$$

This gives

$$\iint\limits_{\Omega} dx\, dy = \int_0^a \int_{(\sqrt{a}-\sqrt{x})^2}^{a-x} dy\, dx$$

$$= \int_0^a [(a - x) - (\sqrt{a} - \sqrt{x})^2]\, dx$$

$$= \int_0^a (-2x + 2\sqrt{a}\sqrt{x})\, dx = \left[-x^2 + \tfrac{4}{3}\sqrt{a}\, x^{3/2} \right]_0^a = -a^2 + \tfrac{4}{3}a^2 = \tfrac{1}{3}a^2.$$

We can also project Ω onto the y-axis and write the boundaries as functions of y:

$$x = (\sqrt{a} - \sqrt{y})^2 \quad \text{and} \quad x = a - y;$$

Ω is then the set of all (x, y) with

$$0 \le y \le a \quad \text{and} \quad (\sqrt{a} - \sqrt{y})^2 \le x \le a - y.$$

This gives

$$\iint_\Omega dx\, dy = \int_0^a \int_{(\sqrt{a} - \sqrt{y})^2}^{a-y} dx\, dy,$$

which is also $\tfrac{1}{3}a^2$. ❏

Symmetry in Double Integration

First we go back to the one-variable case (see Section 5.7). Let's suppose that g is continuous on an interval that is symmetric about the origin, say $[-a, a]$.

If g is odd, then $\displaystyle\int_{-a}^a g(x)\, dx = 0.$

If g is even, then $\displaystyle\int_{-a}^a g(x)\, dx = 2\int_0^a g(x)\, dx.$

We have similar results for double integrals.

Suppose that Ω is symmetric about the y-axis. If f is odd in x [if $f(-x, y) = -f(x, y)$], then

$$\iint_\Omega f(x, y)\, dx\, dy = 0.$$

If f is even in x [if $f(-x, y) = f(x, y)$], then

$$\iint_\Omega f(x, y)\, dx\, dy = 2 \iint_{\substack{\text{right half} \\ \text{of } \Omega}} f(x, y)\, dx\, dy.$$

Suppose that Ω is symmetric about the x-axis. If f is odd in y [if $f(x, -y) = -f(x, y)$], then

$$\iint_\Omega f(x, y)\, dx\, dy = 0.$$

If f is even in y [if $f(x, -y) = f(x, y)$], then

$$\iint_\Omega f(x, y)\, dx\, dy = 2 \iint_{\substack{\text{upper half} \\ \text{of } \Omega}} f(x, y)\, dx\, dy.$$

As an example take the region Ω of Figure 16.4.8. Suppose we wanted to calculate

$$\iint_\Omega (2x - \sin x^2 y)\, dx\, dy.$$

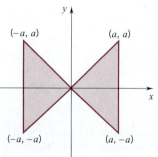

Figure 16.4.8

First,

$$\iint_\Omega (2x - \sin x^2 y) \, dx \, dy = \iint_\Omega 2x \, dx \, dy - \iint_\Omega \sin x^2 y \, dx \, dy.$$

Now, since $f(x, y) = 2x$ is odd in x, the symmetry of Ω about the y-axis gives

$$\iint_\Omega 2x \, dx \, dy = 0.$$

Since $g(x, y) = \sin x^2 y$ is odd in y, the symmetry of Ω about the x-axis gives

$$\iint_\Omega \sin x^2 y \, dx \, dy = 0.$$

Therefore

$$\iint_\Omega (2x - \sin x^2 y) \, dx \, dy = 0. \quad \square$$

As another example we go to Example 1 and reevaluate

$$\iint_\Omega (x^4 - 2y) \, dx \, dy,$$

this time capitalizing on the symmetry of Ω. Note that

$$\underset{\substack{\downarrow \\ \text{symmetry about } x\text{-axis}}}{\iint_\Omega 2y \, dx \, dy} = 0 \quad \text{and} \quad \underset{\substack{\downarrow \\ \text{symmetry about } y\text{-axis}}}{\iint_\Omega x^4 \, dx \, dy} = 2 \underset{\substack{\text{right half} \\ \text{of } \Omega}}{\iint x^4 \, dx \, dy} = \underset{\substack{\downarrow \\ \text{symmetry about } x\text{-axis}}}{4} \underset{\substack{\text{upper part} \\ \text{of right half} \\ \text{of } \Omega}}{\iint x^4 \, dx \, dy}.$$

Therefore

$$\iint_\Omega (x^4 - 2y) \, dx \, dy = 4 \int_0^1 \int_0^{x^2} x^4 \, dy \, dx = 4 \int_0^1 x^6 \, dx = \tfrac{4}{7}. \quad \square$$

Example 5 Calculate the volume within the cylinder $x^2 + y^2 = b^2$ between the planes $y + z = a$ and $z = 0$ given that $a \geq b > 0$.

SOLUTION See Figure 16.4.9. The solid in question is bounded below by the disc

$$\Omega: \quad 0 \leq x^2 + y^2 \leq b^2$$

and above by the plane

$$z = a - y.$$

The volume is given by the double integral

$$\iint_\Omega (a - y) \, dx \, dy.$$

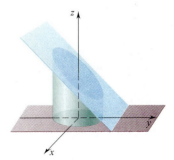

Figure 16.4.9

Since Ω is symmetric about the x-axis,

$$\iint_\Omega y \, dx \, dy = 0.$$

Thus

$$\iint_\Omega (a - y) \, dx \, dy = \iint_\Omega a \, dx \, dy = a \iint_\Omega dx \, dy = a \, (\text{area of } \Omega) = \pi ab^2. \quad \square$$

Concluding Remarks

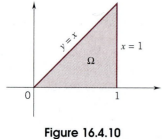

Figure 16.4.10

When two orders of integration are possible, one order may be easy to carry out, while the other may present serious difficulties. Take as an example the double integral

$$\iint_\Omega \cos \tfrac{1}{2}\pi x^2 \, dx \, dy$$

with Ω as in Figure 16.4.10. Projection onto the x-axis leads to

$$\int_0^1 \int_0^x \cos \tfrac{1}{2}\pi x^2 \, dy \, dx.$$

Projection onto the y-axis leads to

$$\int_0^1 \int_y^1 \cos \tfrac{1}{2}\pi x^2 \, dx \, dy.$$

The first double integral is easy to calculate:

$$\int_0^1 \int_0^x \cos \tfrac{1}{2}\pi x^2 \, dy \, dx = \int_0^1 \left(\int_0^x \cos \tfrac{1}{2}\pi x^2 \, dy \right) dx$$

$$= \int_0^1 \left[y \cos \tfrac{1}{2}\pi x^2 \right]_0^x dx = \int_0^1 x \cos \tfrac{1}{2}\pi x^2 \, dx$$

$$= \left[\tfrac{1}{\pi} \sin \tfrac{1}{2}\pi x^2 \right]_0^1 = \tfrac{1}{\pi},$$

but the second is not:

$$\int_0^1 \int_y^1 \cos \tfrac{1}{2}\pi x^2 \, dx \, dy = \int_0^1 \left(\int_y^1 \cos \tfrac{1}{2}\pi x^2 \, dx \right) dy,$$

and $\cos \tfrac{1}{2}\pi x^2$ does not have an elementary antiderivative. One possible way to handle the integral

$$\int_y^1 \cos \tfrac{1}{2}\pi x^2 \, dx$$

is to expand $\cos \tfrac{1}{2}\pi x^2$ as a series in x and then integrate term by term.

Finally, if Ω, the region of integration, is neither of type I nor of type II, it is often possible to break it up into a finite number of regions $\Omega_1, \ldots, \Omega_n$, each of type I or type II. (See Figure 16.4.11.) Since the double integral is additive,

$$\iint_{\Omega_1} f(x, y)\, dx\, dy + \cdots + \iint_{\Omega_n} f(x, y)\, dx\, dy = \iint_{\Omega} f(x, y)\, dx\, dy.$$

Each of the integrals on the left can be evaluated by the methods of this section.

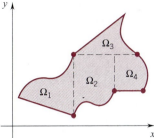

Figure 16.4.11

EXERCISES 16.4

In Exercises 1–3, evaluate taking Ω: $0 \le x \le 1$, $0 \le y \le 3$.

1. $\displaystyle\iint_{\Omega} x^2\, dx\, dy.$ **2.** $\displaystyle\iint_{\Omega} e^{x+y}\, dx\, dy.$

3. $\displaystyle\iint_{\Omega} xy^2\, dx\, dy.$

In Exercises 4–6, evaluate taking Ω: $0 \le x \le 1$, $0 \le y \le x$.

4. $\displaystyle\iint_{\Omega} x^3 y\, dx\, dy.$ **5.** $\displaystyle\iint_{\Omega} xy^3\, dx\, dy.$

6. $\displaystyle\iint_{\Omega} x^2 y^2\, dx\, dy.$

In Exercises 7–9, evaluate taking Ω: $0 \le x \le \frac{1}{2}\pi$, $0 \le y \le \frac{1}{2}\pi$.

7. $\displaystyle\iint_{\Omega} \sin(x + y)\, dx\, dy.$ **8.** $\displaystyle\iint_{\Omega} \cos(x + y)\, dx\, dy.$

9. $\displaystyle\iint_{\Omega} (1 + xy)\, dx\, dy.$

In Exercises 10–18, evaluate the double integral.

10. $\displaystyle\iint_{\Omega} (x + 3y^3)\, dx\, dy,$ Ω: $0 \le x^2 + y^2 \le 1.$

11. $\displaystyle\iint_{\Omega} \sqrt{xy}\, dx\, dy,$ Ω: $0 \le y \le 1$, $y^2 \le x \le y.$

12. $\displaystyle\iint_{\Omega} ye^x\, dx\, dy,$ Ω: $0 \le y \le 1$, $0 \le x \le y^2.$

13. $\displaystyle\iint_{\Omega} (4 - y^2)\, dx\, dy,$ Ω the bounded region between $y^2 = 2x$ and $y^2 = 8 - 2x.$

14. $\displaystyle\iint_{\Omega} (x^4 + y^2)\, dx\, dy,$ Ω the bounded region between $y = x^3$ and $y = x^2.$

15. $\displaystyle\iint_{\Omega} (3xy^2 - y)\, dx\, dy,$ Ω the region between $y = |x|$ and $y = -|x|$, $x \in [-1, 1].$

16. $\displaystyle\iint_{\Omega} e^{-y^2/2}\, dx\, dy,$ Ω the triangle formed by the y-axis, $2y = x$, $y = 1.$

17. $\displaystyle\iint_{\Omega} e^{x^2}\, dx\, dy,$ Ω the triangle formed by the x-axis, $2y = x$, $x = 2.$

18. $\displaystyle\iint_{\Omega} (x + y)\, dx\, dy,$ Ω the region between $y = x^3$ and $y = x^4$, $x \in [-1, 1].$

In Exercises 19–24, sketch the region Ω that gives rise to the repeated integral and change the order of integration.

19. $\displaystyle\int_0^1 \int_{x^4}^{x^2} f(x, y)\, dy\, dx.$ **20.** $\displaystyle\int_0^1 \int_0^{y^2} f(x, y)\, dx\, dy.$

21. $\displaystyle\int_0^1 \int_{-y}^{y} f(x, y)\, dx\, dy.$ **22.** $\displaystyle\int_{1/2}^1 \int_{x^3}^{x} f(x, y)\, dy\, dx.$

23. $\displaystyle\int_1^4 \int_x^{2x} f(x, y)\, dy\, dx.$ **24.** $\displaystyle\int_1^3 \int_{-x}^{x^2} f(x, y)\, dy\, dx.$

In Exercises 25–28, calculate by double integration the area of the bounded region determined by the given pairs of curves.

25. $x^2 = 4y$, $2y - x - 4 = 0.$

26. $y = x$, $x = 4y - y^2.$ **27.** $y = x$, $4y^3 = x^2.$

28. $x + y = 5$, $xy = 6.$

In Exercises 29–32, sketch the region Ω that gives rise to the repeated integral, change the order of integration, and then evaluate.

29. $\displaystyle\int_0^1 \int_{\sqrt{x}}^1 \sin\left(\frac{y^3 + 1}{2}\right) dy\, dx.$

30. $\displaystyle\int_{-1}^0 \int_{-\sqrt{y+1}}^{\sqrt{y+1}} x^2\, dx\, dy.$ **31.** $\displaystyle\int_1^2 \int_0^{\ln y} e^{-x}\, dx\, dy.$

32. $\displaystyle\int_0^1 \int_{x^2}^1 \frac{x^3}{\sqrt{x^4 + y^2}}\, dy\, dx.$

33. Find the area of the first quadrant region bounded by $xy = 2$, $y = 1$, $y = x + 1$.

34. Find the volume of the solid bounded above by $z = x + y$ and below by the triangle with vertices $(0, 0, 0)$, $(0, 1, 0)$, $(1, 0, 0)$.

35. Find the volume of the solid bounded by $\frac{1}{2}x + \frac{1}{3}y + \frac{1}{4}z = 1$ and the coordinate planes.

36. Find the volume of the solid bounded above by the plane $z = 2x + 3y$ and below by the unit square $0 \le x \le 1, 0 \le y \le 1$.

37. Find the volume of the solid bounded above by $z = x^3 y$ and below by the triangle with vertices $(0, 0, 0)$, $(2, 0, 0)$, $(0, 1, 0)$.

38. Find the volume under the paraboloid $z = x^2 + y^2$ within the cylinder $x^2 + y^2 \le 1$.

39. Find the volume of the solid bounded above by the plane $z = 2x + 1$ and below by the disc $(x - 1)^2 + y^2 \le 1$.

40. Find the volume of the solid bounded above by $z = 4 - y^2 - \frac{1}{4}x^2$ and below by the disc $(y - 1)^2 + x^2 \le 1$.

41. Find the volume of the solid in the first octant bounded by $z = x^2 + y^2$, the plane $x + y = 1$ and the coordinate planes.

42. Find the volume of the solid bounded by the circular cylinder $x^2 + y^2 = 1$, the plane $z = 0$ and the plane $x + z = 1$.

43. Find the volume of the solid in the first octant bounded above by $z = x^2 + 3y^2$, below by the xy-plane, and on the sides by the cylinder $y = x^2$ and the plane $y = x$.

44. Find the volume of the solid bounded above by the surface $z = 1 + xy$ and below by the triangle with vertices $(1, 1)$, $(4, 1)$, and $(3, 2)$.

45. Find the volume of the solid in the first octant bounded by the two cylinders $x^2 + y^2 = a^2$, $x^2 + z^2 = a^2$.

46. Find the volume of the tetrahedron bounded by the coordinate planes and the plane

$$\frac{x}{a} + \frac{y}{b} + \frac{z}{c} = 1, \qquad a, b, c > 0.$$

In Exercises 47–50, evaluate the double integral.

47. $\displaystyle\int_0^1 \int_y^1 e^{y/x}\, dx\, dy.$ **48.** $\displaystyle\int_0^1 \int_0^{\cos^{-1} y} e^{\sin x}\, dx\, dy.$

49. $\displaystyle\int_0^1 \int_x^1 x^2\, e^{y^4}\, dy\, dx.$ **50.** $\displaystyle\int_0^1 \int_x^1 e^{y^2}\, dy\, dx.$

In Exercises 51–54, calculate the average value of f over the region Ω.

51. $f(x, y) = x^2 y;$ $\Omega:\ -1 \le x \le 1, \quad 0 \le y \le 4.$

52. $f(x, y) = xy;$ $\Omega:\ 0 \le x \le 1, \quad 0 \le y \le \sqrt{1 - x^2}.$

53. $f(x, y) = \dfrac{1}{xy};$ $\Omega:\ \ln 2 \le x \le 2 \ln 2,$
$\ln 2 \le y \le 2 \ln 2.$

54. $f(x, y) = e^{x+y};$ $\Omega:\ 0 \le x \le 1,$
$x - 1 \le y \le x + 1.$

55. (*Separated variables over a rectangle*) Let R be the rectangle $a \le x \le b, c \le y \le d$. Show that, if f is continuous on $[a, b]$ and g is continuous on $[c, d]$, then

(16.4.3)
$$\iint_R f(x)g(y)\, dx\, dy$$
$$= \left[\int_a^b f(x)\, dx\right] \cdot \left[\int_c^d g(y)\, dy\right].$$

56. Let R be a rectangle symmetric about the x-axis, sides parallel to the coordinate axes. Show that, if f is odd with respect to y, then the double integral of f over R is 0.

57. Let R be a rectangle symmetric about the y-axis, sides parallel to the coordinate axes. Show that, if f is odd with respect to x, then the double integral of f over R is 0.

58. Given that $f(-x, -y) = -f(x, y)$ for all (x, y) in Ω, what form of symmetry in Ω will ensure that the double integral of f over Ω is zero?

59. Let Ω be the triangle with vertices $(0, 0)$, $(0, 1)$, $(1, 1)$. Show that

$$\text{if}\ \int_0^1 f(x)\, dx = 0 \quad \text{then} \quad \iint_\Omega f(x)f(y)\, dx\, dy = 0.$$

60. (*Differentiation under the integral sign*) If f and $\partial f/\partial x$ are continuous, then the function

$$H(t) = \int_a^b \frac{\partial f}{\partial x}(t, y)\, dy$$

can be shown to be continuous. Use the identity

$$\int_0^x \int_a^b \frac{\partial f}{\partial x}(t, y) \, dy \, dt = \int_a^b \int_0^x \frac{\partial f}{\partial x}(t, y) \, dt \, dy$$

to verify that

$$\frac{d}{dx} \left[\int_a^b f(x, y) \, dy \right] = \int_a^b \frac{\partial f}{\partial x}(x, y) \, dy.$$

61. We integrate over regions of type I by means of the formula

$$\iint_\Omega f(x, y) \, dx \, dy = \int_a^b \int_{\phi_1(x)}^{\phi_2(x)} f(x, y) \, dy \, dx.$$

Here f is assumed to be continuous on Ω and ϕ_1, ϕ_2 are assumed to be continuous on $[a, b]$. Show that the function

$$F(x) = \int_{\phi_1(x)}^{\phi_2(x)} f(x, y) \, dy$$

is continuous on $[a, b]$.

■ 16.5 THE DOUBLE INTEGRAL AS THE LIMIT OF RIEMANN SUMS; POLAR COORDINATES

The Double Integral as the Limit of Riemann Sums

In the one-variable case we can write the integral as the limit of Riemann sums:

$$\int_a^b f(x) \, dx = \lim_{\max \Delta x_i \to 0} \sum_{i=1}^n f(x_i^*) \, \Delta x_i.$$

The same approach works with double integrals. To explain it we need to explain what we mean by the *diameter of a set*.

Suppose that S is a bounded closed set (on the line, in the plane, or in three-space). For any two points P and Q of S we can measure their separation, $d(P, Q)$ (the distance between P and Q). The maximum separation between points of S is called the *diameter of S*:

$$\text{diam } S = \max_{P, Q \in S} d(P, Q).$$

NOTE: The distance function $d(P, Q)$ is continuous on the bounded closed set S, and so it has an absolute maximum value.

For a circle, a circular disc, a sphere, or a ball, this sense of diameter agrees with the usual one. Figure 16.5.1 gives some other examples.

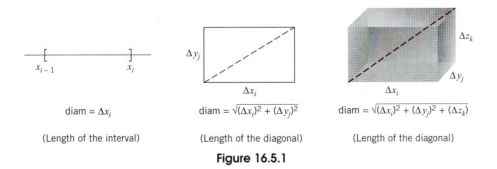

diam $= \Delta x_i$

(Length of the interval)

diam $= \sqrt{(\Delta x_i)^2 + (\Delta y_j)^2}$

(Length of the diagonal)

diam $= \sqrt{(\Delta x_i)^2 + (\Delta y_j)^2 + (\Delta z_k)^2}$

(Length of the diagonal)

Figure 16.5.1

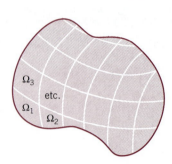

Figure 16.5.2

Now let's start with a basic region Ω and decompose it into a finite number of basic subregions $\Omega_1, \ldots, \Omega_N$. (See Figure 16.5.2.) If f is continuous on Ω, then f is continuous on each Ω_i. Now from each Ω_i we pick an arbitrary point (x_i^*, y_i^*) and form the *Riemann sum*

$$\sum_{i=1}^{N} f(x_i^*, y_i^*)(\text{area of } \Omega_i).$$

As you would expect, the double integral over Ω can be obtained as the limit of such sums; namely, given any $\epsilon > 0$, there exists $\delta > 0$ such that, if the diameters of the Ω_i are all less than δ, then

$$\left| \sum_{i=1}^{N} f(x_i^*, y_i^*)(\text{area of } \Omega_i) - \iint_{\Omega} f(x, y)\, dx\, dy \right| < \epsilon$$

no matter how the (x_i^*, y_i^*) are chosen within the Ω_i. We express this by writing

(16.5.1)
$$\iint_{\Omega} f(x, y)\, dx\, dy = \lim_{\text{diam } \Omega_i \to 0} \sum_{i=1}^{N} f(x_i^*, y_i^*)(\text{area of } \Omega_i).$$

Evaluating Double Integrals Using Polar Coordinates

Here we explain how to calculate double integrals

$$\iint_{\Omega} f(x, y)\, dx\, dy$$

using polar coordinates (r, θ). Throughout we take $r \geq 0$.

We will work with the type of region shown in Figure 16.5.3. The region Ω is then the set of all points (x, y) that have polar coordinates (r, θ) in the set

$$\Gamma: \quad \alpha \leq \theta \leq \beta, \quad \rho_1(\theta) \leq r \leq \rho_2(\theta),$$

where $\beta \leq \alpha + 2\pi$.

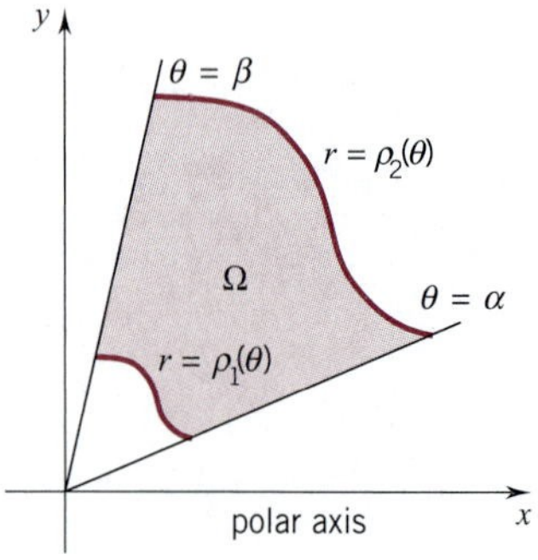

You already know how to calculate the area of Ω. By (9.6.2)

$$\text{area of } \Omega = \int_{\alpha}^{\beta} \tfrac{1}{2}[\rho_2^2(\theta) - \rho_1^2(\theta)]\, d\theta.$$

We can write this as a double integral over Γ:

(16.5.2)
$$\text{area of } \Omega = \iint_{\Gamma} r\, dr\, d\theta.$$

Figure 16.5.3

PROOF Simply note that

$$\tfrac{1}{2}[\rho_2^2(\theta) - \rho_1^2(\theta)] = \int_{\rho_1(\theta)}^{\rho_2(\theta)} r\, dr$$

and therefore

$$\text{area of } \Omega = \int_{\alpha}^{\beta} \int_{\rho_1(\theta)}^{\rho_2(\theta)} r \, dr \, d\theta = \iint_{\Gamma} r \, dr \, d\theta. \quad \square$$

Now let's suppose that f is some function continuous at each point (x, y) of Ω. Then the composition

$$F(r, \theta) = f(r \cos \theta, r \sin \theta)$$

is continuous at each point (r, θ) of Γ. We will show that

(16.5.3)
$$\iint_{\Omega} f(x, y) \, dx \, dy = \iint_{\Gamma} f(r \cos \theta, r \sin \theta) r \, dr \, d\theta. \qquad \text{(note the extra } r\text{)}$$

PROOF Our first step is to place a grid on Ω by using a finite number of rays $\theta = \theta_j$ and a finite number of continuous curves $r = \rho_k(\theta)$ in the manner of Figure 16.5.4. This grid decomposes Ω into a finite number of regions

$$\Omega_1, \ldots \Omega_N$$

with polar coordinates in sets $\Gamma_1, \ldots, \Gamma_N$. Note that by (16.5.2)

$$\text{area of each } \Omega_i = \iint_{\Gamma_i} r \, dr \, d\theta.$$

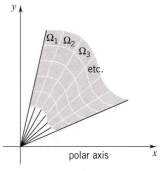

Figure 16.5.4

Writing $F(r, \theta)$ for $f(r \cos \theta, r \sin \theta)$, we have

$$\iint_{\Gamma} F(r, \theta) r \, dr \, d\theta = \sum_{i=1}^{N} \iint_{\Gamma_i} F(r, \theta) r \, dr \, d\theta$$

additivity ————————

for some $(r_i^*, \theta_i^*) \in \Gamma_i$ ————————
(Theorem 16.3.5)

$$= \sum_{i=1}^{N} F(r_i^*, \theta_i^*) \iint_{\Gamma_i} r \, dr \, d\theta$$

$$= \sum_{i=1}^{N} F(r_i^*, \theta_i^*)(\text{area of } \Omega_i)$$

(with $x_i^* = r_i^* \cos \theta_i^*, y_i^* = r_i^* \sin \theta_i^*$) ————————

$$= \sum_{i=1}^{N} f(x_i^*, y_i^*)(\text{area of } \Omega_i).$$

This last expression is a Riemann sum for the double integral

$$\iint_{\Omega} f(x, y) \, dx \, dy$$

and, as such, by (16.5.1), differs from that integral by less than any preassigned positive ϵ provided only that the diameters of all the Ω_i are sufficiently small. This we can guarantee by making our grid sufficiently fine. $\square$

Example 1 Use polar coordinates to evaluate

$$\iint_{\Omega} xy \, dx \, dy,$$

where Ω is the portion of the unit disc that lies in the first quadrant.

SOLUTION Here

$$\Gamma: \quad 0 \le \theta \le \tfrac{1}{2}\pi, \quad 0 \le r \le 1.$$

Therefore

$$\iint_{\Omega} xy \, dx \, dy = \iint_{\Gamma} (r \cos \theta)(r \sin \theta) r \, dr \, d\theta$$

$$= \int_0^{\pi/2} \int_0^1 r^3 \cos \theta \sin \theta \, dr \, d\theta = \tfrac{1}{8}. \quad \square$$

check this ⟶

Example 2 Use polar coordinates to calculate the volume of a sphere of radius R.

SOLUTION In rectangular coordinates

$$V = 2 \iint_{\Omega} \sqrt{R^2 - (x^2 + y^2)} \, dx \, dy$$

where Ω is the disc of radius R centered at the origin. (Verify.) In polar coordinates, the disc of radius R centered at the origin is given by

$$\Gamma: \quad 0 \le \theta \le 2\pi, \quad 0 \le r \le R.$$

Therefore

$$V = 2 \iint_{\Gamma} \sqrt{R^2 - r^2} \, r \, dr \, d\theta = 2 \int_0^{2\pi} \int_0^R \sqrt{R^2 - r^2} \, r \, dr \, d\theta = \tfrac{4}{3}\pi R^3. \quad \square$$

check this ⟶

Example 3 Calculate the volume of the solid bounded above by the cone $z = 2 - \sqrt{x^2 + y^2}$ and bounded below by the disc $\Omega: (x - 1)^2 + y^2 \le 1$. (See Figure 16.5.5.)

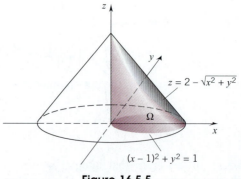

$z = 2 - \sqrt{x^2 + y^2}$

Ω

$(x - 1)^2 + y^2 = 1$

Figure 16.5.5

SOLUTION

$$V = \iint_\Omega (2 - \sqrt{x^2 + y^2}) \, dx \, dy$$

$$= 2 \iint_\Omega dx \, dy - \iint_\Omega \sqrt{x^2 + y^2} \, dx \, dy.$$

The first integral is $2\times$ (area of Ω) $= 2\pi$. We evaluate the second integral by changing to polar coordinates.

The equation $(x - 1)^2 + y^2 = 1$ simplifies to $x^2 + y^2 = 2x$. In polar coordinates this becomes $r^2 = 2r \cos \theta$, which simplifies to $r = 2 \cos \theta$. The disc Ω is the set of all points with polar coordinates in the set

$$\Gamma: \quad -\tfrac{1}{2}\pi \le \theta \le \tfrac{1}{2}\pi, \quad 0 \le r \le 2 \cos \theta.$$

Therefore

$$\iint_\Omega \sqrt{x^2 + y^2} \, dx \, dy = \iint_\Gamma r^2 \, dr \, d\theta = \int_{-\pi/2}^{\pi/2} \int_0^{2\cos\theta} r^2 \, dr \, d\theta = \tfrac{32}{9}.$$

check this ———↑

We then have $V = 2\pi - \tfrac{32}{9} \cong 2.73.$ ❏

Example 4 Evaluate

$$\iint_\Omega \frac{1}{(1 + x^2 + y^2)^{3/2}} \, dx \, dy$$

where Ω is the triangle of Figure 16.5.6.

SOLUTION The vertical side of the triangle is part of the line $x = 1$. In polar coordinates this is $r \cos \theta = 1$, which can be written $r = \sec \theta$. Therefore

$$\iint_\Omega \frac{1}{(1 + x^2 + y^2)^{3/2}} \, dx \, dy = \iint_\Gamma \frac{r}{(1 + r^2)^{3/2}} \, dr \, d\theta$$

where

$$\Gamma: \quad 0 \le \theta \le \pi/4, \quad 0 \le r \le \sec \theta. \qquad \text{(Figure 16.5.7)}$$

The double integral over Γ reduces to

$$\int_0^{\pi/4} \int_0^{\sec\theta} \frac{r}{(1 + r^2)^{3/2}} \, dr \, d\theta = \int_0^{\pi/4} \left[\frac{1}{\sqrt{1 + r^2}} \right]_{\sec\theta}^0 d\theta$$

$$= \int_0^{\pi/4} \left(1 - \frac{1}{\sqrt{1 + \sec^2\theta}} \right) d\theta.$$

For $\theta \in [0, \pi/4]$

$$\frac{1}{\sqrt{1 + \sec^2\theta}} = \frac{\cos \theta}{\sqrt{\cos^2\theta + 1}} = \frac{\cos \theta}{\sqrt{2 - \sin^2\theta}}.$$

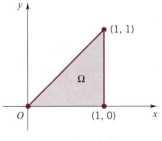

Figure 16.5.6

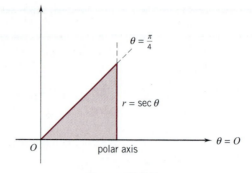

Figure 16.5.7

Therefore the integral can be written

$$\int_0^{\pi/4} \left(1 - \frac{\cos\theta}{\sqrt{2 - \sin^2\theta}} \right) d\theta = \left[\theta - \sin^{-1}\left(\frac{\sin\theta}{\sqrt{2}} \right) \right]_0^{\pi/4} = \frac{\pi}{4} - \frac{\pi}{6} = \frac{\pi}{12}. \quad \square$$

(7.8.4) ————↑

The function $f(x) = e^{-x^2}$ has no elementary antiderivative. Nevertheless, by taking a circuitous route and then using polar coordinates, we can show that

(16.5.4)
$$\int_{-\infty}^{\infty} e^{-x^2}\, dx = \sqrt{\pi}. \quad †$$

PROOF The circular disc D_b: $x^2 + y^2 \leq b^2$ is the set of all (x, y) with polar coordinates (r, θ) in the set Γ: $0 \leq \theta \leq 2\pi$, $0 \leq r \leq b$. Therefore

$$\iint\limits_{D_b} e^{-(x^2+y^2)}\, dx\, dy = \iint\limits_{\Gamma} e^{-r^2}\, r\, dr\, d\theta = \int_0^{2\pi}\int_0^b e^{-r^2} r\, dr\, d\theta$$

$$= \int_0^{2\pi} \tfrac{1}{2}(1 - e^{-b^2})\, d\theta = \pi(1 - e^{-b^2}).$$

Let S_a be the square $-a \leq x \leq a$, $-a \leq y \leq a$. Since $D_a \subseteq S_a \subseteq D_{2a}$ and $e^{-(x^2+y^2)}$ is positive,

$$\iint\limits_{D_a} e^{-(x^2+y^2)}\, dx\, dy \leq \iint\limits_{S_a} e^{-(x^2+y^2)}\, dx\, dy \leq \iint\limits_{D_{2a}} e^{-(x^2+y^2)}\, dx\, dy.$$

It follows that

$$\pi(1 - e^{-a^2}) \leq \iint\limits_{S_a} e^{-(x^2+y^2)}\, dx\, dy \leq \pi(1 - e^{-4a^2}).$$

---

† This integral comes up frequently in probability theory and plays an important role in the branch of physics called "statistical mechanics."

As $a \to \infty$, $\pi(1 - e^{-a^2}) \to \pi$ and $\pi(1 - e^{-4a^2}) \to \pi$. Therefore

$$\lim_{a \to \infty} \iint_{S_a} e^{-(x^2 + y^2)} \, dx \, dy = \pi.$$

But

$$\iint_{S_a} e^{-(x^2 + y^2)} \, dx \, dy = \int_{-a}^{a} \int_{-a}^{a} e^{-(x^2 + y^2)} \, dx \, dy$$

$$= \int_{-a}^{a} \int_{-a}^{a} e^{-x^2} \cdot e^{-y^2} \, dx \, dy$$

$$= \left(\int_{-a}^{a} e^{-x^2} \, dx \right) \left(\int_{-a}^{a} e^{-y^2} \, dy \right) = \left(\int_{-a}^{a} e^{-x^2} \, dx \right)^2.$$

Therefore

$$\lim_{a \to \infty} \int_{-a}^{a} e^{-x^2} \, dx = \lim_{a \to \infty} \left(\iint_{S_a} e^{-(x^2 + y^2)} \, dx \, dy \right)^{1/2} = \sqrt{\pi}. \quad \square$$

EXERCISES 16.5

In Exercises 1–4, evaluate the iterated integral.

1. $\displaystyle\int_0^{\pi/2} \int_0^{\sin \theta} r \cos \theta \, dr \, d\theta.$

2. $\displaystyle\int_0^{\pi/4} \int_0^{\cos 2\theta} r \, dr \, d\theta.$

3. $\displaystyle\int_0^{\pi/2} \int_0^{3 \sin \theta} r^2 \, dr \, d\theta.$

4. $\displaystyle\int_{-\pi/3}^{2\pi/3} \int_0^{2 \cos \theta} r \sin \theta \, dr \, d\theta.$

5. Integrate $f(x, y) = \cos(x^2 + y^2)$ over:
(a) the closed unit disc;
(b) the annular region $1 \le x^2 + y^2 \le 4$.

6. Integrate $f(x, y) = \sin(\sqrt{x^2 + y^2})$ over:
(a) the closed unit disc;
(b) the annular region $1 \le x^2 + y^2 \le 4$.

7. Integrate $f(x, y) = x + y$ over:
(a) $0 \le x^2 + y^2 \le 1, x \ge 0, y \ge 0$;
(b) $1 \le x^2 + y^2 \le 4, x \ge 0, y \ge 0$.

8. Integrate $f(x, y) = \sqrt{x^2 + y^2}$ over the triangle with vertices $(0, 0)$, $(1, 0)$, $(1, \sqrt{3})$.

In Exercises 9–16, calculate by changing to polar coordinates.

9. $\displaystyle\int_{-1}^{1} \int_0^{\sqrt{1-y^2}} \sqrt{x^2 + y^2} \, dx \, dy.$

10. $\displaystyle\int_0^{2} \int_0^{\sqrt{4-x^2}} \sqrt{x^2 + y^2} \, dy \, dx.$

11. $\displaystyle\int_{1/2}^{1} \int_0^{\sqrt{1-x^2}} (x^2 + y^2)^{3/2} \, dy \, dx.$

12. $\displaystyle\int_0^{1/2} \int_0^{\sqrt{1-x^2}} xy\sqrt{x^2 + y^2} \, dy \, dx.$

13. $\displaystyle\int_0^{1} \int_0^{\sqrt{1-x^2}} \sin(x^2 + y^2) \, dy \, dx.$

14. $\displaystyle\int_{-1}^{1} \int_{-\sqrt{1-y^2}}^{\sqrt{1-y^2}} e^{-(x^2 + y^2)} \, dx \, dy.$

15. $\displaystyle\int_0^{2} \int_0^{\sqrt{2x-x^2}} x \, dy \, dx.$

16. $\displaystyle\int_0^{1} \int_{-\sqrt{x-x^2}}^{\sqrt{x-x^2}} (x^2 + y^2) \, dy \, dx.$

In Exercises 17–22, use double integrals to find the area of the given region.

17. One leaf of the petal curve $r = 3 \sin 3\theta$.

18. The region enclosed by the cardioid $r = 2(1 - \cos \theta)$.

19. The region inside the circle $r = 4 \cos \theta$ and outside the circle $r = 2$.

20. The region inside the large loop and outside the small loop of the limaçon $r = 1 + 2 \cos \theta$.

21. The region enclosed by the lemniscate $r^2 = 4 \cos 2\theta$.

22. The region inside the circle $r = 3 \cos \theta$ and outside the cardioid $r = 1 + \cos \theta$.

23. Find the volume of the solid bounded above by the plane $z = y + b$, below by the xy-plane, and on the sides by the circular cylinder $x^2 + y^2 = b^2$.

24. Find the volume of the solid bounded below by the xy-plane and above by the paraboloid $z = 1 - (x^2 + y^2)$.

25. Find the volume of the ellipsoid $x^2/4 + y^2/4 + z^2/3 = 1$.

26. Find the volume of the solid bounded below by the xy-plane and above by the surface $x^2 + y^2 + z^6 = 5$.

27. Find the volume of the solid bounded below by the xy-plane, above by the spherical surface $x^2 + y^2 + z^2 = 4$, and on the sides by the cylinder $x^2 + y^2 = 1$.

28. Find the volume of the solid bounded above by $z = 1 - (x^2 + y^2)$, below by the xy-plane, and on the sides by the cylinder $x^2 + y^2 - x = 0$.

29. Find the volume of the solid bounded above by the plane $z = 2x$ and below by the disc $(x - 1)^2 + y^2 \leq 1$.

30. Find the volume of the solid bounded above by the cone $z^2 = x^2 + y^2$ and below by the region Ω which lies inside the curve $x^2 + y^2 = 2ax$.

31. Find the volume of the solid bounded below by the xy-plane, above by the ellipsoid of revolution $b^2x^2 + b^2y^2 + a^2z^2 = a^2b^2$, and on the sides by the cylinder $x^2 + y^2 - ay = 0$.

32. A cylindrical hole of radius r is drilled through the center of a sphere of radius R, $0 < r < R$.
    (a) Determine the volume of the material that has been removed from the sphere.
    (b) Determine the volume of the ring-shaped solid that remains.

■ 16.6 SOME APPLICATIONS OF DOUBLE INTEGRATION

The Mass of a Plate

Suppose that a thin distribution of matter, called a *plate*, is laid out in the xy-plane in the form of a basic region Ω. If the mass density of the plate (the mass per unit area) is a constant λ, then the total mass M of the plate is simply the density λ times the area of the plate:

$$M = \lambda \times \text{the area of } \Omega.$$

If the density varies continuously from point to point, say $\lambda = \lambda(x, y)$, then the mass of the plate is the average density of the plate times the area of the plate:

$$M = \text{average density} \times \text{the area of } \Omega.$$

This is an integral:

(16.6.1)
$$M = \iint\limits_{\Omega} \lambda(x, y) \, dx \, dy.$$

The Center of Mass of a Plate

The center of mass of a rod x_M is a density-weighted average taken over the interval occupied by the rod:

$$x_M M = \int_a^b x\lambda(x) \, dx. \qquad \text{[This you have seen: (5.9.5).]}$$

The center of mass of a plate (x_M, y_M) is determined by two density-weighted averages, each taken over the region occupied by the plate:

(16.6.2)
$$x_M M = \iint_\Omega x\lambda(x, y)\, dx\, dy, \qquad y_M M = \iint_\Omega y\lambda(x, y)\, dx\, dy.$$

Example 1 A plate is in the form of a half-disc of radius a. Find the mass of the plate and the center of mass given that the mass density of the plate is directly proportional to the distance from the midpoint of the straight edge of the plate.

SOLUTION Place the plate over the region Ω: $-a \le x \le a$, $0 \le y \le \sqrt{a^2 - x^2}$. See Figure 16.6.1. The mass density can then be written $\lambda(x, y) = k\sqrt{x^2 + y^2}$, where $k > 0$ is the constant of proportionality. Now

$$M = \iint_\Omega k\sqrt{x^2 + y^2}\, dx\, dy = \int_0^\pi \int_0^a (kr)r\, dr\, d\theta = k\left(\int_0^\pi 1\, d\theta\right)\left(\int_0^a r^2\, dr\right)$$

change to polar coordinates

$$= k(\pi)(\tfrac{1}{3}a^3) = \tfrac{1}{3}ka^3\pi,$$

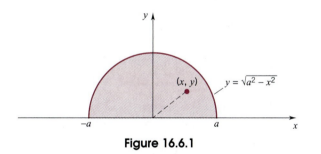

Figure 16.6.1

and

$$x_M M = \iint_\Omega x(k\sqrt{x^2 + y^2})\, dx\, dy = 0$$

(Ω is symmetric with respect to the y-axis and the integrand is odd with respect to x). Thus $x_M = 0$. (This result is also clear from the symmetry of the plate.) Also,

$$y_M M = \iint_\Omega y(k\sqrt{x^2 + y^2})\, dx\, dy = \int_0^\pi \int_0^a (r\sin\theta)(kr)r\, dr\, d\theta$$

$$= k\left(\int_0^\pi \sin\theta\, d\theta\right)\left(\int_0^a r^3\, dr\right)$$

$$= k(2)(\tfrac{1}{4}a^4) = \tfrac{1}{2}ka^4.$$

Since $M = \tfrac{1}{3}ka^3\pi$, we have $y_M = (\tfrac{1}{2}ka^4)/(\tfrac{1}{3}ka^3\pi) = 3a/2\pi$ Thus, the center of mass of the plate is the point $(0, 3a/2\pi) \cong (0, 0.48a)$. ❑

Centroids

If the plate is homogeneous, then the mass density λ is constantly M/A where A is the area of the base region Ω. In this case the center of mass of the plate falls on the *centroid* of the base region (a notion with which you are already familiar). The centroid $(\bar{x}, \bar{y})$ depends only on the geometry of Ω:

$$\bar{x}M = \iint\limits_{\Omega} x(M/A)\,dx\,dy = (M/A)\iint\limits_{\Omega} x\,dx\,dy,$$

$$\bar{y}M = \iint\limits_{\Omega} y(M/A)\,dx\,dy = (M/A)\iint\limits_{\Omega} y\,dx\,dy.$$

Dividing by M and multiplying through by A we have

(16.6.3)
$$\bar{x}A = \iint\limits_{\Omega} x\,dx\,dy, \qquad \bar{y}A = \iint\limits_{\Omega} y\,dx\,dy.$$

Thus $\bar{x}$ is the average x-coordinate on Ω and $\bar{y}$ is the average y-coordinate on Ω. The mass of the plate does not enter into this at all.

Example 2 Find the centroid of the region

$$\Omega: \quad a \leq x \leq b, \quad \phi_1(x) \leq y \leq \phi_2(x). \qquad \text{(Figure 16.6.2.)}$$

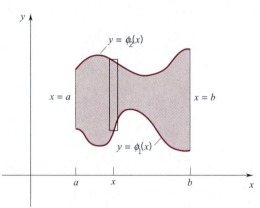

Figure 16.6.2

SOLUTION

$$\bar{x}A = \iint\limits_{\Omega} x\,dx\,dy = \int_a^b \int_{\phi_1(x)}^{\phi_2(x)} x\,dy\,dx = \int_a^b x[\phi_2(x) - \phi_1(x)]\,dx;$$

$$\bar{y}A = \iint\limits_{\Omega} y\,dx\,dy = \int_a^b \int_{\phi_1(x)}^{\phi_2(x)} y\,dy\,dx = \int_a^b \tfrac{1}{2}([\phi_2(x)]^2 - [\phi_1(x)]^2)\,dx.$$

These are the formulas for the centroid that we developed in Section 6.4. Having calculated many centroids at that time, we won't do so here. ❏

Kinetic Energy and Moment of Inertia

A particle of mass m at a distance r from a given line rotates about that line (called the *axis of rotation*) with angular speed ω. The speed v of the particle is then $r\omega$ and the kinetic energy is given by the formula

$$KE = \tfrac{1}{2}mv^2 = \tfrac{1}{2}mr^2\omega^2.$$

Imagine now a rigid body composed of a finite number of point masses m_i located at distances r_i from some fixed line. If the rigid body rotates about that line with angular speed ω, then all the point masses rotate about that same line with that same angular speed ω. The kinetic energy of the body can be obtained by adding up the kinetic energies of all the individual particles:

$$KE = \sum_i \tfrac{1}{2}m_i r_i^2 \omega^2 = \tfrac{1}{2}\left(\sum_i m_i r_i^2\right)\omega^2.$$

The expression in parentheses is called the *moment of inertia* (or *rotational inertia*) of the body and is denoted by the letter I:

(16.6.4)
$$I = \sum_i m_i r_i^2.$$

For a rigid body in straight-line motion

$$KE = \tfrac{1}{2}Mv^2, \qquad \text{where } v \text{ is the speed of the body.}$$

For a rigid body in rotational motion

$$KE = \tfrac{1}{2}I\omega^2, \qquad \text{where } \omega \text{ is the angular speed of the body.}$$

The Moment of Inertia of a Plate

Suppose that a plate in the shape of a basic region Ω rotates about a line. The moment of inertia of the plate about that axis of rotation is given by the formula

(16.6.5)
$$I = \iint_\Omega \lambda(x, y)[r(x, y)]^2 \, dx \, dy$$

where $\lambda = \lambda(x, y)$ is the mass density function and $r(x, y)$ is the distance from the axis to the point (x, y).

Derivation of (16.6.5) Decompose the plate into N pieces in the form of basic regions $\Omega_1, \ldots, \Omega_N$. From each Ω_i choose a point (x_i^*, y_i^*) and view all the mass of the ith piece as concentrated there. The moment of inertia of this piece is then approximately

$$\underbrace{[\lambda(x_i^*, y_i^*)(\text{area of } \Omega_i)]}_{\text{approx. mass of piece}} \underbrace{[r(x_i^*, y_i^*)]^2}_{(\text{approx. distance})^2} = \lambda(x_i^*, y_i^*)[r(x_i^*, y_i^*)]^2 (\text{area of } \Omega_i).$$

The sum of these approximations,

$$\sum_{i=1}^{N} \lambda(x_i^*, y_i^*)[r(x_i^*, y_i^*)]^2 (\text{area of } \Omega_i),$$

is a Riemann sum for the double integral

$$\iint_{\Omega} \lambda(x, y)[r(x, y)]^2 \, dx \, dy.$$

As the maximum diameter of the Ω_i tends to zero, the Riemann sum tends to this integral. ❏

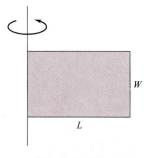

Example 3 A rectangular plate of mass M, length L, and width W rotates about the line shown in Figure 16.6.3. Find the moment of inertia of the plate about that line: **(a)** given that the plate has uniform mass density; **(b)** given that the mass density of the plate is directly proportional to the square of the distance from the rightmost side.

Figure 16.6.3

SOLUTION Coordinatize the plate as in Figure 16.6.4 and call the base region R.
(a) Here $\lambda(x, y) = M/LW$ and $r(x, y) = x$. Thus

$$I = \iint_{R} \frac{M}{LW} x^2 \, dx \, dy = \frac{M}{LW} \int_0^W \int_0^L x^2 \, dx \, dy$$

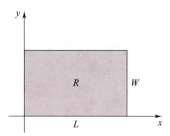

Figure 16.6.4

$$= \frac{M}{LW} W \int_0^L x^2 \, dx = \frac{M}{L} (\tfrac{1}{3} L^3) = \tfrac{1}{3} ML^2.$$

(b) In this case $\lambda(x, y) = k(L - x)^2$, but we still have $r(x, y) = x$. Therefore

$$I = \iint_{R} k(L - x)^2 x^2 \, dx \, dy = k \int_0^W \int_0^L (L - x)^2 x^2 \, dx \, dy$$

$$= kW \int_0^L (L^2 x^2 - 2Lx^3 + x^4) \, dx = \tfrac{1}{30} kL^5 W.$$

We can eliminate the constant of proportionality k by noting that

$$M = \iint_{R} k(L - x)^2 \, dx \, dy = k \int_0^W \int_0^L (L - x)^2 \, dx \, dy$$

$$= kW \left[-\tfrac{1}{3} (L - x)^3 \right]_0^L = \tfrac{1}{3} kWL^3.$$

Therefore,

$$k = \frac{3M}{WL^3} \quad \text{and} \quad I = \tfrac{1}{30} \left(\frac{3M}{WL^3} \right) L^5 W = \tfrac{1}{10} ML^2. \quad ❏$$

Radius of Gyration

If the mass M of an object is all concentrated at a distance r from a given line, then the moment of inertia about that line is given by the product Mr^2.

Suppose now that we have a plate of mass M (actually any object of mass M will do here), and suppose that l is some line. The object has some moment of inertia I about l. Its *radius of gyration* about l is the distance K for which

$$I = MK^2.$$

Namely, the radius of gyration about l is the distance from l at which all the mass of the object would have to be concentrated to effect the same moment of inertia. The formula for radius of gyration is usually written

(16.6.6)
$$K = \sqrt{I/M}.$$

Example 4 A homogeneous circular plate of mass M and radius R rotates about an axle that passes through the center of the plate and is perpendicular to the plate. Calculate the moment of inertia and the radius of gyration.

SOLUTION Take the axle as the z-axis and let the plate rest on the circular region Ω: $x^2 + y^2 \le R^2$ (Figure 16.6.5). The density of the plate is $M/A = M/\pi R^2$ and $r(x, y) = \sqrt{x^2 + y^2}$. Hence

$$I = \iint_{\Omega} \frac{M}{\pi R^2}(x^2 + y^2)\, dx\, dy = \frac{M}{\pi R^2} \int_0^{2\pi} \int_0^R r^3\, dr\, d\theta = \tfrac{1}{2}MR^2.$$

The radius of gyration K is $\sqrt{I/M} = R/\sqrt{2}$.

The circular plate of radius R has the same moment of inertia about the central axle as a circular wire of the same mass with radius $R/\sqrt{2}$. The circular wire is a more efficient carrier of moment of inertia than the circular plate. ❑

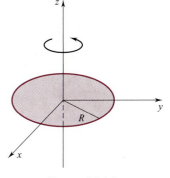

Figure 16.6.5

The Parallel Axis Theorem

Suppose we have an object of mass M and a line l_M that passes through the center of mass of the object. The object has some moment of inertia about that line; call it I_M. If l is any line parallel to l_M, then the object has a certain moment of inertia about l; call that I. The parallel axis theorem states that

(16.6.7)
$$I = I_M + d^2 M$$

where d is the distance between the axes.

We prove the theorem under somewhat restrictive assumptions. Assume that the object is a plate of mass M in the shape of a basic region Ω, and assume that l_M is perpendicular to the plate. Call l the z-axis. Call the plane of the plate the xy-plane.

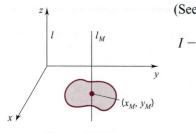

Figure 16.6.6

(See Figure 16.6.6.) Denoting the points of Ω by (x, y) we have

$$I - I_M = \iint_\Omega \lambda(x, y)(x^2 + y^2)\, dx\, dy - \iint_\Omega \lambda(x, y)[(x - x_M)^2 + (y - y_M)^2]\, dx\, dy$$

$$= \iint_\Omega \lambda(x, y)[2x_M x + 2y_M y - (x_M^2 + y_M^2)]\, dx\, dy$$

$$= 2x_M \iint_\Omega x\lambda(x, y)\, dx\, dy + 2y_M \iint_\Omega y\lambda(x, y)\, dx\, dy$$

$$-(x_M^2 + y_M^2) \iint_\Omega \lambda(x, y)\, dx\, dy$$

$$= 2x_M^2 M + 2y_M^2 M - (x_M^2 + y_M^2)M = (x_M^2 + y_M^2)M = d^2 M. \quad \square$$

An obvious consequence of the parallel axis theorem is that $I_M \le I$ for all lines l parallel to l_M. To minimize the moment of inertia we must pass our axis of rotation through the center of mass.

EXERCISES 16.6

In Exercises 1–10, find the mass and center of mass of the plate that occupies the given region Ω and has the given density function λ.

1. Ω: $-1 \le x \le 1$, $0 \le y \le 1$, $\lambda(x, y) = x^2$.

2. Ω: $0 \le x \le 1$, $0 \le y \le \sqrt{x}$, $\lambda(x, y) = x + y$.

3. Ω: $0 \le x \le 1$, $x^2 \le y \le 1$, $\lambda(x, y) = xy$.

4. Ω: $0 \le x \le \pi$, $0 \le y \le \sin x$, $\lambda(x, y) = y$.

5. Ω: $0 \le x \le 8$, $0 \le y \le \sqrt[3]{x}$, $\lambda(x, y) = y^2$.

6. Ω: $0 \le x \le a$, $0 \le y \le \sqrt{a^2 - x^2}$, $\lambda(x, y) = xy$.

7. Ω is the triangle with vertices $(0, 0)$, $(1, 2)$, and $(1, 3)$; $\lambda(x, y) = xy$.

8. Ω is the triangular region in the first quadrant bounded by the lines $x = 0$, $y = 0$, and $3x + 2y = 6$; $\lambda(x, y) = x + y$.

9. Ω is the region bounded by the cardioid $r = 1 + \cos\theta$; λ is the distance to the pole.

10. Ω is the region inside the circle $r = 2\sin\theta$ and outside the circle $r = 1$; $\lambda(x, y) = y$.

In the exercises that follow, I_x, I_y, and I_z denote the moments of inertia about the x, y, and z axes.

11. A rectangular plate of mass M, length L, and width W is placed on the xy-plane with center at the origin, long sides parallel to the x-axis. (We assume here that $L \ge W$.) Find I_x, I_y, I_z if the plate is homogeneous. Determine the corresponding radii of gyration K_x, K_y, K_z.

12. Verify that I_x, I_y, I_z are unchanged if the mass density of

the plate of Exercise 11 varies directly as the distance from the leftmost side.

13. Determine the center of mass of the plate of Exercise 11 if the mass density varies as in Exercise 12.

14. Show that for any plate in the xy-plane

$$I_z = I_x + I_y.$$

How are the corresponding radii of gyration K_x, K_y, K_z related?

15. A homogeneous plate of mass M in the form of a quarter disc of radius R is placed in the xy-plane as in the figure. Find I_x, I_y, I_z and the corresponding radii of gyration.

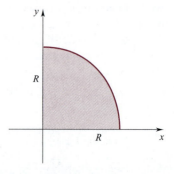

16. A plate in the xy-plane undergoes a rotation in that plane about its center of mass. Show that I_z remains unchanged.

17. A homogeneous disc of mass M and radius R is to be placed on the xy-plane so that it has moment of inertia I_0 about the z-axis. Where should the disc be placed?

18. A homogeneous plate of mass density λ occupies the region under the curve $y = f(x)$ from $x = a$ to $x = b$. Show that

$$I_x = \tfrac{1}{3}\lambda \int_a^b [f(x)]^3 \, dx \qquad \text{and} \qquad I_y = \lambda \int_a^b x^2 f(x) \, dx.$$

19. A homogeneous plate of mass M in the form of an elliptical quadrant is placed on the xy-plane. (See the figure below.) Find I_x, I_y, I_z.

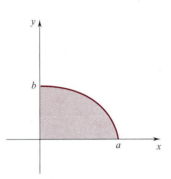

20. Find I_x, I_y, I_z for the plate in Exercise 2.
21. Find I_x, I_y, I_z for the plate in Exercise 3.
22. Find I_x, I_y, I_z for the plate in Exercise 5.
23. Find I_x, I_y, I_z for the plate in Exercise 9.
24. A plate of varying density occupies the region $\Omega = \Omega_1 \cup \Omega_2$ shown in the figure below. Find the center of mass of the plate given that the Ω_1 piece has mass M_1 and center of mass (x_1, y_1), and the Ω_2 piece has mass M_2 and center of mass (x_2, y_2).

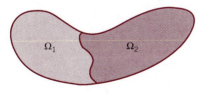

25. A homogeneous plate of mass M is in the form of a

ring. (See the figure below.) Calculate the moment of inertia of the plate:
(a) about a diameter;
(b) about a tangent to the inner circle;
(c) about a tangent to the outer circle.

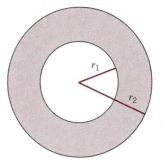

26. Find the moment of inertia of a homogeneous circular wire of mass M and radius r:
(a) about a diameter;
(b) about a tangent. HINT: Use the previous exercise.

27. The plate of Exercise 25 rotates about the axis that is perpendicular to the plate and passes through the center. Find the moment of inertia.

28. Prove the parallel axis theorem for the case where the line through the center of mass lies in the plane of the plate.

29. A plate of mass M has the form of a half disc Ω: $-R \leq x \leq R, \; 0 \leq y \leq \sqrt{R^2 - x^2}$. Find the center of mass given that the mass density varies directly as the distance from the curved boundary.

30. Find I_x, I_y, I_z for the plate of Exercise 29.

31. A plate of mass M is in the form of a disc of radius R. Given that the mass density of the plate varies directly as the distance from a point P on the boundary of the plate, locate the center of mass.

32. A plate of mass M is in the form of a right triangle of base b and height h. Given that the mass density of the plate varies directly as the square of the distance from the vertex of the right angle, locate the center of mass of the plate.

33. Use double integrals to justify an assumption we made about centroids in Chapter 6: Formula (6.4.1).

■ 16.7 TRIPLE INTEGRALS

Now that you are familiar with double integrals

$$\iint_\Omega f(x, y) \, dx \, dy,$$

you will find it easy to understand triple integrals

$$\iiint_T f(x, y, z)\, dx\, dy\, dz.$$

Basically the only difference is this: instead of working with functions of two variables continuous on a plane region Ω, we will be working with functions of three variables continuous on some portion T of three-space.

The Triple Integral over a Box

For double integration we began with a rectangle

$$R: \quad a_1 \leq x \leq a_2, \quad b_1 \leq y \leq b_2.$$

For triple integration we begin with a *box* (a rectangular solid)

$$\Pi: \quad a_1 \leq x \leq a_2, \quad b_1 \leq y \leq b_2, \quad c_1 \leq z \leq c_2. \qquad \text{(Figure 16.7.1)}$$

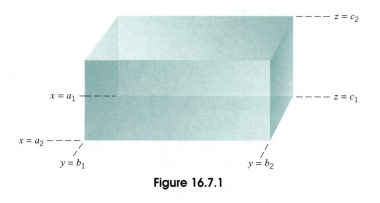

Figure 16.7.1

To partition this box, we first partition the edges.   Taking

$$P_1 = \{x_0, \ldots, x_m\} \quad \text{as a partition of } [a_1, a_2],$$
$$P_2 = \{y_0, \ldots, y_n\} \quad \text{as a partition of } [b_1, b_2],$$
$$P_3 = \{z_0, \ldots, z_q\} \quad \text{as a partition of } [c_1, c_2],$$

we form the set

$$P = P_1 \times P_2 \times P_3 = \{(x_i, y_j, z_k): x_i \in P_1, y_j \in P_2, z_k \in P_3\}\dagger$$

and call this a *partition of* Π.   The partition P breaks up Π into $m \times n \times q$ nonoverlapping boxes

$$\Pi_{ijk}: \quad x_{i-1} \leq x \leq x_i, \quad y_{j-1} \leq y \leq y_j, \quad z_{k-1} \leq z \leq z_k.$$

$\dagger$ $P_1 \times P_2 \times P_3$ is the Cartesian product of P_1, P_2, and P_3.

A typical such box is pictured in Figure 16.7.2.

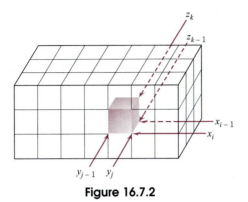

Figure 16.7.2

Let f be continuous on Π. Then, taking

$$M_{ijk} \text{ as the maximum value of } f \text{ on } \Pi_{ijk}$$

and

$$m_{ijk} \text{ as the minimum value of } f \text{ on } \Pi_{ijk},^\dagger$$

we form the *upper sum*

$$U_f(P) = \sum_{i=1}^{m} \sum_{j=1}^{n} \sum_{k=1}^{q} M_{ijk}(\text{volume of } \Pi_{ijk}) = \sum_{i=1}^{m} \sum_{j=1}^{n} \sum_{k=1}^{q} M_{ijk} \, \Delta x_i \, \Delta y_j \, \Delta z_k$$

and the *lower sum*

$$L_f(P) = \sum_{i=1}^{m} \sum_{j=1}^{n} \sum_{k=1}^{q} m_{ijk}(\text{volume of } \Pi_{ijk}) = \sum_{i=1}^{m} \sum_{j=1}^{n} \sum_{k=1}^{q} m_{ijk} \, \Delta x_i \, \Delta y_j \, \Delta z_k.$$

As in the case of functions of one and two variables, it turns out that, with f continuous on Π, there is one and only one number I that satisfies the inequality

$$L_f(P) \le I \le U_f(P) \qquad \text{for all partitions } P \text{ of } \Pi.$$

DEFINITION 16.7.1 THE TRIPLE INTEGRAL OVER A BOX Π

Let f be continuous on the closed box Π. The unique number I that satisfies the inequality

$$L_f(P) \le I \le U_f(P) \qquad \text{for all partitions } P \text{ of } \Pi$$

is called the *triple integral* of f over Π and is denoted by

$$\iiint_{\Pi} f(x, y, z) \, dx \, dy \, dz.$$

---

† A continuous function on a closed bounded set has an absolute maximum and an absolute minimum.

The Triple Integral over a More General Solid

We start with a three-dimensional, bounded, closed, connected set T. We assume that T is a *basic solid*; that is, we assume that the boundary of T consists of a finite number of continuous surfaces $z = \alpha(x, y), y = \beta(x, z), x = \gamma(y, z)$. See, for example, Figure 16.7.3.

Now let's suppose that f is some function continuous on T. To define the triple integral of f over T we first encase T in a rectangular box Π. (Figure 16.7.4.) We then extend f to all of Π by defining f to be zero outside of T. This extended function f is bounded on Π, and it is continuous on all of Π except possibly at the boundary of T. In spite of these possible discontinuities, f is still integrable over Π; that is, there still exists a unique number I such that

$$L_f(P) \leq I \leq U_f(P) \qquad \text{for all partitions } P \text{ of } \Pi.$$

Figure 16.7.3

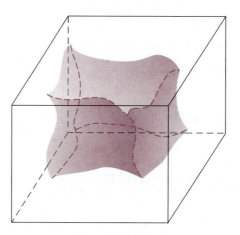

Figure 16.7.4

(We will not attempt to prove this.) The number I is by definition the triple integral

$$\iiint_{\Pi} f(x, y, z)\, dx\, dy\, dz.$$

We define the triple integral over T by setting

(16.7.2)
$$\iiint_{T} f(x, y, z)\, dx\, dy\, dz = \iiint_{\Pi} f(x, y, z)\, dx\, dy\, dz.$$

Volume as a Triple Integral

The simplest triple integral of interest is the triple integral of the function that is constantly one on T. This gives the volume of T:

(16.7.3)
$$\text{volume of } T = \iiint_{T} dx\, dy\, dz.$$

PROOF   Set $f(x, y, z) = 1$ for all (x, y, z) in T.   Encase T in a box Π.   Define f to be zero outside of T.   An arbitrary partition P of Π breaks up T into little boxes Π_{ijk}. Note that

$L_f(P) = $ the sum of the volumes of all the Π_{ijk} that are contained in T

$U_f(P) = $ the sum of the volumes of all the Π_{ijk} that intersect T.

It follows that

$$L_f(P) \le \text{the volume of } T \le U_f(P).$$

The arbitrariness of P gives the formula.   ❑

Some Properties of the Triple Integral

Below we give without proof the salient elementary properties of the triple integral. They are all analogous to what you saw in the one- and two-variable cases.   Assume throughout that T is a basic solid.   The functions f and g are assumed to be continuous on T.

I. Linearity:

$$\iiint_T [\alpha f(x, y, z) + \beta g(x, y, z)] \, dx \, dy \, dz =$$

$$\alpha \iiint_T f(x, y, z) \, dx \, dy \, dz + \beta \iiint_T g(x, y, z) \, dx \, dy \, dz.$$

II. Order:

$$\text{if } f \ge 0 \text{ on } T, \text{ then } \iiint_T f(x, y, z) \, dx \, dy \, dz \ge 0;$$

$$\text{if } f \le g \text{ on } T, \text{ then } \iiint_T f(x, y, z) \, dx \, dy \, dz \le \iiint_T g(x, y, z) \, dx \, dy \, dz.$$

III. Additivity: if T is broken up into a finite number of basic solids $T_1, \ldots, T_n$, then

$$\iiint_T f(x, y, z) \, dx \, dy \, dz = \iiint_{T_1} f(x, y, z) \, dx \, dy \, dz +$$

$$\cdots + \iiint_{T_n} f(x, y, z) \, dx \, dy \, dz.$$

IV. Mean-value condition: there is a point (x_0, y_0, z_0) in T for which

$$\iiint_T f(x, y, z) \, dx \, dy \, dz = f(x_0, y_0, z_0) \cdot (\text{volume of } T).$$

We call $f(x_0, y_0, z_0)$ *the average value of f on T.*

The notion of average given above enables us to write

(16.7.4)

$$\iiint_T f(x, y, z)\, dx\, dy\, dz = \left(\begin{array}{c}\text{the average value}\\ \text{of } f \text{ on } T\end{array}\right) \cdot (\text{volume of } T).$$

We can also take weighted averages: if f and g are continuous and g is nonnegative on T, then there is a point (x_0, y_0, z_0) in T for which

(16.7.5)

$$\iiint_T f(x, y, z)g(x, y, z)\, dx\, dy\, dz = f(x_0, y_0, z_0) \iiint_T g(x, y, z)\, dx\, dy\, dz.$$

As you would expect, we call $f(x_0, y_0, z_0)$ *the g-weighted average of f on T.*

The formulas for mass, center of mass, and moments of inertia derived in the previous section for two-dimensional plates are easily extended to three-dimensional objects.

Suppose that T is an object in the form of a basic solid. If T has constant mass density λ (here density is mass per unit volume), then the mass of T is the density λ times the volume of T:

$$M = \lambda V.$$

If the mass density varies continuously over T, say $\lambda = \lambda(x, y, z)$, then *the mass of T is the average density of T times the volume of T.* Thus, by (16.7.4) the mass of T is given by the triple integral

(16.7.6)

$$M = \iiint_T \lambda(x, y, z)\, dx\, dy\, dz.$$

The coordinates of the center of mass (x_M, y_M, z_M) are density-weighted averages:

(16.7.7)

$$x_M M = \iiint_T x\lambda(x, y, z)\, dx\, dy\, dz, \qquad \text{etc.}$$

If the object T is homogeneous (constant mass density M/V), then the center of mass of T depends only on the geometry of T and falls on the centroid $(\bar{x}, \bar{y}, \bar{z})$ of the space occupied by T. The density is irrelevant. The coordinates of the centroid are simple averages over T:

(16.7.8)

$$\bar{x} V = \iiint_T x \, dx \, dy \, dz, \qquad \text{etc.}$$

The moment of inertia of T about a line is given by the formula

(16.7.9)

$$I = \iiint_T \lambda(x, y, z)[r(x, y, z)]^2 \, dx \, dy \, dz.$$

Here $\lambda(x, y, z)$ is the mass density of T at (x, y, z) and $r(x, y, z)$ is the distance of (x, y, z) from the line in question. The moments of inertia about the x, y, z axes are again denoted by I_x, I_y, I_z.

All of this should be readily understandable. Techniques for evaluating triple integrals are considered in the next three sections.

EXERCISES 16.7

1. Let $f = f(x, y)$ be a function continuous and nonnegative on a basic region Ω and set

$$T = \{(x, y, z): \ (x, y) \in \Omega, \ 0 \le z \le f(x, y)\}.$$

Compare

$$\iiint_T dx \, dy \, dz \qquad \text{to} \qquad \iint_\Omega f(x, y) \, dx \, dy.$$

2. Set $f(x, y, z) = xyz$ on $\Pi: 0 \le x \le 1, 0 \le y \le 1, 0 \le z \le 1$ and take P as the partition $P_1 \times P_2 \times P_3$.
   (a) Find $L_f(P)$ and $U_f(P)$ given that

$$P_1 = \{x_0, \ \ldots \ , x_m\}, \qquad P_2 = \{y_0, \ \ldots \ , y_n\},$$
$$P_3 = \{z_0, \ \ldots \ , z_q\}$$

   are all arbitrary partitions of $[0, 1]$.
   (b) Use your answer to (a) to calculate

$$\iiint_\Pi xyz \, dx \, dy \, dz.$$

3. Let $f(x, y, z) = \alpha$, constant, over the rectangular solid Π: $a_1 \le x \le a_2, b_1 \le y \le b_2, c_1 \le z \le c_2$. Show that

$$\iiint_\Pi \alpha \, dx \, dy \, dz = \alpha(a_2 - a_1)(b_2 - b_1)(c_2 - c_1).$$

4. Find the average value of $f(x, y, z) = xyz$ over the region Π in Exercise 2.

5. Calculate

$$\iiint_\Pi xy \, dx \, dy \, dz \quad \text{where} \quad \Pi: 0 \le x \le a, \ 0 \le y \le b,$$
$$0 \le z \le c.$$

6. Let T be a basic solid of varying mass density $\lambda = \lambda(x, y, z)$. The moment of inertia of T about the xy-plane is defined by setting

$$I_{xy} = \iiint_T \lambda(x, y, z) z^2 \, dx \, dy \, dz.$$

The other plane moments of inertia, I_{xz} and I_{yz}, have comparable definitions. Express I_x, I_y, I_z in terms of the plane moments of inertia.

7. A box Π_1: $0 \le x \le 2a, \ 0 \le y \le 2b, \ 0 \le z \le 2c$ is cut away from a larger box Π_0: $0 \le x \le 2A, 0 \le y \le 2B$, $0 \le z \le 2C$. Locate the centroid of the remaining solid.

8. Show that, if f is continuous and nonnegative on a basic solid T, then the triple integral of f over T is nonnegative.

9. Calculate the mass M of the cube Π: $0 \leq x \leq a$, $0 \leq y \leq a$, $0 \leq z \leq a$ given that the mass density varies directly with the distance from the face on the xy-plane.

10. Locate the center of mass of the cube of Exercise 9.

11. Find the moment of inertia I_z of the cube of Exercise 9.

■ 16.8 REDUCTION TO REPEATED INTEGRALS

In this section we give no proofs. You can assume that all the solids that appear are basic solids and all the functions that you encounter are continuous.

In Figure 16.8.1 we have sketched a solid T. The projection of T onto the xy-plane has been labeled Ω_{xy}. The solid T is then the set of all (x, y, z) with

$$(x, y) \text{ in } \Omega_{xy} \qquad \text{and} \qquad \psi_1(x, y) \leq z \leq \psi_2(x, y).$$

The triple integral over T can be evaluated by setting

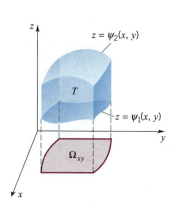

Figure 16.8.1

$$(*) \qquad \iiint_T f(x, y, z)\, dx\, dy\, dz = \iint_{\Omega_{xy}} \left(\int_{\psi_1(x,y)}^{\psi_2(x,y)} f(x, y, z)\, dz \right) dx\, dy.$$

Moving over to Figure 16.8.2 we see that in this case Ω_{xy} is the region

$$a_1 \leq x \leq a_2, \qquad \phi_1(x) \leq y \leq \phi_2(x),$$

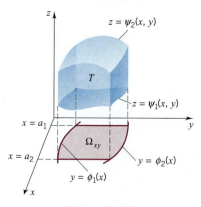

Figure 16.8.2

and T itself is the set of all (x, y, z) with

$$a_1 \leq x \leq a_2, \qquad \phi_1(x) \leq y \leq \phi_2(x), \qquad \psi_1(x, y) \leq z \leq \psi_2(x, y).$$

The triple integral over T can then be expressed by three ordinary integrals:

$$\iiint_T f(x, y, z)\, dx\, dy\, dz = \int_{a_1}^{a_2} \left[\int_{\phi_1(x)}^{\phi_2(x)} \left(\int_{\psi_1(x,y)}^{\psi_2(x,y)} f(x, y, z)\, dz \right) dy \right] dx.$$

It is customary to omit the brackets and parentheses and write

(16.8.1)
$$\iiint_T f(x, y, z)\, dx\, dy\, dz = \int_{a_1}^{a_2} \int_{\phi_1(x)}^{\phi_2(x)} \int_{\psi_1(x,y)}^{\psi_2(x,y)} f(x, y, z)\, dz\, dy\, dx. \qquad †$$

Here we first integrate with respect to z [from $z = \psi_1(x, y)$ to $z = \psi_2(x, y)$], then with respect to y [from $y = \phi_1(x)$ to $y = \phi_2(x)$], and finally with respect to x [from $x = a_1$ to $x = a_2$].

There is nothing special about this order of integration. Other orders of integration are possible and in some cases more convenient. Suppose, for example, that the projection of T onto the xz-plane is a region of the form

$$\Omega_{xz}: \quad a_1 \le z \le a_2, \quad \phi_1(z) \le x \le \phi_2(z).$$

If T is the set of all (x, y, z) with

$$a_1 \le z \le a_2, \quad \phi_1(z) \le x \le \phi_2(z), \quad \psi_1(x, z) \le y \le \psi_2(x, z)$$

then

$$\iiint_T f(x, y, z)\, dx\, dy\, dz = \int_{a_1}^{a_2} \int_{\phi_1(z)}^{\phi_2(z)} \int_{\psi_1(x,z)}^{\psi_2(x,z)} f(x, y, z)\, dy\, dx\, dz.$$

In this case we integrate first with respect to y, then with respect to x, and finally with respect to z. Still four other orders of integration are possible.

Example 1 Evaluate the triple integral

$$\int_0^2 \int_0^x \int_0^{4-x^2} xyz\, dz\, dy\, dx.$$

SOLUTION

$$\int_0^2 \int_0^x \int_0^{4-x^2} xyz\, dz\, dy\, dx = \int_0^2 \int_0^x \left(\int_0^{4-x^2} xyz\, dz \right) dy\, dx$$

$$= \int_0^2 \int_0^x \left(\left[\tfrac{1}{2} xyz^2 \right]_0^{4-x^2} \right) dy\, dx$$

$$= \tfrac{1}{2} \int_0^2 \int_0^x x(4 - x^2)^2 y\, dy\, dx$$

$$= \tfrac{1}{2} \int_0^2 \left(\int_0^x x(4 - x^2)^2 y\, dy \right) dx$$

$$= \tfrac{1}{2} \int_0^2 \left(\left[\tfrac{1}{2} x(4 - x^2)^2 y^2 \right]_0^x \right) dx$$

$$= \tfrac{1}{4} \int_0^2 x^3 (4 - x^2)^2\, dx = \tfrac{1}{4} \int_0^2 x^3 (16 - 8x^2 + x^4)\, dx$$

$$= \tfrac{1}{4} \left[4x^4 - \tfrac{8}{6} x^6 + \tfrac{1}{8} x^8 \right]_0^2 = \tfrac{8}{3}. \quad □$$

---

† This formula is formula (∗) taken one step further. Usually we skip the double-integral stage and go directly to three integrals.

Remark The solid determined by the limits of integration in Example 1 is the region T in the first octant bounded by the parabolic cylinder $z = 4 - x^2$, the plane $z = 0$, the plane $y = x$, and the plane $y = 0$. This solid is shown in Figure 16.8.3. ❑

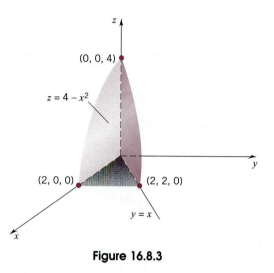

Figure 16.8.3

Example 2 Use triple integration to find the volume of the tetrahedron T shown in Figure 16.8.4, and find the coordinates of the centroid.

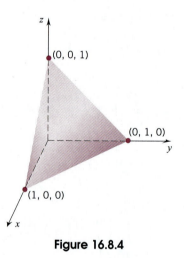

Figure 16.8.4

SOLUTION The volume of T is given by the triple integral

$$V = \iiint_T dx\, dy\, dz.$$

To evaluate this triple integral we can project T onto any one of the three coordinate planes. We will project onto the xy-plane. The base region is then the triangle

$$\Omega_{xy}: \quad 0 \le x \le 1, \quad 0 \le y \le 1 - x. \qquad \text{(Figure 16.8.5)}$$

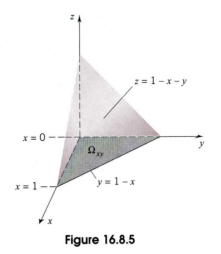

Figure 16.8.5

Since the inclined face is part of the plane $z = 1 - x - y$, we have T as the set of all (x, y, z) with

$$0 \le x \le 1, \quad 0 \le y \le 1 - x, \quad 0 \le z \le 1 - x - y.$$

It follows that

$$V = \iiint\limits_{T} dx \, dy \, dz = \int_0^1 \int_0^{1-x} \int_0^{1-x-y} dz \, dy \, dx$$

$$= \int_0^1 \int_0^{1-x} (1 - x - y) \, dy \, dx$$

$$= \int_0^1 \left[(1 - x)y - \tfrac{1}{2}y^2 \right]_0^{1-x} dx$$

$$= \int_0^1 \tfrac{1}{2}(1 - x)^2 \, dx = \left[-\tfrac{1}{6}(1 - x)^3 \right]_0^1 = \tfrac{1}{6}.$$

By symmetry $\bar{x} = \bar{y} = \bar{z}$. We can calculate $\bar{x}$ as follows:

$$\bar{x} V = \iiint\limits_{T} x \, dx \, dy \, dz = \int_0^1 \int_0^{1-x} \int_0^{1-x-y} x \, dz \, dy \, dx = \tfrac{1}{24}.$$

check this ⟶

Since $V = \tfrac{1}{6}$, we have $\bar{x} = \tfrac{1}{4}$. The centroid is the point $(\tfrac{1}{4}, \tfrac{1}{4}, \tfrac{1}{4})$. See Exercise 37 for a general result concerning the volume and centroid of a tetrahedron. ❑

Example 3 Find the mass of a solid right circular cylinder of radius r and height h given that the mass density is directly proportional to the distance from the lower base.

SOLUTION Call the solid T. In the setup of Figure 16.8.6 we can characterize T by the following inequalities:

$$-r \le x \le r, \quad -\sqrt{r^2 - x^2} \le y \le \sqrt{r^2 - x^2}, \quad 0 \le z \le h.$$

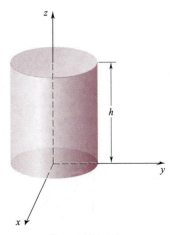

Figure 16.8.6

The first two inequalities define the base region Ω_{xy}. Since the density varies directly with the distance from the lower base, we have $\lambda(x, y, z) = kz$, where $k > 0$ is the constant of proportionality. Then

$$M = \iiint_T kz \, dx \, dy \, dz$$

$$= \int_{-r}^{r} \int_{-\sqrt{r^2 - x^2}}^{\sqrt{r^2 - x^2}} \int_0^h kz \, dz \, dy \, dx$$

$$= \int_{-r}^{r} \int_{-\sqrt{r^2 - x^2}}^{\sqrt{r^2 - x^2}} \tfrac{1}{2} kh^2 \, dy \, dx$$

$$= 4 \int_0^r \int_0^{\sqrt{r^2 - x^2}} \tfrac{1}{2} kh^2 \, dy \, dx \qquad \text{(using the symmetry)}$$

$$= 2kh^2 \int_0^r \sqrt{r^2 - x^2} \, dx.$$

This integral can be evaluated either by using a trigonometric substitution (Section 8.5), or by using Formula 87 in the Table of Integrals. Either way,

$$\int_0^r \sqrt{r^2 - x^2} \, dx = \tfrac{1}{4} \pi r^2.$$

Thus,

$$M = 2kh^2(\tfrac{1}{4}\pi r^2) = \tfrac{1}{2} kh^2 r^2 \pi. \quad \square$$

Remark In Example 3 we would have profited by not skipping the double integral stage; namely, we could have written

$$M = \iint_{\Omega_{xy}} \left(\int_0^h kz \, dz \right) dx \, dy = \iint_{\Omega_{xy}} \tfrac{1}{2} kh^2 \, dx \, dy$$

$$= \tfrac{1}{2} kh^2 (\text{area of } \Omega_{xy}) = \tfrac{1}{2} kh^2 r^2 \pi. \quad \square$$

Example 4 Integrate $f(x, y, z) = yz$ over that part of the first octant $x \geq 0$, $y \geq 0$, $z \geq 0$ that is cut off by the ellipsoid

$$\frac{x^2}{a^2} + \frac{y^2}{b^2} + \frac{z^2}{c^2} = 1.$$

SOLUTION Call the solid T. The upper boundary of T has equation

$$z = \psi(x, y) = \frac{c}{ab} \sqrt{a^2 b^2 - b^2 x^2 - a^2 y^2}.$$

This surface intersects the xy-plane in the curve

$$y = \phi(x) = \frac{b}{a} \sqrt{a^2 - x^2}.$$

We can take

$$\Omega_{xy}: \quad 0 \leq x \leq a, \quad 0 \leq y \leq \phi(x)$$

as the base region (see Figure 16.8.7) and characterize T as the set of all (x, y, z) with

$$0 \le x \le a, \quad 0 \le y \le \phi(x), \quad 0 \le z \le \psi(x, y).$$

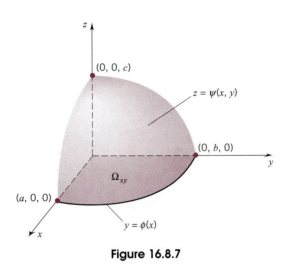

Figure 16.8.7

We can therefore calculate the triple integral by evaluating

$$\int_0^a \int_0^{\phi(x)} \int_0^{\psi(x,y)} yz \, dz \, dy \, dx.$$

A straightforward (but somewhat lengthy) computation that you can carry out yourself gives an answer of $\frac{1}{15} ab^2c^2$. ❏

ANOTHER SOLUTION We return to Example 4 but this time carry out the integration in a different order. In Figure 16.8.8 you can see the same solid projected this time onto the yz-plane. In terms of y and z the curved surface has equation

$$x = \Psi(y, z) = \frac{a}{bc}\sqrt{b^2c^2 - c^2y^2 - b^2z^2}.$$

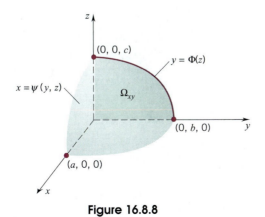

Figure 16.8.8

This surface intersects the yz-plane in the curve

$$y = \Phi(z) = \frac{b}{c}\sqrt{c^2 - z^2}.$$

We can take

$$\Omega_{yz}: \quad 0 \le z \le c, \quad 0 \le y \le \Phi(z)$$

as the base region and characterize T as the set of all (x, y, z) with

$$0 \le z \le c, \quad 0 \le y \le \Phi(z), \quad 0 \le x \le \Psi(y, z).$$

This leads to the repeated integral

$$\int_0^c \int_0^{\Phi(z)} \int_0^{\Psi(y,z)} yz \, dx \, dy \, dz$$

which, as you can check, also gives $\frac{1}{15} ab^2c^2$. ❑

Example 5 Use triple integration to find the volume of the solid T bounded above by the parabolic cylinder $z = 4 - y^2$ and bounded below by the elliptic paraboloid $z = x^2 + 3y^2$.

SOLUTION Solving the two equations simultaneously, we have

$$4 - y^2 = x^2 + 3y^2 \quad \text{and thus} \quad x^2 + 4y^2 = 4.$$

This tells us that the two surfaces intersect in a space curve that lies along the elliptic cylinder $x^2 + 4y^2 = 4$. The projection of this intersection onto the xy-plane is the ellipse $x^2 + 4y^2 = 4$. (See Figure 16.8.9.)

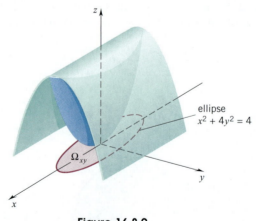

Figure 16.8.9

The projection of T onto the xy-plane is the region

$$\Omega_{xy}: \quad -2 \le x \le 2, \quad -\tfrac{1}{2}\sqrt{4 - x^2} \le y \le \tfrac{1}{2}\sqrt{4 - x^2}.$$

The solid T is then the set of all (x, y, z) with

$$-2 \le x \le 2, \quad -\tfrac{1}{2}\sqrt{4-x^2} \le y \le \tfrac{1}{2}\sqrt{4-x^2}, \quad x^2+3y^2 \le z \le 4-y^2.$$

Its volume is given by

$$V = \int_{-2}^{2} \int_{-\sqrt{4-x^2}/2}^{\sqrt{4-x^2}/2} \int_{x^2+3y^2}^{4-y^2} dz\, dy\, dx$$

$$= 4 \int_{0}^{2} \int_{0}^{\sqrt{4-x^2}/2} \int_{x^2+3y^2}^{4-y^2} dz\, dy\, dx = 4\pi. \quad \square$$

└── check this

EXERCISES 16.8

In Exercises 1–10, evaluate the given repeated integral.

1. $\int_{0}^{a} \int_{0}^{b} \int_{0}^{c} dx\, dy\, dz.$

2. $\int_{0}^{1} \int_{0}^{x} \int_{0}^{y} y\, dz\, dy\, dx.$

3. $\int_{0}^{1} \int_{1}^{2y} \int_{0}^{x} (x+2z)\, dz\, dx\, dy.$

4. $\int_{0}^{1} \int_{1-x}^{1+x} \int_{0}^{xy} 4z\, dz\, dy\, dx.$

5. $\int_{0}^{2} \int_{-1}^{1} \int_{1}^{3} (z-xy)\, dz\, dy\, dx.$

6. $\int_{0}^{2} \int_{-1}^{1} \int_{1}^{3} (z-xy)\, dy\, dx\, dz.$

7. $\int_{0}^{\pi/2} \int_{0}^{1} \int_{0}^{\sqrt{1-x^2}} x\cos z\, dy\, dx\, dz.$

8. $\int_{-1}^{2} \int_{1}^{y+2} \int_{e}^{e^2} \frac{x+y}{z}\, dz\, dx\, dy.$

9. $\int_{1}^{2} \int_{y}^{y^2} \int_{0}^{\ln x} ye^z\, dz\, dx\, dy.$

10. $\int_{0}^{\pi/2} \int_{0}^{\pi/2} \int_{0}^{1} e^z \cos x \sin y\, dz\, dy\, dx.$

11. (*Separated variables over a box*) Set Π: $a_1 \le x \le a_2$, $b_1 \le y \le b_2$, $c_1 \le z \le c_2$. Show that, if f is continuous

on $[a_1, a_2]$, g is continuous on $[b_1, b_2]$ and h is continuous on $[c_1, c_2]$, then

(16.8.2)
$$\iiint_{\Pi} f(x)g(y)h(z)\, dx\, dy\, dx =$$
$$\left(\int_{a_1}^{a_2} f(x)\, dx\right)\left(\int_{b_1}^{b_2} g(y)\, dy\right)\left(\int_{c_1}^{c_2} h(z)\, dz\right).$$

In Exercises 12 and 13, evaluate taking Π: $0 \le x \le 1$, $0 \le y \le 2$, $0 \le z \le 3$.

12. $\iiint_{\Pi} x^3 y^2 z\, dx\, dy\, dz.$

13. $\iiint_{\Pi} x^2 y^2 z^2\, dx\, dy\, dz.$

In Exercises 14–16, the mass density of a box Π: $0 \le x \le a$, $0 \le y \le b$, $0 \le z \le c$ varies directly with the product xyz.

14. Calculate the mass of Π.

15. Locate the center of mass.

16. Determine the moment of inertia of Π about: (a) the vertical line that passes through the point (a, b, c); (b) the vertical line that passes through the center of mass.

In Exercises 17–20, a homogeneous solid T of mass M consists of all points (x, y, z) with $0 \le x \le 1$, $0 \le y \le 1$, $0 \le z \le 1-y$.

17. Sketch T.

18. Find the volume of T.

19. Locate the center of mass.

20. Find the moments of inertia of T about the coordinate axes.

In Exercises 21–26, express the indicated quantity by repeated integrals. Do not evaluate.

21. The mass of a ball $x^2 + y^2 + z^2 \leq r^2$ given that the density varies directly with the distance from the outer shell.

22. The mass of the solid bounded above by $z = 1$ and bounded below by $z = \sqrt{x^2 + y^2}$ given that the density varies directly with the distance from the origin. Identify the solid.

23. The volume of the solid bounded above by the parabolic cylinder $z = 1 - y^2$, below by the plane $2x + 3y + z + 10 = 0$, and on the sides by the circular cylinder $x^2 + y^2 - x = 0$.

24. The volume of the solid bounded above by the paraboloid $z = 4 - x^2 - y^2$ and bounded below by the parabolic cylinder $z = 2 + y^2$.

25. The mass of the solid bounded by the elliptic paraboloids $z = 4 - x^2 - \frac{1}{4}y^2$ and $z = 3x^2 + \frac{1}{4}y^2$ given that the density varies directly with the vertical distance from the lower surface.

26. The mass of the solid bounded by the paraboloid $x = z^2 + 2y^2$ and the parabolic cylinder $x = 4 - z^2$ given that the density varies directly with the distance from the z-axis.

In Exercises 27–32, evaluate the given triple integral.

27. $\iiint_T (x^2z + y) \, dx \, dy \, dz$, where T is the solid bounded by the planes $x = 0$, $x = 1$, $y = 1$, $y = 3$, $z = 0$, $z = 2$.

28. $\iiint_T 2ye^x \, dx \, dy \, dz$, where T is the solid given by $0 \leq y \leq 1, 0 \leq x \leq y, 0 \leq z \leq x + y$.

29. $\iiint_T x^2y^2z^2 \, dx \, dy \, dz$, where T is the solid bounded by the planes $z = y + 1$, $y + z = 1$, $x = 0$, $x = 1$, $z = 0$.

30. $\iiint_T xy \, dx \, dy \, dz$, where T is the solid in the first octant bounded by the coordinate planes and the hemisphere $z = \sqrt{4 - x^2 - y^2}$.

31. $\iiint_T y^2 \, dx \, dy \, dz$, where T is the tetrahedron in the first octant bounded by the coordinate planes and the plane $2x + 3y + z = 6$.

32. $\iiint_T y^2 \, dx \, dy \, dz$, where T is the solid in the first octant bounded by the cylinders $x^2 + y = 1$, $z^2 + y = 1$.

33. Find the volume of the portion of the first octant bounded by the planes $z = x$, $y - x = 2$, and the cylinder $y = x^2$. Where is the centroid?

34. Find the mass of a block in the shape of a unit cube given that the density varies directly with: (a) the distance from one of the faces; (b) the square of the distance from one of the vertices.

35. Find the volume and the centroid of the solid bounded above by the cylindrical surface $x^2 + z = 4$, below by the plane $x + z = 2$, and on the sides by the planes $y = 0$ and $y = 3$.

36. Show that, if $(\bar{x}, \bar{y}, \bar{z})$ is the centroid of a solid T, then

$$\iiint_T (x - \bar{x}) \, dx \, dy \, dz = 0,$$

$$\iiint_T (y - \bar{y}) \, dx \, dy \, dz = 0,$$

$$\iiint_T (z - \bar{z}) \, dx \, dy \, dz = 0.$$

37. Taking a, b, c as positive, find the volume of the tetrahedron with vertices $(0, 0, 0)$, $(a, 0, 0)$, $(0, b, 0)$, and $(0, 0, c)$. Where is the centroid?

38. A homogeneous solid of mass M in the form and position of the tetrahedron of Figure 16.8.4 rotates about the z-axis with moment of inertia I_z. Find I_z.

39. A homogeneous box of mass M has edges a, b, c. Calculate the moment of inertia about:
(a) the edge c;
(b) the line that passes through the center of the box and is parallel to the edge c;
(c) the line that passes through the center of the face bc and is parallel to the edge c.

40. Where is the centroid of the solid bounded above by the plane $z = 1 + x + y$, below by the plane $z = -2$, and on the sides by the planes $x = 1$, $x = 2$, $y = 1$, $y = 2$?

41. Let T be the solid bounded above by the plane $z = y$, below by the xy-plane, and on the sides by the planes $x = 0$, $x = 1$, $y = 0$, $y = 1$. Find the mass of T given that the density varies directly with the square of the distance from the origin. Where is the center of mass?

42. What can you conclude about T given that

$$\iiint_T f(x, y, z) \, dx \, dy \, dz = 0$$

(a) for every continuous function f that is odd in x? (b) for every continuous function f that is odd in y? (c) for every continuous function f that is odd in z? (d) for every continuous function f that satisfies the relation $f(-x, -y, -z) = -f(x, y, z)$?

43. (a) Integrate $f(x, y, z) = x + y^3 + z$ over the unit ball centered at the origin.

(b) Integrate $f(x, y, z) = a_1x + a_2y + a_3z + a_4$ over the unit ball centered at the origin.

44. Integrate $f(x, y, z) = x^2y^2$ over the solid bounded above by the cylinder $y^2 + z = 4$, below by the plane $y + z = 2$, and on the sides by the planes $x = 0$ and $x = 2$.

45. Use triple integrals to find the volume enclosed by the sphere $x^2 + y^2 + z^2 = a^2$.

46. Use triple integrals to find the volume enclosed by the ellipsoid

$$\frac{x^2}{a^2} + \frac{y^2}{b^2} + \frac{z^2}{c^2} = 1.$$

47. Find the mass of the solid Example 5 given that the density varies directly with $|x|$.

48. Find the volume of the solid bounded by the paraboloids $z = 2 - x^2 - y^2$ and $z = x^2 + y^2$.

49. Find the mass and the center of mass of the solid of Exercise 35 given that the density varies directly with $1 + y$.

50. Let T be a solid with volume

$$V = \iiint_T dx\, dy\, dz = \int_0^2 \int_0^{9-x^2} \int_0^{2-x} dz\, dy\, dx.$$

Sketch T and fill in the blanks.

(a) $V = \int_\square^\square \int_\square^\square \int_\square^\square dy\, dx\, dz.$

(b) $V = \int_\square^\square \int_\square^\square \int_\square^\square dy\, dz\, dx.$

(c) $V = \int_0^5 \int_\square^\square \int_\square^\square dz\, dx\, dy + \int_5^9 \int_\square^\square \int_\square^\square dz\, dx\, dy.$

51. Let T be a solid with volume

$$V = \iiint_T dx\, dy\, dz = \int_0^3 \int_x^{6-x} \int_0^{2x} dz\, dy\, dx.$$

Sketch T and fill in the blanks.

(a) $V = \int_\square^\square \int_\square^\square \int_\square^\square dy\, dx\, dz.$

(b) $V = \int_\square^\square \int_\square^\square \int_\square^\square dy\, dz\, dx.$

(c) $V = \int_\square^\square \int_\square^\square \int_\square^\square dx\, dy\, dz + \int_\square^\square \int_\square^\square \int_\square^\square dx\, dy\, dz.$

For the remaining exercises, let V be the volume of the solid T enclosed by the parabolic cylinder $y = 4 - z^2$ and the V-shaped cylinder $y = |x|$. Let $\Omega_{xy}, \Omega_{yz}, \Omega_{xz}$ be the projections of T onto the xy, yz, xz planes, respectively. Fill in the blanks.

52. (a) $V = \iint_{\Omega_{xy}} \boxed{}\, dx\, dy.$

(b) $V = \iint_{\Omega_{xy}} \left(\int_\square^\square dz \right) dx\, dy.$

(c) $V = \int_\square^\square \int_\square^\square \int_\square^\square dz\, dy\, dx.$

(d) $V = \int_\square^\square \int_\square^\square \int_\square^\square dz\, dx\, dy.$

53. (a) $V = \iint_{\Omega_{yz}} \boxed{}\, dy\, dz.$

(b) $V = \iint_{\Omega_{yz}} \left(\int_\square^\square dx \right) dy\, dz.$

(c) $V = \int_\square^\square \int_\square^\square \int_\square^\square dx\, dz\, dy.$

(d) $V = \int_\square^\square \int_\square^\square \int_\square^\square dx\, dy\, dz.$

54. (a) $V = \iint_{\Omega_{xz}} \boxed{}\, dx\, dz.$

(b) $V = \iint_{\Omega_{xz}} \left(\int_\square^\square dy \right) dx\, dz.$

(c) $V = \int_\square^\square \int_\square^\square \int_\square^\square dy\, dx\, dz.$

(d) $V = \int_{-2}^0 \int_\square^\square \int_\square^\square dy\, dz\, dx + \int_0^2 \int_\square^\square \int_\square^\square dy\, dz\, dx.$

■ 16.9 CYLINDRICAL COORDINATES

Introduction to Cylindrical Coordinates

The cylindrical coordinates (r, θ, z) of a point P in xyz-space are shown geometrically in Figure 16.9.1. The first two coordinates, r and θ, are the usual plane polar coordinates, except that r is taken as nonnegative and θ is restricted to the interval $[0, 2\pi)$. The third coordinate is the third rectangular coordinate z. Thus, rectangular coordinates (x, y, z) are related to cylindrical coordinates (r, θ, z) by the equations

$$x = r \cos \theta, \quad y = r \sin \theta, \quad z = z.$$

Conversely, with the obvious exclusions, cylindrical coordinates are related to rectangular coordinates by

$$r = \sqrt{x^2 + y^2}, \quad \tan \theta = \frac{y}{x}, \quad z = z.$$

In rectangular coordinates, the coordinate surfaces

$$x = x_0, \quad y = y_0, \quad z = z_0$$

are three mutually perpendicular planes. In cylindrical coordinates the coordinate surfaces take the form

$$r = r_0, \quad \theta = \theta_0, \quad z = z_0.$$

These surfaces are drawn in Figure 16.9.2.

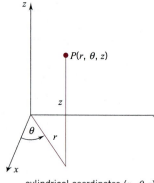

cylindrical coordinates (r, θ, z):
$r \geq 0$, $0 \leq \theta < 2\pi$, z real

Figure 16.9.1

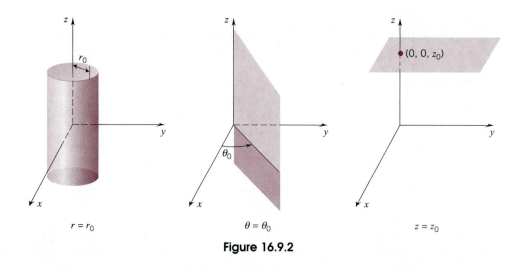

$r = r_0$ $\theta = \theta_0$ $z = z_0$

Figure 16.9.2

The first surface is a right circular cylinder of radius r_0. The central axis of the cylinder is the z-axis. The surface $\theta = \theta_0$ is a vertical half-plane hinged at the z-axis. The plane stands at an angle of θ_0 radians from the positive x-axis. The last coordinate surface is the plane $z = z_0$.

The xyz-solids easiest to describe in cylindrical coordinates are the *cylindrical* wedges. One is pictured in Figure 16.9.3. That wedge consists of all points (x, y, z) that have cylindrical coordinates (r, θ, z) in the box

$$\Pi: \quad a_1 \leq r \leq a_2, \quad b_1 \leq \theta \leq b_2, \quad c_1 \leq z \leq c_2.$$

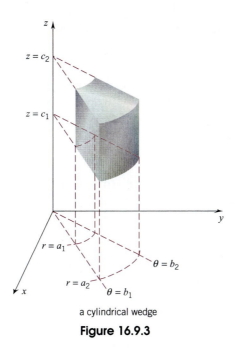

a cylindrical wedge

Figure 16.9.3

Evaluating Triple Integrals Using Cylindrical Coordinates

Suppose that T is some basic solid in xyz-space, not necessarily a wedge. If T is the set of all (x, y, z) with cylindrical coordinates in some basic solid S in $r\theta z$-space, then

(16.9.1)
$$\iiint_T f(x, y, z) \, dx \, dy \, dz = \iiint_S f(r \cos \theta, r \sin \theta, z) r \, dr \, d\theta \, dz.$$

Derivation of (16.9.1) We will carry out the argument on the assumption that T is projectable onto some basic region Ω_{xy} of the xy-plane. (It is for such solids that the formula is most useful.) T has some lower boundary $z = \psi_1(x, y)$ and some upper boundary $z = \psi_2(x, y)$. T is then the set of all (x, y, z) with

$$(x, y) \in \Omega_{xy} \quad \text{and} \quad \psi_1(x, y) \leq z \leq \psi_2(x, y).$$

The region Ω_{xy} has polar coordinates in some set $\Omega_{r\theta}$ (which we assume is a basic region). Then S is the set of all (r, θ, z) with

$$(r, \theta) \in \Omega_{r\theta} \quad \text{and} \quad \psi_1(r \cos \theta, r \sin \theta) \leq z \leq \psi_2(r \cos \theta, r \sin \theta).$$

Therefore

$$\iiint_T f(x, y, z)\, dx\, dy\, dz = \iint_{\Omega_{xy}} \left(\int_{\psi_1(x,y)}^{\psi_2(x,y)} f(x, y, z)\, dz \right) dx\, dy$$

$$(16.5.3) \longrightarrow \; = \iint_{\Omega_{r\theta}} \left(\int_{\psi_1(r\cos\theta,\, r\sin\theta)}^{\psi_2(r\cos\theta,\, r\sin\theta)} f(r\cos\theta,\, r\sin\theta,\, z)\, dz \right) r\, dr\, d\theta$$

$$= \iiint_S f(r\cos\theta,\, r\sin\theta,\, z)\, r\, dr\, d\theta\, dz. \quad \square$$

Volume Formula

If $f(x, y, z) = 1$ for all (x, y, z) in T, then (16.9.1) reduces to

$$\iiint_T dx\, dy\, dz = \iiint_S r\, dr\, d\theta\, dz.$$

The triple integral on the left is the volume of T. In summary, if T is a basic solid in xyz-space and the cylindrical coordinates of T constitute a basic solid S in $r\theta z$-space, then the volume of T is given by the formula

(16.9.2)

$$V = \iiint_S r\, dr\, d\theta\, dz.$$

Calculations

Cylindrical coordinates are particularly useful when an axis of symmetry is present. The axis of symmetry is then taken as the z-axis.

Example 1 Evaluate

$$\int_{-2}^{2} \int_{-\sqrt{4-x^2}}^{\sqrt{4-x^2}} \int_{0}^{4-x^2-y^2} (x^2 + y^2)\, dz\, dy\, dx.$$

SOLUTION This is a triple integral over the solid T given by

$$T: \quad -2 \le x \le 2, \quad -\sqrt{4 - x^2} \le y \le \sqrt{4 - x^2}, \quad 0 \le z \le 4 - x^2 - y^2.$$

This solid is bounded above by the paraboloid of revolution $z = 4 - x^2 - y^2$ and below by the xy-plane (see Figure 16.9.4). Since the solid is symmetric about the z-axis, the integral has a simpler representation in cylindrical coordinates:

$$S: \quad 0 \le r \le 2, \quad 0 \le \theta \le 2\pi, \quad 0 \le z \le 4 - r^2.$$

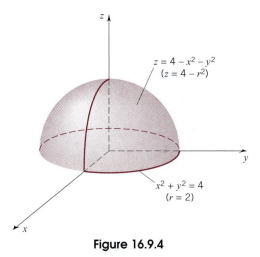

$z = 4 - x^2 - y^2$
$(z = 4 - r^2)$

$x^2 + y^2 = 4$
$(r = 2)$

Figure 16.9.4

Now,

$$\iiint_T (x^2 + y^2)\, dx\, dy\, dz = \iiint_S r^2\, r\, dr\, d\theta\, dz = \int_0^{2\pi} \int_0^2 \int_0^{4-r^2} r^3\, dz\, dr\, d\theta$$

$$= \int_0^{2\pi} \int_0^2 \left[r^3 z \right]_0^{4-r^2} dr\, d\theta = \int_0^{2\pi} \int_0^2 (4r^3 - r^5)\, dr\, d\theta$$

$$= \int_0^{2\pi} \left[r^4 - \tfrac{1}{6} r^6 \right]_0^2 d\theta$$

$$= \tfrac{16}{3} \int_0^{2\pi} d\theta = \frac{32\pi}{3}. \quad \square$$

Example 2 Find the mass of a solid right circular cylinder T of radius R and height h given that the density varies directly with the distance from the axis of the cylinder.

SOLUTION Place the cylinder T on the xy-plane so that the axis of T coincides with the z-axis. The density function then takes the form $\lambda(x, y, z) = k\sqrt{x^2 + y^2}$ and T consists of all points (x, y, z) with cylindrical coordinates (r, θ, z) in the set

$$S: \quad 0 \leq r \leq R, \quad 0 \leq \theta \leq 2\pi, \quad 0 \leq z \leq h.$$

Therefore

$$M = \iiint_T k\sqrt{x^2 + y^2}\, dx\, dy\, dz = \iiint_S (kr)r\, dr\, d\theta\, dz$$

$$= k \int_0^R \int_0^{2\pi} \int_0^h r^2\, dz\, d\theta\, dr = \tfrac{2}{3} k\pi R^3 h. \quad \square$$

└─ check this

Example 3 Use cylindrical coordinates to find the volume of the solid T bounded above by the plane $z = y$ and below by the paraboloid $z = x^2 + y^2$.

SOLUTION In cylindrical coordinates the plane has equation $z = r \sin \theta$ and the paraboloid has equation $z = r^2$. Solving these two equations simultaneously, we have $r = \sin \theta$. This tells us that the two surfaces intersect in a space curve that lies along the circular cylinder $r = \sin \theta$. The projection of this intersection onto the xy-plane is the circle with polar equation $r = \sin \theta$. (See Figure 16.9.5.) The base region Ω_{xy} is thus the set of all (x, y) with polar coordinates in the set

$$0 \le \theta \le \pi, \quad 0 \le r \le \sin \theta.$$

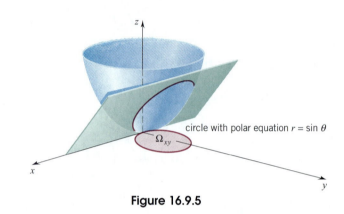

circle with polar equation $r = \sin \theta$

Ω_{xy}

Figure 16.9.5

T itself is the set of all (x, y, z) with cylindrical coordinates in the set

$$S: \quad 0 \le \theta \le \pi, \quad 0 \le r \le \sin \theta, \quad r^2 \le z \le r \sin \theta. \qquad \text{(check this)}$$

Therefore,

$$V = \iiint_T dx\, dy\, dz = \iiint_S r\, dr\, d\theta\, dz$$

$$= \int_0^\pi \int_0^{\sin \theta} \int_{r^2}^{r \sin \theta} r\, dz\, dr\, d\theta$$

$$= \int_0^\pi \int_0^{\sin \theta} (r^2 \sin \theta - r^3)\, dr\, d\theta$$

$$= \int_0^\pi \left[\tfrac{1}{3} r^3 \sin \theta - \tfrac{1}{4} r^4 \right]_0^{\sin \theta} d\theta$$

$$= \tfrac{1}{12} \int_0^\pi \sin^4 \theta\, d\theta = \tfrac{1}{12} (\tfrac{3}{8} \pi) = \tfrac{1}{32} \pi. \quad \square$$

(verify this; Exercise 11, Section 8.3)

Example 4 Locate the centroid of the solid in Example 3.

SOLUTION Since T is symmetric about the yz-plane, we see that $\bar{x} = 0$. To get $\bar{y}$ we begin as usual:

$$\bar{y}V = \iiint_T y\, dx\, dy\, dz = \iiint_S (r\sin\theta)\, r\, dr\, d\theta\, dz$$

$$= \int_0^\pi \int_0^{\sin\theta} \int_{r^2}^{r\sin\theta} r^2\sin\theta\, dz\, dr\, d\theta$$

$$= \int_0^\pi \int_0^{\sin\theta} (r^3\sin^2\theta - r^4\sin\theta)\, dr\, d\theta$$

$$= \int_0^\pi \left[\tfrac{1}{4}r^4\sin^2\theta - \tfrac{1}{5}r^5\sin\theta \right]_0^{\sin\theta} d\theta$$

$$= \tfrac{1}{20} \int_0^\pi \sin^6\theta\, d\theta \;=\; \tfrac{1}{20}\left(\tfrac{5}{16}\pi\right) = \tfrac{1}{64}\pi.$$

⎿ (verify this; Exercise 19, Section 8.3)

Since $V = \tfrac{1}{32}\pi$, we have $\bar{y} = \tfrac{1}{2}$. Now for $\bar{z}$:

$$\bar{z}V = \iiint_T z\, dx\, dy\, dz$$

$$= \iiint_S zr\, dr\, d\theta\, dz$$

$$= \int_0^\pi \int_0^{\sin\theta} \int_{r^2}^{r\sin\theta} zr\, dz\, dr\, d\theta = \cdots = \tfrac{1}{24} \int_0^\pi \sin^6\theta\, d\theta = \tfrac{1}{24}\left(\tfrac{5}{16}\pi\right) = \tfrac{5}{384}\pi.$$

details are left to you ⎯↑

Division by $V = \tfrac{1}{32}\pi$ gives $\bar{z} = \tfrac{5}{12}$. The centroid is thus the point $(0, \tfrac{1}{2}, \tfrac{5}{12})$. ❑

EXERCISES 16.9

In Exercises 1–6, write the given equation in cylindrical coordinates and sketch the graph of the surface.

1. $x^2 + y^2 + z^2 = 9$.   2. $x^2 + y^2 = 4$.

3. $z = 2\sqrt{x^2 + y^2}$.   4. $x = 4z$.

5. $4x^2 + 4y^2 - z^2 = 0$.   6. $y^2 + z^2 = 8$.

In Exercises 7–10, evaluate the repeated integral.

7. $\displaystyle\int_0^\pi \int_0^2 \int_0^{4-r^2} r\, dz\, dr\, d\theta$.

8. $\displaystyle\int_0^{\pi/4} \int_0^1 \int_0^{\sqrt{1-r^2}} r\cos\theta\, dz\, dr\, d\theta$.

9. $\displaystyle\int_0^{\pi/4} \int_0^1 \int_0^{r\cos\theta} r\sec^3\theta\, dz\, dr\, d\theta$.

10. $\displaystyle\int_0^\pi \int_0^{4\cos\theta} \int_0^{\sqrt{16-r^2}} r\, dz\, dr\, d\theta$.

In Exercises 11–13, let T be a solid right circular cylinder of base radius R and height h. Assume that the mass density varies directly with the distance from one of the bases.

11. Use cylindrical coordinates to find the mass M of T.

12. Locate the center of mass of T.

13. Find the moment of inertia of T about the axis of the cylinder.

14. Let T be a homogeneous right circular cylinder of mass M, base radius R, and height h. Find the moment of inertia of the cylinder about: (a) the central axis; (b) a line that lies in the plane of one of the bases and passes through the center of that base; (c) a line that passes through the center of the cylinder and is parallel to the bases.

In Exercises 15–18, let T be a homogeneous solid right circular cone of mass M, base radius R, and height h.

15. Use cylindrical coordinates to verify that the volume of the cone is given by the formula $V = \frac{1}{3}\pi R^2 h$.

16. Locate the center of mass.

17. Find the moment of inertia about the axis of the cone.

18. Find the moment of inertia about a line that passes through the vertex and is parallel to the base.

In Exercises 19–21, let T be the solid bounded above by the paraboloid $z = 1 - (x^2 + y^2)$ and bounded below by the xy-plane.

19. Use cylindrical coordinates to find the volume of T.

20. Find the mass of T if the density varies directly with the distance to the xy-plane.

21. Find the mass of T if the density varies directly with the square of the distance from the origin.

In Exercises 22–27, evaluate using cylindrical coordinates.

22. $\int_{-1}^{1} \int_{0}^{\sqrt{1-x^2}} \int_{\sqrt{x^2+y^2}}^{1} z^3 \, dz \, dy \, dx$.

23. $\int_{0}^{1} \int_{0}^{\sqrt{1-x^2}} \int_{0}^{\sqrt{4-(x^2+y^2)}} dz \, dy \, dx$.

24. $\int_{0}^{1} \int_{0}^{\sqrt{1-x^2}} \int_{0}^{\sqrt{1-x^2-y^2}} z \, dz \, dy \, dx$.

25. $\int_{0}^{3} \int_{0}^{\sqrt{9-y^2}} \int_{0}^{\sqrt{9-x^2-y^2}} \frac{1}{\sqrt{x^2+y^2}} \, dz \, dx \, dy$.

26. $\int_{-1}^{1} \int_{-\sqrt{1-x^2}}^{\sqrt{1-x^2}} \int_{x^2+y^2}^{2-x^2-y^2} \sqrt{x^2+y^2} \, dz \, dy \, dx$.

27. $\int_{0}^{1} \int_{0}^{\sqrt{1-x^2}} \int_{0}^{2} \sin(x^2 + y^2) \, dz \, dy \, dx$.

28. Find the volume of the solid bounded by the paraboloid of revolution $x^2 + y^2 = az$, the xy-plane, and the cylinder $x^2 + y^2 = 2ax$.

29. Find the volume of the solid bounded above by the cone $z^2 = x^2 + y^2$, below by the xy-plane, and on the sides by the cylinder $x^2 + y^2 = 2ax$.

30. Find the volume of the solid bounded above by the plane $2z = 4 + x$, below by the xy-plane, and on the sides by the cylinder $x^2 + y^2 = 2x$.

31. Find the volume of the solid bounded above by $z = a - \sqrt{x^2 + y^2}$, below by the xy-plane, and on the sides by the cylinder $x^2 + y^2 = ax$.

32. Find the volume of the solid that is bounded above by $x^2 + y^2 + z^2 = 25$ and below by $z = \sqrt{x^2 + y^2} + 1$.

33. Find the volume of the solid bounded by the paraboloid $z = x^2 + y^2$ and the plane $z = x$.

34. Find the volume of the solid bounded by the hyperboloid $z^2 = a^2 + x^2 + y^2$ and the upper nappe of the cone $z^2 = 2(x^2 + y^2)$.

35. Find the volume of the "ice cream cone" bounded below by the half-cone $z = \sqrt{3(x^2 + y^2)}$ and above by the unit sphere $x^2 + y^2 + z^2 = 1$.

36. Find the volume of the solid that lies between the cylinders $x^2 + y^2 = 1$ and $x^2 + y^2 = 4$, and is bounded above by the ellipsoid $x^2 + y^2 + 4z^2 = 36$ and below by the xy-plane.

■ **16.10 THE TRIPLE INTEGRAL AS THE LIMIT OF RIEMANN SUMS; SPHERICAL COORDINATES**

The Triple Integral as the Limit of Riemann Sums

You have seen how single integrals and double integrals can be obtained as limits of Riemann sums. The same holds true for triple integrals.

Start with a basic solid T in xyz-space and decompose it into a finite number of basic solids $T_1, \ldots, T_N$. If f is continuous on T, then f is continuous on each T_i. From each T_i pick an arbitrary point (x_i^*, y_i^*, z_i^*) and form the *Riemann sum*

$$\sum_{i=1}^{N} f(x_i^*, y_i^*, z_i^*)(\text{volume of } T_i).$$

As you would expect, the triple integral over T is the limit of such sums; namely, given any $\epsilon > 0$, there exists $\delta > 0$ such that, if the diameters of the T_i are all less than δ, then

$$\left| \sum_{i=1}^{N} f(x_i^*, y_i^*, z_i^*)(\text{volume of } T_i) - \iiint_{T} f(x, y, z) \, dx \, dy \, dz \right| < \epsilon$$

no matter how the (x_i^*, y_i^*, z_i^*) are chosen within the T_i. We express this by writing

(16.10.1)
$$\iiint_T f(x, y, z)\, dx\, dy\, dz = \lim_{\operatorname{diam} T_i \to 0} \sum_{i=1}^{N} f(x_i^*, y_i^*, z_i^*)(\text{volume of } T_i).$$

Introduction to Spherical Coordinates

The spherical coordinates (ρ, θ, ϕ) of a point P in xyz-space are shown geometrically in Figure 16.10.1. The first coordinate, ρ, is the distance from P to the origin; thus $\rho \geq 0$. The second coordinate, the angle marked θ, is the second coordinate of cylindrical coordinates; θ ranges from 0 to 2π. We call θ the *longitude*. The third coordinate, the angle marked ϕ, ranges only from 0 to π. We call ϕ the *colatitude*, or more simply the *polar angle*. (The complement of ϕ would be the *latitude* on a globe.)

The coordinate surfaces

$$\rho = \rho_0, \quad \theta = \theta_0, \quad \phi = \phi_0$$

are shown in Figure 16.10.2. The surface $\rho = \rho_0$ is a sphere; the radius is ρ_0 and the center is the origin. The second surface, $\theta = \theta_0$, is the same as in cylindrical coordinates: the vertical half-plane hinged at the z-axis and standing at an angle of θ_0 radians from the positive x-axis. The surface $\phi = \phi_0$ requires detailed explanation. If $0 < \phi_0 < \tfrac{1}{2}\pi$ or $\tfrac{1}{2}\pi < \phi_0 < \pi$, the surface is a nappe of a cone; it is generated by revolving about the z-axis any ray that emerges from the origin at an angle of ϕ_0 radians from the positive z-axis. The surface $\phi = \tfrac{1}{2}\pi$ is the xy-plane. (The nappe of the cone has opened up completely.) The equation $\phi = 0$ gives the nonnegative z-axis, and the equation $\phi = \pi$ gives the nonpositive z-axis. (When $\phi = 0$ or $\phi = \pi$, the nappe of the cone has closed up completely.)

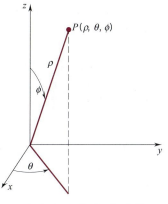

spherical coordinates (ρ, θ, ϕ):
$\rho \geq 0,\ 0 \leq \theta < 2\pi,\ 0 \leq \phi \leq \pi$

Figure 16.10.1

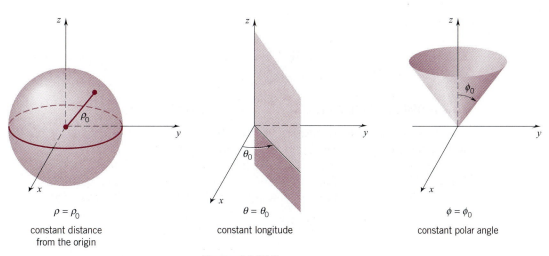

$\rho = \rho_0$
constant distance
from the origin

$\theta = \theta_0$
constant longitude

$\phi = \phi_0$
constant polar angle

Figure 16.10.2

Rectangular coordinates (x, y, z) are related to spherical coordinates (ρ, θ, ϕ) by the following equations:

$$x = \rho \sin \phi \cos \theta, \qquad y = \rho \sin \phi \sin \theta, \qquad z = \rho \cos \phi.$$

You can verify these relations by referring to Figure 16.10.3. (Note that the factor $\rho \sin \phi$ appearing in the first two equations is the r of cylindrical coordinates: $r = \rho \sin \phi$.) Conversely, with obvious exclusions, we have

$$\rho = \sqrt{x^2 + y^2 + z^2}, \qquad \tan \theta = \frac{y}{x}, \qquad \cos \phi = \frac{z}{\sqrt{x^2 + y^2 + z^2}}.$$

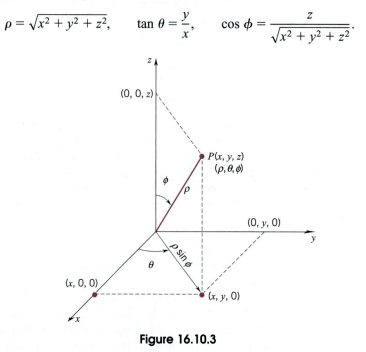

Figure 16.10.3

The Volume of a Spherical Wedge

Figure 16.10.4 shows a *spherical wedge* W in xyz-space. The wedge W consists of all points (x, y, z) that have spherical coordinates in the box

$$\Pi: \quad a_1 \le \rho \le a_2, \quad b_1 \le \theta \le b_2, \quad c_1 \le \phi \le c_2.$$

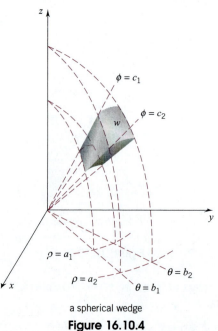

a spherical wedge
Figure 16.10.4

The volume of this wedge is given by the formula

(16.10.2)

$$V = \iiint\limits_{\Pi} \rho^2 \sin \phi \, d\rho \, d\theta \, d\phi.$$

PROOF Note first that W is a solid of revolution. One way to obtain W is to rotate the $\theta = b_1$ face of W, call it Ω, about the z-axis for $b_2 - b_1$ radians. (See Figure 16.10.4.) On that face ρ and $\alpha = \frac{1}{2}\pi - \phi$ play the role of polar coordinates. (See Figure 16.10.5.) In the setup of Figure 16.10.5 the face Ω is the set of all (z, X) with polar coordinates (ρ, α) in the set $\Gamma: a_1 \le \rho \le a_2, \frac{1}{2}\pi - c_2 \le \alpha \le \frac{1}{2}\pi - c_1$. The centroid of Ω is at a distance $\overline{X}$ from the z-axis where

$$\overline{X}(\text{area of } \Omega) = \iint\limits_{\Omega} X \, dX \, dz = \iint\limits_{\Gamma} \rho^2 \cos \alpha \, d\rho \, d\alpha$$

this follows from (16.5.3) together with the fact that here (ρ, α) play the role of polar coordinates

$$= \left(\int_{a_1}^{a_2} \rho^2 \, d\rho \right) \left(\int_{\frac{1}{2}\pi - c_2}^{\frac{1}{2}\pi - c_1} \cos \alpha \, d\alpha \right)$$

$\phi = \frac{1}{2}\pi - \alpha$

$$= \left(\int_{a_1}^{a_2} \rho^2 \, d\rho \right) \left(\int_{c_1}^{c_2} \sin \phi \, d\phi \right).$$

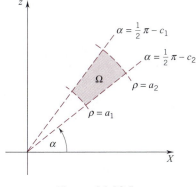

Figure 16.10.5

As the face Ω is rotated from $\theta = b_1$ to $\theta = b_2$, the centroid travels through a circular arc of length

$$s = (b_2 - b_1)\overline{X} = (b_2 - b_1) \frac{1}{\text{area of } \Omega} \left(\int_{a_1}^{a_2} \rho^2 \, d\rho \right) \left(\int_{c_1}^{c_2} \sin \phi \, d\phi \right).$$

From Pappus (see the remark on p. 375) we know that

$$\text{the volume of } W = s(\text{area of } \Omega) = (b_2 - b_1)\left(\int_{a_1}^{a_2} \rho^2 \, d\rho\right)\left(\int_{c_1}^{c_2} \sin\phi \, d\phi\right)$$

$$= \left(\int_{b_1}^{b_2} d\theta\right)\left(\int_{a_1}^{a_2} \rho^2 \, d\rho\right)\left(\int_{c_1}^{c_2} \sin\phi \, d\phi\right)$$

$$= \int_{a_1}^{a_2}\int_{b_1}^{b_2}\int_{c_1}^{c_2} \rho^2\sin\phi \, d\phi \, d\theta \, d\rho$$

$$= \iiint_{\Pi} \rho^2\sin\phi \, d\rho \, d\theta \, d\phi. \quad \square$$

Evaluating Triple Integrals Using Spherical Coordinates

Suppose that T is a basic solid in xyz-space with spherical coordinates in some basic solid S of $\rho\theta\phi$-space. Then

(16.10.3)

$$\iiint_T f(x, y, z) \, dx \, dy \, dz =$$
$$\iiint_S f(\rho\sin\phi\cos\theta, \rho\sin\phi\sin\theta, \rho\cos\phi)\,\rho^2\sin\phi \, d\rho \, d\theta \, d\phi.$$

DERIVATION OF (16.10.3) Assume first that T is a spherical wedge W. The solid S is then a box Π. Now decompose Π into N boxes $\Pi_1, \ldots, \Pi_N$. This induces a subdivision of W into N spherical wedges $W_1, \ldots, W_N$.

Writing $F(\rho, \theta, \phi)$ for $f(\rho\sin\phi\cos\theta, \rho\sin\phi\sin\theta, \rho\cos\phi)$ to save space, we have

$$\iiint_{\Pi} F(\rho, \theta, \phi)\,\rho^2\sin\phi \, d\rho \, d\theta \, d\phi = \sum_{i=1}^{N} \iiint_{\Pi_i} F(\rho, \theta, \phi)\,\rho^2\sin\phi \, d\rho \, d\theta \, d\phi$$

additivity

$$= \sum_{i=1}^{N} F(\rho_i^*, \theta_i^*, \phi_i^*)\iiint_{\Pi_i} \rho^2\sin\phi \, d\rho \, d\theta \, d\phi$$

for some $(\rho_i^*, \theta_i^*, \phi_i^*) \in \Pi_i$
(by 16.7.5)

$$= \sum_{i=1}^{N} F(\rho_i^*, \theta_i^*, \phi_i^*)(\text{volume of } W_i)$$

(16.10.2) applied to Π_i

$x_i^* = \rho_i^*\sin\phi_i^*\cos\theta_i^*,$
$y_i^* = \rho_i^*\sin\phi_i^*\sin\theta_i^*,$
$z_i^* = \rho_i^*\cos\phi_i^*$

$$= \sum_{i=1}^{N} f(x_i^*, y_i^*, z_i^*)(\text{volume of } W_i).$$

This last expression is a Riemann sum for

$$\iiint_W f(x, y, z) \, dx \, dy \, dz$$

and, as such, by (16.10.1), will differ from that integral by less than any preassigned positive number ϵ provided only that the diameters of all the W_i are sufficiently small. This we can guarantee by making the diameters of all the Π_i sufficiently small.

This verifies the formula for the case where T is a spherical wedge. The more general case is left to you. HINT: Encase T in a wedge W and define f to be zero outside of T. ❑

Volume Formula

If $f(x, y, z) = 1$ for all (x, y, z) in T, then the change of variables formula reduces to

$$\iiint_T dx\,dy\,dz = \iiint_S \rho^2 \sin\phi\,d\rho\,d\theta\,d\phi.$$

The integral on the left is the volume of T. It follows that the volume of T is given by the formula

(16.10.4)

$$V = \iiint_S \rho^2 \sin\phi\,d\rho\,d\theta\,d\phi.$$

Calculations

Spherical coordinates are commonly used in applications where there is a center of symmetry. The center of symmetry is then taken as the origin.

Example 1 Calculate the mass M of a solid ball of radius 1 given that the density varies directly with the square of the distance from the center of the ball.

SOLUTION Center the ball at the origin. The ball, call it T, is now the set of all (x, y, z) with spherical coordinates (ρ, θ, ϕ) in the box

$$S: \quad 0 \le \rho \le 1, \quad 0 \le \theta \le 2\pi, \quad 0 \le \phi \le \pi. \qquad \text{(Figure 16.10.6)}$$

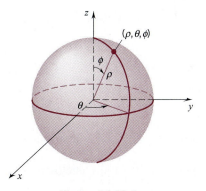

Figure 16.10.6

Therefore

$$M = \iiint_T k(x^2 + y^2 + z^2) \, dx \, dy \, dz$$

$$= \iiint_S (k\rho^2)\rho^2 \sin \phi \, d\rho \, d\theta \, d\phi = k \int_0^\pi \int_0^{2\pi} \int_0^1 \rho^4 \sin \phi \, d\rho \, d\theta \, d\phi$$

$$= k\left(\int_0^\pi \sin \phi \, d\phi\right)\left(\int_0^{2\pi} d\theta\right)\left(\int_0^1 \rho^4 \, d\rho\right) = k(2)(2\pi)(\tfrac{1}{5}) = \tfrac{4}{5}k\pi. \quad \square$$

Example 2   Find the volume of the solid T that is bounded above by the cone $z^2 = x^2 + y^2$, below by the xy-plane, and on the sides by the hemisphere $z = \sqrt{4 - x^2 - y^2}$ (see Figure 16.10.7).

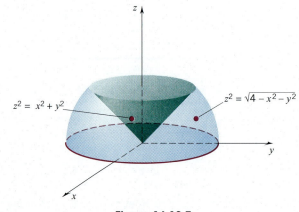

Figure 16.10.7

SOLUTION   The hemisphere is given by $\rho = 2$, $0 \le \phi \le \pi/2$, in spherical coordinates. You can verify that the hemisphere and the cone intersect in a circle which lies in the plane $z = \sqrt{2}$ and has its center on the z-axis. For points on this circle, the angle $\phi = \pi/4$ (verify this). Therefore, it follows that the solid T is the set of all (x, y, z) with spherical coordinates (ρ, θ, ϕ) in the set

$$S: \quad 0 \le \rho \le 2, \quad 0 \le \theta \le 2\pi, \quad \pi/4 \le \phi \le \pi/2.$$

Thus,

$$V = \iiint_T dx \, dy \, dz = \iiint_S \rho^2 \sin \phi \, d\rho \, d\phi \, d\theta$$

$$= \int_{\pi/4}^{\pi/2} \int_0^{2\pi} \int_0^2 \rho^2 \sin \phi \, d\rho \, d\theta \, d\phi = \left(\int_{\pi/4}^{\pi/2} \sin \phi \, d\phi\right)\left(\int_0^{2\pi} d\theta\right)\left(\int_0^2 \rho^2 \, d\rho\right)$$

$$= (\sqrt{2}/2)(2\pi)(\tfrac{8}{3}) = \frac{8\pi\sqrt{2}}{3} \cong 11.85.$$

This is about 70.7 percent of the volume of a hemisphere of radius 2. $\square$

Example 3   Find the volume of the solid T enclosed by the surface

$$(x^2 + y^2 + z^2)^2 = 2z(x^2 + y^2).$$

SOLUTION   In spherical coordinates the bounding surface takes the form

$$\rho = 2 \sin^2\phi \cos \phi. \qquad \text{(check this out)}$$

This equation places no restriction on θ; thus θ can range from 0 to 2π.   Since ρ remains nonnegative, ϕ can range only from 0 to $\frac{1}{2}\pi$.   Thus the solid T is the set of all (x, y, z) with spherical coordinates (ρ, θ, ϕ) in the set

$$S: \quad 0 \le \theta \le 2\pi, \quad 0 \le \phi \le \tfrac{1}{2}\pi, \quad 0 \le \rho \le 2 \sin^2\phi \cos \phi.$$

The rest is straightforward:

$$V = \iiint_T dx\, dy\, dz = \iiint_S \rho^2 \sin \phi \, d\rho \, d\theta \, d\phi$$

$$= \int_0^{2\pi} \int_0^{\pi/2} \int_0^{2\sin^2\phi\cos\phi} \rho^2 \sin \phi \, d\rho \, d\phi \, d\theta = \int_0^{2\pi} \int_0^{\pi/2} \tfrac{8}{3} \sin^7 \phi \cos^3 \phi \, d\phi \, d\theta$$

$$= \tfrac{8}{3} \left(\int_0^{2\pi} d\theta \right) \left(\int_0^{\pi/2} (\sin^7 \phi \cos \phi - \sin^9 \phi \cos \phi) \, d\phi \right)$$

$$= \tfrac{8}{3}(2\pi)(\tfrac{1}{40}) = \tfrac{2}{15}\pi \cong 0.42. \quad \square$$

EXERCISES 16.10

1. Find the spherical coordinates (ρ, θ, ϕ) of the point with rectangular coordinates $(1, 1, 1)$.
2. Find the rectangular coordinates of the point with spherical coordinates $(2, \frac{1}{6}\pi, \frac{1}{4}\pi)$.
3. Find the rectangular coordinates of the point with spherical coordinates $(3, \frac{1}{3}\pi, \frac{1}{6}\pi)$.
4. Find the spherical coordinates of the point with cylindrical coordinates $(2, \frac{2}{3}\pi, 6)$.
5. Find the spherical coordinates of the point with rectangular coordinates $(2, 2, \frac{2}{3}\sqrt{6})$.
6. Find the spherical coordinates of the point with rectangular coordinates $(2\sqrt{2}, -2\sqrt{2}, -4\sqrt{3})$.
7. Find the rectangular coordinates of the point with spherical coordinates $(3, \pi/2, 0)$.
8. Find the spherical coordinates of the point with cylindrical coordinates $(1/\sqrt{2}, \pi/4, 1/\sqrt{2})$.

In Exercises 9–14, equations are given in spherical coordinates.   Interpret each one geometrically.

9. $\rho \sin \phi = 1$.
10. $\sin \phi = 1$.
11. $\cos \phi = -\frac{1}{2}\sqrt{2}$.
12. $\tan \theta = 1$.
13. $\rho \cos \phi = 1$.
14. $\rho = \cos \phi$.

In Exercises 15–18, evaluate the repeated integral.

15. $\displaystyle\int_0^{\pi/3} \int_0^{2\pi} \int_0^1 \rho^2 \sin \phi \, d\rho \, d\theta \, d\phi.$

16. $\displaystyle\int_0^{2\pi} \int_0^{\pi} \int_0^2 \rho^3 \sin \phi \, d\rho \, d\phi \, d\theta.$

17. $\displaystyle\int_0^{\pi/4} \int_0^{\pi} \int_0^{2\cos\phi} \rho^2 \sin \phi \, d\rho \, d\theta \, d\phi.$

18. $\displaystyle\int_0^{\pi/4} \int_0^{2\pi} \int_0^{\sec\phi} \rho^3 \sin \phi \cos \phi \, d\rho \, d\theta \, d\phi.$

In Exercises 19–22, evaluate the repeated integral by changing to spherical coordinates.

19. $\displaystyle\int_0^1 \int_0^{\sqrt{1-x^2}} \int_{\sqrt{x^2+y^2}}^{\sqrt{2-x^2-y^2}} dz \, dy \, dx.$

20. $\displaystyle\int_0^2 \int_0^{\sqrt{4-y^2}} \int_{\sqrt{x^2+y^2}}^{\sqrt{4-x^2-y^2}} (x^2 + y^2 + z^2) \, dz \, dx \, dy.$

21. $\displaystyle\int_0^3 \int_0^{\sqrt{9-y^2}} \int_0^{\sqrt{9-x^2-y^2}} z\sqrt{x^2 + y^2 + z^2} \, dz \, dx \, dy.$

22. $\displaystyle\int_0^1 \int_0^{\sqrt{1-x^2}} \int_0^{\sqrt{1-x^2-y^2}} \frac{1}{x^2 + y^2 + z^2} \, dz \, dy \, dx.$

23. Derive the formula for the volume of a sphere of radius R using spherical coordinates.

24. Express cylindrical coordinates in terms of spherical coordinates.

25. A wedge is cut from a ball of radius R by two planes that meet in a diameter.   Find the volume of the wedge if the angle between the planes is α radians.

26. Find the mass of a ball of radius R given that the density varies directly with the distance from the boundary.

27. Find the mass of a right circular cone of base radius r and height h given that the density varies directly with the distance from the vertex.

28. Use spherical coordinates to derive the formula for the volume of a right circular cone of base radius r and height h.

In Exercises 29 and 30, let T be a homogeneous ball of mass M and radius R.

29. Calculate the moment of inertia about: (a) a diameter; (b) a tangent line.

30. Locate the center of mass of the upper half given that the center of the ball is at the origin.

In Exercises 31 and 32, let T be a homogeneous solid bounded by two concentric spherical shells, an outer shell of radius R_2 and an inner shell of radius R_1.

31. (a) Calculate the moment of inertia about a diameter.
    (b) Use your result in (a) to determine the moment of inertia of a spherical shell of radius R and mass M about a diameter.
    (c) What is the moment of inertia of that same shell about a tangent line?

32. (a) Locate the center of mass of the upper half of T given that the center of T is at the origin.
    (b) Use your result in (a) to locate the center of mass of a homogeneous hemispherical shell of radius R.

33. Find the volume of the solid common to the sphere $\rho = a$ and the cone $\phi = \alpha$. Take $\alpha \in (0, \frac{1}{2}\pi)$.

34. Let T be the solid bounded below by the half-cone $z = \sqrt{x^2 + y^2}$ and above by the spherical surface $x^2 + y^2 + z^2 = 1$. Use spherical coordinates to evaluate

$$\iiint_T e^{(x^2 + y^2 + z^2)^{3/2}} \, dx \, dy \, dz.$$

35. (a) Find an equation for the sphere
    $x^2 + y^2 + (z - R)^2 = R^2$ in spherical coordinates.

(b) Express the upper half of the ball
$x^2 + y^2 + (z - R)^2 \leq R^2$ by inequalities in spherical coordinates.

36. Find the mass of the ball $\rho \leq 2R \cos \phi$ given that the density varies directly with:
    (a) ρ.   (b) $\rho \sin \phi$.
    (c) $\rho \cos^2 \theta \sin \phi$.

37. Find the volume of the solid common to the spheres $\rho = 2\sqrt{2} \cos \phi$ and $\rho = 2$.

38. Find the volume of the solid enclosed by the surface $\rho = 1 - \cos \phi$.

39. Finish the argument in (16.10.3).

40. (*Gravitational Attraction*) Let T be a basic solid and let (a, b, c) be a point not in T. Show that, if T has continuously varying mass density $\lambda = \lambda(x, y, z)$, then T attracts a point mass m at (a, b, c) with a force

$$\mathbf{F} = \iiint_T Gm\lambda(x, y, z)\mathbf{f}(x, y, z) \, dx \, dy \, dz, \text{ where}$$

$$\mathbf{f}(x, y, z) = \frac{[(x - a)\mathbf{i} + (y - b)\mathbf{j} + (z - c)\mathbf{k}]}{[(x - a)^2 + (y - b)^2 + (z - c)^2]^{3/2}}.$$

(Assume that a point mass m_1 at P_1 attracts a point mass m_2 at P_2 with a force $\mathbf{F} = -(Gm_1m_2/r^3)\mathbf{r}$, where $\mathbf{r}$ is the vector $\overrightarrow{P_1P_2}$. Interpret the triple integral component by component.)

41. Let T be the upper half of the ball $x^2 + y^2 + (z - R)^2 \leq R^2$. Given that T is homogeneous and has mass M, find the gravitational force exerted by T on a point mass m located at the origin. (Note Exercise 40.)

42. A point mass m is placed on the axis of a homogeneous solid right circular cylinder at a distance α from the nearest base of the cylinder. Find the gravitational force exerted by the cylinder on the point mass given that the cylinder has base radius R, height h, and mass M. (Note Exercise 40.)

■ 16.11 JACOBIANS; CHANGING VARIABLES IN MULTIPLE INTEGRATION

During the course of the last few sections you have met several formulas for changing variables in multiple integration: to polar coordinates, to cylindrical coordinates, to spherical coordinates. The purpose of this section is to bring some unity into that material and provide a general description for other changes of variable.

We begin with a consideration of area. Figure 16.11.1 shows a basic region Γ in a plane that we are calling the *uv*-plane. (In this plane we denote the abscissa of a point by u and the ordinate by v.) Suppose that

$$x = x(u, v), \qquad y = y(u, v)$$

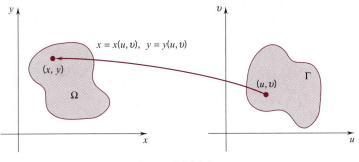

Figure 16.11.1

are functions continuously differentiable on Γ. As (u, v) ranges over Γ, the point $(x, y) = (x(u, v), y(u, v))$ generates a region Ω in the xy-plane. If the mapping

$$(u, v) \longrightarrow (x, y)$$

is one-to-one on the interior of Γ and the *Jacobian J* given by the 2×2 determinant

$$J(u, v) = \begin{vmatrix} \dfrac{\partial x}{\partial u} & \dfrac{\partial y}{\partial u} \\[2ex] \dfrac{\partial x}{\partial v} & \dfrac{\partial y}{\partial v} \end{vmatrix} = \frac{\partial x}{\partial u}\frac{\partial y}{\partial v} - \frac{\partial x}{\partial v}\frac{\partial y}{\partial u}$$

is never zero on the interior of Γ, then

(16.11.1)
$$\text{area of } \Omega = \iint\limits_{\Gamma} |J(u, v)|\; du\; dv.$$

It is very difficult to prove this assertion without making additional assumptions. A proof valid for all cases of practical interest is given in the supplement to Section 17.5. At this point we simply assume this area formula and go on from there.

Suppose now that we want to integrate some continuous function $f = f(x, y)$ over Ω. If this proves difficult to do directly, then we can change variables to u, v and try to integrate over Γ instead. It follows from (16.11.1) that

(16.11.2)
$$\iint\limits_{\Omega} f(x, y)\; dx\; dy = \iint\limits_{\Gamma} f(x(u, v), y(u, v))|J(u, v)|\; du\; dv.$$

The derivation of this formula from (16.11.1) follows the usual lines. Break up Γ into N little basic regions $\Gamma_1, \dots, \Gamma_N$. These induce a decomposition of Ω into N little basic regions $\Omega_1, \dots, \Omega_N$. We can then write

$$\underset{\Gamma}{\int\int} f(x(u, v), y(u, v))|J(u, v)| \, du \, dv \underset{\underset{\text{additivity}}{\uparrow}}{=} \sum_{i=1}^{N} \underset{\Gamma_i}{\int\int} f(x(u, v), y(u, v))|J(u, v)| \, du \, dv$$

$$\underset{\underset{\text{Theorem (16.3.5) applied to } \Gamma_i}{\uparrow}}{=} \sum_{i=1}^{N} f(x(u_i^*, v_i^*), y(u_i^*, v_i^*)) \underset{\Gamma_i}{\int\int} |J(u, v)| \, du \, dv$$

$$\underset{\underset{\text{set } x_i^* = x(u_i^*, v_i^*), \, y_i^* = y(u_i^*, v_i^*)}{\uparrow}}{=} \sum_{i=1}^{N} f(x_i^*, y_i^*) \underset{\Gamma_i}{\int\int} |J(u, v)| \, du \, dv$$

$$\underset{\underset{\text{(16.11.1) applied to } \Gamma_i}{\uparrow}}{=} \sum_{i=1}^{N} f(x_i^*, y_i^*)(\text{area of } \Omega_i).$$

This last expression is a Riemann sum for

$$\underset{\Omega}{\int\int} f(x, y) \, dx \, dy$$

and tends to that integral as the maximum diameter of the Ω_i tends to zero. This we can ensure by letting the maximum diameter of the Γ_i tend to zero. ❑

Example 1 Evaluate

$$\underset{\Omega}{\int\int} (x + y)^2 \, dx \, dy$$

where Ω is the parallelogram bounded by the lines

$$x + y = 0, \qquad x + y = 1, \qquad 2x - y = 0, \qquad 2x - y = 3.$$

SOLUTION The parallelogram is shown in Figure 16.11.2. The boundaries suggest that we set

$$u = x + y, \qquad v = 2x - y.$$

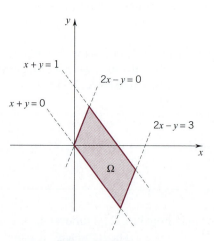

Figure 16.11.2

We want x and y in terms of u and v. Since

$$u + v = (x + y) + (2x - y) = 3x \qquad \text{and} \qquad 2u - v = (2x + 2y) - (2x - y) = 3y,$$

we have

$$x = \frac{u + v}{3}, \qquad y = \frac{2u - v}{3}.$$

This transformation maps the rectangle Γ of Figure 16.11.3 onto Ω. The Jacobian is given by

$$J(u, v) = \begin{vmatrix} \dfrac{\partial}{\partial u}\left(\dfrac{u + v}{3}\right) & \dfrac{\partial}{\partial u}\left(\dfrac{2u - v}{3}\right) \\[2ex] \dfrac{\partial}{\partial v}\left(\dfrac{u + v}{3}\right) & \dfrac{\partial}{\partial v}\left(\dfrac{2u - v}{3}\right) \end{vmatrix} = \begin{vmatrix} \frac{1}{3} & \frac{2}{3} \\[1ex] \frac{1}{3} & -\frac{1}{3} \end{vmatrix} = -\frac{1}{3}.$$

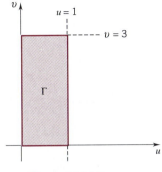

Figure 16.11.3

Therefore

$$\iint_{\Omega} (x + y)^2 \, dx \, dy = \iint_{\Gamma} u^2 \, |J(u, v)| \, du \, dv$$

$$= \tfrac{1}{3} \int_0^3 \int_0^1 u^2 \, du \, dv$$

$$= \tfrac{1}{3} \left(\int_0^3 dv \right) \left(\int_0^1 u^2 \, du \right) = \tfrac{1}{3}(3)\tfrac{1}{3} = \tfrac{1}{3}. \quad \square$$

Example 2 Evaluate

$$\iint_{\Omega} xy \, dx \, dy$$

where Ω is the first-quadrant region bounded by the curves

$$x^2 + y^2 = 4, \qquad x^2 + y^2 = 9, \qquad x^2 - y^2 = 1, \qquad x^2 - y^2 = 4.$$

SOLUTION The region is shown in Figure 16.11.4. The boundaries suggest that we set

$$u = x^2 + y^2, \qquad v = x^2 - y^2.$$

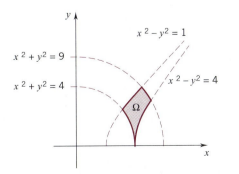

Figure 16.11.4

We want x and y in terms of u and v. Since

$$u + v = 2x^2 \qquad \text{and} \qquad u - v = 2y^2,$$

we have

$$x = \sqrt{\frac{u + v}{2}}, \qquad y = \sqrt{\frac{u - v}{2}}.$$

The transformation maps the rectangle Γ of Figure 16.11.5 onto Ω with Jacobian

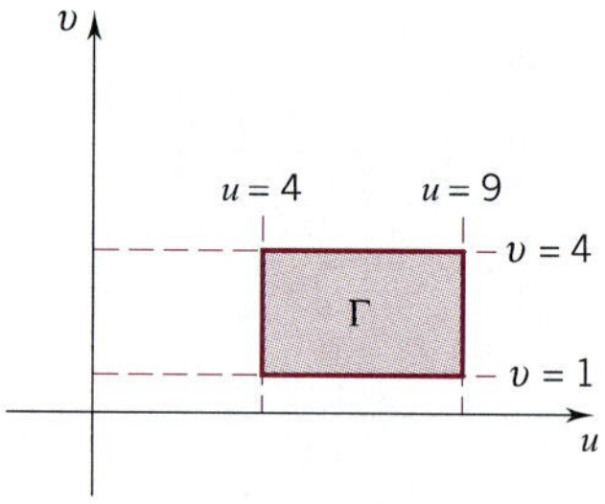

Figure 16.11.5

$$J(u, v) = \begin{vmatrix} \dfrac{\partial}{\partial u}\left(\sqrt{\dfrac{u + v}{2}}\right) & \dfrac{\partial}{\partial u}\left(\sqrt{\dfrac{u - v}{2}}\right) \\[2em] \dfrac{\partial}{\partial v}\left(\sqrt{\dfrac{u + v}{2}}\right) & \dfrac{\partial}{\partial v}\left(\sqrt{\dfrac{u - v}{2}}\right) \end{vmatrix} \overset{\text{check this}}{=} -\frac{1}{4\sqrt{u^2 - v^2}}.$$

Therefore

$$\iint\limits_{\Omega} xy \, dx \, dy = \iint\limits_{\Gamma} \left(\sqrt{\frac{u + v}{2}}\right)\left(\sqrt{\frac{u - v}{2}}\right)\left(\frac{1}{4\sqrt{u^2 - v^2}}\right) du \, dv$$

$$= \iint\limits_{\Gamma} \tfrac{1}{8} \, du \, dv = \tfrac{1}{8}\,(\text{area of } \Gamma) = \tfrac{15}{8}. \quad \square$$

In Section 16.5 you saw the formula for changing variables from rectangular coordinates (x, y) to polar coordinates (r, θ). The formula reads

$$\iint\limits_{\Omega} f(x, y) \, dx \, dy = \iint\limits_{\Gamma} f(r \cos \theta, r \sin \theta) \, r \, dr \, d\theta. \tag{16.5.3}$$

The factor r in the double integral over Γ is the Jacobian of the transformation $x = r \cos \theta, y = r \sin \theta$:

$$J(r, \theta) = \begin{vmatrix} \dfrac{\partial}{\partial r}(r \cos \theta) & \dfrac{\partial}{\partial r}(r \sin \theta) \\[1.5em] \dfrac{\partial}{\partial \theta}(r \cos \theta) & \dfrac{\partial}{\partial \theta}(r \sin \theta) \end{vmatrix} = \begin{vmatrix} \cos \theta & \sin \theta \\ -r \sin \theta & r \cos \theta \end{vmatrix} = r(\cos^2 \theta + \sin^2 \theta) = r.$$

As you can see, (16.5.3) is a special case of (16.11.2).

When changing variables in a triple integral we make three coordinate changes:

$$x = x(u, v, w), \qquad y = y(u, v, w), \qquad z = z(u, v, w).$$

If these functions carry a basic solid Γ onto a solid T, then, under conditions analogous to the two-dimensional case,

$$\text{volume of } T = \iiint\limits_{\Gamma} |J(u, v, w)| \, du \, dv \, dw$$

where now the Jacobian† is a three-by-three determinant:

$$J(u, v, w) = \begin{vmatrix} \dfrac{\partial x}{\partial u} & \dfrac{\partial y}{\partial u} & \dfrac{\partial z}{\partial u} \\[2ex] \dfrac{\partial x}{\partial v} & \dfrac{\partial y}{\partial v} & \dfrac{\partial z}{\partial v} \\[2ex] \dfrac{\partial x}{\partial w} & \dfrac{\partial y}{\partial w} & \dfrac{\partial z}{\partial w} \end{vmatrix}.$$

In this case the change of variables formula reads

$$\iiint_T f(x, y, z)\, dx\, dy\, dz = \iiint_\Gamma f(x(u, v, w), y(u, v, w), z(u, v, w)) \, |J(u, v, w)|\, du\, dv\, dw.$$

---

†A historical note on Jacobians: the study of these functional determinants goes back to a memoir by the German mathematician C. G. Jacobi, 1804–1851.

EXERCISES 16.11

In Exercises 1–11, find the Jacobian of the transformation.

1. $x = au + bv, \quad y = cu + dv.$ (linear transformation)

2. $x = u \cos \theta - v \sin \theta, \quad y = u \sin \theta + v \cos \theta.$ (rotation by θ)

3. $x = uv, \quad y = u^2 + v^2.$

4. $x = u \ln v, \quad y = uv.$

5. $x = uv^2, \quad y = u^2 v.$

6. $x = u - \ln v, \quad y = \ln u + v.$

7. $x = au, \quad y = bv, \quad z = cw.$

8. $x = v + w, \quad y = u + w, \quad z = u + v.$

9. $x = r \cos \theta, \quad y = r \sin \theta, \quad z = z.$ (cylindrical coordinates)

10. $x = \rho \sin \phi \cos \theta, \quad y = \rho \sin \phi \sin \theta, \quad z = \rho \cos \phi.$ (spherical coordinates)

11. $x = (1 + w \cos v) \cos u, \quad y = (1 + w \cos v) \sin u,$ $z = w \sin v.$

12. Every linear transformation

$$x = au + bv, \quad y = cu + dv \quad \text{with} \quad ad - bc \neq 0$$

maps lines of the uv-plane onto lines of the xy-plane. Find the image of:
(a) a vertical line $u = u_0$;
(b) a horizontal line $v = v_0$.

For Exercises 13–15, take Ω as the parallelogram bounded by

$$x + y = 0, \quad x + y = 1, \quad x - y = 0, \quad x - y = 2.$$

Evaluate.

13. $\displaystyle \iint_\Omega (x^2 - y^2)\, dx\, dy.$ 14. $\displaystyle \iint_\Omega 4xy\, dx\, dy.$

15. $\displaystyle \iint_\Omega (x + y) \cos [\pi(x - y)]\, dx\, dy.$

For Exercises 16–18, take Ω as the parallelogram bounded by $x - y = 0, \quad x - y = \pi, \quad x + 2y = 0, \quad x + 2y = \frac{1}{2}\pi.$

Evaluate.

16. $\displaystyle \iint_\Omega (x + y)\, dx\, dy.$

17. $\displaystyle \iint_\Omega \sin (x - y) \cos (x + 2y)\, dx\, dy.$

18. $\displaystyle \iint_\Omega \sin 3x\, dx\, dy.$

19. Let Ω be the first-quadrant region bounded by the curves $xy = 1, xy = 4, y = x, y = 4x.$
(a) Determine the area of Ω and (b) locate the centroid.

20. Show that the ellipse $b^2 x^2 + a^2 y^2 = a^2 b^2$ has area πab setting $x = ar \cos \theta, y = br \cos \theta.$

21. A homogeneous plate in the xy-plane is in the form of a parallelogram. The parallelogram is bounded by the lines $x + y = 0, x + y = 1, 3x - 2y = 0, 3x - 2y = 2.$ Calculate the moments of inertia of the plate about the three coordinate axes. Express your answers in terms of the mass of the plate.

22. Calculate the area of the region Ω bounded by the curves

$$x^2 - 2xy + y^2 + x + y = 0, \qquad x + y + 4 = 0.$$

HINT: Set $u = x - y$, $v = x + y$.

23. Calculate the area of the region Ω bounded by the curves

$$x^2 - 4xy + 4y^2 - 2x - y - 1 = 0, \qquad y = \tfrac{2}{5}.$$

24. Locate the centroid of the region in Exercise 22.

25. Calculate the area of the region Ω enclosed by the curve

$$11x^2 + 4\sqrt{3}xy + 7y^2 - 1 = 0.$$

HINT: Use a rotation $x = u \cos \theta - v \sin \theta$, $y = u \sin \theta + v \cos \theta$ such that the resulting uv-equation has no uv-term.

26. *(Generalized Polar Coordinates)* For a, b, α fixed positive numbers, define

$$x = ar(\cos \theta)^\alpha, \qquad y = br(\sin \theta)^\alpha.$$

(a) Show that the mapping carries the strip

$$0 \le r < \infty, \quad 0 \le \theta \le \tfrac{1}{2}\pi$$

onto the first quadrant of the xy-plane. [HINT: Find a point (r, θ) in the strip that maps onto (x, y) given that $x \ge 0$, $y \ge 0$.]

(b) Show that the mapping is one-to-one on the interior of the strip.

27. Determine the Jacobian of the mapping of Exercise 26.

28. Calculate the area enclosed by the curve $(x/a)^{1/4} + (y/b)^{1/4} = 1$ in the first quadrant by setting $x = ar \cos^8 \theta$, $y = br \sin^8 \theta$.

For Exercises 29–32, let T be the solid ellipsoid $x^2/a^2 + y^2/b^2 + z^2/c^2 \le 1$.

29. Calculate the volume of T by setting

$$x = a\rho \sin \phi \cos \theta, \quad y = b\rho \sin \phi \sin \theta, \quad z = c\rho \cos \phi.$$

30. Locate the centroid of the upper half of T.

31. View the upper half of T as a homogeneous solid of mass M. Find the moments of inertia of this solid about the coordinate axes.

32. Evaluate

$$\iiint_T \left(\frac{x^2}{a^2} + \frac{y^2}{b^2} + \frac{z^2}{c^2} \right) dx\, dy\, dz.$$

33. Evaluate

$$\int_{-\infty}^{\infty} \int_{-\infty}^{\infty} \frac{e^{-(x-y)^2}}{1 + (x+y)^2}\, dx\, dy$$

by integrating over the square S_a: $-a \le x \le a$, $-a \le y \le a$ and taking the limit as $a \to \infty$. HINT: Set $u = x - y$, $v = x + y$ and see (16.5.4).

■ CHAPTER HIGHLIGHTS

16.1 Multiple-Sigma Notation

16.2 The Double Integral Over a Rectangle

partition (p. 1037)   upper sum, lower sum (p. 1038)
double integral over a rectangle (p. 1039)

16.3 The Double Integral Over a Region

basic region (p. 1044)   double integral over a region (p. 1045)
double integral as a volume (p. 1045)   area as a double integral (p. 1046)
properties of the double integral (pp. 1046–1047)
average value over a region (p. 1046)   weighted average over a region (p. 1047)

16.4 The Evaluation of Double Integrals by Repeated Integrals

evaluation formulas (pp. 1049–1050)   symmetry (p. 1056)
separated variables over a rectangle (p. 1060)

16.5 The Double Integral as the Limit of Riemann Sums; Polar Coordinates

diameter of a set (p. 1061)   Riemann sum (p. 1062)

$$\iint_\Omega f(x, y)\, dx\, dy = \lim_{\text{diam } \Omega_i \to 0} \sum_{i=1}^{N} f(x_i^*, y_i^*)(\text{area of } \Omega_i)$$

evaluating double integrals using polar coordinates (p. 1063)
area formula (p. 1062)

16.6 Some Applications of Double Integration

mass of a plate (p. 1068)   center of mass of a plate (p. 1068)
centroid of a region (p. 1070)   moment of inertia of a plate (p. 1071)
radius of gyration (pp. 1073–1074)   parallel axis theorem (p. 1074)

16.7 Triple Integrals

box (p. 1076)   partition of a box (p. 1076)
upper sum, lower sum (p. 1077)   triple integral over a box (p. 1077)
basic solid (p. 1078)   triple integral over a basic solid (p. 1078)
volume as a triple integral (p. 1078)   properties of the triple integral (p. 1079)
average value (p. 1079)   weighted average (p. 1080)
mass (p. 1080)   center of mass (p. 1080)
centroid (p. 1081)   moment of inertia (p. 1081)

16.8 Reduction to Repeated Integrals

possible orders of integration (pp. 1082–1083)
separated variables over a box (p. 1089)

16.9 Cylindrical Coordinates

cylindrical coordinates (r, θ, z) (p. 1092)   cylindrical wedge (p. 1093)
evaluating triple integrals using cylindrical coordinates (p. 1093)
volume in cylindrical coordinates (p. 1094)

16.10 The Triple Integral as the Limit of Riemann Sums; Spherical Coordinates

$$\iiint\limits_{T} f(x, y, z)\, dx\, dy\, dz = \lim_{\text{diam } T_i \to 0} \sum_{i=1}^{N} f(x_i^*, y_i^*, z_i^*)(\text{volume of } T_i)$$

spherical coordinates (ρ, θ, ϕ) (p. 1099)
spherical wedge (p. 1100)
volume of a spherical wedge (p. 1101)
evaluating triple integrals using spherical coordinates (p. 1102)
volume in spherical coordinates (p. 1103)

16.11 Jacobians; Changing Variables in Multiple Integration

Jacobian: two-dimensional case (pp. 1107–1108)
change of variables in double integration (p. 1107)
Jacobian: three-dimensional case (p. 1111)
change of variables in triple integration (p. 1111)

■ PROJECTS AND EXPLORATIONS USING TECHNOLOGY

To do these exercises you will need a graphics calculator or a computer with graphing capability. The majority of these problems are open-ended so different approaches may be used to solve them. You should be aware that different approaches can result in slight variations in the answers. Round your numerical answers to at least four decimal places. The rounding method that your calculator or computer uses also may cause variations in answers.

The methods used to approximate ordinary definite integrals can be extended to multiple integrals, but due to the large number of calculations involved, numerical methods for multiple integrals are much more time-consuming. This is illustrated in the following exercises for integrals of the form

$$I = \int_a^b \int_{g(x)}^{h(x)} f(x, y) \, dy \, dx.$$

16.1 (a) Write a program for your calculator or computer to implement the trapezoidal rule for the integral I.

(b) Test your program on the integral

$$\int_1^5 \int_{\sin x}^{e^x} \cos(e^x + y) \, dy \, dx.$$

(c) Explain why it takes much longer for this program to run than the trapezoidal rule for the definite integral of a function of one variable. Your program might run slow enough that you will be able to hand time it. If so, find an approximate relationship between n, the number of subintervals, and the length of time for your program to run.

(d) If the number of subintervals is doubled when using the trapezoidal rule for a function of one variable, then the error is approximately quartered. See Section 8.8. Show that if J is the value of the integral of a function of one variable and J_k is the value of the trapezoidal approximation for k subintervals, then

$$J \cong \frac{4 J_{2k} - J_k}{3}.$$

(e) Use the values from part (b) and the formula in part (d) to test whether doubling the number of subintervals in the trapezoidal approximation of a double integral also quarters the error.

16.2 (a) Write a program for your calculator or computer to implement Simpson's rule for the integral I.

(b) Test your program on the integral

$$\int_1^5 \int_{\sin x}^{e^x} \cos(e^x + y) \, dy \, dx.$$

(c) Explain why it takes much longer for this program to run than Simpson's rule for a function of one variable. Your program might run slow enough that you will be able to hand time it. If so, find the approximate relationship between n, the number of subintervals, and the length of time for your program to run.

(d) If the number of subintervals is doubled when using Simpson's rule for a function of one variable, then the error is approximately divided by 16. See Section 8.8. Show that if J is the value of the integral of a function of one variable and J_k is the value of Simpson's rule approximation for k subintervals, then

$$J \cong \frac{16 J_{2k} - J_k}{15}.$$

(e) Use the values from part (b) and the formula in part (d) to test whether doubling the number of subintervals in Simpson's rule approximation of a double integral also divides the error by 16.

16.3 (a) Write a program for your calculator or computer to approximate I using random numbers.

(b) Run your program 10 times on the integral

$$\int_1^5 \int_{\sin x}^{e^x} \cos(e^x + y) \, dy \, dx.$$

(c) Compute the mean and standard deviation of the numbers found in part (b).

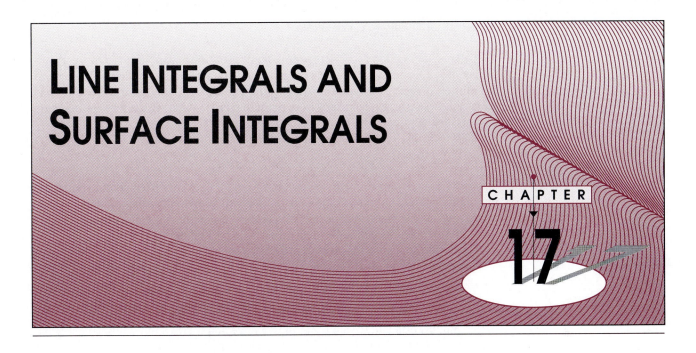

LINE INTEGRALS AND SURFACE INTEGRALS

CHAPTER

17

In this chapter we will study integration along curves and integration over surfaces. At the heart of this subject lie three great integration theorems: *Green's Theorem, Gauss' Theorem* (commonly known as *The Divergence Theorem*), and *Stokes's Theorem.*

All three theorems are ultimately based on *The Fundamental Theorem of Integral Calculus,* and all can be cast in the same general form:

An integral over a set S = a related integral over the boundary of S.

A word about terminology. Suppose that S is some subset of the plane or of three-dimensional space. A function that assigns a scalar to each point of S (say, the temperature at that point or the mass density at that point) is known in science as a *scalar field. A function that assigns a vector to each point of S* (say, the wind-velocity at that point) is called a *vector field.* We will be using this "field" language throughout.

■ 17.1 LINE INTEGRALS

We are led to the definition of *line integral* by the notion of work.

The Work Done by a Varying Force Over a Curved Path

The work done by a constant force $\mathbf{F}$ on an object that moves along a straight line is, by definition, the component of $\mathbf{F}$ in the direction of the displacement multiplied by the length of the displacement vector $\mathbf{d}$ (see Exercises 12.3):

$$W = (\text{comp}_{\mathbf{d}} \mathbf{F}) \, \|\mathbf{d}\|$$

We can write this more briefly as a dot product:

(17.1.1)

$$W = \mathbf{F} \cdot \mathbf{d}$$

1115

This elementary notion of work is useful, but it is not sufficient. Consider, for example, an object that moves through a magnetic field or a gravitational field. The path of the motion is then usually not a straight line but a curve, and the force, rather than remaining constant, tends to vary from point to point. What we want now is a notion of work that applies to this more general situation.

Let's suppose that an object moves along a curve

$$C: \quad \mathbf{r}(u) = x(u)\mathbf{i} + y(u)\mathbf{j} + z(u)\mathbf{k}, \qquad u \in [a, b]$$

subject to a continuous force $\mathbf{F}$. (The vector field $\mathbf{F}$ may vary from point to point, not only in magnitude but also in direction.) We will suppose that the curve is *smooth;* namely, we will suppose that the tangent vector $\mathbf{r}'$ is continuous and never zero. What we want to do here is define the total work done by $\mathbf{F}$ along the curve C.

To decide how to do this, we begin by focusing our attention on what happens over a short parameter interval $[u, u + h]$. As an estimate for the work done over this interval we can use the dot product

$$\mathbf{F}(\mathbf{r}(u)) \cdot [\mathbf{r}(u + h) - \mathbf{r}(u)].$$

In making this estimate we are using the force vector at $\mathbf{r}(u)$ and we are replacing the curved path from $\mathbf{r}(u)$ to $\mathbf{r}(u + h)$ by the line segment from $\mathbf{r}(u)$ to $\mathbf{r}(u + h)$. (See Figure 17.1.1.) If we set

$$W(u) = \text{total work done by } \mathbf{F} \text{ from } \mathbf{r}(a) \text{ to } \mathbf{r}(u), \text{ and}$$

$$W(u + h) = \text{total work done by } \mathbf{F} \text{ from } \mathbf{r}(a) \text{ to } \mathbf{r}(u + h),$$

then the work done by $\mathbf{F}$ from $\mathbf{r}(u)$ to $\mathbf{r}(u + h)$ must be the difference

$$W(u + h) - W(u).$$

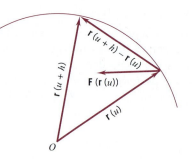

Figure 17.1.1

Bringing our estimate into play, we are led to the approximate equation

$$W(u + h) - W(u) \cong \mathbf{F}(\mathbf{r}(u)) \cdot [\mathbf{r}(u + h) - \mathbf{r}(u)],$$

which, upon division by h, becomes

$$\frac{W(u + h) - W(u)}{h} \cong \mathbf{F}(\mathbf{r}(u)) \cdot \frac{[\mathbf{r}(u + h) - \mathbf{r}(u)]}{h}.$$

The quotients here are average rates of change, and the equation is only an approximate one. The notion of work is made precise by requiring that both sides have exactly the same limit as h tends to zero; in other words, by requiring that

$$W'(u) = \mathbf{F}(\mathbf{r}(u)) \cdot \mathbf{r}'(u).$$

The rest is now determined. Since

$$W(a) = 0 \quad \text{and} \quad W(b) = \text{total work done by } \mathbf{F} \text{ on } C,$$

we have

$$\text{total work done by } \mathbf{F} \text{ on } C = W(b) - W(a) = \int_a^b W'(u)\, du = \int_a^b [\mathbf{F}(\mathbf{r}(u)) \cdot \mathbf{r}'(u)]\, du.$$

In short, we have arrived at the following definition of work:

(17.1.2)

$$W = \int_a^b [\mathbf{F}(\mathbf{r}(u)) \cdot \mathbf{r}'(u)]\, du.$$

Example 1 Determine the work done by the vector field

$$\mathbf{F}(x, y, z) = xy\,\mathbf{i} + 2\mathbf{j} + 4z\,\mathbf{k}$$

along the circular helix C: $\quad \mathbf{r}(u) = \cos u\,\mathbf{i} + \sin u\,\mathbf{j} + u\,\mathbf{k}, \qquad 0 \le u \le 2\pi.$

SOLUTION First we calculate $\mathbf{F}(\mathbf{r}(u))$ and $\mathbf{r}'(u)$:

$$\mathbf{F}(\mathbf{r}(u)) = \cos u \sin u\,\mathbf{i} + 2\mathbf{j} + 4u\,\mathbf{k},$$

$$\mathbf{r}'(u) = -\sin u\,\mathbf{i} + \cos u\,\mathbf{j} + \mathbf{k}.$$

Now,

$$\mathbf{F}(\mathbf{r}(u)) \cdot \mathbf{r}'(u) = -\sin^2 u \cos u + 2 \cos u + 4u$$

and

$$W = \int_0^{2\pi} \mathbf{F}(\mathbf{r}(u)) \cdot \mathbf{r}'(u)\, du = \int_0^{2\pi} (-\sin^2 u \cos u + 2 \cos u + 4u)\, du$$

$$= \left[-\frac{\sin^3 u}{3} + 2 \sin u + 2u^2 \right]_0^{2\pi} = 8\pi^2. \quad \square$$

Line Integrals

The integral on the right of (17.1.2) can be calculated not only for a force function $\mathbf{F}$ but for any vector field $\mathbf{h}$ continuous on C.

DEFINITION 17.1.3 LINE INTEGRAL

Let $\mathbf{h}(x, y, z) = h_1(x, y, z)\mathbf{i} + h_2(x, y, z)\mathbf{j} + h_3(x, y, z)\mathbf{k}$ be a vector field continuous on a smooth curve

$$C: \quad \mathbf{r}(u) = x(u)\mathbf{i} + y(u)\mathbf{j} + z(u)\mathbf{k}, \qquad u \in [a, b].$$

The *line integral* of $\mathbf{h}$ over C is the number

$$\int_C \mathbf{h}(\mathbf{r}) \cdot d\mathbf{r} = \int_a^b [\mathbf{h}(\mathbf{r}(u)) \cdot \mathbf{r}'(u)]\, du.$$

Although we stated this definition in terms of three-dimensional vector fields and curves in space, it also includes the two-dimensional case: $\mathbf{h}(x, y) = h_1(x, y)\mathbf{i} + h_2(x, y)\mathbf{j}$ is a two-dimensional vector field and C: $\mathbf{r}(u) = x(u)\mathbf{i} + y(u)\mathbf{j}$ is a smooth plane curve. Some of the examples which follow are two-dimensional.

Note also that, while we speak of integrating over C, we actually carry out the calculations over the parameter set $[a, b]$. If our definition of line integral is to make sense, the line integral as defined must be independent of the particular parametrization chosen for C. Within the limitations spelled out in the following this is indeed the case:

THEOREM 17.1.4

Let $\mathbf{h}$ be a vector field continuous on a smooth curve C. The line integral

$$\int_C \mathbf{h}(\mathbf{r}) \cdot d\mathbf{r} = \int_a^b [\mathbf{h}(\mathbf{r}(u)) \cdot \mathbf{r}'(u)] \, du$$

is left invariant by every *sense-preserving* change of parameter.

NOTE: Sense-preserving and sense-reversing changes of parameter were discussed in Section 13.3.

PROOF Suppose that ϕ maps $[c, d]$ onto $[a, b]$ and that ϕ' is positive and continuous on $[c, d]$. We must show that the line integral over C as parametrized by

$$\mathbf{R}(w) = \mathbf{r}(\phi(w)), \qquad w \in [c, d]$$

equals the line integral over C as parametrized by $\mathbf{r}$. The argument is straightforward:

$$\int_C \mathbf{h}(\mathbf{R}) \cdot d\mathbf{R} = \int_c^d [\mathbf{h}(\mathbf{R}(w)) \cdot \mathbf{R}'(w)] \, dw$$

$$= \int_c^d [\mathbf{h}(\mathbf{r}(\phi(w))) \cdot \mathbf{r}'(\phi(w)) \, \phi'(w)] \, dw$$

$$\left.\begin{array}{l} \text{Set } u = \phi(w), \, du = \phi'(w) \, dw. \\ \text{At } w = c, \, u = a; \text{ at } w = d, \, u = b. \end{array}\right\} \qquad = \int_c^d [\mathbf{h}(\mathbf{r}(\phi(w))) \cdot \mathbf{r}'(\phi(w))]\phi'(w) \, dw$$

$$= \int_a^b [\mathbf{h}(\mathbf{r}(u)) \cdot \mathbf{r}'(u)] \, du = \int_C \mathbf{h}(\mathbf{r}) \cdot d\mathbf{r}. \quad \square$$

Example 2 Calculate

$$\int_C \mathbf{h}(\mathbf{r}) \cdot d\mathbf{r}$$

given that

$$\mathbf{h}(x, y) = xy\,\mathbf{i} + y^2\mathbf{j} \qquad \text{and} \qquad C: \ \mathbf{r}(u) = u\,\mathbf{i} + u^2\mathbf{j}, \qquad u \in [0, 1].$$

SOLUTION Here

$$x(u) = u, \qquad y(u) = u^2$$

and

$$\begin{aligned}
\mathbf{h}(\mathbf{r}(u)) \cdot \mathbf{r}'(u) &= [x(u)y(u)\mathbf{i} + [y(u)]^2\mathbf{j}] \cdot [x'(u)\mathbf{i} + y'(u)\mathbf{j}] \\
&= x(u)y(u)x'(u) + [y(u)]^2 y'(u) \\
&= u(u^2)1 + u^4(2u) = u^3 + 2u^5.
\end{aligned}$$

If follows that

$$\int_C \mathbf{h}(\mathbf{r}) \cdot d\mathbf{r} = \int_0^1 (u^3 + 2u^5)\, du = \left[\tfrac{1}{4}u^4 + \tfrac{1}{3}u^6 \right]_0^1 = \tfrac{7}{12}. \quad \Box$$

Example 3 Integrate the vector field $\mathbf{h}(x, y, z) = xy\,\mathbf{i} + yz\,\mathbf{j} + xz\,\mathbf{k}$ over the twisted cubic

$$\mathbf{r}(u) = u\,\mathbf{i} + u^2\,\mathbf{j} + u^3\,\mathbf{k}$$

from $(-1, 1, -1)$ to $(1, 1, 1)$.

SOLUTION The path of integration begins at $u = -1$ and ends at $u = 1$. In this case

$$x(u) = u, \qquad y(u) = u^2, \qquad z(u) = u^3.$$

Therefore

$$\begin{aligned}
\mathbf{h}(\mathbf{r}(u)) \cdot \mathbf{r}'(u) &= [x(u)y(u)\mathbf{i} + y(u)z(u)\mathbf{j} + x(u)z(u)\mathbf{k}] \cdot [x'(u)\mathbf{i} + y'(u)\mathbf{j} + z'(u)\mathbf{k}] \\
&= x(u)y(u)x'(u) + y(u)z(u)y'(u) + x(u)z(u)z'(u) \\
&= u(u^2)1 + u^2(u^3)2u + u(u^3)3u^2 \\
&= u^3 + 5u^6
\end{aligned}$$

and

$$\int_C \mathbf{h}(\mathbf{r}) \cdot d\mathbf{r} = \int_{-1}^1 (u^3 + 5u^6)\, du = \left[\tfrac{1}{4}u^4 + \tfrac{5}{7}u^7 \right]_{-1}^1 = \tfrac{10}{7}. \quad \Box$$

If a curve C is not smooth but is made up of a finite number of adjoining smooth pieces $C_1, C_2, \ldots, C_n$, then we define the integral over C as the sum of the integrals over the C_i:

(17.1.5)

$$\int_C = \int_{C_1} + \int_{C_2} + \cdots + \int_{C_n}$$

A curve of this type is said to be *piecewise smooth*. See Figure 17.1.2.

All polygonal paths are piecewise-smooth curves. In the next problem we integrate over a triangle. We do this by integrating over each of the sides and then adding up the results. Observe that the directed line segment that begins at $\mathbf{a} = (a_1, a_2)$ and ends at $\mathbf{b} = (b_1, b_2)$ can be parametrized by setting

$$\mathbf{r}(u) = (1 - u)\mathbf{a} + u\mathbf{b}, \qquad u \in [0, 1].$$

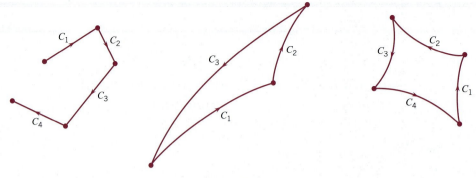

Figure 17.1.2

Example 4 Evaluate the line integral

$$\int_C \mathbf{h(r)} \cdot \mathbf{dr}$$

if $\mathbf{h}(x, y) = e^y \mathbf{i} - \sin \pi x \mathbf{j}$ and C is the triangle with vertices $(1, 0)$, $(0, 1)$, $(-1, 0)$ traversed counterclockwise.

SOLUTION The path C is made up of three line segments:

$$C_1: \quad \mathbf{r}(u) = (1 - u)\mathbf{i} + u\mathbf{j}, \quad u \in [0, 1],$$
$$C_2: \quad \mathbf{r}(u) = (1 - u)\mathbf{j} + u(-\mathbf{i}) = -u\mathbf{i} + (1 - u)\mathbf{j}, \quad u \in [0, 1],$$
$$C_3: \quad \mathbf{r}(u) = (1 - u)(-\mathbf{i}) + u\mathbf{i} = (2u - 1)\mathbf{i}, \quad u \in [0, 1].$$

See Figure 17.1.3.

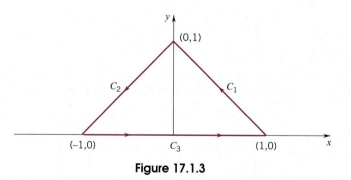

Figure 17.1.3

As you can verify,

$$\int_{C_1} \mathbf{h(r)} \cdot \mathbf{dr} = \int_0^1 \left[e^{y(u)}x'(u) - \sin[\pi x(u)]y'(u) \right] du$$

$$= \int_0^1 \left[-e^u - \sin[\pi(1 - u)] \right] du = 1 - e - \frac{2}{\pi};$$

$$\int_{C_2} \mathbf{h}(\mathbf{r}) \cdot d\mathbf{r} = \int_0^1 \left[e^{y(u)} x'(u) - \sin\left[\pi x(u) \right] y'(u) \right] du$$

$$= \int_0^1 \left[-e^{1-u} + \sin\left(-\pi u \right) \right] du = 1 - e - \frac{2}{\pi};$$

$$\int_{C_1} \mathbf{h}(\mathbf{r}) \cdot d\mathbf{r} = \int_0^1 \left[e^{y(u)} x'(u) - \sin\left[\pi x(u) \right] y'(u) \right] du = \int_0^1 2 \, du = 2.$$

The integral over the entire triangle is the sum of these integrals:

$$\int_C \mathbf{h}(\mathbf{r}) \cdot d\mathbf{r} = \left(1 - e - \frac{2}{\pi} \right) + \left(1 - e - \frac{2}{\pi} \right) + 2 = 4 - 2e - \frac{4}{\pi} \cong -2.71. \quad \square$$

When we integrate over a parametrized curve, we integrate in the direction determined by the parametrization. If we integrate in the opposite direction, our answer is altered by a factor of -1. To be precise, let C be a piecewise-smooth curve and let $-C$ denote the same path traversed in the *opposite direction*. (See Figure 17.1.4.) If C is parameterized by a vector function $\mathbf{r}$ defined on $[a, b]$, then $-C$ can be parameterized by setting

$$R(w) = \mathbf{r}(a + b - w), \qquad w \in [a, b]. \qquad \text{(Section 13.3)}$$

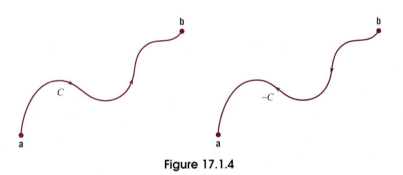

Figure 17.1.4

Our assertion is that

(17.1.6)
$$\int_{-C} \mathbf{h}(\mathbf{R}) \cdot d\mathbf{R} = -\int_C \mathbf{h}(\mathbf{r}) \cdot d\mathbf{r}$$

or more briefly that

(17.1.7)
$$\int_{-C} = -\int_C$$

We leave the proof of this to you. ❏

We were led to the definition of line integral by the notion of work. From (17.1.2) it is clear that, if a force **F** is continually applied to an object that moves over a piecewise-smooth curve C, then the work done by **F** is the line integral of **F** over C:

(17.1.8)

$$W = \int_C \mathbf{F}(\mathbf{r}) \cdot d\mathbf{r}.$$

Example 5 An object, acted upon by various forces, moves along the parabola $y = 3x^2$ from the origin to the point $(1, 3)$. (Figure 17.1.5.) One of the forces acting on the object is $\mathbf{F}(x, y) = x^3\,\mathbf{i} + y\,\mathbf{j}$. Calculate the work done by **F**.

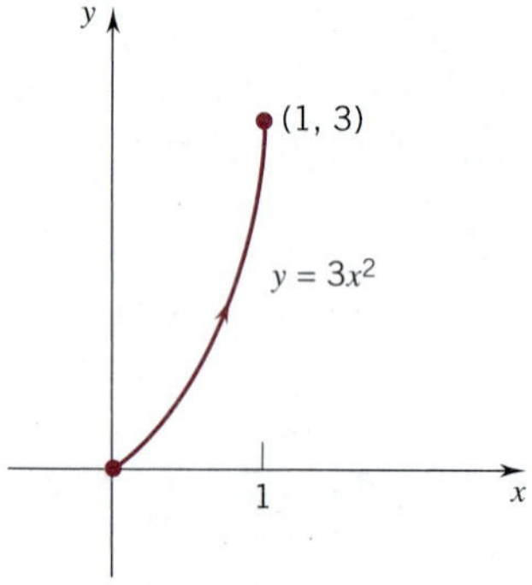

Figure 17.1.5

SOLUTION We can parameterize the path by setting

$$C: \quad \mathbf{r}(u) = u\,\mathbf{i} + 3u^2\,\mathbf{j}, \quad u \in [0, 1].$$

Here

$$x(u) = u, \qquad y(u) = 3u^2$$

and

$$\mathbf{F}(\mathbf{r}(u)) \cdot \mathbf{r}'(u) = [x(u)]^3 x'(u) + y(u)y'(u) = u^3(1) + 3u^2(6u) = 19u^3.$$

If follows that

$$W = \int_C \mathbf{F}(\mathbf{r}) \cdot d\mathbf{r} = \int_0^1 19u^3\, du = \tfrac{19}{4}. \quad \square$$

If an object of mass m moves so that at time t it has position $\mathbf{r}(t)$, then, from Newton's second law $\mathbf{F} = m\mathbf{a}$, the total force acting on the object at time t must be $m\mathbf{r}''(t)$.

Example 6 An object of mass m moves from time $t = 0$ to $t = 1$ so that its position at time t is given by the vector function

$$\mathbf{r}(t) = \alpha t^2\,\mathbf{i} + \sin \beta t\,\mathbf{j} + \cos \beta t\,\mathbf{k}, \qquad \alpha, \beta \text{ constant.}$$

Find the total force acting on the object at time t and calculate the total work done by this force.

SOLUTION Differentiation gives

$$\mathbf{r}'(t) = 2\alpha t\,\mathbf{i} + \beta \cos \beta t\,\mathbf{j} - \beta \sin \beta t\,\mathbf{k}, \qquad \mathbf{r}''(t) = 2\alpha\,\mathbf{i} - \beta^2 \sin \beta t\,\mathbf{j} - \beta^2 \cos \beta t\,\mathbf{k}.$$

The total force on the object at time t is therefore

$$m\mathbf{r}''(t) = m(2\alpha\,\mathbf{i} - \beta^2 \sin \beta t\,\mathbf{j} - \beta^2 \cos \beta t\,\mathbf{k}).$$

We can calculate the total work done by this force by integrating the force over the curve

$$C: \quad \mathbf{r}(t) = \alpha t^2\,\mathbf{i} + \sin \beta t\,\mathbf{j} + \cos \beta t\,\mathbf{k}, \qquad t \in [0, 1].$$

We leave it to you to verify that

$$W = \int_0^1 [m\mathbf{r}''(t) \cdot \mathbf{r}'(t)]\, dt = m\int_0^1 4\alpha^2 t\, dt = 2\alpha^2 m. \quad \square$$

EXERCISES 17.1

1. Integrate $\mathbf{h}(x, y) = y\mathbf{i} + x\mathbf{j}$ over the indicated path:
   (a) $\mathbf{r}(u) = u\mathbf{i} + u^2\mathbf{j}, \quad u \in [0, 1]$.
   (b) $\mathbf{r}(u) = u^3\mathbf{i} - 2u\mathbf{j}, \quad u \in [0, 1]$.

2. Integrate $\mathbf{h}(x, y) = x\mathbf{i} + y\mathbf{j}$ over the paths of Exercise 1.

3. Integrate $\mathbf{h}(x, y) = y\mathbf{i} + x\mathbf{j}$ over the unit circle traversed clockwise.

4. Integrate $\mathbf{h}(x, y) = y^2x\mathbf{i} + 2\mathbf{j}$ over the indicated path:
   (a) $\mathbf{r}(u) = e^u\mathbf{i} + e^{-u}\mathbf{j}, \quad u \in [0, 1]$.
   (b) $\mathbf{r}(u) = (1 - u)\mathbf{i}, \quad u \in [0, 2]$.

5. Integrate $\mathbf{h}(x, y) = (x - y)\mathbf{i} + xy\mathbf{j}$ over the indicated path:
   (a) The line segment from $(2, 3)$ to $(1, 2)$.
   (b) The line segment from $(1, 2)$ to $(2, 3)$.

6. Integrate $\mathbf{h}(x, y) = x^{-1}y^{-2}\mathbf{i} + x^{-2}y^{-1}\mathbf{j}$ over the indicated path:
   (a) $\mathbf{r}(u) = \sqrt{u}\mathbf{i} + \sqrt{1 + u}\mathbf{j}, \quad u \in [1, 4]$.
   (b) The line segment from $(1, 1)$ to $(2, 2)$.

7. Integrate $\mathbf{h}(x, y) = y\mathbf{i} - x\mathbf{j}$ over the triangle with vertices $(-2, 0), (2, 0), (0, 2)$ traversed counterclockwise.

8. Integrate $\mathbf{h}(x, y) = e^{x-y}\mathbf{i} + e^{x+y}\mathbf{j}$ over the line segment from $(-1, 1)$ to $(1, 2)$.

9. Integrate $\mathbf{h}(x, y) = (x + y)\mathbf{i} + (y^2 - x)\mathbf{j}$ over the closed curve that begins at $(-1, 0)$, goes along the x-axis to $(1, 0)$, and returns to $(-1, 0)$ by the upper part of the unit circle.

10. Integrate $\mathbf{h}(x, y) = 3x^2y\mathbf{i} + (x^3 + 2y)\mathbf{j}$ over the square with vertices $(0, 0), (1, 0), (1, 1), (0, 1)$ traversed counterclockwise.

11. Integrate $\mathbf{h}(x, y, z) = yz\mathbf{i} + x^2\mathbf{j} + xz\mathbf{k}$ over the indicated path:
    (a) The line segment from $(0, 0, 0)$ to $(1, 1, 1)$.
    (b) $\mathbf{r}(u) = u\mathbf{i} + u^2\mathbf{j} + u^3\mathbf{k}, \quad u \in [0, 1]$.

12. Integrate $\mathbf{h}(x, y, z) = e^x\mathbf{i} + e^y\mathbf{j} + e^z\mathbf{k}$ over the paths of Exercise 11.

13. Integrate $\mathbf{h}(x, y, z) = \cos x\mathbf{i} + \sin y\mathbf{j} + yz\mathbf{k}$ over the indicated path:
    (a) The line segment from $(0, 0, 0)$ to $(2, 3, -1)$.
    (b) $\mathbf{r}(u) = u^2\mathbf{i} - u^3\mathbf{j} + u\mathbf{k}, \quad u \in [0, 1]$.

14. Integrate $\mathbf{h}(x, y, z) = xy\mathbf{i} + x^2z\mathbf{j} + xyz\mathbf{k}$ over the indicated path:
    (a) The line segment from $(0, 0, 0)$ to $(2, -1, 1)$.
    (b) $\mathbf{r}(u) = e^u\mathbf{i} + e^{-u}\mathbf{j} + u\mathbf{k}, \quad u \in [0,1]$.

15. Calculate the work done by the force $\mathbf{F}(x, y, z) = x\mathbf{i} + xy\mathbf{j} + xyz\mathbf{k}$ applied to an object that moves in a straight line from $(0, 1, 4)$ to $(1, 0, -4)$.

16. A mass m, moving in a force field, traces out a circular arc at constant speed. Show that the force field does no work. Give a physical explanation for this.

17. Find the work done by the force $\mathbf{F}(x, y) = (x + 2)y\mathbf{i} + (2x + y)\mathbf{j}$ applied to an object that moves along the indicated path:
    (a) The parabola from $y = x^2$ from $(0, 0)$ to $(2, 4)$.
    (b) The quarter circle $\mathbf{r}(u) = \cos u\mathbf{i} + \sin u\mathbf{j}, \quad u \in [0, \pi/2]$.

18. Find the work done by the force $\mathbf{F}(x, y) = x \cos y\mathbf{i} - y \sin x\mathbf{j}$ applied to an object that moves along the line segments that connect $(0, 0), (1, 0), (1, 1)$, and $(0, 1)$ in that order.

19. Find the work done by the force $\mathbf{F}(x, y, z) = x^2\mathbf{i} + xy\mathbf{j} + z^2\mathbf{k}$ applied to an object that moves along the circular helix $\mathbf{r}(u) = \cos u\mathbf{i} + \sin u\mathbf{j} + u\mathbf{k}, \quad u \in [0, 2\pi]$.

20. Find the work done by the force $\mathbf{F}(x, y, z) = yz\mathbf{i} + xz\mathbf{j} + xy\mathbf{k}$ applied to an object that moves along the line segments that connect $(0, 0, 0), (1, 0, 0), (1, 1, 0)$, and $(1, 1, 1)$ in that order.

21. Let C: $\mathbf{r} = \mathbf{r}(u), u \in [a, b]$ be a smooth curve and $\mathbf{q}$ a fixed vector. Show that

    $$\int_C \mathbf{q} \cdot d\mathbf{r} = \mathbf{q} \cdot [\mathbf{r}(b) - \mathbf{r}(a)] \qquad \text{and}$$

    $$\int_C \mathbf{r} \cdot d\mathbf{r} = \frac{\|\mathbf{r}(b)\|^2 - \|\mathbf{r}(a)\|^2}{2}.$$

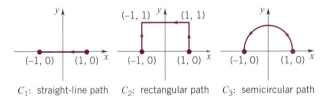

C_1: straight-line path C_2: rectangular path C_3: semicircular path

22. The preceding figure shows three paths from $(1, 0)$ to $(-1, 0)$. Calculate the line integral of

    $$\mathbf{h}(x, y) = x^2\mathbf{i} + y\mathbf{j}$$

    over (a) the straight-line path; (b) the rectangular path; (c) the semicircular path.

23. Let f be a continuous real-valued function of a real variable. Show that, if

    $\mathbf{f}(x, y, z) = f(x)\mathbf{i} \qquad \text{and} \qquad C: \mathbf{r}(u) = u\mathbf{i}, \quad u \in [a, b]$,

    then

    $$\int_C \mathbf{f}(\mathbf{r}) \cdot d\mathbf{r} = \int_a^b f(u)\, du.$$

24. (*Linearity*) Show that, if **f** and **g** are continuous vector fields and C is piecewise smooth, then

(17.1.9)
$$\int_C [\alpha\mathbf{f}(\mathbf{r}) + \beta\mathbf{g}(\mathbf{r})] \cdot d\mathbf{r}$$
$$= \alpha\int_C \mathbf{f}(\mathbf{r}) \cdot d\mathbf{r} + \beta\int_C \mathbf{g}(\mathbf{r}) \cdot d\mathbf{r}$$

for all real α, β.

25. The force $\mathbf{F}(x, y) = -\frac{1}{2}[y\mathbf{i} - x\mathbf{j}]$ is continually applied to an object that orbits an ellipse in standard position. Find a relation between the work done and the area of the ellipse.

26. An object of mass m moves from time $t = 0$ to $t = 1$ so that its position at time t is given by the vector function

$$\mathbf{r}(t) = \alpha t\mathbf{i} + \beta t^2\mathbf{j}, \quad \alpha, \beta \text{ constant}$$

Find the total force acting on the object at time t and calculate the work done by that force during the time interval $[0, 1]$.

27. Exercise 26 with $\mathbf{r}(t) = \alpha t\mathbf{i} + \beta t^2\mathbf{j} + \gamma t^3\mathbf{k}$.

28. (*Important*) The *circulation* of a vector field **v** around an oriented closed curve C is by definition the line integral

$$\int_C \mathbf{v}(\mathbf{r}) \cdot d\mathbf{r}.$$

Let **v** be the velocity field of a fluid in counterclockwise circular motion about the z-axis with constant angular speed ω.
(a) Verify that $\mathbf{v}(\mathbf{r}) = \omega\mathbf{k} \times \mathbf{r}$.
(b) Show that the circulation of **v** around any circle C in the xy-plane with center at the origin is $\pm 2\omega$ times the area of the circle.

29. Let **v** be the velocity field of a fluid that moves radially from the origin: $\mathbf{v} = f(x, y)\mathbf{r}$. What is the circulation of **v** around a circle C centered at the origin?

30. The force exerted on a charged particle at the point $(x, y) \neq (0, 0)$ in the xy-plane by an infinitely long uniformly charged wire lying along the z-axis is given by

$$\mathbf{F}(x, y) = \frac{k(x\mathbf{i} + y\mathbf{j})}{x^2 + y^2}, \quad k > 0, \quad \text{constant}.$$

Find the work done by **F** in moving the particle along the indicated path:
(a) The line segment from $(1, 0)$ to $(1, 2)$.
(b) The line segment from $(0, 1)$ to $(1, 1)$.

31. An inverse-square force field is given by

$$\mathbf{F}(x, y, z) = \frac{k\mathbf{r}}{\|\mathbf{r}\|^3} = \frac{k(x\mathbf{i} + y\mathbf{j} + z\mathbf{k})}{(x^2 + y^2 + z^2)^{3/2}}, \quad k > 0, \text{ constant}.$$

Find the work done by **F** in moving an object along the indicated path:
(a) The line segment from $(1, 0, 2)$ to $(1, 3, 2)$.
(b) From $(1, 0, 0)$ to $(0, \frac{5}{2}\sqrt{2}, \frac{5}{2}\sqrt{2})$ by the line segment from $(1, 0, 0)$ to $(5, 0, 0)$ and then to $(0, \frac{5}{2}\sqrt{2}, \frac{5}{2}\sqrt{2})$ along a path on the surface of the sphere $x^2 + y^2 + z^2 = 25$.

32. Assume that the earth is located at the origin of a rectangular coordinate system. The gravitational force on an object at the point (x, y) is given by

$$\mathbf{F}(x, y) = \frac{-k\mathbf{r}}{\|\mathbf{r}\|^3} = \frac{-k(x\mathbf{i} + y\mathbf{j})}{(x^2 + y^2)^{3/2}}, \quad k > 0, \quad \text{constant}.$$

Find the work done by **F** in moving an object from $(3, 0)$ to $(0, 4)$ along the indicated path:
(a) The first quadrant part of the ellipse $\mathbf{r}(u) = 3\cos u\mathbf{i} + 4\sin u\mathbf{j}, \quad u \in [0, \pi/2]$.
(b) The line segment connecting the two points.

33. An object moves from the point $(0, 0)$ to the point $(1, 0)$ along the curve $y = \alpha x(1 - x)$ in the force field $\mathbf{F}(x, y) = (y^2 + 1)\mathbf{i} + (x + y)\mathbf{j}$. Find α so that the work done by **F** is a minimum.

■ 17.2 THE FUNDAMENTAL THEOREM FOR LINE INTEGRALS

In general, if we integrate a vector field **h** from one point to another, the value of the line integral depends upon the path chosen. There is, however, an important exception. If the vector field is a *gradient field*,

$$\mathbf{h} = \nabla f,$$

then the value of the line integral depends only on the endpoints of the path and not on the path itself. The details are spelled out in the following theorem.

THEOREM 17.2.1   THE FUNDAMENTAL THEOREM FOR LINE INTEGRALS

Let C: $\mathbf{r} = \mathbf{r}(u)$, $u \in [a, b]$, be a piecewise-smooth curve that begins at $\mathbf{a} = \mathbf{r}(a)$ and ends at $\mathbf{b} = \mathbf{r}(b)$. If the scalar field f is continuously differentiable on an open set that contains the curve C, then

$$\int_C \nabla f(\mathbf{r}) \cdot d\mathbf{r} = f(\mathbf{b}) - f(\mathbf{a}).$$

NOTE: It is important to see that this result is an extension of the fundamental theorem of integral calculus: $\int_a^b f'(x)\, dx = f(b) - f(a)$. (See Theorem 5.4.2.)

PROOF   If C is smooth,

$$\int_C \nabla f(\mathbf{r}) \cdot d\mathbf{r} = \int_a^b [\nabla f(\mathbf{r}(u)) \cdot \mathbf{r}'(u)]\, du$$

$$= \int_a^b \frac{d}{du} [f(\mathbf{r}(u))]\, du$$

chain rule (15.4.1) ⟶

$$= f(\mathbf{r}(b)) - f(\mathbf{r}(a))$$
$$= f(\mathbf{b}) - f(\mathbf{a}).$$

If C is not smooth but only piecewise smooth, then we break up C into smooth pieces

$$C = C_1 \cup C_2 \cup \cdots \cup C_n.$$

With obvious notation,

$$\int_C \nabla f(\mathbf{r}) \cdot d\mathbf{r} = \int_{C_1} \nabla f(\mathbf{r}) \cdot d\mathbf{r} + \int_{C_2} \nabla f(\mathbf{r}) \cdot d\mathbf{r} + \cdots + \int_{C_n} \nabla f(\mathbf{r}) \cdot d\mathbf{r}$$

$$= [f(\mathbf{a}_1) - f(\mathbf{a}_0)] + [f(\mathbf{a}_2) - f(\mathbf{a}_1)] + \cdots + [f(\mathbf{a}_n) - f(\mathbf{a}_{n-1})]$$
$$= f(\mathbf{a}_n) - f(\mathbf{a}_0)$$
$$= f(\mathbf{b}) - f(\mathbf{a}). \quad \square$$

The theorem we just proved has an important corollary:

(17.2.2)   If the curve C is *closed* (that is, if $\mathbf{a} = \mathbf{b}$), then $f(\mathbf{b}) = f(\mathbf{a})$ and

$$\int_C \nabla f(\mathbf{r}) \cdot d\mathbf{r} = 0.$$

Example 1   Integrate the vector field $\mathbf{h}(x, y) = y^2 \mathbf{i} + (2xy - e^y)\mathbf{j}$ over the circular arc

$$C: \quad \mathbf{r}(u) = \cos u\, \mathbf{i} + \sin u\, \mathbf{j}, \qquad u \in [0, \tfrac{1}{2}\pi].$$

SOLUTION   First we try to determine whether **h** is a gradient.   We do this by applying Theorem 15.10.2.

Note that $\mathbf{h}(x, y)$ has the form $P(x, y)\mathbf{i} + Q(x, y)\mathbf{j}$ with

$$P(x, y) = y^2 \qquad \text{and} \qquad Q(x, y) = 2xy - e^y.$$

Since P and Q are continuously differentiable everywhere and

$$\frac{\partial P}{\partial y}(x, y) = 2y = \frac{\partial Q}{\partial x}(x, y),$$

we can conclude that **h** is a gradient.   Since the integral depends then only on the endpoints of C and not on C itself, we can simplify the computations by integrating over the line segment C' that joins these same endpoints.   (See Figure 17.2.1.)

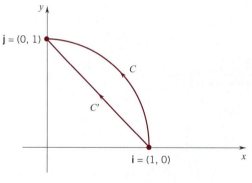

Figure 17.2.1

We parametrize C' by setting

$$\mathbf{r}(u) = (1 - u)\mathbf{i} + u\mathbf{j}, \qquad u \in [0, 1].$$

We then have

$$\int_C \mathbf{h}(\mathbf{r}) \cdot d\mathbf{r} = \int_{C'} \mathbf{h}(\mathbf{r}) \cdot d\mathbf{r}$$

$$= \int_0^1 [\mathbf{h}(\mathbf{r}(u)) \cdot \mathbf{r}'(u)] \, du$$

$$= \int_0^1 \left[[y(u)]^2 x'(u) + [2x(u)y(u) - e^{y(u)}]y'(u) \right] du$$

$$= \int_0^1 [2u - 3u^2 - e^u] \, du = \left[u^2 - u^3 - e^u \right]_0^1$$

$$= 1 - e \cong -1.718 \quad \square$$

ALTERNATIVE SOLUTION   Once we have recognized that $\mathbf{h}(x, y) = y^2\mathbf{i} + (2xy - e^y)\mathbf{j}$ is a gradient $\nabla f(x, y)$, we can determine $f(x, y)$ by the methods of

Section 15.10. Since

$$\frac{\partial f}{\partial x}(x, y) = y^2 \qquad \text{and} \qquad \frac{\partial f}{\partial y}(x, y) = 2xy - e^y,$$

we have

$$f(x, y) = xy^2 + \phi(y) \qquad \text{and therefore} \qquad \frac{\partial f}{\partial y}(x, y) = 2xy + \phi'(y).$$

The two expressions for $\partial f/\partial y$ can be reconciled only if

$$\phi'(y) = -e^y \qquad \text{and thus} \qquad \phi(y) = -e^y + c.$$

This means that

$$f(x, y) = xy^2 - e^y + c.$$

Since the curve C begins at $(1, 0)$ and ends at $(0, 1)$, we see that

$$\int_C \mathbf{h}(\mathbf{r}) \cdot d\mathbf{r} = f(0, 1) - f(1, 0) = (-e + c) - (-1 + c) = 1 - e. \quad \square$$

Example 2 Evaluate the line integral

$$\int_C \mathbf{h}(\mathbf{r}) \cdot d\mathbf{r}$$

where C is the unit circle

$$\mathbf{r}(u) = \cos u\, \mathbf{i} + \sin u\, \mathbf{j}, \qquad u \in [0, 2\pi].$$

and

$$\mathbf{h}(x, y) = (y^2 + y)\mathbf{i} + (2xy - e^y)\mathbf{j}.$$

SOLUTION Although $(y^2 + y)\mathbf{i} + (2xy - e^y)\mathbf{j}$ is not a gradient, part of it,

$$y^2\mathbf{i} + (2xy - e^y)\mathbf{j},$$

is a gradient. (See Example 1.) Thus, we can write $\mathbf{h}$ as

$$\mathbf{h}(x, y) = (y^2 + y)\mathbf{i} + (2xy - e^y)\mathbf{j} = \nabla f(x, y) + \mathbf{g}(x, y),$$

where $\mathbf{g}(x, y) = y\mathbf{i}$, and

$$\int_C \mathbf{h}(\mathbf{r}) \cdot d\mathbf{r} = \int_C \nabla f(\mathbf{r}) \cdot d\mathbf{r} + \int_C \mathbf{g}(\mathbf{r}) \cdot d\mathbf{r}.$$

Now, since we are integrating over a closed curve, the contribution of the gradient part is 0. The contribution of the remaining part is easy to evaluate:

$$\int_C \mathbf{g}(\mathbf{r}) \cdot d\mathbf{r} = \int_0^{2\pi} [\mathbf{g}(\mathbf{r}(u)) \cdot \mathbf{r}'(u)]\, du = \int_0^{2\pi} y(u)x'(u)\, du$$

$$= \int_0^{2\pi} -\sin^2 u\, du = -\pi. \quad \square$$

EXERCISES 17.2

In Exercises 1–11, calculate the line integral of **h** over the indicated curve.

1. $\mathbf{h}(x, y) = x\mathbf{i} + y\mathbf{j};$ $\mathbf{r}(u) = a \cos u\,\mathbf{i} + b \sin u\,\mathbf{j},$ $u \in [0, 2\pi].$

2. $\mathbf{h}(x, y) = (x + y)\mathbf{i} + y\mathbf{j};$ the curve of Exercise 1.

3. $\mathbf{h}(x, y) = \cos \pi y\,\mathbf{i} - \pi x \sin \pi y\,\mathbf{j};$ $\mathbf{r}(u) = u^2\mathbf{i} - u^3\mathbf{j},$ $u \in [0, 1].$

4. $\mathbf{h}(x, y) = (x^2 - y)\mathbf{i} + (y^2 - x)\mathbf{j};$ the curve of Exercise 1.

5. $\mathbf{h}(x, y) = xy^2\mathbf{i} + x^2y\mathbf{j};$ $\mathbf{r}(u) = u \sin \pi u\,\mathbf{i} + \cos \pi u^2\,\mathbf{j},$ $u \in [0, 1].$

6. $\mathbf{h}(x, y) = (1 + e^y)\mathbf{i} + (xe^y - x)\mathbf{j};$ the square with vertices $(-1, -1), (1, -1), (1, 1), (-1, 1)$ traversed counterclockwise.

7. $\mathbf{h}(x, y) = (2xy - y^2)\mathbf{i} + (x^2 - 2xy)\mathbf{j};$ $\mathbf{r}(u) = \cos u\,\mathbf{i} + \sin u\,\mathbf{j},$ $u \in [0, \pi].$

8. $\mathbf{h}(x, y) = 3x(x^2 + y^4)^{1/2}\mathbf{i} + 6y^3(x^2 + y^4)^{1/2}\mathbf{j};$ the circular arc $y = (1 - x^2)^{1/2}$ from $(1,0)$ to $(-1, 0).$

9. $\mathbf{h}(x, y) = 3x(x^2 + y^4)^{1/2}\mathbf{i} + 6y^3(x^2 + y^4)^{1/2}\mathbf{j};$ the arc $y = -(1 - x^2)^{1/2}$ from $(-1, 0)$ to $(1, 0).$

10. $\mathbf{h}(x, y) = 2xy \sinh x^2y\,\mathbf{i} + x^2\sinh x^2y\,\mathbf{j};$ the curve of Exercise 1.

11. $\mathbf{h}(x, y) = (2x \cosh y - y)\mathbf{i} + (x^2\sinh y - y)\mathbf{j};$ the square of Exercise 6.

In Exercises 12–15, verify that the given vector field **h** is a gradient. Then calculate the line integral of **h** over the indicated curve C in two ways: (a) by calculating $\int_C \mathbf{h}(\mathbf{r}) \cdot d\mathbf{r}$ and (b) by finding f such that $\nabla f = \mathbf{h}$ and evaluating f at the endpoints of C.

12. $\mathbf{h}(x, y) = xy^2\mathbf{i} + yx^2\mathbf{j};\ \mathbf{r}(u) = u\mathbf{i} + u^2\mathbf{j},\ u \in [0, 2].$

13. $\mathbf{h}(x, y) = (3x^2y^3 + 2x)\mathbf{i} + (3x^3y^2 - 4y)\mathbf{j};\ \mathbf{r}(u) = u\mathbf{i} + e^u\mathbf{j},\ u \in [0, 1].$

14. $(2x \sin y - e^x)\mathbf{i} + (x^2 \cos y)\mathbf{j};$ $\mathbf{r}(u) = \cos u\,\mathbf{i} + u\,\mathbf{j},\ u \in [0, \pi].$

15. $\mathbf{h}(x, y) = (e^{2y} - 2xy)\mathbf{i} + (2xe^{2y} - x^2 + 1)\mathbf{j};\ \mathbf{r}(u) = ue^u\mathbf{i} + (1 + u)\mathbf{j},\ u \in [0, 1].$

In Exercises 16–18, use the three dimensional analog of Theorem 15.10.2 given in Exercises 15.10 to show that the vector function **h** is a gradient. Then evaluate the line integral of **h** over the indicated curve.

16. $\mathbf{h}(x, y, z) = y^2z^3\mathbf{i} + 2xyz^3\mathbf{j} + 3xy^2z^2\mathbf{k};\ \mathbf{r}(u) = u^2\mathbf{i} + u^4\mathbf{j} + u^6\mathbf{k},\ u \in [0, 1].$

17. $\mathbf{h}(x, y, z) = (2xz + \sin y)\mathbf{i} + x \cos y\,\mathbf{j} + x^2\mathbf{k};\ \mathbf{r}(u) = \cos u\,\mathbf{i} + \sin u\,\mathbf{j} + u\,\mathbf{k},\ u \in [0, 2\pi].$

18. $\mathbf{h}(x, y, z) = e^{-x} \ln y\,\mathbf{i} - [\dfrac{e^{-x}}{y}]\mathbf{j} + 3z^2\mathbf{k};\ \mathbf{r}(u) = (u + 1)\mathbf{i} + e^{2u}\mathbf{j} + (u^2 + 1)\mathbf{k},\ u \in [0, 1].$

19. Calculate the work done by the force $\mathbf{F}(x, y) = (x + e^{2y})\mathbf{i} + (2y + 2xe^{2y})\mathbf{j}$ applied to an object that moves along the curve $\mathbf{r}(u) = 3 \cos u\,\mathbf{i} + 4 \sin u\,\mathbf{j},\ u \in [0, 2\pi].$

20. Calculate the work done by the force $\mathbf{F}(x, y, z) = (2x \ln y - yz)\mathbf{i} + [(x^2/y) - xz]\mathbf{j} - xy\,\mathbf{k}$ applied to an object that moves from the point $(1, 2, 1)$ to the point $(3, 2, 2).$

21. If g is a continuously differentiable real-valued function defined on $[a, b]$, then by the fundamental theorem of integral calculus

$$\int_a^b g'(u)\, du = g(b) - g(a).$$

Show how this result is included in Theorem 17.2.1.

22. Let $\mathbf{r} = x\mathbf{i} + y\mathbf{j} + z\mathbf{k}$ and set $r = \|\mathbf{r}\|.$ The central force field

$$\mathbf{F}(\mathbf{r}) = \frac{K}{r^n}\mathbf{r}, \qquad n \text{ a positive integer}$$

is a gradient field. Find f such that $\nabla f(\mathbf{r}) = \mathbf{F}(\mathbf{r})$ if: (a) $n = 2$; (b) $n \neq 2.$

23. Let $\mathbf{r} = x\mathbf{i} + y\mathbf{j} + z\mathbf{k}$ and set $r = \|\mathbf{r}\|.$ The function

$$\mathbf{F}(\mathbf{r}) = -\frac{mG}{r^3}\mathbf{r} \qquad (G \text{ is a gravitational constant})$$

gives the gravitational force exerted by a unit mass at the origin on a mass m located at $\mathbf{r}$. What is the work done by $\mathbf{F}$ if m moves from $\mathbf{r}_1$ to $\mathbf{r}_2$?

24. Set

$$P(x, y) = \frac{y}{x^2 + y^2} \qquad \text{and} \qquad Q(x, y) = -\frac{x}{x^2 + y^2}$$

on *the punctured unit disc* $\Omega: 0 < x^2 + y^2 < 1.$
(a) Verify that P and Q are continuously differentiable on Ω and that

$$\frac{\partial P}{\partial y}(x, y) = \frac{\partial Q}{\partial x}(x, y) \qquad \text{for all } (x, y) \in \Omega.$$

(b) Verify that, in spite of (a), the vector field $\mathbf{h}(x, y) = P(x, y)\mathbf{i} + Q(x, y)\mathbf{j}$ is not a gradient on Ω.
HINT: Integrate **h** over a circle of radius less than one centered at the origin.
(c) Explain how (b) does not contradict Theorem 15.10.2.

25. The gravitational force acting on an object of mass m at a height z above the surface of the earth is given by

$$\mathbf{F}(x, y, z) = \frac{-mGr_0^2}{(r_0 + z)^2}\mathbf{k},$$

where G is the gravitational constant and r_0 is the radius of the earth. Show that $\mathbf{F}$ is a gradient field and find f such that $\nabla f = \mathbf{F}$.

26. A 4-ton rocket falls to earth from a height of 300 miles. How much work is done by the gravitational force? Use Exercise 25 and assume that the radius of the earth is 4000 miles.

27. An elevator carries 20 people to the fiftieth floor of a building. If the elevator weighs 5000 pounds, the people weigh 3000 lbs, and the fiftieth floor is 500 feet above the ground, find the work done by the elevator against the force of gravity. Use Exercise 25 and assume that the radius of the earth is 4000 miles.

*■ 17.3 WORK–ENERGY FORMULA; CONSERVATION OF MECHANICAL ENERGY

Suppose that a continuous force field $\mathbf{F} = \mathbf{F}(\mathbf{r})$ accelerates a mass m from $\mathbf{r} = \mathbf{a}$ to $\mathbf{r} = \mathbf{b}$ along some smooth curve C. The object undergoes a change in kinetic energy:

$$\tfrac{1}{2}m[\nu(\beta)]^2 - \tfrac{1}{2}m[\nu(\alpha)]^2.$$

The force does a certain amount of work W. How are these quantities related? They are equal:

(17.3.1)
$$\boxed{W = \tfrac{1}{2}m[\nu(\beta)]^2 - \tfrac{1}{2}m[\nu(\alpha)]^2.}$$

This relation is called the *Work–Energy Formula.*

Derivation of the Work–Energy Formula

We parametrize the path of the motion by the time parameter t:

$$C: \quad \mathbf{r} = \mathbf{r}(t), \quad t \in [\alpha, \beta],$$

Where $\mathbf{r}(\alpha) = \mathbf{a}$ and $\mathbf{r}(\beta) = \mathbf{b}$. The work done by $\mathbf{F}$ is given by the formula

$$W = \int_C \mathbf{F}(\mathbf{r}) \cdot d\mathbf{r} = \int_\alpha^\beta [\mathbf{F}(\mathbf{r}(t)) \cdot \mathbf{r}'(t)]\, dt.$$

From Newton's second law of motion, we know that at time t,

$$\mathbf{F}(\mathbf{r}(t)) = m\, \mathbf{a}(t) = m\, \mathbf{r}''(t).$$

It follows that

$$\mathbf{F}(\mathbf{r}(t)) \cdot \mathbf{r}'(t) = m\, \mathbf{r}''(t) \cdot \mathbf{r}'(t)$$
$$= \frac{d}{dt}\left[\frac{1}{2}m(\mathbf{r}'(t) \cdot \mathbf{r}'(t))\right] = \frac{d}{dt}\left[\frac{1}{2}m\|\mathbf{r}'(t)\|^2\right] = \frac{d}{dt}\left[\frac{1}{2}m[\nu(t)]^2\right].$$

Substituting this last expression into the work integral, we see that

$$W = \int_\alpha^\beta \frac{d}{dt}\left(\frac{1}{2}m[\nu(t)]^2\right) dt = \frac{1}{2}m[\nu(\beta)]^2 - \frac{1}{2}m[\nu(\alpha)]^2.$$

as asserted. ❑

Conservative Force Fields

In general, if an object moves from one point to another, the work done (and hence the change in kinetic energy) depends on the path of the motion. There is, however, an important exception: if the force field is a gradient field,

$$\mathbf{F} = \nabla f,$$

then the work done (and hence the change in kinetic energy) depends only on the endpoints of the path and not on the path itself. (This follows directly from the Fundamental Theorem for Line Integrals.) A force field that is a gradient field is called a *conservative field*.

Since the line integral over a closed path is zero, *the work done by a conservative field over a closed path is always zero. An object that passes through a given point with a certain kinetic energy returns to that same point with exactly the same kinetic energy.*

Potential Energy Functions

Suppose that $\mathbf{F}$ is a conservative force field. It is then a gradient field. Then $-\mathbf{F}$ is a gradient field. The functions U for which $\nabla U = -\mathbf{F}$ are called *potential energy functions* for $\mathbf{F}$.

The Conservation of Mechanical Energy

Suppose that $\mathbf{F}$ is a conservative force field: $\mathbf{F} = -\nabla U$. In our derivation of the work–energy formula we showed that

$$\frac{d}{dt}(\tfrac{1}{2}m[v(t)]^2) = \mathbf{F}(\mathbf{r}(t)) \cdot \mathbf{r}'(t).$$

Since

$$\frac{d}{dt}[U(\mathbf{r}(t))] = \nabla U(\mathbf{r}(t)) \cdot \mathbf{r}'(t) = -\mathbf{F}(\mathbf{r}(t)) \cdot \mathbf{r}'(t),$$

we have

$$\frac{d}{dt}[\tfrac{1}{2}m[v(t)]^2 + U(\mathbf{r}(t))] = 0$$

and therefore

$$\underbrace{\tfrac{1}{2}m[v(t)]^2}_{\text{KE}} + \underbrace{U(\mathbf{r}(t))}_{\text{PE}} = \text{a constant.}$$

As an object moves in a conservative force field, its kinetic energy can vary and its potential energy can vary, but the sum of these two quantities remains constant. We call this constant *the total mechanical energy.*

The total mechanical energy is usually denoted by the letter E. The law of conservation of mechanical energy can then be written

(17.3.2.)

$$\tfrac{1}{2}mv^2 + U = E.$$

The conservation of energy is one of the cornerstones of physics. Here we have been talking about mechanical energy. There are other forms of energy and other energy conservation laws.

Differences in Potential Energy

Potential energy at a particular point has no physical significance. Only differences in potential energy are significant:

$$U(\mathbf{b}) - U(\mathbf{a}) = \int_C -\mathbf{F}(\mathbf{r}) \cdot d\mathbf{r}$$

is the work required to move from $\mathbf{r} = \mathbf{a}$ to $\mathbf{r} = \mathbf{b}$ *against* the force field $\mathbf{F}$.

Example 1   A planet moves in the gravitational field of the sun:

$$\mathbf{F}(\mathbf{r}) = -m\rho \frac{\mathbf{r}}{r^3}. \qquad \text{(\textit{m} is the mass of the planet)}$$

Show that the force field is conservative, find a potential energy function, and determine the total energy of the planet. How does the planet's speed vary with the planet's distance from the sun?

SOLUTION   The field is conservative since

$$\mathbf{F}(\mathbf{r}) = -m\rho \frac{\mathbf{r}}{r^3} = \nabla\left(\frac{\rho m}{r}\right). \qquad \text{(check this out)}$$

As a potential energy function we can use

$$U(\mathbf{r}) = -\frac{\rho m}{r}.$$

The total energy of the planet is the constant

$$E = \tfrac{1}{2}mv^2 - \frac{\rho m}{r}.$$

(You met this quantity before: Exercises 2 and 4 of Section 13.6.)
    Solving the energy equation for v we have

$$v = \sqrt{\frac{2E}{m} + \frac{2\rho}{r}}.$$

As r decreases, $2\rho/r$ increases, and v increases; as r increases, $2\rho/r$ decreases, and v decreases. Thus every planet speeds up as it comes near the sun and slows down as it moves away. (The same holds true for Halley's comet. The fact that it slows down as it gets farther away helps explain why it comes back. The simplicity of all this is a testimony to the power of the principle of energy conservation.) ❑

EXERCISES 17.3

1. Let f be a continuous real-valued function of the real variable x. Show that the force field $\mathbf{F}(x, y, z) = f(x)\mathbf{i}$ is conservative and the potential functions for $\mathbf{F}$ are (except for notation) the antiderivatives of $-f$.

2. A particle with electric charge e and velocity $\mathbf{v}$ moves in a magnetic field $\mathbf{B}$, experiencing the force

$$\mathbf{F} = \frac{e}{c}[\mathbf{v} \times \mathbf{B}]. \quad (c \text{ is the velocity of light})$$

$\mathbf{F}$ is not a gradient—it can't be, depending as it does on the *velocity* of the particle. Still, we can find a conserved quantity: the *kinetic energy* $\frac{1}{2}mv^2$. Show by differentiation with respect to t that this quantity is constant. (Assume Newton's second law.)

3. An object is subject to a constant force in the direction of $-\mathbf{k}$: $\mathbf{F} = -c\mathbf{k}$ with $c > 0$. Find a potential energy function for $\mathbf{F}$, and use energy conservation to show that the speed of the object at time t_2 is related to that at time t_1 by the equation

$$v(t_2) = \sqrt{[v(t_1)]^2 + \frac{2c}{m}[z(t_1) - z(t_2)]}$$

where $z(t_1)$ and $z(t_2)$ are the z-coordinates of the object at times t_1 and t_2. (This analysis is sometimes used to model the behavior of an object in the gravitational field near the surface of the earth.)

4. (*Escape velocity*) An object is to be fired straight up from the surface of the earth. Assume that the only force acting on the object is the gravitational pull of the earth and determine the initial speed v_0 necessary to send the object off to infinity. HINT: Appeal to conservation of energy and use the idea that the object is to arrive at infinity with zero speed.

5. (a) Justify the statement that a conservative force field $\mathbf{F}$ always acts so as to encourage motion toward regions of lower potential energy U.
   (b) Evaluate $\mathbf{F}$ at a point where U has a minimum.

6. A harmonic oscillator has a restoring force $\mathbf{F} = -\lambda x\mathbf{i}$. The associated potential is $U(x, y, z) = -\frac{1}{2}\lambda x^2$, and the constant total energy is

$$E = \frac{1}{2}mv^2 + U(x, y, z) = \frac{1}{2}mv^2 + \frac{1}{2}\lambda x^2.$$

Given that $x(0) = 2$ and $x'(0) = 1$, calculate the maximum speed of the oscillator and the maximum value of x.

7. The *equipotential surfaces* of a conservative field $\mathbf{F}$ are the surfaces where the potential energy is constant. Show that: (a) the speed of an object in such a field is constant on every equipotential surface; and (b) at each point of such a surface the force field is perpendicular to the surface.

8. Suppose a force field $\mathbf{F}$ is directed away from the origin with a magnitude that is inversely proportional to the distance from the origin. Show that $\mathbf{F}$ is a conservative field.

9. Let $\mathbf{F}$ be the inverse-square force field:

$$\mathbf{F}(x, y, z) = \frac{k(x\mathbf{i} + y\mathbf{j} + z\mathbf{k})}{(x^2 + y^2 + z^2)^{3/2}}$$

and let C be any curve on the unit sphere $x^2 + y^2 + z^2 = 1$. Show that the work done by $\mathbf{F}$ in moving an object along C is 0. Explain this result.

■ 17.4 ANOTHER NOTATION FOR LINE INTEGRALS; LINE INTEGRALS WITH RESPECT TO ARC LENGTH

If $\mathbf{h}(x, y, z) = P(x, y, z)\mathbf{i} + Q(x, y, z)\mathbf{j} + R(x, y, z)\mathbf{k}$, the line integral

$$\int_C \mathbf{h}(\mathbf{r}) \cdot d\mathbf{r} \text{ can be written } \int_C P(x, y, z)\, dx + Q(x, y, z)\, dy + R(x, y, z)\, dz.$$

The notation arises as follows. With

$$C: \quad \mathbf{r}(u) = x(u)\mathbf{i} + y(u)\mathbf{j} + z(u)\mathbf{k}, \quad u \in [a, b]$$

the line integral

$$\int_C \mathbf{h}(\mathbf{r}) \cdot d\mathbf{r} = \int_a^b [\mathbf{h}(\mathbf{r}(t)) \cdot \mathbf{r}'(t)]\, dt$$

expands to

$$\int_a^b \{P[x(u), y(u), z(u)]x'(u) + Q[x(u), y(u), z(u)]y'(u) + R[x(u), y(u), z(u)]z'(u)\}\, du.$$

Now set

$$\int_C P(x, y, z) \, dx = \int_a^b P[x(u), y(u), z(u)]x'(u) \, du,$$

$$\int_C Q(x, y, z) \, dy = \int_a^b Q[x(u), y(u), z(u)]y'(u) \, du,$$

$$\int_C R(x, y, z) \, dz = \int_a^b R[x(u), y(u), z(u)]z'(u) \, du.$$

Writing the sum of these integrals as

$$\int_C P(x, y, z) \, dx + Q(x, y, z) \, dy + R(x, y, z) \, dz,$$

we have

$$\int_C P(x, y, z) \, dx + Q(x, y, z) \, dy + R(x, y, z) \, dz = \int_C \mathbf{h}(\mathbf{r}) \cdot d\mathbf{r}. \quad \square$$

If C lies in the xy-plane and $\mathbf{h}(x, y) = P(x, y)\mathbf{i} + Q(x, y)\mathbf{j}$, then the line integral reduces to

$$\int_C P(x, y) \, dx + Q(x, y) \, dy.$$

The notation is easy to use; for example, with C as before,

$$\int_C y \, dx + 2z \, dy - x \, dz = \int_a^b [y(u)x'(u) + 2z(u)y'(u) - x(u)z'(u)] \, du,$$

$$\int_C \sin(xy) \, dx - \cos(xy) \, dy = \int_a^b \{\sin[x(u)y(u)]x'(u) - \cos[x(u)y(u)]y'(u)\} \, du.$$

Line Integrals with Respect to Arc Length

Suppose that f is a scalar field continuous on a piecewise-smooth curve

$$C: \quad \mathbf{r}(u) = x(u)\mathbf{i} + y(u)\mathbf{j} + z(u)\mathbf{k}, \quad u \in [a, b].$$

If $s(u)$ is the length of the curve from the tip of $\mathbf{r}(a)$ to the tip of $\mathbf{r}(u)$, then, as you have seen,

$$s'(u) = \|\mathbf{r}'(u)\| = \sqrt{[x'(u)]^2 + [y'(u)]^2 + [z'(u)]^2}.$$

The line integral of f over C *with respect to arc length* s is defined by setting

(17.4.1)

$$\int_C f(\mathbf{r}) \, ds = \int_a^b f(\mathbf{r}(u)) \, s'(u) \, du.$$

In the *xyz*-notation we have

$$\int_C f(x, y, z) \, ds = \int_a^b f(x(u), y(u), z(u)) \, s'(u) \, du,$$

which, in the two-dimensional case, becomes

$$\int_C f(x, y)\, ds = \int_a^b f(x(u), y(u))\, s'(u)\, du.$$

Suppose now that C represents a thin wire (a material curve) of varying mass density $\lambda = \lambda(\mathbf{r})$. [Here mass density is mass per unit length.] The *length* of the wire can be written

(17.4.2)
$$L = \int_C ds.$$

The *mass* of the wire is given by

(17.4.3)
$$M = \int_C \lambda(\mathbf{r})\, ds,$$

and the *center of mass* $\mathbf{r}_M$ can be obtained from the vector equation

(17.4.4)
$$\mathbf{r}_M M = \int_C \mathbf{r}\lambda(\mathbf{r})\, ds.$$

The equivalent scalar equations read

$$x_M M = \int_C x\lambda(\mathbf{r})\, ds, \qquad y_M M = \int_C y\lambda(\mathbf{r})\, ds, \qquad z_M M = \int_C z\lambda(\mathbf{r})\, ds.$$

Finally, the *moment of inertia* about an axis is given by the formula

(17.4.5)
$$I = \int_C \lambda(\mathbf{r})[R(\mathbf{r})]^2\, ds$$

where $R(\mathbf{r})$ is the distance from the axis to the tip of $\mathbf{r}$.

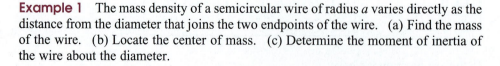

Figure 17.4.1

Example 1 The mass density of a semicircular wire of radius a varies directly as the distance from the diameter that joins the two endpoints of the wire. (a) Find the mass of the wire. (b) Locate the center of mass. (c) Determine the moment of inertia of the wire about the diameter.

SOLUTION Placed as in Figure 17.4.1 the wire can be parameterized by

$$\mathbf{r}(u) = a\cos u\, \mathbf{i} + a\sin u\, \mathbf{j}, \qquad u \in [0, \pi]$$

and the mass density function can be written $\lambda(x, y) = ky$. Since $\mathbf{r}'(u) = -a \sin u \, \mathbf{i} + a \cos u \, \mathbf{j}$, we have

$$s'(u) = \|\mathbf{r}'(u)\| = a.$$

Therefore

$$
\begin{aligned}
M &= \int_C \lambda(x, y) \, ds = \int_C ky \, ds \\
&= \int_0^\pi ky(u)s'(u) \, du \\
&= \int_0^\pi k(a \sin u)a \, du = ka^2 \int_0^\pi \sin u \, du = 2ka^2.
\end{aligned}
$$

By the symmetry of the configuration, $x_M = 0$. To find y_M we have to integrate:

$$
\begin{aligned}
y_M M &= \int_C y\lambda(x, y) \, ds = \int_C ky^2 \, ds \\
&= \int_0^\pi k[y(u)]^2 s'(u) \, du \\
&= \int_0^\pi k(a \sin u)^2 a \, du = ka^3 \int_0^\pi \sin^2 u \, du = \tfrac{1}{2}ka^3 \pi.
\end{aligned}
$$

Since $M = 2ka^2$, we have $y_M = (\tfrac{1}{2}ka^3 \pi)/(2ka^2) = \tfrac{1}{4}a\pi$. The center of mass G lies on the perpendicular bisector of the wire at a distance $\tfrac{1}{4}a\pi$ from the diameter. (See Figure 17.4.1.) Note that in this instance, the center of mass does not lie on the wire.

Now let's find the moment of inertia about the diameter:

$$
\begin{aligned}
I &= \int_C \lambda(x, y)[R(x, y)]^2 \, ds = \int_C (ky)y^2 \, ds \\
&= \int_C k[y(u)]^3 s'(u) \, du \\
&= \int_0^\pi k(a \sin u)^3 a \, du \\
&= ka^4 \int_0^\pi \sin^3 u \, du = \tfrac{4}{3}ka^4.
\end{aligned}
$$

It is customary to express I in terms of M. With $M = 2ka^2$, we have

$$I = \tfrac{2}{3}(2ka^2)a^2 = \tfrac{2}{3}Ma^2. \quad \square$$

EXERCISES 17.4

In Exercises 1–4, evaluate

$$\int_C (x - 2y) \, dx + 2x \, dy$$

along the given path C from $(0, 0)$ to $(1, 2)$.

1. The straight-line path.
2. The parabolic path $y = 2x^2$.
3. The polygonal path $(0, 0)$, $(1, 0)$, $(1, 2)$.
4. The polygonal path $(0, 0)$, $(0, 2)$, $(1, 2)$.

In Exercises 5–8, evaluate

$$\int_C y \, dx + xy \, dy$$

along the given path C from $(0, 0)$ to $(2, 1)$.

5. The parabolic path $x = 2y^2$.
6. The straight-line path.
7. The polygonal path $(0, 0)$, $(0, 1)$, $(2, 1)$.
8. The cubic path $x = 2y^3$.

In Exercises 9–12, evaluate

$$\int_C y^2 \, dx + (xy - x^2) \, dy$$

along the given path C from $(0, 0)$ to $(2, 4)$.

9. The straight-line path.

10. The parabolic path $y = x^2$.

11. The parabolic path $y^2 = 8x$.

12. The polygonal path $(0, 0)$ $(2, 0)$ $(2, 4)$.

In Exercises 13–16, evaluate

$$\int_C x^2 y \, dx + xy \, dy$$

along the given path C from $(1, 0)$ to $(0, 1)$.

13. The straight-line path.

14. The circular path $y = \sqrt{1 - x^2}$, $0 \le x \le 1$.

15. $\mathbf{r}(u) = \cos u \, \mathbf{i} + \sin u \, \mathbf{j}$, $u \in [0, \pi/2]$.

16. The polygonal path $(1, 0)$, $(1, 1)$, $(0, 1)$.

In Exercises 17–20, evaluate

$$\int_C (y^2 + 2x + 1) \, dx + (2xy + 4y - 1) \, dy$$

along the given path C from $(0, 0)$ to $(1, 1)$.

17. The straight-line path.

18. The parabolic path $y = x^2$.

19. The cubic path $y = x^3$.

20. The polygonal path $(0, 0)$, $(4, 0)$, $(4, 2)$, $(1, 1)$.

In Exercises 21–24, evaluate

$$\int_C y \, dx + 2z \, dy + x \, dz$$

along the given path C from $(0, 0, 0)$ to $(1, 1, 1)$.

21. The straight-line path.

22. $\mathbf{r}(u) = u\mathbf{i} + u^2\mathbf{j} + u^3\mathbf{k}$.

23. The polygonal path $(0, 0, 0)$, $(0, 0, 1)$, $(0, 1, 1)$, $(1, 1, 1)$.

24. The polygonal path $(0, 0, 0)$, $(1, 0, 0)$, $(1, 1, 0)$, $(1, 1, 1)$.

In Exercises 25–28, evaluate

$$\int_C xy \, dx + 2z \, dy + (y + z) \, dz$$

along the given path C from $(0, 0, 0)$ to $(2, 2, 8)$.

25. The straight-line path.

26. The polygonal path $(0, 0, 0)$, $(2, 0, 0)$, $(2, 2, 0)$, $(2, 2, 8)$.

27. The parabolic path $\mathbf{r}(u) = t\mathbf{i} + t\mathbf{j} + 2t^2\mathbf{k}$.

28. The polygonal path $(0, 0, 0)$, $(2, 2, 2)$ $(2, 2, 8)$.

29. Evaluate $\int_C x^2 y \, dx + y \, dy + xz \, dz$, where C is the curve of intersection of the cylinder $y - 2z^2 = 1$ and the plane $z = x + 1$ from $(0, 3, 1)$ to $(1, 9, 2)$.

30. Evaluate $\int_C y \, dx + yz \, dy + z(x - 1) \, dz$, where C is the curve of intersection of the sphere $x^2 + y^2 + z^2 = 4$ and the cylinder $(x - 1)^2 + y^2 = 1$ from $(2, 0, 0)$ to $(0, 0, 2)$.

31. Given the vector field
$$\mathbf{F}(x, y) = (x^2 + 6xy - 2y^2)\mathbf{i} + (3x^2 - 4xy + 2y)\mathbf{j}.$$
(a) Show that $\mathbf{F}$ is a gradient field.
(b) Evaluate

$$\int_C (x^2 + 6xy - 2y^2) \, dx + (3x^2 - 4xy + 2y) \, dy$$

along any piecewise smooth curve from $(3, 0)$ to $(0, 4)$.
(c) What is the value of

$$\int_{C'} (x^2 + 6xy - 2y^2) \, dx + (3x^2 - 4xy + 2y) \, dy$$

where C' is any piecewise smooth curve from $(4, 0)$ to $(0, 3)$.

32. Given the vector field
$$\mathbf{F}(x, y, z) = (2xy + z^2)\mathbf{i} + (x^2 - 2yz)\mathbf{j} + (2xz - y^2)\mathbf{k}.$$
(a) Show that $\mathbf{F}$ is a gradient field.
(b) Evaluate

$$\int_C (2xy + z^2) \, dx + (x^2 - 2yz) \, dy + (2xz - y^2) \, dz$$

along any piecewise smooth curve from $(1, 0, 1)$ to $(3, 2, -1)$.
(c) What is the value of

$$\int_{C'} (2xy + z^2) \, dx + (x^2 - 2yz) \, dy + (2xz - y^2) \, dz$$

where C' is any piecewise smooth curve from $(3, 2, -1)$ to $(1, 0, 1)$?

33. A wire in the shape of the quarter-circle

$$C: \quad \mathbf{r}(u) = a(\cos u \, \mathbf{i} + \sin u \, \mathbf{j}), \quad u \in [0, \tfrac{1}{2}\pi]$$

has varying mass density $\lambda(x, y) = k(x + y)$ where k is a positive constant.
(a) Find the total mass of the wire and locate the center of mass.
(b) What is the moment of inertia of the wire about the x-axis?

34. Find the moment of inertia of a homogeneous circular wire of radius a and mass M about: (a) a diameter; (b) the axis through the center that is perpendicular to the plane of the wire.

35. Find the moment of inertia of the wire of Exercise 33 about: (a) the z-axis;   (b) the line $y = x$.

36. A wire of constant mass density k has the form

$$\mathbf{r}(u) = (1 - \cos u)\mathbf{i} + (u - \sin u)\mathbf{j}, \qquad u \in [0, 2\pi].$$

(a) Determine the mass of the wire.
(b) Locate the center of mass.

37. A homogeneous wire of mass M winds around the z-axis as

$$C: \quad \mathbf{r}(u) = a \cos u\,\mathbf{i} + a \sin u\,\mathbf{j} + bu\,\mathbf{k}, \qquad u \in [0, 2\pi].$$

(a) Find the length of the wire.
(b) Locate the center of mass.
(c) Determine the moments of inertia of the wire about the coordinate axes.

38. A homogeneous wire of mass M is of the form

$$C: \quad \mathbf{r}(u) = u\mathbf{i} + u^2\mathbf{j} + \tfrac{2}{3}u^3\mathbf{k}, \quad u \in [0, a].$$

(a) Find the length of the wire.
(b) Locate the center of mass.
(c) Determine the moment of inertia of the wire about the z-axis.

39. Calculate the mass of the wire of Exercise 37 if the mass density varies directly as the square of the distance from the origin.

40. Show that

(17.4.6)
$$\int_C \mathbf{h}(\mathbf{r}) \cdot d\mathbf{r} = \int_C [\mathbf{h}(\mathbf{r}) \cdot \mathbf{T}(\mathbf{r})]\, ds$$

where $\mathbf{T}$ is the unit tangent vector.

■ 17.5 GREEN'S THEOREM

(Green's theorem is the first of the three integration theorems heralded at the beginning of this chapter.)

Recall that a Jordan curve is a plane curve that is both closed and simple. Thus circles, ellipses, and triangles are Jordan curves; figure eights are not. (See Section 15.10, page 1025.)

Figure 17.5.1 depicts a closed region Ω, the total boundary of which is a Jordan curve C. Such a region is called a *Jordan region*.

We know how to integrate over Ω, and, if the boundary C is piecewise smooth, we know how to integrate over C. Green's theorem expresses a double integral over Ω as a line integral over C.

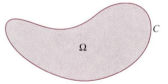

Figure 17.5.1

THEOREM 17.5.1 GREEN'S THEOREM†

Let Ω be a Jordan region with a piecewise-smooth boundary C. If P and Q are scalar fields continuously differentiable on an open set that contains Ω, then

$$\iint_\Omega \left[\frac{\partial Q}{\partial x}(x, y) - \frac{\partial P}{\partial y}(x, y) \right] dx\, dy = \oint_C P(x, y)\, dx + Q(x, y)\, dy$$

where the integral on the right is the line integral over C taken in the counterclockwise direction.

---

† The result was established in 1828 by the English mathematician George Green (1793–1841).

NOTE: As indicated, the symbol $\oint_C$ is used to denote the line integral over a simple closed curve C taken in the counterclockwise direction. A line integral over C taken in the clockwise direction would be written $\oint_C$.

We will prove the theorem only for special cases. First of all let's assume that Ω is an *elementary region*, a region that is both of type I and type II as defined in Section 16.4. For simplicity we take Ω as in Figure 17.5.2.

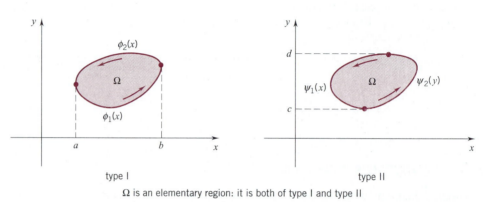

type I   type II

Ω is an elementary region: it is both of type I and type II

Figure 17.5.2

With Ω being of type I, we can show that

(1)
$$\oint_C P(x, y)\, dx = \iint_\Omega -\frac{\partial P}{\partial y}(x, y)\, dx\, dy.$$

In the first place

$$\iint_\Omega -\frac{\partial P}{\partial y}(x, y)\, dx\, dy = -\int_a^b \int_{\phi_1(x)}^{\phi_2(x)} \frac{\partial P}{\partial y}(x, y)\, dy\, dx$$

$$= -\int_a^b \{P[x, \phi_2(x)] - P[x, \phi_1(x)]\}\, dx$$

by The Fundamental Theorem
of Integral Calculus

$$(*) \qquad = \int_a^b P[x, \phi_1(x)]\, dx - \int_a^b P[x, \phi_2(x)]\, dx.$$

The graph of ϕ_1 parametrized from left to right is the curve

$$C_1: \quad \mathbf{r}_1(u) = u\mathbf{i} + \phi_1(u)\mathbf{j}, \qquad u \in [a, b];$$

the graph of ϕ_2, also parametrized from left to right, is the curve

$$C_2: \quad \mathbf{r}_2(u) = u\mathbf{i} + \phi_2(u)\mathbf{j}, \qquad u \in [a, b].$$

Since C traversed counterclockwise consists of C_1 followed by $-C_2$ (C_2 traversed from right to left), you can see that

$$\oint_C P(x, y)\, dx = \int_{C_1} P(x, y)\, dx - \int_{C_2} P(x, y)\, dx$$

$$= \int_a^b P[u, \phi_1(u)]\, du - \int_a^b P[u, \phi_2(u)]\, du.$$

Since u is a dummy variable, it can be replaced by x. Comparison with $(*)$ proves (1).

We leave it to you to verify that

(2)
$$\oint_C Q(x, y)\, dy = \iint_\Omega \frac{\partial Q}{\partial x}(x, y)\, dx\, dy$$

by using the fact that Ω is of type II. This completes the proof of the theorem for Ω as in Figure 17.5.2.

A slight modification of this argument applies to elementary regions which are bordered entirely or in part by line segments parallel to the coordinate axes.

Figure 17.5.3 shows a Jordan region that is not elementary but can be broken up into two elementary regions. (Figure 17.5.4.) Green's theorem applied to the elementary parts tells us that.

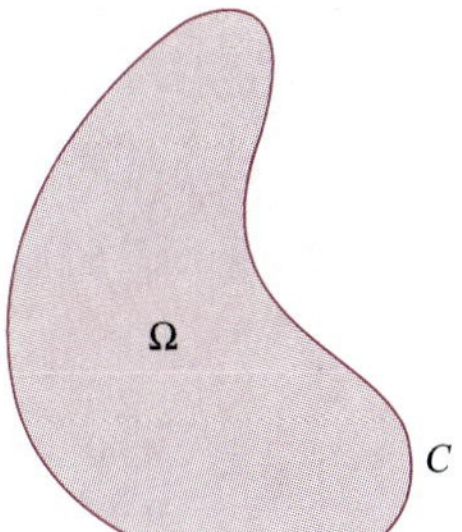

$$\iint_{\Omega_1} \left(\frac{\partial Q}{\partial x} - \frac{\partial P}{\partial y} \right) dx\, dy = \oint_{\text{bdry of } \Omega_1} P\, dx + Q\, dy,$$

$$\iint_{\Omega_2} \left(\frac{\partial Q}{\partial x} - \frac{\partial P}{\partial y} \right) dx\, dy = \oint_{\text{bdry of } \Omega_2} P\, dx + Q\, dy.$$

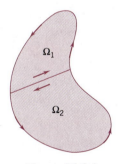

Figure 17.5.3

We now add these equations. The sum of the double integrals is, by additivity, the double integral over Ω. The sum of the line integrals is the integral over C (see the figure) plus the integrals over the crosscut. Since the crosscut is traversed twice and in opposite directions, the total contribution of the crosscut is zero and therefore Green's theorem holds:

$$\iint_\Omega \left(\frac{\partial Q}{\partial x} - \frac{\partial P}{\partial y} \right) dx\, dy = \oint_C P\, dx + Q\, dy.$$

This same argument can be extended to a Jordan region Ω that breaks up into n elementary regions $\Omega_1, \ldots, \Omega_n$. (Figure 17.5.5 gives an example with $n = 4$.)

Figure 17.5.4

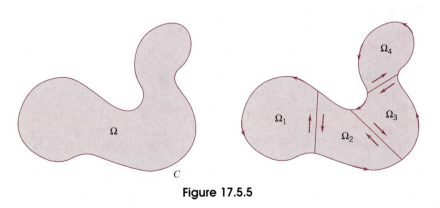

Figure 17.5.5

The double integrals over the Ω_i add up to the double integral over Ω, and, since the line integrals over the crosscuts cancel, the line integrals over the boundaries of the Ω_i add up to the line integral over C. (This is as far as we will carry the proof of Green's theorem. It is far enough to cover all the Jordan regions encountered in practice.) ❏

Example 1 Use Green's theorem to evaluate

$$\oint_C^C (3x^2 + y)\, dx + (2x + y^3)\, dy$$

where C is the circle $x^2 + y^2 = a^2$. (Figure 17.5.6.)

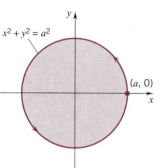

$x^2 + y^2 = a^2$

$(a, 0)$

Figure 17.5.6

SOLUTION Let Ω be the closed disc $0 \le x^2 + y^2 \le a^2$. With

$$P(x, y) = 3x^2 + y \qquad \text{and} \qquad Q(x, y) = 2x + y^3,$$

we have

$$\frac{\partial P}{\partial y} = 1, \qquad \frac{\partial Q}{\partial x} = 2, \qquad \text{and} \qquad \frac{\partial Q}{\partial x} - \frac{\partial P}{\partial y} = 2 - 1 = 1.$$

By Green's theorem

$$\oint_C (3x^2 + y)\, dx + (2x + y^3)\, dy = \iint_\Omega 1\, dx\, dy = \text{area of } \Omega = \pi a^2. \qquad ❏$$

Remark The line integral in Example 1 could have been calculated directly as follows: The circle $x^2 + y^2 = a^2$ is parametrized by

$$x = a \cos u, \qquad y = a \sin u, \qquad 0 \le u \le 2\pi.$$

Thus,

$$\oint_C (3x^2 + y)\, dx + (2x + y^3)\, dy$$

$$= \int_0^{2\pi} [(3a^2\cos^2 u + a \sin u)(-a \sin u) + (2a \cos u + a^3\sin^3 u)(a \cos u)]\, du$$

$$= \int_0^{2\pi} [-3a^3\cos^2 u \sin u - a^2\sin^2 u + 2a^2\cos^2 u + a^4\sin^3 u \cos u]\, du$$

which is obviously a much more difficult integral to evaluate. You can verify, however, that the result is πa^2. ❏

Example 2 Use Green's theorem to evaluate

$$\oint_C (1 + 10xy + y^2)\, dx + (6xy + 5x^2)\, dy$$

where C is the square with vertices $(0, 0)$, $(a, 0)$, (a, a), $(0, a)$. (Figure 17.5.7.)

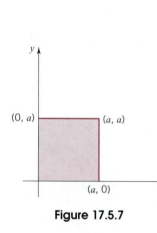

$(0, a)$ (a, a)

$(a, 0)$

Figure 17.5.7

SOLUTION Let Ω be the square region enclosed by C. With

$$P(x, y) = 1 + 10xy + y^2 \qquad \text{and} \qquad Q(x, y) = 6xy + 5x^2,$$

we have

$$\frac{\partial P}{\partial y} = 10x + 2y, \qquad \frac{\partial Q}{\partial x} = 6y + 10x, \qquad \text{and} \qquad \frac{\partial Q}{\partial x} - \frac{\partial P}{\partial y} = 4y.$$

By Green's theorem

$$\oint_C (1 + 10xy + y^2) \, dx + (6xy + 5x^2) \, dy = \iint_\Omega 4y \, dx \, dy$$

$$= \int_0^a \int_0^a 4y \, dx \, dy$$

$$= \left(\int_0^a dx \right) \left(\int_0^a 4y \, dy \right)$$

$$= (a)(2a^2) = 2a^3. \quad \square$$

ALTERNATE SOLUTION By 16.6.3,

$$\iint_\Omega y \, dx \, dy = \bar{y} \, (\text{area of } \Omega)$$

where $\bar{y}$ is the y-coordinate of the centroid of the region Ω. Since $\bar{y} = \tfrac{1}{2}a$, we have,

$$\iint_\Omega 4y \, dx \, dy = 4 \, \bar{y} \, (\text{area of } \Omega) = 4 \left(\tfrac{1}{2}a \right) a^2 = 2a^3. \quad \square$$

Example 3 Use Green's theorem to evaluate

$$\oint_C e^x \sin y \, dx + e^x \cos y \, dy$$

where C is the closed curve consisting of the semicircle $y = \sqrt{1 - x^2}$ and the interval $[-1, 1]$. (Figure 17.5.8.)

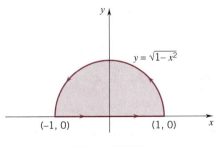

$$y = \sqrt{1 - x^2}$$

$(-1, 0)$ $(1, 0)$

Figure 17.5.8

SOLUTION The curve bounds the closed semicircular disc Ω: $x^2 + y^2 \leq 1$, $y \geq 0$. Here

$$\frac{\partial P}{\partial y} = e^x \cos y, \qquad \frac{\partial Q}{\partial x} = e^x \cos y \qquad \text{and} \qquad \frac{\partial Q}{\partial x} - \frac{\partial P}{\partial y} = 0.$$

By Green's theorem,

$$\oint_C e^x \sin y \; dx + e^x \cos y \; dy = \int\int_\Omega 0 \; dx \; dy = 0.$$

To see the power of Green's theorem, try to evaluate this line integral directly.   ❏

The preceding examples illustrate the use of Green's theorem to convert a line integral along the boundary of a Jordan region Ω into a double integral over Ω. Green's theorem can also be used in reverse. That is, the value of a double integral over a Jordan region can be found by evaluating a line integral along its boundary. A useful application of this involves the computation of the area of a region enclosed by a piecewise-smooth Jordan curve C.

(17.5.2)

$$\text{The area enclosed by } C = \oint_C -y \; dx = \oint_C x \; dy$$

$$= \tfrac{1}{2} \oint_C -y \; dx + x \; dy.$$

PROOF   Let Ω be the region enclosed by C.   In the first integral

$$P(x, y) = -y, \qquad Q(x, y) = 0.$$

Therefore

$$\frac{\partial P}{\partial y} = -1, \qquad \frac{\partial Q}{\partial x} = 0, \qquad \text{and} \qquad \frac{\partial Q}{\partial x} - \frac{\partial P}{\partial y} = 1.$$

Thus by Green's theorem

$$\oint_C -y \; dx = \int\int_\Omega 1 \; dx \; dy = \text{area of } \Omega.$$

That the second integral also gives the area of Ω can be verified in a similar manner.   Finally, the third integral is a consequence of

$$\oint_C -y \; dx + \oint_C x \; dy = \text{twice the area enclosed by } C.   ❏$$

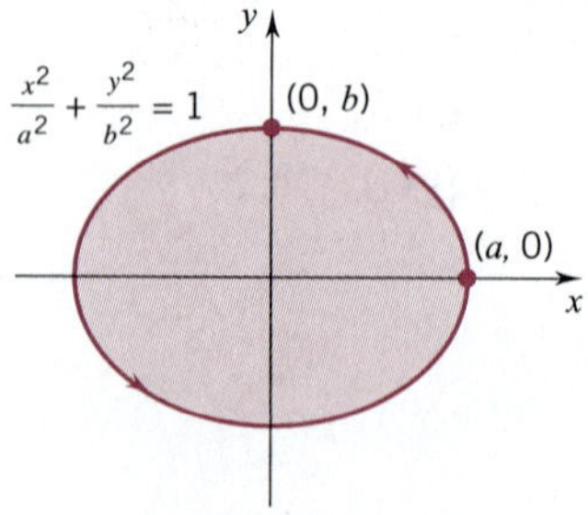

$\frac{x^2}{a^2} + \frac{y^2}{b^2} = 1$ $(0, b)$

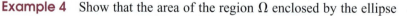

$(a, 0)$

Figure 17.5.9

Example 4   Show that the area of the region Ω enclosed by the ellipse

$$\frac{x^2}{a^2} + \frac{y^2}{b^2} = 1 \qquad\qquad \text{(Figure 17.5.9)}$$

is πab.

SOLUTION   The ellipse is oriented counterclockwise by the parametrization

$$x = a \cos u, \qquad y = b \sin u, \qquad 0 \leq u \leq 2\pi.$$

Although the third integral in (17.5.2) appears to be more complicated than either of the other two integrals, it is actually simpler to use in this case:

$$\text{area of } \Omega = \tfrac{1}{2} \oint_C -y\,dx + x\,dy$$

$$= \tfrac{1}{2} \int_0^{2\pi} [-(b \sin u)(-a \sin u) + (a \cos u)(b \cos u)]\,du$$

$$= \tfrac{1}{2}ab \int_0^{2\pi} du = \pi ab. \quad \square$$

Example 5 Let Ω be a Jordan region of area A with a piecewise-smooth boundary C. Show that the coordinates of the centroid of Ω are given by

$$\bar{x}A = \tfrac{1}{2} \oint_C x^2\,dy, \qquad \bar{y}A = -\tfrac{1}{2} \oint_C y^2\,dx.$$

SOLUTION

$$\tfrac{1}{2} \oint_C x^2\,dy = \tfrac{1}{2} \iint_\Omega 2x\,dx\,dy = \iint_\Omega x\,dx\,dy = \bar{x}A,$$

by Green's theorem

$$-\tfrac{1}{2} \oint_C y^2\,dx = -\tfrac{1}{2} \iint_\Omega (-2y)\,dx\,dy = \iint_\Omega y\,dx\,dy = \bar{y}A. \quad \square$$

Regions Bounded by Two or More Jordan Curves

(All the curves appearing in the following are assumed to be piecewise smooth.)

Figure 17.5.10 shows an annular region Ω. The region is not a Jordan region: the boundary consists of two Jordan curves C_1 and C_2. We cannot apply Green's theorem to Ω directly, but we can break up Ω into two Jordan regions as in Figure 17.5.11 and then apply Green's theorem to each piece. With Ω_1 and Ω_2 as in Figure 17.5.11,

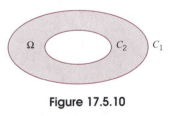

Figure 17.5.10

$$\iint_{\Omega_1} \left(\frac{\partial Q}{\partial x} - \frac{\partial P}{\partial y} \right) dx\,dy = \oint_{\text{bdry of } \Omega_1} P\,dx + Q\,dy,$$

$$\iint_{\Omega_2} \left(\frac{\partial Q}{\partial x} - \frac{\partial P}{\partial y} \right) dx\,dy = \oint_{\text{bdry of } \Omega_2} P\,dx + Q\,dy.$$

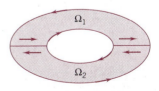

Figure 17.5.11

When we add the double integrals, we get the double integral over Ω. When we add the line integrals, the integrals over the crosscuts cancel and we are left with the *counterclockwise integral* over C_1 and the *clockwise integral* over C_2. (See the figure.) Thus, for the annular region,

(17.5.3)
$$\iint_\Omega \left(\frac{\partial Q}{\partial x} - \frac{\partial P}{\partial y} \right) dx\,dy = \oint_{C_1} P\,dx + Q\,dy + \oint_{C_2} P\,dx + Q\,dy.$$

As a corollary to this we see that, if $\partial Q/\partial x = \partial P/\partial y$ throughout Ω, then the double integral on the left is 0, the sum of the integrals on the right is 0. Therefore

(17.5.4)

$$
\boxed{
\begin{array}{c}
\text{if } \partial Q/\partial x = \partial P/\partial y \text{ throughout } \Omega, \text{ then} \\[4pt]
\oint_{C_1} P\,dx + Q\,dy = \oint_{C_2} P\,dx + Q\,dy.
\end{array}
}
$$

Example 6 Let C_1 be a Jordan curve that does not pass through the origin $(0, 0)$. Show that

$$
\oint_{C_1} -\frac{y}{x^2 + y^2}\,dx + \frac{x}{x^2 + y^2}\,dy = \begin{cases} 0, \text{ if } C_1 \text{ does not enclose the origin} \\ 2\pi, \text{ if } C_1 \text{ does enclose the origin.} \end{cases}
$$

SOLUTION In this case

$$
\frac{\partial P}{\partial y} = \frac{\partial}{\partial y}\left(-\frac{y}{x^2 + y^2}\right) = -\left[\frac{(x^2 + y^2)1 - 2y^2}{(x^2 + y^2)^2}\right] = \frac{y^2 - x^2}{(x^2 + y^2)^2},
$$

$$
\frac{\partial Q}{\partial x} = \frac{\partial}{\partial x}\left(\frac{x}{x^2 + y^2}\right) = \frac{(x^2 + y^2)1 - 2x^2}{(x^2 + y^2)^2} = \frac{y^2 - x^2}{(x^2 + y^2)^2}.
$$

Thus

$$
\frac{\partial Q}{\partial x} = \frac{\partial P}{\partial y} \qquad \text{except at the origin.}
$$

If C_1 does not enclose the origin, then $\partial Q/\partial x - \partial P/\partial y = 0$ throughout the region enclosed by C_1, and, by Green's theorem, the line integral is 0.

If C_1 does enclose the origin, we draw within the inner region of C_1 a small circle centered at the origin

$$
C_2: \quad x^2 + y^2 = a^2. \qquad \text{(See Figure 17.5.12)}
$$

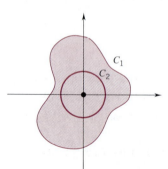

Since $\partial Q/\partial x - \partial P/\partial y = 0$ on the annular region bounded by C_1 and C_2, we know from (17.5.4) that the line integral over C_1 equals the line integral over C_2. All we have to show now is that the line integral over C_2 is 2π. This is straightforward. Parametrizing the circle by

$$
\mathbf{r}(u) = a \cos u \,\mathbf{i} + a \sin u \,\mathbf{j} \qquad \text{with} \quad u \in [0, 2\pi],
$$

we have

$$
\oint_{C_2} -\frac{y}{x^2 + y^2}\,dx + \frac{x}{x^2 + y^2}\,dy = \int_0^{2\pi} (\sin^2 u + \cos^2 u)\,du = \int_0^{2\pi} du = 2\pi. \quad \square
$$

check this

Figure 17.5.12

Figure 17.5.13 shows a region bounded by three Jordan curves: C_2 and C_3 both exterior to one another, both within C_1. For such a region Green's theorem gives

$$
\iint_\Omega \left(\frac{\partial Q}{\partial x} - \frac{\partial P}{\partial y}\right) dx\,dy =
$$

$$
\oint_{C_1} P\,dx + Q\,dy + \oint_{C_2} P\,dx + Q\,dy + \oint_{C_3} P\,dx + Q\,dy.
$$

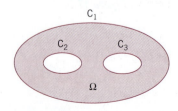

Figure 17.5.13

To see this, break up Ω into two regions by making the crosscuts shown in Figure 17.5.14.

The general formula for configurations of this type reads

$$\iint_\Omega \left(\frac{\partial Q}{\partial x} - \frac{\partial P}{\partial y} \right) dx\, dy = \oint_{C_1} P\, dx + Q\, dy + \sum_{i=2}^{n} \oint_{C_i} P\, dx + Q\, dy.$$

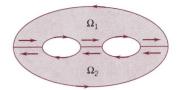

Figure 17.5.14

EXERCISES 17.5

In Exercises 1–4, evaluate the line integral two ways: (a) directly, and (b) using Green's theorem.

1. $\oint_C xy\, dx + x^2\, dy$; where C is the triangle with vertices $(0, 0)$, $(0, 1)$, $(1, 1)$.

2. $\oint_C x^2 y\, dx + 2y^2\, dy$; where C is the square with vertices $(0, 0)$, $(1, 0)$, $(1, 1)$, $(0, 1)$.

3. $\oint_C (3x^2 + y)\, dx + (2x + y^3)\, dy$; $C: x^2 + y^2 = 1$.

4. $\oint_C y^2\, dx + x^2\, dy$; where C is the boundary of the region bounded by the curves $y = x$ and $y = x^2$.

In Exercises 5–16, evaluate by Green's theorem.

5. $\oint_C 3y\, dx + 5x\, dy$; $\quad C: x^2 + y^2 = 1$.

6. $\oint_C 5x\, dx + 3y\, dy$; $\quad C: (x - 1)^2 + (y + 1)^2 = 1$.

7. $\oint_C x^2\, dy$; where C is the rectangle with vertices $(0, 0)$, $(a, 0)$, (a, b), $(0, b)$.

8. $\oint_C y^2\, dx$; where C is the rectangle of Exercise 7.

9. $\oint_C (3xy + y^2)\, dx + (2xy + 5x^2)\, dy$; $C: (x - 1)^2 + (y + 2)^2 = 1$.

10. $\oint_C (xy + 3y^2)\, dx + (5xy + 2x^2)\, dy$; $C: (x - 1)^2 + (y + 2)^2 = 1$.

11. $\oint_C (2x^2 + xy - y^2)\, dx + (3x^2 - xy + 2y^2)\, dy$; $C: \quad (x - a)^2 + y^2 = r^2$.

12. $\oint_C (x^2 - 2xy + 3y^2)\, dx + (5x + 1)\, dy$; $C: x^2 + (y - b)^2 = r^2$.

13. $\oint_C e^x \sin y\, dx + e^x \cos y\, dy$; $C: (x - a)^2 + (y - b)^2 = r^2$.

14. $\oint_C e^x \cos y\, dx + e^x \sin y\, dy$; $\quad C$ the rectangle with vertices $(0, 0)$, $(1, 0)$, $(1, \pi)$, $(0, \pi)$.

15. $\int_C 2xy\, dx + x^2\, dy$; where C is the cardioid $r = 1 - \cos\theta$, $\theta \in [0, 2\pi]$.

16. $\int_C y^2\, dx + 2xy\, dy$; where C is the first quadrant loop of the petal curve $r = 2\sin 2\theta$.

In Exercises 17 and 18, find the area enclosed by the curve by integrating over the curve.

17. The circle $x^2 + y^2 = a^2$.

18. The astroid $x^{2/3} + y^{2/3} = a^{2/3}$.

19. Find the area of the region in the first quadrant bounded by the coordinate axes and the line $x/a + y/b = 1$.

20. Find the area of the region enclosed by the parabola $y = x^2$ and the line $y = 4$.

21. Let C be a piecewise-smooth Jordan curve. Calculate

$$\oint_C (ay + b)\, dx + (cx + d)\, dy$$

given that C encloses a region of area A.

22. Calculate

$$\oint_C \mathbf{F(r)} \cdot \mathbf{dr}$$

given that $\mathbf{F}(x, y) = 2y\,\mathbf{i} - 3x\,\mathbf{j}$ and C is the astroid $x^{2/3} + y^{2/3} = a^{2/3}$.

23. Use Green's theorem to find the area under one arch of the cycloid

$$x(\theta) = R(\theta - \sin\theta), \qquad y(\theta) = R(1 - \cos\theta).$$

24. Find the Jordan curve C that maximizes the line integral

$$\oint_C y^3\,dx + (3x - x^3)\,dy.$$

25. Find the work done by the force $\mathbf{F}(x, y) = (x^2 - y^3)\mathbf{i} + (x^2 + y^2)\mathbf{j}$ in moving an object around the circle $x^2 + y^2 = 1$ in the counterclockwise direction.

26. Find the work done by the force $\mathbf{F}(x, y) = (x^3 - x^2 y)\mathbf{i} + xy^2\mathbf{j}$ in moving an object in the counterclockwise direction around the boundary of the region determined by the parabolas $y = x^2$ and $x = y^2$.

27. Complete the proof of Green's theorem for the elementary region of Figure 17.5.2 by showing that

$$\oint_C Q(x, y)\,dy = \iint_\Omega \frac{\partial Q}{\partial x}(x, y)\,dx\,dy.$$

28. Let Ω be a plate of constant mass density λ in the form of a Jordan region with a piecewise-smooth boundary C. Show that the moments of inertia of the plate about the coordinate axes are given by the formulas

(17.5.5) $\quad I_x = -\frac{\lambda}{3}\oint_C y^3\,dx, \qquad I_y = \frac{\lambda}{3}\oint_C x^3\,dy.$

29. Let P and Q be continuously differentiable functions on the region Ω of Figure 17.5.13. Given that $\partial P/\partial y = \partial Q/\partial x$ on Ω, find a relation between the line integrals

$$\oint_{C_1} P\,dx + Q\,dy, \qquad \oint_{C_2} P\,dx + Q\,dy,$$

$$\oint_{C_3} P\,dx + Q\,dy.$$

30. Show that, if $f = f(x)$ and $g = g(y)$ are continuously differentiable, then

$$\int_C f(x)\,dx + g(y)\,dy = 0$$

for all piecewise-smooth Jordan curves C.

31. Let C be a piecewise-smooth Jordan curve that does not pass through the origin. Evaluate

$$\oint_C \frac{x}{x^2 + y^2}\,dx + \frac{y}{x^2 + y^2}\,dy.$$

(a) If C does not enclose the origin.
(b) If C does enclose the origin.

32. Let C be a piecewise-smooth Jordan curve that does not pass through the origin. Evaluate

$$\oint_C -\frac{y^3}{(x^2 + y^2)^2}\,dx + \frac{xy^2}{(x^2 + y^2)^2}\,dy.$$

(a) If C does not enclose the origin.
(b) If C does enclose the origin.

33. Let $\mathbf{v}$ be a continuously differentiable vector field on the entire plane. Use Green's theorem to verify that if $\mathbf{v}$ is a gradient field $[\mathbf{v} = \nabla\phi]$, then

$$\oint_C \mathbf{v}\cdot d\mathbf{r} = 0$$

for every piecewise-smooth Jordan curve C.

34. Let C be the line segment from the point (x_1, y_1) to the point (x_2, y_2). Show that

$$\int_C -y\,dx + x\,dy = x_1 y_2 - x_2 y_1.$$

35. Let $(x_1, y_1), (x_2, y_2), \ldots, (x_n, y_n)$ be the vertices of a polygon in counterclockwise order. Show that the area of the polygon is

$$A = \tfrac{1}{2}[(x_1 y_2 - x_2 y_1) + (x_2 y_3 - x_3 y_2) + \cdots$$
$$+ (x_{n-1} y_n - x_n y_{n-1}) + (x_n y_1 - x_1 y_n)].$$

36. Use the result in Exercise 35 to find the area of:
(a) The triangle with vertices $(0, 0), (2, 1), (1, 4)$.
(b) The pentagon with vertices $(0, 0), (3, 1), (2, 4), (0, 6), (-1, 2)$.

---

*SUPPLEMENT TO SECTION 17.5

A JUSTIFICATION OF THE JACOBIAN AREA FORMULA

We based the change of variables for double integrals on the Jacobian area formula (Formula 16.10.1). Green's theorem enables us to derive this formula under the conditions spelled out in the following.

Let Γ be a Jordan region in the uv-plane with a piecewise-smooth boundary C_Γ. A vector function $\mathbf{r}(u, v) = x(u, v)\mathbf{i} + y(u, v)\mathbf{j}$ with continuous second partials maps Γ onto a region Ω of the xy-plane. If $\mathbf{r}$ is one-to-one on the interior of Γ and the Jacobian J of the components of $\mathbf{r}$ is different from zero on the interior of Γ, then

$$\text{area of } \Omega = \iint\limits_{\Gamma} |J(u, v)|\, du\, dv.$$

PROOF Suppose that C_Γ is parametrized $u = u(t)$, $v = v(t)$ with $t \in [a, b]$. Then the boundary of Ω is a piecewise-smooth curve C given by

$$\mathbf{r}[u(t), v(t)] = x[u(t), v(t)]\mathbf{i} + y[u(t), v(t)]\mathbf{j}, \qquad t \in [a, b].$$

By Green's theorem, the area of Ω is

$$\oint_C x\, dy = \left| \int_a^b x[u(t), v(t)] \frac{d}{dt} (y[u(t), v(t)])\, dt \right|$$

$$= \left| \int_a^b x[u(t), v(t)] \left(\frac{\partial y}{\partial u}[u(t), v(t)]u'(t) + \frac{\partial y}{\partial v}[u(t), v(t)]v'(t) \right) dt \right|$$

$$= \left| \int_a^b \left(x[u(t), v(t)] \frac{\partial y}{\partial u}[u(t), v(t)]u'(t) + x[u(t), v(t)] \frac{\partial y}{\partial v}[u(t), v(t)]v'(t) \right) dt \right|$$

$$= \left| \int_{C_\Gamma} x \frac{\partial y}{\partial u}\, du + x \frac{\partial y}{\partial v}\, dv \right|$$

— again by Green's theorem

$$= \left| \iint\limits_{\Gamma} \left[\frac{\partial}{\partial u} \left(x \frac{\partial y}{\partial v} \right) - \frac{\partial}{\partial v} \left(x \frac{\partial y}{\partial u} \right) \right] du\, dv \right|.$$

Now

$$\frac{\partial}{\partial u} \left(x \frac{\partial y}{\partial v} \right) - \frac{\partial}{\partial v} \left(x \frac{\partial y}{\partial u} \right) = \frac{\partial x}{\partial u} \frac{\partial y}{\partial v} + x \frac{\partial^2 y}{\partial u\, \partial v} - \frac{\partial x}{\partial v} \frac{\partial y}{\partial u} - x \frac{\partial^2 y}{\partial v\, \partial u}$$

$$= \frac{\partial x}{\partial u} \frac{\partial y}{\partial v} - \frac{\partial x}{\partial v} \frac{\partial y}{\partial u} = J(u, v).$$

Therefore

$$\text{area of } \Omega = \left| \iint\limits_{\Gamma} J(u, v)\, du\, dv \right| = \iint\limits_{\Gamma} |J(u, v)|\, du\, dv,$$

the final equality holding because $J(u, v)$ cannot change sign on Γ. ❑

■ 17.6 PARAMETRIZED SURFACES; SURFACE AREA

You have seen that a space curve C can be parametrized by a vector function $\mathbf{r} = \mathbf{r}(u)$ where u ranges over some interval I of the u-axis (Figure 17.6.1). In an analogous manner we can parametrize a surface S in space by a vector function $\mathbf{r} = \mathbf{r}(u, v)$ where (u, v) ranges over some region Ω of the uv-plane (Figure 17.6.2).

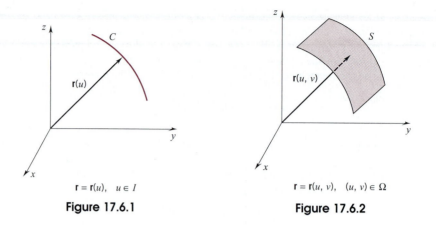

$\mathbf{r} = \mathbf{r}(u), \quad u \in I$

Figure 17.6.1

$\mathbf{r} = \mathbf{r}(u, v), \quad (u, v) \in \Omega$

Figure 17.6.2

Example 1 (*The graph of a function*) As you have seen, the graph of a function

$$y = f(x), \qquad x \in [a, b]$$

can be parametrized by setting

$$\mathbf{r}(u) = u\mathbf{i} + f(u)\mathbf{j}, \qquad u \in [a, b].$$

In the same vein the graph of a function

$$z = f(x, y), \qquad (x, y) \in \Omega$$

is a surface which can be parametrized by setting

$$\mathbf{r}(u, v) = u\mathbf{i} + v\mathbf{j} + f(u, v)\mathbf{k}, \qquad (u, v) \in \Omega.$$

As (u, v) ranges over Ω, the tip of $\mathbf{r}(u, v)$ traces out the graph of f. ❑

Example 2 (*A plane*) If two vectors $\mathbf{a}$ and $\mathbf{b}$ are not parallel, then the set of all linear combinations $u\mathbf{a} + v\mathbf{b}$ generate a plane p_0 that passes through the origin. We can parametrize this plane by setting

$$\mathbf{r}(u, v) = u\mathbf{a} + v\mathbf{b}, \qquad u, v \text{ real numbers.}$$

The plane p that is parallel to p_0 and passes through the tip of $\mathbf{c}$ can be parametrized by setting

$$\mathbf{r}(u, v) = u\mathbf{a} + v\mathbf{b} + \mathbf{c}, \qquad u, v \text{ real numbers.}$$

Note that the plane contains the lines

$$l_1: \quad \mathbf{r}(u, 0) = u\mathbf{a} + \mathbf{c} \qquad \text{and} \qquad l_2: \quad \mathbf{r}(0, v) = v\mathbf{b} + \mathbf{c}. \quad ❑$$

Example 3 (*A sphere*) The sphere of radius a centered at the origin can be parametrized by

$$\mathbf{r}(u, v) = a \cos u \cos v\,\mathbf{i} + a \sin u \cos v\,\mathbf{j} + a \sin v\,\mathbf{k}$$

with (u, v) ranging over the rectangle $R: 0 \leq u \leq 2\pi, \ -\frac{1}{2}\pi \leq v \leq \frac{1}{2}\pi$.

To derive this parametrization we refer to Figure 17.6.3. The points of latitude v (see the figure) form a circle of radius $a \cos v$ on the horizontal plane $z = a \sin v$. This circle can be parametrized by

$$\mathbf{R}(u) = a \cos v(\cos u\,\mathbf{i} + \sin u\,\mathbf{j}) + a \sin v\,\mathbf{k}, \qquad u \in [0, 2\pi].$$

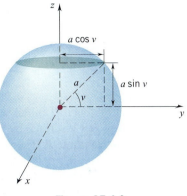

Figure 17.6.3

This expands to give

$$\mathbf{R}(u) = a \cos u \cos v\,\mathbf{i} + a \sin u \cos v\,\mathbf{j} + a \sin v\,\mathbf{k}, \qquad u \in [0, 2\pi].$$

Letting v range from $-\frac{1}{2}\pi$ to $\frac{1}{2}\pi$, we obtain the entire sphere.

The xyz-equation for this same sphere is $x^2 + y^2 + z^2 = a^2$. It is easy to verify that the parametrization satisfies this equation:

$$
\begin{aligned}
x^2 + y^2 + z^2 &= a^2\cos^2 u \cos^2 v + a^2\sin^2 u \cos^2 v + a^2\sin^2 u \\
&= a^2(\cos^2 u + \sin^2 u)\cos^2 u + a^2\sin^2 v \\
&= a^2(\cos^2 v + \sin^2 v) = a^2. \quad \square
\end{aligned}
$$

Example 4 (*A cone*) Figure 17.6.4 shows a right circular cone with apex semiangle α and slant height s. The points of slant height v (see the figure) form a circle of radius $v \sin \alpha$ on the horizontal plane $z = v \cos \alpha$. This circle can be parametrized by

$$
\begin{aligned}
R(u) &= v \sin \alpha\,(\cos u\,\mathbf{i} + \sin u\,\mathbf{j}) + v \cos \alpha\,\mathbf{k} \\
&= v \cos u \sin \alpha\,\mathbf{i} + v \sin u \sin \alpha\,\mathbf{j} + v \cos \alpha\,\mathbf{k}, \qquad u \in [0, 2\pi].
\end{aligned}
$$

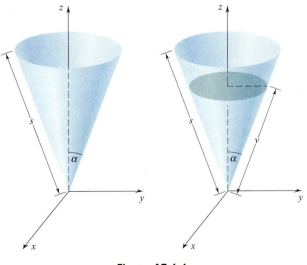

Figure 17.6.4

Since we can obtain the entire cone by letting v range from 0 to s, the cone is parametrized by

$$\mathbf{r}(u, v) = v \cos u \sin \alpha \, \mathbf{i} + v \sin u \sin \alpha \, \mathbf{j} + v \cos \alpha \, \mathbf{k},$$

with $0 \leq u \leq 2\pi$, $0 \leq v \leq s$. ❏

Example 5 (*A spiral ramp*) A rod of length l initially resting on the x-axis and attached at one end to the z-axis sweeps out a surface by rotating about the z-axis at constant rate ω while climbing at a constant rate b. The surface is pictured in Figure 17.6.5.

To parametrize this surface we mark the point of the rod at a distance u from the z-axis ($0 \leq u \leq l$) and ask for the position of this point at time v. At time v the rod will have climbed a distance bv and rotated through an angle ωv. Thus the point will be found at the tip of the vector

$$u(\cos \omega v \, \mathbf{i} + \sin \omega v \, \mathbf{j}) + bv \, \mathbf{k} = u \cos \omega v \, \mathbf{i} + u \sin \omega v \, \mathbf{j} + bv \, \mathbf{k}.$$

The entire surface can be parametrized by

$$\mathbf{r}(u, v) = u \cos \omega v \, \mathbf{i} + u \sin \omega v \, \mathbf{j} + bv \, \mathbf{k} \qquad \text{with } 0 \leq u \leq l, \quad v \geq 0. \quad ❏$$

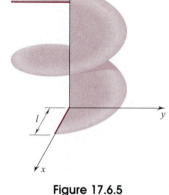

Figure 17.6.5

The Fundamental Vector Product

Let S be a surface parametrized by a differentiable vector function

$$\mathbf{r} = \mathbf{r}(u, v) = x(u, v)\mathbf{i} + y(u, v)\mathbf{j} + z(u, v)\mathbf{k}.$$

For simplicity, let us suppose that (u, v) varies over the open rectangle R: $a < u < b$, $c < v < d$. Since $\mathbf{r}$ is a function of two variables, we can take two partial derivatives:

$$\mathbf{r}'_u = \frac{\partial x}{\partial u}\mathbf{i} + \frac{\partial y}{\partial u}\mathbf{j} + \frac{\partial z}{\partial u}\mathbf{k},$$

the partial of $\mathbf{r}$ with respect to u, and

$$\mathbf{r}'_v = \frac{\partial x}{\partial v}\mathbf{i} + \frac{\partial y}{\partial v}\mathbf{j} + \frac{\partial z}{\partial v}\mathbf{k},$$

the partial of $\mathbf{r}$ with respect to v. Now let (u_0, v_0) be a point of R for which

$$\mathbf{r}'_u(u_0, v_0) \times \mathbf{r}'_v(u_0, v_0) \neq \mathbf{0}.$$

(The reason for this condition will be apparent as we go on.) The vector function

$$\mathbf{r}_1(u) = \mathbf{r}(u, v_0), \qquad u \in (a, b)$$

(here we are keeping v fixed at v_0) traces a differentiable curve C_1 that lies on S (Figure 17.6.6). The vector function

$$\mathbf{r}_2(v) = \mathbf{r}(u_0, v), \qquad v \in (c, d)$$

(this time we are keeping u fixed at u_0) traces a differentiable curve C_2 that also lies on S. Both curves pass through the tip of $\mathbf{r}(u_0, v_0)$:

$$C_1 \text{ with tangent vector } \mathbf{r}'_1(u_0) = \mathbf{r}'_u(u_0, v_0)$$

$$C_2 \text{ with tangent vector } \mathbf{r}'_2(v_0) = \mathbf{r}'_v(u_0, v_0).$$

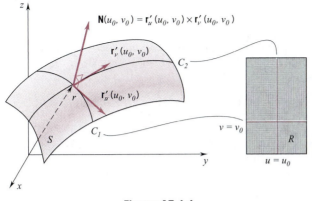

Figure 17.6.6

The cross product $\mathbf{N}(u_0, v_0) = \mathbf{r}'_u(u_0, v_0) \times \mathbf{r}'_v(u_0, v_0)$, which we have assumed to be different from zero, is thus perpendicular to both curves at the tip of $\mathbf{r}(u_0, v_0)$ and can be taken as a normal to the surface at that point. We record the result as follows:

(17.6.1) If S is a surface given by a differentiable function $\mathbf{r} = \mathbf{r}(u, v)$, then the vector $\mathbf{N}(u, v) = \mathbf{r}'_u(u, v) \times \mathbf{r}'_v(u, v)$ is perpendicular to the surface at the point with position vector $\mathbf{r}(u, v)$ and, if different from zero, can be taken as a normal to the surface at that point.

The cross product

$$\mathbf{N} = \mathbf{r}'_u \times \mathbf{r}'_v = \begin{vmatrix} \mathbf{i} & \mathbf{j} & \mathbf{k} \\ \dfrac{\partial x}{\partial u} & \dfrac{\partial y}{\partial u} & \dfrac{\partial z}{\partial u} \\ \dfrac{\partial x}{\partial v} & \dfrac{\partial y}{\partial v} & \dfrac{\partial z}{\partial v} \end{vmatrix}$$

is called the *fundamental vector product* of the surface.

Example 6 For the plane $\mathbf{r}(u, v) = u\mathbf{a} + v\mathbf{b} + \mathbf{c}$ we have

$$\mathbf{r}'_u(u, v) = \mathbf{a}, \qquad \mathbf{r}'_v(u, v) = \mathbf{b} \qquad \text{and therefore} \qquad \mathbf{N}(u, v) = \mathbf{a} \times \mathbf{b}.$$

The vector $\mathbf{a} \times \mathbf{b}$ is normal to the plane. ❑

Example 7 We parametrized the sphere $x^2 + y^2 + z^2 = a^2$ by setting

$$\mathbf{r}(u, v) = a \cos u \cos v\, \mathbf{i} + a \sin u \cos v\, \mathbf{j} + a \sin v\, \mathbf{k}$$

with $0 \le u \le 2\pi$, $-\tfrac{1}{2}\pi \le v \le \tfrac{1}{2}\pi$. In this case

$$\mathbf{r}'_u(u, v) = -a \sin u \cos v\, \mathbf{i} + a \cos u \cos v\, \mathbf{j}$$

and

$$\mathbf{r}'_v(u, v) = -a \cos u \sin v\, \mathbf{i} - a \sin u \sin v\, \mathbf{j} + a \cos v\, \mathbf{k}.$$

Thus

$$N(u, v) = \begin{vmatrix} \mathbf{i} & \mathbf{j} & \mathbf{k} \\ -a \sin u \cos v & a \cos u \cos v & 0 \\ -a \cos u \sin v & -a \sin u \sin v & a \cos v \end{vmatrix}$$

check this ⟶

$$= a \cos v \, (a \cos u \cos v \, \mathbf{i} + a \sin u \cos v \, \mathbf{j} + a \sin v \, \mathbf{k})$$

$$= a \cos v \, \mathbf{r}(u, v).$$

As was to be expected, the fundamental vector product of a sphere is parallel to the radius vector $\mathbf{r}(u, v)$. ❏

The Area of a Parametrized Surface

A linear function

$$\mathbf{r}(u, v) = u\mathbf{a} + v\mathbf{b} + \mathbf{c} \qquad \text{(a and b not parallel)}$$

parametrizes a plane p. Horizontal lines from the uv-plane, lines with equations of the form $v = v_0$, are mapped onto lines parallel to $\mathbf{a}$, and vertical lines, $u = u_0$, are carried onto lines parallel to $\mathbf{b}$:

$$\mathbf{r}(u, v_0) = \underbrace{u\mathbf{a}}_{\text{direction vector}} + \underbrace{v_0\mathbf{b} + \mathbf{c}}_{\text{constant}}, \qquad \mathbf{r}(u_0, v) = \underbrace{v\mathbf{b}}_{\text{direction vector}} + \underbrace{u_0\mathbf{a} + \mathbf{c}}_{\text{constant}}.$$

Thus a rectangle R in the uv-plane with sides parallel to the u and v axes,

$$R: \quad u_1 \leq u \leq u_2, \quad v_1 \leq v \leq v_2,$$

is mapped onto a parallelogram on p with sides parallel to $\mathbf{a}$ and $\mathbf{b}$. What is important to us here is that

the area of the parallelogram $= \|\mathbf{a} \times \mathbf{b}\| \cdot$ (the area of R).

To show this, we refer to Figure 17.6.7. The parallelogram is generated by the vectors

$$\mathbf{r}(u_2, v_1) - \mathbf{r}(u_1, v_1) = (u_2\mathbf{a} + v_1\mathbf{b} + \mathbf{c}) - (u_1\mathbf{a} + v_1\mathbf{b} + \mathbf{c}) = (u_2 - u_1)\mathbf{a},$$

$$\mathbf{r}(u_1, v_2) - \mathbf{r}(u_1, v_1) = (u_1\mathbf{a} + v_2\mathbf{b} + \mathbf{c}) - (u_1\mathbf{a} + v_1\mathbf{b} + \mathbf{c}) = (v_2 - v_1)\mathbf{b}.$$

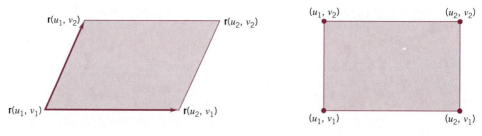

Figure 17.6.7

The area of the parallelogram is thus

$$\|(u_2 - u_1)\mathbf{a} \times (v_2 - v_1)\mathbf{b}\| = \|\mathbf{a} \times \mathbf{b}\|(u_2 - u_1)(v_2 - v_1)$$

$$= \|\mathbf{a} \times \mathbf{b}\| \cdot \text{(area of } R). \quad ❏$$

We can summarize as follows:

(17.6.2)

Let R be a rectangle in the uv-plane with sides parallel to the coordinate axes. If $\mathbf{a}$ and $\mathbf{b}$ are not parallel, the linear function

$$\mathbf{r}(u, v) = u\mathbf{a} + v\mathbf{b} + \mathbf{c}, \qquad (u, v) \in R$$

parametrizes a parallelogram with sides parallel to $\mathbf{a}$ and $\mathbf{b}$ and the area of the parallelogram $= \|\mathbf{a} \times \mathbf{b}\| \cdot$ (the area of R).

More generally, let's suppose that we have a surface S parametrized by a continuously differentiable function

$$\mathbf{r} = \mathbf{r}(u, v), \qquad (u, v) \in \Omega.$$

We assume that Ω is a basic region in the uv-plane and that $\mathbf{r}$ is one-to-one on the interior of Ω. (We don't want $\mathbf{r}$ to cover parts of S more than once.) Also we assume that the fundamental vector product $\mathbf{N} = \mathbf{r}'_u \times \mathbf{r}'_v$ is never zero on the interior of Ω. (We can then use it as a normal.) Under these conditions we call S a *continuously differentiable surface* and define

(17.6.3)

$$\text{area of } S = \iint_\Omega \|\mathbf{N}(u, v)\| \, du \, dv.$$

We show the reasoning behind this definition in the case that Ω is a rectangle R with sides parallel to the coordinate axes. We begin by breaking up R into N little rectangles $R_1, \ldots, R_N$. This induces a decomposition of S into little pieces $S_1, \ldots, S_N$. Taking (u_i^*, v_i^*) as the center of R_i, we have the tip of $\mathbf{r}(u_i^*, v_i^*)$ in S_i. Since the vector $\mathbf{r}'_u(u_i^*, v_i^*) \times \mathbf{r}'_v(u_i^*, v_i^*)$ is normal to the surface at the tip of $\mathbf{r}(u_i^*, v_i^*)$, we can parametrize the tangent plane at this point by the linear function

$$\mathbf{f}(u, v) = u\,\mathbf{r}'_u(u_i^*, v_i^*) + v\,\mathbf{r}'_v(u_i^*, v_i^*) + [\mathbf{r}(u_i^*, v_i^*) - u_i^*\mathbf{r}'_u(u_i^*, v_i^*) - v_i^*\mathbf{r}'_v(u_i^*, v_i^*)].$$

(Check that this linear function gives the right plane.) S_i is the portion of S that corresponds to R_i. The portion of the tangent plane that corresponds to this same R_i is a parallelogram with area

$$\|\mathbf{r}'_u(u_i^*, v_i^*) \times \mathbf{r}'_v(u_i^*, v_i^*)\| \cdot (\text{area of } R_i) = \|\mathbf{N}(u_i^*, v_i^*)\| \cdot (\text{area of } R_i). \quad \text{(by (17.6.2))}$$

Taking this as our estimate for the area of S_i, we have

$$\text{area of } S = \sum_{i=1}^N \text{area of } S_i \cong \sum_{i=1}^N \|\mathbf{N}(u_i^*, v_i^*)\| \cdot (\text{area of } R_i).$$

This is a Riemann sum for

$$\iint_R \|\mathbf{N}(u, v)\| \, du \, dv$$

and tends to this integral as the maximum diameter of the R_i tends to zero. ❑

To make sure that Formula 17.6.3 does not violate our previously established notion of area, we must verify that it gives the expected result both for plane regions and surfaces of revolution. This is done in Examples 9 and 10. By way of introduction we begin with the sphere.

Example 8 (*The surface area of a sphere*) The function

$$\mathbf{r}(u, v) = a \cos u \cos v \, \mathbf{i} + a \sin u \cos v \, \mathbf{j} + a \sin v \, \mathbf{k}$$

with (u, v) ranging over the set $\Omega: 0 \le u \le 2\pi$, $-\frac{1}{2}\pi \le v \le \frac{1}{2}\pi$ parametrizes a sphere of radius a. For this parametrization

$$\mathbf{N}(u, v) = a \cos v \, \mathbf{r}(u, v) \quad \text{and} \quad \|\mathbf{N}(u, v)\| = a^2 |\cos v| = a^2 \cos v.$$

Example 7 ⟶ ↑ $-\frac{1}{2}\pi \le v \le \frac{1}{2}\pi$ ⟶ ↑

According to the new formula

$$\text{area of the sphere} = \iint_\Omega a^2 \cos v \, du \, dv$$

$$= \int_0^{2\pi} \left(\int_{-\frac{1}{2}\pi}^{\frac{1}{2}\pi} a^2 \cos v \, dv \right) du = 2\pi a^2 \int_{-\frac{1}{2}\pi}^{\frac{1}{2}\pi} \cos v \, dv = 4\pi a^2,$$

which, as you know, is correct. ❏

Example 9 (*The area of a region*) If S is a plane region Ω, then S can be parametrized by setting

$$\mathbf{r}(u, v) = u\mathbf{i} + v\mathbf{j}, \quad (u, v) \in \Omega.$$

Here $\mathbf{N}(u, v) = \mathbf{r}'_u(u, v) \times \mathbf{r}'_v(u, v) = \mathbf{i} \times \mathbf{j} = \mathbf{k}$ and $\|\mathbf{N}(u, v)\| = 1$. In this case (17.6.3) reduces to the familiar formula

$$A = \iint_\Omega du \, dv. \quad ❏$$

Example 10 (*The area of a surface of revolution*) Let S be the surface generated by revolving the graph of a function

$$y = f(x), \quad x \in [a, b] \qquad \text{(Figure 17.6.8)}$$

about the x-axis. We assume that f is positive and continuously differentiable. We can parametrize S by setting

$$\mathbf{r}(u, v) = v\mathbf{i} + f(v)\cos u \, \mathbf{j} + f(v)\sin u \, \mathbf{k}$$

with (u, v) ranging over the set $\Omega: 0 \le u \le 2\pi$, $a \le v \le b$. (We leave it to you to verify that this is right.) In this case

$$\mathbf{N}(u, v) = \mathbf{r}'_u(u, v) \times \mathbf{r}'_v(u, v) = \begin{vmatrix} \mathbf{i} & \mathbf{j} & \mathbf{k} \\ 0 & -f(v)\sin u & f(v)\cos u \\ 1 & f'(v)\cos u & f'(v)\sin u \end{vmatrix}$$

$$= -f(v)f'(v)\mathbf{i} + f(v)\cos u \, \mathbf{j} + f(v)\sin u \, \mathbf{k}.$$

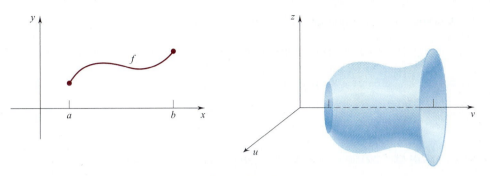

Figure 17.6.8

Therefore $\|\mathbf{N}(u, v)\| = f(v)\sqrt{[f'(v)]^2 + 1}$ and

$$\text{area of } S = \iint_{\Omega} f(v)\sqrt{[f'(v)]^2 + 1} \, du \, dv$$

$$= \int_0^{2\pi} \left(\int_a^b f(v)\sqrt{[f'(v)]^2 + 1} \, dv \right) du = \int_a^b 2\pi f(v)\sqrt{[f'(v)]^2 + 1} \, dv.$$

This is in agreement with Formula 9.10.3. ❑

Example 11 (*Spiral ramp*) One turn of the spiral ramp of Example 5 is the surface

$$S: \quad \mathbf{r}(u, v) = u \cos \omega v \, \mathbf{i} + u \sin \omega v \, \mathbf{j} + bv \, \mathbf{k}$$

with (u, v) ranging over the set Ω: $0 \le u \le l$, $0 \le v \le 2\pi/\omega$. In this case

$$\mathbf{r}'_u(u, v) = \cos \omega v \, \mathbf{i} + \sin \omega v \, \mathbf{j}, \qquad \mathbf{r}'_v(u, v) = -\omega u \sin \omega v \, \mathbf{i} + \omega u \cos \omega v \, \mathbf{j} + b \, \mathbf{k}.$$

Therefore

$$\mathbf{N}(u, v) = \begin{vmatrix} \mathbf{i} & \mathbf{j} & \mathbf{k} \\ \cos \omega v & \sin \omega v & 0 \\ -\omega u \sin \omega v & \omega u \cos \omega v & b \end{vmatrix} = b \sin \omega v \, \mathbf{i} - b \cos \omega v \, \mathbf{j} + \omega u \, \mathbf{k}$$

and

$$\|\mathbf{N}(u, v)\| = \sqrt{b^2 + \omega^2 u^2}.$$

Thus

$$\text{area of } S = \iint_{\Omega} \sqrt{b^2 + \omega^2 u^2} \, du \, dv$$

$$= \int_0^{2\pi/\omega} \left(\int_0^1 \sqrt{b^2 + \omega^2 u^2} \, du \right) dv = \frac{2\pi}{\omega} \int_0^1 \sqrt{b^2 + \omega^2 u^2} \, du.$$

The integral can be evaluated by setting $u = (b/\omega) \tan x$. ❑

The Area of a Surface $z = f(x, y)$

Figure 17.6.9 shows a surface that projects onto a basic region Ω of the xy-plane. Above each point (x, y) of Ω there is one and only one point of S. The surface S is then the graph of a function

$$z = f(x, y), \qquad (x, y) \in \Omega.$$

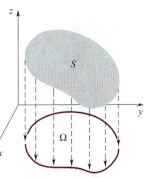

Figure 17.6.9

As we show, if f is continuously differentiable, then

(17.6.4)
$$\text{area of } S = \iint\limits_{\Omega} \sqrt{[f_x(x, y)]^2 + [f_y(x, y)]^2 + 1} \; dx \, dy.$$

DERIVATION OF FORMULA 17.6.4 We can parametrize S by setting

$$\mathbf{r}(u, v) = u\mathbf{i} + v\mathbf{j} + f(u, v)\mathbf{k}, \qquad (u, v) \in \Omega.$$

We may just as well use x and y and write

$$\mathbf{r}(x, y) = x\mathbf{i} + y\mathbf{j} + f(x, y)\mathbf{k}, \qquad (x, y) \in \Omega.$$

Clearly

$$\mathbf{r}'_x(x, y) = \mathbf{i} + f_x(x, y)\mathbf{k} \qquad \text{and} \qquad \mathbf{r}'_y(x, y) = \mathbf{j} + f_y(x, y)\mathbf{k}.$$

Thus

$$\mathbf{N}(x, y) = \begin{vmatrix} \mathbf{i} & \mathbf{j} & \mathbf{k} \\ 1 & 0 & f_x(x, y) \\ 0 & 1 & f_y(x, y) \end{vmatrix} = -f_x(x, y)\mathbf{i} - f_y(x, y)\mathbf{j} + \mathbf{k}.$$

Therefore $\|\mathbf{N}(x, y)\| = \sqrt{[f_x(x, y)]^2 + [f_y(x, y)]^2 + 1}$ and the formula is verified. ❑

Example 12 Find the surface area of that part of the parabolic cylinder $z = y^2$ that lies over the triangle with vertices $(0, 0)$, $(0, 1)$, $(1, 1)$ in the xy-plane.

SOLUTION Here $f(x, y) = y^2$ so that

$$f_x(x, y) = 0, \qquad f_y(x, y) = 2y.$$

The base triangle can be expressed by writing

$$\Omega: \quad 0 \le y \le 1, \quad 0 \le x \le y.$$

The surface has area

$$A = \iint\limits_{\Omega} \sqrt{[f_x(x, y)]^2 + [f_y(x, y)]^2 + 1} \; dx \, dy$$

$$= \int_0^1 \int_0^y \sqrt{4y^2 + 1} \; dx \, dy$$

$$= \int_0^1 y\sqrt{4y^2 + 1} \; dy = \left[\tfrac{1}{12}(4y^2 + 1)^{3/2} \right]_0^1 = \tfrac{1}{12}(5\sqrt{5} - 1) \cong 0.85. \quad ❑$$

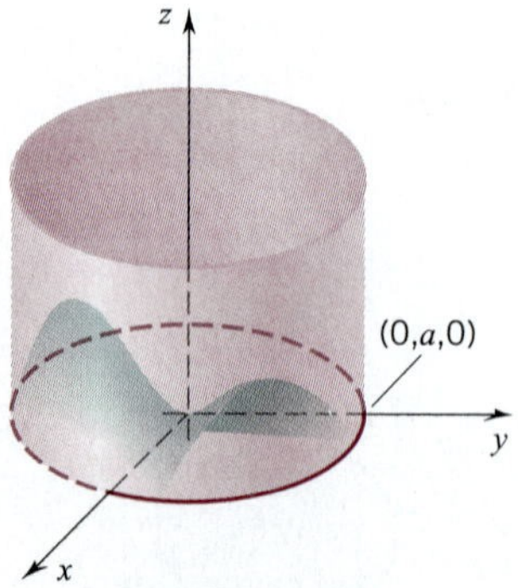

Example 13 Find the surface area of that part of the hyperbolic paraboloid $z = xy$ that lies inside the cylinder $x^2 + y^2 = a^2$. See Figure 17.6.10.

SOLUTION Here $f(x, y) = xy$ so that

$$f_x(x, y) = y, \qquad f_y(x, y) = x.$$

Figure 17.6.10

The formula gives

$$A = \int\int_{\Omega} \sqrt{y^2 + x^2 + 1} \; dx \; dy.$$

In polar coordinates the base region takes the form

$$\Gamma: \quad 0 \le r \le a, \quad 0 \le \theta \le 2\pi.$$

Thus we have

$$A = \int\int_{\Gamma} \sqrt{r^2 + 1} \; r \; dr \; d\theta = \int_0^{2\pi} \int_0^a \sqrt{r^2 + 1} \; r \; dr \; d\theta = \tfrac{2}{3}\pi[(a^2 + 1)^{3/2} - 1]. \quad \square$$

There is an elegant version of this last area formula (Formula 17.6.4) that is geometrically vivid. We know that the vector

$$\mathbf{r}'_x(x, y) \times \mathbf{r}'_y(x, y) = -f_x(x, y)\mathbf{i} - f_y(x, y)\mathbf{j} + \mathbf{k}$$

is normal to the surface at the point $(x, y, f(x, y))$. The unit vector in that direction, the vector

$$\mathbf{n}(x, y) = \frac{-f_x(x, y)\mathbf{i} - f_y(x, y)\mathbf{j} + \mathbf{k}}{\sqrt{[f_x(x, y)]^2 + [f_y(x, y)]^2 + 1}},$$

is called the *upper unit normal.* (It is the unit normal with a nonnegative **k**-component.)

Now let $\gamma(x, y)$ be the angle between $\mathbf{n}(x, y)$ and $\mathbf{k}$ (Figure 17.6.11). Since $\mathbf{n}(x, y)$ and $\mathbf{k}$ are both unit vectors,

$$\cos[\gamma(x, y)] = \mathbf{n}(x, y) \cdot \mathbf{k} = \frac{1}{\sqrt{[f_x(x, y)]^2 + [f_y(x, y)]^2 + 1}}.$$

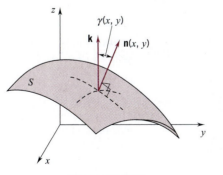

Figure 17.6.11

Taking reciprocals we have

$$\sec[\gamma(x, y)] = \sqrt{[f_x(x, y)]^2 + [f_y(x, y)]^2 + 1}.$$

The area formula can therefore be written

(17.6.5)

$$A = \iint\limits_{\Omega} \sec[\gamma(x, y)] \, dx \, dy.$$

EXERCISES 17.6

In Exercises 1–4, calculate the fundamental vector product.

1. $\mathbf{r}(u, v) = (u^2 - v^2)\mathbf{i} + (u^2 + v^2)\mathbf{j} + 2uv\mathbf{k}$.

2. $\mathbf{r}(u, v) = u \cos v\mathbf{i} + u \sin v\mathbf{j} + \mathbf{k}$.

3. $\mathbf{r}(u, v) = (u + v)\mathbf{i} + (u + v)\mathbf{j} + (u - v)\mathbf{k}$.

4. $\mathbf{r}(u, v) = \cos u \sin v\mathbf{i} + \sin u \cos v\mathbf{j} + u\mathbf{k}$.

In Exercises 5–10, find a parametric representation for the given surface.

5. The upper half of the ellipsoid $4x^2 + 9y^2 + z^2 = 36$.

6. The part of the cylinder $x^2 + y^2 = 4$ that lies between the planes $z = 1$ and $z = 4$.

7. The part of the sphere $x^2 + y^2 + z^2 = 4$ that lies above the plane $z = \sqrt{2}$.

8. The part of the plane $z = x + 2$ that lies inside the cylinder $x^2 + y^2 = 1$.

9. $y = g(x, z)$, $(x, z) \in \Omega$.   10. $x = h(y, z)$, $(y, z) \in \Gamma$.

In Exercises 11–13, find an equation in x, y, z for the given surface and identify the surface.

11. $\mathbf{r}(u, v) = a \cos u \cos v\mathbf{i} + b \sin u \cos v\mathbf{j} + c \sin v\mathbf{k}$;
    $0 \le u \le 2\pi$, $-\frac{1}{2}\pi \le v \le \frac{1}{2}\pi$.

12. $\mathbf{r}(u, v) = au \cos v\mathbf{i} + bu \sin v\mathbf{j} + u^2\mathbf{k}$; $0 \le u$,
    $0 \le v \le 2\pi$.

13. $\mathbf{r}(u, v) = au \cosh v\mathbf{i} + bu \sinh v\mathbf{j} + u^2\mathbf{k}$; u real, v real.

14. Show that the hyperboloid of one sheet

    $$\frac{x^2}{a^2} + \frac{y^2}{b^2} - \frac{z^2}{c^2} = 1 \qquad \text{(Figure 14.2.2)}$$

    can be parametrized by the vector function

    $\mathbf{r}(u, v) = a \cos u \cosh v\mathbf{i} + b \sin u \cosh v\mathbf{j}$
    $\qquad\qquad + c \sinh v\mathbf{k}$; $0 \le u \le 2\pi$, v real.

15. The graph of a continuously differentiable function $y = f(x)$, $x \in [a, b]$ is revolved about the y-axis. Parametrize the surface given that $a \ge 0$.

16. Show that the area of the surface of Exercise 15 is given by the formula

    $$A = \int_a^b 2\pi x \sqrt{1 + [f'(x)]^2} \, dx.$$

17. A plane p intersects the xy-plane at an angle γ. (Draw a figure.) Find the area of a region Γ on p given that the projection of Γ onto the xy-plane is a region Ω of area A_Ω.

18. Determine the area of the portion of the plane $x + y + z = a$ that lies within the cylinder $x^2 + y^2 = b^2$.

19. Find the area of the part of the plane $bcx + acy + abz = abc$ that lies within the first octant.

20. Find the area of the surface $z^2 = x^2 + y^2$ from $z = 0$ to $z = 1$.

21. Find the area of the surface $z = x^2 + y^2$ from $z = 0$ to $z = 4$.

In Exercises 22–28, calculate the area of the surface.

22. $z^2 = 2xy$ with $0 \le x \le a$, $0 \le y \le b$, $z \ge 0$.

23. $z = a^2 - (x^2 + y^2)$ with $\frac{1}{4}a^2 \le x^2 + y^2 \le a^2$.

24. $3z^2 = (x + y)^3$ with $x + y \le 2$, $x \ge 0$, $y \ge 0$.

25. $3z = x^{3/2} + y^{3/2}$ with $0 \le x \le 1$, $0 \le y \le x$.

26. $z = y^2$ with $0 \le x \le 1$, $0 \le y \le 1$.

27. $x^2 + y^2 + z^2 - 4z = 0$ with $0 \le 3(x^2 + y^2) \le z^2$, $z \ge 2$.

28. $x^2 + y^2 + z^2 - 2az = 0$ with $0 \le x^2 + y^2 \le bz$. Assume $a > b > 0$.

29. (a) Find a formula for the area of a surface that is projectable onto a region Ω of the yz-plane; say,

    $$S: \quad x = g(y, z), \quad (y, z) \in \Omega.$$

    Assume that g is continuously differentiable.
    (b) Find a formula for the area of a surface that is projectable onto a region Ω of the xz-plane; say,

    $$S: \quad y = h(x, z), \quad (x, z) \in \Omega.$$

    Assume that h is continuously differentiable.

30. (a) Determine the fundamental vector product for the cylindrical surface

    $\mathbf{r}(u, v) = a \cos u\mathbf{i} + a \sin u\mathbf{j} + v\mathbf{k}$;
    $0 \le u \le 2\pi$, $0 \le v \le l$.

    (b) Use your answer to (a) to find the area of the surface.

31. (a) Determine the fundamental vector product for the cone of Example 4:

$$\mathbf{r}(u, v) = v \cos u \sin \alpha\, \mathbf{i} + v \sin u \sin \alpha\, \mathbf{j} + v \cos \alpha\, \mathbf{k};$$
$$0 \leq u \leq 2\pi, \quad 0 \leq v \leq s.$$

(b) Use your answer to (a) to calculate the area of the cone.

32. Verify that the ellipsoid of revolution

$$\frac{x^2}{a^2} + \frac{y^2}{a^2} + \frac{z^2}{b^2} = 1$$

can be parametrized by

$$\mathbf{r}(u, v) = a \cos u \sin v\, \mathbf{i} + a \sin u \sin v\, \mathbf{j} + b \cos v\, \mathbf{k};$$
$$0 \leq u \leq 2\pi, \quad 0 \leq v \leq \pi.$$

Then show that the surface area of the ellipsoid is given by the formula

$$A = 2\pi a \int_0^\pi \sin v \sqrt{b^2 \sin^2 v + a^2 \cos^2 v}\, dv.$$

33. Let Ω be a plane region in space and let A_1, A_2, A_3 be the areas of the projections of Ω onto the three coordinate planes. Express the area of Ω in terms of A_1, A_2, A_3.

34. Let S be a surface given in cylindrical coordinates by $z = f(r, \theta)$, $(r, \theta) \in \Omega$. Show that, if f is continuously differentiable, then

$$\text{area of } S = \iint_\Omega \sqrt{r^2 [f_r(r, \theta)]^2 + [f_\theta(r, \theta)]^2 + r^2}\, dr\, d\theta$$

provided the integrand is never zero on the interior of Ω.

35. The following surfaces are given in cylindrical coordinates. Find the surface area.
(a) $z = r + \theta$; $0 \leq r \leq 1$, $0 \leq \theta \leq \pi$.
(b) $z = re^\theta$; $0 \leq r \leq a$, $0 \leq \theta \leq 2\pi$.

36. Show that, for a flat surface S that is part of the xy-plane, (17.6.3) gives

$$\text{area of } S = \iint_\Omega |J(u, v)|\, du\, dv$$

where J is the Jacobian of the components of the vector function that parametrizes S. Except for notation this is Formula 16.11.1.

17.7 SURFACE INTEGRALS

The Mass of a Material Surface

Imagine a thin distribution of matter spread out over a surface S. We call this a *material surface.*

If the mass density (the mass per unit area) is a constant λ throughout, then the total mass of the material surface is the density λ times the area of S:

$$M = \lambda(\text{area of } S).$$

If, however, the mass density varies continuously from point to point, $\lambda = \lambda(x, y, z)$, then the total mass must be calculated by integration.

To develop the appropriate integral we suppose that

$$S: \quad \mathbf{r} = \mathbf{r}(u, v) = x(u, v)\mathbf{i} + y(u, v)\mathbf{j} + z(u, v)\mathbf{k}, \quad (u, v) \in \Omega$$

is a continuously differentiable surface, a surface that meets the conditions for area formula (17.6.3).† Our first step is to break up Ω into N little basic regions $\Omega_1, \ldots, \Omega_N$. This decomposes the surface into N little pieces $S_1, \ldots, S_N$. The area of S_i is given by the integral

$$\iint_{\Omega_i} \|\mathbf{N}(u, v)\|\, du\, dv. \qquad \text{(Formula 17.6.3)}$$

---

† We repeat the conditions here: $\mathbf{r}$ is continuously differentiable; Ω is a basic region in the uv-plane; $\mathbf{r}$ is one-to-one on the interior of Ω; $\mathbf{N} = \mathbf{r}_u' \times \mathbf{r}_v'$ is never zero on the interior of Ω.

By the mean-value theorem for double integrals, there exists a point (u_i^*, v_i^*) in Ω_i for which

$$\iint\limits_{\Omega_i} \|\mathbf{N}(u, v)\| \, du \, dv = \|\mathbf{N}(u_i^*, v_i^*)\|(\text{area of } \Omega_i).$$

It follows that

$$\text{area of } S_i = \|\mathbf{N}(u_i^*, v_i^*)\|(\text{area of } \Omega_i).$$

Since the point (u_i^*, v_i^*) is in Ω_i, the tip of $\mathbf{r}(u_i^*, v_i^*)$ is on S_i. The mass density at this point is $\lambda[\mathbf{r}(u_i^*, v_i^*)]$. If S_i is small (which we can guarantee by choosing Ω_i small), then the mass density on S_i is approximately the same throughout. Thus we can estimate M_i, the mass contribution of S_i, by writing

$$M_i \cong \lambda[\mathbf{r}(u_i^*, v_i^*)](\text{area of } S_i) = \lambda[\mathbf{r}(u_i^*, v_i^*)]\|\mathbf{N}(u_i^*, v_i^*)\|(\text{area of } \Omega_i).$$

Adding up these estimates we have an estimate for the total mass of the surface:

$$M \cong \sum_{i=1}^{N} \lambda[\mathbf{r}(u_i^*, v_i^*)]\|\mathbf{N}(u_i^*, v_i^*)\|(\text{area of } \Omega_i)$$

$$= \sum_{i=1}^{N} \lambda[x(u_i^*, v_i^*), y(u_i^*, v_i^*), z(u_i^*, v_i^*)]\|\mathbf{N}(u_i^*, v_i^*)\|(\text{area of } \Omega_i).$$

This last expression is a Riemann sum for

$$\iint\limits_{\Omega} \lambda[x(u, v), y(u, v), z(u, v)]\|\mathbf{N}(u, v)\| \, du \, dv$$

and tends to this integral as the maximum diameter of the Ω_i tends to zero. We can therefore conclude that

(17.7.1)
$$M = \iint\limits_{\Omega} \lambda[x(u, v), y(u, v), z(u, v)]\|\mathbf{N}(u, v)\| \, du \, dv.$$

Surface Integrals

The double integral in (17.7.1) can be calculated not only for a mass density function λ but for any scalar field H continuous over S. We call this integral *the surface integral of H over S* and write

(17.7.2)
$$\iint\limits_{S} H(x, y, z) \, d\sigma = \iint\limits_{\Omega} H[x(u, v), y(u, v), z(u, v)]\|\mathbf{N}(u, v)\| \, du \, dv.$$

Note that, if $H(x, y, z)$ is identically 1, then the right-hand side of (17.7.2) gives the area of S. Thus

(17.7.3)

$$\iint_S d\sigma = \text{area of } S.$$

Example 1 Let $\mathbf{a} = a_1\mathbf{i} + a_2\mathbf{j} + a_3\mathbf{k}$ and $\mathbf{b} = b_1\mathbf{i} + b_2\mathbf{j} + b_3\mathbf{k}$ be vectors, and calculate

$$\iint_S xy \, d\sigma \quad \text{where} \quad S: \quad \mathbf{r}(u, v) = u\mathbf{a} + v\mathbf{b}; \quad 0 \le u \le 1, \quad 0 \le v \le 1.$$

SOLUTION Call the parameter set Ω. Then

$$\iint_S xy \, d\sigma = \iint_\Omega x(u, v) \, y(u, v)\|\mathbf{N}(u, v)\| \, du \, dv.$$

By Example 6, Section 17.6, $\|\mathbf{N}(u, v)\| = \|\mathbf{a} \times \mathbf{b}\|$. Thus

$$\iint_S xy \, d\sigma = \|\mathbf{a} \times \mathbf{b}\| \iint_\Omega x(u, v) \, y(u, v) \, du \, dv.$$

To find $x(u, v)$ and $y(u, v)$ we need the $\mathbf{i}$ and $\mathbf{j}$ components of $\mathbf{r}(u, v)$. We can get these as follows:

$$\mathbf{r}(u, v) = u\mathbf{a} + v\mathbf{b} = u(a_1\mathbf{i} + a_2\mathbf{j} + a_3\mathbf{k}) + v(b_1\mathbf{i} + b_2\mathbf{j} + b_3\mathbf{k})$$
$$= (a_1u + b_1v)\mathbf{i} + (a_2u + b_2v)\mathbf{j} + (a_3u + b_3v)\mathbf{k}.$$

Therefore $x(u, v) = a_1u + b_1v$ and $y(u, v) = a_2u + b_2v$. We can now write

$$\iint_S xy \, d\sigma = \|\mathbf{a} \times \mathbf{b}\| \iint_\Omega (a_1u + b_1v)(a_2u + b_2v) \, du \, dv$$

$$= \|\mathbf{a} \times \mathbf{b}\| \int_0^1 \left(\int_0^1 [a_1a_2u^2 + (a_1b_2 + b_1a_2)uv + b_1b_2v^2] \, du \right) dv$$

$$= \|\mathbf{a} \times \mathbf{b}\|[\tfrac{1}{3}a_1a_2 + \tfrac{1}{4}(a_1b_2 + b_1a_2) + \tfrac{1}{3}b_1b_2]. \quad \square$$

check this ———↑

Example 2 Calculate

$$\iint_S \sqrt{x^2 + y^2} \, d\sigma$$

where S is the spiral ramp of Example 11, Section 17.6:

$$S: \quad \mathbf{r}(u, v) = u \cos \omega v \, \mathbf{i} + u \sin \omega v \, \mathbf{j} + bv \, \mathbf{k}; \quad 0 \le u \le l, \quad 0 \le v \le 2\pi/\omega.$$

SOLUTION Call the parameter set Ω. As we saw in Example 11, Section 17.6,

$$\|N(u, v)\| = \sqrt{b^2 + \omega^2 u^2}.$$

Therefore

$$\iint_S \sqrt{x^2 + y^2}\, d\sigma = \iint_\Omega \sqrt{[x(u, v)]^2 + [y(u, v)]^2}\, \|N(u, v)\|\, du\, dv$$

$$= \iint_\Omega \sqrt{u^2\cos^2\omega v + u^2\sin^2\omega v}\sqrt{b^2 + \omega^2 u^2}\, du\, dv$$

$$= \iint_\Omega u\sqrt{b^2 + \omega^2 u^2}\, du\, dv$$

$u \geq 0 \text{ on } \Omega \longrightarrow$

$$= \int_0^{2\pi/\omega} \left(\int_0^l u\sqrt{b^2 + \omega^2 u^2}\, du \right) dv$$

$$= \frac{2\pi}{\omega} \int_0^l u\sqrt{b^2 + \omega^2 u^2}\, du = \frac{2\pi}{3\omega^3}[(b^2 + \omega^2 l^2)^{3/2} - b^3]. \quad \square$$

Like the other integrals you have studied, the surface integral satisfies a mean-value condition; namely, if the scalar field H is continuous, then there is a point (x_0, y_0, z_0) on S for which

$$\iint_S H(x, y, z)\, d\sigma = H(x_0, y_0, z_0)(\text{area of } S).$$

We call $H(x_0, y_0, z_0)$ *the average value of H on s.* We can thus write

(17.7.4)
$$\iint_S H(x, y, z)\, d\sigma = \left(\begin{array}{c} \text{average value} \\ \text{of } H \text{ on } S \end{array} \right) \cdot (\text{area of } S).$$

We can also take weighted averages: if H and G are continuous on S, then there is a point (x_0, y_0, z_0) on S for which

(17.7.5)
$$\iint_S H(x, y, z)G(x, y, z)\, d\sigma = H(x_0, y_0, z_0) \iint_S G(x, y, z)\, d\sigma.$$

As you would expect, we call $H(x_0, y_0, z_0)$ *the G-weighted average of H on S.*

The coordinates of the centroid $(\bar{x}, \bar{y}, \bar{z})$ of a surface are simply averages over the surface: for a surface S of area A

$$\bar{x}A = \iint_S x\, d\sigma, \qquad \bar{y}A = \iint_S y\, d\sigma, \qquad \bar{z}A = \iint_S z\, d\sigma.$$

In the case of a material surface of mass density $\lambda = \lambda(x, y, z)$, the coordinates of the center of mass (x_M, y_M, z_M) are density-weighted averages: for a surface S of total mass M

$$x_M M = \iint\limits_S x\lambda(x, y, z)\, d\sigma, \quad y_M M = \iint\limits_S y\lambda(x, y, z)\, d\sigma, \quad z_M M = \iint\limits_S z\lambda(x, y, z)\, d\sigma.$$

Example 3   Locate the center of mass of a material surface in the form of a hemispherical shell $x^2 + y^2 + z^2 = a^2$ with $z \geq 0$ given that the mass density is directly proportional to the distance from the xy-plane.   See Figure 17.7.1.

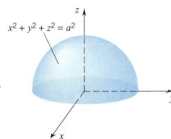

SOLUTION   The surface S can be parametrized by

$$\mathbf{r}(u, v) = a \cos u \cos v\, \mathbf{i} + a \sin u \cos v\, \mathbf{j} + a \sin v\, \mathbf{k}; \quad 0 \leq u \leq 2\pi, 0 \leq v \leq \tfrac{1}{2}\pi.$$

Call the parameter set Ω and recall that $\|\mathbf{N}(u, v)\| = a^2 \cos v$ (Example 8, Section 17.6).   The density function can be written $\lambda(x, y, z) = kz$, where k is the constant of proportionality.   We can calculate the mass as follows:

$$M = \iint\limits_S \lambda(x, y, z)\, d\sigma = k \iint\limits_S z\, d\sigma = k \iint\limits_\Omega z(u, v)\|\mathbf{N}(u, v)\|\, du\, dv$$

$$= k \int_0^{2\pi} \left(\int_0^{\pi/2} (a \sin v)(a^2 \cos v)\, dv \right) du$$

$$= 2\pi k a^3 \int_0^{\pi/2} \sin v \cos v\, dv = \pi k a^3.$$

Figure 17.7.1

By symmetry $x_M = 0$ and $y_M = 0$.   To find z_M we write

$$z_M M = \iint\limits_S z\lambda(x, y, z)\, d\sigma = k \iint\limits_S z^2\, d\sigma$$

$$= k \iint\limits_\Omega [z(u, v)]^2 \|\mathbf{N}(u, v)\|\, du\, dv$$

$$= k \int_0^{2\pi} \left(\int_0^{\pi/2} (a^2 \sin^2 v)(a^2 \cos v)\, dv \right) du$$

$$= 2\pi k a^4 \int_0^{\pi/2} \sin^2 v \cos v\, dv = \tfrac{2}{3}\pi k a^4.$$

Since $M = \pi k a^3$, we see that $z_M = \tfrac{2}{3}\pi k a^4 / M = \tfrac{2}{3}a$.   The center of mass is the point $(0, 0, \tfrac{2}{3}a)$.   ❑

Suppose that a material surface S rotates about an axis.   The moment of inertia of the surface about that axis is given by the formula

(17.7.6)

$$I = \iint\limits_S \lambda(x, y, z)[R(x, y, z)]^2\, d\sigma$$

where $\lambda = \lambda(x, y, z)$ is the mass density function and $R(x, y, z)$ is the distance from the axis to the point (x, y, z). (As with other configurations the moments of inertia about the x, y, z axes are denoted by I_x, I_y, I_z.)

Example 4 A homogeneous material surface with mass density 1 is in the shape of a spherical shell

$$S: \quad x^2 + y^2 + z^2 = a^2. \qquad \text{(Figure 17.7.2)}$$

Calculate the moment of inertia about the z-axis.

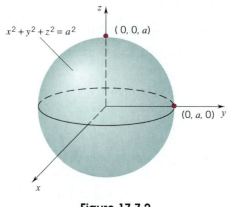

Figure 17.7.2

SOLUTION We parametrize S by setting

$$\mathbf{r}(u, v) = a \cos u \cos v\,\mathbf{i} + a \sin u \cos v\,\mathbf{j} + a \sin v\,\mathbf{k}; \quad 0 \le u \le 2\pi, \; -\tfrac{1}{2}\pi \le v \le \tfrac{1}{2}\pi.$$

Call the parameter set Ω and recall that $\|\mathbf{N}(u, v)\| = a^2 \cos v$. We can calculate the moment of inertia as follows:

$$I_z = \iint_S 1(x^2 + y^2)\, d\sigma = \iint_\Omega ([x(u, v)]^2 + [y(u, v)]^2)\|\mathbf{N}(u, v)\|\, d\sigma$$

$$= \iint_\Omega (a^2 \cos^2 v)(a^2 \cos v)\, du\, dv$$

$$= a^4 \int_0^{2\pi} \left(\int_{-\pi/2}^{\pi/2} \cos^3 v\, dv \right) du$$

$$= 2\pi a^4 \int_{-\pi/2}^{\pi/2} \cos^3 v\, dv = \tfrac{8}{3}\pi a^4.$$

‾‾‾‾‾ check this

Since the surface has mass $M = A = 4\pi a^2$, it follows that $I_z = \tfrac{2}{3}Ma^2$. ❑

A surface

$$S: \quad z = f(x, y), \quad (x, y) \in \Omega$$

can be parametrized by

$$\mathbf{r}(x, y) = x\mathbf{i} + y\mathbf{j} + f(x, y)\mathbf{k}, \qquad (x, y) \in \Omega.$$

As you saw in Section 17.6, $\|\mathbf{N}(x, y)\| = \sec[\gamma(x, y)]$ where $\gamma(x, y)$ is the angle between $\mathbf{k}$ and the upper unit normal. Therefore, for any continuous scalar field H on S,

(17.7.7)
$$\iint_S H(x, y, z) \, d\sigma = \iint_\Omega H(x, y, z) \sec[\gamma(x, y)] \, dx \, dy.$$

In evaluating this last integral we use the fact that
$$\sec[\gamma(x, y)] = \sqrt{[f_x(x, y)]^2 + [f_y(x, y)]^2 + 1}. \qquad \text{(Section 17.6)}$$

Example 5 Calculate

$$\iint_S \sqrt{x^2 + y^2} \, d\sigma \qquad \text{with} \qquad S: \quad z = xy, \quad 0 \le x^2 + y^2 \le 1.$$

SOLUTION The base region Ω is the unit disc. The function $z = f(x, y) = xy$ has partial derivatives $f_x(x, y) = y$, $f_y(x, y) = x$. Therefore
$$\sec[\gamma(x, y)] = \sqrt{y^2 + x^2 + 1} = \sqrt{x^2 + y^2 + 1}$$

and
$$\iint_S \sqrt{x^2 + y^2} \, d\sigma = \iint_\Omega \sqrt{x^2 + y^2} \sqrt{x^2 + y^2 + 1} \, dx \, dy.$$

We evaluate this last integral by changing to polar coordinates. The region Ω is the set of all (x, y) with polar coordinates (r, θ) in the set
$$\Gamma: \quad 0 \le \theta \le 2\pi, \quad 0 \le r \le 1.$$

Therefore
$$\iint_S \sqrt{x^2 + y^2} \, d\sigma = \iint_\Gamma r\sqrt{r^2 + 1} \, r \, dr \, d\theta = \int_0^{2\pi} \left(\int_0^1 r^2\sqrt{r^2 + 1} \, dr \right) d\theta$$
$$= 2\pi \int_0^1 r^2\sqrt{r^2 + 1} \, dr$$
$$= 2\pi \int_0^{\pi/4} \tan^2\phi \sec^3\phi \, d\phi$$
$$= 2\pi \int_0^{\pi/4} [\sec^5\phi - \sec^3\phi] \, d\phi$$
$$= \tfrac{1}{4}\pi[3\sqrt{2} - \ln(\sqrt{2} + 1)] \cong 2.64. \quad \square$$

The Flux of a Vector Field

Suppose that
$$S: \quad \mathbf{r} = \mathbf{r}(u, v), \quad (u, v) \in \Omega$$

is a continuously differentiable surface with a unit normal $\mathbf{n} = \mathbf{n}(x, y, z)$ that is continuous on all of S. Such a surface is called a *smooth surface*. If $\mathbf{v} = \mathbf{v}(x, y, z)$ is a vector field continuous on S, then we can form the surface integral

(17.7.8)
$$\iint_S (\mathbf{v} \cdot \mathbf{n})\, d\sigma = \iint_S [\mathbf{v}(x, y, z) \cdot \mathbf{n}(x, y, z)]\, d\sigma.$$

This surface integral is called *the flux of* $\mathbf{v}$ *across* S *in the direction of* $\mathbf{n}$.

Note that the flux across a surface depends on the choice of unit normal. If $-\mathbf{n}$ is chosen instead of $\mathbf{n}$, the sign of the flux is reversed:

$$\iint_S (\mathbf{v} \cdot -\mathbf{n})\, d\sigma = \iint_S -(\mathbf{v} \cdot \mathbf{n})\, d\sigma = -\iint_S (\mathbf{v} \cdot \mathbf{n})\, d\sigma.$$

Example 6 Calculate the flux of the vector field $\mathbf{v}(x, y, z) = x\mathbf{i} + y\mathbf{j}$ across the sphere S: $x^2 + y^2 + z^2 = a^2$ in the outward direction (Figure 17.7.3).

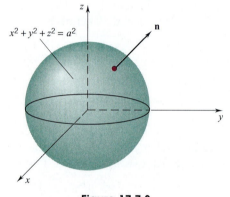

Figure 17.7.3

SOLUTION The outward unit normal is the vector

$$\mathbf{n}(x, y, z) = \frac{1}{a}(x\mathbf{i} + y\mathbf{j} + z\mathbf{k}).$$

Here

$$\mathbf{v} \cdot \mathbf{n} = (x\mathbf{i} + y\mathbf{j}) \cdot \frac{1}{a}(x\mathbf{i} + y\mathbf{j} + z\mathbf{k}) = \frac{1}{a}(x^2 + y^2).$$

Therefore

$$\text{flux out of } S = \frac{1}{a} \iint_S (x^2 + y^2)\, d\sigma.$$

To evaluate this integral we use the usual parametrization

$$\mathbf{r}(u, v) = a\cos u \cos v\, \mathbf{i} + a\sin u \cos v\, \mathbf{j} + a\sin v\, \mathbf{k}; \quad 0 \le u \le 2\pi, \ -\tfrac{1}{2}\pi \le v \le \tfrac{1}{2}\pi.$$

Also, recall that $\|\mathbf{N}(u, v)\| = a^2 \cos v.$ Thus

$$\text{flux out of } S = \frac{1}{a} \iint_\Omega ([x(u, v)]^2 + [y(u, v)]^2) \|\mathbf{N}(u, v)\| \, du \, dv$$

$$= \frac{1}{a} \iint_\Omega (a^2\cos^2 u \, \cos^2 v + a^2\sin^2 u \, \cos^2 v)(a^2\cos v) \, du \, dv$$

$$= a^3 \iint_\Omega \cos^3 v \, du \, dv$$

$$= a^3 \int_0^{2\pi} \left(\int_{-\pi/2}^{\pi/2} \cos^3 v \, dv \right) du = 2\pi a^3 \int_{-\pi/2}^{\pi/2} \cos^3 v \, dv = \tfrac{8}{3} \pi a^3. \quad \square$$

If S is the graph of a function $z = f(x, y)$, $(x, y) \in \Omega$ and $\mathbf{n}$ is the upper unit normal, then the flux of the vector field $\mathbf{v} = v_1 \mathbf{i} + v_2 \mathbf{j} + v_3 \mathbf{k}$ across S in the direction of $\mathbf{n}$ is

(17.7.9)

$$\iint_S (\mathbf{v} \cdot \mathbf{n}) \, d\sigma = \iint_\Omega (-v_1 f_x - v_2 f_y + v_3) \, dx \, dy.$$

PROOF From Section 17.6 we know that

$$\mathbf{n} = \frac{-f_x \mathbf{i} - f_y \mathbf{j} + \mathbf{k}}{\sqrt{(f_x)^2 + (f_y)^2 + 1}} = (-f_x \mathbf{i} - f_y \mathbf{j} + \mathbf{k}) \cos \gamma,$$

where γ is the angle between $\mathbf{n}$ and $\mathbf{k}$. Therefore $\mathbf{v} \cdot \mathbf{n} = (-v_1 f_x - v_2 f_y + v_3) \cos \gamma$ and

$$\iint_S (\mathbf{v} \cdot \mathbf{n}) \, d\sigma = \iint_\Omega (\mathbf{v} \cdot \mathbf{n}) \sec \gamma \, dx \, dy = \iint_\Omega (-v_1 f_x - v_2 f_y + v_3) \, dx \, dy. \quad \square$$

Example 7 Let S be the portion of the paraboloid $z = 1 - (x^2 + y^2)$ that lies above the unit disc Ω. Calculate the flux of $\mathbf{v} = x \mathbf{i} + y \mathbf{j} + z \mathbf{k}$ across this surface in the direction of the upper unit normal (Figure 17.7.4).

SOLUTION

$$f(x, y) = 1 - (x^2 + y^2); \qquad f_x = -2x, \quad f_y = -2y.$$

$$\text{Flux} = \iint_\Omega (-v_1 f_x - v_2 f_y + v_3) \, dx \, dy$$

$$= \iint_\Omega [(-x)(-2x) - y(-2y) + 1 - (x^2 + y^2)] \, dx \, dy$$

$$= \iint_\Omega (1 + x^2 + y^2) \, dx \, dy = \int_0^{2\pi} \left(\int_0^1 (1 + r^2) r \, dr \right) d\theta = \tfrac{3}{2} \pi. \quad \square$$

in polar coordinates

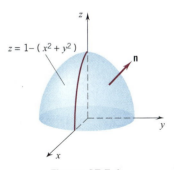

$z = 1 - (x^2 + y^2)$

Figure 17.7.4

The flux through a closed *piecewise-smooth surface* (a closed surface that consists of a finite number of smooth pieces joined together at the boundaries) can be evaluated by integrating over each smooth piece and adding up the results.

Example 8 Calculate the total flux of $\mathbf{v}(x, y, z) = xy\,\mathbf{i} + 4yz^2\mathbf{j} + yz\,\mathbf{k}$ out of the unit cube: $0 \le x \le 1, 0 \le y \le 1, 0 \le z \le 1$. (Figure 17.7.5)

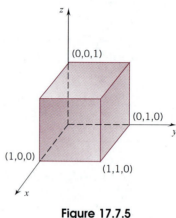

Figure 17.7.5

SOLUTION The total flux is the sum of the fluxes across the faces of the cube:

Face	$\mathbf{n}$	$\mathbf{v} \cdot \mathbf{n}$	Flux
$x = 0$	$-\mathbf{i}$	$-xy = 0$	0
$x = 1$	$\mathbf{i}$	$xy = y$	$\frac{1}{2}$
$y = 0$	$-\mathbf{j}$	$-4yz^2 = 0$	0
$y = 1$	$\mathbf{j}$	$4yz^2 = 4z^2$	$\frac{4}{3}$
$z = 0$	$-\mathbf{k}$	$-yz = 0$	0
$z = 1$	$\mathbf{k}$	$yz = y$	$\frac{1}{2}$

The total flux is $\frac{1}{2} + \frac{4}{3} + \frac{1}{2} = \frac{7}{3}$. ❏

Flux plays an important role in the study of fluid motion. Imagine a surface S within a fluid and choose a unit normal $\mathbf{n}$. Take $\mathbf{v}$ as the velocity of the fluid that passes through S. You can expect $\mathbf{v}$ to vary from point to point, but, for simplicity, let's assume that $\mathbf{v}$ does not change with time. (Such a time-independent flow is said to be steady state.) The flux of $\mathbf{v}$ across S is the average component of $\mathbf{v}$ in the direction of $\mathbf{n}$ times the area of the surface S. This is just the volume of fluid that passes through S in unit time, from the $-\mathbf{n}$ side of S to the $\mathbf{n}$ side of S. If S is a closed surface (such as a cube, a sphere, or an ellipsoid) and $\mathbf{n}$ is chosen as the *outer unit normal*, then the flux across S gives the volume of fluid that flows *out* through S in unit time; if $\mathbf{n}$ is chosen as the *inner unit normal,* then the flux gives the volume of liquid that flows *in* through S in unit time.

EXERCISES 17.7

In Exercises 1–6, evaluate these integrals over the surface S: $z = \frac{1}{2}y^2$; $0 \le x \le 1, 0 \le y \le 1$.

1. $\displaystyle\iint_S d\sigma.$

2. $\displaystyle\iint_S x^2 \, d\sigma.$

3. $\displaystyle\iint_S 3y \, d\sigma.$

4. $\displaystyle\iint_S (x - y) \, d\sigma.$

5. $\displaystyle\iint_S \sqrt{2z} \, d\sigma.$

6. $\displaystyle\iint_S \sqrt{1 + y^2} \, d\sigma.$

In Exercises 7–12, evaluate the given surface integral.

7. $\displaystyle\iint_S xy \, d\sigma;$ S is the first octant part of the plane $x + 2y + 3z = 6$.

8. $\displaystyle\iint_S xyz \, d\sigma;$ S is the first octant part of the plane $x + y + z = 1$.

9. $\displaystyle\iint_S x^2z \, d\sigma;$ S is that part of the cylinder $x^2 + z^2 = 1$, which is between the planes $y = 0$ and $y = 2$, and above the xy-plane.

10. $\displaystyle\iint_S (x^2 + y^2 + z^2) \, d\sigma;$ S is that part of the plane $z = x + 2$ that lies inside the cylinder $x^2 + y^2 = 1$.

11. $\displaystyle\iint_S (x^2 + y^2) \, d\sigma;$ S is the hemisphere $z = \sqrt{1 - x^2 + y^2}$.

12. $\displaystyle\iint_S (x^2 + y^2) \, d\sigma;$ S is that part of the paraboloid $z = 1 - x^2 - y^2$ that lies above the xy-plane.

In Exercises 13–15, find the mass of a material surface in the shape of a triangle $(a, 0, 0), (0, a, 0), (0, 0, a)$ with the given density function.

13. $\lambda(x, y, z) = k.$

14. $\lambda(x, y, z) = k(x + y).$

15. $\lambda(x, y, z) = kx^2.$

16. Locate the centroid of the triangle $(a, 0, 0), (0, a, 0), (0, 0, a), a > 0$.

17. Locate the centroid of the hemisphere $x^2 + y^2 + z^2 = a^2$, $z \ge 0$.

In Exercises 18–19, let S be the parallelogram given by $\mathbf{r}(u, v) = (u + v)\mathbf{i} + (u - v)\mathbf{j} + 2u\,\mathbf{k}; 0 \le u \le 1, 0 \le v \le 1$.

18. Find the area of S.

19. Determine the flux of $\mathbf{v} = x\mathbf{i} - y\mathbf{j}$ across S in the direction of the fundamental vector product.

20. Find the mass of the material surface S: $z = 1 - \frac{1}{2}(x^2 + y^2)$ with $0 \le x \le 1, 0 \le y \le 1$ if the density at each point (x, y, z) is proportional to xy.

In Exercises 21–23, calculate the flux out of the sphere $x^2 + y^2 + z^2 = a^2$.

21. $\mathbf{v} = z\mathbf{k}.$ **22.** $\mathbf{v} = x\mathbf{i} + y\mathbf{j} + z\mathbf{k}.$ **23.** $\mathbf{v} = y\mathbf{i} - x\mathbf{j}.$

24. A homogeneous plate of mass density 1 is in the form of the parallelogram of Exercises 18 and 19. Determine the moments of inertia about the coordinate axes: (a) I_x. (b) I_y. (c) I_z.

In Exercises 25–27, determine the flux across the triangle $(a, 0, 0), (0, a, 0), (0, 0, a), a > 0$, in the direction of the upper unit normal.

25. $\mathbf{v} = x\mathbf{i} + y\mathbf{j} + z\mathbf{k}.$ **26.** $\mathbf{v} = (x + z)\mathbf{k}.$

27. $\mathbf{v} = x^2\mathbf{i} - y^2\mathbf{j}.$

In Exercises 28–30, determine the flux across S: $z = xy$ with $0 \le x \le 1, 0 \le y \le 2$ in the direction of the upper unit normal.

28. $\mathbf{v} = -xy^2\mathbf{i} + z\mathbf{j}.$ **29.** $\mathbf{v} = xz\mathbf{j} - xy\mathbf{k}.$

30. $\mathbf{v} = x^2y\mathbf{i} + z^2\mathbf{k}.$

31. Calculate the flux of $\mathbf{v} = x\mathbf{i} + y\mathbf{j} + z\mathbf{k}$ out of the cylindrical surface

$$S: \quad \mathbf{r}(u, v) = a \cos u\,\mathbf{i} + a \sin u\,\mathbf{j} + v\,\mathbf{k};$$
$$0 \le u \le 2\pi, 0 \le v \le l.$$

32. Calculate the flux of the gravitational force function

$$\mathbf{F}(\mathbf{r}) = G\frac{mM}{r^3}\mathbf{r} \quad \text{out of the sphere } x^2 + y^2 + z^2 = a^2.$$

In Exercises 33–36, find the flux across S: $z = \frac{2}{3}(x^{3/2} + y^{3/2})$ with $0 \le x \le 1, 0 \le y \le 1 - x$ in the direction of the upper unit normal.

33. $\mathbf{v} = x\mathbf{i} - y\mathbf{j} + \frac{3}{2}z\mathbf{k}.$ **34.** $\mathbf{v} = x^2\mathbf{i}.$

35. $\mathbf{v} = y^2\mathbf{j}.$ **36.** $\mathbf{v} = y\mathbf{i} - \sqrt{xy}\,\mathbf{j}.$

37. The cone

$$\mathbf{r}(u, v) = v \cos u \sin \alpha\,\mathbf{i} + v \sin u \sin \alpha\,\mathbf{j} + v \cos \alpha\,\mathbf{k};$$
$$0 \le u \le 2\pi, \quad 0 \le v \le s$$

has area $A = \pi s^2 \sin \alpha$. Locate the centroid.

In Exercises 38–40, the mass density of a material cone $z = \sqrt{x^2 + y^2}$ with $0 \leq z \leq 1$ varies directly as the distance from the z-axis.

38. Find the mass of the cone.

39. Locate the center of mass.

40. Determine the moments of inertia about the coordinate axes: (a) I_x.   (b) I_y.   (c) I_z.

41. You have seen that, if S is a smooth surface immersed in a fluid of velocity $\mathbf{v}$, then the flux

$$\iint_S (\mathbf{v} \cdot \mathbf{n}) \, d\sigma$$

is the volume of fluid that passes through S in unit time from the $-\mathbf{n}$ side of S to the $\mathbf{n}$ side of S. This requires that S be a two-sided surface. There are, however, one-sided surfaces: for example, the Möbius band. To construct a material Möbius band start with a piece of paper in the form of the rectangle in the figure above. Now give the piece of paper a single twist and join the two far edges together so that C coincides with A and D coincides with B. (a) Convince yourself that this surface is one-sided and therefore the notion of flux cannot be applied to it. (b) The surface is not smooth because it is impossible to erect a unit normal $\mathbf{n}$ that varies continuously over the entire surface. Convince yourself of this as follows: erect a unit normal $\mathbf{n}$ at some point P and make a circuit of the surface with $\mathbf{n}$. Note that, as $\mathbf{n}$ returns to P, the direction of $\mathbf{n}$ has been reversed.

In Exercises 42–44, assume that the parallelogram of Exercises 18 and 19 is a material surface with a mass density that varies directly as the square of the distance from the x-axis.

42. Determine the mass.

43. Find the x-coordinate of the center of mass.

44. Find the moment of inertia about the z-axis.

45. Calculate the total flux of $\mathbf{v}(x, y, z) = y\mathbf{i} - x\mathbf{j}$ out of the solid bounded on the sides by the cylinder $x^2 + y^2 = 1$ and above and below by the planes $z = 1$ and $z = 0$.   HINT: Draw a figure.

46. Calculate the total flux of $\mathbf{v}(x, y, z) = y\mathbf{i} - x\mathbf{j}$ out of the solid bounded above by $z = 4$ and below by $z = x^2 + y^2$.

47. Calculate the total flux of $\mathbf{v}(x, y, z) = x\mathbf{i} + y\mathbf{j} + z\mathbf{k}$ out of the solid bounded above by $z = \sqrt{2 - (x^2 + y^2)}$ and below by $z = x^2 + y^2$.

48. Calculate the total flux of $\mathbf{v}(x, y, z) = xz\mathbf{i} + 4xyz^2\mathbf{j} + 2z\mathbf{k}$ out of the unit cube: $0 \leq x \leq 1$, $0 \leq y \leq 1$, $0 \leq z \leq 1$.

■ 17.8 THE VECTOR DIFFERENTIAL OPERATOR ∇

Divergence ∇ · v, Curl ∇ × v

The *vector differential operator* ∇ (this is an inverted delta and it's read *del*) is defined formally by setting

(17.8.1)
$$\nabla = \frac{\partial}{\partial x}\mathbf{i} + \frac{\partial}{\partial y}\mathbf{j} + \frac{\partial}{\partial z}\mathbf{k}.$$

By "formally" we mean that this is not an ordinary vector. Its "components" are differentiation symbols. As the term "operator" suggests, ∇ is to be thought of as something that "operates" on things. What sort of things? Scalar fields and vector fields.

Suppose that f is a differentiable scalar field. Then ∇ operates on f as follows:

$$\nabla f = \left(\frac{\partial}{\partial x}\mathbf{i} + \frac{\partial}{\partial y}\mathbf{j} + \frac{\partial}{\partial z}\mathbf{k} \right) f = \frac{\partial f}{\partial x}\mathbf{i} + \frac{\partial f}{\partial y}\mathbf{j} + \frac{\partial f}{\partial z}\mathbf{k}.$$

This is just the *gradient* of f, with which you are already familiar.

How does ∇ operate on vector fields? In two ways. If $\mathbf{v} = v_1\mathbf{i} + v_2\mathbf{j} + v_3\mathbf{k}$ is a differentiable vector field, then by definition

(17.8.2)

$$\nabla \cdot \mathbf{v} = \frac{\partial v_1}{\partial x} + \frac{\partial v_2}{\partial y} + \frac{\partial v_3}{\partial z}$$

and

(17.8.3)

$$\nabla \times \mathbf{v} = \begin{vmatrix} \mathbf{i} & \mathbf{j} & \mathbf{k} \\ \dfrac{\partial}{\partial x} & \dfrac{\partial}{\partial y} & \dfrac{\partial}{\partial z} \\ v_1 & v_2 & v_3 \end{vmatrix}$$

$$= \left(\frac{\partial v_3}{\partial y} - \frac{\partial v_2}{\partial z}\right)\mathbf{i} + \left(\frac{\partial v_1}{\partial z} - \frac{\partial v_3}{\partial x}\right)\mathbf{j} + \left(\frac{\partial v_2}{\partial x} - \frac{\partial v_1}{\partial y}\right)\mathbf{k}.$$

The first "product," $\nabla \cdot \mathbf{v}$, defined in imitation of the ordinary dot product, is called the *divergence* of $\mathbf{v}$:

$$\nabla \cdot \mathbf{v} = \text{div } \mathbf{v}.$$

The second "product," $\nabla \times \mathbf{v}$, defined in imitation of the ordinary cross product, is called the *curl* of $\mathbf{v}$:

$$\nabla \times \mathbf{v} = \text{curl } \mathbf{v}.$$

Interpretation of Divergence and Curl

Suppose we know the divergence of a field and also the curl. What does that tell us? For definitive answers we must wait for the divergence theorem and Stokes's theorem, but, in a preliminary way, we can give you some rough answers right now.

View $\mathbf{v}$ as the velocity field of some fluid. The divergence of $\mathbf{v}$ at a point P gives us an indication of whether the fluid tends to accumulate near P (negative divergence) or tends to move away from P (positive divergence). In the first case, P is sometimes called a *sink*, and in the second case, it is called a *source*. The curl at P measures the rotational tendency of the fluid.

Example 1 Set

$$\mathbf{v}(x, y, z) = \alpha x\mathbf{i} + \alpha y\mathbf{j} + \alpha z\mathbf{k}. \qquad (\alpha \text{ a constant})$$

Divergence:

$$\nabla \cdot \mathbf{v} = \alpha\frac{\partial x}{\partial x} + \alpha\frac{\partial y}{\partial y} + \alpha\frac{\partial z}{\partial z} = 3\alpha.$$

Curl:

$$\nabla \times \mathbf{v} = \begin{vmatrix} \mathbf{i} & \mathbf{j} & \mathbf{k} \\ \dfrac{\partial}{\partial x} & \dfrac{\partial}{\partial y} & \dfrac{\partial}{\partial z} \\ \alpha x & \alpha y & \alpha z \end{vmatrix} = \alpha \begin{vmatrix} \mathbf{i} & \mathbf{j} & \mathbf{k} \\ \dfrac{\partial}{\partial x} & \dfrac{\partial}{\partial y} & \dfrac{\partial}{\partial z} \\ x & y & z \end{vmatrix} = \mathbf{0}$$

because the partial derivatives that appear in the expanded determinant

$$\frac{\partial y}{\partial x}, \quad \frac{\partial x}{\partial y}, \quad \text{etc.,}$$

are all zero. ❑

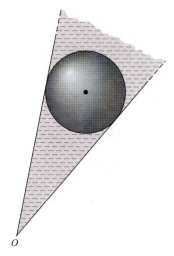

The field of Example 1

$$\mathbf{v}(x, y, z) = \alpha(x\,\mathbf{i} + y\,\mathbf{j} + z\,\mathbf{k}) = \alpha\mathbf{r}$$

can be viewed as the velocity field of a fluid in radial motion—toward the origin if $\alpha < 0$, away from the origin if $\alpha > 0$. Figure 17.8.1 shows a point (x, y, z), a spherical neighborhood of that point, and a cone emanating from the origin that is tangent to the boundary of the neighborhood.

Note two things: all the fluid in the cone stays in the cone, and the speed of the fluid is proportional to its distance from the origin. Therefore, if the divergence 3α is negative, then α is negative, the motion is toward the origin, and the neighborhood *gains fluid* because the fluid coming in is moving more quickly than the fluid going out. (Also the entry area is larger than the exit area.) If, however, the divergence 3α is positive, then α is positive, the motion is away from the origin, and the neighborhood *loses fluid* because the fluid coming in is moving more slowly than the fluid going out. (Also the entry area is smaller than the exit area.)

Since the motion is radial, the fluid has no rotational tendency whatsoever, and we would expect the curl to be identically zero. It is. ❑

O

Figure 17.8.1

Example 2 Set

$$\mathbf{v}(x, y, z) = -\omega y\,\mathbf{i} + \omega x\,\mathbf{j}. \qquad (\omega \text{ a positive constant})$$

Divergence:

$$\nabla \cdot \mathbf{v} = -\omega\frac{\partial y}{\partial x} + \omega\frac{\partial x}{\partial y} = 0 + 0 = 0.$$

Curl:

$$\nabla \times \mathbf{v} = \begin{vmatrix} \mathbf{i} & \mathbf{j} & \mathbf{k} \\ \dfrac{\partial}{\partial x} & \dfrac{\partial}{\partial y} & \dfrac{\partial}{\partial z} \\ -\omega y & \omega x & 0 \end{vmatrix} = \left(\omega\frac{\partial x}{\partial x} + \omega\frac{\partial y}{\partial y}\right)\mathbf{k} = 2\omega\mathbf{k}. \quad ❑$$

The field of Example 2

$$\mathbf{v}(x, y, z) = -\omega y\,\mathbf{i} + \omega x\,\mathbf{j}$$

is the velocity field of uniform counterclockwise rotation about the z-axis with angular speed ω. You can see this by noting that $\mathbf{v}$ is perpendicular to $\mathbf{r} = x\,\mathbf{i} + y\,\mathbf{j} + z\,\mathbf{k}$:

$$\mathbf{v} \cdot \mathbf{r} = (-\omega y\,\mathbf{i} + x\,\mathbf{j}) \cdot (x\,\mathbf{i} + y\,\mathbf{j} + z\,\mathbf{k}) = -\omega yx + \omega xy = 0$$

and the speed at each point is ωR where R is the radius of rotation:

$$v = \sqrt{\omega^2 y^2 + \omega^2 x^2} = \omega\sqrt{x^2 + y^2} = \omega R.$$

How is the curl, $2\omega\mathbf{k}$, related to the rotation? The angular velocity vector (see Exercise 24, Section 13.5) is the vector $\boldsymbol{\omega} = \omega\mathbf{k}$. In this case, then, the curl of $\mathbf{v}$ is twice the angular velocity vector.

With this rotation no neighborhood gains any fluid and no neighborhood loses any fluid. As we saw, the divergence is identically zero. ❏

Basic Identities

For vectors we have $\mathbf{a} \times \mathbf{a} = \mathbf{0}$. Is it true that $\nabla \times \nabla = \mathbf{0}$? Define $(\nabla \times \nabla)f$ by

$$(\nabla \times \nabla)f = \nabla \times (\nabla f).$$

THEOREM 17.8.4 (THE CURL OF A GRADIENT IS ZERO)

If f is a scalar field with continuous second partials, then

$$\nabla \times (\nabla f) = \mathbf{0}.$$

PROOF

$$\nabla \times (\nabla f) = \begin{vmatrix} \mathbf{i} & \mathbf{j} & \mathbf{k} \\ \dfrac{\partial}{\partial x} & \dfrac{\partial}{\partial y} & \dfrac{\partial}{\partial z} \\ \dfrac{\partial f}{\partial x} & \dfrac{\partial f}{\partial y} & \dfrac{\partial f}{\partial z} \end{vmatrix} =$$

$$\left(\frac{\partial^2 f}{\partial y \partial z} - \frac{\partial^2 f}{\partial z \partial y} \right)\mathbf{i} + \left(\frac{\partial^2 f}{\partial z \partial x} - \frac{\partial^2 f}{\partial x \partial z} \right)\mathbf{j} + \left(\frac{\partial^2 f}{\partial x \partial y} - \frac{\partial^2 f}{\partial y \partial x} \right)\mathbf{k} = \mathbf{0}. \quad ❏$$

For vectors we have $\mathbf{a} \cdot (\mathbf{a} \times \mathbf{c}) = 0$. The analogous operator formula, $\nabla \cdot (\nabla \times \mathbf{v}) = 0$, is also valid.

THEOREM 17.8.5 (THE DIVERGENCE OF A CURL IS ZERO)

If the components of the vector field $\mathbf{v} = v_1\mathbf{i} + v_2\mathbf{j} + v_3\mathbf{k}$ have continuous second partials, then

$$\nabla \cdot (\nabla \times \mathbf{v}) = 0.$$

PROOF   Again the key is the equality of the mixed partials:

$$\nabla \cdot (\nabla \times \mathbf{v}) = \frac{\partial}{\partial x}\left(\frac{\partial v_3}{\partial y} - \frac{\partial v_2}{\partial z} \right) + \frac{\partial}{\partial y}\left(\frac{\partial v_1}{\partial z} - \frac{\partial v_3}{\partial x} \right) + \frac{\partial}{\partial z}\left(\frac{\partial v_2}{\partial x} - \frac{\partial v_1}{\partial y} \right) = 0,$$

since for each component v_i the mixed partials cancel. Try it for v_1. ❏

The next two identities are product rules. Here f is a scalar field and $\mathbf{v}$ is a vector field.

(17.8.6)
$$\nabla \cdot (f\mathbf{v}) = (\nabla f) \cdot \mathbf{v} + f(\nabla \cdot \mathbf{v}). \quad [\text{div}(f\mathbf{v}) = (\text{grad} f) \cdot \mathbf{v} + f(\text{div } \mathbf{v})]$$

(17.8.7)
$$\nabla \times (f\mathbf{v}) = (\nabla f) \times \mathbf{v} + f(\nabla \times \mathbf{v}). \quad [\text{curl}(f\mathbf{v}) = (\text{grad} f) \times \mathbf{v} + f(\text{curl } \mathbf{v})]$$

The verification of these identities is left to you in the exercises.

We know from Example 1 that $\nabla \cdot \mathbf{r} = 3$ and $\nabla \times \mathbf{r} = \mathbf{0}$ at all points of space. Now we can show that, if n is an integer, then, for all $\mathbf{r} \neq \mathbf{0}$,

(17.8.8)
$$\nabla \cdot (r^n \mathbf{r}) = (n + 3)r^n \quad \text{and} \quad \nabla \times (r^n \mathbf{r}) = \mathbf{0}. \quad \dagger$$

PROOF Recall that $\nabla r^n = nr^{n-2}\mathbf{r}$. (See 15.1.5.) Using (17.8.6), we have

$$\nabla \cdot (r^n \mathbf{r}) = (\nabla r^n) \cdot \mathbf{r} + r^n(\nabla \cdot \mathbf{r})$$
$$= (nr^{n-2}\mathbf{r}) \cdot \mathbf{r} + r^n(3)$$
$$= nr^{n-2}(\mathbf{r} \cdot \mathbf{r}) + 3r^n = nr^n + 3r^n = (n + 3)r^n.$$

From (15.1.5) you can see that $r^n \mathbf{r}$ is a gradient. Its curl is therefore $\mathbf{0}$ (17.8.4). ❏

The Laplacian

From the operator ∇ we can construct other operators. The most important of these is the Laplacian $\nabla^2 = \nabla \cdot \nabla$. The Laplacian (named after the French mathematician Pierre-Simon Laplace) operates on scalar fields according to the following rule:

(17.8.9)
$$\nabla^2 f = \nabla \cdot (\nabla f) = \frac{\partial^2 f}{\partial x^2} + \frac{\partial^2 f}{\partial y^2} + \frac{\partial^2 f}{\partial z^2}. \quad \dagger\dagger$$

Example 3 If $f(x, y, z) = x^2 + y^2 + z^2$, then

$$\nabla^2 f = \frac{\partial^2}{\partial x^2}(x^2 + y^2 + z^2) + \frac{\partial^2}{\partial y^2}(x^2 + y^2 + z^2) + \frac{\partial^2}{\partial z^2}(x^2 + y^2 + z^2)$$
$$= 2 + 2 + 2 = 6. \quad ❏$$

---

† If n is positive and even, these formulas also hold at $\mathbf{r} = \mathbf{0}$.

†† In some texts, you will see the Laplacian of f written Δf. Unfortunately this can be misread as the increment of f.

Example 4 If $f(x, y, z) = e^{xyz}$, then

$$\nabla^2 f = \frac{\partial}{\partial x^2}(e^{xyz}) + \frac{\partial^2}{\partial y^2}(e^{xyz}) + \frac{\partial^2}{\partial z^2}(e^{xyz})$$

$$= \frac{\partial}{\partial x}(yze^{xyz}) + \frac{\partial}{\partial y}(xze^{xyz}) + \frac{\partial}{\partial z}(xye^{xyz})$$

$$= y^2z^2e^{xyz} + x^2z^2e^{xyz} + x^2y^2e^{xyz}$$

$$= (y^2z^2 + x^2z^2 + x^2y^2)e^{xyz}. \quad \square$$

Example 5 To calculate $\nabla^2(\sin r) = \nabla^2(\sin\sqrt{x^2 + y^2 + z^2})$ we could write

$$\frac{\partial^2}{\partial x^2}(\sin\sqrt{x^2 + y^2 + z^2}) + \frac{\partial^2}{\partial y^2}(\sin\sqrt{x^2 + y^2 + z^2}) + \frac{\partial^2}{\partial z^2}(\sin\sqrt{x^2 + y^2 + z^2})$$

and proceed from there. The calculations are straightforward but lengthy. We will do it in a different way.

Recall that

$$\nabla^2 f = \nabla \cdot \nabla f \quad \text{(definition)} \qquad \nabla \cdot (f\mathbf{v}) = (\nabla f \cdot \mathbf{v}) + f(\nabla \cdot \mathbf{v}) \quad (17.8.6)$$

$$\nabla f(r) = f'(r)r^{-1}\mathbf{r} \quad (15.4.8) \qquad \nabla \cdot (r^n\mathbf{r}) = (n + 3)r^n \quad (17.8.8)$$

Using these relations, we have

$$\nabla^2(\sin r) = \nabla \cdot (\nabla \sin r) = \nabla \cdot [(\cos r)r^{-1}\mathbf{r}]$$
$$= [(\nabla \cos r) \cdot r^{-1}\mathbf{r}] + \cos r(\nabla \cdot r^{-1}\mathbf{r})$$
$$= \{[(-\sin r)r^{-1}\mathbf{r}] \cdot r^{-1}\mathbf{r}\} + (\cos r)(2r^{-1})$$
$$= -\sin r + 2r^{-1}\cos r.$$

We leave it to you to justify each step. $\quad \square$

EXERCISES 17.8

In Exercises 1–12, find $\nabla \cdot \mathbf{v}$ and $\nabla \times \mathbf{v}$.

1. $\mathbf{v}(x, y) = x\mathbf{i} + y\mathbf{j}$.   **2.** $\mathbf{v}(x, y) = y\mathbf{i} + x\mathbf{j}$.

3. $\mathbf{v}(x, y) = \dfrac{x}{x^2 + y^2}\mathbf{i} + \dfrac{y}{x^2 + y^2}\mathbf{j}$.

4. $\mathbf{v}(x, y) = \dfrac{y}{x^2 + y^2}\mathbf{i} + \dfrac{x}{x^2 + y^2}\mathbf{j}$.

5. $\mathbf{v}(x, y, z) = x\mathbf{i} + 2y\mathbf{j} + 3z\mathbf{k}$.

6. $\mathbf{v}(x, y, z) = yz\mathbf{i} + xz\mathbf{j} + xy\mathbf{k}$.

7. $\mathbf{v}(x, y, z) = xyz\mathbf{i} + xz\mathbf{j} + z\mathbf{k}$.

8. $\mathbf{v}(x, y, z) = x^2y\mathbf{i} + y^2z\mathbf{j} + xy^2\mathbf{k}$.

9. $\mathbf{v}(r) = r^{-2}\mathbf{r}$.   **10.** $\mathbf{v}(r) = e^x\mathbf{r}$.

11. $\mathbf{v}(r) = e^{r^2}(\mathbf{i} + \mathbf{j} + \mathbf{k})$.

12. $\mathbf{v}(r) = e^{y^2}\mathbf{i} + e^{z^2}\mathbf{j} + e^{x^2}\mathbf{k}$.

13. Suppose that f is a differentiable function of one variable and $\mathbf{v}(x, y, z) = f(x)\mathbf{i}$. Determine $\nabla \cdot \mathbf{v}$ and $\nabla \times \mathbf{v}$.

14. Show that, if $\mathbf{v}$ is a differentiable field of the form $\mathbf{v}(\mathbf{r}) = f(x)\mathbf{i} + g(y)\mathbf{j} + h(z)\mathbf{k}$, then $\nabla \times \mathbf{v} = \mathbf{0}$.

15. Show that divergence and curl are *linear* operators:

$$\nabla \cdot (\alpha\mathbf{u} + \beta\mathbf{v}) = \alpha(\nabla \cdot \mathbf{u}) + \beta(\nabla \cdot \mathbf{v}) \quad \text{and}$$

$$\nabla \times (\alpha\mathbf{u} + \beta\mathbf{v}) = \alpha(\nabla \times \mathbf{u}) + \beta(\nabla \times \mathbf{v}).$$

16. (*Important*) Show that the gravitational field

$$\mathbf{F}(\mathbf{r}) = -\frac{GmM}{r^3}\mathbf{r}$$

has zero divergence and zero curl at each $\mathbf{r} \neq \mathbf{0}$.

A vector field $\mathbf{F}$ with the property that $\nabla \cdot \mathbf{F} = 0$ is said to be *solenoidal*. Exercise 16 shows that the gravitational field is solenoidal. If $\mathbf{F}$ is the velocity field of some fluid and $\nabla \cdot \mathbf{F} = 0$ in a solid T in three-dimensional space, then $\mathbf{F}$ has no sources or sinks within T.

17. Show that the vector field $\mathbf{F}(x, y, z) = (2x + y + 2z)\mathbf{i} + (x + 4y - 3z)\mathbf{j} + (2x - 3y - 6z)\mathbf{k}$ is solenoidal.

18. Show that the vector field $\mathbf{F}(x, y, z) = 3x^2\mathbf{i} - y^2\mathbf{j} + (2yz - 6xz)\mathbf{k}$ is solenoidal.

A vector field $\mathbf{F}$ with the property that $\nabla \times \mathbf{F} = \mathbf{0}$ is said to be *irrotational.* Exercise 16 shows that the gravitational field is irrotational. If $\mathbf{F}$ is the velocity field of some fluid, then $\nabla \times \mathbf{F}$ measures the tendency of the fluid to "curl" or rotate about an axis. Thus $\nabla \times \mathbf{F} = \mathbf{0}$ in some solid T in three-dimensional space can be interpreted to mean that the fluid is tending to move in a straight line.

19. Show that the vector field $\mathbf{F}(x, y, z) = x\mathbf{i} + y\mathbf{j} - 2z\mathbf{k}$ is irrotational.

20. Show that the vector field of Exercise 17 is irrotational.

In Exercises 21–26, calculate the Laplacian $\nabla^2 f$.

21. $f(x, y, z) = x^4 + y^4 + z^4$.

22. $f(x, y, z) = xyz$.

23. $f(x, y, z) = x^2 y^3 z^4$.   **24.** $f(\mathbf{r}) = \cos r$.

25. $f(\mathbf{r}) = e^r$.   **26.** $f(\mathbf{r}) = \ln r$.

27. Given a vector field $\mathbf{u}$, the operator $\mathbf{u} \cdot \nabla$ is defined by setting

$$(\mathbf{u} \cdot \nabla)f = \mathbf{u} \cdot \nabla f = u_1 \frac{\partial f}{\partial x} + u_2 \frac{\partial f}{\partial y} + u_3 \frac{\partial f}{\partial z}.$$

Calculate $(\mathbf{r} \cdot \nabla)f$ for: (a) $f(\mathbf{r}) = r^2$;   (b) $f(\mathbf{r}) = 1/r$.

28. (Based on Exercise 27) We can also apply $\mathbf{u} \cdot \nabla$ to a vector field $\mathbf{v}$ by applying it to each component. By definition

$$(\mathbf{u} \cdot \nabla)\mathbf{v} = (\mathbf{u} \cdot \nabla v_1)\mathbf{i} + (\mathbf{u} \cdot \nabla v_2)\mathbf{j} + (\mathbf{u} \cdot \nabla v_3)\mathbf{k}.$$

(a) Calculate $(\mathbf{u} \cdot \nabla)\mathbf{r}$ for an arbitrary vector field $\mathbf{u}$.
(b) Calculate $(\mathbf{r} \cdot \nabla)\mathbf{u}$ given that $\mathbf{u} = yz\mathbf{i} + xz\mathbf{j} + xy\mathbf{k}$.

29. Show that, if $f(\mathbf{r}) = g(r)$ and g is twice differentiable, then

$$\nabla^2 f = g''(r) + 2r^{-1}g'(r).$$

30. Verify the following identities.
(a) $\nabla \cdot (f\mathbf{v}) = (\nabla f) \cdot \mathbf{v} + f(\nabla \cdot \mathbf{v})$.
(b) $\nabla \times (f\mathbf{v}) = (\nabla f) \times \mathbf{v} + f(\nabla \times \mathbf{v})$.
(c) $\nabla \times (\nabla \times \mathbf{v}) = \nabla(\nabla \cdot \mathbf{v}) - \nabla^2 \mathbf{v}$ where
$\nabla^2 \mathbf{v} = (\nabla^2 v_1)\mathbf{i} + (\nabla^2 v_2)\mathbf{j} + (\nabla^2 v_3)\mathbf{k}$.
HINT: Begin (c) by writing out the ith-component of each side.

As you saw in Exercises 14.6, the equation

$$\nabla^2 f = \frac{\partial^2 f}{\partial x^2} + \frac{\partial^2 f}{\partial y^2} + \frac{\partial^2 f}{\partial z^2} = 0$$

is called *Laplace's equation in three dimensions.* A scalar field $f = f(x, y, z)$ with continuous second partial derivatives is said to be *harmonic* on a solid T if it is a solution of Laplace's equation.

31. Show that the scalar field

$$f(x, y, z) = x^2 + 2y^2 - 3z^2 + xy + 2xz - 3yz$$

is harmonic.

32. Show that the scalar field

$$f(x, y, z) = \frac{1}{\sqrt{x^2 + y^2 + z^2}}$$

is harmonic on every solid T that excludes the origin. Except for a constant multiplier, f is a potential function for the gravitational field.

33. For what nonzero integers n is $f(\mathbf{r}) = r^n$ harmonic on every solid T that excludes the origin? Here $\mathbf{r} = x\mathbf{i} + y\mathbf{j} + z\mathbf{k}$ and $r = \|\mathbf{r}\| = \sqrt{x^2 + y^2 + z^2}$.

34. Show that if $f = f(x, y, z)$ satisfies Laplace's equation, then its gradient field is both solenoidal and irrotational.

■ 17.9 THE DIVERGENCE THEOREM

Let Ω be a Jordan region with a piecewise smooth boundary C, and let P and Q be continuously differentiable scalar fields on an open set containing Ω. Green's theorem allows us to express a double integral over Ω as a line integral over C:

$$\iint_\Omega \left(\frac{\partial Q}{\partial x} - \frac{\partial P}{\partial y} \right) dx\, dy = \oint_C P\, dx + Q\, dy.$$

In vector terms Green's theorem can be written:

(17.9.1)
$$\iint_\Omega (\nabla \cdot \mathbf{v})\, dx\, dy = \oint_C (\mathbf{v} \cdot \mathbf{n})\, ds.$$

Here **n** is the outer unit normal and the integral on the right is taken with respect to arc length. (Section 17.4)

PROOF Set $\mathbf{v} = Q\mathbf{i} - P\mathbf{j}$. Then

$$\iint_{\Omega} (\nabla \cdot \mathbf{v}) \, dx \, dy = \iint_{\Omega} \left(\frac{\partial Q}{\partial x} - \frac{\partial P}{\partial y} \right) dx \, dy.$$

All we have to show then is that

$$\oint_C (\mathbf{v} \cdot \mathbf{n}) \, ds = \oint_C P \, dx + Q \, dy.$$

For C traversed counterclockwise, $\mathbf{n} = \mathbf{T} \times \mathbf{k}$ where **T** is the unit tangent vector. (Draw a figure.) Thus

$$\mathbf{v} \cdot \mathbf{n} = \mathbf{v} \cdot (\mathbf{T} \times \mathbf{k}) = (-\mathbf{v}) \cdot (\mathbf{k} \times \mathbf{T}) = (-\mathbf{v} \times \mathbf{k}) \cdot \mathbf{T}.$$
$$\mathbf{a} \cdot (\mathbf{b} \times \mathbf{c}) = (\mathbf{a} \times \mathbf{b}) \cdot \mathbf{c}$$

Since $-\mathbf{v} \times \mathbf{k} = (P\mathbf{j} - Q\mathbf{i}) \times \mathbf{k} = P\mathbf{i} + Q\mathbf{j}$, we have $\mathbf{v} \cdot \mathbf{n} = (P\mathbf{i} + Q\mathbf{j}) \cdot \mathbf{T}$. Therefore

$$\oint_C (\mathbf{v} \cdot \mathbf{n}) \, ds = \oint_C [(P\mathbf{i} + Q\mathbf{j}) \cdot \mathbf{T}] \, ds = \oint_C (P\mathbf{i} + Q\mathbf{j}) \cdot d\mathbf{r} = \oint_C P \, dx + Q \, dy.$$
$$(17.4.6)$$

❑

Green's theorem expressed as (17.9.1) has a higher dimensional analog that is known as the divergence theorem.†

THEOREM 17.9.2 THE DIVERGENCE THEOREM

Let T be a solid bounded by a closed surface S which, if not smooth, is piecewise smooth. If the vector field $\mathbf{v} = \mathbf{v}(x, y, z)$ is continuously differentiable throughout T, then

$$\iiint_T (\nabla \cdot \mathbf{v}) \, dx \, dy \, dz = \iint_S (\mathbf{v} \cdot \mathbf{n}) \, d\sigma$$

where **n** is the outer unit normal.

PROOF We will carry out the proof under the assumption that S is smooth and that any line parallel to a coordinate axis intersects S at most twice.

Our first step is to express the outer unit normal **n** in terms of its direction cosines:

$$\mathbf{n} = \cos \alpha_1 \mathbf{i} + \cos \alpha_2 \mathbf{j} + \cos \alpha_3 \mathbf{k}.$$

Then, for $\mathbf{v} = v_1 \mathbf{i} + v_2 \mathbf{j} + v_3 \mathbf{k}$,

$$\mathbf{v} \cdot \mathbf{n} = v_1 \cos \alpha_1 + v_2 \cos \alpha_2 + v_3 \cos \alpha_3.$$

† Also called Gauss' theorem after the German mathematician Karl Friedrich Gauss (1777–1855). Often referred to as "The Prince of Mathematicians," Gauss is regarded by many as one of the greatest geniuses of all time.

The idea of the proof is to show that

(1)
$$\iint_S v_1 \cos \alpha_1 \, d\sigma = \iiint_T \frac{\partial v_1}{\partial x} \, dx \, dy \, dz,$$

(2)
$$\iint_S v_2 \cos \alpha_2 \, d\sigma = \iiint_T \frac{\partial v_2}{\partial y} \, dx \, dy \, dz,$$

(3)
$$\iint_S v_3 \cos \alpha_3 \, d\sigma = \iiint_T \frac{\partial v_3}{\partial z} \, dx \, dy \, dz.$$

All three equations can be verified in much the same manner. We will carry out the details only for the third equation.

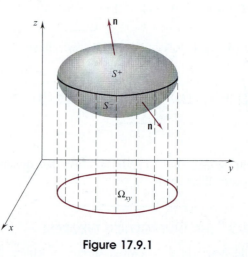

Figure 17.9.1

Let Ω_{xy} be the projection of T onto the xy-plane. (See Figure 17.9.1.) If $(x, y) \in \Omega_{xy}$, then, by assumption, the vertical line through (x, y) intersects S in at most two points, an upper point P^+ and a lower point P^-. (If the vertical line intersects S at only one point P, we set $P = P^+ = P^-$.) As (x, y) ranges over Ω_{xy}, the upper point P^+ describes a surface

$$S^+: \quad z = f^+(x, y), \quad (x, y) \in \Omega_{xy} \qquad \text{(see the figure)}$$

and the lower point describes a surface

$$S^-: \quad z = f^-(x, y), \quad (x, y) \in \Omega_{xy}.$$

By our assumptions, f^+ and f^- are continuously differentiable, $S = S^+ \cup S^-$, and the solid T is the set of all points (x, y, z) with

$$f^-(x, y) \leq z \leq f^+(x, y), \quad (x, y) \in \Omega_{xy}.$$

Now let γ be the angle between the positive z-axis and the upper unit normal. On S^+ the outer unit normal $\mathbf{n}$ is the upper unit normal. Thus on S^+

$$\gamma = \alpha_3 \qquad \text{and} \qquad \cos \alpha_3 \sec \gamma = 1.$$

On S^- the outer unit normal $\mathbf{n}$ is the lower unit normal. In this case

$$\gamma = \pi - \alpha_3 \qquad \text{and} \qquad \cos \alpha_3 \sec \gamma = -1.$$

Thus,

$$\iint_{S^-} v_3 \cos \alpha_3 \, d\sigma = \iint_{\Omega_{xy}} v_3 \cos \alpha_3 \sec \gamma \, dx \, dy = \iint_{\Omega_{xy}} v_3[x, y, f^+(x, y)] \, dx \, dy$$

and

$$\iint_{S^-} v_3 \cos \alpha_3 \, d\sigma = \iint_{\Omega_{xy}} v_3 \cos \alpha_3 \sec \gamma \, dx \, dy = -\iint_{\Omega_{xy}} v_3[x, y, f^-(x, y)] \, dx \, dy.$$

It follows that

$$\iint_S v_3 \cos \alpha_3 \, d\sigma = \iint_{S^+} v_3 \cos \alpha_3 \, d\sigma + \iint_{S^-} v_3 \cos \alpha_3 \, d\sigma$$

$$= \iint_{\Omega_{xy}} (v_3[x, y, f^+(x, y)] - v_3[x, y, f^-(x, y)]) \, dx \, dy$$

$$= \iint_{\Omega_{xy}} \left(\int_{f^-(x, y)}^{f^+(x, y)} \frac{\partial v_3}{\partial z}(x, y, z) \, dz \right) dx \, dy$$

$$= \iiint_T \frac{\partial v_3}{\partial z}(x, y, z) \, dx \, dy \, dz.$$

This confirms Equation (3). Equation (2) can be confirmed by projection onto the xz plane; Equation (1) can be confirmed by projection onto the yz-plane.   ❏

Divergence as Outward Flux per Unit Volume

Choose a point P and surround it by a closed ball N_ϵ of radius ϵ. According to the divergence theorem

$$\iiint_{N_\epsilon} (\nabla \cdot \mathbf{v}) \, dx \, dy \, dz = \text{flux of } \mathbf{v} \text{ out of } N_\epsilon.$$

Thus

$$(\text{average divergence of } \mathbf{v} \text{ on } N_\epsilon) \, (\text{volume of } N_\epsilon) = \text{flux of } \mathbf{v} \text{ out of } N_\epsilon$$

and

$$\text{average divergence of } \mathbf{v} \text{ on } N_\epsilon = \frac{\text{flux of } \mathbf{v} \text{ out of } N_\epsilon}{\text{volume of } N_\epsilon}.$$

Taking the limit of both sides as ϵ shrinks to 0, we have

$$\text{divergence of } \mathbf{v} \text{ at } P = \lim_{\epsilon \to 0^+} \frac{\text{flux of } \mathbf{v} \text{ out of } N_\epsilon}{\text{volume of } N_\epsilon}.$$

In this sense *divergence is outward flux per unit volume.*

Think of $\mathbf{v}$ as the velocity of a fluid. As suggested in Section 17.8, negative divergence at P signals an accumulation of fluid near P:

$$\nabla \cdot \mathbf{v} < 0 \text{ at } P \Rightarrow \text{flux out of } N_\epsilon < 0 \Rightarrow \text{net flow into } N_\epsilon.$$

Positive divergence at P signals a flow of liquid away from P:

$$\nabla \cdot \mathbf{v} > 0 \text{ at } P \Rightarrow \text{flux out of } N_\epsilon > 0 \Rightarrow \text{net flow out of } N_\epsilon.$$

Points at which the divergence is negative are called *sinks*; points at which the divergence is positive are called *sources*. If the divergence of $\mathbf{v}$ is 0 throughout, then the flow has no sinks and no sources and $\mathbf{v}$ is called *solenoidal* (see Exercises 17.8).

Solids Bounded by Two or More Closed Surfaces

The divergence theorem, stated for solids bounded by a single closed surface, can be extended to solids bounded by several closed surfaces. Suppose, for example, that we start with a solid bounded by a closed surface S_1 and extract from the interior of that solid a solid bounded by a closed surface S_2. The remaining solid, call it T and see Figure 17.9.2, has a boundary S that consists of two pieces: an outer piece S_1 and an inner piece S_2. The key here is to note that the *outer* normal for T points *out* of S_1 but *into* S_2. The divergence theorem can be proved for T by slicing T into two pieces T_1 and T_2 as in Figure 17.9.3 and applying the divergence theorem to each piece:

$$\iiint_{T_1} (\nabla \cdot \mathbf{v})\, dx\, dy\, dz = \iint_{\text{bdry of } T_1} (\mathbf{v} \cdot \mathbf{n})\, d\sigma,$$

$$\iiint_{T_2} (\nabla \cdot \mathbf{v})\, dx\, dy\, dz = \iint_{\text{bdry of } T_2} (\mathbf{v} \cdot \mathbf{n})\, d\sigma.$$

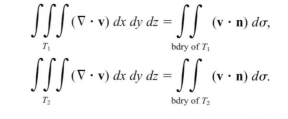

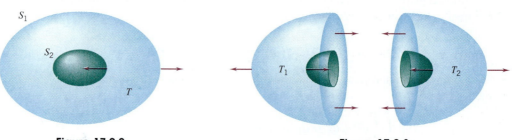

Figure 17.9.2   **Figure 17.9.3**

The triple integrals over T_1 and T_2 add up to the triple integral over T. When the surface integrals are added together, the integrals along the common cut cancel (because the normals are in opposite directions), and therefore only the integrals over S_1 and S_2 remain. Thus the surface integrals add up to the surface integral over $S = S_1 \cup S_2$ and the divergence theorem still holds:

$$\iiint_{T} (\nabla \cdot \mathbf{v})\, dx\, dy\, dz = \iint_{S} (\mathbf{v} \cdot \mathbf{n})\, d\sigma.$$

An Application to Static Charges

Consider a point charge q somewhere in space. This charge creates around itself an *electric field* $\mathbf{E}$, which in turn exerts an electric force on every other nearby charge. If we center our coordinate system at q, then the electric field at $\mathbf{r}$ can be written

$$\mathbf{E}(\mathbf{r}) = q \, \frac{\mathbf{r}}{r^3}.$$

(This is found experimentally.)   Note that this field has exactly the same form as the gravitational field: a constant multiple of r^{-3} **r**.   It follows from (17.8.8) that

$$\nabla \cdot \mathbf{E} = 0 \qquad \text{for all } \mathbf{r} \neq \mathbf{0}.$$

We are interested in the flux of **E** out of a closed surface S.   We assume that S does not pass through q.

If the charge q is outside of S, then **E** is continuously differentiable on the region T bounded by S, and, by the divergence theorem.

$$\text{flux of } \mathbf{E} \text{ out of } S = \iiint_T (\nabla \cdot \mathbf{E}) \, dx \, dy \, dz = \iiint_T 0 \, dx \, dy \, dz = 0.$$

If q is inside of S, then the divergence theorem does not apply to T directly because **E** is not differentiable on all of T.   We can circumvent this difficulty by surrounding q by a small sphere S_a of radius a and applying the divergence theorem to the region T' bounded on the outside by S and on the inside by S_a.   (Figure 17.9.4)

Since **E** is continuously differentiable on T',

$$\iiint_{T'} (\nabla \cdot \mathbf{E}) \, dx \, dy \, dz = \text{flux of } \mathbf{E} \text{ out of } S + \text{flux of } \mathbf{E} \text{ into } S_a$$

$$= \text{flux of } \mathbf{E} \text{ out of } S - \text{flux of } \mathbf{E} \text{ out of } S_a.$$

Since $\nabla \cdot \mathbf{E} = 0$ on T', the triple integral on the left is zero and therefore

$$\text{flux of } \mathbf{E} \text{ out of } S = \text{flux of } \mathbf{E} \text{ out of } S_a.$$

The quantity on the right is easy to calculate: on S_a, $\mathbf{n} = \mathbf{r}/r$ and therefore

$$\mathbf{E} \cdot \mathbf{n} = q \frac{\mathbf{r}}{r^3} \cdot \frac{\mathbf{r}}{r} = \frac{q}{r^2} = \frac{q}{a^2}.$$

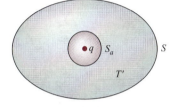

Figure 17.9.4

Thus

$$\text{flux of } \mathbf{E} \text{ out of } S_a = \iint_{S_a} (\mathbf{E} \cdot \mathbf{n}) \, d\sigma = \iint_{S_a} \frac{q}{a^2} d\sigma = \frac{q}{a^2} (\text{area of } S_a) = 4 \pi q.$$

$$\text{area of } S_a = 4\pi a^2 \underline{\qquad} \uparrow$$

It follows that the flux of **E** out of $S = 4\pi q$.

In summary, if **E** is the electric field of a point charge q and S is a closed surface that does not pass through q, then

$$\text{flux of } \mathbf{E} \text{ out of } S = \begin{cases} 0, & \text{if } q \text{ is outside of } S \\ 4\pi q, & \text{if } q \text{ is inside of } S. \end{cases}$$

EXERCISES 17.9

In Exercises 1–4, calculate the flux **v** out of the unit ball $x^2 + y^2 + z^2 \leq 1$ by applying the divergence theorem.

1. $\mathbf{v}(x, y, z) = x\mathbf{i} + y\mathbf{j} + z\mathbf{k}$.
2. $\mathbf{v}(x, y, z) = (1 - x)\mathbf{i} + (2 - y)\mathbf{j} + (3 - z)\mathbf{k}$.
3. $\mathbf{v}(x, y, z) = x^2\mathbf{i} + y^2\mathbf{j} + z^2\mathbf{k}$.
4. $\mathbf{v}(x, y, z) = (1 - x^2)\mathbf{i} - y^2\mathbf{j} + z\mathbf{k}$.

In Exercises 5–8, verify the divergence theorem on the unit cube $0 \leq x \leq 1, 0 \leq y \leq 1, 0 \leq z \leq 1$ for the following vector fields.

5. $\mathbf{v}(x, y, z) = x\mathbf{i} + y\mathbf{j} + z\mathbf{k}$.
6. $\mathbf{v}(x, y, z) = xy\mathbf{i} + yz\mathbf{j} + xz\mathbf{k}$.
7. $\mathbf{v}(x, y, z) = x^2\mathbf{i} - xz\mathbf{j} + z^2\mathbf{k}$.

8. $\mathbf{v}(x, y, z) = x\mathbf{i} + xy\mathbf{j} + xyz\mathbf{k}.$

In Exercises 9–14, use the divergence theorem to find the total flux out of the given solid.

9. $\mathbf{v}(x, y, z) = x\mathbf{i} + 2y^2\mathbf{j} + 3z^2\mathbf{k};\ 0 \le x^2 + y^2 \le 9,$
$0 \le z \le 1.$

10. $\mathbf{v}(x, y, z) = xy\mathbf{i} + yz\mathbf{j} + xz\mathbf{k};\ 0 \le x \le 1,$
$0 \le y \le 1 - x,\quad 0 \le z \le 1 - x - y.$

11. $\mathbf{v}(x, y, z) = x^2\mathbf{i} + xy\mathbf{j} - 2xz\mathbf{k};\ 0 \le x \le 1,$
$0 \le y \le 1 - x,\quad 0 \le z \le 1 - x - y.$

12. $\mathbf{v}(x, y, z) = (2xy + 2z)\mathbf{i} + (y^2 + 1)\mathbf{j} - (x + y)\mathbf{k};$
$0 \le x \le 4,\quad 0 \le y \le 4 - x,\quad 0 \le z \le 4 - x - y.$

13. $\mathbf{v}(x, y, z) = x^2\mathbf{i} + y^2\mathbf{j} + z^2\mathbf{k};$ the cylinder
$0 \le x^2 + y^2 \le 4,\quad 0 \le z \le 4,$ including the top and base.

14. $\mathbf{v}(x, y, z) = 2x\mathbf{i} + xy\mathbf{j} + xz\mathbf{k};$ the ball
$x^2 + y^2 + z^2 \le 4.$

In Exercises 15 and 16 calculate the total flux of $\mathbf{v}(x, y, z) = 2xy\mathbf{i} + y^2\mathbf{j} + 3yz\mathbf{k}$ out of the given solid.

15. The ball: $0 \le x^2 + y^2 + z^2 \le a^2.$

16. The cube: $0 \le x \le a,\ 0 \le y \le a,\ 0 \le z \le a.$

17. What is the flux of $\mathbf{v}(x, y, z) = Ax\mathbf{i} + By\mathbf{j} + Cz\mathbf{k}$ out of a solid of volume V?

18. Let T be a basic solid with a piecewise-smooth boundary. Show that, if f is harmonic on T (defined in Exercises 17.8), then the flux of ∇f out of T is zero.

19. Let S be a closed smooth surface with continuous unit normal $\mathbf{n} = \mathbf{n}(x, y, z)$. Show that

$$\iint_S \mathbf{n}\, d\sigma = \left(\iint_S n_1\, d\sigma\right)\mathbf{i} + \left(\iint_S n_2\, d\sigma\right)\mathbf{j}$$
$$+ \left(\iint_S n_3\, d\sigma\right)\mathbf{k} = \mathbf{0}.$$

20. Let T be a solid with a piecewise-smooth boundary S and let $\mathbf{n}$ be the outer unit normal.
(a) Verify the identity $\nabla \cdot (f\nabla f) = \|\nabla f\|^2 + f(\nabla^2 f)$ and show that, if f is harmonic on T, then

$$\iint_S (ff'_\mathbf{n})\, d\sigma = \iiint_T \|\nabla f\|^2\, dx\, dy\, dz$$

where $f'_\mathbf{n}$ is the directional derivative $\nabla f \cdot \mathbf{n}$.
(b) Show that, if g is continuously differentiable on T, then

$$\iint_S (gf'_\mathbf{n})\, d\sigma =$$
$$\iiint_T [(\nabla g \cdot \nabla f) + g(\nabla^2 f)]\, dx\, dy\, dz.$$

21. Let T be a solid with a piecewise-smooth boundary. Show that, if f and g have continuous second partials, then the flux of $\nabla f \times \nabla g$ out of T is zero.

22. Let T be a solid with a piecewise-smooth boundary S. Express the volume of T as a surface integral over S.

23. Suppose that a solid T (boundary S, outer unit normal $\mathbf{n}$) is immersed in a fluid. The fluid exerts a pressure $p = p(x, y, z)$ at each point of S, and therefore the solid T experiences a force. The total force on the solid due to the pressure distribution is given by the surface integral

$$\mathbf{F} = -\iint_S p\mathbf{n}\, d\sigma.$$

(The formula says that the force on the solid is the average pressure against S times the area of S.) Now choose a coordinate system with the z-axis vertical and assume that the fluid fills a region of space to the level $z = c$. The depth of a point (x, y, z) is then $c - z$ and we have $p(x, y, z) = \rho(c - z)$, where ρ is the weight density of the fluid (the weight per unit volume). Apply the divergence theorem to each component of $\mathbf{F}$ to show that $\mathbf{F} = W\mathbf{k}$, where W is the weight of the fluid displaced by the solid. We call this the *buoyant force* on the solid. (This shows that the object is not pushed from side to side by the pressure and verifies the *principle of Archimedes*: that the buoyant force on an object at rest in a fluid equals the weight of the fluid displaced.)

24. If $\mathbf{F}$ is a force applied at the tip of a radius vector $\mathbf{r}$, then the *torque,* or twisting strength, of $\mathbf{F}$ about the origin is given by the cross product $\boldsymbol{\tau} = \mathbf{r} \times \mathbf{F}$. From physics we learn that the total torque on the solid T of Exercise 23 is given by the formula

$$\boldsymbol{\tau}_{\text{Tot}} = -\iint_S [\mathbf{r} \times \rho(c - z)\mathbf{n}]\, d\sigma.$$

Use the divergence theorem to find the components of $\boldsymbol{\tau}_{\text{Tot}}$ and show that $\boldsymbol{\tau}_{\text{Tot}} = \bar{\mathbf{r}} \times \mathbf{F}$ where $\mathbf{F}$ is the buoyant force $W\mathbf{k}$ and $\bar{\mathbf{r}}$ is the centroid of T.
(This indicates that for calculating the twisting effect of the buoyant force we can view this force as being applied at the centroid. This is very important in ship design. Imagine, for example, a totally submerged submarine. While the buoyant force acts upward through the centroid of the submarine, gravity acts downward through the center of mass. Suppose the submarine should tilt a bit to one side as depicted in the figure. If the centroid lies above the center of mass, then the buoyant force acts to restore the submarine to an upright position. If, however, the centroid lies below the center of mass, then disaster. Once the submarine has tilted a bit, the buoyant

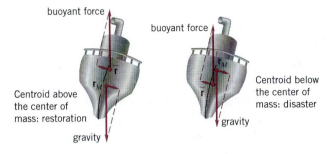

buoyant force

buoyant force

Centroid above
the center of
mass: restoration

gravity

Centroid below
the center of
mass: disaster

gravity

force will make it tilt further. This kind of analysis also applies to surface ships, but in this case the buoyant force acts upward through *the centroid of the portion of the ship that is submerged.* This point is called the *center of flotation.* One must design and load a ship to keep the center of flotation above the center of mass. Putting a lot of heavy cargo on the deck, for instance, tends to raise the center of mass and destabilize the ship.)

■ 17.10 STOKES'S THEOREM

We return to Green's theorem

$$\iint_\Omega \left(\frac{\partial Q}{\partial x} - \frac{\partial P}{\partial y} \right) dx \, dy = \oint_C P \, dx + Q \, dy.$$

Setting $\mathbf{v} = P\mathbf{i} + Q\mathbf{j} + R\mathbf{k}$, we have

$$(\nabla \times \mathbf{v}) \cdot \mathbf{k} = \begin{vmatrix} \mathbf{i} & \mathbf{j} & \mathbf{k} \\ \dfrac{\partial}{\partial x} & \dfrac{\partial}{\partial y} & \dfrac{\partial}{\partial z} \\ P & Q & R \end{vmatrix} \cdot \mathbf{k} = \frac{\partial Q}{\partial x} - \frac{\partial P}{\partial y}.$$

Thus in terms of $\mathbf{v}$, Green's theorem can be written

$$\iint_\Omega [(\nabla \times \mathbf{v}) \cdot \mathbf{k}] \, dx \, dy = \oint_C \mathbf{v}(\mathbf{r}) \cdot d\mathbf{r}.$$

Since any plane can be coordinatized as the xy-plane, this result can be phrased as follows: Let S be a flat surface in space bounded by a Jordan curve C: If $\mathbf{v}$ is continuously differentiable on S, then

$$\iint_S [(\nabla \times \mathbf{v}) \cdot \mathbf{n}] \, d\sigma = \oint_C \mathbf{v}(\mathbf{r}) \cdot d\mathbf{r}$$

where $\mathbf{n}$ is a unit normal for S and the line integral is taken in the *positive sense,* meaning, in the direction of the unit tangent $\mathbf{T}$ for which $\mathbf{T} \times \mathbf{n}$ points away from the surface. See Figure 17.10.1. (An observer marching along C with the same

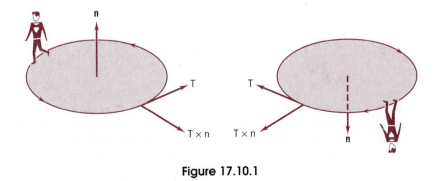

Figure 17.10.1

orientation as **n** keeps the surface to his left.)   The symbol $\displaystyle\oint_C$ denotes this line integral.

Figure 17.10.2 shows a *polyhedral surface S* bounded by a closed polygonal path *C*. The surface *S* consists of a finite number of flat faces $S_1, \ldots, S_N$ with polygonal boundaries $C_1, \ldots, C_N$ and unit normals $\mathbf{n}_1, \ldots, \mathbf{n}_N$. We choose these unit

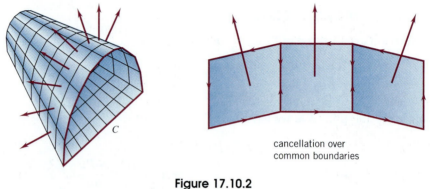

cancellation over
common boundaries

Figure 17.10.2

normals in a consistent manner; that is, they emanate from the same side of the surface. Now let $\mathbf{n} = \mathbf{n}(x, y, z)$ be a vector function of norm 1 which is $\mathbf{n}_1$ on S_1, $\mathbf{n}_2$ on S_2, $\mathbf{n}_3$ on S_3, etc. It is immaterial how **n** is defined on the line segments that join the different faces. Suppose now that $\mathbf{v} = \mathbf{v}(x, y, z)$ is a vector function continuously differentiable on an open set that contains *S*.   Then

$$\iint_S [(\nabla \times \mathbf{v}) \cdot \mathbf{n}] \, d\sigma = \sum_{i=1}^{N} \iint_{S_i} [(\nabla \times \mathbf{v}) \cdot \mathbf{n}_i] \, d\sigma = \sum_{i=1}^{N} \oint_{C_i} \mathbf{v}(\mathbf{r}) \cdot d\mathbf{r},$$

the integral over C_i being taken in the positive sense with respect to $\mathbf{n}_i$.   Now when we add these line integrals, we find that all the line segments that make up the C_i but are not part of *C* are traversed twice and in opposite directions.   (See the figure.) Thus these line segments contribute nothing to the sum of the line integrals and we are left with the integral around *C*.   It follows that for a polyhedral surface *S* with boundary *C*

$$\iint_S [(\nabla \times \mathbf{v}) \cdot \mathbf{n}] \, d\sigma = \oint_C \mathbf{v}(\mathbf{r}) \cdot d\mathbf{r}.$$

This result can be extended to smooth surfaces with smooth bounding curves (see Figure 17.10.3) by approximating these configurations by polyhedral configurations of the type considered and using a limit process.   In an admittedly informal way we have arrived at Stokes's theorem.

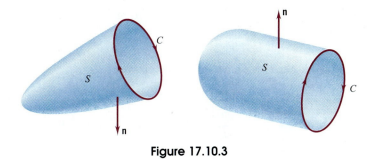

Figure 17.10.3

> **THEOREM 17.10.1 STOKES'S THEOREM†**
>
> Let S be a smooth surface with a smooth bounding curve C. If $\mathbf{v} = \mathbf{v}(x, y, z)$ is a continuously differentiable vector field on an open set that contains S, then
>
> $$\iint\limits_{S} [(\nabla \times \mathbf{v}) \cdot \mathbf{n}]\, d\sigma = \oint_{C} \mathbf{v}(\mathbf{r}) \cdot d\mathbf{r}$$
>
> where $\mathbf{n} = \mathbf{n}(x, y, z)$ is a unit normal that varies continuously on S and the line integral is taken in the positive sense with respect to $\mathbf{n}$.

Example 1 Verify Stokes's theorem for

$$\mathbf{v} = -3y\,\mathbf{i} + 3x\,\mathbf{j} + z^4\,\mathbf{k}$$

taking S as the portion of the ellipsoid $2x^2 + 2y^2 + z^2 = 1$ that lies above the plane $z = 1/\sqrt{2}$ (Figure 17.10.4).

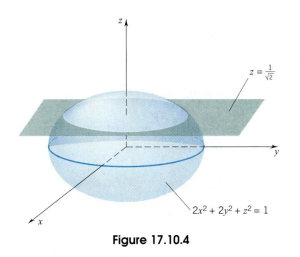

$z = \dfrac{1}{\sqrt{2}}$

$2x^2 + 2y^2 + z^2 = 1$

Figure 17.10.4

---

† The result was announced publicly for the first time by George Gabriel Stokes (1819–1903), an Irish mathematician and physicist who, like Green, was a Cambridge professor.

SOLUTION   A little algebra shows that S is the graph of

$$f(x, y) = \sqrt{1 - 2(x^2 + y^2)}$$

with (x, y) restricted to the disc $\Omega: x^2 + y^2 \leq \frac{1}{4}$. Now

$$\nabla \times \mathbf{v} = \begin{vmatrix} \mathbf{i} & \mathbf{j} & \mathbf{k} \\ \dfrac{\partial}{\partial x} & \dfrac{\partial}{\partial y} & \dfrac{\partial}{\partial z} \\ -3y & 3x & z^4 \end{vmatrix} = \left[\dfrac{\partial}{\partial x}(3x) + \dfrac{\partial}{\partial y}(3y) \right] \mathbf{k} = 6\mathbf{k}.$$

Taking $\mathbf{n}$ as the upper unit normal (i.e., $\mathbf{n} = -f_x \mathbf{i} - f_y \mathbf{j} + \mathbf{k}$), we have

$$\iint_S [(\nabla \times \mathbf{v}) \cdot \mathbf{n}]\, d\sigma = \iint_S (6\mathbf{k} \cdot \mathbf{n})\, d\sigma$$

(17.7.9) _____↑
$$= \iint_\Omega (-(0)f_x - (0)f_y + 6)\, dx\, dy$$

$$= \iint_\Omega 6\, dx\, dy = 6(\text{area of } \Omega) = 6(\tfrac{1}{4}\pi) = \tfrac{3}{2}\pi.$$

The bounding curve C is the set of all (x, y, z) with $x^2 + y^2 = \frac{1}{4}$ and $z = 1/\sqrt{2}$. We can parametrize C by setting

$$\mathbf{r}(u) = \tfrac{1}{2}\cos u\, \mathbf{i} + \tfrac{1}{2}\sin u\, \mathbf{j} + \frac{1}{\sqrt{2}}\mathbf{k}, \qquad u \in [0, 2\pi].$$

Since $\mathbf{n}$ is the upper unit normal, this parametrization gives C in the positive sense. Thus

$$\oint_C \mathbf{v}(\mathbf{r}) \cdot d\mathbf{r} = \int_0^{2\pi} (-\tfrac{3}{2}\sin u\, \mathbf{i} + \tfrac{3}{2}\cos u\, \mathbf{j} + \tfrac{1}{4}\mathbf{k}) \cdot (-\tfrac{1}{2}\sin u\, \mathbf{i} + \tfrac{1}{2}\cos u\, \mathbf{j})\, du$$

$$= \int_0^{2\pi} (\tfrac{3}{4}\sin^2 u + \tfrac{3}{4}\cos^2 u)\, du = \int_0^{2\pi} \tfrac{3}{4}\, du = \tfrac{3}{2}\pi$$

This is the value we obtained for the surface integral.   ❏

Example 2   Verify Stokes's theorem for

$$\mathbf{v} = z^2 \mathbf{i} - 2x\mathbf{j} + y^3 \mathbf{k}$$

taking S as the upper half of the unit sphere $x^2 + y^2 + z^2 = 1$.

SOLUTION   We use the upper unit normal $\mathbf{n} = x\mathbf{i} + y\mathbf{j} + z\mathbf{k}$. Now

$$\nabla \times \mathbf{v} = \begin{vmatrix} \mathbf{i} & \mathbf{j} & \mathbf{k} \\ \dfrac{\partial}{\partial x} & \dfrac{\partial}{\partial y} & \dfrac{\partial}{\partial z} \\ z^2 & -2x & y^3 \end{vmatrix} = 3y^2 \mathbf{i} + 2z\mathbf{j} - 2\mathbf{k}.$$

Therefore

$$\iint\limits_{S} [(\nabla \times \mathbf{v}) \cdot \mathbf{n}]\, d\sigma = \iint\limits_{S} [(3y^2\mathbf{i} + 2z\mathbf{j} - 2\mathbf{k}) \cdot (x\mathbf{i} + y\mathbf{j} + z\mathbf{k})\, d\sigma$$

$$= \iint\limits_{S} (3xy^2 + 2yz - 2z)\, d\sigma$$

$$= \iint\limits_{S} 3xy^2\, d\sigma + \iint\limits_{S} 2yz\, d\sigma - \iint\limits_{S} 2z\, d\sigma.$$

The first integral is zero because S is symmetric about the yz-plane and the integrand is odd with respect to x. The second integral is zero because S is symmetric about the xz-plane and the integrand is odd with respect to y. Thus

$$\iint\limits_{S} [(\nabla \times \mathbf{v}) \cdot \mathbf{n}]\, d\sigma = -\iint\limits_{S} 2z\, d\sigma = -2\bar{z}\ (\text{area of } S) = -2(\tfrac{1}{2})2\,\pi = -2\,\pi.$$

S Exercise 17, Section17.7 _____↑

This is also the value of the integral along the bounding base circle taken in the positive sense: $\mathbf{r}(u) = \cos u\,\mathbf{i} + \sin u\,\mathbf{j}, \quad u \in [2, \pi]$, and

$$\oint_{C} \mathbf{v}(\mathbf{r}) \cdot d\mathbf{r} = \oint_{C} z^2\, dx - 2x\, dy = -2 \oint_{C} x\, dy$$

$$= -2 \int_{0}^{2\pi} \cos^2 u\, du = -2\,\pi. \quad \square$$

Earlier you saw that the curl of a gradient is zero. Using Stokes's theorem we can prove a partial converse.

(17.10.2) | If a vector field $\mathbf{v} = \mathbf{v}(x, y, z)$ is continuously differentiable on an open convex† set U and $\nabla \times \mathbf{v} = 0$ on all of U, then $\mathbf{v}$ is the gradient of some scalar field ϕ defined on U.

PROOF Choose a point $\mathbf{a}$ in U and for each point $\mathbf{x}$ in U define

$$\phi(\mathbf{x}) = \int_{\mathbf{a}}^{\mathbf{x}} \mathbf{v}(\mathbf{r}) \cdot d\mathbf{r}.$$

(This is the line integral from $\mathbf{a}$ to $\mathbf{x}$ taken along the line segment that joins these two points. We know that this line segment lies in U because U is convex.)

Since U is open, $\mathbf{x} + \mathbf{h}$ is in U for all $\mathbf{h}$ sufficiently small. Assume then that $\mathbf{h}$ is sufficiently small for $\mathbf{x} + \mathbf{h}$ to be in U. Since U is convex, the triangular region with vertices at $\mathbf{a}, \mathbf{x}, \mathbf{x} + \mathbf{h}$ lies in U. (See Figure 17.10.5.) Since $\nabla \times \mathbf{v} = \mathbf{0}$ on U, we can conclude from Stokes's theorem that

$$\int_{\mathbf{a}}^{\mathbf{x}} \mathbf{v}(\mathbf{r}) \cdot d\mathbf{r} + \int_{\mathbf{x}}^{\mathbf{x}+\mathbf{h}} \mathbf{v}(\mathbf{r}) \cdot d\mathbf{r} + \int_{\mathbf{x}+\mathbf{h}}^{\mathbf{a}} \mathbf{v}(\mathbf{r}) \cdot d\mathbf{r} = 0.$$

---

† Recall that the set U is *convex* if for each pair of points $p, q \in U$, the line segment $\overline{pq}$ lies entirely in U.

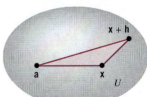

Figure 17.10.5

Therefore

$$\int_{\mathbf{x}}^{\mathbf{x}+\mathbf{h}} \mathbf{v}(\mathbf{r}) \cdot d\mathbf{r} = -\int_{\mathbf{x}+\mathbf{h}}^{\mathbf{a}} \mathbf{v}(\mathbf{r}) \cdot d\mathbf{r} - \int_{\mathbf{a}}^{\mathbf{x}} \mathbf{v}(\mathbf{r}) \cdot d\mathbf{r}$$

$$= \int_{\mathbf{a}}^{\mathbf{x}+\mathbf{h}} \mathbf{v}(\mathbf{r}) \cdot d\mathbf{r} - \int_{\mathbf{a}}^{\mathbf{x}} \mathbf{v}(\mathbf{r}) \cdot d\mathbf{r}.$$

By our definition of ϕ we have

$$\phi(\mathbf{x} + \mathbf{h}) - \phi(\mathbf{x}) = \int_{\mathbf{x}}^{\mathbf{x}+\mathbf{h}} \mathbf{v}(\mathbf{r}) \cdot d\mathbf{r}.$$

We can parametrize the line segment from $\mathbf{x}$ to $\mathbf{x} + \mathbf{h}$ by $\mathbf{r}(u) = \mathbf{x} + u\mathbf{h}$ with $u \in [0, 1]$. Therefore

$$\phi(\mathbf{x} + \mathbf{h}) - \phi(\mathbf{x}) = \int_0^1 [\mathbf{v}(\mathbf{r}(u)) \cdot \mathbf{r}'(u)] \, du$$

$$= \int_0^1 [\mathbf{v}(\mathbf{r}(u)) \cdot \mathbf{h}] \, du$$

$$\text{Theorem 5.9.1} \longrightarrow = \mathbf{v}(\mathbf{r}(u_0)) \cdot \mathbf{h} \qquad \text{for some } u_0 \text{ in } [0, 1]$$

$$= \mathbf{v}(\mathbf{x} + u_0\mathbf{h}) \cdot \mathbf{h} = \mathbf{v}(\mathbf{x}) \cdot \mathbf{h} + [\mathbf{v}(\mathbf{x} + u_0\mathbf{h}) - \mathbf{v}(\mathbf{x})] \cdot \mathbf{h}.$$

The fact that $\mathbf{v} = \nabla\phi$ follows from observing that $[\mathbf{v}(\mathbf{x} + u_0\mathbf{h}) - \mathbf{v}(\mathbf{x})] \cdot \mathbf{h}$ is $o(\mathbf{h})$:

$$\frac{|[\mathbf{v}(\mathbf{x} + u_0\mathbf{h}) - \mathbf{v}(\mathbf{x})] \cdot \mathbf{h}|}{\|\mathbf{h}\|} \leq \frac{\|\mathbf{v}(\mathbf{x} + u_0\mathbf{h}) - \mathbf{v}(\mathbf{x})\| \, \|\mathbf{h}\|}{\|\mathbf{h}\|}$$

$$= \|\mathbf{v}(\mathbf{x} + u_0\mathbf{h}) - \mathbf{v}(\mathbf{x})\| \to 0$$

as $\mathbf{h} \to \mathbf{0}$. ❏

The Normal Component of $\nabla \times \mathbf{v}$ as Circulation per Unit Area; Irrotational Flow

Interpret $\mathbf{v} = \mathbf{v}(x, y, z)$ as the velocity of a fluid flow. In Section 17.8 we stated that $\nabla \times \mathbf{v}$ measures the rotational tendency of the fluid. Now we can be more precise.

Take a point P within the flow and choose a unit vector $\mathbf{n}$. Let D_ϵ be the ϵ-disc that is centered at P and is perpendicular to $\mathbf{n}$. Let C_ϵ be the circular boundary of D_ϵ directed in the positive sense with respect to $\mathbf{n}$. (See Figure 17.10.6) By Stokes's theorem

$$\iint_{D_\epsilon} [(\nabla \times \mathbf{v}) \cdot \mathbf{n}] \, d\sigma = \oint_{C_\epsilon} \mathbf{v}(\mathbf{r}) \cdot d\mathbf{r}.$$

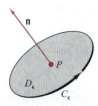

Figure 17.10.6

The line integral on the right is called *the circulation* of $\mathbf{v}$ around C_ϵ. Thus we can say that

$$\left(\begin{array}{c}\text{The average } \mathbf{n}\text{-component} \\ \text{of } \nabla \times \mathbf{v} \text{ on } D_\epsilon\end{array}\right) (\text{the area of } D_\epsilon) = \text{the circulation of } \mathbf{v} \text{ around } C_\epsilon.$$

It follows that

$$\text{The average } \mathbf{n}\text{-component of } \nabla \times \mathbf{n} \text{ on } D_\epsilon = \frac{\text{the circulation of } \mathbf{v} \text{ around } C_\epsilon}{\text{the area of } D_\epsilon}$$

Taking the limit as ϵ shrinks to 0, you can see that

$$\text{The } \mathbf{n}\text{-component of } \nabla \times \mathbf{v} \text{ at } P = \lim_{\epsilon \to 0^+} \frac{\text{the circulation of } \mathbf{v} \text{ around } C_\epsilon}{\text{the area of } D_\epsilon}.$$

At each point P the component of $\nabla \times \mathbf{v}$ in any direction $\mathbf{n}$ is the circulation of $\mathbf{v}$ per unit area in the plane normal to $\mathbf{n}$. If $\nabla \times \mathbf{v} = \mathbf{0}$ identically, the fluid has no rotational tendency and the flow is called irrotational.

EXERCISES 17.10

In Exercises 1–4, let S be the upper half of the unit sphere $x^2 + y^2 + z^2 = 1$ and take $\mathbf{n}$ as the upper unit normal. Find

$$\iint_S [(\nabla \times \mathbf{v}) \cdot \mathbf{n}] \, d\sigma$$

(a) By direct calculation.   (b) By Stokes's theorem.

1. $\mathbf{v}(x, y, z) = x\mathbf{i} + y\mathbf{j} + z\mathbf{k}$.
2. $\mathbf{v}(x, y, z) = y\mathbf{i} - x\mathbf{j} + z\mathbf{k}$.
3. $\mathbf{v}(x, y, z) = z^2\mathbf{i} + 2x\mathbf{j} - y^3\mathbf{k}$.
4. $\mathbf{v}(x, y, z) = 6xz\mathbf{i} - x^2\mathbf{j} - 3y^2\mathbf{k}$.

In Exercises 5–7, let S be the triangular surface with vertices $(2, 0, 0)$, $(0, 2, 0)$, $(0, 0, 2)$ and take $\mathbf{n}$ as the upper unit normal. Find

$$\iint_S [(\nabla \times \mathbf{v}) \cdot \mathbf{n}] \, d\sigma$$

(a) By direct calculation.   (b) By Stokes's theorem.

5. $\mathbf{v}(x, y, z) = 2z\mathbf{i} - y\mathbf{j} + x\mathbf{k}$.
6. $\mathbf{v}(x, y, z) = (x^2 + y^2)\mathbf{i} + y^2\mathbf{j} + (x^2 + z^2)\mathbf{k}$.
7. $\mathbf{v}(x, y, z) = x^4\mathbf{i} + xy\mathbf{j} + z^4\mathbf{k}$.
8. Show that, if $\mathbf{v} = \mathbf{v}(x, y, z)$ is continuously differentiable everywhere and its curl is identically zero, then

$$\int_C \mathbf{v}(\mathbf{r}) \cdot d\mathbf{r} = 0 \qquad \text{for every smooth closed curve } C.$$

9. Let $\mathbf{v} = y\mathbf{i} + z\mathbf{j} + x^2 y^2\mathbf{k}$ and let S be the surface $z = x^2 + y^2$ from $z = 0$ to $z = 4$. Calculate the flux of $\nabla \times \mathbf{v}$ in the direction of the lower unit normal $\mathbf{n}$.
10. Let $\mathbf{v} = \frac{1}{2}y\mathbf{i} + 2xz\mathbf{j} - 3x\mathbf{k}$ and let S be the surface $y = 1 - (x^2 + z^2)$ from $y = -8$ to $y = 1$. Calculate the flux of $\nabla \times \mathbf{v}$ in the direction of the unit normal $\mathbf{n}$ with positive $\mathbf{j}$-component.
11. Let $\mathbf{v} = 2x\mathbf{i} + 2y\mathbf{j} + x^2 y^2 z^2\mathbf{k}$ and let S be the lower half of the ellipsoid

$$\frac{x^2}{4} + \frac{y^2}{9} + \frac{z^2}{27} = 1.$$

Calculate the flux of $\nabla \times \mathbf{v}$ in the direction of the upper unit normal $\mathbf{n}$.

12. Let S be a smooth closed surface and let $\mathbf{v} = \mathbf{v}(x, y, z)$ be a vector field with second partials continuous on an open convex set that contains S. Show that

$$\iint_S [(\nabla \times \mathbf{v}) \cdot \mathbf{n}] \, d\sigma = 0$$

where $\mathbf{n} = \mathbf{n}(x, y, z)$ is the outer unit normal.
13. The upper half of the ellipsoid $\frac{1}{2}x^2 + \frac{1}{2}y^2 + z^2 = 1$ intersects the cylinder $x^2 + y^2 - y = 0$ in a curve C. Calculate the circulation of $\mathbf{v} = y^3\mathbf{i} + (xy + 3xy^2)\mathbf{j} + z^4\mathbf{k}$ around C by using Stokes's theorem.
14. The sphere $x^2 + y^2 + z^2 = a^2$ intersects the plane $x + 2y + z = 0$ in a curve C. Calculate the circulation of $\mathbf{v} = 2y\mathbf{i} - z\mathbf{j} + 2x\mathbf{k}$ about C by using Stokes's theorem.
15. The paraboloid $z = x^2 + y^2$ intersects the plane $z = y$ in a curve C. Calculate the circulation of $\mathbf{v} = 2z\mathbf{i} + x\mathbf{j} + y\mathbf{k}$ about C using Stokes's theorem.
16. The cylinder $x^2 + y^2 = b^2$ intersects the plane $y + z = a^2$ in a curve C. Assume $a^2 \geq b > 0$. Calculate the circulation of $\mathbf{v} = xy\mathbf{i} + yz\mathbf{j} + xz\mathbf{k}$ about C using Stokes's theorem.
17. Let S be a smooth surface with a smooth bounding curve C and let $\mathbf{a}$ be a fixed vector. Show that

$$\iint_S (2\mathbf{a} \cdot \mathbf{n}) \, d\sigma = \oint_C (\mathbf{a} \times \mathbf{r}) \cdot d\mathbf{r}$$

where $\mathbf{n} = \mathbf{n}(x, y, z)$ is a unit vector that varies continuously over S and the line integral is taken in the positive sense with respect to $\mathbf{n}$.
18. Let S be a smooth surface with smooth bounding curve C. Show that, if ϕ and ψ are sufficiently differentiable scalar fields, then

$$\iint_S [(\nabla\phi \times \nabla\psi) \cdot \mathbf{n}] \, d\sigma = \oint_C (\phi\nabla\psi) \cdot d\mathbf{r}$$

where $\mathbf{n} = \mathbf{n}(x, y, z)$ is a unit normal that varies continuously on S and the line integral is taken in the positive sense with respect to $\mathbf{n}$.

19. Let S be a smooth surface with a smooth plane bounding curve C and let $\mathbf{v} = \mathbf{v}(x, y, z)$ be a vector field with second partials continuous on an open convex set that con-

tains S. If S does not cross the plane of C, then Stokes's theorem for S follows readily from the divergence theorem and Green's theorem. Carry out the argument.

20. Our derivation of Stokes's theorem was admittedly nonrigorous. The following version of Stokes's theorem lends itself more readily to rigorous proof.

THEOREM 17.10.3

Let Γ be a Jordan region in the uv-plane with a piecewise-smooth boundary C_Γ given in a counterclockwise sense by a pair of functions $u = u(t)$, $v = v(t)$ with $t \in [a, b]$. Let $\mathbf{R}(u, v) = x(u, v)\mathbf{i} + y(u, v)\mathbf{j} + z(u, v)\mathbf{k}$ be a vector function with continuous second partials on Γ. Assume that $\mathbf{R}$ is one-to-one on Γ and that the fundamental vector product $\mathbf{N} = \mathbf{R}'_u \times \mathbf{R}'_v$ is never zero. The surface S: $\mathbf{R} = \mathbf{R}(u, v)$, $(u, v) \in \Gamma$ is a smooth surface bounded by the oriented space curve C: $\mathbf{r}(t) = \mathbf{R}(u(t), v(t))$, $t \in [a, b]$. If $\mathbf{v} = \mathbf{v}(x, y, z)$ is a vector field continuously differentiable on S, then

$$\iint_S [(\nabla \times \mathbf{v}) \cdot \mathbf{n}] \, d\sigma = \oint_C \mathbf{v}(\mathbf{r}) \cdot d\mathbf{r}$$

where $\mathbf{n}$ is the unit normal in the direction of the fundamental vector product.

Give a detailed proof of the theorem.   HINT: Set $\mathbf{v} = v_1\mathbf{i} + v_2\mathbf{j} + v_3\mathbf{k}$. Then

$$\iint_S [(\nabla \times \mathbf{v}) \cdot \mathbf{n}] \, d\sigma = \iint_S [(\nabla \times v_1\mathbf{i}) \cdot \mathbf{n}] \, d\sigma$$
$$+ \iint_S [(\nabla \times v_2\mathbf{j}) \cdot \mathbf{n}] \, d\sigma + \iint_S [(\nabla \times v_3\mathbf{k}) \cdot \mathbf{n}] \, d\sigma$$

and

$$\oint_C \mathbf{v}(\mathbf{r}) \cdot d\mathbf{r} = \int_C v_1 \, dx + \int_C v_2 \, dy + \int_C v_3 \, dz.$$

Show that

$$\iint_S [(\nabla \times v_1\mathbf{i}) \cdot \mathbf{n}] \, d\sigma = \int_C v_1 \, dx$$

by showing that both integrals can be written

$$\iint_\Gamma \left[\frac{\partial v_1}{\partial u} \frac{\partial x}{\partial v} - \frac{\partial v_1}{\partial v} \frac{\partial x}{\partial u} \right] du \, dv.$$

A similar argument (no need to carry it out) equates the integrals for v_2 and v_3 and proves the theorem.

■ CHAPTER HIGHLIGHTS

17.4   Another Notation for Line Integrals; Line Integrals with Respect to Arc Length

$$\int_C P\,dx + Q\,dy + R\,dz \qquad \int_C f(\mathbf{r})\,ds$$

mass, center of mass, moment of inertia of a wire (pp. 1133–1134)

17.5   Green's Theorem

Jordan region (p. 1137)   Green's theorem for a Jordan region (p. 1137)
area enclosed by a plane curve (p. 1142)
Green's theorem for regions bounded by several Jordan curves (pp. 1143–1145)
a justification of the Jacobian area formula (supplement) (p. 1146)

17.6   Parametrized Surfaces; Surface Area

parametrized surface: $\mathbf{r} = \mathbf{r}(u, v), \quad (u, v) \in \Omega$
fundamental vector product: $\mathbf{N} = \mathbf{r}'_u \times \mathbf{r}'_v$
$\mathbf{N}$ as a normal (p. 1151)   area of a parametrized surface (p. 1152)
area of a surface $z = f(x, y)$ (p. 1155)   upper unit normal (p. 1157)
secant area formula (p. 1158)

17.7   Surface Integrals

mass of a material surface (p. 1159)   surface integral (p. 1160)
average value on a surface, weighted average (p. 1162)   centroid (p. 1162)
center of mass of a material surface (p. 1163)   moment of inertia (p. 1163)
smooth surface (p. 1166)   flux of a vector field (p. 1166)
closed, piecewise-smooth surface (p. 1168)
one sided surface, Möbius band (p. 1170)

17.8   The Vector Differential Operator ∇

the operator del: $\nabla = \dfrac{\partial}{\partial x}\mathbf{i} + \dfrac{\partial}{\partial y}\mathbf{j} + \dfrac{\partial}{\partial z}\mathbf{k}$ gradient of f: ∇f

divergence of $\mathbf{v}$: $\nabla \cdot \mathbf{v}$ curl of $\mathbf{v}$: $\nabla \times \mathbf{v}$
curl of a gradient is zero (p. 1173)   divergence of a curl is zero (p. 1173)
$\nabla \cdot (f\mathbf{v}) = (\nabla f) \cdot \mathbf{v} + f(\nabla \cdot \mathbf{v})$ $\nabla \times (f\mathbf{v}) = (\nabla f) \times \mathbf{v} + f(\nabla \times \mathbf{v})$
$\nabla \cdot (r^n \mathbf{r}) = (n + 3)r^n$ $\nabla \times (r^n \mathbf{r}) = \mathbf{0}$

Laplacian: $\nabla^2 f = \nabla \cdot \nabla f = \dfrac{\partial^2 f}{\partial x^2} + \dfrac{\partial^2 f}{\partial y^2} + \dfrac{\partial^2 f}{\partial z^2}$

17.9   The Divergence Theorem

divergence theorem for a solid bounded by a single closed surface (p. 1177)
divergence as outward flux per unit volume (p. 1179)   sinks and sources (p. 1180)
divergence theorem for solids bounded by two or more closed surfaces (p. 1180)
an application to static charges (p. 1180)
buoyant force, principle of Archimedes (p. 1182)

17.10   Stokes's Theorem

positive sense along a curve bounding an open surface (p. 1183)
Stokes's theorem (p. 1185)
conditions under which $\nabla \times \mathbf{v} = \mathbf{0}$ implies $\mathbf{v} = \nabla f$
normal component of $\nabla \times \mathbf{v}$ as circulation per unit area
irrotational flow (p. 1189)

■ PROJECTS AND EXPLORATIONS USING TECHNOLOGY

To do these exercises you will need a graphics calculator or a computer with graphing capability. The majority of these problems are open-ended so different approaches may be used to solve them. You should be aware that different approaches can result in slight variations in the answers. Round your numerical answers to at least four decimal places. The rounding method that your calculator or computer uses also may cause variations in answers.

17.1 This exercise uses technology to approximate line integrals.

   (a) A line integral is defined in terms of a single variable definite integral. Thus, any existing technology for approximating definite integrals can be used. Set up a procedure for manipulating the integrand of a line integral to use definite integral software. Test your procedure on the following examples from the text

     (i) $\displaystyle\int_C \mathbf{f}(\mathbf{r}) \cdot d\mathbf{r}$ where $\mathbf{f}(x, y) = y\,\mathbf{i} + x\,\mathbf{j}$; $\mathbf{r}(u) = u\,\mathbf{i} + u^2\,\mathbf{j}$, $u \in [0, 1]$.

     (ii) $\displaystyle\int_C \mathbf{f}(\mathbf{r}) \cdot d\mathbf{r}$ where $\mathbf{f}(x, y, z) = y^2 z^3\,\mathbf{i} + 2xyz^3\,\mathbf{j} + 3xy^2 z^2\,\mathbf{k}$;

$$\mathbf{r}(u) = u^2\,\mathbf{i} + u^4\,\mathbf{j} + u^6\,\mathbf{k}, \quad u \in [0, 1].$$

   (b) Prepare a trapezoidal rule procedure for approximating line integrals and test it on the two examples in part (a).

   (c) We have approximated definite integrals by averaging random function values over the domain (see, for example, Technology Exercise 2 at the end of Chapter 8). Prepare such a procedure for approximating line integrals and test it on the two examples in part (a).

   (d) Theorem 17.1.4 states that, under certain conditions, line integrals are independent of the parametrization used for the curve. Check this result by calculating the length of a spiral on a sphere of radius π using the following two parametrizations of the spiral.

     (i) $\mathbf{r}(u) = u \sin u\,\mathbf{i} + u \cos u\,\mathbf{j} + \sqrt{\pi^2 - u^2}\,\mathbf{k}$, $u \in [0, \pi]$,

     (ii) $\mathbf{R}(u) = u^2 \cos\left(\dfrac{\pi}{2} - u^2\right)\mathbf{i} + u^2 \sin\left(\dfrac{\pi}{2} - u^2\right)\mathbf{j} + \sqrt{\pi^2 - u^4}\,\mathbf{k}$, $u \in [0, \sqrt{\pi}]$.

    Use any one of the approximation methods developed above.

   (e) Using the paths in part (d), calculate

$$\int_C \mathbf{f}(\mathbf{r}) \cdot d\mathbf{r} \text{ and } \int_C \mathbf{f}(\mathbf{R}) \cdot d\mathbf{R}, \quad \text{where } \mathbf{f}(x, y, z) = x^3\,\mathbf{i} - xz\,\mathbf{j} + yz\,\mathbf{k}.$$

   (f) Approximating line integrals using random function values over the domain [part (c)] may seem to give different results for different parametrizations of the curve. Explain why an arc length parametrization should give the best result.

17.2 This exercise uses technology to approximate a surface integral.

   (a) Surface integrals can be expressed in terms of double integrals. If you have technology that evaluates double integrals, determine a method for entering a surface integral. Test your method on the following integral:

$$\iint_S H(x, y, z)\, d\sigma, \quad \text{where } H(x, y, z) + x^2 + y^3 + z^4 \quad \text{and}$$

$$S: \mathbf{r}(u, v) = \cos u \cosh v\,\mathbf{i} + \sin u \cosh v\,\mathbf{j} + \sinh v\,\mathbf{k}, \quad u \in [0, 2\pi], \quad v \in [0, 1].$$

(b) Exercise 17.1, part (b), involved a trapezoidal rule for approximating a line integral. Modify that procedure to allow surface integrals to be the input. Then test it on the integral in part (a).

(c) Prepare a procedure for approximating a surface integral using random points on the surface and test it on the integral in part (a). What parametrization of the surface do you think you should use?

(d) Use the procedures from this exercise and Exercise 17.1 to verify Green's theorem for the following problem: Find the work done by the force $\mathbf{F}(x, y) = (x^3 - x^2y)\,\mathbf{i} + xy^2\,\mathbf{j}$ in moving an object in a counterclockwise direction around the boundary of the region enclosed by the parabolas $y = x^2$ and $x = y^2$.

17.3 This exercise uses technology to simplify calculations with the vector differential operator ∇.

(a) Test the capabilities of your technology by calculating the divergence and curl of the vector functions

(i) $\mathbf{v}(x, y, z) = xyz\,\mathbf{i} + \ln(xyz)\,\mathbf{j} - \sin(2x + 3y - 4z)\,\mathbf{k}$,

(ii) $\mathbf{v}(x, y, z) = (x^2 + y^3 - 3z^7)\,\mathbf{i} - \tan^{-1}(x/y)\,\mathbf{j} + e^{x-z}\,\mathbf{k}$,

and the gradient of the function

$$f(x, y, z) = \cos(\sqrt{x^2 + 3\sin y + 5z^4}).$$

If you use symbolic manipulation software, use it on these functions directly. If your technology has only numeric capabilities, then evaluate your results at $x = 2, y = -3$, and $z = 5$.

(b) As you have seen, round off error is a problem in computation. Use the following identities to test the accuracy of the technology that you used in part (a):

$$\nabla \times (\nabla f) = 0 \quad \text{and} \quad \nabla \cdot (\nabla \times \mathbf{v}) = 0.$$

(c) Use the results of the preceding exercises to rework Exercise 13, Section 17.10: The upper half of the ellipsoid $\frac{1}{2}x^2 + \frac{1}{2}y^2 + z^2 = 1$ intersects the cylinder $x^2 + y^2 - y = 0$ in a curve C. Calculate the circulation of $\mathbf{v}(x, y, z) = y^3\,\mathbf{i} + (xy + 3xy^2)\,\mathbf{j} + z^4\,\mathbf{k}$ around C using Stokes's theorem.

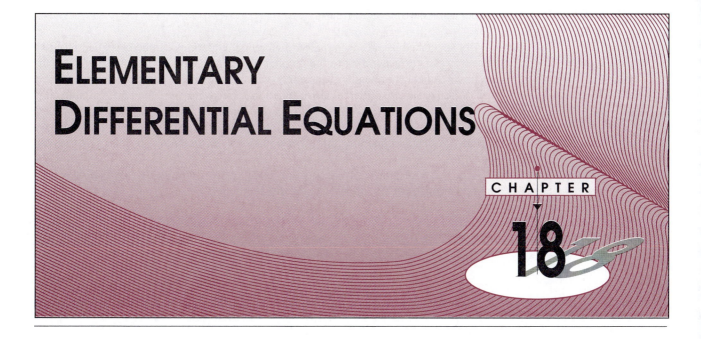

ELEMENTARY DIFFERENTIAL EQUATIONS

CHAPTER

18

■ 18.1 INTRODUCTION

From time to time throughout this text we have spoken about differential equations. In Chapter 7, for instance, we considered the differential equations which modeled exponential growth and decay, and saw that the only functions that satisfy the equation

$$y' = ky \qquad (k \text{ a constant})$$

have the form $y = Ce^{kx}$, where C is an arbitrary constant. Differential equations involving Newton's second law of motion

$$F = ma = m\frac{d^2r}{dt^2},$$

were studied in several different contexts, including the motion of freely falling bodies and planetary motion (see Sections 3.4 and 13.5).

Many problems in the physical and life sciences, engineering, and the social sciences involve "change" (change in size, shape, motion, and so forth). Since a derivative is a "rate of change," the mathematical formulations of such problems are usually described by differential equations. This chapter is intended as an elementary introduction to a study of differential equations. We begin with some terminology.

First, recall that a *differential equation* is an equation which contains an unknown function together with one or more of its derivatives. Some examples are

$$x\frac{dy}{dx} - y = e^x, \qquad y''' + xy'' - yy' = y \sin x,$$

$$\frac{\partial^2 u}{\partial x^2} + \frac{\partial^2 u}{\partial y^2} = 0 \qquad \text{(Laplace's equation, see Exercises 14.6).}$$

In the first two equations, x is the independent variable and $y = y(x)$ is the unknown function. In the third equation, x and y are independent variables and $u = u(x, y)$ is the unknown function.

As suggested by these examples, differential equations can be classified into two general types corresponding to the number of independent variables involved. If the unknown function depends on a single variable, then the equation is an *ordinary differential equation*; if the unknown function depends on more than one variable, then the equation is a *partial differential equation* since the derivatives that appear in the equation must be partial derivatives. Thus, for example, Laplace's equation is a partial differential equation; the other equations just given are ordinary differential equations.

The *order* of a differential equation is the order of the highest derivative (of the unknown function) that appears in the equation

$$y' + ay = b, \qquad y' = x^2 y^2, \qquad \frac{\partial u}{\partial x} = k \frac{\partial u}{\partial y}$$

are first-order equations;

$$x^2 y'' + xy' + x^2 y = 0, \qquad y'' + \omega^2 y = \sin 2x, \qquad \frac{\partial^2 u}{\partial x^2} - \frac{\partial^2 u}{\partial y^2} = 0$$

are equations of the second order; and

$$y''' + ay'' + by' + cy = d$$

is a third-order differential equation.

The obvious question that we want to consider in this chapter is that of "solving" a given differential equation. By a *solution* of a differential equation we mean a function f defined on some interval (in the case of an ordinary differential equation), or on some region in n-dimensional space, $n \geq 2$ (in the case of a partial differential equation), with the property that f and its derivatives satisfy the equation. To determine whether a given function satisfies a certain differential equation, we simply substitute the function and its derivatives into the equation. If the result is an identity, then the function is a solution; otherwise it is not.

Example 1 In each of the following, determine whether the given function satisfies the corresponding differential equation:

(a) $y(x) = x^2 - 5x$; $\quad x\dfrac{dy}{dx} - y = x^2$.

(b) $u(x, y) = (\sin x)(\sinh y)$; $\quad \dfrac{\partial^2 u}{\partial x^2} + \dfrac{\partial^2 u}{\partial y^2} = 0$.

(c) $y(x) = e^x - x^2$; $\quad (x - 1)y'' - xy' + y = (x - 1)^2$.

SOLUTION **(a)** Substituting $y(x) = x^2 - 5x$, $y'(x) = 2x - 5$ into the differential equation

$$x\frac{dy}{dx} - y = x^2,$$

we have

$$x(2x - 5) - (x^2 - 5x) \overset{?}{=} x^2,$$

$$2x^2 - 5x - x^2 + 5x \overset{?}{=} x^2,$$

$$x^2 = x^2.$$

Thus, $y(x) = x^2 - 5x$ is a solution of the differential equation. It can be verified in the same manner that $y = x^2 - Cx$ is a solution of the differential equation for *any* constant C.

(b) Let $u(x, y) = (\sin x)(\sinh y)$. Then

$$\frac{\partial u}{\partial x} = (\cos x)(\sinh y) \qquad \text{and} \qquad \frac{\partial u}{\partial y} = (\sin x)(\cosh y),$$

$$\frac{\partial^2 u}{\partial x^2} = -(\sin x)(\sinh y) \qquad \text{and} \qquad \frac{\partial^2 u}{\partial y^2} = (\sin x)(\sinh y).$$

Since $-(\sin x)(\sinh y) + (\sin x)(\sinh y) = 0$, it follows that $u(x, y) = (\sin x)(\sinh y)$ is a solution of the differential equation $\partial^2 u/\partial x^2 + \partial^2 u/\partial y^2 = 0$.

(c) Substituting $y(x) = e^x - x^2$ and its derivatives $y'(x) = e^x - 2x$, $y''(x) = e^x - 2$, into the differential equation

$$(x - 1)y'' - xy' + y = (x - 1)^2,$$

we have

$$(x - 1)(e^x - 2) - x(e^x - 2x) + (e^x - x^2) \stackrel{?}{=} (x - 1)^2,$$

$$xe^x - e^x - 2x + 2 - xe^x + 2x^2 + e^x - x^2 \stackrel{?}{=} (x - 1)^2,$$

$$x^2 - 2x + 2 \neq (x - 1)^2.$$

Therefore, $y(x) = e^x - x^2$ is not a solution of the differential equation. ❑

From your experience in calculus, you know that the study of functions of several variables (graphing, differentiation, integration) is more complicated than the corresponding study of functions of a single variable. You would probably suspect, therefore, that partial differential equations are typically more complicated than ordinary differential equations. This is indeed the case. Since the intent of this chapter is simply to introduce some of the basic theory and methods for differential equations, we shall confine ourselves to ordinary differential equations. Hereafter, the term *differential equation* will mean *ordinary differential equation.*

In Chapter 5 we saw that the set of all solutions of the first-order differential equation

(1) $$\frac{dy}{dx} = f(x)$$

is given by $y(x) = F(x) + C$, where F is an antiderivative of f and C is an arbitrary constant. Also, as we noted at the beginning of this section, the set of functions $y(x) = Ce^{kx}$, C an arbitrary constant, represents all solutions of

(2) $$y' = ky \qquad (k \text{ a constant}).$$

The set of solutions of the second-order differential equation

(3) $$\frac{d^2y}{dt^2} = -32$$

governing the motion of a freely falling body is given by

$$y(t) = -16t^2 + C_1 t + C_2,$$

where C_1 and C_2 are independent arbitrary constants (in Section 3.4 these constants were denoted v_0, the initial velocity, and y_0, the initial height). The arbitrary constants involved in these sets of solutions are often called *parameters*; Equations (1) and (2) are said to have *one-parameter families of solutions* while Equation (3) has a *two-parameter family of solutions*.

Intuitively, to find a set of solutions of an *n*th-order differential equation, we would expect to integrate *n* times, with each integration step producing (independently) an arbitrary constant. Thus, we would expect an *n*th-order equation to have an *n-parameter family of solutions*. To *solve* a differential equation of order *n* means to find an *n*-parameter family of solutions. In this chapter we study equations of orders one and two, and so we will be looking for one- and two-parameter families of solutions.

In the examples just given, the families of solutions actually included *all* solutions of the corresponding differential equations. In such cases, the term *general solution* is often used in place of *n-parameter family of solutions*. We comment further on this in the sections which follow.

If specific values are assigned to each of the arbitrary constants in an *n*-parameter family of solutions, then the resulting solution is called a *particular solution*. For example, the function

(4) $$y(t) = -16t^2 + 40t + 144$$

is a particular solution of Equation (3) (it represents the path of an object which is projected upward with an initial velocity of 40 ft/sec from an initial height of 144 ft). Specific values for the constants are usually determined by imposing additional conditions, called *side conditions* or *initial conditions*. The function given in (4) is the solution of (3) which satisfies the side conditions

$$y(0) = 144, \qquad y'(0) = 40.$$

Example 2 Show that $y(x) = -\frac{3}{2} + Ce^{x^2}$ is a one-parameter family of solutions of

$$\frac{dy}{dx} - 2xy = 3x$$

and determine a solution which satisfies the side condition $y(1) = 0$.

SOLUTION First, we substitute

$$y = -\frac{3}{2} + Ce^{x^2}, \qquad y' = 2Cxe^{x^2}$$

into the differential equation:

$$2Cxe^{x^2} - 2x\left[-\frac{3}{2} + Ce^{x^2} \right] = 2Cxe^{x^2} + 3x - 2Cxe^{x^2} = 3x.$$

Thus, $y(x) = -\frac{3}{2} + Ce^{x^2}$ is a one-parameter family of solutions. Imposing the side condition $y(1) = 0$, we have

$$y(1) = -\frac{3}{2} + Ce = 0,$$

which gives

$$C = \frac{3}{2e}.$$

Therefore, $y(x) = -\dfrac{3}{2} + \dfrac{3}{2e}e^{x^2}$ is a particular solution which satisfies the given side condition. ❏

Example 3   Show that $y = C_1e^{2x} + C_2e^{-x}$ is a two-parameter family of solutions of

$$y'' - y' - 2y = 0$$

and determine a particular solution that satisfies the side conditions

$$y(0) = 2, \quad y'(0) = 1.$$

SOLUTION   Substituting

$$y = C_1e^{2x} + C_2e^{-x}, \qquad y' = 2C_1e^{2x} - C_2e^{-x}, \qquad y'' = 4C_1e^{2x} + C_2e^{-x}$$

into the differential equation, we have

$$4C_1e^{2x} + C_2e^{-x} - (2C_1e^{2x} - C_2e^{-x}) - 2(C_1e^{2x} + C_2e^{-x})$$
$$= (4C_1 - 2C_1 - 2C_1)e^{2x} + (C_2 + C_2 - 2C_2)e^{-x} = 0,$$

which shows that $y(x) = C_1e^{2x} + C_2e^{-x}$ is a two-parameter family of solutions. Now, imposing the side conditions leads to the pair of equations

$$y(0) = C_1 + C_2 = 2$$
$$y'(0) = 2C_1 - C_2 = 1.$$

This system has the solution $C_1 = 1$, $C_2 = 1$. Thus $y(x) = e^{2x} + e^{-x}$ is a particular solution that satisfies the side conditions. ❏

Remark   Since an nth order differential equation will have an n-parameter family of solutions, we will need to impose n side conditions to evaluate the n arbitrary constants. Typically, the n side conditions lead to a system of n equations in the n "unknowns" $C_1, C_2, \ldots, C_n$. ❏

EXERCISES 18.1

In Exercises 1–8, classify the given differential equation with respect to type (ordinary or partial) and give the order.

1. $xy' - y = e^{2x}$.

2. $(y')^3 + xy'' = x^2 - 1$.

3. $\dfrac{\partial u}{\partial x} = k\dfrac{\partial u}{\partial y}$, k constant.

4. $y''' + xy' + (\sin x)y = \cos x$.

5. $y'' + e^{xy} = \tan x$.

6. $\alpha^2\dfrac{\partial^2 u}{\partial x^2} = \dfrac{\partial u}{\partial t}$, α constant.

7. $\dfrac{\partial^2 u}{\partial x^2} + \dfrac{\partial^2 u}{\partial y^2} = 0$.

8. $y^{(4)} + 3y''' + 3y'' + y' = xe^{-x}$.

In Exercises 9–18, determine whether the given differential equation has the indicated functions as solutions.

9. $2y' - y = 0$; $y_1(x) = e^{x/2}$, $y_2(x) = x^2 + 2e^{x/2}$.

10. $y' + xy = x$; $y_1(x) = e^{-x^2/2}$, $y_2(x) = 1 + Ce^{-x^2/2}$, C any constant.

11. $y' + y = y^2$; $y_1(x) = \dfrac{1}{e^x + 1}$, $y_2(x) = \dfrac{1}{Ce^x + 1}$, C any constant.

12. $y'' + 4y = 0$; $y_1(x) = 2\sin 2x$, $y_2(x) = 2\cos x$.

13. $y'' - 4y = 0$; $y_1(x) = e^{2x}$, $y_2(x) = C\sinh 2x$, C any constant.

14. $y'' - 2y' - 3y = 7e^{3x}$; $y_1(x) = e^{-x} + 2e^{3x}$, $y_2(x) = \tfrac{7}{4}xe^{3x}$.

15. $a^2\dfrac{\partial^2 u}{\partial x^2} = \dfrac{\partial^2 u}{\partial t^2}$, $u_1(x, t) = \cos \lambda x \sin \lambda at$, λ any constant, $u_2(x, t) = \sin(x - at)$.

16. $\dfrac{\partial^2 u}{\partial x^2} + \dfrac{\partial^2 u}{\partial y^2} = 0$; $u_1(x, y) = \sin x \cosh y$, $u_2(x, y) = \ln (x^2 + y^2)$.

17. $4x^2y'' - 12xy' - 9y = 0$; $y_1(x) = x^{-1/2}$,
$y_2(x) = C_1 x^{-1/2} + C_2 x^{9/2}$.

18. $a^2 \dfrac{\partial^2 u}{\partial x^2} = \dfrac{\partial u}{\partial t}$, $u_1(x, t) = e^{-a^2 t} \sin x$,
$u_2(x, t) = e^{-a^2 t} \sin \lambda x$.

In Exercises 19–26, an n-parameter family of functions is given. Show that the members of the family are solutions of the differential equation and find a member of the family that satisfies the side conditions.

19. $y = Ce^{5x}$; $y' = 5y$, $y(0) = 2$.

20. $y = \dfrac{x^2}{3} + \dfrac{C}{x}$; $xy' + y = x^2$, $y(3) = 2$.

21. $y = \dfrac{1}{Ce^x + 1}$; $y' + y = y^2$, $y(1) = -1$.

22. $y = x \ln \dfrac{C}{x}$; $y' = \dfrac{y - x}{x}$, $y(2) = 4$.

23. $y = C_1 x + C_2 x^{1/2}$; $2x^2 y'' - xy' + y = 0$,
$y(4) = 1$, $y'(4) = -2$.

24. $y = C_1 \sin 3x + C_2 \cos 3x$; $y'' + 9y = 0$,
$y(\pi/2) = y'(\pi/2) = 1$.

25. $y = C_1 x^2 + C_2 x^2 \ln x$; $x^2 y'' - 3xy' + 4y = 0$,
$y(1) = 0$, $y'(1) = 1$.

26. $y = C_1 + C_2 e^x + C_3 e^{2x} + \frac{1}{4}x^2 + \frac{3}{4}x - xe^x$;
$y''' - 3y'' + 2y' = x + e^x$,
$y(0) = 1$, $y'(0) = -\frac{1}{4}$, $y''(0) = -\frac{3}{2}$.

In Exercises 27–30, determine the values of r, if any, such that $y = e^{rx}$ is a solution of the given differential equation.

27. $y' + 3y = 0$.

28. $y'' - 5y' + 6y = 0$.

29. $y'' + 6y' + 9y = 0$.

30. $y''' - 3y' + 2y = 0$.

In Exercises 31–34, determine values of r, if any, such that $y = x^r$ is a solution of the given differential equation.

31. $xy'' + y' = 0$.

32. $x^2 y'' + xy' - y = 0$.

33. $4x^2 y'' - 4xy' + 3y = 0$.

34. $x^3 y''' - 2x^2 y'' - 2xy' + 8y = 0$.

An nth-order differential equation together with conditions that are specified at two, or more, points on an interval I is usually called a *boundary-value problem*. In particular, if conditions are specified at two points, the problem is called a *two-point boundary-value problem*.

35. Each member of the two-parameter family $y = C_1 \sin 4x + C_2 \cos 4x$ is a solution of the differential equation $y'' + 16y = 0$.

(a) Find all the members of this family that satisfy the boundary conditions:

$$y(0) = 0, \quad y(\pi/2) = 0.$$

(b) Find all the members of this family that satisfy the boundary conditions:

$$y(0) = 0, \quad y(\pi/8) = 0.$$

36. For each real number r, each member of the two-parameter family $y = C_1 \sin rx + C_2 \cos rx$ is a solution of the differential equation $y'' + r^2 y = 0$.

(a) Determine the numbers r such that the two-point boundary-value problem

$$y'' + r^2 y = 0, \quad y(0) = 0, \quad y(\pi) = 0$$

has a nonzero solution.

(b) Determine the numbers r such that the two-point boundary-value problem

$$y'' + r^2 y = 0, \quad y(0) = 0, \quad y(\pi/2) = 0$$

has a nonzero solution.

■ 18.2 FIRST ORDER LINEAR DIFFERENTIAL EQUATIONS

A first-order differential equation is *linear* if it can be written in the form

(18.2.1)
$$y' + p(x)y = q(x),$$

where p and q are continuous functions on some interval I.

To solve (18.2.1), we first calculate

$$H(x) = \int p(x)\, dx$$

(let the constant of integration equal 0) and form $e^{H(x)}$. Note that

$$\frac{d}{dx}[e^{H(x)}] = e^{H(x)} \frac{d}{dx}[H(x)] = p(x)\, e^{H(x)}.$$

Now multiply the differential equation by $e^{H(x)}$ to obtain the equivalent equation (since $e^{H(x)} \neq 0$)

$$e^{H(x)}y' + e^{H(x)}p(x)y = e^{H(x)}q(x).$$

The left side of this equation has the form

$$\frac{d}{dx}[e^{H(x)}y],$$

and this was our motivation for multiplying by $e^{H(x)}$; it converted the left-hand side of the equation into the derivative of a product. Now the equation can be written

$$\frac{d}{dx}[e^{H(x)}y] = e^{H(x)}q(x),$$

and integration gives

$$e^{H(x)}y = \int e^{H(x)}q(x)\, dx + C.$$

Thus

(18.2.2)

$$y = e^{-H(x)} \left\{ \int e^{H(x)}q(x)\, dx + C \right\}$$

is a one-parameter family of solutions. In Exercise 37 you are asked to show that *every* solution can be written in this form, and so (18.2.2) is actually the *general solution* of Equation (18.2.1). Different choices of C give different *particular solutions*.

Remark The key step in solving the first-order linear differential equation

$$y' + p(x)y = q(x)$$

is multiplication by $e^{H(x)}$, where

$$H(x) = \int p(x)\, dx.$$

It is multiplication by $e^{H(x)}$ that enables us to write the left side of the equation in a form that we can integrate directly. The expression $e^{H(x)}$ is called an *integrating factor*. Integrating factors are discussed again in Section 18.4. ❏

Example 1   Find the general solution of the linear differential equation

$$y' + ay = b, \qquad a, b \text{ constants and } a \neq 0.$$

SOLUTION   Here $p(x) = a$ and $q(x) = b$ are continuous for all x, and so we can take the interval I to be $(-\infty, \infty)$. An integrating factor is

$$e^{\int a \, dx} = e^{ax}.$$

Multiplication by e^{ax} gives

$$e^{ax} y' + e^{ax} ay = e^{ax} b.$$

Since the left-hand side is the derivative of $e^{ax} y$, we have

$$\frac{d}{dx}[e^{ax} y] = e^{ax} b.$$

Integrating this equation, we get

$$e^{ax} y = \frac{b}{a} e^{ax} + C$$

and so

$$y = \frac{b}{a} + Ce^{-ax}. \quad \square$$

Example 2   Find the general solution of the linear differential equation

$$y' + 2xy = x$$

and determine a particular solution that satisfies the side condition $y(0) = 2$.

SOLUTION   The functions $p(x) = 2x$ and $q(x) = x$ are continuous for all x, and so we can take $I = (-\infty, \infty)$. An integrating factor is

$$e^{\int (2x) \, dx} = e^{x^2}.$$

Multiplication by e^{x^2} gives

$$e^{x^2} y' + 2xe^{x^2} y = xe^{x^2},$$

which is the same as

$$\frac{d}{dx}[e^{x^2} y] = xe^{x^2}.$$

Integrating both sides of this equation, we get

$$e^{x^2} y = \tfrac{1}{2} e^{x^2} + C.$$

Thus, the general solution of the equation is

$$y = \tfrac{1}{2} + Ce^{-x^2}.$$

To determine a solution that satisfies the given side condition, we set $x = 0$ and $y = 2$ in the general solution and solve for C:

$$2 = \tfrac{1}{2} + C \qquad \text{and so} \qquad C = \tfrac{3}{2}.$$

Therefore

$$y = \tfrac{1}{2} + \tfrac{3}{2} e^{-x^2}$$

is a particular solution that satisfies $y(0) = 2$. $\quad \square$

Remark   Note that the particular solution in Example 2 was uniquely determined by the side condition $y(0) = 2$.   In Exercise 42 you are asked to show that this situation holds in general.   That is, if p and q are continuous functions on the interval I, and if $x_0 \in I$ and y_0 is any real number, then the linear differential equation

$$y' + p(x)y = q(x) \qquad \text{with the side condition} \qquad y(x_0) = y_0$$

has a unique solution.   ❏

Example 3   Find the general solution of

$$xy' - 2y = 2x^2 + x.$$

SOLUTION   This equation is not in the form of (18.2.1), but it can be put in that form by dividing through by x:

$$y' - \frac{2}{x}y = 2x + 1.$$

The functions $p(x) = -2/x$ and $q(x) = 2x + 1$ are continuous on $(0, \infty)$ and on $(-\infty, 0)$.   We take $I = (0, \infty)$.   Now,

$$H(x) = \int -\frac{2}{x}dx = -2 \ln x = \ln x^{-2},$$

and so

$$e^{\ln x^{-2}} = x^{-2}$$

is an integrating factor.   Multiplication by x^{-2} gives

$$x^{-2}y' - 2x^{-3}y = 2x^{-1} + x^{-2}.$$

Therefore,

$$\frac{d}{dx}[x^{-2}y] = 2x^{-1} + x^{-2},$$

and

$$x^{-2}y = 2 \ln x - x^{-1} + C.$$

Solving for y, we get

$$y = 2x^2 \ln x - x + Cx^2, \qquad (x > 0)$$

and this is the general solution.   ❏

Applications

Example 4   A new company is being formed.   A market analysis projects that the rate of growth of the company's income at any future time t will be proportional to the difference between the actual income at that time and an upper limit of $10 million. Further, it is estimated that the income after 3 years will be $4 million; the income is $0 initially.   Find an expression for the company's income at any time t, and determine how long it will take for the income to reach $8 million.

SOLUTION   Let $A(t)$ denote the company's income at time t.   According to the market analysis, the rate of growth of A is proportional to the difference between A and $10 million.   That is,

$$\frac{dA}{dt} = k(10 - A),$$

where k is the constant of proportionality and A is measured in millions. Rewriting this equation, we have

$$\frac{dA}{dt} + kA = 10k,$$

a first-order linear equation in which $p(x) = k$ and $q(x) = 10k$ are constants. By the result in Example 1,

$$A(t) = \frac{10k}{k} + Ce^{-kt} = 10 + Ce^{-kt}.$$

Since $A(0) = 0$, we have

$$0 = 10 + Ce^0 = 10 + C \qquad \text{and} \qquad C = -10.$$

Therefore,

$$A(t) = 10 - 10\,e^{-kt}.$$

To determine k, we use the fact that $A = 4$ when $t = 3$:

$$4 = 10 - 10e^{-3k},$$

$$e^{-3k} = \tfrac{3}{5} = 0.6,$$

$$-3k = \ln(0.6) \cong -0.511,$$

$$k \cong 0.17.$$

Thus, the company's income at any time t is projected to be (approximately)

$$A(t) = 10 - 10\,e^{-0.17t}.$$

To find out how long it will take for the income to reach \$8 million, we solve

$$8 = 10 - 10e^{-0.17t}$$

for t:

$$e^{-0.17t} = \tfrac{1}{5} = 0.2,$$

$$-0.17t = \ln(0.2) \cong -1.609,$$

$$t \cong 9.47.$$

Thus, it will take approximately $9\tfrac{1}{2}$ years for the company's income to reach \$8 million. ❑

Example 5 A chemical manufacturing company has a 1000-gallon holding tank which it uses to control the release of pollutants into a sewage system. Initially the tank has 360 gallons of water containing 2 pounds of pollutant per gallon. Water containing 3 pounds of pollutant per gallon enters the tank at the rate of 80 gallons per hour and is uniformly mixed with the water already in the tank. Simultaneously, water is released from the tank at the rate of 40 gallons per hour. Determine how many pounds of pollutant are in the tank at any time t. Also, determine the rate (lbs/gal) at which the pollutant is being released after 10 hours.

SOLUTION   Let $P(t)$ be the amount of pollutant (in pounds) in the tank at time t. The rate of change of pollutant in the tank, dP/dt, is given by

$$\frac{dP}{dt} = (\text{rate in}) - (\text{rate out}).$$

The pollutant is entering the tank at the rate of $3 \times 80 = 240$ pounds per hour ($=$ rate in).

Water is entering the tank at the rate of 80 gallons per hour and is leaving at the rate of 40 gallons per hour. Therefore the amount of water in the tank is increasing at the rate of 40 gallons per hour, and so there are $360 + 40t$ gallons of water in the tank at time t, where $0 \le t \le 16$ (the tank will be full at time $t = 16$ hours). We can now conclude that the amount of pollutant per gallon in the tank at time t is

$$\frac{P(t)}{360 + 40t},$$

and the rate at which pollutant is leaving the tank is

$$\frac{40P(t)}{360 + 40t} = \frac{P(t)}{9 + t}(= \text{rate out}).$$

Therefore, our differential equation is

$$\frac{dP}{dt} = 240 - \frac{P}{9 + t}$$

or

$$\frac{dP}{dt} + \frac{1}{9 + t}P = 240,$$

a first-order, linear differential equation. Here we have

$$p(t) = \frac{1}{9 + t} \quad \text{and} \quad H(t) = \int \frac{1}{9 + t}dt = \ln|9 + t| = \ln(9 + t) \quad (9 + t > 0).$$

Thus, an integrating factor is

$$e^{H(t)} = e^{\ln(9 + t)} = 9 + t.$$

Multiplying the equation by $9 + t$ gives

$$(9 + t)\frac{dP}{dt} + P = 240(9 + t),$$

$$\frac{d}{dt}[(9 + t)P] = 240(9 + t),$$

$$(9 + t)P = 120(9 + t)^2 + C,$$

and

$$P(t) = 120(9 + t) + \frac{C}{9 + t}.$$

Since the amount of pollutant in the tank initially is $2 \times 360 = 720$ (pounds), we have

$$P(0) = 120(9) + \frac{C}{9} = 720, \quad \text{which implies} \quad C = -3240.$$

Thus, the amount of pollutant in the tank at any time t is

$$P(t) = 120(9 + t) - \frac{3240}{9 + t} \quad \text{(pounds)}.$$

After 10 hours there are $360 + 40(10) = 760$ gallons of water in the tank, and there are

$$P(10) = 120(19) - \frac{3240}{19} \cong 2109$$

pounds of pollutant. Therefore, the rate at which pollutant is being released into the sewage system after 10 hours is

$$\frac{2109}{760} \cong 2.78 \quad \text{(pounds per gallon)}. \quad \square$$

EXERCISES 18.2

In Exercises 1–16, find the general solution.

1. $y' - 2y = 1.$

2. $xy' - 2y = -x.$

3. $2y' + 5y = 2.$

4. $y' - y = -2e^{-x}.$

5. $y' - 2y = 1 - 2x.$

6. $xy' - 2y = -3x.$

7. $xy' - 4y = -2nx.$

8. $y' + y = 2 + 2x.$

9. $y' - e^x y = 0.$

10. $y' - y = e^x.$

11. $(1 + e^x)y' + y = 1.$

12. $xy' + y = (1 + x)e^x.$

13. $y' + 2xy = xe^{-x^2}.$

14. $xy' - y = 2x \ln x.$

15. $y' + \dfrac{2}{x + 1} y = 0.$

16. $y' + \dfrac{2}{x + 1} y = (x + 1)^{5/2}.$

In Exercises 17–24, find the particular solution determined by the given side condition.

17. $y' + y = x, \quad y(0) = 1.$

18. $y' - y = e^{2x}, \quad y(1) = 1.$

19. $y' + y = \dfrac{1}{1 + e^x}, \quad y(0) = e.$

20. $y' + y = \dfrac{1}{1 + 2e^x}, \quad y(0) = e.$

21. $xy' - 2y = x^3 e^x, \quad y(1) = 0.$

22. $x^2 y' + 2xy = 1, \quad y(1) = 2.$

23. $y' + 2xy = 0, \quad y(x_0) = y_0.$

24. $xy' + 2y = xe^{-x}, \quad y(1) = -1.$

25. A 200-liter tank initially full of water develops a leak at the bottom. Given that 20% of the water leaks out in the first 5 minutes, find the amount of water left in the tank t minutes after the leak develops

(a) if the water drains off at a rate that is proportional to the amount of water present;

(b) if the water drains off at a rate that is proportional to the product of the time elapsed and the amount of water present.

26. An object falling in air is subject not only to gravitational force but also to air resistance. Find $v(t)$, the velocity of the object at time t, given that

$$v'(t) + Kv(t) = 32 \quad \text{and} \quad v(0) = 0 \quad \text{(take } K > 0\text{)}.$$

Show that $v(t)$ cannot exceed $32/K$ ($32/K$ is called the *terminal velocity*).

27. At a certain moment a 100-gallon mixing tank is full of brine containing 0.25 pounds of salt per gallon. Find the amount of salt present t minutes later if the brine is being continuously drawn off at the rate of 3 gallons per minute and replaced by brine containing 0.2 pound of salt per gallon.

28. The current i in an electric circuit varies with time t according to the formula

$$L\frac{di}{dt} + Ri = E$$

where E (the voltage), L (the inductance), and R (the resistance) are positive constants. Measure time in seconds, current in amperes, and suppose that the initial current is 0.

(a) Find a formula for the current at each subsequent time t.

(b) What upper limit does the current approach as t increases?

(c) In how many seconds will the current reach 90% of its upper limit?

29. *Newton's Law of Cooling* states that the rate of change of the temperature T of an object is proportional to the difference between T and the temperature τ of the surrounding medium:

$$\frac{dT}{dt} = -k(T - \tau)$$

where k is a positive constant.

(a) Solve this equation for T. (Find the general solution.)

(b) Solve this equation for T given that $T(0) = T_0$.

(c) Take $T_0 > \tau$. What is the limiting temperature to which the object cools as t increases? What happens if $T_0 < \tau$?

(d) A cup of coffee is served to you at $185°F$ in a room where the temperature is $65°F$. Two minutes later the temperature of the coffee has dropped to $155°F$. How many more minutes would you expect to wait for the coffee to cool to $105°F$?

30. A drug is fed intravenously into a patient's blood stream at the constant rate r. Simultaneously, the drug diffuses into the patient's body at a rate proportional to the amount of the drug present.

(a) Determine the differential equation that describes the amount $Q(t)$ of the drug in the patient's bloodstream at time t.

(b) Determine the solution $Q = Q(t)$ of the differential equation from part (a) that satisfies the initial condition $Q(0) = 0$.

(c) Find $\lim_{t \to \infty} Q(t)$.

31. An advertising company is trying to expose a new product in a certain metropolitan area by advertising on television. Suppose that the rate of exposure to new people is proportional to the number of people who have not seen the product out of a total population of M viewers. Let $P(t)$ denote the number of viewers who have been exposed to the product at time t. The company has determined that no one was aware of the product at the start of the advertising campaign ($P(0) = 0$) and that 30% of the viewers were aware of the product after 10 days.

(a) Determine the differential equation that describes the number of viewers who are aware of the product at time t.

(b) Determine the solution of the differential equation from part (a) that satisfies the side condition $P(0) = 0$.

(c) How long will it take for 90% of the population to be aware of the product.

32. Let k be an integer different from 0 and 1. Transform *Bernoulli's equation*

$$y' + a(x)y = b(x)y^k$$

into the linear equation

$$v' + (1 - k)a(x)v = (1 - k)b(x)$$

by setting $v = y^{1-k}$.

In Exercises 33–36, use the result in Exercise 32 to solve the given differential equation.

33. $y' + y = 3e^x y^3$.

34. $y' - y = xy^2$.

35. $y' + \dfrac{y}{x} = x^3 y^2$.

36. $y' + xy = \dfrac{x}{y}$.

37. Show that if $y = y(x)$ is a solution of the differential equation (18.2.1), then y is a member of the one-parameter family (18.2.2). Thus, the one-parameter family of solutions (18.2.2) is the general solution of (18.2.1). HINT: If $y(x)$ is a solution of (18.2.1), then

$$y'(x) + p(x)y(x) = q(x).$$

Multiply by $e^{H(x)}$, where $H(x) = \int p(x)\,dx$, and continue as in the derivation of (18.2.2).

First Order, Linear, Homogeneous Differential Equations

If the function q in equation (18.2.1) is identically zero on I, then the equation is said to be *homogeneous*; if q is not identically zero on I, then equation (18.2.1) is said to be *nonhomogeneous*. The first order, linear homogeneous differential equation is:

(18.2.3) $\qquad\qquad y' + p(x)y = 0$

where p is continuous on I.

38. Show that the one-parameter family of solutions of equation (18.2.3) is given by

$$y(x) = Ce^{-H(x)}, \qquad \text{where} \qquad H(x) = \int p(x)\,dx.$$

39. Let $y = y_1(x)$ and $y = y_2(x)$ be any two solutions of (18.2.3), and let c be any real number.

(a) Prove that $y = y_1(x) + y_2(x)$ is a solution of (18.2.3).

(b) Prove that $y = cy_1(x)$ is a solution of (18.2.3).

40. Suppose that the coefficient function p in equation (18.2.3) is continuously differentiable. Show that each solution of (18.2.3) has a continuous second derivative.

41. (a) Prove that if $y = y(x)$ is a solution of equation (18.2.3) and $y(a) = 0$ for some $a \in I$, then $y(x) = 0$ for all $x \in I$.

(b) Suppose that $y = y_1(x)$ and $y = y_2(x)$ are solutions of (18.2.3). Prove that if $y_1(a) = y_2(a)$ at some $a \in I$, then $y_1(x) = y_2(x)$ for all $x \in I$.

42. Let $y = u_1(x)$ and $y = u_2(x)$ be solutions of equation (18.2.1).

(a) Prove that $y = u_1(x) - u_2(x)$ is a solution of the homogeneous equation (18.2.3).

(b) Prove that if $u_1(a) = u_2(a)$ at some $a \in I$, then $u_1(x) = u_2(x)$ for all $x \in I$. Note that this result implies that equation (18.2.1) with a side condition $y(x_0) = y_0$ has a unique solution.

43. Consider the linear homogeneous equation $y' + ry = 0$, where r is a real number, and let $y = y(x)$ be any solution.

(a) Show that if $r > 0$, then $\lim\limits_{x\to\infty} y(x) = 0$.

(b) Show that if $r < 0$, then the solution $y = y(x)$ is unbounded.

44. Suppose that the coefficient function p in equation (18.2.3) is positive on $[0, \infty)$.

(a) Show that all solutions of (18.2.3) are bounded on $[0, \infty)$.

(b) Under what conditions will every solution $y = y(x)$ have the property $\lim\limits_{x\to\infty} y(x) = 0$?

■ 18.3 SEPARABLE EQUATIONS; HOMOGENEOUS EQUATIONS

A first-order differential equation can always be written in the form

$$(18.3.1) \qquad F(x, y, y') = 0$$

by gathering all of the terms on the left side of the equation. For example, the first-order linear equation (18.2.1) written in this form is

$$y' + p(x)y - q(x) = 0.$$

Of course, in order for (18.3.1) to be a differential equation, y' must appear among the terms of F. We assume, in addition, that the equation can be solved for y'. This means that the equation can be expressed as

$$(18.3.2) \qquad P(x, y) + Q(x, y)y' = 0.$$

It will also be convenient to write this equation in the equivalent differential form

$$P(x, y)\, dx + Q(x, y)\, dy = 0.$$

The first-order equation (18.3.2) is *separable* iff it can be put in the form

$$(18.3.3) \qquad p(x) + q(y)y' = 0$$

or in the equivalent differential form

$$(18.3.4) \qquad p(x)\, dx + q(y)\, dy = 0.$$

Separable equations were introduced in Section 7.6. We provide the details of the solution method here.

To solve such an equation, we begin by writing (18.3.3) as

$$q(y)y' = -p(x).$$

This separates the variables: y's on the left, x's on the right. If p and q are continuous, they have antiderivatives. Setting

$$P(x) = \int p(x)\, dx \quad \text{and} \quad Q(y) = \int q(y)\, dy,$$

we have

$$Q'(y)y' = -P'(x).$$

Since, by the chain rule,

$$Q'(y)y' = \frac{d}{dx}[Q(y)],$$

we have

$$\frac{d}{dx}[Q(y)] = -P'(x).$$

Integration with respect to x gives

$$Q(y) = -P(x) + C.$$

Back in terms of p and q, we have

$$\int q(y)\, dy = -\int p(x)\, dx + C,$$

which is usually written

(18.3.5)
$$\int p(x)\, dx + \int q(y)\, dy = C.$$

This is a one-parameter family of solutions of (18.3.3). It may or may not be the general solution of the equation, as we illustrate below. Now that we have established the form of the solutions, note that (18.3.5) can be obtained simply by integrating the differential form (18.3.4) term-by-term.

If

$$Q(y) = \int q(y)\, dy$$

is one-to-one, then Q has an inverse and (18.3.5) can be solved for y:

$$y = Q^{-1}\left(C - \int p(x)\, dx\right).$$

Usually this is not the case and we have to be satisfied with (18.3.5) as it stands. For each choice of C, (18.3.5) gives a curve in the xy-plane. Such a curve is called an *integral curve* of the differential equation. An integral curve is a *solution* of the differential equation in the sense that at each point (x, y) of the curve, the slope y' is related to x and y by the equation

$$p(x) + q(y)y' = 0.$$

Example 1 The differential equation

$$x + yy' = 0$$

is separable. We can solve the equation (that is, we can obtain the integral curves) by writing the equation in differential form and integrating term-by-term:

$$x\,dx + y\,dy = 0,$$

$$\int x\,dx + \int y\,dy = C.$$

Carrying out the integrations, we have

$$\tfrac{1}{2}x^2 + \tfrac{1}{2}y^2 = C,$$

and thus

$$x^2 + y^2 = 2C.$$

Since C is an arbitrary constant, $2C$ is also arbitrary, and so we simply write the one-parameter family of solutions as

$$x^2 + y^2 = C.$$

This equation defines y implicitly as a function of x; the integral curves are circles centered at the origin. See Figure 18.3.1.

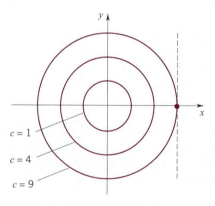

Figure 18.3.1

We can find the integral curve that passes through a specified point by adjusting the constant C. The integral curve that passes through the point (x_0, y_0) is the circle

$$x^2 + y^2 = x_0^2 + y_0^2.$$

It can be obtained by setting

$$C = x_0^2 + y_0^2. \quad \square$$

Let's return briefly to the general form

$$p(x) + q(y)y' = 0.$$

At a point (x, y) where $q(y) = 0$ but $p(x) \neq 0$, y' is infinite, and thus the tangent line, if it exists, must be vertical. In the case of

$$x + yy' = 0$$

there are two points on each circle where $y = 0$ and $x \neq 0$ (see Figure 18.3.1). At each of these points the tangent line is vertical.

Example 2 Show that the differential equation

$$y' = -2x(y - 1)^2$$

is separable and find a one-parameter family of solutions.

SOLUTION We try to write the equation in the form of (18.3.3):

$$2x(y - 1)^2 + y' = 0$$

and for $y \neq 1$,

$$2x + \frac{1}{(y - 1)^2} y' = 0.$$

Thus, the equation is separable. To solve, we write the equation in differential form and integrate term-by-term:

$$2x \, dx + \frac{1}{(y - 1)^2} \, dy = 0,$$

$$\int 2x \, dx + \int (y - 1)^{-2} \, dy = C,$$

and

$$x^2 - (y - 1)^{-1} = C$$

is a one-parameter family of solutions. In this case, we can solve for y explicitly:

$$\frac{1}{(y - 1)} = x^2 - C \qquad \text{and} \qquad y = 1 + \frac{1}{x^2 - C}.$$

Since y cannot be 1, no integral curve can cross the line $y = 1$. We can, however, take any point (x_0, y_0) with $y_0 \neq 1$ and adjust the constant C so that the curve passes through that point: the integral curve

$$y = 1 + \frac{1}{x^2 - C}$$

passes through the point (x_0, y_0) iff $C = x_0^2 - (y_0 - 1)^{-1}$. Representative graphs of the integral curves are shown in Figure 18.3.2. ❑

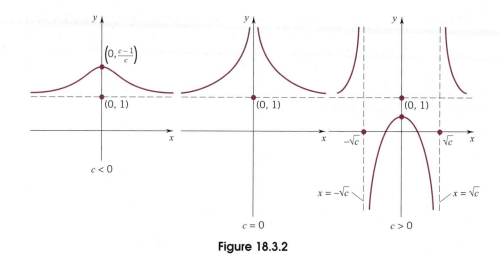

Figure 18.3.2

Remark   It is easy to verify that the constant function $y = 1$ is a solution of the differential equation in Example 2 which is not a member of the one-parameter family of solutions (we had to exclude this function when we divided through by $(y - 1)^2$). Thus, the one-parameter family is not the *general solution* of the equation.   A solution of a first-order differential equation which is not included in a one-parameter family of solutions is called a *singular solution*.   ❏

Example 3   Find a solution of the differential equation

$$y' = \frac{xy - y}{y + 1},$$

which satisfies the side condition $y(2) = 1$.

SOLUTION   We show first that the equation is separable:

$$y - xy + (y + 1)y' = 0,$$

$$1 - x + \frac{y + 1}{y}y' = 0 \qquad (y \neq 0)$$

or

$$1 - x + \left(1 + \frac{1}{y}\right)y' = 0.$$

Now,

$$(1 - x)\,dx + \left(1 + \frac{1}{y}\right)dy = 0$$

$$\int (1 - x)\,dx + \int \left(1 + \frac{1}{y}\right)dy = C,$$

and

$$x - \tfrac{1}{2}x^2 + y + \ln|y| = C$$

is a one-parameter family of solutions which defines y implicitly as a function of x. The constant function $y = 0$ is not included in the one-parameter family; it is a singular solution of the equation.

To find a solution which satisfies the side condition, we set $x = 2$ and $y = 1$ in the one-parameter family:

$$2 - \tfrac{1}{2}(2)^2 + 1 + \ln(1) = C \quad \text{and} \quad C = 1.$$

Thus, the particular solution

$$x - \tfrac{1}{2}x^2 + y + \ln|y| = 1$$

satisfies the side condition $y(2) = 1$. ❏

An Application

In the mid-nineteenth century the Belgian biologist P. F. Verhulst used the differential equation

(18.3.6)

$$\frac{dy}{dt} = ky(M - y),$$

where k and M are positive constants, to study the population growth of various countries. This equation is now known as the *logistic equation* and its solutions are called *logistic functions*. Life scientists have used this equation to model the spread of a disease through a population, and social scientists have used it to study the flow of information. In the case of a disease, if M denotes the total number of people in the population and $y(t)$ is the number of infected people at time t, then the differential equation states that the rate of growth of the disease is proportional to the product of the number of people who are infected and the number who are not.

The differential equation is separable since it can be written in the form

$$\frac{1}{y(M - y)}y' - k = 0.$$

Writing the equation in differential form and integrating term-by-term, we have

$$\frac{1}{y(M - y)}dy - k\,dt = 0,$$

$$\int \frac{1}{y(M - y)}dy - \int k\,dt = C_1,$$

$$\int \left[\frac{1/M}{y} + \frac{1/M}{M - y}\right] dy - kt = C_1, \qquad \text{(partial fraction decomposition)}$$

and

$$\frac{1}{M}\ln|y| - \frac{1}{M}\ln|M - y| = kt + C_1$$

is a one-parameter family of solutions. We can solve this equation for y as follows:

$$\frac{1}{M}\ln\left|\frac{y}{M-y}\right| = kt + C_1$$

$$\ln\left|\frac{y}{M-y}\right| = Mkt + MC_1 = Mkt + C_2, \qquad (C_2 = MC_1)$$

$$\left|\frac{y}{M-y}\right| = e^{(C_2+Mkt)} = e^{C_2}e^{Mkt} = Ce^{Mkt}. \qquad (C = e^{C_2})$$

Now, in the context of this application, $y = y(t)$ satisfies $0 < y < M$. Therefore $y/(M - y) > 0$ and we have

$$\frac{y}{M-y} = Ce^{Mkt}.$$

Solving this equation for y we get

$$y(t) = \frac{CM}{C + e^{-Mkt}}.$$

If $y(t)$ satisfies the side condition $y(0) = R$, $R < M$, then

$$R = \frac{CM}{C+1}, \qquad \text{which implies} \quad C = \frac{R}{M-R}$$

and

(18.3.7)
$$y(t) = \frac{MR}{R + (M - R)e^{-Mkt}}.$$

The graph of this particular solution is shown in Figure 18.3.3. Note that y is increasing for all $t \geq 0$. In Exercise 38 you are asked to show that the graph is concave up on $[0, t_1]$ and concave down on $[t_1, \infty)$. This means that the disease is spreading at an increasing rate up to time $t = t_1$; after t_1, the disease is still spreading, but at a decreasing rate. Note, also, that $y(t) \to M$ as $t \to \infty$.

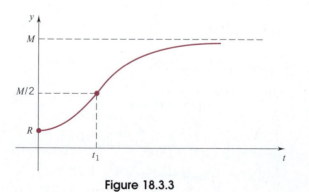

Figure 18.3.3

Example 4 A rumor spreads through a population of 5000 people at a rate proportional to the product of the number of people who have heard it and the number who have not. Suppose that 100 people initiate the rumor and that a total of 500 people know the rumor after two days. How long will it take for half the population to hear the rumor?

SOLUTION Let $y(t)$ denote the number of people who know the rumor at time t. Then y satisfies the differential equation (18.3.6) with $M = 5000$ and the side condition $R = y(0) = 100$. Thus, by (18.3.7),

$$y(t) = \frac{100(5000)}{100 + 4900e^{-5000kt}} = \frac{5000}{1 + 49e^{-5000kt}},$$

where the constant of proportionality k can be determined from the condition $y(2) = 500$. We have

$$500 = \frac{5000}{1 + 49e^{-10,000k}},$$

$$1 + 49e^{-10,000k} = 10,$$

$$e^{-10,000k} = \frac{9}{49},$$

$$-10,000k = \ln\left[\frac{9}{49}\right] \quad \text{and} \quad k \cong 0.00017.$$

Therefore,

$$y(t) = \frac{5000}{1 + 49e^{-0.85t}}.$$

To determine how long it will take for half the population to hear the rumor, we solve the equation

$$2500 = \frac{5000}{1 + 49e^{-0.85t}}$$

for t. We get

$$1 + 49e^{-0.85t} = 2, \quad e^{-0.85t} = \frac{1}{49}, \quad \text{and} \quad t = \frac{\ln(1/49)}{-0.85} \cong 4.58.$$

Thus, it will take slightly more than $4\frac{1}{2}$ days for half the population to hear the rumor. ❑

Homogeneous Equations

If we solve equation (18.3.2) for y', we get

$$y' = \frac{-P(x, y)}{Q(x, y)},$$

which we can write as

(18.3.8)
$$\boxed{y' = f(x, y).}$$

The first-order differential equation (18.3.8) is *homogeneous* iff the function f has the property

$$f(tx, ty) = f(x, y)$$

for all $t \neq 0$. This condition is equivalent to saying that f can be expressed as a function of y/x. For if f is homogeneous, then

$$f(x, y) = f\left(x, x\left(\frac{y}{x}\right)\right) = f\left(1, \frac{y}{x}\right) = g\left(\frac{y}{x}\right) \qquad (x \neq 0).$$

A homogeneous equation can be transformed into a separable equation by setting

$$v = \frac{y}{x}.$$

To see this, write

$$xv = y$$

and differentiate with respect to x:

$$v + xv' = y'.$$

Substituting y and y' into (18.3.8), gives

$$v + xv' = f(x, xv) = f(1, v) = g(v).$$

Thus, we have

$$v + xv' = g(v)$$

and

(18.3.9)
$$\boxed{\frac{1}{x} + \left[\frac{1}{v - g(v)}\right]v' = 0.}$$

This equation is separable.

If Equation (18.3.8) is homogeneous, then we solve it by solving the transformed equation, Equation (18.3.9), and then substitute y/x back in for v.

Example 5 Show that the differential equation

$$y' = \frac{3x^2 + y^2}{xy}$$

is homogeneous and find a one-parameter family of solutions.

SOLUTION The equation is homogeneous since

$$f(tx, ty) = \frac{3(tx)^2 + (ty)^2}{(tx)(ty)} = \frac{t^2(3x^2 + y^2)}{t^2(xy)} = \frac{3x^2 + y^2}{xy} = f(x, y).$$

Note that f can be written as

$$\frac{3 + (y/x)^2}{(y/x)} \left[= g\left(\frac{y}{x}\right)\right],$$

which is a function of y/x as previously indicated.

Now, set $vx = y$. Then, $v + xv' = y'$ and

$$v + xv' = \frac{3x^2 + v^2x^2}{vx^2} = \frac{3 + v^2}{v} \ [= g(v)].$$

Thus,

$$v - \frac{3 + v^2}{v} + xv' = 0,$$

$$-3 + xvv' = 0,$$

$$-\frac{3}{x} dx + v \, dv = 0,$$

$$-\int \frac{3}{x} dx + \int v \, dv = C_1$$

$$-3 \ln|x| + \frac{v^2}{2} = C_1 \quad \text{or} \quad \ln|x| - \tfrac{1}{6}v^2 = C. \quad (C = -\tfrac{1}{3}C_1)$$

Substituting y/x back in for v, we have

$$\ln|x| - \frac{1}{6}\left(\frac{y}{x}\right)^2 = C.$$

The integral curves take the form

$$y^2 = 6x^2 \, (\ln|x| - C). \quad \square$$

EXERCISES 18.3

In Exercises 1–10, solve the differential equation.

1. $y' = y \sin(2x + 3)$.

2. $y' = (x^2 + 1)(y^2 + y)$.

3. $y' = (xy)^3$.

4. $y' = 3x^2(1 + y^2)$.

5. $y' = -\dfrac{\sin 1/x}{x^2 y \cos y}$.

6. $y' = \dfrac{y^2 + 1}{y + yx}$.

7. $y' = xe^{y-x}$.

8. $y' = xy^2 - x - y^2 + 1$.

9. $(y \ln x)y' = \dfrac{(y + 1)^2}{x}$.

10. $e^y \sin 2x \, dx + \cos x(e^{2y} - y) \, dy = 0$.

In Exercises 11–16, find the particular solution determined by the side condition.

11. $y' = x\sqrt{\dfrac{1 - y^2}{1 - x^2}}, \quad y(0) = 0$.

12. $y' = \dfrac{e^{x-y}}{1 + e^x}, \quad y(1) = 0$.

13. $y' = \dfrac{x^2 y - y}{y + 1}, \quad y(3) = 1$.

14. $x^2 y' = y - xy, \quad y(-1) = -1$.

15. $(xy^2 + y^2 + x + 1) \, dx + (y - 1) \, dy = 0, \quad y(2) = 0$.

16. $\cos y \, dx + (1 + e^{-x})\sin y \, dy = 0, \quad y(0) = \pi/4$.

In Exercises 17–24, verify that the equation is homogeneous, and then solve the equation.

17. $y' = \dfrac{x^2 + y^2}{2xy}$.

18. $y' = \dfrac{y^2}{xy + x^2}$.

19. $y' = \dfrac{x - y}{x + y}$.

20. $y' = \dfrac{x + y}{x - y}$.

21. $y' = \dfrac{x^2(e^y)^{1/x} + y^2}{xy}$.

22. $y' = \dfrac{x^2 + 3y^2}{4xy}$.

23. $y' = \dfrac{y}{x} + \sin\left(\dfrac{y}{x}\right)$.

24. $x \, dy = y\left[1 + \ln\left(\dfrac{y}{x}\right)\right] dx$.

In Exercises 25 and 26, find the particular solution determined by the side condition

25. $y' = \dfrac{y^3 - x^3}{xy^2}, \quad y(1) = 2$.

26. $x \sin\left(\dfrac{y}{x}\right) dy = \left[x + y \sin\left(\dfrac{y}{x}\right)\right] dx,$ $\quad y(1) = 0.$

27. (a) Given that

$$x'(t) = a_1 x - a_2 xy \quad \text{and} \quad y'(t) = -a_3 y + a_4 xy,$$

find a differential equation that links y to x independently of the parameter t.
HINT: $dy/dx = y'(t)/x'(t)$.
(b) Solve the equation obtained in (a).

28. Given that

$$x'(t) = -Ax \quad \text{and} \quad y'(t) = -Bxy,$$

(a) Express y as a function of x.
(b) Express x as a function of t with $x(0) = x_0$.
(c) Express y as a function of t with $y(0) = y_0$.

Orthogonal trajectories If two curves intersect at right angles, one with slope m_1 and the other with slope m_2, then $m_1 m_2 = -1$. A curve that intersects every member of a family of curves at right angles is called an *orthogonal trajectory* for that family of curves. A differential equation of the form

$$y' = f(x, y)$$

generates a family of curves: all the integral curves of that particular equation. The orthogonal trajectories of this family are the integral curves of the differential equation

$$y' = -\dfrac{1}{f(x, y)}.$$

The figure shows the family of parabolas $y = Cx^2$. The orthogonal trajectories of this family of parabolas are ellipses.

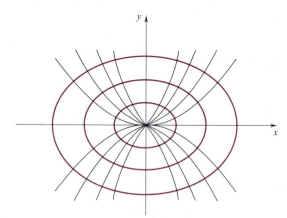

We can establish this by starting with the equation

$$y = Cx^2$$

and differentiating. Differentiation gives

$$y' = 2Cx.$$

Since $y = Cx^2$, we have $C = y/x^2$, and thus

$$y' = 2\left(\dfrac{y}{x^2}\right) x = \dfrac{2y}{x}.$$

The orthogonal trajectories are the integral curves of the differential equation

$$y' = -\dfrac{x}{2y}.$$

As you can check, the solutions of this equation are all of the form

$$x^2 + 2y^2 = K. \quad \square$$

In Exercises 29–34, find the orthogonal trajectories for the following families of curves. In each case draw several of the curves and several of the orthogonal trajectories.

29. $2x + 3y = C.$ **30.** $y = Cx.$

31. $xy = C.$ **32.** $y = Cx^3.$

33. $y = Ce^x.$ **34.** $x = Cy^4.$

▷ In Exercises 35 and 36, show that the given family is *self-orthogonal*. Use a graphing utility to graph at least four members of the family.

35. $y^2 = 4C(x + C).$ **36.** $\dfrac{x^2}{C^2} + \dfrac{y^2}{C^2 - 4} = 1.$

▷ **37.** A flu virus is spreading rapidly through a small town with a population of 25,000. The disease is spreading at a rate proportional to the product of the number of people who have it and the number who don't. Suppose that 100 people had the disease initially, and 400 had it after 10 days.
(a) How many people will have the disease at an arbitrary time t? How many will have it after 20 days?
(b) How long it will take for half the population to have the flu?
(c) Use a graphing utility to sketch the graph of your solution in part (a).

38. Consider the logistic equation (18.3.6). Show that dy/dt is increasing if $y < M/2$ and is decreasing if $y > M/2$. What can you conclude about dy/dt at the instant $y = M/2$? Explain.

▷ **39.** A rescue package whose mass is 100 kilograms is dropped from an airplane at a height of 4000 meters. As it falls, the air resistance is equal to twice its velocity. After 10 seconds, the package's parachute opens and the air resistance is now four times the square of its velocity.
(a) What is the package's velocity at the instant the parachute opens?
(b) Determine an expression for the velocity of the package at time t after the parachute opens.
(c) What is the terminal velocity of the package?

HINT: There are two differential equations that govern the package's velocity and position; one for the free-fall period and one for the period after the parachute opens.

40. It is known that m parts of chemical A combine with n parts of chemical B to produce a compound C. Suppose that the rate at which C is produced varies directly with the product of the amounts of A and B present at that instant. Find the amount of C produced in t minutes from an initial mixing of A_0 pounds of A with B_0 pounds of B, given that:

(a) $n = m$, $A_0 = B_0$, and A_0 pounds of C are produced in the first minute.

(b) $n = m$, $A_0 = \frac{1}{2}B_0$, and A_0 pounds of C are produced in the first minute.

(c) $n \neq m$, $A_0 = B_0$, and A_0 pounds of C are produced in the first minute.

HINT: Denote by $A(t)$, $B(t)$, and $C(t)$ the amounts of A, B, and C present at time t. Observe that $C'(t) = kA(t)B(t)$. Then note that

$$A_0 - A(t) = \frac{m}{m + n}C(t) \quad \text{and} \quad B_0 - B(t) = \frac{n}{m + n}C(t)$$

and thus

$$C'(t) = k\left[A_0 - \frac{m}{m + n}C(t)\right]\left[B_0 - \frac{n}{m + n}C(t)\right].$$

■ 18.4 EXACT EQUATIONS; INTEGRATING FACTORS

We continue to consider the first-order differential equation

(18.4.1)
$$P(x, y) + Q(x, y)y' = 0$$

and its equivalent differential form

(18.4.2)
$$P(x, y)\,dx + Q(x, y)\,dy = 0.$$

Throughout this section, we assume that the functions $P = P(x, y)$ and $Q = Q(x, y)$ are continuously differentiable on a simply connected region Ω in the xy-plane.

Equation (18.4.1) (or (18.4.2)) is said to be exact on Ω iff

(18.4.3)
$$\frac{\partial P}{\partial y}(x, y) = \frac{\partial Q}{\partial x}(x, y) \qquad \text{for all } (x, y) \in \Omega.$$

The reason for this terminology is as follows. If the equation

$$P(x, y) + Q(x, y)y' = 0$$

is exact, then (by Theorem 15.10.2) the vector-valued function

$$P(x, y)\mathbf{i} + Q(x, y)\mathbf{j}$$

is "exactly" in the form of a gradient and there is a function f defined on Ω such that

$$\frac{\partial f}{\partial x} = P \quad \text{and} \quad \frac{\partial f}{\partial y} = Q.$$

We can therefore write (18.4.1) as

(1)
$$\frac{\partial f}{\partial x}(x, y) + \frac{\partial f}{\partial y}(x, y)y' = 0.$$

Now, if we regard y as a differentiable function of x, then, by the chain rule,

$$\frac{d}{dx}[f(x, y)] = \frac{\partial f}{\partial x}(x, y) + \frac{\partial f}{\partial y}(x, y)y'$$

and equation (1) can be written

$$\frac{d}{dx}[f(x, y)] = 0.$$

Integrating with respect to x, we have

$$f(x, y) = C.$$

The level curves of f are thus the integral curves of equation (18.4.1).

A procedure for calculating f from P and Q is shown in the following example. This procedure was explained in detail in Section 15.10.

Example 1   Show that the differential equation

$$(xy^2 - x^3 + y) + (x^2y + x - 2y + 3)y' = 0$$

is exact on $\Omega = $ the xy-plane, and find a one-parameter family of solutions.

SOLUTION   The coefficients

$$P(x, y) = xy^2 - x^3 + y \qquad \text{and} \qquad Q(x, y) = (x^2y + x - 2y + 3)$$

are continuously differentiable on Ω, and for all (x, y) we have

$$\frac{\partial P}{\partial y} = 2xy + 1 = \frac{\partial Q}{\partial x}.$$

To find the integral curves

$$f(x, y) = C$$

we set

$$\frac{\partial f}{\partial x}(x, y) = xy^2 - x^3 + y \qquad \text{and} \qquad \frac{\partial f}{\partial y}(x, y) = x^2y + x - 2y + 3.$$

Integrating $\partial f/\partial x$ with respect to x, holding y constant, we have

$$f(x, y) = \tfrac{1}{2}x^2y^2 - \tfrac{1}{4}x^4 + xy + \phi(y),$$

where $\phi(y)$ is a function of y which is to be determined. Differentiation with respect to y gives

$$\frac{\partial f}{\partial y}(x, y) = x^2y + x + \phi'(y).$$

It now follows that

$$x^2y + x - 2y + 3 = x^2y + x + \phi'(y),$$

and so

$$\phi'(y) = -2y + 3.$$

Thus,

$$\phi(y) = -y^2 + 3y, \qquad \text{(we omit the arbitrary constant)}$$

and

$$f(x, y) = \tfrac{1}{2}x^2y^2 - \tfrac{1}{4}x^4 + xy - y^2 + 3y.$$

Finally, a one-parameter family of solutions is

$$\tfrac{1}{2}x^2y^2 - \tfrac{1}{4}x^4 + xy - y^2 + 3y = C.$$

You can verify this by differentiating implicitly. ❑

If the equation

$$P(x, y) + Q(x, y)y' = 0$$

is not exact, it may be possible to find a function $\mu = \mu(x, y)$ not identically zero such that the equation

$$\mu(x, y) P(x, y) + \mu(x, y) Q(x, y)y' = 0$$

is exact. If μ is never zero or is zero only where P and Q are both zero, then any solution of this second equation gives a solution of the first equation. We call $\mu(x, y)$ an *integrating factor*.

Example 2 The equation

(*) $$(e^x - \sin y) + (\cos y)y' = 0$$

is not exact:

$$\frac{\partial}{\partial y}(e^x - \sin y) = -\cos y \quad \text{but} \quad \frac{\partial}{\partial x}(\cos y) = 0.$$

Multiplication by e^{-x} gives

(**) $$(1 - e^{-x}\sin y) + (e^{-x}\cos y)y' = 0.$$

This equation is exact:

$$\frac{\partial}{\partial y}(1 - e^{-x}\sin y) = -e^{-x}\cos y = \frac{\partial}{\partial x}(e^{-x}\cos y).$$

Thus $\mu(x, y) = e^{-x}$ is an integrating factor, and we can solve equation (*) by solving (**).

To solve (**) we set

$$\frac{\partial f}{\partial x} = 1 - e^{-x}\sin y, \qquad \frac{\partial f}{\partial y} = e^{-x}\cos y.$$

Thus

$$f(x, y) = \int (1 - e^{-x}\sin y)\, dx = x + e^{-x}\sin y + \phi(y)$$

so that

$$\frac{\partial f}{\partial y} = e^{-x} \cos y + \phi'(y).$$

This gives $\phi'(y) = 0$ and shows that we can set $\phi(y) = 0$. The integral curves take the form

$$x + e^{-x} \sin y = C.$$

You can verify this by differentiation. ❏

Let's return to the general situation and write our equation in the form

(1) $$P + Qy' = 0.$$

How can we find an integrating factor if this equation is not exact?
Observe first of all that the equation

$$\mu P + \mu Q y' = 0$$

is exact iff

$$\frac{\partial}{\partial y}(\mu P) = \frac{\partial}{\partial x}(\mu Q)$$

and this occurs iff

(2) $$\mu \frac{\partial P}{\partial y} + \frac{\partial \mu}{\partial y} P = \mu \frac{\partial Q}{\partial x} + \frac{\partial \mu}{\partial x} Q.$$

Thus μ is an integrating factor for (1) iff μ satisfies (2).

In theory, all we have to do now to find an integrating factor for (1) is to solve equation (2) for μ. Unfortunately, Equation (2) is a partial differential equation which is, in general, much more difficult to solve than Equation (1). However, there are two special cases where an integrating factor can be determined.

(18.4.4)

If $v = \dfrac{1}{Q}\left(\dfrac{\partial P}{\partial y} - \dfrac{\partial Q}{\partial x}\right)$ is a function of x only, then

$\mu(x) = e^{\int v(x)\,dx}$ is an integrating factor for equation (1);

if $w = \dfrac{1}{P}\left(\dfrac{\partial P}{\partial y} - \dfrac{\partial Q}{\partial x}\right)$ is a function of y only, then

$\mu(y) = e^{\int -w(y)\,dy}$ is an integrating factor for equation (1).

PROOF We prove the first part and leave the proof of the second part as an exercise. Assume that

$$v = \frac{1}{Q}\left(\frac{\partial P}{\partial y} - \frac{\partial Q}{\partial x}\right)$$

is a function of x only and multiply (1) by $e^{\int v}$. This gives

$$P e^{\int v(x)\,dx} + Q e^{\int v(x)\,dx} y' = 0.$$

All we have to do is show that

$$\frac{\partial}{\partial y}\left(Pe^{\int v(x)\,dx}\right) = \frac{\partial}{\partial x}\left(Qe^{\int v(x)\,dx}\right).$$

This can be seen as follows:

$$\frac{\partial}{\partial x}\left(Qe^{\int v(x)\,dx}\right) = Qve^{\int v(x)\,dx} + \frac{\partial Q}{\partial x}e^{\int v(x)\,dx}$$

$$= \left[Qv + \frac{\partial Q}{\partial x}\right]e^{\int v(x)\,dx}$$

$$= \left[\left(\frac{\partial P}{\partial y} - \frac{\partial Q}{\partial x}\right) + \frac{\partial Q}{\partial x}\right]e^{\int v(x)\,dx}$$

$$= \frac{\partial P}{\partial y}e^{\int v(x)\,dx} = \frac{\partial}{\partial y}\left(Pe^{\int v(x)\,dx}\right). \quad ❏$$

Example 3 Earlier we showed that the equation

$$(e^x - \sin y) + (\cos y)y' = 0$$

is not exact, but that e^{-x} is an integrating factor. We can discover this integrating factor by using (18.4.4). In this case,

$$\frac{1}{Q}\left(\frac{\partial P}{\partial y} - \frac{\partial Q}{\partial x}\right) = \frac{1}{\cos y}(-\cos y - 0) = -1,$$

and

$$e^{\int v(x)\,dx} = e^{\int(-1)\,dx} = e^{-x}. \quad ❏$$

EXERCISES 18.4

In Exercises 1–10, verify that the equation is exact and solve the equation.

1. $(xy^2 - y) + (x^2y - x)y' = 0.$

2. $e^x \sin y + (e^x \cos y)y' = 0.$

3. $(e^y - ye^x) + (xe^y - e^x)y' = 0.$

4. $\sin y + (x \cos y + 1)y' = 0.$

5. $(\ln y + 2xy) + \left(\dfrac{x}{y} + x^2\right)y' = 0.$

6. $(2x \tan^{-1} y) + \left(\dfrac{x^2}{1 + y^2}\right)y' = 0.$

7. $\left(\dfrac{y}{x} + 6x\right)dx + (\ln x - 2)\,dy = 0.$

8. $e^x + \ln y + \dfrac{y}{x} + \left(\dfrac{x}{y} + \ln x + \sin y\right)y' = 0.$

9. $(y^3 - y^2 \sin x - x)\,dx +$
$\qquad (3xy^2 + 2y \cos x + e^{2y})\,dy = 0.$

10. $(e^{2y} - y \cos xy) + (2x\,e^{2y} - x \cos xy + 2y)\,y' = 0.$

11. Let p and q be functions of one variable everywhere continuously differentiable.

(a) Is the equation $p(x) + q(y)y' = 0$ necessarily exact?

(b) Show that the equation $p(y) + q(x)y' = 0$ is not necessarily exact. Then find an integrating factor.

12. Prove the second part of (18.4.4).

In Exercises 13–18, solve each of the following equations using an integrating factor if necessary.

13. $(e^{y-x} - y) + (xe^{y-x} - 1)y' = 0.$

14. $(x + e^y) - \frac{1}{2}x^2y' = 0.$

15. $(3x^2y^2 + x + e^y) + (2x^3y + y + xe^y)y' = 0.$

16. $(\sin 2x \cos y) - (\sin^2 x \sin y)y' = 0.$

17. $(y^3 + x + 1) + (3y^2)y' = 0.$

18. $(e^{2x+y} - 2y) + (xe^{2x+y} + 1)y' = 0.$

In Exercises 19–25, find the integral curve that passes through (x_0, y_0). Use an integrating factor if necessary.

19. $(x^2 + y) + (x + e^y)y' = 0; \quad (x_0, y_0) = (1, 0).$

20. $(3x^2 - 2xy + y^3) + (3xy^2 - x^2)y' = 0$;
$(x_0, y_0) = (1, -1)$.

21. $(2y^2 + x^2 + 2) + (2xy)y' = 0$; $(x_0, y_0) = (1, 0)$.

22. $(x^2 + y) + (3x^2y^2 - x)y' = 0$; $(x_0, y_0) = (1, 1)$.

23. $y^3 + (1 + xy^2)y' = 0$; $(x_0, y_0) = (-2, -1)$.

24. $(x + y)^2 + (2xy + x^2 - 1)y' = 0$; $(x_0, y_0) = (1, 1)$.

25. $[\cosh(x - y^2) + e^{2x}] \, dx + y[1 - 2\cosh(x - y^2)]$
$dy = 0; (x_0, y_0) = (2, \sqrt{2})$.

26. In Section 18.2 we solved the linear differential equation

$$y' + p(x)y = q(x)$$

by using the integrating factor

$$e^{\int p(x)\,dx}.$$

Show that this integrating factor is obtainable by the methods of this section.

27. (a) Find a value of k, if possible, such that the differential equation

$$(xy^2 + k\,x^2y + x^3)\,dx + (x^3 + x^2y + y^2)\,dy = 0$$

is exact.

(b) Find a value of k, if possible, such that the differential equation

$$ye^{2xy} + 2x + (k\,xe^{2xy} - 2y)y' = 0$$

is exact.

28. (a) Find functions f and g, not both identically zero, such that the differential equation

$$g(y)\sin x \, dx + y^2 f(x)\, dy = 0$$

is exact.

(b) Find all functions g such that the differential equation $g(y)\,e^y + xy\,y' = 0$ is exact.

In Exercises 29–34, solve the given differential equation by using a suitable method from Sections 18.2–18.4.

29. $y' = y^2x^3$.

30. $y' = \dfrac{y}{x + \sqrt{xy}}$.

31. $y' + \dfrac{4y}{x} = x^4$.

32. $y' + 2xy - 2x^3 = 0$.

33. $(y\,e^{xy} - 2x)\,dx + \left(\dfrac{2}{y} + xe^{xy}\right)dy = 0$.

34. $y\,dx + (2xy - e^{-2y})\,dy = 0$.

■ 18.5 THE EQUATION $y'' + ay' + by = 0$

A differential equation of the form

$$y'' + ay' + by = \phi(x),$$

where a and b are real numbers and ϕ is a continuous function on some interval I, is a *second-order linear differential equation with constant coefficients.* By a solution of such an equation we mean a function $y = y(x)$ that satisfies the equation for all x in I. In this section we set $\phi = 0$ and consider the *reduced equation*

(18.5.1)

$$\boxed{y'' + ay' + by = 0} \quad †$$

Linear Combinations of Solutions; The Characteristic Equation

Observe, first of all, that *if y_1 and y_2 are both solutions of the reduced equation, then every linear combination*

$$C_1y_1 + C_2y_2$$

is also a solution.

---

† Often called the *homogeneous equation;* see Exercises 18.2.

PROOF Set

$$u = C_1 y_1 + C_2 y_2$$

and observe that

$$u' = C_1 y_1' + C_2 y_2' \quad \text{and} \quad u'' = C_1 y_1'' + C_2 y_2''.$$

If

$$y_1'' + ay_1' + by_1 = 0 \quad \text{and} \quad y_2'' + ay_2' + by_2 = 0,$$

then

$$\begin{aligned} u'' + au' + bu &= (C_1 y_1'' + C_2 y_2'') + a(C_1 y_1' + C_2 y_2') + b(C_1 y_1 + C_2 y_2) \\ &= C_1(y_1'' + ay_1' + by_1) + C_2(y_2'' + ay_2' + by_2) \\ &= C_1(0) + C_2(0) = 0. \quad \square \end{aligned}$$

Earlier you saw that the function $y = e^{-ax}$ satisfies the differential equation

$$y' + ay = 0.$$

This suggests that the differential equation

$$y'' + ay' + by = 0$$

might also have solutions of the form $y = e^{rx}$.

If $y = e^{rx}$, then

$$y' = re^{rx} \quad \text{and} \quad y'' = r^2 e^{rx}.$$

Substitution into the differential equation gives

$$r^2 e^{rx} + are^{rx} + be^{rx} = 0$$

and, since $e^{rx} \neq 0$,

$$r^2 + ar + b = 0.$$

This shows that *the function $y = e^{rx}$ satisfies the differential equation iff*

$$r^2 + ar + b = 0.$$

This quadratic equation in r is called the *characteristic equation* of (18.5.1).†
By the quadratic formula, the roots of the characteristic equation are

$$r = \frac{-a \pm \sqrt{a^2 - 4b}}{2}.$$

The nature of the solutions of the differential equation

$$y'' + ay' + by = 0$$

depends on the nature of these roots.

Case 1. $a^2 > 4b$

If $a^2 > 4b$, the characteristic equation has two distinct real roots:

$$r_1 = -\tfrac{1}{2}a + \tfrac{1}{2}\sqrt{a^2 - 4b}, \qquad r_2 = -\tfrac{1}{2}a - \tfrac{1}{2}\sqrt{a^2 - 4b}.$$

---

† Sometimes called the *auxiliary equation*.

Both

$$y_1 = e^{r_1 x} \quad \text{and} \quad y_2 = e^{r_2 x}$$

are solutions of the differential equation. ❏

Case 2. $a^2 = 4b$

If $a^2 = 4b$, the characteristic equation has only one root

$$r_1 = -\tfrac{1}{2}a.$$

In this case, the characteristic equation must be

$$(r - r_1)^2 = r^2 - 2r_1 r + r_1^2 = 0 \qquad \text{(verify this)}$$

and the corresponding differential equation is

(1) $$y'' - 2r_1 y' + r_1^2 y = 0.$$

We know that $y_1 = e^{r_1 x}$ is one solution of this equation; we need to find a second solution. To motivate our choice for the second solution, consider the special case $r_1 = 0$. Setting $r_1 = 0$ in equation (1), we get

$$y'' = 0,$$

which has

$$y = C_1 + C_2 x$$

as a two-parameter family of solutions. Since $e^{0x} = 1$, we can write this family of solutions in the form

$$y = C_1 e^{0x} + C_2 x e^{0x},$$

and $y_1 = e^{0x}$ and $y_2 = xe^{0x}$ are particular solutions. Thus, we are led to consider $y_2 = xe^{r_1 x}$ as a possible second solution of equation (1). The function y_2 and its first two derivatives are

$$y_2 = xe^{r_1 x}$$
$$y_2' = e^{r_1 x} + r_1 xe^{r_1 x}$$
$$y_2'' = 2r_1 e^{r_1 x} + r_1^2 xe^{r_1 x}.$$

Substituting y_2 and its derivatives into equation (1), we have

$$y_2'' - 2r_1 y_2' + r_1^2 y_2 =$$
$$(2 r_1 e^{r_1 x} + r_1^2 xe^{r_1 x}) - 2r_1(e^{r_1 x} + r_1 xe^{r_1 x}) + r_1^2 xe^{r_1 x} = 0,$$

and so $y_2(x) = xe^{r_1 x}$ is also a solution of (1). ❏

Case 3. $a^2 < 4b$

If $a^2 < 4b$, the characteristic equation has two complex conjugate roots:

$$r_1 = -\tfrac{1}{2}a + i\tfrac{1}{2}\sqrt{4b - a^2}, \qquad r_2 = -\tfrac{1}{2}a - i\tfrac{1}{2}\sqrt{4b - a^2} \qquad (i^2 = -1).$$

Setting

$$\alpha = -\tfrac{1}{2}a \quad \text{and} \quad \beta = \tfrac{1}{2}\sqrt{4b - a^2},$$

we can write

$$r_1 = \alpha + i\beta, \qquad r_2 = \alpha - i\beta.$$

The complex-valued functions

$$z_1 = e^{(\alpha + i\beta)x}, \qquad z_2 = e^{(\alpha - i\beta)x}$$

both satisfy the differential equation.

We can obtain real solutions by forming suitable linear combinations of z_1 and z_2. In Exercise 41 you are asked to verify Euler's formulas

$$e^{i\theta} = \cos\theta + i\sin\theta \qquad \text{and} \qquad e^{-i\theta} = \cos\theta - i\sin\theta.$$

Now, using these formulas, we have

$$z_1 = e^{\alpha x + i\beta x} = e^{\alpha x}\, e^{i\beta x} = e^{\alpha x}[\cos\beta x + i\sin\beta x]$$

and

$$z_2 = e^{\alpha x - i\beta x} = e^{\alpha x}\, e^{-i\beta x} = e^{\alpha x}[\cos\beta x - i\sin\beta x].$$

Adding and subtracting these equations, we have

$$z_1 + z_2 = 2e^{\alpha x}\cos\beta x, \qquad z_1 - z_2 = 2ie^{\alpha x}\sin\beta x$$

and thus

$$\frac{1}{2}(z_1 + z_2) = e^{\alpha x}\cos\beta x, \qquad \frac{1}{2i}(z_1 - z_2) = e^{\alpha x}\sin\beta x.$$

The functions

$$y_1 = e^{\alpha x}\cos\beta x, \qquad y_2 = e^{\alpha x}\sin\beta x$$

are real-valued and, being linear combinations of solutions, are themselves solutions. ❏

Remark In our manipulations of complex exponential functions we have implicitly assumed that the basic properties of real exponents (for example, $e^a e^b = e^{a+b}$) carry over to complex exponents. This is indeed true, but the treatment needed to verify these properties is beyond the scope of our discussion here. ❏

Existence and Uniqueness of Solutions; Wronskians, Fundamental Solutions

You have seen how to obtain solutions of the differential equation

$$y'' + ay' + by = 0$$

from the characteristic equation

$$r^2 + ar + b = 0.$$

We can form more solutions by taking linear combinations of these solutions. Question: Are there still other solutions or do all solutions arise in this manner? Answer: All solutions of the reduced equation are linear combinations of the solutions that we have already found.

To show this, we have to go a little deeper into the theory. Our point of departure is a result proved in more advanced texts.

THEOREM 18.5.2

Let x_0, y_0, y_1 be arbitrary real numbers. The reduced equation

$$y'' + ay' + by = 0$$

has a unique solution $y = y(x)$ that satisfies the side conditions

$$y(x_0) = y_0, \qquad y'(x_0) = y_1.$$

Geometrically, the theorem says that there is one and only one solution whose graph passes through a prescribed point (x_0, y_0) with a prescribed slope y_1. We assume this result and go on from there.

DEFINITION 18.5.3

Let y_1 and y_2 be two solutions of

$$y'' + ay' + by = 0.$$

The *Wronskian*† of y_1 and y_2 is the function W defined for all real x by

$$W(x) = y_1(x)y_2'(x) - y_2(x)y_1'(x).$$

It follows from this definition that the Wronskian can be written as the 2×2 determinant (see Appendix A.2).

$$W(x) = \begin{vmatrix} y_1(x) & y_2(x) \\ y_1'(x) & y_2'(x) \end{vmatrix}.$$

THEOREM 18.5.4

If both y_1 and y_2 are solutions of the reduced equation

$$y'' + ay' + by = 0,$$

then their Wronskian W is either identically zero or never zero.

PROOF   Assume that both y_1 and y_2 are solutions of the reduced equation and set

$$W = y_1 y_2' - y_1' y_2.$$

Differentiation gives

$$W' = y_1 y_2'' + y_1' y_2' - y_1' y_2' - y_1'' y_2 = y_1 y_2'' - y_1'' y_2.$$

---

† Named in honor of Count Hoëné Wronski, a Polish mathematician (1778–1853).

Since y_1 and y_2 are solutions, we know that

$$y_1'' + ay_1' + by_1 = 0$$

and

$$y_2'' + ay_2' + by_2 = 0.$$

Multiplying the first equation by $-y_2$ and the second equation by y_1, we have

$$-y_1''y_2 - ay_1'y_2 - by_1y_2 = 0$$

$$y_2''y_1 + ay_2'y_1 + by_2y_1 = 0.$$

We now add these two equations and obtain

$$(y_1y_2'' - y_2y_1'') + a(y_1y_2' - y_2y_1') = 0.$$

In terms of the Wronskian, we have

$$W' + aW = 0.$$

This is a first-order linear differential equation with general solution

$$W = Ce^{-ax}.$$

If $C = 0$, then W is identically 0; if $C \neq 0$, then W is never zero. ❏

Two solutions y_1, y_2 of the reduced equation

$$y'' + ay' + by = 0$$

are said to be *fundamental* iff their Wronskian is not identically zero (in which case, by Theorem 18.5.4, the Wronskian is never zero).

THEOREM 18.5.5

Every solution of the reduced equation

$$y'' + ay' + by = 0$$

can be expressed in a unique manner as the linear combination of any two fundamental solutions.

PROOF Let u be any solution of the reduced equation and let y_1, y_2 be any pair of fundamental solutions. Choose a number x_0 and form the equations

(1)
$$C_1 y_1(x_0) + C_2 y_2(x_0) = u(x_0)$$
$$C_1 y_1'(x_0) + C_2 y_2'(x_0) = u'(x_0).$$

Since y_1, y_2 are fundamental solutions, the Wronskian at x_0,

$$W(x_0) = y_1(x_0)y_2'(x_0) - y_1'(x_0)y_2(x_0),$$

is different from zero. This guarantees that equations (1) can be solved for C_1, C_2. By elementary algebra

$$C_1 = \frac{u(x_0)y_2'(x_0) - u'(x_0)y_2(x_0)}{y_1(x_0)y_2'(x_0) - y_1'(x_0)y_2(x_0)}, \qquad C_2 = \frac{y_1(x_0)u'(x_0) - y_1'(x_0)u(x_0)}{y_1(x_0)y_2'(x_0) - y_1'(x_0)y_2(x_0)},$$

The denominator in both cases is $W(x_0)$, which, since y_1 and y_2 are fundamental, is different from zero.

Our work is finished. The function $C_1y_1 + C_2y_2$ is a solution of the reduced equation which by (1) has the same value as u at x_0 and the same derivative. Thus, by the uniqueness part of Theorem 18.5.2, $C_1y_1 + C_2y_2$ and u cannot be different functions. It shows that

$$u = C_1y_1 + C_2y_2$$

and proves the theorem. ❏

The arbitrary linear combination

$$y = C_1y_1 + C_2y_2$$

of any two fundamental solutions is the *general solution*. By Theorem 18.5.5 we can obtain any *particular solution* by adjusting C_1 and C_2.

We now return to the solutions obtained earlier and prove the final result.

THEOREM 18.5.6

Given the reduced equation

(1) $$y'' + ay' + by = 0,$$

we form the characteristic equation

$$r^2 + ar + b = 0.$$

I. If the characteristic equation has two distinct real roots r_1 and r_2, then the general solution of (1) takes the form

$$y = C_1e^{r_1 x} + C_2e^{r_2 x}.$$

II. If the characteristic equation has only one real root r_1, then the general solution takes the form

$$y = C_1e^{r_1 x} + C_2xe^{r_1 x} = (C_1 + C_2x)e^{r_1 x}.$$

III. If the characteristic equation has two complex roots

$$r_1 = \alpha + i\beta \quad \text{and} \quad r_2 = \alpha - i\beta, \quad \beta \neq 0,$$

then the general solution takes the form

$$y = C_1e^{\alpha x}\cos \beta x + C_2e^{\alpha x}\sin \beta x = e^{\alpha x}(C_1\cos \beta x + C_2\sin \beta x).$$

PROOF To prove this theorem it is enough to show that the three solution pairs

$$e^{r_1 x}, e^{r_2 x} \qquad e^{r_1 x}, xe^{r_1 x} \qquad e^{\alpha x}\cos \beta x, e^{\alpha x}\sin \beta x$$

all have nonzero Wronskians. The Wronskian of the first pair is the function

$$W(x) = e^{r_1 x}\frac{d}{dx}(e^{r_2 x}) - \frac{d}{dx}(e^{r_1 x})e^{r_2 x}$$

$$= e^{r_1 x}r_2e^{r_2 x} - r_1e^{r_1 x}e^{r_2 x} = (r_2 - r_1)e^{(r_1 + r_2)x}.$$

$W(x)$ is different from zero since by assumption $r_2 \neq r_1$.

We leave it to you to verify that the other two pairs also have nonzero Wronskians. ❏

It's time to look at specific examples.

Example 1 Find the general solution of the equation

$$y'' + 2y' - 15y = 0.$$

Then find the particular solution that satisfies the side conditions

$$y(0) = 0, \qquad y'(0) = -1.$$

SOLUTION The characteristic equation is the quadratic

$$r^2 + 2r - 15 = 0.$$

Factoring the left side, we have

$$(r + 5)(r - 3) = 0.$$

There are two real roots: -5 and 3. The general solution takes the form

$$y = C_1 e^{-5x} + C_2 e^{3x}.$$

Differentiating the general solution, we have

$$y' = -5C_1 e^{-5x} + 3C_2 e^{3x}.$$

The conditions

$$y(0) = 0, \qquad y'(0) = -1$$

are satisfied iff

$$C_1 + C_2 = 0 \qquad \text{and} \qquad -5C_1 + 3C_2 = -1.$$

Solving these two equations simultaneously, we find that

$$C_1 = \tfrac{1}{8}, \qquad C_2 = -\tfrac{1}{8}.$$

The solution that satisfies the prescribed side conditions is the function

$$y = \tfrac{1}{8} e^{-5x} - \tfrac{1}{8} e^{3x}.$$

You should verify this. ❏

Example 2 Find the general solution of the equation

$$y'' + 4y' + 4y = 0.$$

Then find the particular solution that satisfies the side conditions

$$y(1) = 1, \qquad y'(1) = 2.$$

SOLUTION The characteristic equation takes the form

$$r^2 + 4r + 4 = 0,$$

which can be written

$$(r + 2)^2 = 0.$$

The number -2 is the only root. The general solution can be written

$$y = C_1 e^{-2x} + C_2 x e^{-2x}.$$

Differentiating the general solution, we have

$$y' = -2C_1 e^{-2x} - 2C_2 x e^{-2x} + C_2 e^{-2x} = -2C_1 e^{-2x} + C_2(1 - 2x)e^{-2x}.$$

The conditions

$$y(1) = 1, \qquad y'(1) = 2$$

require that

$$C_1 e^{-2} + C_2 e^{-2} = 1 \qquad \text{and} \qquad -2C_1 e^{-2} - C_2 e^{-2} = 2.$$

Solve these equations simultaneously for C_1, C_2 and you will find that

$$C_1 = -3e^2, \qquad C_2 = 4e^2.$$

The solution that satisfies the prescribed side conditions is the function

$$y = -3e^2 e^{-2x} + 4e^2 x e^{-2x} = -3e^{2(1-x)} + 4xe^{2(1-x)}.$$

We leave the checking to you. ❏

Example 3 Find the general solution of the equation

$$y'' + y' + 3y = 0.$$

SOLUTION The characteristic equation is

$$r^2 + r + 3 = 0.$$

The quadratic formula shows that there are two complex roots:

$$r_1 = -\tfrac{1}{2} + i\tfrac{1}{2}\sqrt{11}, \qquad r_2 = -\tfrac{1}{2} - i\tfrac{1}{2}\sqrt{11}.$$

The general solution takes the form

$$y = e^{-\frac{1}{2}x}\left[C_1 \cos\left(\tfrac{1}{2}\sqrt{11}\,x\right) + C_2 \sin\left(\tfrac{1}{2}\sqrt{11}\,x\right)\right] \quad ❏$$

Example 4 Find the general solution of the equation

$$y'' + \omega^2 y = 0, \qquad (\omega \neq 0).$$

SOLUTION The characteristic equation is

$$r^2 + \omega^2 = 0$$

and the roots are

$$r_1 = \omega i, \qquad r_2 = -\omega i.$$

Thus, the general solution is

$$y = C_1 \cos \omega x + C_2 \sin \omega x. \quad ❏$$

Remark The equation in Example 4 describes the oscillatory motion of an object suspended by a spring under the assumption that there are no forces acting on the spring–mass system other than the restoring force of the spring. This application and some generalizations are considered in Section 18.7. In the Exercises you are asked to show that the general solution can also be written as

$$y = A \sin(\omega x + \phi_0),$$

where A and ϕ_0 are constants with $A > 0$ and $\phi_0 \in [0, 2\pi)$.

EXERCISES 18.5

In Exercises 1–18, find the general solution of the given equation.

1. $y'' + 2y' - 15y = 0$. 2. $y'' - 13y' + 42y = 0$.
3. $y'' + 8y' + 16y = 0$. 4. $y'' + 7y' + 3y = 0$.
5. $y'' + 2y' + 5y = 0$. 6. $y'' - 3y' + 8y = 0$.
7. $2y'' + 5y' - 3y = 0$. 8. $y'' - 12y = 0$.
9. $y'' + 12y = 0$. 10. $y'' - 3y' + \frac{9}{4}y = 0$.
11. $5y'' + \frac{11}{4}y' - \frac{3}{4}y = 0$. 12. $2y'' + 3y' = 0$.
13. $y'' + 9y = 0$. 14. $y'' - y' - 30y = 0$.
15. $2y'' + 2y' + y = 0$. 16. $y'' - 4y' + 4y = 0$.
17. $8y'' + 2y' - y = 0$. 18. $5y'' - 2y' + y = 0$.

In Exercises 19–24, solve the given initial value problems.

19. $y'' - 5y' + 6y = 0$, $y(0) = 1$, $y'(0) = 1$.
20. $y'' + 2y' + y = 0$, $y(2) = 1$, $y'(2) = 2$.
21. $y'' + \frac{1}{4}y = 0$, $y(\pi) = 1$, $y'(\pi) = -1$.
22. $y'' - 2y' + 2y = 0$, $y(0) = -1$, $y'(0) = -1$.
23. $y'' + 4y' + 4y = 0$, $y(-1) = 2$, $y'(-1) = 1$.
24. $y'' - 2y' + 5y = 0$, $y(\pi/2) = 0$, $y'(\pi/2) = 2$.

25. Find all solutions of the equation $y'' - y' - 2y = 0$ that satisfy the given side conditions:
   (a) $y(0) = 1$. (b) $y'(0) = 1$.
   (c) $y(0) = 1$, $y'(0) = 1$.

26. Prove that the general solution of the differential equation

$$y'' - \omega^2 y = 0 \qquad (\omega > 0)$$

can be written

$$y = C_1 \cosh \omega x + C_2 \sinh \omega x.$$

27. Suppose that the roots r_1 and r_2 of the characteristic equation of (18.5.1) are real and distinct. Then they can be written as $r_1 = \alpha + \beta$ and $r_2 = \alpha - \beta$, where α and β are real. Show that the general solution of the equation (18.5.1) can be expressed in the form

$$y = e^{\alpha x}(C_1 \cosh \beta x + C_2 \sinh \beta x).$$

28. Show that the general solution of the differential equation

$$y'' + \omega^2 y = 0$$

can be written

$$y = A \sin(\omega x + \phi_0)$$

where A and ϕ_0 are constants with $A > 0$ and $\phi_0 \in [0, 2\pi)$.

29. Complete the proof of Theorem 18.5.6 by showing that the following solutions are fundamental:
   (a) $y_1 = e^{r_1 x}$, $y_2 = xe^{r_1 x}$. (one root case)
   (b) $y_1 = e^{\alpha x}\cos \beta x$, $y_2 = e^{\alpha x}\sin \beta x$. (complex root case)

30. In the absence of any external electromotive force, the current i in a simple electrical circuit varies with time t according to the formula

$$L\frac{d^2 i}{dt^2} + R\frac{di}{dt} + \frac{1}{C}i = 0. \qquad (L, R, C \text{ constants})\dagger$$

Find the general solution of this equation given that $L = 1$, $R = 10^3$, and
   (a) $C = 5 \times 10^{-6}$. (b) $C = 4 \times 10^{-6}$.
   (c) $C = 2 \times 10^{-6}$.

31. Determine a differential equation $y'' + ay' + by = 0$ that has the given functions as solutions:
   (a) $y_1 = e^{2x}$, $y_2 = e^{-4x}$.
   (b) $y_1 = 3e^{-x}$, $y_2 = 4e^{5x}$.
   (c) $y_1 = 2e^{3x}$, $y_2 = xe^{3x}$.

32. Determine a differential equation $y'' + ay' + by = 0$ that has the given functions as solutions:
   (a) $y_1 = 2\cos 2x$, $y_2 = -\sin 2x$.
   (b) $y_1 = e^{-2x}\cos 3x$, $y_2 = 2e^{-2x}\sin 3x$.

33. Let y_1, y_2 be two solutions of the reduced equation with y_2 never zero. Show that the Wronskian of y_1, y_2 is zero iff y_1 is a constant multiple of y_2.

Exercises 34 and 35 concern the differential equation $y'' + ay' + by = 0$ where a and b are nonnegative constants.

34. Prove that if a and b are both positive, then $y(x) \to 0$ as $x \to \infty$ for all solutions y of the equation.

35. (a) Prove that if $a = 0$ and $b > 0$, then all solutions of the equation are bounded.
   (b) Suppose that $a > 0$ and $b = 0$, and let $y = y(x)$ be a solution of the equation. Prove that

$$\lim_{x \to \infty} y(x) = k$$

for some constant k. Determine k for the solution that satisfies the side conditions: $y(0) = y_0$, $y'(0) = y_1$.

Euler Equations An equation of the form

$$(*) \qquad x^2 y'' + \alpha xy' + \beta y = 0,$$

where α and β are real numbers, is called an *Euler equation*.

36. Show that the Euler equation $(*)$ can be transformed into an equation of the form

$$\frac{d^2 y}{dz^2} + a\frac{dy}{dz} + by = 0,$$

---

† L is inductance, R is resistance, and C is capacitance. If L is given in henrys, R in ohms, C in farads, and t in seconds, then the current i is given in amperes.

where a and b are real numbers, by means of the change of variable $z = \ln x$. HINT: If $z = \ln x$, then, by the chain rule,

$$\frac{dy}{dx} = \frac{dy}{dz}\frac{dz}{dx} = \frac{dy}{dz}\frac{1}{x}.$$

Now calculate $\dfrac{d^2 y}{dx^2}$ and substitute into the differential equation.

In Exercises 37–40, use the change of variable indicated in Exercise 36 to transform the given equation into an equation with constant coefficients. Find the general solution of that equation, and then express it in terms of x.

37. $x^2 y'' - xy' - 8y = 0.$ **38.** $x^2 y'' - 2xy' + 2y = 0.$
39. $x^2 y'' - 3xy' + 4y = 0.$ **40.** $x^2 y'' - xy' + 5y = 0.$
41. (a) Expand the function $f(x) = e^x$ in a Taylor series in powers of x (see Section 11.6).
   (b) Assuming that the manipulations can be justified, replace x in the series for e^x by $i\theta$, where $i^2 = -1$, and derive Euler's formula
$$e^{i\theta} = \cos\theta + i\sin\theta.$$
   (c) Now replace x by $-i\theta$ and conclude that
$$e^{-i\theta} = \cos\theta - i\sin\theta.$$

■ 18.6 THE EQUATION $y'' + ay' + by = \phi(x)$

In Section 18.5, we solved the reduced equation

$$y'' + ay' + by = 0.$$

In this section we go on to the *complete equation*†

(18.6.1)
$$y'' + ay' + by = \phi(x).$$

The function $\phi = \phi(x)$ appearing on the right side of this equation is often called the *nonhomogeneous term* or the *forcing function*. We assume that ϕ is continuous on some interval I.

We begin by proving three simple but important results.

(18.6.2) If y_1 and y_2 are solutions of the complete equation, then their difference $u = y_1 - y_2$ is a solution of the reduced equation.

PROOF Suppose the y_1 and y_2 are solutions of the complete equation. Then

$$y_1'' + ay_1' + by_1 = \phi(x) \quad \text{and} \quad y_2'' + ay_2' + by_2 = \phi(x).$$

Now let $u = y_1 - y_2$. Then we have

$$u'' + au' + bu = (y_1'' - y_2'') + a(y_1' - y_2') + b(y_1 - y_2)$$
$$= (y_1'' + ay_1' + by_1) - (y_2'' + ay_2' + by_2)$$
$$= \phi(x) - \phi(x) = 0. \quad \square$$

(18.6.3) If y_p is a particular solution of the complete equation, then every solution of the complete equation can be written as a solution of the reduced equation plus y_p.

† Sometimes called the *nonhomogeneous equation*.

PROOF Let y_p be a particular solution of the complete equation. If y is also a solution of the complete equation, then, by (18.6.2), $y - y_p$ is a solution of the reduced equation. Obviously

$$y = (y - y_p) + y_p. \quad \square$$

It follows from (18.6.3) that we can obtain *the general solution* of the complete equation by starting with the general solution of the reduced equation and adding a particular solution of the complete equation. The general solution of the complete equation can thus be written

(18.6.4)
$$\boxed{y = C_1 u_1 + C_2 u_2 + y_p,}$$

where u_1, u_2 are any two fundamental solutions of the reduced equation and y_p is any particular solution of the complete equation.

The third result is called *the superposition principle*. It allows us to reduce a given complete equation to a set of simpler complete equations whose forcing functions have only one term.

(18.6.5)

Given the complete equation

$$(*) \qquad y'' + ay' + by = \phi_1(x) + \phi_2(x).$$

If y_1 is a solution of

$$y'' + ay' + by = \phi_1(x)$$

and y_2 is a solution of

$$y'' + ay' + by = \phi_2(x),$$

then $y_p = y_1 + y_2$ is a solution of $(*)$.

The proof of the superposition principle is left as an exercise.

We now describe two methods for finding a particular solution of the complete equation.

Variation of Parameters

The method that we present here gives (at least in theory) particular solutions to all equations of the form

$$(1) \qquad y'' + ay' + by = \phi(x).$$

The general solution of the reduced equation

$$y'' + ay' + by = 0$$

can be written

$$y = C_1 u_1(x) + C_2 u_2(x),$$

where u_1, u_2 are any fundamental solutions and the coefficients C_1 and C_2 are arbitrary constants. In the method called *variation of parameters* we replace the constants C_1, C_2 by functions

$$z_1 = z_1(x), \qquad z_2 = z_2(x),$$

which are to be determined so that

(2) $$y_p = z_1 u_1 + z_2 u_2$$

is a solution of Equation (1). Since we have two "unknowns," $z_1(x)$ and $z_2(x)$, we can impose two conditions on them. One condition is that (2) satisfies (1). The second condition is still at our disposal, and we shall choose it in a way that will simplify our calculations.

Differentiating (2), we have

$$y_p' = z_1 u_1' + z_1' u_1 + z_2 u_2' + z_2' u_2 = (z_1 u_1' + z_2 u_2') + (z_1' u_1 + z_2' u_2).$$

We now impose the second condition: we require that

(3) $$z_1' u_1 + z_2' u_2 = 0.$$

With this additional restriction, we have

$$y_p' = z_1 u_1' + z_2 u_2',$$

and, differentiating once more,

$$y_p'' = z_1 u_1'' + z_1' u_1' + z_2 u_2'' + z_2' u_2'.$$

Substituting y_p and its derivatives into the differential equation (1), we have

$$y_p'' + a y_p' + b y_p = (z_1 u_1'' + z_1' u_1' + z_2 u_2'' + z_2' u_2') + a(z_1 u_1' + z_2 u_2') + b(z_1 u_1 + z_2 u_2)$$
$$= \phi(x).$$

Rearranging the terms, we obtain

$$(u_1'' + a u_1' + b u_1)z_1 + (u_2'' + a u_2' + b u_2)z_2 + (z_1' u_1' + z_2' u_2') = \phi(x).$$

Since u_1 and u_2 are solutions of the reduced equation, this equation reduces to

(4) $$z_1' u_1' + z_2' u_2' = \phi(x).$$

Equations (3) and (4) can now be solved simultaneously for z_1' and z_2'. As you can verify yourself, the unique solutions are

(5) $$z_1' = -\frac{u_2 \phi}{W} \quad \text{and} \quad z_2' = \frac{u_1 \phi}{W},$$

where the denominator W is the Wronskian of u_1, u_2. The functions z_1, z_2 are now determined up to an additive constant:

$$z_1 = -\int \frac{u_2 \phi}{W} dx, \qquad z_2 = \int \frac{u_1 \phi}{W} dx.$$

It follows then that the function

(18.6.6) $$y_p(x) = \left(-\int \frac{u_2(x)\phi(x)}{W} dx\right) u_1(x) + \left(\int \frac{u_1(x)\phi(x)}{W} dx\right) u_2(x)$$

is a particular solution of the complete equation

$$y'' + ay' + by = \phi(x). \quad \square$$

Example 1 Use variation of parameters to find a solution of the equation

$$y'' + y = \tan x, \qquad -\frac{1}{2}\pi < x < \frac{1}{2}\pi.$$

Also give the general solution of this equation.

SOLUTION The reduced equation

$$y'' + y = 0$$

has fundamental solutions

$$u_1 = \cos x, \qquad u_2 = \sin x.$$

In this instance the Wronskian W is 1:

$$W = u_1 u_2' - u_1' u_2 = (\cos x)(\cos x) - (-\sin x)\sin x = \cos^2 x + \sin^2 x = 1.$$

Since $\phi(x) = \tan x$, we can set

$$z_1 = -\int \frac{u_2 \phi}{W}\, dx$$

$$= -\int \frac{\sin x \tan x}{1}\, dx$$

$$= -\int \frac{\sin^2 x}{\cos x}\, dx$$

$$= \int \frac{\cos x^2 - 1}{\cos x}\, dx = \int (\cos x - \sec x)\, dx = \sin x - \ln|\sec x + \tan x|$$

and

$$z_2 = \int \frac{u_1 \phi}{W}\, dx = \int \frac{\cos x \tan x}{1}\, dx = \int \sin x\, dx = -\cos x.$$

By (18.6.6) the function

$$y_p = (\sin x - \ln|\sec x + \tan x|)\cos x + (-\cos x)\sin x$$
$$= -(\ln|\sec x + \tan x|)\cos x$$

is a particular solution of the complete equation.

The general solution of the complete equation is

$$y = C_1 \cos x + C_2 \sin x - (\ln|\sec x + \tan x|)\cos x \quad \square$$

Example 2 Find the general solution of

$$y'' - 5y' + 6y = 4e^{2x}.$$

SOLUTION The reduced equation

$$y'' - 5y' + 6y = 0$$

has characteristic equation

$$r^2 - 5r + 6 = (r - 2)(r - 3) = 0.$$

Thus, $u_1(x) = e^{2x}$, $u_2(x) = e^{3x}$ are fundamental solutions. Their Wronskian W is given by

$$W = u_1 u_2' - u_2 u_1' = e^{2x}3e^{3x} - e^{3x}2e^{2x} = e^{5x}.$$

Since $\phi(x) = 4e^{2x}$, we have

$$z_1 = -\int \frac{u_2 \phi}{W} dx = -\int \frac{e^{3x} 4e^{2x}}{e^{5x}} dx = -\int 4 \, dx = -4x$$

and

$$z_2 = \int \frac{u_1 \phi}{W} dx = \int \frac{e^{2x} 4e^{2x}}{e^{5x}} dx = \int 4e^{-x} \, dx = -4e^{-x}.$$

Therefore, by (18.6.6),

$$y_p = -4xe^{2x} - 4e^{-x}e^{3x} = -4xe^{2x} - 4e^{2x}$$

is a particular solution of the complete equation.

The general solution of the complete equation is given by

$$y = A_1 u_1 + A_2 u_2 + y_p = A_1 e^{2x} + A_2 e^{3x} - 4xe^{2x} - 4e^{2x}$$
$$= (A_1 - 4)e^{2x} + A_2 e^{3x} - 4xe^{2x}$$
$$= C_1 e^{2x} + C_2 e^{3x} - 4xe^{2x}, \qquad (C_1 = A_1 - 4, \, C_2 = A_2). \quad \square$$

The Method of Undetermined Coefficients

In contrast to variation of parameters, the method of undetermined coefficients only works for equations

$$(1) \qquad\qquad y'' + ay' + by = \phi(x),$$

where the forcing function ϕ has a special form. To motivate the method, consider the expression on the left side of (1):

$$(2) \qquad\qquad y'' + ay' + by.$$

If we calculate (2) for an exponential function $y = Ae^{rx}$, we have

$$y = Ae^{rx}, \qquad y' = Are^{rx}, \qquad y'' = Ar^2 e^{rx}$$

and

$$y'' + ay' + by = Ar^2 e^{rx} + aAre^{rx} + bAe^{rx} = (Ar^2 + aAr + bA)e^{rx}$$
$$= Ke^{rx}, \qquad \text{where } K = Ar^2 + aAr + bA.$$

Thus, the expression (2) "transforms" $y = Ae^{rx}$ into a constant multiple of itself.

Similarly, if we calculate (2) for $y = A \cos \beta x$, $y = A \sin \beta x$, or $y = A \cos \beta x + B \sin \beta x$, we will get

$$y'' + ay' + by = K \cos \beta x + M \sin \beta x,$$

where K and M are constants which are given in terms of A, B, β, a, and b. For example, if we let $y = A \cos \beta x + B \sin \beta x$ and substitute into (2), we will get

$$y'' + ay' + by = (-A\beta^2 + aB\beta + bA)\cos \beta x + (-B\beta^2 - aA\beta + bB)\sin \beta x$$
$$= K \cos \beta x + M \sin \beta x,$$

where $K = -A\beta^2 + aB\beta + bA$ and $M = -B\beta^2 - aA\beta + bB$.

Finally, if we substitute $y = Ae^{\alpha x} \cos \beta x$, $y = Ae^{\alpha x} \sin \beta x$, or $y = Ae^{\alpha x} \cos \beta x + Be^{\alpha x} \sin \beta x$ into (2), we will obtain

$$y'' + ay' + by = Ke^{\alpha x} \cos \beta x + Me^{\alpha x} \sin \beta x,$$

where K and M are constants which are given in terms of A, B, α, β, a, and b.

The point of these observations is that if the forcing function ϕ in Equation (1) has one of the forms

$$\phi(x) = \begin{cases} ce^{rx}, \\ c \cos \beta x, \quad c \sin \beta x, \quad \text{or} \quad c_1 \cos \beta x + c_2 \sin \beta x \\ ce^{\alpha x} \cos \beta x, \quad ce^{\alpha x} \sin \beta x, \quad \text{or} \quad c_1 e^{\alpha x} \cos \beta x + c_2 e^{\alpha x} \sin \beta x, \end{cases}$$

then we can assume that a particular solution of equation (1) will have the corresponding form

$$y_p = \begin{cases} Ae^{rx}, \\ A \cos \beta x + B \sin \beta x, \\ Ae^{\alpha x} \cos \beta x + Be^{\alpha x} \sin \beta x, \end{cases}$$

where the coefficient A, or the coefficients A and B, are to be determined. We then substitute y_p into equation (1) and solve for the *undetermined coefficients*. The following example illustrates the method.

Example 3 Use the method of undetermined coefficients to find a particular solution of each of the following differential equations:

(a) $y'' + 2y' + 5y = 10e^{-2x}$.   **(b)** $y'' + 2y' + y = 10 \cos 3x + 6 \sin 3x$.

SOLUTION **(a)** We assume that the equation has a solution of the form $y_p = Ae^{-2x}$. Then

$$y_p' = -2Ae^{-2x} \quad \text{and} \quad y_p'' = 4Ae^{-2x}.$$

Substituting y_p and its derivatives into the equation, we have

$$4Ae^{-2x} + 2(-2Ae^{-2x}) + 5Ae^{-2x} = 10e^{-2x}$$

$$5Ae^{-2x} = 10e^{-2x}.$$

Therefore,

$$5A = 10 \quad \text{and} \quad A = 2.$$

Thus, a particular solution of the equation is $y_p = 2e^{-2x}$. You can verify that the general solution of the complete equation is

$$y = C_1 e^{-x} \cos 2x + C_2 e^{-x} \sin 2x + 2e^{-2x}.$$

(b) We assume that the equation has a solution of the form $y_p = A \cos 3x + B \sin 3x$. Then

$$y_p' = -3A \sin 3x + 3B \cos 3x \quad \text{and} \quad y_p'' = -9A \cos 3x - 9B \sin 3x.$$

Substituting y_p and its derivatives into the equation, we have

$$(-9A \cos 3x - 9B \sin 3x) + 2(-3A \sin 3x + 3B \cos 3x)$$
$$+ (A \cos 3x + B \sin 3x) = 10 \cos 3x + 6 \sin 3x$$

and

$$(-8A + 6B) \cos 3x + (-6A - 8B)\sin 3x = 10 \cos 3x + 6 \sin 3x.$$

This equation will be satisfied for all x iff

$$\begin{aligned} -8A + 6B &= 10 \\ -6A - 8B &= 6 \end{aligned} \quad \text{or} \quad \begin{aligned} -4A + 3B &= 5 \\ 3A + 4B &= -3. \end{aligned}$$

The solution of this pair of equations is $A = -\frac{29}{25}$, $B = \frac{3}{25}$. Thus

$$y_p = -\frac{29}{25}\cos 3x + \frac{3}{25}\sin 3x$$

is a particular solution. You can verify that

$$y = C_1 e^{-x} + C_2 x e^{-x} - \frac{29}{25}\cos 3x + \frac{3}{25}\sin 3x$$

is the general solution of the complete equation. ❏

There is a difficulty that can arise in trying to use the method of undetermined coefficients. Consider the differential equation in Example 2:

$$y'' - 5y' + 6y = 4e^{2x}.$$

To use the method of undetermined coefficients, we would assume that a particular solution has the form $y = Ae^{2x}$ since $\phi(x) = 4e^{2x}$. Substituting y and its derivatives

$$y' = 2Ae^{2x}, \qquad y'' = 4Ae^{2x}$$

into the equation, we have

$$4Ae^{2x} - 5(2Ae^{2x}) + 6Ae^{2x} = 4e^{2x}$$

$$0 \cdot Ae^{2x} = 4e^{2x}$$

and

$$0 \cdot A = 4.$$

Clearly, this equation has no solutions. The problem is that our assumed solution $y = Ae^{2x}$ is a solution of the reduced equation

$$y'' - 5y' + 6y = 0.$$

From Example 2 we note that the equation has the particular solution $y = -4xe^{2x}$. This suggests that we should try $y_p = Axe^{2x}$ instead of $y = Ae^{2x}$. The derivatives of y_p are

$$y_p' = Ae^{2x} + 2Axe^{2x}, \qquad y_p'' = 4Ae^{2x} + 4Axe^{2x}.$$

Substituting into the differential equation, we have

$$4Ae^{2x} + 4Axe^{2x} - 5(Ae^{2x} + 2Axe^{2x}) + 6Axe^{2x} = 4e^{2x}$$

$$-Ae^{2x} = 4e^{2x}$$

and

$$A = -4.$$

Thus $y_p = -4xe^{2x}$ is a particular solution as we saw in Example 2.

In a similar manner, you can verify that the form of a particular solution of

$$y'' - 4y' + 4y = e^{2x}$$

is $y_p = Ax^2 e^{2x}$ since $u_1 = e^{2x}$ and $u_2 = xe^{2x}$ are fundamental solutions of the reduced equation:

$$y'' - 4y' + 4y = 0.$$

The form of a particular solution of

$$y'' + 9y = \cos 3x$$

is $y_p = Ax \cos 3x + Bx \sin 3x$ since $u_1 = \cos 3x$ and $u_2 = \sin 3x$ are fundamental solutions of

$$y'' + 9y = 0.$$

Table 18.6.1 summarizes our discussion up to this point.

■ **Table 18.6.1** A Particular Solution of $y'' + ay' + by = \phi(x)$

If $\phi(x)$ is	Try $y_p =$
ce^{rx}	Ae^{rx}
$c_1 \cos \beta x + c_2 \sin \beta x,$	$A \cos \beta x + B \sin \beta x$
$c_1 e^{\alpha x} \cos \beta x + c_2 e^{\alpha x} \sin \beta x,$	$Ae^{\alpha x} \cos \beta x + Be^{\alpha x} \sin \beta x$

Note: If y_p satisfies the reduced equation $y'' + ay' + by = 0$, try xy_p; if xy_p also satisfies the reduced equation, then $x^2 y_p$ will give a particular solution.

Remark   In practice it is a good idea to solve the reduced equation before selecting the trial particular solution of the complete equation.   In this way, you will not make the mistake of selecting a y_p which satisfies the reduced equation.   ❏

Example 4   Find the general solution of

$$y'' + 3y' + 2y = 2e^{-x} + 4e^{-x} \cos x.$$

SOLUTION   First consider the reduced equation

$$y'' + 3y' + 2y = 0.$$

The characteristic equation is

$$r^2 + 3r + 2 = (r + 2)(r + 1) = 0,$$

and it follows that $u_1 = e^{-2x}$ and $u_2 = e^{-x}$ are fundamental solutions.

Now we need to find a particular solution of the complete equation.   By the superposition principle (18.6.5), if y_1 is a particular solution of

(∗) $$y'' + 3y' + 2y = 2e^{-x}$$

and y_2 is a particular solution of

(∗∗) $$y'' + 3y' + 2y = 4e^{-x} \cos x,$$

then $y_p = y_1 + y_2$ is a particular solution of the complete equation.

Since $y = e^{-x}$ is a solution of the reduced equation, a particular solution of (∗) has the form $y_1 = Axe^{-x}$.   We then have

$$y_1' = Ae^{-x} - Axe^{-x} \quad \text{and} \quad y_1'' = -2Ae^{-x} + Axe^{-x}.$$

Substituting y_1 and its derivatives into (∗) gives

$$y_1'' + 3y_1' + 2y_1 = -2Ae^{-x} + Axe^{-x} + 3(Ae^{-x} - Axe^{-x}) + 2Axe^{-x} = 2e^{-x},$$

$$Ae^{-x} = 2e^{-x}.$$

Thus, $A = 2$ and $y_1 = 2xe^{-x}$.

A particular solution of (∗∗) has the form $y_2 = Be^{-x} \cos x + Ce^{-x} \sin x$. The derivatives of y_2 are

$$y_2' = (-B + C)e^{-x} \cos x - (B + C)e^{-x} \sin x,$$

$$y_2'' = -2Ce^{-x} \cos x + 2Be^{-x} \sin x. \qquad \text{(verify these).}$$

Substituting y_2 and its derivatives into equation (∗∗), we get

$$[-2C + 3(-B + C) + 2B]e^{-x} \cos x + [2B - 3(B + C) + 2C]e^{-x} \sin x$$
$$= 4e^{-x} \cos x,$$

$$(-B + C)e^{-x} \cos x - (B + C)e^{-x} \sin x = 4e^{-x} \cos x.$$

It follows that

$$-B + C = 4$$

$$B + C = 0$$

and so $B = -2$, $C = 2$. Thus, $y_2 = -2e^{-x} \cos x + 2e^{-x} \sin x$ is a particular solution of (∗∗).

Now, a particular solution of the complete equation is

$$y_p = 2xe^{-x} - 2e^{-x} \cos x + 2e^{-x} \sin x$$

and

$$y = C_1 e^{-2x} + C_2 e^{-x} + 2xe^{-x} - 2e^{-x} \cos x + 2e^{-x} \sin x$$

is the general solution. ❏

Remark It was not actually necessary to solve for the two particular solutions y_1 and y_2 separately. We could have substituted

$$y_p = Axe^{-x} + Be^{-x} \sin x + Ce^{-x} \cos x$$

and its derivatives directly into the complete equation and solved for the three coefficients A, B, C, simultaneously. ❏

We have illustrated how the method of undetermined coefficients can be applied to the complete equation

$$y'' + ay' + by = \phi(x),$$

where the forcing function ϕ has one of the forms ce^{rx}, $c \cos \beta x$, $c \sin \beta x$, $ce^{\alpha x} \cos \beta x$, $ce^{\alpha x} \sin \beta x$, or is a sum of such forms. However, the method actually applies in the more general situation where ϕ has one of the forms:

$$p(x)e^{rx}, \quad p(x) \cos \beta x, \quad p(x) \sin \beta x, \quad p(x)e^{\alpha x} \cos \beta x, \quad p(x)e^{\alpha x} \sin \beta x,$$

where $p(x)$ is a polynomial, or is a sum of such forms. This follows from the fact that the expression

$$y'' + ay' + by$$

will transform $y_p = (A_0 + A_1 x + A_2 x^2 + \cdots + A_n x^n)e^{rx}$ into $P(x)e^{rx}$, where P is a polynomial of degree n (or less); it will transform

$$y_p = (A_0 + A_1 x + A_2 x^2 + \cdots + A_n x^n) \cos \beta x$$

into

$$P(x) \cos \beta x + Q(x) \sin \beta x,$$

where P and Q are polynomials of degree n (or less); and so on. For example, if we calculate

$$y_p'' + 3y_p$$

for $y_p = (A + Bx + Cx^2)e^{2x}$, we get

$$y_p' = 2(A + Bx + Cx^2)e^{2x} + (B + 2Cx)e^{2x},$$

$$y_p'' = 4(A + Bx + Cx^2)e^{2x} + 4(B + 2Cx)e^{2x} + 2Ce^{2x},$$

and

$$y_p'' + 3y_p = 4(A + Bx + Cx^2)e^{2x} + 4(B + 2Cx)e^{2x} + 2Ce^{2x} + 3(A + Bx + Cx^2)e^{2x}$$
$$= [(7A + 4B + 2C) + (7B + 8C)x + 7Cx^2]e^{2x}.$$

The general version of the method of undetermined coefficients can be summarized as follows:

(1) If $\phi(x) = p(x) e^{rx}$, where p is a polynomial of degree n, then try

$$y_p(x) = (A_0 + A_1 x + \cdots + A_n x^n) e^{rx}.$$

(2) If $\phi(x) = p_1(x) \cos \beta x + p_2(x) \sin \beta x$, where p_1 and p_2 are polynomials with maximum degree $(p_1, p_2) = n$, then try

$$y_p(x) = (A_0 + A_1 x + \cdots + A_n x^n) \cos \beta x + (B_0 + B_1 x + \cdots + B_n x^n) \sin \beta x.$$

(3) If $\phi(x) = p_1(x) e^{rx} \cos \beta x + p_2(x) e^{rx} \sin \beta x$, where p_1 and p_2 are polynomials with maximum degree $(p_1, p_2) = n$, then try

$$y_p(x)$$
$$= (A_0 + A_1 x + \cdots + A_n x^n) e^{rx} \cos \beta x + (B_0 + B_1 x + \cdots + B_n x^n) e^{rx} \sin \beta x.$$

Note: If any term in y_p satisfies the reduced equation $y'' + ay' + by = 0$, then try $x \cdot y_p$; if any term in $x \cdot y_p$ satisfies the reduced equation then $x^2 \cdot y_p$ will give a particular solution.

Example 5 Find a particular solution of

$$y'' + 4y = (3 + 2x)e^{-x}.$$

SOLUTION You can verify that $u_1 = \cos 2x$ and $u_2 = \sin 2x$ are fundamental solutions of the reduced equation. Therefore we should try to find a particular solution of the form

$$y_p = (A + Bx)e^{-x}.$$

The derivatives of y_p are

$$y_p' = -(A + Bx)e^{-x} + Be^{-x}, \qquad y_p'' = (A + Bx)e^{-x} - 2Be^{-x}.$$

Substituting y_p and its derivatives into the differential equation, we have

$$y_p'' + 4y_p = (A + Bx)e^{-x} - 2Be^{-x} + 4(A + Bx)e^{-x}$$
$$= (3 + 2x)e^{-x},$$

$$[(5A - 2B) + 5Bx]e^{-x} = (3 + 2x)e^{-x},$$

and it follows that

$$5B = 2$$

$$5A - 2B = 3.$$

The solution of this pair of equations is $A = \frac{19}{25}$, $B = \frac{2}{5}$. Thus,

$$y_p = (\tfrac{19}{25} + \tfrac{2}{5}x)e^{-x}$$

is a particular solution of the complete equation.   ❏

EXERCISES 18.6

In Exercises 1–16, find a particular solution.

1. $y'' + 5y' + 6y = 3x + 4$.

2. $y'' - 3y' - 10y = 5$.

3. $y'' + 2y' + 5y = x^2 - 1$.

4. $y'' + y' - 2y = x^3 + x$.

5. $y'' + 6y' + 9y = e^{3x}$.   **6.** $y'' + 6y' + 9y = e^{-3x}$.

7. $y'' + 2y' + 2y = e^x$.   **8.** $y'' + 4y' + 4y = xe^{-x}$.

9. $y'' - y' - 12y = \cos x$.   **10.** $y'' - y' - 12y = \sin x$.

11. $y'' + 7y' + 6y = 3\cos 2x$.

12. $y'' + y' + 3y = \sin 3x$.

13. $y'' - 2y' + 5y = e^{-x}\sin 2x$.

14. $y'' + 4y' + 5y = e^{2x}\cos x$.

15. $y'' + 6y' + 8y = 3e^{-2x}$.

16. $y'' - 2y' + 5y = e^x \sin x$.

In Exercises 17–24, find the general solution.

17. $y'' + y = e^x$.

18. $y'' - 2y' + y = -25\sin 2x$.

19. $y'' - 3y' - 10y = -x - 1$.

20. $y'' + 4y = x\cos 2x$.   **21.** $y'' + 3y' - 4y = e^{-4x}$.

22. $y'' + 2y' = 4\sin 2x$.   **23.** $y'' + y' - 2y = 3xe^x$.

24. $y'' + 4y' + 4y = x e^{-2x}$.

25. Prove the superposition principle (18.6.5).

26. Use (18.6.5) to find a particular solution.
   (a) $y'' + 2y' - 15y = x + e^{2x}$.
   (b) $y'' - 7y' + 12y = e^{-x} + \sin 2x$.

27. Find the general solution of the equation

$$y'' + 4y' + 3y = \cosh x.$$

In Exercises 28–35, use variation of parameters to find a particular solution.

28. $y'' + y = 3\sin x \sin 2x$.

29. $y'' - 2y' + y = xe^x \cos x$.

30. $y'' + y = \csc x$, $0 < x < \pi$.

31. $y'' - 4y' + 4y = \frac{1}{3}x^{-1}e^{2x}$, $x > 0$.

32. $y'' + 4y = \sec^2 2x$.   **33.** $y'' + 4y' + 4y = \dfrac{e^{-2x}}{x^2}$.

34. $y'' + 2y' + y = e^{-x}\ln x$.

35. $y'' - 2y' + 2y = e^x \sec x$.

36. Show that the change of variable $y = ve^{kx}$ transforms the equation

$$y'' + ay' + by = (c_n x^n + \cdots + c_1 x + c_0)e^{kx}$$

into

$$v'' + (2k + a)v' + (k^2 + ak + b)v$$
$$= c_n x^n + \cdots + c_1 x + c_0.$$

37. In Exercise 30 of Section 18.5 we introduced a differential equation for electrical current in a simple circuit. In the presence of an external electromotive force $F(t)$, the equation takes the form

$$L\frac{d^2 i}{dt^2} + R\frac{di}{dt} + \frac{1}{C}i = F(t).$$

Find the current i given that $F(t) = F_0$, $i(0) = 0$, $i'(0) = F_0/L$ and
(a) $CR^2 = 4L$.   (b) $CR^2 < 4L$.

38. (a) Show that $y_1 = x$, $y_2 = x \ln x$ are solutions of the Euler equation (see Exercises 18.5)

$$x^2 y'' - xy' + y = 0$$

and that their Wronskian is nonzero on $(0, \infty)$.
   (b) Use variation of parameters to find a particular solution of

$$x^2 y'' - xy' + y = 4x \ln x.$$

39. (a) Show that $y_1 = \sin(\ln x^2)$, $y_2 = \cos(\ln x^2)$ are solutions of the Euler equation

$$x^2 y'' + xy' + 4y = 0$$

and that their Wronskian is nonzero on $(0, \infty)$.
   (b) Use variation of parameters to find a particular solution of

$$x^2 y'' + xy' + 4y = \sin(\ln x).$$

■ 18.7 MECHANICAL VIBRATIONS

An object moves along a straight line. Instead of continuing in one direction, it moves back and forth, oscillating about a central point. Call the central point $x = 0$ and denote by $x(t)$ the displacement of the object at time t. If the acceleration is a constant negative multiple of the displacement,

$$a(t) = -kx(t), \qquad k > 0,$$

then the object is said to be in *simple harmonic motion*.

Since, by definition,

$$a(t) = x''(t),$$

in simple harmonic motion, we have

$$x''(t) = -kx(t),$$

which is the same as

$$x''(t) + kx(t) = 0.$$

To emphasize that k is positive, we set $k = \omega^2$, where $\omega = \sqrt{k} > 0$. The equation of motion then takes the form

(18.7.1)
$$x''(t) + \omega^2 x(t) = 0.$$

This is a second-order, linear differential equation with constant coefficients. The characteristic equation is

$$r^2 + \omega^2 = 0,$$

and the roots are $\pm\omega i$. Therefore, the general solution of Equation (18.7.1) is

$$x(t) = C_1 \cos \omega t + C_2 \sin \omega t.$$

A routine calculation shows that the general solution can be written (Exercise 28, Section 18.5)

(18.7.2)
$$x(t) = A \sin(\omega t + \phi_0),$$

where A and ϕ_0 are constants with $A > 0$ and $\phi_0 \in [0, 2\pi)$.

Now let's analyze the motion measuring t in seconds. By adding $2\pi/\omega$ to t we increase $\omega t + \phi_0$ by 2π:

$$\omega\left(t + \frac{2\pi}{\omega}\right) + \phi_0 = \omega t + \phi_0 + 2\pi.$$

This means that the motion is *periodic* with *period T* given by:

$$T = \frac{2\pi}{\omega}.$$

A complete oscillation takes $2\pi/\omega$ seconds. The reciprocal of the period gives the number of complete oscillations per second. This is called the *frequency f*:

$$f = \frac{\omega}{2\pi}.$$

The number ω is called the *angular frequency*. Since $\sin(\omega t + \phi_0)$ oscillates between -1 and 1,

$$x(t) = A \sin(\omega t + \phi_0)$$

oscillates between $-A$ and A. The number A is called the *amplitude* of the motion.

In Figure 18.7.1 we have plotted x against t. The oscillations along the x-axis are now waves in the xt-plane. The period of the motion, $2\pi/\omega$, is the t distance (the time separation) between consecutive wave crests. The amplitude of the motion, A, is the height of the waves measured in x units from $x = 0$. The number ϕ_0 is known as the *phase constant,* or *phase shift*. The phase constant determines the initial displacement (in the xt-plane the height of the wave at time $t = 0$). If $\phi_0 = 0$, the object starts at the center of the interval of motion (the wave starts at the origin of the xt-plane).

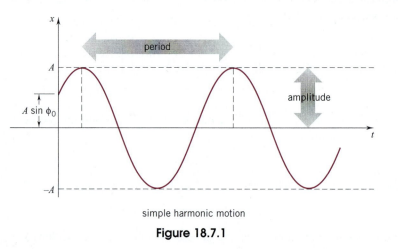

simple harmonic motion

Figure 18.7.1

Example 1 Find an equation for the oscillatory motion of an object, given that the period is $2\pi/3$ and, at time $t = 0$, $x(0) = 1$, $v(0) = x'(0) = 3$.

SOLUTION We begin by setting

$$x(t) = A \sin(\omega t + \phi_0).$$

In general the period is $2\pi/\omega$, so that here

$$\frac{2\pi}{\omega} = \frac{2\pi}{3} \quad \text{and thus} \quad \omega = 3.$$

The equation of motion takes the form

$$x(t) = A \sin(3t + \phi_0).$$

By differentiation

$$v(t) = 3A \cos(3t + \phi_0).$$

The conditions at $t = 0$ give

$$1 = x(0) = A \sin \phi_0, \qquad 3 = v(0) = 3A \cos \phi_0$$

and therefore

$$1 = A \sin \phi_0, \qquad 1 = A \cos \phi_0.$$

Adding the squares, we have

$$2 = A^2 \sin^2 \phi_0 + A^2 \cos^2 \phi_0 = A^2.$$

Since $A > 0$, $A = \sqrt{2}$.

To find ϕ_0 we note that

$$1 = \sqrt{2} \sin \phi_0, \qquad 1 = \sqrt{2} \cos \phi_0.$$

These equations are satisfied by setting $\phi_0 = \frac{1}{4}\pi$. The equation of motion can be written

$$x(t) = \sqrt{2} \sin(3t + \tfrac{1}{4}\pi). \qquad ❏$$

Undamped Vibrations

A coil spring hangs naturally to a length l_0. When a bob of mass m is attached to it, the spring stretches l_1 inches. The bob is later pulled down an additional x_0 inches and then released. What is the resulting motion? Throughout we refer to Figure 18.7.2, taking the downward direction as positive.

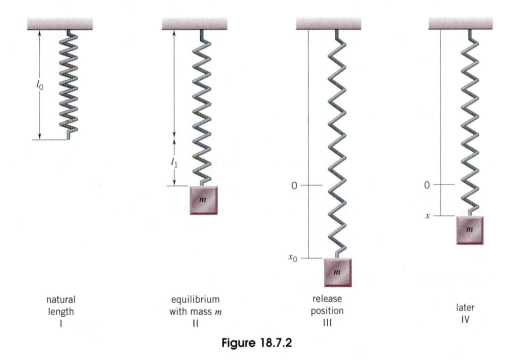

Figure 18.7.2

We begin by analyzing the forces acting on the bob at general position x (stage IV). First there is the weight of the bob:

$$F_1 = mg.$$

This is a downward force, and by our choice of coordinate system, positive. Then there is the restoring force of the spring. This force, by Hooke's law, is proportional to the total displacement $l_1 + x$ and acts in the opposite direction:

$$F_2 = -k(l_1 + x) \qquad \text{with } k > 0.$$

If we neglect resistance, then these are the only forces acting on the bob. Under these conditions the total force is

$$F = F_1 + F_2 = mg - k(l_1 + x),$$

which we rewrite as

(1) $$F = (mg - kl_1) - kx.$$

At stage II (Figure 18.7.2) there was equilibrium. The force of gravity, mg, plus the force of the spring, $-kl_1$, must have been 0:

$$mg - kl_1 = 0.$$

Equation (1) can therefore be simplified to

$$F = -kx.$$

Using Newton's second law

$$F = ma \qquad\qquad (\text{force} = \text{mass} \times \text{acceleration})$$

we have

$$ma = -kx \quad \text{and} \quad \text{thus} \quad a = -\frac{k}{m}x.$$

At any time t,

$$x''(t) = -\frac{k}{m}x(t).$$

Since $k/m > 0$, we can set $\omega = \sqrt{k/m}$ and write

$$x''(t) = -\omega^2 x(t).$$

The motion of the bob is simple harmonic motion with period $T = 2\pi/\omega$. ❏

There is something remarkable about harmonic motion that we have not yet specifically pointed out; namely, that the frequency $f = \omega/2\pi$ is completely independent of the amplitude of the motion. The oscillations of the bob occur with frequency

$$f = \frac{\sqrt{k/m}}{2\pi}, \qquad\qquad (\text{here } \omega = \sqrt{k/m}).$$

By adjusting the spring constant k and the mass of the bob m, we can calibrate the spring–bob system so that the oscillations take place exactly once a second (at least almost exactly). We then have a primitive timepiece (a first cousin of the windup clock). With the passing of time, friction and air resistance reduce the amplitude of the oscillations but not their frequency. By giving the bob a little push or pull once in a while (by rewinding our clock), we can restore the amplitude of the oscillations and thus maintain the steady "ticking."

Damped Vibrations

We derived the equation of motion

$$x'' + \frac{k}{m}x = 0$$

from the force equation

$$F = -kx.$$

Unless the spring is frictionless and the motion takes place in a vacuum, there is a resistance to the motion that tends to dampen the vibrations. Experiment shows that the resistance force R is approximately proportional to the velocity x':

$$R = -cx'. \qquad\qquad (c > 0)$$

Taking this resistance term into account, the force equation reads

$$F = -kx - cx'.$$

Newton's law $F = ma = mx''$ then gives

$$mx'' = -cx' - kx,$$

which we can write as

(18.7.3)

$$x'' + \frac{c}{m}x' + \frac{k}{m}x = 0.$$

This is the equation of motion in the presence of a *damping* factor. To study the motion we analyze this equation.

The characteristic equation

$$r^2 + \frac{c}{m}r + \frac{k}{m} = 0$$

has roots

$$r = \frac{-c \pm \sqrt{c^2 - 4km}}{2m}.$$

There are three possibilities:

$$c^2 - 4km < 0, \qquad c^2 - 4km > 0, \qquad c^2 - 4km = 0.$$

Case 1. $c^2 - 4km < 0$

In this case the characteristic equation has two complex conjugate roots:

$$r_1 = -\frac{c}{2m} + i\omega, \qquad r_2 = -\frac{c}{2m} - i\omega \qquad \text{where} \quad \omega = \frac{\sqrt{4km - c^2}}{2m}.$$

The general solution of (18.7.3),

$$x = e^{-(c/2m)t}(c_1 \cos \omega t + c_2 \sin \omega t),$$

can be written

(18.7.4)

$$x(t) = Ae^{(-c/2m)t} \sin(\omega t + \phi_0),$$

where, as before, A and ϕ_0 are constants, $A > 0$, $\phi_0 \in [0, 2\pi)$. This is called the *underdamped case*. The motion is similar to simple harmonic motion, except that the damping term $e^{(-c/2m)t}$ ensures that $x \to 0$ as $t \to \infty$. The vibrations continue indefinitely with constant frequency $\omega/2\pi$ but diminishing amplitude $Ae^{(-c/2m)t}$. As $t \to \infty$, the amplitude of the vibrations tends to zero; the vibrations die down. The motion is illustrated in Figure 18.7.3. ❏

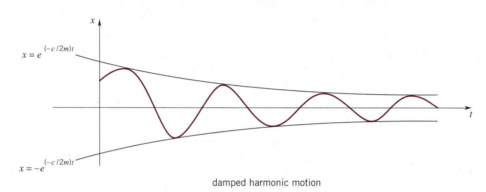

damped harmonic motion

Figure 18.7.3

Case 2. $c^2 - 4km > 0$

In this case the characteristic equation has two distinct real roots:

$$r_1 = \frac{-c + \sqrt{c^2 - 4km}}{2m}, \qquad r_2 = \frac{-c - \sqrt{c^2 - 4km}}{2m}.$$

The general solution takes the form

(18.7.5)

$$x = C_1 e^{r_1 t} + C_2 e^{r_2 t}.$$

This is called the *overdamped case*. The motion is nonoscillatory. Since $\sqrt{c^2 - 4km} < \sqrt{c^2} = c$, both r_1 and r_2 are negative. As $t \to \infty$, $x \to 0$. ❏

Case 3. $c^2 - 4km = 0$

In this case the characteristic equation has only one root

$$r_1 = -\frac{c}{2m}$$

and the general solution takes the form

(18.7.6)
$$x = C_1 e^{-(c/2m)t} + C_2 t e^{-(c/2m)t}.$$

This is called the *critically damped case*. Once again the motion is nonoscil-latory. Moreover, as $t \to \infty$, $x \to 0$. ❏

In both the overdamped and critically damped cases, the mass moves slowly back to its equilibrium position ($x \to 0$ as $t \to \infty$). Depending upon the side conditions, the mass may move through the equilibrium once, but only once; there is no oscillatory motion. Two typical examples of the motion are shown in Figure 18.7.4.

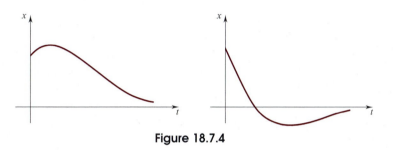

Figure 18.7.4

Forced Vibrations

The vibrations that we have been considering result from the interplay of three forces: the force of gravity, the elastic force of the spring, and the retarding force of the surrounding medium. Such vibrations are called *free vibrations*.

The application of an external force to a freely vibrating system modifies the vibrations and results in what are called *forced vibrations*. In what follows we exam-ine the effect of a pulsating force $F_0 \cos \gamma t$. Without loss of generality we can take both F_0 and γ as positive.

In an undamped system the force equation reads

$$F = -kx + F_0 \cos \gamma t$$

and the equation of motion takes the form

(18.7.7)
$$x'' + \frac{k}{m} x = \frac{F_0}{m} \cos \gamma t.$$

As usual we set $\omega = \sqrt{k/m}$ and write

(18.7.8)
$$x'' + \omega^2 x = \frac{F_0}{m} \cos \gamma t.$$

As you'll see, the nature of the vibrations depends on the relation between the *applied frequency*, $\gamma/2\pi$, and the *natural frequency* of the system, $\omega/2\pi$.

Case 1. $\gamma \neq \omega$

In this case the method of undetermined coefficients gives the particular solution

$$x_p = \frac{F_0/m}{\omega^2 - \gamma^2} \cos \gamma t.$$

The general equation of motion can thus be written

(18.7.9)
$$x = A \sin(\omega t + \phi_0) + \frac{F_0/m}{\omega^2 - \gamma^2} \cos \gamma t.$$

If ω/γ is rational, the vibrations are periodic. If, on the other hand, ω/γ is not rational, then the vibrations are not periodic and the motion, though bounded by

$$|A| + \left| \frac{F_0/m}{\omega^2 - \gamma^2} \right|,$$

can be highly irregular. ❏

Case 2. $\gamma = \omega$

In this case the method of undetermined coefficients gives

$$x_p = \frac{F_0}{2\omega m} t \sin \omega t$$

and the general solution takes the form

(18.7.10)
$$x = A \sin(\omega t + \phi_0) + \frac{F_0}{2\omega m} t \sin \omega t.$$

The undamped system is said to be in *resonance*. The motion is oscillatory, but, because of the extra t present in the second summand, it is far from periodic. As $t \to \infty$, the amplitude of vibration increases without bound. The motion is illustrated in Figure 18.7.5. ❏

Undamped systems and unbounded vibrations are mathematical fictions. No real mechanical system is totally undamped, and unbounded vibrations do not occur in nature. Nevertheless a form of resonance can occur in a real mechanical system. (See Exercises 24–28.) A periodic external force applied to a mechanical system that is insufficiently damped can set up vibrations of very large amplitude. Such vibrations have caused the destruction of some formidable man-made structures. In 1850 the suspension bridge at Angers, France, was destroyed by vibrations set up by the unified step of a column of marching soldiers. More than two hundred French soldiers were killed in that catastrophe. (Soldiers today are told to break ranks before crossing a

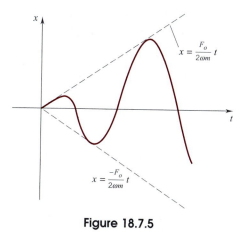

Figure 18.7.5

bridge.) The collapse of the bridge at Tacoma, Washington, is a more recent event. Slender in construction and graceful in design, the Tacoma bridge was opened to traffic on July 1, 1940. The third longest suspension bridge in the world, with a main span of 2800 feet, the bridge attracted many admirers. On November 1 of that same year, after less than five months of service, the main span of the bridge broke loose from its cables and crashed into the water below. (Luckily only one person was on the bridge at the time, and he was able to crawl to safety.) A driving wind had set up vibrations in resonance with the natural vibrations of the roadway, and the stiffening girders of the bridge had not provided sufficient damping to keep the vibrations from reaching destructive magnitude.

EXERCISES 18.7

1. An object is in simple harmonic motion. Find an equation for the motion given that the period is $\frac{1}{4}\pi$ and, at time $t = 0$, $x = 1$ and $v = 0$. What is the amplitude? What is the frequency?

2. An object is in simple harmonic motion. Find an equation for the motion given that the frequency is $1/\pi$ and, at time $t = 0$, $x = 0$ and $v = -2$. What is the amplitude? What is the period?

3. An object is in simple harmonic motion with period T and amplitude A. What is the velocity at the central point $x = 0$?

4. An object is in simple harmonic motion with period T. Find the amplitude given that $v = \pm v_0$ at $x = x_0$.

5. An object in simple harmonic motion passes through the central point $x = 0$ at time $t = 0$ and every 3 seconds thereafter. Find the equation of motion given that $v(0) = 5$.

6. Show that simple harmonic motion $x(t) = A \sin(\omega t + \phi_0)$ can just as well be written: (a) $x(t) = A \cos(\omega t + \phi_1)$; (b) $x(t) = B \sin \omega t + C \cos \omega t$.

Exercises 7–12 are concerned with the motion of the bob depicted in Figure 18.7.2.

7. What is $x(t)$ for the bob of mass m?

8. Find the positions of the bob where the bob attains: (a) maximum speed; (b) zero speed; (c) maximum acceleration; (d) zero acceleration.

9. Where does the bob take on half of its maximum speed?

10. Find the maximal kinetic energy obtained by the bob. (Remember: $KE = \frac{1}{2}mv^2$ where m is the mass of the object and v is the speed.)

11. Find the time average of the kinetic energy of the bob during one period T.

12. Express the velocity of the bob in terms of k, m, x_0, and $x(t)$.

13. Given that $x''(t) = 9 - 4x(t)$ with $x(0) = 0$ and $x'(0) = 0$, show that the motion is simple harmonic motion centered at $x = 2$. Find the amplitude and the period.

14. The figure shows a pendulum of mass m swinging on an arm of length L. The angle θ is measured counterclockwise. Neglecting friction and the weight of the arm, we can describe the motion by the equation

$$mL\theta''(t) = -mg \sin \theta(t)$$

which reduces to

$$\theta''(t) = -\frac{g}{L}\sin\theta(t).$$

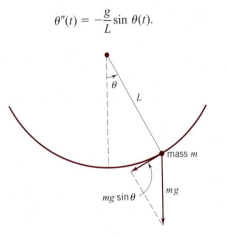

mass m

$mg\sin\theta$

mg

(a) For small angles we replace $\sin\theta$ by θ and write

$$\theta''(t) \cong -\frac{g}{L}\theta(t).$$

Justify this step.

(b) Solve the approximate equation in (a) given that the pendulum
   (i) is held at an angle $\theta_0 > 0$ and released at time $t = 0$.
   (ii) passes through the vertical position at time $t = 0$ and $\theta'(0) = -\sqrt{g/L}\,\theta_0$.

(c) Find L given that the motion repeats itself every 2 seconds.

15. A cylindrical buoy of mass m and radius r centimeters floats with its axis vertical in a liquid of density ρ kilograms per cubic centimeter. Suppose that the buoy is pushed x_0 centimeters down into the liquid. See the figure.

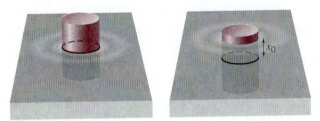

x_0

(a) Neglecting friction and given that the buoyancy force is equal to the weight of the liquid displaced, show that the buoy bobs up and down in simple harmonic motion by finding the equation of motion.

(b) Solve the equation obtained in (a). Specify the amplitude and the period.

16. Explain in detail the connection between uniform circular motion and simple harmonic motion.

17. What is the effect of an increase in the resistance constant c on the amplitude and frequency of the vibrations given by Equation (18.7.4).

18. Prove that the motion given by (18.7.5) can pass through the equilibrium point at most once. How many times can the motion change directions?

19. Prove that the motion given by (18.7.6) can pass through the equilibrium point at most once. How many times can the motion change directions?

20. Show that, if $\gamma \neq \omega$, then the method of undetermined coefficients applied to (18.7.8) gives

$$x_p = \frac{F_0/m}{\omega^2 - \gamma^2}\cos\gamma t.$$

21. Show that, if ω/γ, is rational, then the vibrations given by (18.7.9) are periodic.

22. Show that, if $\gamma = \omega$, then the method of undetermined coefficients applied to (18.7.8) gives

$$x_p = \frac{F_0}{2\omega m}t\sin\omega t.$$

Forced vibrations in a damped system   Write the equation

$$x'' + \frac{c}{m}x' + \frac{k}{m}x = \frac{F_0}{m}\cos\gamma t$$

as

$$(*) \qquad x'' + 2\alpha x' + \omega^2 x = \frac{F_0}{m}\cos\gamma t.$$

We will assume throughout that $0 < \alpha < \omega$. (For large α the resistance is large and the motion is not as interesting.)

23. Find the general solution of the reduced equation $x'' + 2\alpha x' + \omega^2 x = 0$.

24. Verify that the function

$$x_p = \frac{F_0/m}{(\omega^2 - \gamma^2)^2 + 4\alpha^2\gamma^2}[(\omega^2 - \gamma^2)\cos\gamma t + 2\alpha\gamma\sin\gamma t]$$

is a particular solution of (*).

25. Determine x_p if $\omega = \gamma$. Show that the amplitude of the vibrations is very large if the resistance constant c is very small.

26. Show that the solution x_p in Exercise 24 can be written

$$x_p = \frac{F_0/m}{\sqrt{(\omega^2 - \gamma^2)^2 + 4\alpha^2\gamma^2}}\sin(\gamma t + \phi).$$

27. Show that, if $2\alpha^2 \geq \omega^2$, then the amplitude of vibration of the solution x_p in Exercise 26 decreases as γ increases.

28. Suppose now that $2\alpha^2 \leq \omega^2$.
   (a) Find the value of γ that maximizes the amplitude of the solution x_p in Exercise 26.
   (b) Determine the frequency that corresponds to this value of γ. (This is called the *resonant frequency* of the damped system.)
   (c) What is the *resonant amplitude* of the system? (In other words, what is the amplitude of vibration when the applied force is at resonant frequency?)
   (d) Show that, if c, the constant of resistance, is very small, then the resonant amplitude is very large.

■ CHAPTER HIGHLIGHTS

18.1 Introduction

Differential equation (p. 1195)   Order (p. 1196)
Solution (p. 1196) N-parameter family of solutions (p. 1198)
General solution (p. 1198)   Side conditions/initial conditions (p. 1198)

18.2 First Order Linear Differential Equations

A first order differential equation is *linear* if it can be written in the form

$$y' + p(x)y = q(x), \quad \text{where } p \text{ and } q \text{ are continuous on some interval } I.$$

The general solution is $y = e^{-H(x)}\left\{\int e^{H(x)} q(x)\,dx + C\right\}$ (p. 1201).

Integrating factor (p. 1201)

18.3 Separable Equations; Homogeneous Equations

A first order differential equation is *separable* if it can be written in the form

$$p(x) + q(y)y' = 0 \quad \text{or} \quad p(x)\,dx + q(y)\,dy = 0.$$

A one-parameter family of solutions is $\int p(x)\,dx + \int q(y)\,dy = C$ (p. 1209).

The logistic equation (p. 1213)
The first order differential equation $y' = f(x, y)$ is *homogeneous* if $f(tx, ty) = f(x, y)$ for all $t \neq 0$ (p. 1216).
The transformation $v = y/x$ transforms a homogeneous equation into a separable equation (p. 1216).

18.4 Exact Equations; Integrating Factors

The first order differential equation

$P(x, y) + Q(x, y)y' = 0$, where P and Q are continuously differentiable in a region Ω,

is *exact* iff $\partial P/\partial y = \partial Q/\partial x$ for all $(x, y) \in \Omega$.
The function $\mu = \mu(x, y)$ is an *integrating factor* for $P(x, y) + Q(x, y)y' = 0$ if

$$\mu(x, y)P(x, y) + \mu(x, y)Q(x, y)y' = 0$$

is exact (p. 1221).

18.5 The Equation $y'' + ay' + by = 0$

Reduced equation (homogeneous equation) (p. 1224)
Linear combinations of solutions (p. 1224)   The characteristic equation (p. 1225)
Existence and uniqueness of solutions (p. 1228)
The Wronskian (p. 1228)   Fundamental solutions (p. 1229)
Let r_1, r_2 be the roots of the characteristic equation. The general solution of

$$y'' + ay' + by = 0 \text{ is}$$

I. $y = C_1 e^{r_1 x} + C_2 e^{r_2 x}$ if r_1 and r_2 are real and unequal.

II. $y = C_1 e^{rx} + C_2 x e^{rx}$ if $r_1 = r_2 = r$ are real and equal.

III. $y = e^{\alpha x}(C_1 \cos \beta x + C_2 \sin \beta x)$ if $r_1 = \alpha + i\beta$ and $r_2 = \alpha - i\beta$ are complex.

18.6 The Equation $y'' + ay' + by = \phi(x)$

Complete equation (nonhomogeneous equation) (p. 1234)
Nonhomogeneous term (forcing function) (p. 1234)
The general solution of the complete equation is $y = C_1 u_1(x) + C_2 u_2(x) + y_p(x)$, where u_1 and u_2 are fundamental solutions of the reduced equation and y_p is a particular solution of the complete equation.

Superposition principle (p. 1235)   Variation of parameters (p. 1235)

If u_1 and u_2 are fundamental solutions of the reduced equation and W is their Wronskian, then

$$y_p = \left(-\int \frac{u_2(x)\phi(x)}{W}\,dx \right) u_1 + \left(\int \frac{u_1(x)\alpha(x)}{W}\,dx \right) u_2$$

is a particular solution of the complete equation.

Undetermined coefficients, see Table 18.6.1 (p. 1241).

18.7 Mechanical Vibrations

Simple harmonic motion: $x'' + \omega^2 x = 0$ (p. 1245)   Undamped vibrations (p. 1247)

Damped vibrations: $x'' + \dfrac{c}{m}x' + \dfrac{k}{m}x = 0$ (p. 1249)   Damping factor (p. 1249)

Underdamped (p. 1250)   Overdamped (p. 1250)   Critically damped (p. 1251)

Forced vibrations: $x'' + \omega^2 x = \dfrac{F_0}{m}\cos \gamma t$ (p. 1251)

Natural frequency (p. 1252)   Applied frequency (p. 1252)

■ PROJECTS AND EXPLORATIONS USING TECHNOLOGY

To do these exercises you will need a graphics calculator or a computer with graphing capability. The majority of these problems are open-ended so different appraoches may be used to solve them. You should be aware that different approaches can result in slight variations in the answers. Round your numerical answers to at least four decimal places. The rounding method that your calculator or computer uses also may cause variations in answers.

18.1 The problem of solving a first order differential equation $y' = f(x, y)$ subject to an initial condition $y(a) = \alpha$ is called an *initial value problem*. Recall that a function $y = g(x)$ defined on an interval I is a solution of the initial value problem

(1) $\qquad\qquad y' = f(x, y), \quad y(a) = \alpha, \quad \text{where } a \in I,$

if $g(a) = \alpha$ and $g'(x) = f(x, g(x))$ for all $x \in I$. For our purposes in these exercises, we assume that (1) has a unique solution. For example, it can be shown that if f and $\partial f/\partial y$ are continuous functions on a region Ω in the x,y-plane and $(a, \alpha) \in \Omega$, then (1) has a unique solution that is defined on some interval I containing a.

There are no general methods for solving initial value problems such as (1). Consequently, there is a long history of methods for approximating solutions. Numerical methods for constructing approximate solutions of differential equations were among the first applications of digital computers.

Euler's method is a relatively simple method for constructing an approximate solution of (1) on a closed interval $[a, b]$. It is based on the first order Taylor (or tangent line) approximation. If $y = g(x)$ is the solution of (1) and $h \neq 0$ is fixed, then

$$g(x + h) \cong g(x) + g'(x)h = g(x) + f(x, g(x))h.$$

Euler's method uses this approximation to calculate a sequence of points (x_i, w_i), $i = 0, 1, 2, \ldots, n$, where $x_i = a + ih$, $h = (b - a)/n$ and $w_i \cong g(x_i)$. The values w_i are calculated recursively by

$$w_0 = \alpha, \quad \text{and} \quad w_{i+1} = w_i + h f(x_i, w_i), \quad i = 0, 1, 2, \ldots, n - 1.$$

These ideas are illustrated in the figure below.

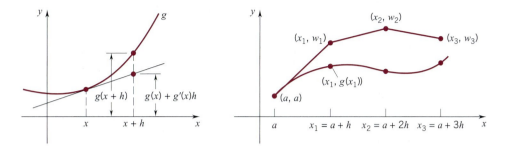

(a) Write a program for your calculator or computer to implement Euler's method. Test your program on the following problems:

(i) $y' = y$, $y(0) = 1$; $x \in [0, 2]$.

(ii) $y' = x + 2y$, $y(0) = 1$; $x \in [0, 1]$.

(iii) $y' = 2xy$, $y(1) = 1$; $x \in [1, 2]$.

(b) The initial value problems in part (a) can be solved exactly. Using your results in (a), plot the points (x_j, w_j), connect them with straight line segments, and compare with the graph of the exact solution for each of the problems.

(c) We would expect the computed sequence (x_j, w_j) to get closer to the exact solution as n increases. We want to determine how quickly this happens. For each of the problems in part (a) calculate $w_n \cong g(b)$ for $n = 1, 2, 4, 8, 16, 32, 64, 128$, and plot $|w_n - g(b)|$ versus n. Then try to find curves that fit each of the data sets.

(d) We would expect to find better approximations if we used a second order Taylor approximation instead of the first order (tangent line) approximation. Fix $h \ne 0$. Show that if the function f in (1) is differentiable, then the solution $y = g(x)$ of the initial value problem (1) is twice differentiable and

$$g(x + h) \cong g(x) + g'(x)h + g''(x)\frac{h^2}{2}$$

$$= g(x) + f(x, g(x))h + \left[\frac{\partial f}{\partial x}(x, g(x)) + \frac{\partial f}{\partial y}(x, g(x))f(x, g(x)) \right] \frac{h^2}{2}.$$

(e) The approximation in (d) leads to the recurrence relation

$$w_0 = \alpha, \text{ and } w_{i+1} = w_i + h f(x_i, w_i) + \frac{h^2}{2} \left[\frac{\partial f}{\partial x}(x_i, w_i) + \frac{\partial f}{\partial y}(x_i, w_i) f(x_i, w_i) \right],$$

$$i = 0, 1, \ldots, n - 1.$$

Modify the procedures written for part (a) to use this recurrence relation. Then repeat part (c) and compare your results with the results of Euler's method.

18.2 A geometric method for studying the first order differential equation

(2) $$y' = f(x, y)$$

involves the use of *slope fields* (also called *direction fields*). Slope fields are used to give qualitative information about the solutions (are solutions increasing? decreasing? concave up? and so forth).

According to the differential equation (2), if y is a solution, then its slope y' at a point (x, y) is simply $f(x, y)$. A slope field for (2) is constructed as follows. Select a set of grid points (x_i, y_i), $i = 1, 2, \ldots, n$ in the plane, calculate $f(x_i, y_i)$ at each grid point, and then draw a short line segment through (x_i, y_i) with slope $f(x_i, y_i)$. For example, a slope field for

$$y' = x - y$$

on the rectangle $-3 \leq x \leq 3$, $-3 \leq y \leq 3$, is shown in Figure A. To sketch the graph of a solution of an initial value problem, start at the initial point and follow the line segments. A sketch of the solution to $y' = x - y$, $y(0) = 0$, is shown in Figure B.

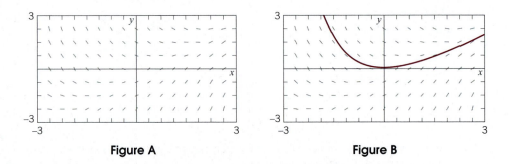

Figure A   **Figure B**

(a) Use technology to draw slope fields for the differential equations in Exercise 18.1 (a). Then use the slope fields to sketch the graphs of the solutions with the given initial conditions. (Some software has these capabilities programmed in, your instructor may provide special software, or you may need to write your own programs.)

(b) Use your technology from part (a) to draw a slope field for the differential equation $y' = x^4 - y^5$. Select grid points in the rectangle $-10 \leq x \leq 10$, $-10 \leq y \leq 10$.

(c) The initial value problem $y' = x^4 - y^5$, $y(0) = \alpha$ has a unique solution. Using your slope field in part (b), sketch the solution curves for the initial value problems corresponding to $\alpha = -1$, $\alpha = 0$, $\alpha = 1$, $\alpha = 5$, and $\alpha = 10$. If possible, compare these sketches with the solution curves drawn by technology.

(d) Explain why each solution is decreasing to a minimum, and then increasing.

(e) Show that every solution is eventually less than $y = \sqrt{x}$. Explain why the solutions get close together as $x \to \infty$.

(f) Let $y = g(x)$ be the solution of the initial value problem $y' = x^4 - y^5$, $y(0) = 10$, on the interval $[0, 2]$. Use Euler's method from Exercise 18.1 to see how well it does with this problem.

SOME ADDITIONAL TOPICS

APPENDIX

A

■ A.1 ROTATION OF AXES; EQUATIONS OF SECOND DEGREE

Rotation of Axes

We begin by referring to Figure A.1.1. From the figure,

$$\cos \theta = \frac{x}{r}, \qquad \sin \theta = \frac{y}{r}.$$

Thus

(A.1.1)

$$x = r \cos \theta, \qquad y = r \sin \theta.$$

Figure A.1.1

Equations (A.1.1) come up repeatedly in calculus. In particular, these are the equations that you use to convert polar coordinates to rectangular coordinates (see Section 9.2).

Consider now a rectangular coordinate system Oxy. If we rotate this system counterclockwise α radians about the origin, we obtain a new coordinate system OXY. See Figure A.1.2.

A point P will now have two pairs of rectangular coordinates:

$$(x, y) \text{ in the } Oxy \text{ system} \qquad \text{and} \qquad (X, Y) \text{ in the } OXY \text{ system}.$$

Here we investigate the relation between (x, y) and (X, Y). With P as in Figure A.1.3,

$$x = r \cos (\alpha + \beta), \qquad y = r \sin (\alpha + \beta)$$

and

$$X = r \cos \beta, \qquad Y = r \sin \beta.$$

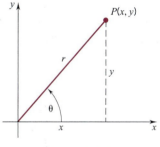

Figure A.1.2

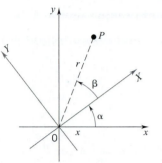

Figure A.1.3

Since

$$\cos (\alpha + \beta) = \cos \alpha \cos \beta - \sin \alpha \sin \beta,$$
$$\sin (\alpha + \beta) = \sin \alpha \cos \beta + \cos \alpha \sin \beta,$$

we have

$$x = r \cos (\alpha + \beta) = (\cos \alpha) r \cos \beta - (\sin \alpha) r \sin \beta,$$
$$y = r \sin (\alpha + \beta) = (\sin \alpha) r \cos \beta + (\cos \alpha) r \sin \beta,$$

and therefore

(A.1.2)

$$x = (\cos \alpha)X - (\sin \alpha)Y, \qquad y = (\sin \alpha)X + (\cos \alpha)Y.$$

These formulas give the algebraic consequences of a counterclockwise rotation of α radians.

Equations of Second Degree

Equations of the form

$$(1) \qquad\qquad ax^2 + cy^2 + dx + ey + f = 0,$$

where a, c, d, e, f are constants and a and c are not both zero, are studied in Sections 1.4 and 9.1, and they occur throughout the text in the examples and exercises. Except for degenerate cases (for example, $x^2 + y^2 + 1 = 0$ or $x^2 - y^2 = 0$), the graph of (1) is a conic section: circle, ellipse, parabola, or hyperbola.

The *general equation of second degree in x and y* is an equation of the form

(A.1.3)

$$ax^2 + bxy + cy^2 + dx + ey + f = 0,$$

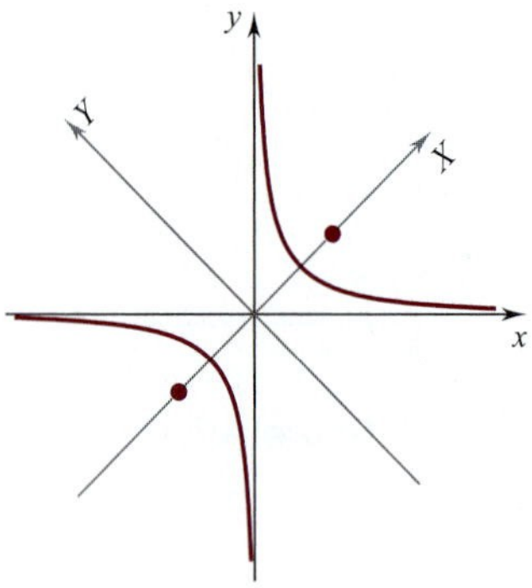

where a, b, c, d, e, f are constants and a, b, c are not all zero. The graph of such an equation is still a conic section (again, except for degenerate cases). For example, the graph of the equation

$$xy - 2 = 0$$

Figure A.1.4

is the hyperbola shown in Figure A.1.4.

Eliminating the xy-Term

Rotations of the coordinate system enable us to simplify equations of the second degree by eliminating the xy-term. That is, if in the Oxy coordinate system, a curve S has an equation of the form

$$(2) \qquad ax^2 + bxy + cy^2 + dx + ey + f = 0 \qquad \text{with} \qquad b \ne 0,$$

then there exists a coordinate system OXY, differing from Oxy by a rotation α, where $0 < \alpha < \pi/2$, such that in the OXY system S has an equation of the form

$$(3) \qquad\qquad AX^2 + CY^2 + DX + EY + F = 0,$$

where A and C are not both zero. To see this, substitute

$$x = (\cos \alpha)X - (\sin \alpha)Y, \qquad y = (\sin \alpha)X + (\cos \alpha)Y$$

in equation (2). This will give you a second-degree equation in X and Y in which the coefficient of XY is

$$-2a \cos \alpha \sin \alpha + b(\cos^2 \alpha - \sin^2 \alpha) + 2c \cos \alpha \sin \alpha.$$

This can be simplified to

$$(c - a) \sin 2\alpha + b \cos 2\alpha.$$

To eliminate the XY term we must have this coefficient equal to zero, that is, we must have

$$b \cos 2\alpha = (a - c)\sin 2\alpha$$

or

$$\cot 2\alpha = \frac{a - c}{b} \qquad \text{(recall } b \neq 0\text{)}.$$

Therefore,

$$2\alpha = \cot^{-1}\left(\frac{a - c}{b}\right)$$

and

$$\alpha = \frac{1}{2}\cot^{-1}\left(\frac{a - c}{b}\right).$$

Since the range of the inverse cotangent function is $(0, \pi)$, it follows that $0 < \alpha < \pi/2$.

We have shown that an equation of the form (2) can be transformed into an equation of the form (3) by rotating the axes through the angle α given by

(A.1.5)

$$\alpha = \frac{1}{2}\cot^{-1}\left(\frac{a - c}{b}\right)$$

We leave it as an exercise to show that the coefficients A and C in (3) are not both zero.

Example 1 In the case of

$$xy - 2 = 0,$$

we have $a = c = 0$, $b = 1$, and $\alpha = \frac{1}{2}\cot^{-1}(0) = \frac{1}{4}\pi$. Setting

$$x = (\cos \tfrac{1}{4}\pi)X - (\sin \tfrac{1}{4}\pi)Y = \tfrac{1}{2}\sqrt{2}\,(X - Y),$$
$$y = (\sin \tfrac{1}{4}\pi)X + (\cos \tfrac{1}{4}\pi)Y = \tfrac{1}{2}\sqrt{2}\,(X + Y),$$

we find that $xy - 2 = 0$ becomes

$$\tfrac{1}{2}(X^2 - Y^2) - 2 = 0,$$

which can be written

$$\frac{X^2}{4} - \frac{Y^2}{4} = 1.$$

This is the equation of a hyperbola in standard position in the OXY system. The hyperbola is shown in Figure A.1.4. ❏

Example 2 In the case of

$$11x^2 + 4\sqrt{3}xy + 7y^2 - 1 = 0,$$

we have $a = 11$, $b = 4\sqrt{3}$, and $c = 7$. Thus we can choose

$$\alpha = \tfrac{1}{2}\cot^{-1}\left(\frac{11-7}{4\sqrt{3}}\right) = \tfrac{1}{2}\cot^{-1}\left(\frac{1}{\sqrt{3}}\right) = \tfrac{1}{6}\pi$$

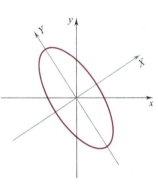

Setting

$$x = (\cos \tfrac{1}{6}\pi)X - (\sin \tfrac{1}{6}\pi)Y = \tfrac{1}{2}(\sqrt{3}X - Y),$$
$$y = (\sin \tfrac{1}{6}\pi)X + (\cos \tfrac{1}{6}\pi)Y = \tfrac{1}{2}(X + \sqrt{3}Y),$$

we find that our initial equation simplifies to $13X^2 + 5Y^2 - 1 = 0$, which we can write as

$$\frac{X^2}{(1/\sqrt{13})^2} + \frac{Y^2}{(1/\sqrt{5})^2} = 1.$$

Figure A.1.5

This is the equation of an ellipse. The ellipse is pictured in Figure A.1.5. ❏

The Discriminant

It is possible to draw general conclusions about the graph of a second-degree equation

$$ax^2 + bxy + cy^2 + dx + ey + f = 0, \qquad a, b, c \text{ not all } 0,$$

just from the *discriminant* $\Delta = b^2 - 4ac$. There are three cases:

Case 1. If $\Delta < 0$, the graph is an ellipse, a circle, a point, or empty.

Case 2. If $\Delta > 0$, the graph is a hyperbola or a pair of intersecting lines.

Case 3. If $\Delta = 0$, the graph is a parabola, a line, a pair of lines, or empty.

Below we outline how these assertions can be verified. A useful first step is to rotate the coordinate system so that the equation takes the form

(4) $$AX^2 + CY^2 + DX + EY + F = 0.$$

An elementary but time-consuming computation shows that the discriminant is unchanged by a rotation, so that in this instance we have

$$\Delta = b^2 - 4ac = -4AC.$$

Moreover, A and C cannot both be zero. If $\Delta < 0$, then $AC > 0$ and we can rewrite (4) as

$$\frac{X^2}{C} + \frac{D}{AC}X + \frac{Y^2}{A} + \frac{E}{AC}Y + \frac{F}{AC} = 0.$$

By completing the squares, we obtain an equation of the form

$$\frac{(X - \alpha)^2}{(\sqrt{|C|})^2} + \frac{(Y - \beta)^2}{(\sqrt{|A|})^2} = K.$$

If $K > 0$, we have an ellipse or a circle. If $K = 0$, we have the point (α, β). If $K < 0$, the set is empty.

If $\Delta > 0$, then $AC < 0$. Proceeding as before, we obtain an equation of the form

$$\frac{(X - \alpha)^2}{(\sqrt{|C|})^2} - \frac{(Y - \beta)^2}{(\sqrt{|A|})^2} = K.$$

If $K \neq 0$, we have a hyperbola. If $K = 0$, the equation becomes

$$\left(\frac{X - \alpha}{\sqrt{|C|}} - \frac{Y - \beta}{\sqrt{|A|}} \right) \left(\frac{X - \alpha}{\sqrt{|C|}} + \frac{Y - \beta}{\sqrt{|A|}} \right) = 0,$$

so that we have a pair of lines intersecting at the point (α, β).

If $\Delta = 0$, then $AC = 0$, so that either $A = 0$ or $C = 0$. Since A and C are not both zero, there is no loss in generality in assuming that $A \neq 0$ and $C = 0$. In this case equation (4) reduces to

$$AX^2 + DX + EY + F = 0.$$

Dividing by A and completing the square we have an equation of the form

$$(X - \alpha)^2 = \beta Y + K.$$

If $\beta \neq 0$, we have a parabola. If $\beta = 0$ and $K = 0$, we have a line. If $\beta = 0$ and $K > 0$, we have a pair of parallel lines. If $\beta = 0$ and $K < 0$, the set is empty. ❑

EXERCISES A.1

In Exercises 1–8, (a) use the discriminant to give a possible identification of the graph of the equation; (b) find a rotation $\alpha \in (0, \pi/2)$ that eliminates the xy-term; (c) rewrite the equation in terms of the new coordinate system; (d) sketch the graph displaying both coordinate systems.

1. $xy = 1.$ **2.** $xy - y + x = 1.$

3. $11x^2 + 10\sqrt{3}xy + y^2 - 4 = 0.$

4. $52x^2 - 72xy + 73y^2 - 100 = 0.$

5. $x^2 - 2xy + y^2 + x + y = 0.$

6. $3x^2 + 2\sqrt{3}xy + y^2 - 2x + 2\sqrt{3}y = 0.$

7. $x^2 + 2\sqrt{3}xy + 3y^2 + 2\sqrt{3}x - 2y = 0.$

8. $2x^2 + 4\sqrt{3}xy + 6y^2 + (8 - \sqrt{3})x + (8\sqrt{3} + 1)y + 8 = 0.$

In Exercises 9 and 10, find a rotation $\alpha \in (0, \pi/2)$ that eliminates the xy-term. Then find $\cos \alpha$ and $\sin \alpha$.

9. $x^2 + xy + Kx + Ly + M = 0.$

10. $5x^2 + 24xy + 12y^2 + Kx + Ly + M = 0.$

11. Show that after a rotation of axes through an angle α, the coefficients in the equation

$$AX^2 + BXY + CY^2 + DX + EY + F = 0$$

are related to the coefficients in the equation

$$ax^2 + bxy + cy^2 + dx + ey + f = 0, \quad a, b, c \text{ not all } 0,$$

as follows:

$A = a \cos^2\alpha + b \cos \alpha \sin \alpha + c \sin^2\alpha,$

$B = 2(c - a)\cos \alpha \sin \alpha + b(\cos^2\alpha - \sin^2\alpha),$

$C = a \sin^2\alpha - b \cos \alpha \sin \alpha + c \cos^2\alpha,$

$D = d \cos \alpha + e \sin \alpha,$

$E = e \cos \alpha - d \sin \alpha,$

$F = f.$

12. Use the results of Exercise 11 to show:
(a) $B^2 - 4AC = b^2 - 4ac.$
(b) If $B = 0$, then A and C cannot both be 0.

■ A.2 DETERMINANTS

By a *matrix* we mean a rectangular arrangement of numbers enclosed in parentheses. For example,

$$\begin{pmatrix} 2 & 4 \\ 3 & 1 \end{pmatrix} \qquad \begin{pmatrix} 1 & 6 & 3 \\ 5 & 2 & 2 \end{pmatrix} \qquad \begin{pmatrix} 2 & 4 & 0 \\ 4 & 7 & 1 \\ 0 & 1 & 1 \end{pmatrix}$$

are all matrices. The numbers occurring in a matrix are called the *entries*.

Each matrix has a certain number of rows and a certain number of columns. A matrix with m rows and n columns is called an $m \times n$ *matrix*. Thus the first matrix above is a 2×2 matrix, the second a 2×3 matrix, the third a 3×3 matrix. The first and third matrices are called *square*; they have the same number of rows as columns. Here we will be working with square matrices as these are the only ones that have determinants.

We could give a definition of determinant that is applicable to all square matrices, but the definition is complicated and would serve little purpose at this point. Our interest here is in the 2×2 case and the 3×3 case. We begin with the 2×2 case.

(A.2.1)

> The *determinant* of the matrix
>
> $$\begin{pmatrix} a_1 & a_2 \\ b_1 & b_2 \end{pmatrix}$$
>
> is the number $a_1 b_2 - a_2 b_1$.

We have a special notation for the determinant. We change the parentheses of the matrix to vertical bars:

$$\text{Determinant of } \begin{pmatrix} a_1 & a_2 \\ b_1 & b_2 \end{pmatrix} = \begin{vmatrix} a_1 & a_2 \\ b_1 & b_2 \end{vmatrix} = a_1 b_2 - a_2 b_1.$$

Thus, for example,

$$\begin{vmatrix} 5 & 8 \\ 4 & 2 \end{vmatrix} = (5 \cdot 2) - (8 \cdot 4) = 10 - 32 = -22$$

and

$$\begin{vmatrix} 4 & 0 \\ 0 & \frac{1}{4} \end{vmatrix} = (4 \cdot \tfrac{1}{4}) - (0 \cdot 0) = 1.$$

We remark on three properties of 2×2 determinants:

1. If the rows or columns of a 2×2 determinant are interchanged, the determinant changes sign:

$$\begin{vmatrix} b_1 & b_2 \\ a_1 & a_2 \end{vmatrix} = - \begin{vmatrix} a_1 & a_2 \\ b_1 & b_2 \end{vmatrix}, \qquad \begin{vmatrix} a_2 & a_1 \\ b_2 & b_1 \end{vmatrix} = - \begin{vmatrix} a_1 & a_2 \\ b_1 & b_2 \end{vmatrix}.$$

PROOF Just note that

$$b_1 a_2 - b_2 a_1 = -(a_1 b_2 - a_2 b_1) \quad \text{and} \quad a_2 b_1 - a_1 b_2 = -(a_1 b_2 - a_2 b_1). \quad \square$$

2. A common factor can be removed from any row or column and placed as a multiplier in front of the determinant:

$$\begin{vmatrix} \lambda a_1 & \lambda a_2 \\ b_1 & b_2 \end{vmatrix} = \lambda \begin{vmatrix} a_1 & a_2 \\ b_1 & b_2 \end{vmatrix}, \qquad \begin{vmatrix} \lambda a_1 & a_2 \\ \lambda b_1 & b_2 \end{vmatrix} = \lambda \begin{vmatrix} a_1 & a_2 \\ b_1 & b_2 \end{vmatrix}.$$

PROOF   Just note that

$$(\lambda a_1)b_2 - (\lambda a_2)b_1 = \lambda(a_1 b_2 - a_2 b_1)$$
$$\text{and} \quad (\lambda a_1)b_2 - a_2(\lambda b_1) = \lambda(a_1 b_2 - a_2 b_1). \quad \square$$

3. If the rows or columns of a 2 × 2 determinant are the same, the determinant is 0.

PROOF

$$\begin{vmatrix} a_1 & a_2 \\ a_1 & a_2 \end{vmatrix} = a_1 a_2 - a_2 a_1 = 0, \qquad \begin{vmatrix} a_1 & a_1 \\ b_1 & b_1 \end{vmatrix} = a_1 b_1 - a_1 b_1 = 0. \quad \square$$

The determinant of a 3 × 3 matrix is harder to define. One definition is this:

$$\begin{vmatrix} a_1 & a_2 & a_3 \\ b_1 & b_2 & b_3 \\ c_1 & c_2 & c_3 \end{vmatrix} = a_1 b_2 c_3 - a_1 b_3 c_2 + a_2 b_3 c_1 - a_2 b_1 c_3 + a_3 b_1 c_2 - a_3 b_2 c_1.$$

The problem with this definition is that it is hard to remember. What saves us is that the expansion on the right can be conveniently written in terms of 2 × 2 determinants; namely, the expression on the right can be written

$$a_1(b_2 c_3 - b_3 c_2) - a_2(b_1 c_3 - b_3 c_1) + a_3(b_1 c_2 - b_2 c_1),$$

which turns into

$$a_1 \begin{vmatrix} b_2 & b_3 \\ c_2 & c_3 \end{vmatrix} - a_2 \begin{vmatrix} b_1 & b_3 \\ c_1 & c_3 \end{vmatrix} + a_3 \begin{vmatrix} b_1 & b_2 \\ c_1 & c_2 \end{vmatrix}.$$

We then have

(A.2.2)
$$\begin{vmatrix} a_1 & a_2 & a_3 \\ b_1 & b_2 & b_3 \\ c_1 & c_2 & c_3 \end{vmatrix} = a_1 \begin{vmatrix} b_2 & b_3 \\ c_2 & c_3 \end{vmatrix} - a_2 \begin{vmatrix} b_1 & b_3 \\ c_1 & c_3 \end{vmatrix} + a_3 \begin{vmatrix} b_1 & b_2 \\ c_1 & c_2 \end{vmatrix}.$$

We will take this as our definition. It is called the *expansion of the determinant along the first row.* Note that the coefficients are the entries a_1, a_2, a_3 of the first row, that they occur alternately with + and − signs, and that each is multiplied by a determinant. You can remember which determinant goes with which entry a_i as follows: in the original matrix, mentally cross out the row and column in which the entry a_i is found, and take the determinant of the remaining 2 × 2 matrix. For example, the determinant that goes with a_3 is

$$\begin{vmatrix} a_1 & a_2 & a_3 \\ b_1 & b_2 & b_3 \\ c_1 & c_2 & c_3 \end{vmatrix} = \begin{vmatrix} b_1 & b_2 \\ c_1 & c_2 \end{vmatrix}.$$

When first starting to work with specific 3×3 determinants, it is a good idea to set up the formula with blank 2×2 determinants:

$$\begin{vmatrix} a_1 & a_2 & a_3 \\ b_1 & b_2 & b_3 \\ c_1 & c_2 & c_3 \end{vmatrix} = a_1 \begin{vmatrix} & \\ & \end{vmatrix} - a_2 \begin{vmatrix} & \\ & \end{vmatrix} + a_3 \begin{vmatrix} & \\ & \end{vmatrix}$$

and then fill in the 2×2 determinants by the "crossing out" rule explained above.

Example 1

$$\begin{vmatrix} 1 & 2 & 1 \\ 0 & 3 & 4 \\ 6 & 2 & 5 \end{vmatrix} = 1 \begin{vmatrix} 3 & 4 \\ 2 & 5 \end{vmatrix} - 2 \begin{vmatrix} 0 & 4 \\ 6 & 5 \end{vmatrix} + 1 \begin{vmatrix} 0 & 3 \\ 6 & 2 \end{vmatrix}$$

$$= 1(15 - 8) - 2(0 - 24) + 1(0 - 18)$$
$$= 7 + 48 - 18 = 37. \quad \square$$

A straightforward (but somewhat laborious) calculation shows that 3×3 determinants have the same three properties we proved earlier for 2×2 determinants:

1. If two rows or columns are interchanged, the determinant changes sign.
2. A common factor can be removed from any row or column and placed as a multiplier in front of the determinant.
3. If two rows or columns are the same, the determinant is 0.

EXERCISES A.2

Evaluate the following determinants.

1. $\begin{vmatrix} 1 & 2 \\ 3 & 4 \end{vmatrix}$.

2. $\begin{vmatrix} 1 & -1 \\ -1 & 1 \end{vmatrix}$.

3. $\begin{vmatrix} 1 & 1 \\ a & a \end{vmatrix}$.

4. $\begin{vmatrix} a & b \\ b & d \end{vmatrix}$.

5. $\begin{vmatrix} 1 & 0 & 3 \\ 2 & 4 & 1 \\ 0 & 1 & 0 \end{vmatrix}$.

6. $\begin{vmatrix} 1 & 0 & 0 \\ 0 & 2 & 0 \\ 0 & 0 & 3 \end{vmatrix}$.

7. $\begin{vmatrix} 0 & 0 & 1 \\ 0 & 2 & 0 \\ 3 & 0 & 0 \end{vmatrix}$.

8. $\begin{vmatrix} a & 0 & 0 \\ b & c & 0 \\ d & e & f \end{vmatrix}$.

9. If A is a matrix, its *transpose* A^T is obtained by interchanging the rows and columns. Thus

$$\begin{pmatrix} a_1 & a_2 \\ b_1 & b_2 \end{pmatrix}^T = \begin{pmatrix} a_1 & b_1 \\ a_2 & b_2 \end{pmatrix}$$

and

$$\begin{pmatrix} a_1 & a_2 & a_3 \\ b_1 & b_2 & b_3 \\ c_1 & c_2 & c_3 \end{pmatrix}^T = \begin{pmatrix} a_1 & b_1 & c_1 \\ a_2 & b_2 & c_2 \\ a_3 & b_3 & c_3 \end{pmatrix}.$$

Show that the determinant of a matrix equals the determinant of its transpose: (a) the 2×2 case. (b) the 3×3 case.

Verify the assertions in Exercises 10–14.

10. $\begin{vmatrix} 1 & 2 & 3 \\ 4 & 5 & 6 \\ 7 & 8 & 9 \end{vmatrix} + \begin{vmatrix} 4 & 5 & 6 \\ 1 & 2 & 3 \\ 7 & 8 & 9 \end{vmatrix} = 0$.

11. $\begin{vmatrix} 1 & 2 & 3 \\ 4 & 5 & 6 \\ 7 & 8 & 9 \end{vmatrix} = \begin{vmatrix} 4 & 5 & 6 \\ 7 & 8 & 9 \\ 1 & 2 & 3 \end{vmatrix}$.

12. $\begin{vmatrix} 1 & 2 & 3 \\ 4 & 5 & 6 \\ 7 & 8 & 9 \end{vmatrix} + \begin{vmatrix} 1 & 2 & 3 \\ 1 & 2 & 3 \\ 7 & 8 & 9 \end{vmatrix} = \begin{vmatrix} 1 & 2 & 3 \\ 4 & 5 & 6 \\ 7 & 8 & 9 \end{vmatrix}$.

13. $\frac{1}{2} \begin{vmatrix} 1 & 0 & 7 \\ 3 & 4 & 5 \\ 2 & 4 & 6 \end{vmatrix} = \begin{vmatrix} 1 & 0 & 7 \\ 3 & 4 & 5 \\ 1 & 2 & 3 \end{vmatrix}$.

14. $\begin{vmatrix} 1 & 2 & 3 \\ x & 2x & 3x \\ 4 & 5 & 6 \end{vmatrix} = 0$.

15. (a) Verify that the equations

$$3x + 4y = 6$$

$$2x - 3y = 7$$

are solved by the prescription

$$x = \frac{\begin{vmatrix} 6 & 4 \\ 7 & -3 \end{vmatrix}}{\begin{vmatrix} 3 & 4 \\ 2 & -3 \end{vmatrix}}, \qquad y = \frac{\begin{vmatrix} 3 & 6 \\ 2 & 7 \end{vmatrix}}{\begin{vmatrix} 3 & 4 \\ 2 & -3 \end{vmatrix}}.$$

(b) More generally, verify that the equations

$$a_1 x + a_2 y = d$$

$$b_1 x + b_2 y = e$$

are solved by the prescription

$$x = \frac{\begin{vmatrix} d & a_2 \\ e & b_2 \end{vmatrix}}{\begin{vmatrix} a_1 & a_2 \\ b_1 & b_2 \end{vmatrix}}, \qquad y = \frac{\begin{vmatrix} a_1 & d \\ b_1 & e \end{vmatrix}}{\begin{vmatrix} a_1 & a_2 \\ b_1 & b_2 \end{vmatrix}}$$

provided that the determinant in the denominator is different from 0.

(c) Conjecture an analogous rule for solving three linear equations in three unknowns.

16. Show that a 3×3 determinant can be "expanded along the bottom row" as follows:

$$\begin{vmatrix} a_1 & a_2 & a_3 \\ b_1 & b_2 & b_3 \\ c_1 & c_2 & c_3 \end{vmatrix} = c_1 \begin{vmatrix} a_2 & a_3 \\ b_2 & b_3 \end{vmatrix} - c_2 \begin{vmatrix} a_1 & a_3 \\ b_1 & b_3 \end{vmatrix} + c_3 \begin{vmatrix} a_1 & a_2 \\ b_1 & b_2 \end{vmatrix}.$$

HINT: You can check this directly by writing out the values of the determinants on the right, or you can interchange rows twice to bring the bottom row to the top and then use expansion along the top row.

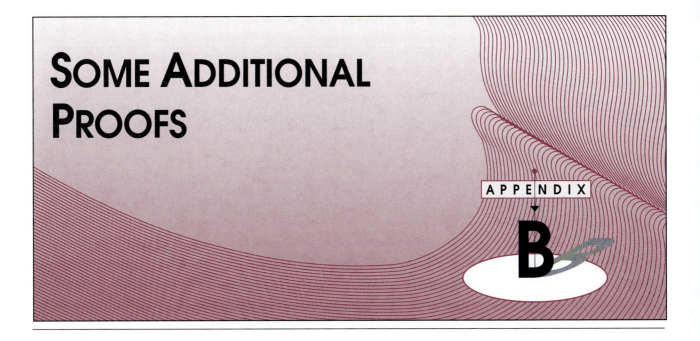

SOME ADDITIONAL PROOFS

In this appendix we present some proofs that many would consider too advanced for the main body of the text. Some details are omitted. These are left to you.

The arguments presented in Sections B.1, B.2, and B.4 require some familiarity with the *least upper bound axiom.* This is discussed in Section 10.1. In addition, Section B.4 requires some understanding of *sequences,* for which we refer you to Sections 10.2 and 10.3.

■ B.1 THE INTERMEDIATE-VALUE THEOREM

> **LEMMA B.1.1**
>
> Let f be continuous on $[a, b]$. If $f(a) < 0 < f(b)$ or $f(b) < 0 < f(a)$, then there is a number c between a and b for which $f(c) = 0$.

PROOF Suppose that $f(a) < 0 < f(b)$. (The other case can be treated in a similar manner.) Since $f(a) < 0$, we know from the continuity of f that there exists a number ξ such that f is negative on $[a, \xi)$. Let

$$c = \text{lub } \{\xi: f \text{ is negative on } [a, \xi)\}.$$

Clearly, $c \leq b$. We cannot have $f(c) > 0$, for then f would be positive on some interval extending to the left of c, and we know that, to the left of c, f is negative. Incidentally this argument excludes the possibility $c = b$ and means that $c < b$. We cannot have $f(c) < 0$, for then there would be an interval $[a, t)$, with $t > c$, on which f is negative, and this would contradict the definition of c. It follows that $f(c) = 0$. ❏

THEOREM B.1.2 THE INTERMEDIATE-VALUE THEOREM

If f is continuous on $[a, b]$ and C is a number between $f(a)$ and $f(b)$, then there is at least one number c between a and b for which $f(c) = C$.

PROOF Suppose for example that

$$f(a) < C < f(b).$$

(The other possibility can be handled in a similar manner.) The function

$$g(x) = f(x) - C$$

is continuous on $[a, b]$. Since

$$g(a) = f(a) - C < 0 \qquad \text{and} \qquad g(b) = f(b) - C > 0,$$

we know from the lemma that there is a number c between a and b for which $g(c) = 0$. Obviously, then, $f(c) = C$. ❑

■ B.2 THE MAXIMUM-MINIMUM THEOREM

LEMMA B.2.1

If f is continuous on $[a, b]$, then f is bounded on $[a, b]$.

PROOF Consider

$$\{x: x \in [a, b] \text{ and } f \text{ is bounded on } [a, x]\}.$$

It is easy to see that this set is nonempty and bounded above by b. Thus we can set

$$c = \text{lub } \{x: f \text{ is bounded on } [a, x]\}.$$

Now we argue that $c = b$. To do so, we suppose that $c < b$. From the continuity of f at c, it is easy to see that f is bounded on $[c - \epsilon, c + \epsilon]$ for some $\epsilon > 0$. Being bounded on $[a, c - \epsilon]$ and on $[c - \epsilon, c + \epsilon]$, it is obviously bounded on $[a, c + \epsilon]$. This contradicts our choice of c. We can therefore conclude that $c = b$. This tells us that f is bounded on $[a, x]$ for all $x < b$. We are now almost through. From the continuity of f, we know that f is bounded on some interval of the form $[b - \epsilon, b]$. Since $b - \epsilon < b$, we know from what we have just proved that f is bounded on $[a, b - \epsilon]$. Being bounded on $[a, b - \epsilon]$ and bounded on $[b - \epsilon, b]$, it is bounded on $[a, b]$. ❑

THEOREM B.2.2 THE MAXIMUM-MINIMUM THEOREM

If f is continuous on $[a, b]$, then f takes on both a maximum value M and a minimum value m on $[a, b]$.

PROOF By the lemma, f is bounded on $[a, b]$. Set

$$M = \text{lub } \{f(x): x \in [a, b]\}.$$

We must show that there exists c in $[a, b]$ such that $f(c) = M$. To do this, we set

$$g(x) = \frac{1}{M - f(x)}.$$

If f does not take on the value M, then g is continuous on $[a, b]$ and thus, by the lemma, bounded on $[a, b]$. A look at the definition of g makes it clear that g cannot be bounded on $[a, b]$. The assumption that f does not take on the value M has led to a contradiction. (That f takes on a minimum value m can be proved in a similar manner.) ❏

■ B.3 INVERSES

> **THEOREM B.3.1 CONTINUITY OF THE INVERSE**
>
> Let f be a one-to-one function defined on an interval (a, b). If f is continuous, then its inverse f^{-1} is also continuous.

PROOF If f is continuous, then, being one-to-one, f either increases throughout (a, b) or it decreases throughout (a, b). The proof of this assertion we leave to you.

Let's suppose now that f increases throughout (a, b). Let's take c in the domain of f^{-1} and show that f^{-1} is continuous at c.

We first observe that $f^{-1}(c)$ lies in (a, b) and choose $\epsilon > 0$ sufficiently small so that $f^{-1}(c) - \epsilon$ and $f^{-1}(c) + \epsilon$ also lie in (a, b). We seek $\delta > 0$ such that

$$\text{if } c - \delta < x < c + \delta, \quad \text{then} \quad f^{-1}(c) - \epsilon < f^{-1}(x) < f^{-1}(c) + \epsilon.$$

This condition can be met by choosing δ to satisfy

$$f(f^{-1}(c) - \epsilon) < c - \delta \quad \text{and} \quad c + \delta < f(f^{-1}(c) + \epsilon)$$

for then, if $c - \delta < x < c + \delta$, then

$$f(f^{-1}(c) - \epsilon) < x < f(f^{-1}(c) + \epsilon),$$

and, since f^{-1} also increases,

$$f^{-1}(c) - \epsilon < f^{-1}(x) < f^{-1}(c) + \epsilon.$$

The case where f decreases throughout (a, b) can be handled in a similar manner. ❏

> **THEOREM B.3.2 DIFFERENTIABILITY OF THE INVERSE**
>
> Let f be a one-to-one function defined on an interval (a, b). If f is differentiable and its derivative does not take on the value 0, then f^{-1} is differentiable and
>
> $$(f^{-1})'(x) = \frac{1}{f'(f^{-1}(x))}.$$

PROOF (Here we use the characterization of derivative spelled out in Theorem 3.5.8.) Let x be in the domain of f^{-1}. We take $\epsilon > 0$ and show that there exists $\delta > 0$ such that

$$\text{if } 0 < |t - x| < \delta, \quad \text{then} \quad \left| \frac{f^{-1}(t) - f^{-1}(x)}{t - x} - \frac{1}{f'(f^{-1}(x))} \right| < \epsilon.$$

Since f is differentiable at $f^{-1}(x)$ and $f'(f^{-1}(x)) \neq 0$, there exists $\delta_1 > 0$ such that

$$\text{if } 0 < |y - f^{-1}(x)| < \delta_1, \quad \text{then} \quad \left| \frac{1}{\dfrac{f(y) - f(f^{-1}(x))}{y - f^{-1}(x)}} - \frac{1}{f'(f^{-1}(x))} \right| < \epsilon$$

and therefore

$$\left| \frac{y - f^{-1}(x)}{f(y) - f(f^{-1}(x))} - \frac{1}{f'(f^{-1}(x))} \right| < \epsilon.$$

By the previous theorem, f^{-1} is continuous at x and therefore there exists $\delta > 0$ such that

$$\text{if } 0 < |t - x| < \delta, \quad \text{then} \quad 0 < |f^{-1}(t) - f^{-1}(x)| < \delta_1.$$

It follows from the special property of δ_1 that

$$\left| \frac{f^{-1}(t) - f^{-1}(x)}{t - x} - \frac{1}{f'(f^{-1}(x))} \right| < \epsilon. \quad \square$$

■ B.4 THE INTEGRABILITY OF CONTINUOUS FUNCTIONS

The aim here is to prove that, if f is continuous on $[a, b]$, then there is one and only one number I that satisfies the inequality

$$L_f(P) \leq I \leq U_f(P) \qquad \text{for all partitions } P \text{ of } [a, b].$$

DEFINITION B.4.1

A function f is said to be *uniformly continuous* on $[a, b]$ iff for each $\epsilon > 0$ there exists $\delta > 0$ such that

$$\text{if } x, y \in [a, b] \text{ and } |x - y| < \delta, \quad \text{then } |f(x) - f(y)| < \epsilon.$$

For convenience, let's agree to say that *the interval $[a, b]$ has the property P_ϵ* iff there exist sequences $\{x_n\}, \{y_n\}$ satisfying

$$x_n, y_n \in [a, b], \qquad |x_n - y_n| < 1/n, \qquad |f(x_n) - f(y_n)| \geq \epsilon.$$

LEMMA B.4.2

If f is not uniformly continuous on $[a, b]$, then $[a, b]$ has the property P_ϵ for some $\epsilon > 0$.

PROOF If f is not uniformly continuous on $[a, b]$, then there is no $\delta > 0$ such that

$$\text{if } x, y \in [a, b] \text{ and } |x - y| < \delta, \quad \text{then } |f(x) - f(y)| < \epsilon.$$

The interval $[a, b]$ has the property P_ϵ for that choice of ϵ. The details of the argument are left to you. ❑

LEMMA B.4.3

Let f be continuous on $[a, b]$. If $[a, b]$ has the property P_ϵ, then at least one of the subintervals $[a, \frac{1}{2}(a + b)]$, $[\frac{1}{2}(a + b), b]$ has the property P_ϵ.

PROOF Let's suppose that the lemma is false. For convenience, we let $c = \frac{1}{2}(a + b)$, so that the halves become $[a, c]$ and $[c, b]$. Since $[a, c]$ fails to have the property P_ϵ, there exists an integer p such that

$$\text{if } x, y \in [a, c] \text{ and } |x - y| < 1/p, \quad \text{then } |f(x) - f(y)| < \epsilon.$$

Since $[c, b]$ fails to have the property P_ϵ, there exists an integer q such that

$$\text{if } x, y \in [c, b] \text{ and } |x - y| < 1/q, \quad \text{then} \quad |f(x) - f(y)| < \epsilon.$$

Since f is continuous at c, there exists an integer r such that, if $|x - c| < 1/r$, then $|f(x) - f(c)| < \frac{1}{2}\epsilon$. Set $s = \max \{p, q, r\}$ and suppose that

$$x, y \in [a, b], \quad |x - y| < 1/s.$$

If x, y are both in $[a, c]$ or both in $[c, b]$, then

$$|f(x) - f(y)| < \epsilon.$$

The only other possibility is that $x \in [a, c]$ and $y \in [c, b]$. In this case we have

$$|x - c| < 1/r, \quad |y - c| < 1/r,$$

and thus

$$|f(x) - f(c)| < \tfrac{1}{2}\epsilon, \quad |f(y) - f(c)| < \tfrac{1}{2}\epsilon.$$

By the triangle inequality, we again have

$$|f(x) - f(y)| < \epsilon.$$

In summary, we have obtained the existence of an integer s with the property that

$$x, y \in [a, b], |x - y| < 1/s \quad \text{implies} \quad |f(x) - f(y)| < \epsilon.$$

Hence $[a, b]$ does not have the property P_ϵ. This is a contradiction and proves the lemma. ❑

THEOREM B.4.4

If f is continuous on $[a, b]$, then f is uniformly continuous on $[a, b]$.

PROOF   We suppose that f is not uniformly continuous on $[a, b]$ and base our argument on a mathematical version of "bisection."

By the first lemma of this section, we know that $[a, b]$ has the property P_ϵ for some $\epsilon > 0$. We bisect $[a, b]$ and note by the second lemma that one of the halves, say $[a_1, b_1]$, has the property P_ϵ. We then bisect $[a_1, b_1]$ and note that one of the halves, say $[a_2, b_2]$, has the property P_ϵ. Continuing in this manner, we obtain a sequence of intervals $[a_n, b_n]$, each with the property P_ϵ. Then for each n, we can choose x_n, $y_n \in [a_n, b_n]$ such that

$$|x_n - y_n| < 1/n \qquad \text{and} \qquad |f(x_n) - f(y_n)| \geq \epsilon.$$

Since

$$a \leq a_n \leq a_{n+1} < b_{n+1} \leq b_n \leq b,$$

we see that the sequences $\{a_n\}$ and $\{b_n\}$ are both bounded and monotonic. Thus they are convergent. Since $b_n - a_n \to 0$, we see that $\{a_n\}$ and $\{b_n\}$ both converge to the same limit, say L. From the inequality

$$a_n \leq x_n \leq y_n \leq b_n,$$

we conclude that

$$x_n \to L \quad \text{and} \quad y_n \to L.$$

This tells us that

$$|f(x_n) - f(y_n)| \to |f(L) - f(L)| = 0,$$

which contradicts the statement that $|f(x_n) - f(y_n)| \geq \epsilon$ for all n.   ❑

LEMMA B.4.5

If P and Q are partitions of $[a, b]$, then $L_f(P) \leq U_f(Q)$.

PROOF $P \cup Q$ is a partition of $[a, b]$ that contains both P and Q. It is obvious then that

$$L_f(P) \leq L_f(P \cup Q) \leq U_f(P \cup Q) \leq U_f(Q).   ❑$$

From the last lemma it follows that the set of all lower sums is bounded above and has a least upper bound L. The number L satisfies the inequality

$$L_f(P) \leq L \leq U_f(P) \qquad \text{for all partitions } P$$

and is clearly the least of such numbers. Similarly, we find that the set of all upper sums is bounded below and has a greatest lower bound U. The number U satisfies the inequality

$$L_f(P) \leq U \leq U_f(P) \qquad \text{for all partitions } P$$

and is clearly the greatest of such numbers.

We are now ready to prove the basic theorem.

THEOREM B.4.6 THE INTEGRABILITY THEOREM

If f is continuous on $[a, b]$, then there exists one and only one number I that satisfies the inequality

$$L_f(P) \le I \le U_f(P) \qquad \text{for all partitions } P \text{ of } [a, b].$$

PROOF We know that

$$L_f(P) \le L \le U \le U_f(P) \qquad \text{for all } P,$$

so that existence is no problem. We will have uniqueness if we can prove that

$$L = U.$$

To do this, we take $\epsilon > 0$ and note that f, being continuous on $[a, b]$, is uniformly continuous on $[a, b]$. Thus there exists $\delta > 0$ such that, if

$$x, y \in [a, b] \text{ and } |x - y| < \delta, \qquad \text{then} \qquad |f(x) - f(y)| < \frac{\epsilon}{b - a}.$$

We now choose a partition $P = \{x_0, x_1, \ldots, x_n\}$ for which max $\Delta x_i < \delta$. For this partition P, we have

$$U_f(P) - L_f(P) = \sum_{i=1}^{n} M_i \, \Delta x_i - \sum_{i=1}^{n} m_i \, \Delta x_i$$

$$= \sum_{i=1}^{n} (M_i - m_i) \, \Delta x_i$$

$$< \sum_{i=1}^{n} \frac{\epsilon}{b - a} \Delta x_i = \frac{\epsilon}{b - a} \sum_{i=1}^{n} \Delta x_i = \frac{\epsilon}{b - a}(b - a) = \epsilon.$$

Since

$$U_f(P) - L_f(P) < \epsilon \qquad \text{and} \qquad 0 \le U - L \le U_f(P) - L_f(P),$$

you can see that

$$0 \le U - L < \epsilon.$$

Since ϵ was chosen arbitrarily, we must have $U - L = 0$ and $L = U$. ❑

■ B.5 THE INTEGRAL AS THE LIMIT OF RIEMANN SUMS

For the notation we refer to Section 5.10.

THEOREM B.5.1

If f is continuous on $[a, b]$, then

$$\int_a^b f(x) \, dx = \lim_{\|P\| \to 0} S^*(P).$$

PROOF   Let $\epsilon > 0$.   We must show that there exists $\delta > 0$ such that

$$\text{if } \|P\| < \delta, \quad \text{then} \quad \left| S^*(P) - \int_a^b f(x)\, dx \right| < \epsilon.$$

From the proof of Theorem B.4.6 we know that there exists $\delta > 0$ such that

$$\text{if } \|P\| < \delta, \quad \text{then} \quad U_f(P) - L_f(P) < \epsilon.$$

For such P we have

$$U_f(P) - \epsilon < L_f(P) \le S^*(P) \le U_f(P) < L_f(P) + \epsilon.$$

This gives

$$\int_a^b f(x)\, dx - \epsilon < S^*(P) < \int_a^b f(x)\, dx + \epsilon,$$

and therefore

$$\left| S^*(P) - \int_a^b f(x)\, dx \right| < \epsilon. \quad \square$$

CHAPTER 10

SECTION 10.1

1. lub = 2; glb = 0   **3.** no lub; glb = 0   **5.** lub = 2; glb = −2   **7.** no lub; glb = 2   **9.** lub = $2\frac{1}{2}$; glb = 2

11. lub = 1; glb = 0.9   **13.** lub = e; glb = 0   **15.** lub = $\frac{1}{2}(-1 + \sqrt{5})$; glb = $\frac{1}{2}(-1 - \sqrt{5})$ **17.** no lub; no glb

19. no lub; no glb   **21.** glb $S = 0$, $0 \le (\frac{1}{11})^3 < 0 + 0.001$ **23.** glb $S = 0$, $0 \le (\frac{1}{10})^{2n-1} < 0 + (\frac{1}{10})^k$, $n > \frac{1}{2}(k + 1)$

25. Let $\epsilon > 0$. The condition $m \le s$ is satisfied by all numbers s in S. All we have to show therefore is that there is some number s in S such that $s < m + \epsilon$. Suppose on the contrary that there is no such number in S. We then have $m + \epsilon \le x$ for all $x \in S$, so that $m + \epsilon$ becomes a lower bound for S. But this cannot happen, for it makes $m + \epsilon$ a lower bound that is *greater* than m, and by assumption, m is the *greatest* lower bound.

27. Let $c = \text{lub } S$. Since $b \in S$, $b \le c$. Since b is an upper bound for S, $c \le b$. Thus $b = c$.

29. (a) Any upper bound for S is an upper bound for T; any lower bound for S is a lower bound for T.
(b) Let $a = \text{glb } S$. Then $a \le t$ for all $t \in T$. Therefore $a \le \text{glb } T$. Similarly, if $b = \text{lub } S$, then $t \le b$ for all $t \in T$, so $\text{lub } T \le b$. It now follows that glb $S \le$ glb $T \le$ lub $T \le$ lub S.

31. Let M be any positive number and consider M/c. Since the set of positive integers is not bounded above, there exists a positive integer k such that $k \ge M/c$. This implies $kc \ge M$. Since $kc \in S$, it follows that S is not bounded above.

33. (a)

a_1	a_2	a_3	a_4	a_5	a_6	a_7	a_8	a_9	a_{10}
1.4142	1.6818	1.8340	1.9152	1.9571	1.9785	1.9892	1.9946	1.9973	1.9986

(b) Let S be the set of positive integers for which $a_n < 2$. Then $1 \in S$ since $a_1 = \sqrt{2} \cong 1.4142 < 2$. Assume that $k \in S$. Now $a_{k+1}^2 = 2a_k < 4$, which implies $a_{k+1} < 2$. Thus $k + 1 \in S$ and S is the set of positive integers.
(c) yes   (d) The number you chose is the least upper bound of the corresponding set S.

SECTION 10.2

1. $a_n = 2 + 3(n - 1)$, $n = 1, 2, 3, \ldots$ **3.** $a_n = \dfrac{(-1)^{n-1}}{2n - 1}$, $n = 1, 2, 3, \ldots$ **5.** $a_n = \dfrac{n^2 + 1}{n}$, $n = 1, 2, 3, \ldots$

7. $a_n = \begin{cases} n & \text{if } n = 2k - 1, \\ \frac{1}{n} & \text{if } n = 2k, \end{cases}$ where $k = 1, 2, 3, \ldots$ **9.** decreasing; bounded below by 0 and above by 2

11. increasing; bounded below by 1 but not bounded above   **13.** not monotonic; bounded below by 0 and above by $\frac{3}{2}$

15. decreasing; bounded below by 0 and above by 0.9   **17.** increasing; bounded below by $\frac{1}{2}$ but not bounded above

19. increasing; bounded below by $\frac{4}{5}\sqrt{5}$ and above by 2   **21.** increasing; bounded below by $\frac{2}{5!}$ but not bounded above

23. decreasing; bounded below by 0 and above by $\frac{1}{2}(10^{10})$ **25.** increasing; bounded below by 0 and above by ln 2

27. decreasing; bounded below by 1 and above by 4   **29.** increasing; bounded below by $\sqrt{3}$ and above by 2

31. decreasing; bounded above by -1 but not bounded below   **33.** increasing; bounded below by $\frac{1}{2}$ and above by 1

35. decreasing; bounded below by 0 and above by 1   **37.** decreasing; bounded below by 0 and above by $\frac{5}{6}$

39. decreasing; bounded below by 0 and above by $\frac{1}{2}$ **41.** decreasing; bounded below by 0 and above by $\frac{1}{3} \ln 3$

43. increasing; bounded below by $\frac{3}{4}$ but not bounded above   **45.** For $n \ge 5$

$$\frac{a_{n+1}}{a_n} = \frac{5^{n+1}}{(n + 1)!} \cdot \frac{n!}{5^n} = \frac{5}{n + 1} < 1 \qquad \text{and thus} \qquad a_{n+1} < a_n.$$

Sequence is not nonincreasing: $a_1 = 5 < \frac{25}{2} = a_2$.

47. boundedness: $0 < (c^n + d^n)^{1/n} < (2d^n)1/n = 2^{1/n} d \le 2d$.
monotonicity: $a_{n+1}^{n+1} = c^{n+1} + d^{n+1} = cc^n + dd^n < (c^n + d^n)^{1/n} c^n + (c^n + d^n)^{1/n} d^n$
$= (c^n + d^n)^{1 + (1/n)} = (c^n + d^n)^{(n+1)/n} = a_n^{n+1}.$
Taking the $n + 1$-th root of each side we have $a_{n+1} < a_n$. The sequence is monotonic decreasing.

49. $a_1 = 1, a_2 = \frac{1}{2}, a_3 = \frac{1}{6}, a_4 = \frac{1}{24}, a_5 = \frac{1}{120}, a_6 = \frac{1}{720};$ $a_n = 1/n!$ **51.** $a_1 = a_2 = a_3 = a_4 = a_5 = a_6 = 1;$ $a_n = 1$

53. $a_1 = 1, a_2 = 3, a_3 = 5, a_4 = 7, a_5 = 9, a_6 = 11;$ $a_n = 2n - 1$ **55.** $a_1 = 1, a_2 = 4, a_3 = 13, a_4 = 40, a_5 = 121, a_6 = 364;$ $a_n = \frac{1}{2}(3^n - 1)$

57. $a_1 = 1, a_2 = 4, a_3 = 9, a_4 = 16, a_5 = 25, a_6 = 36;$ $a_n = n^2$ **59.** $a_1 = 1, a_2 = 3, a_3 = 4, a_4 = 8, a_5 = 16, a_6 = 32;$ $a_n = 2^{n-1}$ $(n \ge 3)$

61. $a_1 = 2, a_2 = 1, a_3 = 2, a_4 = 1, a_5 = 2, a_6 = 1;$ $a_n = \frac{1}{2}[3 - (-1)^n]$ **63.** $a_1 = 1, a_2 = 3, a_3 = 5, a_4 = 7, a_5 = 9, a_6 = 11;$ $a_n = 2n - 1$

65. First $a_1 = 2^1 - 1 = 1$. Next suppose $a_k = 2^k - 1$ for some $k \geq 1$. Then

$$a_{k+1} = 2a_k + 1 = 2(2^k - 1) + 1 = 2^{k+1} - 1.$$

67. First $a_1 = \dfrac{1}{2^0} = 1$. Next suppose $a_k = \dfrac{k}{2^{k-1}}$ for some $k \geq 1$. Then $a_{k+1} = \dfrac{k+1}{2k} a_k = \dfrac{k+1}{2k} \dfrac{k}{2^{k-1}} = \dfrac{k+1}{2^k}$.

69. (a) n (b) $\dfrac{1 - r^n}{1 - r}$ **71.** (a) $150(\tfrac{3}{4})^{n-1}$ (b) $\dfrac{5\sqrt{3}}{2} \left(\dfrac{3}{4}\right)^{\frac{n-1}{2}}$

73. (a) $a_2 = 1 + \sqrt{a_1} = 2 > 1 = a_1$. Assume $a_k = 1 + \sqrt{a_{k-1}} > a_{k-1}$. Then $a_{k+1} = 1 + \sqrt{a_k} > 1 + \sqrt{a_{k-1}} = a_k$. Thus $\{a_n\}$ is an increasing
sequence.

(b) $a_n = 1 + \sqrt{a_{n-1}} < 1 + \sqrt{a_n}$, since $a_{n-1} < a_n$.

$a_n - \sqrt{a_n} - 1 < 0$, or $(\sqrt{a_n})^2 - \sqrt{a_n} - 1 < 0$, which implies (solve the inequality) that $\sqrt{a_n} < \dfrac{1 + \sqrt{5}}{2}$, hence $a_n < \dfrac{3 + \sqrt{5}}{2}$ for all n.

(c) lub $\{a_n\} \cong 2.6180$.

SECTION 10.3

1. diverges   **3.** converges to 0   **5.** converges to 1   **7.** converges to 0   **9.** converges to 0   **11.** diverges   **13.** converges to 0

15. converges to 1   **17.** converges to $\tfrac{4}{9}$ **19.** converges to $\tfrac{1}{2}\sqrt{2}$ **21.** diverges   **23.** converges to 1   **25.** diverges

27. converges to 0   **29.** converges to $\tfrac{1}{2}$ **31.** converges to e^2 **33.** diverges   **35.** converges to 0

37. Use $|(a_n + b_n) - (L + M)| \leq |a_n - L| + |b_n - M|$.   **39.** $\left(1 + \dfrac{1}{n}\right)^{n+1} = \left(1 + \dfrac{1}{n}\right)^n \left(1 + \dfrac{1}{n}\right)$.   Note that $\left(1 + \dfrac{1}{n}\right)^n \to e$

and $\left(1 + \dfrac{1}{n}\right) \to 1$.

41. Imitate the proof given for the nondecreasing case in Theorem 10.3.6.   **43.** Let $\epsilon > 0$. Choose k so that, for $n \geq k$,

$$L - \epsilon < a_n < L + \epsilon, \quad L - \epsilon < c_n < L + \epsilon \quad \text{and} \quad a_n \leq b_n \leq c_n.$$

For such n,

$$L - \epsilon < b_n < L + \epsilon.$$

45. Let $\epsilon > 0$. Since $a_n \to L$, there exists a positive integer N such that $L - \epsilon < a_n < L + \epsilon$ for all $n \geq N$. Now $a_n \leq M$ for all n, so $L - \epsilon < M$, or $L < M + \epsilon$. Since ϵ is arbitrary, $L \leq M$.

47. By the continuity of f, $f(L) = f(\lim_{n \to \infty} a_n) = \lim_{n \to \infty} f(a_n) = \lim_{n \to \infty} a_{n+1} = L$.   **49.** Use Theorem 10.3.12 with $f(x) = x^{1/p}$.   **51.** converges to 0

53. converges to 0   **55.** diverges   **57.** converges to 2   **59.** $L = 0, n = 32$ **61.** $L = 0, n = 4$ **63.** $L = 0, n = 7$

65. $L = 0, n = 65$ **67.** (a) $\dfrac{3 + \sqrt{5}}{2}$ (b) 3

69. (a)

a_2	a_3	a_4	a_5	a_6	a_7	a_8	a_9	a_{10}
0.540302	0.857553	0.654290	0.793480	0.701369	0.763960	0.722102	0.750418	0.731404

(b) 0.739085; it is the fixed point of $f(x) = \cos x$.

71. (a)

a_2	a_3	a_4	a_5	a_6	a_7	a_8
2.000000	1.750000	1.732143	1.732051	1.732051	1.732051	1.732051

(b) $L = \dfrac{1}{2}\left(L + \dfrac{3}{L}\right)$, which implies $L^2 = 3$, or $L = \sqrt{3}$.

SECTION 10.4

1. converges to 1   **3.** converges to 0   **5.** converges to 0   **7.** converges to 0   **9.** converges to 1   **11.** converges to 0

13. converges to 1   **15.** converges to 1   **17.** converges to π **19.** converges to 1   **21.** converges to 0   **23.** diverges

25. converges to 0   **27.** converges to e^{-1} **29.** converges to 0   **31.** converges to 0   **33.** converges to e^x **35.** converges to 0

37. $\sqrt{n+1} - \sqrt{n} = \dfrac{\sqrt{n+1} - \sqrt{n}}{\sqrt{n+1} + \sqrt{n}} (\sqrt{n+1} + \sqrt{n}) = \dfrac{1}{\sqrt{n+1} + \sqrt{n}} \to 0$

39. (b) $2\pi r$. As $n \to \infty$, the perimeter of the polygon tends to the circumference of the circle.   **41.** $\tfrac{1}{2}$ **43.** $\tfrac{1}{8}$

45. (a) $m_{n+1} - m_n = \dfrac{1}{n+1}(a_1 + \cdots + a_n + a_{n+1}) - \dfrac{1}{n}(a_1 + \cdots + a_n)$

$\qquad = \dfrac{1}{n(n+1)}\left[na_{n+1} - (a_1 + \cdots + a_n) \right] > 0$ since $\{a_n\}$ is increasing.

(b) We begin with the hint $m_n < \dfrac{|a_1 + \cdots + a_j|}{n} + \dfrac{\epsilon}{2}\left(\dfrac{n-j}{n}\right)$. Since j is fixed, $\dfrac{|a_1 + \cdots + a_j|}{n} \to 0$, and therefore for n sufficiently

large $\dfrac{|a_1 + \cdots + a_j|}{n} < \dfrac{\epsilon}{2}$. Since $\dfrac{\epsilon}{2}\left(\dfrac{n-j}{n}\right) < \dfrac{\epsilon}{2}$, we see that, for n sufficiently large, $|m_n| < \epsilon$. This shows that $m_n \to 0$.

47. (a) Let S be the set of positive integers n ($n \geq 2$) for which the inequalities hold. Since $(\sqrt{b})^2 - 2\sqrt{ab} + (\sqrt{a})^2 = (\sqrt{b} - \sqrt{a})^2 > 0$, it follows that

$\dfrac{a+b}{2} > \sqrt{ab}$ and $a_1 > b_1$. Now $a_2 = \dfrac{a_1 + b_1}{2} < a_1$ and $b_2 = \sqrt{a_1 b_1} < b_1$. Also, by the argument above, $a_2 = \dfrac{a_1 + b_1}{2} > \sqrt{a_1 b_1} = b_2$, and

so $a_1 > a_2 > b_2 > b_1$. Thus $2 \in S$. Assume that $k \in S$. Then $a_{k+1} = \dfrac{a_k + b_k}{2} < \dfrac{a_k + a_k}{2} = a_k$, $b_{k+1} = \sqrt{a_k b_k} > \sqrt{b_k^2} = b_k$, and $a_{k+1} =$

$\dfrac{a_k + b_k}{2} > \sqrt{a_k b_k} = b_{k+1}$. Thus $k+1 \in S$. Therefore the inequalities hold for all $n \geq 2$.

(b) $\{a_n\}$ is a decreasing sequence which is bounded below.

$\{b_n\}$ is an increasing sequence which is bounded above.

Let $L_a = \lim\limits_{n\to\infty} a_n$, $L_b = \lim\limits_{n\to\infty} b_n$. Then $a_n = \dfrac{a_{n-1} + b_{n-1}}{2}$ implies $L_a = \dfrac{L_a + L_b}{2}$ and $L_a = L_b$.

49. The numerical work suggests $L \cong 1$. Justification: Set $f(x) = \sin x - x^2$. Note that $f(0) = 0$ and for x close to 0, $f'(x) = \cos x - 2x > 0$.
Therefore $\sin x - x^2 > 0$ for x close to 0 and $\sin(1/n) - 1/n^2 > 0$ for n large. Thus, for n large,

$$\frac{1}{n^2} < \sin\frac{1}{n} < \frac{1}{n}$$

$|\sin x| \leq |x|$ for all x

$$\left(\frac{1}{n^2}\right)^{1/n} < \left(\sin\frac{1}{n}\right)^{1/n} < \left(\frac{1}{n}\right)^{1/n}$$

$$\left(\frac{1}{n^{1/n}}\right)^2 < \left(\sin\frac{1}{n}\right)^{1/n} < \frac{1}{n^{1/n}}.$$

As $n \to \infty$ both bounds tend to 1 and therefore the middle term also tends to 1.

51. (a)

a_3	a_4	a_5	a_6	a_7	a_8	a_9	a_{10}
2	3	5	8	13	21	34	55

(b)

r_1	r_2	r_3	r_4	r_5	r_6
1	2	1.5	1.6667	1.6000	1.625

(c) $L = \dfrac{1 + \sqrt{5}}{2} \cong 1.618033989$

SECTION 10.5

1. 0   **3.** 1   **5.** $\frac{1}{2}$ **7.** $\ln 2$ **9.** $\frac{1}{4}$ **11.** 2   **13.** $\dfrac{1+\pi}{1-\pi}$ **15.** $\frac{1}{2}$ **17.** π **19.** $-\frac{1}{2}$ **21.** -2 **23.** $\frac{1}{3}\sqrt{6}$

25. $-\frac{1}{8}$ **27.** 4   **29.** $\frac{1}{2}$ **31.** $\frac{1}{2}$ **33.** 1   **35.** 1   **37.** $\lim\limits_{x\to 0}(2 + x + \sin x) \neq 0$, $\lim\limits_{x\to 0}(x^3 + x - \cos x) \neq 0$

39. $a = \pm 4$, $b = 1$ **41.** $f(0)$ **43.** $\frac{3}{4}$ **45.** (a) $f(x) \to \infty$ as $x \to \pm\infty$ (b) 10   **47.** (b) $\ln 2 \cong 0.6931$

SECTION 10.6

1. ∞ **3.** -1 **5.** ∞ **7.** $\frac{1}{5}$ **9.** 1   **11.** 0   **13.** ∞ **15.** $\frac{1}{3}$ **17.** e **19.** 1   **21.** $\frac{1}{2}$ **23.** 0   **25.** 1

27. e^3 **29.** e **31.** e^2 **33.** 0   **35.** $-\frac{1}{2}$ **37.** 0   **39.** 1   **41.** 1   **43.** 0

45. y-axis vertical asymptote   **47.** x-axis horizontal asymptote   **49.** x-axis horizontal asymptote

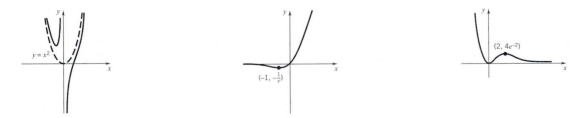

51. $\dfrac{b}{a}\sqrt{x^2-a^2}-\dfrac{b}{a}x=\dfrac{\sqrt{x^2-a^2}+x}{\sqrt{x^2-a^2}+x}\left(\dfrac{b}{a}\right)(\sqrt{x^2-a^2}-x)=\dfrac{-ab}{\sqrt{x^2-a^2}+x}\to 0$ as $x\to\infty$ **53.** Example: $f(x)=x^2+\dfrac{(x-1)(x-2)}{x^3}$

55. $\lim\limits_{x\to0^+}\cos x\neq 0$

57. (a) Let S be the set of positive integers for which the statement is true. Since $\lim\limits_{x\to\infty}\dfrac{\ln x}{x}=0$, $1\in S$. Assume that $k\in S$. By L'Hospital's rule,

$$\lim_{x\to\infty}\frac{(\ln x)^{k+1}}{x}\overset{*}{=}\lim_{x\to\infty}\frac{(k+1)(\ln x)^k}{x}=0\qquad(\text{since }k\in S).$$

Thus $k+1\in S$, and S is the set of positive integers.
(b) Choose any positive number α. Let $k-1$ and k be positive integers such that $k-1\leq\alpha\leq k$. Then, for $x>e$,

$$\frac{(\ln x)^{k-1}}{x}\leq\frac{(\ln x)^\alpha}{x}\leq\frac{(\ln x)^k}{x}$$

and the result follows by the pinching theorem.

59. (a) L'Hospital's rule applied to the given limit results in $\lim\limits_{x\to0}\dfrac{2e^{-1/x^2}}{x^3}$. Rewrite the quotient as $\dfrac{1/x}{e^{1/x^2}}\to0$ as $x\to0$.   (b) $f'(0)=0$.

61. $\lim\limits_{x\to0^+}(1+x^2)^{1/x}=1$.   **63.** $\lim\limits_{x\to\infty}g(x)=-5/3$.

SECTION 10.7

1. 1   **3.** $\frac{1}{4}\pi$ **5.** diverges   **7.** 6   **9.** $\frac{1}{2}\pi$ **11.** 2   **13.** diverges   **15.** $-\frac{1}{4}$ **17.** π **19.** diverges   **21.** $\ln 2$

23. 4   **25.** diverges   **27.** diverges   **29.** diverges   **31.** $\frac{1}{2}$ **33.** $2e-2$ **35.** $\dfrac{\pi}{2}-1$ **37.** π

39. (a)   (b) 2 (c) $V=\displaystyle\int_0^1\pi\left(\dfrac{1}{\sqrt{x}}\right)^2dx=\pi\int_0^1\dfrac{1}{x}\,dx,$ diverges   **41.** (b) 1   (a)

(c) $\frac{1}{2}\pi$
(d) 2π
(e) $\pi[\sqrt{2}+\ln(1+\sqrt{2})]$

43. (a) The interval $[0,1]$ causes no problem. For $x\geq1$, $e^{-x^2}\leq e^{-x}$ and $\displaystyle\int_1^\infty e^{-x}\,dx$ is finite.

   (b) $V_y=\displaystyle\int_0^\infty 2\pi xe^{-x^2}\,dx=\pi$

45. (a)   (b) $\frac{4}{3}$ **47.** converges by comparison with $\displaystyle\int_0^\infty\dfrac{dx}{x^{3/2}}$

(c)2π
(d)$\frac{8}{7}\pi$

49. diverges since for x large the integrand is greater than $\dfrac{1}{x}$ and $\displaystyle\int_1^\infty\dfrac{1}{x}\,dx$ diverges   **51.** converges by comparison with $\displaystyle\int_1^\infty\dfrac{dx}{x^{3/2}}$

53. $L=(a\sqrt{1+c^2}/c)e^{c\theta_1}$ **55.** $\dfrac{1}{s}$; $\ \mathrm{dom}(F)=(0,\infty)$ **57.** $\dfrac{s}{s^2+4}$; $\ \mathrm{dom}(F)=(0,\infty)$

59. $f(x)\geq0$ for all x and $\displaystyle\int_{-\infty}^\infty f(x)\,dx=\int_0^\infty\dfrac{6x}{(1+3x^2)^2}\,dx=1$ **61.** $\dfrac{1}{k}$

CHAPTER 11
SECTION 11.1

1. 12   **3.** 15   **5.** -10 **7.** $\frac{15}{16}$ **9.** $-\frac{2}{15}$ **11.** $\frac{85}{64}$ **13.** $\displaystyle\sum_{n=1}^{11}2n-1$ **15.** $\displaystyle\sum_{n=1}^{25}2n$ **17.** $\displaystyle\sum_{n=1}^{81}(-1)^{n-1}\sqrt{n}$ **19.** $\displaystyle\sum_{k=1}^{n}m_k\Delta x_k$

21. $\displaystyle\sum_{k=1}^{n}f(x_k^*)\Delta x_k$ **23.** $\displaystyle\sum_{k=0}^{5}(-1)^ka^{5-k}b_k$ **25.** $\displaystyle\sum_{k=0}^{4}a_kx^{4-k}$ **27.** $\displaystyle\sum_{k=0}^{4}(-1)^k(k+1)x^k$ **29.** $\displaystyle\sum_{k=3}^{10}\dfrac{1}{2^k},\ \sum_{i=0}^{7}\dfrac{1}{2i+3}$

31. $\displaystyle\sum_{k=3}^{10} (-1)^{k+1} \frac{k}{k+1}$, $\displaystyle\sum_{i=0}^{7} (-1)^i \frac{i+3}{i+4}$ **33.** let $k = n + 3$ **35.** let $k = n - 3$

37. (a) $\displaystyle(1-x)\sum_{k=0}^{n} x^k = \sum_{k=0}^{n} (x^k - x^{k+1})$ (b) converges to $\frac{3}{2}$

$$= (1-x) + (x - x^2) + (x^2 - x^3) + \cdots + (x^n - x^{n+1}) = 1 - x^{n+1}.$$

39. $|a_n - L| < \epsilon$ for $n \geq k$ iff $|a_{n-p} - L| < \epsilon$ for $n \geq k + p$.

41. True for $n = 1$: $\displaystyle\sum_{k=1}^{1} k = 1 = \frac{1}{2}(1)(2)$. Suppose true for $n = p$. Then

$$\sum_{k=1}^{p+1} k = \sum_{k=1}^{p} k + (p+1) = \frac{1}{2}(p)(p+1) + (p+1) = \frac{1}{2}(p+1)(p+2) = \frac{1}{2}(p+1)[(p+1)+1]$$

and thus true for $n = p + 1$.

43. True for $n = 1$: $\displaystyle\sum_{k=1}^{1} k^2 = 1 = \frac{1}{6}(1)(2)(3)$. Suppose true for $n = p$. Then   **45.** 140   **47.** 680

$$\sum_{k=1}^{p+1} k^2 = \sum_{k=1}^{p} k^2 + (p+1)^2 = \frac{1}{6}(p)(p+1)(2p+1) + (p+1)^2$$

$$= \frac{1}{6}(p+1)(2p^2 + 7p + 6) = \frac{1}{6}(p+1)(p+2)(2p+3) = \frac{1}{6}(p+1)[(p+1)+1][2(p+1)+1]$$

and thus true for $n = p + 1$.

SECTION 11.2

1. $\frac{1}{4}$ **3.** $\frac{1}{2}$ **5.** $\frac{11}{18}$ **7.** $\frac{10}{3}$ **9.** $\frac{67000}{999}$ **11.** 4   **13.** $-\frac{3}{2}$ **15.** $\frac{1}{2}$ **17.** 24   **19.** $\displaystyle\sum_{k=1}^{\infty} \frac{7}{10^k} = \frac{7}{9}$ **21.** $\displaystyle\sum_{k=1}^{\infty} \frac{24}{100^k} = \frac{8}{33}$

23. $\displaystyle\sum_{k=1}^{\infty} \frac{112}{1000^k} = \frac{112}{999}$ **25.** $\displaystyle\frac{62}{100} + \frac{1}{100}\sum_{k=1}^{\infty} \frac{45}{100^k} = \frac{687}{1100}$

27. Let $x = .\overline{a_1 a_2 \cdots a_n} \overline{a_1 a_2 \cdots a_n} \cdots$. Then

$$x = \sum_{k=1}^{\infty} \frac{a_1 a_2 \cdots a_n}{(10^n)^k} = a_1 a_2 \cdots a_n \sum_{k=1}^{\infty} \left(\frac{1}{10^n}\right)^k = a_1 a_2 \cdots a_n \left[\frac{1}{1 - \frac{1}{10^n}} - 1\right] = \frac{a_1 a_2 \cdots a_n}{10^n - 1}.$$

29. $\displaystyle\frac{1}{1+x} = \frac{1}{1-(-x)} = \sum_{k=0}^{\infty} (-x)^k = \sum_{k=0}^{\infty} (-1)^k x^k$ **31.** $\displaystyle\sum_{k=0}^{\infty} x^{k+1}$ **33.** $\displaystyle\sum_{k=0}^{\infty} (-1)^k x^{2k+1}$ **35.** $\displaystyle\sum_{k=0}^{\infty} (-1)^k (2x)^{2k}$

37. $\displaystyle\sum_{n=0}^{\infty} \left(\frac{3}{2}\right)^n$; geometric series with $r = \frac{3}{2} > 1$ **39.** $\displaystyle\lim_{k\to\infty}\left(\frac{k}{k+1}\right)^k = \frac{1}{e} \neq 0$ **41.** $4 + \frac{1}{3}\displaystyle\sum_{k=0}^{\infty}\left(\frac{1}{12}\right)^k = 4 + \frac{4}{11}$ o'clock   **43.** 12 ft

45. $\displaystyle\sum_{k=1}^{\infty} n_k \left(1 + \frac{r}{100}\right)^{-k}$ **47.** \$9   **49.** 32   **51.** $\displaystyle\lim_{n\to\infty} s_n = L = \sum_{k=0}^{\infty} a_k$. Thus $\displaystyle\lim_{n\to\infty} R_n = \lim_{n\to\infty}\left(s_n - \sum_{k=0}^{\infty} a_k\right) = \lim_{n\to\infty} s_n - L = 0$.

53. $\displaystyle\frac{1 + (-1)^n}{2}$, $n = 0, 1, 2, \ldots$ **55.** $\displaystyle s_n = \sum_{k=1}^{n} \ln\left(\frac{k+1}{k}\right) = \sum_{k=1}^{n} [\ln(k+1) - \ln k] = \ln(n+1) \to \infty$

57. (a) $\displaystyle s_n = \sum_{k=1}^{n} (d_k - d_{k+1}) = d_1 - d_{n+1} \to d_1$

(b) (i) $\displaystyle\sum_{k=1}^{\infty} \frac{\sqrt{k+1} - \sqrt{k}}{\sqrt{k(k+1)}} = \sum_{k=1}^{\infty}\left(\frac{1}{\sqrt{k}} - \frac{1}{\sqrt{k+1}}\right) = 1$ (ii) $\displaystyle\sum_{k=1}^{\infty} \frac{2k+1}{2k^2(k+1)^2} = \sum_{k=1}^{\infty} \frac{1}{2}\left(\frac{1}{k^2} - \frac{1}{(k+1)^2}\right) = \frac{1}{2}$

59. $N = 6$ **61.** $N = 9999$ **63.** $N = \left[\!\left[\dfrac{\ln(\epsilon[1 - x])}{\ln|x|}\right]\!\right]$, where $[\![\ \ ]\!]$ denotes the greatest integer function.

SECTION 11.3

1. converges; comparison $\Sigma\, 1/k^2$ **3.** converges; comparison $\Sigma\, 1/k^2$ **5.** diverges; comparison $\Sigma\, 1/(k+1)$ **7.** diverges; limit comparison $\Sigma\, 1/k$

9. converges; integral test   **11.** diverges; p-series with $p = \frac{2}{3} \leq 1$ **13.** diverges; $a_k \not\to 0$ **15.** diverges; comparison $\Sigma\, 1/k$

17. diverges; $a_k \not\to 0$ **19.** converges; limit comparison $\Sigma\, 1/k^2$ **21.** diverges; integral test   **23.** converges; limit comparison $\Sigma\, 1/k^2$

25. diverges; limit comparison $\Sigma\, 1/k$ **27.** converges; limit comparison $\Sigma\, 1/k^{3/2}$ **29.** converges; integral test

31. converges; comparison $\Sigma\, 3/k^2$ **33.** converges; comparison $\Sigma\, 2/k^2$ **35.** $p > 1$

37. (a) The improper integral $\int_0^\infty e^{-\alpha x}\,dx = \dfrac{1}{\alpha}$ converges. (b) The improper integral $\int_0^\infty x e^{-\alpha x}\,dx = \dfrac{1}{\alpha^2}$ converges.

(c) The improper integral $\int_0^\infty x^n e^{-\alpha x}\,dx = \dfrac{1}{\alpha^{n+1}}$ converges.

39. (a) 1.1777 (b) $0.02 < R_4 < 0.0313$ (c) $1.1977 < \sum_{k=1}^\infty \dfrac{1}{k^3} < 1.209$ **41.** (a) $1/101 < R_{100} < 1/100$ (b) 10,001   **43.** (a) 15 (b) 1.082

45. (a) If $a_k/b_k \to 0$, then $a_k/b_k < 1$ for all $k \geq K$ for some K. But then $a_k < b_k$ for all $k \geq K$ and, since Σb_k converges, Σa_k converges.
[The basic comparison test, Theorem 11.3.5.]
  (b) Similar to (a) except that this time we appeal to part (ii) of Theorem 11.3.5.

(c) $\Sigma a_k = \Sigma \dfrac{1}{k^2}$ converges, $\Sigma b_k = \Sigma \dfrac{1}{k^{3/2}}$ converges, $\dfrac{1/k^2}{1/k^{3/2}} = \dfrac{1}{\sqrt{k}} \to 0$

  $\Sigma a_k = \Sigma \dfrac{1}{k^2}$ converges, $\Sigma b_k = \Sigma \dfrac{1}{\sqrt{k}}$ diverges, $\dfrac{1/k^2}{1/\sqrt{k}} = \dfrac{1}{k^{3/2}} \to 0$

(d) $\Sigma b_k = \Sigma \dfrac{1}{\sqrt{k}}$ diverges, $\Sigma a_k = \Sigma \dfrac{1}{k^2}$ converges, $\dfrac{1/k^2}{1/\sqrt{k}} = \dfrac{1}{k^{3/2}} \to 0$

  $\Sigma b_k = \Sigma \dfrac{1}{\sqrt{k}}$ diverges, $\Sigma a_k = \Sigma \dfrac{1}{k}$ diverges, $\dfrac{1/k}{1/\sqrt{k}} = \dfrac{1}{\sqrt{k}} \to 0$

47. (a) Since Σa_k converges, $\lim\limits_{k \to \infty} a_k = 0$. Therefore there exists a positive integer N such that $0 < a_k < 1$ for $k \geq N$.
  Thus, for $k \geq N$, $a_k^2 < a_k$ and so Σa_k^2 converges by the comparison test.
  (b) Σa_k may either converge or diverge.
  $\Sigma 1/k^4$ converges and $\Sigma 1/k^2$ converges; $\Sigma 1/k^2$ converges and $\Sigma 1/k$ diverges.

49. $0 < L - \sum_{k=1}^n f(k) = L - s_n = \sum_{k=n+1}^\infty f(k) < \int_n^\infty f(x)\,dx$ **51.** $N = 3$.

53. (a) Set $f(x) = x^{1/4} - \ln x$. Then $f'(x) = \dfrac{1}{4}x^{-3/4} - \dfrac{1}{x} = \dfrac{1}{4x}(x^{1/4} - 4)$. Since $f(e^{12}) = e^3 - 12 > 0$ and $f'(x) > 0$ for $x > e^{12}$, we know that $n^{1/4} >$
  $\ln n$ and therefore $\dfrac{1}{n^{5/4}} > \dfrac{\ln n}{n^{3/2}}$ for sufficiently large n. Since $\Sigma \dfrac{1}{n^{5/4}}$ is a convergent p-series, $\Sigma \dfrac{\ln n}{n^{3/2}}$ converges by the basic comparison test.
  (b) By L'Hospital's rule $\lim\limits_{x \to \infty} \left[\left(\dfrac{\ln x}{x^{3/2}} \right) \Big/ \left(\dfrac{1}{x^{5/4}} \right) \right] = 0$

SECTION 11.4

1. converges; ratio test   **3.** converges; root test   **5.** diverges; $a_k \not\to 0$ **7.** diverges; limit comparison $\Sigma 1/k$ **9.** converges; root test

11. diverges; limit comparison $\Sigma 1/\sqrt{k}$ **13.** diverges; ratio test   **15.** converges; comparison $\Sigma 1/k^{3/2}$ **17.** converges; comparison $\Sigma 1/k^2$

19. diverges; integral test   **21.** diverges; $a_k \to e^{-100} \neq 0$ **23.** diverges; limit comparison $\Sigma 1/k$ **25.** converges; ratio test

27. converges; comparison $\Sigma 1/k^{3/2}$ **29.** converges; ratio test   **31.** converges; ratio test: $a_{k+1}/a_k \to \frac{4}{27}$ **33.** converges; ratio test

35. converges; root test   **37.** converges; root test   **39.** converges; ratio test   **41.** $\frac{10}{81}$

43. The series $\Sigma \dfrac{k!}{k^k}$ converges. Therefore $\lim\limits_{k \to \infty} \dfrac{k!}{k^k} = 0$ by Theorem 11.2.6.   **45.** $p \geq 2$ **47.** $|x| < 1$ **49.** converges for all x

51. Set $b_k = a_k r^k$. If $(a_k)^{1/k} \to \rho$ and $\rho < 1/r$, then
  $(b_k)^{1/k} = (a_k r^k)^{1/k} = (a_k)^{1/k} r \to \rho r < 1$
  and thus, by the root test, $\Sigma b_k = \Sigma a_k r^k$ converges.

SECTION 11.5

1. diverges; $a_k \not\to 0$ **3.** diverges; $a_k \not\to 0$ **5.** (a) does not converge absolutely; integral test (b) converges conditionally; Theorem 11.5.4

7. diverges; limit comparison $\Sigma 1/k$ **9.** (a) does not converge absolutely; limit comparison $\Sigma 1/k$ (b) converges conditionally; Theorem 11.5.4

11. diverges; $a_k \not\to 0$ **13.** (a) does not converge absolutely; comparison $2 \Sigma 1/\sqrt{k+1}$ (b) converges conditionally; Theorem 11.5.4

15. converges absolutely (terms already positive); $\Sigma \sin\left(\dfrac{\pi}{4k^2} \right) \leq \Sigma \dfrac{\pi}{4k^2} = \dfrac{\pi}{4} \Sigma \dfrac{1}{k^2}$ $(|\sin x| \leq |x|)$ **17.** converges absolutely; ratio test

19. (a) does not converge absolutely; limit comparison $\Sigma 1/k$ (b) converges conditionally; Theorem 11.5.4   **21.** diverges; $a_k \not\to 0$

23. diverges; $a_k \not\to 0$ **25.** converges absolutely; ratio test   **27.** diverges; $a_k = \dfrac{1}{k}$ for all k **29.** converges absolutely; comparison $\Sigma 1/k^2$

31. diverges; $a_k \nrightarrow 0$ **33.** 0.1104   **35.** 0.001   **37.** $\frac{10}{11}$ **39.** $N = 39{,}998$ **41.** (a) 4 (b) 6

43. No. For instance, set $a_{2k} = 2/k$ and $a_{2k+1} = 1/k$.

45. (a) Since $\Sigma |a_k|$ converges, $\Sigma |a_k|^2 = \Sigma a_k^2$ converges (Exercise 47, Section 11.3). (b) $\Sigma\, 1/k^2$ is convergent, $\Sigma \frac{(-1)^k}{k}$ is not absolutely convergent.

47. See the proof of Theorem 11.8.2.   **49.** (a) $\displaystyle\sum_{k=1}^{\infty} \frac{(-1)^{k-1}(a+b) + (a-b)}{2k}$ (b) if $a = b = 0$, absolutely convergent (vacuously)
if $a = b \neq 0$, conditionally convergent
if $a \neq b$, divergent

SECTION 11.6

1. $-1 + x + \frac{1}{2}x^2 - \frac{1}{24}x^4$ **3.** $-\frac{1}{2}x^2 - \frac{1}{12}x^4$ **5.** $1 - x + x^2 - x^3 + x^4 - x^5$ **7.** $x + \frac{1}{3}x^3 + \frac{2}{15}x^5$

9. $P_0(x) = 1$, $P_1(x) = 1 - x$, $P_2(x) = 1 - x + 3x^2$, $P_3(x) = 1 - x + 3x^2 + 5x^3$

11. $\displaystyle\sum_{k=0}^{n} (-1)^k \frac{x^k}{k!}$ **13.** $\displaystyle\sum_{k=0}^{m} \frac{x^{2k}}{(2k!)}$ where $m = \frac{n}{2}$ and n is even   **15.** $\displaystyle\sum_{k=0}^{n} \frac{r^k}{k!} x^k$ **17.** 79/48 $(79/48 \cong 1.646)$

19. 5/6 $(5/6 \cong 0.833)$ **21.** 13/24 $(13/24 \cong 0.542)$ **23.** 0.17   **25.** $\dfrac{4e^{2c}}{15}x^5$, $|c| < |x|$ **27.** $\dfrac{-4\sin 2c}{15}x^5$, $|c| < |x|$

29. $\dfrac{3\sec^4 c - 2\sec^2 c}{3}x^3$, $|c| < |x|$ **31.** $\dfrac{3c^2 - 1}{3(1 + c^2)^3}x^3$, $|c| < |x|$ **33.** $\dfrac{(-1)^{n+1}e^{-c}}{(n+1)!}x^{n+1}$, $|c| < |x|$ **35.** $\dfrac{1}{(1 - c)^{n+2}}x^{n+1}$, $|c| < |x|$

37. (a) 4 (b) 2 (c) 999   **39.** (a) 1.649 (b) 0.368   **41.** For $0 \leq k \leq n$, $P^{(k)}(0) = k!a_k$; for $k > n$, $P^{(k)}(0) = 0$. Thus $P(x) = \displaystyle\sum_{k=0}^{\infty} P^{(k)}(0)\frac{x^k}{k!}$.

43. $\dfrac{d^{2k}(\sinh x)}{dx^{2k}}\bigg|_{x=0} = \sinh(0) = 0$; $\dfrac{d^{2k+1}(\sinh x)}{dx^{2k+1}}\bigg|_{x=0} = \cosh(0) = 1$

Therefore $\sinh x = x + \dfrac{x^3}{3!} + \dfrac{x^5}{5!} + \cdots = \displaystyle\sum_{k=0}^{\infty} \dfrac{1}{(2k+1)!}x^{2k+1}$

45. $\displaystyle\sum_{k=0}^{\infty} \frac{a^k}{k!}x^k$, $(-\infty, \infty)$ **47.** $\displaystyle\sum_{k=0}^{\infty} \frac{(-1)^k a^{2k}}{(2k)!}x^{2k}$, $(-\infty, \infty)$ **49.** $\ln a + \displaystyle\sum_{k=1}^{\infty} \frac{(-1)^{k-1}}{ka^k}x^k$, $(-a, a]$

51. $\ln 2 = \ln\left(\dfrac{1 + \frac{1}{3}}{1 - \frac{1}{3}}\right) \cong 2\left[\dfrac{1}{3} + \dfrac{1}{3}\left(\dfrac{1}{3}\right)^3 + \dfrac{1}{5}\left(\dfrac{1}{3}\right)^5\right] = \dfrac{842}{1215}$ $\left(\dfrac{842}{1215} \cong 0.693\right)$ **53.** routine; use $u = (x - t)^k$ and $dv = f^{(k+1)}(t)\,dt$

55. (b) $f(x) = \dfrac{x^{-n}}{e^{1/x^2}}$ and $\lim_{x\to 0} f(x)$ has the form ∞/∞. Successive applications of L'Hospital's rule will finally produce a quotient of the form

$\dfrac{cx^k}{e^{1/x^2}}$, where k is a nonnegative integer and c is a constant. It follows that $\lim_{x\to 0} f(x) = 0$.

(c) $f'(0) = \lim_{x\to 0} \dfrac{e^{-1/x^2} - 0}{x} = 0$ by part (b). Assume that $f^{(k)}(0) = 0$. Then

$f^{(k+1)}(0) = \lim_{x\to 0} \dfrac{f^{(k)}(x) - 0}{x} = \lim_{x\to 0} \dfrac{f^{(k)}(x)}{x}$.

Now, $\dfrac{f^{(k)}(x)}{x}$ is a sum of terms of the form $\dfrac{ce^{-1/x^2}}{x^n}$, where n is a positive integer and c is a constant. By part (b), $f^{(k+1)}(0) = 0$.

Therefore $f^{(n)}(0) = 0$ for all n.

(d) 0 (e) $x = 0$

57.

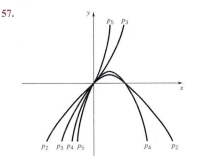

SECTION 11.7

1. $P_3(x) = 2 + \frac{1}{4}(x-4) - \frac{1}{64}(x-4)^2 + \frac{1}{512}(x-4)^3$
$R_4(x) = \frac{-5}{128c^{7/2}}x^4, \quad |c| < |x|$

3. $P_4(x) = \frac{\sqrt{2}}{2} + \frac{\sqrt{2}}{2}\left(x - \frac{\pi}{4}\right) - \frac{\sqrt{2}}{4}\left(x - \frac{\pi}{4}\right)^2 - \frac{\sqrt{2}}{12}\left(x - \frac{\pi}{4}\right)^3 + \frac{\sqrt{2}}{48}\left(x - \frac{\pi}{4}\right)^4$
$R_5(x) = \frac{\cos c}{120}\left(x - \frac{\pi}{4}\right)^5, \quad \left|c - \frac{\pi}{4}\right| < \left|x - \frac{\pi}{4}\right|$

5. $P_3(x) = \frac{\pi}{4} + \frac{1}{2}(x-1) - \frac{1}{4}(x-1)^2 + \frac{1}{12}(x-1)^3$ $\quad R_4(x) = \frac{c(1-c^2)}{(1+c^2)^3}(x-1)^4, \quad |c-1| < |x-1|$

7. $6 + 9(x-1) + 7(x-1)^2 + 3(x-1)^3, \quad (-\infty, \infty)$ **9.** $-3 + 5(x+1) - 19(x+1)^2 + 20(x+1)^3 - 10(x+1)^4 + 2(x+1)^5, \quad (-\infty, \infty)$

11. $\sum_{k=0}^{\infty} (-1)^k \left(\frac{1}{2}\right)^{k+1}(x-1)^k, \quad (-1, 3)$ **13.** $\frac{1}{5}\sum_{k=0}^{\infty} \left(\frac{2}{5}\right)^k (x+2)^k, \quad \left(-\frac{9}{2}, \frac{1}{2}\right)$ **15.** $\sum_{k=0}^{\infty} \frac{(-1)^{k+1}}{(2k+1)!}(x-\pi)^{2k+1}, \quad (-\infty, \infty)$

17. $\sum_{k=0}^{\infty} \frac{(-1)^{k+1}}{(2k)!}(x-\pi)^{2k}, \quad (-\infty, \infty)$ **19.** $\sum_{k=0}^{\infty} \frac{(-1)^k}{(2k)!}\left(\frac{\pi}{2}\right)^{2k}(x-1)^{2k}, \quad (-\infty, \infty)$ **21.** $\ln 3 + \sum_{k=1}^{\infty} \frac{(-1)^{k+1}}{k}\left(\frac{2}{3}\right)^k(x-1)^k, \quad \left(-\frac{1}{2}, \frac{5}{2}\right]$

23. $2\ln 2 + (1 + \ln 2)(x-2) + \sum_{k=2}^{\infty} \frac{(-1)^k}{k(k-1)2^{k-1}}(x-2)^k$ **25.** $\sum_{k=0}^{\infty} \frac{(-1)^k}{(2k+1)!}x^{2k+2}$ **27.** $\sum_{k=0}^{\infty} (k+2)(k+1)\frac{2^{k-1}}{5^{k+3}}(x+2)^k$

29. $1 + \sum_{k=1}^{\infty} \frac{(-1)^k 2^{2k-1}}{(2k)!}(x-\pi)^{2k}$ **31.** $\sum_{n=0}^{\infty} \frac{n!}{(n-k)!\, k!}(x-1)^k$

33. (a) $\frac{e^x}{e^a} = e^{x-a} = \sum_{k=0}^{\infty} \frac{(x-a)^k}{k!}, \quad e^x = e^a \sum_{k=0}^{\infty} \frac{(x-a)^k}{k!}$ (b) $e^{a+(x-a)} = e^x = e^a \sum_{k=0}^{\infty} \frac{(x-a)^k}{k!}, \quad e^{x_1+x_2} = e^{x_1}\sum_{k=0}^{\infty} \frac{x_2^k}{k!} = e^{x_1}e^{x_2}$
(c) $e^{-a}\sum_{k=0}^{\infty} (-1)^k \frac{(x-a)^k}{k!}$

35. (a) $P_2(x) = \frac{1}{2} + \frac{\sqrt{3}}{2}\left(x - \frac{\pi}{6}\right) - \frac{1}{4}\left(x - \frac{\pi}{6}\right)^2 - \frac{\sqrt{3}}{12}\left(x - \frac{\pi}{6}\right)^3$ (b) 0.5736

37. $P_2(x) = 6 + \frac{1}{12}(x-36) - \frac{1}{1728}(x-36)^2; \quad \sqrt{38} \cong 6.164$

SECTION 11.8

1. $(-1, 1)$ **3.** $(-\infty, \infty)$ **5.** $\{0\}$ **7.** $[-2, 2)$ **9.** $\{0\}$ **11.** $[-\frac{1}{2}, \frac{1}{2})$ **13.** $(-1, 1)$ **15.** $(-10, 10)$ **17.** $(-\infty, \infty)$
19. $(-\infty, \infty)$ **21.** $(-3/2, 3/2)$ **23.** converges only at $x = 1$ **25.** $(-4, 0)$ **27.** $(-\infty, \infty)$ **29.** $(-1, 1)$ **31.** $(0, 4)$
33. $(-\frac{5}{2}, \frac{1}{2})$ **35.** $(-2, 2)$ **37.** $\left[-\frac{1}{\sqrt{3}}, \frac{1}{\sqrt{3}}\right]$

39. Examine the convergence of $\Sigma \, |a_k x^k|$; for (a) use the root test and for (b) use the ratio test. **41.** $\Sigma \, |a_k(-r)^k| = \Sigma \, |a_k r^k|$

SECTION 11.9

1. $1 + 2x + 3x^2 + \cdots + nx^{n-1} + \cdots$ **3.** $1 + kx + \frac{(k+1)k}{2!}x^2 + \cdots + \frac{(n+k-1)!}{n!(k-1)!}x^n + \cdots$

5. $\ln(1-x^2) = -x^2 - \frac{1}{2}x^4 - \frac{1}{3}x^6 - \cdots - \frac{1}{n+1}x^{2n+2} - \cdots$ **7.** $1 + x^2 + \frac{2}{3}x^4 + \frac{17}{45}x^6 + \cdots$ **9.** -72 **11.** $\sum_{k=0}^{\infty} \frac{(-1)^k}{(2k+1)!}x^{4k+2}$

13. $\sum_{k=0}^{\infty} \frac{3^k}{k!}x^{3k}$ **15.** $2\sum_{k=0}^{\infty} x^{2k+1}$ **17.** $\sum_{k=0}^{\infty} \frac{(k!+1)}{k!}x^k$ **19.** $\sum_{k=1}^{\infty} \frac{(-1)^{k+1}}{k}x^{3k+1}$ **21.** $\sum_{k=0}^{\infty} \frac{(-1)^k}{k!}x^{3k+3}$ **23.** $\frac{1}{2}$ **25.** $-\frac{1}{2}$

27. $\sum_{k=1}^{\infty} \frac{(-1)^{k-1}}{k^2}x^k, \quad -1 \le x \le 1$ **29.** $\sum_{k=1}^{\infty} \frac{(-1)^{k-1}}{(2k-1)^2}x^{2k-1}$ **31.** $0.804 \le I \le 0.808$ **33.** $0.600 \le I \le 0.603$ **35.** $0.294 \le I \le 0.304$

37. 0.9461 **39.** 0.4485 **41.** e^{x^3} **43.** $3x^2 e^{x^3}$ **45.** (a) $\sum_{n=0}^{\infty} \frac{1}{n!}x^{n+1}$ (b) $\int_0^1 xe^x\, dx = 1 = \int_0^1 \left(\sum_{n=0}^{\infty} \frac{1}{n!}x^{n+1}\right) dx = \sum_{n=0}^{\infty} \frac{1}{n!(n+2)}$

47. Let $f(x)$ be the sum of these series; a_k and b_k are both $f^{(k)}(0)/k!$.

49. (a) If f is even, then $f^{(2k-1)}$ is odd for $k = 1, 2, \ldots$. This implies that $f^{(2k-1)}(0) = (0)$, and so $a_{2k-1} = \frac{f^{(2k-1)}(0)}{(2k-1)!} = 0$ for all k.

(b) If f is odd, then $f^{(2k)}$ is odd for $k = 1, 2, \ldots$, which implies $a_{2k} = 0$ for all k.

51. $0.0352 \le I \le 0.0359; \quad I = \frac{3}{6} - \frac{3}{8}\ln 1.5 \cong 0.0354505$ **53.** $0.2640 \le I \le 0.2643; \quad I = 1 - 2/e \cong 0.2642411$

SECTION 11.10

1. $1 + \frac{1}{2}x - \frac{1}{8}x^2 + \frac{1}{16}x^3 - \frac{5}{128}x^4$ **3.** $1 + \frac{1}{2}x^2 - \frac{1}{8}x^4$ **5.** $1 - \frac{1}{2}x + \frac{3}{8}x^2 - \frac{5}{16}x^3 + \frac{35}{128}x^4$ **7.** $1 - \frac{1}{4}x - \frac{3}{32}x^2 - \frac{7}{128}x^3 - \frac{77}{2048}x^4$

9. $8 + 3x + \frac{3}{16}x^2 - \frac{1}{128}x^3 + \frac{3}{4096}x^4$ **11.** (a) $\sum_{k=0}^{\infty} (-1)^k \binom{-1/2}{k}x^{2k}$ (b) $\sum_{k=0}^{\infty} (-1)^k \binom{-1/2}{k}\frac{1}{2k+1}x^{2k+1}, \quad R = 1$ **13.** 9.8995

15. 2.0799 **17.** 0.4925 **19.** 0.3349 **21.** 0.4815

CHAPTER 12
SECTION 12.1

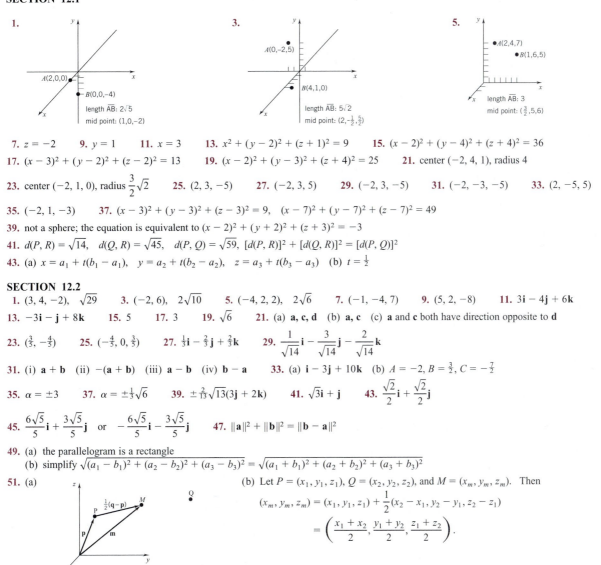

1. length $\overline{AB}$: $2\sqrt{5}$, mid point: $(1,0,-2)$, $A(2,0,0)$, $B(0,0,-4)$

3. length $\overline{AB}$: $5\sqrt{2}$, mid point: $(2,-\frac{1}{2},\frac{5}{2})$, $A(0,-2,5)$, $B(4,1,0)$

5. length $\overline{AB}$: 3, mid point: $(\frac{3}{2},5,6)$, $A(2,4,7)$, $B(1,6,5)$

7. $z=-2$ **9.** $y=1$ **11.** $x=3$ **13.** $x^2+(y-2)^2+(z+1)^2=9$ **15.** $(x-2)^2+(y-4)^2+(z+4)^2=36$

17. $(x-3)^2+(y-2)^2+(z-2)^2=13$ **19.** $(x-2)^2+(y-3)^2+(z+4)^2=25$ **21.** center $(-2,4,1)$, radius 4

23. center $(-2,1,0)$, radius $\dfrac{3}{2}\sqrt{2}$ **25.** $(2,3,-5)$ **27.** $(-2,3,5)$ **29.** $(-2,3,-5)$ **31.** $(-2,-3,-5)$ **33.** $(2,-5,5)$

35. $(-2,1,-3)$ **37.** $(x-3)^2+(y-3)^2+(z-3)^2=9$, $(x-7)^2+(y-7)^2+(z-7)^2=49$

39. not a sphere; the equation is equivalent to $(x-2)^2+(y+2)^2+(z+3)^2=-3$

41. $d(P,R)=\sqrt{14}$, $d(Q,R)=\sqrt{45}$, $d(P,Q)=\sqrt{59}$, $[d(P,R)]^2+[d(Q,R)]^2=[d(P,Q)]^2$

43. (a) $x=a_1+t(b_1-a_1)$, $y=a_2+t(b_2-a_2)$, $z=a_3+t(b_3-a_3)$ (b) $t=\frac{1}{2}$

SECTION 12.2

1. $(3,4,-2)$, $\sqrt{29}$ **3.** $(-2,6)$, $2\sqrt{10}$ **5.** $(-4,2,2)$, $2\sqrt{6}$ **7.** $(-1,-4,7)$ **9.** $(5,2,-8)$ **11.** $3\mathbf{i}-4\mathbf{j}+6\mathbf{k}$

13. $-3\mathbf{i}-\mathbf{j}+8\mathbf{k}$ **15.** 5   **17.** 3   **19.** $\sqrt{6}$ **21.** (a) $\mathbf{a},\mathbf{c},\mathbf{d}$ (b) $\mathbf{a},\mathbf{c}$ (c) $\mathbf{a}$ and $\mathbf{c}$ both have direction opposite to $\mathbf{d}$

23. $(\frac{3}{5},-\frac{4}{5})$ **25.** $(-\frac{4}{5},0,\frac{3}{5})$ **27.** $\frac{1}{3}\mathbf{i}-\frac{2}{3}\mathbf{j}+\frac{2}{3}\mathbf{k}$ **29.** $\dfrac{1}{\sqrt{14}}\mathbf{i}-\dfrac{3}{\sqrt{14}}\mathbf{j}-\dfrac{2}{\sqrt{14}}\mathbf{k}$

31. (i) $\mathbf{a}+\mathbf{b}$ (ii) $-(\mathbf{a}+\mathbf{b})$ (iii) $\mathbf{a}-\mathbf{b}$ (iv) $\mathbf{b}-\mathbf{a}$ **33.** (a) $\mathbf{i}-3\mathbf{j}+10\mathbf{k}$ (b) $A=-2,B=\frac{3}{2},C=-\frac{7}{2}$

35. $\alpha=\pm3$ **37.** $\alpha=\pm\frac{1}{3}\sqrt{6}$ **39.** $\pm\frac{2}{13}\sqrt{13}(3\mathbf{j}+2\mathbf{k})$ **41.** $\sqrt{3}\mathbf{i}+\mathbf{j}$ **43.** $\dfrac{\sqrt{2}}{2}\mathbf{i}+\dfrac{\sqrt{2}}{2}\mathbf{j}$

45. $\dfrac{6\sqrt{5}}{5}\mathbf{i}+\dfrac{3\sqrt{5}}{5}\mathbf{j}$ or $-\dfrac{6\sqrt{5}}{5}\mathbf{i}-\dfrac{3\sqrt{5}}{5}\mathbf{j}$ **47.** $\|\mathbf{a}\|^2+\|\mathbf{b}\|^2=\|\mathbf{b}-\mathbf{a}\|^2$

49. (a) the parallelogram is a rectangle
(b) simplify $\sqrt{(a_1-b_1)^2+(a_2-b_2)^2+(a_3-b_3)^2}=\sqrt{(a_1+b_1)^2+(a_2+b_2)^2+(a_3+b_3)^2}$

51. (a)

(b) Let $P=(x_1,y_1,z_1)$, $Q=(x_2,y_2,z_2)$, and $M=(x_m,y_m,z_m)$. Then
$$(x_m,y_m,z_m)=(x_1,y_1,z_1)+\frac{1}{2}(x_2-x_1,y_2-y_1,z_2-z_1)$$
$$=\left(\frac{x_1+x_2}{2},\frac{y_1+y_2}{2},\frac{z_1+z_2}{2}\right).$$

$$\mathbf{m}=\mathbf{p}+\tfrac{1}{2}(\mathbf{q}-\mathbf{p})$$

53. $\mathbf{F}_1\cong-153.21\mathbf{i}-128.56\mathbf{j}$; $\|\mathbf{F}_1\|=200.02$; $\mathbf{F}_2\cong153.21\mathbf{i}+71.44\mathbf{j}$; $\|\mathbf{F}_2\|=169.05$

55. $\mathbf{R}=(300+25\sqrt{2})\mathbf{i}+(300\sqrt{3}-25\sqrt{2})\mathbf{j}\cong335.36\mathbf{i}+484.26\mathbf{j}$.   True course: N34.70°E; ground speed 589.05 miles/hr.

57. (a) $\|\mathbf{r}-\mathbf{a}\|=3$ where $\mathbf{a}=a_1\mathbf{i}+a_2\mathbf{j}+a_3\mathbf{k}$ (b) $\|\mathbf{r}\|\le2$ (c) $\|\mathbf{r}-\mathbf{a}\|\le1$ where $\mathbf{a}=a_1\mathbf{i}+a_2\mathbf{j}+a_3\mathbf{k}$
(d) $\|\mathbf{r}-\mathbf{a}\|=\|\mathbf{r}-\mathbf{b}\|$ (e) $\|\mathbf{r}-\mathbf{a}\|+\|\mathbf{r}-\mathbf{b}\|=k$

SECTION 12.3

1. -1 **3.** 0   **5.** -1 **7.** $\mathbf{a}\cdot\mathbf{b}$ **9.** $\mathbf{a}\cdot(\mathbf{b}+\mathbf{c})$

11. (a) $\mathbf{a}\cdot\mathbf{b}=5$, $\mathbf{a}\cdot\mathbf{c}=8$, $\mathbf{b}\cdot\mathbf{c}=18$ (b) $\cos\sphericalangle(\mathbf{a},\mathbf{b})=\frac{1}{14}\sqrt{70}$, $\cos\sphericalangle(\mathbf{a},\mathbf{c})=\frac{8}{25}\sqrt{5}$, $\cos\sphericalangle(\mathbf{b},\mathbf{c})=\frac{9}{35}\sqrt{14}$
(c) $\text{comp}_\mathbf{b}\,\mathbf{a}=\frac{5}{14}\sqrt{14}$, $\text{comp}_\mathbf{c}\,\mathbf{a}=\frac{8}{5}$ (d) $\text{proj}_\mathbf{b}\,\mathbf{a}=\frac{5}{14}(3\mathbf{i}-\mathbf{j}+2\mathbf{k})$, $\text{proj}_\mathbf{c}\,\mathbf{a}=\frac{8}{25}(4\mathbf{i}+3\mathbf{k})$

13. $\frac{1}{2}\mathbf{i}+\frac{1}{2}\sqrt{2}\mathbf{j}-\frac{1}{2}\mathbf{k}$ **15.** $\frac{1}{3}\pi$ **17.** $\frac{1}{3}\pi,\frac{2}{3}\pi,\frac{1}{4}\pi$ **19.** 2.2 radians, or 126.3°   **21.** 2.5 radians, or 145.3°

23. $\cos\alpha=\frac{1}{3}$, $\cos\beta=\frac{2}{3}$, $\cos\gamma=\frac{2}{3}$; $\alpha\cong70.5°$, $\beta\cong48.2°$, $\gamma\cong48.2°$

25. $\cos \alpha = \frac{3}{13}$, $\cos \beta = \frac{12}{13}$, $\cos \gamma = \frac{4}{13}$; $\alpha \cong 76.7°$, $\beta \cong 22.6°$, $\gamma \cong 72.1°$

27. (a) $\mathbf{proj}_b\, \alpha a = (\alpha a \cdot \mathbf{u}_b)\mathbf{u}_b = \alpha(a \cdot \mathbf{u}_b)\mathbf{u}_b = \alpha \mathbf{proj}_b\, a$

(b) $\mathbf{proj}_b(a + c) = [(a + c) \cdot \mathbf{u}_b]\mathbf{u}_b$
$= (a \cdot \mathbf{u}_b + c \cdot \mathbf{u}_b)\mathbf{u}_b$
$= (a \cdot \mathbf{u}_b)\mathbf{u}_b + (c \cdot \mathbf{u}_b)\mathbf{u}_b = \mathbf{proj}_b\, a + \mathbf{proj}_b\, c$

29. (a) for $\mathbf{a} \neq 0$ the following statements are equivalent:

$$\mathbf{a} \cdot \mathbf{b} = \mathbf{a} \cdot \mathbf{c}, \qquad \mathbf{b} \cdot \mathbf{a} = \mathbf{c} \cdot \mathbf{a},$$

$$\mathbf{b} \cdot \frac{\mathbf{a}}{\|\mathbf{a}\|} = \mathbf{c} \cdot \frac{\mathbf{a}}{\|\mathbf{a}\|}, \qquad \mathbf{b} \cdot \mathbf{u}_a = \mathbf{c} \cdot \mathbf{u}_a,$$

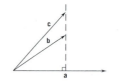

$$(\mathbf{b} \cdot \mathbf{u}_a)\mathbf{u}_a = (\mathbf{c} \cdot \mathbf{u}_a)\mathbf{u}_a, \qquad \mathbf{proj}_a\, \mathbf{b} = \mathbf{proj}_a\, \mathbf{c}. \qquad \mathbf{a} \cdot \mathbf{b} = \mathbf{a} \cdot \mathbf{c} \text{ but } \mathbf{b} \neq \mathbf{c}.$$

(b) $\mathbf{b} = (\mathbf{b} \cdot \mathbf{i})\mathbf{i} + (\mathbf{b} \cdot \mathbf{j})\mathbf{j} + (\mathbf{b} \cdot \mathbf{k})\mathbf{k} = (\mathbf{c} \cdot \mathbf{i})\mathbf{i} + (\mathbf{c} \cdot \mathbf{j})\mathbf{j} + (\mathbf{c} \cdot \mathbf{k})\mathbf{k} = \mathbf{c}$.

31. (a) Express the norms as dot products.

(b) The following statements are equivalent:

$$\mathbf{a} \perp \mathbf{b}, \qquad \mathbf{a} \cdot \mathbf{b} = 0, \qquad \|\mathbf{a} + \mathbf{b}\|^2 - \|\mathbf{a} - \mathbf{b}\|^2 = 0, \qquad \|\mathbf{a} + \mathbf{b}\| = \|\mathbf{a} - \mathbf{b}\|.$$

(c) By (b), the relation $\|\mathbf{a} + \mathbf{b}\| = \|\mathbf{a} - \mathbf{b}\|$ gives $\mathbf{a} \perp \mathbf{b}$. The relation $\mathbf{a} + \mathbf{b} \perp \mathbf{a} - \mathbf{b}$ gives

$$0 = (\mathbf{a} + \mathbf{b}) \cdot (\mathbf{a} - \mathbf{b}) = \|\mathbf{a}\|^2 - \|\mathbf{b}\|^2 \qquad \text{and thus} \qquad \|\mathbf{a}\| = \|\mathbf{b}\|.$$

The parallelogram is a square since it has two adjacent sides of equal length that meet at right angles.

33. $\|\mathbf{a} + \mathbf{b}\|^2 = (\mathbf{a} + \mathbf{b}) \cdot (\mathbf{a} + \mathbf{b}) = \mathbf{a} \cdot \mathbf{a} + 2\mathbf{a} \cdot \mathbf{b} + \mathbf{b} \cdot \mathbf{b}$ **35.** $\pi - \alpha$, $\pi - \beta$, $\pi - \gamma$
$\|\mathbf{a} - \mathbf{b}\|^2 = (\mathbf{a} - \mathbf{b}) \cdot (\mathbf{a} - \mathbf{b}) = \mathbf{a} \cdot \mathbf{a} - 2\mathbf{a} \cdot \mathbf{b} + \mathbf{b} \cdot \mathbf{b}$
and the result follows.

37. If $\mathbf{a} \perp \mathbf{b}$ and $\mathbf{a} \perp \mathbf{c}$, then $\mathbf{a} \cdot \mathbf{b} = 0$ and $\mathbf{a} \cdot \mathbf{c} = 0$, so that

$$\mathbf{a} \cdot (\alpha \mathbf{b} + \beta \mathbf{c}) = \alpha(\mathbf{a} \cdot \mathbf{b}) + \beta(\mathbf{a} \cdot \mathbf{c}) = 0.$$

Thus $\mathbf{a} \perp (\alpha \mathbf{b} + \beta \mathbf{c})$.

39. Existence of decomposition: $\mathbf{a} = (\mathbf{a} \cdot \mathbf{u}_b)\mathbf{u}_b + [\mathbf{a} - (\mathbf{a} \cdot \mathbf{u}_b)\mathbf{u}_b]$. Uniqueness of decomposition: suppose that $\mathbf{a} = \mathbf{a}_\| + \mathbf{a}_\perp = \mathbf{A}_\| + \mathbf{A}_\perp$. Then the vector $\mathbf{a}_\| - \mathbf{A}_\| = \mathbf{A}_\perp - \mathbf{a}_\perp$ is both parallel to $\mathbf{b}$ and perpendicular to $\mathbf{b}$. (Exercises 37 and 38.) Therefore it is zero. Consequently $\mathbf{A}_\| = \mathbf{a}_\|$ and $\mathbf{A}_\perp = \mathbf{a}_\perp$.

41. $x = 0$, $x = 4$ **43.** $\theta = \cos^{-1}(1 - 3\sqrt{3}) \cong 0.96$ radians

45. (a) The direction angles of a vector always satisfy $\cos^2 \alpha + \cos^2 \beta + \cos^2 \gamma = 1$, and, as you can check, $\cos^2 \frac{1}{4}\pi + \cos^2 \frac{1}{6}\pi + \cos^2 \frac{2}{3}\pi \neq 1$.

(b) The relation $\cos^2 \alpha + \cos^2 \frac{1}{4}\pi + \cos^2 \frac{1}{4}\pi = 1$ gives

$$\cos^2 \alpha + \frac{1}{2} + \frac{1}{2} = 1, \quad \cos \alpha = 0, \quad a_1 = \|\mathbf{a}\| \cos \alpha = 0.$$

47. $\mathbf{u} = \pm\frac{1}{165}\sqrt{165}\,(8\mathbf{i} + \mathbf{j} - 10\mathbf{k})$ **49.** Place center of sphere at the origin.

$$\overrightarrow{P_1Q} \cdot \overrightarrow{P_2Q} = (-\mathbf{a} + \mathbf{b}) \cdot (\mathbf{a} + \mathbf{b})$$
$$= -\|\mathbf{a}\|^2 + \|\mathbf{b}\|^2$$
$$= 0.$$

51. (a) $\mathbf{W} = \mathbf{F} \cdot \mathbf{r}$ (b) 0 (c) $\|\mathbf{F}\|\mathbf{i} \cdot (b - a)\mathbf{i} = \|\mathbf{F}\|(b - a)$ **53.** (a) 614.5 joules (b) 614.29 joules

55. (a) $W_1 = W_2$ (b) $W_1 = \frac{1}{2}\|\mathbf{F}\|\|\mathbf{r}\|$, $W_2 = \frac{\sqrt{3}}{2}\|\mathbf{F}\|\|\mathbf{r}\|$, $W_2 = \sqrt{3}W_1$

SECTION 12.4

1. $-2\mathbf{k}$ **3.** $\mathbf{i} + \mathbf{j} + \mathbf{k}$ **5.** $-3\mathbf{i} - \mathbf{j} - 2\mathbf{k}$ **7.** -1 **9.** 0 **11.** 1 **13.** $3\mathbf{i} - 2\mathbf{j} - 3\mathbf{k}$ **15.** $\mathbf{i} + \mathbf{j} - 2\mathbf{k}$ **17.** -3

19. $5\mathbf{i} - 4\mathbf{j} - \mathbf{k}$ **21.** $\left(\frac{1}{\sqrt{6}}, \frac{-1}{\sqrt{6}}, \frac{-2}{\sqrt{6}}\right)$, $\left(\frac{-1}{\sqrt{6}}, \frac{1}{\sqrt{6}}, \frac{2}{\sqrt{6}}\right)$ **23.** $\mathbf{N} = 3\mathbf{j}$; area $= \frac{3}{2}$ **25.** $\mathbf{N} = 8\mathbf{i} + 4\mathbf{j} + 4\mathbf{k}$; area $= 2\sqrt{6}$ **27.** 1

29. 2 **31.** $-2(\mathbf{a} \times \mathbf{b})$ **33.** $\mathbf{a} = 0$ **35.** $\begin{vmatrix} \alpha & \beta \\ \gamma & \delta \end{vmatrix}(\mathbf{a} \times \mathbf{b})$ **37.** $\mathbf{a} \cdot (\mathbf{b} \times \mathbf{c}) = (\mathbf{a} \times \mathbf{b}) \cdot \mathbf{c} = (\mathbf{c} \times \mathbf{a}) \cdot \mathbf{b} = (\mathbf{b} \times \mathbf{c}) \cdot \mathbf{a} = (\mathbf{a} \times -\mathbf{c}) \cdot \mathbf{b}$
$\mathbf{a} \cdot (\mathbf{c} \times \mathbf{b}) = \mathbf{c} \cdot (\mathbf{b} \times \mathbf{a}) = (-\mathbf{a} \times \mathbf{b}) \cdot \mathbf{c}$

39. $\mathbf{a} \times \mathbf{b}$ is perpendicular to the plane determined by $\mathbf{a}$ and $\mathbf{b}$;   **41.** $\mathbf{a} \cdot \mathbf{b} = \mathbf{a} \cdot \mathbf{c}$ implies $\mathbf{a} \cdot (\mathbf{b} - \mathbf{c}) = 0$; $\mathbf{a}$ is perpendicular to $\mathbf{b} - \mathbf{c}$.
$\mathbf{c}$ is in this plane iff $\mathbf{a} \times \mathbf{b} \cdot \mathbf{c} = 0$. $\mathbf{a} \times \mathbf{b} = \mathbf{a} \times \mathbf{c}$ implies $\mathbf{a} \times (\mathbf{b} - \mathbf{c}) = 0$; $\mathbf{a}$ is parallel to $\mathbf{b} - \mathbf{c}$.
Since $\mathbf{a} \neq 0$, it follows that $\mathbf{b} - \mathbf{c} = 0$, or $\mathbf{b} = \mathbf{c}$.

43. $\mathbf{c} \times \mathbf{a} = \|\mathbf{a}\|^2 \mathbf{b}$ **47.** 12.77 ft-lb; bolt moves into the plane of the paper.

SECTION 12.5

1. P and Q **3.** $\mathbf{r}(t) = (3\mathbf{i} + \mathbf{j}) + t\mathbf{k}$ **5.** $\mathbf{r}(t) = t(x_1\mathbf{i} + y_1\mathbf{j} + z_1\mathbf{k})$ **7.** $x(t) = 1 + t$, $y(t) = -t$, $z(t) = 3 + t$

9. $x(t) = 2$, $y(t) = t$, $z(t) = 3$ **11.** $\mathbf{r}(t) = (-\mathbf{i} + 2\mathbf{j} - 3\mathbf{k}) + t(2\mathbf{i} + \mathbf{j} + 4\mathbf{k})$ **13.** $P(1, 2, 0)$, $\frac{1}{4}\pi$ rad

15. $P(1, 3, 1)$, $\cos^{-1}(\frac{1}{6}\sqrt{3}) \cong 1.28$ rad   **17.** $(x_0 - [d_1/d_3]z_0, \ y_0 - [d_2/d_3]z_0, \ 0)$ **19.** The lines are parallel.

21. $\mathbf{r}(t) = (2\mathbf{i} + 7\mathbf{j} - \mathbf{k}) + t(2\mathbf{i} - 5\mathbf{j} + 4\mathbf{k})$, $0 \le t \le 1$ **23.** $\mathbf{u} = -\frac{2}{3}\mathbf{i} + \frac{1}{3}\mathbf{j} + \frac{2}{3}\mathbf{k}$, $9 \le t \le 15$

25. triples of the form $X(u) = 3 + au$, $Y(u) = -1 + bu$, $Z(u) = 8 + cu$ with $2a - 4b + 6c = 0$ **27.** 1   **29.** $\sqrt{69/14} \cong 2.22$

31. $\sqrt{3} \cong 1.73$ **33.** (a) 1   (b) $\sqrt{3}$ **35.** $\mathbf{r}(t) = \frac{1}{11}(7\mathbf{i} + 4\mathbf{j} - \mathbf{k}) \pm t[\frac{1}{11}\sqrt{11}(\mathbf{i} - \mathbf{j} + 3\mathbf{k})]$ **37.** $0 < t < s$

SECTION 12.6

1. Q **3.** $x - 4y + 3z - 2 = 0$ **5.** $3x - 2y + 5z - 9 = 0$ **7.** $y - z - 2 = 0$ **9.** $x_0(x - x_0) + y_0(y - y_0) + z_0(z - z_0) = 0$

11. $\dfrac{1}{\sqrt{30}}(2\mathbf{i} - \mathbf{j} + 5\mathbf{k})$, $-\dfrac{1}{\sqrt{30}}(2\mathbf{i} - \mathbf{j} + 5\mathbf{k})$ **13.** $\dfrac{1}{15}x + \dfrac{1}{12}y - \dfrac{1}{10}z = 1$ **15.** $\frac{1}{2}\pi$ **17.** $\cos\theta = \frac{2}{21}\sqrt{42} \cong 0.617$, $\theta \cong 0.91$ rad

19. coplanar   **21.** not coplanar   **23.** $\dfrac{2}{\sqrt{21}}$ **25.** $\frac{22}{5}$ **27.** $x + z = 2$ **29.** $3x - 4z - 5 = 0$ **31.** $\dfrac{x - x_0}{A} = \dfrac{y - y_0}{B} = \dfrac{z - z_0}{C}$

33. $(x - x_0)/d_1 = (y - y_0)/d_2$, $(y - y_0)/d_2 = (z - z_0)/d_3$ **35.** $x(t) = t$, $y(t) = t$, $z(t) = -t$ **37.** $P(-\frac{19}{14}, \frac{15}{7}, \frac{17}{14})$

39. $10x - 7y + z = 0$ **41.** circle centered at P with radius $\|\mathbf{N}\|$ **43.** if $\alpha > 0$, then P_1 lies on the same side of the plane as the tip of $\mathbf{N}$; if $\alpha < 0$, then P_1 and the tip of $\mathbf{N}$ lie on opposite sides of the plane

45. $\mathbf{a} \cdot \mathbf{b} \times \mathbf{c} = 0$ **47.** 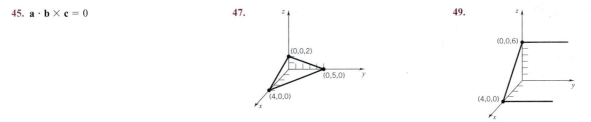   **49.**

51. $10x + 4y + 5z = 20$ **53.** $5x + 3y = 15$

CHAPTER 13

SECTION 13.1

1. $\mathbf{f}'(t) = 2\mathbf{i} - \mathbf{j} + 3\mathbf{k}$ **3.** $\mathbf{f}'(t) = -\dfrac{1}{2\sqrt{1-t}}\mathbf{i} + \dfrac{1}{2\sqrt{1+t}}\mathbf{j} + \dfrac{1}{(1-t)^2}\mathbf{k}$ **5.** $\mathbf{f}'(t) = \cos t\,\mathbf{i} - \sin t\,\mathbf{j} + \sec^2 t\,\mathbf{k}$ **7.** $-\dfrac{1}{1-t}\mathbf{i} - \sin t\mathbf{j} + 2t\mathbf{k}$

9. $12t\mathbf{j} + 2\mathbf{k}$ **11.** $-2e^t\sin t\mathbf{i} + 2e^t\cos t\mathbf{j}$ **13.** $\mathbf{i} + 3\mathbf{j}$ **15.** $(e - 1)\mathbf{i} + (1 - 1/e)\mathbf{k}$ **17.** $\dfrac{\pi}{4}\mathbf{i} + \tan(1)\mathbf{j}$ **19.** $\frac{1}{2}\mathbf{i} + \mathbf{j}$

21. $0\mathbf{i} + 0\mathbf{j} + 0\mathbf{k} = \mathbf{0}$ **23.** does not exist

25. 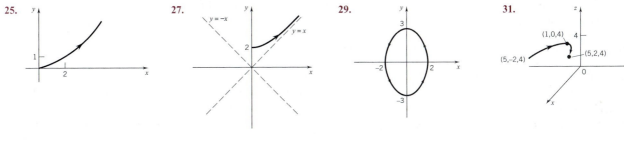   **27.**   **29.**   **31.**

33. 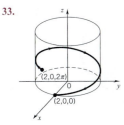   **35.** 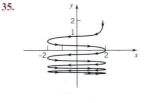   **37.**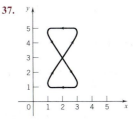

39. (a) $\mathbf{f}(t) = 3\cos t\mathbf{i} + 2\sin t\mathbf{j}$ (b) $\mathbf{f}(t) = 3\cos t\mathbf{i} - 2\sin t\mathbf{j}$ **41.** (a) $\mathbf{f}(t) = t\mathbf{i} + t^2\mathbf{j}$ (b) $\mathbf{f}(t) = -t\mathbf{i} + t^2\mathbf{j}$

43. $\mathbf{f}(t) = (1 + 2t)\mathbf{i} + (4 + 5t)\mathbf{j} + (-2 + 8t)\mathbf{k}, 0 \le t \le 1$ **45.** $\mathbf{f}'(t_0) = \mathbf{i} + m\mathbf{j}$, $\displaystyle\int_a^b \mathbf{f}(t)\,dt = \tfrac{1}{2}(b^2 - a^2)\mathbf{i} + A\mathbf{j}$, $\displaystyle\int_a^b \mathbf{f}'(t)\,dt = (b - a)\mathbf{i} + (d - c)\mathbf{j}$

47. $\mathbf{f}(t) = t\mathbf{i} + (\tfrac{1}{3}t^3 + 1)\mathbf{j} - \mathbf{k}$ **49.** $\mathbf{f}(t) = e^{\alpha t}\mathbf{c}$

51. (a) if $\mathbf{f}'(t) = \mathbf{0}$ on an interval, then the derivative of each component is 0 on that interval, each component is constant on that interval, and therefore $\mathbf{f}$ itself is constant on that interval
(b) set $\mathbf{h}(t) = \mathbf{f}(t) - \mathbf{g}(t)$ and apply part (a)

53. set $\mathbf{f}(t) = f_1(t)\mathbf{i} + f_2(t)\mathbf{j} + f_3(t)\mathbf{k}$, and apply (3.1.4) to f_1, f_2, f_3 **55.** no; as a counterexample set $\mathbf{f}(t) = \mathbf{i} = \mathbf{g}(t)$

57. $\|\mathbf{f}(t)\|^2 = \mathbf{f}(t) \cdot \mathbf{f}(t)$

$2\|\mathbf{f}(t)\|\dfrac{d\|\mathbf{f}(t)\|}{dt} = 2\mathbf{f}(t) \cdot \mathbf{f}'(t)$

$\dfrac{d\|\mathbf{f}(t)\|}{dt} = \dfrac{\mathbf{f}(t) \cdot \mathbf{f}'(t)}{\|\mathbf{f}(t)\|}$

SECTION 13.2

1. $\mathbf{f}'(t) = \mathbf{b}$, $\mathbf{f}''(t) = \mathbf{0}$ **3.** $\mathbf{f}'(t) = 2e^{2t}\mathbf{i} - \cos t\mathbf{j}$, $\mathbf{f}''(t) = 4e^{2t}\mathbf{i} + \sin t\mathbf{j}$ **5.** $\mathbf{f}'(t) = (3t^2 - 8t^3)\mathbf{j}$, $\mathbf{f}''(t) = (6t - 24t^2)\mathbf{j}$

7. $\mathbf{f}'(t) = -2t\mathbf{i} + e^t(t + 1)\mathbf{k}$, $\mathbf{f}''(t) = -2\mathbf{i} + e^t(t + 2)\mathbf{k}$

9. $\mathbf{f}'(t) = (\sin t + t\cos t + 2\sin 2t)\mathbf{i} + (2\cos 2t - \cos t + t\sin t)\mathbf{j} - 3\sin 3t\mathbf{k}$
$\mathbf{f}''(t) = (2\cos t - t\sin t + 4\cos 2t)\mathbf{i} + (-4\sin 2t + 2\sin t + t\cos t)\mathbf{j} - 9\cos 3t\mathbf{k}$

11. $\mathbf{f}'(t) = \tfrac{1}{2}\sqrt{t}\,\mathbf{g}'(\sqrt{t}) + \mathbf{g}(\sqrt{t})$, $\mathbf{f}''(t) = \tfrac{1}{4}\mathbf{g}''(\sqrt{t}) + \tfrac{3}{4}(1/\sqrt{t})\mathbf{g}'(\sqrt{t})$ **13.** $-\sin t\,e^{\cos t}\,\mathbf{i} + \cos t\,e^{\sin t}\,\mathbf{j}$ **15.** $4e^{2t} - 4e^{-2t}$

17. $(\mathbf{a} \times \mathbf{d}) + (\mathbf{b} \times \mathbf{c}) + 2t(\mathbf{b} \times \mathbf{d})$ **19.** $(\mathbf{a} \cdot \mathbf{d}) + (\mathbf{b} \cdot \mathbf{c}) + 2t(\mathbf{b} \cdot \mathbf{d})$ **21.** $\mathbf{r}(t) = \mathbf{a} + t\mathbf{b}$ **23.** $\mathbf{r}(t) = \tfrac{1}{2}t^2\mathbf{a} + \tfrac{1}{6}t^3\mathbf{b} + t\mathbf{c} + \mathbf{d}$

25. $\mathbf{r}''(t) = -\sin t\mathbf{i} - \cos t\mathbf{j} = -\mathbf{r}(t)$; no. **27.** $\mathbf{r}(t) \cdot \mathbf{r}'(t) = 0$, $\mathbf{r}(t) \times \mathbf{r}'(t) = \mathbf{k}$

29. $\dfrac{d}{dt}[\mathbf{f}(t) \times \mathbf{f}'(t)] = [\mathbf{f}(t) \times \mathbf{f}''(t)] + \underbrace{[\mathbf{f}'(t) \times \mathbf{f}'(t)]}_{\mathbf{0}} = \mathbf{f}(t) \times \mathbf{f}''(t)$

31. $[\mathbf{f} \cdot \mathbf{g} \times \mathbf{h}]' = \mathbf{f}' \cdot (\mathbf{g} \times \mathbf{h}) + \mathbf{f} \cdot (\mathbf{g} \times \mathbf{h})' = \mathbf{f}' \cdot (\mathbf{g} \times \mathbf{h}) + \mathbf{f} \cdot [\mathbf{g}' \times \mathbf{h} + \mathbf{g} \times \mathbf{h}']$ and the result follows.

33. The following four statements are equivalent: $\|\mathbf{r}(t)\| = \sqrt{\mathbf{r}(t) \cdot \mathbf{r}(t)}$ is constant, $\mathbf{r}(t) \cdot \mathbf{r}(t)$ is constant, $d/dt\,[\mathbf{r}(t) \cdot \mathbf{r}(t)] = 2[\mathbf{r}(t) \cdot \mathbf{r}'(t)] = 0$ identically, $\mathbf{r}(t) \cdot \mathbf{r}'(t) = 0$ identically.

35. $\dfrac{[\mathbf{f}(t + h) \times \mathbf{g}(t + h)] - [\mathbf{f}(t) \times \mathbf{g}(t)]}{h} = \left(\mathbf{f}(t + h) \times \left[\dfrac{\mathbf{g}(t + h) - \mathbf{g}(t)}{h}\right]\right) + \left(\left[\dfrac{\mathbf{f}(t + h) - \mathbf{f}(t)}{h}\right] \times \mathbf{g}(t)\right)$. Now take the limit as $h \to 0$.
(Appeal to Theorem 13.1.3.)

SECTION 13.3

1. $\pi\mathbf{j} + \mathbf{k}$, $\mathbf{R}(u) = (\mathbf{i} + 2\mathbf{k}) + u(\pi\mathbf{j} + \mathbf{k})$ **3.** $\mathbf{b} - 2\mathbf{c}$, $\mathbf{R}(u) = (\mathbf{a} - \mathbf{b} + \mathbf{c}) + u(\mathbf{b} - 2\mathbf{c})$

5. $4\mathbf{i} - \mathbf{j} + 4\mathbf{k}$, $\mathbf{R}(u) = (2\mathbf{i} + 5\mathbf{k}) + u(4\mathbf{i} - \mathbf{j} + 4\mathbf{k})$ **7.** $-\sqrt{2}\mathbf{i} + \dfrac{3\sqrt{2}}{2}\mathbf{j} + \mathbf{k}$, $\mathbf{R}(u) = \left(\sqrt{2}\mathbf{i} + \dfrac{3\sqrt{2}}{2}\mathbf{j} + \dfrac{\pi}{4}\mathbf{k}\right) + u\left(-\sqrt{2}\mathbf{i} + \dfrac{3\sqrt{2}}{2}\mathbf{j} + \mathbf{k}\right)$

9. The scalar components $x(t) = at$ and $y(t) = bt^2$ satisfy the equation $a^2y(t) = a^2(bt^2) = b(at)^2 = b[x(t)]^2$ and generate the parabola $a^2y = bx^2$.

11. (a) $P(0, 1)$ (b) $P(1, 2)$ (c) $P(-1, 2)$

13. The tangent line at $t = t_0$ has the form $\mathbf{R}(u) = \mathbf{r}(t_0) + u\mathbf{r}'(t_0)$. If $\mathbf{r}'(t_0) = \alpha\mathbf{r}(t_0)$, then **15.** $\pi/2 \cong 1.57$, or $90°$

$\mathbf{R}(u) = \mathbf{r}(t_0) + u\alpha\mathbf{r}(t_0) = (1 + u\alpha)\mathbf{r}(t_0)$.

The tangent line passes through the origin at $u = -1/\alpha$.

17. $P(1, 2, -2)$; $\cos^{-1}(\tfrac{1}{3}\sqrt{5}) \cong 1.11$ rad **19.** (a) $\mathbf{r}(t) = a\cos t\mathbf{i} + b\sin t\mathbf{j}$ (b) $\mathbf{r}(t) = a\cos t\mathbf{i} - b\sin t\mathbf{j}$
(c) $\mathbf{r}(t) = a\cos 2t\mathbf{i} + b\sin 2t\mathbf{j}$ (d) $\mathbf{r}(t) = a\cos 3t\mathbf{i} - b\sin 3t\mathbf{j}$

21. $\mathbf{r}'(t) = t^3\mathbf{i} + 2t\mathbf{j}$ **23.** $\mathbf{r}'(t) = 2e^{2t}\mathbf{i} - 4e^{-4t}\mathbf{j}$ **25.** $\mathbf{r}'(t) = -2\sin t\mathbf{i} + 3\cos t\mathbf{j}$

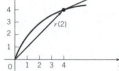

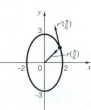

27. $\mathbf{r}(t) = (t^2 + 1)\mathbf{i} + t\mathbf{j}, \quad t \geq 1;$ or, $\mathbf{r}(t) = \sec^2 t\mathbf{i} + \tan t\mathbf{j}, \quad t \in [\frac{1}{4}\pi, \frac{1}{2}\pi)$ **29.** $\mathbf{r}(t) = \cos t \sin 3t\mathbf{i} + \sin t \sin 3t\mathbf{j}, \quad t \in [0, \pi]$

31. 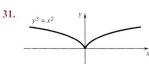   **33.** $(2, 4, 8); \quad \cos^{-1}\left(\dfrac{24}{\sqrt{21}\sqrt{161}}\right) \cong 1.15$ rad

There is no tangent vector at the origin.

35. $\mathbf{T}(1) = \dfrac{1}{\sqrt{2}}\mathbf{j} + \dfrac{1}{\sqrt{2}}\mathbf{k}, \quad \mathbf{N}(1) = \dfrac{1}{\sqrt{2}}\mathbf{j} + \dfrac{1}{\sqrt{2}}\mathbf{k}, \quad x - 1 = 0$ **37.** $\frac{1}{5}\sqrt{5}\,(-2\mathbf{i} + \mathbf{k}), \quad -\mathbf{j}, \quad x + 2z = \frac{1}{2}\pi$

39. $\frac{1}{14}\sqrt{14}\,(\mathbf{i} + 2\mathbf{j} + 3\mathbf{k}), \quad \frac{1}{266}\sqrt{266}\,(-11\mathbf{i} - 8\mathbf{j} + 9\mathbf{k}), \quad 3x - 3y + z = 1$

41. $\mathbf{T}(0) = \dfrac{1}{\sqrt{3}}\mathbf{i} + \dfrac{1}{\sqrt{3}}\mathbf{j} + \dfrac{1}{\sqrt{3}}\mathbf{k}, \quad \mathbf{N}(0) = \dfrac{1}{\sqrt{2}}\mathbf{i} - \dfrac{1}{\sqrt{2}}\mathbf{j}, \quad x + y - 2z + 1 = 0$

43. $\mathbf{T}_1 = \dfrac{\mathbf{R}'(u)}{\|\mathbf{R}'(u)\|} = -\dfrac{\mathbf{r}'(a+b-u)}{\|\mathbf{r}'(a+b-u)\|} = -\mathbf{T}.$ Then $\mathbf{T}_1'(u) = \mathbf{T}'(a+b-u)$ and thus $\mathbf{N}_1 = \mathbf{N}.$

45. Let $\mathbf{T}$ be the unit tangent at the tip of $\mathbf{R}(u) = \mathbf{r}(\phi(u))$ as calculated from the parametrization $\mathbf{r}$ and let $\mathbf{T}_1$ be the unit tangent at the same point as calculated from the parametrization $\mathbf{R}$. Then

$$\mathbf{T}_1 = \frac{\mathbf{R}'(u)}{\|\mathbf{R}'(u)\|} = \frac{\mathbf{r}'(\phi(u))\,\phi'(u)}{\|\mathbf{r}'(\phi(u))\,\phi'(u)\|} = \frac{\mathbf{r}'(\phi(u))}{\|\mathbf{r}'(\phi(u))\|} = \mathbf{T}.$$

$$\phi'(u) > \underline{}$$

This shows the invariance of the unit tangent. The invariance of the principal normal and the osculating plane follows directly from the invariance of the unit tangent.

47. (a) Let $t = \Psi(v) = 2\pi - v^2$. When t increases from 0 to 2π, v decreases from $\sqrt{2\pi}$ to 0.

(b) $\mathbf{T}_r\left(\dfrac{\pi}{4}\right) = -\dfrac{1}{\sqrt{10}}\mathbf{i} + \dfrac{1}{\sqrt{10}}\mathbf{j} + \dfrac{2}{\sqrt{5}}\mathbf{k}, \quad \mathbf{T}_R\left(\dfrac{\sqrt{7\pi}}{2}\right) = \dfrac{1}{\sqrt{10}}\mathbf{i} - \dfrac{1}{\sqrt{10}}\mathbf{j} - \dfrac{2}{\sqrt{5}}\mathbf{k}$

$\mathbf{N}_r\left(\dfrac{\pi}{4}\right) = -\dfrac{\sqrt{2}}{2}\mathbf{i} - \dfrac{\sqrt{2}}{2}\mathbf{j}; \quad \mathbf{N}_R\left(\dfrac{\sqrt{7\pi}}{2}\right) = -\dfrac{\sqrt{2}}{2}\mathbf{i} - \dfrac{\sqrt{2}}{2}\mathbf{j}$

SECTION 13.4

1. $\frac{52}{3}$ **3.** $2\pi\sqrt{a^2 + b^2}$ **5.** $\ln(1 + \sqrt{2})$ **7.** $\frac{1}{27}(13\sqrt{13} - 8)$ **9.** $\sqrt{2}(e^\pi - 1)$ **11.** $6 + \frac{1}{2}\sqrt{2}\ln(2\sqrt{2} + 3)$ **13.** e^2

15. differentiate $s(t) = \displaystyle\int_a^t \sqrt{[x'(u)]^2 + [y'(u)]^2 + [z'(u)]^2}\,du$ **17.** see Exercise 16, differentiate $s(x) = \displaystyle\int_a^x \sqrt{1 + [f'(t)]^2}\,dt$

19. Let L be the length as computed from $\mathbf{r}$ and L^* the length as computed from $\mathbf{R}$. Then

$$L^* = \int_c^d \|\mathbf{R}'(u)\|\,du = \int_c^d \|\mathbf{r}'(\phi(u))\|\,\phi'(u)\,du = \int_a^b \|\mathbf{r}'(t)\|\,dt = L.$$

$$t = \phi(u) \underline{}$$

21. (a) $s = 5t$ (b) $\mathbf{R}(s) = 3\cos\left(\dfrac{s}{5}\right)\mathbf{i} + 3\sin\left(\dfrac{s}{5}\right)\mathbf{j} + \dfrac{4s}{5}\mathbf{k}$ (c) $Q(-3, 0, 4\pi)$

(d) $\mathbf{R}'(s) = \dfrac{-3}{5}\sin\left(\dfrac{s}{5}\right)\mathbf{i} + \dfrac{3}{5}\cos\left(\dfrac{s}{5}\right)\mathbf{j} + \dfrac{4}{5}\mathbf{k}, \quad \|\mathbf{R}'(s)\| = 1$

23. 0.5077 **25.** 22.0939

SECTION 13.5

1. $v/r, \ v^2/r$ **3.** $\|\mathbf{r}''(t)\| = a^2|b\sin at| = a^2|y(t)|$ **5.** $y = \cos \pi x, \quad 0 \leq x \leq 2$ **7.** $x = \sqrt{1 + y^2}, \quad y \geq -1$

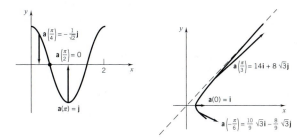

9. (a) (x_0, y_0, z_0) (b) $\alpha \cos \theta \mathbf{j} + \alpha \sin \theta \mathbf{k}$ (c) $|\alpha|$ (d) $-32\mathbf{k}$
(e) arc from parabola $z = z_0 + (\tan \theta)(y - y_0) - 16(y - y_0)^2/(\alpha^2 \cos^2 \theta)$ in the plane $x = x_0$

11. (a) $\mathbf{r}'(0) = b\omega\mathbf{j}$ (b) $\mathbf{r}''(t) = \omega^2\mathbf{r}(t)$ (c) The torque is $\mathbf{0}$ and the angular momentum is constant.

13. (a) $\mathbf{v}(t) = 2\mathbf{j} + (\alpha/m)t\mathbf{k}$ (b) $v(t) = (1/m)\sqrt{4m^2 + \alpha^2 t^2}$ (c) $\mathbf{p}(t) = 2m\mathbf{j} + \alpha t\mathbf{k}$
(d) $\mathbf{r}(t_1) = [2t + y_0]\mathbf{j} + [(\alpha/2m)t^2 + z_0]\mathbf{k}$, $t \ge 0$ $z = (\alpha/8m)(y - y_0)^2 + z_0$, $y \ge y_0$, $x = 0$

15. $\mathbf{F}(t) = 2m\mathbf{k}$ 17. (a) $\pi b\mathbf{j} + \mathbf{k}$ (b) $\sqrt{\pi^2 b^2 + 1}$ (c) $-\pi^2 a\mathbf{i}$ (d) $m(\pi b\mathbf{j} + \mathbf{k})$
(e) $m[b(1 - \pi)\mathbf{i} - 2a\mathbf{j} + 2\pi ab\mathbf{k}]$ (f) $-m\pi^2 a[\mathbf{j} - b\mathbf{k}]$

19. We have $m\mathbf{v} = m\mathbf{v}_1 + m\mathbf{v}_2$ and $\frac{1}{2}mv^2 = \frac{1}{2}mv_1^2 + \frac{1}{2}mv_2^2$. Therefore $\mathbf{v} = \mathbf{v}_1 + \mathbf{v}_2$ and $v^2 = v_1^2 + v_2^2$. Since

$$v^2 = \mathbf{v} \cdot \mathbf{v} = (\mathbf{v}_1 + \mathbf{v}_2) \cdot (\mathbf{v}_1 + \mathbf{v}_2) = v_1^2 + v_2^2 + 2(\mathbf{v}_1 \cdot \mathbf{v}_2),$$

we have $\mathbf{v}_1 \cdot \mathbf{v}_2 = \mathbf{0}$ and $\mathbf{v}_1 \perp \mathbf{v}_2$.

21. Here $\mathbf{r}''(t) = \mathbf{a}$, $\mathbf{r}'(t) = \mathbf{v}(0) + t\mathbf{a}$, $\mathbf{r}(t) = \mathbf{r}(0) + t\mathbf{v}(0) + \frac{1}{2}t^2\mathbf{a}$. If neither $\mathbf{v}(0)$ nor $\mathbf{a}$ is zero, the displacement $\mathbf{r}(t) - \mathbf{r}(0)$ is a linear combination of $\mathbf{v}(0)$ and $\mathbf{a}$ and thus remains on the plane determined by these vectors. The equation of this plane can be written

$$[\mathbf{a} \times \mathbf{v}(0)] \cdot [\mathbf{r} - \mathbf{r}(0)] = 0.$$

(If either $\mathbf{v}(0)$ or $\mathbf{a}$ is zero, the motion is restricted to a straight line; if both of these vectors are zero, the particle remains at its initial position $\mathbf{r}(0)$.)

23. $\mathbf{r}(t) = \mathbf{i} + t\mathbf{j} + (qE_0/2m)t^2\mathbf{k}$ 25. $\mathbf{r}(t) = (1 + t^3/6m)\mathbf{i} + (t^4/12m)\mathbf{j} + \mathbf{k}$.

27. $\dfrac{d}{dt}(\frac{1}{2}mv^2) = mv\dfrac{dv}{dt} = m\left(\mathbf{v} \cdot \dfrac{d\mathbf{v}}{dt}\right) = m\dfrac{d\mathbf{v}}{dt} \cdot \mathbf{v} = \mathbf{F} \cdot \dfrac{d\mathbf{r}}{dt} = 4r^2\left(\mathbf{r} \cdot \dfrac{d\mathbf{r}}{dt}\right) = 4r^2\left(r\dfrac{dr}{dt}\right) = 4r^3\dfrac{dr}{dt} = \dfrac{d}{dt}(r^4)$. Therefore $d/dt\,(\frac{1}{2}mv^2 - r^4) = 0$
and $\frac{1}{2}mv^2 - r^4$ is a constant E. Evaluating E from $t = 0$, we find that $E = 2m$. Thus $\frac{1}{2}mv^2 - r^4 = 2m$ and $v = \sqrt{4 + (2/m)r^4}$.

SECTION *13.6

1. about 61.5% of an earth year   3. set $x = r \cos \theta$, $y = r \sin \theta$

5. Substitute

$$r = \frac{a}{1 + e\cos\theta}, \qquad \left(\frac{dr}{d\theta}\right)^2 = \frac{(ae\sin\theta)^2}{(1 + e\cos\theta)^4}$$

into the right side of the equation and you will see that, with a and e^2 as given, the expression reduces to E.

SECTION 13.7

1. $\dfrac{e^{-x}}{(1 + e^{-2x})^{3/2}}$ 3. $\dfrac{2}{(1 + 4x)^{3/2}}$ 5. $|\cos x|$ 7. $\dfrac{|\sin x|}{(1 + \cos^2 x)^{3/2}}$ 9. $\dfrac{30x^4}{(9 + 4x^5)^{3/2}}$ 11. $\frac{5}{2}\sqrt{5}$ 13. $5\sqrt{5}$ 15. $\dfrac{10\sqrt{10}}{3}$

17. $\frac{4}{3}$ 19. $(\frac{1}{2}\sqrt{2}, \frac{1}{2}\ln\frac{1}{2})$ 21. $\dfrac{1}{(1 + t^2)^{3/2}}$ 23. $\dfrac{12|t|}{(4 + 9t^4)^{3/2}}$ 25. $\frac{1}{2}\sqrt{2}\,e^{-t}$ 27. $\dfrac{2 + t^2}{(1 + t^2)^{3/2}}$ 29. $\sqrt{2}$ 31. $\dfrac{a^4 b^4}{(b^4 x^2 + a^4 y^2)^{3/2}}$

33. follow the hint   35. $k = \frac{1}{3}\sqrt{2}\,e^{-t}$, $a_\mathbf{T} = \sqrt{3}\,e^t$, $a_\mathbf{N} = \sqrt{2}\,e^t$ 37. $k = 1$, $a_\mathbf{T} = 0$, $a_\mathbf{N} = 4$

39. $k = \dfrac{\sqrt{t^4 + 4t^2 + 1}}{(t^4 + t^2 + 1)^{3/2}}$, $a_\mathbf{T} = \dfrac{2t^3 + t}{\sqrt{t^4 + t^2 + 1}}$, $a_\mathbf{N} = \dfrac{\sqrt{t^4 + 4t^2 + 1}}{\sqrt{t^4 + t^2 + 1}}$ 41. $\dfrac{e^{-a\theta}}{\sqrt{1 + a^2}}$ 43. $\dfrac{3}{2\sqrt{2a^2(1 - \cos\theta)}} = \dfrac{3}{2\sqrt{2ar}}$

45. (a) $s(\theta) = 4R|\cos\frac{1}{2}\theta|$ (b) $\rho(\theta) = 4R\sin\frac{1}{2}\theta$ (c) $\rho^2 + s^2 = 16R^2$ 47. $9\rho^2 + s^2 = 16a^2$

CHAPTER 14

SECTION 14.1

1. dom (f) = the first and third quadrants, including the axes; ran $(f) = [0, \infty)$

3. dom (f) = the set of all points (x, y) not on the line $y = -x$; ran $(f) = (-\infty, 0) \cup (0, \infty)$ 5. dom (f) = the entire plane; ran $(f) = (-1, 1)$

7. dom (f) = the first and third quadrants, excluding the axes; ran $(f) = (-\infty, \infty)$

9. dom (f) = the set of all points (x, y) with $x^2 < y$; in other words, the set of all points of the plane above the parabola $y = x^2$; ran $(f) = (0, \infty)$

11. dom (f) = the set of all points (x, y) with $-3 \le x \le 3$, $-2 \le y \le 2$ (a rectangle); ran $(f) = [-2, 3]$

13. dom (f) = the set of all points (x, y, z) not on the plane $x + y + z = 0$; ran $(f) = \{-1, 1\}$

15. dom (f) = the set of all points (x, y, z) with $|y| < |x|$; ran $(f) = (-\infty, 0]$

17. dom (f) = the set of all points (x, y) such that $x^2 + y^2 < 9$; in other words, the set of all points of the plane inside the circle $x^2 + y^2 = 9$; ran $(f) = [\frac{2}{3}, \infty)$

19. dom (f) = the set of all points (x, y, z) with $x + 2y + 3z > 0$; in other words, the set of all points in space that lie on the same side of the plane $x + 2y + 3z = 0$ as the point $(1, 1, 1)$; ran $(f) = (-\infty, \infty)$

21. dom (f) = all of space; ran $(f) = (0, \infty)$

23. dom $(f) = \{x: x \geq 0\}$; range $(f) = [0, \infty)$
dom $(g) = \{(x, y): x \geq 0, y \text{ real}\}$; range $(g) = [0, \infty)$
dom $(h) = \{(x, y, z): x \geq 0, y, z \text{ real}\}$; range $(h) = [0, \infty)$

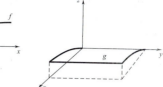

25. $\lim\limits_{h \to 0} \dfrac{f(x + h, y) - f(x, y)}{h} = 4x; \quad \lim\limits_{h \to 0} \dfrac{f(x, y + h) - f(x, y)}{h} = -1$

27. $\lim\limits_{h \to 0} \dfrac{f(x + h, y) - f(x, y)}{h} = 3 - y; \quad \lim\limits_{h \to 0} \dfrac{f(x, y + h) - f(x, y)}{h} = -x + 4y$

29. $\lim\limits_{h \to 0} \dfrac{f(x + h, y) - f(x, y)}{h} = -y \sin(xy); \quad \lim\limits_{h \to 0} \dfrac{f(x, y + h) - f(x, y)}{h} = -x \sin(xy)$

31. (a) $f(x, y) = x^2 y$ (b) $f(x, y) = \pi x^2 y$ (c) $f(x, y) = 2|y|$ **33.** $V = \dfrac{lw(10 - lw)}{l + w}$ **35.** $V = \pi r^2 h + \frac{4}{3} \pi r^3$

SECTION 14.2

1. a quadric cone **3.** a parabolic cylinder **5.** a hyperboloid of one sheet **7.** sphere of radius 2 centered at the origin

9. an elliptic paraboloid **11.** a hyperbolic paraboloid

13.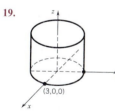

15.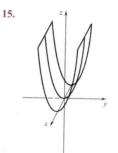

17.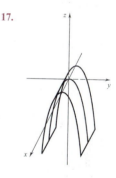

19.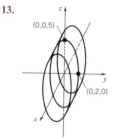

21.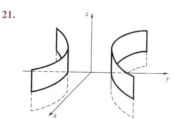

23.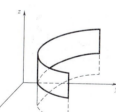

25. elliptic paraboloid, xy-trace: the origin, xz-trace: the parabola $x^2 = 4z$, yz-trace: the parabola $y^2 = 9z$, surface has the form of Figure 14.2.5

27. a quadric cone, xy-trace: the origin, xz-trace: the lines $x = \pm 2z$, yz-trace: the lines $y = \pm 3z$, surface has the form of Figure 14.2.4

29. a hyperboloid of two sheets, xy-trace: none, xz-trace: the hyperbola $4z^2 - x^2 = 4$, yz-trace: the hyperbola $9z^2 - y^2 = 9$, surface has the form of Figure 14.2.3

31. hyperboloid of two sheets, xy-trace: the hyperbola $\dfrac{x^2}{4} - \dfrac{y^2}{9} = 1$, xz-trace: the hyperbola $\dfrac{x^2}{4} - z^2 = 1$, yz-trace: none, see Figure 14.2.3 for an example

33. elliptic paraboloid, xy-trace: the origin, xz-trace: the parabola $x^2 = 4z$, yz-trace: the parabola $y^2 = 9z$, surface has the form of Figure 14.2.5

35. hyperboloid of two sheets, xy-trace: the hyperbola $\dfrac{y^2}{4} - \dfrac{x^2}{9} = 1$, xz-trace: none, yz-trace: the hyperbola $\dfrac{y^2}{4} - z^2 = 1$, see Figure 14.2.3 for an example

37. paraboloid of revolution, xy-trace: the origin, xz-trace: the parabola $x^2 = 4z$, yz-trace: the parabola $y^2 = 4z$, surface has the form of Figure 14.2.5.

39. (a) an elliptic paraboloid (vertex down if A and B are both positive, vertex up if A and B are both negative)
(b) a hyperbolic paraboloid   (c) the xy-plane if A and B are both zero; otherwise, a parabolic cylinder

41. $x^2 + y^2 - 4z = 0$ (paraboloid of revolution)   **43.** (a) a circle   (b) (i) $\sqrt{x^2 + y^2} = -3z$ (ii) $\sqrt{x^2 + z^2} = \frac{1}{3}y$

45. the line $5x + 7y = 30$ **47.** the circle $x^2 + y^2 = \frac{5}{4}$ **49.** the ellipse $x^2 + 2y^2 = 2$ **51.** the parabola $x^2 = -4(y - 1)$

SECTION 14.3

1. lines of slope 1: $y = x - c$

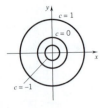

3. parabolas, $y = x^2 - c$

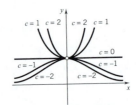

5. the y-axis and the lines $y = [(1 - c)/c]x$,
the origin omitted throughout

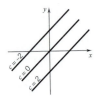

7. the cubics $y = x^3 - c$

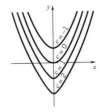

9. the lines $y = \pm x$ and the hyperbolas $x^2 - y^2 = c$

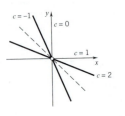

11. pairs of horizontal lines $y = \pm\sqrt{c}$ and the x-axis

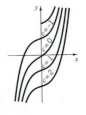

13. the circles $x^2 + y^2 = e^c$ with c real

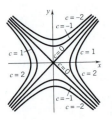

15. the curves $y = e^{cx^2}$ with the point $(0, 1)$ omitted

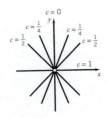

17. the coordinate axes and pairs of lines $y = \pm(\sqrt{1 - c}/\sqrt{c})x$, the origin omitted throughout

19. $x + 2y + 3z = 0$, plane through the origin   **21.** $z = \sqrt{x^2 + y^2}$, the upper nappe of the circular cone $z^2 = x^2 + y^2$; Figure 14.2.4

23. the elliptic paraboloid $\dfrac{x^2}{(3\sqrt{2})^2} + \dfrac{y^2}{(2\sqrt{2})^2} = z$; Figure 14.2.5

25. (i) hyperboloid of two sheets; Figure 14.2.3   (ii) circular cone; Figure 14.2.4   (iii) hyperboloid of one sheet; Figure 14.2.2

27. $4x^2 + y^2 = 1$ **29.** $y^2\tan^{-1}x = \pi$ **31.** $x^2 + 2y^2 - 2xyz = 13$

33. $x^2 + y^2 + z^2 = \dfrac{GmM}{c}$. The surfaces of constant gravitational force are concentric spheres.

35. (a) $T(x, y) = \dfrac{k}{x^2 + y^2}$ (b) $x^2 + y^2 = \dfrac{k}{c}$, concentric circles   (c) $10°$ **37.** F   **39.** A   **41.** E

SECTION 14.4

1. $\dfrac{\partial f}{\partial x} = 6x - y, \quad \dfrac{\partial f}{\partial y} = 1 - x$ **3.** $\dfrac{\partial \rho}{\partial \phi} = \cos \phi \cos \theta, \quad \dfrac{\partial \rho}{\partial \theta} = -\sin \phi \sin \theta$ **5.** $\dfrac{\partial f}{\partial x} = e^{x-y} + e^{y-x}, \quad \dfrac{\partial f}{\partial y} = -e^{x-y} - e^{y-x}$

7. $\dfrac{\partial g}{\partial x} = \dfrac{(AD - BC)y}{(Cx + Dy)^2}, \quad \dfrac{\partial g}{\partial y} = \dfrac{(BC - AD)x}{(Cx + Dy)^2}$ **9.** $\dfrac{\partial u}{\partial x} = y + z, \quad \dfrac{\partial u}{\partial y} = x + z, \quad \dfrac{\partial u}{\partial z} = x + y$

11. $\dfrac{\partial f}{\partial x} = z \cos (x - y), \quad \dfrac{\partial f}{\partial y} = -z \cos (x - y), \quad \dfrac{\partial f}{\partial z} = \sin (x - y)$

13. $\dfrac{\partial \rho}{\partial \theta} = e^{\theta + \phi} [\cos (\theta - \phi) - \sin(\theta - \phi)], \quad \dfrac{\partial \rho}{\partial \phi} = e^{\theta + \phi} [\cos (\theta - \phi) + \sin (\theta - \phi)]$

15. $\dfrac{\partial f}{\partial x} = 2xy \sec(xy) + x^2 y^2 \sec(xy)\tan (xy), \quad \dfrac{\partial f}{\partial y} = x^2\sec(xy) + x^3 y \sec(xy)\tan (xy)$ **17.** $\dfrac{\partial h}{\partial x} = \dfrac{y^2 - x^2}{(x^2 + y^2)^2}, \quad \dfrac{\partial h}{\partial y} = -\dfrac{2xy}{(x^2 + y^2)^2}$

19. $\dfrac{\partial f}{\partial x} = \dfrac{\sin y (\cos x + x \sin x)}{y \cos^2 x}, \quad \dfrac{\partial f}{\partial y} = \dfrac{x (y \cos y - \sin y)}{y^2 \cos x}$ **21.** $\dfrac{\partial h}{\partial x} = 2f(x)f'(x) g(y), \quad \dfrac{\partial h}{\partial y} = [f(x)]^2 g'(y)$

23. $\dfrac{\partial f}{\partial x} = (y^2 \ln z)z^{xy^2}, \quad \dfrac{\partial f}{\partial y} = (2xy \ln z)z^{xy^2}, \quad \dfrac{\partial f}{\partial z} = xy^2 z^{xy^2 - 1}$

25. $\dfrac{\partial h}{\partial r} = 2re^{2t}\cos (\theta - t), \quad \dfrac{\partial h}{\partial \theta} = -r^2 e^{2t}\sin (\theta - t), \quad \dfrac{\partial h}{\partial t} = r^2 e^{2t}[2 \cos (\theta - t) + \sin (\theta - t)]$

27. $\dfrac{\partial f}{\partial x} = -\dfrac{yz}{x^2 + y^2}, \quad \dfrac{\partial f}{\partial y} = \dfrac{xz}{x^2 + y^2}, \quad \dfrac{\partial f}{\partial z} = \tan^{-1}(y/x)$ **29.** $f_x(0, e) = 1, \quad f_y(0, e) = e^{-1}$ **31.** $f_x(1, 2) = \frac{2}{9}, \quad f_y(1, 2) = -\frac{1}{9}$

33. $f_x(x, y) = 2xy, \quad f_y(x, y) = x^2$ **35.** $f_x(x, y) = \dfrac{2}{x}, \quad f_y(x, y) = \dfrac{1}{y}$

37. $f_x(x, y) = -\dfrac{1}{(x - y)^2}, \quad f_y(x, y) = \dfrac{1}{(x - y)^2}$ **39.** $f_x(x, y, z) = y^2 z, \quad f_y(x, y, z) = 2xyz, \quad f_z(x, y, z) = xy^2$

41. (b) $x = x_0, \quad z - z_0 = f_y(x_0, y_0)(y - y_0)$ **43.** $x = 2, \quad z - 5 = 2(y - 1)$ **45.** $y = 2, \quad z - 9 = 6(x - 3)$

47. $u_x = v_y = 2x, \quad u_y = -v_x = -2y$ **49.** $u_x = v_y = \dfrac{x}{x^2 + y^2}, \quad u_y = -v_x = \dfrac{y}{x^2 + y^2}$

51. (a) f depends only on y (b) f depends only on x **53.** (a) $50\sqrt{3}$ in.²   (b) $5\sqrt{3}$ (c) 50   (d) $\frac{5}{18} \pi$ in.²   (e) -2

55. (a) y_0-section: $\mathbf{r}(x) = x\mathbf{i} + y_0\mathbf{j} + f(x, y_0)\mathbf{k}$

tangent line: $\mathbf{R}(t) = [x_0\mathbf{i} + y_0\mathbf{j} + f(x_0, y_0)\mathbf{k}] + t \left[\mathbf{i} + \dfrac{\partial f}{\partial x} (x_0, y_0)\mathbf{k}\right]$

(b) x_0-section: $\mathbf{r}(y) = x_0\mathbf{i} + y\mathbf{j} + f(x_0, y)\mathbf{k}$

tangent line: $\mathbf{R}(t) = [x_0\mathbf{i} + y_0\mathbf{j} + f(x_0, y_0)\mathbf{k}] + t \left[\mathbf{j} + \dfrac{\partial f}{\partial y} (x_0, y_0)\mathbf{k}\right]$

(c) For (x, y, z) in the plane

$[(x - x_0)\mathbf{i} + (y - y_0)\mathbf{j} + (z - f(x_0, y_0))\mathbf{k}] \cdot \left[\left(\mathbf{i} + \dfrac{\partial f}{\partial x} (x_0, y_0)\mathbf{k}\right) \times \left(\mathbf{j} + \dfrac{\partial f}{\partial y} (x_0, y_0)\mathbf{k}\right)\right] = 0.$

From this it follows that

$z - f(x_0, y_0) = (x - x_0)\dfrac{\partial f}{\partial x} (x_0, y_0) + (y - y_0)\dfrac{\partial f}{\partial y} (x_0, y_0).$

57. (a) Set $u = ax + by$. Then $\dfrac{\partial w}{\partial x} = ag'(u)$ and $\dfrac{\partial w}{\partial y} = bg'(u)$. (b) Set $u = x^m y^n$. Then $\dfrac{\partial w}{\partial x} = mx^{m-1}y^n g'(u)$ and $\dfrac{\partial w}{\partial y} = nx^m y^{n-1} g'(u)$.

59. $x\dfrac{\partial u}{\partial x} + y\dfrac{\partial u}{\partial y} = x(4Ax^3 + 4Bxy^2) + y(4Bx^2y + 4Cy^3) = 4(Ax^4 + 2Bx^2y^2 + Cy^4) = 4u$ **61.** r

SECTION 14.5

1. interior $= \{(x, y): 2 < x < 4, 1 < y < 3\}$ (the inside of the rectangle)
boundary $=$ the union of the four line segments that bound the rectangle
set is closed

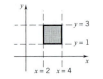

3. interior $=$ the entire set   (region between two concentric circles)
boundary $= \{(x, y): x^2 + y^2 = 1 \text{ or } x^2 + y^2 = 4\}$ (the two circles)
set is open

5. interior $= \{(x, y): 1 < x^2 < 4\} = \{(x, y): -2 < x < 1\} \cup \{(x, y): 1 < x < 2\}$
(two vertical strips without the boundary lines)
boundary $= \{(x, y): x = -2, x = -1, x = 1, \text{ or } x = 2\}$ (four vertical lines)
set is neither open nor closed

7. interior $= \{(x, y): y < x^2\}$ (region below the parabola)
boundary $= \{(x, y): y = x^2\}$ (the parabola)
set is closed

9. interior $= \{(x, y, z): x^2 + y^2 < 1, 0 < z < 4\}$
(the inside of a cylinder)
boundary $=$ the total surface of the cylinder
(the curved part, the top, the bottom)
set is closed

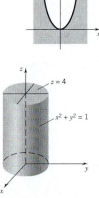

11. (a) ϕ (b) S (c) closed   **13.** interior $= \{x: 1 < x < 3\}$, boundary $= \{1, 3\}$; set is closed

15. interior $=$ the entire set, boundary $= \{1\}$; set is open   **17.** interior $= \{x: |x| > 1\}$, boundary $= \{1, -1\}$; set is neither open nor closed

19. interior $= \phi$, boundary $= \{\text{the entire set}\} \cup \{0\}$; the set is neither open nor closed

SECTION 14.6

1. $\dfrac{\partial^2 f}{\partial x^2} = 2A, \quad \dfrac{\partial^2 f}{\partial y^2} = 2C, \quad \dfrac{\partial^2 f}{\partial y \partial x} = \dfrac{\partial^2 f}{\partial x \partial y} = 2B$ **3.** $\dfrac{\partial^2 f}{\partial x^2} = Cy^2 e^{xy}, \quad \dfrac{\partial^2 f}{\partial y^2} = Cx^2 e^{xy}, \quad \dfrac{\partial^2 f}{\partial y \partial x} = \dfrac{\partial^2 f}{\partial x \partial y} = Ce^{xy}(xy + 1)$

5. $\dfrac{\partial^2 f}{\partial x^2} = 2, \quad \dfrac{\partial^2 f}{\partial y^2} = 4(x + 3y^2 + z^3), \quad \dfrac{\partial^2 f}{\partial z^2} = 6z(2x + 2y^2 + 5z^3)$

$\dfrac{\partial^2 f}{\partial x \partial y} = \dfrac{\partial^2 f}{\partial y \partial x} = 4y, \quad \dfrac{\partial^2 f}{\partial z \partial x} = \dfrac{\partial^2 f}{\partial x \partial z} = 6z^2, \quad \dfrac{\partial^2 f}{\partial z \partial y} = \dfrac{\partial^2 f}{\partial y \partial z} = 12yz^2$

7. $\dfrac{\partial^2 f}{\partial x^2} = \dfrac{1}{(x + y)^2} - \dfrac{1}{x^2}, \quad \dfrac{\partial^2 f}{\partial y^2} = \dfrac{1}{(x + y)^2}, \quad \dfrac{\partial^2 f}{\partial y \partial x} = \dfrac{\partial^2 f}{\partial x \partial y} = \dfrac{1}{(x + y)^2}$

9. $\dfrac{\partial^2 f}{\partial x^2} = 2(y + z), \quad \dfrac{\partial^2 f}{\partial y^2} = 2(x + z), \quad \dfrac{\partial^2 f}{\partial z^2} = 2(x + y)$; the second mixed partials are all $2(x + y + z)$

11. $\dfrac{\partial^2 f}{\partial x^2} = y(y-1)x^{y-2}, \quad \dfrac{\partial^2 f}{\partial y^2} = (\ln x)^2\, x^y, \quad \dfrac{\partial^2 f}{\partial y\,\partial x} = \dfrac{\partial^2 f}{\partial x\,\partial y} = x^{y-1}(1 + y \ln x)$

13. $\dfrac{\partial^2 f}{\partial x^2} = ye^x, \quad \dfrac{\partial^2 f}{\partial y^2} = xe^y, \quad \dfrac{\partial^2 f}{\partial y\,\partial x} = e^y + e^x = \dfrac{\partial^2 f}{\partial x\,\partial y}$ **15.** $\dfrac{\partial^2 f}{\partial x^2} = \dfrac{y^2 - x^2}{(x^2 + y^2)^2}, \quad \dfrac{\partial^2 f}{\partial y^2} = \dfrac{x^2 - y^2}{(x^2 + y^2)^2}, \quad \dfrac{\partial^2 f}{\partial y\,\partial x} = -\dfrac{2xy}{(x^2 + y^2)^2} = \dfrac{\partial^2 f}{\partial x\,\partial y}$

17. $\dfrac{\partial^2 f}{\partial x^2} = -2y^2\cos 2xy, \quad \dfrac{\partial^2 f}{\partial y^2} = -2x^2\cos 2xy, \quad \dfrac{\partial^2 f}{\partial y\,\partial x} = -[\sin 2xy + 2xy \cos 2xy] = \dfrac{\partial^2 f}{\partial x\,\partial y}$

19. $\dfrac{\partial^2 f}{\partial x^2} = 0, \quad \dfrac{\partial^2 f}{\partial y^2} = xz \sin y, \quad \dfrac{\partial^2 f}{\partial z^2} = -xy \sin z$

$\dfrac{\partial^2 f}{\partial y\,\partial x} = \sin z - z \cos y = \dfrac{\partial^2 f}{\partial x\,\partial y}$

$\dfrac{\partial^2 f}{\partial z\,\partial x} = y \cos z - \sin y = \dfrac{\partial^2 f}{\partial x\,\partial z}$

$\dfrac{\partial^2 f}{\partial z\,\partial y} = x \cos z - x \cos y = \dfrac{\partial^2 f}{\partial y\,\partial z}$

21. $x^2\dfrac{\partial^2 u}{\partial x^2} + 2xy\dfrac{\partial^2 u}{\partial x\,\partial y} + y^2\dfrac{\partial^2 u}{\partial y^2} = x^2\left(\dfrac{-2y^2}{(x+y)^3}\right) + 2xy\left(\dfrac{2xy}{(x+y)^3}\right) + y^2\left(\dfrac{-2x^2}{(x+y)^3}\right) = 0$ **23.** $\dfrac{\partial^2 f}{\partial x^2} + \dfrac{\partial^2 f}{\partial y^2} = e^x\sin y - e^x\sin y = 0$

25. $\dfrac{\partial^2 f}{\partial x^2} + \dfrac{\partial^2 f}{\partial y^2} = \dfrac{y^2 - x^2}{x^2 + y^2} + \dfrac{x^2 - y^2}{x^2 + y^2} = 0$ **27.** $\dfrac{\partial^2 f}{\partial x^2} = e^{x+y}\cos \sqrt{2}z, \quad \dfrac{\partial^2 f}{\partial y^2} = e^{x+y}\cos \sqrt{2}z, \quad \dfrac{\partial^2 f}{\partial z^2} = -2e^{x+y}\cos \sqrt{2}z$

29. $\dfrac{\partial^2 f}{\partial t^2} = -5c^2\sin (x + ct)\cos (2x + 2ct) - 4c^2\cos (x + ct)\sin (2x + 2ct)$

$\dfrac{\partial^2 f}{\partial x^2} = -5 \sin (x + ct)\cos (2x + 2ct) - 4 \cos (x + ct)\sin (2x + 2ct),$ so $\dfrac{\partial^2 f}{\partial t^2} - c^2\dfrac{\partial^2 f}{\partial x^2} = 0$

31. $\dfrac{\partial^2 f}{\partial t^2} = c^2k^2(Ae^{kx} + Be^{-kx})(Ce^{ckt} + De^{-ckt}), \quad \dfrac{\partial^2 f}{\partial x^2} = k^2(Ae^{kx} + Be^{-kx})(Ce^{ckt} + De^{-ckt}),$ so $\dfrac{\partial^2 f}{\partial t^2} - c^2\dfrac{\partial^2 f}{\partial x^2} = 0$

33. (a) mixed partials are 0 (b) mixed partials are $g'(x)\, h'(y)$ (c) by the hint mixed partials for each term $x^m y^n$ are $mn x^{m-1}y^{n-1}$

35. (a) no, since $\dfrac{\partial^2 f}{\partial x\,\partial y} \neq \dfrac{\partial^2 f}{\partial y\,\partial x}$ (b) no, since $\dfrac{\partial^2 f}{\partial x\,\partial y} \neq \dfrac{\partial^2 f}{\partial y\,\partial x}$ for $x \neq y$

37. $\underset{\text{by definition}}{\dfrac{\partial^3 f}{\partial x^2\,\partial y}} = \dfrac{\partial}{\partial x}\left(\dfrac{\partial^2 f}{\partial x\,\partial y}\right)$

$\underset{(14.6.5)}{} = \underset{\text{by definition}}{\dfrac{\partial}{\partial x}\left(\dfrac{\partial^2 f}{\partial y\,\partial x}\right)} = \dfrac{\partial^2}{\partial x\,\partial y}\left(\dfrac{\partial f}{\partial x}\right) \underset{(14.6.5)}{=} \dfrac{\partial^2}{\partial y\,\partial x}\left(\dfrac{\partial f}{\partial x}\right) = \dfrac{\partial}{\partial y}\left(\dfrac{\partial^2 f}{\partial x^2}\right) = \underset{\text{by definition}}{\dfrac{\partial^3 f}{\partial y\,\partial x^2}}$

39. (a) 0 (b) 0 (c) $\dfrac{m}{1 + m^2}$ (d) 0 (e) $\dfrac{f'(0)}{1 + [f'(0)]^2}$ (f) $\tfrac{1}{4}\sqrt{3}$ (g) does not exist

41. (a) $\dfrac{\partial g}{\partial x}(0, 0) = \lim_{h\to 0}\dfrac{g(h, 0) - g(0, 0)}{h} = \lim_{h\to 0} 0 = 0, \quad \dfrac{\partial g}{\partial y}(0, 0) = \lim_{h\to 0}\dfrac{g(0, h) - g(0, 0)}{h} = \lim_{h\to 0} 0 = 0$

(b) as (x, y) tends to $(0, 0)$ along the x-axis, $g(x, y) = g(x, 0) = 0$ tends to 0;

as (x, y) tends to $(0, 0)$ along the line $y = x$, $g(x, y) = g(x, x) = \tfrac{1}{2}$ tends to $\tfrac{1}{2}$

43. For $y \neq 0$, $\dfrac{\partial f}{\partial x}(0, y) = \lim_{h\to 0}\dfrac{f(h, y) - f(0, y)}{h} = \lim_{h\to 0}\dfrac{y(y^2 - h^2)}{h^2 + y^2} = y.$ Since $\dfrac{\partial f}{\partial x}(0, 0) = \lim_{h\to 0}\dfrac{f(h, 0) - f(0, 0)}{h} = \lim_{h\to 0} 0 = 0,$

we have $\dfrac{\partial f}{\partial x}(0, y) = y$ for all y. For $x \neq 0$, $\dfrac{\partial f}{\partial y}(x, 0) = \lim_{h\to 0}\dfrac{f(x, h) - f(x, 0)}{h} = \lim_{h\to 0}\dfrac{x(h^2 - x^2)}{x^2 + h^2} = -x.$

Since $\dfrac{\partial f}{\partial y}(0, 0) = \lim_{h\to 0}\dfrac{f(0, h) - f(0, 0)}{h} = \lim_{h\to 0} 0 = 0,$ we have $\dfrac{\partial f}{\partial y}(x, 0) = -x$ for all x.

Therefore $\dfrac{\partial^2 f}{\partial y\,\partial x}(0, y) = 1$ for all y and $\dfrac{\partial^2 f}{\partial x\,\partial y}(x, 0) = -1$ for all x. In particular, $\dfrac{\partial^2 f}{\partial y\,\partial x}(0, 0) = 1,$ while $\dfrac{\partial^2 f}{\partial x\,\partial y}(0, 0) = -1.$

CHAPTER 15

SECTION 15.1

1. $\nabla f = e^{xy}[(xy + 1)\mathbf{i} + x^2\mathbf{j}]$ **3.** $\nabla f = (6x - y)\mathbf{i} + (1 - x)\mathbf{j}$ **5.** $\nabla f = 2xy^{-2}\mathbf{i} - 2x^2y^{-3}\mathbf{j}$

7. $\nabla f = z \cos (x - y)\mathbf{i} - z \cos (x - y)\mathbf{j} + \sin (x - y)\mathbf{k}$ **9.** $\nabla f = (y + z)\mathbf{i} + (x + z)\mathbf{j} + (x + y)\mathbf{k}$

11. $\nabla f = e^{x-y}[(1 + x + y)\mathbf{i} + (1 - x - y)\mathbf{j}]$ **13.** $\nabla f = e^{x}[\ln y\mathbf{i} + y^{-1}\mathbf{j}]$ **15.** $\nabla f = \dfrac{AD - BC}{(Cx + Dy)^2}[y\mathbf{i} - x\mathbf{j}]$

17. $\nabla f = (ye^x + xye^x - ze^y \cos xz)\mathbf{i} + (xe^x + e^z - e^y\sin xz)\mathbf{j} + (ye^z - xe^y\cos xz)\mathbf{k}$

19. $\nabla f = e^{x+2y}\cos(z^2 + 1)\mathbf{i} + 2e^{x+2y}\cos(z^2 + 1)\mathbf{j} - 2ze^{x+2y}\sin(z^2 +)\mathbf{k}$ **21.** $\nabla f = -\mathbf{i} + 18\mathbf{j}$ **23.** $\frac{4}{5}\mathbf{i} + \frac{2}{5}\mathbf{j}$ **25.** $\mathbf{i}$

27. $\nabla f = -\frac{1}{2}\sqrt{2}(\mathbf{i} + 2\mathbf{j} + \mathbf{k})$ **29.** $\mathbf{i} + \frac{3}{5}\mathbf{j} - \frac{4}{5}\mathbf{k}$ **31.** $(6x - y)\mathbf{i} + (1 - x)\mathbf{j}$ **33.** $(2xy + z^2)\mathbf{i} + (2yz + x^2)\mathbf{j} + (2xz + y^2)\mathbf{k}$

35. $f(x, y) = x^2y + y$ **37.** $f(x, y) = \dfrac{x^2}{2} + x\sin y - y^2$ **39.** (a) $(1/r^2)\mathbf{r}$ (b) $[(\cos r)/r]\mathbf{r}$ (c) $(e^r/r)\mathbf{r}$

41. (a) $(0, 0)$ (b)   (c) f has an absolute minimum at $(0, 0)$

(0,0,1)

43. (a) Let $\mathbf{c} = c_1\mathbf{i} + c_2\mathbf{j} + c_3\mathbf{k}$. First, we take $\mathbf{h} = h\mathbf{i}$. Since $\mathbf{c} \cdot \mathbf{h}$ is $o(\mathbf{h})$,

$$0 = \lim_{h \to 0} \frac{\mathbf{c} \cdot \mathbf{h}}{\|\mathbf{h}\|} = \lim_{h \to 0} \frac{c_1 h}{h} = c_1.$$

Similarly, $c_2 = 0$ and $c_3 = 0$.
(b) $(\mathbf{y} - \mathbf{z}) \cdot \mathbf{h} = [f(\mathbf{x} + \mathbf{h}) - f(\mathbf{x}) - \mathbf{z} \cdot \mathbf{h}] + [\mathbf{y} \cdot \mathbf{h} - f(\mathbf{x} + \mathbf{h}) + f(\mathbf{x})] = o(\mathbf{h}) + o(\mathbf{h}) = o(\mathbf{h})$, so that, by part (a), $\mathbf{y} - \mathbf{z} = \mathbf{0}$.

45. (a) In Section 14.6 we showed that f was not continuous at $(0, 0)$. It is therefore not differentiable at $(0, 0)$.

(b) For $(x, y) \neq (0, 0)$, $\dfrac{\partial f}{\partial x} = \dfrac{2y(y^2 - x^2)}{(x^2 + y^2)^2}$. As (x, y) tends to $(0, 0)$ along the y-axis, $\partial f/\partial x = 2/y$ tends to ∞.

47. (a), (b) $\left(\dfrac{\sqrt{2}}{2}, \dfrac{\sqrt{2}}{2}\right)$ maximum, $\left(-\dfrac{\sqrt{2}}{2}, \dfrac{\sqrt{2}}{2}\right)$ minimum, $\left(\dfrac{\sqrt{2}}{2}, -\dfrac{\sqrt{2}}{2}\right)$ minimum, $\left(-\dfrac{\sqrt{2}}{2}, -\dfrac{\sqrt{2}}{2}\right)$ maximum

SECTION 15.2

1. $-2\sqrt{2}$ **3.** $\frac{1}{5}(7 - 4e)$ **5.** $\frac{1}{4}\sqrt{2}(a - b)$ **7.** $\dfrac{2}{\sqrt{65}}$ **9.** $\frac{2}{3}\sqrt{6}$ **11.** $-3\sqrt{2}$ **13.** $\dfrac{\sqrt{3}\pi}{12}$ **15.** $-(x^2 + y^2)^{-1/2}$

17. (a) $\sqrt{2}[a(B - A) + b(C - B)]$ (b) $\sqrt{2}[a(A - B) + b(B - C)]$ **19.** $-\frac{7}{5}\sqrt{5}$ **21.** $\dfrac{18}{\sqrt{14}}$ or $\dfrac{-18}{\sqrt{14}}$

23. increases most rapidly in the direction of $\dfrac{1}{\sqrt{2}}\mathbf{i} + \dfrac{1}{\sqrt{2}}\mathbf{j}$,   rate of change $2\sqrt{2}$; decreases most rapidly in the direction of $-\dfrac{1}{\sqrt{2}}\mathbf{i} - \dfrac{1}{\sqrt{2}}\mathbf{j}$,   rate of change $-2\sqrt{2}$

25. increases most rapidly in the direction of $\dfrac{1}{\sqrt{6}}\mathbf{i} - \dfrac{2}{\sqrt{6}}\mathbf{j} + \dfrac{1}{\sqrt{6}}\mathbf{k}$,   rate of change 1; decreases most rapidly in the direction of $-\dfrac{1}{\sqrt{6}}\mathbf{i} + \dfrac{2}{\sqrt{6}}\mathbf{j} - \dfrac{1}{\sqrt{6}}\mathbf{k}$,   rate of change -1

27. $\nabla f = f'(x_0)\mathbf{i}$. If $f'(x_0) \neq 0$, the gradient points in the direction in which f increases: to the right if $f'(x_0) > 0$, to the left if $f'(x_0) < 0$.

29. (a) $\lim_{h \to 0} \dfrac{f(h, 0) - f(0, 0)}{h} = \lim_{h \to 0} \dfrac{\sqrt{h^2}}{h} = \lim_{h \to 0} \dfrac{|h|}{h}$ does not exist   (b) no; by Theorem 15.2.5 f cannot be differentiable at $(0, 0)$

31. (a) $-\frac{2}{3}\sqrt{97}$ (b) $-\frac{8}{3}$ (c) $-\frac{26}{3}\sqrt{2}$ **33.** (a) its projection onto the xy-plane is the curve $y = x^3$ from $(1, 1)$ to $(0, 0)$
(b) its projection onto the xy-plane is the curve $y = -2x^3$ from $(1, -2)$ to $(0, 0)$

35. its projection onto the xy-plane is the curve $(b^2)^{a^2}x^{b^2} = (a^2)^{b^2}y^{a^2}$ from (a^2, b^2) to $(0, 0)$

37. the curve $y = \ln |\sqrt{2}\sin x|$ in the direction of decreasing x

39. (a) 16   (b) 4   (c) $\frac{16}{17}\sqrt{17}$
(d) The limits computed in (a) and (b) are not directional derivatives. In (a) and (b) we have, in essence, computed $\nabla f(2, 4) \cdot \mathbf{r}_0$ taking $\mathbf{r}_0 = \mathbf{i} + 4\mathbf{j}$ in (a) and $\mathbf{r}_0 = \frac{1}{4}\mathbf{i} + \mathbf{j}$ in (b). In neither case is $\mathbf{r}_0$ a unit vector.

41. (b) $\dfrac{2\sqrt{3} - 3}{2}$ **43.** $\nabla(fg) = \left(f\dfrac{\partial g}{\partial x} + g\dfrac{\partial f}{\partial x}\right)\mathbf{i} + \left(f\dfrac{\partial g}{\partial y} + g\dfrac{\partial f}{\partial y}\right)\mathbf{j} = f\nabla g + g\nabla f$

45. $\nabla f^n = \dfrac{\partial f^n}{\partial x}\mathbf{i} + \dfrac{\partial f^n}{\partial y}\mathbf{j} = nf^{n-1}\dfrac{\partial f}{\partial x}\mathbf{i} + nf^{n-1}\dfrac{\partial f}{\partial y}\mathbf{j} = nf^{n-1}\nabla f$

SECTION 15.3

1. $C = (\frac{1}{3}, \frac{5}{3})$ **3.** (a) $f(x, y, z) = a_1 x + a_2 y + a_3 z + C$ (b) $f(x, y, z) = g(x, y, z) + a_1 x + a_2 y + a_3 z + C$

5. (a) U is not connected (b) (i) $g(\mathbf{x}) = f(\mathbf{x}) - 1$ (ii) $g(\mathbf{x}) = -f(\mathbf{x})$ (c) U is not connected

7. Since f is continuous at $\mathbf{a}$ and $f(\mathbf{a}) = A$, there exists $\delta > 0$ such that

$$\text{if} \quad \|\mathbf{x} - \mathbf{a}\| < \delta \quad \text{and} \quad x \in \Omega, \qquad \text{then} \quad |f(\mathbf{x}) - A| < \epsilon.$$

Whether $\mathbf{a}$ is on the boundary of Ω or in the interior of Ω, there exists $\mathbf{x}_1$ in the interior of Ω within δ of $\mathbf{a}$. That implies that $|f(\mathbf{x}_1) - A| < \epsilon$. A similar argument shows the existence of $\mathbf{x}_2$ with the desired property.

SECTION 15.4

1. e^t **3.** $\dfrac{-2 \sin 2t}{1 + \cos^2 2t}$ **5.** $t^t\left[\dfrac{1}{t} + \ln t + (\ln t)^2\right] + \dfrac{1}{t}$ **7.** $3t^2 - 5t^4$ **9.** $2\omega(b^2 - a^2)\sin \omega t \cos \omega t + b\omega$ **11.** $\sin 2t - 3 \cos 2t$

13. $e^{t/2}(\frac{1}{2} \sin 2t + 2 \cos 2t) + e^{2t}(2 \sin \frac{1}{2}t + \frac{1}{2} \cos \frac{1}{2}t)$ **15.** $e^{t^2}\left[2t \sin \pi t + \pi \cos \pi t\right]$ **17.** $1 - 4t + 6t^2 - 4t^3$

19. increasing $\frac{1288}{3}\pi$ in.^{3}/sec **21.** 41.34 sq in./sec **23.** $\dfrac{\partial u}{\partial s} = 2s \cos^2 t - t \sin s \cos t - st \cos s \cos t$

$$\dfrac{\partial u}{\partial t} = -2s^2 \sin t \cos t + st \sin s \sin t - s \sin s \cos t$$

25. $\dfrac{\partial u}{\partial s} = 4s^3 t^2 \tan (s + t^2) + s^4 t^2 \sec^2(s + t^2);$ **27.** $\dfrac{\partial u}{\partial s} = 2s \cos^2 t - \sin (t - s) \cos t + s \cos t \cos (t - s) + 2t^2 \sin s \cos s$

$$\dfrac{\partial u}{\partial t} = 2s^4 t \tan (s + t^2) + 2s^4 t^3 \sec^2 (s + t^2) \qquad \dfrac{\partial u}{\partial t} = -2s^2 \sin t \cos t + s \sin (t - s) \sin t - s \cos t \cos (t - s) + 2t \sin^2 s$$

29. $\dfrac{d}{dt}[f(\mathbf{r}(t))] = \left[\nabla f(\mathbf{r}(t)) \cdot \dfrac{\mathbf{r}'(t)}{\|\mathbf{r}'(t)\|}\right]\|\mathbf{r}'(t)\| = f'_{\mathbf{u}(t)}(\mathbf{r}(t)) \|\mathbf{r}'(t)\|$ where $\mathbf{u}(t) = \dfrac{\mathbf{r}'(t)}{\|\mathbf{r}'(t)\|}$

31. (a) $(\cos r)\dfrac{\mathbf{r}}{r}$ (b) $(r \cos r + \sin r)\dfrac{\mathbf{r}}{r}$ **33.** (a) $(r \cos r - \sin r)\dfrac{\mathbf{r}}{r^3}$ (b) $\left(\dfrac{\sin r - r \cos r}{\sin^2 r}\right)\dfrac{\mathbf{r}}{r}$

35. (a) See the figure

(b) $\dfrac{\partial u}{\partial r} = \dfrac{\partial u}{\partial x}\left(\dfrac{\partial x}{\partial w}\dfrac{\partial w}{\partial r} + \dfrac{\partial x}{\partial t}\dfrac{\partial t}{\partial r}\right) + \dfrac{\partial u}{\partial y}\left(\dfrac{\partial y}{\partial w}\dfrac{\partial w}{\partial r} + \dfrac{\partial y}{\partial t}\dfrac{\partial t}{\partial r}\right) + \dfrac{\partial u}{\partial z}\left(\dfrac{\partial z}{\partial w}\dfrac{\partial w}{\partial r} + \dfrac{\partial z}{\partial t}\dfrac{\partial t}{\partial r}\right)$

To obtain $\dfrac{\partial u}{\partial s}$, replace each r by s.

37. $\dfrac{du}{dt} = \dfrac{\partial u}{\partial x}\dfrac{dx}{dt} + \dfrac{\partial u}{\partial y}\dfrac{dy}{dt}$

$$\dfrac{d^2u}{dt^2} = \dfrac{\partial u}{\partial x}\dfrac{d^2x}{dt^2} + \dfrac{dx}{dt}\left[\dfrac{\partial^2 u}{\partial x^2}\dfrac{dx}{dt} + \dfrac{\partial^2 u}{\partial y \partial x}\dfrac{dy}{dt}\right] + \dfrac{\partial u}{\partial y}\dfrac{d^2 y}{dt^2} + \dfrac{dy}{dt}\left[\dfrac{\partial^2 u}{\partial x \partial y}\dfrac{dx}{dt} + \dfrac{\partial^2 u}{\partial y^2}\dfrac{dy}{dt}\right] \text{ and the result follows.}$$

39. (a) $\dfrac{\partial u}{\partial r} = \dfrac{\partial u}{\partial x}\dfrac{\partial x}{\partial r} + \dfrac{\partial u}{\partial y}\dfrac{\partial y}{\partial r} = \dfrac{\partial u}{\partial x}\cos \theta + \dfrac{\partial u}{\partial y}\sin \theta,\quad \dfrac{\partial u}{\partial \theta} = \dfrac{\partial u}{\partial x}\dfrac{\partial x}{\partial \theta} + \dfrac{\partial u}{\partial y}\dfrac{\partial y}{\partial \theta} = \dfrac{\partial u}{\partial x}(-r \sin \theta) + \dfrac{\partial u}{\partial y}(r \cos \theta)$

(b) $\left(\dfrac{\partial u}{\partial r}\right)^2 = \left(\dfrac{\partial u}{\partial x}\right)^2 \cos^2\theta + 2\dfrac{\partial u}{\partial x}\dfrac{\partial u}{\partial y}\cos \theta \sin \theta + \left(\dfrac{\partial u}{\partial y}\right)^2 \sin^2\theta,\quad \dfrac{1}{r^2}\left(\dfrac{\partial u}{\partial \theta}\right)^2 = \left(\dfrac{\partial u}{\partial x}\right)^2 \sin^2\theta - 2\dfrac{\partial u}{\partial x}\dfrac{\partial u}{\partial y}\cos \theta \sin \theta + \left(\dfrac{\partial u}{\partial y}\right)^2 \cos^2\theta,$

$\left(\dfrac{\partial u}{\partial r}\right)^2 + \dfrac{1}{r^2}\left(\dfrac{\partial u}{\partial \theta}\right)^2 = \left(\dfrac{\partial u}{\partial x}\right)^2 (\cos^2\theta + \sin^2\theta) + \left(\dfrac{\partial u}{\partial y}\right)^2 (\sin^2\theta + \cos^2\theta) = \left(\dfrac{\partial u}{\partial x}\right)^2 + \left(\dfrac{\partial u}{\partial y}\right)^2$

41. Solve the equations in Exercise 39 (a) for $\dfrac{\partial u}{\partial x}, \dfrac{\partial u}{\partial y}$: **43.** $\nabla u = r(2 - \sin 2\theta)\mathbf{e}_r + r \cos 2\theta \, \mathbf{e}_\theta$

$$\dfrac{\partial u}{\partial x} = \dfrac{\partial u}{\partial r}\cos \theta - \dfrac{1}{r}\dfrac{\partial u}{\partial \theta}\sin \theta;\quad \dfrac{\partial u}{\partial y} = \dfrac{\partial u}{\partial r}\sin \theta + \dfrac{1}{r}\dfrac{\partial u}{\partial \theta}\cos \theta$$

Then $\nabla f = \dfrac{\partial u}{\partial x}\mathbf{i} + \dfrac{\partial u}{\partial y}\mathbf{j} = \dfrac{\partial u}{\partial r}(\cos \theta \mathbf{i} + \sin \theta \mathbf{j}) + \dfrac{1}{r}\dfrac{\partial u}{\partial \theta}(-\sin \theta \mathbf{i} + \cos \theta \mathbf{j})$

45. From Exercise 39 (a),

$$\frac{\partial^2 u}{\partial r^2} = \frac{\partial^2 u}{\partial x^2}\cos^2\theta + 2\frac{\partial^2 u}{\partial y \partial x}\sin\theta\cos\theta + \frac{\partial^2 u}{\partial y^2}\sin^2\theta$$

$$\frac{\partial^2 u}{\partial \theta^2} = \frac{\partial^2 u}{\partial x^2}r^2\sin^2\theta - 2\frac{\partial^2 u}{\partial y \partial x}r^2\sin\theta\cos\theta + \frac{\partial^2 u}{\partial y^2}r^2\cos^2\theta - r\left(\frac{\partial u}{\partial x}\cos\theta + \frac{\partial u}{\partial y}\sin\theta\right).$$

The term in parentheses is just $\dfrac{\partial u}{\partial r}$, and the result follows.

47. $\dfrac{dy}{dx} = -\dfrac{e^y + ye^x - 4xy}{xe^y + e^x - 2x^2}$

49. $\dfrac{dy}{dx} = \dfrac{\cos xy - xy\sin xy - y\sin x}{x^2\sin xy - \cos x}$

51. $\dfrac{\partial z}{\partial x} = -\dfrac{2x - yz(x^2 + y^2 + z^2)\sin xyz}{2z - xy(x^2 + y^2 + z^2)\sin xyz}$; $\dfrac{\partial z}{\partial y} = -\dfrac{2y - xz(x^2 + y^2 + z^2)\sin xyz}{2z - xy(x^2 + y^2 + z^2)\sin xyz}$

53. $\dfrac{\partial z}{\partial x} = -\dfrac{z\sec^2 x - y^2 z^3 - 2yz}{\tan x - 3xy^2 z^2 - 2xy}$; $\dfrac{\partial z}{\partial y} = \dfrac{2xyz^3 + 2xz}{\tan x - 3xy^2 z^2 - 2xy}$

55. $\dfrac{\partial \mathbf{u}}{\partial s} = \dfrac{\partial \mathbf{u}}{\partial x}\dfrac{\partial x}{\partial s} + \dfrac{\partial \mathbf{u}}{\partial y}\dfrac{\partial y}{\partial s}$, $\dfrac{\partial \mathbf{u}}{\partial t} = \dfrac{\partial \mathbf{u}}{\partial x}\dfrac{\partial x}{\partial t} + \dfrac{\partial \mathbf{u}}{\partial y}\dfrac{\partial y}{\partial t}$

where

$$\frac{\partial \mathbf{u}}{\partial x} = \frac{\partial u_1}{\partial x}\mathbf{i} + \frac{\partial u_2}{\partial x}\mathbf{j}, \quad \frac{\partial \mathbf{u}}{\partial y} = \frac{\partial u_1}{\partial y}\mathbf{i} + \frac{\partial u_2}{\partial y}\mathbf{j}$$

57. (a) $f(x, y) = f\left(x, x\dfrac{y}{x}\right) = x^n f(1, \nu)$, where $\nu = y/x$

$$\frac{\partial f}{\partial x} = nx^{n-1}f(1, \nu) + x^n\frac{df}{d\nu}\left(\frac{-y}{x^2}\right)$$

Multiplying through by x and noting that $x^{n-1}\dfrac{df}{d\nu} = \dfrac{\partial f}{\partial y}$, we get

$$x\frac{\partial f}{\partial x} = nx^n f(1, \nu) - yx^{n-1}\frac{df}{d\nu} = nf(x, y) - y\frac{\partial f}{\partial y}$$

and the result follows

(b) Differentiate the result in part (a) with respect to x and multiply the resulting equation by x; then do the same with y. Add the two results.

SECTION 15.5

1. normal vector $\mathbf{i} + \mathbf{j}$; tangent vector $\mathbf{i} - \mathbf{j}$
tangent line $x + y + 2 = 0$; normal line $x - y = 0$

3. normal vector $\sqrt{2}\mathbf{i} - 5\mathbf{j}$; tangent vector $5\mathbf{i} + \sqrt{2}\mathbf{j}$
tangent line $\sqrt{2}x - 5y + 3 = 0$; normal line $5x + \sqrt{2}y - 6\sqrt{2} = 0$

5. normal vector $7\mathbf{i} - 17\mathbf{j}$; tangent vector $17\mathbf{i} + 7\mathbf{j}$
tangent line $7x - 17y + 6 = 0$; normal line $17x + 7y - 82 = 0$

7. normal vector $\mathbf{i} - \mathbf{j}$; tangent vector $\mathbf{i} + \mathbf{j}$
tangent line $x - y - 3 = 0$; normal line $x + y + 1 = 0$

9. 0

11. $4x - 5y + 4z = 0$ **13.** $x + ay - z - 1 = 0$ **15.** $2x + 2y - z = 0$ **17.** $b^2 c^2 x_0 x - a^2 c^2 y_0 y - a^2 b^2 z_0 z - a^2 b^2 c^2 = 0$

19. $(a^2/b, b^2/a, 3ab)$ **21.** $(0, 0, 0)$ **23.** $(\frac{1}{3}, \frac{11}{6}, -\frac{1}{12})$ **25.** $\dfrac{x - x_0}{\partial f/\partial x\,(x_0, y_0, z_0)} = \dfrac{y - y_0}{\partial f/\partial y\,(x_0, y_0, z_0)} = \dfrac{z - z_0}{\partial f/\partial z\,(x_0, y_0, z_0)}$

27. the tangent planes meet at right angles and therefore the normals ∇F and ∇G must meet at right angles:

$$\frac{\partial F}{\partial x}\frac{\partial G}{\partial x} + \frac{\partial F}{\partial y}\frac{\partial G}{\partial y} + \frac{\partial F}{\partial z}\frac{\partial G}{\partial z} = 0$$

29. $\frac{9}{2}a^3$ $(V = \frac{1}{3}Bh)$ **31.** approx. 0.528 rad **33.** $3x + 4y + 6z = 22$, $6x + y - z = 11$

35. $(1, 1, 2)$ lies on both surfaces and the normals at this point are perpendicular

37. (a) $3x - 4y + 6 = 0$ (b) $\mathbf{r}(t) = (4t - 2)\mathbf{i} - 3t\mathbf{j} + (43t^2 - 16t + 6)\mathbf{k}$ (c) $\mathbf{R}(s) = (2\mathbf{i} - 3\mathbf{j} + 33\mathbf{k}) + s(4\mathbf{i} - 3\mathbf{j} + 70\mathbf{k})$
(d) $4x - 18y - z = 29$ (e) $\mathbf{r}(t) = t\mathbf{i} - (\frac{3}{4}t + \frac{3}{2})\mathbf{j} + (\frac{35}{2}t - 2)\mathbf{k}; l = l'$

SECTION 15.6

1. $(1, 0)$ gives a local max of 1 **3.** $(1, -1)$ is a saddle point **5.** $(-2, 1)$ gives a local min of -2

7. no stationary points and no local extreme values **9.** $(1, 1)$ gives a local min of -3 **11.** $(-2, 2)$ is a saddle point

13. $(0, 0)$ is a saddle point **15.** $(3, 4)$ gives absolute min of $\frac{1}{5}$; $(1, 1)$ gives absolute max of $\frac{1}{2}\sqrt{2}$

17. absolute max of 4 occurs at $(-1, 0)$ and $(1, 0)$; there is no absolute min

19. $(1, 1)$ gives absolute min of 0; $(-\sqrt{2}, -\sqrt{2})$ gives absolute max of $6 + 4\sqrt{2}$

21. absolute min of 0 occurs at each point (x, x) with $0 \le x \le 4$; absolute max of 144 occurs at $(0, 12)$

23. $(1, 1)$ gives absolute min of 10; $(2, 8)$ gives absolute max of 68 **25.** $(\frac{32}{9}, -\frac{16}{9}, \frac{32}{9})$; $\frac{16}{3}$ **27.** $x = 6, y = 6, z = 6$; maximum $= 216$

29. length $=$ width $= 2\sqrt[3]{3}$ feet; height $= \sqrt[3]{3}$ feet

31. (a) $(0, 0)$ (b) no local extremes as $(0, 0)$ is a saddle point
   (c) $(1, 0)$ and $(-1, 0)$ give absolute max of $\frac{1}{4}$; $(0, 1)$ and $(0, -1)$ give absolute min of $-\frac{1}{9}$

33. $\left(\dfrac{x_1 + x_2 + x_3}{3}, \dfrac{y_1 + y_2 + y_3}{3} \right)$

35. $1 - \frac{1}{2}\sqrt{3}$ **37.** \$1 per dozen blades, \$2 per razor   **39.** $(0, 0)$ saddle point; $(1, 1)$ local maximum

41. $(1, 0)$ local minimum; $(-1, 0)$ local maximum

SECTION 15.7

1. $(4, -2)$ gives a local min of -10 **3.** $(\frac{10}{3}, \frac{8}{3})$ gives a local max of $\frac{28}{3}$ **5.** $(0, 0)$ is a saddle point; $(2, 2)$ gives a local min of -8

7. $(1, \frac{3}{2})$ is a saddle point; $(5, \frac{27}{2})$ gives a local min of $-\frac{117}{4}$ **9.** no stationary points and no local extreme values

11. $(0, n\pi)$ for integral n are saddle points; no local extreme values   **13.** $(1, -1)$ and $(-1, 1)$ are saddle points; no local extreme values

15. $(\frac{1}{2}, 4)$ gives a local min of 6   **17.** $(1, 1)$ gives a local min of 3   **19.** $(1, 0)$ gives a local min of -1; $(-1, 0)$ gives a local max of 1

21. $(0, 0)$ is a saddle point; $(1, 0)$ and $(-1, 0)$ give a local min of -3 **23.** $(0, 0)$ is a saddle point; $(\frac{\pi}{3}, \frac{\pi}{3})$ and $(-\frac{\pi}{3}, -\frac{\pi}{3})$ give a local max of $\frac{3}{2}$

25. (a) $f_x = 2x + ky, f_y = 2y + kx$; $f_x(0, 0) = f_y(0, 0) = 0$ independent of k.   (b) $|k| > 2$ (c) $|k| < 2$ (d) $|k| = 2$

27. $(-\frac{5}{7}, \frac{4}{7}, \frac{20}{7})$ **29.** $\frac{1}{27}$ **31.** $\dfrac{\sqrt{114}}{36}$ **33.** (a) saddle point, $f(x, y) = 0$ along the plane
   curve $y = x^{2/3}$ (see figure)
   (b) $(0, 0)$ gives a local max of 3

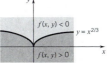

35. $\theta = \frac{1}{6}\pi, x = (2 - \sqrt{3})P, y = \frac{1}{6}(3 - \sqrt{3})P$ **37.** $\frac{2}{3}\sqrt{6}$ **39.** $4 \times 4 \times 6$ m

41. $V = \dfrac{xy(S - 2xy)}{2(x + y)}$ has a maximum value when $x = y = z = \sqrt{\dfrac{S}{6}}$.   **43.** (a) $y = x - \frac{2}{3}$ (b) $y = \frac{14}{13}x^2 - \frac{19}{13}$

45. (a) cross section: 18×18 inches; length: 36 inches   **47.** $x = 4$ in., $\theta = \dfrac{\pi}{3}$
   (b) radius of cross section: $36/\pi$ inches; length: 36 inches.

SECTION 15.8

1. 2   **3.** $-\frac{1}{2}ab$ **5.** $\frac{2}{9}\sqrt{3}ab^2$ **7.** 1   **9.** $\frac{1}{9}\sqrt{3}abc$ **11.** $19\sqrt{2}$ **13.** $\frac{1}{27}abc$ **15.** 1

17. closest point $(\frac{2}{3}, \frac{1}{3}, \frac{2}{3})$; furthest point $(-\frac{2}{3}, -\frac{1}{3}, -\frac{2}{3})$ **19.** $f(3, -2, 1) = 14$ **21.** $|D|(A^2 + B^2 + C^2)^{-1/2}$

23. $4A^2(a^2 + b^2 + c^2)^{-1}$, where A is the area of the triangle and a, b, c, are the sides   **25.** $(2^{-1/3}, -2^{-1/3})$ **27.** hint is given

29. (a) $f(\frac{k}{2}, \frac{k}{2}) = \frac{k}{2}$ is the maximum value   (b) $(xy)^{1/2} = f(x, y) \le f(\frac{k}{2}, \frac{k}{2}) = \frac{k}{2} = \dfrac{x + y}{2}$

31. Same argument as Exercises 29 and 30: $f\left(\dfrac{k}{n}, \dfrac{k}{n}, \ldots, \dfrac{k}{n} \right) = \dfrac{k}{n}$ is the maximum value of $f(x_1, x_2, \ldots, x_n) = (x_1 x_2 \cdots x_n)^{1/n}$

33. radius: $\sqrt[3]{\dfrac{V}{2\pi}}$; height: $\sqrt[3]{\dfrac{4V}{\pi}}$ **35.** length = width = $2\sqrt[3]{3}$ feet; height = $\sqrt[3]{3}$ feet

37. (a) cross section: 18×18 inches; length: 36 inches   (b) radius of cross-section: $36/\pi$ inches; length: 36 inches

39. $Q_1 = 10,000, Q_2 = 20,000, Q_3 = 30,000$ **41.** $\dfrac{\sqrt{2}}{2}$

43. closest point $(-1 + \dfrac{\sqrt{2}}{2}, -1 + \dfrac{\sqrt{2}}{2}, -1 + \sqrt{2})$; furthest point $(-1 - \dfrac{\sqrt{2}}{2}, -1 - \dfrac{\sqrt{2}}{2}, -1 - \sqrt{2})$.

SECTION 15.9

1. $df = (3x^2y - 2xy^2)\,\Delta x + (x^3 - 2x^2y)\,\Delta y$ **3.** $df = (\cos y + y \sin x)\,\Delta x - (x \sin y + \cos x)\,\Delta y$ **5.** $df = \Delta x - (\tan z)\,\Delta y - (y \sec^2 z)\,\Delta z$

7. $df = \dfrac{y(y^2 + z^2 - x^2)}{(x^2 + y^2 + z^2)^2}\,\Delta x + \dfrac{x(x^2 + z^2 - y^2)}{(x^2 + y^2 + z^2)^2}\,\Delta y - \dfrac{2xyz}{(x^2 + y^2 + z^2)^2}\,\Delta z$ **9.** $df = [\cos (x + y) + \cos (x - y)]\,\Delta x + [\cos (x + y) - \cos (x - y)]\,\Delta y$

11. $df = (y^2 z e^{xz} + \ln z)\,\Delta x + 2ye^{xz}\,\Delta y + \left(xy^2 e^{xz} + \dfrac{x}{z} \right)\Delta z$ **13.** $\Delta u = -7.15, du = -7.5$ **15.** $\Delta u = 2.896; du = 2.5$

17. $22\frac{249}{352}$ taking $u = x^{1/2}y^{1/4}, \quad x = 121, \quad y = 16, \quad \Delta x = 4, \quad \Delta y = 1$

19. $\frac{1}{14}\sqrt{2}\pi$ taking $u = \sin x \cos y, \quad x = \pi, \quad y = \frac{1}{4}\pi, \quad \Delta x = -\frac{1}{7}\pi, \quad \Delta y = -\frac{1}{20}\pi$ **21.** $f(2.9, 0.01) \cong 8.67$

23. $f(2.94, 1.1, 0.92) \cong 23.392$ **25.** $dz = -\frac{1}{90}, \quad \Delta z = -\frac{1}{93}$ **27.** decreases about 13.6π in.^{2} **29.** $S \cong 246.8$

31. (a) $dV = 0.24$ (b) $\Delta V = 0.22077$ **33.** $dT = 2.9$ **35.** $dA = x \tan\theta\, \Delta x + \frac{1}{2}x^2\sec^2\theta\, \Delta\theta$; more sensitive to a change in θ

37. (a) $\Delta h = -\dfrac{(2r + \Delta r)h}{(r + \Delta r)^2}\Delta r, \quad \Delta h \cong -\left(\dfrac{2h}{r}\right)\Delta r$ (b) $\Delta h = -\dfrac{(2r + h + \Delta r)}{r + \Delta r}\Delta r, \quad \Delta h \cong -\left(\dfrac{2r + h}{r}\right)\Delta r$

39. (a) 1.962 mm   (b) 1.275 cm^2 **41.** $2.23 \le s \pm |\Delta s| \le 2.27$

SECTION 15.10

1. $f(x, y) = \frac{1}{2}x^2y^2 + C$ **3.** $f(x, y) = xy + C$ **5.** not a gradient   **7.** $f(x, y) = \sin x + y\cos x + C$ **9.** $f(x, y) = e^x\cos y^2 + C$

11. $f(x, y) = xye^x + e^{-y} + C$ **13.** not a gradient   **15.** $f(x, y) = x + xy^2 + \frac{1}{2}x^2y^2 + \frac{1}{2}y^2 + y + C$ **17.** $f(x, y) = \sqrt{x^2 + y^2} + C$

19. $f(x, y) = \frac{1}{3}x^3\sin^{-1}y + y - y\ln y + C$ **21.** $f(x, y) = Ce^{x+y}$ **23.** (a), (b), (c) routine; (d) $f(x, y, z) = x^2 + yz + C$

25. $f(x, y, z) = x^2 + y^2 - z^2 + xy + yz + C$ **27.** $f(x, y, z) = xy^2z^3 + x + \frac{1}{2}y^2 + z + C$ **29.** $\mathbf{F(r)} = \nabla\left(G\dfrac{mM}{r}\mathbf{r}\right)$

CHAPTER 16

SECTION 16.1

1. 819   **3.** 0   **5.** $a_2 - a_1$ **7.** $(a_2 - a_1)(b_2 - b_1)$ **9.** $a_2^2 - a_1^2$ **11.** $(a_2^2 - a_1^2)(b_2 - b_1)$

13. $2n(a_2 - a_1) - 3m(b_2 - b_1)$ **15.** $(a_2 - a_1)(b_2 - b_1)(c_2 - c_1)$ **17.** $a_{111} + a_{222} + \cdots + a_{nnn} = \displaystyle\sum_{p=1}^{n} a_{ppp}$

19. $\displaystyle\sum_{i=1}^{m}\sum_{j=1}^{n}\alpha a_{ij} = \sum_{i=1}^{m}\alpha\left(\sum_{j=1}^{n}a_{ij}\right) = \alpha\sum_{i=1}^{m}\sum_{j=1}^{n}a_{ij}$ **21.** $\displaystyle\sum_{i=1}^{m}\sum_{j=1}^{n}b_ic_j = \sum_{i=1}^{m}b_i\left(\sum_{j=1}^{n}c_j\right) = \left(\sum_{j=1}^{n}c_j\right)\sum_{i=1}^{m}b_i = \left(\sum_{i=1}^{m}b_i\right)\left(\sum_{j=1}^{n}c_j\right)$

SECTION 16.2

1. $L_f(P) = 2\frac{1}{4}, \quad U_f(P) = 5\frac{3}{4}$ **3.** (a) $L_f(P) = \displaystyle\sum_{i=1}^{m}\sum_{j=1}^{n}(x_{i-1} + 2y_{i-1})\,\Delta x_i\,\Delta y_j, \quad U_f(P) = \sum_{i=1}^{m}\sum_{j=1}^{n}(x_i + 2y_j)\,\Delta x_i\,\Delta y_j$

(b) $I = 4$; the volume of the prism bounded above by the plane $z = x + 2y$ and below by R

5. $L_f(P) = -\frac{7}{24}, \quad U_f(P) = \frac{7}{24}$ **7.** (a) $L_f(P) = \displaystyle\sum_{i=1}^{m}\sum_{j=1}^{n}4x_{i-1}\,y_{j-1}\,\Delta x_i\,\Delta y_j, \quad U_f(P) = \sum_{i=1}^{m}\sum_{j=1}^{n}4x_iy_j\Delta x_i\Delta y_j$ (b) $I = b^2d^2$

9. (a) $L_f(P) = \displaystyle\sum_{i=1}^{m}\sum_{j=1}^{n}3(x_{i-1}^2 - y_j^2)\,\Delta x_i\,\Delta y_j, \quad U_f(P) = \sum_{i=1}^{m}\sum_{j=1}^{n}3(x_i^2 - y_{j-1}^2)\,\Delta x_i\,\Delta y_j$ (b) $I = bd(b^2 - d^2)$

11. $0 \le \sin(x + y) \le 1$ for all $(x, y) \in R$. Thus $0 \le \displaystyle\iint_R \sin(x + y)\,dx\,dy \le \iint_R dx\,dy = 1$.

SECTION 16.3

1. $\displaystyle\iint_\Omega dx\,dy = \int_a^b \phi(x)\,dx$

3. Suppose $f(x_0, y_0) \ne 0$. Assume $f(x_0, y_0) > 0$. Since f is continuous, there exists a disc Ω_ϵ with radius ϵ centered at (x_0, y_0) such that $f(x, y) > 0$ on Ω_ϵ. Let R be a rectangle contained in Ω_ϵ. Then $\displaystyle\iint_R f(x, y)\,dx\,dy > 0$, which contradicts the hypothesis.

5. 6

7. By Theorem 16.3.5, there exists a point (x_1, y_1) in D_r such that

$$\iint_{D_r} f(x, y)\,dx\,dy = f(x_1, y_1)\iint_{D_r} dx\,dy = f(x_1, y_1)\pi r^2$$

Thus $f(x_1, y_1) = \dfrac{1}{\pi r^2}\displaystyle\iint_{D_r} f(x, y)\,dx\,dy$. As $r \to 0$, $(x_1, y_1) \to (x_0, y_0)$ and $f(x_1, y_1) \to f(x_0, y_0)$ since f is continuous. The result follows.

SECTION 16.4

1. 1   **3.** $\frac{9}{2}$ **5.** $\frac{1}{24}$ **7.** -2 **9.** $\frac{1}{4}\pi^2 + \frac{1}{64}\pi^4$ **11.** $\frac{2}{27}$ **13.** $\frac{512}{15}$ **15.** 0   **17.** $\frac{1}{4}(e^4 - 1)$

19. $\int_0^1 \int_{y^{1/2}}^{y^{1/4}} f(x, y)\, dx\, dy$

21. $\int_{-1}^0 \int_{-x}^1 f(x, y)\, dy\, dx + \int_0^1 \int_x^1 f(x, y)\, dy\, dx$

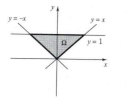

23. $\int_1^2 \int_1^y f(x, y)\, dx\, dy + \int_2^4 \int_{y/2}^y f(x, y)\, dx\, dy + \int_4^8 \int_{y/2}^4 f(x, y)\, dx\, dy$ **25.** 9   **27.** $\frac{1}{160}$

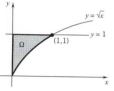

29. $\frac{2}{3}(\cos \frac{1}{2} - \cos 1)$ **31.** $1 - \ln 2$ **33.** $\ln 4 - \frac{1}{2}$ **35.** 4   **37.** $\frac{2}{15}$

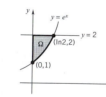

39. 3π **41.** $\frac{1}{6}$ **43.** $\frac{11}{70}$ **45.** $\frac{2}{3}a^3$ **47.** $\frac{1}{2}(e - 1)$ **49.** $\frac{1}{12}(e - 1)$ **51.** $\frac{2}{3}$ **53.** 1

55. $\displaystyle\iint_R f(x)g(y)\, dx\, dy = \int_c^d \int_a^b f(x)g(y)\, dx\, dy = \int_c^d \left(\int_a^b f(x)g(y)\, dx \right) dy = \int_c^d g(y) \left(\int_a^b f(x)\, dx \right) dy$

$$= \left(\int_a^b f(x)\, dx \right) \left(\int_c^d g(y)\, dy \right)$$

57. We have R: $-a \le x \le a,\, c \le y \le d$. Set $f(x, y) = g_y(x)$. For each fixed $y \in [c, d]$, g_y is an odd function. Thus

$$\int_{-a}^a g_y(x)\, dx = 0$$

Therefore

$$\iint_R f(x, y)\, dx\, dy = \int_c^d \int_{-a}^a f(x, y)\, dx\, dy = \int_c^d \int_{-a}^a g_y(x)\, dx\, dy = \int_c^d 0\, dy = 0.$$

59. Note that $\Omega = \{(x, y):\ 0 \le x \le y,\ 0 \le y \le 1\}$. Set $\Omega' = \{(x, y):\ 0 \le y \le x,\ 0 \le x \le 1\}$.

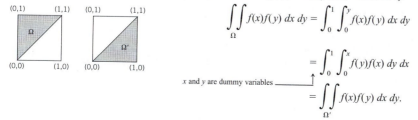

$$\iint_\Omega f(x)f(y)\, dx\, dy = \int_0^1 \int_0^y f(x)f(y)\, dx\, dy$$

$$= \int_0^1 \int_0^x f(y)f(x)\, dy\, dx$$

x and y are dummy variables

$$= \iint_{\Omega'} f(x)f(y)\, dx\, dy.$$

Note that Ω and Ω' don't overlap and their union is the unit square R: $\{(x, y): 0 \le x \le 1, \ 0 \le y \le 1\}$. If $\int_0^1 f(x)\, dx = 0$, then

$$0 = \left(\int_0^1 f(x)\, dx\right)\left(\int_0^1 f(y)\, dy\right) = \iint_R f(x)f(y)\, dx\, dy$$

by Exercise 55

$$= \iint_\Omega f(x)f(y)\, dx\, dy + \iint_{\Omega'} f(x)f(y)\, dx\, dy$$

$$= 2\iint_\Omega f(x)f(y)\, dx\, dy$$

and therefore $\displaystyle\iint_\Omega f(x)f(y)\, dx\, dy = 0$.

61. Let M be the maximum value of $|f(x, y)|$ on Ω.

$$\int_{\phi_1(x+h)}^{\phi_2(x+h)} = \int_{\phi_1(x+h)}^{\phi_1(x)} + \int_{\phi_1(x)}^{\phi_2(x)} + \int_{\phi_2(x)}^{\phi_2(x+h)}$$

$$|F(x+h) - F(x)| = \left|\int_{\phi_1(x+h)}^{\phi_2(x+h)} f(x, y)\, dy - \int_{\phi_1(x)}^{\phi_2(x)} f(x, y)\, dy\right|$$

$$= \left|\int_{\phi_1(x+h)}^{\phi_1(x)} f(x, y)\, dy + \int_{\phi_2(x)}^{\phi_2(x+h)} f(x, y)\, dy\right| \le \left|\int_{\phi_1(x+h)}^{\phi_1(x)} f(x, y)\, dy\right| + \left|\int_{\phi_2(x)}^{\phi_2(x+h)} f(x, y)\, dy\right|$$

$$\le |\phi_1(x) - \phi_1(x+h)|M + |\phi_2(x+h) - \phi_2(x)|M.$$

The expression on the right tends to 0 as h tends to 0 since ϕ_1 and ϕ_2 are continuous.

SECTION 16.5

1. $\frac{1}{6}$ **3.** 6   **5.** (a) $\pi \sin 1 \cong 0.84\pi$ (b) $\pi(\sin 2 - \sin 1) \cong 0.07\pi$ **7.** (a) $\frac{2}{3}$ (b) $\frac{14}{3}$ **9.** $\frac{1}{3}\pi$ **11.** $\frac{1}{15}\pi$ **13.** $\frac{\pi}{4}(1 - \cos 1)$

15. $\frac{\pi}{2}$ **17.** $\frac{3\pi}{4}$ **19.** $\frac{4\pi}{3} + 2\sqrt{3}$ **21.** 4   **23.** $b^3\pi$ **25.** $\frac{16}{3}\sqrt{3}\pi$ **27.** $\frac{2}{3}(8 - 3\sqrt{3})\pi$ **29.** 2π **31.** $\frac{1}{3}\pi a^2 b$

SECTION 16.6

1. $M = \frac{2}{3}$, $x_M = 0$, $y_M = \frac{1}{2}$ **3.** $M = \frac{1}{6}$, $x_M = \frac{12}{21}$, $y_M = \frac{3}{4}$ **5.** $M = \frac{32}{3}$, $x_M = \frac{16}{3}$, $y_M = \frac{9}{7}$ **7.** $M = \frac{5}{8}$, $x_M = \frac{4}{5}$, $y_M = \frac{152}{75}$

9. $M = \frac{5\pi}{3}$, $x_M = \frac{21}{20}$, $y_M = 0$

11. $I_x = \frac{1}{12}MW^2$, $I_y = \frac{1}{12}ML^2$, $I_z = \frac{1}{12}M(L^2 + W^2)$; $K_x = \frac{1}{6}\sqrt{3}W$, $K_y = \frac{1}{6}\sqrt{3}L$, $K_z = \frac{1}{6}\sqrt{3}\sqrt{L^2 + W^2}$ **13.** $x_M = \frac{1}{6}L$, $y_M = 0$

15. $I_x = I_y = \frac{1}{4}MR^2$, $I_z = \frac{1}{2}MR^2$; $K_x = K_y = \frac{1}{2}R$, $K_z = \frac{1}{2}\sqrt{2}R$ **17.** center the disc at a distance $\sqrt{I_0 - \frac{1}{2}MR^2}/\sqrt{M}$ from the origin

19. $I_x = \frac{1}{4}Mb^2$, $I_y = \frac{1}{4}Ma^2$, $I_z = \frac{1}{4}M(a^2 + b^2)$ **21.** $I_x = \frac{1}{10}$, $I_y = \frac{1}{16}$, $I_z = \frac{13}{80}$ **23.** $I_x = \frac{33\pi}{40}$, $I_y = \frac{93\pi}{40}$, $I_z = \frac{63\pi}{20}$

25. (a) $\frac{1}{4}M(r_2^2 + r_1^2)$ (b) $\frac{1}{4}M(r_2^2 + 5r_1^2)$ (c) $\frac{1}{4}M(5r_2^2 + r_1^2)$ **27.** $\frac{1}{2}M(r_2^2 + r_1^2)$ **29.** $x_M = 0$, $y_M = R/\pi$

31. on the diameter through P at a distance $\frac{6}{5}R$ from P

33. Suppose Ω, a basic region of area A, is broken up into n basic regions $\Omega_1, \ldots, \Omega_n$ with areas $A_1, \ldots, A_n$. Then

$$\bar{x}A = \iint_\Omega x\, dx\, dy = \sum_{i=1}^n \left(\iint_{\Omega_i} x\, dx\, dy\right) = \sum_{i=1}^n \bar{x}_i A_i = \bar{x}_1 A_1 + \cdots + \bar{x}_n A_n.$$

The second formula follows in the same manner.

SECTION 16.7

1. they are equal   **3.** $\displaystyle\iiint_\Pi \alpha\, dx\, dy\, dz = \alpha\iiint_\Pi dx\, dy\, dz = \alpha(\text{volume of } \Pi) = \alpha(a_2 - a_1)(b_2 - b_1)(c_2 - c_1)$ **5.** $\frac{1}{8}a^2b^2c$

7. $\bar{x} = \dfrac{A^2BC - a^2bc}{ABC - abc}$, $\bar{y} = \dfrac{AB^2C - ab^2c}{ABC - abc}$, $\bar{z} = \dfrac{ABC^2 - abc^2}{ABC - abc}$

9. $M = \frac{1}{2}ka^4$ where k is the constant of proportionality for the density function   **11.** $I_z = \frac{2}{3}Ma^2$

SECTION 16.8

1. abc **3.** $\frac{2}{3}$ **5.** 16   **7.** $\frac{1}{3}$ **9.** $\frac{47}{24}$ **11.**
$$\iiint_{\Pi} f(x)g(y)h(z)\,dx\,dy\,dz = \int_{c_1}^{c_2}\left[\int_{b_1}^{b_2}\left(\int_{a_1}^{a_2}f(x)g(y)h(z)\,dx\right)dy\right]dz$$
$$= \int_{c_1}^{c_2}\left[\int_{b_1}^{b_2}g(y)h(z)\left(\int_{a_1}^{a_2}f(x)\,dx\right)dy\right]dz$$
$$= \int_{c_1}^{c_2}\left[h(z)\left(\int_{a_1}^{a_2}f(x)\,dx\right)\left(\int_{b_1}^{b_2}g(y)\,dy\right)dz\right]$$
$$= \left(\int_{a_1}^{a_2}f(x)\,dx\right)\left(\int_{b_1}^{b_2}g(y)\,dy\right)\left(\int_{c_1}^{c_2}h(z)\,dz\right)$$

13. 8

15. $(\frac{2}{3}a, \frac{2}{3}b, \frac{2}{3}c)$ **17.**

19. $(\frac{1}{2}, \frac{1}{3}, \frac{1}{3})$

21. $\displaystyle\int_{-r}^{r}\int_{-\sqrt{r^2-x^2}}^{\sqrt{r^2-x^2}}\int_{-\sqrt{r^2-(x^2+y^2)}}^{\sqrt{r^2-(x^2+y^2)}}k(r-\sqrt{x^2+y^2+z^2})\,dz\,dy\,dx$ **23.** $\displaystyle\int_{0}^{1}\int_{-\sqrt{x-x^2}}^{\sqrt{x-x^2}}\int_{-2x-3y-10}^{1-y^2}dz\,dy\,dx$

25. $\displaystyle\int_{-1}^{1}\int_{-2\sqrt{2-2x^2}}^{2\sqrt{2-2x^2}}\int_{3x^2+y^2/4}^{4-x^2-y^2/4}k(z-3x^2-\frac{1}{4}y^2)\,dz\,dy\,dx$ **27.** $\frac{28}{3}$ **29.** $\frac{1}{270}$ **31.** $\frac{12}{5}$ **33.** $V=\frac{8}{3}$, $(\frac{11}{10}, \frac{9}{4}, \frac{11}{20})$

35. $V=\frac{27}{2}$, $(\frac{1}{2}, \frac{3}{2}, \frac{12}{5})$ **37.** $V=\frac{1}{6}abc$, $(\frac{1}{4}a, \frac{1}{4}b, \frac{1}{4}c)$ **39.** (a) $\frac{1}{3}M(a^2+b^2)$ (b) $\frac{1}{12}M(a^2+b^2)$ (c) $\frac{1}{3}Ma^2+\frac{1}{12}Mb^2$

41. $M=\frac{1}{2}k$, $(\frac{7}{12}, \frac{34}{45}, \frac{37}{90})$ **43.** (a) 0 by symmetry (b) $\frac{4}{3}\pi a_4$ **45.** $8\displaystyle\int_{0}^{a}\int_{0}^{\sqrt{a^2-x^2}}\int_{0}^{\sqrt{a^2-x^2-y^2}}dz\,dy\,dz=\frac{4}{3}\pi a^3$ **47.** $M=\frac{128}{15}k$

49. $M=\frac{135}{4}k$, $(\frac{1}{2}, \frac{9}{5}, \frac{12}{5})$ **51.** (a) $V=\displaystyle\int_{0}^{6}\int_{z/2}^{3}\int_{x}^{6-x}dy\,dx\,dz$

(b) $V=\displaystyle\int_{0}^{3}\int_{0}^{2x}\int_{x}^{6-x}dy\,dz\,dx$

(c) $V=\displaystyle\int_{0}^{6}\int_{z/2}^{3}\int_{z/2}^{y}dx\,dy\,dz+\int_{0}^{6}\int_{3}^{(12-z)/2}\int_{z/2}^{6-y}dx\,dy\,dz$

53. (a) $V=\displaystyle\iint_{\Omega_{yz}}2y\,dy\,dz$ (b) $V=\displaystyle\iint_{\Omega_{yz}}\left(\int_{-y}^{y}dx\right)dy\,dz$ (c) $V=\displaystyle\int_{0}^{4}\int_{-\sqrt{4-y}}^{\sqrt{4-y}}\int_{-y}^{y}dx\,dz\,dy$ (d) $V=\displaystyle\int_{-2}^{2}\int_{0}^{4-z^2}\int_{-y}^{y}dx\,dy\,dz$

SECTION 16.9

1. $r^2+z^2=9$ **3.** $z=2r$ **5.** $4r^2=z^2$ **7.** 4π **9.** $\frac{1}{3}$

11. $M=\frac{1}{2}k\pi R^2h^2$ **13.** $\frac{1}{2}MR^2$ **15.** Inverting the cone and placing the vertex at the origin, we have

$$V=\int_{0}^{h}\int_{0}^{2\pi}\int_{0}^{(R/h)z}r\,dr\,d\theta\,dz=\frac{1}{3}\pi R^2h.$$

17. $\frac{3}{10}MR^2$ **19.** $\frac{1}{2}\pi$ **21.** $\frac{1}{4}k\pi$ **23.** $\frac{1}{6}(8-3\sqrt{3})\pi$ **25.** $\dfrac{9\pi^2}{8}$ **27.** $\dfrac{\pi}{2}(1-\cos 1)\cong 0.7221$ **29.** $\frac{32}{9}a^3$

31. $\frac{1}{36}a^2(9\pi - 16)$ **33.** $\frac{1}{32}\pi$ **35.** $\frac{1}{3}\pi(2 - \sqrt{3})$

SECTION 16.10

1. $(\sqrt{3}, \frac{1}{4}\pi, \cos^{-1}[\frac{1}{3}\sqrt{3}])$ **3.** $(\frac{3}{4}, \frac{3}{4}\sqrt{3}, \frac{3}{2}\sqrt{3})$ **5.** $(\rho, \theta, \phi) = \left(\frac{4\sqrt{6}}{3}, \frac{\pi}{4}, \frac{\pi}{3}\right)$ **7.** $(x, y, z) = (0, 0, 3)$

9. the circular cylinder $x^2 + y^2 = 1$; the radius of the cylinder is 1 and the axis is the z-axis   **11.** the lower nappe of the cone $z^2 = x^2 + y^2$

13. horizontal plane one unit above the xy-plane   **15.** $\frac{\pi}{3}$ **17.** $\frac{\pi}{2}$ **19.** $\frac{\pi}{3}(\sqrt{2} - 1)$ **21.** $\frac{243\pi}{20}$ **23.** $V = \frac{4}{3}\pi R^3$ **25.** $V = \frac{2}{3}\alpha R^3$

27. $M = \frac{1}{6}k\pi h[(r^2 + h^2)^{3/2} - h^3]$ **29.** (a) $\frac{2}{5}MR^2$ (b) $\frac{7}{5}MR^2$ **31.** (a) $\frac{2}{5}M\left(\frac{R_2^5 - R_1^5}{R_2^3 - R_1^3}\right)$ (b) $\frac{2}{3}MR^2$ (c) $\frac{5}{3}MR^2$

33. $V = \frac{2}{3}\pi(1 - \cos\alpha)a^3$ **35.** (a) $\rho = 2R\cos\phi$ (b) $0 \le \theta \le 2\pi, 0 \le \phi \le \frac{1}{4}\pi$, $R\sec\phi \le \rho \le 2R\cos\phi$ **37.** $V = \frac{1}{3}(16 - 6\sqrt{2})\pi$

39. Encase T in a spherical wedge W. W has spherical coordinates in a box Π that contains S. Define f to be zero outside of T. Then $F(\rho, \theta, \phi) = f(\rho\sin\phi\cos\theta, \rho\sin\phi\sin\theta, \rho\cos\phi)$ is zero outside of S and

$$\iiint_T f(x, y, z)\, dx\, dy\, dz = \iiint_W f(x, y, z)\, dx\, dy\, dz$$

$$= \iiint_\Pi F(\rho, \theta, \phi)\rho^2\sin\phi\, d\rho\, d\theta\, d\phi = \iiint_S F(\rho, \theta, \phi)\rho^2\sin\phi\, d\rho\, d\theta\, d\phi.$$

41. $\mathbf{F} = \dfrac{GmM}{R^2}(\sqrt{2} - 1)\mathbf{k}$

SECTION 16.11

1. $ad - bc$ **3.** $2(v^2 - u^2)$ **5.** $u^2v^2 - 4uv$ **7.** abc **9.** r **11.** $w(1 + w\cos v)$ **13.** $\frac{1}{2}$ **15.** 0   **17.** $\frac{2}{3}$

19. (a) $A = 3\ln 2$ (b) $\bar{x} = \dfrac{7}{9\ln 2}, \bar{y} = \dfrac{14}{9\ln 2}$ **21.** $I_x = \frac{4}{75}M, I_y = \frac{14}{75}M, I_z = \frac{18}{75}M$ **23.** $A = \frac{32}{15}$ **25.** $A = \pi/\sqrt{65}$

27. $J = ab\alpha r\cos^{\alpha-1}\theta\sin^{\alpha-1}\theta$ **29.** $V = \frac{4}{3}\pi abc$ **31.** $I_x = \frac{1}{3}M(b^2 + c^2), I_y = \frac{1}{3}M(a^2 + c^2), I_z = \frac{1}{3}M(a^2 + b^2)$ **33.** $\frac{1}{2}\pi^{3/2}$

CHAPTER 17

SECTION 17.1

1. (a) 1 (b) -2 **3.** 0   **5.** (a) $-\frac{17}{6}$ (b) $\frac{17}{6}$ **7.** -8 **9.** $-\pi$ **11.** (a) 1 (b) $\frac{23}{21}$

13. (a) $2 + \sin 2 - \cos 3$ (b) $\frac{4}{3} + \sin 1 - \cos 1$ **15.** $\frac{1}{3}$ **17.** (a) 28 (b) $\frac{1}{6}$ **19.** $\dfrac{8\pi^3}{3}$

21. $\displaystyle\int_C \mathbf{q} \cdot d\mathbf{r} = \int_a^b [\mathbf{q} \cdot \mathbf{r}'(u)]\, du + \int_a^b \frac{d}{du}[\mathbf{q} \cdot \mathbf{r}(u)]\, du = [\mathbf{q} \cdot \mathbf{r}(b)] - [\mathbf{q} \cdot \mathbf{r}(a)] = \mathbf{q} \cdot [\mathbf{r}(b) - \mathbf{r}(a)]$

$\displaystyle\int_C \mathbf{r} \cdot d\mathbf{r} = \int_a^b [\mathbf{r}(u) \cdot \mathbf{r}'(u)]\, du = \frac{1}{2}\int_a^b \frac{d}{du}[\mathbf{r}(u) \cdot \mathbf{r}(u)]\, du = \frac{1}{2}\int_a^b \frac{d}{du}(\|\mathbf{r}(u)\|^2)\, du = \frac{1}{2}(\|\mathbf{r}(b)\|^2 - \|\mathbf{r}(a)\|^2)$

23. $\displaystyle\int_C \mathbf{f}(\mathbf{r}) \cdot d\mathbf{r} = \int_a^b [\mathbf{f}(\mathbf{r}(u)) \cdot \mathbf{r}'(u)]\, du = \int_a^b [f(u)\mathbf{i} \cdot \mathbf{i}]\, du = \int_a^b f(u)\, du$ **25.** $|W| = $ area of ellipse

27. force at time $t = m\mathbf{r}''(t) = m(2\beta\mathbf{j} + 6\gamma t\mathbf{k})$; $W = (2\beta^2 + \frac{9}{2}\gamma^2)m$ **29.** 0   **31.** (a) $\left(\dfrac{1}{\sqrt{5}} - \dfrac{1}{\sqrt{14}}\right)\mathbf{k}$ (b) $\dfrac{4}{5}\mathbf{k}$ **33.** $\alpha = \dfrac{15}{6}$

SECTION 17.2

1. 0   **3.** -1 **5.** 0   **7.** 0   **9.** 0   **11.** 4   **13.** $e^3 - 2e^2 + 3$ **15.** $e^5 - 2e^2 + 1$ **17.** 2π **19.** 0

21. Set $f(x, y, z) = g(x)$ and $C: \mathbf{r}(u) = u\mathbf{i}, \quad u \in [a, b]$. In this case, $\nabla f(\mathbf{r}(u)) = g'(x(u))\mathbf{i} = g'(u)\mathbf{i}$ and $\mathbf{r}'(u) = \mathbf{i}$, so that

$$\int_C \nabla f(\mathbf{r}) \cdot d\mathbf{r} = \int_a^b [\nabla f(\mathbf{r}(u)) \cdot \mathbf{r}'(u)]\, du = \int_a^b g'(u)\, du$$

and

$$f(\mathbf{r}(b)) - f(\mathbf{r}(a)) = g(b) - g(a).$$

The statement $\displaystyle\int_C \nabla f(\mathbf{r}) \cdot d\mathbf{r} = f(\mathbf{r}(b)) - f(\mathbf{r}(a))$ reduces to $\displaystyle\int_a^b g'(u)\, du = g(b) - g(a).$

23. $W = mG\left(\dfrac{1}{r_2} - \dfrac{1}{r_1}\right)$ **25.** $f(x, y, z) = \dfrac{mGr_0^2}{r_0 + z}$ **27.** $7575.6G$

SECTION 17.3

1. If f is continuous, then $-f$ is continuous and has antiderivatives u. The scalar fields $U(x, y, z) = u(x)$ are potential functions for $\mathbf{F}$:

$$\nabla U = \frac{\partial U}{\partial x}\mathbf{i} + \frac{\partial U}{\partial y}\mathbf{j} + \frac{\partial U}{\partial z}\mathbf{k} = \frac{du}{dx}\mathbf{i} = -f\mathbf{i} = -\mathbf{F}.$$

3. The scalar field $U(x, y, z) = cz + d$ is a potential energy function for $\mathbf{F}$. We know that the total mechanical energy remains constant. Thus, for any times t_1 and t_2,

$$\tfrac{1}{2}m[v(t_1)]^2 + U(\mathbf{r}(t_1)) = \tfrac{1}{2}m[v(t_2)]^2 + U(\mathbf{r}(t_2)).$$

This gives

$$\tfrac{1}{2}m[v(t_1)]^2 + cz(t_1) + d = \tfrac{1}{2}m[v(t_2)]^2 + cz(t_2) + d.$$

Solve this equation for $v(t_2)$ and you have the desired formula.

5. (a) We know that $-\nabla U$ points in the direction of maximum decrease of U. Thus $\mathbf{F} = -\nabla U$ attempts to drive objects toward a region where U has lower values.   (b) At a point where U has a minimum, $\nabla U = \mathbf{0}$ and therefore $\mathbf{F} = \mathbf{0}$.

7. (a) By conservation of energy $\tfrac{1}{2}mv^2 + U = E$. Since E is constant and U is constant, v is constant.
(b) ∇U is perpendicular to any surface where U is constant. Obviously so is $\mathbf{F} = -\nabla U$.

9. $f(x, y, z) = -\dfrac{k}{\sqrt{x^2 + y^2 + z^2}}$ is a potential function for $\mathbf{F}$. The work done by $\mathbf{F}$ moving an object along C is $W = \displaystyle\int_C \mathbf{F}(\mathbf{r}) \cdot d\mathbf{r} = \int_a^b \nabla f \cdot d\mathbf{r} = f(\mathbf{r}(b)) - f(\mathbf{r}(a))$. Since $\mathbf{r}(a) = (x_0, y_0, z_0)$ and $\mathbf{r}(b) = (x_1, y_1, z_1)$ are points on the unit sphere, $f(\mathbf{r}(b)) = f(\mathbf{r}(a)) = -k$, and so $W = 0$.

SECTION 17.4

1. $\tfrac{1}{2}$ **3.** $\tfrac{9}{2}$ **5.** $\tfrac{11}{6}$ **7.** 2   **9.** 16   **11.** $\tfrac{104}{5}$ **13.** $\tfrac{1}{12}$ **15.** $\tfrac{1}{3} - \tfrac{\pi}{16}$ **17.** 4   **19.** 4   **21.** 2   **23.** 3   **25.** $\tfrac{176}{3}$

27. 32   **29.** $\tfrac{1177}{30}$ **31.** (a) $\dfrac{\partial P}{\partial y} = 6x - 4y = \dfrac{\partial Q}{\partial x}$ (b) 7 (c) $\dfrac{-37}{3}$

33. (a) $M = 2ka^2$, $x_M = y_M = \tfrac{1}{8}a(\pi + 2)$ (b) $I_x = ka^4 = \tfrac{1}{2}Ma^2$ **35.** (a) $I_z = 2ka^4 = Ma^2$ (b) $I = \tfrac{1}{3}ka^4 = \tfrac{1}{6}Ma^2$

37. (a) $L = 2\pi\sqrt{a^2 + b^2}$ (b) $x_M = y_M = 0$, $z_M = \pi b$ (c) $I_x = I_y = \tfrac{1}{6}M(3a^2\pi + 8b^2\pi^2)$, $I_z = Ma^2$ **39.** $M = \tfrac{2}{3}\pi k\sqrt{a^2 + b^2}(3a^2 + 4\pi^2 b^2)$

SECTION 17.5

1. $\tfrac{1}{6}$ **3.** π **5.** 2π **7.** $a^2 b$ **9.** 7π **11.** $5a\pi r^2$ **13.** 0   **15.** 0   **17.** πa^2 **19.** $\dfrac{ab}{2}$ **21.** $(c - a)A$

23. $3\pi r^2$ **25.** $\dfrac{3}{4}\pi$

27. Taking Ω to be of type II, we have

$$\iint_\Omega \frac{\partial Q}{\partial x}(x, y)\, dx\, dy = \int_c^d \int_{\phi_3(y)}^{\phi_4(y)} \frac{\partial Q}{\partial x}(x, y)\, dx\, dy = \int_c^d \{Q[\phi_4(y), y] - Q[\phi_3(y), y]\}\, dy$$

$$* = \int_c^d Q[\phi_4(y), y]\, dy - \int_c^d Q[\phi_3(y), y]\, dy.$$

Set C_3: $\mathbf{r}_3(u) = \phi_3(u)\mathbf{i} + u\mathbf{j}$, $u \in [c, d]$ and C_4: $\mathbf{r}_4(u) = \phi_4(u)\mathbf{i} + u\mathbf{j}$, $u \in [c, d]$. Then

$$\oint_C Q(x, y)\, dy = \int_{C_4} Q(x, y)\, dy - \int_{C_3} Q(x, y)\, dy = \int_c^d Q[\phi_4(u), u]\, du - \int_c^d Q[\phi_3(u), u]\, du.$$

Comparison with $*$ proves the result.

29. $\displaystyle\oint_{C_1} = \oint_{C_2} + \oint_{C_3}$ **31.** (a) 0 (b) 0   **33.** If Ω is the region bounded by C, then

$$\oint_C \mathbf{v} \cdot d\mathbf{r} = \oint_C \frac{\partial\phi}{\partial x}\, dx + \frac{\partial\phi}{\partial y}\, dy = \iint_\Omega \left\{ \frac{\partial}{\partial x}\left(\frac{\partial\phi}{\partial y}\right) - \frac{\partial}{\partial y}\left(\frac{\partial\phi}{\partial x}\right) \right\}\, dx\, dy$$

is zero by equality of mixed partials.

35. $A = \dfrac{1}{2}\oint_C (-y\,dx + x\,dy)$

$= \dfrac{1}{2}\left[\displaystyle\int_{C_1} + \int_{C_2} + \cdots + \int_{C_n}\right]$

Now

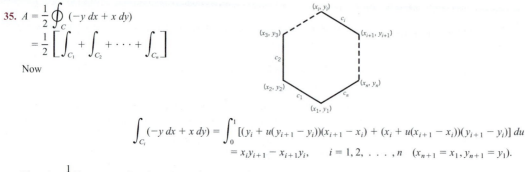

$$\int_{C_i}(-y\,dx + x\,dy) = \int_0^1 [(y_i + u(y_{i+1} - y_i))(x_{i+1} - x_i) + (x_i + u(x_{i+1} - x_i))(y_{i+1} - y_i)]\,du$$

$$= x_i y_{i+1} - x_{i+1}y_i, \qquad i = 1, 2, \ldots, n \quad (x_{n+1} = x_1, y_{n+1} = y_1).$$

Thus $A = \dfrac{1}{2}[(x_1y_2 - x_2y_1) + (x_2y_3 - x_3y_2) + \cdots + (x_ny_1 - x_1y_n)]$.

SECTION 17.6

1. $4[(u^2 - v^2)\mathbf{i} - (u^2 + v^2)\mathbf{j} + 2uv\mathbf{k}]$ **3.** $2(\mathbf{j} - \mathbf{i})$ **5.** $\mathbf{r}(u, v) = 3\cos u \cos v\,\mathbf{i} + 2\sin u \cos v\,\mathbf{j} + 6\sin v\,\mathbf{k}$, $u \in [0, 2\pi]$, $v \in [0, \pi/2]$

7. $\mathbf{r}(u, v) = 2\cos u \cos v\,\mathbf{i} + 2\sin u \cos v\,\mathbf{j} + 2\sin v\,\mathbf{k}$, $u \in [0, 2\pi]$, $v \in (\pi/4, \pi/2]$ **9.** $\mathbf{r}(u, v) = u\,\mathbf{i} + g(u, v)\mathbf{j} + v\,\mathbf{k}$, $(u, v) \in \Omega$

11. $x^2/a^2 + y^2/b^2 + z^2/c^2 = 1$; ellipsoid   **13.** $x^2/a^2 - y^2/b^2 = z$; hyperbolic paraboloid

15. $\mathbf{r}(u, v) = v\cos u\,\mathbf{i} + v\sin u\,\mathbf{j} + f(v)\mathbf{k}$; $0 \le u \le 2\pi$, $a \le v \le b$ **17.** area of $\Gamma = A_\Omega \sec\gamma$ **19.** $\frac{1}{2}\sqrt{a^2b^2 + a^2c^2 + b^2c^2}$

21. $\frac{1}{6}\pi(17\sqrt{17} - 1)$ **23.** $\frac{1}{6}\pi[(4a^2 + 1)^{3/2} - (a^2 + 1)^{3/2}]$ **25.** $\frac{1}{15}(36\sqrt{6} - 50\sqrt{5} + 32)$ **27.** 4π

29. (a) $\displaystyle\iint_\Omega \sqrt{\left[\dfrac{\partial g}{\partial y}(y, z)\right]^2 + \left[\dfrac{\partial g}{\partial z}(y, z)\right]^2 + 1}\,dy\,dz = \iint_\Omega \sec\,[\alpha(y, z)]\,dy\,dz$

where α is the angle between the unit normal with positive $\mathbf{i}$ component and the x-axis

(b) $\displaystyle\iint_\Omega \sqrt{\left[\dfrac{\partial h}{\partial x}(x, z)\right]^2 + \left[\dfrac{\partial h}{\partial z}(x, z)\right]^2 + 1}\,dx\,dz = \iint_\Omega \sec\,[\beta(x, z)]\,dx\,dz$

where β is the angle between the unit normal with positive $\mathbf{j}$ component and the y-axis

31. (a) $\mathbf{N}(u, v) = v\cos u\sin\alpha\cos\alpha\,\mathbf{i} + v\sin u\sin\alpha\cos\alpha\,\mathbf{j} - v\sin^2\alpha\,\mathbf{k}$ (b) $A = \pi s^2\sin\alpha$

33. $A = \sqrt{A_1^2 + A_2^2 + A_3^2}$ **35.** (a) $\frac{1}{4}\sqrt{2}\pi[\sqrt{6} + \ln(\sqrt{2} + \sqrt{3})]$ (b) $\frac{1}{2}a^2[\sqrt{2e^{4\pi} + 1} - \sqrt{3} + 2\pi + \ln(1 + \sqrt{3}) - \ln(1 + \sqrt{2e^{4x} + 1})]$

SECTION 17.7

1. $\frac{1}{2}[\sqrt{2} + \ln(1 + \sqrt{2})]$ **3.** $2\sqrt{2} - 1$ **5.** $\frac{1}{3}[2\sqrt{2} - 1]$ **7.** $\dfrac{9\sqrt{14}}{2}$ **9.** $\dfrac{4}{3}$ **11.** $\dfrac{4\pi}{3}$ **13.** $\frac{1}{2}\sqrt{3}a^2k$ **15.** $\frac{1}{12}\sqrt{3}a^4k$

17. $(0, 0, \frac{1}{2}a)$ **19.** 2   **21.** $\frac{4}{3}\pi a^3$ **23.** 0   **25.** $\frac{1}{2}a^3$ **27.** 0   **29.** $-\frac{3}{2}$ **31.** $2\pi la^2$ **33.** $\frac{8}{35}$ **35.** $-\frac{4}{63}$

37. $\bar{x} = \bar{y} = 0$, $\bar{z} = \frac{2}{3}s\cos\alpha$ **39.** $x_M = y_M = 0$, $z_M = \frac{3}{4}$ **41.** no answer required   **43.** $x_M = \frac{11}{9}$

45. Total flux out of the solid is 0. It is clear from a diagram that the outer unit normal to the cylindrical side of the solid is given by $\mathbf{n} = x\mathbf{i} + y\mathbf{j}$, in which case $\mathbf{v} \cdot \mathbf{n} = 0$. The outer unit normals to the top and bottom of the solid are $\mathbf{k}$ and $-\mathbf{k}$, respectively. So, here as well, $\mathbf{v} \cdot \mathbf{n} = 0$ and the total flux is 0.

47. $(4\sqrt{2} - \frac{7}{2})\pi$

SECTION 17.8

1. $\nabla \cdot \mathbf{v} = 2$, $\nabla \times \mathbf{v} = \mathbf{0}$ **3.** $\nabla \cdot \mathbf{v} = 0$, $\nabla \times \mathbf{v} = \mathbf{0}$ **5.** $\nabla \cdot \mathbf{v} = 6$, $\nabla \times \mathbf{v} = \mathbf{0}$

7. $\nabla \cdot \mathbf{v} = yz + 1$, $\nabla \times \mathbf{v} = -x\mathbf{i} + xy\mathbf{j} + (1 - x)z\mathbf{k}$ **9.** $\nabla \cdot \mathbf{v} = 1/r^2$, $\nabla \times \mathbf{v} = \mathbf{0}$

11. $\nabla \cdot \mathbf{v} = 2(x + y + z)e^{r^2}$, $\nabla \times \mathbf{v} = 2e^{r^2}[(y - z)\mathbf{i} - (x - z)\mathbf{j} + (x - y)\mathbf{k}]$ **13.** $\nabla \cdot \mathbf{v} = f'(x)$, $\nabla \times \mathbf{v} = \mathbf{0}$ **15.** use components

17. $\nabla \cdot \mathbf{F} = \dfrac{\partial P}{\partial x} + \dfrac{\partial Q}{\partial y} + \dfrac{\partial R}{\partial z} = 2 + 4 - 6 = 0$ **19.** $\nabla \times \mathbf{F} = \begin{vmatrix} \mathbf{i} & \mathbf{j} & \mathbf{k} \\ \dfrac{\partial}{\partial x} & \dfrac{\partial}{\partial y} & \dfrac{\partial}{\partial z} \\ x & y & -2z \end{vmatrix} = \mathbf{0}$ **21.** $\nabla^2 f = 12(x^2 + y^2 + z^2)$

23. $\nabla^2 f = 2y^3z^4 + 6x^2yz^4 + 12x^2y^3z^2$ **25.** $\nabla^2 f = e^r(1 + 2r^{-1})$ **27.** (a) $2r^2$ (b) $-1/r$

29. $\nabla^2 f = \nabla^2 g(r) = \nabla \cdot (\nabla g(r)) = \nabla \cdot (g'(r)r^{-1}\mathbf{r})$

$= [(\nabla g'(r)) \cdot r^{-1}\mathbf{r}] + g'(r)(\nabla \cdot r^{-1}\mathbf{r})$

$= \{[g''(r)r^{-1}\mathbf{r}] \cdot r^{-1}\mathbf{r}\} + g'(r)(2r^{-1}) = g''(r) + 2r^{-1}g'(r)$

33. $n = -1$

SECTION 17.9

1. $\displaystyle\iint_S (\mathbf{v} \cdot \mathbf{n}) \, d\sigma = \iiint_T (\mathbf{\nabla} \cdot \mathbf{v}) \, dx \, dy \, dz = \iiint_T 3 \, dx \, dy \, dz = 3V = 4\pi$

3. $\displaystyle\iint_S (\mathbf{v} \cdot \mathbf{n}) \, d\sigma = \iiint_T (\mathbf{\nabla} \cdot \mathbf{v}) \, dx \, dy \, dz = \iiint_T 2(x + y + z) \, dx \, dy \, dz.$

The flux is zero since the function $f(x, y, z) = 2(x + y + z)$ satisfies the relation $f(-x, -y, -z) = -f(x, y, z)$ and T is symmetric about the origin.

5.

Face	n	v · n	Flux
$x = 0$	$-\mathbf{i}$	0	0
$x = 1$	$\mathbf{i}$	1	1
$y = 0$	$-\mathbf{j}$	0	0
$y = 1$	$\mathbf{j}$	1	1
$z = 0$	$-\mathbf{k}$	0	0
$z = 1$	$\mathbf{k}$	1	1

total flux = 3

7.

Face	n	v · n	Flux
$x = 0$	$-\mathbf{i}$	0	0
$x = 1$	$\mathbf{i}$	1	1
$y = 0$	$-\mathbf{j}$	xz	
$y = 1$	$\mathbf{j}$	$-xz$	
$z = 0$	$-\mathbf{k}$	0	0
$z = 1$	$\mathbf{k}$	1	1

fluxes added up to 0   total flux = 2

$\displaystyle\iiint_T (\mathbf{\nabla} \cdot \mathbf{v}) \, dx \, dy \, dz = \iiint_T 3 \, dx \, dy \, dz = 3V = 3$

$\displaystyle\iiint_T (\mathbf{\nabla} \cdot \mathbf{v}) \, dx \, dy \, dz = \iiint_T 2(x + z) \, dx \, dy \, dz = 2(\bar{x} + \bar{z})V = 2(\tfrac{1}{2} + \tfrac{1}{2})1 = 2$

9. flux $= \displaystyle\iiint_T (1 + 4y + 6z) \, dx \, dy \, dz = (1 + 4\bar{y} + 6\bar{z})V = (1 + 0 + 3)9\pi = 36\pi$ **11.** $\frac{1}{24}$ **13.** 64π **15.** 0   **17.** $(A + B + C)V$

19. Let T be the solid enclosed by S and set $n = n_1\mathbf{i} + n_2\mathbf{j} + n_3\mathbf{k}$.

$$\iint_S n_1 d\sigma = \iint_S (\mathbf{i} \cdot \mathbf{n}) \, d\sigma = \iiint_T (\mathbf{\nabla} \cdot \mathbf{i}) \, dx \, dy \, dz = \iiint_T 0 \, dx \, dy \, dz = 0.$$

Similarly $\displaystyle\iint_S n_2 d\sigma = 0$ and $\displaystyle\iint_S n_3 d\sigma = 0.$

21. A routine computation shows that $\mathbf{\nabla} \cdot (\nabla f \times \nabla g) = 0.$ Therefore

$$\iint_S [(\nabla f \times \nabla g) \cdot \mathbf{n}] \, d\sigma = \iiint_T [\mathbf{\nabla} \cdot (\nabla f \times \nabla g)] \, dx \, dy \, dz = 0.$$

23. Set $\mathbf{F} = F_1\mathbf{i} + F_2\mathbf{j} + F_3\mathbf{k}.$

$$F_1 = \iint_S [\rho(z - c)\mathbf{i} \cdot \mathbf{n}] \, d\sigma = \iiint_T [\mathbf{\nabla} \cdot \rho(z - c)\mathbf{i}] \, dx \, dy \, dz = \iiint_T \underbrace{\frac{\partial}{\partial x}[\rho(z - c)]}_{0} \, dx \, dy \, dz = 0.$$

Similarly $F_2 = 0.$

$$F_3 = \iint_S [\rho(z - c)\mathbf{k} \cdot \mathbf{n}] \, d\sigma = \iiint_T [\mathbf{\nabla} \cdot \rho(z - c)\mathbf{k}] \, dx \, dy \, dz$$

$$= \iiint_T \frac{\partial}{\partial z}[(\rho(z - c)] \, dx \, dy \, dz = \iiint_T \rho \, dx \, dy \, dz = W.$$

SECTION 17.10

For Exercises 1 and 3: $\mathbf{n} = x\mathbf{i} + y\mathbf{j} + z\mathbf{k}$ and C: $\mathbf{r}(u) = \cos u\mathbf{i} + \sin u\mathbf{j}$, $u \in [0, 2\pi]$.

1. (a) $\displaystyle\iint_S [(\mathbf{\nabla} \times \mathbf{v}) \cdot \mathbf{n}] \, d\sigma = \iint_S (\mathbf{0} \cdot \mathbf{n}) \, d\sigma = 0.$

(b) S is bounded by the unit circle C: $\mathbf{r}(u) = \cos u\mathbf{i} + \sin u\mathbf{j}$, $u \in [0, 2\pi]$.

$\displaystyle\oint_C \mathbf{v}(\mathbf{r}) \cdot \mathbf{dr} = 0$ since $\mathbf{v}$ is a gradient.

3. (a) $\displaystyle\iint_S [(\nabla \times \mathbf{v}) \cdot \mathbf{n}]\,d\sigma = \iint_S [(-3y^2\mathbf{i} + 2z\mathbf{j} + 2\mathbf{k}) \cdot (x\mathbf{i} + y\mathbf{j} + z\mathbf{k})]\,d\sigma$

$\displaystyle = \iint_S (-3xy^2 + 2yz + 2z)\,d\sigma$

$\displaystyle = \underbrace{\iint_S (-3xy^2)\,d\sigma}_{\substack{0 \text{ by symmetry}}} + \underbrace{\iint_S 2yz\,d\sigma}_{\substack{0 \text{ by} \\ \text{symmetry}}} + \underbrace{\iint_S 2z\,d\sigma = 2\pi}_{\substack{2\bar{z}A = 2\pi \\ \text{by Ex. 17,} \\ \text{Section 17.7}}}$

(b) $\displaystyle\oint_C \mathbf{v}(\mathbf{r}) \cdot d\mathbf{r} = \oint_C z^2\,dx + 2x\,dy = \oint_C 2x\,dy = \int_0^{2\pi} 2\cos^2 u\,du = 2\pi$

For Exercises 5 and 7 take $S: z = 2 - x - y$ with $0 \le x \le 2, 0 \le y \le 2 - x$ and C as the triangle $(2, 0, 0)$, $(0, 2, 0)$, $(0, 0, 2)$. Then $C = C_1 \cup C_2 \cup C_3$ with

$C_1: \mathbf{r}_1(u) = 2(1 - u)\mathbf{i} + 2u\mathbf{j}, u \in [0, 1]$,
$C_2: \mathbf{r}_2(u) = 2(1 - u)\mathbf{j} + 2u\mathbf{k}, u \in [0, 1]$,
$C_3: \mathbf{r}_3(u) = 2(1 - u)\mathbf{k} + 2u\mathbf{i}, u \in [0, 1]$.

$\mathbf{n} = \tfrac{1}{3}\sqrt{3}(\mathbf{i} + \mathbf{j} + \mathbf{k})$.

5. (a) $\displaystyle\iint_S [(\nabla \times \mathbf{v}) \cdot \mathbf{n}]\,d\sigma = \iint_S \tfrac{1}{3}\sqrt{3}\,d\sigma = \tfrac{1}{3}\sqrt{3}A = 2$

(b) $\displaystyle\oint_C \mathbf{v}(r) \cdot d\mathbf{r} = \left(\int_{C_1} + \int_{C_2} + \int_{C_3}\right)\mathbf{v}(r) \cdot d\mathbf{r} = -2 + 2 + 2 = 2$

7. (a) $\displaystyle\iint_S [(\nabla \times \mathbf{v}) \cdot \mathbf{n}]\,d\sigma = \iint_S [y\mathbf{k} \cdot \tfrac{1}{3}\sqrt{3}(\mathbf{i} + \mathbf{j} + \mathbf{k})]\,d\sigma = \tfrac{1}{3}\sqrt{3}\iint_S y\,d\sigma = \tfrac{1}{3}\sqrt{3}\bar{y}A = \tfrac{4}{3}$ **9.** 4π **11.** 0 **13.** $\pm\tfrac{1}{8}\pi$ **15.** $\pm\tfrac{1}{4}\pi$

(b) $\displaystyle\oint_C \mathbf{v}(\mathbf{r}) \cdot d\mathbf{r} = \left(\int_{C_1} + \int_{C_2} + \int_{C_3}\right)\mathbf{v}(\mathbf{r}) \cdot d\mathbf{r} = (\tfrac{4}{3} - \tfrac{32}{5}) + \tfrac{32}{5} + 0 = \tfrac{4}{3}$

17. Straightforward calculation shows that
$$\nabla \times (\mathbf{a} \times \mathbf{r}) = \nabla \times [(a_2z - a_3y)\mathbf{i} + (a_3x - a_1z)\mathbf{j} + (a_1y - a_2x)\mathbf{k}] = 2\mathbf{a}.$$

19. In the plane of C, the curve C bounds some Jordan region that we call Ω. The surface $S \cup \Omega$ is a piecewise-smooth surface that bounds a solid T. Note that $\nabla \times \mathbf{v}$ is continuously differentiable on T. Thus, by the divergence theorem,

$$\iiint_T [\nabla \cdot (\nabla \times \mathbf{v})]\,dx\,dy\,dz = \iint_{S\cup\Omega} [(\nabla \times \mathbf{v}) \cdot \mathbf{n}]\,d\sigma$$

where $\mathbf{n}$ is the outer unit normal. Since the divergence of a curl is identically zero, we have

$$\iint_{S\cup\Omega} [(\nabla \times \mathbf{v}) \cdot \mathbf{n}]\,d\sigma = 0.$$

Now $\mathbf{n}$ is $\mathbf{n}_1$ on S and $\mathbf{n}_2$ on Ω. Thus

$$\iint_S [(\nabla \times \mathbf{v}) \cdot \mathbf{n}_1]\,d\sigma + \iint_\Omega [(\nabla \times \mathbf{v}) \cdot n_2]\,d\sigma = 0.$$

This gives

$$\iint_S [(\nabla \times \mathbf{v}) \cdot \mathbf{n}_1]\,d\sigma = \iint_\Omega [(\nabla \times \mathbf{v}) \cdot (-\mathbf{n}_2)]\,d\sigma = \oint_C \mathbf{v}(\mathbf{r}) \cdot d\mathbf{r}$$

where C is traversed in a positive sense with respect to $-\mathbf{n}_2$ and therefore in a positive sense with respect to $\mathbf{n}_1$. ($-\mathbf{n}_2$ points toward S)

CHAPTER 18

SECTION 18.1
1. first-order ordinary   **3.** first-order partial   **5.** second-order ordinary   **7.** second-order partial   **9.** y_1 is; y_2 is not
11. both y_1 and y_2 are solutions   **13.** both y_1 and y_2 are solutions   **15.** both u_1 and u_2 are solutions   **17.** both y_1 and y_2 are solutions

19. $y = 2e^{5x}$ **21.** $y = \dfrac{1}{-2e^{x-1} + 1}$ **23.** $y = -\frac{17}{4}x + 9x^{1/2}$ **25.** $y = x^2 \ln x$ **27.** $r = -3$ **29.** $r = -3$ **31.** $r = 0$

33. $r = \frac{1}{2}$ or $r = \frac{3}{2}$ **35.** (a) $y = C_1 \sin 4x$ (b) $y = 0$

SECTION 18.2

1. $y = -\frac{1}{2} + Ce^{2x}$ **3.** $y = \frac{2}{5} + Ce^{-(5/2)x}$ **5.** $y = x + Ce^{2x}$ **7.** $y = \frac{2}{3}nx + Cx^4$ **9.** $y = Ce^{e^x}$ **11.** $y = 1 + C(e^{-x} + 1)$

13. $y = e^{-x^2}(\frac{1}{2}x^2 + C)$ **15.** $y = C(x+1)^{-2}$ **17.** $y = 2e^{-x} + x - 1$ **19.** $y = e^{-x}[\ln(1 + e^x) + e - \ln 2]$ **21.** $y = x^2(e^x - e)$

23. $y = y_0 e^{x_0^2 - x^2}$ **25.** (a) $200(\frac{4}{5})^{t/5}$ (b) $200(\frac{4}{5})^{t^2/25}$ **27.** $20 + 5e^{-0.03t}$ lb

29. (a) $T(t) = \tau + Ce^{-kt}$ (b) $T(t) = \tau + (T_0 - \tau)e^{-kt}$ (c) $\lim_{t\to\infty} T(t) = \tau$ (d) $\dfrac{2\ln 3}{\ln(\frac{4}{3})} - 2 \approx 5.64$ min

31. (a) $\dfrac{dP}{dt} = k(M - P)$ (b) $P(t) = M(1 - e^{-0.0357t})$ (c) 65 days   **33.** $y^2 = \dfrac{1}{6e^x + Ce^{2x}}$ **35.** $y = \dfrac{1}{Cx - \frac{1}{3}x^4}$

37. If $y = y(x)$ is a solution of (18.2.1), then

$$y'(x) + p(x)y(x) = q(x)$$

and

$$e^{h(x)}y'(x) + p(x)e^{h(x)}y(x) = e^{h(x)}q(x),$$

where $h(x) = \displaystyle\int p(x)\,dx$.   This implies that

$$e^{h(x)}y(x) = \int e^{h(x)}q(x)\,dx + C$$

for some constant C.   It now follows that $y = y(x)$ is a member of the one-parameter family (18.2.2).

39. Let y_1 and y_2 be solutions of (18.2.3) and let $z = y_1 + y_2$.   Then

$z' = (y_1 + y_2)' = y_1' + y_2' = -py_1 - py_2 = -pz$.   Thus $z' + p(x)z = 0$ and z is a solution of (18.2.3).

41. (a) From Exercise 38, $y(x) = Ce^{-h(x)}$.   If $y(a) = 0$, then $0 = Ce^{-h(a)}$, which implies $C = 0$ since $e^{-h(a)} \neq 0$.   Thus $y(x) = 0$ for all $x \in I$.
(b) Put $u(x) = y_1(x) - y_2(x)$.   Then u is a solution of (18.2.3) by Exercise 39.   Since $u(a) = y_1(a) - y_2(a) = 0$, $u(x) = 0$ for all $x \in I$ by part (a).   Thus $y_1(x) = y_2(x)$ for all $x \in I$.

43. $y(x) = Ce^{-rx}$
(a) If $r > 0$, then $\lim_{x\to\infty} y(x) = \lim_{x\to\infty} Ce^{-rx} = 0$ (b) If $r < 0$, then $\lim_{x\to\infty} y(x) = \lim_{x\to\infty} Ce^{-rx} = \infty$

SECTION 18.3

1. $y = Ce^{-(1/2)\cos(2x+3)}$ **3.** $x^4 + \dfrac{2}{y^2} = C$ **5.** $y\sin y + \cos y = \cos\left(\dfrac{1}{x}\right) + C$ **7.** $e^{-y} = xe^{-x} + e^{-x} + C$

9. $\ln|y+1| + \dfrac{1}{y+1} = \ln|\ln x| + C$ **11.** $\sin^{-1}y = 1 - \sqrt{1-x^2}$ **13.** $y + \ln|y| = \dfrac{x^3}{3} - x - 5$

15. $\dfrac{x^2}{2} + x + \dfrac{1}{2}\ln(y^2+1) - \tan^{-1}y = 4$ **17.** $y^2 - x^2 = Cx$ **19.** $x^2 - 2xy - y^2 = C$ **21.** $y + x = xe^{y/x}(C - \ln x)$

23. $x\left(1 + \cos\dfrac{y}{x}\right) = C\sin\dfrac{y}{x}$ **25.** $y^3 + 3x^3\ln|x| = 8x^3$ **27.** (a) $y' = \dfrac{y(a_4x - a_3)}{x(a_1 - a_2y)}$ (b) $x^{a_3}y^{a_1} = Ce^{a_4x + a_2y}$

29. $y = \frac{3}{2}x + C$ **31.** $x^2 - y^2 = C$ **33.** $y^2 = -2x + C$

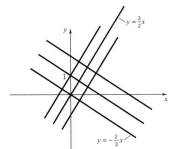

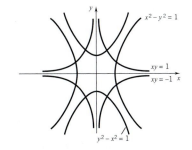

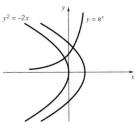

35. A differential equation for the given family is $y^2 = 2xyy' + y^2(y')^2$. Now replace y' by $-\dfrac{1}{y'}$. Then the resulting differential equation is

$y^2 = -\dfrac{2xy}{y'} + \dfrac{y^2}{(y')^2}$, which simplifies to $y^2 = 2xyy' + y^2(y')^2$. Thus the given family is self-orthogonal.

37. (a) $y(t) = \dfrac{25{,}000}{1 + 249e^{-0.1382t}}$, $y(20) \cong 1498$ (b) 40 days **39.** (a) 88.82 m/sec (b) $v(t) = \dfrac{15.65(1 + 0.48e^{-1.25t})}{1 - 0.48e^{-1.25t}}$ (c) 15.65 m/sec

SECTION 18.4

1. $\dfrac{x^2y^2}{2} - xy = C$ **3.** $xe^y - ye^x = C$ **5.** $x \ln y + x^2y = C$ **7.** $y \ln x - 3x^2 - 2y = C$ **9.** $xy^3 + y^2\cos x - \tfrac{1}{2}x^2 + \tfrac{1}{2}e^{2y} = C$

11. (a) yes (b) $\dfrac{1}{p(y)q(x)}$ $(p(y)q(x) \neq 0)$ **13.** $xe^y - ye^x = C$ **15.** $x^3y^2 + \tfrac{1}{2}x^2 + xe^y + \tfrac{1}{2}y^2 = C$ **17.** $y^3e^x + xe^x = C$

19. $x^3 + 3xy + 3e^y = 4$ **21.** $4x^2y^2 + x^4 + 4x^2 = 5$ **23.** $xy - \dfrac{1}{y} = 3$ **25.** $\sinh(x - y^2) + \tfrac{1}{2}e^{2x} + \tfrac{1}{2}y^2 = \tfrac{1}{2}e^4 + 1$

27. (a) $k = 3$ (b) $k = 1$ **29.** $y = \dfrac{-4}{x^4 + C}$ **31.** $y = \tfrac{1}{9}x^5 + Cx^{-4}$ **33.** $e^{xy} - x^2 + 2 \ln|y| = C$

SECTION 18.5

1. $y = C_1e^{3x} + C_2e^{-5x}$ **3.** $y = C_1e^{-4x} + C_2xe^{-4x}$ **5.** $y = e^{-x}(C_1\cos 2x + C_2\sin 2x)$ **7.** $y = C_1e^{(1/2)x} + C_2e^{-3x}$

9. $y = C_1\cos 2\sqrt{3}x + C_2\sin 2\sqrt{3}x$ **11.** $y = C_1e^{(1/5)x} + C_2e^{-(3/4)x}$ **13.** $y = C_1\cos 3x + C_2\sin 3x$ **15.** $y = e^{-(1/2)x}(C_1\cos \tfrac{1}{2}x + C_2\sin \tfrac{1}{2}x)$

17. $y = C_1e^{(1/4)x} + C_2e^{-(1/2)x}$ **19.** $y = 2e^{2x} - e^{3x}$ **21.** $y = 2\cos \dfrac{x}{2} + \sin \dfrac{x}{2}$ **23.** $y = 7e^{-2(x+1)} + 5xe^{-2(x+1)}$

25. (a) $y = Ce^{2x} + (1 - C)e^{-x}$ (b) $y = Ce^{2x} + (2C - 1)e^{-x}$ (c) $y = \tfrac{2}{3}e^{2x} + \tfrac{1}{3}e^{-x}$

27. $\alpha = \dfrac{r_1 + r_2}{2};\ \beta = \dfrac{r_1 - r_2}{2}$

$y = k_1e^{r_1x} + k_2e^{r_2x} = e^{\alpha x}(C_1\cosh \beta x + C_2\sinh \beta x)$, where $k_1 = \dfrac{C_1 + C_2}{2}$, $k_2 = \dfrac{C_1 - C_2}{2}$.

29. (a) The Wronskian of $y_1 = e^{r_1x}$, $y_2 = xe^{r_1x}$ is

$\quad W(x) = e^{r_1x}(e^{r_1x} + r_1xe^{r_1x}) - xe^{r_1x}(r_1e^{r_1x}) = e^{2r_1x} \neq 0.$

(b) The Wronskian of $y_1 = e^{\alpha x}\cos \beta x$, $y_2 = e^{\alpha x}\sin \beta x$, $\beta \neq 0$, is

$\quad W(x) = e^{\alpha x}\cos \beta x[\alpha e^{\alpha x}\sin \beta x + \beta e^{\alpha x}\cos \beta x] - e^{\alpha x}\sin \beta x[\alpha e^{\alpha x}\cos \beta x - \beta e^{\alpha x}\sin \beta x] = \beta e^{2\alpha x} \neq 0.$

31. (a) $y'' + 2y' - 8y = 0$ (b) $y'' - 4y' - 5y = 0$ (c) $y'' - 6y' + 9y = 0$

33. If $y_1(x) = ky_2(x)$, then $y_2(x)y_1'(x) - y_1(x)y_2'(x) = ky_2(x)y_2'(x) - ky_2(x)y_2'(x) = 0$. If $y_2(x)y_1'(x) - y_1(x)y_2'(x) = 0$, then $\left[\dfrac{y_1(x)}{y_2(x)}\right]' = 0$, which implies

$y_1(x) = ky_2(x)$ for some constant k.

35. (a) If $a = 0$, $b > 0$, then the general solution is

$\quad y = C_1\cos \sqrt{b}\,x + C_2\sin \sqrt{b}\,x = A\cos(\sqrt{b}\,x + \phi)$

where A and ϕ are constants. Clearly $|y(x)| \leq |A|$ for all x.

(b) If $a > 0$, $b = 0$, then the general solution is $y = C_1 + C_2e^{-ax}$ and $\lim\limits_{x \to \infty} y(x) = C_1$. The solution which satisfies the conditions $y(0) = y_0$,

$\quad y'(0) = y_1$ is

$\quad y = y_0 + \dfrac{y_1}{a} - \dfrac{y_1}{a}e^{-ax};$

$\quad k = y_0 + \dfrac{y_1}{a}.$

37. $y = C_1x^4 + C_2x^{-2}$ **39.** $y = C_1x^2 + C_2x^2\ln x$

41. (a) $e^x = 1 + x + \dfrac{x^2}{2!} + \dfrac{x^3}{3!} + \cdots + \dfrac{x^n}{n!} + \cdots = \sum\limits_{k=0}^{\infty} \dfrac{1}{k!}x^k$ (b) $e^{i\theta} = 1 + i\theta - \dfrac{1}{2!}\theta^2 - \dfrac{i}{3!}\theta^3 + \cdots + \dfrac{(i)^n}{n!}\theta^n + \cdots = \cos \theta + i \sin \theta$

(c) $e^{-i\theta} = 1 - i\theta - \dfrac{1}{2!}\theta^2 + \dfrac{i}{3!}\theta^3 + \cdots + \dfrac{(-i)^n}{n!}\theta^n + \cdots = \cos \theta - i \sin \theta$

SECTION 18.6

1. $y = \tfrac{1}{2}x + \tfrac{1}{4}$ **3.** $y = \tfrac{1}{5}x^2 - \tfrac{4}{25}x - \tfrac{27}{125}$ **5.** $y = \tfrac{1}{36}e^{3x}$ **7.** $y = \tfrac{1}{3}e^x$ **9.** $y = -\tfrac{13}{170}\cos x - \tfrac{1}{170}\sin x$ **11.** $y = \tfrac{3}{100}\cos 2x + \tfrac{21}{100}\sin 2x$

13. $y = \tfrac{1}{20}e^{-x}\sin 2x + \tfrac{1}{10}e^{-x}\cos 2x$ **15.** $y = \tfrac{3}{2}xe^{-2x}$ **17.** $y = C_1\cos x + C_2\sin x + \tfrac{1}{2}e^x$ **19.** $y = C_1e^{5x} + C_2e^{-2x} + \tfrac{1}{10}x + \tfrac{7}{100}$

21. $y = C_1 e^x + C_2 e^{-4x} - \frac{1}{5} x e^{-4x}$ **23.** $y = C_1 e^{-2x} + C_2 e^x + \frac{1}{2} x^2 e^x - \frac{1}{3} x e^x$

25. Let $z = y_1 + y_2$. Then   **27.** $y = C_1 e^{-3x} + C_2 e^{-x} + \frac{1}{4} x e^{-x} + \frac{1}{16} e^x$

$$z'' + az' + bz = (y_1'' + y_2'') + a(y_1' + y_2') + b(y_1 + y_2)$$
$$= (y_1'' + ay_1' + by_1) + (y_2'' + ay_2' + by_2) = \phi_1 + \phi_2$$

29. $y = 2e^x \sin x - x e^x \cos x$ **31.** $y = \frac{1}{3} x \ln |x| \, e^{2x}$ **33.** $y = -\ln |x| \, e^{-2x}$ **35.** $y = e^x(x \sin x + \cos x \ln |\cos x|)$

37. (a) $i(t) = -CF_0 e^{-(R/2L)t} + \dfrac{F_0}{2L}(2 - RC) t e^{-(R/2L)t} + CF_0$

(b) $i(t) = e^{-(R/2L)t} \left[\dfrac{F_0(2 - RC)}{2L\beta} \sin \beta t - CF_0 \cos \beta t \right] + CF_0,$ where $\beta = \sqrt{\dfrac{4L - CR^2}{C}}$

39. (a) $y_1 y_2' - y_2 y_1' = -\dfrac{2}{x} \neq 0$ (b) $y = \frac{1}{3} \sin (\ln x)$

SECTION 18.7

1. $x(t) = \sin(8t + \frac{1}{2}\pi)$; $A = 1$, $f = 4/\pi$ **3.** $\pm 2\pi A/T$ **5.** $x(t) = (15/\pi) \sin \frac{1}{3}\pi t$ **7.** $x(t) = x_0 \sin (\sqrt{k/m}\, t + \frac{1}{2}\pi)$

9. at $x = \pm\frac{1}{2}\sqrt{3}\, x_0$ **11.** $\frac{1}{4} k x_0^2$

13. Set $y(t) = x(t) - 2$. Equation $x''(t) = 8 - 4x(t)$ can be written $y''(t) + 4y(t) = 0$. This is simple harmonic motion centered at $y = 0$, which is $x = 2$.

$$y(t) = A \sin(2t + \phi_0).$$

The condition $x(0) = 0$ gives $y(0) = -2$ and thus

$$A \sin \phi_0 = -2. \qquad (*)$$

Since $y'(t) = x'(t)$ and $y'(t) = 2A \cos(2t + \phi_0)$, the condition $x'(0) = 0$ gives $y'(0) = 0$, and thus

$$2A \cos \phi_0 = 0. \qquad (**)$$

Equations $(*)$ and $(**)$ are satisfied by $A = 2$, $\phi_0 = \frac{3}{2}\pi$. The equation of motion can therefore be written

$$y(t) = 2 \sin(2t + \tfrac{3}{2}\pi).$$

The amplitude is 2 and the period is π.

15. (a) $x''(t) + \omega^2 x(t) = 0$ with $\omega = r\sqrt{\pi\rho/m}$ (b) $x(t) = x_0 \sin (r\sqrt{\pi\rho/m}\, t + \frac{1}{2}\pi)$, taking downward as positive; $A = x_0$, $T = (2/r)\sqrt{m\pi/\rho}$

17. amplitude and frequency both decrease   **19.** at most once; at most once   **21.** if $\omega/\gamma = m/n$, then $m/\omega = n/\gamma$ is a period

23. $x = e^{-\alpha t}[c_1 \cos \sqrt{\omega^2 - \alpha^2}\, t) + c_2 \sin (\sqrt{\omega^2 - \alpha^2}\, t)]$ or equivalently $x = C_1 e^{-\alpha t}[\sin(\sqrt{\omega^2 - \alpha^2}\, t) + C_2]$

25. $x_P = \dfrac{F_0}{2\alpha\gamma m} \sin \gamma t$;   as $c = 2\alpha m \to 0^+$, the amplitude $\dfrac{F_0}{2\alpha\gamma m} = \dfrac{F_0}{cm} \to \infty$

27. $(\omega^2 - \gamma^2)^2 + 4\alpha^2 \gamma^2 = \omega^2 + \gamma^4 + 2\gamma^2(2\alpha^2 - \omega^2)$ increases as γ increases

APPENDIX A

SECTION A.1

1. hyperbola, $\alpha = \dfrac{\pi}{4}$, $\frac{1}{2}(X^2 - Y^2) = 1$ **3.** hyperbola, $\alpha = \dfrac{\pi}{6}$, $4X^2 - Y^2 = 1$ **5.** parabola, $\alpha = \dfrac{\pi}{4}$, $2Y^2 + \sqrt{2}X = 0$

7. parabola, $\alpha = \dfrac{\pi}{3}$, $X^2 - Y = 0$ **9.** $\alpha = \dfrac{\pi}{8}$, $\cos \alpha = \dfrac{1}{2}\sqrt{2 + \sqrt{2}}$, $\sin \alpha = \dfrac{1}{2}\sqrt{2 - \sqrt{2}}$

SECTION A.2

1. -2 **3.** 0   **5.** 5   **7.** -6 **9.** $\begin{vmatrix} 4 & 5 & 6 \\ 1 & 2 & 3 \\ 7 & 8 & 9 \end{vmatrix} = - \begin{vmatrix} 1 & 2 & 3 \\ 4 & 5 & 6 \\ 7 & 8 & 9 \end{vmatrix}$ (interchanged two rows)

11. The second determinant is zero because two rows are the same.   **13.** $\begin{vmatrix} 1 & 2 & 3 \\ x & 2x & 3x \\ 4 & 5 & 6 \end{vmatrix} = x \begin{vmatrix} 1 & 2 & 3 \\ 1 & 2 & 3 \\ 4 & 5 & 6 \end{vmatrix}$ ⎯ two rows are the same $= 0$

15. The expression on the left can be written

$$\lambda^2 \begin{vmatrix} 1 & 1 \\ a & b \end{vmatrix} - 2\lambda \begin{vmatrix} 1 & 1 \\ a & b \end{vmatrix} + 1 \begin{vmatrix} 1 & 1 \\ a & b \end{vmatrix} = (\lambda^2 - 2\lambda + 1) \begin{vmatrix} 1 & 1 \\ a & b \end{vmatrix} = (\lambda - 1)^2 \begin{vmatrix} 1 & 1 \\ a & b \end{vmatrix}.$$

17. $\begin{vmatrix} a_1 & a_2 & a_3 \\ b_1 & b_2 & b_3 \\ c_1 & c_2 & c_3 \end{vmatrix} = - \begin{vmatrix} a_1 & a_2 & a_3 \\ c_1 & c_2 & c_3 \\ b_1 & b_2 & b_3 \end{vmatrix} = \begin{vmatrix} c_1 & c_2 & c_3 \\ a_1 & a_2 & a_3 \\ b_1 & b_2 & b_3 \end{vmatrix} = c_1 \begin{vmatrix} a_2 & a_3 \\ b_2 & b_3 \end{vmatrix} - c_2 \begin{vmatrix} a_1 & a_3 \\ b_1 & b_3 \end{vmatrix} + c_3 \begin{vmatrix} a_1 & a_2 \\ b_1 & b_2 \end{vmatrix}$

INDEX

A

Abel, Niels Henrik, 759
Absolute convergence, 720–721
Absolute extreme values, functions of several variables, 996–998
Acceleration:
    components of, 893–894
    as displacement, 780
    as vector, 871
Alternating series, 721–722
Alternating series test, 722–724
Angular frequency, 1246
Angular momentum, 874–876
Angular velocity vector, 882
Antipodal points, 804
Applied frequency, 1252
Arc length:
    additivity of, 869–870
    line integrals with respect to, 1133–1135
    space curves, 863–870
        additivity of arc length, 869–870
        definition, 865
        formula, 865
Area:
    of parametrized surface, 1152–1158
        parallelogram, 1152–1153
        region, 1154
        sphere, 1154
        spiral ramp, 1155
        surface of revolution, 1154–1155
        $z = f(x,y)$, 1155–1158
Auxiliary equation, $1225n$

B

Basic comparison test, 710–711
Basic region, 1044
Bernoulli, Jacob, $668n$
Bernoulli's equation, 1207
Binomial coefficients, 766–767
Binomial series, 766–767
Boundary (of set), 929
Boundary-value problems, 1200
Bounded above, 637, 644
Bounded below, 640, 644
Boundedness:
    of function of several variables, 901
    of sequences, 644, 653–654
    of sets of real numbers, 637–641
Box, triple integral over a, 1076–1077
Brahe, Tycho, 882

C

Cantor middle third set, 705
Cartesian product, $1037n$
Cartesian space coordinates, 773–777
    displacements in, 778–782
    distance formula for, 775–776
    symmetry in, 776–777
Cauchy, A. L., $670n$
Cauchy mean-value theorem, 670–671
Cauchy-Riemann equations, 926
Center of curvature, 890
Center of mass, 1068–1069
Central (radial) force, 876
Centroid(s), 1070–1071
Chain rule(s):
    for functions of several variables, 968–976
        along curve, 968–972
        implicit differentiation, 974–975
        on surface, 972–974
Changes of parameter, sense-preserving, 861
Circle(s), unit, 899
Circular helix, 840, 859
Circular motion about origin, 872–873
Circulation per unit area, 1188–1189
Closed region, continuity on, 966–967
Closed sets, 929
Closed unit ball, 900
Closed unit disc, 899
Colatitude, 1099
Collinear points, 827
Collinear vectors, 827
Components (of vector), 799–801
Compositions, continuity of, 935
Compound interest, 900
Computer-generated graphs, 914
Conditional convergence, 721
Cone:
    as parametrized surface, 1149–1150
    quadric, 905–906
Conservation of mechanical energy, 1130–1131
Conservation of momentum, 874
Conservative force fields, 1130
Constants of motion, 874
Continuity:
    on closed region, 966–967
    and differentiability, 137–139

of functions of several variables, 934–939
    and differentiability, 950
    in each variable separately, 936
    examples, 934–935
    and partial differentiability, 936–937
    of vector functions, 843
Continuously differentiable surface, 1153
Convergent improper integral, 686
Convergent sequences, 653–654
Convergent series, 696, 700, 702–704
    absolute convergence, 720–721
    conditional convergence, 721
    nonnegative terms, tests for series with, 706–713, 715–719
    power series, 746–755
Convex set of points, 968
Cooling, Newton's law of, 1206–1207
Coordinate planes, 773–774
Copernicus, 882
Coplanar vectors, 828
Critical points, 992
Critically damped case, 1250–1251
Cross product, 805–812
    components of, 809–812
    definition, 805–806
    distributive laws with, 808
    identities for, 812
    in physics and engineering, 812
    properties of, 806–807
    replacement of Lagrange condition by, 1012–1013
    right-handed triples, properties of, 806–807
    and scalar triple product, 807–808
Curl, 1171–1174
Curvature:
    center of, 890
    of plane, 888–892
    radius of, 890
    of space, 892–893
Curve(s):
    integral, 1209–1212
    level, 911–913
    piecewise smooth, 1119
    plane, 888–892
    reversing sense of, 860
    space, 892–893
    vector function, parametrization by, 853–861

Integrating factors, 1201–1203, 1222–1223
Integration, of vector functions, 844–846
of power series, 752, 756–762
Interest, compound, 900
Interior (of set), 928–929
Interior points, 928–929
Intermediate-value theorem(s), 641
for functions of several variables, 966–967
Intersecting curves, 857–858
Intersecting lines, 819–821
Intersecting planes, 829–831
Interval of convergence (power series), 748–751, 761–762
Irrotational flow, 1189

J

Jacobian area formula, 1146–1147
Jacobians, 1107
Jordan, Camille, 1025n
Jordan curve theorem, 1025n
Jordan curves, 1025n
regions bounded by two or more, 1143–1145
Jordan regions, 1137
Joule (unit), 804

K

Kepler, Johannes, 882–883
Kepler's laws, 884–887
Kinetic energy, 1071

L

Lagrange, Joseph-Louis, 731, 1009n
Lagrange formula for the remainder, 731–733, 741–742
Lagrange multiplier, 1009–1111
Laplace, Pierre-Simon, 939, 1174
Laplace transforms, 689
Laplace's equation, 939–940
Laplacian, 1174
Law of cosines, 796
Least upper bound, 637–641
of sequence, 644
Least upper bound axiom, 638–641
Length (of vector), 785–786
Level curves, 911–913
Level surfaces, 914–916
L'Hospital, G. F. A., 668n
L'Hospital's rule:
∞/∞, 674, 675, 677
0/0, 668–672, 677
Limit comparison test, 712–713
Limit(s). *See also under* Sequence(s) of real numbers

of functions of several variables, 931–934
of vector function, 841–843
Linear first-order differential equations, 1200–1201
Linear second-order differential equations, 1224
Line integrals, 1115–1122
alternative notation for, 1132–1133
and conservation of mechanical energy, 1130–1131
definition of, 1117–1118
fundamental theorem for, 1124–1127
with respect to arc length, 1133–1135
and sense-preserving changes of parameter, 1118
and work done by varying force over curved path, 1115–1117
and work-energy formula, 1129
Line(s):
distance from point to, 822–823
normal, 982
scalar parametric equations of, 817–819
tangent, 856, 982
vectors, parametrized by, 814–817, 819–823
intersecting lines, 819–821
motion along straight line, 872
parallel lines, 820
perpendicular lines, 821
symmetric form of equations, 817–818
Little-o(h), 945–946
Local extreme values:
functions of several variables, 991–996
and second-partials test, 999–1005
Logistic equations, 1213–1215
Logistic functions, 1213–1215
Longitude, 1099
Lower bound, 640
Lower sum, 1038
Lower unit normal, 990

M

Machin, John, 765n
Maclaurin, Colin, 733
Maclaurin series, 733
Magnetic field lines, spiraling around, 880
Magnitude:
of displacement, 778
of vector, 785–786
Mass:
center of, 1068–1069
of material surface, 1159–1160

Maximum-minimum theorem, 641
Mean-value theorem(s):
for double integrals, 1047–1048
for functions of several variables, 963–966
Mechanical energy, conservation of, 1130–1131
Mechanical vibrations, 1245–1253
damped vibrations, 1248–1251
forced vibrations, 1251–1253
undamped vibrations, 1247–1248
Mechanics, 873–880
angular momentum, 874–876
initial-value problems, 877–880
momentum, 874
statistical, 1066n
torque, 876
Midpoint formula, 777
Minimax (of elliptic paraboloid), 906
Minor axis (of ellipse), 556
Mirrors, parabolic, 562–563
Mixed partials, equality of, 937–939
Mixing problems, 451–453
Moment of inertia, 1071
in plate, 1071–1072
Momentum, 874
angular, 874–876
Monotonic sequences, 644–645
Multiple integration, changing variables in, 1107–1111
Multiple sigma notation, 1033–1035

N

Nappes, 906
Natural frequency, 1252
Neighborhood of a point, 928
Newton, Sir Isaac, 883
Newton-meter (unit), 804
Newton's law of cooling, 1206–1207
Newton's Second Law of Motion, 780, 870–871, 873–874
for extended three-dimensional objects, 883–884
and Kepler's laws, 884–885
Nondecreasing sequences, 644
boundedness of, 654
Nonhomogeneous equation, 1207, 1234
Nonhomogeneous function, 1234
Nonincreasing sequences, 644
boundedness of, 654–655
Norm (of vector), 785–786
Normal line, 982
Normal vectors, 824, 981
n-parameter family of solutions, 1198